Lecture Notes in Computer Science 9389

Commenced Publication in 1973
Founding and Former Series Editors:
Gerhard Goos, Juris Hartmanis, and Jan van Leeuwen

More information about this series at http://www.springer.com/series/7412

Frank Nielsen · Frédéric Barbaresco (Eds.)

Geometric Science of Information

Second International Conference, GSI 2015
Palaiseau, France, October 28–30, 2015
Proceedings

Springer

Editors
Frank Nielsen
École Polytechnique, LIX
Palaiseau
France

and

Sony Computer Science Laboratories, Inc.
Tokyo
Japan

Frédéric Barbaresco
Thales Land and Air Systems
Limours
France

ISSN 0302-9743   ISSN 1611-3349   (electronic)
Lecture Notes in Computer Science
ISBN 978-3-319-25039-7   ISBN 978-3-319-25040-3   (eBook)
DOI 10.1007/978-3-319-25040-3

Library of Congress Control Number: 2015950451

LNCS Sublibrary: SL6 – Image Processing, Computer Vision, Pattern Recognition, and Graphics

Springer Cham Heidelberg New York Dordrecht London

Cover page painting: Woman teaching Geometry, from French medieval edition of Euclid's Elements (14th century) © The British Library, used with granted permission.

Printed on acid-free paper

Springer International Publishing AG Switzerland is part of Springer Science+Business Media
(www.springer.com)

Preface

On behalf of both the Organizing and the Scientific Committees, it is our great pleasure to welcome you to the proceedings of the Second International SEE Conference on "Geometric Science of Information" (GSI 2015), hosted by École Polytechnique (Palaiseau, France), during October 28–30, 2015 (http://www.gsi2015.org/).

GSI 2015 benefited from the scientific sponsorship of Société de Mathématique Appliquées et Industrielles (SMAI, smai.emath.fr/) and the financial sponsorship of:

- CNRS
- École Polytechnique
- Institut des Systèmes Complexes
- Inria (http://www.inria.fr/en/)
- Telecom ParisTech
- THALES (www.thalesgroup.com)

GSI 2015 was also supported by CNRS Federative Networks MIA and ISIS.

The 3-day conference was organized in the framework of the relations set up between SEE (http://www.see.asso.fr/) and the following scientific institutions or academic laboratories: École Polytechnique, École des Mines de Paris, INRIA, Supélec, Université Paris-Sud, Institut Mathématique de Bordeaux, Sony Computer Science Laboratories, Telecom SudParis, and Telecom ParisTech.

We would like to express our thanks to the Computer Science Department LIX of École Polytechnique for hosting this second scientific event at the interface between geometry, probability, and information geometry. In particular, we warmly thank Evelyne Rayssac of LIX, École Polytechnique, for her kind administrative support that helped us book the auditorium and various resources at École Polytechnique, and Olivier Bournez (LIX Director) for providing financial support.

The GSI conference cycle was initiated by the Brillouin Seminar Team (http://repmus.ircam.fr/brillouin/home). The 2015 event was motivated in continuing the first initiatives launched in 2013 (see LNCS proceedings 8085, http://www.springer.com/us/book/9783642400193). We mention that in 2011, we organized an Indo-French workshop on "Matrix Information Geometry" that yielded an edited book in 2013 (http://www.springer.com/us/book/9783642302312).

The technical program of GSI 2015 covered all the main topics and highlights in the domain of "geometric science of information" including information geometry manifolds of structured data/information and their advanced applications. These proceedings consist solely of *original research papers* that were carefully *peer-reviewed* by two or three experts and revised before acceptance.

The program included the renown invited speaker Professor Charles-Michel Marle (UPMC, Université Pierre et Marie Curie, Paris, France), who gave a talk on "Actions of Lie Groups and Lie Algebras on Symplectic and Poisson Manifolds," and three distinguished keynote speakers:

- Professor Marc Arnaudon (Bordeaux University, France): "Stocastic Euler-Poincaré Reduction"
- Professor Tudor Ratiu (EPFL, Switzerland): "Symmetry Methods in Geometric Mechanics"
- Professor Matilde Marcolli (Caltech, USA): "From Geometry and Physics to Computational Linguistics"

A short course was given by Professor Dominique Spehner (Grenoble University, France) on the "Geometry on the Set of Quantum States and Quantum Correlations" chaired by Roger Balian (CEA, France).

The collection of papers have been arranged into the following 17 thematic sessions, illustrating the richness and versatility of the field:

- Dimension Reduction on Riemannian Manifolds
- Optimal Transport
- Optimal Transport and Applications in Imagery/Statistics
- Shape Space and Diffeomorphic Mappings
- Random Geometry and Homology
- Hessian Information Geometry
- Topological Forms and Information
- Information Geometry Optimization
- Information Geometry in Image Analysis
- Divergence Geometry
- Optimization on Manifold
- Lie Groups and Geometric Mechanics/Thermodynamics
- Computational Information Geometry
- Lie Groups: Novel Statistical and Computational Frontiers
- Geometry of Time Series and Linear Dynamical Systems
- Bayesian and Information Geometry for Inverse Problems
- Probability Density Estimation

Historical Background

As for the first edition of GSI (2013) and in past publications (https://www.see.asso.fr/node/11950), GSI 2015 addressed inter-relations between different mathematical domains such as shape spaces (geometric statistics on manifolds and Lie groups, deformations in shape space), probability/optimization and algorithms on manifolds (structured matrix manifold, structured data/information), relational and discrete metric spaces (graph metrics, distance geometry, relational analysis), computational and Hessian information geometry, algebraic/infinite dimensional/Banach information manifolds, divergence geometry, tensor-valued morphology, optimal transport theory, and manifold and topology learning, as well as applications such as geometries of audio-processing, inverse problems, and signal processing.

At the turn of the century, new and fruitful interactions were discovered between several branches of science: information science (information theory, digital communications, statistical signal processing), mathematics (group theory, geometry and topology, probability, statistics), and physics (geometric mechanics, thermodynamics, statistical physics, quantum mechanics).

From Statistics to Geometry

In the middle of the last century, a new branch in the geometric approach of statistical problems was initiated independently by Harold Hotelling and Calyampudi Radhakrishna Rao, who introduced a metric space in the parameter space of probability densities. The metric tensor was proved to be equal to the Fisher information matrix. This result was axiomatized by Nikolai Nikolaevich Chentsov in the framework of category theory. This idea was also latent in the work of Maurice Fréchet, who had noticed that the "distinguished densities" that reach lower bounds of statistical estimators are defined by a function that is given by a solution of the Legendre–Clairaut equation (cornerstone equation of "information geometry"), and in the works of Jean-Louis Koszul with a generalized notion of characteristic function.

From Probability to Geometry

Probability is again the subject of a new foundation to apprehend new structures and generalize the theory to more abstract spaces (metric spaces, shape space, homogeneous manifolds, graphs). An initial attempt to probability generalization in metric spaces was made by Maurice Fréchet in the middle of the last century, in the framework of abstract spaces topologically affine and "distance space" ("espace distancié"). More recently, Misha Gromov, at IHES (Institute of Advanced Scientific Studies), indicated the possibilities for (non-)homological linearization of basic notions of probability theory and also the replacement of real numbers as values of probabilities by objects of suitable combinatorial categories. In parallel, Daniel Bennequin, from Institut mathématique de Jussieu, observed that entropy is a universal co-homological class in a theory associated with a family of observable quantities and a family of probability distributions.

From Groups Theory to Geometry

As observed by Gaston Bachelard, "The group provides evidence of a mathematic closed on itself. Its discovery closes the era of conventions, more or less independent, more or less coherent." About Elie Cartan's work on group theory, Henri Poincaré said that "The problems addressed by Elie Cartan are among the most important, most abstract, and most general dealing with mathematics; group theory is, so to speak, the whole mathematics, stripped of its material and reduced to pure form. This extreme level of abstraction has probably made my presentation a little dry. To assess each of the results, I would have had to virtually render it the material of which it had been stripped; but this refund can be made in a thousand different ways; and this is the only form that can be found as well as a host of various garments, which is the common link between mathematical theories whose proximity is often surprising."

From Mechanics to Geometry

The last elaboration of geometric structure on information is emerging at the inter-relations between "geometric mechanics" and "information theory" that was largely debated at the GSI 2015 conference with invited speakers including C.M. Marle, T. Ratiu, and M. Arnaudon. Elie Cartan, the master of geometry during the last century, said: "distinguished service that has rendered and will make even the absolute differential calculus of Ricci and Levi–Civita should not prevent us from avoiding too exclusively formal calculations, where debauchery indices often mask a very simple geometric fact. It is this reality that I have sought to put in evidence everywhere." Elie Cartan was the son of Joseph Cartan, who was the village blacksmith, and Elie recalled that his childhood had passed under "blows of the anvil, which started every morning from dawn." One can imagine that the hammer blows made by Joseph on the anvil, giving shape and curvature to the metal, influenced Elie's mind with germinal intuition of fundamental geometric concepts. The alliance between geometry and mechanics is beautifully illustrated by the image of Forge, in the painting of Velasquez about the Vulcan God (see Figure 1). This concordance of meaning is also confirmed by the etymology of the word "forge," which comes from late fourteenth century, "a smithy," from the Old French *forge* "forge, smithy" (twelfth century), earlier *faverge*, from the Latin *fabrica* "workshop, smith's shop," from faber (genitive fabri) "workman in hard materials, smith."

Fig. 1. Into the Flaming Forge of Vulcan, into the Ninth Sphere, Mars descends in order to retemper his flaming sword and conquer the heart of Venus (Diego Velázquez, Museo Nacional del Prado). Public domain image, courtesy of https://en.wikipedia.org/wiki/Apollo_in_the_Forge_of_Vulcan

As Henri Bergson said in his book *The Creative Evolution* in 1907: "As regards human intelligence, there is not enough [acknowledgment] that mechanical invention was first its essential approach … we should say perhaps not *Homo sapiens*, but *Homo faber*. In short, intelligence, considered in what seems to be its original feature, is the faculty of manufacturing artificial objects, especially tools to make tools, and of indefinitely varying the manufacture."

Geometric Science of Information: A new Grammar of Sciences

Henri Poincaré said that "mathematics is the art of giving the same name to different things" ("La mathématique est l'art de donner le même nom à des choses différentes" in *Science et méthode*, 1908). By paraphrasing Henri Poincaré, we could claim that the "geometric science of information" is the art of giving the same name to different sciences. The rules and the structures developed at the GSI 2015 conference comprise a kind of new grammar for these sciences.

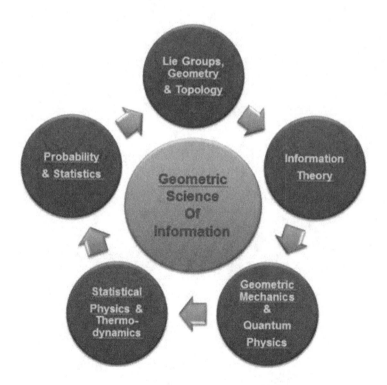

We give our thanks to all the authors and co-authors for their tremendous effort and scientific contribution. We would also like to acknowledge all the Organizing and Scientific Committee members for their hard work in evaluating the submissions. We warmly thank Jean Vieille, Valerie Alidor, and Flore Manier from the SEE for their kind support.

As with GSI 2013, a selected number of contributions focusing on a core topic were invited to contribute a chapter without page restriction to the edited book *Geometric Theory of Information* (http://www.springer.com/us/book/9783319053165) in 2014. Similarly, for GSI 2015, we invite prospective authors to submit their original work to a special issue on "advances in differential geometrical theory of statistics" of the MDPI *Entropy* journal (http://www.mdpi.com/journal/entropy/special_issues/entropy-statistics).

It is our hope that the fine collection of peer-reviewed papers presented in these LNCS proceedings will be a valuable resource for researchers working in the field of information geometry and for graduate students.

July 2015

Frank Nielsen
Frédéric Barbaresco

Organization

Program Chairs

Frédéric Barbaresco   Thales Air Systems, France
Frank Nielsen   École Polytechnique, France and Sony CSL, Japan

Scientific Committee

Pierre-Antoine Absil   University of Louvain, Belgium
Bijan Afsari   John Hopkins University, USA
Stéphanie Allassonnière   École Polytechnique, France
Jésus Angulo   MINES ParisTech, France
Marc Arnaudon   Institut de Mathématiques de Bordeaux, France
Michael Aupetit   Qatar Computing Research Institute, Quatar
Roger Balian   Academy of Sciences, France
Barbara Trivellato   Politecnico di Torino, Italy
Pierre Baudot   Max Planck Institute of Leipzig, Germany
Daniel Bennequin   University of Paris Diderot, France
Yannick Berthoumieu   École Nationale d'Electronique, Informatique et
     Radiocommunications de Bordeaux, France
Jérémie Bigot   Institut de Mathématiques de Bordeaux, France
Silvère Bonnabel   Mines ParisTech, France
Michel Boyom   University of Montpellier, France
Michel Broniatowski   University of Pierre and Marie Curie, France
Martins Bruveris   Brunel University London, UK
Charles Cavalcante   Universidade Federal do Ceara, Brazil
Frédéric Chazal   Inria, France
Arshia Cont   IRCAM, France
Gery de Saxcé   Université des Sciences et des Technologies de Lille,
     France
Laurent Decreusefond   Telecom ParisTech, France
Michel Deza   ENS Paris, France
Stanley Durrleman   Inria, France
Patrizio Frosini   University of Bologna, Italy
Alfred Galichon   New York University, USA
Alexander Ivanov   Imperial College, UK
Jérémie Jakubowicz   Institut Mines-Telécom, France
Hongvan Le   Mathematical Institute of ASCR, Czech Republik
Nicolas Le Bihan   University of Melbourne, Australia
Luigi Malagò   Shinshu University, Japan

Jonathan Manton	University of Melbourne, Australia
Jean-François Marcotorchino	Thales, France
Bertrand Maury	Université Paris-Sud, France
Ali Mohammad-Djafari	Supelec, France
Richard Nock	NICTA, Australia
Yann Ollivier	CNRS, France
Xavier Pennec	Inria, France
Michel Petitjean	Université Paris 7, France
Gabriel Peyré	CNRS, France
Giovanni Pistone	Collegio Carlo Alberto de Castro Statistics Initiative, Italy
Olivier Rioul	Télécom ParisTech, France
Said Salem	IMS Bordeaux, France
Olivier Schwander	University of Geneva, Switzerland
Rodolphe Sepulchre	Cambridge University, UK
Hichem Snoussi	University of Technology of Troyes, France
Alain Trouvé	ENS Cachan, France
Claude Vallée	Poitiers University, France
Geert Verdoolaege	Ghent University, Belgium
Rui Vigelis	Federal University of Ceara, Brazil
Susan Holmes	Stanford University, USA
Martin Kleinsteuber	Technische Universität München, Germany
Shiro Ikeda	ISM, Japan
Martin Bauer	University of Vienna, Austria
Charles-Michel Marle	Université Pierre et Marie Curie, France
Mathilde Marcolli	Caltech, USA
Jean-Philippe Ovarlez	Onera, France
Jean-Philippe Vert	Mines ParisTech, France
Allessandro Sarti	École des hautes études en sciences sociales, Paris
Jean-Paul Gauthier	University of Toulon, France
Wen Huang	University of Louvain, Belgium
Antonin Chambolle	École Polytechnique, France
Jean-François Bercher	ESIEE, France
Bruno Pelletier	University of Rennes, France
Stephan Weis	Universidade Estadual de Campinas, Brazil
Gilles Celeux	Inria, France
Jean-Michel Loubes	Toulouse University, France
Anuj Srivastana	Florida State University, USA
Johannes Rauh	Leibniz Universität Hannover, Germany
Joan Alexis Glaunes	Mines ParisTech, France
Quentin Mérigot	Université Paris-Dauphine, Paris
K.S. Subrahamanian Moosath	University of Calicut, India
K.V. Harsha	Indian Institute of Space Science and Technology, India

Emmanuel Trelat	UPMC, France
Lionel Bombrun	IMS Bordeaux, France
Olivier Cappé	Telecom Paris, France
Stephan Huckemann	Institut für Mathematische Stochastik; Göttingen, Germany
Piotr Graczyk	University of Angers, France
Fernand Meyer	Mines ParisTech
Corinne Vachier	Université Paris Est Créteil, France
Tudor Ratiu	EPFL, Swiss
Klas Modin	Chalmers University of Technology, Göteborg, Sweden
Hervé Lombaert	Inria, France
Michèle Basseville	IRISA, France
Juliette Matiolli	Thales, France
Peter D. Grünwald	CWI, Amsterdam, The Netherlands
François-Xavier Viallard	CEREMADE, Paris, France
Guido Francisco Montúfar	Max Planck Institute for Mathematics in the Sciences, Leipzig, Germany
Emmanuel Chevallier	Mines ParisTech, France
Christian Leonard	École Polytechnique, France
Nikolaus Hansen	Inria, France
Laurent Younes	John Hopkins University, USA
Sylvain Arguillère	John Hopkins University, USA
Shun-Ichi Amari	RIKEN, Japan
Julien Rabin	ENSICAEN, France
Dena Asta	Carnegie Mellon University, USA
Pierre-Yves Gousenbourger	École Polytechnique de Louvain, Belgium
Nicolas Boumal	Inria and ENS Paris, France
Jun Zhang	University of Michigan, Ann Arbor, USA
Jan Naudts	University of Antwerp, Belgium
Alexis Decurninge	Huawei Technologies, Paris
Roman Belavkin	Middlesex University, UK
Hugo Boscain	École Polytecnique, France
Eric Moulines	Telecom ParisTech, France
Udo Von Toussaint	Max-Planck-Institut fuer Plasmaphysik, Garching, Germany
Jean-Philippe Anker	University of Orléans, France
Charles Bouveyron	University Paris Descartes, France
Michael Blum	IMAG, France
Sylvain Chevallier	IUT of Vélizy, France
Jeremy Bensadon	LRI, France
Philippe Cuvillier	IRCAM, France
Frédéric Barbaresco	Thales, France
Frank Nielsen	École Polytechnique, France and Sony CSL, Japan

Sponsors and Organizer

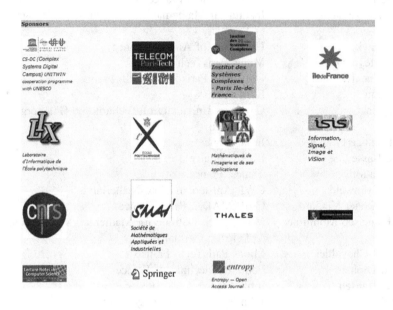

Contents

Shape Space and Diffeomorphic Mappings

Random Geometry/Homology

Hessian Information Geometry

Topological Forms and Information

Information Geometry Optimization

Information Geometry in Image Analysis

Divergence Geometry

Optimization on Manifold

Lie Groups and Geometric Mechanics/Thermodynamics

Geometry of Time Series and Linear Dynamical systems

Bayesian and Information Geometry for Inverse Problems

Probability Density Estimation

Dimension Reduction on Riemannian Manifolds

Evolution Equations with Anisotropic Distributions and Diffusion PCA

Stefan Sommer[✉]

Department of Computer Science, University of Copenhagen, Copenhagen, Denmark
sommer@di.ku.dk

Abstract. This paper presents derivations of evolution equations for the family of paths that in the Diffusion PCA framework are used for approximating data likelihood. The paths that are formally interpreted as most probable paths generalize geodesics in extremizing an energy functional on the space of differentiable curves on a manifold with connection. We discuss how the paths arise as projections of geodesics for a (non bracket-generating) sub-Riemannian metric on the frame bundle. Evolution equations in coordinates for both metric and cometric formulations of the sub-Riemannian geometry are derived. We furthermore show how rank-deficient metrics can be mixed with an underlying Riemannian metric, and we use the construction to show how the evolution equations can be implemented on finite dimensional LDDMM landmark manifolds.

1 Introduction

The diffusion PCA framework [1,2] models data on non-linear manifolds as samples from distributions generated by anisotropic diffusion processes. These processes are mapped from Euclidean space to the manifold by stochastic development in the frame bundle [3]. The construction is connected to a (non bracket-generating) sub-Riemannian metric on the bundle of linear frames of the manifold, the frame bundle.

Velocity vectors and length of geodesics are conventionally used for estimation and statistics in Riemannian manifolds, i.e. for Principal Geodesic Analysis [4] or tangent space statistics [5]. In contrast to this, the anisotropic nature of the distributions considered for Diffusion PCA makes geodesics for the sub-Riemannian metric the natural vehicle for estimation and statistics. These paths were presented in [2] and formally interpreted as most probable paths for the driving diffusion processes that are mapped from $\mathbb{R}^n$ to M by stochastic development.

In this paper, we present derivations of the evolution equations for the paths. We discuss the role of frames as representing either metrics or cometrics and how the sub-Riemannian metric is related to the Sasaki-Mok metric on FM. We then develop a construction that allows the sub-Riemannian metric to be defined as a sum of a rank-deficient generator and an underlying Riemannian metric. Finally, we show how the evolution equations manifest themselves in a specific case, the finite dimensional manifolds arising in the LDDMM landmark matching problem.

© Springer International Publishing Switzerland 2015
F. Nielsen and F. Barbaresco (Eds.): GSI 2015, LNCS 9389, pp. 3–11, 2015.
DOI: 10.1007/978-3-319-25040-3_1

1.1 Diffusion PCA

Diffusion PCA (DPCA, [1,2]) provides a generalization of the Euclidean Principal Component Analysis (PCA) procedure to Riemannian manifolds or, more generally, differentiable manifolds with connection. In contrast to procedures such as Principal Geodesic Analysis (PGA, [4]), Geodesic PCA (GPCA, [6]), and Horizontal Component Analysis (HCA, [7]), DPCA does not employ explicit representations of low-dimensional subspaces. Instead of generalizing the maximum variance/minimum residual formulation of PCA, it is based on a formulation of PCA as a maximum likelihood fit of a Gaussian distribution to data [8,9]. Through the process of stochastic development [3], a class of anisotropic distributions are defined that generalizes normal distributions to the manifold situation. DPCA is thereby a maximum likelihood fit in this family of distributions.

2 Anisotropic Diffusions, Frame Bundles and Development

Development and stochastic development provides an invertible map $\varphi_{(x,X_\alpha)}$ from paths in $\mathbb{R}^n$ starting at the origin to paths on the manifold M starting at a given point $x \in M$. The development map $\varphi_{(x,X_\alpha)}$ is dependent on both the starting point x and a frame X_α for $T_x M$. Through $\varphi_{(x,X_\alpha)}$, diffusion processes in $\mathbb{R}^n$ map to processes on M. This construction is called the Eells-Elworthy-Malliavin construction of Brownian motion [10]. We here outline the process of development and stochastic development and describe its use in Diffusion PCA.

Let (M, ∇) be a differentiable manifold of dimension n with connection ∇. For each point $x \in M$, let $F_x M$ be the set of frames X_α, i.e. ordered bases of $T_x M$. The set $\{F_x M\}_{x \in M}$ can be given a natural differential structure as a fiber bundle on M called the frame bundle FM. It can equivalently be defined as the principal bundle $\mathrm{GL}(\mathbb{R}^n, TM)$. We let the map $\pi_{FM} : FM \to M$ denote the canonical projection. For a differentiable curve x_t in M with $x = x_0$, a frame $X_\alpha = X_{\alpha,0}$ for $T_{x_0} M$ can be parallel transported along x_t thus giving a path $(x_t, X_{\alpha,t})$ in FM. Such paths are called horizontal, and their derivatives form n-dimensional subspaces of the $n + n^2$-dimensional tangent spaces $T_{(x,X_\alpha)} FM$. This horizontal subspace HFM and the vertical subspace VFM of vectors tangent to the fibers $\pi^{-1}(x)$ together split the tangent spaces, i.e. $T_{(x,X_\alpha)} FM = H_{(x,X_\alpha)} FM \oplus V_{(x,X_\alpha)FM}$. The split induces a map $\pi_* : HFM \to TM$ and isomorphisms $\pi_{*,(x,X_\alpha)} : H_{(x,X_\alpha)} FM \to T_x M$ with inverses $\pi^*_{(x,X_\alpha)}$, see Fig. 1. Using $\pi^*_{(x,X_\alpha)}$, horizontal vector fields H_e on FM are defined for vectors $e \in \mathbb{R}^n$ by $H_e(u) = (ue)^*$. In particular, the standard basis $(e_1, \ldots, e_n)$ on $\mathbb{R}^n$ gives n globally defined horizontal vector fields $H_i \in HFM$, $i = 1, \ldots, n$ by $H_i = H_{e_i}$. A *horizontal lift* of x_t is a curve in FM tangent to HFM that projects to x_t. Horizontal lifts are unique up to the choice of initial frame $X_{\alpha,0}$.

Let W_t be an $\mathbb{R}^n$ valued semimartingale. A solution to the stochastic differential equation $dU_t = \sum_{i=1}^{d} H_i(U_t) \circ dW_t^i$ in FM is called a stochastic development of W_t. The solution projects to a stochastic development $X_t = \pi_{FM}(U_t)$ in M.

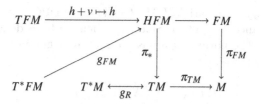

Fig. 1. Commutative diagram for the manifold, frame bundle, the horizontal subspace HFM of TFM, a Riemannian metric g_R and the sub-Riemannian metric g_{FM} defined below. The connection provides the splitting $TFM = HFM \oplus VFM$. The restrictions $\pi_*|_{H_{(x,X_\alpha)}M}$ are invertible maps $H_{(x,X_\alpha)}M \to T_x M$.

We call the process W_t in $\mathbb{R}^n$ that through φ maps to X_t for the driving process of X_t. Since a normal distribution $W \sim \mathcal{N}(0, \Sigma)$ can be obtained as the transition probability of a diffusion process W_t stopped at e.g. $t = 1$, a general class of distributions on the manifold M can be defined by stochastic development of processes W_t resulting in random variables $X = X_1$.

Diffusion PCA uses the map $\int_{\text{Diff}} : FM \to \text{Dens}(M)$ that for each point $(x, X_\alpha) \in FM$ sends a Brownian motion in $\mathbb{R}^n$ to a distribution X_1 by starting a diffusion U_t at (x, X_α) and letting $X_1 = \pi_{FM}(U_1)$ after normalization. The pair (x, X_α) is analogous to the parameters (μ, Σ) for a Euclidean normal distribution: the point $x \in M$ represents the starting point of the diffusion, and X_α represents the square root covariance $\Sigma^{1/2}$. Diffusion PCA fits distributions obtained through $\int_{\text{Diff}}$ by maximum likelihood to observed data, i.e. it optimizes for the most probable parameters (x, X_α) for the anisotropic diffusion process.

3 Evolution Equations

For a Euclidean stationary driftless diffusion process with stochastic generator Σ, the log-probability of a sample path can formally be written

$$\ln \tilde{p}_\Sigma(x_t) \propto - \int_0^1 \|\dot{x}_t\|_\Sigma^2 dt + c_\Sigma \tag{1}$$

with the norm $\| \cdot \|_\Sigma$ given by the inner product $\langle v, w \rangle_\Sigma = \left\langle \Sigma^{-1/2} v, \Sigma^{-1/2} w \right\rangle$. Though only formal as the sample paths are almost surely nowhere differentiable, the interpretation can be given a precise meaning by taking limits of piecewise linear curves [11]. Turning to the manifold situation with the processes mapped to M by stochastic development, the probability of observing a path can either be defined in the manifold by taking limits of small tubes around the curve, or in $\mathbb{R}^n$ trough its anti-development. With the former formulation, a scalar curvature correction term must be added to (1) giving the Onsager-Machlup functional ([12]). The latter formulation corresponds to finding probabilities of paths for the driving process W_t. Taking the maximum of (1) gives geodesics as most probable paths for the driving process when Σ is unitary.

Let now $(x_t, X_{\alpha,t})$ be a path in FM. Recall that in DPCA, $X_{\alpha,t}$ represents the square root covariance $\Sigma^{1/2}$ at x_t. Since $X_{\alpha,t}$ being a basis defines an invertible map $\mathbb{R}^n \to T_{x_t}M$, the norm $\|\cdot\|_\Sigma$ has a direct analogue in the norm $\|\cdot\|_{X_{\alpha,t}}$ defined by the inner product

$$\langle v, w \rangle_{X_{\alpha,t}} = \left\langle X_{\alpha,t}^{-1} v, X_{\alpha,t}^{-1} w \right\rangle_{\mathbb{R}^n} \tag{2}$$

for vectors $v, w \in T_{x_t}M$. The transport of the frame along paths in effect defines a transport of inner product along sample paths: the paths carry with them the inner product defined by the square root covariance $X_{\alpha,0}$ at x_0.

The inner product can equivalently be defined as a metric $g_{X_\alpha} : T_x^*M \to T_xM$. Again using that $X_{\alpha,t}$ can be considered a map $\mathbb{R}^n \to T_{x_t}$, g_{X_α} is defined by $\xi \mapsto X_\alpha((\xi \circ X_\alpha)^\sharp)$ where $\sharp$ is the standard identification $(\mathbb{R}^n)^* \to \mathbb{R}^n$. The sequence of mappings defining g_{X_α} is illustrated below:

$$\begin{array}{ccccccc} T_{x_t}^*M & \to & (\mathbb{R}^n)^* & \to & \mathbb{R}^n & \to & T_{x_t}M \\ \xi & \mapsto & \xi \circ X_\alpha & \mapsto & (\xi \circ X_\alpha)^\sharp & \mapsto & X_\alpha(\xi \circ X_\alpha)^\sharp. \end{array} \tag{3}$$

This definition uses the $\mathbb{R}^n$ inner product in the definition of $\sharp$. Its inverse gives the cometric $g_{X_\alpha}^{-1} : T_{x_t}M \to T_{x_t}^*M$, i.e. $v \mapsto (X_\alpha^{-1}v)^\flat \circ X_\alpha^{-1}$.

Formally, extremal paths for (2) can be interpreted as most probable paths for the driving process W_t when $X_{\alpha,0}$ defines an anisotropic diffusion. Below, we will identify the extremal paths as geodesics for a sub-Riemannian metric, and we use this to find coordinate expressions for evolutions of the paths.

3.1   Sub-Riemannian Metric on the Horizontal Distribution

We now lift the path-dependent metric defined above to a sub-Riemannian metric on HFM. For any $w, \tilde{w} \in H_{(x,X_\alpha)}FM$, the lift of (2) by π_* is the inner product

$$\langle w, \tilde{w} \rangle = \left\langle X_\alpha^{-1}\pi_*w, X_\alpha^{-1}\pi_*\tilde{w} \right\rangle_{\mathbb{R}^n}.$$

The inner product induces a sub-Riemannian metric $g_{FM} : TFM^* \to HFM \subset TFM$ by

$$\langle w, g_{FM}(\xi) \rangle = (\xi | w), \quad \forall w \in H_xFM \tag{4}$$

with $(\xi | w)$ denoting the evaluation $\xi(w)$. The metric g_{FM} gives FM a (non bracket-generating) sub-Riemannian structure [13], see also Fig. 1. It is equivalent to the lift

$$\xi \mapsto \pi_{(x,X_\alpha)}^*(g_{X_\alpha}(\xi \circ \pi_{(x,X_\alpha)}^*)), \quad \xi \in T_{(x,X_\alpha)}FM \tag{5}$$

of the metric g_{X_α} above. The metric is related to the Sasaki-Mok metric on FM [14] that extends the Sasaki metric on TM. The Sasaki-Mok metric allows paths in FM to have derivatives in the vertical space VFM while g_{FM} restricts paths to only have derivatives in HFM. This constraint is nonholonomic thus giving the sub-Riemannian structure.

Following [14], we let (x^i, X^i_α) be coordinates on FM with X^i_α satisfying $X_\alpha = X^i_\alpha \frac{\partial}{\partial x^i}$. The horizontal distribution is then spanned by the n linearly independent vector fields $D_j = \frac{\partial}{\partial x^j} - \Gamma^{h\gamma}_j \frac{\partial}{\partial X^h_\gamma}$ where $\Gamma^{h\gamma}_j = \Gamma^h_{ji} X^i_\gamma$ and Γ^h_{ij} are the Christoffel symbols for the connection ∇. We denote this adapted frame D. The vertical distribution is correspondingly spanned by $D_{j_\beta} = \partial_{X^j_\beta}$, and $D^h = dx^h$, $D^{h\gamma} = \Gamma^{h\gamma}_j dx^j + dX^h_\gamma$ constitutes a dual coframe D^*. The map $\pi_* : HFM \to TM$ is in coordinates $\pi_*(w^j D_j) = w^j \frac{\partial}{\partial x^j}$.

For $(x, X_\alpha) \in FM$, the map $X_\alpha : \mathbb{R}^n \to T_x M$ is in coordinates given by the matrix $[X^i_\alpha]$ so that $X(v) = X^i_\alpha v^\alpha \frac{\partial}{\partial x^i} = X_\alpha v^\alpha$. With $w = w^j D_j$ and $\tilde{w} = \tilde{w}^j D_j$, we have

$$
\begin{aligned}
\langle w, \tilde{w} \rangle &= \langle w^i D_i, \tilde{w}^j D_j \rangle = \left\langle X^{-1} w^i \frac{\partial}{\partial x^i}, X^{-1} \tilde{w}^j \frac{\partial}{\partial x^j} \right\rangle \\
&= \langle w^i X^\alpha_i, \tilde{w}^j X^\alpha_j \rangle_{\mathbb{R}^n} = \delta_{\alpha\beta} w^i X^\alpha_i \tilde{w}^j X^\beta_j = W_{ij} w^i \tilde{w}^j
\end{aligned}
$$

where $[X^\alpha_i]$ is the inverse of $[X^i_\alpha]$ and $W_{ij} = \delta_{\alpha\beta} X^\alpha_i X^\beta_j$. Define now $W^{kl} = \delta^{\alpha\beta} X^k_\alpha X^l_\beta$ so that $W^{ir} W_{rj} = \delta^i_j$ and $W_{ir} W^{rj} = \delta^j_i$. We can then write the sub-Riemannian metric g_{FM} in terms of the adapted frame D,

$$
g_{FM}(\xi_h D^h + \xi_{h_\gamma} D^{h\gamma}) = W^{ih} \xi_h D_i, \tag{6}
$$

because $\langle w, g_{FM}(\xi) \rangle = \langle w, W^{jh} \xi_h D_j \rangle = W_{ij} w^i W^{jh} \xi_h = w^i \xi_i = \xi_h D^h(w^j D_j) = \xi(w)$. The component matrix of the adapted frame D in the coordinates (x^i, X^i_α) is

$$
_{(x, X_\alpha)} L_D = \begin{bmatrix} I & 0 \\ -\Gamma & I \end{bmatrix} \quad \text{and therefore} \quad _D L_{(x, X_\alpha)} = \begin{bmatrix} I & 0 \\ \Gamma & I \end{bmatrix}
$$

with $\Gamma = [\Gamma^{h\gamma}_j]$. Similarly, for the component matrices of the dual frame D^*,

$$
_{(x, X_\alpha)^*} L_{D^*} = \begin{bmatrix} I & \Gamma^T \\ 0 & I \end{bmatrix} \quad \text{and} \quad _{D^*} L_{(x, X_\alpha)^*} = \begin{bmatrix} I & -\Gamma^T \\ 0 & I \end{bmatrix}.
$$

From (6), g_{FM} has D, D^* components

$$
_D g_{FM, D^*} = \begin{bmatrix} W^{-1} & 0 \\ 0 & 0 \end{bmatrix}.
$$

Therefore, g_{FM} has the following components in the coordinates (x^i, X^i_α):

$$
_{(x, X_\alpha)} g_{FM, (x, X_\alpha)^*} = {}_{(x, X_\alpha)} L_D \, {}_D g_{FM, D^*} \, {}_{D^*} L_{(x, X_\alpha)^*} = \begin{bmatrix} W^{-1} & -W^{-1} \Gamma^T \\ -\Gamma W^{-1} & \Gamma W^{-1} \Gamma^T \end{bmatrix}
$$

or $g^{ij}_{FM} = W^{ij}$, $g^{ij_\beta}_{FM} = -W^{ih} \Gamma^{j_\beta}_h$, $g^{i_\alpha j}_{FM} = -\Gamma^{i_\alpha}_h W^{hj}$, and $g^{i_\alpha j_\beta}_{FM} = \Gamma^{i_\alpha}_k W^{kh} \Gamma^{j_\beta}_h$.

3.2 Geodesics for g_{FM}

Geodesics in sub-Riemannian manifolds satisfy the Hamilton-Jacobi equations [13]. Since g_{FM} is a lift of g_{X_α} and geodesics are energy minimizing, the extremal paths for (2) will exactly be geodesics for g_{FM}. In the present case, the Hamiltonian $H(x,\xi) = \frac{1}{2}(\xi|g_{FM}(\xi))$ gives the equations

$$\dot{y}^i = g^{ij}_{FM,y}\xi_j, \quad \dot{\xi}_i = -\frac{1}{2}\frac{\partial}{\partial y^i}g^{pq}_{FM,y}\xi_p\xi_q.$$

We write (x^i, X^i_α) for coordinates on FM as above, and (ξ_i, ξ_{i_α}) for cotangent vectors in T^*FM. This gives

$$\dot{x}^i = g^{ij}\xi_j + g^{ij\beta}\xi_{j_\beta} = W^{ij}\xi_j - W^{ih}\Gamma^{j\beta}_h\xi_{j_\beta}$$

$$\dot{X}^i_\alpha = g^{i\alpha j}\xi_j + g^{i\alpha j\beta}\xi_{j_\beta} = -\Gamma^{i\alpha}_h W^{hj}\xi_j + \Gamma^{i\alpha}_k W^{kh}\Gamma^{j\beta}_h\xi_{j_\beta}$$

$$\dot{\xi}_i = -\frac{1}{2}\left(\frac{\partial}{\partial y^i}g^{hk}_y\xi_h\xi_k + \frac{\partial}{\partial y^i}g^{hk\delta}_y\xi_h\xi_{k_\delta} + \frac{\partial}{\partial y^i}g^{h_\gamma k}_y\xi_{h_\gamma}\xi_k + \frac{\partial}{\partial y^i}g^{h_\gamma k_\delta}_y\xi_{h_\gamma}\xi_{k_\delta}\right)$$

$$\dot{\xi}_{i_\alpha} = -\frac{1}{2}\left(\frac{\partial}{\partial y^{i_\alpha}}g^{hk}_y\xi_h\xi_k + \frac{\partial}{\partial y^{i_\alpha}}g^{hk\delta}_y\xi_h\xi_{k_\delta} + \frac{\partial}{\partial y^{i_\alpha}}g^{h_\gamma k}_y\xi_{h_\gamma}\xi_k + \frac{\partial}{\partial y^{i_\alpha}}g^{h_\gamma k_\delta}_y\xi_{h_\gamma}\xi_{k_\delta}\right)$$

writing $\Gamma^{h_\gamma}_{k,i}$ for $\frac{\partial}{\partial y^i}\Gamma^{h_\gamma}_k$ and where $\frac{\partial}{\partial y^l}g^{ij} = 0$, $\frac{\partial}{\partial y^l}g^{ij\beta} = -W^{ih}\Gamma^{j\beta}_{h,l}$, $\frac{\partial}{\partial y^l}g^{i\alpha j} = -\Gamma^{i\alpha}_{h,l}W^{hj}$, $\frac{\partial}{\partial y^l}g^{i\alpha j\beta} = \Gamma^{i\alpha}_{k,l}W^{kh}\Gamma^{j\beta}_h + \Gamma^{i\alpha}_k W^{kh}\Gamma^{j\beta}_{h,l}$ and $\frac{\partial}{\partial y^{l_\zeta}}g^{ij} = W^{ij}_{,l_\zeta}$, $\frac{\partial}{\partial y^{l_\zeta}}g^{ij\beta} = -W^{ih}_{,l_\zeta}\Gamma^{j\beta}_h - W^{ih}\Gamma^{j\beta}_{h,l_\zeta}$, $\frac{\partial}{\partial y^{l_\zeta}}g^{i\alpha j} = -\Gamma^{i\alpha}_{h,l_\zeta}W^{hj} - \Gamma^{i\alpha}_h W^{hj}_{,l_\zeta}$, $\frac{\partial}{\partial y^{l_\zeta}}g^{i\alpha j\beta} = \Gamma^{i\alpha}_{k,l_\zeta}W^{kh}\Gamma^{j\beta}_h + \Gamma^{i\alpha}_k W^{kh}_{,l_\zeta}\Gamma^{j\beta}_h + \Gamma^{i\alpha}_k W^{kh}\Gamma^{j\beta}_{h,l_\zeta}$ with $\Gamma^{i\alpha}_{h,l_\zeta} = \frac{\partial}{\partial y^{l_\zeta}}(\Gamma^i_{hk}X^k_\alpha) = \delta^{\zeta\alpha}\Gamma^i_{hl}$ and $W^{ij}_{,l_\zeta} = \delta^{il}X^j_\zeta + \delta^{jl}X^i_\zeta$. Combining these expressions, we obtain the flow equations

$$\dot{x}^i = W^{ij}\xi_j - W^{ih}\Gamma^{j\beta}_h\xi_{j_\beta}, \quad \dot{X}^i_\alpha = -\Gamma^{i\alpha}_h W^{hj}\xi_j + \Gamma^{i\alpha}_k W^{kh}\Gamma^{j\beta}_h\xi_{j_\beta}$$

$$\dot{\xi}_i = W^{hl}\Gamma^{k\delta}_{l,i}\xi_h\xi_{k_\delta} - \frac{1}{2}\left(\Gamma^{h_\gamma}_{k,i}W^{kh}\Gamma^{k\delta}_h + \Gamma^{h_\gamma}_k W^{kh}\Gamma^{k\delta}_{h,i}\right)\xi_{h_\gamma}\xi_{k_\delta}$$

$$\dot{\xi}_{i_\alpha} = \Gamma^{h\delta}_{k,i_\alpha}W^{kh}\Gamma^{k\delta}_h\xi_{h_\gamma}\xi_{k_\delta} - \left(W^{hl}_{,i_\alpha}\Gamma^{k\delta}_l + W^{hl}\Gamma^{k\delta}_{l,i_\alpha}\right)\xi_h\xi_{k_\delta}$$

$$\qquad - \frac{1}{2}\left(W^{hk}_{,i_\alpha}\xi_h\xi_k + \Gamma^{h\delta}_k W^{kh}_{,i_\alpha}\Gamma^{k\delta}_h\xi_{h_\gamma}\xi_{k_\delta}\right).$$

4   Cometric Formulation and Low-Rank Generator

We now investigate a cometric $g_{F^dM} + \lambda g_R$ where g_R is Riemannian, g_{F^dM} is a rank d positive semi-definite inner product arising from d linearly independent tangent vectors, and $\lambda > 0$. We assume that g_{F^dM} is chosen so that $g_{F^dM} + \lambda g_R$ is invertible even though g_{F^dM} is rank-deficient. The practical implication of this construction is that a numerical implementation need not transport a full $n \times n$ matrix for the frame but a potentially much lower dimensional $n \times d$ matrix. This situation corresponds to extracting the first d eigenvectors in Euclidean space PCA. When using the frame bundle to model covariances, the sum formulation is more natural for a cometric than a metric because, with the cometric formulation, $g_{F^dM} + \lambda g_R$ represents a sum of covariance matrices instead of a sum of inverse covariance matrices. Thus $g_{F^dM} + \lambda g_R$ can be intuitively thought of as adding isotropic noise of variance λ to the covariance represented by g_{F^dM}.

To pursue this, let F^dM denote the bundle of rank d linear maps $\mathbb{R}^d \to T_xM$. We define a cometric by

$$\langle \xi, \tilde{\xi} \rangle = \delta^{\alpha\beta}(\xi|\pi_*^{-1}X_\alpha)(\tilde{\xi}|\pi_*^{-1}X_\beta) + \lambda\langle \xi, \tilde{\xi}\rangle_{g_R}$$

for $\xi, \tilde{\xi} \in T^*F^dM$. The sum over α, β is for $\alpha, \beta = 1, \ldots, d$. The first term is equivalent to the lift (5) of the cometric $\langle \xi, \tilde{\xi}\rangle = \left(\xi|g_{X_\alpha}(\hat{\xi})\right)$ given $X_\alpha : \mathbb{R}^d \to T_xM$. Note that in the definition (3) of g_{X_α}, the map X_α is not inverted, thus the definition of the metric immediately carries over to the rank-deficient case.

Let (x^i, X_α^i), $\alpha = 1, \ldots, d$ be a coordinate system on F^dM. The vertical distribution is in this case spanned by the nd vector fields $D_{j_\beta} = \partial_{X_\beta^j}$. Except for index sums being over d instead of n terms, the situation is thus similar to the full-rank case. Note that $(\xi|\pi_*^{-1}w) = (\xi|w^j D_j) = w^i\xi_i$. The cometric in coordinates is

$$\langle \xi, \tilde{\xi}\rangle = \delta^{\alpha\beta}X_\alpha^i\xi_i X_\beta^j\tilde{\xi}_j + \lambda g_R^{ij}\xi_i\tilde{\xi}_j = \xi_i\left(\delta^{\alpha\beta}X_\alpha^i X_\beta^j + \lambda g_R^{ij}\right)\tilde{\xi}_j = \xi_i W^{ij}\tilde{\xi}_j$$

with $W^{ij} = \delta^{\alpha\beta}X_\alpha^i X_\beta^j + \lambda g_R^{ij}$. We can then write the corresponding sub-Riemannian metric g_{F^dM} in terms of the adapted frame D

$$g_{F^dM}(\xi_h D^h + \xi_{h_\gamma} D^{h_\gamma}) = W^{ih}\xi_h D_i \tag{7}$$

because $(\xi|g_{F^dM}(\tilde{\xi})) = \left\langle \xi, \tilde{\xi}\right\rangle = \xi_i W^{ij}\tilde{\xi}_j$. That is, the situation is analogous to (6) except the term λg_R^{ij} is added to W^{ij}.

The geodesic system is again given by the Hamilton-Jacobi equations. As in the full-rank case, the system is specified by the derivatives of g_{F^dM}: $\frac{\partial}{\partial y^l}g_{F^dM}^{ij} = W_{,l}^{ij}$, $\frac{\partial}{\partial y^l}g_{F^dM}^{ij_\beta} = -W_{,l}^{ih}\Gamma_h^{j\beta} - W^{ih}\Gamma_{h,l}^{j\beta}$, $\frac{\partial}{\partial y^l}g_{F^dM}^{i_\alpha j} = -\Gamma_h^{i_\alpha}W^{hj} - \Gamma_{h,l}^{i_\alpha}W_{,l}^{hj}$, $\frac{\partial}{\partial y^l}g_{F^dM}^{i_\alpha j_\beta} = \Gamma_{k,l}^{i_\alpha}W^{kh}\Gamma_h^{j_\beta} + \Gamma_k^{i_\alpha}W_{,l}^{kh}\Gamma_h^{j_\beta} + \Gamma_k^{i_\alpha}W^{kh}\Gamma_{h,l}^{j_\beta}$ and $\frac{\partial}{\partial y^{l_\zeta}}g_{F^dM}^{ij} = W_{,l_\zeta}^{ij}$, $\frac{\partial}{\partial y^{l_\zeta}}g_{F^dM}^{ij_\beta} = -W_{,l_\zeta}^{ih}\Gamma_h^{j_\beta} - W^{ih}\Gamma_{h,l_\zeta}^{j_\beta}$, $\frac{\partial}{\partial y^{l_\zeta}}g_{F^dM}^{i_\alpha j} = -\Gamma_h^{i_\alpha}W^{hj} - \Gamma_{h,l_\zeta}^{i_\alpha}W_{,l_\zeta}^{hj}$, $\frac{\partial}{\partial y^{l_\zeta}}g_{F^dM}^{i_\alpha j_\beta} = \Gamma_{k,l_\zeta}^{i_\alpha}W^{kh}\Gamma_h^{j_\beta} + \Gamma_k^{i_\alpha}W_{,l_\zeta}^{kh}\Gamma_h^{j_\beta} + \Gamma_k^{i_\alpha}W^{kh}\Gamma_{h,l_\zeta}^{j_\beta}$ with $\Gamma_{h,l_\zeta}^{i_\alpha} = \frac{\partial}{\partial y^{l_\zeta}}$ $(\Gamma_{hk}^i X_\alpha^k) = \delta^{\zeta\alpha}\Gamma_{hl}^i$, $W_{,l}^{ij} = \lambda g_{R_{,l}^{ij}}$, and $W_{,l_\zeta}^{ij} = \delta^{il}X_\zeta^j + \delta^{jl}X_\zeta^i$. Note that the introduction of the Riemannian metric g_R implies that W^{ij} are now dependent on the manifold coordinates x^i.

5   LDDMM Landmark Equations

The cometric formulation applies immediately to the finite dimensional manifolds that arise when matching N landmarks with the LDDMM framework

[15]. We here use this to provide a concrete example of the flow equations. The LDDMM metric is naturally expressed as a cometric, and, using a rank-deficient inner product g_{F^dM}, we can obtain a reduction of the system of equations to $2(2N + 2Nd)$ compared to $2(2N + (2N)^2)$ when the landmarks are points in $\mathbb{R}^2$. For ease of notation, we consider only the $\mathbb{R}^2$ case here. Please see [2] for illustrations of the generated diffeomorphism flows.

The manifold $M = \{(x_1^1, x_1^2, \ldots, x_m^1, x_m^2) | (x_i^1, x_i^2) \in \mathbb{R}^2\}$ can be represented in coordinates by letting i^1, i^2 denote the first and second indices of the ith landmark. The landmark manifold is in LDDMM given the cometric $g_x(v, w) = \sum_{i,j=1}^m v^i K(x_i, x_j) w^j$ and thus $g_x^{i^k j^l} = K(x_i, x_j)_k^l$. The Christoffel symbols can be written in terms of derivatives of the cometric g^{ij} (recall that $\delta_j^i = g^{ik} g_{kj} = g_{jk} g^{ki}$) [16]

$$\Gamma_{ij}^k = \frac{1}{2} g_{ir} \left(g^{kl} g_{,l}^{rs} - g^{sl} g_{,l}^{rk} - g^{rl} g_{,l}^{ks} \right) g_{sj}. \tag{8}$$

This relation comes from the fact that $g_{jm,k} = -g_{jr} g_{,k}^{rs} g_{sm}$ gives the derivative of the metric. The derivatives of the cometric is simply $g_{,r^q}^{i^k j^l} = (\delta_r^i + \delta_r^j) \partial_{x_r^q} K(x_i, x_j)_k^l$. Using (8), derivatives of the Christoffel symbols can be computed

$$\Gamma_{ij,\xi}^k = \frac{1}{2} g_{ir,\xi} \left(g^{kl} g_{,l}^{rs} - g^{sl} g_{,l}^{rk} - g^{rl} g_{,l}^{ks} \right) g_{sj} + \frac{1}{2} g_{ir} \left(g^{kl} g_{,l}^{rs} - g^{sl} g_{,l}^{rk} - g^{rl} g_{,l}^{ks} \right) g_{sj,\xi}$$

$$+ \frac{1}{2} g_{ir} \left(g_{,\xi}^{kl} g_{,l}^{rs} + g^{kl} g_{,l\xi}^{rs} - g_{,\xi}^{sl} g_{,l}^{rk} - g^{sl} g_{,l\xi}^{rk} - g_{,\xi}^{rl} g_{,l}^{ks} - g^{rl} g_{,l\xi}^{ks} \right) g_{sj}.$$

This provides the full data for numerical integration of the evolution equations on F^dM. An implementation using the above system can be found at http://github.com/stefansommer/dpca.

Acknowledgement. The author wishes to thank Peter W. Michor and Sarang Joshi for suggestions for the geometric interpretation of the sub-Riemannian metric on FM and discussions on diffusion processes on manifolds. This work was supported by the Danish Council for Independent Research and the Erwin Schrödinger Institute in Vienna.

References

1. Sommer, S.: Diffusion Processes and PCA on Manifolds, Mathematisches Forschungsinstitut Oberwolfach (2014)
2. Sommer, S.: Anisotropic distributions on manifolds: template estimation and most probable paths. In: Ourselin, S., Alexander, D.C., Westin, C.-F., Cardoso, M.J. (eds.) IPMI 2015. LNCS, vol. 9123, pp. 193–204. Springer, Heidelberg (2015)
3. Hsu, E.P.: Stochastic Analysis on Manifolds. American Mathematical Soc, Providence (2002)
4. Fletcher, P., Lu, C., Pizer, S., Joshi, S.: Principal geodesic analysis for the study of nonlinear statistics of shape. IEEE Trans. Med. Imaging **23**(8), 995–1005 (2004)
5. Vaillant, M., Miller, M., Younes, L., Trouvé, A.: Statistics on diffeomorphisms via tangent space representations. NeuroImage **23**(Supplement 1), S161–S169 (2004)

6. Huckemann, S., Hotz, T., Munk, A.: Intrinsic shape analysis: geodesic PCA for riemannian manifolds modulo isometric lie group actions. Stat. Sin. **20**(1), 1–100 (2010)
7. Sommer, S.: Horizontal dimensionality reduction and iterated frame bundle development. In: Nielsen, F., Barbaresco, F. (eds.) GSI 2013. LNCS, vol. 8085, pp. 76–83. Springer, Heidelberg (2013)
8. Tipping, M.E., Bishop, C.M.: Probabilistic principal component analysis. J. Roy. Stat. Soc. Ser. B **61**(3), 611–622 (1999)
9. Zhang, M., Fletcher, P.: Probabilistic principal geodesic analysis. In: NIPS, pp. 1178–1186 (2013)
10. Elworthy, D.: Geometric aspects of diffusions on manifolds. In: Hennequin, P.L. (ed.) École d'Été de Probabilités de Saint-Flour XV-XVII, 1985–87. Lecture Notes in Mathematics, vol. 1362, pp. 277–425. Springer, Heidelberg (1988)
11. Andersson, L., Driver, B.K.: Finite dimensional approximations to wiener measure and path integral formulas on manifolds. J. Funct. Anal. **165**(2), 430–498 (1999)
12. Fujita, T., Kotani, S.I.: The onsager-machlup function for diffusion processes. J. Math. Kyoto Univ. **22**(1), 115–130 (1982)
13. Strichartz, R.S.: Sub-riemannian geometry. J. Differ. Geom. **24**(2), 221–263 (1986)
14. Mok, K.P.: On the differential geometry of frame bundles of riemannian manifolds. J. Fur Die Reine Und Angew. Math. **1978**(302), 16–31 (1978)
15. Younes, L.: Shapes and Diffeomorphisms. Springer, Heidelberg (2010)
16. Micheli, M.: The differential geometry of landmark shape manifolds: metrics, geodesics, and curvature. Ph.D. thesis, Brown University, Providence, USA (2008)

Barycentric Subspaces and Affine Spans in Manifolds

Xavier Pennec[✉]

Inria Sophia-Antipolis and Côte d'Azur University (UCA), Sophia Antipolis, France
xavier.pennec@inria.fr

Abstract. This paper addresses the generalization of Principal Component Analysis (PCA) to Riemannian manifolds. Current methods like Principal Geodesic Analysis (PGA) and Geodesic PCA (GPCA) minimize the distance to a "Geodesic subspace". This allows to build sequences of nested subspaces which are consistent with a forward component analysis approach. However, these methods cannot be adapted to a backward analysis and they are not symmetric in the parametrization of the subspaces. We propose in this paper a new and more general type of family of subspaces in manifolds: *barycentric subspaces* are implicitly defined as the locus of points which are weighted means of $k + 1$ reference points. Depending on the generalization of the mean that we use, we obtain the Fréchet/Karcher barycentric subspaces (FBS/KBS) or the affine span (with exponential barycenter). This definition restores the full symmetry between all parameters of the subspaces, contrarily to the geodesic subspaces which intrinsically privilege one point. We show that this definition defines locally a submanifold of dimension k and that it generalizes in some sense geodesic subspaces. Like PGA, barycentric subspaces allow the construction of a forward nested sequence of subspaces which contains the Fréchet mean. However, the definition also allows the construction of backward nested sequence which may not contain the mean. As this definition relies on points and do not explicitly refer to tangent vectors, it can be extended to non Riemannian geodesic spaces. For instance, principal subspaces may naturally span over several strata in stratified spaces, which is not the case with more classical generalizations of PCA.

1 Introduction

For Principal Component Analysis (PCA) in a Euclidean space, one can equivalently define the principal k-dimensional affine subspace using the minimization of the variance of the residuals (the projection of the data point to the subspace) or the maximization of the explained variance within that affine subspace. This is due to the Pythagorean theorem, which does not hold in more general manifolds. A second important observation is that principal components of different orders are nested, which allows to build forward and backward estimation methods by iteratively adding or removing principle components.

Generalizing affine subspaces to manifolds is not so obvious. For the zero-dimensional subspace, intrinsic generalization of the mean on manifolds naturally

© Springer International Publishing Switzerland 2015
F. Nielsen and F. Barbaresco (Eds.): GSI 2015, LNCS 9389, pp. 12–21, 2015.
DOI: 10.1007/978-3-319-25040-3_2

comes into mind: the Fréchet mean is the set of global minima of the variance, as defined by Fréchet in general metric spaces [5]. The set of local minima of the variance was named Karcher mean by W.S Kendall [10] after the work of Karcher et al. on Riemannian centers of mass ([8] see [9] for a discussion of the naming and earlier works).

The one-dimensional component is then quite naturally a geodesic which should passe through the mean point. Higher-order components are more difficult to define. The simplest intrinsic generalization of PCA is tangent PCA (tPCA), which amounts to unfold the whole distribution in the tangent space at the mean using the pullback of the Riemannian exponential map, and to compute the principle components of the covariance matrix in the tangent space. The method is thus based on the maximization of the explained variance. tPCA is often used on manifolds because it is simple and efficient. However, if it is good for analyzing data which are sufficiently centered around a central value (unimodal or Gaussian-like data), it is often not sufficient for multimodal or large support distributions (e.g. uniform on close compact subspaces).

Fletcher et al. proposed in [4] to rely on the least square distance to subspaces which are totally geodesic at one point. These Geodesic Subspaces (GS) are spanned by the geodesics going through one point with tangent vector restricted to a linear subspace of the tangent space. These subspaces are only locally a manifold as they are generally not smooth at the cut locus of the mean point. The procedure was coined Principle Geodesic Analysis (PGA). However, the least-square procedure was computationally expensive, so that the authors implemented in practice a classical tangent PCA. A real implementation of the original PGA procedure was only provided recently by Sommer et al. [16]. PGA is intrinsic and allows to build a sequences of embedded principal geodesic subspaces in a forward component analysis approach by building iteratively the components from dimension 0 (the mean point), dimension 1 (a geodesic), etc. Higher dimensions are obtained iteratively by selecting the direction in the tangent space at the mean that optimally reduce the square distance of data point to the geodesic subspace. However, the mean always belong to geodesic subspaces even when it is not part of the support of the distribution.

Huckemann et al. [14] proposed to start at the first order component by fitting a geodesic to the data, not necessarily through the mean. The second principle geodesic is chosen orthogonally to the first one, and higher order components are added orthogonally at the crossing point to build a geodesic subspace. The method was named Geodesic PCA (GPCA). Sommer [15] proposed a method called horizontal component analysis (HCA) which uses the parallel transport of the 2nd direction along the first principle geodesic to define the second coordinates, and iteratively define higher order coordinates through horizontal development along the previous modes. Other principle decompositions have been proposed, like Principle Graphs [6], extending the idea of k-means.

All the cited methods are intrinsically forward methods that build successively larger approximation spaces for the data. A notable exception is Principle Nested Spheres (PNS), proposed by Jung, et al. [7] as a general framework for non-geodesic decomposition of high-dimensional spheres or high-dimensional planar landmarks shape spaces. Subsphere or radius 0 to 1 are obtained by slicing a higher dimensional sphere by an affine hyperplane. The backward analysis approach, determining a decreasing family of subspace, has been generalized to more general manifold with the help of a nested sequence of relations [3]. However, up to know, such sequences of relationships are only known for spheres, Euclidean spaces or quotient spaces of Lie groups by isometric actions [14].

In this paper, we keep the principle of minimizing the unexplained information. However, we propose to replace Geodesic Subspaces by new and more general types of family of subspaces in manifolds: Barycentric Subspaces (BS). BS are defined as the locus of points which are weighted means of $k + 1$ reference points. Depending on the generalization of the mean that we use on manifolds, Fréchet mean, Karcher mean or exponential barycenter, we obtain the Fréchet/Karcher barycentric subspaces (FBS/KBS) or the affine span. We show that these definition are related and locally define a submanifold of dimension k, and that they generalize in some sense the geodesic subspaces. Like PGA, Barycentric Subspace Analysis (BSA) allows the construction of forward nested subspaces which contains the Fréchet mean. However, it also allows a backward analysis which may not contain the mean. As this definition relies on points and do not explicitly refer to tangent vectors to parametrize geodesics, a very interesting side effect is that it can also be extended to more general geodesic spaces that are not Riemannian. For instance, in stratified spaces, it naturally allows to have principle subspaces that span over several strata. The paper is divided in three parts. We recall in Sect. 2 the background knowledge. Then, we define in Sect. 3 the notions of barycentric subspaces in metric spaces and the affine spans in manifolds. Section 4 finally establishes important properties and relationships between these subspaces.

2   Background Knowledge on Riemannian Manifolds

2.1   Computing in Riemannian Manifolds

We consider an embedding Riemannian manifold $\mathcal{M}$ of dimension n. The Riemannian metric is denoted $\langle\,.\mid.\,\rangle_x$ on each tangent space $T_x\mathcal{M}$ of the manifold. The expression of the the underlying norm in a chart is $\|v\|_x^2 = v^{\mathrm{T}} G(x) v = v^i v^j g_{ij}(x)$ using Einstein notations for tensor contractions. We assume the manifold to be geodesically complete (no boundary nor any singular point that we can reach in a finite time). As an important consequence, the Hopf-Rinow-De Rham theorem states that there always exists at least one minimizing geodesic between any two points of the manifold.

We denote by $\exp_x(v)$ the *exponential map* at point x which associate to each tangent vector $v \in T_x\mathcal{M}$ the point of $\mathcal{M}$ reached by the geodesic starting at x with this tangent vector after a unit time. This map is a local diffeomorphism from $0 \in T_x\mathcal{M}$ to $\mathcal{M}$, and we denote $\overrightarrow{xy} = \log_x(y)$ its inverse: it may be defined as the smallest vector of $T_x\mathcal{M}$ that allows to shoot a geodesic from x to y. A geodesic $\exp_x(tv)$ is minimizing up to a certain cut time t_0 and not anymore after. When t_0 is finite, $t_0 v$ is called a tangential cut-point and $\exp_x(t_0 v)$ a cut point. The domain of injectivity $D(x) \in T_x\mathcal{M}$ of the exponential map can be maximally extended up to the tangential cut-locus $\partial D(x) = C(x)$. It covers all the manifold $\mathcal{M}$ except the *cut locus* $C(x) = \exp_x(C(x))$ which has null measure for the Riemannian measure.

When the tangent space is provided with an orthonormal basis, the Riemannian exponential and logarithmic maps provide a *normal coordinate systems at x*. A set of normal coordinate systems at each point of the manifold realize an atlas which allows to work very easily on the manifold. The implementation of exp and log maps is the basis of programming on Riemannian manifolds, and most the geometric operations needed for statistics or image processing can be rephrased based on them [12,13].

2.2 Taylor Expansions in Normal Coordinate Systems

We consider a normal coordinate system centered at x and $x_v = \exp_x(v)$ a variation of the point x. We denote by $R^i_{jkl}(x)$ the coefficients of the Riemannian curvature tensor at x and by ϵ a conformal gauge scale that encodes the size of the path (in terms of $\|v\|_x$ and $\|\overrightarrow{xy}\|_x$) normalized by the curvature. Following [2], the Taylor expansion of the metric is $g^a_b(v) = \delta^a_b - \frac{1}{3}R^a_{cbd}v^c v^d - \frac{1}{6}\nabla_e R^a_{cbd}v^e v^c v^d + O(\epsilon^4)$, and a geodesic joining x_v to y has initial tangent vector:

$$\left[\log_{x_v}(y)\right]^a = \overrightarrow{xy}^a - v^a + \frac{1}{3}R^a_{cbd}v^b\overrightarrow{xy}^c\overrightarrow{xy}^d + \frac{1}{12}\nabla_c R^a_{dbe}v^b\overrightarrow{xy}^c\overrightarrow{xy}^d\overrightarrow{xy}^e + O(\epsilon^4).$$

Combining these two expansions, we get the expansion of the Riemannian distance: $d^2_{xy}(v) = \mathrm{dist}^2(\exp_x(v), y) = \|\overrightarrow{xy}\|^2_x + (\nabla d^2_{xy})^T v + \frac{1}{2}v^T \nabla^2 d^2_{xy}v + O(\epsilon^3)$, where the gradient $\nabla d^2_{xy} = -2\overrightarrow{xy}$ is -2 times the log and the Hessian is the opposite of the differential of the log:

$$(\nabla^2 d^2_{xy})^a_b = -\left[D_x \log_x(y)\right]^a_b = \delta^a_b - \frac{1}{3}\overrightarrow{xy}^c\overrightarrow{xy}^d R^a_{cbd} - \frac{1}{12}\overrightarrow{xy}^c\overrightarrow{xy}^d\overrightarrow{xy}^e \nabla_c R^a_{dbe} + O(\epsilon^3).$$

2.3 Moments of Point Distributions

Let $\mu(x) = \sum_i \lambda_i \delta_{x_i}(x)$ be a singular distribution of $k+1$ points on $\mathcal{M}$ with weights $(\lambda_0, \dots \lambda_k)$ that do not sum up to zero. To define the moments of that distribution, we have to take care that the Riemannian log and distance functions are not smooth at the cut-locus of the points $\{x_i\}$.

Definition 1 ($(k+1)$-Pointed Riemannian Manifold).
Let $\{x_0, \dots x_k\} \in \mathcal{M}^{k+1}$ be a set of $k+1$ distinct points in the Riemannian manifold $\mathcal{M}$ and $C(x_0, \dots x_k) = \cup_{i=0}^k C(x_i)$ be the union of the cut loci of these points. We call $(k+1)$-pointed manifold $\mathcal{M}^*(x_0, \dots x_k) = \mathcal{M}/C(x_0, \dots x_k)$ the submanifold of the non-cut points of the points.

Since the cut locus of each point is closed and has null measure, $\mathcal{M}^*(x_0, \dots x_k)$ is open and dense in $\mathcal{M}$. Thus, it is a submanifold of $\mathcal{M}$ (not necessarily connected). On this submanifold $\mathcal{M}^*(x_0, \dots x_k)$, the distance to the points x_i and the Riemannian log function $\overrightarrow{xx_i} = \log_x(x_i)$ are smooth.

Definition 2 (Weighted Moments of a $(k+1)$-Pointed Manifold).
Let $(\lambda_0, \ldots \lambda_k) \in \mathbb{R}^{k+1}$ such that $\sum_i \lambda_i \neq 0$. The weighted n-order moment of a $(k+1)$-pointed Riemannian manifold $\mathcal{M}^*(x_0, \ldots x_k)$ is the smooth $(n, 0)$ tensor:

$$\mathfrak{M}_n(x, \lambda) = \sum_i \lambda_i \underbrace{\overrightarrow{xx_i} \otimes \overrightarrow{xx_i} \ldots \otimes \overrightarrow{xx_i}}_{n \; times} \tag{1}$$

The 0-th order moment (the mass) $\mathfrak{M}_0(\lambda) = \sum_i \lambda_i = \mathbb{1}^T \lambda$ is constant. All other moment are homogeneous of degree 1 in λ and can be normalized by dividing by the mass $\mathfrak{M}_0(\lambda)$. The first order moment $\mathfrak{M}_1(x, \lambda) = \sum_i \lambda_i \overrightarrow{xx_i}$ is a smooth vector field on the manifold $\mathcal{M}^*(x_0, \ldots x_k)$. The second and higher order moments are smooth $(n, 0)$ tensor fields that will be used through their contraction with the Riemannian curvature tensor.

3   Barycentric Subspaces

In a Euclidean space, an affine subspace of dimension k is generated by a point and k non-collinear vectors: $\mathrm{Aff}(x_0, v_1 \ldots v_k) = \left\{ x = x_0 + \sum_{i=1}^k \lambda_i v_i, \; \lambda \in \mathbb{R}^k \right\}$. Alternatively, one could also generate the affine span of $k+1$ points in general linear position using the implicit equation $\sum_i \lambda_i (x_i - x) = 0$ where $\sum_{i=0}^k \lambda_i = 1$. The two definitions are equivalent when $x_i = x_0 + v_i$. The last parametrization of k-dimensional affine submanifolds is relying on barycentric coordinates which live in the projective space $\mathcal{P}_k$ minus the orthogonal of the line element $\mathbb{1} = (1 : 1 : \ldots 1)$:

$$\mathcal{P}_k^* = \left\{ (\lambda_0 : \ldots : \lambda_k) \in \mathbb{R}^{k+1} \text{ s.t. } \sum_i \lambda_i \neq 0 \right\}.$$

Standard charts of this space are given either by the intersection of the line elements with the "upper" unit sphere S_k of $\mathbb{R}^{k+1}$ with north pole $\mathbb{1}/\sqrt{k}$ (unit weights) or by the k-plane of $\mathbb{R}^{k+1}$ passing through the point $\mathbb{1}/k$ and orthogonal to this vector. We call normalized weights $\underline{\lambda}_i = \lambda_i / (\sum_{j=0}^k \lambda_j)$ this last projection.

3.1   Fréchet and Karcher Barycentric Subspaces in a Metric Space

The two above definitions of the affine span turn out to have different generalizations in manifolds: the first definition leads to geodesic subspaces, as defined in PGA and GPCA [4,14,16], while the second definition using the affine span suggests a generalization to manifolds either using the Fréchet/Karcher weighted mean or using an exponential barycenter.

Definition 3 (Fréchet/Karcher Barycentric Subspaces of $k+1$ Points).
Let $(\mathcal{M}, \mathrm{dist})$ be a metric space and $(x_0, \ldots x_k) \in \mathcal{M}^k$ be $k+1$ distinct reference points. The (normalized) weighted variance at point x with weight $\lambda \in \mathcal{P}_k^*$ is: $\sigma^2(x, \lambda) = \frac{1}{2} \sum_{i=0}^k \underline{\lambda}_i \, \mathrm{dist}^2(x, x_i) = \frac{1}{2} \sum_{i=0}^k \lambda_i \, \mathrm{dist}^2(x, x_i)/(\sum_{j=0}^k \lambda_j)$. The Fréchet barycentric subspace is the locus of weighted Fréchet means of

these points, i.e. the set of absolute minima of the weighted variance: $FBS(x_0, \ldots x_k) = \{\arg\min_{x \in \mathcal{M}} \sigma^2(x, \lambda), \ \lambda \in \mathcal{P}_k^*\}$. *The Karcher barycentric subspace* $KBS(x_0, \ldots x_k)$ *is defined similarly with local minima instead of global ones.*

This definition restores the full symmetry of all the parameters defining the subspaces, contrarily to the geodesic subspaces which privilege one point. Here, we defined the notion on general metric spaces to show that it works in spaces more general than smooth Riemannian manifolds. In a stratified space for instance, the barycentric subspace spanned by points belonging to different strata naturally maps over all these strata. This is a significant improvement over geodesic subspaces used in PGA which can only be defined within a regular strata.

3.2 Affine Spans as Exponential Barycentric Subspaces

A second way to generalize the affine span to manifolds is to see directly the implicit barycentric coordinates equation as a weighted exponential barycenter:

Definition 4 (Affine Span of a $(k+1)$-Pointed Riemannian Manifold).
A point $x \in \mathcal{M}^*(x_0, \ldots x_k)$ *has barycentric coordinates* $\lambda \in \mathcal{P}_k^*$ *if*

$$\mathfrak{M}_1(x, \lambda) = \sum_{i=0}^{k} \lambda_i \overrightarrow{xx_i} = 0. \tag{2}$$

The affine span of the points $(x_0, \ldots x_k) \in \mathcal{M}^k$ *is the set of weighted exponential barycenters of the reference points in* $\mathcal{M}^*(x_0, \ldots x_k)$:

$$\mathrm{Aff}(x_0, \ldots x_k) = \{x \in \mathcal{M}^*(x_0, \ldots x_k) | \exists \lambda \in \mathcal{P}_k^* : \mathfrak{M}_1(x, \lambda) = 0\}.$$

This definition is only valid on $\mathcal{M}^*(x_0, \ldots x_k)$ and may hide some discontinuities of the affine span on the union of the cut locus of the reference points. Outside this null measure set, one recognizes that Eq. (2) defines nothing else than the critical points of the variance $\sigma^2(x, \lambda) = \frac{1}{2} \sum_i \lambda_i \, \mathrm{dist}^2(x, x_i)$. The affine span is thus a superset of the barycentric subspaces in $\mathcal{M}^*(x_0, \ldots x_k)$. However, we notice that the variance may also have local minima on the cut-locus of the reference points.

Let us consider field of $n \times (k+1)$ matrices $Z(x) = [\overrightarrow{xx_0}, \ldots \overrightarrow{xx_k}]$. We can rewrite Eq. (2) in matrix form: $\mathfrak{M}_1(x, \lambda) = Z(x)\lambda = 0$. Thus, we see that the affine span is controlled by the kernel of the matrix field $Z(x)$:

Theorem 1 (SVD Characterization of the Affine Span).
Let $Z(x) = U(x).S(x).V(x)^{\mathrm{T}}$ *be a singular decomposition of the matrix fields* $Z(x) = [\overrightarrow{xx_0}, \ldots \overrightarrow{xx_k}]$ *on* $\mathcal{M}^*(x_0, \ldots x_k)$ *(with singular values sorted in decreasing order). The barycentric subspace* $\mathrm{Aff}(x_0, \ldots x_k)$ *is the zero level-set of the* $k+1$ *singular value* $s_{k+1}(x)$ *and the subspace of valid barycentric weights is spanned by the right singular vectors corresponding to the* l *vanishing singular values:* $\mathrm{Span}(v_{k-l}, \ldots v_k)$ *(it is void if* $l = 0$).

4   Properties of Barycentric Subspaces in Manifolds

In this section, we restrict the analysis to $\mathcal{M}^*(x_0, \ldots x_k)$ so that all quantities are smooth.

4.1   Karcher Barycentric Subspaces and Affine Span

In $\mathcal{M}^*(x_0, \ldots x_k)$, the critical points of the weighted variance are the points of the affine span. Among these points, the local minima may be characterized by the Hessian $H(x, \lambda) = -\sum_i \lambda_i D_x \log_x(x_i)$ of the weighted variance. Using the Taylor expansion of the differential of the log of Sect. 2.2), we obtain:

$$[H(x,\lambda)]_b^a = \delta_b^a - \tfrac{1}{3} R_{cbd}^a [\mathfrak{M}_2(x,\underline{\lambda})]^{cd} - \tfrac{1}{12} \nabla_c R_{dbe}^a [\mathfrak{M}_3(x,\underline{\lambda})]^{cde} + O(\epsilon^4), \quad (3)$$

The key factor is the contraction of the curvature with the dispersion of the reference points: when the typically distance from x to all the reference points x_i is smaller than the inverse of the curvature, then $H(x, \lambda)$ is essentially close to the identity. In the limit of null curvature, (e.g. for a Euclidean space), $H(x, \lambda)$ is simply the unit matrix. In general Riemannian manifolds, Eq. (3) only gives a qualitative behavior. In order to obtain hard bounds on the spectrum of $H(x, \lambda)$, one has to investigate bounds on Jacobi fields, as is done for the proof of uniqueness of the Karcher and Fréchet means [1,8,10,11,17]. Thanks to these proofs, we can in fact establish that the Karcher barycentric submanifold is locally well defined around the Karcher mean.

When the Hessian is degenerated, we cannot conclude on the local minimality without going to higher order differentials. This leads us to stratify the affine span by the index of the Hessian of the weighted variance.

Definition 5 (Regular and Positive Points of $\mathcal{M}^*(x_0, \ldots x_k)$).
A point $x \in \mathcal{M}^*(x_0, \ldots x_k)$ is said regular (resp. positive) if the Hessian matrix $H(x, \lambda)$ is invertible (resp. positive definite) for all λ in the right singular space of the smallest singular value of $Z(x)$. The set of regular (resp. positive) points is denoted $Reg(\mathcal{M}^*(x_0, \ldots x_k))$ (resp. $Reg^+(\mathcal{M}^*(x_0, \ldots x_k))$). The set of positive points of the affine span is called the positive span $\mathrm{Aff}^+(x_0, \ldots x_k)$.

Positive points are obviously regular, and in Euclidean spaces all the points are positive and regular. However, in Riemannian manifolds, we may have non-regular points and regular points which are non-positive.

Theorem 2 (Karcher Barycentric Subspace and Positive Span).
The positive span $\mathrm{Aff}^+(x_0, \ldots x_k)$ is the set of regular points of the Karcher barycentric subspace $KBS(x_0, \ldots x_k)$ on $\mathcal{M}^*(x_0, \ldots x_k)$.

One generalization of the Fréchet (resp. Karcher) mean is the use of the power α of the metric instead of the square. For instance, one defines the median ($\alpha = 1$) and the modes ($\alpha \to 0$) as the minima of the α-variance $\sigma^\alpha(x) = \tfrac{1}{\alpha} \sum_{i=0}^k \mathrm{dist}^\alpha(x, x_i)$. Following this idea, one could think of generalizing barycentric subspaces to the α-Fréchet (resp. α-Karcher) barycentric subspaces.

In fact, it turns out that the critical points of the α-variance are just elements of the affine span with weights $\lambda'_i = \lambda_i \, \text{dist}^{\alpha-2}(x, x_i)$. Thus, changing the power of the metric just amounts to reparametrizing the barycentric weights, which shows the notion of affine span is really central.

4.2 Dimension of the Barycentric Subspace

We can locally parametrize the affine span thanks to a Taylor expansion of the constraint $Z(x)\lambda = 0$: a change of coordinates $\delta\lambda$ induces a change of position δx verifying $H(x, \lambda)\delta x + Z(x)\delta\lambda = 0$. At the positive points, the Hessian is invertible and the SVD characterization leads us to conclude that:

Theorem 3 (Dimension of the Barycentric Subspaces at Regular Points).
The positive span $\text{Aff}^+(x_0, \ldots, x_k)$ (i.e. the regular KBS), is a stratified space of dimension k on $\text{Reg}(\mathcal{M}^*(x_0, \ldots x_k))$. On the m-dimensional strata, $Z(x)$ has exactly $k - m + 1$ vanishing singular values.

4.3 Geodesic Subspaces as Limit of Barycentric Subspaces

By analogy with Euclidean spaces, one would expects the affine span to be close to the geodesic subspace

$$GS(x, w_1, \ldots w_k) = \left\{ \exp_x \left(\sum_{i=1}^{k} \alpha_i w_i \right) \in \mathcal{M} \text{ for } \alpha \in \mathbb{R}^k \right\}$$

generated by the k independent vectors $w_1, \ldots w_k$ at x when all the points $\{x_i = \exp_{x_0}(\epsilon w_i)\}_{1 \le i \le k}$ are converging to x_0 at first order.

In order to investigate that, we first need to restrict the definition of the geodesic subspaces. Indeed, although the above classical definition is implicitly used in most of the works using PGA, it may not define a k-dimensional submanifold when there is a cut-locus. For instance, it is well known that geodesics of a flat square torus are either periodic or everywhere dense in a flat torus submanifold depending on whether the components of the initial velocity field have rational or irrational ratios. Thus, it makes sense to restrict to the part of the GS which is limited by the cut-locus.

Definition 6 (Restricted Geodesic Submanifolds).
Let $x \in \mathcal{M}$ be a point of a Riemannian manifold and $W_x = \{\sum_{i=1}^{k} \alpha_i w_i, \alpha \in \mathbb{R}^k\}$ the k-dimensional linear subspace of $T_x\mathcal{M}$ generated a k-uplet $\{w_i\}_{1 \le i \le k} \in (T_x\mathcal{M})^k$ of tangent vectors at x. Recall that $D(x) \subset T_x\mathcal{M}$ is the maximal definition domain on which the exponential map is diffeomorphic.

We call restricted geodesic submanifold $GS^*(W_x)$ at x generated by the vector subspace W_x the submanifold of $\mathcal{M}$ generated by the geodesics starting at x with tangent vectors $w \in W_x$, but up to the first cut-point of x only:

$$GS^*(W_x) = GS^*(x, w_1, \ldots w_k) = \{\exp_x(w), w \in W_x \cap D(x)\}$$

This restricted definition correctly defines a k-dimensional submanifold of $\mathcal{M}$, whose completion may be a manifold with boundary.

Let $x = \exp_{x_0}(w) \in GS^*(W_x)$. Thanks to the symmetry of geodesics, we can show that this point is solution of the barycentric equation $\sum_{i=0}^k \lambda_i \log_x(x_i) = O(\eta^2)$ with non-normalized homogeneous coordinates $\lambda_i = \alpha_i$ for $1 \leq i \leq k$ and $\lambda_0 = \eta - (\sum_i \alpha_i)$. These coordinates obviously sum up to zero when η goes to zero, which is a point at infinity in $\mathcal{P}_k^*$. In that sense, points of the restricted geodesic submanifold $GS^*(W)$ are points at infinity of the affine span $\mathrm{Aff}(x, x_1, \ldots x_k)$ when the points $x_i = \exp_x(\eta w_i)$ are converging to x at first order along the tangent vectors w_i.

Theorem 4 (Restricted GS as Limit Case of the Affine Span).
Points of the restricted geodesic submanifold $GS^*(W_x) = \{\exp_x(w), w \in W_x \cap D(x)\}$ are points at infinity in $\mathcal{P}_k^*$ of the affine span $\mathrm{Aff}(x, x_1, \ldots x_k)$ when the points $x_i = \exp_x(\eta w_i)$ are converging to x at first order along the tangent vectors w_i defining the k-dimensional subspace $W_x \subset T_x\mathcal{M}$.

5 Perspectives

We proposed in this paper three generalization of the affine span of $k+1$ points in a manifold. These barycentric subspaces are implicitly defined as the locus of points which are weighted (Fréchet/Karcher/exponential barycenter) means of $k+1$ reference points. In generic conditions, barycentric subspaces are stratified spaces that are locally submanifolds of dimension k. Their singular set of dimension $k - l$ corresponds to the case where l of the reference point belongs to the barycentric subspace defined by the $k - l$ other reference points.

In non-generic conditions, points may coalesce along certain directions, defining non local jets instead of a regular k-tuple. Geodesic subspaces, which are defined by $k - 1$ tangent vectors at a point, do correspond (in some restricted sense) to the limit of the affine span when the k-tuple converges towards that jet. We conjecture that this can be generalized to higher order derivatives using techniques from sub-Riemannian geometry. This way, some non-geodesic decomposition schemes such as loxodromes, splines and principle nested spheres could also be seen as limit cases of barycentric subspaces.

Investigating simple manifolds like spheres and symmetric spaces will provide useful guidelines in that direction. For instance, the closure of the barycentric subspace of $k + 1$ different reference points on the n-dimensional sphere is the k-dimensional great subsphere that contains the reference points. It is noticeable that the closure of the affine span generated by any $k + 1$-tuple of points of a great k-dimensional subsphere generate the same space, which is also a geodesic subspace. This coincidence of spaces is due to the very high symmetry of the sphere. For second order jets, we conjecture that we obtain subspheres of different radii as used in principle nested spheres (PNS) analysis.

Barycentric subspaces can be naturally nested, by defining an ordering of the reference points, which makes is suitable for a generalization of Principal Component Analysis (PCA) to Riemannian manifolds. Several problems however remain to be investigated to use Barycentric Subspace Analysis (BSA) in practice. First, the optimization on k-tuple might have multiple solutions, as in the case of spheres. Here, we need to find a suitable quotient space similar to the quotient definition of Grassmanians. Second, the optimization might converge towards a non-local jet instead on a k-tuple, and good renormalization techniques need to be designed to guaranty the numerical stability. Third, one theoretically needs to define a proper criterion to be optimized by all k-tuple for $k = 0 \dots n$ together and not just a greedy approach as done by the classical forward and backward approaches.

References

1. Afsari, B.: Riemannian l^p center of mass: existence, uniqueness, and convexity. Proc. AMS **180**(2), 655–673 (2010)
2. Brewin, L.: Riemann normal coordinate expansions using cadabra. Class. Quantum Gravity **26**(17), 175017 (2009)
3. Damon, J., Marron, J.S.: Backwards principal component analysis and principal nested relations. J. Math. Imaging Vis. **50**(1–2), 107–114 (2013)
4. Fletcher, P., Lu, C., Pizer, S., Joshi, S.: Principal geodesic analysis for the study of nonlinear statistics of shape. IEEE Trans. Med. Imaging **23**(8), 995–1005 (2004)
5. Fréchet, M.: Les éléments aléatoires de nature quelconque dans un espace distancié. Annales de l'Institut Henri Poincaré **10**, 215–310 (1948)
6. Gorban, A.N., Zinovyev, A.Y.: Principal graphs and manifolds. In: Handbook of Research on Machine Learning Applications and Trends: Algorithms, Methods and Techniques, Chap. 2, pp. 28–59 (2009)
7. Jung, S., Dryden, I.L., Marron, J.S.: Analysis of principal nested spheres. Biometrika **99**(3), 551–568 (2012)
8. Karcher, H.: Riemannian center of mass and mollifier smoothing. Commun. Pure Appl. Math. **30**, 509–541 (1977)
9. Karcher, H.: Riemannian Center of Mass and so called Karcher mean, July 2014. arXiv:1407.2087 [math]
10. Kendall, W.: Probability, convexity, and harmonic maps with small image I: uniqueness and fine existence. Proc. Lond. Math. Soc. **61**(2), 371–406 (1990)
11. Le, H.: Estimation of Riemannian barycenters. LMS J. Comput. Math **7**, 193–200 (2004)
12. Pennec, X.: Intrinsic statistics on Riemannian manifolds: basic tools for geometric measurements. J. Math. Imaging Vis. **25**(1), 127–154 (2006). A preliminary appeared as INRIA RR-5093, January 2004
13. Pennec, X., Fillard, P., Ayache, N.: A Riemannian framework for tensor computing. Int. J. Comput. Vis. **66**(1), 41–66 (2006). A preliminary version appeared as INRIA Research. Report **5255**, (July 2004)
14. Huckemann, A.M.S., Hotz, T.: Intrinsic shape analysis: geodesic principal component analysis for Riemannian manifolds modulo Lie group actions. Statistica Sin. **20**, 1–100 (2010)
15. Sommer, S.: Horizontal dimensionality reduction and iterated frame bundle development. In: Nielsen, F., Barbaresco, F. (eds.) GSI 2013. LNCS, vol. 8085, pp. 76–83. Springer, Heidelberg (2013)
16. Sommer, S., Lauze, F., Nielsen, M.: Optimization over geodesics for exact principal geodesic analysis. Adv. Comput. Math. **40**(2), 283–313 (2013)
17. Yang, L.: Medians of probability measures in Riemannian manifolds and applications to radar target detection. Ph.D. thesis, Poitier University, December 2011

Dimension Reduction on Polyspheres
with Application to Skeletal Representations

Benjamin Eltzner[1]([⊠]), Sungkyu Jung[2], and Stephan Huckemann[1]

[1] Institute for Mathematical Stochastics, University of Göttingen,
Göttingen, Germany
beltzne@uni-goettingen.de
[2] Department of Statistics, University of Pittsburgh, Pittsburgh, USA

Abstract. We present a novel method that adaptively deforms a poly-sphere (a product of spheres) into a single high dimensional sphere which then allows for principal nested spheres (PNS) analysis. Applying our method to skeletal representations of simulated bodies as well as of data from real human hippocampi yields promising results in view of dimension reduction. Specifically in comparison to composite PNS (CPNS), our method of principal nested deformed spheres (PNDS) captures essential modes of variation by lower dimensional representations.

1 Introduction

In data analysis, it is one of the big challenges to discover major modes of variation. For data in a Euclidean space this can be done by principal component analysis (PCA) where the modes are determined by an eigendecomposition of the covariance matrix. Notably, this is equivalent to determining a sequence of nested affine subspaces minimizing residual variance. Inspired by the eigendecomposition, [6,7] proposed PCA in the tangent space of a suitably defined mean, the notion of covariance has been generalized by [3] cf. also [2], and inspired by minimizing residual variances, [9] proposed to find a sequence of orthogonal best approximating geodesics. Taking into account parallel transport, [14] proposed to build a nested sequence of subspaces spanned by geodesics. These methods apply to general manifolds and to some extent also to stratified spaces, e.g. to shape spaces due to isometric (not necessarily free) actions of Lie-groups on manifolds (cf. [8]). For spherical data, it is possible to almost entirely mimic the second characterization of PCA by backward principal nested sphere (PNS) analysis, proposed by [10]. Here in every step, a codimension one small hypersphere is determined, best approximating the data orthogonally projected to the previous small hypersphere. This method hinges on the very geometry of the sphere and cannot be easily generalized to other spaces. For data on polyspheres (products of spheres), which naturally occur in skeletal representations for modeling and analysis of body organs, in composite PNS (CPNS) by [11], PNS is performed in every factor.

In order to make PNS more directly available for polyspheres, in this communication we propose *principal nested deformed spheres* (PNDS) where we first

© Springer International Publishing Switzerland 2015
F. Nielsen and F. Barbaresco (Eds.): GSI 2015, LNCS 9389, pp. 22–29, 2015.
DOI: 10.1007/978-3-319-25040-3_3

deform a polysphere into a single high dimensional sphere, in a data adaptive way, and then perform PNS on this sphere. In Sect. 2 we describe the proposed deformation of the data space and in Sect. 3 we apply the method to simulated and real data from skeletal representations and compare results to those of CPNS. A thorough introduction and analysis of the method is deferred to a future publication in preparation.

2   Polysphere Deformation

We assume in the following that the data space is a polysphere $Q = S_{r_i}^{d_i} \times \ldots \times S_{r_I}^{d_I}$ and that on each individual sphere of dimension $d_i \in \mathbb{N}$ and radius $r_i > 0$ the data are confined to a half sphere, $1 \leq i \leq I$. Notably, then [1] guarantees the existence of a unique spherical mean $\mu_i \in S_{r_i}^{d_i}$ of the data on each individual sphere. In the following we will deform Q stepwise to a single higher-dimensional sphere S^D, $D = d_1 + \ldots + d_I$ where the mapping $P : Q \longrightarrow S^D$ is data-adaptive, i.e. P is as faithful as possible in terms of data variation.

2.1   The Construction for Unit Spheres

For equal radii, the explicit mapping is given below in (3), for varying radii the modification is found further down in (4) and (5).

For the following motivation, we use polar coordinates

$$\forall 1 \leq k \leq d : \quad x_k = \left(\prod_{j=1}^{k-1} \sin \phi_j \right) \cos \phi_k, \quad x_{d+1} = \left(\prod_{j=1}^{d} \sin \phi_j \right) \tag{1}$$

for the embedding $S^d \subset \mathbb{R}^{d+1}$ of the d-dimensional unit sphere. We will formulate the construction of $P = P_I$ recursively, first for two unit spheres,

$$P_1 : S^{d_2} \times S^{d_1} \longrightarrow S^{d_2+d_1}, \quad P_2 : S^{d_3} \times S^{d_2+d_1} \longrightarrow S^{d_3+d_2+d_1}, \quad \ldots$$

where we embed $S^{d_2} \times S^{d_1}$ into $\mathbb{R}^{d_2+d_1+2}$ denoting coordinates as $x_{1,1}, \ldots, x_{1,d_1+1}$, $x_{2,1}, \ldots, x_{2,d_2+1}$. Then the squared line elements of the two spheres are given by

$$ds_1^2 = \sum_{k=1}^{d_1} \left(\prod_{j=1}^{k-1} \sin^2 \phi_{1,j} \right) d\phi_{1,k}^2, \quad ds_2^2 = \sum_{k=1}^{d_2} \left(\prod_{j=1}^{k-1} \sin^2 \phi_{2,j} \right) d\phi_{2,k}^2$$

and the polysphere's squared line element is given by $ds^2 = ds_2^2 + ds_1^2$. The line element of the sphere, i. e. the image of P_1 is then formally defined as

$$ds^2 = ds_2^2 + \left(\prod_{j=1}^{d_2} \sin^2 \phi_{2,j} \right) ds_1^2 \tag{2}$$

which can easily be checked to be a squared line element of a sphere of dimension $D = d_1 + d_2$.

In the next step we give a data-driven choice of coordinates and ordering of the two unit spheres. The transformation of the line element in Eq. (2) amounts to multiplying ds_1^2 by $x_{2,d_2+1}^2 = 1 - x_{2,1}^2 - \ldots - x_{2,d_2}^2$. This yields

$$\left\{ x \in \mathbb{R}^{d_2+d_1+2} \;\middle|\; \sum_{k=1}^{d_2+1} x_{2,k}^2 = 1 = \sum_{k=1}^{d_2} x_{2,k}^2 + \sum_{k=1}^{d_1+1} (x_{2,d_2+1}x_{1,k})^2 \right\}$$

Since we assumed that the data projections to each individual sphere are contained in half-spheres, we may choose coordinates for each individual sphere such that $x_{i,d_i+1} > 0$ for the projections of all data points to the i-th sphere $(2 \leq i \leq I)$. Often the projections are confined to a half sphere centered at the spherical mean μ_i on the i-the sphere. Then the positive x_{i,d_i+1}-unit direction can be chosen as μ_i. As the other coordinates are equally deformed, their choice is arbitrary. Thus, the coordinates of the $S^{d_2+d_1}$ are given by

$$\forall 1 \leq k \leq d_2 : \quad y_j = x_{2,k}, \quad \forall 1 \leq k \leq d_1 + 1 : \quad y_{d_2+k} = x_{2,d_2+1}x_{1,k} \qquad (3)$$

from which angular coordinates can be calculated by inverting the relation (1). Using $x_{2,d_2+1} > 0$ for all data, the data space is thus

$$S^{d_2+d_1} = \left\{ y \in \mathbb{R}^{d_2+d_1+1} \;\middle|\; \sum_{k=1}^{d_2} y_k^2 + \sum_{k=1}^{d_1+1} y_{d_2+k}^2 = 1 \right\}.$$

As the line element of the sphere S^{d_1} in Eq. (2) is multiplied by a factor ≤ 1 for each data point, we call this sphere the "inner" sphere and note that data variation on this sphere is reduced. High data variation on the "outer" sphere S^{d_2} would lead to a greater and more uneven reduction of variation for the inner sphere. In order to prevent this, we first sort the spheres $(i = 1, \ldots, I)$ such that data variation is lowest on the last (outermost, $i = I$) sphere and highest on the first (innermost, $i = 1$) sphere. Indeed, if the data vary little on the i-th sphere then x_{i,d_i+1} is nearly one, causing little deformation.

2.2 Spheres of Different Radii

In general, the spheres in a polysphere of interest will have different radii. In fact, the radius for each sphere and each datum will often be unique. Recall that a logarithmic scale is well suited for lengths, linearizing ratios, such that the geometric mean corresponds to the arithmetic mean on a logarithmic scale. Hence, it is natural to define the mean radius of each individual sphere by the geometric mean of the radii of the data points, cf. [11].

There is yet another subtlety to be dealt with. In Eq. (3), simply multiplying the coordinates of S^{d_2} and S^{d_1} by the corresponding radii, implies that all coordinates of the sphere $S^{d_2+d_1}$ are in particular scaled with the radius of the outer sphere S^{d_2}. This implies that the relative scaling of the spheres will only depend on the radius of the inner S^{d_1}, clearly an unwanted feature. Hence, we normalize the radii with their geometric mean

$$R_i := r_i \left(\prod_{j=1}^{K} r_j \right)^{-\frac{1}{I}}$$

$(i = 1, \ldots, I)$, rescale all coordinates of the first unit sphere

$$\forall 1 \leq k \leq d_1 + 1 : \quad x_{1,k} \mapsto \tilde{x}_{1,k} = R_1 x_{1,k}, \tag{4}$$

only the first d_i coordinates of the i-th unit sphere $(i = 2, \ldots I)$

$$\forall 1 \leq k \leq d_i : \quad x_{i,k} \mapsto \tilde{x}_{i,k} = R_i x_{i,k} \tag{5}$$

and then apply the recursive operations defined in Eq. (3), using now $\tilde{x}$ instead of x. In particular for two spheres only, we thus start with the ellipsoid

$$\left\{ \tilde{x} \in \mathbb{R}^{d_2+d_1+1} \middle| \sum_{k=1}^{d_2} R_2^{-2} \tilde{x}_{2,k}^2 + \sum_{k=1}^{d_1+1} R_1^{-2} (\tilde{x}_{2,d_2+1} \tilde{x}_{1,k})^2 = 1 \right\}.$$

and only in the final step project $\tilde{y}$ to a unit sphere. Now the ordering of the spheres is determined by decreasing rescaled data variance where the data variance on the i-th unit sphere is rescaled by multiplication with R_i $(i = 1, \ldots, I)$.

One of the referees pointed out that radii normalizations could also be left variable to allow for more general optimizations. We will gratefully explore this in further research.

3   Application to Skeletal Representations

Our method is well-suited for application to skeletal representations (s-reps), as these contain data on a product of several spheres. An in-depth exposition of s-reps can be found in [13], cf. also [4,5]. For the s-reps used here, we now give a very brief review from [11].

3.1   The S-Rep Parameter Space

The basic building block of an s-rep is a two-dimensional mesh of $m \times n$ skeletal points which are embedded as medial as possible in the body to be described by the s-rep so that the surface of the body splits into three parts, the *northern sheet* above the mesh, the *southern sheet* below the mesh and the *crest* where the two sheets meet. From each of the skeletal points emerges a spoke to a point on the northern sheet and one to a point on the southern sheet. From each skeletal point on the boundary of the mesh an additional spoke emerges pointing to a point on the crest. This yields a total of $K = 2mn + 2m + 2n - 4$ spokes. Figure 1(b) shows an s-rep of a bent ellipsoid with 9×3 skeletal points (yellow), northern spokes (magenta), southern spokes (blue) and crest spokes (red). S-Reps are frequently used to model body organs in which case the spoke directions are restricted to half spheres due to the limited flexibility of organs.

An s-rep is represented in the following product space giving the size of its centered mesh, the lengths of the spokes, the normalized mesh-points and the spoke directions

$$Q = \mathbb{R}_+ \times \mathbb{R}_+^K \times S^{3mn-1} \times \left(S^2\right)^K. \tag{6}$$

Applying the polysphere deformation for spheres with different radii we obtain the data space

$$Q' = S^{5mn+2m+2n-5}. \tag{7}$$

3.2   PNS, CPNS and PNDS

In PNS (cf. [10]), for data on a unit sphere S^L a nested sequence of l-dimensional small-spheres M_l $(l = 1, \ldots, L - 1)$ is determined that approximates the data best with respect to the least sum of squares of spherical residuals:

$$S^L \supset M_{L-1} \supset \cdots \supset M_2 \supset M_1 \supset \{\mu\}$$

At each reduction step, the residuals are recorded as signed distances from the subsphere. In CPNS (cf. [11]), PNS is applied to every sphere occurring in the product (7) yielding a Euclidean vector of residuals. This vector, appended by the vector of logarithms of the sizes, is then subjected to classical PCA.

In PNDS, as proposed here, PNS is applied to the single polysphere (7) which has been obtained by polysphere deformation for spheres with the different radii given by the $\mathbb{R}_+ \times \mathbb{R}_+^K$ factors. In particular, no further PCA step as in CPNS is necessary.

3.3   Results

We compare the performance of PNDS to that of CPNS in terms of dimension reduction for the following data sets.

- $\mathrm{Hip}_{\mathrm{full}}$ contains s-reps fitted to MRI images of 51 human hippocampi, cf. [11]; $\mathrm{Hip}_{\mathrm{sp}}$ contains only the 66 spokes of variable length.
- Two data sets of simulated ellipsoids that have been twisted ($\mathrm{Sim}_{66,1}$) as well as bent and twisted ($\mathrm{Sim}_{66,2}$) from [12] consisting of 66 unit-length spokes, cf. Fig. 1.

Overall PNDS requires fewer dimensions than CPNS to explain data variation. For the full hippocampi data, PNDS explains 90 % of the variation by 8 dimensions, CPNS by 18 dimensions (9 vs. 20 for the spokes data only). The same effect, although far less prominent is visible for the simulated data, which is far less noisy than the real data, cf. Fig. 2 and Table 1.

Figure 3 elucidates a key difference between PNDS and CPNS. The data producing the **V** shape visible in components 1 and 2 of CPNS (b) obviously is spread along several spoke spheres in (6). Because in (7) they are mapped to a single sphere, that **V** shape can be explained by a single component via PNDS (c). The residual

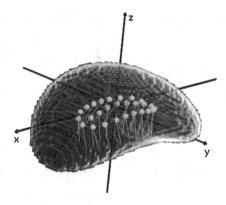

Fig. 1. S-rep with 9×3 skeletal points and 66 spokes fitted to a simulated bent ellipsoid, from [12] (Color figure online).

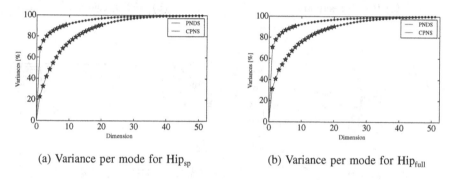

(a) Variance per mode for Hip_{sp}

(b) Variance per mode for Hip_{full}

Fig. 2. PNDS vs. CPNS: displaying scree plots of cumulative variances for s-reps of 51 hippocampi from [11]. Right: the full data set Hip_{full}, left: only the spoke information Hip_{sp}.

Table 1. PNDS vs. CPNS: percentage of variances explained by lower dimensional subsets that are required to explain at least 90 % of the respective total data variance.

		1	2	3	4	5	6	7	8	9	10	11	12	13	14	15	16	17	18	19	20	
$\text{Sim}_{66,1}$	PNDS	92.0																				92.0
	CPNS	62.7	32.1																			94.8
$\text{Sim}_{66,2}$	PNDS	88.7	7.4																			96.1
	CPNS	76.2	15.5																			91.7
Hip_{sp}	PNDS	68.5	7.2	4.3	2.9	2.2	1.9	1.5	1.2	1.1												90.7
	CPNS	22.8	10.0	9.1	6.9	6.1	5.2	4.8	3.8	3.0	2.9	2.5	2.2	2.1	1.8	1.6	1.5	1.4	1.3	1.1	1.0	90.9
Hip_{full}	PNDS	70.4	6.6	4.2	3.1	2.2	1.5	1.3	1.2													90.5
	CPNS	31.2	9.6	8.4	5.1	4.6	4.2	4.1	3.5	3.0	2.3	2.1	2.0	1.8	1.3	1.2	1.2	1.0	0.9			90.5
Dimension		1	2	3	4	5	6	7	8	9	10	11	12	13	14	15	16	17	18	19	20	

data distances to that small circle on the two-dimensional PNDS in (c) have the shape of a **3**, as visible in the first two components of PNDS (a) and the second and third component of CPNS (b). Higher dimensional components (already component 3 in PNDS) only explain low variance noise as seen in (c).

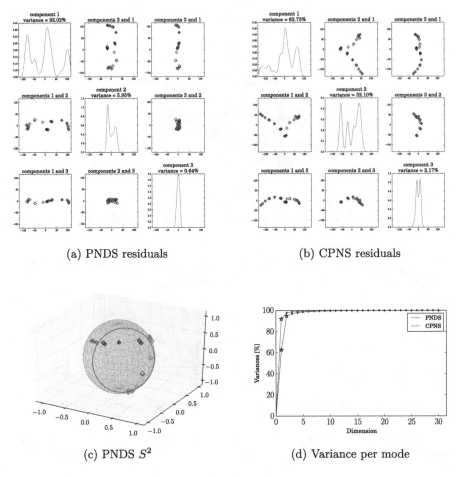

(a) PNDS residuals

(b) CPNS residuals

(c) PNDS S^2

(d) Variance per mode

Fig. 3. PNDS vs. CPNS for simulated twisted ellipsoids ($Sim_{66,1}$): scatter plots of residual signed distances for the first three components in (a) and (b). The data projected to the second component (a small two-sphere) in PNDS with first component (a small circle) inside, is visualized in (c). As in Fig. 2, subfigure (d) shows cumulative variances over dimension.

4   Conclusion and Outlook

We have shown that the deformation of a polysphere data space into a single high dimensional sphere may yield considerable enhancement in terms of dimension reduction. For the application to skeletal representations presented here, this is a crucial step towards a simple parametric model of body organ shapes, which allows for better fits and thus more successful automated localization of organs in MRI images. Applications range from minimizing tissue damage in radiation therapy or surgery to various diagnostic opportunities.

References

1. Afsari, B.: Riemannian L^p center of mass: existence, uniqueness, and convexity. Proc. Am. Math. Soc. **139**, 655–773 (2011)
2. Arsigny, V., Commowick, O., Pennec, X., Ayache, N.: A log-Euclidean framework for statistics on diffeomorphisms. In: Larsen, R., Nielsen, M., Sporring, J. (eds.) MICCAI 2006. LNCS, vol. 4190, pp. 924–931. Springer, Heidelberg (2006)
3. Boisvert, J., Pennec, X., Labelle, H., Cheriet, F., Ayache, N.: Principal spine shape deformation modes using Riemannian geometry and articulated models. In: Perales, F.J., Fisher, R.B. (eds.) AMDO 2006. LNCS, vol. 4069, pp. 346–355. Springer, Heidelberg (2006)
4. Damon, J.: Smoothness and geometry of boundaries associated to skeletal structures I: sufficient conditions for smoothness. Ann. Inst. Fourier **53**(6), 1941–1985 (2003)
5. Damon, J.: Global geometry of regions and boundaries via skeletal and medial integrals. Commun. Anal. Geom. **15**(2), 307–358 (2007)
6. Fletcher, P.T., Lu, C., Pizer, S.M., Joshi, S.C.: Principal geodesic analysis for the study of nonlinear statistics of shape. IEEE Trans. Med. Imaging **23**(8), 995–1005 (2004)
7. Gower, J.C.: Generalized Procrustes analysis. Psychometrika **40**, 33–51 (1975)
8. Huckemann, S., Hotz, T., Munk, A.: Intrinsic shape analysis: geodesic principal component analysis for Riemannian manifolds modulo Lie group actions (with discussion). Stat. Sinica **20**(1), 1–100 (2010)
9. Huckemann, S., Ziezold, H.: Principal component analysis for Riemannian manifolds with an application to triangular shape spaces. Adv. App. Probab. (SGSA) **38**(2), 299–319 (2006)
10. Jung, S., Dryden, I.L., Marron, J.S.: Analysis of principal nested spheres. Biometrika **99**(3), 551–568 (2012)
11. Pizer, S.M., Jung, S., Goswami, D., Vicory, J., Zhao, X., Chaudhuri, R., Damon, J.N., Huckemann, S., Marron, J.: Nested sphere statistics of skeletal models. In: Breuß, M., Bruckstein, A., Maragos, P. (eds.) Innovations for Shape Analysis, pp. 93–115. Springer, Heidelberg (2013)
12. Schulz, J., Jung, S., Huckemann, S., Pierrynowski, M., Marron, J.S., Pizer, S.M.: Analysis of rotational deformations from directional data. J. Comput. Graph. Stat. **24**(2), 539–560 (2015). doi:10.1080/10618600.2014.914947
13. Siddiqi, K., Pizer, S.: Medial Representations: Mathematics, Algorithms and Applications. Springer, The Netherlands (2008)
14. Sommer, S.: Horizontal dimensionality reduction and iterated frame bundle development. In: Nielsen, F., Barbaresco, F. (eds.) GSI 2013. LNCS, vol. 8085, pp. 76–83. Springer, Heidelberg (2013)

Affine-Invariant Riemannian Distance Between Infinite-Dimensional Covariance Operators

Hà Quang Minh[⊠]

Pattern Analysis and Computer Vision (PAVIS),
Istituto Italiano di Tecnologia (IIT), Genova, Italy
minh.haquang@iit.it

Abstract. This paper studies the affine-invariant Riemannian distance on the Riemann-Hilbert manifold of positive definite operators on a separable Hilbert space. This is the generalization of the Riemannian manifold of symmetric, positive definite matrices to the infinite-dimensional setting. In particular, in the case of covariance operators in a Reproducing Kernel Hilbert Space (RKHS), we provide a closed form solution, expressed via the corresponding Gram matrices.

1 Introduction

Symmetric Positive Definite (SPD) matrices play an important role in numerous areas of mathematics, statistics, machine learning, and their applications. Some examples of applications of SPD matrices in practice include brain imaging [3,9,22], object detection [24,25] and image retrieval [8] in computer vision, and radar signal processing [5,11].

It is well-known that for a fixed $n \in \mathbb{N}$, the set of all SPD matrices of size $n \times n$, denote by $\mathrm{Sym}^{++}(n)$, forms a Riemannian manifold with nonpositive curvature. Consequently, many computational methods have been proposed to exploit the Riemannian structure of $\mathrm{Sym}^{++}(n)$. Two of the most commonly applied Riemmannian metrics on $\mathrm{Sym}^{++}(n)$ are the classical affine-invariant metric [6,7,15,17,18,25] and the recently introduced Log-Euclidean metric [3,4]. The current work focuses on the infinite-dimensional generalization of the affine-invariant Riemannian metric.

Generalizations to the Infinite-Dimensional Setting. The affine-invariant Riemannian metric on $\mathrm{Sym}^{++}(n)$ has recently been generalized to the infinite-dimensional setting by [2,12,13] from a purely mathematical viewpoint. In the infinite-dimensional case, this metric measures the distances between positive definite unitized Hilbert-Schmidt operators, which are scalar perturbations of Hilbert-Schmidt operators on a Hilbert space and which are infinite-dimensional generalizations of positive definite matrices. As shown in [2,12,13], these operators form an infinite-dimensional Riemann-Hilbert manifold. A key point that needs to be emphasized is that while the general Hilbert-Schmidt formulation includes $\mathrm{Sym}^{++}(n)$ as a special case, the infinite-dimensional formulation is significantly different from its corresponding finite-dimensional version. In particular, in general

© Springer International Publishing Switzerland 2015
F. Nielsen and F. Barbaresco (Eds.): GSI 2015, LNCS 9389, pp. 30–38, 2015.
DOI: 10.1007/978-3-319-25040-3_4

one *cannot* obtain the infinite-dimensional formulas from the finite-dimensional ones by letting the dimension approach infinity.

Contributions of this Work. In the present paper, we derive the closed form solution for the infinite-dimensional affine-invariant distances between covariance operators on a reproducing kernel Hilbert space (RKHS) induced by a positive definite kernel. From the perspective of kernel methods in machine learning, this is motivated by the fact that covariance operators defined on nonlinear features, which are obtained by mapping the original input data into a high-dimensional feature space, can better capture input correlations than covariance matrices defined on the original input data, see e.g. KernelPCA [23].

Related Work. The present paper is a parallel contribution to [16]. In [16], we introduced the Log-Hilbert-Schmidt metric between positive definite unitized Hilbert-Schmidt operators, which is the infinite-dimensional version of the Log-Euclidean metric on $\mathrm{Sym}^{++}(n)$. In particular, for covariance operators on an RKHS induced by a positive definite kernel, we derived a closed-form solution for the Log-Hilbert-Schmidt distance via the corresponding Gram matrices. Empirical experiments performed on the task of multi-category image classification in [16] with the Log-Hilbert-Schmidt metric show significant improvements in classification accuracies with respect to the Log-Euclidean metric. In the context of functional data analysis, the generalization of the metrics on $\mathrm{Sym}^{++}(n)$ between covariance matrices to metrics between infinite-dimensional covariance operators has also been a problem of particular recent interest [20,21]. In [20], the authors discussed the difficulty of generalizing the affine-invariant and Log-Euclidean metrics to the infinite-dimensional setting and proposed several other metrics instead. As we show in [16] and below, this difficulty is due to the sharp differences between the finite and infinite-dimensional cases and is resolved by the formulation of the Riemann-Hilbert manifold of positive definite operators. Together with [2,12,13], the present work and [16] thus successfully resolve the problems of extending the affine-invariant and Log-Euclidean metrics to the infinite-dimensional setting.[1]

2 Background

The Riemannian manifold of positive definite matrices $\mathrm{Sym}^{++}(n)$ has been studied extensively in the literature, both mathematically and computationally [6,7,15,17,18]. The most commonly used Riemannian metric on $\mathrm{Sym}^{++}(n)$ is the affine-invariant metric, in which the geodesic distance between two positive definite matrices A and B is given by

$$d_{\mathrm{aiE}}(A, B) = || \log(A^{-1/2} B A^{-1/2})||_F, \qquad (1)$$

where log denotes the principal matrix logarithm operation and F is an Euclidean norm on the space of symmetric matrices $\mathrm{Sym}(n)$, the tangent space of

---

[1] The generalization of the Bregman divergences on $\mathrm{Sym}^{++}(n)$ to the infinite-dimensional setting will be presented in a separate paper.

$\text{Sym}^{++}(n)$. In this work we take F to be the Frobenius norm, which is induced by the standard inner product on $\text{Sym}(n)$, given by $\langle A, B \rangle_F = \text{tr}(A^*B)$. The metric (1) has recently been generalized to the infinite-dimensional setting by [2,12,13]. The main goal of this paper is to provide an explicit formulation in the infinite-dimensional case to compute the affine-invariant distances between covariance operators on an RKHS.

Covariance Operators in Reproducing Kernel Hilbert Spaces. Let $\mathcal{X}$ be an arbitrary non-empty set. Let $\mathbf{x} = [x_1, \ldots, x_m]$ be a data matrix sampled from $\mathcal{X}$, where $m \in \mathbb{N}$ is the number of observations. Let K be a positive definite kernel on $\mathcal{X} \times \mathcal{X}$ and $\mathcal{H}_K$ its induced reproducing kernel Hilbert space (RKHS). Let $\Phi : \mathcal{X} \to \mathcal{H}_K$ be the corresponding feature map, so that $K(x, y) = \langle \Phi(x), \Phi(y) \rangle_{\mathcal{H}_K}$ for all pairs $(x, y) \in \mathcal{X} \times \mathcal{X}$. The feature map Φ gives the (potentially infinite) mapped data matrix $\Phi(\mathbf{x}) = [\Phi(x_1), \ldots, \Phi(x_m)]$ of size $\dim(\mathcal{H}_K) \times m$ in the feature space $\mathcal{H}_K$. The corresponding empirical covariance operator for $\Phi(\mathbf{x})$ is defined to be

$$C_{\Phi(\mathbf{x})} = \frac{1}{m}\Phi(\mathbf{x})J_m\Phi(\mathbf{x})^T : \mathcal{H}_K \to \mathcal{H}_K, \qquad (2)$$

where J_m is the centering matrix, defined by $J_m = I_m - \frac{1}{m}\mathbf{1}_m\mathbf{1}_m^T$ with $\mathbf{1}_m = (1, \ldots, 1)^T \in \mathbb{R}^m$. The covariance operator $C_{\Phi(\mathbf{x})}$ can be viewed as a covariance matrix in the feature space $\mathcal{H}_K$, with rank at most $m - 1$. If $\mathcal{X} = \mathbb{R}^n$ and $K(x, y) = \langle x, y \rangle_{\mathbb{R}^n}$, then we recover the standard $n \times n$ empirical covariance matrix $C_{\Phi(\mathbf{x})} = C_{\mathbf{x}} = \frac{1}{m}\sum_{i=1}^m (x_i - \bar{x})(x_i - \bar{x})^T$, where $\bar{x} = \frac{1}{m}\sum_{i=1}^m x_i$.

Regularization. In the finite-dimensional case, since covariance matrices may not be full-rank and thus may only be positive semi-definite, in order to apply the Riemannian structure of $\text{Sym}^{++}(n)$, *empirically* one often needs to employ regularization. One of the most widely used regularizations is $(C_{\mathbf{x}} + \gamma I_{\mathbb{R}^n})$, for some regularization parameter $\gamma > 0$, a technique also known as diagonal loading (see e.g. [1,10] for more general forms of regularization). This simple, yet powerful, regularization ensures that both the matrix inversion and matrix logarithm operations are well-defined and stable, and is readily generalizable to the infinite-dimensional setting. When $\dim(\mathcal{H}_K) = \infty$, regularization is always necessary, both *theoretically* and *empirically*, as discussed in [16] and below.

3   The Riemann-Hilbert Manifold of Positive Definite Unitized Hilbert-Schmidt Operators

Throughout the paper, let $\mathcal{H}$ be a separable Hilbert space of arbitrary dimension. Let $\mathcal{L}(\mathcal{H})$ be the Banach space of bounded linear operators on $\mathcal{H}$, $\text{Sym}(\mathcal{H})$ be the subspace of self-adjoint bounded operators, and $\text{Sym}^+(\mathcal{H})$ and $\text{Sym}^{++}(\mathcal{H})$ be the subsets of positive and strictly positive operators in $\text{Sym}(\mathcal{H})$, respectively. As we now discuss, the infinite-dimensional generalization of $\text{Sym}^{++}(n)$ is *not* $\text{Sym}^{++}(\mathcal{H})$, however, but the set of *positive definite unitized Hilbert-Schmidt operators* on $\mathcal{H}$ under the *extended Hilbert-Schmidt inner product* (see also [16]).

Positive Definite Operators. Let $A \in \text{Sym}^{++}(n)$, then $\log(A) = UDU^T = U\text{diag}(\log\lambda_1, \ldots, \log\lambda_n)U^T$, where $A = UDU^T$ denotes the spectral decomposition for A, with $\{\lambda_k\}_{k=1}^n$ being the eigenvalues of A. Assume now that

$\dim(\mathcal{H}) = \infty$. Let $A \in \mathrm{Sym}^{++}(\mathcal{H})$ be compact, then A has a countable spectrum of positive eigenvalues $\{\lambda_k\}_{k=1}^{\infty}$, counting multiplicities, with $\lim_{k \to \infty} \lambda_k = 0$. Thus $\lim_{k \to \infty} \log \lambda_k = -\infty$ and consequently $\log(A)$ is unbounded. For $\log(A)$ to be well-defined and bounded in the case $\dim(\mathcal{H}) = \infty$, it is thus not sufficient to assume that $A \in \mathrm{Sym}^{++}(\mathcal{H})$. Instead, it is necessary to assume that A is *positive definite* (see e.g. [19]) , that is there exists a constant $M_A > 0$ such that

$$\langle Ax, x \rangle \geq M_A ||x||^2 \quad \text{for all} \ x \in \mathcal{H}, \tag{3}$$

which is equivalent to requiring that A be both *strictly positive* and *invertible*, with $A^{-1} \in \mathcal{L}(\mathcal{H})$. The eigenvalues of A, if they exist, are bounded below by M_A. We denote the set of positive definite, bounded operators on $\mathcal{H}$ by

$$\mathbb{P}(\mathcal{H}) = \{A \in \mathcal{L}(\mathcal{H}), A^* = A, \exists M_A > 0 \ \text{s.t.} \ \langle Ax, x \rangle \geq M_A ||x||^2 \ \forall x \in \mathcal{H}\}, \tag{4}$$

and we write $A > 0 \Longleftrightarrow A \in \mathbb{P}(\mathcal{H})$. The simplest positive definite operators in our setting have the form $(A + \gamma I)$, for any $A \in \mathrm{Sym}^+(\mathcal{H})$ and any $\gamma > 0, \gamma \in \mathbb{R}$.

Extended Hilbert-Schmidt Algebra. Let $\mathrm{HS}(\mathcal{H})$ denote the two-sided ideal of Hilbert-Schmidt operators on $\mathcal{H}$ in $\mathcal{L}(\mathcal{H})$, which is a Banach algebra without unit, under the Hilbert-Schmidt norm, defined by

$$||A||_{\mathrm{HS}}^2 = \mathrm{tr}(A^*A) = \sum_{k=1}^{\dim(\mathcal{H})} \lambda_k(A^*A). \tag{5}$$

This is the natural generalization of the Frobenius norm in (1) to the case $\dim(\mathcal{H}) = \infty$. For $A \in \mathrm{HS}(\mathcal{H}) \cap \mathrm{Sym}^+(\mathcal{H})$ and $\gamma > 0, \gamma \in \mathbb{R}$, we have $(A + \gamma I) \in \mathbb{P}(\mathcal{H})$ and $\log(A + \gamma I)$ is well-defined and bounded. However,

$$|| \log(A + \gamma I)||_{\mathrm{HS}}^2 = \sum_{k=1}^{\infty} [\log(\lambda_k + \gamma)]^2 = \infty \quad \text{for all} \ \gamma \neq 1. \tag{6}$$

This is because the identity operator I is not Hilbert-Schmidt when $\dim(\mathcal{H}) = \infty$ and, consequently, a direct generalization of Eq. (1) would give

$$d(\gamma I, \mu I) = || \log(\gamma/\mu)I||_{\mathrm{HS}} = |\log(\gamma/\mu)| \ ||I||_{\mathrm{HS}} = \infty \quad \text{if} \ \gamma \neq \mu. \tag{7}$$

To resolve this problem, consider the *extended (or unitized) Hilbert-Schmidt algebra* [2,12,13] consisting of all scalar perturbations of Hilbert-Schmidt operators, defined by

$$\mathcal{H}_{\mathbb{R}} = \{A + \gamma I : A^* = A, \ A \in \mathrm{HS}(\mathcal{H}), \gamma \in \mathbb{R}\}, \tag{8}$$

Endowed with the *extended Hilbert-Schmidt inner product*

$$\langle A + \gamma I, B + \mu I \rangle_{\mathrm{eHS}} = \mathrm{tr}(A^*B) + \gamma\mu = \langle A, B \rangle_{\mathrm{HS}} + \gamma\mu, \tag{9}$$

under which the scalar operators γI are orthogonal to the Hilbert-Schmidt operators, the set $\mathcal{H}_{\mathbb{R}}$ is then a Banach algebra with identity I. The corresponding *extended Hilbert-Schmidt norm* is given by

$$||(A + \gamma I)||_{\mathrm{eHS}}^2 = ||A||_{\mathrm{HS}}^2 + \gamma^2, \quad \text{with} \ ||I||_{\mathrm{eHS}} = 1. \tag{10}$$

If $\dim(\mathcal{H}) < \infty$, then we set $|| \ ||_{\mathrm{eHS}} = || \ ||_{\mathrm{HS}}$, with $||(A + \gamma I)||_{\mathrm{eHS}} = ||A + \gamma I||_{\mathrm{HS}}$.

The Riemann-Hilbert Manifold of Positive Definite Unitized Hilbert-Schmidt Operators. As shown recently in [13], the generalization of the finite-dimensional Riemannian manifold $\text{Sym}^{++}(n)$ to the case $\dim(\mathcal{H}) = \infty$ is the infinite-dimensional Riemann-Hilbert manifold $\Sigma(\mathcal{H})$ defined by

$$\Sigma(\mathcal{H}) = \mathbb{P}(\mathcal{H}) \cap \mathcal{H}_{\mathbb{R}} = \{A + \gamma I > 0 \ : A^* = A, \ A \in \text{HS}(\mathcal{H}), \gamma \in \mathbb{R}\}. \tag{11}$$

This is the manifold consisting of all positive definite operators which are scalar perturbations of Hilbert-Schmidt operators on $\mathcal{H}$. At every point $P \in \Sigma(\mathcal{H})$, the tangent space to the manifold at P is $T_P(\Sigma(\mathcal{H})) \cong \mathcal{H}_{\mathbb{R}}$.

Let $(A + \gamma I) \in \Sigma(\mathcal{H})$, then it possesses a countable set of eigenvalues $\{\lambda_k + \gamma\}_{k=1}^{\infty}$. Let $\{\phi_k\}_{k=1}^{\infty}$ denote the corresponding normalized eigenvectors, then $(A + \gamma I)$ admits the spectral decomposition

$$(A + \gamma I) = \sum_{k=1}^{\infty} (\lambda_k + \gamma)\phi_k \otimes \phi_k, \tag{12}$$

where $\phi_k \otimes \phi_k : \mathcal{H} \rightarrow \mathcal{H}$ is a rank-one operator defined by $(\phi_k \otimes \phi_k)w = \langle w, \phi_k \rangle \phi_k$, $w \in \mathcal{H}$. The logarithm map $\log : \Sigma(\mathcal{H}) \rightarrow \mathcal{H}_{\mathbb{R}}$ is a diffeomorphism, with $\log(A + \gamma I)$ defined by

$$\log(A + \gamma I) = \sum_{k=1}^{\dim(\mathcal{H})} \log(\lambda_k + \gamma)\phi_k \otimes \phi_k \in \mathcal{H}_{\mathbb{R}}. \tag{13}$$

On $\Sigma(\mathcal{H})$, the **geodesic distance** between two operators $(A + \gamma I), (B + \mu I) \in \Sigma(\mathcal{H})$ is given by

$$d_{\text{aiHS}}[(A + \gamma I), (B + \mu I)] = ||\log[(A + \gamma I)^{-1/2}(B + \mu I)(A + \gamma I)^{-1/2}]||_{\text{eHS}}, \tag{14}$$

This is the infinite-dimensional version of the affine-invariant distance (1). In particular, for $A = B = 0$, we have

$$d_{\text{aiHS}}[\gamma I, \mu I] = ||\log(\gamma/\mu)I||_{\text{eHS}} = |\log(\gamma/\mu)| < \infty \quad \forall \gamma, \mu > 0, \tag{15}$$

which resolves the problem encountered in Eq. (7).

4   Affine-Invariant Distance Between Positive Definite Operators

In this section, we give concrete formulas for the affine-invariant distance as defined in Eq. (14). In particular, for the case $\dim(\mathcal{H}) = \infty$, we show that the square of the affine-invariant distance decomposes into two components, namely a square Hilbert-Schmidt norm plus a scalar term. The formulas obtained in this section are used to derive closed-form solution for the affine-invariant distance between RKHS covariance operators in Sect. 5.

Consider the operator exponential operation, which is well-defined for all bounded operators $A \in \mathcal{L}(\mathcal{H})$, given by

$$\exp(A) = \sum_{k=0}^{\infty} \frac{A^k}{k!}, \quad \text{with operator norm} || \exp(A)|| \leq \exp(||A||). \tag{16}$$

As shown in [13], the map $\exp : \mathcal{H}_{\mathbb{R}} \to \Sigma(\mathcal{H})$ and its inverse $\log : \Sigma(\mathcal{H}) \to \mathcal{H}_{\mathbb{R}}$ are diffeomorphisms. If $B \in \mathcal{L}(\mathcal{H})$ is an invertible operator, then it follows directly from the definition of $\exp$ that

$$\exp(BAB^{-1}) = B\exp(A)B^{-1}. \tag{17}$$

From the above property of the exponential, it follows that for $A \in \Sigma(\mathcal{H})$ and an invertible operator $B \in \mathcal{L}(\mathcal{H})$, $\log(BAB^{-1})$ is well-defined and is given by

$$\log(BAB^{-1}) = B\log(A)B^{-1}. \tag{18}$$

The operator BAB^{-1} is not necessarily self-adjoint, but its eigenvalues are the same as those of A and hence are all positive. Let $A, B \in \Sigma(\mathcal{H})$, then

$$A^{-1}B = A^{-1/2}(A^{-1/2}BA^{-1/2})A^{1/2}. \tag{19}$$

Thus $\log(A^{-1}B)$ is well-defined and is given by

$$\log(A^{-1}B) = A^{-1/2}\log(A^{-1/2}BA^{-1/2})A^{1/2}. \tag{20}$$

We can now state the following results.

Theorem 1. *Let $\mathcal{H}$ be a separable Hilbert space with $\dim(\mathcal{H}) = \infty$. Assume that $(A + \gamma I), (B + \mu I) \in \Sigma(\mathcal{H})$. Then*

$$d_{\mathrm{aiHS}}^2[(A + \gamma I), (B + \mu I)]$$

$$= \left\| \log\left[\left(\frac{A}{\gamma} + I\right)^{-1/2} \left(\frac{B}{\mu} + I\right) \left(\frac{A}{\gamma} + I\right)^{-1/2} \right] \right\|_{\mathrm{HS}}^2 + \left(\log\frac{\gamma}{\mu}\right)^2 \tag{21}$$

$$= \mathrm{tr}\left\{ \log\left[\left(\frac{A}{\gamma} + I\right)^{-1} \left(\frac{B}{\mu} + I\right) \right] \right\}^2 + \left(\log\frac{\gamma}{\mu}\right)^2. \tag{22}$$

We note that a different but equivalent statement to Eq. (21) has also been given in [14] (Corollary 2.6). Our approach here is more constructive and the formulas we give are more explicit, leading to the closed-form distances between RKHS covariance operators below.

Finite-Dimensional Case. There is a sharp break between the cases $\dim(\mathcal{H}) = \infty$ and $\dim(\mathcal{H}) < \infty$, as shown in the formula for the affine-invariant distance in the case $\dim(\mathcal{H}) < \infty$ below, which depends explicitly on the dimension $\dim(\mathcal{H})$.

Theorem 2. *Assume that $\dim(\mathcal{H}) < \infty$. Let $A, B \in \mathrm{Sym}^+(\mathcal{H})$ and $\gamma, \mu > 0$. Then*

$$d_{\mathrm{aiHS}}^2[(A + \gamma I), (B + \mu I)] = \mathrm{tr}\left\{ \log\left[\left(\frac{A}{\gamma} + I\right)^{-1} \left(\frac{B}{\mu} + I\right) \right] \right\}^2$$

$$- 2\left(\log\frac{\gamma}{\mu}\right)\mathrm{tr}\left\{ \log\left[\left(\frac{A}{\gamma} + I\right)^{-1} \left(\frac{B}{\mu} + I\right) \right] \right\} + \left(\log\frac{\gamma}{\mu}\right)^2 \dim(\mathcal{H}). \tag{23}$$

In particular, in (23), for $A = B = 0$, we have $d_{\text{aiHS}}^2(\gamma I, \mu I) = \left(\log \frac{\gamma}{\mu}\right)^2 \dim(\mathcal{H}) \to \infty$ as $\dim(\mathcal{H}) \to \infty$ if $\gamma \neq \mu$, illustrating that, in general, the infinite-dimensional case cannot be obtained from the finite-dimensional case by letting $\dim(\mathcal{H}) \to \infty$.

Let $\mathcal{H}_1, \mathcal{H}_2$ be two separable Hilbert spaces. We are particularly interested in Hilbert-Schmidt operators of the form $AA^* : \mathcal{H}_2 \to \mathcal{H}_2$, where $A : \mathcal{H}_1 \to \mathcal{H}_2$ is a compact operator such that $A^*A : \mathcal{H}_1 \to \mathcal{H}_1$ is Hilbert-Schmidt. For these operators, we have the following result.

Proposition 1. *Let* $\mathcal{H}_1, \mathcal{H}_2$ *be separable Hilbert spaces and* $A, B : \mathcal{H}_1 \to \mathcal{H}_2$ *be compact operators such that* $A^*A, B^*B : \mathcal{H}_1 \to \mathcal{H}_1$ *are Hilbert-Schmidt operators. Then* $AA^*, BB^* : \mathcal{H}_2 \to \mathcal{H}_2$ *are Hilbert-Schmidt operators and*

$$\text{tr}\left\{\log\left[(AA^* + I_{\mathcal{H}_2})^{-1}(BB^* + I_{\mathcal{H}_2})\right]\right\} = \tag{24}$$

$$\text{tr}\left\{\log\left[\begin{pmatrix} B^*B & -B^*A(I_{\mathcal{H}_1} + A^*A)^{-1} & -B^*A(I_{\mathcal{H}_1} + A^*A)^{-1}A^*B \\ A^*B & -A^*A(I_{\mathcal{H}_1} + A^*A)^{-1} & -A^*A(I_{\mathcal{H}_1} + A^*A)^{-1}A^*B \\ B^*B & -B^*A(I_{\mathcal{H}_1} + A^*A)^{-1} & -B^*A(I_{\mathcal{H}_1} + A^*A)^{-1}A^*B \end{pmatrix} + I_{\mathcal{H}_1} \otimes I_3\right]\right\}.$$

$$\text{tr}\left\{\log\left[(AA^* + I_{\mathcal{H}_2})^{-1}(BB^* + I_{\mathcal{H}_2})\right]\right\}^2 = \tag{25}$$

$$\text{tr}\left\{\log\left[\begin{pmatrix} B^*B & -B^*A(I_{\mathcal{H}_1} + A^*A)^{-1} & -B^*A(I_{\mathcal{H}_1} + A^*A)^{-1}A^*B \\ A^*B & -A^*A(I_{\mathcal{H}_1} + A^*A)^{-1} & -A^*A(I_{\mathcal{H}_1} + A^*A)^{-1}A^*B \\ B^*B & -B^*A(I_{\mathcal{H}_1} + A^*A)^{-1} & -B^*A(I_{\mathcal{H}_1} + A^*A)^{-1}A^*B \end{pmatrix} + I_{\mathcal{H}_1} \otimes I_3\right]\right\}^2.$$

5   Affine-Invariant Distance Between Regularized RKHS Covariance Operators

Let $\mathcal{X}$ be an arbitrary non-empty set. We now apply the formulas of Sect. 4 to compute the affine-invariant distance between covariance operators on an RKHS induced by a positive definite kernel K on $\mathcal{X} \times \mathcal{X}$. In this case, we have explicit formulas for d_{aiHS} via the corresponding Gram matrices. Let $\mathbf{x} = [x_i]_{i=1}^m$, $\mathbf{y} = [y_i]_{i=1}^m$, $m \in \mathbb{N}$, be two data matrices sampled from $\mathcal{X}$ and $C_{\Phi(\mathbf{x})}, C_{\Phi(\mathbf{y})}$ be the corresponding covariance operators induced by the kernel K, as defined in Sect. 2. Let $K[\mathbf{x}], K[\mathbf{y}]$, and $K[\mathbf{x}, \mathbf{y}]$ be the $m \times m$ Gram matrices defined by $(K[\mathbf{x}])_{ij} = K(x_i, x_j)$, $(K[\mathbf{y}])_{ij} = K(y_i, y_j)$, $(K[\mathbf{x}, \mathbf{y}])_{ij} = K(x_i, y_j)$, $1 \leq i, j \leq m$. Let $A = \frac{1}{\sqrt{\gamma m}}\Phi(\mathbf{x})J_m : \mathbb{R}^m \to \mathcal{H}_K$, $B = \frac{1}{\sqrt{\mu m}}\Phi(\mathbf{y})J_m : \mathbb{R}^m \to \mathcal{H}_K$, so that

$$AA^* = \frac{1}{\gamma}C_{\Phi(\mathbf{x})}, \quad BB^* = \frac{1}{\mu}C_{\Phi(\mathbf{y})}, \tag{26}$$

$$A^*A = \frac{1}{\gamma m}J_m K[\mathbf{x}]J_m, \quad B^*B = \frac{1}{\mu m}J_m K[\mathbf{y}]J_m, \quad A^*B = \frac{1}{\sqrt{\gamma\mu}m}J_m K[\mathbf{x}, \mathbf{y}]J_m. \tag{27}$$

Theorem 3. *Assume that* $\dim(\mathcal{H}_K) = \infty$. *Let* $\gamma > 0, \mu > 0$. *Then*

$$d_{\text{aiHS}}^2[(C_{\Phi(\mathbf{x})} + \gamma I_{\mathcal{H}_K}), (C_{\Phi(\mathbf{y})} + \mu I_{\mathcal{H}_K})] = \text{tr}\left\{\log\left[\begin{pmatrix} C_{11} & C_{12} & C_{13} \\ C_{21} & C_{22} & C_{23} \\ C_{11} & C_{12} & C_{13} \end{pmatrix} + I_{3m}\right]\right\}^2$$

$$+ \left(\log \frac{\gamma}{\mu}\right)^2, \tag{28}$$

where the $m \times m$ matrices C_{ij}, $i, j = 1, 2, 3$, are given by

$$C_{11} = \frac{1}{\mu m} J_m K[\mathbf{y}] J_m,$$

$$C_{12} = -\frac{1}{\sqrt{\gamma \mu} m} J_m K[\mathbf{y}, \mathbf{x}] J_m \left(I_m + \frac{1}{\gamma m} J_m K[\mathbf{x}] J_m \right)^{-1},$$

$$C_{13} = -\frac{1}{\gamma \mu m^2} J_m K[\mathbf{y}, \mathbf{x}] J_m \left(I_m + \frac{1}{\gamma m} J_m K[\mathbf{x}] J_m \right)^{-1} J_m K[\mathbf{x}, \mathbf{y}] J_m,$$

$$C_{21} = \frac{1}{\sqrt{\gamma \mu} m} J_m K[\mathbf{x}, \mathbf{y}] J_m,$$

$$C_{22} = -\frac{1}{\gamma m} J_m K[\mathbf{x}] J_m \left(I_m + \frac{1}{\gamma m} J_m K[\mathbf{x}] J_m \right)^{-1},$$

$$C_{23} = -\frac{1}{\gamma m} J_m K[\mathbf{x}] J_m \left(I_m + \frac{1}{\gamma m} J_m K[\mathbf{x}] J_m \right)^{-1} \frac{1}{\sqrt{\gamma \mu} m} J_m K[\mathbf{x}, \mathbf{y}] J_m.$$

Theorem 4. *Assume that* $\dim(\mathcal{H}_K) < \infty$. *Let* $\gamma > 0, \mu > 0$. *Then*

$$d_{\mathrm{aiHS}}^2[(C_{\Phi(\mathbf{x})} + \gamma I_{\mathcal{H}_K}), (C_{\Phi(\mathbf{y})} + \mu I_{\mathcal{H}_K})] = \mathrm{tr} \left\{ \log \left[\begin{pmatrix} C_{11} & C_{12} & C_{13} \\ C_{21} & C_{22} & C_{23} \\ C_{11} & C_{12} & C_{13} \end{pmatrix} + I_{3m} \right] \right\}^2$$

$$- 2 \left(\log \frac{\gamma}{\mu} \right) \mathrm{tr} \left\{ \log \left[\begin{pmatrix} C_{11} & C_{12} & C_{13} \\ C_{21} & C_{22} & C_{23} \\ C_{11} & C_{12} & C_{13} \end{pmatrix} + I_{3m} \right] \right\} + \left(\log \frac{\gamma}{\mu} \right)^2 \dim(\mathcal{H}_K), \qquad (29)$$

where the $m \times m$ matrices C_{ij}'s, $i, j = 1, 2, 3$, are as in Theorem 3.

Remark 1. Let $m \in \mathbb{N}$ be fixed. Since the $m \times m$ matrices C_{ij} in Theorem 4 are all finite, it follows that in Eq. (29), $d_{\mathrm{aiHS}}^2[(C_{\Phi(\mathbf{x})} + \gamma I_{\mathcal{H}_K}), (C_{\Phi(\mathbf{y})} + \mu I_{\mathcal{H}_K})] \to \infty$ as $\dim(\mathcal{H}_K) \to \infty$ if $\gamma \neq \mu$.

Special Case. With $\mathcal{X} = \mathbb{R}^n$ and the linear kernel $K(x, t) = \langle x, t \rangle$ (29) gives $d_{\mathrm{aiHS}}[(C_{\Phi(\mathbf{x})} + \gamma I_{\mathcal{H}_K}), (C_{\Phi(\mathbf{y})} + \mu I_{\mathcal{H}_K})] = d_{\mathrm{aiE}}[(C_\mathbf{x} + \gamma I_n), (C_\mathbf{y} + \mu I_n)]$, the affine-invariant distance between the covariance matrices $C_\mathbf{x}$ and $C_\mathbf{y}$.

Remark 2. The proofs for all mathematical results and numerical experiments will be provided in the longer version of the paper.

References

1. Arsenin, V.I., Tikhonov, A.N.: Solutions of Ill-Posed Problems. Winston, Washington (1977)
2. Andruchow, E., Varela, A.: Non positively curved metric in the space of positive definite infinite matrices. Revista de la Union Mat. Argent. **48**(1), 7–15 (2007)
3. Arsigny, V., Fillard, P., Pennec, X., Ayache, N.: Fast and simple calculus on tensors in the log-euclidean framework. In: Duncan, J.S., Gerig, G. (eds.) MICCAI 2005. LNCS, vol. 3749, pp. 115–122. Springer, Heidelberg (2005)
4. Arsigny, V., Fillard, P., Pennec, X., Ayache, N.: Geometric means in a novel vector space structure on symmetric positive-definite matrices. SIAM J. Matrix An. App. **29**(1), 328–347 (2007)

5. Barbaresco, F.: Information geometry of covariance matrix: Cartan-Siegel homogeneous bounded domains, Mostow/Berger fibration and Frechet median. In: Nielsen, F., Bhatia, R. (eds.) Matrix Information Geometry, pp. 199–255. Springer, Heidelberg (2013)
6. Bhatia, R.: Positive Definite Matrices. Princeton University Press, Princeton (2007)
7. Bini, D.A., Iannazzo, B.: Computing the Karcher mean of symmetric positive definite matrices. Linear Algebra Appl. **438**(4), 1700–1710 (2013)
8. Cherian, A., Sra, S., Banerjee, A., Papanikolopoulos, N.: Jensen-Bregman LogDet divergence with application to efficient similarity search for covariance matrices. TPAMI **35**(9), 2161–2174 (2013)
9. Dryden, I.L., Koloydenko, A., Zhou, D.: Non-Euclidean statistics for covariance matrices, with applications to diffusion tensor imaging. Ann. Appl. Stat. **3**, 1102–1123 (2009)
10. Engl, H.W., Hanke, M., Neubauer, A.: Regularization of Inverse Problems. Mathematics and Its Applications, vol. 375. Springer, Netherlands (1996)
11. Formont, P., Ovarlez, J.-P., Pascal, F.: On the use of matrix information geometry for polarimetric SAR image classification. In: Nielsen, F., Bhatia, R. (eds.) Matrix Information Geometry, pp. 257–276. Springer, Heidelberg (2013)
12. Larotonda, G.: Geodesic Convexity, Symmetric Spaces and Hilbert-Schmidt Operators. Ph.D. thesis, Universidad Nacional de General Sarmiento, Buenos Aires, Argentina (2005)
13. Larotonda, G.: Nonpositive curvature: a geometrical approach to Hilbert-Schmidt operators. Differ. Geom. Appl. **25**, 679–700 (2007)
14. Lawson, J., Lim, Y.: The least squares mean of positive Hilbert-Schmidt operators. J. Math. Anal. Appl. **403**(2), 365–375 (2013)
15. Lawson, J.D., Lim, Y.: The geometric mean, matrices, metrics, and more. Am. Math. Monthly **108**(9), 797–812 (2001)
16. Minh, H.Q., San Biagio, M., Murino, V.: Log-Hilbert-Schmidt metric between positive definite operators on Hilbert spaces. In: Advances in Neural Information Processing Systems 27 (NIPS 2014), pp. 388–396 (2014)
17. Mostow, G.D.: Some new decomposition theorems for semi-simple groups. Memoirs Am. Math. Soc. **14**, 31–54 (1955)
18. Pennec, X., Fillard, P., Ayache, N.: A Riemannian framework for tensor computing. Int. J. Comput. Vis. **66**(1), 41–66 (2006)
19. Petryshyn, W.V.: Direct and iterative methods for the solution of linear operator equations in Hilbert spaces. Trans. Am. Math. Soc. **105**, 136–175 (1962)
20. Pigoli, D., Aston, J., Dryden, I.L., Secchi, P.: Distances and inference for covariance operators. Biometrika **101**(2), 409–422 (2014)
21. Pigoli, D., Aston, J., Dryden, I.L., Secchi, P.: Permutation tests for comparison of covariance operators. In: Contributions in infinite-dimensional statistics and related topics, pp. 215–220. Società Editrice Esculapio (2014)
22. Qiu, A., Lee, A., Tan, M., Chung, M.K.: Manifold learning on brain functional networks in aging. Med. Image Anal. **20**(1), 52–60 (2015)
23. Schölkopf, B., Smola, A., Müller, K.-R.: Nonlinear component analysis as a kernel eigenvalue problem. Neural Comput. **10**(5), 1299–1319 (1998)
24. Tosato, D., Spera, M., Cristani, M., Murino, V.: Characterizing humans on Riemannian manifolds. TPAMI **35**(8), 1972–1984 (2013)
25. Tuzel, O., Porikli, F., Meer, P.: Pedestrian detection via classification on Riemannian manifolds. TPAMI **30**(10), 1713–1727 (2008)

A Sub-Riemannian Modular Approach
for Diffeomorphic Deformations

Barbara Gris[1,2]([✉]), Stanley Durrleman[2], and Alain Trouvé[1]

[1] CMLA, UMR 8536, École normale supérieure de Cachan, Cachan, France
barbara.gris@outlook.com
[2] Inria Paris-Rocquencourt, Sorbonne Universités, UPMC Univ Paris 06 UMR
S1127, Inserm U1127, CNRS UMR 7225, ICM, 75013 Paris, France

Abstract. We develop a generic framework to build large deformations from a combination of base modules. These modules constitute a dynamical dictionary to describe transformations. The method, built on a coherent sub-Riemannian framework, defines a metric on modular deformations and characterises optimal deformations as geodesics for this metric. We will present a generic way to build local affine transformations as deformation modules, and display examples.

1 Introduction

A central aspect of Computational Anatomy is the comparison of different shapes, which are encoded as meshes or images. A common approach is the study of deformations matching one shape onto another, so that the differences between the two shapes are encoded by the deformation parameters [9,11,17]. In order to study differences between subjects on a particular structure, it should be useful to constrain locally the deformation, to favour realistic anatomic deformations, or to introduce some anatomical priors. For instance, for cortical surfaces with different sulci topography, one can prefer to favour lateral displacement over the creation of new sulci. Large deformations are commonly obtained through the integration of a vector field [4,10,13,18] and a natural route is to introduce the constraints in the vector fields instead of the final diffeomorphism [19]. The vector field could be restricted, via a finite dimensional control variable, to a state dependent finite dimensional subspace generated by a finite basis and conceptualized in structures called hereafter *deformation modules*. Deformation modules should create interpretable deformations, and several modules should be allowed to combine to form more complex compound deformation modules in the spirit of Grenander's Pattern Theory [7].

Preliminary instantiations of the concept of deformation modules can be found in several early works. In the poly-affine framework [3,14,20], deformations are created by the integration of a poly-affine stationary vector field. This vector field is a sum of few local affine transformations, which are then easily interpretable and share some features of deformation modules even if regions of each affine component are not updated during the deformation. In the LDDMM

© Springer International Publishing Switzerland 2015
F. Nielsen and F. Barbaresco (Eds.): GSI 2015, LNCS 9389, pp. 39–47, 2015.
DOI: 10.1007/978-3-319-25040-3_5

framework, a Riemannian structure is defined on the group of diffeomorphisms and optimal matchings are geodesics for this metric [2,10]. Several discretization schemes based on landmarks induce examples of finite dimensional state dependent representations of the velocity fields updated along the deformation and could be rephrased inside our definition of deformation modules. Note that the discretization scheme are thought of as approximations of the unconstrained non-parametrized infinite dimensional case. In a recently developed sparse-LDDMM framework [6,12,15], the vector field is constrained to be a sum of a fixed number of local translations, carried by control points, with a more clearer focus on finite-dimensionality and local interpretability. Extension to locally more complex transformation are considered in [8,16].

However, a consistent and general mathematical framework able to handle a large body of modular based large deformations is still missing in computational anatomy. A useful theory should not only provide a clear definition of deformation modules, but also explains how a hierarchy of deformation modules can be induced from basic one and how a Riemannian (or sub-Riemannian) setting can be defined underlying the computation of optimal large deformations and organizing the action of the different modules in the deformation process. This paper is a first attempt in that direction and presents a mathematical and computational sub-Riemannian framework to build large deformations from well defined deformation modules.

We will present several instances of deformation modules generating multi-scale and locally affine vector fields as simple illustrative examples. We will show trajectories that can be built from the combination of such modules, and how the component of a particular module can be recovered and followed through the integration.

2   Definition of a Deformation Module

Intuitively, a deformation module creates a vector field parametrized in low dimension, describing a distinctive aspect of a larger deformation pattern. This notion should embrace at least the notion of a sum of local translations, scalings or other local affine transformations as simple examples. In the following, $C_0^1(\mathbb{R}^d)$ will be the set on C^1 continuous mapping v vanishing at infinity equipped with the usual supremum norm on v and its first derivative, and $\mathrm{Diff}_0^1(\mathbb{R}^d)$ the open subset of $\mathrm{Id} + C_0^1(\mathbb{R}^d)$ of C^1 diffeomorphisms. We recall that for any curve $v \in L^1 \doteq L^1([0,1], C_0^1(\mathbb{R}^d))$ there exists a unique curve $t \mapsto \phi_t^v \in \mathrm{Diff}_0^1(\mathbb{R}^d)$ solution of the flow equation $\dot{\phi}_t^v = v_t \circ \phi_t^v$, with $\phi_0^v = \mathrm{id}$.

Let O be a finite dimensional manifold and $(\phi, o) \mapsto \phi.o$ a C^1 action of $\mathrm{Diff}_0^1(\mathbb{R}^d)$ on O in the sense that $(\phi, o) \mapsto \phi.o$ is continuous and there exists a continuous mapping $\xi : O \times C_0^1(\mathbb{R}^d) \to TO$ called the infinitesimal action, so that $v \mapsto \xi_o(v) \doteq \xi(o, v)$ is linear continous, $o \mapsto \xi_o$ is locally Lipschitz and for any $v \in L^1$, the curve $t \mapsto o_t \doteq \phi_t^v.o_0$ is absolutely continuous (a.c.) and satisfies $\dot{o}_t = \xi_{o_t}(v_t)$ for almost every $t \in [0,1]$.

Remark 1. *In fact O is a shape space as defined by S. Arguillère in [2].*

Definition 1 (Deformation module). *We say that* $M = (O, H, V, \zeta, \xi, c)$ *is a deformation module with geometrical descriptors in O, controls in H, infinitesimal action ξ, field generator ζ and cost c if H is a finite dimensional Euclidean space, V is an Hilbert space with $V \overset{C^o}{\hookrightarrow} C_0^1(\mathbb{R}^d)$, $\zeta : O \times H \to O \times V$ is a continuous mapping such that $h \mapsto \zeta_o(h)$ is linear where $\zeta(o, h) = (o, \zeta_o(h))$, $o \mapsto \zeta_o$ is locally Lipschitz and $c : O \times H \to \mathbb{R}^+$ is a continuous mapping such that $h \mapsto c_o(h) \doteq c(o, h)$ is a positive quadratic form on H and there exists $C > 0$ such that for each o, h:*

$$|\zeta_o(h)|_V^2 \leq C c_o(h). \tag{1}$$

Let us explain how a deformation modules induces large deformations.

Definition 2 (Finite energy controled paths on O). *Let $a, b \in O$. We denote $\Omega_{a,b}$ the set of mesurable curves $t \mapsto (o_t, h_t) \in O \times H$ where o_t is a.c., starting from a and ending at b, such that $\dot{o}_t = \xi_{o_t}(v_t)$ for $v_t \doteq \zeta_{o_t}(h_t)$ and $E(o, h) \doteq \int_0^1 c_{o_t}(h_t) dt < \infty$ where $E(o, h)$ is called the energy of (o, h).*

Thanks to (1) and the smoothness condition for deformation modules one get the following construction of flows:

Proposition 1 (Flows generated by a deformation module). *Let $(o, h) \in \Omega_{a,b}$ and $v = \zeta_o(h)$. Then $\int_0^1 |v_t|_V^2 dt \leq C \int_0^1 c_{o_t}(h_t) dt < \infty$ so that $v \in L^2([0,1], V) \subset L^1$ and $o_t = \phi_t^v.o_0$. Moreover, $\int_0^1 |h_t|_H^2 dt \leq (\sup_t \|c_{o_t}^{-1}\|) \int_0^1 c_{o_t}(h_t) dt < \infty$ (with $\|c_o^{-1}\| \doteq \sup_{|h|_H = 1} c_o(h)^{-1}$) so that $h \in L^2([0,1], H)$.*

A more geometrical point of view on deformation modules would be to identify ζ (resp. ξ) as a continuous morphisms between the two vector bundles $O \times H$ and $O \times V$ (resp. $O \times V \to TO$) and c as a metric on $O \times H$. Now, $\rho \doteq \xi \circ \zeta : O \times H \to TO$ and c induce a sub-Riemannian structure on O (as defined in [1]). Moreover, indexed by the choice of $a \in O$, we can induce a sub-Riemannian structure on $\text{Diff}_0^1(\mathbb{R}^d)$ by considering $\rho^a : \text{Diff}_0^1(\mathbb{R}^d) \times H \to T\text{Diff}_0^1(\mathbb{R}^d) = C_0^1(\mathbb{R}^d)$ such that $\rho_\phi^a(h) \doteq \zeta_\phi.a(h)$ and the metric on $\text{Diff}_0^1(\mathbb{R}^d) \times H$ given by $c_\phi^a(h) \doteq c_{\phi.a}(h)$.

During the trajectory, the geometrical descriptor o_t creates the vector field $\zeta_o(h)$ which acts back on o_t through the infinitesimal action ξ. Then, as explained in Fig. 1, ξ can be seen as a feedback action, allowing geometrical descriptors to evolve with the vector field.

2.1   First Example: Sum of Local Translations

This first example explains how the construction of [6] can be seen as a deformation module. Let $\sigma \in \mathbb{R}^+$, and $D \in \mathbb{N}$, we want to build a module M that would generate a sum of D local translations acting at scale σ. We set $O = (\mathbb{R}^d)^D$ (families of D points), $H = (\mathbb{R}^d)^D$ (families of D vectors) and $V = V_\sigma$ the scalar Gaussian RKHS of scale σ. For $o = (z_i) \in O$, we define $\zeta_o : h = (\alpha_i) \in H \mapsto \sum_{i=1}^D K_\sigma(z_i, \cdot)\alpha_i$, $\xi_o : v \in V \mapsto (v(z_i))_i$ (application of the vector field at each point) and $c_o : h = (\alpha_i) \in H \mapsto |\sum_i K_\sigma(z_i, \cdot)\alpha_i|_{V_\sigma}^2$.

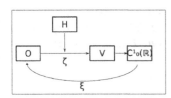

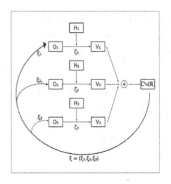

Fig. 1. Schematic view of a deformation module.

Fig. 2. Schematic view of a combination of deformation modules.

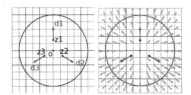

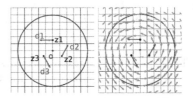

Fig. 3. Local scaling. **Left**: Geometrical descriptor o (in blue) and intermediate tools (in black and red). **Right**: Plot of the resulting vector field in green (Color figure online).

Fig. 4. Local rotation. **Left**: Geometrical descriptor o (in blue) and intermediate tools (in black and red). **Right**: Plot of the resulting vector field in green (Color figure online).

2.2 Second Example: Constrained Local Affine Transformations

We present here a generic way to build a particular kind of local affine transformation as a deformation module. Let us first start by an illustrative example of a local scaling in dimension 2 parametrized by a scale σ, a center o and a scaling ratio h. From σ and o, we build 3 points z_j and 3 unit vectors d_j as described in Fig. 3. We can then define the vector field generated by the geometrical descriptor o and the control h by $\zeta_o(h) \doteq \sum_{j=1}^{3} K_\sigma(z_j(o,\sigma),\cdot)d_j(o,\sigma)h$. We emphasize here that points z_j and vectors d_j are intermediate tools used to build the vector field but that the latter is only parametrized by σ, o and h. We can then define the module M by the following spaces : $V = V_\sigma$, $O = \mathbb{R}^d$, $H = \mathbb{R}$ and applications, for $o \in O$: $\zeta_o, \xi_o : v \in C_0^1(\mathbb{R}^d) \mapsto v(o)$ and $c_o : h \in H \mapsto |\zeta_o(h)|^2_{V_\sigma} = \sum_{j,j'} K_\sigma(z_j, z_{j'})d_j^T d_{j'}h^2$.

If we change the rule to build vectors d_j, we can build a local rotation, see Fig. 4. More generally, we can set any kind of rules to build points z_j and unit vectors d_j from a geometrical descriptor o to create another type of local transformation. We have here defined a generic way to build a module that generates a vector field based on user's assumptions. It is the way to incorporate anatomical prior in the deformation.

2.3   Third Example: Unconstrained Local Affine Transformations

Some more complex local affine transformations can be needed. Any local affine deformation in dimension d can be approximated by a sum of $d+1$ local translations carried by points close to each other with respect to the scale. In this spirit we can build another type of module, defining local affine transformations in P different areas of size defined by the same scale σ, by summing $D \doteq P \times (d+1)$ local translations, whose centres would be assembled in groups of $d+1$ points. We can build a module corresponding to the sum of D local translations, as in Sect. 2.1. This example differs from [6] as we suppose that $(d+1)$ centres of translations are pooled together. This construction allows to build modules that generate a vector field which is locally an affine transformation, without prior constraints. This module class differs from the poly-affine framework in that the neighbourhood which is affected by the local module is transported by the global transformation (via ξ).

3   Combination of Modules

We want to define the combination of L modules so that it remains a module in the sense of Definition 1. Modules are defined by spaces O^l, H^l, V^l, and applications ζ^l, ξ^l and c^l for each $l = 1 \cdots L$. We can define $\pi : w = (w_1, ..., w_L) \in W \doteq \prod_l V^l \mapsto \sum_i w_i \in C_0^1(\mathbb{R}^d)$. One can show that $V \doteq \pi(W)$ is a Hilbert space and is continuously embedded in $C_0^1(\mathbb{R}^d)$, with for $v \in V$, $|v|_V^2 = \inf\{\sum_i |v_i|_{V^i}^2 \mid \pi((v_i)_i) = v\}$. We are then able to define the compound module M by spaces: $O \doteq \prod_l O^l$, $H \doteq \prod_l H^l$, $V = Im(\pi)$ and applications, for $o = (o^l)_l \in O$: $\zeta_o : h = (h^l) \in H \mapsto \pi((\zeta_{o^l}^l(h^l))_l) = \sum_l \zeta_{o^l}^l(h^l) \in V$, $\xi_o : v \in C_0^1(\mathbb{R}^d) \mapsto (\xi^l(o^l, v))_l \in T_oO$ and $c_o : h = (h^l) \in H \mapsto \sum_l c_{o^l}^l(h^l)$. Then for any $h = (h^l) \in H$ we have

$$|\zeta_o(h)|_V^2 \leq \sum_{l=1}^{L} |\zeta_{o^l}^l(h^l)|_{V_l} \leq \sum_{l=1}^{L} C_l c_{o^l}^l(h^l) \leq (\max_{l \in L} C_l) c_o(h).$$

All necessary hypotheses to build a module are satisfied. A schematic view of this combination can be seen on Fig. 2. Note that even if the cost of the elementary module for each l is given by $c_{o^l}^l(h^l) = |\zeta_{o^l}^l(h^l)|_{V^l}^2$, as in our following examples, the cost of the compound module is then $c_o(h) = \sum_l |\zeta_{o^l}^l(h^l)|_{V^l}^2 \neq |\zeta(h)|_V^2$ so that that generically (when π is not one to one) $C > 1$ and c is not the pull-back metric on $O \times H$ of the metric on $O \times V$. In alternative extensions of sparse-LDDMM to locally more complex transformation ([8,16]) the norm of V has been a natural choice for the cost. Then the cost depends on the built vector field but not on the way it is built, unlike in our construction. Here minimizing the cost corresponds to selecting one way of building the vector field, and then choosing a particular cost enables to favour certain decomposition over others.

3.1   An Example of Combination: Sum of Multi-scale Translations

Let us build a module M which would be a sum of local translations, acting at different scales σ_l. For each σ_l can be built a module of type defined in Sect. 2.1, let $D_l \in \mathbb{N}$ be the number of local translations acting at this scale. The multi-scale module is then the combination of these modules. In particular the cost is, for $o = (z_j^l)$ and $h = (\alpha_j^l)$, $c_o(h) = \sum_l \sum_j K_{\sigma_l}(z_j^l, z_{j'}^l) \alpha_j^{lT} \alpha_{j'}^l$. It is clear here that it is not derived from the norm of the vector field $\zeta_o(h) = \sum_{l=1}^N \sum_{j=1}^{D_l} K_{\sigma_l}(z_j^l, \cdot) \alpha_j^l$ in V, which is here the RKHS of kernel $\sum_l K_{\sigma_l}$, as in [12] but from the way it is built as sum of elements of $V_{o^l}^l = \zeta_{o^l}^l(H^l)$.

4   Shooting

Let us consider a generic deformation module M and fix two values $a, b \in O$. For each trajectory $(o, h) \in \Omega_{a,b}$ (see Definition 2) we get a flow ϕ^v with $v = \zeta_o(h)$ and $\int_0^1 |v_t|_V dt \leq CE(o, h) < \infty$ (see Proposition 1).

We assume that for a distance d_O compatible with the topology on O, there exists $\gamma > 0$ and $K \subset \mathbb{R}^d$ such that $d_O(\phi.a, \phi'.a) \leq \gamma \|\phi - \phi'\|_{\infty, K}$ where $\| \ \|_{\infty, K}$ is the uniform C^1 norm on K. Note that this property is verified in our examples.

Theorem 1. *If $\Omega_{a,b}$ is not empty, then E reaches its minimum on $\Omega_{a,b}$.*

Proof. (Sketch) One consider a minimizing sequence $(o^n, h^n) \in \Omega_{a,b}$ and the associated flows ϕ^{v_n} for $v_n \doteq \zeta_{o^n}(h^n)$. Since $\int_0^1 |v_t^n|_V^2 \leq CE(o^n, h^n)$ we can assume (up to the extraction of a subsequence) that v^n weakly converges to v^∞ so that $t \mapsto \phi_t^{v^n}$ converges uniformly for the $\| \ \|_{\infty, K}$ norm and o^n converges uniformly to o^∞. Thus there exists a compact $K' \subset O$ such that o^∞ and the o^n stay in K'. Now, we have that $\int_0^1 |h_t^n|_H^2 dt \leq \sup_{K'} \|c_o^{-1}\| E(o^n, h^n)$ so that (up to the the the extraction of a subsequence) we can assume that h^n weakly converges to h^∞. Hence, for any $w \in L^2([0, 1], V)$ we have $|\int_0^1 \langle v_t^\infty - \zeta_{o_t^\infty}(h_t^\infty), w_t \rangle_V dt| \leq \overline{\lim} |\int_0^1 \langle (\zeta_{o_t^\infty} - \zeta_{o_t^n})(h_t^n), w_t \rangle_V dt| \leq \overline{\lim}(\int_0^1 |w_t|_V^2 dt \int_0^1 \|\zeta_{o_t^\infty} - \zeta_{o_t^n}\|^2 |h_t^n|_H^2 dt)^{1/2} = 0$. Since w is arbitrary, $v^\infty = \zeta_{o_t^\infty}(h_t^\infty)$ and $\dot{o}_t^\infty = \xi_{o_t}(v_t)$ so that $(o^\infty, h^\infty) \in \Omega_{a,b}$. Now, $\int_0^1 c_{o_t^\infty}(h_t^\infty) dt \leq \underline{\lim}(\int_0^1 c_{o_t^\infty}(h_t^n) dt \int_0^1 c_{o_t^\infty}(h_t^\infty) dt)^{1/2}$ where we have used that $c_o(h)$ can be written as $\langle C_o h, h \rangle_H$ where $o \mapsto C_o$ is continous and that h^n is weakly converging in $L^2([0, 1], H)$. Since $|\int c_{o_t^\infty}(h_t^n) - c_{o_t^n}(h_t^n) dt| \leq (\sup_t \|C_{o_t^\infty} - C_{o_t^n}\|) \int_0^1 |h_t^n|_H dt \to 0$ we get $\int_0^1 c_{o_t^\infty}(h_t^\infty) dt \leq \underline{\lim} \int_0^1 c_{o_t^n}(h_t^n) dt.$ □

A trajectory of $\Omega_{a,b}$ minimizing E can be obtained from the Hamiltonian and optimal control point of view [2] that we briefly describe below. Let us define the dual variable $\eta \in (T_o O)^*$ and introduce the Hamiltonian $H(\eta, o, h) = \langle \eta, \xi \circ \zeta(o, h) \rangle \rangle - \frac{1}{2} c_o(h)$. It can be shown that trajectories of $\Omega_{a,b}$ minimizing E can be separated in two categories: normal and abnormal geodesics, we will concentrate on normal geodesics as justified at the end of this section. Normal geodesics are such that there exists a trajectory (o, η) in T^*O such that (in a local chart)

$$\begin{cases} \dot{o} = \xi(o, \zeta_o(h)) \\ \dot{\eta} = -\frac{\partial H}{\partial o}(\eta, o, h) \\ h = A(o)\eta \end{cases} \tag{2}$$

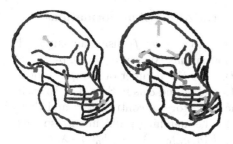

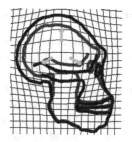

Fig. 5. Initial positions for modules: in red rotations, in green translation and in cyan scaling. In blue is the initial shape, in black the shape at $t = 1$. **Left:** parametrization of the vector field: initial geometrical descriptors (in black) and momenta. **Right:** Initial geometrical descriptors, controls and intermediate tools (color figure online).

Fig. 6. In blue the initial shape, in cyan are intermediate tools useful to build the vector field at $t = 0$, in black is the transported shape while following the largest scaling (color figure online).

where $A(o)$ is a matrix depending on o. The whole geodesic trajectory is then parametrized by initial values of (o, η) (of dimension twice the dimension of o). We have obtained a geodesic shooting from an element (a, η_0), with $\eta_0 \in (T_a O)^*$, to a geodesic path (o, h) and then to a trajectory ϕ^v with $v = \zeta_o(h)$.

Remark 2. *In practice we do not minimize E with fixed a end-point but minimize $J(o_{t=0}, h) = E(o, h) + g(o_{t=1})$ (with o such that $\dot{o} = \xi_o(\zeta_o(h))$). Trajectories minimizing J are normal geodesics following Eq. 2 (see [2]).*

5 Examples

5.1 Shooting with Constrained Types of Local Affine Transformations

We present here an example of geodesic trajectory created by the combination of different modules (in dimension 2): one translation (scale 100), two rotations (scales 20 and 60), two scaling (scales 50 and 20). Initial values of the geometrical descriptor (5 points: dimension 10) and the momentum (same dimension as o) define the total trajectory: it is parametrized in dimension 20. These parameters are displayed in Fig. 5 as well as initial controls (lengths of directions d_j for scaling and rotation), intermediate tools (z_j, d_j) and the transported shape (Fig. 6). We can follow the action of a particular module M_l by fixing the trajectory of its controls h^l, and integrating the new trajectory $(\tilde{v}, \tilde{o}^l)$ such that: $\tilde{o}^l(t = 0) = o^l(t = 0)$ and for each t : $\tilde{v}(t) = \zeta^l_{\tilde{o}^l(t)}(h(t))$, $\dot{\tilde{o}}^l(t) = \xi^l_{\tilde{o}^l(t)}(\tilde{v}(t))$.

5.2 Matching with Unconstrained Local Affine Transformations

We present here the matching from a shape f_0 to another f_1, using a compound module of unconstrained local affine transformations. We set $\sigma_1 = 60$, $\sigma_2 = 20$, $\sigma_3 = 8$ three scales and $P_1 = 1$, $P_2 = 9$, $P_3 = 20$ the number of groups of local translations at each scale. We define O, H, V, ζ, ξ and c as in Sect. 3.1. To each $(a, \eta) \in TO^*$, can be associated a geodesic trajectory of controls $h^{a,\eta}$ and then a diffeomorphism $\phi^{a,\eta}$ as defined in Sect. 4. The matching problem corresponds then to finding values of a and η minimizing $E(a, \eta, f_0, f_1) = c_a(h_0^{a,\eta}) + \lambda D(\phi_1^{a,\eta} \cdot f_0, f_1)$ where D is the varifold distance [5]. Our first implementation is adapted from the software Deformetrica where we introduce constraints to remain control points pooled during optimization. An example of result is shown on Fig. 7.

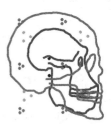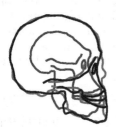

Fig. 7. Matching from f_0 (blue) to f_1 (red), with parameters per scale (σ_1: black, σ_2: green, σ_3: magenta). **Left:** initialization geometrical descriptors. **Middle:** Optimized initial geometrical descriptors and controls. **Right:** Match at $t = 1$ (black) (Color figure online).

6 Conclusion

We constructed a mathematical framework for generic deformation modules, which is stable under combination. Large deformations can be built then by integrating vector fields generated by these modules. By defining a cost on module we allow optimal deformations to come from geodesic paths in a sub-Riemannian manifold. We presented several examples of modules and geodesics. The construction allows an easy interpretation of the computed deformation and the incorporation of anatomical prior, so that this work may have important applications for the analysis of biological shapes.

References

1. Agrachev, A., Boscain, U., Charlot, G., Ghezzi, R., Sigalotti, M.: Two-dimensional almost-riemannian structures with tangency points. In: Proceedings of the 48th IEEE Conference on Decision and Control, 2009 held Jointly with the 2009 28th Chinese Control Conference, CDC/CCC 2009, pp. 4340–4345. IEEE (2009)

2. Arguillere, S.: Géométrie sous-riemannienne en dimension infinie et applications à l'analyse mathématique des formes. Ph.D. thesis, Paris 6 (2014)
3. Arsigny, V., Commowick, O., Ayache, N., Pennec, X.: A fast and log-euclidean polyaffine framework for locally linear registration. J. Math. Imaging Vis. **33**(2), 222–238 (2009)
4. Ashburner, J.: A fast diffeomorphic image registration algorithm. Neuroimage **38**(1), 95–113 (2007)
5. Charon, N., Trouvé, A.: The varifold representation of non-oriented shapes for diffeomorphic registration (2013). arXiv preprint arXiv:1304.6108
6. Durrleman, S., Prastawa, M., Gerig, G., Joshi, S.: Optimal data-driven sparse parameterization of diffeomorphisms for population analysis. In: Székely, G., Hahn, H.K. (eds.) IPMI 2011. LNCS, vol. 6801, pp. 123–134. Springer, Heidelberg (2011)
7. Grenander, U.: Elements of Pattern Theory. JHU Press, Baltimore (1996)
8. Jacobs, H.: Symmetries in LDDMM with higher order momentum distributions (2013). arXiv preprint arXiv:1306.3309
9. Joshi, S., Lorenzen, P., Gerig, G., Bullitt, E.: Structural and radiometric asymmetry in brain images. Med. Image Anal. **7**(2), 155–170 (2003)
10. Miller, M.I., Trouvé, A., Younes, L.: On the metrics and euler-lagrange equations of computational anatomy. Ann. Rev. Biomed. Eng. **4**(1), 375–405 (2002)
11. Miller, M.I., Younes, L., Trouvé, A.: Diffeomorphometry and geodesic positioning systems for human anatomy. Technology **2**(01), 36–43 (2014)
12. Risser, L., Vialard, F., Wolz, R., Murgasova, M., Holm, D.D., Rueckert, D.: Simultaneous multi-scale registration using large deformation diffeomorphic metric mapping. IEEE Trans. Med. Imaging **30**(10), 1746–1759 (2011)
13. Rueckert, D., Aljabar, P., Heckemann, R.A., Hajnal, J.V., Hammers, A.: Diffeomorphic registration using B-Splines. In: Larsen, R., Nielsen, M., Sporring, J. (eds.) MICCAI 2006. LNCS, vol. 4191, pp. 702–709. Springer, Heidelberg (2006)
14. Seiler, C., Pennec, X., Reyes, M.: Capturing the multiscale anatomical shape variability with polyaffine transformation trees. Med. Image Anal. **16**(7), 1371–1384 (2012)
15. Sommer, S., Lauze, F., Nielsen, M., Pennec, X.: Sparse multi-scale diffeomorphic registration: the kernel bundle framework. J. Math. Imaging Vis. **46**(3), 292–308 (2013)
16. Sommer, S., Nielsen, M., Darkner, S., Pennec, X.: Higher-order momentum distributions and locally affine LDDMM registration. SIAM J. Imaging Sci. **6**(1), 341–367 (2013)
17. Vaillant, M., Miller, M.I., Younes, L., Trouvé, A.: Statistics on diffeomorphisms via tangent space representations. NeuroImage **23**, S161–S169 (2004)
18. Vercauteren, T., Pennec, X., Perchant, A., Ayache, N.: Symmetric log-domain diffeomorphic registration: a demons-based approach. In: Metaxas, D., Axel, L., Fichtinger, G., Székely, G. (eds.) MICCAI 2008, Part I. LNCS, vol. 5241, pp. 754–761. Springer, Heidelberg (2008)
19. Younes, L.: Constrained diffeomorphic shape evolution. Found. Comput. Math. **12**(3), 295–325 (2012)
20. Zhang, W., Noble, J.A., Brady, J.M.: Adaptive non-rigid registration of real time 3D ultrasound to cardiovascular MR images. In: Karssemeijer, N., Lelieveldt, B. (eds.) IPMI 2007. LNCS, vol. 4584, pp. 50–61. Springer, Heidelberg (2007)

Optimal Transport

The Nonlinear Bernstein-Schrödinger Equation in Economics

Alfred Galichon[1]([✉]), Scott Duke Kominers[2], and Simon Weber[3]

[1] Economics Department, Sciences Po, Paris, France and New York University,
New York, USA
alfred.galichon@sciencespo.fr

[2] Society of Fellows, Harvard University, Cambridge, USA
kominers@fas.harvard.edu

[3] Department of Economics, Sciences Po, 27 rue Saint-Guillaume, 75007 Paris, France
simon.weber@sciencespo.fr

Abstract. In this paper we relate the Equilibrium Assignment Problem (EAP), which is underlying in several economics models, to a system of nonlinear equations that we call the "nonlinear Bernstein-Schrödinger system", which is well-known in the linear case, but whose nonlinear extension does not seem to have been studied. We apply this connection to derive an existence result for the EAP, and an efficient computational method.

In this note, we will review and extend some results from our previous work [8] where we introduced a novel approach to imperfectly transferable utility and unobserved heterogeneity in tastes, based on a nonlinear generalization of the Bernstein-Schrödinger equation. We consider an assignment problem where agents from two distinct populations may form pairs, which generates utility to each agent. Utility may be transfered across partners, possibly with frictions. This general framework hence encompasses both the classic Non-Tranferable Utility (NTU) model of Gale and Shapley [6], sometimes called the "stable marriage problem", where there exists no technology to allow transfers between matched partners; and the Transferable Utility (TU) model of Becker [1] and Shapley-Shubik [14], a.k.a. the "optimal assignment problem," where utility (money) is additively transferable across partners.

If the NTU assumption seems natural for many markets (including school choices), TU models are more appropriate in most settings where there can be bargaining (labour and marriage markets for example). However, even in those markets, there can be transfer frictions. For example, in marriage markets, the transfers between partners might take the form of favor exchange (rather than

A.Galichon—gratefully acknowledges funding from the European Research Council under the European Union's Seventh Framework Programme (FP7/2007–2013)/ERC grant agreements no 313699 and 295298, and FiME.

S.D.Kominers—gratefully acknowledges the support of NSF grants CCF-1216095 and, as well as the Harvard Milton Fund.

© Springer International Publishing Switzerland 2015
F. Nielsen and F. Barbaresco (Eds.): GSI 2015, LNCS 9389, pp. 51–59, 2015.
DOI: 10.1007/978-3-319-25040-3_6

cash), and the cost of a favor to one partner may not exactly equal the benefit
to the other.

In [8], we developed a general Imperfectly Transferable Utility model with
unobserved heterogeneity, which includes as special cases the classic fully- and
non-transferable utility models, but also extends to collective models, and settings with taxes on transfers, deadweight losses, and risk aversion. As we argue
in the present note, the models we consider in [8] obey a particularly simple
system of equations we dub "Nonlinear Bernstein-Schrödinger equations". The
present contribution gives a general result for the latter equation, and also derives
several consequences. The main result is derived in Sect. 1. Sections 2 and 3 consider equilibrium assignment problems with and without heterogeneity. Finally,
we provide a discussion of our results in Sect. 4.

1 The Main Result

For $x \in \mathcal{X}$ and $y \in \mathcal{Y}$, we consider a function $M_{xy} : \mathbb{R}^2 \to \mathbb{R}$, and $M_{x0} : \mathbb{R} \to \mathbb{R}$
and $M_{0y} : \mathbb{R} \to \mathbb{R}$. Let $(n_x)_{x \in \mathcal{X}}$ and $(m_y)_{y \in \mathcal{Y}}$ be vectors of positive numbers.
Consider the *nonlinear Bernstein-Schrödinger system*, which consists in looking
for two vectors $u \in \mathbb{R}^{\mathcal{X}}$ and $v \in \mathbb{R}^{\mathcal{Y}}$ such that

$$\begin{cases} M_{x0}(u_x) + \sum_{y \in \mathcal{Y}} M_{xy}(u_x, v_y) = n_x \\ M_{0y}(v_y) + \sum_{x \in \mathcal{X}} M_{xy}(u_x, v_y) = m_y \end{cases}.$$

Theorem 1. *Assume M_{xy} satisfies the following three conditions:*

*(i) Continuity. The maps $M_{xy} : (u_x, v_y) \longmapsto M_{xy}(u_x, v_y)$, $M_{x0} : (u_x) \longmapsto$
$M_{x0}(u_x)$ and $M_{0y} : (v_y) \longmapsto M_{0y}(v_y)$ are continuous.*

*(ii) Monotonicity. The map $M_{xy} : (u_x, v_y) \longmapsto M_{xy}(u_x, v_y)$ is monotonically
decreasing, i.e. if $u_x \leq u'_x$ and $u_y \leq u'_y$, then $M_{xy}(u_x, v_y) \geq M(u'_x, v'_y)$. The
maps $M_{x0} : (u_x) \longmapsto M_{x0}(u_x)$ and $M_{0y} : (v_y) \longmapsto M_{0y}(v_y)$ are monotonically
decreasing.*

*(iii) Limits. For each v_y, $\lim_{u_x \to \infty} M_{xy}(u_x, v_y) = 0$ and $\lim_{u_x \to -\infty} M_{xy}$
$(u_x, v_y) = +\infty$, and for each u_x, $\lim_{u_y \to \infty} M_{xy}(u_x, v_y) = 0$ and $\lim_{u_y \to -\infty} M_{xy}$
$(u_x, v_y) = +\infty$. Additionally, $u_x \to \infty M_{x0}(u_x) = 0$ and $\lim_{u_y \to \infty} M_{0y}(u_y) = 0$,
and $\lim_{u_x \to -\infty} M_{x0}(u_x) = +\infty$ and $\lim_{u_y \to -\infty} M_{0y}(u_y) = +\infty$.*

*Then there exists a solution to the nonlinear Bernstein-Schrödinger system.
Further, if the maps M are C^1, the solution is unique.*

This theorem appears in [8] under a slightly different form. The proof is
interesting as it provides an algorithm of determination of u and v. It is an
important generalization of the Iterated Projection Fitting Procedure (see [5,12,
13]), which has been rediscovered and utilized many times under different names
for various applied purposes: "RAS algorithm" [10], "biproportional fitting",
"Sinkhorn Scaling" [4], etc. However, all these techniques and their variants can
be recast as particular cases of the method described in the proof of Theorem 1.
For convenience, we recall the algorithm used to provide a constructive proof of
existence.

Algorithm 2.

Step 0	*Fix the initial value of v_y at $v_y^0 = +\infty$.*
Step $2t+1$	*Keep the values v_y^{2t} fixed. For each $x \in \mathcal{X}$, solve for the value, u_x^{2t+1} such that equality $\sum_{y \in \mathcal{Y}} M_{xy}(u_x, v_y^{2t}) + M_{x0}(u_x) = n_x$ holds.*
Step $2t+2$	*Keep the values u_x^{2t+1} fixed. For each $y \in \mathcal{Y}$, solve for which is the value, v_y^{2t+2} such that equality $\sum_{x \in \mathcal{X}} M_{xy}(u_x^{2t+1}, v_y) + M_{0y}(v_y) = m_y$ holds.*

Then u^t and v^t converge monotonically to a solution of the nonlinear Bernstein-Schrödinger system.

In practice a precision level $\epsilon > 0$ will be chosen, and the algorithm will terminate when $\sup_y |v_y^{2t+2} - v_y^{2t}| < \epsilon$.

Proof. (i) Existence. The proof of existence is an application of Tarski's fixed point theorem and relies on the previous Algorithm. We need to prove that the construction of u_x^{2t+1} and v_y^{2t+2} at each step is well defined. Consider step $2t+1$. For each $x \in \mathcal{X}$, the equation to solve is

$$\sum_{y \in \mathcal{Y}} M_{xy}(u_x, v_y) + M_{x0}(u_x) = n_x$$

but the right-hand side is a continuous and decreasing function of u_x, tends to 0 when $u_x \to +\infty$, and tends to $+\infty$ when $u_x \to -\infty$. Note that by letting $v_y \to +\infty$, the terms in the sum tends to 0, providing a lower bound for u_x. Hence u_x^{2t+1} is well defined and belongs in $\left(M_{x0}^{-1}(n_x), +\infty\right)$, and let us denote

$$u_x^{2t+1} = F_x(v^{2t})$$

and clearly, F is anti-isotone, meaning that $v_y^{2t} \leq \tilde{v}_y^{2t}$ for all $y \in \mathcal{Y}$ implies $F_x(\tilde{v}^{2t}) \leq F_x(v^{2t})$ for all $x \in \mathcal{X}$. By the same token, at step $2t+2$, v_y^{2t+2} is well defined in $\left(M_{0y}^{-1}(m_y), +\infty\right)$, and let us denote

$$v_y^{2t+2} = G_y(u^{2t+1})$$

where, similarly, G is anti-isotone. Thus

$$v^{2t+2} = G \circ F\left(v^{2t}\right)$$

where $G \circ F$ is isotone. But $v_y^2 < \infty = v_y^0$ implies that $v^{2t+2} \leq G \circ F\left(v^{2t}\right)$. Hence $\left(v^{2t+2}\right)_{t \in \mathbb{N}}$ is a decreasing sequence, bounded from below by 0. As a result v^{2t+2} converges. Letting $\bar{v}$ its limit, and letting $\bar{u} = F(\bar{v})$, one can see that $(\bar{u}, \bar{v})$ is a solution to the nonlinear Bernstein-Schrödinger system. (ii) Unicity. Introduce map ζ defined by

$$\zeta : (u_x, v_y) \to \begin{pmatrix} \zeta_x = \sum_{y \in \mathcal{Y}} M_{xy}(u_x, v_y) + M_{x0}(u_x) \\ \zeta_y = \sum_{x \in \mathcal{X}} M_{xy}(u_x, v_y) + M_{0y}(v_y) \end{pmatrix}.$$

One has

$$D\zeta(u_x, v_y) = \left(\begin{array}{c|c} A & B \\ \hline C & D \end{array}\right)$$

where:

- $A = (\partial \zeta_x / \partial u_{x'})_{xx'} = \sum_{y' \in \mathcal{Y}} \partial_{u_x} M_{xy'}(u_x, v_{y'}) + 1$ if $x = x'$, 0 otherwise,
- $B = (\partial \zeta_x / \partial v_y)_{xy} = \partial_{v_y} M_{xy}(u_x, v_y)$
- $C = (\partial \zeta_y / \partial u_x)_{yx} = \partial_{u_x} M_{xy}(u_x, v_y)$
- $D = (\partial \zeta_y / \partial v_{y'})_{yy'} = \sum_{x' \in \mathcal{X}} \partial_{v_y} M_{x'y}(u_{x'} v_y) + 1$ if $y = y'$, 0 otherwise.

It is straightforward to show that the matrix $D\zeta$ is dominant diagonal. A result from [11] states that a dominant diagonal matrix with positive diagonal entries is a P-matrix. Hence $D\zeta(u_x, v_y)$ is a P-matrix. Applying Theorem 4 in [9] it follows that ζ is injective.

2 Equilibrium Assignment Problem

In this section, we consider the equilibrium assignment problem, which is a far-reaching generalization of the optimal assignment problem. To describe this framework, consider two finite populations $\mathcal{I}$ and $\mathcal{J}$, and a two-sided matching framework (for simplicity, we will call the two sides of this market "men" and "women") with imperfect transfers and without heterogeneity. Agents $i \in \mathcal{I}$ and $j \in \mathcal{J}$ get respectively utility u_i and v_j at equilibrium. If either i or j remains unmatched, that agent gets utility 0; however, if they match together, they may get any respective utilities u_i and v_j such that the feasibility constraint

$$\Psi_{ij}(u_i, v_j) \leq 0, \tag{1}$$

is satisfied, where the transfer function Ψ_{ij} is assumed to be continuous and isotone with respect to its arguments. Note that at equilibrium, $u_i \geq 0$ and $v_j \geq 0$ as the agents always have the option to remain unassigned; by the same token, for any pair i, j (matched or not), one cannot have a strict inequality in (1), otherwise i and j would have an incentive to form a blocking pair, and achieve a higher payoff than their equilibrium payoff. Thus the stability condition $\Psi_{ij}(u_i, v_j) \geq 0$ holds in general. Let $\mu_{ij} = 1$ if i and j are matched, and 0 otherwise; we have therefore that $\mu_{ij} > 0$ implies that $\Psi_{ij}(u_i, v_j) = 0$. This allows us to define an equilibrium outcome.

Definition 1. *The equilibrium assignment problem defined by Ψ has an equilibrium outcome (μ_{ij}, u_i, v_j) whenever the following conditions are met: (i) $\mu_{ij} \geq 0$, $u_i \geq 0$ and $v_j \geq 0$ (ii) $\sum_j \mu_{ij} \leq 1$ and $\sum_i \mu_{ij} \leq 1$ (iii) $\Psi_{ij}(u_i, v_j) \geq 0$ (iv) $\mu_{ij} > 0$ implies $\Psi_{ij}(u_i, v_j) = 0$.*

Note that, by the Birkhoff-von Neumann theorem, the existence of an equilibrium in this problem leads to the existence of an equilibrium satisfying the stronger integrality requirement $\mu_{ij} \in \{0, 1\}$. This general framework allow us to express the optimal assignment problem (matching with Transferable Utility), as the case where

$$\Psi_{ij}(u_i, v_j) = u_i + v_j - \Phi_{ij},$$

while in the NTU case

$$\Psi_{ij}(u_i, v_j) = \max(u_i - \alpha_{ij}, v_j - \gamma_{ij}).$$

Other interesting cases are considered in [8]. For instance, the *Linearly Transferable Utility (LTU)* model, where

$$\Psi_{ij}(u_i, v_j) = \lambda_{ij}(u_i - \alpha_{ij}) + \zeta_{ij}(v_j - \gamma_{ij})$$

with $\lambda_{ij}, \zeta_{ij} > 0$, and the *Exponentially Transferable Utility (ETU)* model, in which Ψ_{ij} takes the form

$$\Psi_{ij}(u_i, v_j) = \tau \log \left(\frac{\exp(u_i/\tau) + \exp(v_j/\tau)}{2} \right).$$

In the ETU model, the parameter τ_{ij} is defined as the *degree of transferability*, since $\tau \to +\infty$ recovers the TU case and $\tau \to 0$ recovers the NTU framework.

Theorem 3. *Assume Ψ is such that: (a) For any $x \in \mathcal{X}$ and $y \in \mathcal{Y}$, we have $\Psi_{xy}(\cdot, \cdot)$ continuous. (b) For any $x \in \mathcal{X}$, $y \in \mathcal{Y}$, $t \leq t'$ and $r \leq r'$, we have $\Psi_{xy}(t, r) \leq \Psi_{xy}(t', r')$; furthermore, when $t < t'$ and $r < r'$, we have $\Psi_{xy}(t, r) < \Psi_{xy}(t', r')$. (c) For any sequence (t_n, r_n), if (r_n) is bounded and $t_n \to +\infty$, then $\liminf \Psi_{xy}(t_n, r_n) > 0$. Analogously, if (t_n) is bounded and $r_n \to +\infty$, then $\liminf \Psi_{xy}(t_n, r_n) > 0$. (d) For any sequence (t_n, r_n) such that if $(t_n - r_n)$ is bounded and $t_n \to -\infty$ (or equivalently, $r_n \to +\infty$), we have that $\limsup \Psi_{xy}(t_n, r_n) < 0$. Then the equilibrium assignment problem defined by Ψ_{ij} has an equilibrium outcome.*

Proof. Consider $T > 0$ and let

$$M_{ij}(u_i, v_j) = \exp \left(-\frac{\Psi_{ij}(u_i, v_j)}{T} \right)$$

$$M_{i0}(u_i) = \exp \left(-\frac{u_i}{T} \right)$$

$$M_{0j}(v_j) = \exp \left(-\frac{v_j}{T} \right)$$

and consider the Bernstein-Schrödinger system

$$\begin{cases} M_{i0}(u_i) + \sum_{j \in \mathcal{J}} M_{ij}(u_i, v_j) = 1 \\ M_{0j}(v_j) + \sum_{i \in \mathcal{I}} M_{ij}(u_i, v_j) = 1 \end{cases}.$$

We need to show that $M_{xy}(.,.)$, $M_{x0}(.)$ and $M_{0y}(.)$ satisfy the properties stated in Theorem 1. It is straightforward to show that conditions (i) and (ii) in Theorem 1 follow directly from assumptions (a) and (b) on Ψ. Moreover, letting $T \to 0^+$, assumptions (c) and (d) together imply that condition (iii) is satisfied. Hence we can apply Theorem 1, and it follows that a solution u_i^T, v_j^T to the system exists. Note that $M_{i0}(u_i^T) \leq 1$ and $M_{0j}(v_j^T) \leq 1$ imply that $u_i^T \geq 0$ and $v_j^T \geq 0$. Now, consider the sequence obtained by taking $T = k$, $k \in \mathbb{N}$. Then, up to a subsequence extraction, we may assume $u_i^k \to \bar{u}_i \in \mathbb{R}^+ \cup \{+\infty\}$ and $v_j^k \to \bar{v}_j \in \mathbb{R}^+ \cup \{+\infty\}$. It follows that $\Psi_{ij}\left(u_i^k, v_j^k\right)$ converges in $\mathbb{R}^+ \cup \{+\infty\}$, hence $\mu_{ij}^k = M_{ij}\left(u_i^k, v_j^k\right)$ converges toward $\bar{\mu}_{ij} \in [0, 1]$. Similarly, the limits $\bar{\mu}_{i0}$ and $\bar{\mu}_{0j}$ exist in $[0, 1]$. Hence (i) in Definition 1 is met. By continuity, $\bar{\mu}$ satisfies

$$\begin{cases} \bar{\mu}_{i0} + \sum_{j \in \mathcal{J}} \bar{\mu}_{ij} = 1, \\ \bar{\mu}_{0j} + \sum_{i \in \mathcal{I}} \bar{\mu}_{ij} = 1 \end{cases},$$

which established (ii). Let us show that (iii) holds, that is, that $\Psi_{ij}\left(u_i, v_j\right) \geq 0$ for any i and j. Assume otherwise. Then there exists $\epsilon > 0$ such that for k large enough, $\Psi_{ij}\left(u_i^k, v_j^k\right) < -\epsilon$, so that $\mu_{ij}^k > \exp\left(\epsilon/T\right) \to +\infty$, contradicting $\bar{\mu}_{ij} \leq 1$. Thus, we have established (iii). Finally, we show that (iv) holds. Assume otherwise. Then there is i and j such that $\bar{\mu}_{ij} > 0$ and $\Psi_{ij}\left(\bar{u}_i, \bar{v}_j\right) > 0$. This implies that there exists $\epsilon > 0$ such that for k large enough, $\mu_{ij}^k > \epsilon$, thus $\Psi_{ij}\left(u_i^k, v_j^k\right) < -T \log \epsilon \to 0$, hence $\Psi_{ij}\left(\bar{u}_i, \bar{v}_j\right) \leq 0$, a contradiction. Hence (iv) holds; this completes the proof and establishes that $(\bar{\mu}, \bar{u}, \bar{v})$ is an equilibrium assignment.

3 ITU Matching with Heterogeneity

Following [8], we now assume that individuals may be gathered into groups of agents of similar observable characteristics, or types, but heterogeneous tastes. We let $\mathcal{X}$ and $\mathcal{Y}$ be the sets of *types* of men and women. An individual man $i \in \mathcal{I}$ has type $x_i \in \mathcal{X}$; similarly, an individual woman $j \in \mathcal{J}$ has type $y_j \in \mathcal{Y}$. We assume that there is a mass n_x of men of type x and m_y of women of type y, respectively. Assume further that

$$\Psi_{ij}\left(u_i, v_j\right) = \Psi_{x_i y_j}\left(u_i - T\varepsilon_{iy}, v_j - T\eta_{xj}\right),$$

where ϵ and η are i.i.d. random vectors drawn from a Gumbel distribution, and where $T > 0$ is a temperature parameter. Unassigned agents get $T\varepsilon_{i0}$ and $T\eta_{0j}$. For all i such that $x_i = x$ and $y_j = y$, the stability condition implies

$$\Psi_{x_i y_j}\left(u_i - T\varepsilon_{iy}, v_j - T\eta_{xj}\right) \geq 0.$$

Hence,

$$\min_{\substack{i:x_i=x \\ j:y_j=y}} \Psi_{x_i y_j}\left(u_i - T\varepsilon_{iy}, v_j - T\eta_{xj}\right) \geq 0.$$

Thus, letting

$$U_{xy} = \min_{i:x_i=x}\{u_i - T\varepsilon_{iy}\} \text{ and } V_{xy} = \min_{j:y_j=y}\{v_j - T\eta_{xj}\},$$

we have $\Psi_{xy}\left(U_{xy}, V_{xy}\right) \geq 0$, and with

$$\mu_{xy} = \sum_{\substack{i:x_i=x \\ j:y_j=y}} \mu_{ij}$$

we have that $\mu_{xy} > 0$ implies $\Psi_{xy}(U_{xy}, V_{xy}) = 0$, and by a standard argument (the random vectors ϵ and η are drawn from distributions with full support, hence there will be at least a man i of type x and a woman j of type y such that i prefers type y and j prefers type x, that is, $\mu_{xy} > 0$ for all x and y)

$$\Psi_{xy}(U_{xy}, V_{xy}) = 0 \ \forall x \in \mathcal{X}, y \in \mathcal{Y}.$$

Note that this is an extension to the ITU case of the analysis in Galichon and Salanié [7], building on Choo and Siow [3]. We have that

$$u_i = \max_{y} \{U_{xy} + T\varepsilon_{iy}, T\varepsilon_{i0}\} \text{ and } v_j = \max_{x} \{V_{xy} + T\eta_{xj}, T\eta_{0j}\},$$

thus a standard result from Extreme Value Theory (see Choo and Siow [3] for a derivation) yields

$$U_{xy} = T \log \frac{\mu_{xy}}{\mu_{x0}} \text{ and } V_{xy} = T \log \frac{\mu_{xy}}{\mu_{0y}},$$

so we see that μ_{xy} satisfies

$$\begin{cases} \mu_{x0} + \sum_{y \in \mathcal{Y}} \mu_{xy} = n_x \\ \mu_{0y} + \sum_{x \in \mathcal{X}} \mu_{xy} = m_y \\ \Psi_{xy}\left(T \log \frac{\mu_{xy}}{\mu_{x0}}, T \log \frac{\mu_{xy}}{\mu_{0y}}\right) = 0 \end{cases}. \tag{2}$$

The various cases of interest discussed above, namely TU, NTU, LTU, and ETU cases yield, respectively,

$$\mu_{xy} = \mu_{x0}^{1/2} \mu_{0y}^{1/2} \exp \frac{\Phi_{xy}}{2} \text{ (TU)}$$

$$\mu_{xy} = \min\left(\mu_{x0} e^{\alpha_{xy}}, \mu_{0y} e^{\gamma_{xy}}\right) \text{ (NTU)}$$

$$\mu_{xy} = e^{(\lambda_{xy}\alpha_{xy} + \zeta_{xy}\gamma_{xy})/(\lambda_{xy} + \zeta_{xy})} \mu_{x0}^{\lambda_{xy}/(\lambda_{xy} + \zeta_{xy})} \mu_{0y}^{\zeta_{xy}/(\lambda_{xy} + \zeta_{xy})} \text{ (LTU)}$$

$$\mu_{xy} = \left(\frac{e^{-\alpha_{xy}/\tau_{xy}} \mu_{x0}^{-1/\tau_{xy}} + e^{-\gamma_{xy}/\tau_{xy}} \mu_{0y}^{-1/\tau_{xy}}}{2}\right)^{-\tau_{xy}} \text{ (ETU)},$$

(see [3]). To apply Theorem 1, we let $M_{xy}(u_x, v_y)$ be the value m solution to (for a proof of existence and uniqueness of such a solution, see Lemma 1 of [8])

$$\Psi_{xy}(T \log m + u_x, T \log m + v_y) = 0,$$

and let

$$M_{x0}(u_x) = \exp\left(\frac{-u_x}{T}\right) \text{ and } M_{0y}(v_y) = \exp\left(\frac{-v_y}{T}\right).$$

In [8], we rewrote system (2) as a nonlinear Bernstein-Schrödinger system, namely

$$\begin{cases} M_{x0}(u_x) + \sum_{y \in \mathcal{Y}} M_{xy}(u_x, v_y) = n_x \\ M_{0y}(v_y) + \sum_{x \in \mathcal{X}} M_{xy}(u_x, v_y) = m_y \end{cases}.$$

Theorem 4. *The nonlinear Bernstein-Schrödinger system in (2) has a unique solution.*

Proof. The proof follows directly from the application of Theorem 1. It is easy to check that the conditions on $M_{x0}(.)$ and $M_{0y}(.)$ required by Theorem 1 are met in this case. Lemma 1 in [8] provides a proof that M_{xy} satisfies these conditions. $\blacksquare$

4 Discussion

In this note, we have argued how matching problems may be formulated as a system of nonlinear equations, also known as the Bernstein-Schrödinger equation or Schrödinger's problem [15]. We have shown existence and uniqueness of a solution under certain conditions, and have explicated the link with various matching problems, with or without heterogeneity. Solving such a system of equations requires an algorithm that we call the Iterative Projection Fitting Procedure (IPFP); in practice, this algorithm converges very quickly. Our setting can be extended in several ways. One of them is to consider the case with unassigned agents. In that case, we have the additional constraint that $\sum_x n_x = \sum_y m_y$, thus the nonlinear Bernstein-Schrödinger system, which in this case writes as

$$\begin{cases} \sum_{y \in \mathcal{Y}} M_{xy}(u_x, v_y) = n_x \\ \sum_{x \in \mathcal{X}} M_{xy}(u_x, v_y) = m_y \end{cases}$$

has a degree of freedom, as the sum over $x \in \mathcal{X}$ of the first set of equations coincides with the sum over $y \in \mathcal{Y}$ of the second one. The one-dimensional manifold of solutions of this problem is studied in [2].

References

1. Becker, G.S.: A theory of marriage: part I. J. Polit. Econ. **81**(4), 46–813 (1973)
2. Carlier, G., Galichon, A.: A remark on the manifold of solutions of a nonlinear Schrödinger system. Working Paper (2014)
3. Choo, E., Siow, A.: Who marries whom and why. J. Polit. Econ. **114**(1), 175–201 (2006)
4. Cuturi, M.: Sinkhorn distances: lightspeed computation of optimal transport. Adv. Neural Inf. Process. Syst. **26**, 2292–2300 (2013)
5. Deming, W.E., Stephan, F.F.: On a least squares adjustment of a sampled frequency table when the expected marginal totals are known. Ann. Math. Stat. **11**(4), 427–444 (1940)
6. Gale, D., Shapley, L.S.: College admissions and the stability of marriage. Am. Math. Monthly **69**(1), 9–15 (1962)
7. Galichon, A., Salanié, B.: Cupid's invisible hand: social surplus and identification in matching models. Working Paper (2014)
8. Galichon, A., Kominers, S.D., Weber, S.: An empirical framework for matching with imperfectly transferable utility. Working Paper (2014)
9. Gale, D., Nikaido, H.: The Jacobian matrix and global univalence of mappings. Mathematische Annalen **159**(2), 81–93 (1965)

10. Kruithof, J.: Telefoonverkeersrekening. De Ingenieur **52**, E15–E25 (1937)
11. McKenzie, L.W.: Matrices with Dominant Diagonals and Economic Theory. Mathematical Methods in the Social Sciences 1959. Stanford University Press, Stanford (1960)
12. Ruschendorf, L.: Convergence of the iterative proportional fitting procedure. Ann. Stat. **23**(4), 1160–1174 (1995)
13. Rüschendorf, L., Thomsen, W.: Closedness of sum spaces and the generalized Schrödinger problem. Theory Probab. Appl. **42**(3), 483–494 (1998)
14. Shapley, L., Shubik, M.: The assignment game I: the core. Int. J. Game Theory **1**, 111–130 (1972)
15. Schrödinger, E.: Über die Umkehrung der Naturgesetze. In: Akad. Wiss. Berlin. Phys. Math., vol. 144, pp. 144–153. Sitzungsberichte Preuss (1931)

Some Geometric Consequences
of the Schrödinger Problem

Christian Léonard$^{(\boxtimes)}$

Modal-X, Université Paris Ouest,
Bât. G, 200 av. de la République, 92001 Nanterre, France
christian.leonard@u-paris10.fr

Abstract. This note presents a short review of the Schrödinger problem and of the first steps that might lead to interesting consequences in terms of geometry. We stress the analogies between this entropy minimization problem and the renowned optimal transport problem, in search for a theory of lower bounded curvature for metric spaces, including discrete graphs.

Keywords: Schrödinger problem · Entropic interpolations · Optimal transport · Displacement interpolations · Lower bounded curvature of metric spaces · Lott-Sturm-Villani theory

Introduction

This note presents a short review of the Schrödinger problem and of the first steps that might lead to interesting consequences in terms of geometry. It doesn't contain any new result, but is aimed at introducing this entropy minimization problem to the community of geometric sciences of information.

We briefly describe Schrödinger's problem, see [12] for a recent review. It is very similar to an optimal transport problem. Several analogies with the Lott-Sturm-Villani theory about lower bounded curvature on geodesic spaces, which has been thoroughly investigated recently with great success, will be emphasized. The results are presented in the setting of a Riemannian manifold.

As a conclusion, some arguments are put forward that advocate for replacing the optimal transport problem by the Schrödinger problem when seeking for a theory of lower bounded curvature on discrete graphs.

For any measurable space Y, $\mathrm{M}(Y)$ is the set of all positive measures and $\mathrm{P}(Y)$ is the subset of all probability measures on Y.

1 Optimal Transport

Let $\mathcal{X}$ be some state space equipped with a σ-field so that we can consider measures on $\mathcal{X}$ and $\mathcal{X}^2$. The Monge problem amounts to find a mapping $T : \mathcal{X} \to \mathcal{X}$ that solves the minimizing problem

$$\int_{\mathcal{X}} c(x, Tx)\, \mu_0(dx) \to \min; \qquad T : \mathcal{X} \to \mathcal{X} \text{ such that } T_{\#}\mu_0 = \mu_1, \qquad (1)$$

© Springer International Publishing Switzerland 2015
F. Nielsen and F. Barbaresco (Eds.): GSI 2015, LNCS 9389, pp. 60–68, 2015.
DOI: 10.1007/978-3-319-25040-3_7

where $c : \mathcal{X}^2 \to [0, \infty)$ is a given measurable function, $\mu_0, \mu_1 \in P(\mathcal{X})$ are prescribed probability measures on $\mathcal{X}$ and $T_\# \mu_0(dy) := \mu_0(T^{-1}(dy))$ is the image of μ_0 by the measurable mapping T. One interprets $c(x, y)$ as the cost for transporting a unit mass from $x \in \mathcal{X}$ to $y \in \mathcal{X}$. Hence the integral $\int_\mathcal{X} c(x, Tx) \mu_0(dx)$ represents the global cost for transporting the mass profile $\mu_0 \in P(\mathcal{X})$ onto $T_\# \mu_0 \in P(\mathcal{X})$ by means of the transport mapping T. A solution of the Monge problem is a mapping T from $\mathcal{X}$ to $\mathcal{X}$ which transports the mass distribution μ_0 onto the target mass distribution μ_1 at a minimal cost.

The most efficient way to solve Monge's problem is to consider the following relaxed version which was introduced by Kantorovich during the 40's. The Monge-Kantorovich problem is

$$\int_{\mathcal{X}^2} c(x, y)\, \pi(dxdy) \to \min; \qquad \pi \in P(\mathcal{X}^2) : \pi_0 = \mu_0, \pi_1 = \mu_1, \qquad (2)$$

where $\pi_0(dx) := \pi(dx \times \mathcal{X})$ and $\pi_1(dy) := \pi(\mathcal{X} \times dy)$ are the marginals of the joint distribution π on the product space $\mathcal{X}^2$. It consists of finding a coupling $\pi \in P(\mathcal{X}^2)$ of the mass distributions μ_0 and μ_1 which minimizes the average cost $\int_{\mathcal{X}^2} c(x, y)\, \pi(dxdy)$. Considering the so-called deterministic coupling $\pi^T(dxdy) := \mu_0(dx)\delta_{Tx}(dy)$ of μ_0 and $T_\# \mu_0$ where δ_y stands for the Dirac measure at y, we see that (2) extends (1): if (2) admits π^T as a solution, then T solves (1). In contrast with the highly nonlinear problem (1), (2) enters the well-understood class of convex minimization problems. The interest of the Monge-Kantorovich problem goes over its tight relation with the Monge transport problem. It is a source of fertile connections. For instance, it leads to the definition of many useful distances on the set $P(\mathcal{X})$ of probability measures. Other connections are sometimes more surprising at first sight. We shall invoke below a few links with the geometric notion of curvature.

A key reference for the optimal transport theory is Villani's textbook [19].

2 Schrödinger Problem

In 1931, that is ten years before Kantorovich discovered (2), Schrödinger [16,17] addressed a new statistical physics problem motivated by its amazing similarity with several aspects of the time reversal symmetry in quantum mechanics. In modern terms, the Schrödinger problem is expressed as follows

$$H(\pi|\rho) \to \min; \pi \in P(\mathcal{X}^2) : \pi_0 = \mu_0, \pi_1 = \mu_1, \qquad (3)$$

where $\rho \in M(\mathcal{X}^2)$ is some reference positive measure on the product space $\mathcal{X}^2$ and

$$H(p|r) := \int_Y \log(dp/dr)\, dp \in (-\infty, \infty], \quad p, r \in P(Y)$$

denotes the relative entropy of the probability measure p on Y with respect to the reference positive measure r on the same space Y and it is understood that $H(p|r) = \infty$ when p is not absolutely continuous with respect to r. For a survey of basic results about the Schrödinger problem, see [12].

Schrödinger considers a large collection of N independent particles living in a Riemannian manifold $\mathcal{X}$ (he takes $\mathcal{X} = \mathbb{R}^n$, but we need a manifold for further developments) and moving according to a Brownian motion. It is supposed that at time $t = 0$ they are spatially distributed according to a profile $\mu_0 \in \mathrm{P}(\mathcal{X})$ and that at time $t = 1$, we observe that they are distributed according to a profile $\mu_1 \in \mathrm{P}(\mathcal{X})$ which is far away from the expected configuration. One asks what is the most likely behavior of the whole system of particles which performs this very unlikely event. It is a large deviation problem which is solved by means of Sanov's theorem (see Föllmer's lecture notes [7] for the first rigorous derivation of Schrödinger's problem) and leads to (3) with the reference measure

$$\rho^\epsilon(dxdy) = \mathrm{vol}(dx)\,(2\pi\epsilon)^{-n/2}\exp\left(-\frac{d(x,y)^2}{2\epsilon}\right)\mathrm{vol}(dy)$$

where d is the Riemannian distance. It is the joint law of the endpoint position (X_0, X_1) of a Brownian motion $(X_t)_{0 \leq t \leq 1}$ on the unit time interval $[0,1]$ with variance ϵ which starts at time $t = 0$ uniformly at random according to the volume measure on $\mathcal{X}$ (this process is reversible). For any probability π on $\mathcal{X}^2$ with a finite entropy we easily see that

$$\lim_{\epsilon \to 0^+} \epsilon H(\pi|\rho^\epsilon) = \int_{\mathcal{X}^2} \frac{d(x,y)^2}{2}\,\pi(dxdy)$$

which is an average cost as in (2) with respect to the so-called quadratic cost

$$c(x,y) = d(x,y)^2/2. \tag{4}$$

Therefore the Monge-Kantorovich problem with the quadratic cost

$$\int_{\mathcal{X}^2} \frac{d(x,y)^2}{2}\,\pi(dxdy) \to \min; \qquad \pi \in \mathrm{P}(\mathcal{X}^2) : \pi_0 = \mu_0, \pi_1 = \mu_1, \tag{5}$$

appears as the limit as the fluctuation parameter ϵ tends to zero of a family of Schrödinger problems associated with a Gaussian reference measure. This is a specific instance of a general phenomenon that has been discovered by Mikami [14] and explored in detail in [11].

3 Dynamical Analogues

Many aspects of the static problems (2) and (3) are easier to clarify by means of their dynamical analogues. To keep things easy, we stick to the quadratic cost (4) on a Riemannian manifold.

Notation. We introduce some useful notation. The path space on $\mathcal{X}$ is denoted by $\Omega \subset \mathcal{X}^{[0,1]}$. The canonical process $(X_t)_{t \in [0,1]}$ is defined for each $t \in [0,1]$ and $\omega \in \Omega$ by $X_t(\omega) = \omega_t \in \mathcal{X}$. For any $Q \in \mathrm{M}(\Omega)$ and $0 \leq t \leq 1$, we denote $Q_t := (X_t)_\# Q := Q(X_t \in \cdot) \in \mathrm{M}(\mathcal{X})$ the law of X_t under Q. We denote the endpoint distribution $Q_{01}(dxdy) := Q(X_0 \in dx, X_1 \in dy) \in \mathrm{M}(\mathcal{X}^2)$ and use the probabilistic notation E_P for $\int_\Omega dP$.

Displacement Interpolations. We introduce the dynamical analogue of (5). It consists of minimizing the average kinetic action

$$E_P \int_{[0,1]} |\dot{X}_t|^2_{X_t}/2\, dt \to \min; \quad P \in \mathrm{P}(\Omega) : P_0 = \mu_0, P_1 = \mu_1 \tag{6}$$

under the constraint that the initial and final marginals P_0 and P_1 of P are equal to the prescribed probability measures μ_0 and $\mu_1 \in \mathrm{P}(\mathcal{X})$ on $\mathcal{X}$.

Suppose for simplicity that there is a unique solution P to this problem. Then P has the form $P(\cdot) = \int_{\mathcal{X}^2} \delta_{\gamma^{xy}}(\cdot)\, \pi(dxdy)$ where $\delta_{\gamma^{xy}}$ is the Dirac mass at γ^{xy}: the unique geodesic between x and y, and its endpoint projection $P_{01} = \pi \in \mathrm{P}(\mathcal{X}^2)$ is the unique solution of the optimal transport problem (5).

Definition 1. *The* displacement interpolation *between μ_0 and μ_1 is the flow of marginals* $[\mu_0, \mu_1] := (P_t)_{0 \le t \le 1}$ *of the solution P of (6).*

This notion has been introduced by McCann in his PhD Thesis [13]. It is the basis of the development of the theory of lower bounds for the Ricci curvature of geodesic spaces, see the textbooks [1,19].

Entropic Interpolations. Now, we introduce the dynamical analogue of (3). It consists of minimizing the relative entropy

$$H(P|R) := E_P \log(dP/dR) \to \min; \quad P \in \mathrm{P}(\Omega) : P_0 = \mu_0, P_1 = \mu_1 \tag{7}$$

with respect to the reference path measure $R \in \mathrm{M}(\Omega)$ under the same marginal constraints as in (6). Following Schrödinger, if we choose R to be the law of the reversible Brownian motion on $\mathcal{X}$, we obtain with Girsanov's theory that

$$H(P|R) = H(P_0|\mathrm{vol}) + E_P \int_{[0,1]} |v_t^P(X_t)|^2_{X_t}/2\, dt$$

where vol denotes the volume measure and v_t^P is the Nelson forward velocity field of the diffusion law P, [15]. When $\mathcal{X} = \mathbb{R}^n$, denoting $E_P[\cdot|\cdot]$ the conditional expectation,

$$v_t^P(x) = \lim_{h \to 0, h > 0} \frac{1}{h} E_P[X_{t+h} - X_t \mid X_t = x]. \tag{8}$$

Definition 2. *The* entropic interpolation *between μ_0 and μ_1 is the flow of marginals* $[\mu_0, \mu_1]^R := (P_t)_{0 \le t \le 1}$ *of the unique solution P of (7).*

If $P \in \mathrm{P}(\Omega)$ solves the dynamical problem (7), then $P_{01} \in \mathrm{P}(\mathcal{X}^2)$ solves the static problem

$$H(\pi|R_{01}) \to \min; \pi \in \mathrm{P}(\mathcal{X}^2) : \pi_0 = \mu_0, \pi_1 = \mu_1, \tag{9}$$

where the reference measure $R_{01} \in \mathrm{M}(\mathcal{X}^2)$ is the endpoint projection of the reference path measure $R \in \mathrm{M}(\Omega)$.

Slowing Down. As already seen in the static case, the analogy between (6) and (7) is not only formal. Considering the slowed down process $R^\epsilon = (X^\epsilon)_\# R$ which is the law of $X_t^\epsilon = X_{\epsilon t}, 0 \le t \le 1$, it is known that

$$\epsilon H(P|R^\epsilon) \to \min; \quad P \in \mathrm{P}(\Omega) : P_0 = \mu_0, P_1 = \mu_1$$

Γ-converges to (6), see [11]. In particular, the entropic interpolation $[\mu_0, \mu_1]^{R^\epsilon}$ is a smooth approximation of the displacement interpolation $[\mu_0, \mu_1]$.

This kind of convergence also holds for optimal L^1-transport on graphs [9] and Finsler manifolds (instead of optimal L^2-transport on a Riemannian manifold) where diffusion processes must be replaced by random processes with jumps (work in progress).

4   Dynamics of the Interpolations

Unlike entropic interpolations, displacement interpolations lack regularity. Already known results about the dynamics of the displacement interpolations in the so-called RCD geodesic spaces with a Ricci curvature bounded from below can be found in [8]. Understanding the dynamics of entropic interpolations could be a first step (before letting ϵ tend to zero) to recover such results.

Dynamics of the Displacement Interpolations. A *formal* representation of the displacement interpolation is given by

$$\dot{X}_t = \nabla \psi(t, X_t), \quad P\text{-a.s.}$$

where P is a solution of (6), ψ is the viscosity solution of the Hamilton-Jacobi equation

$$\begin{cases} \partial_t \psi + |\nabla \psi|^2/2 = 0 \\ \psi_{t=1} = \psi_1 \end{cases} \tag{10}$$

and ψ_1 is in accordance with the endpoint data μ_0 and μ_1. Note that

$$\ddot{X}_t = \nabla[\partial_t \psi + |\nabla \psi|^2/2](t, X_t) = 0, \quad P\text{-a.s.} \tag{11}$$

fitting the standard geodesic picture.

Dynamics of the Entropic Interpolations. Similarly, a *rigorous* representation of the entropic interpolation is given by

$$v_t^P = \nabla \psi(t, X_t), \quad P\text{-a.s.}$$

where v^P is defined at (8), P is the solution of (7) and ψ is the classical solution of the Hamilton-Jacobi-Bellman equation

$$\begin{cases} \partial_t \psi + \Delta \psi/2 + |\nabla \psi|^2/2 = 0, \\ \psi_{t=1} = \psi_1. \end{cases} \tag{12}$$

Iterating time derivations in the spirit of (8) in both directions of time allows to define a relevant notion of stochastic acceleration a^P, see for instance [5,15]. We obtain the following analogue of (11)

$$a_t^P = \frac{1}{2}\nabla[\partial_t\psi + \Delta\psi/2 + |\nabla\psi|^2/2] + \frac{1}{2}\nabla[-\partial_t\varphi + \Delta\varphi/2 + |\nabla\varphi|^2/2](t, X_t) = 0, \quad P\text{-a.s.}$$

where φ solves some HJB equation $\begin{cases} -\partial_t\varphi + \Delta\varphi/2 + |\nabla\varphi|^2/2 = 0 \\ \varphi_{t=0} = \varphi_0 \end{cases}$ in the other direction of time.

5 Interpolations are Sensitive to Ricci Curvature

On a Riemannian manifold $\mathcal{X}$, one says that the Ricci curvature is bounded below by some constant $\kappa \in \mathbb{R}$, when

$$\mathrm{Ric}_x(v, v) \geq \kappa g_x(v, v), \quad \forall (x, v) \in \mathrm{T}\mathcal{X}$$

where Ric is the Ricci tensor and g is the Riemannian metric defined on the tangent bundle $\mathrm{T}\mathcal{X}$.

Displacement Interpolations. Ten years ago, Sturm and von Renesse [18] have discovered that this lower bound holds if and only if along any displacement interpolation $(\mu_t)_{0\leq t\leq 1}$, the entropy

$$t \in [0, 1] \mapsto H(\mu_t|\mathrm{vol}) \in (-\infty, \infty]$$

is κ-convex with respect to W_2, i.e.

$$H(\mu_t|\mathrm{vol}) \leq (1 - t)H(\mu_0|\mathrm{vol}) + tH(\mu_1|\mathrm{vol}) - \kappa\frac{t(1 - t)}{2}W_2^2(\mu_0, \mu_1), \quad \forall t \in [0, 1].$$
(13)

The Wasserstein distance W_2 of order 2 is defined by means of the quadratic optimal transport problem by $W_2^2(\mu_0, \mu_1) := \inf(5)$. It plays the role of a Riemannian distance on the set $\mathrm{P}_2(\mathcal{X}) := \{\mu \in \mathrm{P}(\mathcal{X}); \int_{\mathcal{X}} d(x_o, x)^2 \mu(dx) < \infty\}$ of all probability measures on $\mathcal{X}$ with a finite second moment. Accordingly, the displacement interpolations are similar to geodesics. Unfortunately $(\mathrm{P}_2(\mathcal{X}), W_2)$ is not a Riemannian manifold and the displacement interpolations are not regular enough to be differentiable in time. In particular, the expected equivalent local statement

$$\frac{d^2}{dt^2}H(\mu_t|\mathrm{vol}) \geq \kappa W_2^2(\mu_0, \mu_1), \quad \forall 0 \leq t \leq 1$$
(14)

of the convex inequality (13) is meaningless.

However, this remarkable result of Sturm and von Renesse was the basic step for developing the Lott-Sturm-Villani theory of lower bounded Ricci curvature of geodesic spaces, see [19]. The program of this theory is to extend the notion of lower bounded Ricci curvature from Riemannian manifolds to geodesic spaces (a special class of metric spaces) by taking advantage of the *almost* Riemannian structure of $(P_2(\mathcal{X}), W_2)$ and in particular of the dynamical properties the corresponding almost geodesics: the displacement interpolations. The heuristic formula obtained with Otto's heuristic calculus, see [19, Chap. 15], for the second derivative of the entropy along a displacement interpolations (μ_t) is

$$\frac{d^2}{dt^2} H(\mu_t|\mathrm{vol}) = \Gamma_2(\psi_t), \quad 0 \le t \le 1, \tag{15}$$

where ψ solves the Hamilton-Jacobi equation (10). We see that it formally implies (14) under the Γ_2-criterion

$$\Gamma_2(\psi) \ge \kappa g(\nabla\psi, \nabla\psi), \quad \forall\psi$$

where the Bakry-Émery operator Γ_2 is given by

$$\Gamma_2(\psi) = \mathrm{Ric}(\nabla\psi) + \sum_{i,j}(\partial_i\partial_j\psi)^2.$$

Entropic Interpolations. As an interesting consequence of the dynamical properties of the entropic interpolations, we obtain in [10] that along any entropic interpolation $(\mu_t)_{0\le t\le 1}$ on a Riemannian manifold, we have

$$\frac{d^2}{dt^2} H(\mu_t|\mathrm{vol}) = \frac{1}{2}\left\{\Gamma_2(\varphi_t) + \Gamma_2(\psi_t)\right\}$$

where φ and ψ are the solutions of the above HJB equations (12) in both directions of time. This formula is a rigorous (in the sense that the second derivative is well defined) analogue of the heuristic identity (15).

Conclusion

As a conclusion we sketch a research program and cite a few recent publications related to the Schrödinger problem in the domains of numerical analysis and engineering sciences.

A Research Program

In view of the analogies between the optimal transport problem and the Schrödinger problem on a Riemannian manifold, one can hope that the program of the Lott-Sturm-Villani theory can be transferred successfully from geodesic spaces to a larger class of metric spaces. As a guideline, one should consider the Schrödinger problem as the basic "geodesic" problem instead of the Monge-Kantorovich problem. We see several advantages to this strategy:

1. Unlike the displacement interpolations, the entropic interpolations are regular enough for their second derivative in time to be considered without any trouble.
2. Slowing down the reference process, which might be a diffusion process on a RCD space (see [8]) or a random walk on a graph (see [9,10]), one retrieves displacement interpolations as limits of entropic interpolations.
3. As shown in [9], the entropic interpolations are well defined on discrete graphs. They also lead to natural displacement interpolations. Remark that discrete graphs are not geodesic and as a consequence, are ruled out by the Lott-Sturm-Villani approach.

This program remains to be investigated ...

Recent Literature

A recent resurgence of the use of the Schrödinger problem arises in applied and numerical sciences. In [6], the Schrödinger problem is solved using the Sinkhorn algorithm. This appears to be very competitive with respect to other optimal transport solvers because of its simplicity, parallelism and convergence speed (at the expense of an extra smoothing).

A notion of interpolation quite similar to the entropic interpolation might be defined by means of entropic barycenters as introduced in [2]. It would be interesting to investigate their curvature properties.

Motivated both by engineering problems and theoretical physics, in the spirit of [14] the recent papers [3,4] look at the entropic interpolations with a stochastic control viewpoint.

References

1. Ambrosio, L., Gigli, N., Savaré, G.: Gradient Flows in Metric Spaces and in the Space of Probability Measures. Lectures in Mathematics ETH Zürich, 2nd edn. Birkhäuser, Basel (2008)
2. Benamou, J.-D., Carlier, G., Cuturi, M., Nenna, L., Peyré, G.: Iterative Bregman projections for regularized transportation problems. SIAM J. Sci. Comput. **37**(2), A1111–A1138 (2015)
3. Chen, Y., Georgiou, T., Pavon, M.: On the relation between optimal transport and schrödinger bridges: a stochastic control viewpoint. Preprint, arXiv:1412.4430
4. Chen, Y., Georgiou, T., Pavon, M.: Optimal transport over a linear dynamical system. Preprint, arXiv:1502.01265
5. Chung, K.L., Zambrini, J.-C.: Introduction to Random Time and Quantum Randomness. World Scientific, Hackensack (2008)
6. Cuturi, M.: Sinkhorn distances: lightspeed computation of optimal transport. NIPS 2013, arXiv:1306.0895 (2013)
7. Föllmer, H.: Random fields and diffusion processes. In: Hennequin, P.-L. (ed.) École d'été de Probabilités de Saint-Flour XV-XVII-1985-87. Lecture Notes in Mathematics, vol. 1362, pp. 101–203. Springer, Berlin (1988)

8. Gigli, N.: Nonsmooth differential geometry. An approach tailored for spaces with Ricci curvature bounded from below. Preprint. http://cvgmt.sns.it/paper/2468/

9. Léonard, C.: Lazy random walks and optimal transport on graphs. Preprint. arXiv:1308.0226, to appear in Ann. Probab

10. Léonard, C.: On the convexity of the entropy along entropic interpolations. Preprint, arXiv:1310.1274

11. Léonard, C.: From the Schrödinger problem to the Monge-Kantorovich problem. J. Funct. Anal. **262**, 1879–1920 (2012)

12. Léonard, C.: A survey of the Schrödinger problem and some of its connections with optimal transport. Discrete Contin. Dyn. Syst. A **34**(4), 1533–1574 (2014)

13. McCann, R.: A convexity theory for interacting gases and equilibrium crystals. Ph.D. thesis, Princeton Univ. (1994)

14. Mikami, T.: Monge's problem with a quadratic cost by the zero-noise limit of h-path processes. Probab. Theor. Relat. Fields **129**, 245–260 (2004)

15. Nelson, E.: Dynamical Theories of Brownian Motion. Princeton University Press, Princeton (1967)

16. Schrödinger, E.: Über die umkehrung der naturgesetze. Sitzungsberichte Preuss. Akad. Wiss. Berlin. Phys. Math. **144**, 144–153 (1931)

17. Schrödinger, E.: Sur la théorie relativiste de l'électron et l'interprétation de la mécanique quantique. Ann. Inst. H. Poincaré **2**, 269–310 (1932). http://archive.numdam.org/ARCHIVE/AIHP/

18. Sturm, K.-T., von Renesse, M.-K.: Transport inequalities, gradient estimates, entropy, and Ricci curvature. Comm. Pure Appl. Math. **58**(7), 923–940 (2005)

19. Villani, C.: Optimal Transport. Old and New. Grundlehren der mathematischen Wissenschaften, vol. 338. Springer, Heidelberg (2009)

Optimal Transport, Independance Versus Indetermination Duality, Impact on a New Copula Design

Benoit Huyot[1], Yves Mabiala[1], and J. F. Marcotorchino[2]([✉])

[1] Thales Communications and Security, CENTAI Lab, Gennevilliers, France
[2] Thales R&T Directorate and LSTA Lab, UPMC Paris VI University, Paris, France
jeanfrancois.marcotorchino@thalesgroup.com

Abstract. This article leans on some previous results already presented in [10], based on the Fréchet's works, Wilson's entropy and Minimal Trade models in connection with the MKP transportation problem (MKP, stands for Monge-Kantorovich Problem). Using the duality between "independance" and "indetermination" structures, shown in this former paper, we are in a position to derive a novel approach to design a copula, suitable and efficient for anomaly detection in IT systems analysis.

Keywords: Optimal transport · MKP problem · Indetermination and independance structures · Condorcet and relational analysis · Copula theory

1 Introduction

The main purpose of this article is to link the Optimal Transport Theory to a special use of Copula Theory devoted to Anomaly Detection. Relying on MKP approaches through Wilson's entropy and Minimal Trade models variants, we derive a new copula function which gave very good and efficient results for practical and real life applications, dedicated to the prevention of Cyber-attacks.

2 Optimizing the Transportation Problem

In this section we consider two particular cost functions of the discrete transport problem: the *Alan Wilson's Entropy Model* and *the Minimal Trade Model*. In both cases, the cost is a function of the unkown joint distribution $h(\pi_{uv})$. The optimal solutions will be given in the next paragraphs. It is quite important to see that those two optimization problems induce a new duality: the duality between statistical "Independance" and logical "Indetermination" (see [1,9] and [10]).

© Springer International Publishing Switzerland 2015
F. Nielsen and F. Barbaresco (Eds.): GSI 2015, LNCS 9389, pp. 69–76, 2015.
DOI: 10.1007/978-3-319-25040-3_8

2.1 Alan Wilson's Entropy Model

The "Flows Entropy Model" of Alan Wilson was introduced in [19] and [15] for Spatial Interaction Modeling. His purpose is to determine the distribution of the normalized frequency flows π_{uv} (supposing $\pi_{uv} > 0 \ \forall u, v$) which maximizes the entropy of the system. The objective function to be maximized is based upon the Boltzman's or Shannon's Entropies:

$$max_\pi - \sum_{u=1}^{p} \sum_{v=1}^{q} \pi_{uv} ln(\pi_{uv}) \tag{1}$$

The optimal solution is obtained by using the Lagrange's multipliers to maximize the MKP problem. The explicit expression of the optimal solution is shown in Table 1. The flow maximizing entropy reveals a "statistical independance" between the p suppliers and the q clients.

2.2 The Minimal Trade Model

In the Minimal Trade Model, the criterion is a quadratic function measuring the squared deviation of the cells values from the no information situation (the uniform joint distribution) in order to get a smooth ventilation[1] of the origins-destinations π_{uv} values subject to the balanced marginals and mass preserving (see [9, 10]).

$$min_\pi \sum_{u=1}^{p} \sum_{v=1}^{q} \left(\pi_{uv} - \frac{1}{pq} \right)^2 \tag{2}$$

We solve this problem by using the Lagrange multipliers, since the function to optimize is convex, we are looking for a minimum. The optimal solution is shown in Table 1[2]. The optimal solution reveals an "indetermination structure" between the p suppliers and the q clients. This concept of "indetermination structure", studied in ([9] and [10]), is related to the relational aspect of the Condorcet's Voting Theory (see [1] and [12]), when "votes for" = "votes against".

Table 1. Variants of the MKP problem

Model	Objective function	Optimal solution
Alan wilson's Entropy Model	$max_\pi - \sum_{u=1}^{p} \sum_{v=1}^{q} \pi_{uv} ln \pi_{uv}$	$\pi_{uv}^* = \mu_u \nu_v \forall (u, v)$ $n_{uv}^* = \frac{n_{u.} n_{.v}}{N}$
The Minimal Trade Model	$min_\pi \sum_{u=1}^{p} \sum_{v=1}^{q} \left(\pi_{uv} - \frac{1}{pq} \right)^2$	$\pi_{uv}^* = \frac{\mu_u}{q} + \frac{\nu_v}{p} - \frac{1}{pq}$ $n_{uv}^* = \frac{n_{u.}}{q} + \frac{n_{.v}}{p} - \frac{N}{pq}$

---

[1] This explains the term: "Minimal Trade Model".

[2] There exist some constraints to satisfy for garanteeing the positivity of the optimal values π_{uv} (see [9]).

3   Continuous Variants of the Discrete Optimal Solution

3.1   The Density Solution for Alan Wilson's Problem

It can be simply shown that the continuous solutions of the Alan Wilson's problem is given by: $\pi^*(x, y) = f(x)g(y)$, where $\pi : [a, b] \times [c, d] \to [0, 1]$ is defined on the product of the two closed intervals of the cartesian plan with lengths: A and B. μ and ν have densities f and g respectively. The cumulative distribution function associated to the density solution is given by:

$$\Pi^*(x, y) = F(x)G(y)$$

3.2   The Density Solution for the Minimal Trade Problem

The optimal solution of the continuous version of the "Minimal Trade Problem" obtain by considering the Kantorovich's duality is given (see [11]) by:

$$\pi^*(x, y) = \frac{f(x)}{B} + \frac{g(y)}{A} - \frac{1}{AB} \tag{3}$$

where $\pi^* : [a, b] \times [c, d] \to [0, 1]$ is defined on the product of two closed intervals of the cartesian plan with lengths: A and B. f and g are density function assumed both be absolutely continuous and square-integrable. We have also the cumulative distribution function as follows:

$$\Pi^*(x, y) = y\frac{F(x)}{B} + x\frac{G(y)}{A} - \frac{xy}{AB} \tag{4}$$

4   Relationship with the Copula Theory

4.1   Some Basic Definitions About Copula

Copula have been introduced by M. Fréchet in 1951 (see [4]) as a function of cross-dependancy between random variables. Those initial definitions has been improved by the addition of a very important theorem, originated by A. Sklar in 1959 (see [16]) providing the existence, and in some case, the unicity of a copula.

A copula[3] is a function defined as a map $C : [0, 1] \times [0, 1] \to [0, 1]$ where :

- $C(u, 0) = C(0, v) = 0$, C is grounded
- $C(u, 1) = C(1, u) = u$, $\forall u \in [0, 1]$, marginal uniformity

Let us define $\Pi(x, y) = P(X \leq x, Y \leq y)$ as a joint cumulative distribution function of two random variables. Then we can present the Sklar's Theorem:

---

[3] $C(u, v) - C(u, v') - C(u', v) + C(u', v') \geq 0$ $\forall 0 \leq u \leq u' \leq 1$ $\forall 0 \leq v \leq v' \leq 1$ is known as the 2-increasing property. It is nothing but the so called Monge's condition which was coined by Alan Hoffmann in 1963 (see [6]), this is an additional link between optimal transport and copula theory.

Theorem 1 (Sklar's Theorem). *If X and Y are two continuous random variables then the joint cumulative distribution function could be written as a function of each cumulative distribution function, and this function is a copula:*

$$P(X \leq x, Y \leq y) = C(P(X \leq x), P(Y \leq y)) \Rightarrow \Pi(x,y) = C(F(x), G(y))$$

where Π is the joint cumulative distribution function, F (resp. G) is the cumulative distribution function of X (resp. Y).

Theorem 2 (Fréchet-Hoeffding bounds). $\forall (u,v) \in [0,1] \times [0,1]$ *if C is a copula function then:* $Max(u + v - 1, 0) \leq C(u,v) \leq Min(u,v)$

4.2 Illustration of the Sklar's Theorem on the Previous Alan Wilson's and Minimal Trade Solution

In the case of Alan Wilson's problem the corresponding copula is given by:

$$C^*(u,v) = uv \tag{5}$$

For the minimal trade problem the copula associated to the bivariate solution Π^* is given by:

$$C^*(u,v) = v\frac{F^{-1}(u)}{A} + u\frac{G^{-1}(v)}{B} - \frac{F^{-1}(u)G^{-1}(v)}{AB} \tag{6}$$

C^* verifies the characteristic properties of a copula since $F^{-1}(0) = 0, G^{-1}(0) = 0, F^{-1}(1) = A$ and $G^{-1}(1) = B$. With these bound assumptions we got:

- $C^*(u,0) = C^*(0,v) = 0$
- $C^*(1,v) = v$ and $C^*(u,1) = u$

4.3 Some Introductive Notations

$Y \in \{0,1\}$ is a binary random variable representing the abnormality status as:

- $Y = 0$ if event is abnormal, $Y = 1$ if it is not.
- $\hat{Y} = 0$ if event is detected as abnormal, $\hat{Y} = 1$ else.

By definition we put $P_0 = P(Y = 0)$. With this inverse notation we could write $G(y) = P_0 \ \forall \ 0 \leq y < 1$. If X is a random vector we will use the following definitions.[4]

$$P(X \leq x) = P(X_1 \leq x_1, ..., X_p \leq x_p) = F(x)$$

---

[4] In the last paragraph we suppose X to be at least bivariate.

4.4   Using Fréchet's Upper Bound as Copula Anomaly Detector

According to the definition of a conditional probability we have: $P(Y = 0|X) = P(Y = 0, X)/P(X)$. Using the upper bound of Theorem 2 and assuming: $P(X) \geq P_0$, we obtain the following inequality: $P(Y = 0|X) \leq P_0/P(X)$. According to the maximum of an a posteriori Bayes' rule, alarms can be raised only if the probability of this event is lower than twice the a priori probability of a targeted class.

$$P(Y = 0|X) \geq \frac{1}{2} \Rightarrow P(X) \leq 2P_0 \tag{7}$$

Consequently we obtain an upper bound for alarm activation condition on $P(X)$. If $0 \leq P(X) \leq 2P_0$, it is possible to activate an alarm. In the following we will see that under some conditions the more unfrequent an event is, the more likely abnormal it is.

4.5   Copula as a Tool for Detecting Unfrequent Events

So, an upper bound exists allowing to define a decision region as a trigger to detect an attack (abnormal event). Here we want to determine the limits of the conditional probability of very unfrequent event. Using a copula approach we can turn (using Theorem 1) the scoring function into:

$$P(Y = 0|X < x) = \frac{C(P_0, F(x))}{F(x)} = \frac{C(P_0, v)}{v}$$

Using simultaneously the 1-lipschitzian property of C, the 2-increasing property applied to a copula function the *de L'Hôpital's rule* to prove the differentiability at point 0 and relying on the definition of the "Lower Tail" dependance limit which is given by $\lambda_L = \lim_{v \to 0} C(v, v)/v$, we get:

$$\lim_{v \to 0} \frac{C(P_0, v)}{v} \geq \lambda_L \quad \forall P_0 \geq v. \tag{8}$$

This result gives a lower bound for: $C(P_0, v)/v$, which from now on will be considered as our scoring decision function. We are interested in measuring its variation on the decision region $[0, 2P_0]$.

5   Measure of the Classification Capability Through a Copula Approach

In this paragraph, we focus on the classification capability of the copula based model. The choice of the notion of *Receiver Operating Curve* (ROC) to represent the different compromises between *detection* and *false alarms* is determined by the fact that this ROC Curve allows to simultaneously compare two different classifiers. And consequently *"best"* model is the one for which the area under the ROC curve is the greatest possible. This decision region is called AUC (*Area Under ROC Curve*), finding its optimal value (the best classifier) amounts to solve a variational problem. In this section we show the link between AUC and the density function c, where $c(u, v) = \frac{\partial^2}{\partial u \partial v} C(u, v)$, from the Sklar's theorem and a variable change process we can derive:

$$\pi(x, y) = c(F(x), G(y)) f(x) g(y) \tag{9}$$

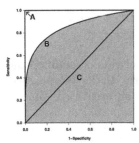

A "good" model optimizes the compromise between both the following criteria written according to a threshold[5] "s".

- **Sensitivity:** this measure counts how many events are properly classified as *anomalies*. This criterion is defined as the *true positive rate*. It represents the following quantity $P(s) = P(\hat{Y} = 0 | Y = 0) = \frac{C(P_0, s)}{P_0}$
- **1-Specificity (anti-Specificity):** it measures the misclassifications. 1-Specificity is the simple *false positive rate*. Mathematically we have: $Q(s) = 1 - P(\hat{Y} = 1 | Y = 1) = \frac{s}{1 - P_0}(1 - C(P_0, s))$

Choosing the *best* model for all potential thresholds "s" remains to maximize AUC. This area corresponds to the area under the function which associates for each given anti-specificity the related sensitivity given by the model. Classically the AUC is defined by:

$$AUC = \int_0^1 P(Q) dQ = \int_0^1 P(t) \cdot \frac{\partial}{\partial t} Q(t) dt \tag{10}$$

where P refers to Sensitivity and Q to the 1-Specificity. As $[0, s] \in [0, 1]$ and $X = (X_1, X_2)$, to measure the model performance according to the variation of each component of X, and its consequences on the scalar s, we introduce the link between X_1 and X_2 through the copula trick $s = C(F^{-1}(x_1), G^{-1}(x_2)) = C(s_1, s_2)$. We have shown in [7] that after developments and grouping (10) could be turned into:

$$AUC = -\frac{1}{2P_0(1 - P_0)} \int_0^1 (C(P_0, s) - 1)^2 \, ds + \frac{1 - P_0^2}{2P_0(1 - P_0)}$$

Using simultaneously a bounded limit on P_0 and the Fréchef's bounds, maximizing AUC is equivalent to minimizing the quantity under the integral.

$$Argmax_C AUC \sim Argmin_C \int_0^1 \int_0^1 (C(s_1, s_2) - 1)^2 ds_1 ds_2 \tag{11}$$

The problem remains to test different copulas according to their capabilities to deliver the best AUC criterion value. We have practically tested some usual

---

[5] Where $\{\hat{Y} = 0\} = \{P(X \le x) \le s\}$.

copulas (Clayton, Gumbel, Carlie-Gumbel Morgernstein and the upper Frechet's copula plus the copula we have defined in formula (6)) and surprisingly the minimal trade ones, gives very impressive results in real-life applications. This seems to indicate that this minimal trade or relational copula could be "the optimal solution" of problem (11) at least a very good candidate to. Although we are not handling the same types of entities, intuitively the criterion is very similar to the minimal trade problem presented in (2). And we are currently working to solve this research problem: providing the optimality of the minimal trade copula.

6   Anomaly Detection Based on This New Copula

Cyber-attacks which occurred during the last years, have shown the limitations of existing intrusion detection systems. Those systems are often based on expert rules (i.e. known signatures) which make them very tractable for detecting known attacks but unable at discovering new patterns. The algorithmic genericity of this present approach allows to use it in many real-life contexts. The algorithm has been designed to be mainly applied in cybersecurity but it is also quite suitable for other general purpose applications. Event structure is captured through the cross dependancy function, and summarized in the lower tail dependancy ratio. Moreover, in this context, the online learning capability has a major impact on the performances improvements. In cybersecurity, our algorithm has been tested to support until 3.5 GB of data per second (about 300 TB per day). Our benchmarking tests were performed on the DARPA intrusion detection datasets. On those datasets we obtained 73.86 % of detected attacks with 2.32 % of false alarms, which is quite satisfactory result which can be compared favorably with the other existing attempts and in the context of the copulas family, our proposed copula gives, by far, the best results compared to the usual ones: Clayton, Gumbel etc. (see [5] for other Archimedian copulas) (Table 2).

Table 2. Benchmark of different copulas on the DARPA dataset

Quantile level used for copula benchmark					
Quantile level	10^{-4}	5.10^{-4}	10^{-3}	5.10^{-3}	10^{-2}
"Minimal Trade" or Relational Copula					
Detection rate	18.64 %	73.86 %	74.32 %	74.82 %	75.09 %
False alarms rate	23.15 %	2.32 %	4.38 %	3.72 %	4.71 %
Gumbel copula					
Detection rate	27.05 %	33.19 %	38.50 %	57.69 %	62.53 %
False alarms rate	16.51 %	18.97 %	15.28 %	24.39 %	42.95 %
Clayton copula					
Detection rate	0.0 %	0.0 %	19.28 %	71.73 %	79.86 %
False alarms rate	0.0 %	0.0 %	0.63 %	36.76 %	34.20 %

References

1. Ah-Pine, J., Marcotorchino, J.F.: Overview of the relational analysis approach in data-mining and multi-criteria decision making. In: Web Intelligence and Intelligent Agents, pp. 325–346. InTech publishing (2010)

2. Carlier, G.: Optimal transport and economic applications. New Mathematical Models in Economics and Finance, pp. 1–82. Lecture Notes IMA (2010)

3. Evans, L.C.: Partial differential equations and monge-kantorovich mass transfer. In: Yau, S.T. (ed.) Current Developments in Mathematics. International Press, Cambridge (1997)

4. Fréchet, M.: Sur les tableaux de corrélations dont les marges sont données. Section A n° 14, pp. 53–77, Annales de l'Université de Lyon (1951)

5. Genest, C., Rivest, L.P.: Statistical inference procedures for bivariate Archimedean copulas. J. Am. Stat. Assoc. **88**(423), 1034–1043 (1993)

6. Hoffman, A.J.: On simple linear programming problems. In: Klee V. (ed.) Proceedings of Symposia in Pure Mathematics, vol. VII, pp. 317–327, AMS, Providence (1963)

7. Huyot, B., Mabiala, Y.: Online Unsupervised Anomaly Detection in large information systems using copula theory. In: Proceedings of the IEEE International Conference on Cloud Computing and Intelligence System, Hong-Kong (2014)

8. Kantorovich, L.: On the translocation of masses. Comptes Rendus (Doklady), n°37, pp. 199–201, Acad. Sci. URSS (1942)

9. Marcotorchino, J.F.: Utilisation des Comparaisons par Paires en Statistique des Contingences, Cahiers du Séminaire Analyse des Données et Processus Stochastiques Université Libre de Bruxelles, publication Brussel's University (1984)

10. Marcotorchino, F., Céspedes, P.C.: Optimal transport and minimal trade problem, impacts on relational metrics and applications to large graphs and networks modularity. In: Nielsen, F., Barbaresco, F. (eds.) GSI 2013. LNCS, vol. 8085, pp. 169–179. Springer, Heidelberg (2013)

11. Marcotorchino, J.F.: Optimal Transport, Spatial Interaction models and related problems impacts on Relational Metrics, adapted to Large Graphs and Networks Modularity, Thales internal publication presented to IRCAM Léon Brillouin's Seminar (2013)

12. Michaud, P.: Condorcet, a man of the avant-garde. ASMDA J. 3(2) (1997). Wiley

13. Monge, G.: Mémoire sur la théorie des déblais et de remblais. Histoire de l'Académie Royale des Sciences de Paris, avec les Mémoires de Mathématique et de Physique pour la même année, pp. 666–704 (1781)

14. Rachev, S.T., Ruschendorf, L.: Mass Transportation Problem. Theory, vol. 1. Springer, New York (1998)

15. Schrodinger, E.: "Uber die Umkehrung der Naturgesetze" Sitzungsberichte Preuss. Akad. Wiss Berlin. Phys. Math **144**, 144–153 (1931)

16. Sklar, A.: Fonctions de répartition à n dimensions et leurs marges. Université Paris 8 (1959)

17. Villani, C.: Topics in Optimal Transportation. Graduate Studies in Mathematics, vol. 58. The American Mathematical Society, Providence (2003)

18. Villani, C.: Optimal Transport Old and New. Springer, New York (2009)

19. Wilson, A.G.: The use of entropy maximising models. J. Transport Economies Policy **3**, 108–126 (1969)

Optimal Mass Transport over Bridges

Yongxin Chen[1], Tryphon Georgiou[1], and Michele Pavon[2]($\boxtimes$)

[1] Department of Electrical and Computer Engineering,
University of Minnesota, 200 Union Street S.E., Minneapolis, MN 55455, USA
{chen2468,tryphon}@umn.edu
[2] Dipartimento di Matematica, Università di Padova,
via Trieste 63, 35121 Padova, Italy
pavon@math.unipd.it

Abstract. We present an overview of our recent work on implementable solutions to the Schrödinger bridge problem and their potential application to optimal transport and various generalizations.

1 Introduction

In a series of papers, Mikami, Thieullen and Léonard [21, 22, 24–26] have investigated the connections between the optimal mass transport problem (OMT) and the Schrödinger bridge problem (SBP). The former may be shown to be the Γ-limit of a sequence of the latter, and thereby, SBP can be seen as a regularization of the OMT. Since OMT is well-known to be challenging from a computational viewpoint, this observation leads to the question of whether we can get approximate solutions to OMT via solving a sequence of SBPs. Both types of problem admit a control, fluid-dynamic formulation and it is in this setting that the connection between the two becomes apparent. There are, however, several difficulties in carrying out this program:

(i) The solution of the SBP is usually not given in *implementable form*;
(ii) SBP has been studied only for *non degenerate, constant diffusion coefficient* processes with control and noise entering through *identical channels*;
(iii) No SBP *steady-state* theory;
(iv) No OMT problem with *nontrivial prior*.

Notice that (ii) and (iii) exclude most engineering applications. In the past year, we have set out to partially remedy this situation [5,12]. We present here an overview of this work.

2 Background

2.1 Optimal Transport

Consider the Monge-Kantorovich (OMT) problem [1,29,30]

$$\inf_{\pi \in \Pi(\mu,\nu)} \int_{\mathbb{R}^n \times \mathbb{R}^n} c(x,y) d\pi(x,y),$$

© Springer International Publishing Switzerland 2015
F. Nielsen and F. Barbaresco (Eds.): GSI 2015, LNCS 9389, pp. 77–84, 2015.
DOI: 10.1007/978-3-319-25040-3_9

where $\Pi(\mu, \nu)$ are "couplings" of μ and ν, and $c(x, y) = \frac{1}{2}\|x - y\|^2$.

If μ does not give mass to sets of dimension $\leq n - 1$, by Brenier's theorem, there exists a unique optimal transport plan π (Kantorovich) induced by a map T (Monge), where $T = \nabla\varphi$, φ is a *convex* function, $\pi = (I \times \nabla\varphi)\#\mu$, and $\nabla\varphi\#\mu = \nu$ where $\#$ indicates "push-forward". Assume from now on $\mu(dx) = \rho_0(x)dx$, $\nu(dy) = \rho_1(y)dy$. The static OMT above was given a dynamical formulation by Benamou-Brenier in [2]:

$$\inf_{(\rho, v)} \int_{\mathbb{R}^n} \int_0^1 \frac{1}{2}\|v(x, t)\|^2 \rho(x, t)dtdx, \tag{1}$$

$$\frac{\partial \rho}{\partial t} + \nabla \cdot (v\rho) = 0, \tag{2}$$

$$\rho(x, 0) = \rho_0(x), \quad \rho(y, 1) = \rho_1(y). \tag{3}$$

Proposition 1. *Let $\rho^*(x, t)$ with $t \in [0, 1]$ and $x \in \mathbb{R}^n$, satisfy*

$$\frac{\partial \rho^*}{\partial t} + \nabla \cdot (\nabla\psi\rho^*) = 0, \quad \rho^*(x, 0) = \rho_0(x), \tag{4}$$

where ψ is a (viscosity) solution of the Hamilton-Jacobi equation

$$\frac{\partial \psi}{\partial t} + \frac{1}{2}\|\nabla\psi\|^2 = 0 \tag{5}$$

for some boundary condition $\psi(x, 1) = \psi_1(x)$. If $\rho^*(x, 1) = \rho_1(x)$, then the pair (ρ^*, v^*) with $v^*(x, t) = \nabla\psi(x, t)$ is a solution of (1)–(3).

2.2 Schrödinger Bridges

The ingredients of the classical Schrödinger bridge problem are the following:

– a cloud of N independent Brownian particles,
– an initial and a final marginal density $\rho_0(x)dx$ and $\rho_1(y)dy$, resp.,
– ρ_0 and ρ_1 are not compatible with the transition mechanism

$$\rho_1(y) \neq \int_0^1 p(0, x, 1, y)\rho_0(x)dx,$$

where

$$p(s, y, t, x) = [2\pi(t - s)]^{-\frac{n}{2}} \exp\left[-\frac{\|x - y\|^2}{2(t - s)}\right], \quad s < t.$$

In view of the law of large numbers, particles have been transported in an *unlikely way* (N being large). Then, Schrödinger in (1931) posed the following question: *Of the many unlikely ways in which this could have happened, which one is the most likely?* Föllmer in 1988 observed that this is a problem of *large deviations of the empirical distribution* [15] on path space connected through Sanov's theorem to a *maximum entropy problem.*

Schrödinger's solution (*bridge from ρ_0 to ρ_1 over Brownian motion*) has at each time a density ρ that factors as $\rho(x,t) = \varphi(x,t)\hat{\varphi}(x,t)$, where φ and $\hat{\varphi}$ solve the *Schrödinger's system*

$$\varphi(x,t) = \int p(t,x,1,y)\varphi(y,1)dy, \quad \varphi(x,0)\hat{\varphi}(x,0) = \rho_0(x), \qquad (6)$$

$$\hat{\varphi}(x,t) = \int p(0,y,t,x)\hat{\varphi}(y,0)dy, \quad \varphi(x,1)\hat{\varphi}(x,1) = \rho_1(x). \qquad (7)$$

The new evolution has drift field $b(x,t) = \nabla\varphi(x,t)$. His result extends to the case when the "prior" evolution is a general Markov diffusion process possibly with creation and killing [32]. Existence and uniqueness for the Schrödinger's system has been studied in particular by Beurling, Fortet, Jamison and Föllmer [3,18–20], see [22,32] for a survey.

The *maximum entropy* formulation of the Schrödinger bridge problem (SBP) with "prior" P is

$$\text{Minimize} \quad H(Q,P) = \mathbb{E}_Q\left[\log \frac{dQ}{dP}\right] \quad \text{over} \quad \mathcal{D}(\rho_0,\rho_1),$$

where $\mathcal{D}$ is the family of distributions on $\Omega := C([0,1],\mathbb{R}^n)$ that are equivalent to stationary Wiener measure $W = \int W_x\,dx$. It can be turned, thanks to *Girsanov's theorem*, into a stochastic control problem see [4,13,14,17,27] with fluid dynamic counterpart. Here $P = W^\epsilon$, namely stationary Wiener measure with variance ϵ, in which case the problem assumes a form similar to (1)–(3)

$$\inf_{(\rho,v)} \int_{\mathbb{R}^n} \int_0^1 \frac{1}{2\epsilon}\|v(x,t)\|^2\rho(x,t)dtdx,$$

$$\frac{\partial\rho}{\partial t} + \nabla\cdot(v\rho) - \frac{\epsilon}{2}\Delta\rho = 0,$$

$$\rho(x,0) = \rho_0(x), \quad \rho(y,1) = \rho_1(y).$$

3   Gauss-Markov Bridges

Consider the problem in the case where the prior evolution and the marginals are *Gaussian*. In [6,8], the following two problems have been addressed:

Problem 1: Find a control u, adapted to X_t and minimizing

$$J(u) := \mathbb{E}\left\{\int_0^1 \frac{1}{2}\|u(t)\|^2\,dt\right\},$$

among those which achieve the transfer

$$dX_t = A(t)X_t dt + B(t)u(t)dt + B_1(t)dW_t,$$
$$X_0 \sim \mathcal{N}(0,\Sigma_0), \quad X_1 \sim \mathcal{N}(0,\Sigma_1).$$

If the pair (A, B) is controllable (for constant A and B, this amounts to the matrix $(B, AB, ..., A^{n-1}B)$ having full row rank), Problem 1 turns out to be always *feasible* (this result is highly nontrivial as the control may be "handicapped" with respect to the effects of the noise).

Problem 2: Find $u = -Kx$ minimizing $J_{\text{power}}(u) := \mathbb{E}\{\frac{1}{2}\|u\|^2\}$ and such that

$$dX_t = (A - BK)X_t dt + B_1 dW_t$$

has

$$\rho(x) = (2\pi)^{-n/2} \det(\Sigma)^{-1/2} \exp\left(-\frac{1}{2}x' \Sigma^{-1} x\right)$$

as *invariant probability density.*

Problem 2 may not have a solution (not all values for Σ can be maintained by state feedback).

Sufficient conditions for optimality have been provided in [6,8] in terms of:

- a system of two *matrix Riccati equations* (*Lyapunov equations* if $B = B_1$) in the finite horizon case. The Riccati equations are nonlinearly coupled through the boundary conditions. In the case where $B \neq B_1$, which falls outside the classical maximum entropy problem, the two equations are also *dynamically coupled.*
- in terms of *algebraic conditions* for the stationary case.

Optimal controls may be computed via semidefinite programming in both cases.

4   Cooling for Stochastic Oscillators

Cooling for micro and macro-mechanical systems consists in implementing via feedback a frictional force to steer the state of a thermodynamical system to a non equilibrium steady state with *effective temperature* that is lower than that of the heat bath. Important applications of such *Brownian motors* [28] are found in molecular dynamics, Atomic Force Microscopy and gravitational wave detectors [16,23,31], to name a few.

The basic model is provided by a *controlled stochastic oscillator* deriving from the Nyquist-Johnson model of RLC electrical network with noisy resistor (1928) and the Ornstein-Uhlenbeck model of physical Brownian motion (1930):

$$dx(t) = v(t)\, dt, \tag{8}$$

$$dv(t) = -\beta v(t)\, dt - \frac{1}{m}\nabla V(x(t))dt + u(x(t), v(t), t) + \sigma dW_t, \tag{9}$$

$$\sigma^2 = \frac{2k\beta T}{m}, \quad Einstein's\ fluctuation\text{-}dissipation\ relation. \tag{10}$$

Here $u(x, v, t)$ is a feedback control law and V is such that the initial value problem is well-posed on bounded time intervals. For $u \equiv 0$,

$$\rho(x, v, t) \rightarrow \rho_{MB}(x, v) = Z^{-1} \exp\left[-\frac{H(x, v)}{kT}\right], \quad H(x, v) = \frac{1}{2}m\|v\|^2 + V(x).$$

Let $\bar{\rho}(x,v) = \bar{Z}^{-1} \exp\left[-\frac{H(x,v)}{kT_{\text{eff}}}\right]$ and let $T_{\text{eff}} < T$ be a desired *steady state* effective temperature. In [9], we have studied the following two problems:

- Efficient asymptotic steering of the system to $\bar{\rho}$;
- Efficient steering of the system from ρ_0 to $\bar{\rho}$ at a finite time $t = 1$.

In both cases, we get a solution for a general system of nonlinear stochastic oscillators, where we allow for both potential and dissipative interactions between the particles, by extending the theory of the Schrödinger bridges accordingly.

Consider the case of a scalar oscillator in a quadratic potential with Gaussian marginals. For a suitable choice of constants, the model is

$$
\begin{aligned}
dx(t) &= v(t)dt, \\
dv(t) &= -v(t)dt - x(t)dt + u(t)dt + dW_t.
\end{aligned}
$$

Using the results in [6,8], through velocity feedback control the system is first efficiently steered to the desired state $\bar{\rho}$ at time $t = 1$ and then maintained efficiently in $\bar{\rho}$. This is illustrated by Fig. 1 that depicts some sample paths and a transparent tube outlining the "3σ region" of the one-time densities.

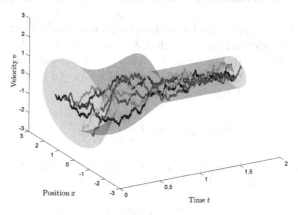

Fig. 1. Inertial particles: trajectories in phase space.

5   OMT with Prior

In [10,11], we have formulated and studied a generalization of optimal transport problem that includes prior dynamics. It is the natural candidate for the zero-noise limit of SBP where the prior is a general Markovian evolution and not just stationary Wiener measure. In particular, in [11] we have studied the case where there are fewer control than state variables and Gaussian marginals and derived the corresponding limiting transport problem. The latter can be put in the form of a classical OMT with cost deriving from a Lagrangian action, where, however, the Lagrangian is not strictly convex with respect to the $\dot{x}$ variable. Convergence of solutions is proven directly. Simulations confirm that in the zero-noise limit the "entropic interpolation" provided by the (generalized) Schrödinger bridge converges to the "displacement interpolation" of the limiting OMT problem.

In conclusion, in [6,7,9–11], we have worked out a number of cases where an implementable form of the solution of a (possibly generalized) Schrödinger bridge problem can be obtained. We have also explored to some extent the connection between zero-noise limits of SBP and suitable reformulations of OMT problems. These cases include degenerate, hypoelliptic diffusions like the Ornstein-Uhlenbeck model (8)–(10). The case of differing noise and control channels which does not have a classical SBP counterpart has also been studied. Finally, in [8], we have extended the fluid-dynamic SBP theory to the case of anisotropic diffusions with killing, a situation where again no probabilistic counterpart is available in general. The new evolution is obtained by solving a suitable generalization of the Schrödinger bridge system. How can we solve this generalized Schrödinger system as well as those corresponding to problems not covered in [6,7,9–11]? An alternative powerful tool is given by iterative schemes which contract Birkhoff's version of Hilbert's metric. This is discussed in the next section.

6   Positive Contraction Mappings for Schrödinger systems

Let S be a real Banach space and K a closed solid cone in S. That is, K is closed with nonempty interior and is such that $K + K \subseteq K$, $K \cap -K = \{0\}$ as well as $\lambda K \subseteq K$ for all $\lambda \geq 0$. Define $x \preceq y \Leftrightarrow y - x \in K$, and for $x, y \in K\backslash\{0\}$, $M(x, y) := \inf\{\lambda \mid x \preceq \lambda y\}$ and $m(x, y) := \sup\{\lambda \mid \lambda y \preceq x\}$. The *Hilbert metric* is the projective metric defined on $K\backslash\{0\}$ by

$$d_H(x, y) := \log\left(\frac{M(x, y)}{m(x, y)}\right).$$

A map $\mathcal{E}$ from S to S is said to be *positive* provided it takes the interior of K into itself. For such a map define its *projective diameter*

$$\Delta(\mathcal{E}) := \sup\{d_H(\mathcal{E}(x), \mathcal{E}(y)) \mid x, y \in K\backslash\{0\}\}$$

and the *contraction ratio*

$$\|\mathcal{E}\|_H := \inf\{\lambda \mid d_H(\mathcal{E}(x), \mathcal{E}(y)) \leq \lambda d_H(x, y), \text{ for all } x, y \in K\backslash\{0\}\}.$$

Theorem 1. *(Garrett Birkhoff 1957, P. Bushell 1973) Let $\mathcal{E}$ be a positive map. If $\mathcal{E}$ is monotone and homogeneous of degree m ($\mathcal{E}(\lambda x) = \lambda^m \mathcal{E}(x)$), then*

$$\|\mathcal{E}\|_H \leq m.$$

If $\mathcal{E}$ is also linear, the (possibly stronger) bound also holds

$$\|\mathcal{E}\|_H = \tanh(\frac{1}{4}\Delta(\mathcal{E})).$$

Consider now a Markov chain with T-step transition probabilities π_{x_0,x_T} (prior) and consider two marginal distributions $\mathbf{p}_0$ and $\mathbf{p}_T$, where x_0, x_T are indices corresponding to initial and final states. An adaptation of Schrödinger's question to this setting leads to the following Schrödinger system:

$$\varphi(x_0,0) = \sum_{x_T} \pi_{x_0,x_T}\varphi(x_T,T) = \mathcal{E}\left(\varphi(x_T,T)\right), \quad \varphi(x_0,0)\hat{\varphi}(x_0,0) = \mathbf{p}_0(x_0),$$

$$\hat{\varphi}(x_T,T) = \sum_{x_0} \pi_{x_0,x_T}\hat{\varphi}(x_0,0) = \mathcal{E}^\dagger\left(\hat{\varphi}(x_0,0)\right), \quad \varphi(x_T,T)\hat{\varphi}(x_T,T) = \mathbf{p}_T(x_T).$$

It turns out that the composition of the four maps

$$\hat{\varphi}(x_0,0) \longrightarrow \hat{\varphi}(x_T,T) := \mathcal{E}^\dagger(\hat{\varphi}(x_0,0)) \longrightarrow \varphi(x_T,T) := \frac{\mathbf{p}_T(x_T)}{\hat{\varphi}(x_T,T)}$$

$$\longrightarrow \varphi(x_0,0) := \mathcal{E}\left(\varphi(x_T,T)\right) \longrightarrow \left(\hat{\varphi}(x_0,0)\right)_{\text{next}} := \frac{\mathbf{p}_0(x_0)}{\varphi(x_0,0)}$$

where division of vectors is performed componentwise, is *contractive in the Hilbert metric*. Indeed, the linear maps are non-expansive with $\mathcal{E}$ strictly contractive, whereas componentwise divisions are isometries (and contractive when the marginals have zero entries). In [5], we have obtained similar results for Kraus maps of statistical quantum mechanics with pure states or uniform marginals. The case of diffusion processes is studied in [12]. Applications include interpolation of 2D images to construct a 3D model (MRI).

References

1. Ambrosio, L., Gigli, N., Savaré, G.: Gradient Flows in Metric Spaces and in the Space of Probability Measures. Lectures in Mathematics ETH Zürich. Birkhäuser Verlag, Basel (2008)
2. Benamou, J., Brenier, Y.: A computational fluid mechanics solution to the Monge-Kantorovich mass transfer problem. Num. Math. **84**(3), 375–393 (2000)
3. Beurling, A.: An automorphism of product measures. Ann. Math. **72**, 189–200 (1960)
4. Blaquière, A.: Controllability of a Fokker-Planck equation, the Schrödinger system, and a related stochastic optimal control. Dyn. Control **2**(3), 235–253 (1992)
5. Georgiou, T.T., Pavon, M.: Positive contraction mappings for classical and quantum Schrödinger systems, May 2014, arXiv:1405.6650v2. J. Math. Phys. to appear
6. Chen, Y., Georgiou, T.T., Pavon, M.: Optimal steering of a linear stochastic system to a final probability distribution, Part I, arXiv:1408.2222v1. IEEE Trans. Aut. Control, to appear
7. Chen, Y., Georgiou, T.T., Pavon, M.: Optimal steering of inertial particles diffusing anisotropically with losses, arXiv:1410.1605v1, ACC Conf. (2015)
8. Chen, Y., Georgiou, T.T., Pavon, M.: Optimal steering of a linear stochastic system to a final probability distribution, Part II, arXiv:1410.3447v1. IEEE Trans. Aut. Control, to appear
9. Chen, Y., Georgiou, T.T., Pavon, M.: Fast cooling for a system of stochastic oscillators, Nov 2014, arXiv:1411.1323v1

84   Y. Chen et al.

10. Chen, Y., Georgiou, T.T., Pavon, M.: On the relation between optimal transport and Schrödinger bridges: A stochastic control viewpoint, arXiv:1412.4430v1
11. Chen, Y., Georgiou, T.T., Pavon, M.: Optimal transport over a linear dynamical system, arXiv:1502.01265v1
12. Chen, Y., Georgiou, T.T., Pavon, M.: A computational approach to optimal mass transport via the Schrödinger bridge problem (2015, in preparation)
13. Dai Pra, P.: A stochastic control approach to reciprocal diffusion processes. Appl. Math. Optim. 23(1), 313–329 (1991)
14. Dai Pra, P., Pavon, M.: On the Markov processes of Schroedinger, the Feynman-Kac formula and stochastic control. In: Kaashoek, M.A., van Schuppen, J.H., Ran, A.C.M. (eds.) Realization and Modeling in System Theory - Proceedings of the 1989 MTNS Conference, pp. 497–504. Birkaeuser, Boston (1990)
15. Dembo, A., Zeitouni, O.: Large Deviations Techniques and Applications. Applied Math., vol. 38, 2nd edn. Springer, New York (1998)
16. Doi, M., Edwards, S.F.: The Theory of Polymer Dynamics. Oxford University Press, New York (1988)
17. Fillieger, R., Hongler, M.-O., Streit, L.: Connection between an exactly solvable stochastic optimal control problem and a nonlinear reaction-diffusion equation. J. Optimiz. Theory Appl. 137, 497–505 (2008)
18. Föllmer, H.: Random fields and diffusion processes. In: Hennequin, P.L. (ed.) École d'Ètè de Probabilitès de Saint-Flour XV-XVII. Lecture Notes in Mathematics, vol. 1362, pp. 102–203. Springer, New York (1988)
19. Fortet, R.: Résolution d'un système d'equations de M. Schrödinger. J. Math. Pure Appl. IX, 83–105 (1940)
20. Jamison, B.: The Markov processes of Schrödinger. Z. Wahrscheinlichkeitstheorie verw. Gebiete 32, 323–331 (1975)
21. Léonard, C.: From the Schrödinger problem to the Monge-Kantorovich problem. J. Funct. Anal. 262, 1879–1920 (2012)
22. Léonard, C.: A survey of the Schroedinger problem and some of its connections with optimal transport. Discrete Contin. Dyn. Syst. A 34(4), 1533–1574 (2014)
23. Liang, S., Medich, D., Czajkowsky, D.M., Sheng, S., Yuan, J., Shao, Z.: Ultramicroscopy 84, 119 (2000)
24. Mikami, T.: Monge's problem with a quadratic cost by the zero-noise limit of h-path processes. Probab. Theory Relat. Fields 129, 245–260 (2004)
25. Mikami, T., Thieullen, M.: Duality theorem for the stochastic optimal control problem. Stoch. Proc. Appl. 116, 1815–1835 (2006)
26. Mikami, T., Thieullen, M.: Optimal transportation problem by stochastic optimal control. SIAM J. Control Opt. 47(3), 1127–1139 (2008)
27. Pavon, M., Wakolbinger, A.: On free energy, stochastic control, and Schroedinger processes. In: Di Masi, G.B., Gombani, A., Kurzhanski, A. (eds.) Modeling, Estimation and Control of Systems with Uncertainty, pp. 334–348. Birkauser, Boston (1991)
28. Reimann, P.: Brownian motors: noisy transport far from equilibrium. Phys. Rep. 361, 57 (2002)
29. Villani, C.: Topics in Optimal Transportation, vol. 58. AMS, Providence (2003)
30. Villani, C.: Optimal Transport: Old and New, vol. 338. Springer, Berlin (2008)
31. Vinante, A., Bignotto, M., Bonaldi, M., et al.: Feedback cooling of the normal modes of a massive electromechanical system to submillikelvin temperature. Phys. Rev. Lett. 101, 033601 (2008)
32. Wakolbinger, A.: Schroedinger bridges from 1931 to 1991. In: Proceedings of the 4th Latin American Congress in Probability and Mathematical Statistics, Mexico City 1990, Contribuciones en probabilidad y estadistica matematica, 3, 61–79 (1992)

Optimal Transport and Applications in Imagery/Statistics

Non-convex Relaxation of Optimal Transport for Color Transfer Between Images

Julien Rabin[1] and Nicolas Papadakis[2]([⊠])

[1] GREYC, UMR 6072, Université de Caen, 14050 Caen, France
julien.rabin@unicaen.fr
[2] CNRS, IMB, UMR 5251, 33400 Talence, France
nicolas.papadakis@math.u-bordeaux.fr

Abstract. Optimal transport (OT) is a major statistical tool to measure similarity between features or to match and average features. However, OT requires some relaxation and regularization to be robust to outliers. With relaxed methods, as one feature can be matched to several ones, important interpolations between different features arise. This is not an issue for comparison purposes, but it involves strong and unwanted smoothing for transfer applications. We thus introduce a new regularized method based on a non-convex formulation that minimizes transport dispersion by enforcing the one-to-one matching of features. The interest of the approach is demonstrated for color transfer purposes.

Keywords: Optimal transport · Relaxation · Color transfer

1 Introduction

Many image processing applications require the modification or the prescription of some characteristics (colors, frequencies or wavelet coefficients) of a given image, while preserving other features. Statistics to be prescribed may come from prior knowledge, or more generally, are learned from an example. In such a case, another image is selected from a database to define a template. Such a framework arises for image enhancement, inpainting, colorization of grayscale or infrared images, tone mapping, color grading or color transfer. In this paper, we will focus on this last application through histogram transfer between images.

Color transfer consists in modifying an image to match the color palette of another one, while preserving its geometry. In the literature, the different interpretations and definitions of color palettes have led to various algorithms. In the following, we only consider unsupervised approaches.

Parametric and Histogram Modeling. Since the seminal work of [17], methods have been designed to transfer some simple color statistics (*i.e.* the mean and

A preliminary version of this work has been presented at the NIPS 2014 Workshop on Optimal Transport and Machine Learning (pdf).

F. Nielsen and F. Barbaresco (Eds.): GSI 2015, LNCS 9389, pp. 87–95, 2015.
DOI: 10.1007/978-3-319-25040-3_10

standard deviation [9]) in any color space [20]. More general approaches match the complete histogram of features from two images. When considering grayscale images, the problem is known as 1-D *histogram specification*. This framework has been extended for color histograms, using for instance 3-D cumulative histograms [10] or 1-D ΔE-color index [6].

Histogram Matching via *Optimal Transportation*. As pointed out by [11], there exist strong links between histogram specification and the Monge-Kantorovich Optimal Transport (OT) problem. The OT problem consists in estimating the map that transfers a source probability distribution onto a target one, while minimizing a given cost function. The transport cost is also referred to as the Wasserstein distance or the Earth Mover's distance. The associated OT map is the key element to perform the transfer of colors. Some approaches to find fast approximate solution of OT were investigated in [11, 15].

Spatial Information. The exact transfer of color palette is generally not satisfying for practical applications in image processing [12]. The color distributions to be matched may have very different shapes, so that outliers generally appear in the processed image. Moreover, as the process is performed in the color space, it may not transfer coherent colors to neighboring pixels, resulting in undesirable artifacts, such as JPEG compression blocks, enhanced noise, saturation or contrast inversion [11, 13, 18]. As a consequence, various models have been designed to incorporate some regularity priors on the image domain, such as Total Variation [7]. Color transfer may be formalized [10, 14] as a variational problem in the image domain, in order to directly incorporate a spatial regularization of colors.

Approximate and Regularized Matching. While spatial regularization suppresses small artifacts due to exact histogram specification, it cannot handle strong artifacts due to an irregular OT map [14]. Ferradans *et al.* [4] thus proposed to regularize the optimal assignment between point clouds. The exact matching constraint is relaxed to enforce robustness to outliers. Instead of providing one-to-one assignment, the use of capacity variables makes it possible one-to-many correspondences. The additional introduction of regularity priors on the OT map produces smoother and more robust transport maps for image processing applications. This work has been extended in [16] for histogram purposes: spatial information is used to drive the regularization of color transfer, while the capacity variables are automatically estimated. Notice that a fast method based on entropy prior for estimating smooth OT between histograms has been proposed in [3]. It also yields one-to-many correspondences between histogram bins.

Color Dispersion. Due to the one-to-many relaxation and the fact that only the gradient of the *average transport* flow is penalized, the regularization does not prevent the transport map to associate very different colors to a single pixel or cluster. This leads to undesirable results such as color mixing or color inconsistencies in the modified image.

Contributions and Outline. In this paper, we propose a new model that takes into account the aforementioned issues. We also relax and regularize the transport map and introduce a non-convex constraint that minimizes the variance of colors assigned to each cluster. We rely on a fast proximal splitting algorithm in order to compute the transport map that is finally used for color transfer purposes. The organization of the paper is as follows. Background on OT is given in Sect. 2. The proposed model is introduced in Sect. 3 and experimented in Sect. 4.

2   Color Transfer *via* Optimal Transport

We refer to u as the input image to be modified, and to v as an exemplar image v provided by the user. We consider clustered feature distributions, which may be seen as multi-dimensional histograms or discrete probability distribution (i.e. color palettes, wavelet coefficient histograms). We refer to $h_u = \sum_{i=1}^{n} h_u[i]\delta_{X_i}$ as the histogram of features $X := \{X_i \in \mathbb{R}^d\}_{i \leq n} \in \mathbb{R}^{n \times d}$ from the input image, so that $\sum_{i=1}^{n} h_u[i] = 1$. The histogram is thus composed of n features, each of them being of dimension d. In the same way, $h_v = \sum_{j=1}^{m} h_v[j]\delta_{Y_j}$ is the target distribution of features $Y := \{Y_j \in \mathbb{R}^d\}_{j \leq m} \in \mathbb{R}^{m \times d}$. We have no assumption on the way those histograms are built (uniform quantization, k-means, *etc*). Hence, we have a quantized version of our image, for instance using Nearest-Neighbor interpolation w.r.t a metric d (that is the quadratic L_2 distance in the following): $\tilde{u}_i = X_{I(i)}$ where $I(i) = \mathrm{argmin}_I \, d(u(i), X_I)$.

2.1   Optimal Transport of Histogram (Histogram Specification)

The OT of h_u onto h_v is obtained by estimating the transport matrix:

$$P^\star \in \underset{P \in \mathcal{P}_{h_u, h_v}}{\mathrm{argmin}} \; C(P) := \langle P, C_{X,Y} \rangle = \sum_{i,j} P_{i,j}(C_{X,Y})_{i,j} \tag{1}$$

within the set $\mathcal{P}_{h_u, h_v} = \{P \in \mathbb{R}_+^{n \times m}, P\mathbf{1}_m = h_u, P^T\mathbf{1}_n = h_v\}$ where $\mathbf{1}_N \in \mathbb{R}^N$ is the unit vector, and the cost matrix $C_{X,Y}$ is generally defined from quadratic distances $(C_{X,Y})_{i,j} = \|X_i - Y_j\|^2$ in applications where the transport map is expected to be regular. In order to change the statistical distribution of u accordingly to the transport matrix P, the transfer map is defined as: $T : X_i \mapsto \{Y_j \text{ s.t. } P_{i,j} > 0\}$. In order to avoid quantization artifacts, it has been proposed in [19] to incorporate spatial information in a multivariate Gaussian mixture model $S : u_i \mapsto \frac{1}{W} \sum_i w_i(u_i) T(X_{I(i)})$, which average a mapping T using adaptive weights functions $w_i(\cdot) = \exp\left(-\frac{1}{2}\| \cdot - X_i\|_{V_i}^2\right)$, and a normalization factor $W = \sum_i w_i$. This method uses estimated covariance matrices V_i of clusters in spatial and color product space. While this post-processing removes small quantization artifacts, it can barely attenuate large irregularities of the transport map T.

2.2 Optimal Transport Relaxation

One main limitation of OT comes from the exact matching constraint which is not robust to outliers. To address this issue, it has been proposed to make use of relaxed constraints for the optimal assignment problem, defining min/max capacities on the optimal flow [4]. In the context of histogram matching, such relaxed OT problem corresponds to solve problem (1) onto the relaxed constraint set $\mathcal{K}_{h_u,h_v} = \{P \in \mathbb{R}_+^{n \times m}, \ P\mathbf{1}_m = h_u, \ \kappa_{min} h_v \leq P^T \mathbf{1}_n \leq \kappa_{max} h_v\}$. This model includes two vectorial parameters, $\kappa_{min} \leq 1$ and $\kappa_{max} \geq 1$ defined in $\mathbb{R}^m$, that control the proportion of the target histogram's bins that can be used by the color transfer. Outliers can be taken into account by taking $\kappa_{min} < 1$. By setting $\kappa_{max} > 1$, some colors of the source palette will be used more frequently than in the example image.

However, there is no statistical control of how "close" the transported histogram is over the source one and it is very difficult to tune so many parameters by hand. In [16], the following extended model has been shown to tackle these two limitations, by including the calibration of the capacity parameters $\{\kappa_j\}_{j \leq m}$ within the model through the penalization of their distance to 1:

$$(P, \kappa)^* \in \underset{\substack{P \in \mathbb{R}_+^{n \times m} \ s.t. \ P\mathbf{1}_m = h_u \\ \kappa \in \mathbb{R}_+^m \ s.t. \ \langle \kappa, h_v \rangle \geq 1}}{\operatorname{argmin}} C(P) + \rho \|\kappa - \mathbf{1}_m\|_1. \tag{2}$$

2.3 Optimal Transport Regularization

Another limitation of the OT framework is the lack of control over the regularity of the solution. In [4], the authors propose to measure the regularity of the average transfer mapping. First, one considers the following definition of *Posterior mean* to define a one-to-one mapping T from a transfer matrix P:

$$T(X_i) = \overline{Y}_i = \frac{1}{\sum_{j=1}^m P_{ij}} \sum_{j=1}^m P_{ij} Y_j = \frac{1}{h_u(i)} \sum_j P_{i,j} Y_j = (D_{h_u} PY)_i, \tag{3}$$

where the normalization matrix D_{h_u} is diagonal: $(D_{h_u})_{ii} = h_u(i)^{-1}$.

The regularity of this average transfer map is then evaluated on a graph $\mathcal{G}_X = (I_X, E_X)$, built from the set of input features $\{X_i\}_i$. Denoting as $I_X = \{1, \cdots, I_n\}$ the set of nodes representing the features $\{X_1, \ldots X_n\}$, and $E_X \subset I_X^2$ the set of edges, the gradient $\mathcal{G}_X V \in \mathbb{R}^{n \times n \times d}$ of a multi-valued function $V = \{V_i^l\} \in \mathbb{R}^{n \times d}$ on $\mathcal{G}_X$ is computed at point X_i as $(\mathcal{G}_X V)_i = (w_{ij}(V_i - V_j))_{j \in E_X(i)} \in \mathbb{R}^{n \times d}$, where the weight w_{ij} between X_i and X_j relies on their similarit: $w_{ij} \propto \exp -d(X_i, X_j)$. The OT matrix now solves the following problem

$$P^* \in \underset{P \in \mathcal{K}_{h_u,h_v}}{\operatorname{argmin}} C(P) + \lambda \|\mathcal{G}_X (D_{h_u} PY - X)\|, \tag{4}$$

where $\|\mathcal{G}_X V\|$ can be interpreted as the TV norm of field V on the graph $\mathcal{G}_X$. The flow is taken as and $V = T(X) - X$ so that color translation are not penalized. With such regularization, artifacts or contrast inversion are also avoided.

Together with functional (4), the relaxed formulation (2) yields smooth transport maps. that enforces transport between one cluster to many. For color transfer purposes, the obtained prescribed colors are then defined from linear combination of the target color palette, resulting in false colors artifacts and a lost of color contrast. In this paper, we propose to solve this issue by incorporating information on the color transfer dispersion.

3   Non Convex Relaxation of Color Palette Transport

The relaxation considered here is different from the one proposed in [5] where a capacity relaxation of the target histogram is considered. The closeness to the target histogram is imposed through a data fidelity term which makes easier the control of the color transfer result, while simplifying the projection onto the set of acceptable transport matrices. By using linear programming to optimize the regularized problem (2) as in [4,16], the dimension of the variables to estimate is greatly increased. Simplifications of the regularization term (through the mean transport and the use of divergence) are thus needed to reduce the complexity. Such regularizers limit the inter-cluster color dispersion but they induce the creation of new drab colors since an important interpolation of the target color palette may occur (*i.e.* the intra-cluster color variance may be large with the one-to-many assignment). To cope with these issues, we propose to penalize the dispersion of assigned colors with a non-convex energy term. We also rely on a different optimization tool which decreases the dimension of the problem w.r.t linear programming, as we only deal with the estimation of the OT matrix.

3.1   Optimization Problem

In order to deal with the aforementioned limitations, we propose the following relaxed and regularized OT problem:

$$P^* \in \underset{P \in \mathcal{P}_{h_u}}{\operatorname{argmin}} \left\{ \mathcal{E}(P) := C(P) + \rho F(P) + \lambda R(P) + \alpha D(P) \right\}, \tag{5}$$

where $C(P) = \langle C_{X,Y}, P \rangle$ is the linear cost matching function. The set $\mathcal{P}_{h_u} = \{\mathbb{R}^{n \times m}, P \geq 0, P\mathbf{1}_m = h_u\}$ is the convex set of right stochastic matrices (where each row sums to the corresponding bin value in h_u). The constraint $P \in \mathcal{P}_{h_u}$ is incorporated using an indicator function $\iota_{\mathcal{P}_{h_u}}(P) = 0$ if $P \in \mathcal{P}_{h_u}$ and $+\infty$ otherwise. The orthogonal projector $\operatorname{Proj}_{\mathcal{P}_{h_u}}(P)$ is done by projecting onto the corresponding simplex for each row of P. In order to solve the problem (5) with a fast projected gradient descent, we used differentiable functions for the other terms. Observe that other choice would lead to different optimization algorithms.

Fidelity Term. As the set of acceptable transport matrices $\mathcal{P}_{h_u}$ does not anymore take into account the target distribution, we have to make sure that the transported histogram $P^T\mathbf{1}_n$ is close enough to the target histogram h_v. To do so, we define the fidelity term $F(P)$ w.r.t the target histogram h_v by relying on the Pearson's χ^2 statistics. For any bistochastic matrix $P \in \mathcal{P}_{h_u,h_v}$, this term reads

$$F(P) = \tfrac{1}{2}\chi_{h_v}^2\left(P^T\mathbf{1}_n\right) = \tfrac{1}{2}\left\|D_{h_v}^{1/2}(P^T\mathbf{1}_n - h_v)\right\|^2 = \tfrac{1}{2}\left\|\mathbf{1}_n^T P D_{h_v}^{1/2}\right\|^2 - \tfrac{1}{2}. \qquad (6)$$

We assume therefore from here, without loss of generality, that h_v has non empty bins. Observe that the corresponding fidelity term can be interpreted as a weighted L^2 metric, which further penalizes bins of the target histogram that has small values. This will prevent the model from using very rare features from the exemplar image. The gradient then reads $\nabla F(P) = \mathbf{1}_n \cdot (\mathbf{1}_n^T P D_{h_v}) = \mathbf{1}_{n \times n} P D_{h_v}$.

Regularity Term. We consider a Tikhonov regularization of the gradient of a flow V, incorporating spatial information from the input feature distribution in the gradient operator G_X defined on the graph of clusters $\mathcal{G}_X$. We measure the gradient of the average transport flow $V = D_{h_u} PY - X$ and therefore define

$$R(P) = \frac{1}{2}\left\|D_{h_u}^{-1}G_X(D_{h_u}PY - X)\right\|_2^2, \qquad (7)$$

where the gradient norm is weighted by the corresponding histogram bin value: $(D_{h_u}^{-1})_{ii} = h_u[i]$. Its derivative is related to the graph-Laplacian $G_X^T G_X$ and writes $\nabla R(P) = D_{h_u} G_X^T D_{h_u}^{-2} G_X(D_{h_u}PY - X)Y^T$.

Dispersion Term. The regularized transfer induces a high variability of color assigned to each input color. We then propose to minimize the variance of the flow. Denoting $\overline{Y} = D_{h_u}PY$ from Eq. 3, the intra-cluster variance of the color assigned to X_i is: $\mathrm{Var}(Y)_i = \left(\overline{Y^2}\right)_i - \overline{Y}_i^2 = \frac{1}{h_u[i]}\sum_j P_{i,j}\|Y_j\|^2 - \left\|\frac{1}{h_u[i]}\sum_j P_{i,j}Y_j\right\|^2$. We therefore penalize this variance with respect to each cluster weight and thus obtain the functional term:

$$D(P) := \sum_i h_u[i]\,\mathrm{Var}(Y)_i = \langle P,\ \mathbf{1}_n\,\mathrm{Diag}(YY^T)^T - D_{h_u}^{-1}PYY^T\rangle \qquad (8)$$

where $\mathrm{Diag} : \mathbb{R}^{n \times n} \mapsto \mathbb{R}^n$ is the diagonal extraction of a square matrix, that is used here to compute the norm vector $\mathrm{Diag}(YY^T) = (Y^T \odot Y^T)\mathbf{1}_d$. This dispersion term measures the "sparsity" of average transport map. The derivative writes $\nabla D(P) = \mathbf{1}_n\,\mathrm{Diag}(YY^T)^T - 2D_{h_u}^{-1}PYY^T$.

3.2   Algorithm

From the penalization of the variance of the flow, the objective function (5) is non-convex. We rely on a Forward-Backward (FB) algorithm to find a critical point of this non-convex problem [2], as it contains a convex non-smooth term (the linear constraint $\iota_{\mathcal{P}_{h_u}}(P)$) and the sum of differentiable terms $G(P) = \langle C_{XY}, P\rangle + \lambda R(P) + \rho F(P) + \alpha D(P)$, where the gradient of G is L-lipschitz and L is proportional to ρ, λ, α and m. A raw estimation gives us $L \leq \tilde{L} = \lambda\|G_X^T G_X\|_2\|YY^T\|_2 + \rho m + 2\alpha\|D_{h_u}^{-1}\|_2\|YY^T\|_2$. but the constant can also be estimated empirically using a few random normalized matrices P. In practice, we use the inertial FB algorithm proposed in [8]:

$$P^{k+1} = \underset{\mathcal{P}_{h_u}}{\mathrm{Proj}}\left(P^k - \tau\left(C + \rho\nabla F(P^k) + \lambda\nabla R(P^k) + \alpha\nabla D(P^k)\right) + \beta(P^k - P^{k-1})\right).$$

which converges to a local minima of (5) by taking $\beta \in [0, 1[$ and $\tau < 2(1 - \beta)/L$.

4   Experiments

4.1   Regularized 1-D Histogram Matching

The first step of our color transfer process consists in defining the source and target sets X, Y, which involves spatio-color clustering on the input image u and the exemplar one v, respectively. Here, the clustering is performed using the fast super-pixels method [1], with the default regularization parameter 0.02 and a raw 20×20 seed initialization. These clusters are then used to build a weighted graph $(\omega_{i,j})$ and define the transport cost matrix C_{XY} that are involved in the minimization of the non-convex functional (5). The color transfer is finally applied using the estimated relaxed and regularized transport map.

As we work at a super-pixel scale to speed-up the OT computation, the last step of the proposed approach is to synthesize a new image w from the source image u using the new color palette. Like [19], we use maximum likelihood estimation to incorporate geometrical information from the source image u into the synthesis process. In order to restore the sharp details from the original image that may have been lost in the process, we run a post-processing filter, as detailed in [16].

As illustrated in Fig. 1, when the histograms of the two images have very different shapes, the classic OT color transfer create a lot of artifacts (Fig. 1c). The original colors are better recovered with increasing penalization of the color variance (Fig. 1e, f, g and h). Such property is illustrated in other examples in Fig. 2. When no penalization is applied to the color variance (*i.e.* $\alpha = 0$),

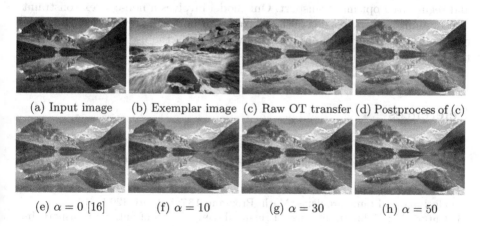

  (a) Input image   (b) Exemplar image (c) Raw OT transfer (d) Postprocess of (c)

   (e) $\alpha = 0$ [16]   (f) $\alpha = 10$ (g) $\alpha = 30$ (h) $\alpha = 50$

Fig. 1. Illustration of the penalization of transport variance for color transfer preservation. Colors of image (a) are modified using image (b) as a template. (c): Result with optimal transfer, without any regularization. (d): Post-processing of (c) to remove small artifacts, large color inconsistencies still occur. (e): adaptive approach [16], which mixes capacity relaxation and spatial regularization yields better results but final colors may be washed-out due to the mixing of colors. (f), (g) and (h): Proposed model. The parameter α directly controls the amount of transport dispersion.

(a) Input   (b) $\alpha = 0$ [16]   (c) $\alpha = 10$ (d) $\alpha = 100$ (e) Exemplar

Fig. 2. Color transfer. The colors of the exemplar images (column (e)) are transfered to the input images (column (a)) with different dispersion parameters.

it corresponds to the model of [16]. By monitoring the capacity of the target histogram and regularizing the average flow [16], the synthesized images look more plausible (Fig. 2b) but they contain new drab colors (that do not exist in the target image) and they are over-smoothed. On the other hand, the transfer is visually far better when the color variance is penalized (with high values of α). In this case, the final images only contain the colors of the target images.

5   Conclusion and Future Work

We have proposed a method for transferring color between images using relaxed and regularized optimal transport. Our model involves a non-convex constraint that minimizes the dispersion of the relaxed transport and prevents from creating new drab colors. Further improvements will concern the use of faster optimization tools and the incorporation of high-order moments (i.e. covariances of the transfered clusters) into the final synthesis.

References

1. Achanta, R., Shaji, A., Smith, K., Lucchi, A., Fua, P., Süsstrunk, S.: SLIC superpixels compared to state-of-the-art superpixel methods. IEEE TPAMI **34**(11), 2274–2282 (2012)
2. Attouch, H., Bolte, J., Svaiter, B.: Convergence of descent methods for semi-algebraic and tame problems. Math. Program. **137**(1–2), 91–129 (2013)
3. Cuturi, M.: Sinkhorn distances: lightspeed computation of optimal transport. In: NIPS 2013, pp. 2292–2300 (2013)
4. Ferradans, S., Papadakis, N., Peyré, G., Aujol, J.F.: Regularized discrete optimal transport. SIAM J. Imaging Sci. **7**(3), 1853–1882 (2014)
5. Ferradans, S., Papadakis, N., Rabin, J., Peyré, G., Aujol, J.F.: Blind deblurring using a simplified sharpness index. In: Kuijper, A., Bredies, K., Pock, T., Bischof, H. (eds.) SSVM 2013. LNCS, vol. 7893, pp. 86–97. Springer, Heidelberg (2013)
6. Morovic, J., Sun, P.L.: Accurate 3d image colour histogram transformation. Pattern Recogn. Lett. **24**(11), 1725–1735 (2003)

7. Nikolova, M., Wen, Y.W., Chan, R.H.: Exact histogram specification for digital images using a variational approach. JMIV **46**(3), 309–325 (2013)

8. Ochs, P., Chen, Y., Brox, T., Pock, T.: ipiano: inertial proximal algorithm for nonconvex optimization. SIAM J. Imaging Sci. **7**(2), 1388–1419 (2014)

9. Papadakis, N., Bugeau, A., Caselles, V.: Image editing with spatiograms transfer. IEEE TIP **21**(5), 2513–2522 (2012)

10. Papadakis, N., Provenzi, E., Caselles, V.: A variational model for histogram transfer of color images. IEEE TIP **20**(6), 1682–1695 (2011)

11. Pitié, F., Kokaram, A.C., Dahyot, R.: Automated colour grading using colour distribution transfer. CVIU **107**, 123–137 (2007)

12. Pouli, T., Reinhard, E.: Progressive color transfer for images of arbitrary dynamic range. Comput. Graph. **35**(1), 67–80 (2011)

13. Rabin, J., Delon, J., Gousseau, Y.: Removing artefacts from color and contrast modifications. IEEE TIP **20**(11), 3073–3085 (2011)

14. Rabin, J., Peyré, G.: Wasserstein regularization of imaging problem. In: IEEE ICIP 2011, pp. 1541–1544 (2011)

15. Rabin, J., Peyré, G., Delon, J., Bernot, M.: Wasserstein barycenter and its application to texture mixing. In: Bruckstein, A.M., ter Haar Romeny, B.M., Bronstein, A.M., Bronstein, M.M. (eds.) SSVM 2011. LNCS, vol. 6667, pp. 435–446. Springer, Heidelberg (2012)

16. Rabin, J., Ferradans, S., Papadakis, N.: Adaptive color transfer with relaxed optimal transport. In: IEEE ICIP 2014 (2014)

17. Reinhard, E., Adhikhmin, M., Gooch, B., Shirley, P.: Color transfer between images. IEEE Trans. Comput. Graphics Appl. **21**(5), 34–41 (2001)

18. Su, Z., Zeng, K., Liu, L., Li, B., Luo, X.: Corruptive artifacts suppression for example-based color transfer. IEEE Trans. Multimedia **16**(4), 988–999 (2014)

19. Tai, Y.W., Jia, J., Tang, C.K.: Local color transfer via probabilistic segmentation by expectation-maximization. In: CVPR 2005, pp. 747–754 (2005)

20. Xiao, X., Ma, L.: Color transfer in correlated color space. In: ACM VRCIA 2006, pp. 305–309 (2006)

Generalized Pareto Distributions, Image Statistics and Autofocusing in Automated Microscopy

Reiner Lenz$^{(\boxtimes)}$

Linköping University, 60174 Norrköping, Sweden
reiner.lenz@liu.se

Abstract. We introduce the generalized Pareto distributions as a statistical model to describe thresholded edge-magnitude image filter results. Compared to the more common Weibull or generalized extreme value distributions these distributions have at least two important advantages, the usage of the high threshold value assures that only the most important edge points enter the statistical analysis and the estimation is computationally more efficient since a much smaller number of data points have to be processed. The generalized Pareto distributions with a common threshold zero form a two-dimensional Riemann manifold with the metric given by the Fisher information matrix. We compute the Fisher matrix for shape parameters greater than -0.5 and show that the determinant of its inverse is a product of a polynomial in the shape parameter and the squared scale parameter. We apply this result by using the determinant as a sharpness function in an autofocus algorithm. We test the method on a large database of microscopy images with given ground truth focus results. We found that for a vast majority of the focus sequences the results are in the correct focal range. Cases where the algorithm fails are specimen with too few objects and sequences where contributions from different layers result in a multi-modal sharpness curve. Using the geometry of the manifold of generalized Pareto distributions more efficient autofocus algorithms can be constructed but these optimizations are not included here.

1 Introduction

Analyzing the statistical properties of images is a fundamental problem in vision science and low-level image and signal processing. In vision science one is analyzing the properties of biological vision systems and their relation to the statistical properties of their environment. A typical technical example is transform coding in which correlations between different pixels or color channels are used to reduce the size of image files. Very often non-parametric models and numerical values like the empirical means and variances are used. In the cases where parametric models are available it is possible to use the additional information to design more effective processing methods, for example to control the system.

© Springer International Publishing Switzerland 2015
F. Nielsen and F. Barbaresco (Eds.): GSI 2015, LNCS 9389, pp. 96–103, 2015.
DOI: 10.1007/978-3-319-25040-3_11

In the case of image data a popular class of distributions are the Weibull distributions which are used to describe the distribution of the magnitude values of the results of difference based filtering. They have been used to analyze natural image statistics and in [14] it was shown how models based on the Weibull distributions can be used to construct autofocus algorithms. In [9] it was shown how the more general class of generalized extreme value distributions (GEV) can be used for auto-focusing in automated microscopy. In this paper we will use the same database of microscope images as in [9] but we will show how to use the class of generalized Pareto distributions (GPD) instead of the GEVs. The first advantage of this approach is related to a threshold step involved in the distribution fitting. In practice the measured data always follows a mixture distribution, where the majority of the data has very low-values and usually the GEVs are fitted to thresholded data. In the case of the GEVs one wants to use a low threshold so that only the low-level part of the mixture is eliminated but the GEV part is well represented. The GPDs define the distribution of the data over a relatively high threshold and the data entering the fitting consists therefore only of the tail of the data following the GEV. Another advantage of GPD-based methods is computational efficiency. Practically the GEVs distribution is fitted to the data by a maximum likelihood estimation which in turn is numerically an optimization procedure. In the case of image data the number of measurement points is very large and the distribution fitting very slow. Using the GPD model is much more efficient since the optimization uses only a relatively small number (we often use five percent) of the datapoints.

In the following we will first give a brief overview over the low-level filter systems used. These filters are based on the representation theory of the dihedral group D(4), the symmetry group of the square grid. It can be shown that the resulting filters filter vectors transform like the underlying data under the dihedral transforms, that (under fairly general conditions) they are good approximations to the eigenvectors obtained by principal component analysis and that there are fast implementations similar to the discrete Fourier transform (more details can be found in [8,10]).

Next we summarize some results describing the relation between the GEVs and the GPDs and compute the Fisher matrix and its determinant for the GPD. The (inverse) of the determinant of the Fisher matrix as later used as a sharpness function in the auto-focus application. A heuristical motivation for using the determinant as a sharpness measure is based on the observation that the Fisher information matrix describes the local geometry on the manifold of the GPD distributions. Now consider the visual change of the a defocused image under a change of the focal plane of the camera. Intuitively the appearance will not change too much since the image was blurred from the beginning. Now if the image was in focus and we change the focal plane either before or after the correct focal plane then the appearance will change more than in the previous case. The position with the best focal position should therefore correspond to a critical point in the focus sequence. The hypothesis is that this behavior is somehow reflected in the properties of the statistical distributions.

Computing the determinant of the GPDs shows that it is the product of two factors, one depending only on the shape parameter and one only depending on the scale parameter. The scale part motivates a popular choice to use the variance as a sharpness function. Using the GPD-model provides an extra term given by the shape. It shows that estimates based on distributions with small (negative) shape parameters are not very reliable since most of the information is located in datapoints with very small filter responses. In the experimental part of the paper we describe first the database of microscopy images used and then we summarize the results of our experiments which show that the tails of the distributions contain sufficient information for the auto-focus application.

2   Dihedral Filters

Almost all digital images are functions defined on a square grid. The symmetry group of this grid is the dihedral group $D(4)$ consisting of eight elements, four rotations and for rotations combined with one reflection on a symmetry axis of the square. The first step in low level image processing is often a linear filtering and results from the representation theory show that filter functions given by the irreducible representations of the group $D(4)$ are eigenfunctions of group-invariant correlation functions and thus principal components, they share the transformation properties of $D(4)$ and they are computationally very efficient since they can be implemented using a reduced number of additions and subtractions only. The two-dimensional irreducible representation corresponds to a pair of gradient filters. In the following we will only use filter kernels of size 5×5 pixels. We divide the 5×5 window in so-called orbits which are the smallest $D(4)$ subsets. There is one orbit consisting of the center pixel, four orbits consisting of four points each (located on the axes and the corners of a square) and one orbit with the remaining eight points. This results in six pairs of filter results given by pairs of filter kernels of the form (in Matlab notation) $[-1 - 1; 11]$ and $[1 - 1; -11]$. The lengths of the corresponding two-dimensional filter vectors is invariant under all transformations of the underlying symmetry group. In the following we will only consider the lengths of these filter vectors and ignore the orientation information contained in the relative size of the two components of the vectors. More details can be found in [7,8].

3   Extreme Value and Pareto Distributions

For most pixels in an image the filter response to such a filter pair will be very small since neighboring pixels usually have similar intensity values and positions with large filter responses are most interesting. The statistical distribution of such edge-type filter systems has previously been investigated in the framework of the Weibull- or more generally in the framework of the generalized extreme value distributions (GEV) (see, for example [3,4,6,12–14]). From the construction of the filter functions follows that the filter results follow a mixture distribution consisting of near-zero filter results and the distribution of the significant edge magnitude values. This means that a threshold process is required before the extreme value distributions can be fitted. This can be avoided if we use the generalized Pareto distributions instead of the generalized extreme value distributions. The following selection of results from the theory of extreme value distributions may give a heuristic explanation why these distributions may be relevant in the current application.

The Three Types Theorem (originally formulated by Fisher and Tippett [2] and later proved by Gnedenko [5]) states the following: if we have an i.i.d. sequence of random variables $X_1, X_2 \ldots$ and if $M_n = \max(X_1, X_2 \ldots X_n)$ is the sample maximum then we find that if there exists (after a suitable renormalization) a non-degenerate limit distribution then this limit distribution must be of one of three different types. These types are known as the Gumbel-, Weibull- and Frechet distributions. One can combine these three distributions in a single generalized extreme value distribution (GEV). In the following we basically use the maximum likelihood estimators from the Matlab Statistical toolbox and we will therefore also use the definition and notations used there. The probability density function of the GEV is defined as

$$f(x; k, \mu, \sigma) = \frac{1}{\sigma} e^{-\left(1+k\frac{x-\mu}{\sigma}\right)^{-1/k}} \left(1 + k\frac{x-\mu}{\sigma}\right)^{-1-1/k}$$

where μ is the location, σ is the scale- and k is the shape parameter. In our implementation we don't consider the distributions with $k = 0$ since we assume that in the case of real measured data we will encounter this case relatively seldom. Related to the GEV-distributions are the generalized Pareto distributions with probability density functions defined as

$$f(x; k, \mu, \sigma, \theta) = \frac{1}{\sigma} \left(1 + k\frac{x-\theta}{\sigma}\right)^{-1-1/k}$$

where k and σ are again the shape and scale and θ is the threshold parameter. For positive k the support of the distribution is given by the half-axis $\theta < x < \infty$ and for negative k by $\theta < x < -k/\sigma$. It was shown in [11] that the GPDs are obtained as limit distributions of thresholding over some high threshold (instead of the maximum value as in the GEV case).

Both the GEV and the GPD depend on three parameters. In our application we are only interested in the shape and the scale of the GPD since the location parameter is given by the threshold parameter. Therefore we consider the GPD as a class of distributions depending on the two parameters k, σ. For these distributions we can compute the 2×2 Fisher information matrix $\mathcal{G}(\eta) = (g_{kl})$ with elements g_{kl} and parameters $\eta_1 = k, \eta_2 = \sigma$ defined as

$$g_{kl} = \int_x \frac{\partial \log f(x; \eta)}{\partial \eta_k} \frac{\partial \log f(x; \eta)}{\partial \eta_l} f(x; \eta) \, dx \tag{1}$$

For the GPD we used Mathematica to compute the following entries:

$$g_{11} = \frac{2}{2k^2 + 3k + 1}; \quad g_{12} = \frac{1}{2k^2\sigma + 3k\sigma + \sigma}; \quad g_{22} = \frac{1}{(2k + 1)\sigma^2} \tag{2}$$

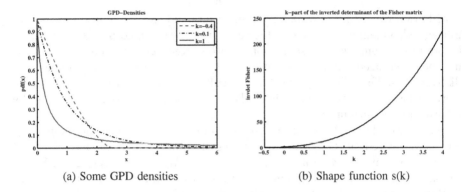

(a) Some GPD densities   (b) Shape function s(k)

Fig. 1. Generalized Pareto distributions

For the inverse of the determinant of the Fisher information matrix we find the simple expression (see Fig. 1b for a plot of S):

$$S(k,\sigma) = 1/\det \mathcal{G}(k,\sigma) = (k+1)^2(2k+1)\sigma^2 = s(k) \cdot \sigma^2$$

We see that this sharpness function consists of two factors, the shape factor $s(k)$ and the squared scale factor σ^2. The scale factor corresponds to the often used variance-based sharpness functions but the new shape term $s(k)$ gives extra information about the reliability of the scale-estimate. For low values of k the distribution is concentrated on finite intervals near zero, indicating a very weak edge content. For high values of $s(k)$ the tails become more significant and the sharpness estimate more reliable. The density functions of the three GPDs with parameters $\theta = 0, \sigma = 1, k = -0.4, 0.1, 4$ are shown in Fig. 1a

4   Database and Implementation

In our experiments we use the microscopy images from set BBBC006v1 in the Broad Bioimage Benchmark Collection which is described in [1]. The images are available at http://www.broadinstitute.org/bbbc/BBBC006. The database contains 52224 images from 384 cells, measured at two positions and prepared with two different types of staining. For each position and each cell a focus sequence consisting of 34 images was recorded. Each image consists of 696×520 pixels in 16-bit TIF format. The images show stained human cells and the imaging process is described as follows: *For each site, the optimal focus was found using laser auto-focusing to find the well bottom. The automated microscope was then programmed to collect a z-stack of 32 image sets. Planes between z = 11 - 23 are considered ground truth as in-focus images.*

As mentioned above we first filtered an image with dihedral filters of size 5×5. Since these filters basically express the 25-dimensional vectors in a new basis we get as a result a sequence of 25 filtered images. These filter results come in different types characterized by their transformation properties under the operations in D(4) (as given by the irreducible representations of D(4)). Here we only consider those filter results that are invariant under all transformations (corresponding to the trivial representation, consisting of the sum of the pixel values on an orbit) and edge-type filter pair results (corresponding to the two-dimensional irreducible representation). A 5×5 grid consists of six orbits (with one, four and eight elements) and after the filtering we first select only those points for further processing where the value at the center pixel (one pixel orbit) is greater than the mean value of the pixels on the other orbits (recall that the images are fluorescent microscope images where object points act as tiny light sources). In this way only positions with a local maximum intensity value are entering the statistical estimation. For these points we now select the twelve components of the filter results that transform like the two-dimensional representation of D(4) and we compute the length of this vector as a measure of the 'edge' strength in this point. In the group representation framework the length of this subvector should be computed as the usual Euclidean norm but we usually use the sum of the absolute values instead. This is faster and the results are comparable. For these length values we compute the Q-quantile and select all samples with a value greater than this quantile. A typical value for Q is 0.95, so that about 5 % of the selected local maxima points enter the GPD-fitting process. From the selected data we compute the minimum value θ and shift the data by subtracting θ. Next we use the Matlab function **gpfit** which implements a maximum-likelihood estimator. For the first image in a focus sequence we use the standard startvalues in the optimization process and for the following images we use the result of the previous fitting as start values, thus reducing the execution time of the computationally intensiv fitting process. For every focus sequence we selected the image with the maximum value of the inverse determinant as the detected optimal focal plane.

5 Results

In the procedure described above there is only one parameter the user can choose: the Q-value of the quantile based threshold parameter. A typical value we used is Q=0.95. We first analysed for how many sequences the detected focal plane lies in the range z = 11 - 23 defined as the ground truth. There are $4 \cdot 384 = 1536$ sequences and the number of wrong estimates (per combination of site/well) are collected in Table 1.

Analyzing the cases where this global estimation process gives results outside the ground truth interval of slices 11 to 23 we can roughly distinguish the following types of errors. The first case consists of cases where the distribution

Table 1. Detected focal planes outside the ground truth interval

Type 1 staining	Type 2 staining
31	20

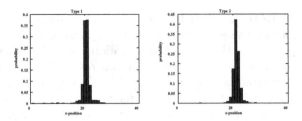

Fig. 2. Detected focal planes

fitting did not converge. This is typically the case for images with very little useful information, i.e. slices containing very few object points. In the second case the fitting was successful but the value of the estimated shape parameter was very low and the estimate is therefore not very reliable. Also in this case the number of useful object points was very small. In the third case we find multiple maxima of the sharpness function. Usually the first one lies in the ground truth interval but the second maximum comes later in the sequence. Visual inspection shows that this is often the case when high intensity cells lie in layers above or below the 'correct' focal plane. Such cases could be excluded when the search process found the first significant maximum in the sharpness function (Fig. 2).

6   Conclusions

We used two fundamental results from the theory of extreme value distributions to select the GPDs as statistical models for the edge-like filter magnitudes over a high threshold. Removing the influence of the threshold parameter (defined by the quantile of the measured filter magnitudes of an image) we obtain distributions in the two-parameter manifold of GPDs with fixed location parameter. For this manifold we computed the Fisher information matrix describing the local geometry of this manifold. We then introduced the determinant of the inverse of the Fisher matrix as a sharpness function and showed that for a vast majority of sequences the autofocus process detected one of the slices in the ground truth focal region. The cases where it missed the correct region where either characterized by poor data quality or multiple maxima detection caused by contributions from neighboring slices. The results obtained show that GPD-based models provide enough information for the control of autofocus procedures. The form of the determinant clarifies the role of the shape of the distribution and the variance in the determination of the focal plane. As a by-product we also gain additional computational efficiency since most of the pixels do not enter the sharpness computation. Another advantage of parametric models is the fact that they provide information about the analytical form of the sharpness function. This can be use in the construction of faster optimization methods where the sequence of an already measured set of parameters controls the step-length of the microscopes focus mechanism before the next image is collected.

Acknowledgements. This research is funded by the The Swedish Research Council through a framework grant for the project Energy Minimization for Computational Cameras (2014-6227) and by the Swedish Foundation for Strategic Research through grant IIS11-0081.

We used the image set BBBC006v1 from the Broad Bioimage Benchmark Collection (Ljosa, et al. "Annotated high- throughput microscopy image sets for validation," Nature Methods, vol. 9, no. 7, p. 637, 2012)

References

1. Bray, M.A., Fraser, A.N., Hasaka, T.P., Carpenter, A.E.: Workflow and metrics for image quality control in large-scale high-content screens. J. Biomol. Screen. **17**(2), 266–274 (2012)
2. Fisher, R., Tippett, L.: Limiting forms of the frequency distribution of the largest or smallest member of a sample. Proc. Camb. Philos. Soc. **24**, 180–190 (1928)
3. Geusebroek, J.-M.: The stochastic structure of images. In: Kimmel, R., Sochen, N.A., Weickert, J. (eds.) Scale-Space 2005. LNCS, vol. 3459, pp. 327–338. Springer, Heidelberg (2005)
4. Geusebroek, J.M., Smeulders, A.W.M.: Fragmentation in the vision of scenes. In: Proceedings of ICCV, pp. 130–135 (2003)
5. Gnedenko, B.: Sur la distribuion limite du terme maximum d'une série aléatoire. Ann. Math. **44**, 423–453 (1943)
6. Jia, Y., Darrell, T.: Heavy-tailed distances for gradient based image descriptors. In: Advances in Neural Information Systems, pp. 1–9 (2011)
7. Lenz, R.: Group Theoretical Methods in Image Processing. LNCS, vol. 413. Springer, Heidelberg (1990)
8. Lenz, R.: Investigation of receptive fields using representations of dihedral groups. J. Vis. Commun. Image Represent. **6**(3), 209–227 (1995)
9. Lenz, R.: Generalized extreme value distributions, information geometry and sharpness functions for microscopy images. In: Proceedings of ICASSP, pp. 2867–2871 (2014)
10. Lenz, R., Zografos, V., Solli, M.: Dihedral color filtering. In: Fernandez-Maloigne, C. (ed.) Advanced Color Image Processing and Analysis, pp. 119–145. Springer, New York (2013)
11. Pickands, J.: Statistical-inference using extreme order statistics. Ann. Statistics **3**(1), 119–131 (1975)
12. Scholte, H.S., Ghebreab, S., Waldorp, L., Smeulders, A.W.M., Lamme, V.A.F.: Brain responses strongly correlate with Weibull image statistics when processing natural images. J. Vis. **9**(4), 29:1–29:15 (2009)
13. Yanulevskaya, V., Geusebroek, J.M.: Significance of the Weibull distribution and its sub-models in natural image statistics. In: Proceedings of International Conference Computer Vision Theory and Application, pp. 355–362 (2009)
14. Zografos, V., Lenz, R., Felsberg, M.: The Weibull manifold in low-level image processing: an application to automatic image focusing. Im. Vis. Comp. **31**(5), 401–417 (2013)

Barycenter in Wasserstein Spaces: Existence and Consistency

Thibaut Le Gouic$^{(\boxtimes)}$ and Jean-Michel Loubes

École Centrale de Marseille, Marseille, France
tle-gouic@centrale-marseille.fr
http://tle-gouic.perso.centrale-marseille.fr

Abstract. We study barycenters in the Wasserstein space $\mathcal{P}_p(E)$ of a locally compact geodesic space (E, d). In this framework, we define the barycenter of a measure $\mathbb{P}$ on $\mathcal{P}_p(E)$ as its Fréchet mean. The paper establishes its existence and states consistency with respect to $\mathbb{P}$. We thus extends previous results on $\mathbb{R}^d$, with conditions on $\mathbb{P}$ or on the sequence converging to $\mathbb{P}$ for consistency.

Keywords: Barycenter · Wasserstein space · Geodesic spaces

1 Introduction

The Fréchet mean of a Borel probability measure μ, defined on a metric space (E, d), as the minimizer of

$$x \mapsto \mathbb{E}d^2(x, X), X \sim \mu$$

provides a natural extension of the barycenter as it coincides on $\mathbb{R}^d$ with the barycenter $\sum_{i=1}^n \lambda_i x_i$ of the points $(x_i)_{1 \le i \le n}$, with weights $(\lambda_i)_{1 \le i \le n}$ if

$$\mu = \sum_{i=1}^n \lambda_i x_i.$$

Its existence is a straightforward consequence of the local compactness of a geodesic space (E, d) when assumed. But it is not obvious in more general cases.

Mimicking the Fréchet mean, for any $p \ge 1$, we define a p-barycenter (or simply a barycenter) of a measure μ, as any minimizer of $x \mapsto \mathbb{E}d^p(x, X)$, where $X \sim \mu$.

The Wasserstein space $\mathcal{P}_p(E)$ of a locally compact geodesic space (E, d) is the set of all Borel probability measure on (E, d) such that $\mathbb{E}d^p(x, X) < \infty$, for some $x \in E$, endowed with the p-Wasserstein metric defined between two measures μ, ν as

$$W_p^p(\mu, \nu) = \inf_{\pi \in \Gamma(\mu, \nu)} \int d^p(x, y) d\pi(x, y), \tag{1}$$

© Springer International Publishing Switzerland 2015
F. Nielsen and F. Barbaresco (Eds.): GSI 2015, LNCS 9389, pp. 104–108, 2015.
DOI: 10.1007/978-3-319-25040-3_12

where $\Gamma(\mu, \nu)$ is the set of measures on $E \times E$ with marginals μ and ν. Since the Wasserstein space of a locally compact space geodesic space is geodesic but not locally compact (unless (E, d) is compact), the existence of a barycenter is not as straightforward.

This paper presents its existence and study consistency properties. Several works has already been achieved in this field. An important one is the demonstration of existence and uniqueness of the barycenter of measures $\mathbb{P}$ on $\mathcal{P}_2(\mathbb{R}^d)$, with $d \in \mathbb{N}^*$, and $\mathbb{P}$ finitely supported on Dirac masses:

$$\mathbb{P} = \sum_{i=1}^{n} \lambda_i \delta_{\mu_i},$$

such that $\mu_i \in \mathcal{P}_p(\mathbb{R}^d)$, for $1 \leq i \leq n$, with one μ_i *vanishing on small sets*. In this case (when the measure $\mathbb{P}$ is finitely supported on Dirac masses), the barycenter of $\mathbb{P}$ is also the minimizer of

$$\nu \mapsto \sum i = 1^n \lambda_i W_p^p(\nu, \mu_i),$$

which is how the problem is more classically posed.

This *vanishing* property is said to be satisfied for a probability measure if it gives probability 0 to sets of Hausdorff dimension less than $d - 1$. Any measure absolutely with respect to the Lebesgue measure *vanishes on small sets*. This work of [AC] has been extended to compact Riemannian manifolds, with the condition to *vanish on small sets* being replaced by absolute continuity with respect to the volume measure by [KP]. Since the Wasserstein space of a compact space is also compact, the existence in this setting can be easily obtained, but their work provides, among other results, an interesting extension, to our concern, to the work of [AC], by showing a dual problem called the *multidimensional* problem, for any $\mathbb{P}$ of the form

$$\sum_{i=1}^{n} \lambda_i \delta_{\mu_i}.$$

The same dual problem has been used in a previous work to show existence of barycenter whenever there exists a measurable (not necessarily unique) barycenter application on (E^n, d^n) that associate the barycenter of $\sum_{i=1}^{n} \lambda_i \delta_{x_i}$ to every n-uplets $(x_1, ..., x_n)$. It is a first step toward the proof of existence of barycenter for any $\mathbb{P}$.

Two statistical problems arise from the notion of barycenter. The first one can be stated as follows. Given $(\mu_i)_{i \geq 1}$, and $(\lambda_j^J)_{1 \leq j \leq J}$, it would be useful that the (or any) barycenter of $\mathbb{P}_J = \sum_{j=1}^{J} \lambda_j^J \delta_{\mu_j}$ converges to a barycenter of a limit measure of the sequence $(\mathbb{P}_J)_{J \geq 1}$. [BK] studied this problem in the case where $(\mu_j)_{j \geq 1}$ have compact support, are absolutely continuity with respect to the Lebesgue measure and are indexed on a compact set Θ of $\mathbb{R}^d$. They state more precisely that given a probability measure on Θ, one can induce a probability measure $\mathbb{P}$ on $\mathcal{P}_p(\mathbb{R}^d)$, and if the $(\mu_j)_{j \geq 1}$ are chosen randomly under $\mathbb{P}^{\otimes \infty}$, the (unique) barycenter of $\frac{1}{J}\sum_{j=1}\delta_{\mu_j}$ converges to the barycenter of $\mathbb{P}$, $\mathbb{P}$-almost surely.

In a previous work [BLGL], the authors produced a similar result under the assumptions that the $(\mu_j)_{j\geq 1}$ are *admissible deformations*, which is a similar condition.

The second statistical problem rising from this framework is the following. Given $(\lambda_i)_{i\geq 1}$ and $(\mu_j^n)_{1\leq j\leq J}$ converging to some $(\mu_j)_{1\leq j\leq J}$, a question of our interest is whether the barycenter of $\sum_{j=1}^{J}\lambda_j\delta_{\mu_j^n}$ converges to a barycenter of the limit $(\mu_j)_{1\leq j\leq J}$. The problem has been answered positively in [BLGL], up to a subsequence, since the barycenter is not unique. Although these two problems are presented differently, they can be formulated into one problem. Does the (or any) barycenter of $\mathbb{P}_n$ converges to the barycenter of $\mathbb{P}$ when $\mathbb{P}_n$ converges to $\mathbb{P}$?

This paper presents a positive result of [LGL], that implies, in particular, existence of the barycenter for any Borel probability measure $\mathbb{P} \in \mathcal{P}_p(\mathcal{P}_p(E))$.

2   Existence of Barycenter

Let (E, d) be a geodesic locally compact space. For any $p \geq 1$, define by $\mathcal{P}_p(E)$ the Wasserstein space of E, by the space of all Borel probability measures such that for all $x \in E$, $\mathbb{E}d^p(x, X) < \infty$, endowed with the Wasserstein metric defined in (1). Denote thus by $\mathcal{P}_p(\mathcal{P}_p(E))$ the p-Wasserstein space of Borel measures on $\mathcal{P}_p(E)$ endowed with the Wasserstein metric.

For a measure $\mathbb{P} \in \mathcal{P}_p(\mathcal{P}_p(E))$, we define a *barycenter* as a minimizer of the function

$$\mu \mapsto \mathbb{E}(W_p^p(\tilde{\mu}, \mu)),$$

where $\tilde{\mu} \sim \mathbb{P}$. Remark that $\mathbb{E}(W_p^p(\tilde{\mu}, \mu)) = W_p^p(\mathbb{P}, \delta_\mu)$ where the two notations W_p refer to the Wasserstein distance but on different spaces respectively $\mathcal{P}_p(E)$ and $\mathcal{P}_p(\mathcal{P}_p(E))$.

Then [LGL] proves the following result.

Theorem 1. *Let $\mathbb{P} \in \mathcal{P}_p(\mathcal{P}_p(E))$ be a measure on $\mathcal{P}_p(E)$. Then there exists a barycenter of $\mathbb{P}$.*

This result is a consequence of the existence of barycenter for $\mathbb{P}$ finitely supported, showed in [BLGL] or [LG], and the consistency result of [LGL] presented above.

3   Consistency of the Barycenter

Since the barycenter is not necessarily unique for a given $\mathbb{P}$, the continuity of the barycenter with respect to $\mathbb{P}$ does not make sense. However, it is interesting to know for a sequence of measure $(\mathbb{P}_n)_{n\geq 1} \subset \mathcal{P}_p(E)$ converging in $\mathcal{P}_p(\mathcal{P}_p(E))$ to $\mathbb{P}$, a sequence of their barycenter converges to a barycenter of $\mathbb{P}$. [LGL] provides a positive answer.

Theorem 2. *Let $(\mathbb{P}_n)_{n\geq 1} \subset \mathcal{P}_p(\mathcal{P}_p(E))$ be a sequence of measures on $\mathcal{P}_p(E)$, and set μ_n a barycenter of $\mathbb{P}_n$, for all $n \in \mathbb{N}$. Suppose that $W_p(\mathbb{P}, \mathbb{P}_n) \to 0$. Then, the sequence $(\mu_n)_{n\geq 1}$ is compact in $\mathcal{P}_p(E)$ and any limit is a barycenter of $\mathbb{P}$.*

Proof (Main ideas). The proof is in three steps.

The first step is to show that the sequence $(\mu_n)_{n\geq 1}$ is tight. It is indeed a consequence of the fact that balls on (E, d) are compact together with applying a Markov inequality to these balls.

The second step uses Skorokhod representation theorem and lower semicontinuity of $\nu \mapsto W_p(\mu, \nu)$ for any μ, to show that any weak limit of the sequence $(\mu_n)_{n\geq 1}$ is a barycenter of $\mathbb{P}$.

The final step shows that the convergence of the $(\mu_n)_{n\geq 1}$ holds actually in $\mathcal{P}_p(E)$.

Applying this result to a constant sequence provides the following corollary.

Corollary 1. *The set of all barycenters of a given measure $\mathbb{P} \in \mathcal{P}_p(\mathcal{P}_p(E))$ is compact.*

An interesting and immediate corollary follows from the assumption that $\mathbb{P}$ has a unique barycenter.

Corollary 2. *Suppose $\mathbb{P} \in \mathcal{P}_p(\mathcal{P}_p(E))$ has a unique barycenter. Then for any sequence $(\mathbb{P}_n)_{n\geq 1} \subset \mathcal{P}_p(\mathcal{P}_p(E))$ converging to $\mathbb{P}$, any sequence $(\mu_n)_{n\geq 1}$ of their barycenters converges to the barycenter of $\mathbb{P}$.*

On $E = \mathbb{R}^d$ and for $p = 2$, there exists a simple condition that ensures that the barycenter is unique.

Proposition 1. *Let $\mathbb{P} \in \mathcal{P}_2(\mathcal{P}_2(\mathbb{R}^d))$ such that there exists a set $A \in \mathcal{P}_2(\mathbb{R}^d)$ of measures such that for all $\mu \in A$,*

$$B \in \mathcal{B}(\mathbb{R}^d), \dim(B) \leq d - 1 \implies \mu(B) = 0, \tag{2}$$

and $\mathbb{P}(A) > 0$, then, $\mathbb{P}$ admits a unique barycenter.

Proof. It is a consequence of the fact that if ν satisfies (2), then $\mu \mapsto W_2(\mu, \nu)$ is strictly convex and this so is $\mu \mapsto \mathbb{E}W_2^2(\mu, \tilde{\mu})$.

4   Statistical Applications

Previous results imply that the two statistical problems mentioned in the introduction have positive answers. Define

$$\mathbb{P}_J = \sum_{i=1}^{J} \lambda_i^J \delta_{\mu_j}$$

with measure $\mu_j \in \mathcal{P}_p(E)$ and weights λ_j so that $\mathbb{P}_J$ converges to some measure $\mathbb{P}$, then Theorem 2 states that the barycenter (or any barycenter if not unique) of $\mathbb{P}_J$ converges to the barycenter of $\mathbb{P}$ (provided $\mathbb{P}$ has a unique barycenter).

Also, given

$$\mathbb{P}_n = \sum_{j=1}^{J} \lambda_j \delta_{\mu_j^n}$$

with positive weights λ_j and measures $(\mu_j^n)_{1 \le j \le J, n \ge 1} \subset \mathcal{P}_p(E)^J$ converging to some limit measures $(\mu_j)_{1 \le j \le J} \in \mathcal{P}_p(E)^J$, then, Theorem 2 states that the barycenter (or any if not unique) converges to the barycenter of $\sum_{j=1}^{J} \lambda_j \delta_{\mu_j^n}$ (if unique). These applications are further developed in [LGL].

Acknowledgments. The author would like to thank Anonymous Referee #2 for the detailed review.

References

[AC] Agueh, M., Carlier, G.: Barycenters in the Wasserstein space. IAM J. Math. Anal. **43**, 904–924 (2011)

[KP] Kim, Y.H., Pass, B.: Wasserstein Barycenters over Riemannian manifolds (2014)

[BK] Bigot, J., Klein, T.: Consistent estimation of a population barycenter in the Wasserstein space (2012)

[BLGL] Boissard, E., Le Gouic, T., Loubes, J.-M.: Distribution's template estimate with Wasserstein metrics. Bernoulli J. **21**(2), 740–759 (2015)

[LGL] Le Gouic, T., Loubes, J.-M: Barycenter in Wasserstein spaces: existence and consistency (2015)

[LG] Le Gouic, T.: Localisation de masse et espaces de Wasserstein, thesis manuscript (2013)

Multivariate L-Moments Based on Transports

Alexis Decurninge[✉]

Mathematical and Algorithmic Sciences Lab, France Research Center, Huawei
Technologies Co. Ltd., Boulogne-Billancourt, France
alexis.decurninge@huawei.com

Abstract. Univariate L-moments are expressed as projections of the
quantile function onto an orthogonal basis of univariate polynomials.
We present multivariate versions of L-moments expressed as collections
of orthogonal projections of a multivariate quantile function on a basis
of multivariate polynomials. We propose to consider quantile functions
defined as transports from the uniform distribution on $[0;1]^d$ onto the
distribution of interest and present some properties of the subsequent
L-moments. The properties of estimated L-moments are illustrated for
heavy-tailed distributions.

1 Motivations and Notations

Univariate L-moments are either expressed as sums of order statistics or as
projections of the quantile function onto an orthogonal basis of polynomials
in $L_2([0;1], \mathbb{R})$. Both concepts of order statistics and of quantile are specific to
dimension one which makes non immediate a generalization to multivariate data.

Let $r \in \mathbb{N}_* := \mathbb{N}\backslash\{0\}$. For an identically distributed sample $X_1, ..., X_r$ on $\mathbb{R}$,
we note $X_{1:r} \leq ... \leq X_{r:r}$ its order statistics. It should be noted that $X_{1:r}, ..., X_{r:r}$
are still random variables.

Then, if $\mathbb{E}[|X|] < \infty$, the r-th L-moment is defined by:

$$\lambda_r = \frac{1}{r} \sum_{k=0}^{r-1} (-1)^k \binom{r-1}{k} \mathbb{E}[X_{r-k:r}]. \tag{1.1}$$

If we use F to denote the cumulative distribution function (cdf) and define the
quantile function for $t \in [0;1]$ as the generalized inverse of F i.e. $Q(t) = \inf\{x \in \mathbb{R} \text{ s.t. } F(x) > t\}$, this definition can be written:

$$\lambda_r = \int_0^1 Q(t) L_r(t) dt \tag{1.2}$$

where the L_r's are the shifted Legendre polynomials which are a Hilbert orthogonal basis for $L^2([0;1], \mathbb{R})$.

L-moments were introduced by Hosking [6] in 1990 as alternative descriptors to central moments for a univariate distribution especially for the study of
heavy-tailed distributions. They have some properties that we wish to keep for

© Springer International Publishing Switzerland 2015
F. Nielsen and F. Barbaresco (Eds.): GSI 2015, LNCS 9389, pp. 109–117, 2015.
DOI: 10.1007/978-3-319-25040-3_13

the analysis of multivariate data. Serfling and Xiao [8] listed the following key features of univariate L-moments which are desirable for a multivariate generalization:

- The existence of the r-th L-moment for all r if the expectation of the underlying random variable is finite
- A distribution is characterized by its infinite series of L-moments (if the expectation is finite)
- A scalar product representation with mutually orthogonal weight functions (Eq. (1.2))
- Sample L-moments are more stable than classical moments, increasingly with higher order: the impact of each outlier is linear in the L-moment case whereas it is in the order of $(x - \bar{x})^k$ for classical moments of k order

Serfling and Xiao proposed a multivariate extension of L-moments for a vector $(X_1, ..., X_d)^T$, based on matrices built from the conditional distribution of X_j given X_i for all $(i,j) \in \{1, ..., d\}^2$. The coefficients of their L-moment matrices are defined through Eq. (1.2) for diagonal elements and for $i \neq j$ through

$$\lambda_r^{(ij)} = \int_0^1 Q_{j|i}(t_j|Q_i(t_i))L_r(t_i)dt_i dt_j \tag{1.3}$$

where $Q_{j|i}(.|x)$ is the conditional quantile of X_j knowing $X_i = x$.

Their definition satisfies most of the properties of the univariate L-moments, but for the characterization of the multivariate distributions by the family of its L-moments. We generalize their approach by a slightly shift in perspective. Indeed, we define L-moments as projections of a multivariate quantile onto an orthogonal polynomial basis. As multivariate quantile, we will consider a transport of the uniform measure on $[0;1]^d$ onto the measure of interest (see for example Galichon and Henry [4]). Let us recall that T is said to be a transport map between μ and ν if $T\#\mu = \nu$ i.e. if

$$\nu(B) = \mu(T^{-1}(B)) \text{ for every Borel subset B of } \mathbb{R}^d \tag{1.4}$$

Let us now introduce some notation. In the following, we will consider a random variable or vector X with measure ν and $\overset{d}{=}$ means the equality in distribution. The scalar product between x and y in $\mathbb{R}^d$ will be noted $x.y$ or $\langle x, y \rangle$.

All proofs of presented results can be found in a longer version [2,3].

2   General Definition of Multivariate L-Moments

Let X be a random vector in $\mathbb{R}^d$. We wish to exploit the representation given by the Eq. (1.2) in order to define multivariate L-moments. Recall that we chose quantiles as mappings between $[0;1]^d$ and $\mathbb{R}^d$.

Let $\alpha = (i_1, ..., i_d) \in \mathbb{N}^d$ be a multi-index and $L_\alpha(t_1, ..., t_d) = \prod_{k=1}^d L_{i_k}(t_k)$ (where the L_{i_k}'s are univariate shifted Legendre polynomials) the natural multivariate extension of the Legendre polynomials. Indeed, it holds

Lemma 1. *The L_α family is orthogonal and complete in the Hilbert space $L^2([0;1]^d, \mathbb{R})$ equipped with the usual scalar product:*

$$\forall f, g \in L^2([0;1]^d), \quad \langle f, g \rangle = \int_{[0;1]^d} f(u).g(u)du \qquad (2.1)$$

We can finally define the multivariate L-moments.

Definition 2. *Let $Q : [0;1]^d \to \mathbb{R}^d$ be a transport between the uniform distribution on $[0;1]^d$ and ν. Then, if $\mathbb{E}[\|X\|] < \infty$, the L-moment λ_α of multi-index α associated to the transport Q are defined by:*

$$\lambda_\alpha := \int_{[0;1]^d} Q(t_1, ..., t_d) L_\alpha(t_1, ..., t_d) dt_1...dt_d \in \mathbb{R}^d. \qquad (2.2)$$

With this definition, there are as many L-moments as ways to transport $unif$ onto ν. For example, if we consider Rosenblatt transport, Definition (2.2) coincide with Serfling and Xiao L-moment matrices (see [3]).

Remark 3. Given the degree δ of $\alpha = (i_1, ..., i_d)$ that we define by $\delta = \sum_{k=1}^{d}(i_k - 1) + 1$, we may define all L-moments with degree δ, each one associated with a given corresponding α leading to the same δ.

For example, the L-moment of degree 1 is

$$\lambda_1(= \lambda_{1,1,...,1}) = \int_{[0;1]^d} Q(t_1, ..., t_d) dt_1...dt_d = \mathbb{E}[X]. \qquad (2.3)$$

The L-moments of degree 2 can be grouped in a matrix:

$$\Lambda_2 = \left[\int_{[0;1]^d} Q_i(t_1, ..., t_d)(2t_j - 1) dt_1...dt_d \right]_{1 \le i,j \le d}. \qquad (2.4)$$

In Eq. (2.3) we noted $Q(t_1, ..., t_d) = \begin{pmatrix} Q_1(t_1, ..., t_d) \\ \vdots \\ Q_d(t_1, ..., t_d) \end{pmatrix}$.

Proposition 4. *Let ν and ν' be two Borel probability measures. We suppose that Q and Q' respectively transport $unif$ onto ν and ν'.*

Assume that Q and Q' have same multivariate L-moments $(\lambda_\alpha)_{\alpha \in \mathbb{N}_*^d}$ given by the Eq. (2.2).

Then $\nu = \nu'$. Moreover:

$$Q(t_1, ..., t_d) = \sum_{(i_1,...,i_d) \in \mathbb{N}_*^d} \left(\prod_{k=1}^{d} (2i_k + 1) \right) L_{(i_1,...,i_d)}(t_1, ..., t_d) \lambda_{(i_1,...,i_d)} \in \mathbb{R}^d \qquad (2.5)$$

3   Monotone Transport from the Standard Gaussian Distribution and Associated L-Moments

3.1   Monotone Transport

Let us define the particular monotone transport. We consider source measures μ that give no mass to "small sets". For a precise definition of the term "small set", we use the Hausdorff dimension (see [7]). It then holds

Proposition 5 (*McCann/Brenier's Theorem*). *Let μ, ν be two probability measures on $\mathbb{R}^d$, such that μ does not give mass to sets of Hausdorff dimension at most $d - 1$. Then, there is exactly one measurable map T such that $T \# \mu = \nu$ and $T = \nabla\varphi$ for some convex function φ, in the sense that any two such maps coincide $d\mu$-almost everywhere.*

The gradient of convex potentials are called monotone by analogy with the univariate case. We can see this gradient as the solution of a potential differential equation. By abuse of language, we will refer at this transport as monotone transport in the sequel.

3.2   Gaussian L-Moments

The major drawback of the uniform law on $[0; 1]^d$ is its non-invariance by rotation which is a desirable property in order to more easily compute the monotone transports. We then propose a transport leading to the following L-moments:

$$\lambda_\alpha = \int_{[0;1]^d} T_0 \circ Q_{\mathcal{N}}(t_1, ..., t_d) L_\alpha(t_1, ..., t_d) dt_1...dt_d \qquad (3.1)$$

where $Q_{\mathcal{N}}$ is the transport of *unif*, the uniform distribution on $[0; 1]^d$, onto the multivariate standard distribution $\mathcal{N}(0, I_d)$ defined by

$$Q_{\mathcal{N}}(t_1, .., t_d) = \begin{pmatrix} \mathcal{N}^{-1}(t_1) \\ \vdots \\ \mathcal{N}^{-1}(t_d) \end{pmatrix}$$

and T_0 the transport of $\mathcal{N}(0, I_d)$ onto the considered distribution:

$$([0; 1]^d, du) \overset{Q_{\mathcal{N}}}{\rightsquigarrow} (\mathbb{R}^d, d\mathcal{N}) \overset{T_0}{\rightsquigarrow} (\mathbb{R}^d, d\nu) \qquad (3.2)$$

Indeed, $T_0 \circ Q_{\mathcal{N}}$ is then a transport of *unif* onto the considered distribution.

Example 6 (L-moments of multivariate Gaussian).
Let us consider $m \in \mathbb{R}^d$, a positive matrix A and the convex quadratic potential:

$$\varphi(x) = m.x + \frac{1}{2}x^T A x \qquad \text{for } x \in \mathbb{R}^d. \qquad (3.3)$$

The monotone transport associated to this potential is:

$$T_0(x) = \nabla\varphi(x) = m + Ax \qquad \text{for } x \in \mathbb{R}^d. \tag{3.4}$$

Furthermore, $T_0(\mathcal{N}_d(0, I_d)) \overset{d}{=} \mathcal{N}_d(m, A^T A)$. The L-moments of a multivariate Gaussian of mean m and covariance $A^T A$ are:

$$\lambda_\alpha = \int_{[0;1]^d} [m + A\mathcal{N}_d(t_1, ..., t_d)] L_\alpha(t_1, ..., t_d) dt_1...dt_d$$

$$= \mathbb{1}_{\alpha=(1,...,1)} m + \mathbb{1}_{\alpha\neq(1,...,1)} A\lambda_\alpha(\mathcal{N}_d(0, I_d))$$

with the notation $\lambda_\alpha(\mathcal{N}_d(0, I_d))$ denoting the α-th L-moments of the standard multivariate Gaussian, which is easy to compute since it is a random vector with independent components.

In particular, the L-moment matrix of degree 2:

$$\Lambda_2 = (\lambda_{2,1...,1} ... \lambda_{1,...,1,2}) = \frac{1}{\sqrt{\pi}} A. \tag{3.5}$$

Example 7 (Linear combinations of independent variables).
Let $(e_1, ..., e_d)$ be an orthonormal basis of $\mathbb{R}^d$ and $(b_1, ..., b_d)$ the canonical basis. We consider the potential defined by:

$$\varphi(x) = \sum_{i=1}^d \sigma_i \varphi_i(x^T e_i) \tag{3.6}$$

with each function φ_i derivable and convex and $\sigma_i > 0$. Then

$$\nabla\varphi(x) = \sum_{i=1}^d \sigma_i \varphi_i'(x^T e_i) e_i \tag{3.7}$$

Then, if we denote by $P = \sum_{i=1}^d e_i b_i^T$ and $D = \sum_{i=1}^d \sigma_i b_i b_i^T$, this potential generates the random vector

$$Y \overset{d}{=} P^T D \begin{pmatrix} \varphi_1'(X^T e_1) \\ \vdots \\ \varphi_d'(X^T e_d) \end{pmatrix}. \tag{3.8}$$

Let us note that P is orthogonal i.e. $PP^T = P^T P = I_d$ and D is diagonal.

As $e_1,...,e_d$ is an orthonormal family, $X^T e_1, ..., X^T e_d$ are independent Gaussian random variables. Then if we write the increasing functions $\varphi_i'(x) = Q_i(\mathcal{N}_1(x))$ with Q_i the quantile of a random variable Z_i, then

$$Y \overset{d}{=} P^T \begin{pmatrix} \sigma_1 Z_1 \\ \vdots \\ \sigma_d Z_d \end{pmatrix} \tag{3.9}$$

with $Z_1,...,Z_d$ independent. The parameters σ_i are meant to represent a scale parameter for each Z_i but can be absorbed in the function φ_i'.

The L-moments of Y are then for $\alpha \in \mathbb{N}_*^d$:

$$\lambda_\alpha = P^T D \begin{pmatrix} \int_{\mathbb{R}^d} \varphi_1'(\langle x, e_1 \rangle) L_\alpha(\mathcal{N}_d(x)) d\mathcal{N}_d(x) \\ \vdots \\ \int_{\mathbb{R}^d} \varphi_d'(\langle x, e_d \rangle) L_\alpha(\mathcal{N}_d(x)) d\mathcal{N}_d(x) \end{pmatrix}$$

4   Estimation of a Monotone Transport

Let $x_1, ..., x_n$ be an iid sample drawn from a common measure ν. We present the construction of the monotone transport of an absolutely continuous measure μ defined on $\mathbb{R}^d$ with finite expectation onto $\nu_n = \sum_{i=1}^n \delta_{x_i}$. If μ is the standard Gaussian measure on $\mathbb{R}^d$, the resulting transport is an estimate of the transport of $\mathcal{N}(0, I_d)$ onto ν.

4.1   Discrete Monotone Transport

For this purpose, we will present a variational approach initially proposed by Aurenhammer [1]. Let $\phi_h : \mathbb{R}^d \to \mathbb{R}$ be the piecewise linear function defined by

$$\text{for any } u \in \mathbb{R}^d, \quad \phi_h(u) = \max_{1 \leq i \leq n} \{u.x_i + h_i\}. \tag{4.1}$$

Theorem 8. (see [5]) *Let us suppose that $x_1, ..., x_n$ are distinct points of $\mathbb{R}^d$.*

Then $\nabla \phi_h$ is piecewise constant and is a monotone transport of μ onto ν_n for a particular $h = h^*$, unique up to a constant $(b, ..., b)$, which is the minimizer of an energy function E

$$h^* = \arg \min_{h \in \mathbb{R}^n} E(h) = \arg \min_{h \in \mathbb{R}^n} \int_{\mathbb{R}^d} \phi_h(u) d\mu - \frac{1}{n} \sum_{i=1}^n h_i. \tag{4.2}$$

The gradient of the energy function is simply given by:

$$\nabla E(h) = \begin{pmatrix} \int_{W_1(h)} d\mu(x) - \frac{1}{n} \\ \vdots \\ \int_{W_n(h)} d\mu(x) - \frac{1}{n} \end{pmatrix}. \tag{4.3}$$

with $(W_i(h))_{1 \leq i \leq n}$ a collection of sets (called power diagram) defined by

$$W_i(h) = \{y \in \mathbb{R}^d \text{ s.t. } \nabla \phi_h(y) = x_i\}.$$

The computation of optimal h^* can then be performed by a gradient descent:

$$h_{t+1} = h_t - \gamma \nabla E(h_t).$$

Moreover, in order to compute the gradient of E for the standard Gaussian, we use a Monte-Carlo method.

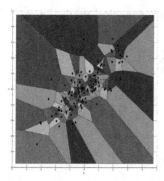
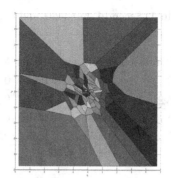

(a) Voronoi cells of the sample   (b) Power diagram corresponding to the optimal transport (the transport maps each cell into one sample i.e. is piecewise linear)

Fig. 1. Discrete optimal transport for a sample of size 100 drawn from a Gaussian distribution with covariance $\begin{pmatrix} 1 & 0.8 \\ 0.8 & 1 \end{pmatrix}$ into the standard Gaussian

4.2   Consistency of the Optimal Transport Estimator

We define two transports, say T and T_n expressed as the gradient of two convex functions, say φ and φ_n, so that $T = \nabla\varphi$ and $T_n = \nabla\varphi_n$.

T and T_n respectively transport μ onto ν and ν_n. Let us recall that, if $\mu = \mathcal{N}(0, I_d)$, we defined the quantiles of ν and ν_n by $Q = T \circ Q_{\mathcal{N}}$ and $Q_n = T_n \circ Q_{\mathcal{N}}$ (Fig. 1).

Theorem 9. *If ν satisfies $\int \|x\| d\nu(x) < +\infty$, then*

$$\|T - T_n\|_1 = \int_\Omega \|T(x) - T_n(x)\| d\mu(x) \overset{a.s.}{\to} 0. \tag{4.4}$$

Moreover, we have for $\alpha \in \mathbb{N}_*^d$

$$\hat{\lambda}_\alpha = \int_\Omega Q_n(u) L_\alpha(u) du \overset{a.s.}{\to} \lambda_\alpha = \int_\Omega Q(u) L_\alpha(u) du \tag{4.5}$$

5   Numerical Applications

We will present some numerical results for the estimation of L-moments issued from the monotone transport. For that purpose, we simulate a linear combination of independent vectors in $\mathbb{R}^2$

$$Y = P \begin{pmatrix} Z_1 \\ Z_2 \end{pmatrix} \quad \text{with} \quad P = \frac{1}{\sqrt{2}} \begin{pmatrix} -1 & 1 \\ 1 & 1 \end{pmatrix}$$

Table 1. Second L-moments and covariance numerical results for $\nu = 0.5$

Parameter	True value	$n = 30$			$n = 100$		
		Mean	Median	CV	Mean	Median	CV
$\Lambda_{2,11}$	0.38	0.28	0.27	0.30	0.38	0.37	0.18
$\Lambda_{2,12}$	0.19	0.14	0.13	0.65	0.20	0.20	0.33
Σ_{11}	0.69	0.70	0.48	1.23	0.69	0.59	0.55
Σ_{12}	0.55	0.55	0.29	1.62	0.54	0.47	0.67

Z_1, Z_2 are drawn from a symmetrized Weibull distribution of shape parameter ν and scale parameter 1.

We perform $N = 100$ estimations of the second L-moment matrix Λ_2 and the covariance matrix Σ for a sample of size $n = 30$ or 100. We present the results in Table 1 through the following features

- The mean of the different estimates
- The median of the different estimates
- The coefficient of variation of the estimates $\hat{\theta}_1, ..., \hat{\theta}_N$ (for an arbitrary parameter θ)

$$CV = \frac{\left(\sum_{i=1}^N \left(\theta_i - \frac{1}{N} \sum_{i=1}^N \theta_i \right)^2 \right)^{1/2}}{\frac{1}{N} \sum_{i=1}^N \theta_i}$$

Table 1 illustrates the fact that the L-moment estimator are more stable than classical covariance estimates for heavy-tailed distributions. The effects should be even more visible for moments of higher order. However, our sampled L-moments introduces a bias for small n contrary to classical empirical covariance.

Acknowledgements. This work was performed during the PhD of A. Decurninge supported by the DGA/MRIS and Thales.

References

1. Aurenhammer, F.: Power diagrams: properties, algorithms and applications. SIAM J. Comput. **16**(1), 78–96 (1987)
2. Decurninge, A.: Multivariate quantiles and multivariate L-moments. arXiv:1409.6013 (2014)
3. Decurninge, A.: Univariate and multivariate quantiles, probabilistic and statistical approaches; radar applications. Ph.D. Dissertation (2015)
4. Galichon, A., Henry, M.: Dual theory of choice with multivariate risks. J. Econ. Theor. **147**(4), 1501–1516 (2012)
5. Gu, X., Luo, F., Sun, J., Yau, S.-T.: Variational principles for Minkowski type problems, discrete optimal transport, and discrete Monge-Ampere equations. arXiv:1302.5472 (2013)

6. Hosking, J.R.: L-moments: analysis and estimation of distributions using linear combinations of order statistics. J. Roy. Stat. Soc. **52**(1), 105–124 (1990)
7. Villani, C.: Topics in optimal transportation. In: Graduate Studies in Mathematics, vol. 58. American Mathematical Society (2003)
8. Serfling, R., Xiao, P.: A contribution to multivariate L-moments: L-comoment matrices. J. Multivariate Anal. **98**(9), 1765–1781 (2007)

Shape Space and Diffeomorphic Mappings

Spherical Parameterization for Genus Zero Surfaces Using Laplace-Beltrami Eigenfunctions

Julien Lefèvre[1,2](✉) and Guillaume Auzias[1,2]

[1] Aix-Marseille Université, LSIS, CNRS UMR 7296, Marseille, France
[2] Institut de Neurosciences de la Timone UMR 7289,
Aix Marseille Université CNRS, Marseille, France
{julien.lefevre,guillaume.auzias}@univ-amu.fr

Abstract. In this work, we propose a fast and simple approach to obtain a spherical parameterization of a certain class of closed surfaces without holes. Our approach relies on empirical findings that can be mathematically investigated, to a certain extent, by using Laplace-Beltrami Operator and associated geometrical tools. The mapping proposed here is defined by considering only the three first non-trivial eigenfunctions of the Laplace-Beltrami Operator. Our approach requires a topological condition on those eigenfunctions, whose nodal domains must be 2. We show the efficiency of the approach through numerical experiments performed on cortical surface meshes.

Keywords: Riemannian manifold · Laplace-Beltrami Operator · Surface parameterization · Nodal domains

1 Introduction

Spherical parametrization of 3D closed (genus-0) meshes is a classical approach in computer graphics for texture mapping, remeshing and morphing [11]. Neuroimaging data analysis is an important field of applications since representing the brain as a closed surface is increasingly popular in the community, in line with the specific advantages for e.g. visualization and inter-subjects mapping [5]. Several works [1,7] translated most recent methodological advances from computer graphics to brain mapping, and put emphasis on two properties that are particularly desirable in this field [8]: the spherical parameterization must be fold-free to ensure the validity of neuroimaging data analysis that rely on the spherical representation, and computationally efficient in order to be applicable to large number of individual meshes whose sub-milimetric resolution involves typically more than 100 K vertices.

In this work, we propose a fast and simple approach to obtain a spherical parameterization of a certain class of genus 0 surfaces. Our approach is a particular case of the one defined by Bérard in [3], by considering only the three

---

This work is funded by the Agence Nationale de la Recherche (ANR-12-JS03-001-01, "Modegy").

F. Nielsen and F. Barbaresco (Eds.): GSI 2015, LNCS 9389, pp. 121–129, 2015.
DOI: 10.1007/978-3-319-25040-3_14

first non-trivial eigenfunctions of the Laplace-Beltrami Operator. A comparable approach has also been proposed in [8] but implies to solve a highly non linear partial differential equation. We note also an empirical approach based on the spatial regularity of the first eigenfunctions in [10]. We complement these papers with preliminary theoretical contributions, which echo with empirical findings that make it very appealing for neuroimaging studies. In particular, our contributions support the C^∞ diffeomorphic nature of our mapping, under specific condition on the considered surface. We show through numerical experiments performed on 138 cortical surface meshes that this condition is systematically met in practice. Moreover, the resulting mapping is computed in few seconds and almost diffeomorphic with less than 1 % of folded triangles on average.

2   Background and Main Results

We recall first classical results on Spectral theory and Laplace-Beltrami eigenfunctions. The interested reader can refer to [12].

2.1   Laplace-Beltrami Eigenfunctions

Definition 1. *Given $(\mathcal{M}, g)$ a compact 2-Riemannian manifold without boundary, denoting $x(p) = (x_1, x_2)$ a local coordinate system, i.e. a local diffeomorphism $\mathcal{M} \to \mathbb{R}^2$ around a point p, the Laplace-Beltrami operator acting on C^∞ functions is*

$$\Delta f = \frac{1}{\sqrt{det(g)}} \sum_{i,j} \partial_{x_j} \left(\sqrt{det(g)} g^{i,j} \partial_{x_i} f \right) \tag{1}$$

Remark 1. In the applications $\mathcal{M}$ will be a closed surface in $\mathbb{R}^3$ and the metric g will correspond to the euclidean inner product.

If we consider the eigenvalue problem:

$$\Delta f = -\lambda f \tag{2}$$

we know that it has eigenvalues $0 = \lambda_0 < \lambda_1 \leq \lambda_2...$ and corresponding eigenfunctions $\Phi_1, \Phi_2, ...$ (See Fig. 1 for a visual intuition of some Φ_i). The eigenfunctions are orthogonal in the sense of the scalar product $< u, v >_{\mathcal{M}} = \int_{\mathcal{M}} uv d\mu$, where the volume form $d\mu$ is given by $\sqrt{det(g)}dx_1 dx_2$.

Definition 2. *Given an eigenfunction Φ of the Laplace-Beltrami Operator, we call **nodal set** the set of points where Φ vanishes. We denote it $N(\Phi)$ in the following. The **nodal domains** correspond to the connected components of the complementary of the nodal set.*

We have some qualitative results on the nodal domains of eigenfunctions:

Theorem 1 (Courant's Nodal Domain Theorem). *The number of nodal domains for the n-th eigenfunction is inferior or equal to $n + 1$.*

There is a global result on the dimension of nodal sets when $\mathcal{M}$ is a 2-manifold:

Theorem 2 (S.H. Cheng [4]). *Except on a closed set of points, the nodal set of an eigenfunction Φ is a C^{∞} 1-manifold, i.e. a line in our applications.*

Last we recall the Green formula for an open set $D \in \mathcal{M}$ that will be of great use in the following:

$$\int_D \Phi \Delta \Psi d\mu = -\int_D g(\nabla \Phi, \nabla \Psi) d\mu + \int_{\partial D} \Phi(\nabla \Psi \cdot n) d\tilde{\mu} \tag{3}$$

$d\tilde{\mu}$ is the induced metric on the boundary. For simplicity, we use the notation $\cdot$ instead of the riemannian metric $g(.,.)$ and remove the volume forms in the following.

Fig. 1. From left to right: eigenfunctions Φ_1, Φ_2 and Φ_3. Colormap goes from blue (negative) to red/yellow (positive). Each nodal sets are in green (Color figure online).

2.2 Main Conjecture

Based on empirical findings we suggest the following result, which allows to define a natural spherical parameterization:

Conjecture 3. *Let $\mathcal{M}$ be a genus zero surface in $\mathbb{R}^3$. Let Φ_1, Φ_2 and Φ_3 be three orthogonal eigenfunctions of the Laplace-Beltrami operator. We assume they have only two nodal domains. Then the mapping*

$$\Phi : \mathcal{M} \longrightarrow \qquad\qquad \mathbb{S}^2$$
$$p \longmapsto \left(\sqrt{\Phi_1(p)^2 + \Phi_2(p)^2 + \Phi_3(p)^2}\right)^{-1} \left(\Phi_1(p), \Phi_2(p), \Phi_3(p)\right)$$

is well defined and it is a C^{∞} diffeomorphism.

Remark 2. The orthogonality condition is more general than assuming different eigenvalues and allows applying the conjecture on the sphere itself. In that case Φ is exactly the identity (up to a multiplicative constant). Namely, given a point p and its spherical coordinates $p = (\sin \theta \cos \phi, \sin \theta \sin \phi, \cos \theta)$, a choice of three first normalized eigenfunctions is $\sqrt{\frac{3}{4\pi}} \cos \theta$, $\sqrt{\frac{3}{4\pi}} \sin \theta \cos \phi$ and $\sqrt{\frac{3}{4\pi}} \sin \theta \sin \phi$.

Remark 3. The mapping $\mathbf{\Phi}$ is a particular case of the mapping $\mathbf{\Phi}_\lambda : \mathcal{M} \to \mathbb{S}^{N(\lambda)}$ proposed by Pierre Bérard in [3] where $N(\lambda)$ is the number of eigenvalues inferior to λ. But it is important to note that the proper definition of $\mathbf{\Phi}$ is easier in [3] when λ is large enough because the denominator is guaranteed to never vanish. It was also shown that $\mathbf{\Phi}_\lambda$ is an embedding for λ large enough. In our case the restrictions on the topology of the eigenfunctions could guarantee the diffeomorphic aspect for only 3 eigenfunctions.

Remark 4. We first proposed a sketch of proof for the injectivity of $\mathbf{\Phi}$ by using properties of the mapping $\mathbf{F} : p \to \big(\varPhi_1(p), \varPhi_2(p), \varPhi_3(p)\big)$. In particular, we used the formula $\varDelta \mathbf{F}(p) = 2H(p)\mathbf{N}(p)$ linking laplacian of coordinates and mean curvature for hypersurfaces, combined with $\varDelta \mathbf{F} = (-\lambda_1 \varPhi_1, -\lambda_2 \varPhi_2, -\lambda_3 \varPhi_3)$. But it is important to see that the first formula holds if $\varDelta$ is the Laplace-Beltrami operator of $\mathbf{F}(\mathcal{M})$ (provided it is a submanifold !) which makes the second equation not true anymore.

Nevertheless our initial flaw was at the origin of experimental observations that yield the conjecture:

Conjecture 4. *With the previous notations and hypotheses, $\mathbf{F}(\mathcal{M})$ is a genus-zero surface whose mean curvature has a constant sign.*

3   Preliminary Results

Our initial strategy to tackle the first conjecture was:

- to prove first that intersection points of two nodal sets exist, thanks to global arguments.
- to characterize those intersections in terms of the angle between the two iso-lines (equivalently the gradient of the eigenfunctions), by using local results on eigenfunctions. That sort of results are known for auto-intersection of nodal sets.

Those two first steps would allow a correct definition of the mapping $\mathbf{\Phi}$ but the diffeomorphic aspect remains the most difficult part.

Proposition 1. *Let $\mathcal{M}$ be a genus zero surface in $\mathbb{R}^3$. We consider two eigenfunctions $\varPhi$ and $\varPsi$ with only two nodal domains and different associated eigenvalues. Then their nodal sets have at least one intersection point.*

Proof. Let λ and λ' be the two eigenvalues associated to $\varPhi$ and $\varPsi$. Since $\varPhi$ and $\varPsi$ have two nodal domains, their nodal sets divide $\mathcal{M}$ in two parts respectively. If we assume that $N(\varPhi) \cap N(\varPsi) = \emptyset$, we have a partition of $\mathcal{M}$ in three connected domains D_1, D_2, D_3 and the two nodal sets (of measure 0).

where

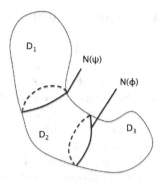

$$D_1 = \{p|\Phi(p) > 0, \Psi(p) > 0\}$$
$$D_2 = \{p|\Phi(p) > 0, \Psi(p) < 0\}$$
$$D_3 = \{p|\Phi(p) < 0, \Psi(p) < 0\}$$

The figure on the right provides a schematic illustration of the configuration. Inspired by this figure, we define $B = D_2$ and use the Green formula in two different ways:

$$\lambda \int_B \Phi\Psi = \int_B -\Psi\Delta\Phi = \int_B \nabla\Psi \cdot \nabla\Phi - \int_{\partial B} \Phi(\nabla\Psi \cdot \mathbf{n})$$

$$\lambda' \int_B \Phi\Psi = \int_B -\Phi\Delta\Psi = \int_B \nabla\Psi \cdot \nabla\Phi - \int_{\partial B} \Psi(\nabla\Phi \cdot \mathbf{n})$$

to obtain

$$(\lambda' - \lambda) \int_B \Phi\Psi = \int_{\partial B} \Phi(\nabla\Psi \cdot \mathbf{n}) - \int_{\partial B} \Psi(\nabla\Phi \cdot \mathbf{n}) \qquad (4)$$

The boundary ∂B equals $N(\Phi) \cup N(\Psi)$. On $N(\Phi)$ (resp. $N(\Psi)$) we have $\Phi = 0$ (resp. $\Psi = 0$) and on $N(\Psi)$ (resp. $N(\Phi)$) we have also $\nabla\Psi\cdot\mathbf{n} = 0$ (resp. $\nabla\Phi\cdot\mathbf{n} = 0$) since $N(\Psi)$ is a level set of Ψ (resp. Φ). Then the two integrals in the right term of (4) vanish which leads to $\int_B \Phi\Psi = 0$. It is a contradiction since both Φ and Ψ have a constant sign on B. ∎

Remark 5. We can find simple examples where this proposition fails when the nodal domains are more than 2. For very elongated ellipsoids the second and third eigenfunctions have respectively 3 and 4 nodal domains. Numerical simulations (not shown here) revealed that the three first eigenfunctions have isolines that are all parallel and nodal sets have no intersection points.

Remark 6. It is probably harder to find examples of surfaces for which there is only one intersection point. This singular configuration implies a colinearity of $\nabla\Phi$ and $\nabla\Psi$ at the crossing point. Conversely if there is no colinearity at an intersection point, one can intuitively conclude that there is at least a second intersection point by an argument *à la Jordan*.

Next our initial attempt to characterize the local behavior at the intersection point followed ideas exposed in [4]. In particular Theorem 2.5 says that when the nodal lines of a given eigenfunction meet, they form an equiangular system. The proof relies on local approximations of solutions of elliptic partial differential equations close to the origin in the C^∞ case thanks to a theorem by Lipman Bers. Nevertheless this strategy appeared to be too general in our case and we were not able to obtain a relationship linking angle between nodal sets of two eigenfunctions and other quantities such as the local mean curvature, even if numerical computations may reveal interesting behaviors (see Fig. 3 right).

4   Experimental Results

Data. We implemented our approach on 138 triangular meshes of cortical surfaces from the OASIS database that were segmented through FreeSurfer software. The number of vertices ranges from 106914 to 167230 vertices. Laplace-Beltrami eigenfunctions were computed through a variational formulation of Equation (2) and a discretization with Finite Element Methods [6]. On Fig. 1 we displayed the three first eigenfunctions and the nodal set in green for a given surface. We observed that the three first eigenfunctions had only 2 nodal domains (yellow and blue in each case). The codes were implemented in MATLAB on a Mac with a 2.6 GHZ processor. CPU time ranges from 3.76 s to 6.71 s.

Diffeomorphic Aspects. On Fig. 2 we illustrated our approach for a brain mesh at left on which the mean curvature was computed following [6] and was represented as an image. The surface $\mathbf{F}(\mathcal{M})$ was represented on the middle with the initial curvature. The orientation followed the one of the brain at left. On the right we showed the sphere $\mathbf{\Phi}(\mathcal{M})$ and the inital mean curvature with the same orientation as previously.

Next we verified that for all the meshes, the number of nodal domains of the three first non-trivial eigenfunctions were exactly 2. We evaluated the number of flipped faces in percentage, ranging from 0.008 % to 7.01 % (average: 0.29 ± 0.7 %). Figure 4 left showed an example of such faces and we observed that they correspond to local defects in the mesh, probably affecting a proper approximation of eigenfunctions.

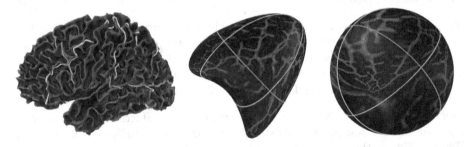

Fig. 2. Left: Cortical surface $\mathcal{M}$ and its mean curvature. Colormap goes from blue (negative) to red/yellow (positive). Middle: $\mathbf{F}(\mathcal{M})$. Right: spherical surface with the initial curvature (Color figure online).

Reproducibility. On Fig. 3 left and middle we represented the intersection points of the 3 first nodal sets for all the meshes superimposed in the Talairach space, a classical reference system in neuroimaging. We observed first that those intersection points are only 6 for each brain and secondly that they are consistently distributed in 3D. This result is of course related to the stability of the first eigenfunctions across the different brains and is more directly interpretable by looking at points than functions. On Fig. 3 right we displayed the distribution of angles between nodal sets. The distribution is unimodal in one case (intersection of Φ_1 and Φ_3) and bimodal for the other cases.

Distorsions. Finally we obtained evaluations of geometric distorsions through angular errors and relative error on lengths across all the meshes (Fig. 4 middle and right). Those errors are discrepencies between the values measured on the initial mesh and on the spherical representation. Even if the errors are larger than for methods that explicitly minimize distorsions [1,5], the average values are of the same order of magnitude.

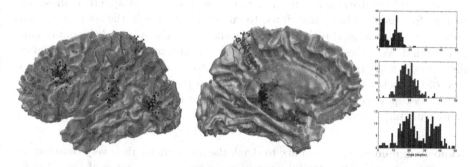

Fig. 3. Left and middle: The 6 intersection points of the 3 first nodal sets for all the meshes superimposed in the Talairach space. A slightly transparent cortical mesh is shown to illustrate the reproducibility. Right: Distribution of angles between nodal sets of 1st and 2nd, 1st and 3rd, 2nd and 3rd eigenfunctions. On the three figures the colors are matched (Color figure online).

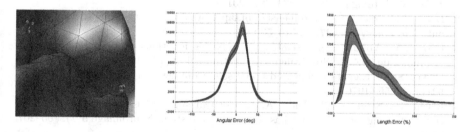

Fig. 4. From left to right: Examples of flipped triangles in green, distributions of angular error (in degree) and relative error on distances (%) (Color figure online).

5 Discussion and Perspectives

In this work we have proposed a spherical mapping with preliminary mathematical and empirical results in favor of a C^∞ diffeomorphic property. Cortical surfaces were very suitable in our case because they satisfy the property proposed in our mathematical part, with a number of nodal domains equal to 2. In practice we obtained (a limited number of) flipped faces due to mesh irregularities. Moreover the computation time was 5 s in average, which outperforms the recent approach in [8]. Our approach offers a correct stability across a group of brain shapes, in the sense that interesting features such as intersection of nodal sets are quite consistent. Our mapping is probably suboptimal when it comes to match brain meshes with respect to strategies using more eigenfunctions [10]. Nevertheless our method would be very appropriate for a faster initialization of a spherical mapping instead of a Gauss map and Tutte map [7]. It can also be used "on the fly" for extensive simulations [9].

We could try to generalize our results for any kind of spherical mesh by looking for eigenfunctions associated to larger eigenvalues in the spectrum, with only 2 nodal domains. But there is no guarantee that there exist 3 eigenfunctions of this kind. We could think also to possible extensions for manifolds of higher dimension. In [2] results have been obtained to bound the embedding dimension of Laplacian eigenfunctions map thanks to Ricci curvature and injectivity radius. A topological condition on the number of nodal domains could produce complementary views on this question.

Aknowledgments. We would like to thank the reviewers for their very constructive comments. In particular a careful observation by one the reviewer is at the origin of the Remark 4 and of important modifications in the structure of the manuscript.

References

1. Auzias, G., Lefevre, J., Le Troter, A., Fischer, C., Perrot, M., Régis, J., Coulon, O.: Model-driven harmonic parameterization of the cortical surface: Hip-hop. IEEE Trans. Med. Imaging **32**(5), 873–887 (2013)
2. Bates, J.: The embedding dimension of laplacian eigenfunction maps. Appl. Comput. Harmonic Anal. **37**(3), 516–530 (2014)
3. Bérard, P.: Volume des ensembles nodaux des fonctions propres du Laplacien. Séminaire de Théorie Spectrale et Géométrie **3**, 1–9 (1984)
4. Cheng, S.-Y.: Eigenfunctions and nodal sets. Commentarii Mathematici Helvetici **51**(1), 43–55 (1976)
5. Fischl, B., Sereno, M.I., Dale, A.M.: Cortical surface-based analysis: II: Inflation, flattening, and a surface-based coordinate system. Neuroimage **9**(2), 195–207 (1999)
6. Germanaud, D., Lefèvre, J., Toro, R., Fischer, C., Dubois, J., Hertz-Pannier, L., Mangin, J.-F.: Larger is twistier: spectral analysis of gyrification (spangy) applied to adult brain size polymorphism. NeuroImage **63**(3), 1257–1272 (2012)
7. Gu, X., Wang, Y., Chan, T.F., Thompson, P.M., Yau, S.T.: Genus zero surface conformal mapping and its application to brain surface mapping. IEEE Trans. Med. Imaging **23**(8), 949–958 (2004)
8. Lai, R., Wen, Z., Yin, W., Gu, X., Lui, L.M.: Folding-free global conformal mapping for genus-0 surfaces by harmonic energy minimization. J. Sci. Comput. **58**(3), 705–725 (2014)

9. Lefèvre, J., Intwali, V., Hertz-Pannier, L., Hüppi, P.S., Mangin, J.-F., Dubois, J., Germanaud, D.: Surface smoothing: a way back in early brain morphogenesis. In: Mori, K., Sakuma, I., Sato, Y., Barillot, C., Navab, N. (eds.) MICCAI 2013, Part I. LNCS, vol. 8149, pp. 590–597. Springer, Heidelberg (2013)

10. Lombaert, H., Sporring, J., Siddiqi, K.: Diffeomorphic spectral matching of cortical surfaces. In: Gee, J.C., Joshi, S., Pohl, K.M., Wells, W.M., Zöllei, L. (eds.) IPMI 2013. LNCS, vol. 7917, pp. 376–389. Springer, Heidelberg (2013)

11. Sheffer, A., Praun, E., Rose, K.: Mesh parameterization methods and their applications. Found. Trends® Comput. Graph. Vision 2(2), 105–171 (2006)

12. Zelditch, S.: Local and global analysis of eigenfunctions. arXiv preprint arXiv:0903.3420 (2009)

Biased Estimators on Quotient Spaces

Nina Miolane[✉] and Xavier Pennec

INRIA, Asclepios Project-Team,
2004 Route des Lucioles, BP 93, 06902 Sophia Antipolis Cedex, France
nina.miolane@inria.fr

Abstract. Usual statistics are defined, studied and implemented on Euclidean spaces. But what about statistics on other mathematical spaces, like manifolds with additional properties: Lie groups, Quotient spaces, Stratified spaces etc? How can we describe the *interaction between statistics and geometry*? The structure of Quotient space in particular is widely used to model data, for example every time one deals with shape data. These can be shapes of constellations in Astronomy, shapes of human organs in Computational Anatomy, shapes of skulls in Palaeontology, etc. Given this broad field of applications, statistics on shapes -and more generally on observations belonging to quotient spaces- have been studied since the 1980's. However, most theories model the variability in the shapes but do not take into account the noise on the observations themselves. In this paper, we show that statistics on quotient spaces are biased and even inconsistent when one takes into account the noise. In particular, some algorithms of template estimation in Computational Anatomy are biased and inconsistent. Our development thus gives a first theoretical geometric explanation of an experimentally observed phenomenon. A biased estimator is not necessarily a problem. In statistics, it is a general rule of thumb that a bias can be neglected for example when it represents less than 0.25 of the variance of the estimator. We can also think about neglecting the bias when it is low compared to the signal we estimate. In view of the applications, we thus characterize geometrically the situations when the bias can be neglected with respect to the situations when it must be corrected.

Introduction

In Quantum Field theory, one can roughly consider a relativistic quantum field of quarks as a function from space-time to the space of colors: $\phi : \mathbb{R}^4 \mapsto \mathbb{C}^3$. The function ϕ associates to each point of space-time an elementary particle: the quark living in the space of colors. The space of quarks fields carries a right action of the group of symmetry of space-time and a left action of the group of quark symmetry. These group actions represent what does not change the laws of Physics.

In Computational Anatomy when we study anatomical shapes, we encounter the same mathematical structures [4]. In others words, we have often two groups acting on the same anatomical object ϕ respectively on the right and on the left. These group actions represent what does not change the shape of the object.

© Springer International Publishing Switzerland 2015
F. Nielsen and F. Barbaresco (Eds.): GSI 2015, LNCS 9389, pp. 130–139, 2015.
DOI: 10.1007/978-3-319-25040-3_15

A first example is a set of K anatomical landmarks [7,10], which is a function from the space of labels to the real 3D space: $\phi : [1, \ldots, K] \mapsto \mathbb{R}^3$. In this case, the manifold $\mathcal{M}$ of landmarks sets carries a right action of the group of permutations of the labels and a left action of the group of rotations and translations of the 3D space $SE(3)$. Relabeling the landmarks, or rotating and translating the whole set of landmarks, does not change the anatomical shape described. A second example is a 1D-signal [9] which is a function: $\phi : [0, 1] \mapsto \mathbb{R}$. The manifold $\mathcal{M}$ of 1D-signals carries a right action of the group of diffeomorphisms of $[0, 1]$: this action represents the reparameterizations of the signal, which do not change its shape. A third example is a parameterized surface, or more generally immersed submanifold, in $\mathbb{R}^d$ [3] which is a function: $\phi : \mathcal{N} \mapsto \mathbb{R}^d$. The manifold $\mathcal{M}$ of parameterized surfaces (resp. of immersed submanifolds) carries a right action of the group of diffeomorphisms of $\mathcal{N}$ and a left action of the group of rotations and translations of the $\mathbb{R}^d$ space $SE(d)$ (poses). Again, reparameterizing, rotating or translating the surface or the submanifold does not change its shape.

The anatomical data that we observe belong to the manifold $\mathcal{M}$. We are interested in the statistical analysis of their *shapes*. The corresponding shape data now belong to a quotient space $\mathcal{Q}$, which is the manifold $\mathcal{M}$ quotiented by the group actions that leave the shape invariant. Thus we perform statistics on $\mathcal{Q}$ rather than on $\mathcal{M}$, even if the data originally belong to $\mathcal{M}$. But $\mathcal{Q}$ not a differentiable manifold in the general case. To be more precise, $\mathcal{Q}$ is a *manifold with singularities* or a *stratified space*.

Statistics on quotient spaces or shape spaces $\mathcal{Q}$ have been studied since the 1980's. Regarding shapes of landmarks, the theory has been first introduced by Kendall [8]. Regarding shapes of curves one can refer to [6,9] and, in the more general setting of manifold quotiented by isometric group actions, to [5]. In all cases when one performs statistics on shape spaces, the estimation of the *mean shape* is central. A way of doing it is via parametric statistical estimation. By doing so, studies model the variability in the shapes of the objects (meaning that they assume some *variability in* $\mathcal{Q}$) and the variability in reparameterization or in objects pose (meaning that they assume some variability in the "orbit" of the group actions).

However, the effect of noise on the objects themselves (which is a *noise in the ambient manifold $\mathcal{M}$*) has not been thoroughly investigated yet. The noise on the objects in $\mathcal{M}$ exists as it comes from imperfect measure instruments or human lack of precision, for example while positioning the landmarks on a medical image. Usual estimators were proven to be consistent in the theory without noise. But experiments have shown that they have bias and even inconsistency in the presence of noise [2]. For example, the standard estimator of the mean shape of the data, the Fréchet mean computed in the quotient space with the max-max algorithm, is biased and even inconsistent. This bias is experimentally observed and shows to be dependent on the noise level. But it is not theoretically understood so far. There is a need of an extended statistical theory on quotient spaces that takes the noise into account and quantifies the induced bias.

This paper is a first step in this direction. We suggest a geometric interpretation of the usual estimator of the mean shape. It enables to show that noise on the observations in $\mathcal{M}$ induces bias on the estimator in $\mathcal{Q}$. We work in the case where the observations belong to a finite dimensional flat manifold $\mathcal{M}$, which is quotiented by a proper isometric action of a finite dimensional Lie group $\mathcal{G}$. We describe how the bias on the *statistical* estimator is controlled by the *geometry* of the quotient space $\mathcal{Q}$, in particular by its singularities at the scale of the noise. Even if we work in finite dimension, we provide the intuition of the behavior for infinite dimension.

In the first section, we present notions of differential geometry on quotient spaces. In the second section, we present the statistical framework and the geometric interpretation of the estimator usually computed. In the third section, we demonstrate the geometric origin of the bias on this estimator. In the fourth section, we show explicit computations of the bias for the case of $\mathbb{R}^m$ quotiented by $SO(m)$, which is the common example of finite dimensional flat manifold quotiented by the isometric action of a finite dimensional Lie group.

1 Differential Geometry of Quotient Spaces $\mathcal{Q}$

We consider $\mathcal{M}$ a finite dimensional flat Riemannian manifold and $\mathcal{G}$ a finite dimensional Lie group. For more details on this section, we refer to [1,13,14].

Basis on Quotient Spaces. A *Lie group action of* $\mathcal{G}$ on $\mathcal{M}$ is a differentiable map $\rho : \mathcal{G} \times \mathcal{M} \ni (g, X) \mapsto g \cdot X \in \mathcal{M}$, such that $e \cdot X = X$ and $g' \cdot (g \cdot X) = (g'g) \cdot X$. If $\mathcal{G} \times \mathcal{M} \ni (g, X) \mapsto (g \cdot X, X) \in \mathcal{M} \times \mathcal{M}$, is a proper mapping, i.e. if the inverse image of every compact is compact, the action is *proper*. If $d\rho_g : T_X \mathcal{M} \to T_{g \cdot X} \mathcal{M}$ with $\rho_g = \rho(g, .)$ leaves the metric of $\mathcal{M}$ invariant, the action is *isometric*. We only consider proper isometric actions.

The *orbit of* $X \in \mathcal{M}$ is defined as $\mathcal{O}_X = \{g \cdot X \mid g \in \mathcal{G}\}$. The orbit of X represents all the points that can be reached by moving X through the action of $\mathcal{G}$. The *isotropy group of* X, also called the *stabilizer of* X, is defined as $\mathcal{G}_X = \{g \in \mathcal{G} \mid g \cdot X = X\}$. The stabilizer of X represents the elements of $\mathcal{G}$ that fail at moving X. The orbit $\mathcal{O}_X$ is a submanifold of $\mathcal{M}$ and the stabilizer $\mathcal{G}_X$ is a Lie subgroup of $\mathcal{G}$. They are related by the orbit-stabilizer theorem as follows: $\mathcal{O}_X \sim \mathcal{G}/\mathcal{G}_X$. The orbits form a partition of $\mathcal{M}$ and the set of orbits is called the *quotient space* $\mathcal{Q} = \mathcal{M}/\mathcal{G}$. Because the action is proper, $\mathcal{Q}$ is Hausdorff and inherits a Riemannian structure from $\mathcal{M}$. This is precisely why we consider a proper action.

The isotropy groups $\mathcal{G}_X$ and $\mathcal{G}_{X'}$ of elements X and X' in the same orbit are conjugate groups in $\mathcal{G}$. This property enables to define the *orbit type* of an orbit as the conjugacy class (H) of the isotropy groups of its elements. From the orbit-stabilizer theorem, we observe that a "large" orbit leads to a "small" isotropy group, so a "small" orbit type. The smallest orbit type is called the *principal orbit type*. The corresponding largest orbits are called *principal orbits*. In contrast, non principal orbits are called *singular orbits*. The corresponding points in $\mathcal{Q}$ are called *singularities*. The action of $\mathcal{G}$ on $\mathcal{M}$ is said to be *free* if there

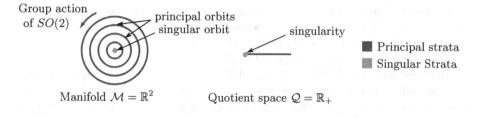

Fig. 1. Stratification and foliation of $\mathcal{M}$ and stratification of $\mathcal{Q}$ for $\mathcal{M} = \mathbb{R}^2$ quotiented by the action of $SO(2)$ so that $\mathcal{Q} = \mathbb{R}^2/SO(2) = \mathbb{R}_+$ (Color figure online).

is no singular orbits. In this case, $\mathcal{Q}$ has no singularities and is a differentiable manifold. Otherwise $\mathcal{Q}$ is a *manifold with singularities*.

We investigate how the singular orbits are distributed in $\mathcal{M}$, or equivalently how the singularities are distributed in $\mathcal{Q}$. The partition of $\mathcal{M}$ into orbits is a singular Riemannian foliation called the *canonical foliation of $\mathcal{M}$*. We can gather the orbits of same orbit type (H) and define $\mathcal{M}_{(H)} = \{X \in \mathcal{M}|\ \mathcal{G}_X$ conjugate to $H\}$. The decomposition of $\mathcal{M}$ into the connected components of the $\mathcal{M}_{(H)}$ is a stratification called the *orbit type stratification of $\mathcal{M}$*. In this stratification, the principal type component is open and dense.

The decomposition of $\mathcal{Q}$ into the corresponding components $\mathcal{Q}_{(H)} = \mathcal{M}_{(H)}/\mathcal{G}$ also forms a stratification, whose principal type component is also dense. Thus, singularities are "sparsely distributed" in $\mathcal{Q}$ in the sense that they are of null Lebesgue measure.

These foliation and stratifications are illustrated in Fig. 1 in the case of $\mathcal{M} = \mathbb{R}^2$ quotiented by the action of $SO(2)$ so that $\mathcal{Q} = \mathbb{R}^2/SO(2) = \mathbb{R}_+$. We use this 2D example throughout the paper as it is convenient for the illustrations. But of course, the theory applies to the general case.

Orbits as immersed submanifolds. An orbit $\mathcal{O}$ is an immersed submanifold in $\mathcal{M}$, i.e. a differentiable manifold together with an immersion $I : \mathcal{O} \mapsto \mathcal{M}$. We identify $\mathcal{O}$ with $I(\mathcal{O})$ and we denote $T_X\mathcal{O}$, resp. $N_X\mathcal{O}$, the tangent space, resp. the normal space with respect to the metric of $\mathcal{M}$, of $\mathcal{O}$ at X. The *first fundamental form* of $\mathcal{O}$ is defined to be the induced metric on $\mathcal{O}$, i.e. the pull-back of the ambient (flat) metric of $\mathcal{M}$ by I. The *second fundamental form* of $\mathcal{O}$ is the quadratic form h defined by $T_X\mathcal{O} \times T_X\mathcal{O} \ni (u,v) \mapsto h(v,w) = (\nabla_v w)^\perp \in N_X\mathcal{O}$, where ∇ is the Levi-Civita connection on $\mathcal{M}$ and $(\nabla_v w)^\perp$ the orthogonal projection on $N_X\mathcal{O}$.

We can write $\mathcal{O}$ locally around X as the graph G of a smooth function from $T_X\mathcal{O}$ to $N_X\mathcal{O}$. Let Z be a point on $\mathcal{O}$ near X. On $T_X\mathcal{O}$ we parameterize Z, and thus $\mathcal{O}$, by its Riemannian Logarithm $z = \mathrm{Log}_X(Z)$. The coordinates in $T_X\mathcal{O}$ are labeled with i,j, the coordinates in $N_X\mathcal{O}$ are labeled with a,b. Locally around X, and equivalently around $z = 0$, we have the graph equation of $\mathcal{O}$:

$$G^a(z) = -\frac{1}{2}h^a_{ij}(0)z^i z^j + O(|z|^3) \tag{1}$$

where we use Einstein summation convention: there is an implicit sum on indices that are repeated up and down. The second fundamental form h at 0 is thus the best approximation of $\mathcal{O}$ around X as the graph of a quadratic function. It quantifies the external curvature of $\mathcal{O}$ in $\mathcal{M}$.

Here we have considered a fixed orbit $\mathcal{O}$. But later in the paper we will consider different orbits. As the space of orbits is by definition the quotient space $\mathcal{Q}$, we will label the orbits by the corresponding point in $\mathcal{Q}$. So we will write $\mathcal{O}_y$ for $y \in \mathcal{Q}$. We will thus have expressions such as $h_{ij}^a(y, 0)$, for the second fundamental form of the orbit represented by y.

2   Geometric Interpretation of the Template Estimation

Here we interpret the statistical template estimation in Computational Anatomy with the quotient space framework. Such an estimation is usually parametric. It means that we assume a parametric model that has produced the observations, in our case: the anatomies. The (parametric) statistical estimation amounts to infer the parameters of the model from the observations, in our case: infer the template, or the mean anatomical shape, from the observed anatomies.

More precisely, one usually assumes that there is an underlying probability density of anatomical shapes, i.e. a probability density on $\mathcal{Q}$, whose mean is called the template. We consider here the simplest generative model with a dirac distribution in $\mathcal{Q}$, i.e. no variability in shapes $\mathbf{y} \sim \delta(y_0)$. Here y_0 is the template. As we do not observe the anatomical shapes in $\mathcal{Q}$, but rather the anatomies in $\mathcal{M}$, one has to model the poses of the shape. One usually assumes that there is a probability density in the orbits. We assume here a uniform distribution on the orbits, i.e. maximal variability in poses: $\mathbf{z} \sim \mathcal{U}(\mathcal{O})$. Finally, we model the noise on the observations by a Riemannian Gaussian in $\mathcal{M}$ with isotropic noise σ: $\mathbf{x} \sim \mathcal{N}_\mathcal{M}((\mathbf{y}, \mathbf{z}), \sigma)$ [12]. As we have an isometric action and a isotropic noise, this whole generative model is equivalent to the even simpler one: $\mathbf{y} \sim \delta(y_0)$ and $\mathbf{z} \sim \delta(z_0)$ and $\mathbf{x} \sim \mathcal{N}_\mathcal{M}((y_0, z_0), \sigma)$.

Figure 2 illustrates the geometric picture for the generative model. The green dot in $\mathcal{Q} = \mathbb{R}_+$ is the template, or the mean shape. The green circle in $\mathcal{M} = \mathbb{R}^2$ is the orbit of y_0. We choose the pose z_0 on this orbit to get the anatomical object

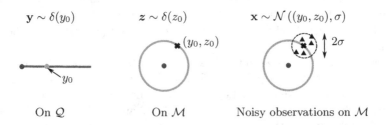

$\mathbf{y} \sim \delta(y_0)$ $\mathbf{z} \sim \delta(z_0)$ $\mathbf{x} \sim \mathcal{N}((y_0, z_0), \sigma)$

(y_0, z_0) 2σ

y_0

On $\mathcal{Q}$ On $\mathcal{M}$ Noisy observations on $\mathcal{M}$

Fig. 2. Geometric interpretation of the three steps of the generative model (Color figure online).

Registration of the observations   Fréchet mean of the registered observations

From $\mathcal{M}$ to $\mathcal{Q}$ Biased statistics on $\mathcal{Q}$

Fig. 3. Geometric interpretation of the two steps of the statistical estimation (Color figure online).

represented by the black dot $x_0 = (y_0, z_0)$ on the green circle. The observed anatomical objects are noisy observations of (y_0, z_0). They are represented by the black triangles generated by the bivariate Gaussian of isotropic noise σ.

The goal of the statistical estimation is to compute an estimator $\hat{y}_0$ of the template y_0. A standard method is the max-max algorithm with unimodal approximation, whose two steps can be geometrically interpreted as follows. First, the registration step amounts to the projection of the observations in $\mathcal{Q}$. Then, one computes the Fréchet mean $\hat{y}_0$ (a generalization of the notion of mean in Riemannian manifolds) of the observations in $\mathcal{Q}$ in order to estimate y_0.

Figure 3 illustrates the geometric picture for the estimation. First, the observations are registered: the triangle follows the black curve on their orbits (blue circles) in order to be "projected" on $\mathcal{Q} = \mathbb{R}_+$. Ultimately, the Fréchet mean is computed from the projected triangles on $\mathcal{Q} = \mathbb{R}_+$.

The algorithm produces an estimator $\hat{y}_0$ of y_0. How good is this estimator? To answer this question, we can study its unbiasedness and its consistency. We generalize the usual definitions on linear spaces to Riemannian manifolds. Thus, the *bias of an estimator* $\hat{y}_0$, relative to the parameter y_0 it is designed to estimate, is defined as: $\text{Bias}_{y_0}[\hat{y}_0] = \text{Log}_{y_0}(\text{E}_{y_0}[\hat{y}_0])$, where Log is the Riemannian logarithm. An *unbiased estimator* has null bias. A *consistent estimator* converges in probability to the estimator is designed to estimate, as the number of observations increases. We show that $\hat{y}_0$ is biased and inconsistent as an estimator of y_0.

3   Geometric Foundations of Bias and Inconsistency

We consider observations in $\mathcal{M}$ generated with the probability density F of our model. We show that they are equivalent to observations in $\mathcal{Q}$ generated with a probability density f which we compute.

The probability density on $\mathcal{M}$ writes as follows: $F(x) = \frac{1}{C} \exp(-\frac{d_{\mathcal{M}}^2(x, x_0)}{2\sigma^2})$ where C is the normalizing constant. We write $\mathcal{O}_y$ for the orbit corresponding to $y \in \mathcal{Q}$. For an isometric action we have: $d\mathcal{M}(x) = d\mathcal{O}_y(x)d\mathcal{Q}(y)$. Thus the probability of having an observation x projecting within the interval of quotient

coordinates $[y, y + dy]$ is: $\mathbb{P}(\mathbf{y} \in [y, y+dy]) = \int_y^{y+dy} \left(\int_{\mathcal{O}_y} F(x) d\mathcal{O}_y(x) \right) dQ(y)$.
By taking $dy \to 0$, we have the induced probability density of the quotient space:

$$f(y) = \int_{\mathcal{O}_y} F(x) d\mathcal{O}_y(x) = \frac{1}{c} \int_{\mathcal{O}_y} \exp(-\frac{d_{\mathcal{M}}(x, x_0)^2}{2\sigma^2}) d\mathcal{O}_y(x) \qquad (2)$$

To compute an approximation of the density f, we make the assumption that the noise is low, i.e. $\sigma << 1$. The low noise assumption enables to use Taylor expansions in the orbit coordinates and then integrate in $\mathcal{O}_y$ locally on the ball $\mathcal{B}_y = \mathcal{B}(x_0, \sigma) \cap \mathcal{O}_y \subset \mathcal{O}_y$. To this aim, we introduce momenta defined for the uniform measure $\mu_y(z)$ of the ball $\mathcal{B}_y$ as:

$$\mathfrak{M}_{\mathcal{O}_y, p}^{i_1 \ldots i_p} = \int_{\mathcal{O}_y} z^{i_1} \ldots z^{i_p} \mu(dz) = \int_{\mathcal{B}_y} z^{i_1} \ldots z^{i_p} dz \qquad (3)$$

We get in coordinates centered at $y_0 \in Q$:

$$f(y) = \frac{1}{2\sigma^2} \exp(-\frac{y_a y^a}{2\sigma^2}) \left(\mathfrak{M}_{\mathcal{O}_y, 0} + S(y)_{ij} \mathfrak{M}_{\mathcal{O}_y, 2}^{ij} + O(|\mathfrak{M}_{\mathcal{O}_y, 3}|) \right) \qquad (4)$$

where: $S_{ij}(y) = -\delta_{ij} + y_a h_{ij}^a(y, 0) + \frac{1}{2} h_{ki}^a(y, 0) h_{kj}^a(y, 0)$.

The probability density f on Q differs from a normal distribution because of the y-dependent term: $\mathfrak{M}_{\mathcal{O}_y, 0} + S(y)_{ij} \mathfrak{M}_{\mathcal{O}_y, 2}^{ij} + O(|\mathfrak{M}_{\mathcal{O}_y, 3}|)$. This term can be interpreted as a *Taylor expansion of the bias with respect to the local geometry of the orbits*. The first order of the bias is $\mathfrak{M}_{\mathcal{O}_y, 0}$ and corresponds to the area of $\mathcal{B}_y$, i.e. the area of $\mathcal{O}_y$ seen at the scale of the noise. The second order of the bias is $S(y)_{ij} \mathfrak{M}_{\mathcal{O}_y, 2}^{ij}$ and we recognize a contraction of the second momentum with the matrix $S(y)$ that depends on the external curvature of $\mathcal{O}_y$. We expect the higher order terms to be also such contractions between higher order momenta and higher order derivatives of the external curvature of the orbits.

The expectation of f on the ball $B(y_0, \sigma) \subset Q$ computed at the tangent space of the mean shape $T_{y_0} Q$ gives: $\text{Log}_{y_0}(\hat{y}_0)^a = \int_{B(y_0, \sigma)} y^a f(y) dQ(y)$. We recognize the bias of the estimator $\hat{y}_0$ in the case of an infinite number of observations. It differs from 0 because the f distribution on Q is not symmetric. Because we are in the case of an infinite number of observations, this also shows the inconsistency of the estimator $\hat{y}_0$. Given the expression of f, we see that bias and inconsistency depend on the external curvature of the orbits and its first derivatives at the scale of the noise. As the external curvature of orbits generally increases when we approach singularities [11], the nearer we are from a singularity of Q, the larger is the bias. All in all, when one performs usual statistics on Q from observations on $\mathcal{M}$, the singularities in Q induce bias and inconsistency.

4   An Illustration on the Quotient $Q = \mathbb{R}^m / SO(m)$

We consider the case $\mathbb{R}^m$ quotiented by $SO(m)$, which is a common example of a finite dimensional flat manifold quotiented by an isometric Lie group action. We perform the computations globally without the low noise assumption.

Generating observations in $\mathcal{M} = \mathbb{R}^m$ with a multivariate normal law and then projecting to $\mathcal{Q} = \mathbb{R}_+$ is equivalent to generating observations directly in $\mathcal{Q}$ with the following probability density:

$$f_m(y) = \frac{2^{1-m/2}}{\sigma^m y^{m-1}} \exp\left(-\frac{y^2 + y_0{}^2}{2\sigma^2}\right) {}_0\tilde{F}_1\left(\frac{m}{2}, \frac{y^2 y_0{}^2}{4\sigma^2}\right) \tag{5}$$

where ${}_0\tilde{F}_1$ is a regularized hypergeometric function. Figure 4 shows that an increase in the noise σ induces that the expectation of f is shifted away from y_0. The expectation is precisely the estimator $\hat{y}_0$ for the case of an infinite number of observations. Thus we see the bias increasing with the noise level. We also note that the probability density and therefore the bias depend on the dimension m.

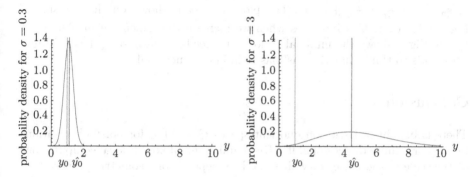

Fig. 4. Induced distributions on $\mathcal{Q} = \mathbb{R}_+$ for $m = 3$ and noise level $\sigma = 0.3$ (left) and $\sigma = 3$ (right). In green, the original mean. In red, the estimator of the mean (Color figure online).

The expectation of the probability density f_m on the quotient $\mathbb{R}_+$ is $\mathbb{E}_m(\alpha) = y_0.e_m(\alpha)$ where $\alpha^2 = \frac{y_0{}^2}{2\sigma^2}$ and $e_m(\alpha) = \frac{1}{\alpha}\Gamma\left(\frac{m+1}{2}\right) {}_1\tilde{F}_1\left(-\frac{1}{2}, \frac{m}{2}, -\alpha^2\right)$. Because $\mathbb{R}_+$ is linear, the bias writes:

$$\text{Bias}_{y_0}(\hat{y}_0) = \mathbb{E}[\hat{y}_0] - y_0 = y_0.(e_m(\alpha) - 1) = y_0.\text{bias}_m(\alpha) \tag{6}$$

The function $\text{bias}_m(\alpha)$ is more precisely the bias of any estimator $\hat{y}_0$ in units of the parameter y_0 it is designed to estimate. It entirely depends on the variable α. First, α can be seen as a signal over noise ratio (SNR): we interpret y_0 as the signal we seek to recover. But we can also interpret α geometrically: it is the ratio of the distance y_0 to the singularity 0 of $\mathcal{Q} = \mathbb{R}_+$ over the noise σ. The most favorable conditions are when there is no noise with respect to the signal, or equivalently when we are far away from the singularity at the scale of the noise: $\alpha \to \infty$. In this case, we have: $\text{bias}_m(\alpha) \underset{\alpha\to\infty}{\to} 0$. In contrast, the less favorable conditions are when the noise is preponderant with respect to the signal, or equivalently when we are close to the singularity at the scale of the noise: $\alpha \to 0$. In this case, we have: $\text{bias}_m(\alpha) \underset{\alpha\to 0}{\to} \infty$ and more precisely

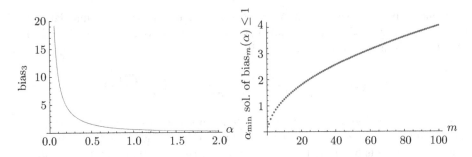

Fig. 5. Bias function of the SNR α (left). Influence of the dimension on the bias (right).

$\text{bias}_m(\alpha) = \frac{\Gamma(\frac{m+1}{2})}{\Gamma(\frac{m}{2})} \frac{1}{\alpha} + O(\alpha)$. The plot of bias_3 is shown on the left side of Fig. 5. Moreover, bias increases when we increase the dimension m. The right side of Fig. 5 shows the minimal ratio α one needs if $\text{bias}_m(\alpha) \leq 1$ is required. This leads to think that there is bias in infinite dimension!

Conclusion

There is bias in statistics on quotient spaces $\mathcal{Q} = \mathcal{M}/\mathcal{G}$ for noisy observations in $\mathcal{M}$. For a finite dimensional flat Riemannian manifold $\mathcal{M}$, local computations at the scale of the noise show that the bias depends on geometric properties of $\mathcal{Q}$, more precisely on its singularities. Global computations on $\mathbb{R}^m$ quotiented by $SO(m)$ further emphasize that bias cannot be neglected as soon as signal and noise are of the same order, or equivalently as soon as we are close to a singularity at the scale of the noise. Additionally, the increase of the bias with the dimension leads to think that the same phenomenon exists in infinite dimension. Further developments will involve computations for non flat manifolds together with an algorithm to correct the bias. Ultimately, one should generalize the study to the infinite dimensional case.

References

1. Alekseevsky, D., Kriegl, A., Losik, M., Michor, P.W.: The Riemannian geometry of orbit spaces. The metric, geodesics, and integrable systems (2001)
2. Allassonnière, S., Amit, Y., Trouvé, A.: Towards a coherent statistical framework for dense deformable template estimation. J. Roy. Stat. Soc. **69**(1), 3–29 (2007)
3. Bauer, M., Bruveris, M., Michor, P.: Overview of the geometries of shape spaces and diffeomorphism groups. J. Math. Imaging Vis. **50**(1–2), 60–97 (2014)
4. Grenander, U., Miller, M.: Computational anatomy: an emerging discipline. Q. Appl. Math. **LVI**(4), 617–694 (1998)
5. Huckemann, S., Hotz, T., Munk, A.: Intrinsic shape analysis: geodesic principal component analysis for riemannian manifolds modulo lie group actions. Stat. Sin. **20**, 1–100 (2010)

6. Joshi, S., Kaziska, D., Srivastava, A., Mio, W.: Riemannian structures on shape spaces: a framework for statistical inferences. In: Krim, H., Yezzi Jr., A. (eds.) Statistics and Analysis of Shapes, pp. 313–333. Birkhäuser, Boston (2006)
7. Kendall, D.G.: Shape manifolds, procrustean metrics, and complex projective spaces. Bull. Lond. Math. Soc. **16**(2), 81–121 (1984)
8. Kendall, D.G.: A survey of the statistical theory of shape. Stat. Sci. **4**(2), 87–99 (1989)
9. Kurtek, S.A., Srivastava, A., Wu, W.: Signal estimation under random time-warpings and nonlinear signal alignment. In: Advances in Neural Information Processing Systems 24, 675–683 (2011)
10. Le, H., Kendall, D.G.: The Riemannian structure of Euclidean shape spaces: a novel environment for statistics. Ann. Stat. **21**(3), 1225–1271 (1993)
11. Lytchak, A., Thorbergsson, G.: Curvature explosion in quotients and applications. J. Differ. Geom. **85**(1), 117–140 (2010)
12. Pennec, X.: Intrinsic statistics on Riemannian manifolds: basic tools for geometric measurements. J. Math. Imaging Vis. **25**(1), 127–154 (2006)
13. Postnikov, M.: Geometry VI: riemannian geometry. Encyclopaedia of Mathematical Sciences. Springer (2001)
14. Small, C.: A survey of the statistical theory of shape. Stat. Sci. (4), 105–108 (1989). The Institute of Mathematical Statistics

Reparameterization Invariant Metric
on the Space of Curves

Alice Le Brigant[1,2]([✉]), Marc Arnaudon[1], and Frédéric Barbaresco[2]

[1] Institut Mathématique de Bordeaux, UMR 5251,
Université de Bordeaux and CNRS, Talence, France
alice.lebrigant@gmail.com
[2] Thales Air System, Surface Radar Domain, Technical Directorate,
Voie Pierre-Gilles de Gennes, 91470 Limours, France

Abstract. This paper focuses on the study of open curves in a manifold M, and its aim is to define a reparameterization invariant distance on the space of such paths. We use the square root velocity function (SRVF) introduced by Srivastava et al. in [11] to define a reparameterization invariant metric on the space of immersions $\mathcal{M} = \mathrm{Imm}([0,1], M)$ by pullback of a metric on the tangent bundle $T\mathcal{M}$ derived from the Sasaki metric. We observe that such a natural choice of Riemannian metric on $T\mathcal{M}$ induces a first-order Sobolev metric on $\mathcal{M}$ with an extra term involving the origins, and leads to a distance which takes into account the distance between the origins and the distance between the image curves by the SRVF parallel transported to a same vector space, with an added curvature term. This provides a generalized theoretical SRV framework for curves lying in a general manifold M.

1 Introduction

Computing distances between shapes of open or closed curves is of interest in many fields that require shape analysis, from medical imaging to video surveillance, to radar detection. While the shape of an organ or a human contour can be modeled by a closed plane curve, some applications require the manipulation of curves lying in a non flat manifold, such as S^2-valued curves representing trajectories on the earth or curves in the space of hermitian positive definite matrices, where the values represent covariance matrices of Gaussian processes. The shape space of planar curves has been widely studied [1,7,8,13], and the more general setting of shapes lying in any manifold M has recently met great interest [3,5,12,14]. Here we consider open oriented curves in a Riemannian manifold M, more precisely the space of immersions $c : [0,1] \to M$,

$$\mathcal{M} = \mathrm{Imm}([0,1], M).$$

Reparameterizations will be represented by increasing diffeomorphisms $\phi : [0,1] \to [0,1]$ (so that they preserve the end points of the curves), and their set is denoted by $\mathrm{Diff}^+([0,1])$. Then, one way to describe a shape is as the equivalence class of all

© Springer International Publishing Switzerland 2015
F. Nielsen and F. Barbaresco (Eds.): GSI 2015, LNCS 9389, pp. 140–149, 2015.
DOI: 10.1007/978-3-319-25040-3_16

the curves that are identical modulo reparameterization, and the shape space as the associated quotient space,

$$\mathscr{S} = \mathrm{Imm}([0,1], M) / \mathrm{Diff}^+([0,1]).$$

The formal principal bundle structure $\pi : \mathscr{M} \to \mathscr{S}$ induces a decomposition of the tangent bundle $T\mathscr{M} = V\mathscr{M} \oplus H\mathscr{M}$ into a vertical subspace $V\mathscr{M} = \ker(T\pi)$ consisting of all vectors tangent to the fibers of $\mathscr{M}$ over $\mathscr{S}$, and a horizontal subspace $H\mathscr{M} = (V\mathscr{M})^{\perp_G}$ defined as the orthogonal complement of $V\mathscr{M}$ according to the metric G that we put on $\mathscr{M}$. We say formal because the manifold structure of the space $\mathrm{Imm}([0,1], M)$ has not yet been thoroughly studied to our knowledge. We require that G be reparameterization invariant, that is to say that the action of $\mathrm{Diff}^+([0,1])$ be isometric for G

$$G_{c\circ\phi}(h \circ \phi, k \circ \phi) = G_c(h,k), \tag{1}$$

for any curve $c \in \mathscr{M}$, reparameterization $\phi \in \mathrm{Diff}^+([0,1])$, and infinitesimal deformations $h, k \in T_c\mathscr{M}$ – h and k can also be seen as vector fields along the curve c in M. If this property is satisfied, then G induces a Riemannian metric $\hat{G}$ on the shape space,

$$\hat{G}_{\pi(c)}(T_c\pi(h), T_c\pi(k)) = G_c(h^H, k^H),$$

in the sense that the above expression does not depend on the choice of the representatives c, h and k. Here h^H, k^H denote the horizontal parts of h and k according to the previously mentioned decomposition, as well as the horizontal lifts of $T_c\pi(h)$ and $T_c\pi(k)$, respectively. The geodesic distances d on $\mathscr{M}$ and $\hat{d}$ on $\mathscr{S}$ are then simply linked by

$$\hat{d}([c_0], [c_1]) = \inf\{ d(c_0, c_1 \circ \phi) \mid \phi \in \mathrm{Diff}^+([0,1]) \},$$

where $[c_0]$ and $[c_1]$ denote the shapes of two given curves c_0 and c_1. The most natural candidate for a reparameterization invariant metric G on $\mathscr{M}$ is the L^2-metric with integration over arc length, but Michor and Mumford have shown in [6] that the induced metric $\hat{G}$ on the shape space always vanishes. This has motivated the study of Sobolev metrics [1,2,8], and particularly of a first-order Sobolev metric on the space of plane curves,

$$G_c(h,k) = \int \langle D_s h^\perp, D_s k^\perp \rangle + \frac{1}{4} \langle D_s h^\parallel, D_s k^\parallel \rangle \, ds, \tag{2}$$

where we integrate according to arc length $ds = \|c'\| \, dt$ and $\langle \cdot, \cdot \rangle$ denotes the euclidean metric on $\mathbb{R}^2$, $D_s h = \frac{1}{\|c'\|} h'$ is the derivation of h according to arc length, $D_s h^\parallel = \langle D_s h, v \rangle v$ is the projection of $D_s h$ on the unit length tangent vector field $v = \frac{1}{\|c'\|} c'$ along c, and $D_s h^\perp = \langle D_s h, n \rangle n$ is the projection of $D_s h$ on the unit length normal vector field n along c. This particular first-order Sobolev metric is of interest because it can be studied via the square root velocity (SRV) framework, introduced by Srivastava et al. in [11] and used in several applications [4,12]. This framework can be extended to curves in a general manifold by using parallel transport, in a way which allows us to move the computations to the tangent plane to the origin of one of the two curves under comparison, see [5,14]. In [5] the transformation used is a generalization of the SRV function introduced by Bauer et al. in [1] as a tool to study a more general form of the Sobolev metric (2). In [14] a Riemannian framework is given, including the associated Riemannian metric and the geodesic equations. While our approach in this paper is similar, we feel that the distance we introduce here will be more directly dependent on the "relief" of the manifold, since it is computed in the manifold itself rather than in one tangent plane as in [5,14]. This enables us to take into account a greater amount of information on the space separating two curves.

2   New Metric on the Space of Parameterized Curves

We consider the square root velocity function (SRVF) introduced in [11] on the space of curves in M,

$$R : \mathcal{M} \to T\mathcal{M}, \quad c \mapsto \frac{c'}{\sqrt{\|c'\|}},$$

where $\|\cdot\|$ is the norm associated to the Riemannian metric on M. This function will allow us to define a metric G on $\mathcal{M}$ by pullback of a metric $\tilde{G}$ on $T\mathcal{M}$. First, we define the following projections from TTM to TM. Let $\xi \in T_{(p,u)}TM$ and (x, U) be a curve in TM that passes through (p, u) at time 0 at speed ξ. Then we define the vertical and horizontal projections

$$\mathrm{vp}_{(p,u)} : T_{(p,u)}TM \to T_pM, \quad \xi \mapsto \xi_V := \nabla_{x'(0)}U,$$
$$\mathrm{hp}_{(p,u)} : T_{(p,u)}TM \to T_pM, \quad \xi \mapsto \xi_H := x'(0).$$

The horizontal and vertical projections live in the tangent bundle TM and are not to be confused with the horizontal and vertical parts which live in the double tangent bundle TTM and will be denoted by ξ^H, ξ^V. Furthermore, let us point out that the horizontal projection is simply the differential of the natural projection $TM \to M$, and that according to these definitions, the Sasaki metric [9, 10] can be written

$$g^S_{(p,u)}(\xi, \eta) = \langle \xi_H, \eta_H \rangle + \langle \xi_V, \eta_V \rangle,$$

where $\langle \cdot, \cdot \rangle$ is the Riemannian metric on M. Now we can define the metric that we put on $T\mathcal{M}$. Let us consider $h \in T\mathcal{M}$ and $\xi, \eta \in T_h T\mathcal{M}$. We define

$$\tilde{G}_h(\xi, \eta) = \langle \xi(0)_H, \eta(0)_H \rangle + \int_0^1 \langle \xi(t)_V, \eta(t)_V \rangle \, dt, \tag{3}$$

where $\xi(t)_H = \mathrm{hp}(\xi(t))$ and $\xi(t)_V = \mathrm{vp}(\xi(t))$ are the horizontal and vertical projections of $\xi(t) \in TTM$ for all t. Then we have the following result.

Proposition 1. *The pullback of the metric $\tilde{G}$ by the square root velocity function R is given by*

$$G_c(h, k) = \langle h(0), k(0) \rangle + \int \langle \nabla_s h^\perp, \nabla_s k^\perp \rangle + \frac{1}{4} \langle \nabla_s h^\|, \nabla_s k^\| \rangle \, ds, \tag{4}$$

for any curve $c \in \mathcal{M}$ and vectors $h, k \in T_c\mathcal{M}$, where we integrate according to arc length, $\nabla_s h = \frac{1}{\|c'\|} \nabla_{c'} h$ is the covariant derivative of h according to arc length, and $\nabla_s h^{\|} = \langle \nabla_s h, v, v \rangle v$ and $\nabla_s h^{\perp} = \nabla_s h - \nabla_s h^{\|}$ are its tangential and normal components respectively, if $v = \frac{1}{\|c'\|} c'$ is the unit tangent vector field along c in M.

Remark 1. In the case of curves in a flat space, G is the first-order Sobolev metric (2), studied in [11], with an added term involving the origins. This extra term guaranties that the induced distance is always greater than the distance between the starting points of the curves in M.

Proof. For any $c \in \mathcal{M}$, and $h, k \in T_c\mathcal{M}$, the metric G is defined by

$$G_c(h, k) = \bar{G}_{R(c)}(T_c R(h), T_c R(k)).$$

For any $t \in [0, 1]$, we have $T_c R(h)(t)_H = h(t)$ and $T_c R(h)_V = \nabla_h R(c)(t)$. To prove this proposition, we just need to compute the latter. Let $a \mapsto C(a, \cdot)$ be a curve in $\mathcal{M}$ such that $C(0, \cdot) = c$ et $\partial_a C(0, \cdot) = h$. Then

$$
\begin{aligned}
\nabla_h R(c)(t) &= \frac{1}{\|c'\|^{1/2}} \nabla_h c' + h\left(\|c'\|^{-1/2}\right) c' \\
&= \frac{1}{\|\partial_t C\|^{1/2}} \nabla_{\partial_a C} \partial_t C + \partial_a \langle \partial_t C, \partial_t C \rangle^{-1/4} \partial_t C \\
&= \frac{1}{\|\partial_t C\|^{1/2}} \nabla_{\partial_t C} \partial_a C - \frac{1}{2} \langle \partial_t C, \partial_t C \rangle^{-5/4} \langle \nabla_a \partial_t C, \partial_t C \rangle \partial_t C \\
&= \|c'\|^{1/2} \left((\nabla_s h)^{\perp} + \frac{1}{2} \langle \nabla_s h, v \rangle v \right),
\end{aligned}
$$

where in the last step we use again the inversion $\nabla_{\partial_a} \partial_t C = \nabla_{\partial_t C} \partial_a C$.

3 Fiber Bundle Structures

Principal Bundle Over the Shape Space. We already know that we have a formal principal bundle structure over the shape space

$$\pi : \mathcal{M} = \text{Imm}([0, 1], M) \to \mathcal{S} = \mathcal{M}/\text{Diff}^+([0, 1]).$$

which induces a decomposition $T\mathcal{M} = V\mathcal{M} \overset{\perp}{\oplus} H\mathcal{M}$. Just as in the planar case, the fact that the square root velocity function R verifies the equivariance property

$$R(c \circ \phi) = \sqrt{\phi'} \left(R(c) \circ \phi \right)$$

for all $c \in \mathcal{M}$, $h, k \in T_c\mathcal{M}$ and $\phi \in \text{Diff}^+([0, 1])$, guaranties that the integral part of G is reparameterization invariant. Remembering that the reparameterizations $\phi \in \text{Diff}^+([0, 1])$ preserve the origins of the curves, we notice that G is constant along the fibers, as expressed in Eq. (1), and so there exists a Riemannian metric $\hat{G}$ on the shape space $\mathcal{S}$ such that π is (formally) a Riemannian submersion from $(\mathcal{M}, G)$ to $(\mathcal{S}, \hat{G})$

$$G_c(h^H, k^H) = \hat{G}_{\pi(c)}(T_c\pi(h), T_c\pi(k)),$$

where h^H and k^H are the horizontal parts of h and k respectively.

Fiber Bundle Over the Starting Points. The special role that plays the starting point in the metric G induces another formal fiber bundle structure, where the base space is the manifold M, seen as the set of starting points of the curves, and the fibers are the set of curves with the same origin. The projection is then

$$\pi^{(*)} : \mathcal{M} \to M, \quad c \mapsto c(0).$$

It induces another decomposition of the tangent bundle in vertical and horizontal bundles

$$V_c^{(*)}\mathcal{M} = \ker T\pi^{(*)} = \{ h \in T_c\mathcal{M} \mid h(0) = 0 \},$$
$$H_c^{(*)}\mathcal{M} = \left(V_c^{(*)}\mathcal{M} \right)^{\perp_G}.$$

Proposition 2. *We have the usual decomposition $T\mathcal{M} = V^{(*)}\mathcal{M} \overset{\perp}{\oplus} H^{(*)}\mathcal{M}$, the horizontal bundle $H_c^{(*)}\mathcal{M}$ consists of parallel vector fields along c, and $\pi^{(*)}$ is (formally) a Riemannian submersion for $(\mathcal{M}, G)$ and $(M, \langle \cdot, \cdot \rangle)$.*

Proof. Let h be a tangent vector. Consider h_0 the parallel vector field along c with initial value $h_0(0) = h(0)$. It is a horizontal vector, since its vanishing covariant derivative along c assures that for any vertical vector l we have $G_c(h_0, l) = 0$. The difference $\bar{h} = h - h_0$ between those two horizontal vectors has initial value 0 and so it is a vertical vector, which gives a decomposition of h into a horizontal vector and a vertical vector. The definition of $H^{(*)}\mathcal{M}$ as the orthogonal complement of $V^{(*)}\mathcal{M}$ guaranties that their sum is direct. Now if k is another tangent vector, then the scalar product between their horizontal parts is

$$G_c(h^H, k^H) = \langle h^H(0), k^H(0) \rangle_{c(0)} = \langle h(0), k(0) \rangle_{c(0)} = \langle T_c\pi^{(*)}(h^H), T_c\pi^{(*)}(k^H) \rangle_{\pi^{(*)}},$$

and this completes the proof.

4   Induced Distance on the Space of Curves

Here we will give an expression for the geodesic distance induced by the metric G. Let us consider two curves $c_0, c_1 \in \mathcal{M}$, and a path of curves $a \mapsto c(a, \cdot)$ linking them in $\mathcal{M}$

$$c(0, t) = c_0(t), \quad c(1, t) = c_1(t),$$

for all $t \in [0, 1]$. We denote by $f(a, \cdot) = R(c(a, \cdot))$ the image of this path of curves by the SRVF R. Note that f is a vector field along the surface c in M. Let now $\tilde{f}$ be the raising of f in the tangent plane $T_{c(0,0)}M$ in the following way

$$\tilde{f}(a, t) = P_{c(\cdot, 0)}^{a, 0} \circ P_{c(a, \cdot)}^{t, 0} \left(f(a, t) \right),$$

where we denote by $P_{\gamma}^{s, t} : T_{\gamma(s)}M \to T_{\gamma(t)}M$ the parallel transport along a curve γ from $\gamma(s)$ to $\gamma(t)$. Notice that $\tilde{f}$ is a surface in a vector space, as illustrated in Fig. 1. Lastly, we introduce a vector field $(b, s) \mapsto \omega^{a,t}(b, s)$ in M, which parallel translates $f(a, t)$

along $c(a, \cdot)$ to its origin, then along $c(\cdot, 0)$ and back down again, as shown in Fig. 1. More precisely

$$\omega^{a,t}(b,s) = P^{0,s}_{c(b,\cdot)} \circ P^{a,b}_{c(\cdot,0)} \circ P^{t,0}_{c(a,\cdot)} \left(f(a,t)\right)$$

for all b, s. That way the quantity $\nabla_{\partial_a c}\omega^{a,t}$ measures the holonomy along the rectangle of infinitesimal width shown in Fig. 1. For convenience, we will adopt the following notations for a vector field ω along a surface $a \mapsto c(a,t)$

$$\nabla_a \omega := \nabla_{\partial_a c}\omega, \quad \nabla_t \omega := \nabla_{\partial_t c}\omega.$$

We can now formulate our result.

Proposition 3. *With the above notations, the geodesic distance induced by the Riemannian metric G between two curves c_0 and c_1 on the space $\mathcal{M} = Imm([0,1], M)$ of parameterized curves is given by*

$$dist(c_0, c_1) = \inf_c \int_0^1 \sqrt{\|\gamma'(a)\|^2 + \int_0^1 \|\nabla_a f(a,t)\|^2 \, dt} \, da,$$

where $\gamma = c(\cdot, 0)$ is the curve linking the origins, $f = R(c)$ and the norm is the one associated to the Riemannian metric on M. It can also be written

$$dist(c_0, c_1) = \inf_c \int_0^1 \sqrt{\|\gamma'(a)\|^2 + \int_0^1 \|\partial_a \tilde{f}(a,t) + \Omega(a,t)\|^2 \, dt} \, da, \qquad (5)$$

where the curvature term Ω is given by

$$\Omega(a,t) = P^{a,0}_{c(\cdot,0)} \circ P^{t,0}_{c(a,\cdot)} \left(\nabla_a \omega^{a,t}(a,t)\right)$$

$$= P^{a,0}_{c(\cdot,0)} \circ P^{t,0}_{c(a,\cdot)} \left(\int_0^t P^{s,t}_{c(a,\cdot)} \left(\mathcal{R}(\partial_s c, \partial_a c) P^{t,s}_{c(a,\cdot)} f(a,t)\right) ds\right),$$

if $\mathcal{R}$ denotes the curvature tensor of the manifold M.

Remark 2. Our original motivation for this work was to find a geodesic distance (that is, a distance induced by a Riemannian metric) that resembled the product distance introduced in [5]. In the first term under the square root of expression (5) we can see the velocity vector of the curve γ linking the two origins, and in the second the velocity vector of the curve $\tilde{f}$ linking the TSRVF-images of the curves – Transported Square Root Velocity Function, as introduced by Su et al. in [12]. However there is also a curvature term Ω which, as previously mentioned, measures the holonomy along the rectangle of infinitesimal width shown in Fig. 1. If instead we equip the tangent bundle $T\mathcal{M}$ with the metric

$$\tilde{G}_h(\xi, \xi) = \|\xi_h(0)\|^2 + \int_0^1 \left\|\xi_v(t) - \int_0^t P^{s,t}_c \left(\mathcal{R}(c', \xi_h) P^{t,s}_c h(t)\right) ds\right\|^2 dt,$$

for $h \in T\mathcal{M}$ and $\xi, \eta \in T_h T\mathcal{M}$, then the curvature term Ω vanishes and the geodesic distance on $\mathcal{M}$ becomes

$$dist(c_0, c_1) = \inf_c \int_0^1 \sqrt{\|\gamma'(a)\|^2 + \|\partial_a \tilde{f}(a,\cdot)\|^2} \, da, \qquad (6)$$

where the norm of the second term under the square root is the L^2-norm, and which corresponds exactly to the geodesic distance associated to the metric on the space $\mathbb{C} = \cup_{p \in M} L^2([0,1], T_p M)$ introduced by Zhang et al. in [14]. Indeed, if

$$q(a,t) = P^{t,0}_{c(a,\cdot)}\left(f(a,t)\right) = P^{0,a}_{c(\cdot,0)}\left(\tilde{f}(a,t)\right),$$

then $a \mapsto (\gamma(a), q(a,\cdot))$ is a curve in $\mathbb{C}$, and the squared norm of its tangent vector according to the metric of [14] is given by

$$\|\gamma'(a)\|^2 + \int_0^1 \|\nabla_a q(a,t)\|^2 \, dt$$

$$= \|\gamma'(a)\|^2 + \int_0^1 \left\|\nabla_a \left(P^{0,a}_{c(\cdot,0)} \tilde{f}(a,t)\right)\right\|^2 \, dt$$

$$= \|\gamma'(a)\|^2 + \int_0^1 \left\|\partial_a \tilde{f}(a,t)\right\|^2 \, dt.$$

The difference between the two distances (5) and (6) resides in the curvature term Ω, which translates the fact that in the first one, we compute the distance in the manifold, whereas in the second, it is computed in the tangent space to one of the origins of the curves. Therefore, the first one takes more directly into account the "relief" of the manifold between the two curves under comparison. For example, if there is a "bump" between two curves in an otherwise relatively flat space, the second distance (6) might not see it, whereas the first one (5) will thanks to the curvature term.

Remark 3. Let us briefly consider the flat case : if the manifold M is flat, the two distances (5) and (6) coincide. If two curves c_0 and c_1 in a flat space have the same starting point p, the first summand under the square root vanishes and the distance becomes the L^2-distance between the two image curves $R(c_0)$ and $R(c_1)$. If two curves in a flat space differ only by a translation, then the distance is simply the distance between their origins.

Proof. Since G is defined by pullback of $\tilde{G}$ by the SRVF R, we know that the lengths of c in $\mathcal{M}$ and of $f = R(c)$ in $T\mathcal{M}$ are equal and so that

$$\text{dist}(c_0, c_1) = \inf_c \int_0^1 \sqrt{\tilde{G}\left(\partial_a f(a,\cdot), \partial_a f(a,\cdot)\right)} \, da,$$

with

$$\tilde{G}\left(\partial_a f(a,\cdot), \partial_a f(a,\cdot)\right) = \|\partial_a c(a,0)\|^2 + \int_0^1 \|\nabla_a f(a,t)\|^2 \, dt.$$

Now let us fix $t \in [0,1]$. Then $a \mapsto P^{t,0}_{c(a,\cdot)}\left(f(a,t)\right)$ is a vector field along $c(\cdot,0)$, and so

$$\nabla_a \left(P^{t,0}_{c(a,\cdot)} f(a,t)\right) = P^{0,a}_{c(\cdot,0)}\left(\frac{\partial}{\partial a} P^{a,0}_{c(\cdot,0)} \circ P^{t,0}_{c(a,\cdot)}\left(f(a,t)\right)\right) = P^{0,a}_{c(\cdot,0)}\left(\partial_a \tilde{f}(a,t)\right).$$

We consider the vector field v along the surface $(a,s) \mapsto c(a,s)$ that is parallel along all curves $c(a,\cdot)$ and takes value $v(a,t) = f(a,t)$ in $s = t$ for any $a \in [0,1]$, that is

$$v(a,s) = P^{t,s}_{c(a,\cdot)}\left(f(a,t)\right),$$

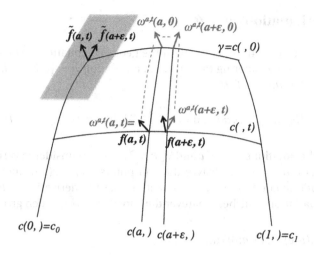

Fig. 1. Illustration of the distance between two curves c_0 and c_1 in the space of curves $\mathcal{M}$

for all $a \in [0,1]$ and $s \in [0,1]$. That way we know that

$$\nabla_a v(a, t) = \nabla_a f(a, t),$$
$$\nabla_a v(a, 0) = P_{c(\cdot,0)}^{0,a} \left(\partial_a \tilde{f}(a, t) \right),$$
$$\nabla_s v(a, s) = 0,$$

for all $a, s \in [0,1]$. Then we can express its covariant derivative in the following way

$$\nabla_a v(a, t) = P_{c(a,\cdot)}^{0,t} \left(\nabla_a v(a, 0) \right) + \int_0^t P_{c(a,\cdot)}^{s,t} \left(\nabla_s \nabla_a v(a, s) \right) ds$$

$$= P_{c(a,\cdot)}^{0,t} \circ P_{c(\cdot,0)}^{0,a} \left(\partial_a \tilde{f}(a, t) \right) + \int_0^t P_{c(a,\cdot)}^{s,t} \left(\mathscr{R}(\partial_s c, \partial_a c) P_{c(a,\cdot)}^{t,s} f(a, t) \right) ds. \quad (7)$$

Now let us fix $a \in [0,1]$ as well. Notice that the vector field $\omega^{a,t}$ defined above verifies

$$\omega^{a,t}(a, t) = f(a, t),$$
$$\nabla_s \omega^{a,t}(b, s) = 0,$$
$$\nabla_b \omega^{a,t}(b, 0) = 0,$$

for all $b, s \in [0,1]$. Note that unlike v, we do *not* have $\nabla_a \omega^{a,t}(a, t) = \nabla_a f(a, t)$ because $\omega^{a,t}(b, t) = f(b, t)$ is only true for $b = a$. It is easy to verify that the last term of Eq. (7) is precisely the covariant derivative of the vector field $\omega^{a,t}$

$$\nabla_a \omega^{a,t}(a, t) = \int_0^t P_{c(a,\cdot)}^{s,t} \left(\mathscr{R}(\partial_s c, \partial_a c) P_{c(a,\cdot)}^{t,s} f(a, t) \right) ds,$$

since for any $s \in [0,1]$, $\omega^{a,t}(a, s) = P_{c(a,\cdot)}^{t,s} f(a, t)$, and finally by composing by $P_{c(\cdot,0)}^{a,0} \circ P_{c(a,\cdot)}^{t,0}$, we obtain the second expression (5), which completes the proof.

5   Geodesic Equation on $T\mathcal{M}$

In the same way as in [14], we can obtain the geodesic equation associated to our metric $\tilde{G}$ on $T\mathcal{M}$ by considering the energy of a variation $b \mapsto \big(\hat{c}(b,\cdot,\cdot), \hat{h}(b,\cdot,\cdot)\big)$ of a curve $a \mapsto (c(a,\cdot), h(a,\cdot))$ of $T\mathcal{M}$

$$E(b) = \int_0^1 \langle \hat{c}_a(b,a,0), \hat{c}_a(b,a,0) \rangle \, da + \int_0^1 \int_0^1 \langle \nabla_a \hat{h}(b,a,t), \nabla_a \hat{h}(b,a,t) \rangle \, dt \, da,$$

where we use the notations $c_a = \partial_a c$ and $\nabla_a h = \nabla_{x_a} h$. The considered variation $(\hat{c}, \hat{h})$ takes value (c, h) in $b = 0$ and leaves the end points $c(0,\cdot)$, $c(1,\cdot)$ unchanged. We obtain the geodesic equation in $T\mathcal{M}$ by writing that the derivative in $b = 0$ of the energy of this variation vanishes whatever the choice of $(\hat{c}, \hat{h})$, which gives

$$\int_0^1 \langle \nabla_a c_a(a,0), \hat{c}_b(0,a,0) \rangle \, da$$

$$+ \int_0^1 \int_0^1 \langle \mathscr{R}(h, \nabla_a h) c_a(a,t), \hat{c}_b(0,a,t) \rangle + \langle \nabla_a \nabla_a h(a,t), \nabla_b \hat{h}(0,a,t) \rangle \, dt \, da = 0.$$

The geodesic equation in $R(\mathcal{M})$ is obtained by putting $h = R(c) = \frac{c_t}{\sqrt{\|c_t\|}}$. Unfortunately we must distinguish the two spaces since R is not bijective from $\mathcal{M}$ to $T\mathcal{M}$.

Remark 4. Let us just point out that if we consider instead a third metric

$$\tilde{G}_h(\xi, \eta) = \int_0^1 \langle \xi(t)_H, \eta(t)_H \rangle \, \|h(t)\|^2 + \langle \xi(t)_V, \eta(t)_V \rangle \, dt,$$

then the derivative in $b = 0$ of the energy of the variation $(\hat{c}, \hat{h})$ is given by

$$E'(0) = -2 \int_0^1 \int_0^1 \langle \|h\|^2 \nabla_a c_a + 2 \langle \nabla_a h, h \rangle c_a + \mathscr{R}(h, \nabla_a h) c_a, \hat{c}_b(0) \rangle$$

$$+ \langle \nabla_a \nabla_a h - \|c_a\|^2 h, \nabla_b \hat{h}(0) \rangle \, dt \, da,$$

and since the two tangent vectors $\hat{c}_b(0)$, $\nabla_b \hat{h}(0)$ can be chosen independently, we get the following geodesic equations in $T\mathcal{M}$

$$\begin{cases} \|h\|^2 \nabla_a c_a + 2 \langle \nabla_a h, h \rangle c_a + \mathscr{R}(h, \nabla_a h) c_a = 0, \\ \nabla_a \nabla_a h = \|x_a\|^2 h, \end{cases}$$

where all terms are considered at any point $(a, t) \in [0, 1]^2$. In $R(\mathcal{M})$ however, we only have an integral form, for we cannot choose $\hat{c}_b(0)$ and $\nabla_b R(\hat{c}(0))$ independently.

6   Conclusion

In the same way that the first-order Sobolev metric (2) on the space of plane curves can be obtained as the pullback of the L^2-metric by the square root velocity function [11], our metric G can be obtained as the pullback of a metric $\tilde{G}$ on the tangent bundle T$\mathcal{M}$ derived from the Sasaki metric, by the same SRVF. As such it is reparameterization invariant, and induces a Riemannian metric $\hat{G}$ on the shape space $\mathcal{S}$ for which the fiber bundle projection is formally a Riemannian submersion. On the other hand, the special role that G gives to the starting points of the curves induces another formal fiber bundle structure, this time over the manifold M seen as the set of starting points of the curves, for which the projection is formally also a Riemannian submersion. In the flat case, the geodesic distance induced by G is a product metric, and when the manifold M is not flat, there is an added curvature term. We can modify the metric $\tilde{G}$ so that this curvature term in the distance induced by its pullback G disappears, but the first option seems preferable since the induced distance takes into account a greater amount of information on the geometry of the manifold.

Acknowledgments. This research was supported by Thales Air Systems and the french MoD DGA.

References

1. Bauer, M., Bruveris, M., Marsland, S., Michor, P.W.: Constructing reparameterization invariant metrics on spaces of plane curves. Differ. Geom. Appl. **34**, 139–165 (2014)
2. Bauer, M., Bruveris, M., Michor, P.: Why use Sobolev metrics on the space of curves. In: Riemannian Computing in Computer Vision. Springer (to appear). arXiv:1502.03229
3. Bauer, M., Harms, P., Michor, P.: Sobolev metrics on shape space of surfaces. J. Geom. Mech. **3**(4), 389–438 (2011)
4. Laga, H., Kurtek, S., Srivastava, A., Miklavcic, S.J.: Landmark-free statistical analysis of the shape of plant leaves. J. Theoret. Biol. **363**, 41–52 (2014)
5. Le Brigant, A., Arnaudon, M., Barbaresco, F.: Reparameterization invariant distance on the space of curves in the hyperbolic plane. AIP Conf. Proc. **1641**, 504 (2015)
6. Michor, P., Mumford, D.: Vanishing geodesic distance on spaces of submanifolds and diffeomorphisms. Documenta Math. **10**, 217–245 (2005)
7. Michor, P., Mumford, D.: Riemannian geometries on spaces of plane curves. J. Eur. Math. Soc. (JEMS) **8**, 1–48 (2006)
8. Michor, P., Mumford, D.: An overview of the Riemannian metrics on spaces of curves using the Hamiltonian approach. Appl. Comput. Harmonic Anal. **23**, 74–113 (2007)
9. Sasaki, S.: On the differential geometry of tangent bundles of Riemannian manifolds. Tohoku Math. J. **10**, 338–354 (1958)
10. Sasaki, S.: On the differential geometry of tangent bundles of Riemannian manifolds II. Tohoku Math. J. **14**, 146–155 (1962)
11. Srivastava, A., Klassen, E., Joshi, S.H., Jermyn, I.H.: Shape analysis of elastic curves in Euclidean spaces. IEEE T. Pattern Anal. **33**(7), 1415–1428 (2011)
12. Su, J., Kurtek, S., Klassen, E., Srivastava, A.: Statistical analysis of trajectories on Riemannian manifolds: bird migration, hurricane tracking and video surveillance. Ann. Appl. Stat. **8**(1), 530–552 (2014)
13. Younes, L., Michor, P., Shah, J., Mumford, D.: A metric on shape space with explicit geodesics. Rend. Lincei Mat. Appl. **9**, 25–57 (2008)
14. Zhang, Z., Su, J., Klassen, E., Le, H., Srivastava, A.: Video-based action recognition using rate-invariant analysis of covariance trajectories (2015). arXiv:1503.06699 [cs.CV]

Invariant Geometric Structures
on Statistical Models

Lorenz Schwachhöfer[1]([✉]), Nihat Ay[2], Jürgen Jost[2], and Hông Vân Lê[3]

[1] Technische Universität Dortmund, Vogelpothsweg 87, 44221 Dortmund, Germany
[2] Max-Planck-Institut Für Mathematik in den Naturwissenschaften,
Inselstrasse 22, 04103 Leipzig, Germany
nay@mis.mpg.de
[3] Mathematical Institute of ASCR, Zitna 25, 11567 Praha, Czech Republic

Abstract. We review the notion of parametrized measure models and
tensor fields on them, which encompasses all statistical models considered
by Chentsov [6], Amari [3] and Pistone-Sempi [10]. We give a complete
description of n-tensor fields that are invariant under sufficient statistics.
In the cases $n = 2$ and $n = 3$, the only such tensors are the Fisher metric
and the Amari-Chentsov tensor. While this has been shown by Chentsov
[7] and Campbell [5] in the case of finite measure spaces, our approach
allows to generalize these results to the cases of infinite sample spaces and
arbitrary n. Furthermore, we give a generalisation of the monotonicity
theorem and discuss its consequences.

1 General Definition of Parametrised Measure Models

Let (Ω, Σ) be a measurable space. We consider the Banach space $\mathcal{S}(\Omega)$ of all
signed finite measures on Ω with the total variation $\| \cdot \|_{TV}$ as Banach norm.
More precisely, the total variation of such a measure μ is defined as

$$\|\mu\|_{TV} := \sup \sum_{i=1}^{n} |\mu(A_i)|$$

where the supremum is taken over all finite partitions $\Omega = A_1 \dot\cup \ldots \dot\cup A_n$ with
disjoint sets $A_i \in \Sigma$. We consider the subsets $\mathcal{M}(\Omega) \subset \mathcal{S}(\Omega)$ of all finite non-
negative measures, and $\mathcal{P}(\Omega) \subset \mathcal{M}(\Omega)$ of all probability measures on Ω.

For a fixed σ-finite non-negative measure μ_0, we also consider the subspace

$$\mathcal{S}(\Omega, \mu_0) := \{\mu = \phi \mu_0 \; : \; \phi \in L^1(\Omega, \mu_0)\}$$

of signed measures dominated by μ_0. This space can be identified in terms of
the canonical map $i_{can} : \mathcal{S}(\Omega, \mu_0) \to L^1(\Omega, \mu_0)$, $\mu \mapsto \frac{d\mu}{d\mu_0}$. Note that

$$\|\mu\|_{TV} = \left\| \frac{d\mu}{d\mu_0} \right\|_{L^1(\Omega, \mu_0)},$$

L. Schwachhöfer—speaker
J. Jost—partially supported by ERC Advanced Grant FP7-267087
H.V. Lê—partially supported by RVO: 6798584.

© Springer International Publishing Switzerland 2015
F. Nielsen and F. Barbaresco (Eds.): GSI 2015, LNCS 9389, pp. 150–158, 2015.
DOI: 10.1007/978-3-319-25040-3_17

which implies that i_{can} is a Banach space isomorphism. Therefore, we refer to the topology of $\mathcal{S}(\Omega, \mu_0)$ also as L^1-*topology*. This is independent of the particular choice of the reference measure μ_0, because if $\phi \in L^1(\Omega, \mu_0)$ and $\psi \in L^1(\Omega, \phi\mu_0)$, then $\psi\phi \in L^1(\Omega, \mu_0)$.

Definition 1. *(Parametrized measure model)*
   Let Ω be a measure space.

1. *A* signed parametrized measure model *is a triple (M, Ω, p) where M is a (finite or infinite dimensional) Banach manifold and $p : M \rightarrow \mathcal{S}(\Omega)$ is a C^1-map between Banach manifolds in the formal sense given in [9].*
2. *Such a triple (M, Ω, p) is called a* parametrized measure model *if it consists only of non-negative measures, i.e., such that the image of p is contained in $\mathcal{M}(\Omega)$.*
3. *The triple (M, Ω, p) is called a* statistical model *if it consists only of probability measures, i.e., such that the image of p is contained in $\mathcal{P}(\Omega)$.*
4. *We call such a model is* dominated by μ_0 *if the image of p is contained in $\mathcal{S}(\Omega, \mu_0)$. In this case, we use the notation (M, Ω, μ_0, p) for this model.*

Remark 1. Evidently, for the applications we have in mind, we are interested mainly in statistical models. However, we can take the point of view that $\mathcal{P}(\Omega)$ is the projectivisation of $\mathcal{P}(\Omega) = \mathbb{P}(\mathcal{M}(\Omega)\backslash 0)$ via rescaling. Thus, given a parametrized measure model (M, p, Ω), normalisation yields a statistical model (M, p_0, Ω) defined by

$$p_0(\xi) := \frac{p(\xi)}{\|p(\xi)\|_{TV}}.$$

which is again a C^1-map.

   Furthermore, regarding general signed parametrized measure models enables us to take the point of view that $p : M \rightarrow \mathcal{S}(\Omega)$ is a C^1-map, since $\mathcal{S}(\Omega)$ is a Banach space, whereas the subsets $\mathcal{M}(\Omega)$ and $\mathcal{P}(\Omega)$ do not carry a canonical manifold structure, as for infinite measure space Ω these are *not* open subsets of $\mathcal{S}(\Omega)$ in the L^1-topology.

   If a (signed) parametrized measure model (M, Ω, μ_0, p) is dominated by μ_0, then there is a *density function* $p : \Omega \times M \rightarrow \mathbb{R}$ such that

$$p(\xi) = p(.; \xi)\mu_0. \tag{1}$$

   From the context, i.e., from the number of arguments, it will be clear which map p is meant, whence we will denote both maps by the same symbol. Evidently, we must have $p(.; \xi) \in L^1(\Omega, \mu_0)$ for all ξ. In particular, for fixed ξ, $p(.; \xi)$ is defined only up to changes on a μ_0-null set.

Definition 2. *(Regular density function)*
   Let (M, Ω, μ_0, p) be a (signed) parametrized measure model dominated by μ_0. We say that this model has a regular density function *if the density function $p : \Omega \times M \rightarrow \mathbb{R}$ satisfying (1) can be chosen such that for all $V \in T_\xi M$ the partial derivative $\partial_V p(.; \xi)$ exists and lies in $L^1(\Omega, \mu_0)$.*

Remark 2. The standard notion of statistical models always assumes that it is dominated by some measure and has a regular density function (e.g. [3, §2, p. 25], [4, §2.1], [11]). In fact, we shall assume regularity of the density function whenever it is convenient.

However, let us point out why for a statistical model (or, more generally, for a signed parametrized measure model), the regularity of the density function is indeed an additional requirement.

The formal definition of differentiability of p implies that for each C^1-path $\xi(t) \in M$ with $\xi(0) = \xi$, $\dot{\xi}(0) =: V \in T_\xi M$, the curve $t \mapsto p(.;\xi(t)) \in L^1(\Omega, \mu_0)$ is differentiable. This implies that there is a $d_\xi p(V) \in L^1(\Omega, \mu_0)$ such that

$$\left\| \frac{p(.;\xi(t)) - p(.;\xi)}{t} - d_\xi p(V)(.) \right\|_1 \xrightarrow{t \to 0} 0.$$

If this is a *pointwise* convergence, then $d_\xi p(V) = \partial_V p(.;\xi)$ is the partial derivative and whence, $\partial_V p(.;\xi)$ lies in $L^1(\Omega, \mu_0)$, so that the density function is regular.

However, in general convergence in $L^1(\Omega, \mu_0)$ does *not* imply pointwise convergence, whence there are parametrized measure models without a regular density function, cf. the example below. Nevertheless, for simplicity we shall frequently use the notation $\partial_V p(\cdot;\xi)$ instead of $d_\xi p(V)(.)$, even if the density function is *not* regular.

By this convention, for a signed parametrized measure model (M, Ω, μ_0, p) we can describe its derivative in the direction of $V \in T_\xi M$ as

$$d_\xi p(V) = \partial_V p(.;\xi)\, \mu_0. \tag{2}$$

Example 1. 1. The family of normal distributions on $\mathbb{R}$

$$p(\mu, \sigma) := \frac{1}{\sqrt{2\pi}\sigma} \exp(-\frac{(x-\mu)^2}{2\sigma^2})\, dx$$

   is a statistical model with regular density function on the upper half plane $H = \{(\mu, \sigma) \; : \; \mu, \sigma \in \mathbb{R}, \sigma > 0\}$.
2. To see that our notion is indeed more general, consider the family of measures on $\Omega = (0, \pi)$

$$p(\xi) := \begin{cases} \left(1 + \xi\,(\sin^2(t - 1/\xi))^{1/\xi^2}\right)\, dt & \text{for } \xi \neq 0 \\ dt & \text{for } \xi = 0. \end{cases}$$

This model is dominated by the Lebesgue measure dt, with density function $p(t;\xi) = 1 + \xi\,(\sin^2(t - 1/\xi))^{1/\xi^2}$ for $\xi \neq 0$, $p(t;0) = 1$. Thus, the partial derivative $\partial_\xi p$ does not exist at $\xi = 0$, whence the density function is not regular.

On the other hand, $p : \mathbb{R} \to \mathcal{M}(\Omega, dt)$ is differentiable in the above sense at $\xi = 0$ with $d_0 p(\partial_\xi) = 0$. To see this, we calculate

$$
\left\| \frac{p(\xi) - p(0)}{\xi} \right\|_1 = \left\| (\sin^2(t - 1/\xi))^{1/\xi^2} \, dt \right\|_1
$$
$$
= \int_0^\pi (\sin^2(t - 1/\xi))^{1/\xi^2} \, dt
$$
$$
= \int_0^\pi (\sin^2 t)^{1/\xi^2} \, dt \xrightarrow{\xi \to 0} 0.
$$

which shows the claim. Here, we used the π-periodicity of the integrand for fixed ξ and dominated convergence in the last step.

In general, for a signed parametrized measure model (M, Ω, p), there is no relation between the measures $p(\xi) \in \mathcal{S}(\Omega)$ and that of the directional derivative $\partial_V p = d_\xi p(V)$. However, we make the following definition.

Definition 3. *A signed parametrized measure model (M, Ω, p) is said to have a logarithmic derivative if $d_\xi p(V)$ is dominated by $p(\xi)$ for all $\xi \in M$, $V \in T_\xi M$.*

If such a model is dominated by μ_0 and has a regular density function p for which (1) holds, then we can calculate the Radon-Nikodym derivative as

$$
\frac{d\{d_\xi p(V)\}}{dp(\xi)} = \frac{d\{d_\xi p(V)\}}{d\mu_0} \cdot \left(\frac{dp(\xi)}{d\mu_0} \right)^{-1}
$$
$$
= \partial_V p(.; \xi)(p(.; \xi))^{-1} = \partial_V \log |p(.; \xi)|,
$$

where we use the convention $\log 0 = 0$. This calculation motivates to *define*

$$
\partial_V \log |p(\xi)| := \frac{d\{d_\xi p(V)\}}{dp(\xi)} \in L^1(\Omega, |p(\xi)|) \tag{3}
$$

for any signed parametrized measure model with logarithmic derivative. That is, for such a model, $\partial_V \log |p(\xi)|$ may be rigorously defined, even though $\log |p(.; \xi)|$ itself is *not* defined unless $p(.; \xi) \neq 0$.

As the following proposition shows, the assumption that a model has a logarithmic derivative is quite natural, since we are mainly interested in (non-signed) parametrized measure models. This proposition holds without any further regularity assumptions on the model.

Proposition 1. [1,2] *Any (non-signed) parametrized measure model and, in particular, any statistical model (M, Ω, p) has a logarithmic derivative.*

2   Roots of Measures and k-integrability

For any $k \geq 1$, we define the *space of k-th roots of measures dominated by μ* as

$$
\mathcal{S}^{1/k}(\Omega, \mu) := \{ \phi \mu^{1/k} \ : \ \phi \in L^k(\Omega, \mu) \}.
$$

This is a Banach space isomorphic to $L^k(\Omega, \mu)$, and there is a canonical bijective map, called the *k-th signed power*, defined as

$$\pi^k : \mathcal{S}^{1/k}(\Omega, \mu) \longrightarrow \mathcal{S}(\Omega, \mu), \qquad \pi^k(\phi\mu^{1/k}) := sign(\phi)|\phi|^k\mu.$$

By taking the directed limit over all positive measures, we can also define the Banach space $\mathcal{S}^{1/k}(\Omega)$ and the k-th power $\pi^k : \mathcal{S}^{1/k}(\Omega) \to \mathcal{S}(\Omega)$, without making use of a fixed reference measure.

As it turns out, π^k is differentiable and a homeomorphism (but not a diffeomorphism), and we denote its continuous inverse by $\pi^{1/k} : \mathcal{S}(\Omega) \to \mathcal{S}^{1/k}(\Omega)$.

Definition 4. *(k-integrable parametrized measure model)*

Let (M, Ω, p) be a statistical model or, more general, a parametrized measure model. Then it is called k-integrable for $k \geq 1$ if for all $\xi \in M$ and $V \in T_\xi M$ we have

$$\partial_V \log p(\xi) = \frac{d\{d_\xi p(V)\}}{dp(\xi)} \in L^k(\Omega, p(\xi)),$$

and moreover, the map

$$TM \longrightarrow \mathcal{S}^{1/k}(\Omega), \qquad \xi \longmapsto \frac{d\{d_\xi p(V)\}}{dp(\xi)} p(\xi)^{1/k}$$

is continuous.

Example 2. [1, Example 2.5.1] Let Ω_n be a finite set of n elements and μ_n a measure of maximal support on Ω_n. It is evident that $\mathcal{M}_+(\Omega_n, \mu_n)$ is diffeomorphic to $\mathbb{R}^n$. Let S be a C^1-submanifold in $\mathcal{P}_+(\Omega_n, \mu_n)$ and $i_S : S \to \mathcal{P}_+(\Omega_n, \mu_n)$ the canonical embedding. Then $(S, \Omega_n, \mu_n, i_S)$ is an immersed k-integrable statistical model for all $k \geq 1$.

Example 3. [1, Proposition 5.11] Non-parametric statistical models in sense of Pistone-Sempi [10] are k-integrable parametrized measure models for any $k \geq 1$.

Example 4. Let $\Omega = (0, 1)$ and $\mu_0 = dt$ be the Lebesgue measure. For $\xi \in (0, \infty)$ we define

$$p(\xi) := (2 + \sin(\xi t^{-\alpha}))\, dt,$$

where $\alpha \in (0, 1)$ is fixed. Then this model is k-integrable if and only if $k < \alpha^{-1}$. In particular, it does not define a model in the sense of Pistone and Sempi.

Definition 5. *(Canonical n-tensor)* *For $n \in \mathbb{N}$, the canonical n-tensor is the covariant n-tensor on $S^{1/n}(\Omega)$, given by*

$$L_\Omega^n(\nu_1, \ldots, \nu^n) = n^n \int_\Omega d(\nu_1 \cdots \nu_n), \qquad where \ \nu_i \in \mathcal{S}^{1/n}(\Omega). \tag{4}$$

The purpose of defining the notion of k-integrability of a model is the following result. If the parametrized measure model $p : M \to \mathcal{M}(\Omega)$ is a n-integrable, then the map

$$p^{1/n} : M \longrightarrow \mathcal{S}^{1/n}(\Omega), \qquad \xi \longmapsto \pi^{1/n}(p(\xi))$$

is differentiable, so that we can define the *canonical n-tensor* $\tau^n_{(M,p,\Omega)}$ as the pull-back of L^n_Ω. That is, we define for $V_1, \ldots, V_n \in T_\xi M$

$$
\begin{aligned}
\tau^n_{(M,p,\Omega)}(V_1, \ldots, V_n) &:= L^n_\Omega(d_\xi p^{1/n}(V_1, \ldots, d_\xi p^{1/n}(V_1)) \\
&= \int_\Omega \partial_{V_1} \log p(\xi) \cdots \partial_{V_n} \log p(\xi)\, dp(\xi).
\end{aligned}
\tag{5}
$$

Example 5.   1. For $n = 1$, the canonical 1-form is given as

$$
\tau^1_{(M,p,\Omega)}(V) := \int_\Omega \partial_V \ln p(\xi)\, dp(\xi) = \partial_V \|p(\xi)\|.
\tag{6}
$$

Thus, it vanishes if and only if $\|p(\xi)\|$ is constant, e.g., if (M, p, Ω) is a *statistical* model.

2. For $n = 2$, $\tau^2_{(M,p,\Omega)}$ coincides with the *Fisher metric*

$$
\mathbf{g}^F(V, W)_\xi := \int_\Omega \partial_V \ln p(\xi)\, \partial_W \ln p(\xi)\, dp(\xi)
\tag{7}
$$

3. For $n = 3$, $\tau^3_{(M,p,\Omega)}$ coincides with the *Amari-Chentsov 3-symmetric tensor*

$$
T^{AC}(V, W, X)_\xi := \int_\Omega \partial_V \ln p(\xi)\, \partial_W \ln p(\xi)\, \partial_X \ln p(\xi)\, dp(\xi).
\tag{8}
$$

3   Sufficient Statistics and Main Results

Given two measure spaces Ω, Ω', a measurable map $\kappa : \Omega \to \Omega'$ is called a *statistic*. Such a statistic induces a map $\kappa_* : \mathcal{M}(\Omega) \to \mathcal{M}(\Omega')$ by

$$
\kappa_*(\mu)(A') := \mu(\kappa^{-1}(A')).
$$

Evidently, κ_* maps probability measures to probability measures, and it extends to a bounded lineear map $\kappa_* : \mathcal{S}(\Omega) \to \mathcal{S}(\Omega')$. Thus, given a parametrized measure model (M, Ω, p), a statistic induces the parametrized measure model (M, Ω', p') given as

$$
p'(\xi) := \kappa_*(p(\xi)).
$$

Proposition 2. *Let (M, Ω, p) and (M, Ω', p') with $p' = \kappa(p)$ for some statistic $\kappa : \Omega \to \Omega'$. If (M, Ω, p) is k-integrable, then so is (M, Ω', p').*

Again, the advantage of our approach is that this proposition is valid without any regularity assumptions on the model or on κ. In general, the parametrized measure model (M, Ω', p') carries less information than the model (M, Ω, p). However, there are situations where this is not the case.

Definition 6. *(Sufficient statistic)*
Let (M, Ω, p) be a *(signed)* parametrized measure model. Then $\kappa : \Omega \to \Omega'$ is called a *sufficient statistic for p if there is a $\mu \in \mathcal{M}(\Omega)$ such that*

$$
p(\xi) = \phi'(\kappa(\cdot); \xi)\mu
\tag{9}
$$

for some $\phi'(\cdot;\xi) \in L^1(\Omega',\mu)$. *In this case,*

$$p'(\xi) = \kappa_* p(\xi) = \phi'(\cdot;\xi)\mu',$$

where $\mu' = \kappa_*(\mu)$.

We now can show the following

Theorem 1. *(Monotonicity theorem) (cf. [1,4])*

Let (M,Ω,p) be a 2-integrable parametrized measure model, let $\kappa : \Omega \to \Omega'$ be a statistic, so that the induced parametrized measure model (M,Ω',p') with $p'(\xi) = \kappa_*(p(\xi))$ is also 2-integrable by Proposition 2. Moreover, let $\mathfrak{g}$ and $\mathfrak{g}'$ denote the Fisher metric of (M,Ω,p) and (M,Ω',p'), respectively. Then

$$\mathfrak{g}(V,V) \geq \mathfrak{g}'(V,V) \qquad \text{for all } V \in T_\xi M \text{ and } \xi \in M. \tag{10}$$

Moreover, if $p(\xi) = p(\cdot;\xi)\mu_0$ with regular and positive density function $p : M \times \Omega \to (0,\infty)$, and M is connected, then equality in (10) holds for all V if and only if κ is a sufficient statistic for the model (M,Ω,p).

Remark 3. The difference $\mathfrak{g}(V,V) - \mathfrak{g}'(V,V) \geq 0$ is called the *information loss* of the model (M,Ω,p) *under* κ. Thus, the interpretation of the monotonicity theorem is that every statistic produces some (non-negative) information loss which vanishes if and only if this statistic is sufficient.

Let (M,Ω,p) be a n-integrable parametrized measure model, let $\kappa : \Omega \to \Omega'$ be a statistic and (M,Ω',p') the induced model, which is also n-integrable by Proposition 2. If κ is a sufficient statistic for the model, then it is not hard to show that

$$\tau^n_{(M,p',\Omega')} = \tau^n_{(M,p,\Omega)}$$

for the canonical n-tensors defined in (5). Remarkably, the converse is also true. Namely, we have the following result.

Theorem 2. *(c.f. [2])* Let $\Theta^n_{(M,p,\Omega)}$ be a family of n-tensors on M for all parametrized measure models (M,p,Ω). Then the following are equivalent.

1. *The family is invariant under sufficient statistics; that is, for a sufficient statistic* $\kappa : \Omega \to \Omega'$ *we have* $\Theta^n_{(M,p,\Omega)} = \Theta^n_{(M,p',\Omega')}$.
2. *The family is invariant under sufficient statistics to finite sets; that is, for a sufficient statistics* $\kappa : \Omega \to I$ *with* I *finite we have* $\Theta^n_{(M,p,\Omega)} = \Theta^n_{(M,p',I)}$.
3. $\Theta^n_{(M,p,\Omega)}$ *is algebraically generated by the canonical tensors* $\tau^n_{(M,p,\Omega)}$ *defined in (5).*

For instance, for $n = 4$, Theorem 2 implies that any invariant family of 4-tensors for statistical models (M,p,Ω) is of the form

$$\Theta^4_{(M,p,\Omega)}(V_1,\ldots,V_4) = c_0\,\tau^4_{M,p,\Omega}(V_1,\ldots,V_4) + c_1\,\mathfrak{g}^F(V_1,V_2)\mathfrak{g}^F(V_3,V_4)$$
$$+ c_2\,\mathfrak{g}^F(V_1,V_3)\mathfrak{g}^F(V_2,V_3) + c_3\,\mathfrak{g}^F(V_1,V_4)\mathfrak{g}^F(V_2,V_3)$$

for constants $c_0, \ldots, c_3$, so that the space of invariant families on statistical models is 4-dimensional for $n = 4$. Evidently, this dimension rapidly increases with n.

For $n = 2, 3$, however, we obtain the following.

Corollary 1. *(Generalisation of Chentsov's and Campbells theorem)*

1. *Let $(\Theta^2_{(M,p,\Omega)})$ be a family of 2-tensors which is invariant under sufficient statistics. Then there are continuous functions $a, b : [0, \infty) \to \mathbb{R}$ such that*

$$(\Theta^2_{(M,p,\Omega)})_\xi(V, W) = a(\|p(\xi)\|)\mathfrak{g}^F(V, W) + b(\|p(\xi)\|)\partial_V \|p(\xi)\| \cdot \partial_W \|p(\xi)\|.$$

   In particular, for a statistical model (i.e., where $\|p(\xi)\| \equiv 1$), $\Theta^2_{(M,p,\Omega)} = c \, \mathfrak{g}^F$ with the constant $c = a(1)$.

2. *Let $(\Theta^3_{(M,p,\Omega)})$ be a family of 3-tensors which is invariant under sufficient statistics. Then it is a constant multiple of there the Amari-Chentsov tensor from (8), i.e., for some $c \in \mathbb{R}$ and any model we have*

$$\Theta^3_{(M,p,\Omega)} = c \, T^{AC}$$

4 Concluding Remarks

The results of Chentsov [7] and Campbell [5] on the characterisation of the Fisher metric and the Amari-Chentsov tensor by their invariance was given for finite sample spaces Ω only. In this case, the issues of convergence are negligable, and the notion of differentiability of the map $p : M \to \mathcal{P}(\Omega)$ is obvious. But for arbitrary sample spaces Ω, most generalisations need further assumptions.

For instance, Theorem 1 was previously known only in case that the statistic $\kappa : \Omega \to \Omega'$ admits transversal measures, e.g., if Ω, Ω' are topological spaces with their Borel σ-algebra and κ is continuous.

Similarly, in a recent paper [8] it was shown that the Fisher metric is uniquely characterized by its diffeomorphism invariance, provided the measure space Ω is a differentiable manifold and the model consists of smooth densities. On the one hand, this is more general as only the invariance under diffeomorphisms, i.e., a very special class of sufficient statistics, is needed, but on the other hand, assuming Ω to be a manifold and all densities to be smooth is a strong restriction.

In contrast, our approach does not make any assumption on the sample space Ω. Indeed, our only assumption is that the map $p : M \to \mathcal{P}(\Omega)$ is a differentiable map between Banach manifolds – whence we do not even require the measure to be given by a smooth density function (cf. Remark 2). Remarkably, this very weak condition already suffices to show Theorems 1 and 2.

References

1. Ay, N., Jost, J., Lê, H.V., Schwachhöfer, L.: Information geometry and sufficient statistics. Probab. Theor. Relat. Fields **162**(1–2), 327–364 (2015)
2. Ay, N., Jost, J., Lê, H.V., Schwachhöfer, L.: Information geometry, book in preparation
3. Amari, S.: Differential Geometrical Theory of Statistics. In: Amari, S.-I., Barndorff-Nielsen, O.E., Kass, R.E., Lauritzen, S.L., Rao, C.R. (eds.) Differential Geometry in Statistical Inference. Lecture Note-Monograph Series, vol. 10. Institute of Mathematical Statistics, California (1987)
4. Amari, S., Nagaoka, H.: Methods of Information Geometry. Translations of Mathematical Monographs, vol. 191. American Mathematical Society, Providence (2000)
5. Campbell, L.L.: An extended Chentsov characterization of a Riemannian metric. Proc. AMS **98**, 135–141 (1986)
6. Chentsov, N.: Category of mathematical statistics. Dokl. Acad. Nauk USSR **164**, 511–514 (1965)
7. Chentsov, N.: Statistical Decision Rules and Optimal Inference. Translation of Mathematical Monograph, vol. 53. AMS, Providence (1982)
8. Bauer, M., Bruveris, M., Michor, P.: Uniqueness of the Fisher-Rao metric on the space of smooth densities (2014). arXiv:1411.5577
9. Lang, S.: Fundamentals Of Differential Geometry. Springer, NewYork (1999)
10. Pistone, G., Sempi, C.: An infinite-dimensional structure on the space of all the probability measures equivalent to a given one. Ann. Stat. **5**, 1543–1561 (1995)
11. Rao, C.R.: Information and the accuracy attainable in the estimation of statistical parameters. Bull. Calcutta Math. Soc. **37**, 81–89 (1945)

The Abstract Setting for Shape Deformation Analysis and LDDMM Methods

Sylvain Arguillère[✉]

Johns Hopkins University, Baltimore, MD, USA
sarguil1@johnshopkins.edu

Abstract. This paper aims to define a unified setting for shape registration and LDDMM methods for shape analysis. This setting turns out to be sub-Riemannian, and not Riemannian. An abstract definition of a space of shapes in $\mathbb{R}^d$ is given, and the geodesic flow associated to any reproducing kernel Hilbert space of sufficiently regular vector fields is showed to exist for all time.

1 Introduction

The purpose of this paper is to define and study abstract shape spaces in $\mathbb{R}^d$ in order to unify and generalize the LDDMM algorithms that have been developed in the past few years. They consist in fixing a Hilbert space V of smooth vector fields in $\mathbb{R}^d$ with reproducing kernel K, and studying the deformations of an initial shape (a *template*) induced by flows of elements of V [8,10,17–21]. This allows to measure the "energy" of this flow by integrating the squared norm of the vector field. One then tries to get as close as possible to a target shape while keeping the energy small. This induces a length structure on the shape space and the problem can be reformulated as a geodesic search for this structure.

However, these methods have some flaws from a theoretical point of view. First of all, the notion of "shape space" has always been ambiguous. While it usually refers to a space of embeddings of a compact surface in $\mathbb{R}^3$ (or, in numerical simulations, to spaces of landmarks), more general spaces are sometimes needed and therefore require a case by case analysis. For example, when studying the movement of a muscle, one needs to take into account the direction of that muscle's fibers, which are not part of the embedding itself.

The second problem only appears for a shape space $\mathcal{S}$ of infinite dimension. Contrarily to what is described in most papers, the length structure induced by the flow of vector fields in V yields a *sub-Riemannian structure* on $\mathcal{S}$, not a Riemannian one. While this raises several difficulties from a theoretical viewpoint, it does not change the optimization algorithms, since those are mainly concerned with finite dimensional shape spaces, for which the structure is indeed Riemannian.

The purpose of this paper is to address both of these issues. In the first section, we briefly summarize the results of [4] on the Hamiltonian geodesic flow of the space of Sobolev diffeomorphisms of $\mathbb{R}^d$ for the right-invariant sub-Riemannian structure induced by a fixed arbitrary Hilbert space V of smooth

© Springer International Publishing Switzerland 2015
F. Nielsen and F. Barbaresco (Eds.): GSI 2015, LNCS 9389, pp. 159–167, 2015.
DOI: 10.1007/978-3-319-25040-3_18

enough vector fields. In the second part of this paper, we define abstract shape spaces in $\mathbb{R}^d$ as Banach manifolds on which the group of diffeomorphisms of $\mathbb{R}^d$ acts in a way that is compatible with its particular topological group structure. We then define the sub-Riemannian structure induced on $\mathcal{S}$ by this action and by V, and see that it admits a global Hamiltonian geodesic flow.

2   Sub-Riemannian Structures on Groups of Diffeomorphisms

The purpose of this section is to give a brief summary of the results of [4].

Fix $d \in \mathbb{N}$. For an integer $s > d/2+1$, let $\mathcal{D}^s(\mathbb{R}^d) = e + H^s(\mathbb{R}^d, \mathbb{R}^d) \cap \mathrm{Diff}(\mathbb{R}^d)$ be the connected component of $e = \mathrm{Id}_M$ in the space of diffeomorphisms of class H^s. It is an open subset of the affine Hilbert space $e + H^s(\mathbb{R}^d, \mathbb{R}^d)$, and therefore a Hilbert manifold. It is also a group for the composition $(\varphi, \psi) \mapsto \varphi \circ \psi$. This group law satisfies the following properties:

1. **Continuity:** $(\varphi, \psi) \mapsto \varphi \circ \psi$ is continuous.
2. **Smoothness on the left:** For every $\psi \in \mathcal{D}^s(\mathbb{R}^d)$, the mapping $R_\psi : \varphi \mapsto \varphi \circ \psi$ is smooth.
3. **Smoothness on the right:** For every $k \in \mathbb{N} \setminus \{0\}$, the mappings

$$\mathcal{D}^{s+k}(\mathbb{R}^d) \times \mathcal{D}^s(\mathbb{R}^d) \longrightarrow \mathcal{D}^s(\mathbb{R}^d) \quad H^{s+k}(\mathbb{R}^d, \mathbb{R}^d) \times \mathcal{D}^s(\mathbb{R}^d) \longrightarrow H^s(\mathbb{R}^d, \mathbb{R}^d)$$
$$(\varphi, \psi) \longmapsto \varphi \circ \psi \qquad\qquad\qquad (X, \psi) \longmapsto X \circ \psi$$

(1)

are of class C^k.
4. **Regularity:** For any $\varphi_0 \in \mathcal{D}^s(\mathbb{R}^d)$ and $X(\cdot) \in L^2(0, 1; H^s(\mathbb{R}^d, \mathbb{R}^d))$, there is a unique curve $\varphi(\cdot) \in H^1(0, 1; \mathcal{D}^s(\mathbb{R}^d))$ such that $\varphi(0) = \varphi_0$ and $\dot\varphi(t) = X(t) \circ \varphi(t)$ almost everywhere on $[0, 1]$.

See [9,12,15,16] for more on this structure.

Sub-Riemannian structures on $\mathcal{D}^s(\mathbb{R}^d)$.

Definition 1. *We define a strong right-invariant structure on $\mathcal{D}^s(\mathbb{R}^d)$ as follows: fix V an arbitrary Hilbert space of vector fields with Hilbert product $\langle \cdot, \cdot \rangle_V$ and norm $\| \cdot \|_V$ and continuous inclusion in $H^{s+k}(\mathbb{R}^d, \mathbb{R}^d)$, $k \in \mathbb{N} \setminus \{0\}$. The* ***sub-Riemannian structure*** *induced by V on $\mathcal{D}^s(\mathbb{R}^d)$ is the one for which horizontal curves satisfy $\dot\varphi(t) = X(t) \circ \varphi(t)$, with $X \in L^2(0, 1; V)$, and have total action $A(\varphi) = A(X) = \frac{1}{2} \int_0^1 \|X(t)\|_V^2 dt$.*

Define $K_V : V^* \to V$ the canonical isometry: for $P \in V^*$, $P = \langle K_V P, \cdot \rangle_V$.

Such a space V admits a reproducing kernel: a matrix-valued mapping $(x, y) \mapsto K(x, y)$ defined on $\mathbb{R}^d \times \mathbb{R}^d$ such that, for any $P \in H^s(\mathbb{R}^d, \mathbb{R}^d)^* = H^{-s}(\mathbb{R}^d, \mathbb{R}^{d*})$, the vector field $K_V P$ is given by convolution (in the distributional sense) of P with K:

$$K_V P(x) = \int_{\mathbb{R}^d} K(x, y) P(y) dy.$$

Geodesics on $\mathcal{D}^s(\mathbb{R}^d)$. We keep the framework and notations used in the previous section, with $V \hookrightarrow H^{s+k}(\mathbb{R}^d, \mathbb{R}^d)$ and $k \geqslant 1$. Define the *endpoint map* from e by end $: L^2(0,1;V) \to \mathcal{D}^s(\mathbb{R}^d)$ such that $\text{end}(X) = \varphi^X(1)$. It is of class $\mathcal{C}^k$. A *geodesic* $\varphi^X(\cdot)$ from e is a critical point of the action $A(X(\cdot))$ among all horizontal curves $\varphi^Y(\cdot)$ from e with the same endpoint $\varphi^Y(1) = \varphi_1$. In other words, for every $\mathcal{C}^1$ variation $a \in (-\varepsilon, \varepsilon) \mapsto X_a(\cdot) \in L^2(0,1;V)$ such that $\text{end}(X_a) = \varphi_1$, we have $\partial_a(A(X_a(\cdot)))_{|a=0} = 0$.

Normal geodesics. It is easy to see that for any such curve, the couple of linear maps

$$(\mathrm{d}A(X(\cdot)), \mathrm{d}\,\text{end}(X(\cdot))) : L^2(0,1;V) \to \mathbb{R} \times T_{\varphi_1}\mathcal{D}^s(\mathbb{R}^d)$$

is **not** onto. A sufficient condition for this to be true is that there exists $P_{\varphi_1} \in T^*_{\varphi_1}\mathcal{D}^s(\mathbb{R}^d) = H^{-s}(\mathbb{R}^d, \mathbb{R}^{d*})$ such that $(\mathrm{d}\,\text{end}(X(\cdot)))^*.P_{\varphi_1} = \mathrm{d}A(X(\cdot))$. If such a P_1 exists, the curve induced by X is called a *normal geodesic*. This is not the only possibility [1,3,14], but it is the one we will focus on, as it is enough for inexact matching problems.

Define the normal Hamiltonian $H : T^*\mathcal{D}^s(\mathbb{R}^d) \to \mathbb{R}$ by

$$H(\varphi, P) = \frac{1}{2}P(dR_\varphi K_V dR_\varphi^* P) = \frac{1}{2}\iint_{\mathbb{R}^d \times \mathbb{R}^d} P(x)K(\varphi(x), \varphi(y))P(y)\mathrm{d}y\mathrm{d}x,$$

with K_V the isometry $V^* \to V$ and $dR_\varphi(\cdot) = \cdot \circ \varphi$ on $H^s(\mathbb{R}^d, \mathbb{R}^d)$. H is of class at least $\mathcal{C}^k$. Its symplectic gradient $\nabla^\omega H(\varphi, P) = (\partial_P H(\varphi, P), -\partial_\varphi H(\varphi, P))$ is of class $\mathcal{C}^{k-1}$.

We have the following theorem.

Theorem 1. *If* $k \geqslant 1$, $\varphi(\cdot)$ *is a geodesic if and only if it is the projection to* $\mathcal{D}^s(\mathbb{R}^d)$ *of an integral curve of* $\nabla^\omega H(\varphi, P)$. *In this case, the corresponding* $P(\cdot)$ *is the associated* **normal covector**.

If $k \geqslant 2$, *then the symplectic gradient of the Hamiltonian admits a well-defined global flow of class* $\mathcal{C}^{k-1}$, *called the Hamiltonian geodesic flow. In other words, for every* $(\varphi_0, P_0) \in T^*\mathcal{D}^s(\mathbb{R}^d)$, *there is a unique solution* $(\varphi(\cdot), P(\cdot)) :$ $\mathbb{R} \to T^*\mathcal{D}^s(\mathbb{R}^d)$ *to the Cauchy problem* $(\varphi(0), P(0)) = (\varphi_0, P_0)$, $(\dot{\varphi}(t), \dot{P}(t)) = (\partial_P H(\varphi(t), P(t)), -\partial_\varphi H(\varphi(t), P(t)))$ *a.e.* $t \in [0,1]$. *Moreover, any subarc of this solution projects to a normal geodesic on* $\mathcal{D}^s(\mathbb{R}^d)$ *and, conversely, any normal geodesic is the projection of such a solution.*

Momentum formulation. We define the *momentum map* $\mu : T^*\mathcal{D}^s(\mathbb{R}^d) \to H^s(\mathbb{R}^d)^* = H^{-s}(\mathbb{R}^{d*})$ by $\mu(\varphi, P) = dR_\varphi^* P$.

Proposition 1. *We assume that* $k \geqslant 1$. *Then a horizontal curve* $\varphi(\cdot) \in H^1(0,1;\mathcal{D}^s(\mathbb{R}^d))$, *flow of* $X(\cdot) \in L^2(0,1;V)$, *is a normal geodesic with normal covector* $P(\cdot)$ *if and only if the corresponding momentum* $\mu(t) = \mu(\varphi(t), P(t))$ *along the curve satisfies, for almost every time* t, $\dot{\mu}(t) = \text{ad}^*_{X(t)}\mu(t) = -\mathcal{L}_{X(t)}\mu(t)$. *Here,* $\text{ad}_X : H^{s+1}(\mathbb{R}^d, \mathbb{R}^d) \to H^s(\mathbb{R}^d, \mathbb{R}^d)$, *with*

$\mathrm{ad}_X Y = [X, Y]$, and $\mathcal{L}_X$ the Lie derivative with respect to X. In particular, this equation integrates as

$$\mu(t) = \varphi(t)_* \mu(0),$$

for every $t \in [0, 1]$ in the sense of distributions.

We recognize the usual EPDiff equations [6,7,13]. See [2,4] for further results on such sub-Riemannian structures on $\mathcal{D}^s(\mathbb{R}^d)$, and [3,11] for more general infinite dimensional sub-Riemannian structures.

3 Shape Spaces

3.1 Definition

Throughout the section, fix a positive integer d and let s_0 be the smallest integer such that $s_0 > d/2$. A shape space in $\mathbb{R}^d$ is a Banach manifold acted upon by $\mathcal{D}^s(\mathbb{R}^d)$ for some s in a way that is compatible with its particular topological group structure. The following definition is adapted from that of [5].

Definition 2. Let $\mathcal{S}$ be a Banach manifold and $\ell \in \mathbb{N} \setminus \{0\}$, and $s = s_0 + \ell$. Assume that $\mathcal{D}^s(\mathbb{R}^d)$ acts on $\mathcal{S}$, according to the action $(\varphi, q) \mapsto \varphi \cdot q = R_q(\varphi)$. We say that $\mathcal{S}$ is a **shape space** of order ℓ in M if the following conditions are satisfied:

1. **Continuity:** $(\varphi, q) \mapsto \varphi \cdot q$ is continuous.
2. **Smoothness on the left:** For every $q \in \mathcal{S}$, the mapping $R_q : \varphi \mapsto \varphi \cdot q$ is smooth. Its differential at e is denoted ξ_q, and is called the **infinitesimal action** of $H^s(\mathbb{R}^d, \mathbb{R}^d)$.
3. **Smoothness on the right:** For every $k \in \mathbb{N}$, the mappings

$$\begin{aligned} R_q : \mathcal{D}^{s+k}(\mathbb{R}^d) \times \mathcal{S} &\longrightarrow \mathcal{S} & \text{and} \quad \xi : H^{s+k}(\mathbb{R}^d, \mathbb{R}^d) \times \mathcal{S} &\longrightarrow T\mathcal{S} \\ (\varphi, q) &\longmapsto \varphi \cdot q & (X, q) &\longmapsto \xi_q X \end{aligned} \quad (2)$$

are of class $\mathcal{C}^k$.
4. **Regularity:** For every $X(\cdot) \in L^2(0, 1; H^s(\mathbb{R}^d, \mathbb{R}^d))$ and $q_0 \in \mathcal{S}$, there exists a unique curve $q(\cdot) = q^X(\cdot) \in H^1(0, 1; \mathcal{S})$ such that $q^X(0) = q_0$ and $\dot{q}^X(t) = \xi_{q^X(t)} X(t)$ for almost every t in $[0, 1]$.

A an element q of $\mathcal{S}$ is called a **state** of the shape. We say that $q \in \mathcal{S}$ has **compact support** if there exists a compact subset U of M such that $R_q : \varphi \mapsto \varphi \cdot q$ is continuous with respect to the semi-norm $\| \cdot \|_{H^{s_0+\ell}(U,M)}$ on $\mathcal{D}^s(\mathbb{R}^d)$. In other words, q has a compact support if $\varphi \cdot q$ depends only on the restriction of φ to a compact subset U of M.

Here are some examples of some of the most widely used shape spaces:

1. $\mathcal{D}^{s_0+\ell}(\mathbb{R}^d)$ is a shape space of order ℓ for its action on itself given by composition on the left.

2. Let S be a smooth compact Riemannian manifold, and α_0 be the smallest integer greater than $\dim(S)/2$. Then $S = \mathrm{Emb}^{\alpha_0+\ell}(S, \mathbb{R}^d)$, the manifold of all embeddings $q : S \to M$ of Sobolev class $H^{\alpha_0+\ell}$ are shape spaces of order ℓ. In this case, $\mathcal{D}^{s_0+\ell}(\mathbb{R}^d)$ acts on S by left composition $\varphi \cdot q = \varphi \circ q$, and this action satisfies all the required properties of Definition 2 (see [5] for the proof), with infinitesimal action $\xi_q X = X \circ q$. Every $q \in S$ has compact support.
3. A particularly interesting case is obtained when $\dim(S) = 0$. Then $S = \{s_1, \ldots, s_n\}$ is simply a finite set. In that case, for any ℓ, the shape space $S = \mathcal{C}^\ell(S, \mathbb{R}^d)$ is identified with the space of n landmarks in $\mathbb{R}^d$:

$$\mathrm{Lmk}_n(\mathbb{R}^d) = \{(x_1, \ldots, x_n) \in (\mathbb{R}^d)^n \mid x_i \neq x_j \text{ if } i \neq j\}.$$

For every $q = (x_1, \ldots, x_n)$, the action of $\mathcal{D}^{s_0+1}(\mathbb{R}^d)$ is given by $\varphi \cdot q = (\varphi(x_1), \ldots, \varphi(x_n))$. For a vector field X of class H^{s_0+1} on M, the infinitesimal action of X at q is given by $\xi_q(X) = (X(x_1), \ldots, X(x_n))$. Spaces of landmarks are actually spaces of order 0 (see [5] for a definition).
4. Let S be a shape space of order $\ell \in \mathbb{N}$. Then TS is a shape space of order $\ell+1$, with the action of $\mathcal{D}^{s_0+\ell+1}(\mathbb{R}^d)$ on TS_1 defined by $\varphi \cdot (q, v) = (\varphi \cdot q, \partial_q(\varphi \cdot q)(v))$.

3.2 Sub-Riemannian Structure on Shape Spaces

Let S be a shape space of order $\ell \geqslant 1$ in $\mathbb{R}^d$, and fix $s = s_0 + \ell$ and $k \in \mathbb{N} \setminus \{0\}$. Consider $(V, \langle \cdot, \cdot \rangle)$ an arbitrary Hilbert space of vector fields with continuous inclusion in $H^{s+k}(\mathbb{R}^d, \mathbb{R}^d)$. According to the previous section, we obtain a strong right-invariant sub-Riemannian structure induced by V on $\mathcal{D}^s(\mathbb{R}^d)$.

The framework of shape and image matching. The classical LDDMM algorithms for exact shape matching seek to minimize

$$\frac{1}{2} \int_0^1 \langle X(t), X(t) \rangle dt$$

over every $X \in L^2(0, 1; V)$ such that $\varphi^X(1) \cdot q_0 = q_1$, where the template q_0 and the target q_1 are fixed. Usually, one only wants to get "close" to the target shape, which is accomplished by minimizing

$$\frac{1}{2} \int_0^1 \langle X(t), X(t) \rangle dt + g(\varphi^X(1) \cdot q_0)$$

over every $X \in L^2(0, 1; V)$, where the endpoint constraint has been replaced with the addition of a data attachment term $g(\varphi^X(1) \cdot q_0)$ in the functional (See [5] and references therein). The function g is usually such that it reaches its minimum at q_1.

The sub-Riemannian structure. This leads us to define a sub-Riemannian structure on S as follows.

Definition 3. *The strong sub-Riemannian structure induced by V is the one for which horizontal curves are those that satisfy $\dot{q}(t) = \xi_{q(t)}X(t)$ for almost every $t \in [0,1]$, for some **control** $X(\cdot) \in L^2(0,1;\mathbb{R}^d)$. The curve $q(\cdot)$ is called a **horizontal deformation** of $q(0)$. Note that $q(t) = \varphi^X(t) \cdot q(0)$ for every t.*

Remark 1. If $\xi_q(V) = T_q\mathcal{S}$ for every $q \in \mathcal{S}$, this is actually a Riemannian structure. This is often the case in numerical simulations, where $\mathcal{S}$ is finite dimensional (usually a space of landmarks). However, in the general case, we *do not* obtain a Riemannian structure.

For example, for $d = 2$, take $\mathcal{S} = Emb^2(S^1, \mathbb{R}^2)$, with S^1 the unit circle, and fix the state $q = \mathrm{Id}_{S^1} \in \mathcal{S}$. If the kernel $K(x,y) = e^{-\|x-y\|^2}$ is Gaussian, all elements of V are analytic. Therefore, any $\xi_q(X) : S^1 \to \mathbb{R}^2$ with $X \in V$ is analytic, while $T_q\mathcal{S} = H^2(S^1, \mathbb{R}^2)$.

The *length* and *action* of a horizontal curve is not uniquely defined and depends on the control $X(\cdot)$. They coincide with the length and action of the flow φ^X which were defined in the previous section. The LDDMM algorithm can therefore be formulated as a search for sub-Riemannian geodesics on $\mathcal{S}$ for this structure.

Sub-Riemannian distance. Define the *sub-Riemannian distance* $d_{SR}^{\mathcal{S}}(q_0, q_1)$ as the infimum over the lengths of every horizontal system $(q(\cdot), X(\cdot))$ with $q(0) = q_0$ and $q(1) = q_1$. It is clear that $d_{SR}^{\mathcal{S}}$ is at least a semi-distance.

Sub-Riemannian geodesics on shape spaces. We assume that $\mathcal{S}$ is a shape space in $\mathbb{R}^d$ of order $\ell \geqslant 1$, and that $\mathcal{D}^s(\mathbb{R}^d)$, $s = s_0 + \ell$, is equipped with a strong right-invariant sub-Riemannian structure induced by the Hilbert space $(V, \langle \cdot, \cdot \rangle)$ of vector fields on $\mathbb{R}^d$, with continuous inclusion $V \hookrightarrow H^{s+k}(\mathbb{R}^d, \mathbb{R}^d)$ for some $k \geqslant 1$.

Geodesics. Fix an initial point q_0 and a final point q_1 in $\mathcal{S}$. The endpoint mapping from q_0 is $\mathrm{end}_{q_0}^{\mathcal{S}}(X(\cdot)) = \varphi^X(1) \cdot q_0 = R_{q_0} \circ \mathrm{end}$, where $\mathrm{end}(X(\cdot)) = \varphi^X(1) \in \mathcal{D}^s(\mathbb{R}^d)$. It is of class $\mathcal{C}^k$. A *geodesic* on $\mathcal{S}$ between the states q_0 and q_1 is a horizontal system $(q(\cdot), X(\cdot))$ joining q_0 and q_1 such that for any $\mathcal{C}^1$-family $a \mapsto X_a(\cdot) \in L^2(0,1;V)$ with $\varphi^{X_a}(1) \cdot q_0 = q_1$ for every a and $X_0 = X$, we have $\partial_a A(X_a(\cdot)) = 0$. We will, once again, focus on *normal geodesics*. A curve $q(\cdot)$ is a normal geodesic if for some control X whose flow φ^X yields $q(\cdot) = \varphi^X(\cdot) \circ q(0)$, and for some $p_1 \in T_{q(1)}^*\mathcal{S}$, we have $\mathrm{d}A(X) = \mathrm{d}\,\mathrm{end}_{q_0}^{\mathcal{S}}(X)^*p_1$.

Canonical symplectic form, symplectic gradient. We denote by ω the canonical weak symplectic form on $T^*\mathcal{S}$, given by the formula $\omega(q,p).(\delta q_1, \delta p_1; \delta q_2, \delta p_2) = \delta p_2(\delta q_1) - \delta p_1(\delta q_2)$, with $(\delta q_i, \delta p_i) \in T_{(q,p)}T^*\mathcal{S} \simeq T_q\mathcal{S} \times T_q^*\mathcal{S}$ in a canonical coordinate system (q,p) on $T^*\mathcal{S}$. A function $f : T^*\mathcal{S} \to \mathbb{R}$, differentiable at some point $(q,p) \in T^*\mathcal{S}$, admits a *symplectic gradient* at (q,p) if there exist a vector $\nabla^\omega f(q,p) \in T_{(q,p)}T^*\mathcal{S}$ such that, for every $z \in T_{(q,p)}T^*\mathcal{S}$, $df_{(q,p)}(z) = \omega(\nabla^\omega f(q,p), z)$. In this case, this symplectic gradient $\nabla^\omega f(q,p)$ is unique. Such a gradient exists if and only if $\partial_p f(q,p) \in T_q^{**}\mathcal{S}$ can be identified with a vector in $T_q\mathcal{S}$ through the canonical inclusion $T_q\mathcal{S} \hookrightarrow T_q^{**}\mathcal{S}$. In that case, we have, in canonical coordinates, $\nabla^\omega f(q,p) = (\partial_p f(q,p), -\partial_q f(q,p))$.

The normal Hamiltonian function and geodesic equation. We define the *normal Hamiltonian* of the system $H^S : T^*S \to \mathbb{R}$ by

$$H^S(q,p) = \frac{1}{2}p(K_q p) = \frac{1}{2}p(\xi_q K_V \xi_q^* p),$$

where $K_q = \xi_q K_V \xi_q^* : T_q^*S \to T_qS$. It can usually be computed thanks to the reproducing kernel of V. This is a function of class C^k, that admits as symplectic gradient $\nabla^\omega H^S(q,p) = (K_q p, -\frac{1}{2}\partial_q(K_q p)^* p)$. of class C^{k-1} on T^*S.

Momentum of the action and Hamiltonian flow. Recall that the *momentum map* associated to the group action of $\mathcal{D}^s(\mathbb{R}^d)$ over S is the mapping $\mu^S : T^*S \to H^s(\mathbb{R}^d, \mathbb{R}^d)^* = H^{-s}(\mathbb{R}^d, \mathbb{R}^{d*})$ given by $\mu^S(q,p) = \xi_q^* p$.

Proposition 2. *A curve $(q(\cdot), p(\cdot))$ in T^*S satisfies the normal Hamiltonian equations $(\dot{q}(t), \dot{p}(t)) = \nabla^\omega H^S(q(t), p(t))$ if and only if, for $\mu^S(t) = \mu^S(q(t), p(t))$ and $X(t) = K_V \xi_{q(t)}^* p(t)$, we have*

$$\dot{\mu}^S(t) = \mathrm{ad}_{X(t)}^* \mu(t).$$

In particular, this is also equivalent to having $\varphi^X(\cdot)$ be a normal geodesic on $\mathcal{D}^s(\mathbb{R}^d)$ with initial covector $P(0) = \xi_{q(0)}^* p(0)$ and momentum $\mu(t) = \mu^S(t)$ along the trajectory.

This result allows for the proof of our main result.

Theorem 2. *Assume $k \geqslant 1$. Then a horizontal curve $q(\cdot)$ with control $X(\cdot)$ is a geodesic if and only if it is the projection of an integral curve $(q(\cdot), p(\cdot))$ of $\nabla^\omega H^S$ (that is, $(\dot{q}(t), \dot{p}(t)) = \nabla^\omega H^S(q(t), p(t))$ for almost every t in $[0,1]$), with $X(t) = K_V \xi_{q(t)}^* p(t)$. This is also equivalent to having the flow φ^X of $X(\cdot)$ be a normal geodesic on $\mathcal{D}^s(\mathbb{R}^d)$ with momentum $\mu(t) = \mu^S(t) = \xi_{q(t)}^* p(t)$.*

Assume $k \geqslant 2$. Then $\nabla^\omega H$ admits a **global flow** *on T^*S of class C^{k-1}, called the* **Hamiltonian geodesic flow.** *In other words, for any initial point $(q_0, p_0) \in T^*S$, there exists a unique curve $t \mapsto (q(t), p(t))$ defined on all of $\mathbb{R}$, such that $(q(0), p(0)) = (q_0, p_0)$ and, for almost every t, $(\dot{q}(t), \dot{p}(t)) = \nabla^\omega H^S(q(t), p(t))$. We say that $p(\cdot)$ is the normal covector along the trajectoy.*

Combining those results, we see that solutions of the normal Hamiltonian equations on S are exactly those curves that come from normal geodesics on $\mathcal{D}^s(\mathbb{R}^d)$ with initial momentum of the form $\xi_{q_0}^* p_0$. In particular, for $k \geqslant 2$, the completeness of the normal geodesic flow on $T^*\mathcal{D}^s(\mathbb{R}^d)$ implies that $\nabla^\omega H^S$ is a complete vector field on T^*S.

On inexact matching. It should be emphasized, again, that other geodesics may also exist [4]. However, when performing LDDMM methods and algorithms for inexact matching, one aims to minimize over $L^2(0, 1; V)$ functionals of the form

$$J(X(\cdot)) = A(X(\cdot)) + g(q^X(1)) = A(X(\cdot)) + g \circ \mathrm{end}_{q_0}^S(X(\cdot)).$$

In this case, $X(\cdot)$ is a critical point if and only if $dA(X) = -d\,\mathrm{end}_{q_0}^{\mathcal{S}}(X)^*dg(q^X(1))$. The trajectory induced by such a critical point X is therefore automatically a normal geodesic, whose covector satisfies $p(1) = -dg(q^X(1))$ (or, equivalently, whose momentum satisfies $\mu(1) = -\xi^*_{q(1)}dg(q^X(1))$). This means that one needs only consider normal geodesics when looking for minimizers of J. Consequently, the search for minimizing trajectories can be reduced to the minimization of

$$\frac{1}{2}\int_0^1 p(t)(K_{q(t)}p(t))dt + g(q(1))$$

among all solutions of the control system $\dot{q}(t) = K_{q(t)}p(t)$, where $p(\cdot)$ is any covector along $q(\cdot)$ and is L^2 in time. This leads to the usual LDDMM methods.

This reduction is very useful in practical applications and numerical simulations, since, when $\mathcal{S}$ is finite dimensional, we obtain a finite dimensional control system, for which many optimization methods are available. See [5] for algorithms to minimize such a functional in the abstract framework of shape spaces in $\mathbb{R}^d$.

The case of images. Images are elements I of the functional space $L^2(\mathbb{R}^d, \mathbb{R})$. They are acted upon by $\mathcal{D}^s(\mathbb{R}^d)$ through $(\varphi, I) \mapsto I \circ \varphi^{-1}$. For a fixed template I_0 and target I_1, one aims to minimize a functional of the form

$$J(X(\cdot)) = A(X(\cdot)) + g\left(I(1)^{-1}\right),$$

with $g(I) = c\|I - I_1\|^2_{L^2}$, $c > 0$ fixed, and $I(t) = I_0 \circ \varphi(t)^{-1}$.

However, the action $(\varphi, I) \mapsto I \circ \varphi^{-1}$ *does not* make $L^2(\mathbb{R}^d, \mathbb{R})$ into a shape space, because it is not continuous. To circumvent this difficulty and still apply the framework developed in this paper, one can simply work on the shape space $\mathcal{D}^s(\mathbb{R}^d)$ itself. In this case, as long as the template I_0 belongs to $\mathcal{C}^1(\mathbb{R}^d, \mathbb{R})$, we can easily check that $\varphi \mapsto g\left(I_0 \circ \varphi^{-1}\right)$ is of class $\mathcal{C}^1$, which implies, according to the results of this section and a quick computation, that minimizers of J are those vector fields whose flow are normal geodesics with final momentum given by

$$\mu(1) = (I_1 - I(1))\,dI(1) \in L^2(\mathbb{R}^d, \mathbb{R}^{d*}).$$

References

1. Agrachev, A.A., Sachkov, Y.L.: Control Theory from the Geometric Viewpoint. Encyclopaedia of Mathematical Sciences, vol. 87. Springer, Heidelberg (2004)
2. Agrachev, A.A., Caponigro, M.: Controllability on the group of diffeomorphisms. Ann. Inst. H. Poincaré Anal. Non Linéaire **26**(6), 2503–2509 (2009)
3. Arguillère, S.: Infinite dimensional sub-Riemannian geometry and applications to shape analysis. Ph.D. thesis (2014)
4. Arguillère, S., Trélat, E.: Sub-Riemannian structures on groups of diffeomorphisms. Preprint http://arxiv.org/abs/1409.8378 (2014)

5. Arguillère, S., Trélat, E., Trouvé, A., Younes, L.: Shape deformation analysis from the optimal control viewpoint. J. Math. Pures et Appliquées (2014). arxiv.org/abs/1401.0661

6. Arnold, V.: Sur la géométrie différentielle des groupes de Lie de dimension infinie et ses applications à l'hydrodynamique des fluides parfaits. Ann. Inst, Fourier (Grenoble), 16(1), 319–361 (1966)

7. Bauer, M., Bruveris, M., Harms, P., Michor, P.W.: Geodesic distance for right invariant sobolev metrics of fractional order on the diffeomorphism group. Ann. Glob. Anal. Geom. 44(1), 5–21 (2013)

8. Dupuis, P., Grenander, U., Miller, M.I.: Variational problems on flows of diffeomorphisms for image matching. Quart. Appl. Math. 56(3), 587–600 (1998)

9. Ebin, D.G., Marsden, J.: Groups of diffeomorphisms and the motion of an incompressible fluid. Ann. Math. 2(92), 102–163 (1970)

10. Grenander, U., Miller, M.I.: Computational anatomy: an emerging discipline. Quart. Appl. Math. 56(4), 617–694 (1998). Current and future challenges in the applications of mathematics, Providence, RI (1997)

11. Grong, E., Markina, I., Vasil'ev, A.: Sub-Riemannian geometry on infinite-dimensional manifolds. Preprint arxiv.org/abs/1201.2251 (2012)

12. Kriegl, A., Michor, P.W.: The Convenient Setting of Global Analysis. Mathematical Surveys and Monographs, vol. 53. American Mathematical Society, Providence (1997)

13. Michor, P.W., Mumford, D.: An overview of the Riemannian metrics on spaces of curves using the Hamiltonian approach. Appl. Comput. Harmon. Anal. 23(1), 74–113 (2007)

14. Montgomery, R.: A Tour of Subriemannian Geometries, Their Geodesics and Applications. Mathematical Surveys and Monographs, vol. 91. American Mathematical Society, Providence (2002)

15. Omori, H.: Infinite Dimensional Lie Transformation Groups. Lecture Notes in Mathematics, vol. 427. Springer, Berlin (1974)

16. Schmid, R.: Infinite dimensional Lie groups with applications to mathematical physics. J. Geom. Symmetry Phys. 1, 54–120 (2004)

17. Trouvé, A.: Action de groupe de dimension infinie et reconnaissance de formes. C. R. Acad. Sci. Paris Sér. I Math. 321(8), 1031–1034 (1995)

18. Trouvé, A.: Diffeomorphism groups and pattern matching in image analysis. Int. J. Comput. Vis. 37(1), 17 (2005)

19. Trouvé, A., Younes, L.: Local geometry of deformable templates. SIAM J. Math. Anal. 37(1), 17–59 (2005). (Electronic)

20. Trouvé, A., Younes, L.: Shape spaces. In: Scherzer, O. (ed.) Handbook of Mathematical Methods in Imaging, pp. 1309–1362. Springer, New York (2011)

21. Younes, L.: Shapes and Diffeomorphisms. Applied Mathematical Sciences, vol. 171. Springer, Berlin (2010)

Random Geometry/Homology

The Extremal Index for a Random Tessellation

Nicolas Chenavier[✉]

Université du Littoral Côte D'Opale, 50 Rue Ferdinand Buisson, Calais, France
nicolas.chenavier@lmpa.univ-littoral.fr

Abstract. Let m be a random tessellation in $\mathbf{R}^d$, $d \geq 1$, observed in the window $\mathbf{W}_\rho = \rho^{1/d}[0,1]^d$, $\rho > 0$, and let f be a geometrical characteristic. We investigate the asymptotic behaviour of the maximum of $f(C)$ over all cells $C \in m$ with nucleus in $\mathbf{W}_\rho$ as ρ goes to infinity. When the normalized maximum converges, we show that its asymptotic distribution depends on the so-called extremal index. Two examples of extremal indices are provided for Poisson-Voronoi and Poisson-Delaunay tessellations.

Keywords: Random tessellations · Extreme values · Poisson point process

1   Introduction

Random tessellations A (convex) tessellation of $\mathbf{R}^d$, $d \geq 1$, endowed with its Euclidean norm $|\cdot|$, is a countable collection of nonempty convex compact subsets, called *cells*, with disjoint interiors which subdivides the space and such that the number of cells intersecting any bounded subset of $\mathbf{R}^d$ is finite. The set $\mathbf{T}$ of (convex) tessellations is endowed with the σ-algebra generated by the sets $\left\{ m \in \mathbf{T}, \bigcup_{C \in m} \partial C \cap K = \emptyset \right\}$ where ∂K is the boundary of K for any compact set K in $\mathbf{R}^d$. By a random tessellation m, we mean a random variable with values in $\mathbf{T}$. It is said to be stationary if its distribution is invariant under translations of the cells. For a complete account on random tessellations, we refer to the book [10].

Given a fixed realization of m, we associate with each cell $C \in m$ a point $z(C) = z_m(C)$ in a deterministic way, which is called the *nucleus* of the cell, such that $z_{m+x}(C + x) = z_m(C) + x$ for all $x \in \mathbf{R}^d$. To describe the mean behaviour of the tessellation, the notions of intensity and typical cell are introduced as follows. Let $B \subset \mathbf{R}^d$ be a Borel subset. The *intensity* γ of the tessellation is defined as $\gamma = \frac{1}{\lambda_d(B)} \cdot \mathbf{E}\left[\#\{C \in m, z(C) \in B\} \right]$, where λ_d is the d-dimensional Lebesgue measure. We assume that $\gamma \in (0, \infty)$ and, without loss of generality, we take $\gamma = 1$. The *typical cell* $\mathcal{C}$ is a random polytope whose the distribution is given by

$$\mathbf{E}[f(\mathcal{C})] = \frac{1}{\lambda_d(B)} \cdot \mathbf{E}\left[\sum_{\substack{C \in m, \\ z(C) \in B}} f(C - z(C)) \right] \tag{1}$$

© Springer International Publishing Switzerland 2015
F. Nielsen and F. Barbaresco (Eds.): GSI 2015, LNCS 9389, pp. 171–178, 2015.
DOI: 10.1007/978-3-319-25040-3_19

for all $f : \mathcal{K}_d \to \mathbf{R}$ bounded measurable function on the set of convex bodies $\mathcal{K}_d$, i.e. convex compact sets, endowed with the Hausdorff topology.

Extremes in stochastic geometry We are interested in the following problem: only a part of the tessellation is observed in the window $\mathbf{W}_\rho = \rho^{1/d}[0,1]^d$. Let $f : \mathcal{K}_d \to \mathbf{R}$ be a translation invariant measurable function, i.e. $f(C+x) = f(C)$ for all $C \in \mathcal{K}_d$ and $x \in \mathbf{R}^d$. We denote by $M_{f,\mathbf{W}_\rho}$ the maximum of $f(C)$ over all cells $C \in m$ with nucleus $z(C)$ in $\mathbf{W}_\rho$, i.e.

$$M_{f,\mathbf{W}_\rho} = \max_{\substack{C \in m, \\ z(C) \in \mathbf{W}_\rho}} f(C).$$

In this paper, we investigate the limit behaviour of $M_{f,\mathbf{W}_\rho}$ when ρ goes to infinity.

To the best of our knowledge, one of the first application of extreme value theory in stochastic geometry was given by Penrose (see Chaps. 6, 7 and 8 in Penrose [7]). More recently, Schulte and Thäle [11] established a theorem to derive the order statistics of a general functional, $f_k(x_1, ..., x_k)$ of k points of a homogeneous Poisson point process. Calka and Chenavier [2] went on to provide a series of results for the extremal properties of cells in the Poisson-Voronoi tessellation, which were then extended in [3]. Besides, extremes for the inradius of a Poisson line tessellation are also considered in [4].

A general theorem Before stating our results, we first recall the main theorem in [3]. To do it, we consider a threshold v_ρ such that the mean number of exceedance cells converges to a limit denoted by $\tau \geq 0$, i.e.

$$\rho \cdot \mathbf{P}\left(f(\mathcal{C}) > v_\rho\right) \xrightarrow[\rho \to \infty]{} \tau.$$

Such an assumption is classical in extreme value theory. We also introduce two conditions on m and f.

The first one deals with R-dependence. To introduce this condition we partition $\mathbf{W}_\rho = \rho^{1/d}[0,1]^d$ into a set V_ρ of $N_\rho = \left\lfloor \frac{\rho}{\log \rho} \right\rfloor$ sub-cubes of equal size. These sub-cubes are indexed by the set of $\mathbf{i} = (i_1, \ldots, i_d) \in \left[1, N_\rho^{1/d}\right]^d$. With a slight abuse of notation, we identify a cube with its index. Let us define a distance between sub-cubes $\mathbf{i}$ and $\mathbf{j}$ as $d(\mathbf{i},\mathbf{j}) = \max_{1 \leq r \leq d}\{|i_r - j_r|\}$. Moreover, if A, B are two sets of sub-cubes, we let $\delta(A,B) = \min_{\mathbf{i} \in A, \mathbf{j} \in B} d(\mathbf{i},\mathbf{j})$. For each $\mathbf{i} \in V_\rho$, we denote by

$$M_{f,\mathbf{i}} = \max_{\substack{C \in m, \\ z(C) \in \mathbf{i} \cap \mathbf{W}_\rho}} f(C).$$

When $\{C \in m, z(C) \in \mathbf{i} \cap \mathbf{W}_\rho\}$ is empty, we take $M_{f,\mathbf{i}} = -\infty$. We are now prepared to introduce our first condition which is referred as the finite range condition (FRC):

CONDITION (FRC): *there exists an integer R and an event A_ρ with $\mathbf{P}(A_\rho) \xrightarrow[\rho \to \infty]{} 1$ such that, conditional on A_ρ, the σ-algebras $\sigma\{M_{f,\mathbf{i}}, \mathbf{i} \in A\}$ and $\sigma\{M_{f,\mathbf{i}}, \mathbf{i} \in B\}$ are independent when $\delta(A,B) > R$.*

Our second condition deals with a local property of m and f and is referred as the local correlation condition (LCC).

CONDITION (LCC): *with the same notation as before, we have*

$$N_\rho \mathbf{E} \left[\sum_{\substack{(C_1,C_2)_{\neq} \in m^2, \\ z(C_1), z(C_2) \in \mathbf{W}_{\log \rho}}} \mathbb{1}_{f(C_1)>v_\rho, f(C_2)>v_\rho} \right] \xrightarrow[\rho \to \infty]{} 0,$$

where $(C_1, C_2)_{\neq} \in m^2$ means that (C_1, C_2) is a pair of distinct cells of m.

This (local) condition means that, with high probability, two neighboring cells (in the sense that their nuclei belong to $\mathbf{W}_{\log \rho}$ which is small compared to $\mathbf{W}_\rho$) are not simultaneously exceedances. Under these assumptions, we have the following result (Theorem 1 in [3]):

Theorem 1. *Let m be a stationary random tessellation of intensity 1 such that* CONDITIONS (FRC) *and* (LCC) *hold. Then*

$$\mathbb{P}(M_{f,\mathbf{W}_\rho} \leq v_\rho) \xrightarrow[\rho \to \infty]{} e^{-\tau}.$$

Theorem 1 can be extended in various directions: order statistics with a rate of convergence, random tessellations satisfying some β-mixing property and marked point processes. Numerous examples of this theorem can be derived such as the minimum of the Voronoi flowers, the maximum and minimum of areas of a planar Poisson-Delaunay tessellation and the maximum of inradius of a Gauss-Poisson Voronoi tessellation (see Sects. 3, 4 and 5 in [3]).

The main difficulty is to apply Theorem 1 and to check CONDITION (LCC) since it requires delicate geometric estimates. Our main question is: does Theorem 1 remains true when this condition does not hold?

Extremal index and new results When CONDITION (LCC) does not hold, the exceedance locations can be divided into clusters. More precisely, we show that the behaviour of $M_{f,\mathbf{W}_\rho}$ can be deduced up to a constant according to the following new result:

Proposition 2. *Let m be a stationary random tessellation of intensity 1 such that* CONDITION (FRC) *holds. Let us assume that for all $\tau \geq 0$, there exists a deterministic function $v_\rho(\tau)$ depending on ρ such that $\rho \cdot \mathbf{P}(f(\mathcal{C}) > v_\rho(\tau))$ converges to τ as ρ goes to infinity. Then there exist constants $\theta, \theta', 0 \leq \theta \leq \theta' \leq 1$ such that, for all $\tau \geq 0$,*

$$\limsup_{\rho \to \infty} \mathbf{P}\left(M_{f,\mathbf{W}_\rho} \leq v_\rho(\tau) \right) = e^{-\theta\tau} \quad \text{and} \quad \liminf_{\rho \to \infty} \mathbf{P}\left(M_{f,\mathbf{W}_\rho} \leq v_\rho(\tau) \right) = e^{-\theta'\tau}.$$

In particular, if $\mathbf{P}\left(M_{f,\mathbf{W}_\rho} \leq v_\rho(\tau) \right)$ converges, then $\theta = \theta'$ and

$$\mathbf{P}\left(M_{f,\mathbf{W}_\rho} \leq v_\rho(\tau) \right) \xrightarrow[\rho \to \infty]{} e^{-\theta\tau}.$$

Proposition 2 is similar to a result due to Leadbetter for stationary sequences of real random variables (see Theorem 2.2 in [5]). Its proof relies notably on the adaptation to our setting of several arguments included in [5]. According to Leadbetter, we say that the random tessellation m has *extremal index* θ if, for each $\tau \geq 0$, we have simultaneously $\rho \cdot \mathbf{P}\left(f(\mathcal{C}) > v_\rho(\tau)\right) \xrightarrow[\rho \to \infty]{} \tau$ and $\mathbf{P}\left(M_{f,\mathbf{W}_\rho} \leq v_\rho(\tau)\right) \xrightarrow[\rho \to \infty]{} e^{-\theta\tau}$.

For a sequence of real random variables, the extremal index can be interpreted as the reciprocal of the mean cluster size. Except in specific cases, the extremal index cannot be made explicit. A lot of inferences was considered to estimate this parameter (e.g. [8,12]). For a random tessellation, we think that the extremal index has a similar geometric interpretation. In a future work, we hope to develop a general method to estimate the extremal index.

The paper is organized as follows. In Sect. 2, we prove Proposition 2. As an illustration, we also provide two examples of extremal indices in Sect. 3.

2   Proof of Proposition 2

We only prove Proposition 2 for the limit superior since the limit inferior can be dealt with a similar method. To do it, for each $\tau \geq 0$, we define:

$$\psi(\tau) = \limsup_{\rho \to \infty} \mathbf{P}\left(M_{f,\mathbf{W}_\rho} \leq v_\rho(\tau)\right). \tag{2}$$

The key idea is to establish a functional equation for ψ. More precisely, for each $k \in \mathbf{N}_+$, we will show that:

$$\limsup_{\rho \to \infty} \mathbf{P}\left(M_{f,\mathbf{W}_{\rho/k^d}} \leq v_\rho(\tau)\right) = \psi(\tau/k^d), \tag{3}$$

$$\limsup_{\rho \to \infty} \mathbf{P}\left(M_{f,\mathbf{W}_{\rho/k^d}} \leq v_\rho(\tau)\right) = \psi^{1/k^d}(\tau). \tag{4}$$

The first convergence only depends on the sequence $v_\rho(\tau)$ while the second one is a consequence of CONDITION (FRC).

Proof of (3). Let us assume that $v_\rho(\tau) \geq v_{\rho/k^d}(\tau/k^d)$. Then

$$\left| \mathbf{P}\left(M_{f,\mathbf{W}_{\rho/k^d}} \leq v_\rho(\tau)\right) - \mathbf{P}\left(M_{f,\mathbf{W}_{\rho/k^d}} \leq v_{\rho/k^d}(\tau/k^d)\right) \right|$$

$$\leq \mathbf{P}\left(\bigcup_{\substack{C \in m, \\ z(C) \in \mathbf{W}_{\rho/k^d}}} \{v_{\rho/k^d}(\tau/k^d) \leq f(C) \leq v_\rho(\tau)\} \right)$$

$$\leq \mathbf{E}\left[\sum_{\substack{C \in m, \\ z(C) \in \mathbf{W}_{\rho/k^d}}} \mathbb{1}_{v_{\rho/k^d}(\tau/k^d) \leq f(C) \leq v_\rho(\tau)} \right].$$

This together with the corresponding inequality when $v_\rho(\tau) \leq v_{\rho/k^d}(\tau/k^d)$ shows that

$$\left| \mathbf{P}\left(M_{f,\mathbf{W}_{\rho/k^d}} \leq v_\rho(\tau) \right) - \mathbf{P}\left(M_{f,\mathbf{W}_{\rho/k^d}} \leq v_{\rho/k^d}(\tau/k^d) \right) \right| \tag{5}$$

$$\leq \frac{\rho}{k^d} \left| \mathbf{P}\left(f(\mathcal{C}) > v_{\rho/k^d}(\tau/k^d) \right) - \mathbf{P}\left(f(\mathcal{C}) > v_\rho(\tau) \right) \right|$$

$$= \frac{\rho}{k^d} \left| \frac{\tau/k^d}{\rho/k^d} - \frac{\tau}{\rho} + o\left(\frac{1}{\rho} \right) \right| \xrightarrow[\rho \to \infty]{} 0$$

according to (1) and the fact that $\mathbf{P}\left(f(\mathcal{C}) > v_\rho(\tau) \right)$ converges to τ for each $\tau \geq 0$. Moreover, from (2) we have

$$\limsup_{\rho \to \infty} \mathbf{P}\left(M_{f,\mathbf{W}_{\rho/k^d}} \leq v_{\rho/k^d}(\tau/k^d) \right) = \psi(\tau/k^d).$$

We obtain (3) from the previous equality and (5).   □

Proof of (4). The main idea is to apply the following adaptation of Lemma 4 in [3]:

Lemma 3. *Let $L \geq 1$ and let $B^{(1)}, \ldots, B^{(L)}$ be a L-tuple of Borel subsets included in W. Under the same assumptions as in Proposition 2, we have:*

$$\mathbf{P}\left(M_{f,\mathbf{W}_\rho} \leq v_\rho \right) - \prod_{l=1}^{L} \mathbf{P}\left(M_{f,\mathbf{B}_\rho^{(l)}} \leq v_\rho \right) \xrightarrow[\rho \to \infty]{} 0,$$

where $\mathbf{B}_\rho^{(l)} = \rho^{1/d} B^{(l)}$, $1 \leq l \leq L$.

Partitioning $W = [0,1]^d$ into a set of k^d sub-cubes of equal volume $1/k^d$, say $B^{(1)}, \ldots, B^{(k^d)}$, and applying Lemma 3, we get

$$\mathbf{P}\left(M_{f,\mathbf{W}_\rho} \leq v_\rho(\tau) \right) - \prod_{l=1}^{k^d} \mathbf{P}\left(M_{f,\mathbf{B}_\rho^{(l)}} \leq v_\rho(\tau) \right) \xrightarrow[\rho \to \infty]{} 0.$$

Since $\mathbf{B}_\rho^{(l)} = \rho^{1/d} B^{(l)}$ is a cube of volume ρ/k^d for any $1 \leq l \leq k^d$ and since m is stationary, we have

$$\mathbf{P}\left(M_{f,\mathbf{W}_\rho} \leq v_\rho(\tau) \right) - \mathbf{P}\left(M_{f,\mathbf{W}_{\rho/k^d}} \leq v_\rho(\tau) \right)^{k^d} \xrightarrow[\rho \to \infty]{} 0.$$

We obtain (4) from the previous convergence and (2).   □

Proof of Proposition 2. For each $\tau \geq 0$ and $k \in \mathbf{N}_+$, it follows from (3) and (4) that $\psi(\tau/k^d) = \psi^{1/k^d}(\tau)$. Moreover, in the same spirit as in (5), we have

$$\mathbf{P}\left(M_{f,\mathbf{W}_{\rho/k^d}} \leq v_\rho(\tau) \right) \geq 1 - \frac{\rho}{k^d} \mathbf{P}\left(f(\mathcal{C}) > v_\rho(\tau) \right) \xrightarrow[\rho \to \infty]{} 1 - \frac{\tau}{k^d}.$$

Hence, taking the kth powers and applying Lemma 3, we deduce that

$$\liminf_{\rho \to \infty} \mathbf{P}\left(M_{f,\mathbf{W}_\rho} \le v_\rho(\tau)\right) = \liminf_{\rho \to \infty} \mathbf{P}\left(M_{f,\mathbf{W}_{\rho/k^d}} \le v_\rho(\tau)\right)^{k^d} \ge \left(1 - \frac{\tau}{k^d}\right)^{k^d}.$$

Letting $k \to \infty$, we obtain $\liminf_{\rho \to \infty} \mathbf{P}\left(M_{f,\mathbf{W}_\rho} \le v_\rho(\tau)\right) \ge e^{-\tau}$. In particular, this shows that $\psi(\tau) > 0$. Since ψ is also non-increasing and since the only solution of the functional equation $\psi(\tau/k^d) = \psi^{1/k^d}(\tau)$ which is strictly positive and non-increasing is an exponential function, we have $\psi(\tau) = e^{-\theta\tau}$ for some $\theta \ge 0$. This concludes the proof of Proposition 2.   □

3 Examples

We provide below two examples where the extremal index differs from 1.

The minimum of inradii of a Poisson-Voronoi tessellation Let $\mathbf{X}$ be a Poisson point process in $\mathbf{R}^d$ of intensity 1. For all $x \in \mathbf{X}$, we denote by $C_\mathbf{X}(x)$ the Voronoi cell of nucleus x:

$$C_\mathbf{X}(x) = \{y \in \mathbf{R}^d : |x - y| \le |x' - y|, x' \in \mathbf{X}\}.$$

The family $m_{PVT} = \{C_\mathbf{X}(x), x \in \mathbf{X}\}$ is the so-called Poisson-Voronoi tessellation. Such a model is extensively used in various domains such as astrophysics [14] and telecommunications [1], see also the reference books [6,10].

In this example, for each cell $C = C_\mathbf{X}(x)$, $x \in \mathbf{X}$, we take $z(C_\mathbf{X}(x)) = x$ and $f(C_\mathbf{X}(x)) = r(C_\mathbf{X}(x)) = \max\{r \ge 0 : B(x,r) \subset C_\mathbf{X}(x)\}$ to denote the inradius of the cell. First we notice that the distribution of $r(C)^d$, with $r(C) = r(C_{\mathbf{X} \cup \{0\}}(0))$, is exponential with parameter $2^d \kappa_d$, where κ_d denotes the volume of the unit ball. Indeed, for any $v \ge 0$, we have $r(C) \le v$ if and only if $\mathbf{X} \cap B(0, 2v) \ne \emptyset$. In particular, for any $t \ge 0$, we get

$$\rho \cdot \mathbf{P}\left(r(C)^d \le (2^d \kappa_d \rho)^{-1} t\right) \xrightarrow[\rho \to \infty]{} t.$$

Moreover, according to the convergence (2b) in [2], we know that

$$\mathbf{P}\left(\min_{x \in \mathbf{X} \cap \mathbf{W}_\rho} r(C_\mathbf{X}(x))^d \ge (2^d \kappa_d \rho)^{-1} t\right) \xrightarrow[\rho \to \infty]{} e^{-\frac{1}{2} \cdot t}.$$

Let us notice that the convergence was established in [2] for a fixed window and for a Poisson point process such that the intensity goes to infinity. By scaling property of the Poisson point process, the result of [2] can be re-written as above for a fixed intensity and for a window $\mathbf{W}_\rho$ where $\rho \to \infty$. This allows us to provide a first example of extremal index:

Example 1. The extremal index of the minimum of inradius of a Poisson-Voronoi tessellation exists and is $\theta = 1/2$.

It can be also explained by a trivial heuristic argument. Indeed, if a cell minimizes the inradius, one of its neighbors has to do the same. Hence the mean cluster size of exceedances is 2 which implies that $\theta = 1/2$.

The maximum of circumradii of a Poisson-Delaunay tessellation Let $\mathbf{X}$ be a Poisson point process in $\mathbf{R}^d$ of intensity β^d, where

$$\beta_d = \frac{(d^3 + d^2)\Gamma\left(\frac{d^2}{2}\right)\Gamma^d\left(\frac{d+1}{2}\right)}{\Gamma\left(\frac{d^2+1}{2}\right)\Gamma^d\left(\frac{d+2}{2}\right)2^{d+1}\pi^{\frac{d-1}{2}}}.$$

We connect two points $x, x' \in \mathbf{X}$ by an edge if and only if $C_{\mathbf{X}}(x) \cap C_{\mathbf{X}'}(x) \neq \emptyset$. The set of these edges defines a random tessellation m_{PDT} of $\mathbf{R}^d$ into simplices with intensity 1 (e.g. Theorem 10.2.8 in [10]) which is the so-called Poisson-Delaunay tessellation. Such a model is extensively used in medical image segmentation [13] and is a powerful tool for reconstructing a $3D$ set from a discrete point set [9].

Here we take $z(C)$ and $f(C) = R(C)$ as the circumcenter and the circumradius of any cell $C \in m_{PDT}$ respectively. A Taylor expansion of $\mathbf{P}\left(R(C) > v\right)$ (e.g. Equation (3.14) in [3]), as v goes to infinity, shows that for each $t \in \mathbf{R}$

$$\rho \cdot \mathbf{P}\left(R(C)^d \geq \delta_d^{-1} \cdot \left(\log\left([(d-1)!]^{-1}\rho\log(\beta_d\rho)^{d-1}\right) + t\right)\right) \xrightarrow[\rho\to\infty]{} e^{-t},$$

where $\delta_d = \beta_d\kappa_d$. Moreover, with standard arguments, we easily show that the maximum of circumradii of Delaunay cells $\max_{\substack{C \in m, \\ z(C) \in \mathbf{W}_\rho}} R(C)$ has the same asymptotic behaviour as the maximum of circumradii of the associated Voronoi cells $\max_{x \in \mathbf{X} \cap \mathbf{W}_\rho} R(C_{\mathbf{X}}(x))$. Besides, thanks to (2c) in [2], we know that

$$\mathbf{P}\left(\max_{x \in \mathbf{X} \cap \mathbf{W}_\rho} R(C_{\mathbf{X}}(x))^d \leq \delta_d^{-1}\left(\log\left(\alpha_d\beta_d\rho\log(\beta_d\rho)^{d-1}\right) + t\right)\right) \xrightarrow[\rho\to\infty]{} e^{-e^{-t}},$$

where $\alpha_d := \frac{1}{d!}\left(\frac{\pi^{1/2}\Gamma\left(\frac{d}{2}+1\right)}{\Gamma\left(\frac{d+1}{2}\right)}\right)^{d-1}$. It follows that

$$\mathbf{P}\left(\max_{\substack{C \in m_{PDT}, \\ z(C) \in \mathbf{W}_\rho}} R(C)^d \leq \delta_d^{-1} \cdot \left(\log\left([(d-1)!]^{-1}\rho\log(\beta_d\rho)^{d-1}\right) + t\right)\right) \xrightarrow[\rho\to\infty]{} e^{-\theta_d \cdot e^{-t}},$$

where

$$\theta_d = \alpha_d\beta_d(d-1)! = \frac{(d^3 + d^2)\Gamma\left(\frac{d^2}{2}\right)\Gamma\left(\frac{d+1}{2}\right)}{2^{d+1}d\Gamma\left(\frac{d^2+1}{2}\right)\Gamma\left(\frac{d+2}{2}\right)}.$$

This allows us to provide a second example of extremal index:

Example 2. The extremal index of the maximum of circumradius of a Poisson-Delaunay tessellation exists and $\theta = \theta_d$. In particular, when $d = 1, 2, 3$, the extremal indices are $\theta = 1$, $\theta = 1/2$ and $\theta = 35/128$ respectively.

References

1. Baccelli,F., Blaszczyszyn, B.: Stochastic Geometry and Wireless Networks Volume 2: Applications, Foundations and Trends in Networkin: Vol. 1: Theory, Vol. 2: Applications No 1-2 (2009)
2. Calka, P., Chenavier, N.: Extreme values for characteristic radii of a Poisson-Voronoi tessellation. Extremes **17**, 359–385 (2014)
3. Chenavier, N.: A general study of extremes of stationary tessellations with examples. Stochast. Process. Appl. **124**, 2917–2953 (2014)
4. Chenavier, N., Hemsley, R.: Extremes for the inradius in the Poisson line tessellation. In: Advances in Applied Probability. http://arxiv.org/pdf/1502.00135.pdf (2015, to appear)
5. Leadbetter, M.R.: Extremes and local dependence in stationary sequences. Z. Wahrsch. Verw. Gebiete **65**, 291–306 (1983)
6. Okabe, A., Boot, B., Sugihara, K., Chiu, S.N.: Spatial Tessellations: Concepts and Applications of Voronoi Diagrams, 2nd edn. Wiley, New York (2000). Wiley Series in Probability and Statistics
7. Penrose, M.: Random geometric graphs. Oxford Stud. Probab. **5**, 344 (2003)
8. Robert, C.Y.: Inference for the limiting cluster size distribution of extreme values. Ann. Statist. **37**, 271–310 (2009)
9. Schaap, W.: DTFE: the Delaunay Tessellation Field Estimator, Ph.D thesis (2007)
10. Weil, R., Weil, W.: Stochastic and Integral Geometry. Probability and Its Applications, 1st edn. Springer, Heidelberg (2008)
11. Schulte, M., Thäle, C.: The scaling limit of Poisson-driven order statistics with applications in geometric probability. Stochast. Process. Appl. **122**, 4096–4120 (2012)
12. Smith, R.L., Weissman, I.: Estimating the extremal index. J. Roy. Statist. Soc. Ser. B **56**, 515–528 (1994)
13. Španěl, M., Kršek, P., Švub, M., Štancl, V., Šiler, O.: Delaunay-based vector segmentation of volumetric medical images. In: Kropatsch, W.G., Kampel, M., Hanbury, A. (eds.) CAIP 2007. LNCS, vol. 4673, pp. 261–269. Springer, Heidelberg (2007)
14. Zaninetti, L.: The oscillating behavior of the pair correlation function in galaxies. Appl. Phys. Res. **6**, 35–46 (2014)

A Two-Color Interacting Random Balls Model for Co-localization Analysis of Proteins

F. Lavancier[1,2]([⊠]) and C. Kervrann[1]

[1] Inria, Centre Rennes Bretagne Atlantique, Rennes, France
[2] University of Nantes, Nantes, France
Frederic.Lavancier@univ-nantes.fr

Abstract. A model of two-type (or two-color) interacting random balls is introduced. Each colored random set is a union of random balls and the interaction relies on the volume of the intersection between the two random sets. This model is motivated by the detection and quantification of co-localization between two proteins. Simulation and inference are discussed. Since all individual balls cannot been identified, e.g. a ball may contain another one, standard methods of inference as likelihood or pseudolikelihood are not available and we apply the Takacs-Fiksel method with a specific choice of test functions.

1 Introduction

This study is motivated by an application to the detection and quantification of co-localization between proteins [2]. As an example, Fig. 1 depicts a M10 cell showing respectively from left to right: Langerin proteins (colored in green), Rab11 GTPase proteins (colored in red) and the superposition of the two previous images resulting in some possible yellow spots. These two proteins have been tagged with Yellow Fluorescence Protein and mCherry respectively and the segmentation algorithm of [1] has been applied to get the images of Fig. 1, though other algorithms could have been used, e.g. [13]. The proteins of interest are known to be involved in the traffic of intermediates transport such as vesicles from endosomes to plasma membrane. The problem of co-localization and co-expression concern the detection and the understanding of their interaction in this process. This amounts to characterizing their joint spatial repartition. In Fig. 1, the occurrence of yellow spots in the right hand side image illustrates the correlation between the locations of the green and red spots in the cell, thus showing some co-localization between the two proteins.

In order to quantify the above phenomenon, object-based methods have been applied in the literature, see [7–9] and the references therein, where the spots are reduced to points and their interaction is analyzed by spatial statistics methods. We instead decide to preserve the intrinsic geometrical nature of the spots and we introduce a model of two-type (or two-color) interacting random balls. Each colored random set is a union of random balls and the interaction relies on the volume of the intersection between the two random sets. The remainder

© Springer International Publishing Switzerland 2015
F. Nielsen and F. Barbaresco (Eds.): GSI 2015, LNCS 9389, pp. 179–186, 2015.
DOI: 10.1007/978-3-319-25040-3_20

Fig. 1. M10 cell showing Langerin proteins (left, in green) and Rab11 GTPase proteins (middle, in red). Right: superposition of the two previous images resulting in some possible yellow spots (Color figure online).

of this paper is organized as follows: the model is defined in Sect. 2, a simulation algorithm is described in Sect. 3 and the inference procedure, especially to estimate the interaction parameter, is explained in Sect. 4. Some simulations on synthesized data show the ability of our procedure to detect and quantify co-localization. The application on a bunch of real data is part of an ongoing work.

2   The Model

We briefly recall some background material on point processes and stochastic geometry before introducing our model. The reader is referred to [11] and [3, Sects. 3–4] for more details.

A finite marked point pattern x on a compact set $W \subset \mathbb{R}^d$ is composed of a finite number of marked points (ξ, R), where $\xi \in W$ represents the location of the point and $R > 0$ is the mark associated to ξ. All along this paper, R represents the radius of the euclidean ball centered at ξ, so that x can be viewed as a collection of balls whose centers lie in W.

We consider a random model in that the centres are distributed according to a point process over W and, independently of the locations, the radii follow some probability distribution μ on the interval $[R_{\min}, R_{\max}]$ for some $0 < R_{\min} < R_{\max}$. The standard example is the Boolean model for which the locations come from a Poisson point process over W, see [3]. Moreover, we randomly mark the balls with a color $i \in \{1, 2\}$, yielding a two-type marked random balls process. We denote by x_1, respectively x_2, the collection of balls over W having color $i = 1$, respectively $i = 2$. A ball $B(\xi, R)$ with color i is denoted for short $(\xi, R)_i$.

The reference model is to assume x_1 and x_2 independent, with locations following a unit-rate Poisson distribution over W, while the radii are independently distributed according to μ. This reference model can be viewed either as an independent bivariate Boolean model or as a (univariate) Boolean model with marks in $\{1, 2\}$ associated to probabilities $(\frac{1}{2}, \frac{1}{2})$, the two point of views being equivalent, see Sect. 6.6 in [11].

In order to introduce some interaction between the two colors, we consider a density with respect to the previous reference models. With respect to the bivariate Boolean model, which is the first point of view for the reference process, our density writes

$$\tilde{f}(x_1, x_2) = \frac{1}{\tilde{c}} z_1^{n_1} z_2^{n_2} e^{\theta |\mathcal{U}_1 \cap \mathcal{U}_2|}$$

where for $i = 1, 2$, $n_i = \mathrm{card}(x_i)$, $\mathcal{U}_i = \cup_{(\xi,R) \in x_i} B(\xi, R)$, $z_i > 0$, $\theta \in \mathbb{R}$ and $\tilde{c} > 0$ is a normalizing constant. The parameters z_1 and z_2 rule the mean number of balls of each color while θ is an interaction parameter. If $\theta = 0$, there is no interaction between the two colors and the model is equivalent to the reference model, up to the intensities z_1 and z_2. If $\theta > 0$, then there is an attraction between the two colors in that realizations of (x_1, x_2) having large values of $|\mathcal{U}_1 \cap \mathcal{U}_2|$ will be more likely than independent realizations of x_1 and x_2. If $\theta < 0$, the converse holds and there is some repulsion between the two colors.

Equivalently, we can consider the density with respect to the Boolean model with equiprobable marks in $\{1, 2\}$, corresponding to the second point of view for the reference process. This density writes

$$f(x) = \frac{1}{c} 2^{n_1 + n_2} z_1^{n_1} z_2^{n_2} e^{\theta |\mathcal{U}_1 \cap \mathcal{U}_2|}, \tag{1}$$

where $x = x_1 \cup x_2$, $c > 0$ is a normalizing constant, and the interpretations of the parameters z_1, z_2 and θ are the same.

Although the bivariate point of view might be more natural to motivate the model, the latter point of view is more convenient for its theoretical study including the inference methodology explained in Sect. 4. Henceforth we view our model as having the density (1).

An important characteristic of the model is the Papangelou conditional intensity of a colored ball $(\xi, R)_i$ in x. It is defined by

$$\lambda((\xi, R)_i, x) = \frac{f(x \cup (\xi, R)_i)}{f(x)}$$

which gives

$$\lambda((\xi, R)_i, x) = 2 z_i \, e^{\theta \, (|\mathcal{U}_i \cup B(\xi, R)| - |\mathcal{U}_i| + |\mathcal{U}_1 \cup \mathcal{U}_2| - |\mathcal{U}_1 \cup \mathcal{U}_2 \cup B(\xi, R)|)}. \tag{2}$$

Heuristically, this can be interpreted as the conditional probability to observe the ball $(\xi, R)_i$ given the configuration elsewhere is x.

The Papangelou conditional intensity is at the heart of our inference procedure presented in Sect. 4. This is also a convenient tool to verify the existence of our model on W for any value of the parameters $z_1 > 0$, $z_2 > 0$ and $\theta \in \mathbb{R}$, which amounts to verify that $c \neq 0$ in (1). The latter is indeed ensured by the local stability property: there exists $\kappa > 0$ such that for any $i = 1, 2$, $(\xi, R) \in \mathbb{R}^d \times [R_{\min}, R_{\max}]$, $\lambda((\xi, R)_i, x) < \kappa$. The definition actually extends to the whole space $\mathbb{R}^d$ thanks to the Georgii-Nguyen-Zessin (GNZ) equation [6,12], in which case (1) becomes the conditional density on W given the outside configuration on $\mathbb{R}^d \setminus W$, see [5] for more details.

3 Simulation

To generate a realization following (1), we use a standard birth-death Metropolis-Hastings algorithm as described in [11]. At the initial state we start with a realization of the Boolean model with equiprobable colors and intensity $z_1 + z_2$. At each iteration, we generate a proposal for the birth of a new colored ball or for the death of an existing ball, each proposal occurring with probability $1/2$. If the proposal is a birth, then a color is chosen with probability $1/2$, a new location ξ is drawn uniformly over W and a radius R is sampled from μ. This birth is then accepted with probability $\min\{1, \lambda((\xi, R)_i, x)|W|/(n(x) + 1)\}$ where x denotes the configuration before the proposal and $|W|$ is the volume of W. If the proposal is a death, then a ball is chosen uniformly in x and its deletion is accepted with probability $\min\{1, n(x)/(\lambda((\xi, R)_i, x \setminus (\xi, R)_i)|W|)\}$. The generated Markov Chain converges to the distribution given by (1) and has further interesting properties due to the local stability property of our model, see Sect. 7.3 in [11] for more details.

From a practical point of view, the implementation of the above algorithm requires to be able to compute $\lambda((\xi, R)_i, x)$ and $\lambda((\xi, R)_i, x \setminus (\xi, R)_i)$, which from (2) amounts to be able to compute the area of a union of balls. This can be done by standard image analysis tools or by exploiting the power tessellations associated to $\mathcal{U}_1$ and $\mathcal{U}_2$ as in [10].

Some simulations in dimension $d = 2$ are represented in Fig. 2 where the first row corresponds to the case of no-interaction between the two colors, i.e. $\theta = 0$ (in which case the MCMC algorithm above is not necessary), while in the second row $\theta = 0.2$ corresponding to an attraction between the two colors. For comparison, for these examples, the area of the intersection of the two colors represents less than one percent of the total volume of their union when $\theta = 0$, while it represents 17 %, 26 % and 15 % respectively of the total volume in the case $\theta = 0.2$.

4 Inference

Given a realization as those in Fig. 2, we are interested in the estimation of the parameters, especially of the interaction parameter θ. We assume in the following that μ, including $R_{\min}$ and $R_{\max}$, is known.

In general, it is impossible to identify all individual balls (ξ, R) in W since some spots as in Fig. 2 can be formed by the union of several balls, or a ball may contain another one. For this reason standard inference procedures such as likelihood or pseudo-likelihood, which require the observation of the number of balls in W, are not available. This problem is discussed in detail in [5] for the estimation of a (one-color) interacting random balls model, namely the Quermass model. To overcome this difficulty, following [5], we use the Takacs-Fiksel method for some specific choice of test functions as described below.

Consider, for any non-negative function h,

$$C(z_1, z_2, \theta; h) = S(h) - z_1 I_1(\theta; h) - z_2 I_2(\theta; h) \tag{3}$$

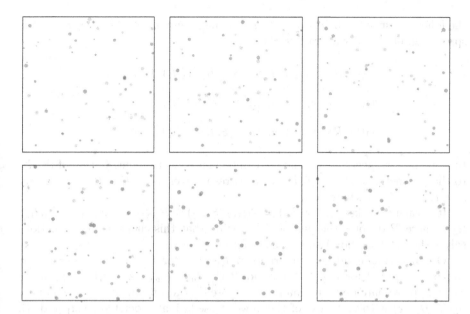

Fig. 2. Samples when $W = [0, 250] \times [0, 280]$, $z_1 = z_2 = 5 \times 10^{-4}$, $R_{\min} = 0.8$, $R_{\max} = 3$, μ is the uniform distribution, $\theta = 0$ (first row) and $\theta = 0.2$ (second row).

where

$$S(h) = \sum_{(\xi, R) \in x} h((\xi, R), x \backslash (\xi, R))$$

and for $i = 1, 2$,

$$I_i(\theta; h) = \int_{R_{\min}}^{R_{\max}} \int_W h((\xi, R)_i, x) \frac{\lambda((\xi, R)_i, x)}{2z_i} \, d\xi \, \mu(dR).$$

Denoting by z_1^*, z_2^* and θ^* the true unknown values of the parameters, we know from the GNZ equation that $E(C(z_1^*, z_2^*, \theta^*; h)) = 0$. Therefore the random variable $C(z_1, z_2, \theta; h)$ should be close to 0 when (z_1, z_2, θ) is close to (z_1^*, z_2^*, θ^*). This remark is at the basis of the Takacs-Fiksel approach. Given K test functions $(h_k)_{1 \le k \le K}$, the Takacs-Fiksel estimator is defined by

$$(\hat{z}_1, \hat{z}_2, \hat{\theta}) := \arg\min_{z_1, z_2, \theta} \sum_{k=1}^{K} C(z_1, z_2, \theta; h_k)^2. \tag{4}$$

The strong consistency and asymptotic normality of $(\hat{z}_1, \hat{z}_2, \hat{\theta})$ when the observation window W grows to $\mathbb{R}^d$ are established in [4], provided $K \ge 3$ and some technical conditions on the test functions h_k.

In our setting, it is in general not possible to compute $S(h)$ since it a priori requires the identification of each individual ball (ξ, R) of x. Nonetheless, for the following specific choices of h, $S(h)$ becomes computable. Let $\mathcal{S}(\xi, R)$ be the sphere centered at ξ with radius R and

$$h_1((\xi, R)_i, x) = \text{Length}\Big(\mathcal{S}(\xi, R) \cap (\mathcal{U}_1)^c\Big) \, \mathbf{1}_{\{i=1\}},$$

$$h_2((\xi, R)_i, x) = \text{Length}\Big(\mathcal{S}(\xi, R) \cap (\mathcal{U}_2)^c\Big) \, \mathbf{1}_{\{i=2\}},$$

$$h_3((\xi, R)_i, x) = \text{Length}\Big(\mathcal{S}(\xi, R) \cap (\mathcal{U}_1 \cup \mathcal{U}_2)^c\Big).$$

In words, h_1, respectively h_2 and h_3, can be viewed as the contribution of $(\xi, R)_i$ to the perimeter of $\mathcal{U}_1 \cup B(\xi, R)$ if $i = 1$, respectively $\mathcal{U}_2 \cup B(\xi, R)$ if $i = 2$ and $\mathcal{U}_1 \cup \mathcal{U}_2 \cup B(\xi, R)$ whatever i.

It is easily checked that $S(h_1) = \mathcal{P}(\mathcal{U}_1)$, $S(h_2) = \mathcal{P}(\mathcal{U}_2)$ and $S(h_3) = \mathcal{P}(\mathcal{U}_1 \cup \mathcal{U}_2)$, where $\mathcal{P}$ denotes the perimeter, proving that this choice of test functions solves the identification issue raised before.

On the other hand the integrand in $I_i(\theta, h_k)$ for $i = 1, 2$ and $k = 1, 2, 3$ can be computed for any $\xi \in W$ and $R \in [R_{\min}, R_{\max}]$ and $I_i(\theta, h_k)$ can be easily approximated by a Riemann sum. Note that a convenient way to compute $h_k((\xi, R)_i, x)$ is to make use of the power tessellation associated to $\mathcal{U}_1$ and $\mathcal{U}_2$, see [10].

From (3) and (2), we notice that the optimisation (4) in z_1 and z_2 can be done explicitly. Using the above simplification for $S(h_k)$, $k = 1, 2, 3$, we deduce that the solution $(\hat{z}_1, \hat{z}_2, \hat{\theta})$ of (4) necessarily belongs to the implicit manifold $(z_1, z_2) = (\tilde{z}_1(\theta), \tilde{z}_2(\theta))$ with

$$\tilde{z}_1(\theta) =$$
$$\Big[\{I_1(\theta, h_1)^2 + I_1(\theta, h_3)^2\} \{I_2(\theta, h_2)^2 + I_2(\theta, h_3)^2\} - I_1(\theta, h_3)^2 I_2(\theta, h_3)^2 \Big]^{-1}$$
$$\times \Big[\{I_1(\theta, h_1)\mathcal{P}(\mathcal{U}_1) + I_1(\theta, h_3)\mathcal{P}(\mathcal{U}_1 \cup \mathcal{U}_2)\} \{I_2(\theta, h_2)^2 + I_2(\theta, h_3)^2\}$$
$$- I_1(\theta, h_3)I_2(\theta, h_3) \{I_2(\theta, h_2)\mathcal{P}(\mathcal{U}_2) + I_2(\theta, h_3)\mathcal{P}(\mathcal{U}_1 \cup \mathcal{U}_2)\} \Big],$$

$$\tilde{z}_2(\theta) =$$
$$\Big[\{I_1(\theta, h_1)^2 + I_1(\theta, h_3)^2\} \{I_2(\theta, h_2)^2 + I_2(\theta, h_3)^2\} - I_1(\theta, h_3)^2 I_2(\theta, h_3)^2 \Big]^{-1}$$
$$\times \Big[\{I_2(\theta, h_2)\mathcal{P}(\mathcal{U}_2) + I_2(\theta, h_3)\mathcal{P}(\mathcal{U}_1 \cup \mathcal{U}_2)\} \{I_1(\theta, h_1)^2 + I_1(\theta, h_3)^2\}$$
$$- I_1(\theta, h_3)I_2(\theta, h_3) \{I_1(\theta, h_1)\mathcal{P}(\mathcal{U}_1) + I_1(\theta, h_3)\mathcal{P}(\mathcal{U}_1 \cup \mathcal{U}_2)\} \Big].$$

Consequently the estimation of θ reduces to the following one-dimensional optimization problem

$$\hat{\theta} = \arg\min_{\theta} \Big[\{\mathcal{P}(\mathcal{U}_1) - \tilde{z}_1(\theta)I_1(\theta, h_1)\}^2 + \{\mathcal{P}(\mathcal{U}_2) - \tilde{z}_2(\theta)I_2(\theta, h_2)\}^2$$
$$+ \{\mathcal{P}(\mathcal{U}_1 \cup \mathcal{U}_2) - \tilde{z}_1(\theta)I_1(\theta, h_3) - \tilde{z}_2(\theta)I_2(\theta, h_3)\}^2 \Big].$$

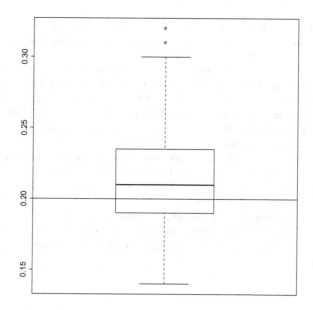

Fig. 3. Repartition of $\hat{\theta}$ computed from 100 samples as in the second row of Fig. 2.

An estimation of the intensities is then $\hat{z}_1 = \tilde{z}_1(\hat{\theta})$ and $\hat{z}_2 = \tilde{z}_2(\hat{\theta})$.

An an illustration, Fig. 3 shows the repartition of $\hat{\theta}$ from 100 replications of our model with the same parameters as in the second row of Fig. 2. For this example, our procedure clearly identifies the attractive behavior between x_1 and x_2, which opens exciting perspectives for the quantification of co-localization in proteins as in Fig. 1. The latter application is part of an ongoing project.

References

1. Basset, A., Boulanger, J., Bouthemy, P., Kervrann, C., Salamero, J.: SLT-LoG: a vesicle segmentation method with automatic scale selection and local threshold-ing applied to tirf microscopy. In: 2014 IEEE 11th International Symposium on Biomedical Imaging (ISBI), Beijing, China, pp. 533–536 (2014)
2. Bolte, S., Cordelieres, F.: A guided tour into subcellular colocalization analysis in light microscopy. J. Microsc. **224**(3), 213–232 (2006)
3. Chiu, S.N., Stoyan, D., Kendall, W.S., Mecke, J.: Stochastic Geometry and Its Applications, 3rd edn. Wiley, Chichester (2013)
4. Coeurjolly, J.F., Dereudre, D., Drouilhet, R., Lavancier, F.: Takacs Fiksel method for stationary marked Gibbs point processes. Scand. J. Stat. **39**(3), 416–443 (2012)
5. Dereudre, D., Lavancier, F., Staňková Helisová, K.: Estimation of the intensity parameter of the germ-grain Quermass-interaction model when the number of germs is not observed. Scand. J. Stat. **41**(3), 809–929 (2014)
6. Georgii, H.: Canonical and grand canonical Gibbs states for continuum systems. Commun. Math. Phy. **48**(1), 31–51 (1976)

7. Helmuth, J.A., Paul, G., Sbalzarini, I.F.: Beyond co-localization: inferring spatial interactions between sub-cellular structures from microscopy images. BMC Bioinform. **11**(1), 372 (2010)
8. Lagache, T., Lang, G., Sauvonnet, N., Olivo-Marin, J.C.: Analysis of the spatial organization of molecules with robust statistics. PLoS One **8**(12), e80914 (2013)
9. Lagache, T., Sauvonnet, N., Danglot, L., Olivo-Marin, J.C.: Statistical analysis of molecule colocalization in bioimaging. Cytom. A **87**(6), 568–579 (2015)
10. Møller, J., Helisová, K.: Power diagrams and interaction processes for union of discs. Adv. Appli. Prob. **40**(2), 321–347 (2008)
11. Møller, J., Waagepetersen, R.: Statistical Inference and Simulation for Spatial Point Processes. Chapman and Hall/CRC, Boca Raton (2003)
12. Nguyen, X., Zessin, H.: Integral and differential characterizations Gibbs processes. Math. Nachr. **88**(1), 105–115 (1979)
13. Pecot, T., Bouthemy, P., Boulanger, J., Chessel, A., Bardin, S., Salamero, J., Kervrann, C.: Background fluorescence estimation and vesicle segmentation in live cell imaging with conditional random fields. IEEE Trans. Image Proc. **24**(2), 667–680 (2015)

Asymptotics of Superposition of Point Processes

L. Decreusefond and A. Vasseur[(✉)]

Institut Mines-Telecom, Telecom ParisTech, LTCI UMR 5141, Paris, France
{laurent.decreusefond,aurelien.vasseur}@telecom-paristech.fr

Abstract. The characteristic independence property of Poisson point processes gives an intuitive way to explain why a sequence of point processes becoming less and less repulsive can converge to a Poisson point process. The aim of this paper is to show this convergence for sequences built by superposing, thinning or rescaling determinantal processes. We use Papangelou intensities and Stein's method to prove this result with a topology based on total variation distance.

Keywords: Stochastic geometry · Ginibre point process · β-Ginibre point process · Poisson point process · Stein's method

1 Motivations

The primary motivation of this work was the following. Consider the locations of base stations (BS), i.e. antennas, of the mobile network in Paris. If we have a look at the global process of all base stations of all operators and for all operating frequencies, we obtain the left picture of Fig. 1. It turns to be compatible with the null hypothesis of being a Poisson process. However, if we look at the positions of base stations deployed by one operator, in one frequency band, we get a picture similar to the right picture of Fig. 1. It was shown in [6] that this deployment is statistically compatible with a point process with repulsion, called β-Ginibre process.

When superposing a large number of independent processes with internal repulsion but few points, it is intuitively clear that the resulting process does not exhibit strong interdependencies between its atoms and should thus resemble a Poisson process. This is this intuition we wanted to quantify by determining how fast does the convergence hold. It is often clear by looking at the Laplace transforms that a superposition of processes converge to a Poisson process, however, this does not yield a convergence rate. We here use the Stein-Dirichlet-Malliavin method, developped in [1,3], to precise this rate. It turns out that the pertinent characteristics of the point processes to be considered is their Papangelou intensity, see [5] and references therein. We show here that the Kantorovitch-Rubinstein between a Poisson point process and any other point process is controlled by the L^1 distance of their Papangelou intensity, thus generalizing the property that the distance between two Poisson processes is controlled by the L^1 distance between their control measure [2]. This result is then applied to several

© Springer International Publishing Switzerland 2015
F. Nielsen and F. Barbaresco (Eds.): GSI 2015, LNCS 9389, pp. 187–194, 2015.
DOI: 10.1007/978-3-319-25040-3_21

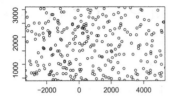
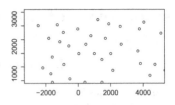

Fig. 1. On the left, positions of all BS in Paris. On the right, locations of BS for one frequency band.

situations involving superpositions and dilations of point processes. This paper is organized as follows: In Sect. 2, we recall the basics of point processes theory and introduce our model of choice, the determinantal point processes. Section 3 is devoted to the explanation of the Stein-Dirichlet-Malliavin method and how we get the main theorem. In Sect. 4, we apply this result to superposition and dilations of determinantal point processes. Due to space limitations, we just give here the main theorems, the proofs are accessible in [4].

2   Preliminaries

2.1   Point Processes

Let Y be a Polish space and $\mathcal{F}_Y$ its Borel algebra, N_Y the space of all locally finite subsets (configurations) in Y, $\widehat{N}_Y$ the space of finite subsets in Y. One may identify an element $\{x_n, \, n \in \mathbb{N}\}$ of N_Y with the atomic measure $\sum_{n \in \mathbb{N}} \delta_{x_n}$, where $\mathbb{N}$ henceforth denotes the set of positive integers. The space N_Y is a Polish space when equipped with the vague convergence (see [7]). A point process Φ is a random variable in N_Y. We denote by ϕ a generic realization, i.e. a configuration, of Φ. The intensity measure of Φ is defined as the measure $A \in \mathcal{F}_Y \mapsto \mathbf{E}[\phi(A)]$. For μ a diffuse measure on Y, the μ-sample measure L is defined by: For any measurable $f : \widehat{N}_Y \to \mathbb{R}_+$,

$$\int_{\widehat{N}_Y} f(\phi) L(\mathrm{d}\phi) = \sum_{k=0}^{+\infty} \frac{1}{k!} \int_{Y^k} f(\{x_1, \ldots, x_k\}) \mu(\,\mathrm{d}x_1) \ldots \mu(\,\mathrm{d}x_k). \qquad (1)$$

Among other characterizations, the distribution $\mathbb{P}_\Phi$ of a point process Φ can be given by its correlation function $\rho : \widehat{N}_Y \to \mathbb{R}_+$ defined by:

$$\int_{N_Y} \sum_{\substack{\alpha \in \widehat{N}_Y \\ \alpha \subset \phi}} f(\alpha) \mathbb{P}_\Phi(d\phi) = \int_{\widehat{N}_Y} f(\alpha) \rho(\alpha) L(d\alpha), \qquad (2)$$

for any measurable bounded function $f : \widehat{N}_Y \to \mathbb{R}_+$. For instance, the correlation function of a Poisson point process (PPP) with control measure $M(dx) = m(x)\mu(dx)$ on Y is given by $\rho(\phi) = \prod_{x \in \phi} m(x)$. Another descriptor of interest of a point process is the so-called Papangelou intensity. It is a function $c : Y \times N_Y \to \mathbb{R}_+$ such that for any measurable function $f : Y \times N_Y \to \mathbb{R}_+$,

$$\int_{N_Y} \sum_{x \in \phi} f(x, \phi \setminus \{x\}) \mathbb{P}_\Phi(d\phi) = \int_Y \int_{N_Y} c(x, \xi) f(x, \xi) \mathbb{P}_\Phi(d\xi) \mu(dx). \quad (3)$$

For a Poisson process of control measure $m(x)\mu(dx)$, $c(x, \phi) = m(x)$ for any (x, ϕ).

2.2 Determinantal Point Processes

For the applications we have in mind, we introduce now the notion of determinantal point processes, for details we refer to [8]. A process of this kind is characterized by a reference measure μ on Y and an Hilbert-Schmidt linear map K from $L^2(Y, \mu; \mathbb{C})$ into itself satisfying the following properties:

- K is positive Hermitian.
- The discrete spectrum of K is included in $[0, 1)$.
- K is a locally trace-class: For any compact $\Lambda \subset Y$, $K_\Lambda = P_\Lambda K P_\Lambda$ (where P_Λ is the orthogonal projection of $L^2(Y, \mu; \mathbb{C})$ to $L^2(\Lambda, \mu; \mathbb{C})$), the restriction of K to $L^2(\Lambda, \mu; \mathbb{C})$, is trace class.

Since K is Hilbert-Schmidt, there exists a kernel, which we still denote by K, from $Y \times Y$ into $\mathbb{C}$, such that for any $x \in Y$,

$$Kf(x) = \int_Y K(x, y) f(y) \mu(dy).$$

Together with K, there is another operator of importance, usually denoted by J and defined as $J = (I - K)^{-1} K$. Since K is hermitian, there exists a complete orthonormal basis $(h_j, j \in \mathbb{N})$ of $L^2(Y, \mu; \mathbb{C})$ and a sequence $(\lambda_j, j \in \mathbb{N}) \subset [0, 1)^{\mathbb{N}}$ such that for all $f \in L^2(Y, \mu; \mathbb{C})$,

$$Kf = \sum_{j=1}^{+\infty} \lambda_j \langle f, h_j \rangle_{L^2(\mu)} h_j, \quad Jf = \sum_{j=1}^{+\infty} \frac{\lambda_j}{1 - \lambda_j} \langle f, h_j \rangle_{L^2(\mu)} h_j,$$

and then, for all $x, y \in Y$,

$$K(x, y) = \sum_{j=1}^{+\infty} \lambda_j h_j(x) h_j(y), \quad J(x, y) = \sum_{j=1}^{+\infty} \frac{\lambda_j}{1 - \lambda_j} h_j(x) h_j(y).$$

The determinantal point process DPP (K, μ) is then defined by its correlation functions (see [8]):

$$\rho(\{x_1, \cdots, x_k\}) = \det(K(x_i, x_j), \ 1 \le i, j \le k).$$

From [5], we know that

$$c(x_0, \{x_1, \cdots, x_k\}) = \frac{\det(J(x_i, x_j), \ 0 \le i, j \le k)}{\det(J(x_i, x_j), \ 1 \le i, j \le k)}.$$

2.3   Superposition, Rescalings and Thinnings

Let Φ be a point process on Y. For $\varepsilon \in [0;1]$, we associate to Φ the ε-thinned point process $t_\varepsilon(\Phi)$ obtained by retaining, independently and with probability ε, each point of ϕ.

If $Y = \mathbb{R}^d$ and γ is a positive real number, we associate to Φ the γ-rescaled point process $r_\gamma(\Phi)$ obtained by applying a dilation of magnitude $\gamma^{1/d}$ to each point of Φ. Note that this modifies the intensity measure of Φ by a factor γ.

For $\beta \in (0;1]$, we associate to Φ the β-point process $r_{\beta^{-1}}(t_\beta(\Phi))$ obtained by combining a β-thinning and a β^{-1}-rescaling, in order to conserve the intensity measure of Φ. Their respective correlation functions are provided by the following proposition.

Theorem 1. *Let Φ be a point process on Y with correlation function ρ_Φ, and $\varepsilon \in [0;1]$. Then, the function correlation of $t_\varepsilon(\Phi)$ is given for any $\alpha \in \widehat{N}_Y$ by*

$$\rho_{t_\varepsilon(\Phi)}(\alpha) = \varepsilon^{|\alpha|}\rho_\Phi(\alpha).$$

Moreover, if $Y = \mathbb{R}^d$ and $\gamma > 0$, the correlation function of $r_\gamma(\Phi)$ is given for any $\alpha \in \widehat{N}_Y$ by

$$\rho_{r_\gamma(\Phi)}(\alpha) = \gamma^{|\alpha|}\rho_\Phi(\gamma^{\frac{1}{d}}\alpha).$$

3   Kantorovitch-Rubinstein Distance and Stein's Method

The total variation distance between two measures ν_1 and ν_2 on Y is defined by

$$d_{\mathrm{TV}}(\nu_1, \nu_2) := \sup_{\substack{A \in \mathcal{F}_Y \\ \nu_1(A), \nu_2(A) < \infty}} |\nu_1(A) - \nu_2(A)|.$$

We say that a measurable map $F : N_Y \to \mathbb{R}$ is 1-Lipschitz if

$$|F(\phi_1) - F(\phi_2)| \leq d_{TV}(\phi_1, \phi_2) \text{ for all } \phi_1, \phi_2 \in N_Y.$$

We denote by Lip_1 the set of bounded 1-Lipschitz maps. The Kantorovich-Rubinstein distance between two probability measures $\mathbb{P}_1$ and $\mathbb{P}_2$ on N_Y is defined by

$$d_{KR}(\mathbb{P}_1, \mathbb{P}_2) = \sup_{F \in \mathrm{Lip}_1} \left| \int_{N_Y} F(\phi)\, \mathbb{P}_1(d\phi) - \int_{N_Y} F(\phi)\, \mathbb{P}_2(d\phi) \right|. \tag{4}$$

According to [3, Proposition 2.1], the topology induced by this distance coincides with the topology of narrow convergence of probability measures on N_Y. Our goal is to evaluate the distance between some probability measure on N_Y and $\mathbb{P}_M$, the distribution of a Poisson point process of control measure M on Y. We assume henceforth that M has a finite mass, i.e. $M(Y) < \infty$. We use the Stein-Dirichlet-Malliavin method which we describe roughly now, for details we refer to [3].

The Glauber process $(G_t, t \geq 0)$ associated to $\mathbb{P}_M$ is the $\widehat{N}_Y$-valued Markov process whose generator is given by

$$LF(\phi) := \int_Y (F(\phi + \delta_y) - F(\phi))\, M(dy) + \sum_{y \in \phi}(F(\phi - \delta_y) - F(\phi)), \quad \phi \in \widehat{N}_Y,$$

where $F : N_Y \to \mathbb{R}$ is a measurable and bounded function. Since M is a finite measure, the dynamics of G are described as follows: Let $G(0) = \phi$ and consider a Poisson process on the half-axis of intensity $M(Y)$. We denote by $(T_n, n \geq 1)$ the arrival times of this process. At each T_n, $G(T_n) = G(T_n^-) + \delta_{Y_n}$ where Y_n is chosen according to M, independently of everything else. All the particles, be they present at the origin or born after, have a lifetime which follows an exponential distribution of parameter 1, independent of everything else. Then, $G(t)$ is the point process of living particles at time t. We denote by $(P_t, t \geq 0)$ its semi-group:

$$P_t F(\phi) = \mathbb{E}\left[F(G(t)) \,|\, G(0) = \phi\right].$$

This Markov process, or at least its semi-group, has two attractive features:

- It is ergodic: $\lim_{t \to \infty} P_t F(\phi) = \int_{N_Y} F(\phi)\mathbb{P}_M(d\phi)$ for all $\phi \in N_Y$.
- If we define the operator D by

$$D_y F(\phi) = F(\phi + \delta_y) - F(\phi),$$

for any $y \in Y$ and $\phi \in N_Y$, we have

$$D_y P_t F(\phi) = e^{-t} P_t D_y F(\phi),$$

for all $t \geq 0$, $y \in Y$ and $\phi \in N_Y$.

As a consequence of the ergodicity and of the markovianity of P, we have the Stein-Dirichlet representation formula, see [3]: For any probability measure $\mathbb{P}$ on N_Y,

$$\int_{N_Y} F(\phi)\,\mathbb{P}(d\phi) - \int_{N_Y} F(\phi)\,\mathbb{P}_M(d\phi) = \int_{N_Y} \int_0^\infty L P_s F(\phi)\, ds\, \mathbb{P}(d\phi).$$

Theorem 2. *Let $\mathbb{P}$ be a finite point process on Y with Papangelou intensity c, and $\mathbb{P}_M$ the distribution of a Poisson point process with finite control measure $M(dy) = m(y)\mu(dy)$ on Y. Then, we have the following upper bound:*

$$d_{KR}(\mathbb{P}, \mathbb{P}_M) \leq \int_Y \int_{N_Y} |m(y) - c(y, \phi)|\nu(d\phi)\mu(dy).$$

Proof. Starting from the expression of L, we have

$$\int_{N_Y} F(\phi)\,\mathbb{P}(d\phi) - \int_{N_Y} F(\phi)\,\mathbb{P}_M(d\phi)$$

$$= \int_{N_Y} \int_0^\infty \int_Y D_y P_s F(\phi) m(y)\mu(dy)\, ds\, \mathbb{P}(d\phi)$$

$$+ \int_{N_Y} \int_0^\infty \sum_{y \in \phi} P_s F(\phi - \delta_y) - P_s F(\phi)\, ds\, \mathbb{P}(d\phi).$$

By the very definition of the Papangelou intensity,

$$\int_{N_Y} \int_0^\infty \sum_{y \in \phi} P_s F(\phi - \delta_y) - P_s F(\phi) \, ds \, \mathbb{P}(d\phi)$$

$$= -\int_{N_Y} \int_0^\infty \int_Y D_y P_s F(\phi) c(y, \phi) \mu(dy) \, ds \, \mathbb{P}(d\phi).$$

Thus,

$$\int_{N_Y} F(\phi) \, \mathbb{P}(d\phi) - \int_{N_Y} F(\phi) \, \mathbb{P}_M(d\phi)$$

$$= \int_{N_Y} \int_0^\infty \int_Y D_y P_s F(\phi)(m(y) - c(y, \phi)) \mu(dy) \, ds \, \mathbb{P}(d\phi).$$

In view of the commutation relationship between D_y and P_s, since F Lipschitz entails that $|D_y F(\phi)| \leq 1$ for any (y, ϕ), we get

$$|D_y P_s F(\phi)| = e^{-s} |P_s D_y F(\phi)| \leq e^{-s} P_s \mathbb{1} = e^{-s}.$$

The result then follows.

4   Applications to Superpositions, Rescalings and Thinnings

Let K be the continuous kernel of a DPP on Y, and Λ a compact subset of Y. For all $n \in \mathbb{N}$, we consider $\Phi_{n,1}, \ldots, \Phi_{n,n}$, n independent point processes and Φ_n their superposition, such that, for all $i \in \{1, \ldots, n\}$, $\Phi_{n,i}$ is a DPP with kernel $K_{n,i}$, supported by Λ and such that there exists a measurable function $m : \Lambda \to \mathbb{C}$ verifying:

$$\lim_{n \to +\infty} \int_\Lambda \Big| \sum_{i=1}^n K_{n,i}(y,y) - m(y) \Big| \mu(dy) = 0.$$

Furthermore, let $c_{n,i}$ be the Papangelou intensity of the point process $\Phi_{n,i}$: by Theorem 3.6 of [5], its existence is guaranteed for all determinantal point process. The Papangelou intensity of the superposition can be deduced from Papangelou intensities $c_{n,i}$ of the $\Phi_{n,i}$:

Lemma 1. *For all $n \in \mathbb{N}$, Φ_n admits a Papangelou intensity c_n given for all $y \in Y$, $\phi = \cup_{i=1}^n \phi_i \in \mathbb{N}_Y$ by:*

$$c_n(y, \phi) = \sum_{i=1}^n c_{n,i}(y, \phi_i).$$

Note that this is not strictly speaking the Papangelou intensity of Φ_n since the right-hand-side depends on the particular construction of Φ_n as the superposition of the $\Phi_{n,i}$.

4.1 Superposition of Rescaled Determinantal Processes

The following theorem states the convergence of a sequence built by superposition of thinned determinantal processes and is a corollary of Theorem 2. The DPP $\Phi_{n,i}$ is obtained by γ_i/n-rescaling the $\beta_{n,i}$-DPP (K,μ) restricted to a compact set $\Lambda \subset Y$. Its kernel $K_{n,i}$ is then defined for $x, y \in Y$ by

$$K_{n,i} : (x,y) \in Y \times Y \mapsto \frac{\gamma_i}{n} K\left(\sqrt{\frac{\gamma_i}{n\beta_{n,i}}} x, \sqrt{\frac{\gamma_i}{n\beta_{n,i}}} y \right) 1_{\Lambda \times \Lambda}(x,y).$$

Let $(R_n, n \in \mathbb{N})$ be the sequence defined for all $n \in \mathbb{N}$ by

$$R_n = \int_\Lambda \left| \sum_{i=1}^{n} K_{n,i}(y,y) - m(y) \right| \mu(dy),$$

where $m : Y \to \mathbb{R}_+$ is a measurable function.

Corollary 1. *Suppose K is such that $\sup_{x,y \in \Lambda} |K(x,y)| < +\infty$. Assume that $\lim_{n \to +\infty} R_n = 0$. Then the sequence $(\Phi_n, n \in \mathbb{N})$ converges strongly to a Poisson point process with control measure $M(dy) = m(y)\mu(dy)$ and*

$$d_{KR}(\mathbb{P}_{\Phi_n}, \mathbb{P}_M) = O\left(\frac{1}{n} + R_n \right).$$

4.2 Superposition of Thinned Determinantal Processes

Consider now a superposition where each DPP $\Phi_{n,i}$ is obtained by γ_i/n-thinning the $\beta_{n,i}$-DPP (K,μ). Its kernel $K_{n,i}$ is then defined by

$$K_{n,i} : (x,y) \in Y \times Y \mapsto \frac{\gamma_i}{n} K\left(\frac{x}{\sqrt{\beta_{n,i}}}, \frac{y}{\sqrt{\beta_{n,i}}} \right) 1_{\Lambda \times \Lambda}(x,y). \tag{5}$$

The next theorem states the convergence of the previously built sequence of superpositions.

Corollary 2. *Suppose $\lim_{n \to +\infty} R_n = 0$. Then the sequence $(\Phi_n, n \in \mathbb{N})$ converges strongly to a Poisson point process with control measure $M(dy) = m(y)\mu(dy)$ and*

$$d_{KR}(\Phi_n, \pi_M) = O\left(\frac{1}{n} \sup_{i=1,\ldots,n} \beta_{n,i} + R_n \right).$$

4.3 Application to β-determinantal Processes

Let K be the kernel of a DPP on Y, assumed to be continuous as a function of x and y. Suppose that the sequence $(\beta_n, n \in \mathbb{N})$ converges to 0 and that Φ_n is the DPP with kernel K_n defined by

$$K_n : (x,y) \in Y \times Y \mapsto K\left(\frac{x}{\sqrt{\beta_n}}, \frac{x}{\sqrt{\beta_n}} \right) 1_{\Lambda \times \Lambda}(x,y). \tag{6}$$

Let $(R_n, n \in \mathbb{N})$ be the sequence defined for all $n \in \mathbb{N}$ by

$$R_n = \int_\Lambda |K_n(y,y) - K(0,0)| \mu(\, dy).$$

Corollary 3. *Suppose* $\lim_{n \to +\infty} R_n = 0$. *Then the sequence* (Φ_n) *converges strongly to an homogeneous Poisson point process with control measure* $M(dy) = K(0,0)\mu(dy)$ *and*

$$d_{KR}(\Phi_n, \pi_M) = O(\beta_n + R_n).$$

References

1. Coutin, L., Decreusefond, L.: Stein's method for Brownian approximations. Commun. Stoch. Anal. **7**, 349–372 (2013)
2. Decreusefond, L., Joulin, A., Savy, N.: Upper bounds on Rubinstein distances on configuration spaces and applications. Commun. Stoch. Anal. **4**(3), 377–399 (2010). http://hal.archives-ouvertes.fr/hal-00347899/fr/
3. Decreusefond, L., Schulte, M., Thäle, C.: Functional Poisson approximation in Rubinstein distance. Annals of Probability (2015). http://de.arxiv.org/abs/1406.5484
4. Decreusefond, L., Vasseur, A.: Asymptotics of superpositions and thinnings of point processes, in preparation
5. Georgii, H.O., Yoo, H.J.: Conditional intensity and gibbsianness of determinantal point processes. J. Stat. Phys. **118**, 55–84 (2004). http://arxiv.org/pdf/math/0401402v2.pdf
6. Gomez, J.S., Vasseur, A., Vergne, A., Decreusefond, L., Martins, P., Chen, W.: A case study on regularity in cellular network deployment. Wireless Communications Letters (2015)
7. Kallenberg, O.: Random Measures, 3rd edn. Academic Press, New York (1983)
8. Shirai, T., Takahashi, Y.: Random point fields associated with certain Fredholm determinants ii: Fermion shifts and their ergodic and Gibbs properties. Ann. Probab. **31**(3), 1533–1564 (2003). http://www.jstor.org/stable/3481499

Asymptotic Properties of Random Polytopes

Pierre Calka [(✉)]

Laboratoire de Mathématiques Raphaël Salem (UMR 6085),
Université de Rouen, Mont-Saint-Aignan, France
pierre.calka@univ-rouen.fr

Abstract. Random polytopes have constituted some of the central objects of stochastic geometry for more than 150 years. They are in general generated as convex hulls of a random set of points in the Euclidean space. The study of such models requires the use of ingredients coming from both convex geometry and probability theory. In the last decades, the study has been focused on their asymptotic properties and in particular expectation and variance estimates. In several joint works with Tomasz Schreiber and J. E. Yukich, we have investigated the scaling limit of several models (uniform model in the unit-ball, uniform model in a smooth convex body, Gaussian model) and have deduced from it limiting variances for several geometric characteristics including the number of k-dimensional faces and the volume. In this paper, we survey the most recent advances on these questions and we emphasize the particular cases of random polytopes in the unit-ball and Gaussian polytopes.

1 Introduction

Stochastic geometry is a branch of probability theory which consists in studying random spatial models embedded in the Euclidean space $\mathbb{R}^d$, $d \geq 2$, or any metric space. Random polytopes are considered as some of the key models of stochastic geometry [20]. The first appearance of a random polytope in the literature goes back to recreational mathematics in the 19th century and in particular to the so-called *Sylvester's four-point problem*. In 1864, J. J. Sylvester has indeed considered 4 random independent points which are uniformly distributed in some fixed convex body K of $\mathbb{R}^2$ and he asked for the probability $p_4(K)$ that these four points are the vertices of a convex quadrilateral. This question and its generalization to n points, i.e. finding the probability $p_n(K)$ that n independent and uniformly distributed points in K are all extreme points of their convex hulls, have drawn attention from both geometers specialized in integral geometry and probabilists specialized in geometric probability. In particular, W. Blaschke [9] showed in 1917 that $p_4(K)$ is maximal when K is a disk and minimal when K is a triangle. This implies that for any planar convex body K, we have

$$\frac{1}{3} \leq p_4(K) \leq \left(1 - \frac{35}{12\pi^2}\right).$$

More recently, I. Bárány [3] has obtained the asymptotic behavior of $p_n(K)$ and P. Valtr [26] has deduced from combinatorial arguments an explicit formula for fixed n when K is a triangle or a parallelogram.

© Springer International Publishing Switzerland 2015
F. Nielsen and F. Barbaresco (Eds.): GSI 2015, LNCS 9389, pp. 195–202, 2015.
DOI: 10.1007/978-3-319-25040-3_22

In parallel, the study of the convex hull of n independent points which are uniformly distributed in a convex body K of $\mathbb{R}^n$ has started more than 50 years ago. In 1962, J. G. Wendel has derived an explicit formula for the probability that the origin is inside the convex hull when the points have a symmetric distribution [29]. B. Efron has obtained a very simple formula which relates the mean number of extreme points of a sample of size n to the mean volume of the convex hull of $(n-1)$ points. Indeed, let K be a convex body in $\mathbb{R}^d$, $d \geq 2$. Denoting by K_n the convex hull of n i.i.d. uniform points in K, by $f_0(K_n)$ the number of extreme points of K_n and by V_d the d-dimensional Lebesgue measure, we have the following equality for every $n \geq (d+1)$:

$$\mathbb{E} f_0(K_n) = n \left(1 - \frac{\mathbb{E} V_d(K_{n-1})}{V_d(K)} \right).$$

Apart from these two works, very few non-asymptotic results have been obtained and focus has quickly turned to the description of the asymptotic behavior of a random convex hull when the size of the input goes to infinity.

In two seminal works published in 1963 and 1964 [21,22], A. Rényi and R. Sulanke obtained explicit formulae for the asymptotics of the mean number of vertices, mean area and mean perimeter of a planar random polytope constructed as the convex hull of n independent and identically distributed points. The three different distributions that they considered were the uniform distribution in a smooth convex body, the uniform distribution in a convex polytope and the standard Gaussian distribution. In the rest of the paper, we will name these three models as the *smooth uniform model*, *polytope uniform model* and *Gaussian model* respectively (see Fig. 1).

The works of A. Rényi and R. Sulanke have revealed completely different behaviors between the three models: in the smooth uniform model, the growth of the number of extreme points is polynomial while it is logarithmic in both the polytope uniform model and the Gaussian model. The study has been extended to higher dimensions in several subsequent works, due notably to R. Schneider and J. A. Wieacker [24], I. Bárány [2] and M. Reitzner [17,19] for the smooth uniform model, I. Bárány and C. Buchta [4] and M. Reitzner [19] for the polytope uniform model, F. Affentranger and R. Schneider [1] and Y. Baryshnikov and R. Vitale [8] for the Gaussian model. The results that are derived from these many references are summarized in the table below (see Fig. 2). Here and below, K_n denotes the convex hull of an input of size n in $\mathbb{R}^d$, K being the mother body in both smooth and polytope uniform models, and $f_k(K_n)$, $0 \leq k \leq d$, is the number of k-dimensional faces of K_n. For two functions f and g, we use the abbreviations $f(n) \sim g(n)$ and $f(n) \approx g(n)$ when $\lim_{n \to \infty} \frac{f(n)}{g(n)} = 1$ and when $\lim_{n \to \infty} \frac{f(n)}{g(n)}$ exists in $(0, \infty)$ respectively. In particular, the multiplicative constants are proportional to the so-called affine area of K and to the number of flags in K in the smooth uniform model and polytope uniform model respectively.

In the more recent years, the works have been focused on second-order results for random polytopes and in particular central limit theorems and variance estimates. After seminal works in the planar case due to P. Groeneboom [15] in

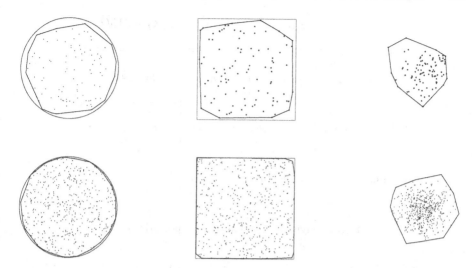

Fig. 1. Simulations of random polytopes in the plane for an input of size 100 (top) and size 500 (bottom). From left to right: smooth uniform model (in the disk), polytope uniform model (in the square), Gaussian model

1988 and T. Hsing [16] in 1994, a new method based on so-called dependency graphs proved successful for deriving central limit theorems for both $f_k(K_n)$ and $V_d(K_n)$ when applied by M. Reitzner [18] and V. H. Vu [28] for the smooth uniform model, by I. Bárány and M. Reitzner [5] for the polytope uniform model and by I. Bárány and V. H. Vu [7] for the Gaussian model.

Showing the existence of limiting variances proved to be more intricate. When proving the central limit theorem, M. Reitzner [18] and I. Bárány and V. H. Vu [7] obtained sharp lower and upper bounds for the smooth uniform model and Gaussian model respectively. Bounds for the polytope uniform model have been obtained by M. Reitzner and I. Bárány [6], independently from the central limit theorem and with a particular use of the so-called floating body of order $1/n$ associated with K.

The goal of this paper is essentially to present several results of existence and calculation of limiting variances. In 2008, T. Schreiber and J. E. Yukich [25] derived for the first time a limiting variance for the number of extreme points in the case of a smooth uniform model in the unit-ball. In [13], we show that the boundary of a random polytope inside the unit-ball converges to a limiting process when properly rescaled and we deduce from this scaling transformation variance asymptotics for several geometric characteristics as well as an invariance principle. In [11], the case of the unit-ball is used to derive variance asymptotics in the general smooth uniform model. Finally, in [12], a new scaling transformation is defined in order to provide results similar to those contained in [13] but for the Gaussian model.

In the rest of the paper, we will concentrate on both the smooth uniform model in the unit-ball and the Gaussian model. Section 2 will be devoted to

	$\mathbb{E}\, f_k(K_n)$	$\mathbb{E}\, V_d(K_n)$	$V_d(K) - \mathbb{E}\, V_d(K_n)$
Smooth uniform	$\approx n^{\frac{d-1}{d+1}}$	$\sim V_d(K)$	$\approx n^{-\frac{2}{d+1}}$
Polytope uniform	$\approx \log^{d-1}(n)$	$\sim V_d(K)$	$\approx n^{-1}\log^{d-1}(n)$
Gaussian	$\approx \log^{\frac{k}{2}}(n)$	$\approx \log^{\frac{d-1}{2}}(n)$	-

Fig. 2. Asymptotic behavior of expectations

the introduction of the scaling transformations on which the variance estimates depend. Section 3 will provide the main results regarding variance asymptotics and describe a few key ingredients of the proofs. In Sect. 4, we provide concluding remarks and suggest several extensions.

The results presented here are based on joint works [11–13] with T. Schreiber and J. E. Yukich.

2   Scaling Transformation

In this section, we focus on two examples: for $n \geq (d+1)$,

– the convex hull K_n of n independents points which are uniformly distributed in the unit-ball of $\mathbb{R}^d$, denoted by $\mathbb{B}^d$;
– the convex hull K_n of n independent points which are distributed according to the standard Gaussian distribution in $\mathbb{R}^d$, i.e. the distribution with density function

$$\phi(x) := (2\pi)^{-d/2} \exp(-\frac{|x|^2}{2}), \quad x \in \mathbb{R}^d,$$

where $|\cdot|$ is the Euclidean norm in $\mathbb{R}^d$.

For sake of simplicity of the exposition, we will keep in the two different examples the same notation for the input, the convex hull, the scaling transform and the limiting process.

Both models are rotation-invariant. The limiting shape of K_n in the uniform model is $\mathbb{B}^d$ itself. In the Gaussian model, thanks to a work due to J. Geffroy in 1961 [14], it is known that the Hausdorff distance between K_n and the ball centered at the origin and of radius $\sqrt{2\log(n)}$ converges to zero. The *sphericity* of the limiting shape of the random polytope in both cases suggests that we have to describe the boundary with spherical coordinates and rescale it in both radial and angle coordinates. Moreover, the rescaling should be done in such a way that the mean number of points per unit volume stays bounded when $n \to \infty$.

Let κ_d be the volume of $\mathbb{B}^d$, $u_0 = (0, \cdots, 0, 1)$ be the north pole of the unit-sphere $\mathbb{S}^{d-1}$ and $\exp_{d-1} : \mathbb{R}^{d-1} \longrightarrow \mathbb{S}^{d-1}$ be the exponential map at u_0. We recall that $\exp_{d-1}$ is injective on the $(d-1)$-dimensional open ball centered at the origin and of radius π and we will denote by $\exp_{d-1}^{-1}$ the inverse of the restriction of $\exp_{d-1}$ to that domain. Moreover, in the Gaussian model, we will use the following critical radius:

$$R_n = \sqrt{2 \log n - \log(2 \cdot (2\pi)^d \cdot \log n)}.$$

The definition of the scaling transform is then

$$T^{(n)} : \begin{cases} \mathbb{B}^d \setminus \{\alpha u_0 : \alpha \le 0\} \longrightarrow \mathbb{R}^{d-1} \times \mathbb{R}_+ \\ x \longmapsto \left(\left(\frac{n}{\kappa_d}\right)^{\frac{1}{d+1}} \exp_{d-1}^{-1}\left(\frac{x}{|x|}\right), \left(\frac{n}{\kappa_d}\right)^{\frac{2}{d+1}} (1 - |x|) \right) \end{cases}$$

in the uniform model and

$$T^{(n)} : \begin{cases} \mathbb{R}^d \setminus \{\alpha u_0 : \alpha \le 0\} \longrightarrow \mathbb{R}^{d-1} \times \mathbb{R}_+ \\ x \longmapsto \left(R_n \exp_{d-1}^{-1}\left(\frac{x}{|x|}\right), R_n^2 (1 - \frac{|x|}{R_\lambda}) \right) \end{cases}$$

in the Gaussian model.

In the next theorem, we describe the asymptotic behavior of the boundary of K_n. In order to do so, we need to introduce a few notation in the rescaled space $\mathbb{R}^{d-1} \times \mathbb{R}$:

- $\mathcal{P}$ is a Poisson point process in $\mathbb{R}^d = \{(v, h) \in \mathbb{R}^{d-1} \times \mathbb{R}\}$ of intensity measure $1_{\mathbb{R}_+}(h) dv dh$ in the case of the uniform model (resp. $e^h dv dh$ in the case of the Gaussian model);
- $\Pi^\uparrow$ and $\Pi^\downarrow$ are respectively the up-paraboloid and down-paraboloid in $\mathbb{R}^{d-1} \times \mathbb{R}$ of equations $(2h = |v|^2)$ and $(2h = -|v|^2)$;
- for any locally finite set $\chi \subset \mathbb{R}^d$,

$$\Psi(\chi) = \bigcup_{x \in \chi} (x + \Pi^\uparrow)$$

and

$$\Phi(\chi) = \bigcup_{x \in \mathbb{R}^d; (x + \Pi^\downarrow) \cap \chi = \emptyset} (x + \Pi^\downarrow);$$

- for any locally finite set $\chi \subset \mathbb{R}^d$, $\mathrm{Ext}(\chi)$ is the set of $x \in \chi$ such that $(x + \Pi^\uparrow) \not\subset \Psi(\chi \setminus \{x\})$.

We then have the two following results.

Theorem 2.1. *As $n \to \infty$,*

(i) *the image by $T^{(n)}$ of the set of extreme points of K_n converges in distribution to $\mathrm{Ext}(\mathcal{P})$;*

(ii) *the re-scaled boundary $T^{(n)}(\partial K_n)$ converges in probability to $\partial(\Phi(\mathcal{P}))$ in the space of continuous functions on $\mathbb{R}^{d-1}$ endowed with the uniform convergence on compact sets (Fig. 3).*

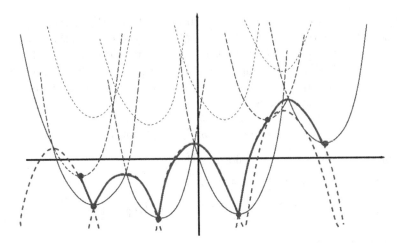

Fig. 3. In blue: point process $\text{Ext}(\mathcal{P})$, in green: the boundary $\partial(\Phi(\mathcal{P}))$ of the parabolic hull process, in red: the boundary $\partial(\Psi(\mathcal{P}))$ of the parabolic growth process (Color figure online).

3   Variance Asymptotics

We are interested in the asymptotic behavior of some of the classical geometric characteristics of the convex hull K_n, i.e. its number $f_k(K_n)$ of k-dimensional faces, $0 \leq k \leq d$, and its volume $V_d(K_n)$. For every $n \geq 1$, let

$$\alpha_n = \begin{cases} n^{\frac{d-1}{d+1}} & \text{for the uniform model} \\ (2\log n)^{\frac{d-1}{2}} & \text{for the Gaussian model} \end{cases}, \text{ and } \beta_n = \begin{cases} n^{-\frac{d+3}{d+1}} & \text{for the uniform model} \\ (2\log n)^{\frac{d-3}{2}} & \text{for the Gaussian model} \end{cases}.$$

We deduce from Theorem 2.1 the following variance asymptotics.

Theorem 3.2. *(i)* $\lim_{n\to\infty} \alpha_n^{-1} Var(f_k(K_n))$ *exists in* $(0,\infty)$ *and can be written explicitly as a function of* $\Phi(\mathcal{P})$;
(ii) $\lim_{n\to\infty} \beta_n^{-1} Var(V_d(K_n))$ *exists in* $(0,\infty)$ *and can be written explicitly as a function of* $\Phi(\mathcal{P})$;

In the few lines below, we provide a few hints to prove Theorem 3.2.

An important step of the reasoning is the Poissonization of the input as it is often the case in stochastic geometry: indeed, we observe that it is easier to study the convex hull of a Poisson point process $\mathcal{P}_\lambda$ of intensity measure $\lambda 1_{\mathbb{B}^d}(x)dx$ ($\lambda\phi(x)dx$ respectively) when $\lambda \to \infty$.

Now the functionals $f_k(K_n)$ and $V_d(K_n)$ can be rewritten as sums over all points x from the Poisson point process of a certain score depending on x and on the point process (for instance, $f_0(K_n)$ is the sum of 1_x is extreme over x).

Using both the scaling transform defined in the previous section and the Mecke-Slivnyak's formula (Corollary 3.2.3 in [23]), the variance can then be rewritten as an integral over the product space $\mathbb{R}^{d-1} \times \mathbb{R}$. The convergence provided by Theorem 2.1 will imply in turn the convergence of the integrand.

The existence of the limiting variance will finally be deduced from a careful use of Lebesgue's dominated convergence.

4 Concluding Remarks

- Results from Theorem 3.2 can be extended to so-called intrinsic volumes of order k, $0 \le k \le (d-1)$, of the convex set K_n. Moreover, the method provides a functional central limit theorem for the volume with an explicit limiting variance.
- Such variance asymptotics in the uniform model also occur when the mother body K is a smooth convex body with a C^3 boundary. The limiting variance is then equal to the affine area of K up to a multiplicative constant [11].
- First large deviation-type results for these functionals have been derived in [27] and [10].
- A different way of defining a random polytope consists in constructing a process of random hyperplanes and considering the intersection of the half-spaces bounded by these hyperplanes and containing for instance the origin. This results in studying cells from a random hyperplane tessellation (see e.g. Chapter 10 from [23]). Some of the preceding results can be translated in this new set-up for cells with a large inradius.

References

1. Affentranger, F., Schneider, R.: Random projections of regular simplices. Discret. Comput. Geom. **7**, 219–226 (1992)
2. Bárány, I.: Random polytopes in smooth convex bodies. Mathematika **39**, 81–92 (1992)
3. Bárány, I.: Sylvesters question: the probability that n points are in convex position. Ann. Probab. **27**, 2020–2034 (1999)
4. Bárány, I., Buchta, C.: Random polytopes in a convex polytope, independence of shape, and concentration of vertices. Math. Ann. **297**, 467–497 (1993)
5. Bárány, I., Reitzner, M.: Poisson polytopes. Ann. Probab. **38**, 1507–1531 (2010)
6. Bárány, I., Reitzner, M.: The variance of random polytopes. Adv. Math. **225**, 1986–2001 (2010)
7. Bárány, I., Vu, H.V.: Central limit theorems for Gaussian polytopes. Ann. Probab. **35**, 1593–1621 (2007)
8. Baryshnikov, Y.M., Vitale, R.A.: Regular simplices and Gaussian samples. Discret. Comput. Geom. **11**, 141–147 (1994)
9. Blaschke, W.: Über affine Geometrie XI: Losung des Vierpunk- tproblemsvon Sylvester aus der Theorie der geometrischen Wahrschein- lichkeiten. Ber. Verh. S achs. Ges. Wiss. Leipzig, Math.-Phys. Kl. **69**, 436–453 (1917)
10. Calka, P., Schreiber, T.: Large deviation probabilities for the number of vertices of random polytopes in the ball. Adv. Appl. Probab. **38**, 47–58 (2006)
11. Calka, P., Yukich, J.E.: Variance asymptotics for random polytopes in smooth convex bodies. Probab. Theor. Relat. Fields **158**, 435–463 (2014)
12. Calka, P., Yukich, J.E.: Variance asymptotics and scaling limits for Gaussian Polytopes. Probability Theory and Related Fields (2015, to appear). http://arxiv.org/abs/1403.1010/

13. Calka, P., Schreiber, T., Yukich, J.E.: Brownian limits, local limits, and variance asymptotics for convex hulls in the ball. Ann. Probab. **41**, 50–108 (2013)
14. Geffroy, J.: Localisation asymptotique du polyèdre d'appui d'un échantillon Laplacien à k dimensions. Publ. Inst. Stat. Univ. Paris **10**, 213–228 (1961)
15. Groeneboom, P.: Limit theorems for convex hulls. Probab. Theor. Relat. Fields **79**, 327–368 (1988)
16. Hsing, T.: On the asymptotic distribution of the area outside a random convex hull in a disk. Ann. Appl. Probab. **4**, 478–493 (1994)
17. Reitzner, M.: Stochastical approximation of smooth convex bodies. Mathematika **51**, 11–29 (2004)
18. Reitzner, M.: Central limit theorems for random polytopes. Probab. Theor. Relat. Fields **133**, 483–507 (2005)
19. Reitzner, M.: The combinatorial structure of random polytopes. Adv. Math. **191**, 178–208 (2005)
20. Reitzner, M.: Random polytopes. In: Kendall, W.S., Molchanov, I. (eds.) New Perspectives in Stochastic Geometry, pp. 45–77. Oxford University Press, Oxford (2010)
21. Rényi, A., Sulanke, R.: Über die konvexe Hülle von n zufällig gewählten Punkten. Z. Wahrscheinlichkeitstheorie und verw. Gebiete **2**, 75–84 (1963)
22. Rényi, A., Sulanke, R.: Über die konvexe Hülle von n zufällig gewählten Punkten II. Z. Wahrscheinlichkeitstheorie und verw. Gebiete **3**, 138–147 (1964)
23. Schneider, R., Weil, W.: Stochastic and Integral Geometry. Springer, Heidelberg (2008)
24. Schneider, R., Wieacker, J.A.: Random polytopes in a convex body. Z. Wahrsch. Verw. Geb. **52**, 69–73 (1980)
25. Schreiber, T., Yukich, J.E.: Variance asymptotics and central limit theorems for generalized growth processes with applications to convex hulls and maximal points. Ann. Probab. **36**, 363–396 (2008)
26. Valtr, P.: The probability that n random points in a triangle are in convex position. Combinatorica **16**, 567–573 (1996)
27. Vu, V.H.: Sharp concentration of random polytopes. Geom. Funct. Anal. **15**, 1284–1318 (2005)
28. Vu, V.H.: Central limit theorems for random polytopes in a smooth convex set. Adv. Math. **207**, 221–243 (2006)
29. Wendel, J.G.: A problem in geometric probability. Math. Scand. **11**, 109–111 (1962)

Asymmetric Topologies on Statistical Manifolds

Roman V. Belavkin[✉]

School of Science and Technology, Middlesex University, London NW4 4BT, UK
r.belavkin@mdx.ac.uk

Abstract. Asymmetric information distances are used to define asymmetric norms and quasimetrics on the statistical manifold and its dual space of random variables. Quasimetric topology, generated by the Kullback-Leibler (KL) divergence, is considered as the main example, and some of its topological properties are investigated.

1 Introduction

It is difficult to overestimate the importance of the Kullback-Leibler (KL) divergence $D_{KL}[p,q] = \mathbb{E}_p\{\ln(p/q)\}$ in probability and information theories, statistics and physics [1]. Not only it plays the role of a non-symmetric squared Euclidean distance on the set $\mathcal{P}(\Omega)$ of all probability measures on measurable set $(\Omega, \mathcal{A})$, satisfying the non-symmetric Pythagorean theorem [2] and the generalized law of cosines (see Theorem 2), but it also possesses a number of other useful and often unique to it properties. Indeed, it is Gâteaux differentiable and strictly convex everywhere where it is finite (the convex cone of finite positive measures). It is unique in the sense of additivity: $D_{KL}[p_1 \otimes p_2, q_1 \otimes q_2] = D_{KL}[p_1,q_1] + D_{KL}[p_2,q_2]$, and its Hessian defines Riemannian metric on the statistical manifold $\mathcal{P} \subset Y_+$ invariant in the category of Markov morphisms. The existence and uniqueness of this Riemannian metric is one of the most celebrated results in information geometry due to Chentsov (Lemma 11.3 in [3] or its infinite-dimensional version Theorem 5.1 in [4]).

Perhaps, the only 'inconvenient' property of the KL-divergence is its asymmetry: $D_{KL}[p,q] \neq D_{KL}[q,p]$ for some p and q. It means that a topology defined on $\mathcal{P}(\Omega)$ in terms of the KL-divergence is not symmetric, and the analysis of asymmetric topological spaces (e.g. quasi-normed, quasi-metric or quasi-uniform spaces) is significantly more difficult than that of normed or metric spaces. Many classical results about completeness, total boundedness or compactness do not hold in asymmetric topologies (e.g. see [5,6]). Perhaps, for this reason previous works have considered statistical manifolds as subsets of Banach spaces, such as the Orlicz spaces [7]. This, of course, requires certain symmetrization. Specifically, the Orlicz norm (or the equivalent Luxemburg norm) is defined using the integral of an even function $\phi(x) = \phi(-x)$ (called the N-function), which usually uses the absolute value $|x| = \max\{-x, x\}$ under the argument of ϕ. Because probability measures are positive functions, the transformation $x \mapsto |x|$ appears

This work was supported in part by BBSRC grant BB/L009579/1.

F. Nielsen and F. Barbaresco (Eds.): GSI 2015, LNCS 9389, pp. 203–210, 2015.
DOI: 10.1007/978-3-319-25040-3_23

to be quite innocent and well-justified, because one can apply the highly developed theory of Banach spaces. However, this may loose asymmetry that is quite natural in some random phenomena. Moreover, symmetrization on the statistical manifold $\mathcal{P}(\Omega)$ also automatically symmetrizes the topology of the dual space containing random variables. When these random variables are used in the context of optimization (e.g. as utility or cost functions), their symmetrization is rather unnatural, and some random variables cannot be used. Let us illustrate this in the following examples.

Example 1 (The St. Petersburg lottery). The lottery is played by tossing a coin repeatedly until the first head appears. The probability of head occurring on the nth toss assuming independent and identically distributed (i.i.d.) tosses of a fair coin is $q(n) = 2^{-n}$. If the payoff is $x(n) = 2^n$, then the lottery has infinite expected payoff (this is historically the first example of unbounded expectation [8]). If the coin is biased towards head, however, such as $p(n) = 2^{-(1+\alpha)n}$, ($\alpha > 0$), then the expected payoff becomes finite. The effective domain of the moment generating function $\mathbb{E}_q\{e^{\beta x}\}$ does not contain the ray $\{\beta x : \beta > 0\}$, but it does contain the ray $\{-\beta x : \beta \geq 0\}$. Thus, random variable $x(n) = 2^n$ belongs to the space, where zero is not in the interior of the effective domain of $\mathbb{E}_q\{e^{\beta x}\}$ or of the cumulant generating function $\Psi_q(\beta x) = \ln \mathbb{E}_q\{e^{\beta x}\}$ (the Legendre-Fenchel transform of $D_{KL}[p, q]$). This implies that sublevel sets $\{p : D_{KL}[p, q] \leq \lambda\}$ in the dual space are unbounded (see Theorem 4 of [9,10]). Note that exponential family distributions $p(x; \beta) = e^{\beta x - \Psi_q(\beta x)}q$ solve the problem of maximization of random variable x on $\{p : D_{KL}[p, q] \leq \lambda\}$, while $p(-x; \beta) = e^{-\beta x - \Psi_q(\beta x)}q$ solve minimization (i.e. maximization of $-x$). This example illustrates asymmetry typical of optimization problems, because random variable $x(n) = 2^n$ has bottom ($x(1) = 2$), but it is topless.

Example 2 (Error minimization). Consider the problem of minimization of the error function $z(a, b)$ (or equivalently maximization of utility $x = -z$), which can be defined using some metric $d : \Omega \times \Omega \to [0, \infty)$ on Ω. For example, using the Hamming metric $d_H(a, b) = \sum_{n=1}^{l} \delta_{b_n}(a_n)$ on finite space $\{1, \ldots, \alpha\}^l$ or using squared Euclidean metric $d_E^2(a, b) = \sum_{n=1}^{l} |a_n - b_n|^2$ on a real space $\mathbb{R}^l$ (i.e. by defining utility $x = -d_H$ or $x = -\frac{1}{2}d_E^2$). Let w be the joint distribution of a and $b \in \Omega$, and let q, p be its marginal distributions. The KL-divergence $D_{KL}[w, q \otimes p] =: I_S(a, b)$ defines the amount of mutual information between a, b. Joint distributions minimizing the expected error subject to constraint $I_S(a, b) \leq \lambda$ belong to the exponential family $w(x; \beta) = e^{\beta x - \Psi_{q \otimes p}(\beta x)}q \otimes p$. With the maximum entropy $q \otimes p$ and Hamming metric $x = -d_H$, this $w(x; \beta)$ corresponds to the binomial distribution, and in the case of squared Euclidean metric $x = -\frac{1}{2}d_E^2$ to the Gaussian distribution. In the finite case of $\Omega = \{1, \ldots, \alpha\}^l$, the random variable $x = -d_H$ can be reflected $x \mapsto -x = d_H$, as both $\Psi_{q \otimes p}(-\beta d_H)$ and $\Psi_{q \otimes p}(\beta d_H)$ are finite (albeit possibly with different values). However, in the infinite case of $\Omega = \mathbb{R}^l$, the unbounded random variable $x = -\frac{1}{2}d_E^2$ cannot be reflected, as maximization of Euclidean distance has no solution, and $\Psi_{q \otimes p}(\beta \frac{1}{2}d_E^2) = \infty$ for any $\beta \geq 0$. As in the previous example, $0 \notin \text{Int}(\text{dom}\,\Psi_{q \otimes p})$.

The examples above illustrate that symmetrization of neighbourhoods on the statistical manifold requires random variables to be considered together with their reflections $x \mapsto -x$. However, this is not always desirable or even possible in the infinite-dimensional case. First, random variables used in optimization problems, such as utilities or cost functions, do not form a linear space, but a wedge. Operations $x \mapsto -x$ and $x \mapsto |x|$ are not monotonic. Second, the wedge of utilities or cost functions may include unbounded functions (e.g. concave utilities $x : \Omega \to \mathbb{R} \cup \{-\infty\}$ and convex cost functions $z : \Omega \to \mathbb{R} \cup \{\infty\}$). In some cases, one of the functions x or $-x$ cannot be absorbed into the effective domain of the cumulant generating function Ψ (i.e. $\beta x \notin \operatorname{dom} \Psi$ for any $\beta > 0$), in which case the symmetrization $x \mapsto |x|$ would leave such random variables out.

In the next section, we shall outline the main ideas for defining dual asymmetric topologies using polar sets and sublinear functions related to them. In Sect. 3, we shall introduce a generalization of Bregman divergence, a generalized law of cosines and define associated asymmetric seminorms and quasi-metrics. In Sect. 4, we shall prove that asymmetric topology defined by the KL-divergence is complete, Hausdorff and contains a separable Orlicz subspace.

2 Topologies Induced by Gauge and Support Functions

Let X and Y be a pair of linear spaces over $\mathbb{R}$ put in duality via a non-degenerate bilinear form $\langle \cdot, \cdot \rangle : X \times Y \to \mathbb{R}$:

$$\langle x, y \rangle = 0, \ \forall x \in X \ \Rightarrow y = 0, \qquad \langle x, y \rangle = 0, \ \forall y \in Y \ \Rightarrow x = 0$$

When x is understood as a random variable and y as a probability measure, then the pairing is just the expected value $\langle x, y \rangle = \mathbb{E}_y\{x\}$. We shall define topologies on X and Y that are compatible with respect to the pairing $\langle \cdot, \cdot \rangle$, but the bases of these topologies will be formed by systems of neighbourhoods of zero that are generally non-balanced sets (i.e. $y \in M$ does not imply $-y \in M$). It is important to note that such spaces may fail to be topological vector spaces, because multiplication by scalar can be discontinuous (e.g. see [11]). Let us first recall some properties that depend only on the pairing $\langle \cdot, \cdot \rangle$.

Each non-zero $x \in X$ is in one-to-one correspondence with a hyperplane $\partial \Pi x := \{y : \langle y, x \rangle = 1\}$ or a closed halfspace $\Pi x := \{y : \langle y, x \rangle \leq 1\}$. The intersection of all Πx containing M is the *convex closure* of M denoted by $\operatorname{co}[M]$. Set M is closed and convex if $M = \operatorname{co}[M]$. The *polar* of $M \subseteq Y$ is

$$M^\circ := \{x \in X : \langle x, y \rangle \leq 1, \ \forall y \in M\}$$

The polar set is always closed and convex and $0 \in M^\circ$. Also, $M^{\circ\circ} = \operatorname{co}[M \cup \{0\}]$, and $M = M^{\circ\circ}$ if and only if M is closed, convex and $0 \in M$. Without loss of generality we shall assume $0 \in M$. The mapping $M \mapsto M^\circ$ has the properties:

$$(M \cup N)^\circ = M^\circ \cap N^\circ \tag{1}$$

$$(M \cap N)^\circ = \operatorname{co}[M^\circ \cup N^\circ] \tag{2}$$

We remind that set $M \subseteq Y$ is called:

Absorbing if $y/\alpha \in M$ for all $y \in Y$ and $\alpha \geq \varepsilon(y)$ for some $\varepsilon(y) > 0$.
Bounded] if $M \subseteq \alpha \Pi x$ for any closed halfspace Πx and some $\alpha > 0$.
Balanced if $M = -M$.

Set M is absorbing if and only if its polar M° is bounded; If M is balanced, then so is M°. If M is closed and convex, then the following are balanced closed and convex sets: $-M \cap M$, $\mathrm{co}\,[-M \cup M]$.

Given set $M \subseteq Y$, $0 \in M$, the set $Y_M := \{y : y/\alpha \in M, \forall \alpha \geq \varepsilon(y) > 0\}$ of elements absorbed into M can be equipped with a topology, uniquely defined by the base of closed neighbourhoods of zero $\mathfrak{M} := \{\alpha M : \alpha > 0\}$. The set $X_M := \{x : \langle x, y \rangle \leq \alpha, \forall y \in M\}$ of hyperplanes bounding M are absorbed into the polar set M°, and the collection $\mathfrak{M}^\circ := \{\alpha^{-1} M^\circ : \alpha^{-1} > 0\}$ is the base of the polar topology on X_M. Note that Y_M (resp. X_M) is a strict subset of Y (resp. X), unless M is absorbing (resp. bounded). Moreover, it may fail to be a topological vector space, unless M (or M°) is balanced. Such polar topologies can be defined using gauge or support functions.

The *gauge* (or *Minkowski* functional) of set $N \subseteq X$ is the mapping $\mu N : X \to \mathbb{R} \cup \{\infty\}$ defined as

$$\mu N(x) := \inf\{\alpha > 0 : x/\alpha \in N\}$$

with $\mu N(0) := 0$ and $\mu N(x) := \infty$ if $x/\alpha \notin N$ for all $\alpha > 0$. Note that $\mu N(x) = 0$ if $x/\alpha \in N$ for all $\alpha > 0$. The following statements are implied by the definition.

Lemma 1. $\mu N(x) < \infty$ *for all* $x \in X$ *if and only if* N *is absorbing;* $\mu N(x) > 0$ *for all* $x \neq 0$ *if and only if* N *is bounded.*

The gauge is positively homogeneous function of the first degree, $\mu N(\beta x) = \beta \mu N(x)$, $\beta > 0$, and if N is convex, then it is also subadditive, $\mu N(x_1 + x_2) \leq \mu N(x_1) + \mu N(x_2)$. Thus, the gauge of an absorbing closed convex set satisfies all axioms of a seminorm apart from symmetry, and therefore it is a *quasiseminorm*. Function $\rho_N(x_1, x_2) = \mu N(x_2 - x_1)$ is a *quasi-pseudometric* on X. If N is bounded, then μN is a *quasi-norm* and d_N is a *quasi-metric*. Symmetry $\mu N(x) = \mu N(-x)$ and $\rho_N(x_1, x_2) = \rho_N(x_2, x_1)$ requires N to be balanced.

The *support* function of set $M \subseteq Y$ is the mapping $sM : X \to \mathbb{R} \cup \{\infty\}$:

$$sM(x) := \sup\{\langle x, y \rangle : y \in M\}$$

Like the gauge, the support function is also positively homogeneous of the first degree, and it is always subadditive. Generally, $\mu N(x) \geq sN^\circ(x)$, with equality if and only if N is convex. In fact, the following equality holds:

Lemma 2. $sM(x) = \mu M^\circ(x)$, $\forall M \subseteq Y$, $0 \in M$.

Proof. $\langle x, y \rangle \leq sM(x)$ for all $y \in M$, $sM(x/\alpha) = \alpha^{-1} sM(x)$, $sM(x) := \inf\{\alpha > 0 : \langle x/\alpha, y \rangle \leq 1, \forall y \in M\} = \inf\{\alpha > 0 : x/\alpha \in M^\circ\}$. $\square$

The following is the asymmetric version of the Hölder inequality:

Lemma 3 (Asymmetric Hölder). $\langle x, y \rangle \leq sM(x)sM^\circ(y)$, $\forall M \subseteq Y$, $0 \in M$.

Proof. $\langle x, y \rangle \leq sM(x)$, $\langle x/sM(x), y \rangle \leq 1$ for all $y \in M$, so that $x/sM(x) \in M^\circ$. Hence $\langle x/sM(x), y \rangle \leq sM^\circ(y)$. □

The support function $sM(x)$ can be symmetrized in two ways:

$$s^s M(x) := s[-M \cup M](x), \qquad s^\circ M(x) := s[-M \cap M](x)$$

Lemma 4. *1.* $s^s M(x) \geq sM(x) \geq s^\circ M(x)$.
2. $s^s M(x) = \sup\{sM(-x), sM(x)\}$.
3. $s^\circ M(x) = \text{co}\,[\inf\{sM(-x), sM(x)\}] = \inf\{sM(z) + sM(z - x) : z \in X\}$.
4. $s^\circ M(x) = \sup\{\langle x, y \rangle : s^s M^\circ(y) \leq 1\}$.

Proof. 1. Follows from set inclusions: $-M \cup M \supseteq M \supseteq -M \cap M$.
2. $s^s M(x) = \mu[-M^\circ \cap M^\circ](x) = \sup\{\mu M^\circ(-x), \mu M^\circ(x)\}$ by Lemma 2 and equation (1).
3. $s^\circ M(x) = \mu\text{co}\,[-M^\circ \cup M^\circ](x) = \text{co}\,[\inf\{\mu M^\circ(-x), \mu M^\circ(x)\}]$ by Lemma 2 and equation (2). The second equation follows from the equivalence of convex closure infimum and infimal convolution for sublinear functions [12].
4. Follows from $sM(x) = sM^{\circ\circ}(x)$, $0 \in M$, and $N^\circ = \{y : sN(y) \leq 1\}$ by substituting $N = (-M \cap M)^\circ = \text{co}\,[-M^\circ \cup M^\circ]$. □

3 Distance Functions and Sublevel Neighbourhoods

A closed neighbourhood of $z \in Y$ can be defined by sublevel set $\{y : D[y, z] \leq \lambda\}$ of a *distance* function $D : Y \times Y \to \mathbb{R} \cup \{\infty\}$ satisfying the following axioms:

1. $D[y, z] \geq 0$.
2. $D[y, z] = 0$ if $y = z$.

Thus, a distance is generally not a metric (i.e. non-degeneracy, symmetry or the triangle inequality are not required). A distance function associated with closed functional $F : Y \to \mathbb{R} \cup \{\infty\}$ can be defined as follows:

$$D_F[y, z] := \inf\{F(y) - F(z) - \langle x, y - z \rangle : x \in \partial F(z)\} \tag{3}$$

The set $\partial F(z) := \{x : \langle x, y - z \rangle \leq F(y) - F(z), \forall y \in Y\}$ is called *subdifferential* of F at z. It follows immediately from the definition of subdifferential that $D_F[y, z] \geq 0$. We shall define $D_F[y, z] := \infty$, if $\partial F(z) = \varnothing$ or $F(y) = \infty$. We note that the notion of subdifferential can be applied to a non-convex function F. However, non-empty $\partial F(z)$ implies $F(z) < \infty$ and $F(z) = F^{**}(z)$, $\partial F(z) = \partial F^{**}(z)$ ([13], Theorem 12). Generally, $F^{**} \leq F$, so that $F(y) - F(z) \geq F^{**}(y) - F^{**}(z)$ if $\partial F(z) \neq \varnothing$. If F is Gâteaux differentiable at z, then $\partial F(z)$ has a single element $x = \nabla F(z)$, called the *gradient* of F at z. Thus, definition (3) is a generalization of the Bregman divergence for the case of a non-convex and non-differentiable F. Note that the dual functional F^* defines dual distance D_F^* on X, which is related to D_F as follows: $D_F[y, z] = D_F^*[\nabla F(z), \nabla F(y)]$.

Theorem 1. $D_F[y, z] = 0 \iff \{y, z\} \subseteq \partial F^*(x)$, $\exists x \in X$.

Proof. If $y = z$, then $D_F[y, z] = 0$ by definition. If $y \neq z$, then $\{y, z\} \subseteq \partial F^*(x) \iff \partial F(y) = \partial F(z) = \{x\}$, which follows from the property of subdifferentials: $y \in \partial F^*(x) \iff \partial F(y) \ni x$ ([13], Corollary to Theorem 12). Thus, $D_F[y, z] = D_{F^*}[\nabla F(z), \nabla F(y)] = D_F^*[x, x] = 0$. □

Corollary 1. D_F *separates points of* $\mathrm{dom}\, F \subseteq Y$ *if and only if* F *is Gâteaux differentiable or* F^* *is strictly convex.*

Let us denote by $\nabla_1 D[y, z]$ and $\nabla_1^2 D[y, z]$ the first and the second Gâteaux differentials of $D[y, z]$ with respect to the first argument. For a twice Gâteaux differentiable F they are $\nabla_1 D_F[y, z] = \nabla F(y) - \nabla F(z)$ and $\nabla_1^2 D_F[y, z] = \nabla^2 F(y)$.

Theorem 2 (Generalized Law of Cosines). *The following statements are equivalent:*

$$D[y, z] = \int_0^1 (1 - t) \Big\langle \nabla_1^2 D[z + t(y - z), y](y - z), y - z \Big\rangle \, dt$$

$$D[y, w] = D[y, z] + D[z, w] - \langle \nabla_1 D[z, w], z - y \rangle$$

Proof. Consider the first order Taylor expansion of $D[\cdot, w]$ at z:

$$D[y, w] = D[z, w] + \langle \nabla_1 D[z, w], y - z \rangle + R_1[z, y]$$

where the remainder is $R_1[z, y] = \int_0^1 (1 - t) \langle \nabla_1^2 D[z + t(y - z), y](y - z), y - z \rangle \, dt$. The result follows from the equality $D[y, z] = R_1[z, y]$. □

An asymmetric seminorm on space X can be defined either by the gauge or support function of sublevel sets of distances $D_F^*[x, 0]$ and $D_F[y, z]$ respectively:

$$\|x|_{F^*} := \inf\{\alpha > 0 : D_F^*[x/\alpha, 0] \leq 1\}, \quad \|x|_F := \sup_y\{\langle x, y - z \rangle : D_F[y, z] \leq 1\}$$

The supremum is achieved at $y(\beta) \in \partial F^*(\beta x)$, $D_F[y(\beta), z] = 1$. A quasi-pseudometric is defined as $\rho_{F^*}(w, x) = \|x - w|_{F^*}$ or $\rho_F(w, x) = \|x - w|_F$. The dual space Y is equipped with asymmetric seminorms and quasi-pseudometrics in the same manner. The following characterization of the topology is known.

Theorem 3 ([14] or see Proposition 1.1.40 in [6]). *An asymmetric seminormed space* X *is:*

T_0 *if and only if* $\|x|_{F^*} > 0$ *or* $\| - x|_{F^*} > 0$ *for all* $x \neq 0$;
T_1 *if and only if* $\|x|_{F^*} > 0$ *for all* $x \neq 0$;
T_2 *(Hausdorff) if and only if* $\|x|_{F^*} > 0$ *and* $\| - x|_{F^*} > 0$ *for all* $x \neq 0$.

These separation properties depend on sublevel set $\{x : D_F^*[x, 0] \leq 1\}$. For T_0 it must not contain any hyperplane; for T_1 it must not contain any ray (i.e. it must be bounded); for T_2 it also must contain zero in the interior (i.e. it must be bounded and absorbing). The following theorem is useful in our analysis.

Theorem 4 ([9,10]). *If* $0 \in \mathrm{Int}(\mathrm{dom}\, F^*) \subset X$, *then sublevel sets* $\{y : F(y) \leq \lambda\}$ *are bounded. Conversely, if one of the sublevel sets for* $\lambda > \inf F$ *is bounded, then* $0 \in \mathrm{Int}(\mathrm{dom}\, F^*)$.

4   Asymmetric Topology Generated by the KL-Divergence

The KL-divergence can be defined as Bregman divergence associated with closed convex functional $KL(y) = \langle \ln y - 1, y \rangle$:

$$D_{KL}[y, z] = \langle \ln y - \ln z, y \rangle - \langle 1, y - z \rangle$$

Note that KL is a proper closed convex functional that is finite for all $y \geq 0$, if we define $(\ln 0) \cdot 0 = 0$ and $KL(y) = \infty$ for $y \not\geq 0$. The dual of KL is the moment generating functional $KL^*(x) = \langle e^x, z \rangle$. The dual divergence of x from 0 is:

$$D_{KL}^*[x, 0] = \langle e^x - 1 - x, z \rangle$$

The above divergence can be written as $D_{KL}^*[x, 0] = \langle \phi^*(x), z \rangle$, where $\phi^*(x) = e^x - 1 - x$. The dual of ϕ^* is the closed convex function $\phi(u) = (1+u)\ln(1+u) - u$. Making the change of variables $y \mapsto u = \frac{y}{z} - 1$, the KL-divergence can be written in terms of $\phi(u)$: $D_{KL}[y, z] = D_{KL}[(1+u)z, z] = \langle (1+u)\ln(1+u) - u, z \rangle$. Sublevel set $M = \{y - z : D_{KL}[y, z] \leq 1\}$ is a closed neighbourhood of $0 \in Y - z$; sublevel set $N = \{x : D_{KL}^*[x, 0] \leq 1\}$ is a closed neighbourhoods of $0 \in X$. Both functions $\phi(u)$ and $\phi^*(x)$ are not even, and these neighbourhoods are not balanced. In the theory of Orlicz spaces the symmetrized functions $\phi(|u|)$ and $\phi^*(|x|)$ are used to define even functionals and norms [15]. This approach has been used in infinite-dimensional information geometry [7]. In particular, because $\phi(|u|)$ belongs to the Δ_2 class [15], the corresponding Orlicz space $Y_{\phi(|\cdot|)}$ (and the statistical manifold it contains) is separable. The dual Orlicz space $X_{\phi^*(|\cdot|)}$ is not separable, because $\phi^*(|x|)$ is not Δ_2. Note, however, that another symmetrization is possible: $\phi(-|u|)$, which is not Δ_2, and $\phi^*(-|x|)$, which is Δ_2. Thus, one can introduce the non-separable Orlicz space $Y_{\phi(-|\cdot|)}$, and the dual separable Orlicz space $X_{\phi^*(-|\cdot|)}$. One can check that the following inequalities hold: $\phi(|u|) \leq \phi(u) \leq \phi(-|u|)$ (resp. $\phi^*(|x|) \geq \phi^*(x) \geq \phi^*(-|x|)$), which corresponds to the following symmetrizations and inclusions of sublevel sets: $\mathrm{co}\,[-M \cup M] \supseteq M \supseteq -M \cap M$ (resp. $-N \cap N \subseteq N \subseteq \mathrm{co}\,[-N \cup N]$). Thus, the asymmetric topology of space Y_ϕ, induced by D_{KL} (resp. of X_{ϕ^*}, induced by D_{KL}^*) is finer than topology of the separable Orlicz space $Y_{\phi(|\cdot|)}$ (resp. $X_{\phi^*(-|\cdot|)}$), and so it is Hausdorff. On the other hand, the diameter $\mathrm{diam}(M) = \sup\{\rho_{KL}(y, z) : y, z \in M\}$ of set $M \subset Y$ (resp. for $\rho_{KL}^*(x, w)$ and $N \subset X$) is the diameter with respect to the metric of the Orlicz space $Y_{\phi(-|\cdot|)}$ (resp. $X_{\phi^*(|\cdot|)}$), which is complete. Therefore, every nested sequence of sets with diameters decreasing to zero has non-empty intersection, so that space Y_ϕ (resp. X_{ϕ^*}) is ρ-sequentially complete ([16], Theorem 10). Thus, we have proven the following theorem, concluding this short paper.

Theorem 5. *Asymmetric seminorm* $\|x\|_{KL}^* := \sup\{\langle x, y - z \rangle : D_{KL}[y, z] \leq 1\}$ *(resp.* $\|y - z\|_{KL} := \inf\{\alpha^{-1} > 0 : D_{KL}[z + \alpha(y - z), z] \leq 1\}$*) induces Hausdorff topology on space* X *(resp. on* $Y = Y_+ - Y_+$*), and therefore it is an asymmetric norm. It is* ρ-*sequentially complete and contains a separable subspace, which is an Orlicz space with the norm* $\|x\|_{\phi^*(-|\cdot|)}$ *(resp.* $\|y - z\|_{\phi(|\cdot|)}$*).*

References

1. Kullback, S., Leibler, R.A.: On information and sufficiency. Ann. Math. Stat. **22**(1), 79–86 (1951)
2. Chentsov, N.N.: A nonsymmetric distance between probability distributions, entropy and the Pythagorean theorem. Math. notes Acad. Sci. USSR **4**(3), 323–332 (1968)
3. Chentsov, N.N.: Statistical decision rules and optimal inference. Nauka, Moscow, U.S.S.R. In: Russian, English translation: Providence, p. 1982. AMS, RI (1972)
4. Morozova, E.A., Chentsov, N.N.: Natural geometry of families of probability laws. In: Prokhorov, Y.V. (ed.) Probability Theory 8. Volume 83 of Itogi Nauki i Tekhniki, pp. 133–265. VINITI, Russian (1991)
5. Fletcher, P., Lindgren, W.F.: Quasi-uniform spaces. Lecture notes in pure and applied mathematics. Vol. 77, Marcel Dekker, New York (1982)
6. Cobzas, S.: Functional Analysis in Asymmetric Normed Spaces. Birkhäuser, Basel (2013)
7. Pistone, G., Sempi, C.: An infinite-dimensional geometric structure on the space of all the probability measures equivalent to a given one. Ann. Stat. **23**(5), 1543–1561 (1995)
8. Bernoulli, D.: Commentarii academiae scientarum imperialis petropolitanae. Econometrica **22**, 23–36 (1954)
9. Asplund, E., Rockafellar, R.T.: Gradients of convex functions. Trans. Am. Math. Soc. **139**, 443–467 (1969)
10. Moreau, J.J.: Functionelles Convexes, Lectrue Notes, Séminaire sur les équations aux derivées partielles. Collége de France, Paris (1967)
11. Borodin, P.A.: The Banach-Mazur theorem for spaces with asymmetric norm. Math. Notes **69**(3–4), 298–305 (2001)
12. Tikhomirov, V.M.: Convex analysis. In: Gamkrelidze, R.V. (ed.) Analysis II. Encyclopedia of Mathematical Sciences. vol. 14, pp. 1–92. Springer-Verlag, Heidelberg (1990)
13. Rockafellar, R.T.: Conjugate duality and optimization. In: CBMS-NSF Regional Conference Series in Applied Mathematics, vol. 16, Society for Industrial and Applied Mathematics, PA (1974)
14. García-Raffi, L.M., Romaguera, S., Sánchez-Pérez, E.A.: On Hausdorff asymmetric normed linear spaces. Houston J. Math. **29**(3), 717–728 (2003)
15. Krasnoselskii, M.A., Rutitskii, Y.B.: Convex functions and Orlicz spaces. Fizmatgiz, Moscow (1958). English translation.: P. Noordhoff, Ltd., Groningen (1961)
16. Reilly, I.L., Subrahmanyam, P.V., Vamanamurthy, M.K.: Cauchy sequences in quasi-pseudo-metric spaces. Monatshefte für Mathematik **93**, 127–140 (1982)

Hessian Information Geometry

Hessian Structures and Non-invariant (F, G)-Geometry on a Deformed Exponential Family

K.V. Harsha and K.S. Subrahamanian Moosath[✉]

Department of Mathematics, Indian Institute of Space Science and Technology,
Valiamala P.O, 695547 Thiruvananthapuram, Kerala, India
{harsha.11,smoosath}@iist.ac.in

Abstract. A deformed exponential family has two kinds of dual Hessian structures, the U-geometry and the χ-geometry. In this paper, we discuss the relation between the non-invariant (F, G)-geometry and the two Hessian structures on a deformed exponential family. A generalized likelihood function called F-likelihood function is defined and proved that the Maximum F-likelihood estimator is a Maximum a posteriori estimator.

1 Introduction

On a statistical manifold $\mathcal{S}$, one can categorize geometries into invariant and non-invariant classes according to their invariance under smooth one to one transformations of the random variable. The invariant α-geometry is relevant in the asymptotic theory of an exponential family [1]. One category of non-invariant geometries on a statistical manifold is the (F, G)-geometry defined using an embedding function F and a positive smooth function G [2]. The (F, G)-geometry is relevant in the study of deformed exponential family. A deformed exponential family has two kinds of Hessian structures; the U-geometry introduced by Naudts [3] (see also [4]) and the χ-geometry defined by Amari et al. [5].

In this paper we present the relation between the (F, G)-geometry and these two Hessian structures. In Sect. 2 the two Hessian structures on a deformed exponential family are given. In Sect. 3 we discuss the role of (F, G)-geometry in the study of Hessian structures of the deformed exponential family. In Sect. 4 a generalized product called F-product, F-independence and F-likelihood function are defined using an increasing concave function F and its inverse Z. Then a generalized maximum likelihood estimator called the maximum F-likelihood estimator (F-MLE) is defined and showed that the F-MLE is a Maximum a posteriori estimator (MAP estimator).

Definition 1. (F, G)-geometry
*Let $\mathcal{S} = \{p(x; \theta) \ / \ \theta \in \mathbb{E}\}$ be a statistical manifold defined on $\mathcal{X} \subset \mathbb{R}^n$.
Let $F : (0, \infty) \longrightarrow \mathbb{R}$ be a function which is atleast twice differentiable. Assume that $F'(u) \neq 0, \ \forall \ u \in (0, \infty)$. Then F is an embedding of $\mathcal{S}$ into the space of*

© Springer International Publishing Switzerland 2015
F. Nielsen and F. Barbaresco (Eds.): GSI 2015, LNCS 9389, pp. 213–221, 2015.
DOI: 10.1007/978-3-319-25040-3_24

random variable $\mathbb{R}_{\mathcal{X}}$ *which takes each* $p(x;\theta) \longmapsto F(p(x;\theta))$. *Let* $G : (0,\infty) \longrightarrow$
$\mathbb{R}$ *be a positive smooth function.*
The **G-metric** g^G *and the* (F,G)-**connection** $\nabla^{F,G}$ *are defined as*

$$g_{ij}^G(\theta) = \int \partial_i \ell\, \partial_j \ell\, G(p)\, p\, dx. \tag{1}$$

$$\Gamma_{ijk}^{F,G} = \int \left(\partial_i \partial_j \ell + (1 + \frac{pF''(p)}{F'(p)})\partial_i \ell\, \partial_j \ell \right) \partial_k \ell\, G(p)\, p\, dx. \tag{2}$$

where $\ell(p) = \log p$.
When $G(p) = 1$ *and* $F(p) = L_\alpha(p)$, (F,G)-*connection reduces to the*
α-*connection and the G-metric reduces to the Fisher information metric.*

Theorem 2. *Let* F, H *be two embeddings of statistical manifold* S *into the space*
$\mathbb{R}_X$ *of random variables. Let* G *be a positive smooth function on* $(0,\infty)$. *Then*
$(F,G)-$*connection* $\nabla^{F,G}$ *and the* $(H,G)-$*connection* $\nabla^{H,G}$ *are dual connections*
with respect to the G-metric iff the functions F *and* H *satisfy*

$$H'(p) = \frac{G(p)}{pF'(p)} \tag{3}$$

We call such an embedding H *as a* G-**dual embedding** *of* F.
The components of the dual connection $\nabla^{H,G}$ *can be written as*

$$\Gamma_{ijk}^{H,G} = \int \left(\partial_i \partial_j \ell + (1 + \frac{pH''(p)}{H'(p)})\partial_i \ell\, \partial_j \ell \right) \partial_k \ell\, G(p)\, p\, dx \tag{4}$$

$$= \int \left(\partial_i \partial_j \ell + (\frac{pG'(p)}{G(p)} - \frac{pF''(p)}{F'(p)})\partial_i \ell\, \partial_j \ell \right) \partial_k \ell\, G(p)\, p\, dx. \tag{5}$$

(Refer [2] for more details).

Definition 3. *Let* (M,g) *be a Riemannian manifold and let* ∇ *be a flat connec-*
tion on M. *Then we say that the pair* (M,g) *is a Hessian structure on* M *or*
(M,g,∇) *is a Hessian manifold if there exist a function* ψ *such that* $S = \nabla d\psi$.
Let ∇^* *be the dual connection of* ∇ *with respect to the metric* g. *Then the con-*
dition (M,g,∇) *is a Hessian manifold is equivalent to that* (M,g,∇,∇^*) *is a*
dually flat space [6].

2   Hessian Structures on a F-Exponential Family

A deformed exponential family which is a generalization of exponential family
was introduced by Naudts [3]. A deformed family has two kinds of Hessian structures or equivalently two dually flat structures; U-geometry given by Naudts [3]
(see also [4]) and χ-geometry given by Amari et al. [5] (see also [6,7,11]).
In this section, we describe these two Hessian structures. For the sake of notational convenience deformed exponential family is formulated using the function
F and we call it as a F-exponential family.

Definition 4. F-Exponential Family *Let* $F : (0, \infty) \longrightarrow \mathbb{R}$ *be any smooth increasing concave function. Let Z be the inverse function of F. Define the standard form of a n-dimensional F-exponential family of distributions as*

$$p(x; \theta) = Z(\sum_{i=1}^{n} \theta^i x_i - \psi(\theta)) \quad or \quad F(p(x; \theta)) = \sum_{i=1}^{n} \theta^i x_i - \psi(\theta) \qquad (6)$$

where $x = (x_1, \cdots, x_n)$ is a set of random variables, $\theta = (\theta^1, \cdots, \theta^n)$ are the canonical parameters and $\psi(\theta)$ is determined from the normalization condition.

2.1 U-geometry on the F-exponential Family

Consider an n-dimensional F-exponential family $S = \{p(x; \theta) \ / \ \theta \in E \subseteq \mathbb{R}^n\}$. Now define a divergence of Bregman type given by Naudts [3] as

$$D^F(p, q) = \int \left(\int_q^p (F(u) - F(q)) du \right) dx \qquad (7)$$

The geometry induced from the divergence D^F called U-**geometry** (refer also [4]) is given by

$$g_{ij}^{D^F}(\theta) = \int \partial_i p \, \partial_j F \, dx \ ; \quad \Gamma_{ijk}^{D^F}(\theta) = 0 \ ; \quad \Gamma_{ijk}^{D^{*F}}(\theta) = \int \partial_k F \, \partial_i \partial_j p \, dx \quad (8)$$

Hence affine connection ∇^{D^F} is flat. Now define a function ([6])

$$v(t) = \int_1^t F(s) \, ds, \quad t > 0. \qquad (9)$$

Assume that $v(o) := \lim_{t \to +0} v(t)$ is finite.

The generalized entropy functional I and generalized Massieu potential Ψ are defined as

$$I(p_\theta) := - \int [v(p(x; \theta)) + (p(x; \theta) - 1)v(o)] \, dx. \qquad (10)$$

$$\Psi(\theta) := \int p(x; \theta) F(p(x; \theta)) \, dx + I(p_\theta) + \psi(\theta). \qquad (11)$$

Theorem 5. *For a F-exponential family S, we have*

1. *(g^{D^F}, ∇^{D^F}) and $(g^{D^F}, \nabla^{D^{*F}})$ are mutually dual Hessian structures on S equivalently, $(S, g^{D^F}, \nabla^{D^F}, \nabla^{D^{*F}})$ is a dually flat space.*
2. *The canonical co-ordinate θ is ∇^{D^F}-affine and Ψ is the potential function corresponding to θ.*
3. *The metric $g_{ij}^{D^F}(\theta) = \partial_i \partial_j \Psi(\theta)$.*
4. *The dual co-ordinate η is given by $\eta_i = \partial_i \Psi(\theta) = E_p[x_i]$ and it is $\nabla^{D^{*F}}$-affine.*
5. *The dual potential function Φ corresponding to the dual co-ordinate η is $\Phi(\eta) = -I(p_\theta)$.*

(Refer [3,6] for more details).

2.2   χ-geometry on the F-Exponential Family

Let $S = \{p(x; \theta) \ / \ \theta \in E \subseteq \mathbb{R}^n\}$ be a F-exponential family.

Definition 6. *For a distribution function p parametrized by θ, let us define a probability distribution*

$$\hat{p}_F(x) = \frac{1}{h_F(\theta)F'(p)}, \quad \text{where } h_F(\theta) = \int \frac{1}{F'(p)} dx \qquad (12)$$

called F-escort probability distribution related to p.

Definition 7. *Using this escort probability distribution $\hat{p}_F$, the $\hat{F}$-expectation of a random variable is defined as*

$$E_{\hat{p}_F}(f(x)) = \frac{1}{h_F(\theta)} \int \frac{1}{F'(p)} f(x) dx \qquad (13)$$

Note 8. Note that the F-potential function $\psi_F(\theta)$ is a convex function of θ and we have

$$\partial_i \partial_j \psi_F(\theta) = \frac{1}{h_F(\theta)} \int \frac{-F''(p)}{(F'(p))^3} (x_i - \partial_i \psi_F(\theta)) (x_j - \partial_j \psi_F(\theta)) \qquad (14)$$

is positive semidefinite. Further we assume that it is positive definite. Then $\psi_F(\theta)$ is a strictly convex function of θ.

Definition 9. *A divergence of Bregman-type D_F is defined using $\psi_F(\theta)$, as*

$$D_F(p(x; \theta_1) : p(x; \theta_2)) = \psi_F(\theta_2) - \psi_F(\theta_1) - \nabla\psi_F(\theta_1).(\theta_2 - \theta_1) \qquad (15)$$

Lemma 10. *The geometry induced from the divergence D_F called the χ-geometry is given by*

$$g_{ij}^{D_F}(\theta) = \partial_i \partial_j \psi_F(\theta) ; \quad \Gamma_{ijk}^{D_F} = \partial_i \partial_j \partial_k \psi_F(\theta) ; \quad \Gamma_{ijk}^{D_F^*} = 0 \qquad (16)$$

Hence the affine connection $\nabla^{D_F^*}$ is flat.

Theorem 11. *For a F-exponential family S, we have*

1. *(g^{D_F}, ∇^{D_F}) and $(g^{D_F}, \nabla^{D_F^*})$ are mutually dual Hessian structures on S equivalently, $(S, g^{D_F}, \nabla^{D_F}, \nabla^{D_F^*})$ is a dually flat space.*
2. *The canonical co-ordinate θ is $\nabla^{D_F^*}$-affine and ψ_F is the potential function corresponding to θ.*
3. *The metric $g_{ij}^{D_F}(\theta) = \partial_i \partial_j \psi_F(\theta)$.*
4. *The dual co-ordinate η is given by $\eta_i = \partial_i \psi_F(\theta) = E_{\hat{p}_F}[x_i]$ and it is ∇^{D_F}-affine.*
5. *The dual potential function ϕ_F corresponding to the dual co-ordinate η is $\phi_F(\eta) = E_{\hat{p}_F}(F(p))$.*

(Refer [5, 6] for more details.)

3 Non-invariant (F,G)-geometry on a Deformed Exponential Family

In this section we describe how the two Hessian structures are related to the non-invariant (F,G)-geometry. First we show that the U-geometry is (F,G)-geometry for suitable choices of F and G. Also we show that the χ-geometry is the conformal flattening of (F,G)-geometry for suitable choices of F and G.

3.1 U-geometry as (F,G)-geometry

Proposition 12. *On a F-exponential family, the U-geometry $(g^{D^F}, \nabla^{D^F}, \nabla^{D^{*F}})$ is the (F,G)-geometry $(g^G, \nabla^{F,G}, \nabla^{H,G})$, with $G(p) = pF'(p)$ and H is the G-dual embedding of F given by $H(p) = p$.*

Proof. Using (1), (2) and (3), Eq. (8) can be written as

$$g_{ij}^{D^F}(\theta) = \int pF'(p)\ \partial_i\ell\ \partial_j\ell\ p\ dx = g^G(\theta) \tag{17}$$

$$\Gamma_{ijk}^{D^F}(\theta) = \int \left(\partial_i\partial_j\ell + (1 + \frac{pF''(p)}{F'(p)})\partial_i\ell\ \partial_j\ell \right) \partial_k\ell\ pF'(p)\ p\ dx \tag{18}$$

$$= \Gamma_{ijk}^{F,G}(\theta) \tag{19}$$

$$\Gamma_{ijk}^{D^{*F}}(\theta) = \int (\partial_i\partial_j\ell + \partial_i\ell\ \partial_j\ell)\ \partial_k\ell\ pF'(p)\ p\ dx \tag{20}$$

$$= \Gamma_{ijk}^{H,G}(\theta) \tag{21}$$

where $G(p) = pF'(p)$ and H is the G-dual embedding of F given by

$$1 + \frac{pH''(p)}{H'(p)} = 1 \Rightarrow \quad H(p) = p. \tag{22}$$

3.2 χ-geometry as the Conformal Flattening of (F,G)-geometry

Now we show that the χ-geometry induced from the divergence D_F is the conformal flattening of the (F,G)-geometry for suitable choices of F and G.

Theorem 13. *On a F-exponential family S, the χ-geometry $(g^{D_F}, \nabla^{D_F}, \nabla^{D_F^*})$ induced from the divergence D_F is obtained by the ±1-conformal transformation of the (F,G)-geometry for suitable choices of F and G. That is*

$$g_{ij}^{D_F}(\theta) = K(\theta)g_{ij}^G \tag{23}$$

$$\Gamma_{ijk}^{D_F} = K(\theta)\Gamma_{ijk}^{H,G} + \partial_j K(\theta)g_{ik}^G(\theta) + \partial_i K(\theta)g_{jk}^G(\theta) \tag{24}$$

$$\Gamma_{ijk}^{D_F^*}(\theta) = K(\theta)\Gamma_{ijk}^{F,G} - \partial_k K(\theta)g_{ij}^G(\theta) \tag{25}$$

where $G(p) = \frac{-pF''(p)}{F'(p)}$, H is the G-dual embedding of F and $K(\theta) = \frac{1}{h_F(\theta)}$.

Proof. For the F-exponential family $\mathcal{S}$, we have

$$\partial_i F(p(x;\theta)) = F'(p)\, \partial_i \ell\, p = x_i - \partial_i \psi_F(\theta) \tag{26}$$

From Eqs. (14), (16) and (26), we get

$$g_{ij}^{D_F}(\theta) = \partial_i \partial_j \psi_F(\theta) = \frac{1}{h_F(\theta)} \int \frac{-F''(p)}{(F'(p))^3} \partial_i F\, \partial_j F dx \tag{27}$$

$$= K(\theta) g_{ij}^{G} \tag{28}$$

where $K(\theta) = \frac{1}{h_F(\theta)}$ and $G(p) = \frac{-pF''(p)}{F'(p)}$. Thus the metric g^{D_F} is obtained as a conformal transformation of the G-metric by a gauge function $K(\theta)$.
Also from the Eq. (16), we have

$$\Gamma_{ijk}^{D_F} = \partial_i \partial_j \partial_k \psi_F(\theta)$$

$$= \frac{1}{h_F(\theta)} \int \left(\frac{-pF''(p)}{F'(p)} - \frac{p^2 F'''(p)}{F'(p)} + \frac{2p^2(F''(p))^2}{(F'(p))^2} \right) \partial_i \ell\, \partial_j \ell\, \partial_k \ell\, p dx$$

$$+ \frac{1}{h_F(\theta)} \int (\frac{-pF''(p)}{F'(p)}) \partial_i \partial_j \ell\, \partial_k \ell\, p dx$$

$$+ \frac{1}{h_F(\theta)} \int \partial_j \partial_k \psi_F(\theta) \frac{pF''(p)}{(F'(p))^2} \partial_i \ell\, dx$$

$$+ \frac{1}{h_F(\theta)} \int \partial_i \partial_k \psi_F(\theta) \frac{pF''(p)}{(F'(p))^2} \partial_j \ell\, dx$$

This can be rewritten as

$$\Gamma_{ijk}^{D_F}(\theta) = K(\theta) \int \left[\partial_i \partial_j \ell\, \partial_k \ell + (1 + \frac{pH''(p)}{H'(p)}) \partial_i \ell\, \partial_j \ell\, \partial_k \ell \right] G(p)\, p\, dx$$

$$+ \partial_j K(\theta) g_{ik}^G(\theta) + \partial_i K(\theta) g_{jk}^G(\theta)$$

$$= K(\theta) \Gamma_{ijk}^{H,G} + \partial_j K(\theta) g_{ik}^G(\theta) + \partial_i K(\theta) g_{jk}^G(\theta)$$

with $G(p) = \frac{-pF''(p)}{F'(p)}$, $K(\theta) = \frac{1}{h_F(\theta)}$ and H is the G-dual embedding of F.
Hence the connection induced by the divergence function D_F is the (-1)-conformal transformation of (H, G)-connection $\nabla^{H,G}$ by a gauge function $K(\theta)$.
Similarly, we can show for $\nabla^{D_F^*}$. $\blacksquare$

(Refer [10] for a detailed proof)

4 F-product, F-independence and $F-$likelihood Estimators

Matsuzoe and Ohara [8] defined a generalized product of numbers called q-product. Here using an increasing concave function F and its inverse function Z we also define a generalized product called F-**product** of two numbers x, y as

$$x \otimes_F y = Z[F(x) + F(y)] \tag{29}$$

The F-product satisfies the following properties

$$Z(x) \otimes_F Z(y) = Z(x+y) \quad ; \quad F(x \otimes_F y) = F(x) + F(y) \tag{30}$$

Definition 14. F-independence *The two random variables X and Y are said to be F-independent with normalization if the joint probability density function $p(x,y)$ is given by the F-product of the marginal probability density functions $p_1(x)$ and $p_2(y)$.*

$$p(x,y) = \frac{p_1(x) \otimes_F p_2(y)}{K_{p_1,p_2}} \tag{31}$$

where K_{p_1,p_2} is the normalization defined by

$$K_{p_1,p_2} = \int \int_{xy} p_1(x) \otimes_F p_2(y) dx dy \tag{32}$$

(Refer [9] for U-independence)

Definition 15. *Let $\mathcal{S} = \{p(x;\theta) \ / \ \theta \in E \subseteq \mathbb{R}^n\}$ be an $n-$dimensional statistical manifold defined on a sample space $\mathcal{X} \subseteq \mathbb{R}$. Let $\{x_1, \cdots, x_N\}$ be N observations from a probability density function $p(x;\theta) \in \mathcal{S}$. Let us define a **$F$-likelihood function** $L_F(\theta)$ ([10]) as*

$$L_F(\theta) = p(x_1;\theta) \otimes_F \cdots \otimes_F p(x_N;\theta) = Z(\sum_{i=1}^{N} F(p(x_i;\theta))) \tag{33}$$

*We say that $\hat{\theta}_F$ is the **Maximum F-likelihood estimator (F-MLE)** if*

$$\hat{\theta}_F = \arg \max_{\theta \in E} L_F(\theta) \tag{34}$$

Definition 16. *Maximum a posteriori estimator (MAP estimator)* *Let $p(x \mid \theta)$ be a distribution of the random variable x which depends on an unobserved population parameter θ. Let $p(\theta)$ be a prior distribution of θ. Then the posterior distribution of θ is given by*

$$p(\theta \mid x) = \frac{p(x \mid \theta)p(\theta)}{p(x)} \tag{35}$$

*where $p(x) = \int_\theta p(x \mid \theta) \, p(\theta) \, d\theta$ is the marginal density function of x.
Let $\{x_1, \cdots, x_N\}$ be N observations. Then the MAP estimator $\hat{\theta}_{MAP}$ for θ is given by*

$$\hat{\theta}_{MAP} = \arg \max_{\theta \in E} p(\theta \mid x_1, \cdots, x_N) = \arg \max_{\theta \in E} p(x_1, \cdots, x_N \mid \theta)p(\theta) \tag{36}$$

Theorem 17. *Let $\{x_1, \cdots, x_N\}$ be F-independent observations from $p(x \mid \theta)$, where F is an increasing concave function other than logarithmic function. Then the F-MLE is a MAP estimator with the prior distribution $p(\theta)$ given by*

$$p(\theta) = \frac{K(\theta)}{K_1} \quad ; \quad K_1 = \int K(\theta) \, d\theta < \infty \tag{37}$$

where

$$K(\theta) = \int \cdots \int p(x_1; \theta) \otimes_F \cdots \otimes_F p(x_N; \theta) dx_1 \cdots dx_N \qquad (38)$$

$$= \int \cdots \int Z(\sum_{i=1}^{N} F(p(x_i; \theta))) dx_1 \cdots dx_N \qquad (39)$$

Proof. Since $\{x_1, \cdots, x_N\}$ are F-independent, from Eqs. (31), (32) and (33) we have the joint density function $p(x_1, \cdots, x_N \mid \theta)$

$$p(x_1, \cdots, x_N \mid \theta) = \frac{p(x_1 \mid \theta) \otimes_F \cdots \otimes_F p(x_N \mid \theta)}{K(\theta)} \qquad (40)$$

where

$$K(\theta) = \int \cdots \int p(x_1 \mid \theta) \otimes_F \cdots \otimes_F p(x_N \mid \theta) dx_1 \cdots dx_N \qquad (41)$$

$$= \int \cdots \int Z(\sum_{i=1}^{N} F(p(x_i \mid \theta))) dx_1 \cdots dx_N \qquad (42)$$

Now let $K_1 = \int K(\theta) d\theta$. The F-MLE $\hat{\theta}_F$ is given by

$$\hat{\theta}_F = \arg \max_{\theta \in E} L_F(\theta) = \arg \max_{\theta \in E} p(x_1 \mid \theta) \otimes_F \cdots \otimes_F p(x_N \mid \theta) \qquad (43)$$

From Eq. (40), we have

$$\frac{L_F(\theta)}{K_1} = \frac{p(x_1 \mid \theta) \otimes_F \cdots \otimes_F p(x_N \mid \theta)}{K_1} \qquad (44)$$

$$= \frac{K(\theta)}{K_1} p(x_1, \cdots, x_N \mid \theta) \qquad (45)$$

Hence we have

$$\hat{\theta}_F = \arg \max_{\theta \in E} L_F(\theta) = \arg \max_{\theta \in E} \frac{L_F(\theta)}{K_1} \qquad (46)$$

$$= \arg \max_{\theta \in E} \frac{K(\theta)}{K_1} p(x_1, \cdots, x_N \mid \theta) \qquad (47)$$

$$= \arg \max_{\theta \in E} p(\theta) \, p(x_1, \cdots, x_N \mid \theta) \qquad (48)$$

$$= \hat{\theta}_{MAP} \qquad (49)$$

with the prior distribution $p(\theta)$ of θ given by

$$p(\theta) = \frac{K(\theta)}{K_1} \quad ; \quad K_1 = \int K(\theta) \, d\theta \qquad (50)$$

That is the F-MLE is a MAP estimator with $p(\theta)$ as prior distribution of θ. ∎

5   Conclusion

In this paper we described the two Hessian structures on a deformed exponential family; the U-geometry and the χ-geometry. Then we explored the relation between these two Hessian structures and the non-invariant (F, G)-geometry. We showed that U-geometry is the (F, G)-geometry and χ-geometry is the conformal flattening of (F, G)-geometry for suitable choices of F and G. Using an increasing concave function F and its inverse Z we defined a generalized product called F-product, F-independence and F-likelihood function. Then we defined a generalized estimator called F-MLE and showed that the F-MLE is a MAP estimator.

References

1. Amari, S.I., Nagaoka, H.: Methods of Information Geometry. Translations of Mathematical Monographs. Oxford University Press, Oxford (2000)
2. Harsha, K.V., Subrahamanian Moosath, K.S.: F-geometry and Amari's α-geometry on a statistical manifold. Entropy 16(5), 2472–2487 (2014)
3. Naudts, J.: Estimators, escort probabilities, and phi-exponential families in statistical physics. J. Ineq. Pure Appl. Math., 5(4) (2004). Article no 102
4. Eguchi, S., Komori, O., Ohara, A.: Duality of maximum entropy and minimum divergence. Entropy 16, 3552–3572 (2014)
5. Amari, S.I., Ohara, A., Matsuzoe, H.: Geometry of deformed exponential families: invariant, dually flat and conformal geometries. Phys. A Stat. Mech. Appl. 391, 4308–4319 (2012)
6. Matsuzoe, H., Henmi, M.: Hessian structures on deformed exponential families. In: Nielsen, F., Barbaresco, F. (eds.) GSI 2013. LNCS, vol. 8085, pp. 275–282. Springer, Heidelberg (2013)
7. Amari, S.I., Ohara, A.: Geometry of q-exponential family of probability distributions. Entropy 13, 1170–1185 (2011)
8. Matsuzoe, H. and Ohara, A. : Geometry for $q-$exponential families. In: Proceedings of the 2nd International Colloquium on Differential Geometry and its Related Fields, Veliko Tarnovo, 6–10 September (2010)
9. Fujimoto, Y., Murata, N.: A generalization of independence in naive bayes model. In: Fyfe, C., Tino, P., Charles, D., Garcia-Osorio, C., Yin, H. (eds.) IDEAL 2010. LNCS, vol. 6283, pp. 153–161. Springer, Heidelberg (2010)
10. Harsha, K.V., Subrahamanian Moosath, K.S.: Geometry of F-likelihood Estimators and F-Max-Ent Theorem. In: AIP Conference Proceedings Bayesian Inference and Maximum Entropy Methods in Science and Engineering (MaxEnt 2014), Amboise, France, 21–26 September 2014, vol. 1641, pp. 263–270 (2015)
11. Harsha, K.V., Subrahamanian Moosath, K.S.: Dually flat geometries of the deformed exponential family. Phys. A 433, 136–147 (2015)

New Metric and Connections
in Statistical Manifolds

Rui F. Vigelis[1]([⊠]), David C. de Souza[2], and Charles C. Cavalcante[3]

[1] Computer Engineering, Campus Sobral,
Federal University of Ceará, Sobral, CE, Brazil
rfvigelis@ufc.br
[2] Federal Institute of Ceará, Fortaleza, CE, Brazil
davidcs@ifce.edu.br
[3] Wireless Telecommunication Research Group,
Department of Teleinformatics Engineering,
Federal University of Ceará, Fortaleza, CE, Brazil
charles@ufc.br

Abstract. We define a metric and a family of α-connections in statistical manifolds, based on φ-divergence, which emerges in the framework of φ-families of probability distributions. This metric and α-connections generalize the Fisher information metric and Amari's α-connections. We also investigate the parallel transport associated with the α-connection for $\alpha = 1$.

1 Introduction

In the framework of φ-families of probability distributions [11], the authors introduced a divergence $\mathcal{D}_\varphi(\cdot \parallel \cdot)$ between probabilities distributions, called φ-divergence, that generalizes the Kullback–Leibler divergence. Based on $\mathcal{D}_\varphi(\cdot \parallel \cdot)$ we can define a new metric and connections in statistical manifolds. The definition of metrics or connections in statistical manifolds is a common subject in the literature [2,3,7]. In our approach, the metric and α-connections are intrinsically related to φ-families. Moreover, they can be recognized as a generalization of the Fisher information metric and Amari's α-connections [1,4].

Statistical manifolds are equipped with the Fisher information metric, which is given in terms of the derivative of $l(t; \theta) = \log p(t; \theta)$. Another metric can be defined if the logarithm $\log(\cdot)$ is replaced by the inverse of a φ-function $\varphi(\cdot)$ [11]. Instead of $l(t; \theta) = \log p(t; \theta)$, we can consider $f(t; \theta) = \varphi^{-1}(p(t; \theta))$. The manifold equipped with this metric, which coincides with the metric derived from $\mathcal{D}_\varphi(\cdot \parallel \cdot)$, is called a *generalized statistical manifold*.

Using the φ-divergence $\mathcal{D}_\varphi(\cdot \parallel \cdot)$, we can define a pair of mutually dual connections $D^{(1)}$ and $D^{(-1)}$, and then a family of α-connections $D^{(\alpha)}$. The connections $D^{(1)}$ and $D^{(-1)}$ corresponds to the exponential and mixture connections in classical information geometry. For example, in parametric φ-families, whose

This work was partially funded by CNPq (Proc. 309055/2014-8).

F. Nielsen and F. Barbaresco (Eds.): GSI 2015, LNCS 9389, pp. 222–229, 2015.
DOI: 10.1007/978-3-319-25040-3_25

definition is found in Sect. 2.1, the connection $D^{(1)}$ is flat (i.e., its torsion tensor T and curvature tensor R vanish identically). As a consequence, a parametric φ-family admits a parametrization in which the Christoffel symbols $\Gamma_{ijk}^{(-1)}$ associated with $D^{(-1)}$ vanish identically. In addition, parametric φ-families are examples of Hessian manifolds [8].

The rest of the paper is organized as follows. In Sect. 2, we define the generalized statistical manifolds. Section 2.1 deals with parametric φ-families of probability distribution. In Sect. 3, α-connections are introduced. The parallel transport associated with $D^{(1)}$ is investigated in Sect. 3.1.

2 Generalized Statistical Manifolds

In this section, we provide a definition of generalized statistical manifolds. We begin with the definition of φ-functions. Let (T, Σ, μ) be a measure space. In the case $T = \mathbb{R}$ (or T is a discrete set), the measure μ is considered to be the Lebesgue measure (or the counting measure). A function $\varphi \colon \mathbb{R} \to (0, \infty)$ is said to be a φ-*function* if the following conditions are satisfied:

(a1) $\varphi(\cdot)$ is convex,
(a2) $\lim_{u \to -\infty} \varphi(u) = 0$ and $\lim_{u \to \infty} \varphi(u) = \infty$.

Moreover, we assume that a measurable function $u_0 \colon T \to (0, \infty)$ can be found such that, for each measurable function $c \colon T \to \mathbb{R}$ such that $\varphi(c(t)) > 0$ and $\int_T \varphi(c(t)) d\mu = 1$, we have

(a3) $\displaystyle \int_T \varphi(c(t) + \lambda u_0(t)) d\mu < \infty$, for all $\lambda > 0$.

The exponential function and the Kaniadakis' κ-exponential function [6] satisfy conditions (a1)–(a3) [11]. For $q \neq 1$, the q-exponential function $\exp_q(\cdot)$ [9] is not a φ-function, since its image is $[0, \infty)$. Notice that if the set T is finite, condition (a3) is always satisfied. Condition (a3) is indispensable in the definition of non-parametric families of probability distributions [11].

A generalized statistical manifold is a family of probability distributions $\mathcal{P} = \{p(t; \theta) : \theta \in \Theta\}$, which is defined to be contained in

$$\mathcal{P}_\mu = \left\{ p \in L^0 : p > 0 \text{ and } \int_T p d\mu = 1 \right\},$$

where L^0 denotes the set of all real-valued, measurable functions on T, with equality μ-a.e. Each $p_\theta(t) := p(t; \theta)$ is given in terms of parameters $\theta = (\theta^1, \ldots, \theta^n) \in \Theta \subseteq \mathbb{R}^n$ by a one-to-one mapping. The family $\mathcal{P}$ is called a *generalized statistical manifold* if the following conditions are satisfied:

(P1) Θ is a domain (an open and connected set) in $\mathbb{R}^n$.
(P2) $p(t; \theta)$ is a differentiable function with respect to θ.
(P3) The operations of integration with respect to μ and differentiation with respect to θ^i commute.

(P4) The matrix $g = (g_{ij})$, which is defined by

$$g_{ij} = -E'_\theta \left[\frac{\partial^2 f_\theta}{\partial \theta^i \partial \theta^j} \right], \tag{1}$$

is positive definite at each $\theta \in \Theta$, where $f_\theta(t) = f(t; \theta) = \varphi^{-1}(p(t; \theta))$ and

$$E'_\theta[\cdot] = \frac{\int_T (\cdot) \varphi'(f_\theta) d\mu}{\int_T u_0 \varphi'(f_\theta) d\mu}.$$

Notice that expression (1) reduces to the Fisher information matrix in the case that φ coincides with the exponential function and $u_0 = 1$. Moreover, the right-hand side of (1) is invariant under reparametrization. The matrix (g_{ij}) can also be expressed as

$$g_{ij} = E''_\theta \left[\frac{\partial f_\theta}{\partial \theta^i} \frac{\partial f_\theta}{\partial \theta^j} \right], \tag{2}$$

where

$$E''_\theta[\cdot] = \frac{\int_T (\cdot) \varphi''(f_\theta) d\mu}{\int_T u_0 \varphi'(f_\theta) d\mu}.$$

Because the operations of integration with respect to μ and differentiation with respect to θ^i are commutative, we have

$$0 = \frac{\partial}{\partial \theta^i} \int_T p_\theta d\mu = \int_T \frac{\partial}{\partial \theta^i} \varphi(f_\theta) d\mu = \int_T \frac{\partial f_\theta}{\partial \theta^i} \varphi'(f_\theta) d\mu, \tag{3}$$

and

$$0 = \int_T \frac{\partial^2 f_\theta}{\partial \theta^i \partial \theta^j} \varphi'(f_\theta) d\mu + \int_T \frac{\partial f_\theta}{\partial \theta^i} \frac{\partial f_\theta}{\partial \theta^j} \varphi''(f_\theta) d\mu. \tag{4}$$

Thus expression (2) follows from (4). In addition, expression (3) implies

$$E'_\theta \left[\frac{\partial f_\theta}{\partial \theta^i} \right] = 0. \tag{5}$$

A consequence of (2) is the correspondence between the functions $\partial f_\theta / \partial \theta^i$ and the basis vectors $\partial / \partial \theta^i$. The inner product of vectors

$$X = \sum_i a^i \frac{\partial}{\partial \theta^i} \qquad \text{and} \qquad Y = \sum_i b^j \frac{\partial}{\partial \theta^j}$$

can be written as

$$g(X, Y) = \sum_{i,j} g_{ij} a^i b^j = \sum_{i,j} E''_\theta \left[\frac{\partial f_\theta}{\partial \theta^i} \frac{\partial f_\theta}{\partial \theta^j} \right] a^i b^j = E''_\theta [\widetilde{X} \widetilde{Y}], \tag{6}$$

where

$$\widetilde{X} = \sum_i a^i \frac{\partial f_\theta}{\partial \theta^i} \qquad \text{and} \qquad \widetilde{Y} = \sum_i b^j \frac{\partial f_\theta}{\partial \theta^j}.$$

As a result, the tangent space $T_{p_\theta} \mathcal{P}$ can be identified with $\widetilde{T}_{p_\theta} \mathcal{P}$, which is defined as the vector space spanned by $\partial f_\theta / \partial \theta^i$, equipped with the inner product $\langle \widetilde{X}, \widetilde{Y} \rangle_\theta = E''_\theta [\widetilde{X} \widetilde{Y}]$. By (5), if a vector $\widetilde{X}$ belongs to $\widetilde{T}_{p_\theta} \mathcal{P}$, then $E'_\theta[\widetilde{X}] = 0$. Independent of the definition of (g_{ij}), the expression in the right-hand side of (6) always defines a semi-inner product in $\widetilde{T}_{p_\theta} \mathcal{P}$.

2.1   Parametric φ-Families of Probability Distribution

Let $c\colon T \to \mathbb{R}$ be a measurable function such that $p := \varphi(c)$ is probability density in $\mathcal{P}_\mu$. We take any measurable functions $u_1, \ldots u_n\colon T \to \mathbb{R}$ satisfying the following conditions:

(i) $\int_T u_i \varphi'(c) d\mu = 0$, and
(ii) there exists $\varepsilon > 0$ such that

$$\int_T \varphi(c + \lambda u_i) d\mu < \infty, \qquad \text{for all } \lambda \in (-\varepsilon, \varepsilon).$$

Define $\Theta \subseteq \mathbb{R}^n$ as the set of all vectors $\theta = (\theta^i) \in \mathbb{R}^n$ such that

$$\int_T \varphi\left(c + \lambda \sum_{k=1}^n \theta^i u_i\right) d\mu < \infty, \qquad \text{for some } \lambda > 1.$$

The elements of the *parametric φ-family* $\mathcal{F}_p = \{p(t; \theta) : \theta \in \Theta\}$ centered at $p = \varphi(c)$ are given by the one-to-one mapping

$$p(t; \theta) := \varphi\left(c(t) + \sum_{i=1}^n \theta^i u_i(t) - \psi(\theta) u_0(t)\right), \qquad \text{for each } \theta = (\theta^i) \in \Theta. \quad (7)$$

where the *normalizing function* $\psi\colon \Theta \to [0, \infty)$ is introduced so that expression (7) defines a probability distribution in $\mathcal{P}_\mu$.

Condition (ii) is always satisfied if the set T is finite. It can be shown that the normalizing function ψ is also convex (and the set Θ is open and convex). Under conditions (i)–(ii), the family $\mathcal{F}_p$ is a submanifold of a non-parametric φ-family. For the non-parametric case, we refer to [10, 11].

By the equalities

$$\frac{\partial f_\theta}{\partial \theta^i} = u_i(t) - \frac{\partial \psi}{\partial \theta^i}, \qquad -\frac{\partial^2 f_\theta}{\partial \theta^i \partial \theta^j} = -\frac{\partial^2 \psi}{\partial \theta^i \partial \theta^j},$$

we get

$$g_{ij} = \frac{\partial^2 \psi}{\partial \theta^i \partial \theta^j}.$$

In other words, the matrix (g_{ij}) is the Hessian of the normalizing function ψ.

For $\varphi(\cdot) = \exp(\cdot)$ and $u_0 = 1$, expression (7) defines a parametric exponential family of probability distributions $\mathcal{E}_p$. In exponential families, the normalizing function is recognized as the Kullback–Leibler divergence between $p(t)$ and $p(t; \theta)$. Using this result, we can define the φ-divergence $\mathcal{D}_\varphi(\cdot \parallel \cdot)$, which generalizes the Kullback–Leibler divergence $\mathcal{D}_{\mathrm{KL}}(\cdot \parallel \cdot)$.

By (7) we can write

$$\psi(\theta) u_0(t) = \sum_{i=1}^n \theta^i u_i(t) + \varphi^{-1}(p(t)) - \varphi^{-1}(p(t; \theta)).$$

From condition (i), this equation yields

$$\psi(\theta) \int_T u_0 \varphi'(c) d\mu = \int_T [\varphi^{-1}(p) - \varphi^{-1}(p_\theta)] \varphi'(c) d\mu.$$

In view of $\varphi'(c) = 1/(\varphi^{-1})'(p)$, we get

$$\psi(\theta) = \frac{\displaystyle\int_T \frac{\varphi^{-1}(p) - \varphi^{-1}(p_\theta)}{(\varphi^{-1})'(p)} d\mu}{\displaystyle\int_T \frac{u_0}{(\varphi^{-1})'(p)} d\mu} =: \mathcal{D}_\varphi(p \,\|\, p_\theta), \tag{8}$$

which defines the φ-divergence $\mathcal{D}_\varphi(p \,\|\, p_\theta)$. Clearly, expression (8) can be used to extend the definition of $\mathcal{D}_\varphi(\cdot \,\|\, \cdot)$ to any probability distributions p and q in $\mathcal{P}_\mu$.

3 α-Connections

We use the φ-divergence $\mathcal{D}_\varphi(\cdot \,\|\, \cdot)$ to define a pair of mutually dual connection in generalized statistical manifolds. Let $\mathcal{D}: M \times M \to [0, \infty)$ be a non-negative, differentiable function defined on a smooth manifold M, such that

$$\mathcal{D}(p \,\|\, q) = 0 \quad \text{if and only if} \quad p = q. \tag{9}$$

The function $\mathcal{D}(\cdot \,\|\, \cdot)$ is called a *divergence* if the matrix (g_{ij}), whose entries are given by

$$g_{ij}(p) = -\left[\left(\frac{\partial}{\partial \theta^i}\right)_p \left(\frac{\partial}{\partial \theta^j}\right)_q \mathcal{D}(p \,\|\, q)\right]_{q=p}, \tag{10}$$

is positive definite for each $p \in M$. Hence a divergence $\mathcal{D}(\cdot \,\|\, \cdot)$ defines a metric in M. A divergence $\mathcal{D}(\cdot \,\|\, \cdot)$ also induces a pair of mutually dual connections D and D^*, whose Christoffel symbols are given by

$$\Gamma_{ijk} = -\left[\left(\frac{\partial^2}{\partial \theta^i \partial \theta^j}\right)_p \left(\frac{\partial}{\partial \theta^k}\right)_q \mathcal{D}(p \,\|\, q)\right]_{q=p} \tag{11}$$

and

$$\Gamma^*_{ijk} = -\left[\left(\frac{\partial}{\partial \theta^k}\right)_p \left(\frac{\partial^2}{\partial \theta^i \partial \theta^j}\right)_q \mathcal{D}(p \,\|\, q)\right]_{q=p}, \tag{12}$$

respectively. By a simple computation, we get

$$\frac{\partial g_{jk}}{\partial \theta^i} = \Gamma_{ijk} + \Gamma^*_{ikj},$$

showing that D and D^* are mutually dual.

In Sect. 2.1, the φ-divergence between two probability distributions p and q in $\mathcal{P}_\mu$ was defined as

$$\mathcal{D}_\varphi(p \,\|\, q) := \frac{\displaystyle\int_T \frac{\varphi^{-1}(p) - \varphi^{-1}(q)}{(\varphi^{-1})'(p)} d\mu}{\displaystyle\int_T \frac{u_0}{(\varphi^{-1})'(p)} d\mu}. \tag{13}$$

Because φ is convex, it follows that $\mathcal{D}_\varphi(p \parallel q) \geq 0$ for all $p, q \in \mathcal{P}_\mu$. In addition, if we assume that $\varphi(\cdot)$ is strictly convex, then $\mathcal{D}_\varphi(p \parallel q) = 0$ if and only if $p = q$. In a generalized statistical manifold $\mathcal{P} = \{p(t; \theta) : \theta \in \Theta\}$, the metric derived from the divergence $\mathcal{D}(q \parallel p) := \mathcal{D}_\varphi(p \parallel q)$ coincides with (1). Expressing the φ-divergence $\mathcal{D}_\varphi(\cdot \parallel \cdot)$ between p_θ and p_ϑ as

$$\mathcal{D}(p_\theta \parallel p_\vartheta) = E'_\vartheta[(f_\vartheta - f_\theta)],$$

after some manipulation, we get

$$g_{ij} = -\left[\left(\frac{\partial}{\partial\theta^i}\right)_p \left(\frac{\partial}{\partial\theta^j}\right)_q \mathcal{D}(p \parallel q)\right]_{q=p}$$
$$= -E'_\theta\left[\frac{\partial^2 f_\theta}{\partial\theta^i \partial\theta^j}\right].$$

As a consequence, expression (13) defines a divergence on statistical manifolds.

Let $D^{(1)}$ and $D^{(-1)}$ denote the pair of dual connections derived from $\mathcal{D}_\varphi(\cdot \parallel \cdot)$. By (11) and (12), the Christoffel symbols $\Gamma^{(1)}_{ijk}$ and $\Gamma^{(-1)}_{ijk}$ are given by

$$\Gamma^{(1)}_{ijk} = E''_\theta\left[\frac{\partial^2 f_\theta}{\partial\theta^i \partial\theta^j}\frac{\partial f_\theta}{\partial\theta^k}\right] - E'_\theta\left[\frac{\partial^2 f_\theta}{\partial\theta^i \partial\theta^j}\right]E''_\theta\left[u_0\frac{\partial f_\theta}{\partial\theta^k}\right] \tag{14}$$

and

$$\Gamma^{(-1)}_{ijk} = E''_\theta\left[\frac{\partial^2 f_\theta}{\partial\theta^i \partial\theta^j}\frac{\partial f_\theta}{\partial\theta^k}\right] + E'''_\theta\left[\frac{\partial f_\theta}{\partial\theta^i}\frac{\partial f_\theta}{\partial\theta^j}\frac{\partial f_\theta}{\partial\theta^k}\right]$$
$$- E''_\theta\left[\frac{\partial f_\theta}{\partial\theta^j}\frac{\partial f_\theta}{\partial\theta^k}\right]E''_\theta\left[u_0\frac{\partial f_\theta}{\partial\theta^i}\right] - E''_\theta\left[\frac{\partial f_\theta}{\partial\theta^i}\frac{\partial f_\theta}{\partial\theta^k}\right]E''_\theta\left[u_0\frac{\partial f_\theta}{\partial\theta^j}\right], \tag{15}$$

where

$$E'''_\theta[\cdot] = \frac{\int_T(\cdot)\varphi'''(f_\theta)d\mu}{\int_T u_0\varphi'(f_\theta)d\mu}.$$

Notice that in parametric φ-families, the Christoffel symbols $\Gamma^{(1)}_{ijk}$ vanish identically. Thus, in these families, the connection $D^{(1)}$ is flat.

Using the pair of mutually dual connections $D^{(1)}$ and $D^{(-1)}$, we can specify a family of α-connections $D^{(\alpha)}$ in generalized statistical manifolds. The Christoffel symbol of $D^{(\alpha)}$ is defined by

$$\Gamma^{(\alpha)}_{ijk} = \frac{1+\alpha}{2}\Gamma^{(1)}_{ijk} + \frac{1-\alpha}{2}\Gamma^{(-1)}_{ijk}. \tag{16}$$

The connections $D^{(\alpha)}$ and $D^{(-\alpha)}$ are mutually dual, since

$$\frac{\partial g_{jk}}{\partial\theta^i} = \Gamma^{(\alpha)}_{ijk} + \Gamma^{(-\alpha)}_{ikj}.$$

For $\alpha = 0$, the connection $D^{(0)}$, which is clearly self-dual, corresponds to the Levi–Civita connection ∇. One can show that $\Gamma^{(0)}_{ijk}$ can be derived from the expression defining the Christoffel symbols of ∇ in terms of the metric:

$$\Gamma_{ijk} = \sum_m \Gamma^m_{ij}g_{mk} = \frac{1}{2}\left(\frac{\partial g_{ki}}{\partial\theta^j} + \frac{\partial g_{kj}}{\partial\theta^i} - \frac{\partial g_{ij}}{\partial\theta^k}\right).$$

The connection $D^{(\alpha)}$ can be equivalently defined by

$$\Gamma_{ijk}^{(\alpha)} = \Gamma_{ijk}^{(0)} - \alpha T_{ijk},$$

where

$$T_{ijk} = \frac{1}{2}E_\theta'''\left[\frac{\partial f_\theta}{\partial \theta^i}\frac{\partial f_\theta}{\partial \theta^j}\frac{\partial f_\theta}{\partial \theta^k}\right] - \frac{1}{2}E_\theta''\left[\frac{\partial f_\theta}{\partial \theta^k}\frac{\partial f_\theta}{\partial \theta^i}\right]E_\theta''\left[u_0\frac{\partial f_\theta}{\partial \theta^j}\right]$$
$$- \frac{1}{2}E_\theta''\left[\frac{\partial f_\theta}{\partial \theta^k}\frac{\partial f_\theta}{\partial \theta^j}\right]E_\theta''\left[u_0\frac{\partial f_\theta}{\partial \theta^i}\right] - \frac{1}{2}E_\theta''\left[\frac{\partial f_\theta}{\partial \theta^i}\frac{\partial f_\theta}{\partial \theta^j}\right]E_\theta''\left[u_0\frac{\partial f_\theta}{\partial \theta^k}\right]. \quad (17)$$

In the case that φ is the exponential function and $u_0 = 1$, Eqs. (14), (15), (16) and (17) reduce to the classical expressions for statistical manifolds.

3.1 Parallel Transport

Let $\gamma\colon I \to M$ be a smooth curve in a smooth manifold M, with a connection D. A vector field V along γ is said to be *parallel* if $D_{d/dt}V(t) = 0$ for all $t \in I$. Take any tangent vector V_0 at $\gamma(t_0)$, for some $t_0 \in I$. Then there exists a unique vector field V along γ, called the *parallel transport* of V_0 along γ, such that $V(t_0) = V_0$.

A connection D can be recovered from the parallel transport. Fix any smooth vectors fields X and Y. Given $p \in M$, define $\gamma\colon I \to M$ to be an integral curve of X passing through p. In other words, $\gamma(t_0) = p$ and $\frac{d\gamma}{dt} = X(\gamma(t))$. Let $P_{\gamma,t_0,t}\colon T_{\gamma(t_0)}M \to T_{\gamma(t)}M$ denote the parallel transport of a vector along γ from t_0 to t. Then we have

$$(D_X Y)(p) = \frac{d}{dt}P_{\gamma,t_0,t}^{-1}(Y(c(t)))\Big|_{t=t_0}.$$

For details, we refer to [5].

To avoid some technicalities, we assume that the set T is finite. In this case, we can consider a generalized statistical manifold $\mathcal{P} = \{p(t;\theta) : \theta \in \Theta\}$ for which $\mathcal{P} = \mathcal{P}_\mu$. The connection $D^{(1)}$ can be derived from the parallel transport

$$P_{q,p}\colon \widetilde{T}_q\mathcal{P} \to \widetilde{T}_p\mathcal{P}$$

given by

$$\widetilde{X} \mapsto \widetilde{X} - E_\theta'[\widetilde{X}]u_0,$$

where $p = p_\theta$. Recall that the tangent space $T_p\mathcal{P}$ can be identified with $\widetilde{T}_p\mathcal{P}$, the vector space spanned by the functions $\partial f_\theta/\partial\theta^i$, equipped with the inner product $\langle \widetilde{X}, \widetilde{Y}\rangle = E_\theta''[\widetilde{X}\widetilde{Y}]$, where $p = p_\theta$. We remark that $P_{q,p}$ does not depend on the curve joining q and p. As a result, the connection $D^{(1)}$ is flat. Denote by $\gamma(t)$ the coordinate curve given locally by $\theta(t) = (\theta^1,\ldots,\theta^i + t,\ldots,\theta^n)$. Observing that $P_{\gamma(0),\gamma(t)}^{-1}$ maps the vector $\frac{\partial f_\theta}{\partial\theta^j}(t)$ to

$$\frac{\partial f_\theta}{\partial\theta^j}(t) - E_{\theta(0)}'\left[\frac{\partial f_\theta}{\partial\theta^j}(t)\right]u_0,$$

we define the connection

$$\widetilde{D}_{\partial f_\theta/\partial\theta_i}\frac{\partial f_\theta}{\partial\theta_j} = \frac{d}{dt}P^{-1}_{\gamma(0),\gamma(t)}\left(\frac{\partial f_\theta}{\partial\theta_j}(\gamma(t))\right)\Big|_{t=0}$$

$$= \frac{d}{dt}\left(\frac{\partial f_{\theta(t)}}{\partial\theta^j} - E'_{\theta(0)}\left[\frac{\partial f_{\theta(t)}}{\partial\theta^j}\right]u_0\right)\Big|_{t=0}$$

$$= \frac{\partial^2 f_\theta}{\partial\theta^i\partial\theta^j} - E'_\theta\left[\frac{\partial^2 f_\theta}{\partial\theta^i\partial\theta^j}\right]u_0.$$

Let us denote by D the connection corresponding to $\widetilde{D}$, which acts on smooth vector fields in $T_p\mathcal{P}$. By this identification, we have

$$g\left(D_{\partial/\partial\theta_i}\frac{\partial}{\partial\theta_j},\frac{\partial}{\partial\theta_k}\right) = \left\langle\widetilde{D}_{\partial f_\theta/\partial\theta_i}\frac{\partial f_\theta}{\partial\theta_j},\frac{\partial f_\theta}{\partial\theta_k}\right\rangle$$

$$= E''_\theta\left[\frac{\partial^2 f_\theta}{\partial\theta^i\partial\theta^j}\frac{\partial f_\theta}{\partial\theta^k}\right] - E'_\theta\left[\frac{\partial^2 f_\theta}{\partial\theta^i\partial\theta^j}\right]E''_\theta\left[u_0\frac{\partial f_\theta}{\partial\theta^k}\right]$$

$$= \Gamma^{(1)}_{ijk},$$

showing that $D = D^{(1)}$.

References

1. Shun-ichi Amari and Hiroshi Nagaoka. Methods of information geometry, vol. 191. Translations of Mathematical Monographs. American Mathematical Society, Providence, RI. Oxford University Press, Oxford (2000). Translated from the 1993 Japanese original by Daishi Harada
2. Amari, S., Ohara, A.: Geometry of q-exponential family of probability distributions. Entropy **13**(6), 1170–1185 (2011)
3. Amari, S., Ohara, A., Matsuzoe, H.: Geometry of deformed exponential families: invariant, dually-flat and conformal geometries. Phys. A **391**(18), 4308–4319 (2012)
4. Calin, O., Udrişte, C.: Geometric Modeling in Probability and Statistics. Springer, Cham (2014)
5. do Carmo, M.P.: Riemannian Geometry, 14th edn. Birkhäuser, Boston (2013)
6. Kaniadakis, G.: Statistical mechanics in the context of special relativity. Phys. Rev. E **66**(5), 056125 (2002). (3), 17
7. Matsuzoe, H.: Hessian structures on deformed exponential families and their conformal structures. Differ. Geom. Appl. **35**, 323–333 (2014)
8. Shima, H.: The geometry of Hessian structures. World Scientific Publishing Co., Pte. Ltd., Hackensack (2007)
9. Tsallis, C.: What are the numbers that experiments provide? Quim. Nova **17**(6), 468–471 (1994)
10. Vigelis, R.F., Cavalcante, C.C.: The Δ_2-condition and φ-families of probability distributions. In: Nielsen, F., Barbaresco, F. (eds.) GSI 2013. LNCS, vol. 8085, pp. 729–736. Springer, Heidelberg (2013)
11. Vigelis, R.F., Cavalcante, C.C.: On φ-families of probability distributions. J. Theoret. Probab. **26**(3), 870–884 (2013)

Curvatures of Statistical Structures

Barbara Opozda$^{(\boxtimes)}$

Department of Mathematics and Computer Science, UJ, Cracow, Poland
barbara.opozda@im.uj.edu.pl

Abstract. Curvature properties for statistical structures are studied. The study deals with the curvature tensor of statistical connections and their duals as well as the Ricci tensor of the connections, Laplacians and the curvature operator. Two concepts of sectional curvature are introduced. The meaning of the notions is illustrated by presenting few exemplary theorems.

Keywords: Affine connection · Curvature tensor · Ricci tensor · Sectional curvature · Laplacian · Bochner's technique

1 Introduction

The curvature tensor is one of the most important tensors in differential geometry. On the base of this tensor many other objects can be defined. In particular, the Ricci tensor, the scalar curvature, the Weyl curvature tensor, the Weitzenböck curvature tensor or the sectional curvature. Some of these notions, however, are attributed only to Riemannian structures with their Levi-Civita connections. For instance, the sectional curvature is such a notion. We claim that some of these, especially strongly attributed to Riemannian geometry, notions can be extended to statistical structures and like in the Riemannian case, provide a lot of information on the structures.

By a statistical structure on a manifold M we mean a pair (g, ∇), where g is a metric tensor field and ∇ is a torsion-free connection for which ∇g as a cubic form is symmetric in all arguments, see [1]. Such a structure is also called a Codazzi structure. One can define (equivalently) a statistical structure by equipping a Riemannian manifold (M, g) with a symmetric $(1, 2)$-tensor field K, for which the cubic form $C(X, Y, Z) = g(K(X, Y), Z)$ is symmetric in all arguments. Having K one defines a torsion-free connection ∇ by the formula $\nabla_X Y = \hat{\nabla}_X Y + K(X, Y)$, where $\hat{\nabla}$ is the Levi-Civita connection of g. The pair (g, ∇) turns out to be a statistical structure. Of course, instead of K one can prescribe a symmetric cubic form C. A manifold endowed with a statistical structure is called a statistical manifold.

In information theory classical examples of statistical manifolds are manifolds of probability distributions equipped with the Fisher information metric and an

Research supported by the NCN grant 2013/11/B/ST1/02889.

© Springer International Publishing Switzerland 2015
F. Nielsen and F. Barbaresco (Eds.): GSI 2015, LNCS 9389, pp. 230–239, 2015.
DOI: 10.1007/978-3-319-25040-3_26

appropriate cubic form. Namely, let $(\mathcal{X}, \mathcal{B})$ be a measurable space with σ-algebra $\mathcal{B}$ over $\mathcal{X}$. Let Λ be a domain in $\mathbf{R}^n$ and

$$p : \mathcal{X} \times \Lambda \ni (x, \lambda) \longrightarrow p(x, \lambda) \in \mathbf{R}$$

be a function smoothly depending on λ. Moreover, we assume that $p_\lambda(x) := p(x, \lambda)$ is a probability measure on $\mathcal{X}$ for each $\lambda \in \Lambda$. Set $\ell(x, \lambda) = \log p(x, \lambda)$. The Fisher information metric g on Λ is given by

$$g_{ij}(\lambda) = \mathbb{E}_\lambda[(\partial_i \ell)(\partial_j \ell)], \tag{1}$$

where $\mathbb{E}_\lambda$ denotes the expectation relative to p_λ, $\partial_i \ell$ stands for $\frac{\partial \ell}{\partial \lambda_i}$ and $\lambda = (\lambda_1, ..., \lambda_n)$. One defines a symmetric cubic form C on Λ by the formula

$$C_{ijk}(\lambda) = \mathbb{E}[(\partial_i \ell)(\partial_j \ell)(\partial_k \ell)].$$

The pair $(g, \alpha C)$ constitutes a statistical structure on Λ for every $\alpha \in \mathbf{R}$.

However, the oldest source of statistical structures is the theory of affine hypersurfaces in $\mathbf{R}^n$ or the geometry of the second fundamental form of hypersurfaces in real space forms. Lagrangian submanifolds of complex space forms are also naturally endowed with statistical structures. Nevertheless, most statistical structures are outside these categories. In general, a statistical structure is not realizable on a hypersurface nor on a Lagrangian submanifold, even locally, see [4].

In this paper we present some ideas of extending Riemannian geometry to the case of statistical structures. We concentrate on the ideas depending on naturally defined curvature tensors for a statistical structure. Exemplary theorems concerning these ideas are provided.

2 Statistical Structures

A statistical structure on a manifold M can be defined in few equivalent ways. First of all M must have a Riemannian structure defined by a metric tensor field g. We assume that g is positive definite, although g can be also indefnite. A statistical structure can be defined as a pair (g, K), where g is a Riemannian metric tensor field and K is a symmetric $(1, 2)$-tensor field on M which is also symmetric relative to g, that is, the cubic form

$$C(X, Y, Z) = g(X, K(Y, Z)) \tag{2}$$

is symmetric relative to X, Y. It is clear that any symmetric cubic form C on a Riemannian manifold (M, g) defines by (2) a $(1, 2)$-tensor field K having the symmetry properties as above. Another definition says that a statistical structure is a pair (g, ∇), where ∇ is a torsion-free affine connection on M and ∇g as a $(0, 3)$-tensor field on M is symmetric in all arguments. ∇ is called a statistical connection. The equivalence of the above definitions is established by taking K

as the difference tensor between the connection ∇ and the Levi-Civita connection $\hat{\nabla}$ for g, that is,

$$\nabla_X Y = \hat{\nabla}_X Y + K(X, Y) \tag{3}$$

for every vector fields X, Y on M. The cubic forms C and ∇g are related by the equality $\nabla g = -2C$.

For any connection ∇ on a Riemannian manifold (M, g) one defines its conjugate connection $\overline{\nabla}$ (relative to g) by the formula

$$g(\nabla_X Y, Z) + g(Y, \overline{\nabla}_X Z) = X g(Y, Z) \tag{4}$$

for any vector fields X, Y, Z on M. The connections ∇ and $\overline{\nabla}$ are simultaneously torsion-free. If (g, ∇) is a statistical structure then so is $(g, \overline{\nabla})$. Moreover, if (g, ∇) is trace-free then so is $(g, \overline{\nabla})$. Recall that a statistical structure (g, ∇) is trace-free if $\mathrm{tr}\,_g(\nabla g)(X, \cdot, \cdot) = 0$ for every X or equivalently $\mathrm{tr}\,_g K = 0$, or equivalently $\mathrm{tr}\, K_X = 0$ for every X, where $K_X Y$ stands for $K(X, Y)$. If R is the curvature tensor for ∇ and $\overline{R}$ is the curvature tensor for $\overline{\nabla}$ then we have, see [3],

$$g(R(X, Y)Z, W) = -g(\overline{R}(X, Y)W, Z) \tag{5}$$

for every X, Y, Z, W. In particular, $R = 0$ if and only if $\overline{R} = 0$. If K is the difference tensor between ∇ and $\hat{\nabla}$ then

$$\overline{\nabla}_X Y = \hat{\nabla}_X Y - K_X Y. \tag{6}$$

We also have, [3],

$$R(X, Y) = \hat{R}(X, Y) + (\hat{\nabla}_X K)_Y - (\hat{\nabla}_Y K)_X + [K_X, K_Y], \tag{7}$$

where $\hat{R}$ is the curvature tensor for $\hat{\nabla}$. Writing the same equality for $\overline{\nabla}$ and adding both equalities we get

$$R(X, Y) + \overline{R}(X, Y) = 2\hat{R}(X, Y) + 2[K_X, K_Y]. \tag{8}$$

The above formulas yield, see [4],

Lemma 1. Let (g, K) be a statistical structure. The following conditions are equivalent:

(1) $R = \overline{R}$,
(2) $\hat{\nabla} K(X, Z, Y)$ is symmetric in all arguments,
(3) $g(R(X, Y)Z, W)$ is skew-symmetric relative to Z, W.

A statistical structure is called a Hessian structure if the connection ∇ is flat, that is, $R = 0$. In this case, by (8), we have

$$\hat{R} = -[K, K]. \tag{9}$$

For a statistical structure one defines the vector field E by

$$E = \mathrm{tr}\,_g K. \tag{10}$$

The dual (relative to g) form will be denoted by τ. We have $\operatorname{tr}_g \nabla g(\cdot, \cdot, Z) = -2\tau(Z)$. If ν_g is the volume form determined by g then $\nabla_Z \nu_g = -\tau(Z)\nu_g$. Therefore, a statistical structure (g, ∇) is trace-free if and only if $\nabla \nu_g = 0$. Trace-free statistical structures are of the greatest importance in the classical theory of affine hypersurfaces of $\mathbf{R}^{n+1}$. In this theory they are called Blaschke structures. In the theory of Lagrangian submanifolds trace-free statistical structures appear on minimal submanifolds.

Denote by Ric, $\overline{Ric}$ and $\widehat{Ric}$ the Ricci tensors for ∇, $\overline{\nabla}$ and $\hat{\nabla}$ respectively. Recall that for any linear connection ∇ with curvature tensor R its Ricci tensor is defined by $Ric(Y, Z) = \operatorname{tr}\{X \to R(X, Y)Z\}$. Note that the Ricci tensor does not have to be symmetric. We have

$$Ric(Y, Z) = \widehat{Ric}(Y, Z) + (div^{\hat{\nabla}} K)(Y, Z) - \hat{\nabla}\tau(Y, Z) + \tau(K(Y, Z)) - g(K_Y, K_Z).$$

It follows that

$$Ric(Y, Z) + \overline{Ric}(Y, Z) = 2\widehat{Ric}(Y, Z) - 2g(K_Y, K_Z) + 2\tau(K(Y, Z)). \tag{11}$$

In particular, if (g, ∇) is trace-free then

$$2\widehat{Ric}(X, X) \geq Ric(X, X) + \overline{Ric}(X, X). \tag{12}$$

The above formulas also yield

$$Ric(Y, Z) - Ric(Z, Y) = -d\tau(Y, Z). \tag{13}$$

Hence ∇ is Ricci-symmetric if and only if $d\tau = 0$. Recall that the Ricci tensor of ∇ is symmetric if and only if there is a (locally defined) volume form ν parallel relative to ∇.

Denote by ρ the scalar curvature for ∇, that is, $\rho = \operatorname{tr}_g Ric(\cdot, \cdot)$. By (5) it is clear that the scalar curvature for $\overline{\nabla}$ is equal to ρ. Taking now the trace relative to g on both sides of (11) we get

$$\hat{\rho} = \rho + |K|^2 - |E|^2. \tag{14}$$

The last formula implies, in particular, that the Riemannian scalar curvature for $\hat{\nabla}$ is maximal among scalar curvatures of connections which are statistical for g. More precisely, we have, see [4],

Proposition 1. *The functional*

$$\mathfrak{scal} : \{statistical\ connections\ for\ g\} \ni \nabla \to \operatorname{tr}_g Ric \in \mathcal{C}^\infty(M)$$

attains its maximum for the Levi-Civita connection at each point of M. Conversely, if ∇ is a statistical connection for g and $\mathfrak{scal}$ attains its maximum for ∇ at each point on M, then ∇ is the Levi-Civita connection for g.

As we have already observed the curvature tensor R for ∇ does not have the same symmetries as the curvature tensor of the Levi-Civita connection. By Lemma 1 we see that the symmetry conditions are fulfilled if $R = \overline{R}$. It turns out that this condition is important in many considerations. For instance, in theorems saying that under some curvature conditions a statistical structure is trivial, that is, $\nabla = \hat{\nabla}$. Proofs of the following theorems can be found in [4].

Theorem 1. *Let M be a connected compact surface and (g, ∇) be a trace-free statistical structure on M. If M is of genus 0 and $R = \overline{R}$ then $\nabla = \hat{\nabla}$ on M. If M is of genus 1 and $K = 0$ at one point of M then $\nabla = \hat{\nabla}$ on M.*

Theorem 2. *Let M be a compact manifold equipped with a trace-free statistical structure (g, ∇) such that $R = \overline{R}$. If the sectional curvature $\hat{k}$ for g is positive then $\nabla = \hat{\nabla}$.*

Although the Ricci tensors for ∇ and $\overline{\nabla}$ differs very much from each other, their integrals over a unit sphere bundle UM are the same. Namely we have

Theorem 3. *Let M be a compact oriented manifold and (g, ∇) be a statistical structure on it. Then*

$$\int_{UM} Ric(U, U) dU = \int_{UM} \overline{Ric}(U, U) dU. \qquad (15)$$

If (g, ∇) is trace-free then

$$\int_{UM} Ric(U, U) dU \leq \int_{UM} \widehat{Ric}(U, U) dU \qquad (16)$$

and the equality holds if and only if $\nabla = \hat{\nabla}$ on M.

3   On Examples

As it was mentioned in the Introduction a natural source of statistical structures is the theory of affine hypersurfaces. Let $\mathbf{f} : M \to \mathbf{R}^{n+1}$ be a locally strongly convex hypersurface. For simplicity assume that M is oriented. Let ξ be a transversal vector field on M. We define the induced volume form ν_ξ on M (compatible with the given orientation) as follows

$$\nu_\xi(X_1, ..., X_n) = \det(\mathbf{f}_* X_1, ..., \mathbf{f}_* X_n, \xi).$$

We also have the induced connection ∇ and the second fundamental form g defined on M by the Gauss formula:

$$D_X \mathbf{f}_* Y = \mathbf{f}_* \nabla_X Y + g(X, Y)\xi,$$

where D is the standard flat connection on $\mathbf{R}^{n+1}$. Since the hypersurface is locally strongly convex, g is definite. By multiplying ξ by -1, if necessary, we can assume that g is positive definite. A transversal vector field is called equiaffine if $\nabla \nu_\xi = 0$. This condition is equivalent to the fact that ∇g is symmetric, i.e. (g, ∇) is a statistical structure. It means, in particular, that for a statistical structure obtained on a hypersurface by a choice of a transversal vector field, the Ricci tensor of ∇ is automatically symmetric. In general, the Ricci tensor of a statistical structure on an abstract manifold does not have to be symmetric. Therefore, all structures with non-symmetric Ricci tensor are non-realizable as the induced structures on hypersurfaces. For statistical structures induced on hypersurfaces the condition $R = \overline{R}$ describes the so called affine spheres. The class of affine spheres is very large, very attractive for geometers and still very misterious. Again, it is easy to find examples of statistical structures on abstract manifolds for which $R = \overline{R}$ and which cannot be realized on affine spheres (although in this case the Ricci tensor of ∇ is symmetric). For a statistical structure a necessary condition for being (locally) realizable on a hypersurface is that the connection $\overline{\nabla}$ is projectively flat. It is a strong condition which is rarely satisfied.

Another source of statistical structures is the theory of Lagrangian submanifolds in almost Hermitian manifolds. In this case K can be regarded as the second fundamental tensor of a sumbanifold. The theory is best developed for Lagrangian submanifolds of complex space forms. In this case $\hat{\nabla}K$ as a $(1,3)$-tensor field is symmetric. Hence, by Lemma 1, we also have $\overline{R} = R$. In this case an obstructive condition (which makes that a statistical structure satisfying $R = \overline{R}$ might be non-realizable on a Lagrangian submanifold) is the Gauss equation.

In analogy with the case of hypersurfaces by an equiaffine statistical structure we mean a triple (g, ∇, ν), where (g, ∇) is a statistical structure and ν is a volume form on M (in most cases different than ν_g) parallel relative to ∇.

For more information on dual connections, affine differential geometry and the geometry of statistical structures we refer to [1–4,6].

4   Sectional Curvatures

Of course we have the ordinary sectional curvature for g. In general, the tensor field R for a statistical connection ∇ is not good enough to produce the sectional curvature. The reason is that, in general, $g(R(X,Y)Z,W)$ is not skew symmetric relative to Z, W. But the tensor field $\mathcal{R} = \frac{1}{2}(R + \overline{R})$ has the property $g((\mathcal{R}(X,Y)Z,W) = -g((\mathcal{R}(X,Y)W,Z)$. Moreover, it satisfies the first Bianchi identity. This allows to define the sectional curvature, which we call the sectional ∇-curvature. Namely, if X, Y is an orthonormal basis of a vector plane $\pi \subset T_x M$ then the sectional ∇-curvature by this plane is defined as $g(\mathcal{R}(X,Y)Y,X)$. But for this sectional curvature Schur's lemma does not hold, in general. It is because there is no appropriate universal second Bianchi identity. We have, however, the following analogue of the second Bianchi identity, see [4],

$$\Xi_{U,X,Y}(\hat{\nabla}_U(R + \overline{R}))(X,Y) = \Xi_{U,X,Y}(K_U(\overline{R} - R))(X,Y), \qquad (17)$$

where $\Xi_{U,X,Y}$ denotes the cyclic permutation sum. It follows that for statistical structures satisfying the condition $R = \overline{R}$ Schur's lemma holds. Another result in which the assumption $R = \overline{R}$ is important is the following analogue of Tachibana's theorem, [4],

Theorem 4. *Let M be a connected compact oriented manifold and (g, ∇) be a statistical structure on M such that $R = \overline{R}$. If the curvature operator $\hat{\mathfrak{R}}$ for $\hat{R}$ is*

non-negative and div $\hat{\nabla} R = 0$ *then* $\hat{\nabla} R = 0$. *If additionally* $\hat{\mathfrak{R}} > 0$ *at some point of* M *then the sectional* ∇-*curvature is constant.*

Since the tensor $\mathcal{R}$ has good symmetry properties, one can also define the curvature operator, say $\mathfrak{R}$, for $\mathcal{R}$ sending 2-vectors into 2-vectors. Namely, we set

$$g(\mathfrak{R}(X \wedge Y), Z \wedge U) = g(\mathcal{R}(Z,U)Y, X), \tag{18}$$

where g denotes here the natural extension of g to tensors. The formula defines a linear, symmetric relative to g operator $\mathfrak{R} : \Lambda^2 TM \to \Lambda^2 TM$. In particular, it is diagonalizable and hence it can be positive, negative (definite) etc.

We have the following analogue of a theorem of Meyer-Gallot for trace-free statistical structures, see [4],

Theorem 5. *Let M be a connected compact oriented manifold and (g, ∇) be a trace-free statistical structure on M. If the curvature operator $\mathfrak{R}$ for $\mathcal{R}$ is non-negative on M then each harmonic form is parallel relative to ∇, $\overline{\nabla}$ and $\hat{\nabla}$. If moreover the curvature operator is positive at some point of M then the Betti numbers $b_1(M) = ... = b_{n-1}(M) = 0$.*

Another sectional curvature for a statistical structure (g, K) can be defined by using the tensor field K. We define a $(1,3)$-tensor field $[K, K]$ by

$$[K,K](X,Y)Z = [K_X, K_Y]Z = K_X K_Y Z - K_Y K_X Z \tag{19}$$

for $X, Y, Z \in T_x M$, $x \in M$. Recall that for a Hessian structure we have $[K, K] = -\hat{R}$. The tensor field $[K, K]$ is skew symmetric in X, Y, skew-symmetric relative to g and satisfies the first Bianchi identity. Therefore one can define the sectional K-curvature by a vector plane π tangent to M as $k(\pi) = g([K, K](X, Y)Y, X)$, where X, Y is an orthonormal basis of π. As in the previous case, for this sectional curvature Schur's lemma holds if $\hat{\nabla} K$ is symmetric as a $(1, 3)$-tensor field. Note that the notion of the sectional K-curvature is purely algebraic. The fact that this curvature is constant implies that K has a special expression. Namely, we have, see [5],

Theorem 6. *Let (g, K) be a statistical structure on an n-dimensional manifold M. If the sectional K-curvature is constant and equal to A for all vector planes in TM then for each $x \in M$ there is an orthonormal basis $e_1, ..., e_n$ of $T_x M$ such that*

$$K(e_1, e_1) = \lambda_1 e_1, \quad K(e_1, e_i) = \mu_1 e_i \tag{20}$$

$$K(e_i, e_i) = \mu_1 e_1 + ... + \mu_{i-1} e_{i-1} + \lambda_i e_i, \tag{21}$$

for $i = 2, ...n$ and

$$K(e_i, e_j) = \mu_i e_j \tag{22}$$

for some numbers λ_i, μ_i for $i = 1, ..., n - 1$ and $j > i$. Moreover

$$\mu_i = \frac{\lambda_i - \sqrt{\lambda_i^2 - 4A_{i-1}}}{2}, \tag{23}$$

$$A_i = A_{i-1} - \mu_i^2, \tag{24}$$

for $i = 1, ..., n-1$ where $A_0 = A$. If additionally the statistical structure (g, K) is trace-free then $A \le 0$, λ_i and μ_i are expressed as follows

$$\lambda_i = (n-i)\sqrt{\frac{-A_{i-1}}{n-i+1}}, \quad \mu_i = -\sqrt{\frac{-A_{i-1}}{n-i+1}}. \tag{25}$$

Note that, in general, it is not possible to find a local frame $e_1, ..., e_n$ around a point of M in which K has expression as in the above theorem.

Below there are few theorems serving as examples of results dealing with the sectional K-curvature. In these theorems the notation $[K, K] \cdot K$, $\hat{R} \cdot K$ means that $[K, K]$ and $\hat{R}$ act on K as differentiations. Details concerning the theorems are providedT in [5].

Theorem 7. Let (g, K) be a statistical structure on a manifold M. If the sectional K-curvature is non-positive on M and $[K, K] \cdot K = 0$ then the sectional K-curvature vanishes on M.

Corollary 1. If (g, K) is a Hessian structure on M with non-positive sectional curvature of g and such that $\hat{R} \cdot K = 0$ then $\hat{R} = 0$.

Theorem 8. If (g, K) is a statistical structure on a manifold M, the sectional K-curvature is negative on M and $\hat{R} \cdot K = 0$ then $\hat{R} = 0$.

Theorem 9. Assume that $[K, K] = 0$ on a statistical manifold (M, g, K), $\hat{\nabla} K$ is symmetric and $\hat{\nabla} E = 0$. If K is non-degenerate, that is, the mapping $T_x M \ni X \to K_X \in HOM(T_x M)$ is a monomorphism at each point of M then $\hat{R} = 0$ and $\hat{\nabla} K = 0$ on M.

Theorem 10. Let (g, K) be a trace-free statistical structure on a manifold M with symmetric $\hat{\nabla} K$. If the sectional K-curvature is constant then either $K = 0$ or $\hat{R} = 0$ and $\hat{\nabla} K = 0$ on M.

5 Bochner-Type Theorems

Bochner's theorems for Riemannian manifolds say, roughly speaking, that under some curvature assumptions harmonic forms must be parallel. This is a converse to the trivial statement that a parallel form is harmonic.

For statistical structures one can prove some analogues of this theorems. First we define a new Laplacian Δ^∇ depending on the statistical connection. If (g, ∇) is a statistical structure on M then we define the codifferential δ^∇ acting on differential forms copying the classical Weitzenböck formula

$$\delta^\nabla \omega = -\text{tr}_g \nabla \omega(\cdot, \cdot, ...)$$

for any differential form ω. We now set

$$\Delta^{\nabla} = d\delta^{\nabla} + \delta^{\nabla}d.$$

If the statistical structure is trace-free then Δ^{∇} is the ordinary Laplacian for g. A form ω is called ∇-harmonic if $\Delta^{\nabla}\omega = 0$. Hodge's theory can be adapted to this definition of a Laplacian and harmonicity. In particular, for an equiaffine statistical structure on a compact manifold M we have $\dim \mathcal{H}^{k,\nabla}(M) = b_k(M)$, where $\mathcal{H}^{k,\nabla}(M)$ is the space of all ∇-harmonic forms and $b_k(M)$ is the k-th Betti number of M.

Below are exemplary analogues of Bochner-type theorems for statistical structures.

Theorem 11. *Let M be a connected compact oriented manifold with an equiaffine statistical structure (g, ∇, ν). If the Ricci tensor Ric for ∇ is non-negative on M then every ∇-harmonic 1-form on M is $\overline{\nabla}$-parallel. In particular, the first Betti number $b_1(M)$ is not greater than $\dim M$. If additionally $Ric > 0$ at some point of M then $b_1(M) = 0$.*

Theorem 12. *Let M be a connected compact oriented manifold. Let (g, ∇) be a trace-free statistical structure on M. If $Ric + \overline{Ric} \geq 0$ on M then each harmonic 1-form on M is parallel relative to the connections ∇, $\overline{\nabla}$ and $\hat{\nabla}$. In particular, $b_1(M) \leq \dim M$. If moreover $Ric + \overline{Ric} > 0$ at some point then $b_1(M) = 0$.*

For any statistical structure (g, ∇) one can define the Weitzenböck curvature operator denoted here by $\mathcal{W}^R$. It depends only on g and the curvature tensor R. More precisely, it can be introduced as follows. Let s be a tensor field of type (l, k), where $k > 0$, on M. One defines a tensor field $\mathcal{W}^R s$ of type (l, k) by the formula

$$(\mathcal{W}^R s)(X_1, ..., X_k) = \sum_{i=1}^{k} \sum_{j=1}^{n} (R(e_j, X_i) \cdot s)(X_1, ..., e_j, ..., X_k), \qquad (26)$$

where $e_1, ..., e_n$ is an arbitrary orthonormal frame, $R(e_j, X_i) \cdot s$ means that $R(e_j, X_i)$ acts as a differentiation on s, and e_j in the last parenthesis is at the i-th place. It is possible to prove appropriate generalizations of Bochner-Weitzenböck's and Lichnerowicz's formulas for the Laplacian acting on differential forms on statistical manifolds. In particular, for a trace-free structure we have the following simple formula

$$\Delta = \nabla^* \nabla + \mathcal{W}^R,$$

where ∇^* is suitably defined formal adjoint for ∇.

Details concerning Hodge's theory and Bochner's technique for statistical structures can be found in [4].

References

1. Lauritzen, S.T.: Statistical Manifolds. IMS Lecture Notes-Monograph Series, vol. 10, pp. 163–216 (1987)
2. Li, A.-M., Simon, U., Zhao, G.: Global Affine Differential Geometry of Hypersurfaces. Walter de Gruyter, Berlin (1993). Geom. Appl. 24, 567–578 (2006)
3. Nomizu, K., Sasaki, T.: Affine Differential Geometry. Cambridge University Press, Cambridge (1994)
4. Opozda, B.: Bochner's technique for statistical manifolds. Ann. Glob. Anal. Geom. doi:10.1007/s10455-015-9475-z
5. Opozda, B.: A sectional curvature for statistical structures. arXiv:1504.01279 [math.DG]
6. Shima, H.: The Geometry of Hessian Structures. World Scientific, Singapore (2007)

The Pontryagin Forms of Hessian Manifolds

J. Armstrong[1]([✉]) and S. Amari[2]

[1] King's College London, London WC2R 2LS, UK
john.1.armstrong@kcl.ac.uk
[2] RIKEN Brain Science Institute, Saitama 351-0198, Japan
amari@brain.riken.jp

Abstract. We show that Hessian manifolds of dimensions 4 and above must have vanishing Pontryagin forms. This gives a topological obstruction to the existence of Hessian metrics. We find an additional explicit curvature identity for Hessian 4-manifolds. By contrast, we show that all analytic Riemannian 2-manifolds are Hessian.

1 Introduction

At GSI2013, S. Amari asked the question of when a given Riemannian metric is a Hessian metric. In other words, for what metrics g do there exist local coordinates at every point such that g can be written as the Hessian of some convex potential function ϕ?

As a first result we will show that:

- All analytic metrics in 2 dimensions are Hessian metrics.
- Not all analytic metrics in 3 dimensions and higher are Hessian metrics.
- In dimensions 4 and above there are restrictions on the possible curvature tensors of Hessian metrics.

We will see that these results are quite simple to prove using Cartan–Kähler theory and were found independently by Robert Bryant [3].

A further question posed by Amari was to find conditions and invariants which characterize the Riemannian metrics which are Hessian. The ultimate goal would be to define a set of tensors such that the metric is Hessian if and only if these tensors vanish. We cannot achieve this goal in full. However, a partial answer that we can demonstrate is that the *Pontryagin forms* of the metric must vanish. By the *Pontryagin forms* we mean the differential forms defined in terms of the curvature that provide representatives of the Pontryagin classes.

We note that this provides a topological obstruction to the existence of a Hessian metric: a compact manifold that admits a Hessian metric must have vanishing Pontryagin classes.

To put this into context we recall that Hessian metrics are be locally equivalent to g-dually flat structures. That is g is Hessian if and only if one can locally find flat affine connections ∇ and ∇^* satisfying:

$$g(\nabla_Z X, Y) = g(X, \nabla_Z^* Y).$$

© Springer International Publishing Switzerland 2015
F. Nielsen and F. Barbaresco (Eds.): GSI 2015, LNCS 9389, pp. 240–247, 2015.
DOI: 10.1007/978-3-319-25040-3_27

In [2] the question of when a manifold admits a global g-dually flat structure is considered and some topological obstructions are found. Our result is related, but distinct. We have found an obstruction to the existence of a metric which is required to be locally g-dually flat but which need not have globally defined connections ∇ and ∇^*.

It is trivial that a manifold that is globally g-dually flat must have vanishing Pontryagin classes: simply consider the Pontryagin forms defined by the flat connection. By the same token, the Euler characteristic must vanish on any manifold which is globally g-dually flat.

On the other hand as mentioned above, all 2-manifolds admit Hessian metrics including those with non-vanishing Euler characteristic. Thus in 2 dimensions there is a large difference between the set of manifolds which admit Hessian metrics and those which admit global g-dually flat structures. One can generalize this example to higher dimensions by considering quotients of hyperbolic space. It is well known that the hyperbolic metric is a Hessian metric, yet quotients of hyperbolic space may have non-vanishing Euler characteristic implying that they cannot admit a flat connection never mind a g-dually flat structure.

Thus this paper provides a first step towards answering the interesting question: which manifolds admit a Hessian metric?

This paper is a summary and update of a joint paper with S. Amari. Full details can be found in [1].

2   A Counting Argument

To define a Hessian metric locally near a point p an a manifold M^n we need to choose a set of coordinates $x : M^n \to \mathbb{R}$ defined in a neighbourhood of p and a strictly convex potential function ϕ. We can then write down a Hessian metric

$$g_{ij} = \frac{\partial^2 \phi}{\partial x_i \partial x_j}.$$

Speaking somewhat loosely we can say that a Hessian metric depends upon $n+1$ real valued functions of n variables: the n coordinate functions and the potential ϕ.

On the other hand to write down a general metric we need to choose the $\frac{n(n+1)}{2}$ tensor components g_{ij} in some neighbourhood of the point p. Thus a general Riemannian metric depends upon $\frac{n(n+1)}{2}$ real valued functions of n variables.

Since $\frac{n(n+1)}{2} > n + 1$ when $n > 2$ this strongly suggests that there are many more Riemannian metrics than Hessian metrics in dimensions greater than 3.

This argument is suggestive but not rigorous. In particular it gives the wrong answer in dimension 1! Our formulae would suggest that there are more Hessian metrics than Riemannian metrics in dimension 1. The reason for this is that we haven't taken into account the diffeomorphism group when counting.

To make the argument rigorous we need to consider jet bundles. If the metric is Hessian we see that the k-jet of the metric depends upon the $k + 2$-jet of the functions x and ϕ. The dimension of the space of $(k+2)$-jets of $(n+1)$ functions of n real variables is:

$$\dim J_{k+2}(x, \phi) := \sum_{i=0}^{k+2} (n+1) \dim(S^i T_p) = \sum_{i=0}^{k+2} (n+1) \binom{n+i-1}{i}.$$

Similarly the dimension of the space of k-jets of g_{ij} is:

$$\dim J_k(g) := \sum_{i=0}^{k} \frac{n(n+1)}{2} \dim(S^i T_p) = \sum_{i=0}^{k} \frac{n(n+1)}{2} \binom{n+i-1}{i}.$$

If we now compare the growth rate of $\dim J_k(g)$ and $\dim J_{k+2}(x, \phi)$ as k increases we see that so long as $n > 2$, $\dim J_k(g) > \dim J_{k+2}(x, \phi)$ for sufficiently large n. To see this note that we can write:

$$\dim J_k(g) - \dim J_{k+2}(x, \phi) = (n+1)(a_{k,n} - b_{k,n})$$

where

$$a_{k,n} := \left(\frac{n}{2} - 1\right) \sum_{i=1}^{k} \binom{n+1-i}{i},$$

$$b_{k,n} := \binom{n+k}{k+1} + \binom{n+k+1}{k+2}.$$

We can now rigorously conclude that in dimensions greater than 2 there really are more Riemannian metrics than Hessian metrics. The growth rate of jet bundles provides a rigorous language for heuristic counting arguments.

3   Dimension 2

In dimension 2 our counting argument fails. It seems conceivable that every Riemannian metric is a Hessian metric. One can go further and explicitly identify the mapping from $(k+2)$ jets of (x, ϕ) to k-jets of metrics. It is not difficult to do so for low values of k. One discovers that the mapping is onto. It is easy to write a computer program that computes the mapping for a given value of k, in which case one will again discover that the mapping is onto.

One would like to be able to find a proof that this mapping is onto for all k and one would like to be able to deduce from this that all Riemannian metrics are Hessian metrics.

Fortunately a toolkit already exists for solving precisely this kind of problem. It is called Cartan–Kähler theory.

A general setting is to consider two vector bundles V and W over some n-manifold M^n. Let $D : \Gamma(V) \longrightarrow \Gamma(W)$ be an order k differential operator mapping sections of V to sections of W. In other words let D map k-jets of V at p to elements of W_p.

The top order term of this mapping is called the symbol σ_D of D.

$$\sigma_D : S^k T_p^* \otimes V_p \longrightarrow W_p$$

The reason that the top order term acts on a symmetric power of the tangent bundle simply comes from the fact that derivatives in different directions commute. The top order term only depends on the k-th derivatives of a section $v \in \Gamma(V)$. If we assume that D is quasilinear then σ will be a linear map.

If σ is onto then the differential equation $Dv = w$ can always be solved up to order k at p. Now consider differentiating the equation $Dv = w$. We will get a $k + 1$-th order differential equation. We can associate a symbol σ_1 to this differential equation. If σ_1 is onto too then we can always solve the equation to $k + 1$-st order. Continuing in this way we can define a sequence of symbols σ_i. If they are all onto then we can solve the differential equation to any desired order. Note that the σ_i can be easily computed directly from σ. Thus requiring that σ_i is onto for all i is just an algebraic condition on σ.

How can one prove that σ_i is onto for all i? The solution is to use *Cartan's test* which we will now describe. Given a basis $\{v_1, v_2, \ldots v_n\}$ for T^*M, define the map:

$$\sigma_{i,m} : S^{k+i} \langle v_1, v_2, \ldots v_m \rangle \otimes V_p \longrightarrow S^i T_p^* \otimes W_p$$

to be the restriction of σ_i. Define $g_{i,m} := \dim \ker \sigma_{i,m}$. If one can find a basis $\{v_1, v_2, \ldots v_n\}$ and a number α such that σ_i is onto for all $i \leqslant \alpha$ and such that $g_{\alpha,n} = \sum_{\beta=0}^{k} g_{\alpha-1\beta}$ then the differential equation is said to be involutive. It can be shown that this implies that $\sigma_{\alpha+i}$ is onto for all i. Moreover, if one is working in the analytic category, one can then prove that solutions to the differential equation exist [4–6].

This gives a strategy for proving that all analytic 2-metrics are locally g-dually flat and hence Hessian. Given a metric g we can interpret the requirement that it is Hessian as requiring that we can locally find a g-dually flat connection. This gives rise to a 1-st order differential equation for a connection A.

To be precise, we define

$$\iota : T^* \otimes T^* \otimes T \longrightarrow T^* \otimes T^* \otimes T^*$$

to be raising the final index using the metric then to find a g-dually flat connection we seek a tensor $A \in \iota^{-1}(S^3 T^*)$ such that the connection $\nabla + A$ has curvature zero. It is well known that such a tensor is equivalent to a g-dually flat connection.

The details are not illuminating. The point is that we have expressed the problem as a differential equation and it is a simple algebraic exercise to check that this equation is involutive. It follows that all analytic 2-metrics are Hessian.

4   Dimensions > 4

Our aim in this section is to find more concrete obstructions to the existence of Hessian metrics. The key result is the following [8]:

Proposition 1. *Let (M, g) be a Riemannian manifold. Let ∇ denote the Levi–Civita connection and let $\overline{\nabla} = \nabla + A$ be a g-dually flat connection. Then*

(i) The tensor A_{ijk} lies in $S^3 T^*$. We shall call it the S^3-tensor of $\overline{\nabla}$.
(ii) The S^3-tensor determines the Riemann curvature tensor as follows:

$$R_{ijkl} = -g^{ab} A_{ika} A_{jlb} + g^{ab} A_{ila} A_{jkb}. \tag{1}$$

Proof. $A \in T^* \otimes T^* \otimes T$. The condition that $\overline{\nabla}$ is torsion free is equivalent to requiring that $A \in S^2 T^* \otimes T$. Using the metric to identify T and T^*, the condition that $\overline{\nabla}$ is dually torsion free can be written as $A \in S^3 T^*$.

Expanding the formula $\overline{R}_{XY} Z = \overline{\nabla}_X \overline{\nabla}_Y Z - \overline{\nabla}_Y \overline{\nabla}_X - \overline{\nabla}_{[X,Y]} Z$ in terms of ∇ and A, one obtains the following curvature identity:

$$\overline{R}_{XY} Z = R_{XY} Z + 2(\nabla_{[X} A)_{Y]} Z + 2A_{[X} A_{Y]} Z. \tag{2}$$

Here $\overline{R} = 0$ is the curvature of $\overline{\nabla}$ and the square brackets denote anti-symmetrization. Since $\overline{\nabla}$ is dually flat $\overline{R} = 0$.

Continuing to use the metric to identify T and T^*, the symmetries of the curvature tensor tell us that $R \in \Lambda^2 T \otimes \Lambda^2 T$. On the other hand, $(\nabla_{[.} A)_{.]} \in \Lambda^2 T \otimes S^2 T$. Thus if one projects equation (2) onto $\Lambda^2 T \otimes \Lambda^2 T$ one obtains the curvature identity (1).

We define a quadratic equivariant map ρ from $S^3 T^* \longrightarrow \Lambda^2 T^* \otimes \Lambda^2 T^*$ by:

$$\rho(A_{ijk}) = -g^{ab} A_{ika} A_{jlb} + g^{ab} A_{ila} A_{jkb}$$

Corollary 1. *In dimensions > 4 the condition that R lies in the image of ρ gives a non-trivial necessary condition for a metric g to be a Hessian metric.*

Proof. $\dim S^3 T = \binom{n+2}{n-1} = \frac{1}{6}n(1 + n)(2 + n)$. The dimension of the space of algebraic curvature tensors, $\mathcal{R}$, is $\dim \mathcal{R} = \frac{1}{12}n^2(n^2 - 1)$. So $\dim \mathcal{R} - \dim S^3 T = \frac{1}{12}n(n - 4)(1 + n)^2$. This is strictly positive if $n > 4$.

5   Dimension 4

Surprisingly the condition that R lies in the image of ρ gives a non-trivial condition in dimension 4. In dimension 4, $\dim S^3 T = \dim \mathcal{R} = 20$, yet the dimension of the image of ρ is only 18. The authors discovered this by computer experiment: we picked a random tensor $A \in S^3 T^*$ and then computed the rank of the derivative ρ_* at A. By Sard's theorem we could be rather confident that ρ is not onto.

To prove this rigorously we wanted to identify the explicit conditions on an algebraic curvature tensor that were required for it to lie in the image of ρ. We found these conditions using a computer search. We assumed that the explicit conditions could be written as $SO(4)$-equivariant polynomials in the curvature R and catalogued the possibilities using the representation theory of SO(4). This was feasible to program due to the simple representation theory of Spin(4) $\cong$ SU(2) $\times$ SU(2). We only had to examine up to cubic polynomials to find the 2-dimensions of curvature obstruction suggested by our numerical experiments.

Theorem 1. *The space of possible curvature tensors for a Hessian 4-manifold is* 18 *dimensional. In particular the curvature tensor must satisfy the identities:*

$$\alpha(R_{ija}^{b}R_{klb}^{a}) = 0 \tag{3}$$

$$\alpha(R_{iajb}R_{kcd}^{b}R_{l}^{dac} - 2R_{iajb}R_{kcd}^{a}R_{l}^{dbc}) = 0 \tag{4}$$

where α denotes antisymmetrization of the i, j, k and l indices.

Proof. Using a symbolic algebra package, write the general tensor in S^3T^* with respect to an orthonormal basis in terms of its 20 components. Compute the curvature tensor using equation (1). One can then directly check the above identities.

The first of these equations is particularly interesting. The tensor defined in Eq. 3 is a closed 4-form. Its de Rham cohomology class is independent of the metric and hence defines a topological invariant of the manifold - the first Pontrjagin class. The integral of this form over the 4-manifold is the signature of the 4-manifold. We have proved that the signature must vanish on a Hessian 4-manifold.

6   Pontryagin Classes of Hessian Manifolds

Let us generalize this last result to higher dimensions. To make the proof as vivid as possible, we introduce a graphical notation that simplifies manipulating symmetric powers of the S^3-tensor A (this is based on the notation given in the appendix of [7]). When using this notation we will always assume that our coordinates are orthonormal at the point where we perform the calculations so we can ignore the difference between upper and lower indices of ordinary tensor notation.

Given a tensor defined by taking the n-th tensor power of the S^3-tensor tensor A followed by a number of contractions we can define an associated graph by:

- Adding one vertex to the graph for each occurrence of A;
- Adding an edge connecting the vertices for each contraction between the vertices;
- Adding a vertex for each tensor index that is not contracted and labelling it with the same symbol used for the index. Join this vertex to the vertex representing the associated occurrence of A.

When two tensors written in the Einstein summation convention are juxtaposed in a formula, we will refer to this as "multiplying" the tensors. This multiplication corresponds graphically to connecting labelled vertices of the graphs according to the contractions that need to be performed when the tensors are juxtaposed. Since this multiplication is commutative, and since the S^3-tensor is symmetric, one sees that there is a one to one correspondence between isomorphism classes of such graphs and equivalently defined tensors.

We can use these graphs in formulae as an alternative notation for the tensor represented by the graph. For example, we can write the curvature identity (1) graphically as

$$
R_{ijkl} = - \;
\begin{array}{cc} i & j \\[-2pt] \vdash\!\!-\!\!\dashv \\[-2pt] k & l \end{array}
\; + \;
\begin{array}{cc} i & j \\[-2pt] \times \\[-2pt] k & l \end{array} \; .
\tag{5}
$$

Theorem 2. *The tensor*

$$
Q^p_{i_1 i_2 \ldots i_{2p}} =
$$

$$
\sum_{\sigma \in S_{2p}} \mathrm{sgn}(\sigma) R^{a_2}_{i_{\sigma(1)} i_{\sigma(2)} a_1} R^{a_3}_{i_{\sigma(3)} i_{\sigma(4)} a_2} R^{a_4}_{i_{\sigma(5)} i_{\sigma(6)} a_3} \cdots R^{a_1}_{i_{\sigma(2p-1)} i_{\sigma(2p)} a_p}
$$

vanishes on a Hessian manifold. Hence all Pontryagin forms vanish on a Hessian manifold.

Proof. We can rewrite the curvature identity (1) as:

$$
R_{i_1 i_2 ab} = \sum_{\sigma \in S_2} - \mathrm{sgn}(\sigma) \;
\begin{array}{cc} i_{\sigma(1)} & i_{\sigma(2)} \\[-2pt] \vdash\!\!-\!\!\dashv \\[-2pt] a & b \end{array} \; .
$$

Thus we can replace each R in the formula for Q^p with an 'H'. The legs of adjacent H's are then connected. The result is:

$$
Q^p_{i_1 i_2 \ldots i_{2p}} =
$$

$$
(-1)^p \sum_{\sigma \in S_{2p}} \mathrm{sgn}(\sigma) \;
\begin{array}{cccccccc} i_{\sigma(1)} & i_{\sigma(2)} & i_{\sigma(3)} & i_{\sigma(4)} & i_{\sigma(5)} & i_{\sigma(6)} & \cdots & i_{\sigma(2p-1)} \; i_{\sigma(2p)} \end{array} \; .
$$

Since the cycle $1 \to 2 \to 3 \ldots \to 2p \to 1$ is an odd permutation, one sees that $Q^p = 0$.

The import of this result is that the Pontryagin forms of the manifold can be expressed in terms of the Q_p tensors. Thus it is a corollary that the Pontryagin forms, and hence the Pontryagin classes, vanish on a Hessian manifold. This result is an easy consequence of the standard definition of the Pontryagin forms combined with standard results on symmetric polynomials.

We have seen that equation (3) generalizes easily to higher dimensions. Equation (4) on the other hand does not hold in dimensions $\geqslant 5$. We list some interesting questions that this raises. Can one efficiently find all the explicit curvature conditions that must be satisfied by a Hessian metric in a fixed dimension $n \geqslant 5$? Can one find all the curvature conditions that hold for all n? For large enough n, is the condition that the curvature lies in the image of ρ a sufficient condition for a metric to be Hessian?

References

1. Amari, S., Armstrong, J.: Curvature of Hessian manifolds. Differ. Geom. Appl. **33**, 1–12 (2014)
2. Ay, N., Tuschmann, W.: Dually flat manifolds and global information geometry. Open. Syst. Inf. Dyn. **9**(2), 195–200 (2002)
3. Bryant, R.: http://mathoverflow.net/questions/122308/ (2013)
4. Cartan, E.: Les systèmes différentiels extérieurs et leur applications géométriques, vol. 994. Hermann & cie, Paris (1945)
5. Goldschmidt, H.: Integrability criteria for systems of nonlinear partial differential equations. J. Differ. Geom. **1**(3–4), 269–307 (1967)
6. Guillemin, V.W., Sternberg, S.: An algebraic model of transitive differential geometry. Bull. Am. Math. Soc. **70**(1), 16–47 (1964)
7. Penrose, R., Rindler, W.: Spinors and Space-Time: Volume 1, Two-Spinor Calculus and Relativistic Fields. Cambridge Monographs on Mathematical Physics. Cambridge University Press, Cambridge (1987)
8. Shima, H.: The Geometry of Hessian Structures, vol. 1. World Scientific, Singapore (2007)

Matrix Realization of a Homogeneous Cone

Hideyuki Ishi[✉]

Graduate School of Mathematics,
Nagoya University, Furo-cho, Nagoya 464-8602, Japan
hideyuki@math.nagoya-u.ac.jp

Abstract. Based on the theory of compact normal left-symmetric algebra (clan), we realize every homogeneous cone as a set of positive definite
real symmetric matrices, where homogeneous Hessian metrics as well as
a transitive group action on the cone are described efficiently.

Keywords: Homogeneous cone · Left-symmetric algebra · Hessian
metric

1 Introduction

An open convex cone containing no straight line is called a homogeneous cone if
a linear Lie group acts on the cone transitively. Since the study of homogeneous
cones was originally motivated by the theory of homogeneous bounded complex
domains (see Introduction of [15]), the research on the cones has been developed in connection with various areas of mathematics. Actually, geometry and
analysis on homogeneous cones are nowadays applied to statistics [1,4,7] and
optimization theory [5,14]. These connections with applied mathematics can be
understood well in the context of Hessian geometry. A Hessian manifold is a Riemannian manifold with a flat structure such that the metric is locally expressed
as a Hessian of a smooth function, and it is an important object in information geometry [13]. A homogeneous cone is a typical example of a homogeneous
Hessian manifold, that is, a Hessian manifold on which the automorphism group
acts transitively. Moreover, the homogeneous cone plays a central role in the
general theory of homogeneous Hessian manifolds [12].

In this article, we realize every homogeneous cone as a set of real positive definite symmetric matrices, where a transitive group action on the cone
is obtained very easily (Theorem 3). Furthermore, all the Hessian metrics invariant under the group action, which exhaust the homogeneous Hessian metrics
on the cone, are described in a simple way (Theorem 4). Similar realizations of
homogeneous cones have been obtained by several authors [3,11,16–18], whereas
our approach is near to that of [17], based on the theory of compact normal left-
symmetric algebra. Roughly speaking, the algebra is defined as a linearization
of the transitive group action (see Sect. 2). The left-symmetric algebra is also
called a Koszul-Vinberg algebra [2] or a pre-Lie algebra [10], and it is studied
actively from various viewpoints. In particular, the left-symmetric algebra serves

© Springer International Publishing Switzerland 2015
F. Nielsen and F. Barbaresco (Eds.): GSI 2015, LNCS 9389, pp. 248–256, 2015.
DOI: 10.1007/978-3-319-25040-3_28

as a main algebraic tool in the study of homogeneous Hessian manifolds [12]. In this respect, the present work on homogeneous cones matches with the Hessian geometry perspective very well.

2   Homogeneous Cones and Compact Normal Left-Symmetric Algebras

We shall review briefly the relation between a homogeneous cone and a compact normal left-symmetric algebra (clan) following Vinberg [15, Chap. 2]. Let V be a finite dimensional real vector space, and $\Omega \subset V$ an open convex cone containing no straight line. We denote by $GL(\Omega)$ the group $\{ f \in GL(V) ; f(\Omega) = \Omega \}$ of linear automorphisms on the cone Ω. Then $GL(\Omega)$ is a closed subgroup of $GL(V)$. In other words, $GL(\Omega)$ forms a linear Lie group. In what follows, we assume that Ω is *homogeneous*, that is, the group $GL(\Omega)$ acts on Ω transitively. The homogeneity of Ω is clearly equivalent to the existence of a linear Lie group acting transitively on the cone Ω.

Let H be a maximal connected triangular subgroup of $GL(\Omega)$, where a linear Lie group on V is said to be *triangular* if it is expressed as a set of triangular matrices with respect to an appropriate basis of V. Because of the maximality, H contains the group $\{ e^t \mathrm{Id}_V \}_{t \in \mathbb{R}}$ of dilations on V. It is shown in [15, Chap. 1, Theorem 1] that the group H acts on Ω simply transitively. Take and fix a point $e \in \Omega$. Then we have a diffeomorphism $H \ni h \mapsto he \in \Omega$, and differentiating this map, we obtain a linear isomorphism $\mathfrak{h} \ni L \mapsto Le \in V \equiv T_e\Omega$, where $\mathfrak{h}$ is the Lie algebra of H. Thus, for each $x \in V$, there exists a unique $L_x \in \mathfrak{h}$ for which $L_x e = x$. Let us define a bilinear map $\triangle : V \times V \to V$ by

$$x \triangle y := L_x y \qquad (x, y \in V).$$

Then $e \in V$ is a unit element of V, that is,

$$x \triangle e = x = e \triangle x \qquad (x \in V).$$

In fact, the first equality is clear by definition, while the second follows from the fact that L_e equals the identity operator Id_V, which is the generator of the dilation group $\{ e^t \mathrm{Id}_V \}_{t \in \mathbb{R}}$. Moreover, the following hold:

(C1) For any $x, y, z \in V$, one has $[x \triangle y \triangle z] = [y \triangle x \triangle z]$, where $[x \triangle y \triangle z] := x \triangle (y \triangle z) - (x \triangle y) \triangle z$ (*left-symmetry*),

(C2) There exists a linear form $\xi \in V^*$ such that $(x|y)_\xi := \xi(x \triangle y)$ $(x, y \in V)$ gives a positive inner product on V (*compactness*),

(C3) For every $x \in V$, the linear operator L_x has only real eigenvalues (*normality*).

In general, a (not necessarily associative) $\mathbb{R}$-algebra $(V, \triangle)$ satisfying (C1) is called *a left-symmetric algebra*, and a left-symmetric algebra with the properties (C2) and (C3) is called *a clan* (*c*ompact *n*ormal *l*eft-symmetric *a*lgebra). Vinberg showed that the construction of the clan $(V, \triangle)$ from the homogeneous cone Ω explained above induces a one-to-one correspondence between homogeneous cones and clans with unit element up to natural equivalence [15, Chap. 2, Theorem 2].

3   Clans Consisting of Real Symmetric Matrices

Let V_n be the vector space of real symmetric matrices of size n, and $\mathcal{P}_n \subset V_n$ the set of positive definite symmetric matrices. Then $\mathcal{P}_n$ is a homogeneous cone on which the group $GL(n, \mathbb{R})$ acts transitively by the action ρ defined by $\rho(A)X := AX\,{}^tA$ $(A \in GL(n, \mathbb{R}), X \in V_n)$. Let H_n be the set of lower triangular matrices of size n with positive diagonals. Then H_n is a maximal connected triangular subgroup of $GL(n, \mathbb{R})$, and acts on the cone $\mathcal{P}_n$ simply transitively by the action ρ. The Lie algebra $\mathfrak{h}_n$ of H_n equals the space of all lower triangular matrices of size n. We choose the unit matrix E_n as a reference point e in $\mathcal{P}_n$. The associated clan structure on the vector space V_n is given by

$$X \triangle Y := \underset{\sim}{X}Y + Y\,{}^t(\underset{\sim}{X}) \quad (X, Y \in V_n),$$

where $\underset{\sim}{X} \in \mathfrak{h}_n$ is defined by

$$(\underset{\sim}{X})_{ij} := \begin{cases} 0 & (i < j), \\ X_{ii}/2 & (i = j), \\ X_{ij} & (i > j). \end{cases}$$

If a linear subspace $\mathcal{Z} \subset V_n$ satisfies $\mathcal{Z} \triangle \mathcal{Z} \subset \mathcal{Z}$, then $(\mathcal{Z}, \triangle)$ forms a clan. We call such $\mathcal{Z}$ a subalgebra of $(V_n, \triangle)$. The following crucial fact is proved in [6].

Theorem 1. *Every clan with a unit element is isomorphic to a subalgebra of* $(V_n, \triangle)$ *containing* E_n.

Now we state one of our main results.

Theorem 2. *Let* $\mathcal{Z} \subset V_n$ *be a subalgebra containing* E_n. *Then there exists a permutation matrix* $w = \sum_{\alpha=1}^{n} E_{\sigma(\alpha),\,\alpha}$ *with* $\sigma \in \mathfrak{S}_n$ *such that* $\rho(w) : \mathcal{Z} \ni X \mapsto wX\,{}^tw \in \rho(w)\mathcal{Z}$ *is an algebra isomorphism, and* $\rho(w)\mathcal{Z}$ *is the space of matrices of the form*

$$X = \begin{pmatrix} X_{11} & {}^tX_{21} & \cdots & {}^tX_{r1} \\ X_{21} & X_{22} & & {}^tX_{r2} \\ \vdots & & \ddots & \\ X_{r1} & X_{r2} & & X_{rr} \end{pmatrix} \quad \begin{pmatrix} X_{ll} = x_{ll}E_{n_l},\ x_{ll} \in \mathbb{R}\ (l = 1, \ldots, r) \\ X_{lk} \in \mathcal{V}_{lk}\ (1 \le k < l \le r) \end{pmatrix}, \quad (1)$$

where $n = n_1 + \cdots + n_r$ *is a partition and* $\mathcal{V}_{lk} \subset \mathrm{Mat}(n_l, n_k; \mathbb{R})$ $(1 \le k < l \le r)$ *are some vector spaces.*

We shall say that a linear space $\mathcal{Z} \subset V_n$ admits a *normal block decomposition* if $\mathcal{Z}$ consists of matrices of the form (1) with appropriate vector spaces $\mathcal{V}_{lk} \subset \mathrm{Mat}(n_l, n_k; \mathbb{R})$. For example, let $\mathcal{Z}$ be the space of matrices of the form

$$\begin{pmatrix} x_1 & x_4 & 0 & 0 \\ x_4 & x_2 & 0 & 0 \\ 0 & 0 & x_1 & x_5 \\ 0 & 0 & x_5 & x_3 \end{pmatrix}.$$

Then it is easy to see that $\mathcal{Z}$ is a subalgebra of $(V_4, \triangle)$. Let $w \in GL(4, \mathbb{R})$ be the permutation matrix corresponding to the transposition $(2\ 3)$. Then

$$\rho(w) \begin{pmatrix} x_1 & x_4 & 0 & 0 \\ x_4 & x_2 & 0 & 0 \\ 0 & 0 & x_1 & x_5 \\ 0 & 0 & x_5 & x_3 \end{pmatrix} = \begin{pmatrix} x_1 & 0 & x_4 & 0 \\ 0 & x_1 & 0 & x_5 \\ x_4 & 0 & x_2 & 0 \\ 0 & x_5 & 0 & x_3 \end{pmatrix},$$

so that $\rho(w)\mathcal{Z}$ admits a normal block decomposition with $n_1 = 2$, $n_2 = n_3 = 1$. In checking whether $\rho(w)$ is an algebra isomorphism, the following lemma is useful.

Lemma 1. *Let $\mathcal{Z}$ be a subalgebra of $(V_n, \triangle)$, and take $A \in O(n)$. Assume that $A X\, {}^tA \in \mathfrak{h}_n$ for all $X \in \mathcal{Z}$. Then $\rho(A) : \mathcal{Z} \to \rho(A)\mathcal{Z} \subset V_n$ gives an algebra isomorphism.*

Proof. Note that X is characterized as a unique lower triangular matrix such that $X = \underline{X} + {}^t(\underline{X})$. Thus the assumption of the statement implies that $(\rho(A)X) = A\underline{X}\, {}^tA$ for $X \in \mathcal{Z}$. Therefore we have

$$(\rho(A)X)\triangle(\rho(A)Y) = A\underline{X}{}^tA(AY^tA) + (AY^tA)^t(A\underline{X}{}^tA)$$
$$= A(\underline{X}Y + Y\,{}^t(\underline{X})){}^tA = \rho(A)(X\triangle Y)$$

for $X, Y \in \mathcal{Z}$. ∎

Before proceeding to the proof of Theorem 2, we recall the structure theorem of clans due to [15, Chap. 2, Sect. 4].

Proposition 1 (Vinberg). *Let $(V, \triangle)$ be a clan with a unit element $e \in V$. If the primitive idempotents $e_1, \ldots, e_r$ of V are suitably labeled, one has*

$$V = \sum_{1 \le k \le l \le r}^{\oplus} V_{lk}, \tag{2}$$

where

$$V_{lk} := \{ x \in V \,;\, e_j \triangle x = (\delta_{lj} + \delta_{kj})x/2, \quad x\triangle e_j = \delta_{kj}x \text{ for all } j = 1, \ldots, r \}.$$

Moreover, V_{kk} equals $\mathbb{R}e_k$ for $k = 1, \ldots, r$, and one has $e = e_1 + \cdots + e_r$.

For the clan $(V_n, \triangle)$, the primitive idempotents are exactly the matrix elements $E_{\alpha\alpha}$ $(\alpha = 1, \ldots, n)$, and we have

$$V_{\beta\alpha} = \begin{cases} \mathbb{R}(E_{\alpha\beta} + E_{\beta\alpha}) & (\alpha \le \beta) \\ \{0\} & (\alpha > \beta). \end{cases} \tag{3}$$

We see from [15, Chap. 2, Proposition 9] that any idempotent of V_n is of the form $E_I := \sum_{\alpha \in I} E_{\alpha\alpha}$ with $I \subset \{1, \ldots, n\}$.

We shall apply Proposition 1 to the subalgebra $\mathcal{Z} \subset V_n$. Then we have a partition

$$\{1,\ldots,n\} = \bigsqcup_{k=1}^{r} I_k$$

for which $E_{I_1},\ldots,E_{I_r}$ are primitive idempotents of $\mathcal{Z}$ labeled as in Proposition 1. The subspace

$$\mathcal{Z}_{lk} := \left\{ x \in V \,;\, E_{I_j}\triangle X = (\delta_{lj} + \delta_{kj})X/2, \quad X\triangle E_{I_j} = \delta_{kj}X \text{ for all } j = 1,\ldots,r \right\}$$

is contained in

$$\sum_{(\alpha,\beta)\in I_k \times I_l,\, \alpha<\beta}^{\oplus} \mathbb{R}(E_{\alpha\beta} + E_{\beta\alpha})$$

for $1 \le k < l \le r$ owing to (3). Thus we have

$$\mathcal{Z} \subset \sum_{1\le k\le r}^{\oplus} \mathbb{R}E_{I_k} \oplus \sum_{1\le k<l\le r}^{\oplus} \sum_{(\alpha,\beta)\in I_k \times I_l,\, \alpha<\beta}^{\oplus} \mathbb{R}(E_{\alpha\beta} + E_{\beta\alpha}). \tag{4}$$

Put $\Lambda := \{ (\alpha,\beta)\,;\, \alpha < \beta \text{ and } \alpha \in I_l,\ \beta \in I_k \text{ with } l > k \}$. By (4), the (α,β)-components of $X \in \mathcal{Z}$ must be zero for $(\alpha,\beta) \in \Lambda$. Since Λ is determined from $\mathcal{Z}$, we sometimes write $\Lambda(\mathcal{Z})$ for Λ.

Now we prove Theorem 2 by induction on the cardinality of the set $\Lambda(\mathcal{Z})$. If $\Lambda(\mathcal{Z}) = \emptyset$, then the subalgebra $\mathcal{Z}$ admits a normal block decomposition. If $\Lambda(\mathcal{Z}) \ne \emptyset$, put

$$\beta_0 := \min \{ \beta \,;\, (\alpha,\beta) \in \Lambda \text{ for some } \alpha \}.$$

The minimality of β_0 tells us that $(\beta_0 - 1, \beta_0) \in \Lambda$, so that the $(\beta_0 - 1, \beta_0)$-component of $X \in \mathcal{Z}$ equals 0. Let w_0 be the permutation matrix corresponding to the transposition $(\beta_0 - 1\ \beta_0)$. Then Lemma 1 tells us that $\rho(w_0) : \mathcal{Z} \to \rho(w_0)\mathcal{Z} =: \mathcal{Z}' \subset V_n$ is an algebra isomorphism. Clearly, $\mathcal{Z}'$ is a subalgebra with $\sharp\Lambda(\mathcal{Z}') = \sharp\Lambda(\mathcal{Z}) - 1$, whence Theorem 2 follows.

Let $\mathcal{Z}$ be a subalgebra of $(V_n, \triangle)$ admitting a normal block decomposition. Then $\mathcal{Z}_{lk}$ is the set of matrices X in (1) whose block components except X_{lk} are zero, so that we have a natural linear isomorphism between $\mathcal{Z}_{lk}$ and $V_{lk} \subset \mathrm{Mat}(n_l, n_k; \mathbb{R})$. Keeping this observation in mind, we can easily verify the following statement.

Proposition 2. *The linear space $\mathcal{Z}$ of the matrices X of the form (1) is a subalgebra of $(V_n, \triangle)$ if and only if the following three conditions are satisfied:*
(V1) $A \in V_{lk}$, $B \in V_{ki} \Rightarrow AB \in V_{li}$ for $1 \le i < k < l \le r$,
(V2) $A \in V_{li}$, $B \in V_{ki} \Rightarrow A\,{}^tB \in V_{lk}$ for $1 \le i < k < l \le r$,
(V3) $A \in V_{lk} \Rightarrow A\,{}^tA \in \mathbb{R}E_{n_k}$.

The homogeneous cone corresponding to a given subalgebra of $(V_n, \triangle)$ and a simply transitive group action on the cone are described in a quite simple way as follows.

Theorem 3. *Let $\mathcal{Z}$ be a subalgebra of $(V_n, \triangle)$ containing E_n.*

(i) *Define $\mathfrak{h}_{\mathcal{Z}} := \left\{ X \,;\, X \in \mathcal{Z} \right\} \subset \mathfrak{h}_n$. Then, for any $T, S \in \mathfrak{h}_{\mathcal{Z}}$ one has $TS \in \mathfrak{h}_{\mathcal{Z}}$.*

(ii) *The set $H_{\mathcal{Z}} := \mathfrak{h}_{\mathcal{Z}} \cap H_n$ forms a linear Lie group whose Lie algebra is $\mathfrak{h}_{\mathcal{Z}}$.*

(iii) *Let $\mathcal{P}_{\mathcal{Z}}$ be the open convex cone $\mathcal{Z} \cap \mathcal{P}_n$ in $\mathcal{Z}$. Then $\mathcal{P}_{\mathcal{Z}}$ is a homogeneous cone on which $H_{\mathcal{Z}}$ acts simply transitively by ρ.*

Proof. First we assume that $\mathcal{Z}$ admits a normal block decomposition. Then the assertion (i) follows from (V1) in Proposition 2. We see that $\mathfrak{h}_{\mathcal{Z}}$ is a subalgebra of the matrix algebra $\mathrm{Mat}(n, \mathbb{R})$, whence (ii) follows. In view of (V2) and (V3), we see that $H_{\mathcal{Z}}$ acts on $\mathcal{Z}$ by ρ. Since $H_{\mathcal{Z}} \subset H_n$ and $\mathcal{P}_{\mathcal{Z}} \subset \mathcal{P}_n$, the action of $H_{\mathcal{Z}}$ on the cone $\mathcal{P}_{\mathcal{Z}}$ has no isotropy. This fact together with $\dim H_{\mathcal{Z}} = \dim \mathcal{Z}$ tells us that all $H_{\mathcal{Z}}$-orbits in $\mathcal{P}_{\mathcal{Z}}$ are open. Therefore, we see from the connectedness of the convex cone $\mathcal{P}_{\mathcal{Z}}$ that $\mathcal{P}_{\mathcal{Z}}$ is an $H_{\mathcal{Z}}$-orbit.

For a general subalgebra $\mathcal{Z}$, the assertions are verified using Theorem 2. $\square$

Let us present examples. We set

$$
\mathcal{Z} := \left\{
\begin{pmatrix}
x_{11}E_8 & {}^tC(a) & \begin{matrix} b_0 \\ \vdots \\ b_7 \end{matrix} \\[1mm]
 & & \begin{matrix} c_0 \\ \vdots \\ c_7 \end{matrix} \\[1mm]
C(a) & x_{22}E_8 & \\[1mm]
b_0 \;\cdots\; b_7 & c_0 \;\cdots\; c_7 & x_{33}
\end{pmatrix}
\;;\;
\begin{matrix}
x_{11}, x_{22}, x_{33} \in \mathbb{R} \\[1mm]
a = (a_i)_{i=0}^7,\ (b_i)_{i=0}^7,\ (c_i)_{i=0}^7 \in \mathbb{R}^8
\end{matrix}
\right\}
$$

$$\subset \mathrm{Sym}(17, \mathbb{R}),$$

where $C(a)$ is the matrix

$$
\begin{pmatrix}
a_0 & a_1 & a_2 & a_3 & a_4 & a_5 & a_6 & a_7 \\
-a_1 & a_0 & -a_3 & a_2 & -a_5 & a_4 & a_7 & -a_6 \\
-a_2 & a_3 & a_0 & -a_1 & -a_6 & -a_7 & a_4 & a_5 \\
-a_3 & -a_2 & a_1 & a_0 & -a_7 & a_6 & -a_5 & a_4 \\
-a_4 & a_5 & a_6 & a_7 & a_0 & -a_1 & -a_2 & -a_3 \\
-a_5 & -a_4 & a_7 & -a_6 & a_1 & a_0 & a_3 & -a_2 \\
-a_6 & -a_7 & -a_4 & a_5 & a_2 & -a_3 & a_0 & a_1 \\
-a_7 & a_6 & -a_5 & -a_4 & a_3 & a_2 & -a_1 & a_0
\end{pmatrix}
\in \mathrm{Mat}(8, \mathbb{R})
$$

arising from the multiplication table of the Cayley numbers. In this case we have $r = 3$, $n_1 = n_2 = 8$, $n_3 = 1$, $\mathcal{V}_{31} = \mathcal{V}_{32} = \mathrm{Mat}(1, 8; \mathbb{R})$ and $\mathcal{V}_{21} = \left\{ C(a) \,;\, a \in \mathbb{R}^8 \right\} \subset \mathrm{Mat}(8, \mathbb{R})$. The cone $\mathcal{P}_{\mathcal{Z}}$ is linearly isomorphic to the exceptional symmetric cone $\mathrm{Herm}^+(3, \mathbb{O})$.

Let us investigate another example. For $\theta \in [0, \pi/4]$, define $\mathcal{Z}_\theta \subset \mathrm{Sym}(7, \mathbb{R})$ by

$$
\mathcal{Z}_\theta := \left\{
\begin{pmatrix}
x_{11}E_4 & {}^tM_\theta(a) & \begin{matrix} b_0 \\ \vdots \\ b_3 \end{matrix} \\[1mm]
M_\theta(a) & x_{22}E_2 & \begin{matrix} c_0 \\ c_1 \end{matrix} \\[1mm]
b_0 \;\cdots\; b_3 & c_0 \; c_1 & x_{33}
\end{pmatrix}
\;;\;
\begin{matrix}
x_{11}, x_{22}, x_{33} \in \mathbb{R} \\[1mm]
a = (a_0, a_1),\ (c_0, c_1) \in \mathbb{R}^2,\ (b_i)_{i=0}^3 \in \mathbb{R}^4
\end{matrix}
\right\},
$$

where

$$M_\theta(a) := \begin{pmatrix} a_0\cos\theta & a_1\cos\theta & a_0\sin\theta & a_1\sin\theta \\ -a_1\cos\theta & a_0\cos\theta & a_1\sin\theta & -a_0\sin\theta \end{pmatrix} \in \mathrm{Mat}(2,4).$$

The corresponding homogeneous cones $\mathcal{P}_{\mathcal{Z}_\theta} := \mathcal{P}_7 \cap \mathcal{Z}_\theta$ are mutually non-isomorphic for different θ [9].

4 Homogeneous Hessian Metrics on a Homogeneous Cone

Let us go back to the setting in Sect. 2. Let $\Omega \subset V$ be a homogeneous cone on which a triangular group $H \subset GL(\Omega)$ acts simply transitively. Let $(V, \triangle)$ be the corresponding clan with a unit element e. It is shown in [8, Proposition 2.1] that the linear form $\xi \in V^*$ in (C2) is of the form

$$\xi(x) = \sum_{k=1}^{r} s_k x_{kk} \quad \left(x = \sum_{k=1}^{r} x_{xx} e_k + \sum_{1 \le k < l \le r} X_{lk}, \; x_{kk} \in \mathbb{R}, \; X_{lk} \in V_{lk} \right).$$

with some positive constants $s_1, \ldots, s_r$. We denote by $\xi_{\underline{s}}$ this linear form, where $\underline{s} = (s_1, \ldots, s_r) \in \mathbb{R}^r_{>0}$, and by $(\cdot|\cdot)_{\underline{s}}$ the corresponding inner product on V, that is, $(x|y)_{\underline{s}} := \xi_{\underline{s}}(x \triangle y)$ for $x, y \in V$. By (C1), we have for $x, y, z \in V$,

$$(x\triangle y|z)_{\underline{s}} - (x|y\triangle z)_{\underline{s}} = (y\triangle x|z)_{\underline{s}} - (y|x\triangle z)_{\underline{s}}.$$

Then [12] tells us that there exists an H-invariant Hessian metric $g_{\underline{s}}$ on the homogeneous cone Ω such that the restriction of $g_{\underline{s}}$ to the tangent space $T_e\Omega \equiv V$ equals $(\cdot|\cdot)_{\underline{s}}$. In general, a Hessian metric g on Ω is said to be *homogeneous* if the Hessian automorphism group $\mathrm{Aut}(\Omega, g) := \{ f \in GL(\Omega) ; f^*g = g \}$ acts on Ω transitively. Clearly, every $g_{\underline{s}}$ is homogeneous. Moreover, for any homogeneous Hessian metric g, there exists $f \in GL(\Omega)$ and $\underline{s} \in \mathbb{R}^r_{>0}$ such that $g = f^*g_{\underline{s}}$ because a maximal connected triangular subgroup of $\mathrm{Aut}(\Omega, g)$ is conjugate to the group H as a subgroup of $GL(\Omega)$ owing to [15, Chap. 1]. In this sense, the family $\{g_{\underline{s}}\}_{\underline{s}\in\mathbb{R}^r_{>0}}$ essentially exhausts all homogeneous Hessian metrics on the homogeneous cone Ω.

Now we turn to the matrix realization of a homogeneous cone. Let $\mathcal{Z}$ be a subalgebra of $(V_n, \triangle)$ admitting a normal block decomposition. We set

$$N_1 := 1, \quad N_k := n_1 + \cdots + n_{k-1} + 1 \quad (k = 2, \ldots, r).$$

For $\underline{s} \in \mathbb{R}^r_{>0}$, define $\sigma_1, \ldots, \sigma_r \in \mathbb{R}$ by

$$\sigma_r := s_r, \quad \sigma_k := s_k - n_k \sum_{l=k+1}^{r} \sigma_l \quad (k = 1, \ldots, r-1).$$

Then, for $X \in \mathcal{Z}$ of the form (1), we have

$$\xi_{\underline{s}}(X) = \sum_{k=1}^{r} s_k x_{kk} = \sum_{k=1}^{r} \sigma_k \mathrm{tr} X^{[N_k]}, \tag{5}$$

where $X^{[N]}$ denotes the left-top corner submatrix of X of size N.

Theorem 4. *The function $\phi_{\underline{s}}$ on $\mathcal{P}_{\mathcal{Z}}$ given by*

$$\phi_{\underline{s}}(X) := -\log\left(\prod_{k=1}^{r}(\det X^{[N_k]})^{\sigma_k}\right) \qquad (X \in \mathcal{P}_{\mathcal{Z}})$$

is a global potential of the Hessian metric $g_{\underline{s}}$. Moreover, for $X \in \mathcal{P}_{\mathcal{Z}}$ and $A, B \in T_X \mathcal{P}_{\mathcal{Z}} \equiv \mathcal{Z}$, one has

$$g_{\underline{s}}(A,B)_X = \sum_{k=1}^{r} \sigma_k \mathrm{tr}\left(A^{[N_k]}(X^{[N_k]})^{-1}B^{[N_k]}(X^{[N_k]})^{-1}\right). \qquad (6)$$

Proof. Let us denote by $h_{\underline{s}}(A,B)_X$ the right-hand side of (6). Since $\phi_{\underline{s}}(X) = -\sum_{k=1}^{r}\sigma_k \log(\det X^{[N_k]})$, we see immediately from the well-known formula for the Hessian of the log-determinant function that $h_{\underline{s}}(A,B)_X$ equals the Hessian of $\phi_{\underline{s}}$. On the other hand, for any $T \in H_{\mathcal{Z}}$, we have $\phi_{\underline{s}}(\rho(T)X) = \phi_{\underline{s}}(X) + C_T$ with $C_T = -\log\left(\prod_{k=1}^{r}(\det T^{[N_k]})^{2\sigma_k}\right)$, so that the Hessian $h_{\underline{s}}$ of $\phi_{\underline{s}}$ is invariant under $\rho(T)$. The invariance also follows from the equality $h_{\underline{s}}(\rho(T)A, \rho(T)B)_{\rho(T)X} = h_{\underline{s}}(A,B)_X$ which can be checked by a direct calculation. Let us observe that

$$h_{\underline{s}}(A,B)_{I_n} = \sum_{k=1}^{r} \sigma_k \mathrm{tr}\, A^{[N_k]}B^{[N_k]} = \sum_{k=1}^{r}\sigma_k \mathrm{tr}\,(A^{[N_k]}\triangle B^{[N_k]})$$

$$= \sum_{k=1}^{r}\sigma_k \mathrm{tr}\,(A\triangle B)^{[N_k]},$$

which together with (5) tells us that

$$h_{\underline{s}}(A,B)_{I_n} = \xi_{\underline{s}}(A\triangle B) = (A|B)_{\underline{s}} = g_{\underline{s}}(A|B)_{I_n}.$$

Therefore, the $\rho(H_{\mathcal{Z}})$-invariance of both $g_{\underline{s}}$ and $h_{\underline{s}}$ completes the proof.

The author is grateful to the anonymous referees for their comments which are helpful for the improvement of the paper.

References

1. Andersson, S.A., Wojnar, G.G.: Wishart distributions on homogeneous cones. J. Theor. Probab. **17**, 781–818 (2004)
2. Boyom, M.N.: The cohomology of Koszul-Vinberg algebra nd related topics. Afr. Diaspora J. Math. (New Ser.) **9**, 53–65 (2010)
3. Chua, C.B.: Relating homogeneous cones and positive definite cones via T-algebras. SIAM J. Optim. **14**, 500–506 (2003)
4. Graczyk, P., Ishi, H.: Riesz measures and Wishart laws associated to quadratic maps. J. Math. Soc. Jpn. **66**, 317–348 (2014)
5. Güler, O., Tunçel, L.: Characterization of the barrier parameter of homogeneous convex cones. Math. Program. A **81**, 55–76 (1998)

6. Ishi, H.: Representation of clans and homogeneous cones. Vestnik Tambov Univ. **16**, 1669–1675 (2011)

7. Ishi, H.: Homogeneous cones and their applications to statistics. In: Modern Methods of Multivariate Statistics, Hermann, Paris, pp. 135–154 (2014)

8. Kai, C., Nomura, T.: A characterization of symmetric cones through pseudoinverse maps. J. Math. Soc. Jpn. **57**, 195–215 (2005)

9. Kaneyuki, S., Tsuji, T.: Classification of homogeneous bounded domains of lower dimension. Nagoya Math. J. **53**, 1–46 (1974)

10. Manchon, D.: A short survey on pre-Lie algebras. In: Noncommutative Geometry and Physics: Renormalisation, Motives, Index Theory, pp. 89–102 (2011). ESI Lect. Math. Phys., Eur. Math. Soc., Zurich

11. Rothaus, O. S.: The construction of homogeneous convex cones. Ann. of Math. **83**, 358–376 (1966). Correction: ibid 87, 399 (1968)

12. Shima, H.: Homogeneous Hessian manifolds. Ann. Inst. Fourier Grenoble **30**, 91–128 (1980)

13. Shima, H.: The geometry of Hessian structures. World Scientific Publishing, Hackensack (2007)

14. Truong, V.A., Tunçel, L.: Geometry of homogeneous convex cones, duality mapping, and optimal self-concordant barriers. Math. Program. **100**, 295–316 (2004)

15. Vinberg, E.B.: The theory of convex homogeneous cones. Trans. Moscow Math. Soc. **12**, 340–403 (1963)

16. Xu, Y.-C.: Theory of Complex Homogeneous Bounded Domains. Kluwer, Dordrecht (2005)

17. Yamasaki, T., Nomura, T.: Realization of homogeneous cones through oriented graphs. Kyushu J. Math. **69**, 11–48 (2015)

18. Zhong, J.-Q., Lu, Q.-K.: The realization of affine homogeneous cones. Acta Math. Sinica **24**, 116–142 (1981)

Multiply CR-Warped Product Statistical Submanifolds of a Holomorphic Statistical Space Form

Michel Nguiffo Boyom[1], Mohammed Jamali[2]([✉]),
and Mohammad Hasan Shahid[3]

[1] Department De Mathematiques, Universite Montpellier II, Montpellier, France
nguiffo.boyom@gmail.com
[2] Department of Mathematics, Al-Falah University, Dhauj, faridabad, India
jamali_dbd@yahoo.co.in
[3] Department of Mathematics, Jamia Millia Islamia, New Delhi, India
hasan_jmi@yahoo.com

Abstract. In this article, we derive an inequality satisfied by the squared norm of the imbedding curvature tensor of Multiply CR-warped product statistical submanifolds N of holomorphic statistical space forms M. Furthermore, we prove that under certain geometric conditions, N and M become Einstein.

1 Introduction

The concern of this short note is to introduce warped products of statistical manifolds and to sketch the study of their geometry. This is actually a wide project because in a warped product, two statistical models (M_1, P_1) and (M_2, P_2), for the same measurable set (Σ, Ω), araise many relevant questions which deserve the attention.

For instance, let us assume that the Fisher informations of both (M_1, P_1) and (M_2, P_2) are Riemannian metrics.

Definition 1. [7] A warped product $M_1 \times_\sigma M_2$ is the product $M_1 \times M_2$ of two Riemannian manifolds (M_1, g_1) and (M_2, g_2) endowed with the Riemannian metric

$$g_\sigma(x, y) = g_1(x) + \sigma(x)g_2(y)$$

where $\sigma \in C^\infty(M_1)$ is a positive function.

By the same way, the σ-warped product of two statistical models (M_1, P_1) and (M_2, P_2) is the manifold $M_1 \times M_2$ with the function

$$P_\sigma : M_1 \times M_2 \times \Sigma \longrightarrow R$$

defined by

$$P_\sigma(x, y, \xi) = e^{-\sigma(x)} P_1(x, \xi) + (1 - e^{-\sigma(x)}) P_2(y, \xi).$$

© Springer International Publishing Switzerland 2015
F. Nielsen and F. Barbaresco (Eds.): GSI 2015, LNCS 9389, pp. 257–268, 2015.
DOI: 10.1007/978-3-319-25040-3_29

This P_σ has the following properties:

(i) $P_\sigma(x, y, \xi)$ is smooth with respect to (x, y);
(ii) $P_\sigma(x, y, \xi) \leq 1$;
(iii) $\int_\Sigma P_\sigma(x, y, \xi) d\xi = 1$.

Thus $(M_1 \times M_2, P_\sigma)$ is a statistical model for the measurable set (Σ, Ω).

Therefore araise many questions regarding the relationships between the Information Geometry of (M_1, P_1), (M_2, P_2) and $(M_1 \times M_2, P_\sigma)$. For instance, does the Fisher information of P_σ is always a Riemannian metric?

We may replace the smoothness by another analysis requirement such as analyticity, holomorphy.

The present note focusses on the last setting i.e. holomorphy. We wish to discuss some interesting items such as the geometry of submanifolds in holomorphic statsitical manifolds. This project is a relevant one in the Information Geometry and its applications [2].

It is known that the notion of CR-warped product manifold was first introduced by B. Y. Chen([5,6]). In these papers, he obtained the certain sharp inequalities involving warping functions and the squared norm of second fundamental form. On the other hand, in [1,8], Amari and Furuhata studied the statisitcal submanifolds and hypersurfaces of statisitcal manifolds in the context of Information Geometry. Further, L. Todgihounde [9] established dualistic structures on warped products which has been the motivation behind the study of this article. In the present article, we first establish the dualistic structure on multiply warped product statistical manifolds. Later, we study multiply CR-warped product statistical submanifolds of holomorphic statistical space forms. We refer [8], for more details of holomorphic space forms.

2   Preliminary

Definition 2. *[1] A statistical manifold is a Riemannian manifold (M, g) endowed with a pair of torsion-free affine connections $\tilde{\nabla}$ and $\tilde{\nabla}^*$ satisfying*

$$Zg(X, Y) = g(\tilde{\nabla}_Z X, Y) + g(X, \tilde{\nabla}_Z^* Y) \tag{1}$$

for $X, Y \in \Gamma(TM)$. It is denoted by $(M, g, \tilde{\nabla}, \tilde{\nabla}^*)$. The connections $\tilde{\nabla}$ and $\tilde{\nabla}^*$ are called dual connections and it is easily shown that $\left(\tilde{\nabla}^*\right)^* = \tilde{\nabla}$. If $(\tilde{\nabla}, g)$ is a statistical structure on $\tilde{M}$, then $(\tilde{\nabla}^*, g)$ is also a statistical structure.

Here it should be noted that we have used the symbol $\tilde{\nabla}$ and $\tilde{\nabla}^*$ to denote the dual connections as we have reserved the notations ∇ and ∇^* for later use in Gauss formula.

Remark 1. *For any triplet $(M, g, \tilde{\nabla}) \; \exists$ a unique $\tilde{\nabla}^*$ called the dual of $\tilde{\nabla}$ in (M, g). Therefore, in other words, we can say that the triplet $(M, g, \tilde{\nabla})$ is statistical if the dual $\tilde{\nabla}^*$ of $\tilde{\nabla}$ in (M, g) is torsion-free.*

Let $N_1, N_2, \ldots, N_k$ be Riemannian manifolds of dimensions $n_1, n_2, n_3, \ldots, n_k$ respectively and let $N = N_1 \times N_2 \times \ldots \times N_k$ be the cartesian product of $N_1, N_2, \ldots, N_k$. For each a, denote by $\pi_a : N \longrightarrow N_a$ the canonical projection of N onto N_a. We denote the horizontal lift of N_a in N via π_a by N_a itself. If $\sigma_2, \ldots, \sigma_k : N_1 \longrightarrow R^+$ are positive valued functions, then

$$g(X, Y) = \langle \pi_{1*}X, \pi_{1*}Y \rangle + \sum_{a=1}^{k} (\sigma_a \circ \pi_1)^2 \langle \pi_{a*}X, \pi_{a*}Y \rangle \qquad (2)$$

defines a metric g on N. The product manifold N endowed with this metric is denoted by $N_1 \times_{\sigma_2} N_2 \times \ldots \times_{\sigma_k} N_k$. This product manifold N is known as multiply warped product manifold.

Let D_a denote the distributions obtained from the vectors tangent to N_a. We have

$$T(N) = D_1 \oplus D_2 \oplus \ldots \oplus D_k.$$

Definition 3. *Let $\{e_1, e_2, \ldots, e_n\}$ be any local orthonormal frame of vector fields and f be any smooth function on a manifold N. The gradient of the function f is defined as*

$$grad f = \sum_{i=1}^{n} e_i(f) e_i.$$

Definition 4. *[7] For a differentiable function f on N, the Laplacian Δf of f is defined as*

$$\Delta f = \sum_{i=1}^{n} \{e_i(e_i f) - \nabla_{e_i} e_i f\}.$$

Finally, we state the Green's theorem as follows :

Theorem 1. *In a compact and orientable Riemannian manifold N without boundary, we have*

$$\int_N \Delta f = 0$$

for any function f on N.

3   Multiply CR-Warped Product Statistical Submanifolds of Holomorphic Statistical Space Form

Let N_i and N_j be two statistical manifolds. From [9], the warped product $N_i \times_{\sigma_i} N_j$ is again a statistical manifold, where σ_i is a smooth positive real valued function on N_i. Generalizing this concept we may obtain multiply warped product statistical manifolds as follows :

Definition 5. *If $N_1, N_2, \ldots, N_k$ be k statistical manifolds, then $N = N_1 \times_{\sigma_2} N_2 \times \ldots \times_{\sigma_k} N_k$ is again a statistical manifold with the metric given by (2) . This manifold N is called multiply warped product statistical manifold.*

Now let us denote the part $\sigma_2 N_2 \times ... \times_{\sigma_k} N_k$ by $N_\perp$ and N_1 by N_T. Then N can be represented as $N = N_T \times N_\perp$. We donote by $\bar{X}, \bar{Y}.... \in \Gamma(M)$ as the vector fields on M and $X, Y, ...$ the induced vector fields on N.

Definition 6. *[8] A 2m dimensional statistical manifold M is said to be a holomorphic statistical manifold if it admits an endomorphism over the tangent bundle $\Gamma(M)$ and a metric g and a fundamental form ω given by $\omega(\bar{X}, \bar{Y}) = g(\bar{X}, J\bar{Y})$ such that the following equations are satisfied :*

$$J^2 = -Id; \quad \tilde{\nabla}\omega = 0$$

for any vector fields $\bar{X}, \bar{Y} \in \Gamma(M)$.

Since ω is skew-symmetric, we have $g(\bar{X}, J\bar{Y}) = -g(J\bar{X}, \bar{Y})$.

Definition 7. *A statistical submanifold N of a holomorphic statistical manifold M is called invariant if the almost complex structure J of M carries each tangent space of N into itself whereas it is said to be anti-invariant if the almost complex structure J of M carries each tangent space of N into its corresponding normal space.*

Definition 8. *A multiply warped product statistical submanifold $N = N_T \times N_\perp$ in an almost complex manifold M is called a multiply CR-warped product statistical submanifold if N_T is an invariant submanifold and $N_\perp$ is an anti-invarinat submanifold of M.*

Definition 9. *[8] A holomorphic statistical manifold M is said to be of constant holomorphic curvature $c \in R$ if the following curvature equation holds :*

$$\tilde{R}(\bar{X}, \bar{Y})\bar{Z} = \frac{c}{4}\left\{ g(\bar{Y}, \bar{Z})\bar{X} - g(\bar{X}, \bar{Z})\bar{Y} + g(J\bar{Y}, \bar{Z})J\bar{X} - g(J\bar{X}, \bar{Z})J\bar{Y} + 2g(\bar{X}, J\bar{Y})J\bar{Z} \right\}$$

for any $\bar{X}, \bar{Y}, \bar{Z} \in \Gamma(M)$.

Let N and M be two statistical manifolds and N be a submanifold of M. Then for any $X, Y \in \Gamma(N)$, the corresponding Gauss formulas are

$$\tilde{\nabla}_X Y = \nabla_X Y + h(X, Y),$$

$$\tilde{\nabla}^*_X Y = \nabla^*_X Y + h^*(X, Y)$$

where $\tilde{\nabla}$ and $\tilde{\nabla}^*$ (respectively ∇ and ∇^*) are the dual connections on M (respectively on N), h and h^* are symmetric and bilinear, called the imbedding curvature tensor of N in M for $\tilde{\nabla}$ and the imbedding curvature tensor of N in M for $\tilde{\nabla}^*$ respectively.

Let $x : N_T \times N_\perp \longrightarrow M$ be an isometric statistical immersion of a multiply warped product statistical submanifold into statistical manifold M. Let us denote the normal bundle on N by $\Gamma(TN^\perp)$. Since h and h^* are bilinear, we have the linear transformations A_ξ and A_{ξ^*} defined by

$$g(A_\xi X, Y) = g(h(X, Y), \xi) \text{ and } g(A^*_\xi X, Y) = g(h^*(X, Y), \xi)$$

for any $\xi \in \Gamma(TN^\perp)$ and $X, Y \in \Gamma(TN)$. The corresponding Weingarten equations are as follows [10]

$$\tilde{\nabla}_X \xi = -A_\xi^* X + \nabla_X^\perp Y,$$

$$\tilde{\nabla}_X^* \xi = -A_\xi X + \nabla_X^{\perp *} \xi$$

for any $\xi \in \Gamma(TN^\perp)$ and $X \in \Gamma(TN)$. The connections $\nabla_X^\perp$ and $\nabla_X^{* \perp}$ are Riemannian dual connections with respect to the induced metric on $\Gamma(TN^\perp)$. The corresponding Gauss, Codazzi and Ricci equations are given by the following [3] :

(i) In terms of $\tilde{\nabla}$ and ∇ :

$$g(\tilde{R}(X,Y)Z,W) = g(R(X,Y)Z,W) + g(h(X,Z), h^*(Y,W)) - g(h^*(X,W), h(Y,Z)),$$

$$(\tilde{R}(X,Y)Z)^\perp = \{\nabla_X^\perp h(Y,Z) - h(\nabla_X Y, Z) - h(Y, \nabla_X Z)\}$$
$$-\{\nabla_Y^\perp h(X,Z) - h(\nabla_Y X, Z) - h(X, \nabla_Y Z)\},$$

$$g(R^\perp(X,Y)\xi, \eta) = g(\tilde{R}(X,Y)\xi, \eta) + g([A_\xi^*, A_\eta]X, Y)$$

where $R^\perp$ is the Riemannian curvature tensor for $\nabla^\perp$ on $TN^\perp$, $\xi, \eta \in \Gamma(TN^\perp)$ and $[A_\xi^*, A_\eta] = A_\xi^* A_\eta - A_\eta A_\xi^*$.

Similar curvature equations can be obtained for $\tilde{\nabla}^*$ and ∇^*.

Let $N = N_1 \times_{\sigma_2} N_2 \times ... \times_{\sigma_k} N_k$ be a multiply CR-warped product statistical submanifold of a holomorphic statistical space form M. Let D be the invariant distribution (i.e. $TN_T = D$) such that its orthogonal complementary distribution $D^\perp$ is anti-invariant (i.e. $TN_\perp = T(N = N_2 \times ... \times N_k) = D^\perp$).

Then we have the following decompositions :

$$TN = D \oplus D^\perp \text{ and } T^\perp N = JD^\perp \oplus \lambda \tag{3}$$

where λ denotes the orthogonal complementary distribution of $JD^\perp$ in $T^\perp N$ and is an invariant normal subbundle of $T^\perp N$. For any vector field $X, Y \in TN$, we put

$$(\tilde{\nabla}_X J)Y = P_X Y + Q_X Y$$

where $P_X Y$ (resp. $Q_X Y$) denotes the tangential (resp. normal) component of $(\tilde{\nabla}_X J)Y$ respectively.

Also, let $p \in N$ and $\{e_1, .., e_n, e_{n+1}, ..., e_{2m}\}$ be an orthonormal basis of the tangent space $T_p M$, such that $e_1, ... e_n$ constitute a basis of the tangent space $T_p N$.

We denote by H and H^*, the mean curvature vectors given by

$$H = \frac{1}{n} \sum_{i=1}^{n} h(e_i, e_i),$$

$$H^* = \frac{1}{n} \sum_{i=1}^{n} h^*(e_i, e_i).$$

We call multiply CR-warped product statistical submanifold N of holomorphic statistical space form M minimal, if H and H^* vanishes identically. We also set

$$\|h\|^2 = \sum_{i,j=1}^{n} g(h(e_i, e_j), h(e_i, e_j)).$$

4 Some Inequalities for Multiply CR-Warped Product Statistical Submanifolds

In this section, we establish an inequality satisfied by the squared norm of imbedding curvature tensor. We further prove the Einsteinian property of N and M under slight geometric conditions.

From the decomposition (3), we may write

$$h(X,Y) = h_{JD^\perp}(X,Y) + h_\lambda(X,Y).$$

Also we have for multiply CR-warped product statistical submanifold N of a statistical manifold [9]

$$\nabla_X Z = \sum_{a=2}^{k}(X(\log \sigma_a))Z^a \text{ and } \nabla_X^* Z = \sum_{a=2}^{k}(X(\log \sigma_a))Z^a \tag{4}$$

for any vector fields $X \in D$ and $Z \in D^\perp$, where Z^a denotes the N_a-component of Z. We shall do calculations for ∇ (similar is the case for ∇^*).

For proving the main inequalites we need the following lemma.

Lemma 1. *Let* $N = N_T \times_{\sigma_2} N_2 \times ... \times_{\sigma_k} N_k$ *be a multiply CR-warped product statistical submanifold of a holomorphic statistical space form M. Then we have*

$$\text{(i)} \ h_{JD^\perp}(JX, Z) = \sum_{a=2}^{k}(X(\log \sigma_a))JZ^a + JP_Z JX$$

$$\text{(ii)} \ g(P_Z JX, W) = g(Q_Z X, JW)$$

$$\text{(iii)} \ g(h(JX, Z), Jh(X, Z)) = \|h_\lambda(Z, X)\|^2 + g(Q_Z X, Jh_\lambda(X, Z))$$

for any vector fields X in D and Z, W in $D^\perp$, where Z^a denotes the N_a-component of Z.

Proof. From Gauss formula we can write

$$\nabla_Z JX + h(JX, Z) = P_Z X + Q_Z X + J\nabla_Z X + Jh(Z, X)$$

or

$$h(JX, Z) = P_Z X + Q_Z X + J(\sum_{a=2}^{k}(X(\log \sigma_a))Z^a) + Jh(Z, X) - \sum_{a=2}^{k}(JX(\log \sigma_a))Z^a.$$

$$\tag{5}$$

Comparing tangential parts in the above equation and then taking inner product with $W \in D^{\perp}$, we get

$$h_{JD^{\perp}}(JX, Z) = \sum_{a=2}^{k}(X(\log \sigma_a))JZ^a + JP_Z JX, \quad \forall \, X \in D, Z \in D^{\perp}.$$

This proves (i) of the lemma.

Now comparing normal parts of (5), we get

$$h(JX, Z) - Jh_{\lambda}(Z, X) = Q_Z X + \sum_{a=2}^{k}(X(\log \sigma_a)JZ^a). \tag{6}$$

By taking inner product of the above equation with JW , we obtain

$$g(h_{JD^{\perp}}(JX, Z), JW) = g(Q_Z X, JW) + \sum_{a=2}^{k}(X(\log \sigma_a)g(JZ^a, JW)).$$

Simplifying the above equation by using part (i) of the lemma we arrive at

$$g(P_Z JX, W) = g(Q_Z X, JW)$$

which proves part (ii).

Taking inner product of (6) by $Jh(X, Z)$, we find

$$g(h(JX, Z), Jh(X, Z)) = \|h_{\lambda}(Z, X)\|^2 + g(Q_Z X, Jh_{\lambda}(X, Z))$$

which is (iii) part of the lemma   ∎

Theorem 2. *Let $N = N_T \times_{\sigma_2} N_2 \times \dots \times_{\sigma_k} N_k$ be multiply CR-warped product statistical submanifold of holomorphic statistical space form M with $P_{D^{\perp}} D \in D$, then the squared norm of imbedding curvature tensor of N in M satisfies the following inequality :*

$$\|h\|^2 \geq \sum_{a=2}^{k} n_a^2 \|\nabla \log \sigma_a\|^2 + \|P_{D^{\perp}} D\|^2.$$

Proof. Let $\{X_1, X_2, .., X_p, X_{p+1} = JX_1, ..., X_{2p} = JX_p\}$ be a local orthonormal frame of vector fields on N_T and $\{Z_1, Z_2, ..., Z_q\}$ be such that Z_{Δ_a} is a basis for some N_a , $a = 2, ..., k$ where $\Delta_2 = \{1, 2, .., n_2\}, ..., \Delta_k = \{n_2 + n_3 + \dots + n_{k-1} + 1, ..., n_1 + n_2 + \dots + n_k\}$ and $n_2 + n_3 + \dots + n_k = q$. Then we have

$$\|h\|^2 = \sum_{i,j=1}^{2p} g(h(X_i, X_j), h(X_i, X_j)) + \sum_{i=1}^{2p}\sum_{a=2}^{k} g(h(X_i, Z_{\Delta_a}), h(X_i, Z_{\Delta_a}))$$

$$+ \sum_{a,b=2}^{k} g(h(Z_{\Delta_a}, Z_{\Delta_b}), h(Z_{\Delta_a}, Z_{\Delta_b})).$$

The above equation implies

$$\|h\|^2 \geq \sum_{i=1}^{2p} \sum_{a=2}^{k} g(h(X_i, Z_{\Delta_a}), h(X_i, Z_{\Delta_a})).$$

Now using part (i) of last lemma we get

$$\|h\|^2 \geq \sum_{i=1}^{2p} \sum_{a=2}^{k} g(n_a(JX_i(\log \sigma_a))JZ_{\Delta_a} + JP_{Z_{\Delta_a}}X_i, n_a(JX_i(\log \sigma_a))JZ_{\Delta_a} + JP_{Z_{\Delta_a}}X_i).$$

In view of the assumption $P_{D^\perp}D \in D$, the above inequality takes the form

$$\|h\|^2 \geq \sum_{a=2}^{k} n_a^2 \|\nabla \log \sigma_a\|^2 \|Z_{\Delta_a}\|^2 + \|P_{D^\perp}D\|^2.$$

By Cauchy-Schwartz inequality, the above equation becomes

$$\sum_{a=2}^{k} n_a^2 \|\nabla \log \sigma_a\|^2 \|Z_{\Delta_a}\|^2 + \|P_{D^\perp}D\|^2 \geq \sum_{a=2}^{k} n_a^2 \|(\nabla \log \sigma_a)Z_{\Delta_a}\|^2 + \|P_{D^\perp}D\|^2.$$

Therefore

$$\|h\|^2 \geq \sum_{a=2}^{k} n_a^2 \|(\nabla \log \sigma_a)\|^2 + \|P_{D^\perp}D\|^2.$$

Hence the lemma.   ∎

Theorem 3. *Let $N = N_T \times_{\sigma_2} N_2 \times \ldots \times_{\sigma_k} N_k$ be a compact orientable multiply CR-warped product statistical submanifold without boundary of holomorphic statistical space form M of constant curvature k. If $P_{D^\perp}D \in D$ and $A_{\nabla_X^{\perp *}JZ}JX = A_{\nabla_{JX}^{\perp *}JZ}X$, then*

$$k \leq 0$$

and the equality holds if and only if $grad_D(\log \sigma_a) = 0$.

Proof. Let $X \in D$, $Z \in D^\perp$, then for holomorphic statistical space form of constant curvature k, we have

$$\tilde{R}(X, JX, Z, JZ) = \frac{k}{4} \left\{ \begin{array}{l} g(JX, Z)g(X, JZ) - g(X, Z)g(JX, JZ) - g(X, Z)g(JX, JZ) \\ +g(JX, Z)g(X, JZ) - 2g(X, X)g(JZ, JZ) \end{array} \right\}$$

which implies that

$$\tilde{R}(X, JX, Z, JZ) = -\frac{k}{2}g(X, X)g(Z, Z). \tag{7}$$

On the other hand from Codazzi equation, we may write

$$\tilde{R}(X, JX, Z, JZ) = g\left(\nabla_X^\perp h(JX, Z), JZ\right) - g\left(h(\nabla_X JX, Z), JZ\right) - g\left(h(JX, \nabla_X Z), JZ\right)$$
$$- g\left(\nabla_{JX}^\perp h(X, Z), JZ\right) + g\left(h(\nabla_{JX} X, Z), JZ\right) + g\left(h(X, \nabla_{JX} Z), JZ\right). \quad (8)$$

We shall calculate each term of (8) above in order to obtain the required inequality.

Now since M is statistical, we get

$$Xg(h(JX, Z), JZ) = g(\tilde{\nabla}_X h(JX, Z), JZ) + g(h(JX, Z), \tilde{\nabla}_X^* JZ).$$

Using Weingarten formula, the above equation gives

$$g(\nabla_X^\perp h(JX, Z), JZ) = Xg(h(JX, Z), JZ) - g(h(JX, Z), \tilde{\nabla}_X^* JZ). \quad (9)$$

We now calculate the first term of the above (9). From (6), we have

$$g(h(JX, Z), JZ) - g(h_\lambda(Z, X), JZ) = g(Q_Z X, JZ) + \sum_{a=2}^{k}(X \log \sigma_a)g(JZ^a, JZ)$$

i.e.

$$g(h(JX, Z), JZ) = g(P_Z JX, Z) + \sum_{a=2}^{k}(X \log \sigma_a)g(Z^a, Z^a).$$

or

$$Xg(h(JX, Z), JZ) = \sum_{a=2}^{k} \left[\left\{X(X \log \sigma_a) + 2(X \log \sigma_a)^2\right\} g(Z^a, Z^a)\right].$$

Therefore (9) is obtained as follows:

$$g(\nabla_X^\perp h(JX, Z), JZ) = \sum_{a=2}^{k} \left[\left\{X(X \log \sigma_a) + 2(X \log \sigma_a)^2\right\} g(Z^a, Z^a)\right] - g(h(JX, Z), \tilde{\nabla}_X^* JZ). \quad (10)$$

Similarly by replacing X by JX in the last equation, we get

$$-g(\nabla_{JX}^\perp h(X, Z), JZ) = \sum_{a=2}^{k} \left[\left\{JX(JX \log \sigma_a) + 2(JX \log \sigma_a)^2\right\} g(Z^a, Z^a)\right] + g(h(X, Z), \tilde{\nabla}_{JX}^* JZ). \quad (11)$$

Again using result (i) of the lemma and $P_{D^\perp} D \in D$, we conclude

$$g(h(JX, \nabla_X Z), JZ) = \sum_{a=2}^{k}(X \log \sigma_a)^2 g(Z^a, Z^a). \quad (12)$$

Similarly by replacing X by JX in the above equation, we find

$$g(h(X, \nabla_{JX} Z), JZ) = - \sum_{a=2}^{k} (JX \log \sigma_a)^2 g(Z^a, Z^a). \tag{13}$$

Also from result (i) of the lemma, we have

$$h_{JD^\perp}(\nabla_{JX} X, Z) = JP_Z \nabla_{JX} X - \sum_{a=2}^{k} (J\nabla_{JX} X \log \sigma_a) JZ^a.$$

Therefore we derive finally from the above equation

$$g(h(\nabla_{JX} X, Z), JZ) = g(JP_Z \nabla_{JX} X, JZ) - \sum_{a=2}^{k} (J\nabla_{JX} X \log \sigma_a) g(JZ^a, JZ).$$

But N_T is totally geodesic in N [4] which implies that $\nabla_{JX} X \in D$. Hence we have $P_Z \nabla_{JX} X \in D$. This makes the first term of the above equation zero and hence we get

$$g(h(\nabla_{JX} X, Z), JZ) = - \sum_{a=2}^{k} (J\nabla_{JX} X \log \sigma_a) g(Z^a, Z^a). \tag{14}$$

Similarly we obtain

$$g(h(\nabla_X JX, Z), JZ) = - \sum_{a=2}^{k} (J\nabla_X JX \log \sigma_a) g(Z^a, Z^a).$$

The last equation may further be simplified to

$$g(h(\nabla_X JX, Z), JZ) = \sum_{a=2}^{k} \{(\nabla_X X \log \sigma_a) g(Z^a, Z^a)\} + \sum_{a=2}^{k} \{(\nabla_{JX} JX \log \sigma_a) g(Z^a, Z^a)\}$$
$$- \sum_{a=2}^{k} \{(J\nabla_{JX} X \log \sigma_a) g(Z^a, Z^a)\}. \tag{15}$$

Put (10)-(15) into (8), we derive

$$\tilde{R}(X, JX, Z, JZ) = \sum_{a=2}^{k} \left[\left\{ X(X \log \sigma_a) + (X \log \sigma_a)^2 \right\} g(Z^a, Z^a) \right] - g(h(JX, Z), \tilde{\nabla}_X^* JZ) \tag{16}$$
$$+ \sum_{a=2}^{k} \left[\left\{ JX(JX \log \sigma_a) + (JX \log \sigma_a)^2 \right\} g(Z^a, Z^a) \right] + g(h(X, Z), \tilde{\nabla}_{JX}^* JZ)$$
$$- \sum_{a=2}^{k} \{(\nabla_X X \log \sigma_a) g(Z^a, Z^a)\} - \sum_{a=2}^{k} \{(\nabla_{JX} JX \log \sigma_a) g(Z^a, Z^a)\}.$$

Now from Gauss equation, we have

$$\tilde{R}(X, JX, Z, JZ) = -\frac{k}{2} \|X\|^2 \sum_{a=2}^{k} \|Z^a\|^2. \tag{17}$$

Combining (16) and (17) and taking summation over the range from 1 to p, we have

$$\left(\frac{pk}{4}\right) \sum_{a=2}^{k} \|Z^a\|^2 = \sum_{a=2}^{k} \Delta(\log \sigma_a) \|Z^a\|^2 - \sum_{a=2}^{k} \|grad_D(\log \sigma_a)\|^2 \|Z^a\|^2$$

$$+ \sum_{i=1}^{p} \left[g(h(Je_i, Z), \nabla_{e_i}^{\perp *} JZ) - g(h(e_i, Z), \nabla_{Je_i}^{\perp *} JZ) \right]. \quad (18)$$

Integrating both sides,Green's theorem and the hypothesis lead to

$$k = \frac{-4 \sum_{a=2}^{k} \|Z^a\|^2 \,_N\{\|grad_D(\log \sigma_a)\|^2\}dv}{p \sum_{a=2}^{k} \|Z^a\|^2 \,_N dv} \leq 0$$

since $\sum_{a=2}^{k} \|Z^a\|^2 \,_N dv > 0$ and $\sum_{a=2}^{k} \|Z^a\|^2 \,_N\{\|grad_D(\log \sigma_a)\|^2\}dv \geq 0$. Further the equality holds if and only if $_N\{\|grad_D(\log \sigma_a)\|^2\}dv = 0$ which implies that the equality holds if and only if $grad_D(\log \sigma_a) = 0$.This proves the theorem. ∎

Theorem 4. *Let $N = N_T \times_{\sigma_2} N_2 \times \dots \times_{\sigma_k} N_k$ be a compact orientable anti-invariant multiply warped product statistical submanifold without boundary of holomorphic statistical space form M of constant curvature k. If $P_{D\perp} D \in D$ and $A_{\nabla_X^{\perp *} JZ} JX = A_{\nabla_{JX}^{\perp *} JZ} X$, then*

$$R(X, Y, X, Y) \geq g(H, H^*)$$

and equality holds if and only if $grad_D(\log \sigma_a) = 0$.

Proof. From the previous theorem we have $k \leq 0$. Since N is anti-invariant, we have $N_T = 0$ and $N = N_\perp$. This implies that N becomes completely totally umbilical submanifold of M. Furthermore, from the expression of ambient curvature we have, for two orthonormal vectors $X, Y \in TN$

$$\tilde{R}(X, Y, X, Y) = -\frac{k}{4}.$$

Furthermore from Gauss equation and totally umbilicity of N, we obtain

$$R(X, Y, X, Y) = -\left(\frac{k}{4} + g(H, H^*)\right)$$

or

$$R(X, Y, X, Y) \geq g(H, H^*)$$

and equality holds if and only if $grad_D(\log \sigma_a) = 0$. ∎

Theorem 5. *Let $N = N_T \times_{\sigma_2} N_2 \times \dots \times_{\sigma_k} N_k$ be a compact orientable anti-invariant multiply warped product statistical submanifold without boundary of holomorphic statistical space form M of constant curvature k. If $P_{D\perp} D \in D$ and $A_{\nabla_X^{\perp *} JZ} JX = A_{\nabla_{JX}^{\perp *} JZ} X$, then M is Einstien and N is Einstien if and only if $\frac{k}{4} + g(H, H^*)$ is constant.*

Proof. The proof is straight from the last theorem and the Gauss equation which combinely give

$$Ric(Y, Z) = (n - 1)\left\{\frac{k}{4} + g(H, H^*)\right\} g(Y, Z).$$

∎

Acknowledgement. The authors are thankful to the referees for their vauable suggestions and comments.

References

1. Amari, S.: Differential Geometric Methods in Statistics. Lecture Notes in Statistics, vol. 28. Springer, New York (1985)
2. Amari, S., Nagaoka, H.: Methods of Information Geometry. Transactions of Mathematical Monographs, vol. 191. American Mathematical Society, Providence (2000)
3. Aydin, M.E., Mihai, A., Mihai, I.: Some inequalities on submanifolds in statistical manifolds of constant curvature. Filomat 9(3), 465–477 (2015)
4. Bishop, R.L., O'neill, B.: Manifolds of negative curvature. Trans. Amer. Math. Soc. 145, 1–49 (1969)
5. Chen, B.Y.: Geometry of warped product CR-submanifolds in Kaehler manifold. Monatsh. Math. 133, 177–195 (2001)
6. Chen, B.Y.: Geometry of warped product CR-submanifolds in Kaehler manifold II. Monatsh. Math. 134, 103–119 (2001)
7. Chen, B.Y., Dillen, F.: Optimal inequalities for multiply warped product submanifolds. Int. Elect. J. Geom. 1(1), 1–11 (2008)
8. Furuhata, H.: Hypersurfaces in statistical manifolds. Diff. Geom. Appl. 27, 420–429 (2009)
9. Todgihounde, L.: Dualistic structures on warped product manifolds. Diff. Geom.-Dynam. Syst. 8, 278–284 (2006)
10. Vos, P.W.: Fundamental equations for statistical submanifolds with applications to the Bartlett connection. Ann. Inst. Statist. Math. 41(3), 429–450 (1989)

Topological Forms and Information

Information Algebras and Their Applications

Matilde Marcolli[✉]

California Institute of Technology, Pasadena, USA
matilde@caltech.edu

Abstract. In this lecture we will present joint work with Ryan Thorngren on thermodynamic semirings and entropy operads, with Nicolas Tedeschi on Birkhoff factorization in thermodynamic semirings, ongoing work with Marcus Bintz on tropicalization of Feynman graph hypersurfaces and Potts model hypersurfaces, and their thermodynamic deformations, and ongoing work by the author on applications of thermodynamic semirings to models of morphology and syntax in Computational Linguistics.

Lecture Outline

This is an abstract for an invited talk in the Session on *Information and Topology* of the 2nd conference on *Geometric Science of Information*. The talk is based on joint work with Ryan Thorngren [20] and Nicolas Tedeschi [19], ongoing work with Marcus Bintz [4], and other ongoing work [18].

Tropical Semiring

The min-plus (or tropical) semiring $\mathbb{T}$ is $\mathbb{T} = \mathbb{R} \cup \{\infty\}$, with the operations $\oplus$ and $\odot$ given by

$$x \oplus y = \min\{x, y\},$$

with ∞ the identity element for $\oplus$ and with

$$x \odot y = x + y,$$

with 0 the identity element for $\odot$. The operations $\oplus$ and $\odot$ satisfy associativity and commutativity and distributivity of the product $\odot$ over the sum $\oplus$.

Thermodynamic Semirings (Information Algebras)

A notion of *thermodynamic semiring* was introduced in [20] and further developed in [19]. Thermodynamic semirings (or Information Algebras) are deformations of the min-plus algebra, where the product $\odot$ is unchanged, but the sum $\oplus$ is deformed to a new operation $\oplus_{\beta,S}$,

$$x \oplus_{\beta,S} y = \min_p \{px + (1-p)y - \frac{1}{\beta} S(p)\}. \tag{1}$$

© Springer International Publishing Switzerland 2015
F. Nielsen and F. Barbaresco (Eds.): GSI 2015, LNCS 9389, pp. 271–276, 2015.
DOI: 10.1007/978-3-319-25040-3_30

according to a binary entropy functional S and a deformation parameter β, which we interpret thermodynamically as an inverse temperature $\beta = 1/T$ (up to the Boltzmann constant which we set equal to 1). At zero temperature (that is, $\beta \to \infty$) one recovers the unperturbed idempotent addition. The algebraic properties (commutativity, left and right identity, associativity) of this operation correspond to properties of the entropy functional (symmetry $S(p) = S(1 - p)$, minima $S(0) = S(1) = 0$, and extensivity $S(pq) + (1 - pq)S(p(1 - q)/(1 - pq)) = S(p) + pS(q)$), namely the Khinchin axioms of the Shannon entropy.

More generally, the entropy functional considered in the deformation need not be the Shannon entropy: thermodynamic semirings associated to Rényi and Tsallis entropies have different algebraic properties: the lack of commutativity and associativity is measured in a way that relates to the corresponding axiomatic properties of these more general entropy functionals. In particular, as shown in [20], the general thermodynamic semirings have a natural interpretation in terms of non-extensive thermodynamics, [1,10].

The case where the deformation is achieved by the Shannon entropy was considered in [5] in relation to absolute arithmetic and $\mathbb{F}_1$-geometry. Thermodynamic semirings are also closely related to Maslov dequantization, see [24], and to statistical mechanics [22]. Applications to multifractals are also described in [20].

Entropy Operad

The theory of thermodynamic semirings was also presented in [20] in terms of a general operadic formulation of entropy functionals. A collection $\mathcal{S} = \{S_n\}_{n \in \mathbb{N}}$ of n-ary entropy functionals S_n satisfies a coherence condition if

$$S_n(p_1, \ldots, p_n) = S_m(p_{i_1}, \ldots, p_{i_m}),$$

whenever, for some $m < n$, we have $p_j = 0$ for all $j \notin \{i_1, \ldots, i_m\}$. Shannon, Rényi, Tsallis entropies satisfy this condition.

A collection $\mathcal{S} = \{S_n\}_{n \in \mathbb{N}}$ of coherent entropy functionals determines n-ary operations $C_{n,\beta,\mathcal{S}}$ on $\mathbb{R} \cup \{\infty\}$,

$$C_{n,\beta,\mathcal{S}}(x_1, \ldots, x_n) = \min_p \{\sum_{i=1}^n p_i x_i - \frac{1}{\beta} S_n(p_1, \ldots, p_n)\}, \tag{2}$$

with the minimum taken over $p = (p_i)$, with $\sum_i p_i = 1$. More generally, one obtains n-ary operations $C_{n,\beta,\mathcal{S},\mathcal{T}}(x_1, \ldots, x_n)$ with $\mathcal{S}$ as above and $\mathcal{T}$ planar rooted trees with n leaves. As shown in [20], these operations can be written as

$$C_{n,\beta,\mathcal{S},\mathcal{T}}(x_1, \ldots, x_n) = \min_p \{\sum_{i=1}^n p_i x_i - \frac{1}{\beta} S_\mathcal{T}(p_1, \ldots, p_n)\}, \tag{3}$$

with the $S_\mathcal{T}(p_1, \ldots, p_n)$ obtained from the S_j, for $j = 2, \ldots, n$. One obtains in this way an algebra over the A_∞-operad of rooted trees.

Birkhoff Factorization in Thermodynamic Semirings

As part of the "renormalization and computation" program developed in [14–16], Manin asked in [14] for an extension of the algebraic renormalization method based on Rota–Baxter algebras ([6–9]) to tropical semirings.

This is achieved in [19], by introducing Rota–Baxter structures of weight λ on min-plus semirings and on their thermodynamic deformation. A Rota–Baxter operator of weight λ is defined as a $\oplus$-additive (monotone) map T satisfying

$$T(x) \odot T(y) = T(T(x) \odot y) \oplus T(x \odot T(y)) \oplus T(x \odot y) \odot \log \lambda$$

when $\lambda > 0$, while for $\lambda < 0$ one has the identity

$$T(x) \odot T(y) \oplus T(x \odot y) \odot \log(-\lambda) = T(T(x) \odot y) \oplus T(x \odot T(y)).$$

In the thermodynamic case, the notion of Rota–Baxter operator is the same, but with $\oplus$ replaced by the deformed $\oplus_{\beta,S}$.

Suppose given a map $\psi : \mathcal{H} \to \mathbb{T}_{\beta,S}$ from a commutative graded Hopf algebra $\mathcal{H}$ to a thermodynamic semiring $\mathbb{T}_{\beta,S}$ with a Rota–Baxter operator of weight $+1$, such that $\psi(xy) = \psi(x) \odot \psi(y)$. It is shown in [19] that ψ has a unique Birkhoff factorization $\psi_+ = \psi_- \star \psi$, with $\psi_- = T(\tilde{\psi})$, where $\tilde{\psi}$ is the Bogolyubov preparation of ψ,

$$\tilde{\psi}(X) = \psi(X) \oplus_{\beta,S} \bigoplus_{\beta,S} \psi_-(X') \odot \psi(X'')$$

where the Hopf algebra coproduct is $\Delta(X) = X \otimes 1 + 1 \otimes X + \sum X' \otimes X''$. The product $\star$ in the Birkhoff factorization is defined as

$$(\psi_1 \star \psi_2)(X) = \bigoplus_{\beta,S} (\psi_1(X^{(1)}) \odot \psi_2(X^{(2)})),$$

where $\oplus_{\beta,S}, \odot$ are the semiring operations and $\Delta(X) = \sum X^{(1)} \otimes X^{(2)} = X \otimes 1 + 1 \otimes X + \sum X' \otimes X''$ is the Hopf algebra coproduct.

The Hopf algebra can be taken to be, for instance, a Hopf algebra of Feynman graphs as in [6] or of flow charts as in [14]. Rota–Baxter operators can be constructed using running time or memory size, in the case of flow charts, or Markov random fields, or the order of polynomial countability of the graph hypersurface (or infinity when not polynomially countable) in the case of Feynman graphs, [19].

Tropical Hypersurfaces

Tropical geometry is a version of algebraic geometry over min-plus (or max-plus) semirings, see [11,12] for a general introduction. In recent years, tropical geometry has also been studied in relation to algebraic statistics, [21].

A tropical polynomial is a function $p : \mathbb{R}^n \to \mathbb{R}$ of the form

$$p(x_1, \ldots, x_n) = \oplus_{j=1}^m a_j \odot x_1^{k_{j1}} \odot \cdots \odot x_n^{k_{jn}} =$$

$$\min\{a_1 + k_{11}x_1 + \cdots + k_{1n}x_n, a_2 + k_{21}x_1 + \cdots + k_{2n}x_n, \cdots, a_m + k_{m1}x_1 + \cdots + k_{mn}x_n\}.$$

A tropical hypersurface is the set of points where the piecewise linear tropical polynomial is non-differentiable.

Thermodynamic Tropicalization

When deforming the tropical semiring to a thermodynamic semiring, one can similarly consider polynomials of the form

$$p_{\beta,\mathcal{S}}(x_1, \ldots, x_n) = \oplus_{\beta,\mathcal{S},j} a_j \odot x_1^{k_{j1}} \odot \cdots \odot x_n^{k_{jn}} =$$

$$\min_{p=(p_j)} \{\sum_j p_j(a_j + k_{j1}x_1 + \cdots + k_{jn}x_n) - \frac{1}{\beta} S_n(p_1, \ldots, p_n)\},$$

where $\mathcal{S} = \{S_n\}$ is a coherent family of entropy functionals, or more generally

$$p_{\beta,\mathcal{S},\mathcal{T}}(x_1, \ldots, x_n) = \min_p \{\sum_j p_j(a_j + k_{j1}x_1 + \cdots + k_{jn}x_n) - \frac{1}{\beta} S_{\mathcal{T}}(p_1, \ldots, p_n)\},$$

$$(4)$$

for $\mathcal{T}$ a rooted tree with n leaves and $S_{\mathcal{T}}(p_1, \ldots, p_n)$ the corresponding entropy functional determined by the S_k in $\mathcal{S}$ with $2 \le k \le n$.

In the case where the entropy function is the Shannon entropy, this deformation of the tropical polynomial can be related to Maslov dequantization, see [24].

Applications to Feynman Graph and Potts Model Hypersurfaces

In perturbative quantum field theory, Feynman integrals can be written as period integrals (up to renormalization of divergences) on the complement of certain hypersurfaces, defined by the vanishing of the graph polynomial

$$\Psi_\Gamma(t) = \sum_T \prod_{e \notin E(T)} t_e,$$

with the sum over spanning trees of the graph and variables t_e assigned to the edges of the graph.

The algebro-geometric and motivic properties of these hypersurfaces have been widely studied in recent years, see [17] for an overview. These algebro-geometric methods were recently extended to other hypersurfaces that arise as zeros of partition functions of Potts models in [2].

We will discuss properties of the tropicalization of graph hypersurfaces and of Potts model hypersurfaces and their thermodynamic deformations, based on ongoing work [4].

Lexicographic Semirings and Geometric Models in Linguistics

Min-plus type semirings are widely used, in the form of "lexicographic semirings", in computational models of morphology and syntax in Linguistics, [13,23]. Another application of thermodynamics semirings that we will discuss is based on ongoing work [18], where entropy deformations of these linguistics models are considered, as a way of introducing an inverse temperature parameter β in deterministic finite-state representations of n-gram models based on the tropical semiring. These are a tropical geometry version of the Viterbi sequence algorithm, as described in [21]. Introducing the deformation parameter β plays a role, in these models, analogous to the thermodynamic formalism of [3]. We will discuss some consequences of this approach.

References

1. Abe, S., Okamoto, Y.: Nonextensive Statistical Mechanics and Its Applications. Lecture Notes in Physics, vol. 560, 1st edn. Springer, Heidelberg (2001)
2. Aluffi, P., Marcolli, M.: A motivic approach to phase transitions in Potts models. J. Geom. Phys. **63**, 6–31 (2013)
3. Beck, C., Schlögl, F.: Thermodynamics of Chaotic Systems: An Introduction. Cambridge University Press, Cambridge (1993)
4. Bintz, M., Marcolli, M.: Thermodynamic tropicalization of Feynman graph and Potts model hypersurfaces (in preparation)
5. Connes, A., Consani, C.: From Monoids to Hyperstructures: in Search of an Absolute Arithmetic, "Casimir Force, Casimir Operators and the Riemann Hypothesis". Walter de Gruyter, Berlin (2010). pp. 147–198
6. Connes, A., Kreimer, D.: Renormalization in quantum field theory and the Riemann-Hilbert problem. I. The Hopf algebra structure of graphs and the main theorem. Commun. Math. Phys. **210**(1), 249–273 (2000)
7. Connes, A., Marcolli, M.: Noncommutative Geometry, Quantum Fields and Motives, vol. 55, p. 785, Colloquium Publications, American Mathematical Society (2008)
8. Ebrahimi-Fard, K., Guo, L.: Rota-Baxter algebras in renormalization of perturbative quantum field theory in "universality and renormalization". Fields Inst. Commun. **50**, 47–105 (2007). American Mathematical Society
9. Ebrahimi-Fard, K., Guo, L., Kreimer, D.: Integrable renormalization II the general case. Ann. Henri Poincaré **6**(2), 369–395 (2005)
10. Gell-Mann, M., Tsallis, C. (eds.): Nonextensive Entropy. Oxford University Press, New York (2004)
11. Itenberg, I., Mikhalkin, G.: Geometry in tropical limit. Math. Semesterber. **59**(1), 57–73 (2012)
12. Itenberg, I., Mikhalkin, G., Shustin, E.: Tropical algebraic geometry, Oberwolfach Seminars, vol. 35. Birkhäuser Verlag, Basel (2007)
13. Kuich, W., Salomaa, A.: Semirings, Automata, Languages. Springer, Heidelberg (1986)
14. Manin, Y.I.: Renormalization and computation, I: motivation and background, in "OPERADS 2009". Sémin Congr. Soc. Math. France **26**, 181–222 (2013)
15. Manin, Y.I.: Renormalization and computation II: time cut-off and the halting problem. Math. Struct. Comput. Sci. **22**(5), 729–751 (2012)

16. Manin, Y.I.: Infinities in quantum field theory and in classical computing: renormalization program. In: Ferreira, F., Löwe, B., Mayordomo, E., Mendes Gomes, L. (eds.) CiE 2010. LNCS, vol. 6158, pp. 307–316. Springer, Heidelberg (2010)
17. Marcolli, M.: Feynman Motives. World Scientific, Singapore (2010)
18. Marcolli, M.: Thermodynamics Semirings in Computational Linguistics (in preparation)
19. Marcolli, M., Tedeschi, N.: Entropy algebras and Birkhoff factorization. J. Geom. Phys. 97, 243–265 (2015)
20. Marcolli, M., Thorngren, R.: Thermodynamic semirings. J. Noncommut. Geom. 8(2), 337–392 (2014)
21. Pachter, L., Sturmfels, B.: Tropical geometry of statistical models. Proc. Nat. Ac. Sci. USA 101(46), 16132–16137 (2004)
22. Quadrat, J.P., Max-Plus Working Group: Min-Plus linearity and statistical mechanics. Markov Process. Relat. Fields 3(4), 565–597 (1997)
23. Roark, B., Sproat, R.: Computational Approaches to Morphology and Syntax. Oxford University Press, Oxford (2007)
24. Viro, O.: Dequantization of real algebraic geometry on logarithmic paper. In: European Congress of Mathematics, Vol. I (Barcelona, 2000), Progr. Math., 201, pp. 135–146. Birkhäuser (2001)

Finite Polylogarithms, Their Multiple Analogues and the Shannon Entropy

Philippe Elbaz-Vincent[1]([✉]) and Herbert Gangl[2]

[1] Institut Fourier, CNRS-Université Grenoble Alpes,
BP 74, 38402 Saint Martin D'hères, France
`Philippe.Elbaz-Vincent@ujf-grenoble.fr`
[2] Department of Mathematical Sciences, University of Durham,
South Road, Durham, UK
`herbert.gangl@durham.ac.uk`

Abstract. We show that the entropy function—and hence the finite 1-logarithm—behaves a lot like certain derivations. We recall its cohomological interpretation as a 2-cocycle and also deduce $2n$-cocycles for any n. Finally, we give some identities for finite multiple polylogarithms together with number theoretic applications.

1 Information Theory, Entropy and Polylogarithms

It is well known that the notion of entropy occurs in many sciences. In thermodynamics, it means a measure of the quantity of disorder, or more accurately, the tendancy of a system to go toward a disordered state. In information theory, the entropy measures (in terms of real positive numbers) the quantity of information of a certain property [17,20]. From a practical viewpoint, entropies play also a key role in the study of random bit generators (deterministic or not) [8], in particular due to the Maurer test [16]. A general definition of entropy has been given by Rényi [18]: let $S = \{s_1, \ldots, s_n\}$ be a set of discrete events for which the probabilities are given by $p_i = P(s = s_i)$ for $i = 1, \ldots, n$. The Rényi entropy S is then defined for $\alpha > 0$ and $\alpha \neq 1$ as

$$H_\alpha(S) = \frac{1}{1 - \alpha} \log \left(\sum_{i=1}^{n} p_i^\alpha \right).$$

The Shannon entropy [20] can be recovered from the one of Rényi when $\alpha \to 1$

$$H_1(S) = \lim_{\alpha \to 1} H_\alpha(S) = - \sum_{i=1}^{n} p_i \log(p_i).$$

Partially supported by the LabEx PERSYVAL-Lab (ANR-11-LABX-0025-01) and by the LabEx AMIES.

© Springer International Publishing Switzerland 2015
F. Nielsen and F. Barbaresco (Eds.): GSI 2015, LNCS 9389, pp. 277–285, 2015.
DOI: 10.1007/978-3-319-25040-3_31

We also often use the minimal entropy which is related to the probability of the most predictable event (while the Shannon entropy gives an averaged measure):

$$H_{\min}(S) = \lim_{\alpha \to \infty} H_\alpha(S) = -\log(\max_{i=1,\dots,n}(p_i)).$$

Those different entropies are related by the following inequalities

$$H_{\min}(S) \leqslant \dots \leqslant H_2(S) \leqslant H_1(S) \leqslant \log(\text{card}(S)) = \lim_{\alpha \to 0} H_\alpha(S).$$

The Shannon entropy can be characterised in the framework of information theory, assuming that the propagation of information follows a Markovian model [17,20]. If H is the Shannon entropy, it fulfills the equation, often called the *Fundamental Equation of Information Theory* (FEITH),

$$H(x) + (1-x)H\left(\frac{y}{1-x}\right) - H(y) - (1-y)H\left(\frac{x}{1-y}\right) = 0. \qquad \text{(FEITH)}$$

In [2](section 5.4, pp.66–69), it is shown that if g is a real function locally integrable on $]0,1[$ and if, moreover, g fulfills FEITH, then there exists $c \in \mathbb{R}$ such that $g = cH$ (we can also restrict the hypothesis to Lebesgue measurable). There are several papers (e.g., [1,7]) on the equation FEITH and the understanding of its structural properties, with the motivation to weaken either the probabilistic hypothesis or the analytical ones. The following generalisation of the equation FEITH has also been considered [14], for β positive and x and y in some admissible range,

$$H(x) + (1-x)^\beta H\left(\frac{y}{1-x}\right) - H(y) - (1-y)^\beta H\left(\frac{x}{1-y}\right) = 0. \qquad (1)$$

It turns out that FEITH can be derived, in a precise formal sense [9], from the 5-term equation of the classical (or p-adic) dilogarithm. Cathelineau [5] found that an appropriate derivative of the Bloch–Wigner dilogarithm coincides with the classical entropy function, and that the five term relation satisfied by the former implies the four term relation of the latter. Kontsevich [13] discovered that the truncated finite logarithm over a finite field $\mathbb{F}_p$, with p prime, defined by

$$\pounds_1(x) = \sum_{k=1}^{p-1} \frac{x^k}{k},$$

satisfies FEITH (or its generalisation for $\beta = 1$ or p). In [9] we showed how one can expand this relationship for "higher analogues" in order to produce and prove similar functional identities for finite polylogarithms from those for classical polylogarithms. It was also shown that functional equations for finite polylogarithms often hold even as polynomial identities over finite fields. In particular, we have shown that the polynomial version of $\mathcal{L}_1$ fulfills (1) with $\beta = p$. Another approach, due to Bloch and Esnault [3], gives a more geometric version in terms of algebraic cycles, and further structural properties have been investigated by Cathelineau [6].

In this paper we propose some new formal characterisations of the entropy from an algebraic viewpoint, using formal derivations and a relation to cohomology (Sect. 2), and we give complementary relations involving multiple analogues of the finite polylogarithms with a few applications to number theory (Sect. 3). The details for this last work will be given in a subsequent paper. In the remainder of the paper, rings are assumed to be commutative. We will denote by $\mathbb{F}_q$ the finite field with q elements. If q is prime, we set $\mathbb{F}_q = \mathbb{Z}/q\mathbb{Z}$.

2   Algebraic Interpretation of the Entropy Function

2.1   Formal Entropy as Formal Derivations

Definition 1. *Let R be a (commutative) ring and let D be a map from R to R. We will say that D is a* unitary derivation *over R if the following axioms hold:*

1. *"Leibniz's rule": for all $x, y \in R$, we have $D(xy) = xD(y) + yD(x)$.*
2. *"Additivity on partitions of unity": for all $x \in R$, we have $D(x) + D(1-x) = 0$.*

We will denote by $Der^u(R)$ the set of unitary derivations over R.

Applying analogous arguments as for derivations (see for instance [15], chap. 9), we have

Proposition 1. *The set of unitary derivations over R, $Der^u(R)$, is an R-module, which has $Der_{\mathbb{Z}}(R)$ as a submodule. If D and D' are two unitary derivations, then the composition $D \circ D'$ and the Lie bracket $[D, D'] = D \circ D' - D' \circ D$ are unitary derivations.*

Let D be a unitary derivation over R.

1. *For all $x \in R$ and all $n \in \mathbb{N}$ we have $D(x^n) = nx^{n-1}D(x)$. Furthermore, if $x \in R^\times$ the rule is also true for $n \in \mathbb{Z}$.*
2. *For all $n \in \mathbb{N}$, $D((n+1)1_R) = \frac{n(n+1)}{2}D(-1)$, and $2D(-1) = 0$.*
3. *If R has no 2-torsion or if $2(-1) = 0$ in R, then for all $x \in R$ and all $n \in \mathbb{Z}$, we have $D(nx) = nD(x)$.*
4. *Suppose that R has no 2-torsion, or that $2(-1) = 0$ in R, and let $m \in \mathbb{Z}$ with $m \in R^\times$, then $D(\frac{1}{m}) = 0$. If moreover $\mathbb{Q} \subset R$, then $D(\mathbb{Q}) = 0$.*

Proof.   1. Works as the classical proof for derivations.
  2. First, using the standard fact that $0 = 0 \cdot 0$, we deduce that $D(0) = 0$, and then $D(1) = 0$. Then we can see that $2D(-1) = 0$ and that $D(n+1) - D(n) = nD(-1)$. Thus an induction argument proves the formula.

3. If R has no 2-torsion, or if $2(-1) = 0$ in R, then $D(-1) = 0$, and using the previous result together with the fact that $D(-n) = -D(1+n)$, we deduce $D(n) = 0$ for all $n \in \mathbb{Z}$. Then the desired formula follows.
4. Direct consequence of the previous rules. $\square$

Remark 1. We can get nicer statements by working in $Der^u(R)/\langle D(-1) \rangle$, where $\langle D(-1) \rangle$ denotes the submodule of $Der^u(R)$ spanned by $D(-1)$.

Corollary 1. *Suppose that $nR = 0$, for a given $n \in \mathbb{N} - \{0\}$. Then if D is a unitary derivation over R and if $\lambda_n : R \longrightarrow R$ is defined by $\lambda_n(x) = x^n$, we then have $D \circ \lambda_n = 0$. In particular if p is a prime number, $\nu \in \mathbb{N} - \{0\}$ and $q = p^\nu$, then $D(\mathbb{F}_q) = 0$.*

Recall the following definition from [12].

Definition 2. *Let R be a commutative ring and k be a natural number. We say that R is k-fold stable if for any family of k unimodular vectors $(a_i, b_i)_{1 \leqslant i \leqslant k} \in R^2$ (i.e. $a_i R + b_i R = R$), there exists $t \in R$, such that $a_i + t b_i \in R^\times$ for all i.*

Proposition 2. "Unitary Derivations are almost Derivations"
Let R be a 2-fold stable ring, and suppose that R is of characteristic 2 (i.e. $2R = 0$) or that R has no 2-torsion. Then $Der_{\mathbb{Z}}(R) = Der^u(R)$.

Proof. According to Proposition 1, we have to show that any unitary derivation is additive. Let $D \in Der^u(R)$ and let $x, y \in R$. Suppose first that x is invertible. Then $x+y = x(1+\frac{y}{x})$, and by Leibniz's rule, we have $D(x+y) = xD(1+\frac{y}{x})+(1+\frac{y}{x})D(x)$. Using the additivity on partitions of unity, $D(1+\frac{y}{x}) = -D(-\frac{y}{x})$ and also $D(-\frac{y}{x}) = -D(\frac{y}{x})$. Hence we deduce $D(x+y) = D(x) + D(y)$. Now suppose that x is not invertible. Then applying the 2-fold stability to the unimodular vectors $(0,1)$, $(x,1)$, we deduce the existence of $t \in R^\times$ such that $x + t$ is invertible. Setting $x' = x+t$ and $y' = y-t$, we have $D(x+y) = D(x'+y')$. Then we can apply the previous arguments to x', y', and deduce that $D(x+y) = D(x+t)+D(y-t)$. Now we again apply the same arguments to x, t, and y, $-t$. Using the rules of Proposition 1, we conclude that $D(x + y) = D(x) + D(y)$, and the claim follows. $\square$

Example 1. As any semilocal ring R such that any of its residue fields has at least 3 elements is 2-fold stable [12], we then deduce that $Der_{\mathbb{Z}}(R) = Der^u(R)$.

2.2   Unitary Derivations and Symmetric Information Function of Degree 1

For more details on this section related to information theory see [14].

Definition 3. *Let R be a commutative ring. We will say that a map $f : R \to R$ is an abstract symmetric information function of degree 1 if the two following conditions hold: for all $x, y \in R$ such that $x, y, 1 - x, 1 - y \in R^\times$, the functional equation FEITH holds and for all $x \in R$, we have $f(x) = f(1 - x)$.*

Denote by $\mathbb{JF}_1(R)$ the set of abstract symmetric information functions of degree 1 over R. Then $\mathbb{JF}_1(R)$ is an R-module. Let $Leib(R)$ be the set of Leibniz functions over R (i.e. which fulfill the "Leibniz rule"), then it is also an R-module (in fact the composition and the Lie bracket still hold in $Leib(R)$). The proof of the following proposition is a straightforward computation.

Proposition 3. *We have a morphism of R-modules $H : Leib(R) \rightarrow \mathbb{JF}_1(R)$, defined by $H(\varphi) = \varphi + \varphi \circ \tau$, with $\tau(x) = 1 - x$. Furthermore, $Ker(H) = Der^u(R)$.*

Remark 2. The morphism H is not necessarily onto. If $R = \mathbb{F}_q$, a finite field, then $Leib(\mathbb{F}_q) = 0$, but $\mathbb{JF}_1(\mathbb{F}_q) \neq 0$.

2.3 Cohomological Interpretation of Formal Entropy Functions

The following results are classical in origin (see [4], pp.58–59, and also the references cited there, and also [13]). We try in this section to render the proofs (for the finite case) more transparent, and also emphasize the derivation aspect of the previous sections.

Theorem 1. *Let F be a finite prime field and $H : F \rightarrow F$ a function which fulfills the following conditions: $H(x) = H(1-x)$, the functional equation (FEITH) holds for H and $H(0) = 0$. Then the function $\varphi : F \times F \rightarrow F$ defined by $\varphi(x, y) = (x+y)H(\frac{x}{x+y})$ if $x + y \neq 0$ and 0 otherwise, is a non-trivial 2-cocycle.*

Proof. The fact that φ is a 2-cocycle is a straightforward consequence of the properties on H. In order to see this, we use the inversion relation, which in turn one can deduce from (FEITH), and the relation $H(x) = H(1 - x)$. By setting $Y = \frac{x}{x+y+z}$ and $X = \frac{y}{x+y+z}$ (assuming some suitable admissibility conditions on x, y and z), and modulo some modifications using the other relations, the 2-cocycle condition is deduced from (FEITH). For the non-triviality, notice that φ is homogeneous and recall that as F is a field we can endow the cochains with a structure of F-vector space. Suppose that φ is a 2-coboundary. Then, there exists a map $Q : F \rightarrow F$, such that $\varphi(x, y) = Q(x+y) - Q(x) - Q(y)$. Notice that $Q(0) = 0$. As φ is homogeneous, we have $\varphi(\lambda x, \lambda y) = \lambda Q(x+y) - \lambda Q(x) - \lambda Q(y)$. Thus the function $\psi_\lambda(x) = Q(\lambda x) - \lambda Q(x)$ is an additive morphism $F \rightarrow F$, hence entirely determined by $\psi_\lambda(1)$. The map $\psi_\lambda(1)$ fulfills the Leibniz chain rule on $F^\times$. Indeed, assuming $F = \mathbb{Z}/p\mathbb{Z}$, if λ, μ are arbitrary elements of F, as $\mu\psi_\lambda(1) = \psi_\lambda(\mu)$, by a straightforward computation we deduce $\psi_{\lambda\mu}(1) = \psi_\lambda(\mu) + \lambda\psi_\mu(1)$. Thus we formally have $\psi_{\lambda^m}(1) = m\lambda^{m-1}\psi_\lambda(1)$. But $F^\times$ is generated by a primitive root, say ω. Let $p = card(F)$. Then $\omega^{p-1} = 1$. Moreover $0 = \psi_1(1) = (p-1)\omega^{p-2}\psi_\omega(1)$. Hence $\psi_\omega(1) = 0$ and then $Q(\lambda x) = \lambda Q(x)$ for all $\lambda, x \in F$. This implies that Q is an additive map and thus $\varphi = 0$, which contradicts the fact that it is a non-zero 2-cochain. □

Remark 3. We should notice that $H(\lambda) = \varphi(\lambda, 1-\lambda) = \psi_\lambda(1) + \psi_{(1-\lambda)}(1)$, which is very similar to the results of Maksa [14].

Corollary 2. *The map $F \to H^2(F, F)$, given by $\lambda \mapsto \lambda\varphi$, is an isomorphism and, up to a constant, $\mathcal{L}_1$ is unique.*

Using the (cup) product structure on the cohomology ring $H^*(F, F)$ (cf. [10], chap. 3), we can check the following property:

Corollary 3. *Let n be a positive integer. The map*

$$\varphi(x_1, \dots, x_{2n}) = \prod_{i=1,\, i \text{ even}}^{2n-1} \varphi(x_i, x_{i+1})$$

induces a non-trivial cocycle in $H^{2n}(F, F)$, which corresponds to the cup product induced by φ. This cocycle corresponds to the product of n functions H, and is unique up to a constant.

3 Finite Multiple Polylogarithms

While *classical* polylogarithms play an important role in the theory of mixed Tate motives over a field, it turns out that it is often preferable to also consider the larger class of *multiple* polylogarithms (e.g., [11]). In a similar way it is useful to investigate their finite analogues. We are mainly concerned with finite double polylogarithms which are given as functions $\mathbb{Z}/p \times \mathbb{Z}/p \to \mathbb{Z}/p$ by

$$\mathcal{L}_{a,b}(x, y) = \sum_{0 < m < n < p} \frac{x^m}{m^a} \frac{y^n}{n^b} \, .$$

3.1 Expressing $\mathcal{L}_{1,1}$ via $\mathcal{L}_2$

Our arguably most interesting result, from which we will deduce a couple of consequences, is the following.

Theorem 2. *The finite $(1,1)$-logarithm $\mathcal{L}_{1,1}(x, y)$ can be expressed in terms of $\mathcal{L}_2$. More precisely, we have*

$$y\mathcal{L}_{1,1}(x, \frac{1}{y}) = \mathcal{L}_2\left(-y^p \left[\frac{x}{y}\right] - (1-y)^p \left[\frac{1-x}{1-y}\right] + [1-x] + [1-y] \right). \tag{2}$$

The proof of this result takes $(1 - y)^p \mathcal{L}_2\left(\frac{1-x}{1-y}\right)$ and decomposes the (triangular) domain over which the summation variables run into an "open" part (a triangle) and three "boundary" parts (one diagonal, a vertical and a horizontal line) and identifies the former with the $\mathcal{L}_{1,1}$-expression and the latter with the three remaining terms in the equation. At a crucial step one uses the binomial identity

$$\sum_{r=0}^{N} \binom{N-r}{s} \binom{r}{t} = \binom{N+1}{s+t+1} \, .$$

3.2 Cathelineau's $\mathcal{L}_2$-identity

Combining the well-known shuffle identity $\mathcal{L}_{a,b}(x,y) + \mathcal{L}_{b,a}(y,x) + \mathcal{L}_{a+b}(xy) = \mathcal{L}_a(x)\mathcal{L}_b(y)$ for $a = b = 1$ with the above we find that the product $\mathcal{L}_1(x)\mathcal{L}_1(y)$ can indeed be expressed as a sum of $\mathcal{L}_2$-terms. In fact, the resulting expression is precisely Cathelineau's "double bracket" $[[x,y]]$ ([5], p.1344, Déf. 4). Now the sum obtained from the four terms of (FEITH) for one of the two arguments and while leaving the second argument fixed kills the products of $\mathcal{L}_1$-terms so we are left with only $\mathcal{L}_2$-terms, hence we have proved a functional equation—in fact, Cathelineau's 22-term equation in ([5], p.1346, (2)).

3.3 Further Identities

We can prove an inversion formula for finite multiple polylogarithms

$$T_1^p \cdots T_\ell^p \, \mathcal{L}_{m_\ell,\ldots,m_1}\left(\frac{1}{T_\ell},\ldots,\frac{1}{T_1}\right) = (-1)^{m_1+\cdots+m_\ell} \, \mathcal{L}_{m_1,\ldots,m_\ell}(T_1,\ldots,T_\ell),$$

and we can also build a four variable identity for $\mathcal{L}_{1,1}$.

Proposition 4. *Define* $[x,y]_s = \mathcal{L}_{1,1}(x,y) + \mathcal{L}_{1,1}(y,x)$ *and consider the following linear combination*

$$K(x,y) = [x,y]_s + x^p\left[\frac{1}{x},y\right]_s - (1-y)^p\left[1-x,\frac{y}{y-1}\right]_s + (1-y)^p\left[1-x,\frac{1}{1-y}\right]_s$$
$$- x^p(1-y)^p\left[1-\frac{1}{x},\frac{y}{y-1}\right]_s + x^p(1-y)^p\left[1-\frac{1}{x},\frac{1}{1-y}\right]_s.$$

Then the following functional equation (purely in $\mathcal{L}_{1,1}$) holds:

$$I(x,y;z,w) - I(x,z;y,w) = 0,$$

where

$$I(x,y;z,w) = (1+z)(1+w)K(x,y) + (1+x)(1+y)K(z,w).$$

3.4 Finite Polylogarithms and Fermat's Last Theorem

Several classical criteria used by Kummer, Mirimanoff and Wieferich to prove certain cases of Fermat's Last Theorem can be rephrased in terms of functional equations and evaluations of finite (multiple) polylogarithms. For example, Mirimanoff was led to the study of (nowadays called) *Mirimanoff polynomials* (cf. [19], VIII, (1.11))

$$\varphi_j(T) = \sum_{j=1}^{p-1} k^{j-1}T^k,$$

which are nothing else but finite polylogarithms:

$$\varphi_j(T) \equiv \mathcal{L}_{p-j}(T) \pmod{p}.$$

(Note that Mirimanoff's original polynomials correspond to $-\varphi_j(-T)$).

Part of the groundwork for Mirimanoff's congruences was formed by the **crucial identity**

$$-\frac{1}{2}\left[\varphi_{p-1}(T)\right]^2 \equiv \varphi_{p-2}(T) + (T-1)^{2p}\varphi_{p-2}\left(\frac{T}{T-1}\right) \quad (\bmod\ p)$$

([19], VIII, (1.29)) which is nothing but the special case product formula $x = y\,(=T)$ in our identity for $\pounds_1(x)\pounds_1(y)$ alluded to in 3.2.

The *Mirimanoff congruences* ([19], VIII, (1B)) can be reformulated as follows: for any solution (x, y, z) of $x^p + y^p + z^p = 0$ in pairwise prime integers not divisible by p (i.e. a *Fermat triple*) and for $t = -\frac{x}{y}$ we have

$$\pounds_1(t) = 0, \quad \pounds_j(t)\pounds_{p-j}(t) = 0 \qquad \left(j = 2, \ldots, \frac{p-1}{2}\right).$$

One can prove these congruences using an identity expressing $\pounds_{p-j-1,j+1}(1, T)$ in terms of $\pounds_n(T)$: denoting the Bernoulli numbers by B_n, we have

$$\pounds_{p-j-1,1+j}(1, T) \equiv \frac{1}{j+1}\sum_{n=0}^{j}\binom{j+1}{n}B_n\pounds_n(T) \qquad j = 1, \ldots, p-2. \tag{3}$$

Also, *Wieferich's criterion* states that if the first case of FLT for the prime p is false then p^2 divides $2^p - 1$ (only two such primes are known for which that latter holds: $p = 1093$ and $p = 3511$). This criterion can be rephrased in terms of finite polylogarithms as saying $\pounds_1(-1) = 0$ for such primes.

Acknowledgement. We would like to express our sincere gratitude to the reviewers for their valuable comments who have helped improve this paper

References

1. Aczél, J.: Entropies old and new (and both new and old) and their characterizations. In: Erickson, G., Zhai, Y. (eds.) CP707, Bayesian Inference and Maximum Entropy Methods in Science and Engineering: 23rd International Workshop, American Institute of Physics (2004)
2. Aczél, J., Dhombres, J.: Functional Equations in Several Variables. Encyclopedia of Mathematics and its Applications, vol. 31. Cambridge University Press, Cambridge (1989)
3. Bloch, S., Esnault, H.: An additive version of higher Chow groups. Ann. Sci. École Norm. Sup. **36**(4), 463–477 (2003)
4. Cathelineau, J.-L.: Sur l'homologie de SL_2 à coefficients dans l'action adjointe. Math. Scand. **63**, 51–86 (1988)
5. Cathelineau, J.-L.: Remarques sur les différentielles des polylogarithmes uniformes. Ann. Inst. Fourier **46**(5), 1327–1347 (1996). Grenoble
6. Cathelineau, J.-L.: The tangent complex to the Bloch-Suslin complex. Bull. Soc. math. France **135**(4), 565–597 (2007)

7. Csiszár, I.: Axiomatic characterizations of information measures. Entropy **10**, 261–273 (2008). doi:10.3390/e10030261
8. De Julis, G.: Analyse d'accumulateurs d'entropie pour les générateurs aléatoires cryptographiques. Ph.D thesis, Université Grenoble Alpes, December 2014 https://tel.archives-ouvertes.fr/tel-01102765v1
9. Elbaz-Vincent, Ph, Gangl, H.: On poly(ana)logs I. Compos. Math. **130**(2), 161–210 (2002)
10. Evens, L.: The Cohomology of Groups. Oxford Math. Monographs. Clarendon Press, Oxford (1991)
11. Goncharov, A.B.: Galois symmetries of fundamental groupoids and noncommutative geometry. Duke Math. J. **128**(2), 209–284 (2005)
12. van der Kallen, W.: The K_2 of rings with many units. Ann. Scient. Ec. Norm. Sup. **10**, 473–515 (1977)
13. Kontsevich, M.: The $1\frac{1}{2}$-logarithm. Appendix to [9]. Compos. Math. **130**(2), 211–214 (2002)
14. Maksa, Gy: The general solution of a functional equation related to the mixed theory of information. Aeq. Math. **22**, 90–96 (1981)
15. Matsumura, H.: Commutative Ring Theory. Cambridge studies in advanced Math., vol. 8. Cambridge University Press, Cambridge (1986)
16. Maurer, U.M.: A universal statistical test for random bit generators. J. Cryptol. **5**(2), 89–105 (1992)
17. Ollivier, Y.: Aspects de l'entropie en mathématiques et en physique, Technical report (2002) http://www.yann-ollivier.org/entropie/entropie.pdf
18. Rényi, A.: On measures of entropy and information. In: Proceeding of 4th Berkeley Symposium on Mathematical Statistics and Probability, vol. 1, pp. 547–561 (1960)
19. Ribenboim, P.: 13 Lectures on Fermat's Last Theorem. Springer, New York (1979)
20. Shannon, C.: A Mathematical theory of communication. Bell Syst. Tech. J., Vol. 27, pp. 379–423, 623–656, July, October (1948)

Heights of Toric Varieties, Entropy and Integration over Polytopes

José Ignacio Burgos Gil[1], Patrice Philippon[2($\boxtimes$)], and Martín Sombra[3]

[1] Instituto de Ciencias Matemáticas (CSIC-UAM-UCM-UCM3),
Calle Nicolás Cabrera 15, Campus UAM, Cantoblanco, 28049 Madrid, Spain
burgos@icmat.es
http://www.icmat.es/miembros/burgos
[2] Institut de Mathématiques de Jussieu – U.M.R. 7586 du CNRS,
Équipe de Théorie des Nombres, BP 247, 4 Place Jussieu, 75005 Paris, France
patrice.philippon@imj-prg.fr
http://webusers.imj-prg.fr/~patrice.philippon
[3] ICREA & Departament d'Àlgebra i Geometria, Universitat de Barcelona,
Gran Via 585, 08007 Barcelona, Spain
sombra@ub.edu
http://atlas.mat.ub.es/personals/sombra

Abstract. We present a dictionary between arithmetic geometry of toric varieties and convex analysis. This correspondence allows for effective computations of arithmetic invariants of these varieties. In particular, combined with a closed formula for the integration of a class of functions over polytopes, it gives a number of new values for the height (arithmetic analog of the degree) of toric varieties, with respect to interesting metrics arising from polytopes. In some cases these heights are interpreted as the average entropy of a family of random processes.

1 Introduction

Toric varieties form a remarkable class of algebraic varieties, endowed with an action of a torus having one Zariski dense open orbit. It is well known that their geometric properties can be described in terms of combinatorial objects such as fans and polytopes having the same dimension, say n, as the toric variety. For instance, the degree of a toric variety with respect to a nef toric divisor is $n!$ times the volume of the corresponding polytope.

In the book [4], we have extended this dictionary by linking the arithmetic geometry of toric varieties defined over a number field to convex analysis. Here, the arithmetic ingredients are given by (semipositive) metrics on the toric line

Burgos was partially supported by the MINECO research project MTM2013-42135-P. Philippon was partially supported by the CNRS project PICS "Géométrie diophantienne et calcul formel" and the ANR research project "Hauteurs, modularité, transcendance". Sombra was partially supported by the MINECO research project MTM2012-38122-C03-02.

F. Nielsen and F. Barbaresco (Eds.): GSI 2015, LNCS 9389, pp. 286–295, 2015.
DOI: 10.1007/978-3-319-25040-3_32

bundle associated to a toric divisor. Each of these metrics correspond to a continuous concave function on the associated polytope, that we call the *local roof function*. These functions combine in a *global* roof function over the polytope. In this context, the arithmetic invariant analogous to the degree is the height which, similarly to the degree, can be expressed as $(n + 1)!$ times the integral over the polytope of the global roof function.

For particular choices of metrics, these heights coincide with the average entropy of certain random processes associated to the polytope. Our toric "dictionary", combined with a closed formula for the integration of a class of functions over polytopes, allows to compute the values of these heights. All the results presented here can be found in more details in [4].

2   Heights and Toric Varieties

In this section we recall some of the basic constructions and results in [4]. Details and more information can be found in this reference. The reader can also consult [5,6] for a background on the algebraic geometry of toric varieties.

Let $N \simeq \mathbb{Z}^n$ be a lattice of rank n and $M := \mathrm{Hom}(N, \mathbb{Z})$ its dual lattice. Set $N_{\mathbb{R}} := N \otimes_{\mathbb{Z}} \mathbb{R} \simeq \mathbb{R}^n$ and $M_{\mathbb{R}} := M \otimes_{\mathbb{Z}} \mathbb{R}$. We denote by $\langle x, u \rangle$ the pairing between $x \in M_{\mathbb{R}}$ and $u \in N_{\mathbb{R}}$.

To a *lattice fan* Σ on $N_{\mathbb{R}}$ we associate a *toric scheme* over the integers, denoted by X_{Σ}. This scheme is flat over $\mathrm{Spec}(\mathbb{Z})$ of relative dimension n. It is equipped with an action of the algebraic torus $\mathbb{T}_N := \mathrm{Spec}(\mathbb{Z}[M]) \simeq \mathbb{G}_{m,S}^n$ extending the natural action of $\mathbb{T}_N$ on itself. This action has a dense orbit, denoted X_{Σ}° and which is canonically isomorphic to $\mathbb{T}_N$. The scheme $X_{\Sigma,S}$ is projective whenever the fan Σ is complete and regular, and it is smooth whenever each cone of Σ is generated by a subset of a basis of N, see [5]. We will assume both properties from now on.

A *virtual support function* is a continuous function $\Psi \colon N_{\mathbb{R}} \to \mathbb{R}$ whose restriction to each of the cones of Σ is an element of M. Such a function defines an invariant Cartier divisor D_{Ψ} of $X_{\Sigma,S}$ or, equivalently, an equivariant line bundle $L_{\Psi,S}$ together with an invariant rational section s_{Ψ} such that $\mathrm{div}(s_{\Psi}) = D_{\Psi}$, see [6, Sects. 3.3 and 3.4]. The divisor D_{Ψ} is relatively ample if and only if Ψ is concave and restricts to different elements of M on each of the maximal cones of Σ. We will also suppose this from now on. Under this assumption, the polyhedron

$$\Delta_{\Psi} := \{x \in M_{\mathbb{R}} : \langle x, y \rangle \geq \Psi(y) \text{ for all } y \in N_{\mathbb{R}}\} \subset M_{\mathbb{R}}$$

is an n-dimensional polytope.

Let $X_{\Sigma}(\mathbb{C})$ and $\mathbb{T}_N(\mathbb{C})$ respectively denote the analytification of the scheme X_{Σ} and of the algebraic torus $\mathbb{T}$. Also let $\mathbb{S} := \{t \in \mathbb{T}_N(\mathbb{C}) \mid |t| = 1\}$ be the *compact subtorus* of $\mathbb{T}_N(\mathbb{C})$. There is a map $\mathrm{val} \colon X_{\Sigma}^{\circ}(\mathbb{C}) \to N_{\mathbb{R}}$, defined, in a given splitting $X_{\Sigma}^{\circ}(\mathbb{C}) = \mathbb{T}_N(\mathbb{C}) \simeq (\mathbb{C}^{\times})^n$, by

$$\mathrm{val}(x_1, \ldots, x_n) = (-\log |x_1|_v, \ldots, -\log |x_n|_v).$$

This map does not depend on the choice of the splitting and the compact torus $\mathbb{S}$ coincides with its fiber over the point $0 \in N_{\mathbb{R}}$.

We furthermore consider a semipositive *toric metric* $\|\cdot\|$ on the analytification of the line bundle L_{Ψ}, that is, a semipositive metric which is invariant under the action of $\mathbb{S}$. We denote by $\overline{L}_{\Psi}$ the line bundle metrized in this way. Such a toric metrized line bundle defines a continuous function $\psi_{\overline{L}} \colon N_{\mathbb{R}} \to \mathbb{R}$ given, for $p \in \mathbb{T}(\mathbb{C})$, by

$$\psi_{\overline{L}}(\mathrm{val}(p)) = \log \|s_{\Psi}(p)\|_v.$$

This function is concave. We can then consider its *Legendre-Fenchel dual* $\psi_{\overline{L}}^{\vee} \colon M_{\mathbb{R}} \to \mathbb{R} \cup \{-\infty\}$ defined by

$$\psi_{\overline{L}}^{\vee}(x) = \inf_{u \in N_{\mathbb{R}}} (\langle u, x \rangle - \psi_{\overline{L}}(u)).$$

The *stability set* of a concave function is the set of points where its Legendre-Fenchel dual is $> -\infty$. It turns out that the stability set of $\psi_{\overline{L}}$ coincides with the polytope Δ_{Ψ} and that the function $\psi_{\overline{L}}^{\vee}$ is continuous and concave on Δ_{Ψ}. The *roof function* of $\overline{L}$, denote $\vartheta_{\overline{L}}$, is defined as the restriction of $\psi_{\overline{L}}^{\vee}$ to the polytope Δ_{Ψ}.

Given a flat projective and smooth scheme X over $\mathrm{Spec}(\mathbb{Z})$ equipped with a semipositive metrized line bundle $\overline{L}$, we can define using arithmetic intersection theory a *height function*, denoted $\mathrm{h}_{\overline{L}}$, for subschemes of X_{Σ}, see [2,8,10]. It is the arithmetic analogue of the notion of degree of subvarieties.

One of the main results in [4] is that the height of a toric scheme with respect to a toric semipositive metrized line bundle can be expressed as the integral of the associated roof function [4, Theorem 5.2.5]. In precise terms,

$$\mathrm{h}_{\overline{L}_{\Psi}}(X_{\Sigma}) = (n+1)! \int_{\Delta_{\Psi}} \vartheta_{\overline{L}} \, \mathrm{d\,vol}_M, \tag{1}$$

where vol_M is the Haar measure on $M_{\mathbb{R}}$ normalized so that the lattice M has covolume 1.

3   Metrics from Polytopes and Entropy

In some cases, the height of a toric variety with respect to a toric semipositive metrized line bundle has an interpretation in terms of the average entropy of a family of random processes.

Let $\Delta \subset \mathbb{R}^n$ be a lattice polytope of dimension n and Γ an arbitrary polytope containing it. For a point x in the interior of Δ, we denote by Π_x the partition of Γ consisting of the cones $\eta_{x,F}$ of vertex x and base the relative interior of each proper face F of Γ. We consider Γ as a probability space endowed with the uniform probability distribution. Let β_x be the random variable that maps a point $y \in \Gamma$ to the base F of the unique cone $\eta_{x,F}$ that contains y. Clearly, the probability that a given face F is returned is the ratio of the volume of the cone based on F to the volume of Γ. We have

$$\mathrm{vol}_n(\eta_{x,F}) = n^{-1}\mathrm{dist}(x, F)\mathrm{vol}_{n-1}(F),$$

where vol_n and vol_{n-1} respectively denote the Euclidean n-th and $(n-1)$-th Euclidean volume of convex subsets of $\mathbb{R}^n$, and $\text{dist}(x, F)$ denotes the distance of the point x to the face F. Hence,

$$P(\beta_x = F) = \begin{cases} \dfrac{\text{dist}(x, F)\text{vol}_{n-1}(F)}{n\text{vol}_n(\Gamma)} & \text{if } \dim(F) = n - 1, \\ 0 & \text{if } \dim(F) \leq n - 2. \end{cases} \qquad (2)$$

The *entropy* of the random variable β_x is

$$\mathcal{E}(x) = -\sum_F P(\beta_x = F) \log(P(\beta_x = F)),$$

where the sum is over the facets F of Γ.

From the polytope Γ, we can construct a concave function on itself as follows. For each facet F of Γ, we denote by $u'_F \in \mathbb{R}^n$ the inner normal vector to F of Euclidean norm $(n-1)!\text{vol}_{n-1}(F)$. Set $\lambda(F) = \inf_{x \in \Gamma}\langle x, u'_F\rangle$ and consider the affine polynomial defined as

$$\ell_F(x) = \langle x, u'_F\rangle - \lambda(F). \qquad (3)$$

Hence,

$$\Gamma = \{x \in M_{\mathbb{R}} \mid \ell_F(x) \geq 0 \text{ for every facet } F\}.$$

In particular, ℓ_F is nonnegative on Γ, and we can consider the function $\vartheta_\Gamma \colon \Gamma \to \mathbb{R}$ defined by

$$\vartheta_\Gamma(x) = -\frac{1}{2}\sum_F \ell_F(x) \log(\ell_F(x)).$$

By [4, Lemma 6.2.1], this function is concave.

Notation 1. Let Σ_Δ and Ψ_Δ be the fan and the support function on $N_{\mathbb{R}} = \mathbb{R}^n$ induced by the polytope Δ, see [4, Example 2.5.13]. Let X_{Σ_Δ} and L_{Ψ_Δ} be the corresponding toric scheme and line bundle. The restriction of the function ϑ_Γ above to Δ is a continuous concave function and so, by [4, Theorem 4.8.1], it corresponds to a semipositive toric metric on L_{Ψ_Δ}. We denote this metric by $\|\cdot\|_{\Delta,\Gamma}$ and we write $\overline{L}_{\Psi_\Delta}$ for the line bundle L_{Ψ_Δ} equipped with this toric metric.

Example 1. Let $\Delta^n = \{(x_1, \ldots, x_n) \mid x_i \geq 0, \sum_i x_i \leq 1\}$ be the standard simplex of $\mathbb{R}^n$ and consider the case when $\Gamma = \Delta = \Delta^n$. The corresponding concave function on Δ^n is given by

$$\vartheta_{\Delta^n}(x_1, \ldots, x_n) = -\frac{1}{2}\left(1 - \sum_i x_i\right)\log\left(1 - \sum_i x_i\right) - \frac{1}{2}\sum_{i=1}^n x_i \log(x_i).$$

From [4, Example 2.4.3 and 4.3.9(1)], we deduce that the corresponding toric metric is the Fubini-Study metric on $\mathcal{O}(1)$, the universal line bundle on the projective space $\mathbb{P}^n$.

Remark 1. This kind of metrics are interesting for the Kähler geometry of toric varieties. Given a Delzant polytope $\Delta \subset \mathbb{R}^n$, Guillemin has constructed a "canonical" Kähler structure on the associated symplectic toric variety, see [7] for details. Following Guillemin, this canonical Kähler structure is codified by a convex function on the polytope, dubbed the "symplectic potential".

With the notation above, when $\Gamma = \Delta$ and u'_F is a primitive vector in N for every facet F, the function $-\vartheta_\Gamma$ coincides with this symplectic potential, see [7, Appendix 2, (3.9)]. In this case, the metric $\| \cdot \|_{\Delta,\Gamma}$ on the line bundle L_{Ψ_Δ} is smooth and positive, and its Chern form gives this canonical Kähler form.

The following result shows that the average entropy of the random variables β_x, $x \in \Delta$, with respect to the uniform distribution on Δ can be expressed in terms of the height of the toric variety X_{Σ_Δ} with respect to $\overline{L}$.

Theorem 1. *With the above notation,*

$$\frac{1}{\mathrm{vol}_n(\Delta)} \int_\Delta \mathcal{E} \ d\mathrm{vol}_n = \frac{1}{n!\mathrm{vol}_n(\Gamma)} \left(\frac{2\,\mathrm{h}_{\overline{L}}(X_{\Sigma_\Delta})}{(n+1)\deg_L(X_{\Sigma_\Delta})} - \lambda(\Gamma)\log(n!\mathrm{vol}_n(\Gamma)) \right)$$

with $\lambda(\Gamma) = \sum_F \lambda(F)$, the sum being over the facets F of Γ. In particular, if $\Gamma = \Delta$,

$$\frac{1}{\mathrm{vol}_n(\Delta)} \int_\Delta \mathcal{E} \ d\mathrm{vol}_n = \frac{2\,\mathrm{h}_{\overline{L}}(X_{\Sigma_\Delta})}{(n+1)\deg_L(X_{\Sigma_\Delta})^2} - \lambda(\Gamma)\frac{\log(\deg_L(X_{\Sigma_\Delta}))}{\deg_L(X_{\Sigma_\Delta})}.$$

Proof. By [9, Lemma 5.1.1], the vectors u'_F satisfy the Minkowski condition $\sum_F u'_F = 0$. Hence

$$\sum_F \ell_F = -\sum_F \lambda(F) = -\lambda(\Gamma).$$

Let x be a point in the interior of Δ and F a facet of Γ. We deduce from (2) that $P(\beta_x = F) = \ell_F(x)/(n!\mathrm{vol}_n(\Gamma))$. Hence,

$$\mathcal{E}(x) = -\sum_F \frac{\ell_F(x)}{n!\mathrm{vol}_n(\Gamma)} \log\left(\frac{\ell_F(x)}{n!\mathrm{vol}_n(\Gamma)} \right)$$

$$= \frac{1}{n!\mathrm{vol}_n(\Gamma)} \left(-\sum_F \ell_F(x)\log(\ell_F(x)) - \lambda(\Gamma)\log(n!\mathrm{vol}_n(\Gamma)) \right)$$

$$= \frac{1}{n!\mathrm{vol}_n(\Gamma)} \left(2\vartheta_\Gamma(x) - \lambda(\Gamma)\log(n!\mathrm{vol}_n(\Gamma)) \right).$$

The result then follows from the expression for the height of X_{Σ_Δ} in (1) and the analogous expression for its degree in [6, page 111, Corollary]. ∎

Example 2. The Fubini-Study metric of $\mathcal{O}(1)$ corresponds to the case when Γ and Δ are the standard simplex Δ^n. In that case, the average entropy of the random variables β_x, $x \in \Delta$, is

$$\frac{1}{n!} \int_{\Delta^n} \mathcal{E} \ d\mathrm{vol}_n = \frac{2\,\mathrm{h}_{\overline{\mathcal{O}(1)}}(\mathbb{P}^n)}{(n+1)}.$$

4   Integration on Polytopes

In this section, we present a closed formula for the integral over a polytope of a function of one variable composed with a linear form, extending in this direction Brion's formula for the case of a simplex [3], see Proposition 1 and Corollary 2 below. This formula allow us to compute the height of toric varieties with respect to the metrics arising from polytopes as in Sect. 3.

We consider the vector space $\mathbb{R}^n$ with its usual scalar product, that we denote $\langle \cdot, \cdot \rangle$, and its Lebesgue measure, that we denote vol_n. We also consider a polytope $\Delta \subset \mathbb{R}^n$ of dimension n.

Definition 1. Let $u \in \mathbb{R}^n$ and $\lambda \in \mathbb{R}$, the *aggregate of* Δ *in the affine subset*

$$L_{u,\lambda} := \{ x \in \mathbb{R}^n \mid \langle x, u \rangle = \lambda \}$$

is the union of all the faces of Δ contained in $L_{u,\lambda}$. An *aggregate* V *of* Δ *in the direction* u is an aggregate in $L_{u,\lambda}$ for some $\lambda \in \mathbb{R}$.

We denote by $\dim(V)$ the maximal dimension of a face of Δ contained in V. In particular, $\dim(\emptyset) = -1$.

We write $\Delta(u)$ for the set of non-empty aggregates of Δ in the direction u. In particular, $\Delta(0) = \{\Delta\}$. Note that, if $V \in \Delta(u)$ and x is a point in the affine space spanned by V, then the value $\langle x, u \rangle$ is independent of x. We denote this common value by $\langle V, u \rangle$.

For any two aggregates $V_1, V_2 \in \Delta(u)$, we have $V_1 = V_2$ if and only if $\langle V_1, u \rangle = \langle V_2, u \rangle$.

Example 3

(1) Every facet of a polytope is an aggregate in the direction orthogonal to the facet.
(2) If u is general enough, the set $\Delta(u)$ agrees with the set of vertices of Δ.
(3) Let $\Delta = \{(x, y) \in \mathbb{R}^2 \mid 0 \le x, y \le 1\}$ be the unit square and $u = (1, 1)$. Then the set of aggregates $\Delta(u)$ contains three elements: $\{(0,0)\}$, $\{(1,0),(0,1)\}$ and $\{(1,1)\}$.

In each facet F of Δ we choose a point m_F. Let L_F be the linear hyperplane defined by F and π_F the orthogonal projection of $\mathbb{R}^n$ onto L_F. Then, $F - m_F$ is a polytope in L_F of full dimension $n - 1$. To ease the notation, we identify $F - m_F$ with F. Observe that, with this identification, for $V \in \Delta(u)$, the intersection $V \cap F$ is an aggregate of F in the direction $\pi_F(u)$. We also denote by u_F the inner normal vector to F of norm 1.

Definition 2 Let $u \in \mathbb{R}^n$ be a vector. For each aggregate V in the direction of u, we define the coefficients $C_k(\Delta, u, V)$, $k \in \mathbb{N}$, recursively. If $u = 0$, then V is either $\emptyset$ or Δ. For both cases, we set

$$C_k(\Delta, 0, V) = \begin{cases} \mathrm{vol}_n(V) & \text{if } k = n, \\ 0 & \text{otherwise.} \end{cases}$$

If $u \neq 0$, we set

$$C_k(\Delta, u, V) = - \sum_F \frac{\langle u_F, u \rangle}{\|u\|^2} C_k(F, \pi_F(u), V \cap F),$$

where the sum is over the facets F of Δ. This recursive formula implies that $C_k(\Delta, u, V) = 0$ for all $k > \dim(V)$.

Let $C^n(\mathbb{R})$ be the space of functions of one real variable which are n-times continuously differentiable. For $f \in C^n(\mathbb{R})$ and $0 \leq k \leq n$, we write $f^{(k)}$ for the k-th derivative of f. We want to give a formula that, for $f \in C^n(\mathbb{R})$, computes

$$\int_\Delta f^{(n)}(\langle x, u \rangle) \, \mathrm{dvol}_n(x)$$

in terms of the values of the function $x \mapsto f(\langle x, u \rangle)$ at the vertices of Δ. However, when u is orthogonal to some faces of Δ of positive dimension, such a formula necessarily depends on the values of the derivatives of f.

Proposition 1 ([4, Proposition 6.1.4]) *Let $\Delta \subset \mathbb{R}^n$ be a polytope of dimension n and $u \in \mathbb{R}^n$. Then, for any $f \in C^n(\mathbb{R})$,*

$$\int_\Delta f^{(n)}(\langle x, u \rangle) \, \mathrm{dvol}_n(x) = \sum_{V \in \Delta(u)} \sum_{k \geq 0} C_k(\Delta, u, V) f^{(k)}(\langle V, u \rangle).$$

The coefficients $C_k(\Delta, u, V)$ are uniquely determined by this identity.

Corollary 1 *Let $\Delta \subset \mathbb{R}^n$ be a polytope of dimension n and $u \in \mathbb{R}^n$. Then,*

$$\sum_{V \in \Delta(u)} \sum_{k=0}^{\min\{i, \dim(V)\}} C_k(\Delta, u, V) \frac{\langle V, u \rangle^{i-k}}{(i-k)!} = \begin{cases} 0 & \text{for } i = 0, \ldots, n-1, \\ \mathrm{vol}_n(\Delta) & \text{for } i = n. \end{cases}$$

Proof This follows from proposition 1 applied to the functions $f(z) = z^i/i!$. □

The following result gives the basic properties of the coefficients associated to the aggregates of a polytope.

Proposition 2 ([4, Proposition 6.1.6]) *Let $\Delta \subset \mathbb{R}^n$ be a polytope of dimension n and $u \in \mathbb{R}^n$. Let $V \in \Delta(u)$ and $k \geq 0$.*

(1) *The coefficient $C_k(\Delta, u, V)$ is homogeneous of weight $k - n$ in the sense that, for $\lambda \in \mathbb{R}^\times$,*

$$C_k(\Delta, \lambda u, V) = \lambda^{k-n} C_k(\Delta, u, V).$$

(2) *The coefficients $C_k(\Delta, u, V)$ satisfy the vector relation*

$$C_k(\Delta, u, V) \cdot u = - \sum_F C_k(F, \pi_F(u), V \cap F) \cdot u_F,$$

where the sum is over the facets F of Δ.

(3) Let $\Delta_1, \Delta_2 \subset \mathbb{R}^n$ be two polytopes of dimension n intersecting along a common facet and such that $\Delta = \Delta_1 \cup \Delta_2$. Then $V \cap \Delta_i = \emptyset$ or $V \cap \Delta_i \in \Delta_i(u)$ and

$$C_k(\Delta, u, V) = C_k(\Delta_1, u, V \cap \Delta_1) + C_k(\Delta_2, u, V \cap \Delta_2).$$

In case Δ is a simplex, the linear system given by Corollary 1 has as many unknowns as equations. In this case, the coefficients corresponding to an aggregate in a given direction are determined by this linear system. The following result gives a closed formula for those coefficients.

Proposition 3 ([4, Proposition 6.1.7]) Let $\Delta \subset \mathbb{R}^n$ be a simplex and $u \in \mathbb{R}^n$. Write $d_W = \dim(W)$ for $W \in \Delta(u)$. Then, for $V \in \Delta(u)$ and $0 \le k \le \dim(V)$,

$$C_k(\Delta, u, V) = (-1)^{d_V - k} \frac{n!}{k!} \mathrm{vol}_n(\Delta) \sum_{\substack{\eta \in \mathbb{N}^{\Delta(u) \setminus \{V\}} \\ |\eta| = d_V - k}} \prod_{W \in \Delta(u) \setminus \{V\}} \frac{\binom{d_W + \eta_W}{d_W}}{\langle V - W, u \rangle^{d_W + \eta_W + 1}}.$$

Remark 2 We can rewrite the formula in Proposition 3 in terms of vertices instead of aggregates as follows:

$$C_k(\Delta, u, V) = (-1)^{d_V - k} \frac{n!}{k!} \mathrm{vol}_n(\Delta) \sum_{|\beta| = d_V - k} \prod_{\nu \notin V} \langle V - \nu, u \rangle^{-\beta_\nu - 1}, \tag{4}$$

where the product is over the vertices ν of Δ not lying in V and the sum is over the tuples β of non negative integers of length $d_V - k$, indexed by those same vertices of Δ that are not in V, that is, $\beta \in \mathbb{N}^{n - d_V}$ and $|\beta| = d_V - k$.

Example 4 Let $\Delta \subset \mathbb{R}^n$ be a simplex and $u \in \mathbb{R}^n$. If a vertex ν_0 of Δ is an aggregate in the direction of u, then formula (4) reduces to

$$C_0(\Delta, u, \nu_0) = n! \mathrm{vol}_n(\Delta) \prod_{\nu \ne \nu_0} \langle \nu_0 - \nu, u \rangle^{-1}, \tag{5}$$

where the product runs over all vertices of Δ different from ν_0. Suppose that the simplex is presented as the intersection of $n + 1$ halfspaces as

$$\Delta = \bigcap_{i=0}^{n} \{x \in \mathbb{R}^n | \langle x, u_i \rangle - \lambda_i \ge 0\}$$

with $u_i \in \mathbb{R}^n \setminus \{0\}$ and $\lambda_i \in \mathbb{R}$. Up to a reordering, we can assume that u_0 is an inner normal vector to the unique face of Δ not containing ν_0. We denote by ε the sign of $(-1)^n \det(u_1, \dots, u_n)$. Then the above coefficient can be alternatively written as

$$C_0(\Delta, u, \nu_0) = \frac{\varepsilon \det(u_1, \dots, u_n)^{n-1}}{\prod_{i=1}^{n} \det(u_1, \dots, u_{i-1}, u, u_{i+1}, \dots, u_n)}.$$

From the equation (5), we obtain the following extension of Brion's "short formula" for the case of a simplex [3, Théorème 3.2], see also [1].

Corollary 2 *Let $\Delta \subset \mathbb{R}^n$ be a simplex of dimension n that is the convex hull of points ν_i, $i = 0, \ldots, n$, and let $u \in \mathbb{R}^n$ such that $\langle \nu_i, u \rangle \neq \langle \nu_j, u \rangle$ for $i \neq j$. Then, for any $f \in \mathcal{C}^n(\mathbb{R})$,*

$$\int_\Delta f^{(n)}(\langle x, u \rangle) \ \mathrm{dvol}_n(x) = n! \mathrm{vol}_n(\Delta) \sum_{i=0}^n \frac{f(\langle \nu_i, u \rangle)}{\prod_{j \neq i} \langle \nu_i - \nu_j, u \rangle}.$$

Proof This follows from Proposition 1 and formula (5).   □

The following result gives the value of the integral over a simplex of a function of the form $\ell(x) \log(\ell(x))$, where ℓ is an affine function.

Proposition 4 *Let $\Delta \subset \mathbb{R}^n$ be a simplex of dimension n and $\ell \colon \mathbb{R}^n \to \mathbb{R}$ an affine function which is non-negative on Δ. Write $\ell(x) = \langle x, u \rangle - \lambda$ for some vector u and constant λ. Then*

$$\frac{1}{\mathrm{vol}_n(\Delta)} \int_\Delta \ell(x) \log(\ell(x)) \ \mathrm{dvol}_n(x) = \sum_{V \in \Delta(u)} \sum_{\beta'} \binom{n}{n - |\beta'|} \frac{\ell(V) \left(\log(\ell(V)) - \sum_{j=2}^{|\beta'|+1} \frac{1}{j} \right)}{(|\beta'| + 1) \prod_{\nu \notin V} \left(-(\frac{\ell(\nu)}{\ell(V)} - 1)^{\beta'_\nu} \right)},$$

where the second sum runs over $\beta' \in (\mathbb{N}^\times)^{n-\dim(V)}$ with $|\beta'| \leq n$ and the product is over the $n - \dim(V)$ vertices ν of Δ not in V.

If $\ell(x)$ is the defining equation of a hyperplane containing a facet F of Δ, then

$$\frac{1}{\mathrm{vol}_n(\Delta)} \int_\Delta \ell(x) \log(\ell(x)) \ \mathrm{d}x = \frac{\ell(\nu_F)}{n+1} \left(\log(\ell(\nu_F)) - \sum_{j=2}^{n+1} \frac{1}{j} \right),$$

where ν_F denotes the unique vertex of Δ not contained in F.

Proof This follows from the formula in proposition 1 and (4) with the function $f^{(n)}(z) = (z - \lambda) \log(z - \lambda)$, a $(n - k)$-th primitive of which is

$$f^{(k)}(z) = \frac{(z - \lambda)^{n-k+1}}{(n - k + 1)!} \left(\log(z - \lambda) - \sum_{j=2}^{n-k+1} \frac{1}{j} \right).$$

We obtain the following formula for the height of a toric variety with respect to the toric metrics considered in §3, in terms of the coefficients $C_k(\Delta, u_i, V)$.

Theorem 2 *Let $\Delta \subset \mathbb{R}^n$ be a lattice polytope of dimension n and Γ an arbitrary polytope containing it. Let X_{Σ_Δ} and $\overline{L} = \overline{L}_{\Psi_\Delta}$ be as in Notation 1, and ℓ_F and u'_F the affine polynomial and the inner normal vector associated to a facet F of Γ as in (3) . Then*

$$\mathrm{h}_{\overline{L}}(X_{\Sigma_\Delta}) = \frac{(n + 1)!}{2} \sum_{F} \sum_{V \in \Delta(u'_F)} \sum_{k=0}^{\dim(V)} C_k(\Delta, u'_F, V) \frac{\ell_F(V)^{n-k+1}}{(n - k + 1)!} \left(\sum_{j=2}^{n-k+1} \frac{1}{j} - \log(\ell_F(V)) \right),$$

the first sum being over the facets F of Γ. Suppose furthermore that $\Delta \subset \mathbb{R}^n$ is a simplex. Then

$$h_{\overline{L}}(X_{\Sigma_\Delta}) = \frac{n!}{2}\mathrm{vol}_M(\Delta)\sum_F \ell_F(\nu_F)\left(\sum_{j=2}^{n+1}\frac{1}{j} - \log(\ell_F(\nu_F))\right), \qquad (6)$$

where ν_F is the unique vertex of Δ not contained in the facet F.

Proof The first statement follows readily from the formula (1) and Proposition 1 applied to the functions

$$f_i(z) = \left(\log(z - \lambda_i) - \sum_{j=2}^{n+1}\frac{1}{j}\right)(z - \lambda_i)^{n+1}/(n+1)!.$$

The second statement follows similarly from Proposition 4.   □

Example 5 Let $\mathcal{O}(1)$ be the universal line bundle of $\mathbb{P}^n$. As we have seen in Example 1, the Fubini-Study metric of $\mathcal{O}(1)$ corresponds to the case of the standard simplex Δ^n. Hence we recover from (6) the well known expression for the height of $\mathbb{P}^n$ with respect to the Fubini-Study metric in [2, Lemma 3.3.1]:

$$h_{\overline{\mathcal{O}(1)}}(\mathbb{P}^n) = \frac{n+1}{2}\sum_{j=2}^{n+1}\frac{1}{j} = \sum_{h=1}^{n}\sum_{j=1}^{h}\frac{1}{2j}.$$

Hence, in this case the average entropy of the random variables β_x, $x \in \Delta^n$, is

$$\frac{1}{n!}\int_{\Delta^n}\mathcal{E}\,\mathrm{dvol}_n = \frac{2\,h_{\overline{\mathcal{O}(1)}}(\mathbb{P}^n)}{(n+1)} = \sum_{j=2}^{n+1}\frac{1}{j}.$$

References

1. Baldoni, V., Berline, N., De Loera, J., Köppe, M., Vergne, M.: How to integrate a polynomial over a simplex. Math. Comput. **80**, 297–325 (2011)
2. Bost, J.-B., Gillet, H., Soulé, C.: Heights of projective varieties and positive Green forms. J. Amer. Math. Soc. **7**, 903–1027 (1994)
3. Brion, M.: Points entiers dans les polyèdres convexes. Ann. Sci. École Norm. Sup. **21**(4), 653–663 (1988)
4. Burgos-Gil, J.I., Philippon, P., Sombra, M.: Arithmetic Geometry of Toric Varieties: Metrics, Measures and Heights. Astérisque 360, Soc. Math., France (2014)
5. Cox, D.A., Little, J.D., Schenck, H.K.: Toric varieties, Grad. Stud. Math. 124, Amer. Math. Soc. (2011)
6. Fulton, W.: Introduction to Toric Varieties. Ann. of Math. Stud., vol. 131. Princeton University Press, Princeton (1993)
7. Guillemin, V.: Moment maps and combinatorial invariants of Hamiltonian $\mathbb{T}^n$-spaces. Progr. Math., vol. 122. Birkhäuser, Boston (1995)
8. Maillot, V.: Géométrie d'Arakelov des variétés toriques et fibrés en droites intégrables. Mém. Soc. Math. France **80** (2000)
9. Schneider, R.: Convex Bodies: The Brunn-Minkowski Theory. Cambridge University Press, Cambridge (1993)
10. Zhang, S.-W.: Small points and adelic metrics. J. Algebraic Geom. **4**, 281–300 (1995)

Characterization and Estimation of the Variations of a Random Convex Set by Its Mean n-Variogram: Application to the Boolean Model

Saïd Rahmani$^{(\boxtimes)}$, Jean-Charles Pinoli, and Johan Debayle

École Nationale Supérieure des Mines de Saint Etienne,
SPIN/LGF UMR CNRS 5307, Saint-Etienne, France
{said.rahmani,pinoli,debayle}@emse.fr

Abstract. In this paper we propose a method to characterize and estimate the variations of a random convex set Ξ_0 in terms of shape, size and direction. The mean n-variogram $\gamma_{\Xi_0}^{(n)} : (u_1 \cdots u_n) \mapsto \mathbb{E}[\nu_d(\Xi_0 \cap (\Xi_0 - u_1) \cdots \cap (\Xi_0 - u_n))]$ of a random convex set Ξ_0 on $\mathbb{R}^d$ reveals information on the nth order structure of Ξ_0. Especially we will show that considering the mean n-variograms of the dilated random sets $\Xi_0 \oplus rK$ by an homothetic convex family $rK_{r>0}$, it's possible to estimate some characteristic of the nth order structure of Ξ_0. If we make a judicious choice of K, it provides relevant measures of Ξ_0. Fortunately the germ-grain model is stable by convex dilatations, furthermore the mean n-variogram of the primary grain is estimable in several type of stationary germ-grain models by the so called n-points probability function. Here we will only focus on the Boolean model, in the planar case we will show how to estimate the nth order structure of the random vector composed by the mixed volumes $^t(A(\Xi_0), W(\Xi_0, K))$ of the primary grain, and we will describe a procedure to do it from a realization of the Boolean model in a bounded window. We will prove that this knowledge for all convex body K is sufficient to fully characterize the so called difference body of the grain $\Xi_0 \oplus \check{\Xi}_0$. we will be discussing the choice of the element K, by choosing a ball, the mixed volumes coincide with the Minkowski's functional of Ξ_0 therefore we obtain the moments of the random vector composed of the area and perimeter $^t(A(\Xi_0), U(\Xi))$. By choosing a segment oriented by θ we obtain estimates for the moments of the random vector composed by the area and the Ferret's diameter in the direction θ, $^t((A(\Xi_0), H_{\Xi_0}(\theta))$. Finally, we will evaluate the performance of the method on a Boolean model with rectangular grain for the estimation of the second order moments of the random vectors $^t(A(\Xi_0), U(\Xi_0))$ and $^t((A(\Xi_0), H_{\Xi_0}(\theta))$.

Keywords: Boolean model · Geometric covariogram · Mixed volumes · Particle size distribution · n points set probability · Random set · Shape variations

© Springer International Publishing Switzerland 2015
F. Nielsen and F. Barbaresco (Eds.): GSI 2015, LNCS 9389, pp. 296–308, 2015.
DOI: 10.1007/978-3-319-25040-3_33

1   Introduction

A random closed set (RACS) denotes a random variable defined on a probability space $(\Omega, \mathfrak{A}, P)$ taking values in $(\mathbb{F}, \mathfrak{F})$ the family of all closed subset of $\mathbb{R}^d$ provided with the σ-algebra $\mathfrak{F} = \sigma\{\{F \in \mathbb{F} \,|\, F \cap X \neq \emptyset\} \, X \in \mathfrak{K}\}$ where $\mathfrak{K}$ denotes the class of compact subsets on $\mathbb{R}^d$. As when we work with random vectors it is necessary to give meaning to the concept of distribution. Choquet and Matheron have shown that a random set is fully characterized by its probability of presence in each place of the space, thus the concept of distribution is replaced by the so called *functional capacity* also called *Choquet capacity* $T_\Xi : \mathfrak{K} \to [0,1]$.

$$T_\Xi(X) = P(\{\Xi \cap X \neq \emptyset\}) \tag{1}$$

Several materials can be modeled by random sets. In fact, the heterogeneity of the materials can be apprehended by a probabilistic approach [1,2]. Especially granular or fibrous media [3,4] can be represented by unions of overlapping particles (the grains) centred on random positions (the germs), thus giving rise to the germ-grain model.

$$\Xi = \bigcup_{x_i \in \Phi} \Xi_i + x_i \tag{2}$$

Where Φ is point process (REF) which generates the germs x_i, and the grains Ξ_i are convex random sets independent and identically distributed. Notice that this definition assumes the independence between the particles Ξ_i and their positions x_i, there is a more general definition authorizing the correlation between germs and grains [5], for more convenience we choose to introduce the model under this hypothesis. There are two types of use of this model. The first one is the simulation of a material, the global characteristics of the model match to the material's ones but the germs and grains have no physical sense: the local characteristics of the model (Φ and Ξ_0) a priori have no connection with the intrinsic structure of the material. Our approach consist in representing the people of crystals by such a model; that is to say that the point process Φ is the repartition of the particles and the convex random sets Ξ_i the particles themselves. The goal is to adjust the model to actual data from measurements acquired by an image acquisition system. Generally, the acquisition is obtained by optical imaging, in other words, we have realizations of $\Xi \cap W$ where W is a bounded window and we want to estimate the characteristics of Φ and Ξ_0. To meet this objective, we focuses on two points: first, estimate the characteristics of Ξ from a realization of $\Xi \cap W$ [6], secondly establish relationships between characteristics of Ξ and the local characteristics of the model (Φ, Ξ_0). This second point raises the problem of non-uniqueness of the representation (2). To remedy this, we introduce an additional assumption: we assume the process Φ comes from a known type (Poisson process, Cox process, ...). We will use the homogeneous Boolean model, a germ-grain model in which Φ is a homogeneous Poisson point process. This model is widely used because we have an analytical formula for the Choquet capacity.

$$T_\Xi(X) = 1 - \exp(-\lambda \mathbb{E}[\nu_d(\Xi_0 \oplus \check{X})]), \forall X \in \mathfrak{K} \tag{3}$$

Several methods are used to connect the global characteristics of the model to the characteristics of the primary grain. In the plane and the space, Miles's formulae [7] or minimum contrast method [8] can estimate the average value of Minkowski's functional of the primary grain. Generally, the primary grain is assumed to have a known and deterministic shape, that is to say, the realizations of the primary grain are homothetic. So as to estimate the variations of the scaling factor from the expectation of the Minkowski functionals. For example, for a disc in the plane, the moments of the first and second orders of the radius of the primary grain are respectively proportional to its average perimeter and its average area. However, if we consider that the shape of the grain can vary, several issues remain unresolved: firstly Minkowski functionals of a random convex set are not enough to characterize its shape and also their average will not provide a sufficient information to characterize its variations. For instance, for a Boolean model whose grain has a shape that depends on several parameters (rectangle, ellipse ...), the estimation of geometric variations of the grain is not direct. The aim of our work is to characterize and estimate the variations of the primary grain of the Boolean model without any assumption concerning its shape, from a realization of the model in a bounded window. The ideal would be to estimate the functional capacity of the primary grain T_{Ξ_0} or an equivalent which completely characterizes the random convex Ξ_0. Applying the Steiner's formula and the linearity of the expectation, for all convex compact set X we have:

$$\mathbb{E}[\nu_d(\Xi_0 \oplus \check{X})] = \sum_{k=0}^{d} \binom{d}{k} \mathbb{E}[W_{d-k}(\Xi_0, \check{X})] \qquad (4)$$

Where $W_{d-k}(\Xi_0, \check{X})$ denotes the $(d-k)^{th}$ mixed volume of Ξ_0 and $\check{X}$. Considering the relationship (3), we understand that the functional capacity of the model evaluated on a convex K depends on the grain only by the expectations of mixed volumes between Ξ_0 and $\check{X}$. In order to reveal the variations of Ξ_0 we need to consider the functional capacity of the model on compacts that are not convex, that is why we are interested in the n-point-probability function [5]: $\mathbb{P}(\{x_1, \cdots x_n\} \subset \Xi)$. For $n = 2$, this quantity is known under the name of covariance and it can be connected to the mean covariogram of the primary grain [9]. For any $n \geq 3$, the n-point-probability function can be used to estimate the mean n-variogram $\gamma_{\Xi_0}^{(n)} : (u_1 \cdots u_n) \mapsto \mathbb{E}[\nu_d(\Xi_0 \cap (\Xi_0 - u_1) \cdots \cap (\Xi_0 - u_n))]$ of Ξ_0, this quantity evaluated on the dilated grain $\Xi_0 \oplus K$ for a convex compact set K reveals the nth order structure of Ξ_0, especially some linear combinations of the expectations $\mathbb{E}[\prod_{k=0}^{d} W_{d,k}(\Xi_0, K)^{p_k}]$ of order $p = \sum_{k=0}^{d} p_k \leq n$ where $W_{d,k}(\Xi, K)$, $k = 0, \cdots d$, denotes the mixed volumes of Ξ_0 by K. First we will discuss the properties of the mean n-variogram of a random convex and how it describes its nth order structure. Secondly we will focus on the planar case, we will discuss the interpretation of the nth order moments of the vector $^t(A(\Xi_0), W(\Xi, K))$, and we will show how they can be estimated for the primary grain of a boolean model. Finally we will test the estimation method by the simulations of a boolean model with rectangular grains.

2 From the Mean n-Variogram of a Random Convex Set to its Variations

In this section we will discuss the properties of the the mean n-variogram of a random convex and how it describes his n^{th} order structure.

2.1 Mean n-Variogram of Random Convex Set

In this paragraph we will define the mean n-variogram of a random convex and discuss its properties. The mean n-variogram is a simple generalization of the concept of mean covariogram introduced by Bruno Galerne in [9] for $n = 2$. The proof of the following results can be easy found by recursion on n from Bruno Galerne's proof [9], that is why we will omit them.

Definition 1. *Let Ξ_0 be a random convex set satisfying $\mathbb{E}[\nu_d(\Xi_0)] < \infty$ and $n \geq 1$ we will call **mean n-variogram** of Ξ_0 the expectation of its n-variogram:*

$$\gamma_{\Xi_0}^{(n)} : \left|\begin{array}{l} \mathbb{R}^{d \times (n-1)} \longrightarrow \mathbb{R}_+ \\ (u_1, \cdots u_{n-1}) \longmapsto \mathbb{E}[\nu_d(\bigcap_{i=1}^{n-1}(\Xi_0 - u_i) \cap \Xi_0)] \end{array}\right.$$

Proposition 1. *Let Ξ_0 be a random convex set satisfying $\mathbb{E}[A(\Xi_0)^n] < \infty$ and $n \geq 3$ (the cases $n < 3$ were treated in [9]). Then the mean n-variogram have the following properties:*

(i) *permutation invariant:*
$$\forall \sigma \in S_n, \ \gamma_{\Xi_0}^{(n)}(u_1, \cdots u_{n-1}) = \gamma_{\Xi_0}^{(n)}(u_{\sigma(1)}, \cdots u_{\sigma(n-1)})$$

(ii) *Reducibility:*
$$\exists i \neq j, u_i = u_j \Rightarrow \gamma_{\Xi_0}^{(n)}(u_1, \cdots u_{n-1}) = \gamma_{\Xi_0}^{(n-1)}(u_1, \cdots u_{i-1}, u_{i+1}, \cdots u_{n-1})$$
$$\text{and } \exists i, u_i = 0 \Rightarrow \gamma_{\Xi_0}^{(n)}(u_1, \cdots u_{n-1}) = \gamma_{\Xi_0}^{(n-1)}(u_1, \cdots u_{i-1}, u_{i+1}, \cdots u_{n-1})$$

(iii) $\forall (u_1, \cdots u_{n-1}) \in \mathbb{R}^{d \times (n-1)}, \ 0 \leq \gamma_{\Xi_0}^{(n-1)}(u_1, \cdots u_{n-1}) \leq \gamma_{\Xi_0}^{(n-1)}(u_2, \cdots u_{n-1})$

(iv) $\forall k \leq n - 1, \ \int_{\mathbb{R}^d} \cdots \int_{\mathbb{R}^d} \gamma_{\Xi_0}^{(n)}(u_1, \cdots u_{n-1}) du_1 \cdots du_k =$
$\ldots \mathbb{E}[\nu_d(\Xi_0)^k] \gamma_{\Xi_0}^{(n-k)}(u_{k+1}, \cdots u_{n-1})$ *especially for $k = n$ we have:*

$$\int_{\mathbb{R}^d} \cdots \int_{\mathbb{R}^d} \gamma_{\Xi_0}^{(n)}(u_1, \cdots u_{n-1}) du_1 \cdots du_{n-1} = \mathbb{E}[\nu_d(\Xi_0)^n] \quad (5)$$

(v) $\forall (u_1, \cdots u_{n-1}) \in \mathbb{R}^{d \times (n-1)}, \ \gamma_{\Xi_0}^{(n)}(-u_1, \cdots - u_n) = \gamma_{-\Xi_0}^{(n)}(u_1, \cdots u_n)$
and $\gamma_{\Xi_0}^{(n)}(-u_1, u_2 \cdots u_{n-1}) = \gamma_{\Xi_0}^{(n)}(u_1, u_2 - u_1, \cdots u_{n-1} - u_1)$.

(vi) *The partial map $u \to \gamma_{\Xi_0}^{(n)}(u_1, \cdots u_{n-2}, u)$ is uniformly continuous and zero limit when $\|u\| \to +\infty$. Furthermore for all $u \in \mathbb{R}^d$ the map*
$r \to \gamma_{\Xi_0}^{(n)}(u_1, \cdots u_{n-2}, ru)$ *is decreasing on $\mathbb{R}$.*

(vii) $\gamma_{\Xi_0}^{(n)}$ *also has an integral formulation:*

$$\gamma_{\Xi_0}^{(n)}(u_1, \cdots u_{n-1}) = \int_{\mathbb{R}^d} \mathbb{P}(\{x, x + u_1, \cdots x + u_{n-1}\} \subset \Xi_0) dx \quad (6)$$

The relationship (5) has a the great advantage of giving access to the nth order moment of the volume of the random set Ξ_0. This relationship is even more important because dilating Ξ_0 by a convex K, it gives access to some linear combinations of the expectations $\mathbb{E}[\prod_{k=0}^{d} W_{d,k}(\Xi_0, K)^{p_k}]$ of order $p = \sum_{k=0}^{d} p_k \leq n$.

2.2 Dilatation of a Random Convex Set

Let Ξ_0 be a convex random set and K a convex compact set, let's recall the Steiner's formula:

$$\forall r \geq 0, \ \nu_d(\Xi_0 \oplus rK) = \sum_{k=0}^{d} \binom{d}{k} W_{d,k}(\Xi_0, K) r^k \tag{7}$$

Where $W_{d,k}(\Xi_0, K)$ denote the mixed volume of k-homogeneity in its first variable $(d-k)$-homogeneity in its second, that is:

$$\forall k = 0, \cdots d, \forall (\alpha, \beta) \in \mathbb{R}_+^2, \ W_{d,k}(\alpha \Xi_0, \beta K) = \alpha^{d-k} \beta^k W_{d,k}(\Xi_0, K) \tag{8}$$

Furthermore $W_{d,d}(\Xi, K) = \nu_d(K)$ and $W_{d,0}(\Xi, K) = \nu_d(\Xi)$. For $r \in \mathbb{R}_+$, $n \geq 0$ and Ξ_0 satisfiying $\mathbb{E}[\nu_d(\Xi_0)^n] < \infty$ we introduce the function:

$$\zeta_{\Xi_0, K}^{(n)} : \begin{vmatrix} \mathbb{R}_+ \longrightarrow \mathbb{R}_+ \\ r \longmapsto \mathbb{E}[\nu_d(\Xi_0 \oplus rK)^n] \end{vmatrix} \tag{9}$$

The existance of $\zeta_{\Xi_0, K}^{(n)}$ is ensured by the existence of $\mathbb{E}[\nu_d(\Xi_0)^n]$ and the convexity of Ξ_0. Notice that for $n \geq 2$ the functional $\zeta_{\Xi_0, K}^{(n)}$ can be connected to the mean n-variogram of $\Xi_0 \oplus rK$ by (5) we have:

$$\zeta_{\Xi_0, K}^{(n)}(r) = \int_{\mathbb{R}^d} \cdots \int_{\mathbb{R}^d} \gamma_{\Xi \oplus rK}^{(n)}(u_1, \cdots u_{n-1}) du_1 \cdots du_{n-1} \tag{10}$$

by injecting (7) in (9) we have:

$$\zeta_{\Xi_0, K}^{(n)}(r) = \mathbb{E}[(\sum_{k=0}^{d} \binom{d}{k} W_{d,k}(\Xi_0, K) r^k)^n] \tag{11}$$

$\zeta_{\Xi_0, K}^{(n)}$ is therefore a polynomial function in r of degree $n \times d$ it can be expressed as:

$$\zeta_{\Xi_0, K}^{(n)} = \sum_{j=0}^{nd} C_{n,j}^{(K)} r^j \tag{12}$$

using the multinomial theorem, each of these coefficients $C_{n,j}^{(K)}$ of degree $j \leq nd$, can be expressed as a linear combination of the interactions $\mathbb{E}[\prod_{k=0}^{d} W_{d,k}(\Xi_0, K)^{p_k}]$ satisfying $j = \sum_{k=0}^{d} k \times p_k$ and $\sum_{k=0}^{d} p_k = n$.

Of course we can extract all coefficients $C_{n,j}^{(K)}$ by searching a polynomial approximation of $\zeta_{\Xi_0,K}^{(n)}$ in a similar way to the minimum contrast method [10]. Unfortunately in the general case it is not sufficient to obtain the interactions $\mathbb{E}[\prod_{k=0}^{d} W_{d,k}(\Xi_0, K)^{p_k}]$ from the coefficients of $\zeta_{\Xi_0,K}^{(n)}$. However in the planar case, we can do this as follows.

Theorem 1. *Let Ξ_0 be a convex random set on the plane $\mathbb{R}^2$, we introduce the polynomial function $\eta_{\Xi_0,K}^{(n)}$ as follows:*

$$\eta_{\Xi_0,K}^{(n)}(r) = \sum_{j=0}^{n} \binom{n}{j}(-1)^j A(K)^j r^{2j} \zeta_{\Xi_0,K}^{(n-j)}(r) \tag{13}$$

Where $A(K) = \nu_2(K)$ denotes the area of K. Then $\eta_{\Xi_0,K}^{(n)}$ is a polynomial function of degree n and if we note $M_{n,k}^{(K)}$ its kth order coefficient for all $n \in \mathbb{N}$ and $k = 1, \cdots n$, we have:

$$M_{n,k}^{(K)} = \sum_{j=0}^{\lfloor \frac{k}{2} \rfloor}(-1)^j A(K)^j \binom{n}{j} C_{2n-2j,k-2j}^{(K)} \tag{14}$$

and

$$\mathbb{E}[A(\Xi)^{n-k}W(\Xi_0,K)^k] = \frac{M_{n,k}^{(K)}}{2^k \binom{n}{k}} \tag{15}$$

Where $W(\Xi_0,K) = W_{2,1}(K)$ denotes the mixed area and $\lfloor \frac{k}{2} \rfloor$ denotes the floor of $\frac{k}{2}$.

Proof. First according to Steiner's formula, $A(\Xi_0) + 2rW(\Xi_0,K) = A(\Xi_0 \oplus rK) - r^2 A(K)$
$\Rightarrow \mathbb{E}[(A(\Xi_0) + 2rW(\Xi_0,K))^n] = \mathbb{E}[(A(\Xi_0 \oplus K) - r^2 A(K))^n]$ by applying the binomial theoerm on each side of the equality and according to linearity of the expectation we have:

$$\eta_{\Xi_0,K}^{(n)}(r) = \sum_{k=0}^{n} 2^k \binom{n}{k} r^k \mathbb{E}[A(\Xi)^{n-k}W(\Xi_0,K)^k] \tag{16}$$

it follows the relationship (15). Injecting (12) in (13) we have:

$$\eta_{\Xi_0,K}^{(n)}(r) = \sum_{j=0}^{n} \sum_{p=0}^{2(n-j)} \binom{n}{j}(-1)^j A(K)^j r^{2j+p} C_{2n-2j,p}^{(K)}$$

applying the change of variable $z = 2j + p$ we have:

$$\eta_{\Xi_0,K}^{(n)}(r) = \sum_{j=0}^{n} \sum_{z=2j}^{2n} \binom{n}{j}(-1)^j A(K)^j r^z C_{2n-2j,z-2j}^{(K)}$$

$$= \sum_{z=0}^{2n} \{ \sum_{j=0}^{\lfloor \frac{z}{2} \rfloor} \binom{n}{j}(-1)^j A(K)^j C_{2n-2j,z-2j}^{(K)} \} r^z$$

by identification with (16) it follows the relationship (14). □

Remark 1. The Theorem 1 shows how the nth order interactions $\mathbb{E}[A(\Xi)^{n-k}W(\Xi_0,K)^k]$, $k = 0, \cdots n$, can be estimated by the knowledge of the functions $\zeta^{(j)}_{\Xi_0,K}$, $j = 1, \cdots n$. We emphasize that for all convex compact K the distribution of the random vector ${}^t(A(\Xi), W(\Xi_0,K))$ is fully characterized and can be reconstructed from its moments: the interactions $\mathbb{E}[A(\Xi)^{n-k}W(\Xi_0,K)^k], n \in \mathbb{N}, k = 1, \cdots n$. There is two way to estimate the nth order interactions: make polynomials approximations of the $\zeta^{(j)}_{\Xi_0,K}$, $j = 1, \cdots n$ to get their coefficients $C^{(K)}_{j,p}$ and using (14) or make directly a polynomial approximation of $\eta^{(n)}_{\Xi_0,K}$ to get the coefficients $M^{(K)}_{n,k}$ and using (15).

2.3   Variation of a Random Convex Set in $\mathbb{R}^2$

Here we will discuss the choice of K, let Ξ_0 be a convex random set of $\mathbb{R}^2$ satisfying $\mathbb{E}[A(\Xi_0)] < \infty$. Let's note B the unit ball of $\mathbb{R}^2$ and S_θ the rotation of the segment $[0,1] \times \{0\}$ with angle $\theta \in [0, 2\pi]$; in other words S_θ is the centred segment directed by θ of length two. We have the well know result [6,11]:

$$W(\Xi_0, B) = \frac{1}{2}U(\Xi_0) \tag{17}$$

$$W(\Xi_0, S_\theta) = H_{\Xi_0}(\theta) \tag{18}$$

Where $U(\Xi_0)$ denotes the perimeter of Ξ_0 and $H_{\Xi_0}(\theta)$ denotes the Ferret's diameter of Ξ_0 in the direction θ. As a direct result, the choice $K = 2B$ provides estimators for all moments of the random vector ${}^t(A(\Xi_0), U(\Xi_0))$, and by choosing $K = S_\theta$ we obtain all moments of the random vector ${}^t(A(\Xi_0), H_{\Xi_0}(\theta))$. Notice that the Ferret's diameter is π-periodic in the variable θ, if Ξ_0 is supposed to be isotropic, then the random variables $H_{\Xi_0}(\theta)$ for $\theta \in [0, \pi]$ are identically distributed. Let's remark that the random process $H_{\Xi_0} = (H_{\Xi_0}(\theta))_{\theta \in [0,\pi]}$ fully characterize the random set $\Xi_0 \oplus \check{\Xi}_0$; in fact, for each $\omega \in \Omega$ the Ferret's diameter $H_{\Xi_0}(\omega) : [0, \pi] \to \mathbb{R}_+$ coincide with the support function of the convex compact set $\Xi_0 \oplus \check{\Xi}_0(\omega)$. It is well known that the support function of the convex compact set fully characterizes the convex compact set concerned [12]. Therefore H_{Ξ_0} fully characterizes the difference body $\Xi_0 \oplus \check{\Xi}_0$ that leads to the following theorem.

Theorem 2 (Characterization of a Random Convex Set by Its Mixed Area). *Let's $\Xi_0^{(1)}$ and $\Xi_0^{(2)}$ be two convex random sets of $\mathbb{R}^2$ satisfying $\mathbb{E}[A(\Xi^{(j)})] < \infty$, $j = 1, 2$ and assume at least one of the distributions of the random variables $A(\Xi^{(j)})$ is M-determinate [13]. Then, the condition*

$$\forall n \geq 1, \ \forall K \in \mathfrak{K}_c, \ \zeta^{(n)}_{\Xi_0^{(1)},K} = \zeta^{(n)}_{\Xi_0^{(2)},K} \tag{19}$$

Implies,

$$\Xi_0^{(1)} \oplus \check{\Xi}_0^{(1)} \overset{\mathcal{L}}{=} \Xi_0^{(2)} \oplus \check{\Xi}_0^{(2)} \tag{20}$$

Where "$\stackrel{\mathcal{L}}{=}$" denotes the so called equality in law, and $\mathfrak{K}_c$ denotes the class of convex compact sets on $\mathbb{R}^2$.

Proof. Let's $\Xi_0^{(1)}$ and $\Xi_0^{(2)}$ be two convex random sets of $\mathbb{R}^2$ satisfying $\mathbb{E}[A(\Xi_0^{(j)})] < \infty$, $j = 1, 2$ and assume at least one of the distributions of the random variables $A(\Xi^{(j)})$ is M-determinate Let's assume the condition (19), according to the Theorem 1 and the M-determinate condition:

$$(19) \Rightarrow \forall n \geq 1, \forall K \in \mathfrak{K}, \mathbb{E}[W(\Xi_0^{(1)}, K)^n] = [W(\Xi_0^{(2)}, K)^n]$$
$$\Rightarrow W(\Xi_0^{(1)}, K) \stackrel{\mathcal{L}}{=} W(\Xi_0^{(2)}, K)$$

For each $k \geq 1$ and for each $(\theta_1, \cdots \theta_k) \in [0, \pi]^k$, let $V_1 =^t (H_{\Xi_0^{(1)}}(\theta_1), \cdots H_{\Xi_0^{(1)}}(\theta_k))$ be a random vector extract of the random process H_{Ξ_1} and $V_2 =^t (H_{\Xi_0^{(2)}}(\theta_1), \cdots H_{\Xi_0^{(2)}}(\theta_k))$ a random vector extract of $H_{\Xi_0^{(2)}}$. We will prove $V_1 \stackrel{\mathcal{L}}{=} V_2$, for this, let's consider a positive linear combination of elements of V_1, $\sum_{i=1}^k \alpha_i H_{\Xi_0^{(1)}}(\theta_i)$ and the convex compact sets $Z = \bigoplus_{i=1}^k \alpha_i S_{\theta_i}$. Notice the following property:

Lemma 1. *Let's X, Y be convex sets, $x \in \mathbb{R}^+$ and $\beta \in [0, \pi]$. Using the Steiner's formula on $A(X \oplus Y \oplus xS_\beta)$ and the properties of the support function, it is easy to show that:*

$$W(X, Y \oplus xS_\beta) = W(X, Y) + xH_X(\beta) \tag{21}$$

Applaying successively this lemma on $W(\Xi_0^{(1)}, Z)$ and on $W(\Xi_0^{(2)}, Z)$ we have: $W(\Xi_0^{(1)}, Z) = \sum_{i=1}^k \alpha_i H_{\Xi_0^{(1)}}(\theta_i)$ and $W(\Xi_0^{(2)}, Z) = \sum_{i=1}^k \alpha_i H_{\Xi_0^{(2)}}(\theta_i)$
Thus,

$$\forall (\alpha_1, \cdots \alpha_k) \in \mathbb{R}_+^k, \sum_{i=1}^k \alpha_i H_{\Xi_0^{(1)}}(\theta_i) = \sum_{i=1}^k \alpha_i H_{\Xi_0^{(2)}}(\theta_i)$$

which implies $V_1 \stackrel{\mathcal{L}}{=} V_2$, we have this result for all $k \geq 1$ and for all $(\theta_1, \cdots \theta_k) \in [0, \pi]^k$, thus $H_{\Xi^{(1)}} \stackrel{\mathcal{L}}{=} H_{\Xi^{(2)}}$ therefore $\Xi_0^{(1)} \oplus \breve{\Xi}_0^{(1)} \stackrel{\mathcal{L}}{=} \Xi_0^{(2)} \oplus \breve{\Xi}_0^{(2)}$. $\square$

Remark 2. First note that the M-determinate condition is not realy restrictive, it can always be assumed in pratical cases [14]. Let's notice that the choice of the convex compact $Z = \bigoplus_{i=1}^k \alpha_i S_{\theta_i}$ provide an explicit expression of the mixed area, it can be use for estimate all characteristics of the random process H_{Ξ_0} in the following way; For each $k \geq 1$ and $n \geq 0$ we define $P_{\Xi_0, (\theta_1, \cdots \theta_k)}^{(k,n)}$, the polynomial function on k variables of degree n as:

$$P_{\Xi_0,(\theta_1,\cdots\theta_k)}^{(k,n)}(\alpha_1,\cdots\alpha_k) = \mathbb{E}[W(\Xi_0,\bigoplus_{i=1}^{k}\alpha_i S_{\theta_i})^n]$$

$$= (\sum_{i=1}^{k}\alpha_i H_{\Xi_0}(\theta_i))^n$$

$$= \sum_{j_1+\cdots j_k=n}\binom{n}{j_1,\cdots j_k}\mathbb{E}[\prod_{i=1}^{k} H_{\Xi_0}(\theta_i)^{j_i}]\prod_{i=1}^{k}\alpha_i^{j_i}$$

Thus all of the expectations $\mathbb{E}[\prod_{i=1}^{k} H_{\Xi_0}(\theta_i)^{j_i}]$ can be estimate by a fit of $P_{\Xi_0,(\theta_1,\cdots\theta_k)}^{(k,n)}$. However, in practice it is difficult to compute and fit $P_{\Xi_0,(\theta_1,\cdots\theta_k)}^{(k,n)}$ for large k, thus we will be more interested in the autocorrelation of the random process H_{Ξ_0} and its marginals moments $\mathbb{E}[H_{\Xi_0}(\theta)^n]$.

3   Application to the Boolean Model

Let Ξ be a Boolean of primary grain Ξ_0 and intensity λ:

$$\Xi = \bigcup_{x_i\in\Phi}\Xi_i + x_i \tag{22}$$

Where Φ is a homogeneous Poisson point process of intensity λ and the Ξ_i are random convex sets identically distributed as Ξ_0.

3.1   The Polynomial $\zeta_{\Xi_0}^{(K)}$ for the Primary Grain of the Boolean Model

Let's recall the fundamental relationship of functional capacity:

$$\forall X \in \mathfrak{K},\ T_\Xi(X) = P(\{\Xi\cap X \neq \emptyset\})$$
$$= 1 - \exp(-\lambda\mathbb{E}[\nu_d(\Xi_0\oplus\check{X})])$$
$$= 1 - \exp(-\Psi_\Xi(X))$$

Where $\Psi_\Xi(X) = -\ln(1 - T_\Xi(X)) = \lambda\mathbb{E}[\nu_d(\Xi_0\oplus\check{X})]$. Let's enunciate a useful lemma the proof is omitted since it can be established by induction.

Lemma 2 (Inclusion-Exclusion Principle). *Let f be a C-additive sets function, and $(A_i)_{1\leq i\leq n}$ be a non degenerate family of subset of $\mathbb{R}^n$, then:*

$$f(\bigcap_{i=1}^{n} A_i) = \sum_{k=1}^{n}\frac{(-1)^{k+1}}{k!}\sum_{(i_1,\cdots i_k)\in I_k} f(\bigcup_{j=1}^{k} A_{i_j})$$

$$f(\bigcup_{i=1}^{n} A_i) = \sum_{k=1}^{n}\frac{(-1)^{k+1}}{k!}\sum_{(i_1,\cdots i_k)\in I_k} f(\bigcap_{j=1}^{k} A_{i_j})$$

Where $I_k = \{(i_1, \cdots i_k) \in \{i_1, \cdots i_k\} \subset \{1, \cdots n\}^k \mid \forall l \leq k, \forall m \leq k, l \neq m \Rightarrow i_l \neq i_m\}$.

For $n \geq 2$ let's note $u_n = 0$, according the Lemma 2 and the expression of Ψ, the mean n-variogram can be expressed as:

$$\gamma_{\Xi_0}^{(n)}(u_1, \cdots u_{n-1}) = \mathbb{E}[\nu_d(\bigcap_{i=1}^{n} \Xi_0 - u_i)]$$

$$= \sum_{k=1}^{n} \frac{(-1)^{k+1}}{k!} \sum_{(i(1),\cdots i(k))\in I_k} \mathbb{E}[\nu_d(\bigcup_{j=1}^{k} \Xi_0 - u_{i(j)})]$$

$$= \sum_{k=1}^{n} \frac{(-1)^{k+1}}{\lambda k!} \sum_{(i(1),\cdots i(k))\in I_k} \Psi_\Xi(\{u_{i(1)}, \cdots u_{i(k)}\})$$

$$\Rightarrow \gamma_{\Xi_0}^{(n)}(u_1, \cdots u_{n-1}) = \sum_{k=1}^{n} \frac{(-1)^{k}}{\lambda k!} \sum_{(i(1),\cdots i(k))\in I_k} \ln(1 - T_\Xi(\{u_{i(1)}, \cdots u_{i(k)}\}))$$

$$(23)$$

Obviously the quantities $\Psi_\Xi(\{u_{i_1}, \cdots u_{i_k}\})$ can be expressed by the n-point probability function $\mathscr{C}_\Xi^{(n)}(x_1, \cdots x_n) = \mathbb{P}(\{x_1, \cdots x_n\} \subset \Xi)$ evaluated on the subsets of $\{u_1, \cdots u_n\}$, we have:

$$\ln(1 - T_\Xi(\{u_{i_1}, \cdots u_{i_k}\})) = \ln(\mathbb{P}(\bigcap_{j=1}^{k} \{u_{i_j} \notin \Xi\}))$$

$$= \ln(\sum_{j=1}^{k} \frac{(-1)^{j+1}}{j!} \sum_{(z_1,\cdots z_j)\in I_j} \mathbb{P}(\bigcup_{l=1}^{j} \{u_{i(z_l)} \notin \Xi\}))$$

$$= \ln(\sum_{j=1}^{k} \frac{(-1)^{j+1}}{j!} \sum_{(z_1,\cdots z_j)\in I_j} (1 - \mathscr{C}_\Xi^{(j)}(u_{i(z_1)}, \cdots u_{i(z_j)})))$$

$$\Rightarrow \gamma_{\Xi_0}^{(n)}(u_1, \cdots u_{n-1}) = \sum_{k=1}^{n} \sum_{(i(1),\cdots i(k))\in I_k} \frac{(-1)^{k}}{\lambda k!} \ln(\sum_{j=1}^{k} \frac{(-1)^{j+1}}{j!}$$

$$\sum_{(z_1,\cdots z_j)\in I_j} (1 - \mathscr{C}_\Xi^{(j)}(u_{i(z_1)}, \cdots u_{i(z_j)}))) \qquad (24)$$

The n-point-probability function $\mathscr{C}_\Xi^{(n)}(x_1, \cdots x_n)$ can be viewed as a volume fraction of $\bigcap_{i=1}^{n}(\Xi + x_i)$, it is easy to see that the stationary of Ξ implies

$\mathbb{P}(\{x_1, \cdots x_n\} \subset \Xi) = \mathbb{P}(\{0, x_1 - x_n, \cdots x_{n-1} - x_n\} \subset \Xi)$, its yield an unbiased estimator for the n-point probability function in bounded windows W:

$$\hat{\mathscr{C}}_{\Xi,W}^{(n)}(x_1, \cdots x_n) = \frac{\nu_d((\Xi \cap W) \ominus \{0, x_1 - x_n, \cdots x_{n-1} - x_n\})}{\nu_d(W \ominus \{0, x_1 - x_n, \cdots x_{n-1} - x_n\})} \qquad (25)$$

Thus, an estimator $\hat{\gamma}^{(n)}_{\Xi_0,W}$ for the mean n-variogram of Ξ_0 can be obtained by (25),(24) associated to an estimator of λ (see [5,7]). We emphasize that the Boolean model is stable by convex dilatation, in other words, for $K \in \mathfrak{K}_c$, the dilated model $\Xi \oplus K$ is also a Boolean model of same intensity λ and of primary grain $\Xi_0 \oplus K$. As a consequence, for each $r \geq 0$ an estimator $\hat{\gamma}^{(n)}_{\Xi_0 \oplus rK}$ can be found. However a precaution must be taken to break the edge effects; if we have a realization of $\Xi \cap W$, the dilated model $\Xi \oplus rK$ is only known within the eroded window $W_{rK} = W \ominus rK$, in fact $((\Xi \cap W) \oplus rK) \cap W_{rK} = (\Xi \oplus rK) \cap W_{rK}$. Therefore it follow from (10) the estimators:

$$\hat{\zeta}^{(n)}_{\Xi_0,K,W}(r) = \int_{\mathbb{R}^d} \cdots \int_{\mathbb{R}^d} \hat{\gamma}^{(n)}_{\Xi \oplus rK,W_{rK}}(u_1,\cdots u_{n-1})du_1 \cdots du_{n-1} \qquad (26)$$

3.2   The Case of the Planar Boolean Model

As a consequence of (26) and the Theorem 1, in the planar case we obtain an estimator for the polynomial $\eta^{(n)}_{\Xi,K}$:

$$\hat{\eta}^{(n)}_{\Xi_0,K,W}(r) = \sum_{j=0}^{n} \binom{n}{j}(-1)^j A(K)^j r^{2j} \hat{\zeta}^{(n-j)}_{\Xi_0,K,W}(r) \qquad (27)$$

Therefore by fitting these quantities, we obtain estimators for the moments of $^t(A(\Xi_0),W(\Xi_0,K))$. Furthermore the polynomial approximation of $\hat{\eta}^{(n)}_{\Xi_0,K,W}$ can be refined by inequality constraints, some of them are probabilistic(inequality between moments, Cauchy-Schwarz inequality). But there are also some morphological constraints like the generalized isoperimetric inequality [15]. Furthermore, if we make additive assumptions concerning the shape of Ξ_0, other constraints can be found [16].

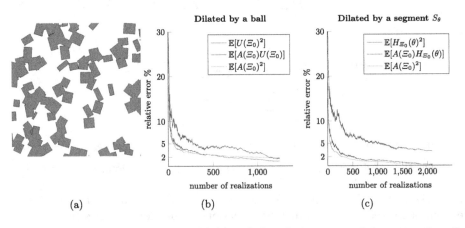

Fig. 1. A realization of the test model (a) and the relatives error of the estimation of the 2nd order moments of $^t(A(\Xi_0),U(\Xi_0))$ on (b) and of $^t(A(\Xi_0),H_{\Xi_0}(\theta))$ on (c).

We have tested this method with $n = 2$ for a Boolean model of rectangular grains: the uncorrelated sides of the primary grain follow $\mathcal{N}(40, 10)$ and $\mathcal{N}(30, 10)$, the intensity parameter $\lambda = \frac{100}{500 \times 500}$ (see Fig. 1). We have simulated several realization of this model in a bounded window $W = 500 \times 500$, and we studies the relative error between the theoretical and estimated moments of $^t(A(\Xi_0), W(\Xi_0, K))$ when K is a ball or a segment (see Fig. 1).

4   Conclusions and Prospects

We have established analytical formulae which allow us to connect the n-variograms of the dilatations of a random convex set with the variations of some of its morphological characteristics. For the Boolean model, we also have shown how the n-variogram of its primary grain can be connected to its n-point probability function. Therefore, it provides estimators of the variations of the primary grain's morphological characteristics. Especially, using dilatation by a disk or a segment, the proposed method can be used to characterize a primary grain whose shape depends on two parameters (rectangle, ellipse,...). We emphasize that our method can be used for any germ-grain model in which we can estimate the mean n-variogram.

In the future we are looking at more complex germ-grain models than the Boolean model. We are also interested in the influence of the model parameters and the observation's window on the accuracy of the estimates. The prospect of describing a convex random set by the characteristics of the random process associated to its Ferret's diameter (see Subsect. 2.3), is even more relevant.

References

1. Torquato, S.: Random Heterogeneous Materials: Microstructure and Macroscopic Properties, vol. 16. Springer, New York (2002)
2. Jeulin, D.: Random texture models for material structures. Stat. Comput. 10(2), 121–132 (2000)
3. Galerne, B.: Modèles d'image aléatoires et synthèse de texture. Ph.D. thesis, Ecole normale supérieure de Cachan-ENS Cachan (2010)
4. Peyrega, C.: Prediction des proprietes acoustiques de materiaux fibreux heterogenes a partir de leur microstructure 3D. Ph.D. thesis, École Nationale Supérieure des Mines de Paris (2010)
5. Chiu, S.N., Stoyan, D., Kendall, W.S., Mecke, J.: Stochastic Geometry and Its Applications. Wiley, Hoboken (2013)
6. Molchanov, I.S.: Statistics of the Boolean Model for Practitioners and Mathematicians. Wiley, Chichester (1997)
7. Miles, R.E.: Estimating aggregate and overall characteristics from thick sections by transmission microscopy. J. Microsc. 107(3), 227–233 (1976)
8. Heinrich, L.: Asymptotic properties of minimum contrast estimators for parameters of Boolean models. Metrika 40(1), 67–94 (1993)
9. Galerne, B.: Computation of the perimeter of measurable sets via their covariogram. Applications to random sets. Image Anal. Stereol. 30(1), 39–51 (2011)

10. Heinrich, L.: Asymptotic properties of minimum contrast estimators for parameters of boolean models. Metrika **40**(1), 67–94 (1993)
11. Michielsen, K., De Raedt, H.: Integral-geometry morphological image analysis. Phys. Rep. **347**(6), 461–538 (2001)
12. Schneider, R.: Convex Bodies: The Brunn-Minkowski Theory, vol. 151. Cambridge University Press, Cambridge (2013)
13. Stoyanov, J.: Krein condition. In: Hazewinkel, M. (ed.) Encyclopedia of Mathematics. Springer, Dordrecht (2001). ISBN 978-1-55608-010-4
14. Akhiezer, N., Kemmer, N.: The Classical Moment Problem and Some Related Questions in Analysis, vol. 5. Oliver & Boyd, Edinburgh (1965)
15. Yau, S.-T.: Institute for Advanced Study (Seminar on differential geometry), no. 102. Princeton University Press, Princeton (1982)
16. Rivollier, S., Debayle, J., Pinoli, J-C.: Analyse morphométrique d'images à tons de gris par diagrammes de forme. In: RFIA 2012 (Reconnaissance des Formes et Intelligence Artificielle) (2012). ISBN 978-2-9539515-2-3

Information Geometry Optimization

Laplace's Rule of Succession
in Information Geometry

Yann Ollivier$^{(\boxtimes)}$

CNRS & LRI, Université Paris-Saclay & INRIA-TAO, Paris, France
yann.ollivier@lri.fr

When observing data $x_1, \ldots, x_t$ modelled by a probabilistic distribution $p_\theta(x)$, the maximum likelihood (ML) estimator $\theta^{\mathrm{ML}} = \arg\max_\theta \sum_{i=1}^t \ln p_\theta(x_i)$ cannot, in general, safely be used to predict x_{t+1}. For instance, for a Bernoulli process, if only "tails" have been observed so far, the probability of "heads" is estimated to 0. (Thus for the standard log-loss scoring rule, this results in infinite loss the first time "heads" appears.)

Bayesian estimators suffer less from this problem, as every value of θ contributes, to some extent, to the Bayesian prediction of x_{t+1} knowing $x_{1:t}$. However, their use can be limited by the need to integrate over parameter space or to use Monte Carlo samples from the posterior distribution.

For Bernoulli distributions, Laplace's famous "add-one" rule of succession (e.g., [CBL06, Grü07]) regularizes θ by adding 1 to the count of "heads" and of "tails" in the observed sequence, thus estimating the Bernoulli parameter p_H by $\hat{p}_H := \frac{n_H+1}{n_H+n_T+2}$ given n_H "heads" and n_T "tails" observations. On the other hand the maximum likelihood estimator is $\frac{n_H}{n_H+n_T}$ so that the two differ at order $O(1/n)$ after $n = n_H + n_T$ observations.

For Bernoulli distributions, Laplace's rule is equivalent to using a uniform Bayesian prior on the Bernoulli parameter [CBL06, Ch. 9.6]. The non-informative Jeffreys prior on the Bernoulli parameter corresponds to Krichevsky and Trofimov's "add-one-half" rule [KT81], namely $\hat{p}_H := \frac{n_H+1/2}{n_H+n_T+1}$. Thus, in this case, some Bayesian predictors have a simple implementation.

We claim (Theorem 1) that for exponential families[1], Bayesian predictors can be approximated by mixing the ML estimator with the *sequential normalized maximum likelihood* (SNML) estimator from universal coding theory [RSKM08, RR08], which is a fully canonical version of Laplace's rule. The weights of this mixture depend on the density of the desired Bayesian prior with respect to the non-informative Jeffreys prior, and are equal to 1/2 for the Jeffreys prior, thus extending Krichevsky and Trofimov's result. The resulting mixture also approximates the "flattened" ML estimator from [KGDR10].

Thus, it is possible to approximate Bayesian predictors without the cost of integrating over θ or sampling from the posterior. The statements below emphasize the special role of the Jeffreys prior and the Fisher information metric. Moreover, the analysis reveals that the direction of the shift from the ML predictor to Bayesian predictors is systematic and given by an intrinsic,

---

[1] For simplicity we only state the results with i.i.d. models. However the ideas extend to non-i.i.d. sequences with $p_\theta(x_{t+1}|x_{1:t})$ in an exponential family, e.g., Markov models.

© Springer International Publishing Switzerland 2015
F. Nielsen and F. Barbaresco (Eds.): GSI 2015, LNCS 9389, pp. 311–319, 2015.
DOI: 10.1007/978-3-319-25040-3_34

information-geometric vector field on statistical manifolds. This could contribute to regularization procedures in statistical learning.

1. Notation and Statement. Let $p_\theta(x)\,\mathrm{d}x$ be a family of distributions on a variable x, smoothly parametrized by θ, with density $p_\theta(x)$ with respect to some reference measure $\mathrm{d}x$ (typically $\mathrm{d}x$ is the counting measure for discrete x, or the Lebesgue measure in $\mathbb{R}^d$).

Let $x_1, \ldots, x_t$ be a sequence of observations to be predicted online using p_θ. The maximum likelihood predictor p^{ML} is given by the probability density

$$p^{\mathrm{ML}}(x_{t+1} = y | x_{1:t}) := p_{\theta_t^{\mathrm{ML}}}(y), \qquad \theta_t^{\mathrm{ML}} := \arg\max_\theta \sum_{i=1}^t \ln p_\theta(x_i) \qquad (1)$$

assuming this arg max is well-defined. Bayesian predictors (e.g., Laplace's rule) usually differ from p^{ML} at order $1/t$.

The *sequential normalized maximum likelihood* predictor ([RSKM08, RR08], see also [TW00]) uses, for each possible value y of x_{t+1}, the parameter $\theta^{\mathrm{ML}+y}$ that would yield the best probability if y had already been observed. Since this increases the probability of every y, it is necessary to renormalize. Define

$$\theta_t^{\mathrm{ML}+y} := \arg\max_\theta \left\{ \ln p_\theta(y) + \sum_{i=1}^t \ln p_\theta(x_i) \right\} \qquad (2)$$

as the ML estimator when adding y at position $t + 1$. For each y, define the *SNML predictor*[2] for time $t + 1$ by the probability density

$$p^{\mathrm{SNML}}(x_{t+1} = y | x_{1:t}) := \frac{1}{Z} p_{\theta_t^{\mathrm{ML}+y}}(y) \qquad (3)$$

where Z is a normalizing constant (assuming $Z < \infty$).

For Bernoulli distributions, p^{SNML} coincides with Laplace's "add-one" rule[3]. For other distributions the two may differ[4]: for instance, defining Laplace's rule for continuous-valued x requires choosing a prior distribution on x, whereas the SNML distribution is completely canonical.

We claim that for exponential families, $\frac{1}{2}(p^{\mathrm{ML}} + p^{\mathrm{SNML}})$ is close to the Bayesian predictor using the Jeffreys prior. This generalizes the "add-one-half" rule.

---

[2] This variant of SNML is SNML-1 in [RSKM08] and CNML-3 in [Grü07].

[3] Note that we describe it in a different way. The usual presentation of Laplace's rule is to define $\theta^{\mathrm{Lap}} := \arg\max_\theta \{\ln p_\theta(\text{"heads"}) + \ln p_\theta(\text{"tails"}) + \sum \ln p_\theta(x_i)\}$ and then use θ^{Lap} to predict x_{t+1}. Here we follow the SNML viewpoint and use a different $\theta^{\mathrm{ML}+y}$ for each possible value y of x_{t+1}.

[4] [HB12, BGH+13] contain a characterization of those one-dimensional exponential families for which the variants of NML predictors coincide exactly between themselves and with a Bayesian prior, which is then necessarily the Jeffreys prior. Here Theorem 1 shows that this happens in some approximate sense for *any* exponential family; further relationship between these results is not obvious.

This extends to any Bayesian prior π by using a *weighted* SNML predictor

$$p^{w\text{-SNML}}(y) := \frac{1}{Z} w(\theta^{\text{ML}+y}) p_{\theta^{\text{ML}+y}}(y) \qquad (4)$$

The weight $w(\theta)$ to be used for a given prior π will depend on the ratio between π and the Jeffreys prior. Recall that the latter is $\pi^{\text{Jeffreys}}(d\theta) := \sqrt{\det \mathcal{I}(\theta)}\, d\theta$ where $\mathcal{I}$ is the *Fisher information matrix* of the family (p_θ),

$$\mathcal{I}(\theta) := -\mathbb{E}_{x \sim p_\theta} \partial^2_\theta \ln p_\theta(x) \qquad (5)$$

where ∂^2_θ stands for the Hessian matrix of a function of θ.

Theorem 1. *Let $p_\theta(x)\, dx$ be an exponential family of probability distributions, and let π be a Bayesian prior on θ. Then, under suitable regularity assumptions, the Bayesian predictor with prior π knowing $x_{1:t}$ has probability density*

$$\frac{1}{2} p^{\text{ML}}(\cdot | x_{1:t}) + \frac{1}{2} p^{\beta^2\text{-SNML}}(\cdot | x_{1:t}) \qquad (6)$$

up to $O(1/t^2)$, where $\beta(\theta)$ is the density of π with respect to the Jeffreys prior, i.e., $\pi(d\theta) = \beta(\theta) \sqrt{\det \mathcal{I}(\theta)}\, d\theta$ with $\mathcal{I}$ the Fisher matrix.

More precisely, both under the prior π and under $\frac{1}{2}(p^{\text{ML}} + p^{\beta^2\text{-SNML}})$, the probability density that $x_{t+1} = y$ given $x_{1:t}$ is asymptotically

$$p_{\theta_t^{\text{ML}}}(y) \left(1 + \frac{1}{2t} \|\partial_\theta \ln p_\theta(y)\|^2_F + \frac{1}{t} \langle \partial_\theta \ln \beta, \partial_\theta \ln p_\theta(y) \rangle_F - \frac{\dim \Theta}{2t} + O(1/t^2) \right) \qquad (7)$$

provided $p_{\theta_t^{\text{ML}}}(y) > 0$, where $\langle \partial_\theta f, \partial_\theta g \rangle_F := (\partial_\theta f)^\top \mathcal{I}^{-1}(\theta) \partial_\theta g$ is the Fisher scalar product and $\|\partial_\theta f\|^2_F = \langle \partial_\theta f, \partial_\theta f \rangle_F$ is the Fisher metric norm of $\partial_\theta f$.

For the Jeffreys prior (constant β), this also coincides up to $O(1/t^2)$ with the "flattened" or "squashed" ML predictor from [KGDR10, GK10] with $n_0 = 0$. In particular, the latter is $O(1/t^2)$ close to the Jeffreys prior, and the optimal regret guarantees in [KGDR10] apply to (7). Note that a multiplicative $1 + O(1/t^2)$ difference between predictors results in an $O(1)$ difference on cumulated log-loss regrets.

Regularity Assumptions. In most of the article we assume that $p_\theta(x_{t+1}|x_{1:t})$ is a non-degenerate exponential family of probability distributions, with θ belonging to an open set of parameters Θ. The key property we need from exponential families is the existence of a parametrization θ in which $\partial^2_\theta \ln p_\theta(x) = -\mathcal{I}(\theta)$ for all x and θ: this holds in the natural parametrization for any exponential family (indeed, $p_\theta(x) = e^{\theta \cdot T(x)}/Z(\theta)$ yields $\partial_\theta \ln p_\theta(x) = T(x) - \partial_\theta \ln Z(\theta)$ so that $\partial^2_\theta \ln p_\theta(x) = -\partial^2_\theta \ln Z(\theta)$ for any x).

For simplicity we assume that the space for x is compact, so that to prove $O(1/t^2)$ convergence of distributions over x it is enough to prove $O(1/t^2)$ convergence for each value of x. We assume that the sequence of observations $(x_t)_{t \in \mathbb{N}}$

is an *ineccsi sequence* [Grü07], namely, that for t large enough, the maximum likelihood estimate is well-defined and stays in a compact subset of the parameter space. We also need to assume the same in a Bayesian sense, namely, that for t large enough, the Bayesian maximum a posteriori using prior π is well-defined and stays in a compact subset of Θ. The Bayesian priors are assumed to be smooth with positive densities. On the other hand we do not assume that the Jeffreys prior or the prior π are proper; it is enough that the posterior given the observations is proper, so that the Bayesian predictor at time t is well-defined.

In some parts of the article we do not need p_θ to be an exponential family, but we still assume that the model p_θ is smooth, that there is a well-defined maximum θ_t^{ML} for any $x_{1:t}$ and no other log-likelihood local maxima.

2. Computing the SNML Predictor. We prove Theorem 1 by proving that both predictors are given by (7). Further proofs are gathered at the end of the text.

We first work on p^{SNML}. Here we do not assume that p_θ is an exponential family. Let J_t be the *observed information matrix*, assumed to be positive-definite,

$$J_t(\theta) := -\frac{1}{t} \sum_{i=1}^{t} \partial_\theta^2 \ln p_\theta(x_i) \tag{8}$$

Proposition 2. *Under suitable regularity assumptions, the maximum likelihood update from t to $t+1$ satisfies*

$$\theta_{t+1}^{\mathrm{ML}} = \theta_t^{\mathrm{ML}} + \frac{1}{t} J_t(\theta_t^{\mathrm{ML}})^{-1} \partial_\theta \ln p_\theta(x_{t+1}) + O(1/t^2) \tag{9}$$

For exponential families, this update is the natural gradient of $\ln p(x_{t+1})$ with learning rate $1/t$ [Ama98], because $J_t(\theta_t^{\mathrm{ML}}) = \mathcal{I}(\theta_t^{\mathrm{ML}})$, the exact Fisher information matrix. (For exponential families *in the natural parametrization*, $J_t(\theta) = \mathcal{I}(\theta)$ for all θ. But since the Hessian of a function f on a manifold is a well-defined tensor at a critical point of f, it follows that at θ_t^{ML} one has $J_t(\theta_t^{\mathrm{ML}}) = \mathcal{I}(\theta_t^{\mathrm{ML}})$ for *any* parametrization of an exponential family.)

Proposition 3. *Under suitable regularity assumptions,*

$$p^{\mathrm{SNML}}(y|x_{1:t}) = \frac{1}{Z} p_{\theta_t^{\mathrm{ML}}}(y) \left(1 + \frac{1}{t} (\partial_\theta \ln p_\theta(y))^\top J_t^{-1} \partial_\theta \ln p_\theta(y) + O(1/t^2) \right) \tag{10}$$

provided $p_{\theta_t^{\mathrm{ML}}}(y) > 0$, where J_t is as above and the derivatives are taken at θ_t^{ML}.

Importantly, the normalization constant Z can be computed without having to sum over y explicitly. Indeed (cf. [KGDR10]), by definition of $\mathcal{I}(\theta)$,

$$\mathbb{E}_{y \sim p_\theta} (\partial_\theta \ln p_\theta(y))^\top J_t^{-1} \partial_\theta \ln p_\theta(y) = \mathrm{Tr}(J_t^{-1} \mathcal{I}(\theta)) \tag{11}$$

so that $Z = 1 + \frac{1}{t} \mathrm{Tr}(J_t^{-1} \mathcal{I}(\theta_t^{\mathrm{ML}})) + O(1/t^2)$. For exponential families, $J_t = \mathcal{I}$ at θ_t^{ML} so that $Z = 1 + \frac{\dim \Theta}{t} + O(1/t^2)$ and

$$p_{\theta_t^{\mathrm{ML}}}(y) \left(1 + \frac{1}{t} (\partial_\theta \ln p_\theta(y))^\top \mathcal{I}^{-1} \partial_\theta \ln p_\theta(y) - \frac{\dim \Theta}{t} \right) \tag{12}$$

is an $O(1/t^2)$ approximation of $p^{\text{SNML}}(y|x_{1:t})$.

For the weighted SNML distribution $p^{w\text{-SNML}}$, a similar argument yields

$$p^{w\text{-SNML}}(y|x_{1:t}) = \frac{1}{Z} p_{\theta_t^{\text{ML}}}(y) \left(1 + \frac{1}{t}(\partial_\theta \ln p_\theta(y))^\top J_t^{-1} (\partial_\theta \ln p_\theta(y) + \partial_\theta \ln w(\theta)) + O(1/t^2) \right) \quad (13)$$

with $Z = 1 + \frac{1}{t} \operatorname{Tr}(J_t^{-1}\mathcal{I}(\theta_t^{\text{ML}})) + O(1/t^2)$ as above. (The $\partial_\theta \ln w$ term does not contribute to Z because $\sum_y p_\theta(y)\partial_\theta \ln p_\theta(y) = 0$).

Computing $\frac{1}{2}p^{\text{ML}} + \frac{1}{2}p^{w\text{-SNML}}$ with $w(\theta) = \beta(\theta)^2$ in (13), and using that $J_t(\theta^{\text{ML}}) = \mathcal{I}$ for exponential families, proves one half of Theorem 1.

3. Computing the Bayesian Posterior. Next, let us establish the asymptotic behavior of the Bayesian posterior. This relies on results from [TK86]. The following proposition may have independent interest.

Proposition 4. *Consider a Bayesian prior $\pi(\mathrm{d}\theta) = \alpha(\theta)\,\mathrm{d}\theta$. Then the posterior mean of a smooth function $f(\theta)$ given data $x_{1:t}$ and prior π is asymptotically*

$$f(\theta_t^{\text{ML}}) + \frac{1}{t}(\partial_\theta f)^\top J_t^{-1} \partial_\theta \left(\ln \frac{\alpha}{\sqrt{\det(-\partial_\theta^2 L)}} \right) + \frac{1}{2t} \operatorname{Tr}(J_t^{-1}\partial_\theta^2 f) + O(1/t^2) \quad (14)$$

where $L(\theta) := \frac{1}{t}\ln p_\theta(x_{1:t})$ is the average log-likelihood function, ∂_θ^2 is the Hessian matrix w.r.t. θ, and $J_t := -\partial_\theta^2 L(\theta_t^{\text{ML}})$ is the observed information matrix.

When p_θ is an exponential family in the natural parametrization, for any $x_{1:t}$, $-\partial_\theta^2 L$ is equal to the Fisher matrix $\mathcal{I}$, so that the denominator in the log is the Jeffreys prior $\sqrt{\det \mathcal{I}}$. In particular, for exponential families in natural coordinates, the first term vanishes if the prior π is the Jeffreys prior.

Corollary 5. *Let p_θ be an exponential family. Consider a Bayesian prior $\beta(\theta)\sqrt{\det \mathcal{I}(\theta)}\,\mathrm{d}\theta$ having density β with respect to the Jeffreys prior. Then the posterior probability that $x_{t+1} = y$ knowing $x_{1:t}$ is asymptotically given by (7) as in Theorem 1.*

This proves the second half of Theorem 1.

4. Intrinsic Viewpoint. When rewritten in intrinsic Riemannian terms, Proposition 4 emphasizes a systematic discrepancy at order $1/t$ between ML prediction and Bayesian prediction, which is often more "centered" as in Laplace's rule.

This is characterized by a canonical vector field on a statistical manifold indicating the direction of the difference between ML and Bayesian predictors, as follows. In intrinsic terms, the posterior mean (14) in Proposition 4 is[5]

$$f(\theta^{\text{ML}}) - \frac{1}{t}(\nabla^2 L)^{-1}\left(\mathrm{d}f, \mathrm{d}\ln\frac{\pi}{\sqrt{\det(-\nabla^2 L)}} \right) - \frac{1}{2t} \operatorname{Tr}\left((\nabla^2 L)^{-1}\nabla^2 f \right) + O(1/t^2)$$

$$(15)$$

---

[5] The equality between (14) and (15) holds only at θ_t^{ML}; the value of (14) is not intrinsic away from θ^{ML}. The equality relies on $\partial_\theta L = 0$ at θ^{ML} to cancel curvature contributions.

where $L(\theta) = \sum_{i=t}^{t} \ln p_\theta(x_i)$ as above and where ∇^2 is the Riemannian Hessian with respect to any Riemannian metric on θ, for instance the Fisher metric. This follows from a direct Riemannian-geometric computation (e.g., in normal coordinates). In this expression both, the prior $\pi(d\theta)$ and $\sqrt{\det(-\nabla^2 L)}$ are volume forms on the tangent space so that their ratio is coordinate-independent[6].

At first order in $1/t$, this is the average of f under a Riemannian Gaussian distribution[7] with covariance matrix $\frac{1}{t}(-\nabla^2 L)^{-1}$, but centered at $\theta^{\mathrm{ML}} - \frac{1}{t}(\nabla^2 L)^{-1} \, d \ln(\pi/\sqrt{\det(-\nabla^2 L)})$ instead of θ^{ML}.

Thus, if we want to approximate the posterior Bayesian distribution by a Gaussian, there is a systematic shift $\frac{1}{t}V(\theta^{\mathrm{ML}})$ between the ML estimate and the center of the Bayesian posterior, where V is the data-dependent vector field

$$V := -(\nabla^2 L)^{-1} \, d \ln \left(\pi/\sqrt{\det(-\nabla^2 L)} \right) \tag{16}$$

A particular case is when π is the Jeffreys prior: then

$$V = \frac{1}{2}(\nabla^2 L)^{-1} \, d \ln \det(-\mathcal{I}^{-1}\nabla^2 L) \tag{17}$$

is an intrinsic vector field defined on any statistical manifold, depending on $x_{1:t}$.

Proposition 6. *When the prior is the Jeffreys prior, the vector V is*

$$V^i = \frac{1}{2}(\nabla_i\nabla_j L)^{-1}(\nabla_k\nabla_l L)^{-1} \, \nabla_j\nabla_k\nabla_l L \tag{18}$$

in Einstein notation, where $L(\theta) = \frac{1}{t}\sum_{s=1}^{t} \ln p_\theta(x_s)$ is the log-likelihood function, and ∇ is the Levi-Civita connection of the Fisher metric[8].

If p_θ is an exponential family with the Jeffreys prior, the value of V at θ^{ML} does not depend on the observations $x_{1:t}$ and is equal to

$$V^i(\theta^{\mathrm{ML}}) = \frac{1}{4}\mathcal{I}^{ij}\mathcal{I}^{kl}T_{jkl} \tag{19}$$

where T is the skewness tensor [AN00, Eq. (2.28)]

$$T_{jkl}(\theta) := \mathbb{E}_{x \sim p_\theta} \frac{\partial \ln p_\theta(x)}{\partial \theta^j} \frac{\partial \ln p_\theta(x)}{\partial \theta^k} \frac{\partial \ln p_\theta(x)}{\partial \theta^l} \tag{20}$$

$V(\theta^{\mathrm{ML}})$ is thus an intrinsic, data-independent vector field for exponential families, which characterizes the discrepancy between maximum likelihood and the "center" of the Jeffreys posterior distribution. Note that V can be computed from log-likelihood derivatives only. This could be useful for regularization of the ML estimator in statistical learning.

---

[6] This is clear when dividing both by the Riemannian volume form $\sqrt{\det g}$: both the prior density $\pi/\sqrt{\det g}$ and $\sqrt{\det(-g^{-1}\nabla^2 L)}$ are intrinsic.

[7] i.e., the image by the exponential map of a Gaussian distribution in a tangent plane.

[8] Note that $\nabla_j\nabla_k\nabla_l L$ is not fully symmetric. Still it is symmetric at θ^{ML}, because the various orderings differ by a curvature term applied to ∇L with vanishes at θ^{ML}.

5. Proofs (Sketch)

Proof of Proposition 2. Minimization of a Taylor expansion of log-likelihood around θ_t^{ML}. This is justified formally by applying the implicit function theorem to $F : (\varepsilon, \theta) \mapsto \partial_\theta \left(\varepsilon \ln p_\theta(x_{t+1}) + \frac{1}{t} \sum_{i=1}^{t} \ln p_\theta(x_t) \right)$ at point $(0, \theta^{\mathrm{ML}})$.

Proof of Proposition 3. Abbreviate $\theta_y := \theta_t^{\mathrm{ML}+y}$. From Proposition 2 we have

$$\theta_y = \theta_t^{\mathrm{ML}} + \frac{1}{t} J_t^{-1} \partial_\theta \ln p_\theta(y) + O(1/t^2) \tag{21}$$

and expanding $\ln p_\theta(y)$ around θ_t^{ML} yields $p_{\theta_y}(y) = p_{\theta_t^{\mathrm{ML}}}(y)(1 + (\theta_y - \theta_t^{\mathrm{ML}})^{\mathsf{T}} \partial_\theta \ln p_\theta(y)) + O((\theta_y - \theta^{\mathrm{ML}})^2)$ and plugging in the value of $\theta_y - \theta_t^{\mathrm{ML}}$ yields the result.

Proof of Proposition 4. The posterior mean is $(\int f(\theta)\alpha(\theta)p_\theta(x_{1:t})\,\mathrm{d}\theta)/(\int \alpha(\theta)p_\theta(x_{1:t})\,\mathrm{d}\theta)$. From [TK86], if $L_1(\theta) = \frac{1}{t} \ln p_\theta(x_{1:t}) + \frac{1}{t} g_1(\theta)$ and $L_2 = \frac{1}{t} \ln p_\theta(x_{1:t}) + \frac{1}{t} g_2(\theta)$ we have

$$\frac{\int e^{tL_2(\theta)} \,\mathrm{d}\theta}{\int e^{tL_1(\theta)} \,\mathrm{d}\theta} = \sqrt{\frac{\det H_1}{\det H_2}} \, e^{t(L_2(\theta_2) - L_1(\theta_1))} (1 + O(1/t^2)) \tag{22}$$

where $\theta_1 = \arg\max L_1$, $\theta_2 = \arg\max L_2$, and H_1 and H_2 are the Hessian matrices of $-L_1$ and $-L_2$ at θ_1 and θ_2, respectively. Here we have $g_1 = \ln \alpha(\theta)$ and $g_2 = g_1 + \ln f(\theta)$ (assuming f is positive; otherwise, add a constant to f).

From a Taylor expansion of L_1 as in Proposition 2 we find $\theta_1 = \theta_t^{\mathrm{ML}} + \frac{1}{t} J_t^{-1} \partial_\theta g_1(\theta_t^{\mathrm{ML}}) + O(1/t^2)$ and likewise for θ_2. So $\theta_1 - \theta_2 = \frac{1}{t} J_t^{-1} \partial_\theta (g_1 - g_2)(\theta_t^{\mathrm{ML}}) + O(1/t^2)$. Since θ_2 maximizes L_2, a Taylor expansion of L_2 around θ_2 gives

$$L_2(\theta_1) = L_2(\theta_2) - \frac{1}{2}(\theta_1 - \theta_2)^{\mathsf{T}} H_2(\theta_1 - \theta_2) + O(1/t^3) \tag{23}$$

so that, using $L_2 = L_1 + \frac{1}{t} \ln f$ we find

$$L_2(\theta_2) - L_1(\theta_1) = L_2(\theta_1) - L_1(\theta_1) + \frac{1}{2}(\theta_1 - \theta_2)^{\mathsf{T}} H_2(\theta_1 - \theta_2) + O(1/t^3) \tag{24}$$

$$= \frac{1}{t} \ln f(\theta_1) + \frac{1}{2t^2}(\partial_\theta \ln f)^{\mathsf{T}} J_t^{-1} H_2 J_t^{-1} \partial_\theta \ln f + O(1/t^3) \tag{25}$$

where the second term is evaluated at θ_t^{ML}. We have $H_2 = J_t + O(1/t)$, so $\exp(t(L_2(\theta_2) - L_1(\theta_1))) = f(\theta_1)(1 + \frac{1}{2t}(\partial_\theta \ln f)^{\mathsf{T}} J_t^{-1} \partial_\theta \ln f + O(1/t^2))$. Meanwhile, by a Taylor expansion of $\ln \det(-\partial_\theta^2 L_2(\theta_2))$ around θ_2,

$$\det H_2 = \det(-\partial_\theta^2 L_2(\theta_2)) = \det(-\partial_\theta^2 L_2(\theta_1)) \left(1 + (\theta_2 - \theta_1)^{\mathsf{T}} \partial_\theta \ln \det(-\partial_\theta^2 L_2) + O(\theta_2 - \theta_1)^2\right) \tag{26}$$

and from $L_2 = L_1 + \frac{1}{t} \ln f$ and $\det(A + \varepsilon B) = \det(A)(1 + \varepsilon \operatorname{Tr}(A^{-1}B) + O(\varepsilon^2))$,

$$\det(-\partial_\theta^2 L_2(\theta_1)) = \det(-\partial_\theta^2 L_1(\theta_1)) \left(1 + \frac{1}{t} \operatorname{Tr}\left((\partial_\theta^2 L_1)^{-1} \partial_\theta^2 (\ln f)\right) + O(1/t^2)\right) \tag{27}$$

$$= (\det H_1) \left(1 - \frac{1}{t} \operatorname{Tr}\left(H_1^{-1} \partial_\theta^2 (\ln f)\right) + O(1/t^2)\right) \tag{28}$$

so, collecting,

$$\sqrt{\frac{\det H_1}{\det H_2}} = 1 - \frac{1}{2}(\theta_2 - \theta_1)^\top \partial_\theta \ln \det(-\partial_\theta^2 L_2) + \frac{1}{2t} \operatorname{Tr}\left(H_1^{-1}\partial_\theta^2(\ln f)\right) + O(1/t^2) \tag{29}$$

but $\theta_2 - \theta_1 = J_t^{-1}\partial_\theta \ln f + O(1/t^2)$, and $L_2 = L + O(1/t)$ and $H_1 = J_t + O(1/t)$, so that

$$\sqrt{\frac{\det H_1}{\det H_2}} = 1 - \frac{1}{2t}(\partial_\theta \ln f)^\top J_t^{-1}\partial_\theta \ln \det(-\partial_\theta^2 L) + \frac{1}{2t}\operatorname{Tr}\left(J_t^{-1}\partial_\theta^2(\ln f)\right) + O(1/t^2) \tag{30}$$

Collecting from (22), expanding $f(\theta_1) = f(\theta_t^{\mathrm{ML}})(1 + \frac{1}{t}(\partial_\theta \ln f)^\top J_t^{-1}\partial_\theta \ln \alpha + O(1/t^2))$, and expanding $\partial_\theta \ln f$ in terms of $\partial_\theta f$ proves Proposition 4.

Proof of Corollary 5. Let us work in natural coordinates for an exponential family (indeed, since the statement is intrinsic, it is enough to prove it in some coordinate system). In these coordinates, for any x, $\partial_\theta^2 \ln p_\theta(x) = -\mathcal{I}(\theta)$ with $\mathcal{I}$ the Fisher matrix, so that $-\partial_\theta^2 L = \mathcal{I}(\theta)$. Apply Proposition 4 to $f(\theta) = p_\theta(y)$, expanding $\partial_\theta f = f\partial_\theta \ln f$ and using $\partial_\theta^2 \ln f = -\mathcal{I}(\theta)$.

Proof of Proposition 6. The Levi-Civita connection on a Riemannian manifold with metric g satisfies $\nabla_l \ln \det A_i^j = (A^{-1})_j^i \nabla_l A_i^j$ thanks to $\partial \ln \det M = \operatorname{Tr}(M^{-1}\partial M)$ and by expanding ∇A. Applying this to $A_i^j = \mathcal{I}^{jk}\nabla_{ki}^2 L$ and using $\nabla \mathcal{I} = 0$ proves the first statement. Moreover, for any function f, *at a critical point of f, $\nabla_l \nabla_j \nabla_k f = \nabla_l \partial_j \partial_k f - \Gamma_{jk}^i \nabla_l \nabla_i f$* and consequently at a critical point of f, with $H_{ij} = \nabla_i \nabla_j f$,

$$\nabla_l \ln \det(g^{ij} H_{jk}) = (H^{-1})^{ij}\nabla_l \partial_i \partial_j f - (H^{-1})^{jk}\Gamma_{jk}^i H_{il} \tag{31}$$

In the natural parametrization of an exponential family, $-\partial^2 L$ is identically equal to the Fisher metric $\mathcal{I}$. Consequently, $\nabla_l \ln \det(-\mathcal{I}^{ij}\nabla_{jk}^2 L) = \mathcal{I}^{ij}\nabla_l \mathcal{I}_{ij} - \mathcal{I}^{jk}\Gamma_{jk}^i \mathcal{I}_{il} = -\mathcal{I}^{jk}\Gamma_{jk}^i \mathcal{I}_{il}$ since $\nabla \mathcal{I} = 0$. So from (17), using $d = \nabla = \partial$ for scalars, and $\nabla^2 L = -\mathcal{I}$ at θ^{ML}, we get in this parametrization

$$V^m = -\frac{1}{2}\mathcal{I}^{ml}\partial_l \ln \det(-\mathcal{I}^{-1}\nabla^2 L) = \frac{1}{2}\mathcal{I}^{ml}\mathcal{I}^{jk}\Gamma_{jk}^i \mathcal{I}_{il} = \frac{1}{2}\mathcal{I}^{jk}\Gamma_{jk}^m \tag{32}$$

The Christoffel symbols Γ in this parametrization can be computed from

$$\partial_i \mathcal{I}_{jk}(\theta) = \partial_i \mathbb{E}_{x \sim p_\theta}\partial_j \ln p_\theta(x)\partial_k \ln p_\theta(x) \tag{33}$$

$$= T_{ijk} - \mathcal{I}_{ij}\mathbb{E}_{x \sim p_\theta}\partial_k \ln p_\theta(x) - \mathcal{I}_{ik}\mathbb{E}_{x \sim p_\theta}\partial_j \ln p_\theta(x) = T_{ijk} \tag{34}$$

because $\partial_i \partial_j \ln p_\theta(x) = -\mathcal{I}_{ij}(\theta)$ for any x in this parametrization, and because $\mathbb{E}\partial \ln p_\theta(x) = 0$. So $\Gamma_{jk}^i = \frac{1}{2}\mathcal{I}^{il}T_{jkl}$ in this parametrization. This ends the proof.

Acknowledgments. I would like to thank Peter Grünwald and the referees for valuable comments and suggestions on this text.

References

Ama98. Amari, S.-I.: Natural gradient works efficiently in learning. Neural Comput. **10**, 251–276 (1998)

AN00. Amari, S.-I., Nagaoka, H.: Methods of Information Geometry. Translations of Mathematical Monographs, vol. 191. American Mathematical Society, Providence (2000). Translated from the 1993 Japanese original by Daishi Harada

BGH+13. Bartlett, P., Grünwald, P., Harremoës, P., Hedayati, F., Kotlowski, W.: Horizon-independent optimal prediction with log-loss in exponential families. In: Conference on Learning Theory (COLT), pp. 639–661 (2013)

CBL06. Cesa-Bianchi, N., Lugosi, G.: Prediction, Learning, and Games. Cambridge University Press, Cambridge (2006)

GK10. Grünwald, P., Kotłowski, W.: Prequential plug-in codes that achieve optimal redundancy rates even if the model is wrong. In: 2010 IEEE International Symposium on Information Theory Proceedings (ISIT), pp. 1383–1387, IEEE (2010)

Grü07. Grünwald, P.D.: The Minimum Description Length Principle. MIT Press, Cambridge (2007)

HB12. Hedayati, F., Bartlett, P.: The optimality of Jeffreys prior for online density estimation and the asymptotic normality of maximum likelihood estimators. In: Conference on Learning Theory (COLT) (2012)

KGDR10. Kotłowski, W., Grünwald, P., De Rooij, S.: Following the flattened leader. In: Conference on Learning Theory (COLT), pp. 106–118, Citeseer (2010)

KT81. Krichevsky, R., Trofimov, V.: The performance of universal encoding. IEEE Trans. Inf. Theor. **27**(2), 199–207 (1981)

RR08. Roos, T., Rissanen, J.: On sequentially normalized maximum likelihood models. In: Proceeedings of 1st Workshop on Information Theoretic Methods in Science and Engineering (WITMSE-2008) (2008)

RSKM08. Roos, T., Silander, T., Kontkanen, P., Myllymäki, P.: Bayesian network structure learning using factorized NML universal models. In: Information Theory and Applications Workshop 2008, pp. 272–276, IEEE (2008)

TK86. Tierney, L., Kadane, J.B.: Accurate approximations for posterior moments and marginal densities. J. Am. Stat. Assoc. **81**(393), 82–86 (1986)

TW00. Takimoto, E., Warmuth, M.K.: The last-step minimax algorithm. In: Arimura, H., Sharma, A.K., Jain, S. (eds.) ALT 2000. LNCS (LNAI), vol. 1968, pp. 279–290. Springer, Heidelberg (2000)

Standard Divergence in Manifold
of Dual Affine Connections

Shun-ichi Amari[1]($^{\boxtimes}$) and Nihat Ay[2]

[1] RIKEN Brain Science Institute, Wako-shi, Saitama 351-0198, Japan
amari@brain.riken.jp
[2] Max-Planck Institute for Mathematics in Science, Leipzig, Germany
nay@mis.mpg.de

Abstract. A divergence function defines a Riemannian metric G and dually coupled affine connections (∇, ∇^*) with respect to it in a manifold M. When M is dually flat, a canonical divergence is known, which is uniquely determined from $\{G, \nabla, \nabla^*\}$. We search for a standard divergence for a general non-flat M. It is introduced by the magnitude of the inverse exponential map, where $\alpha = -(1/3)$ connection plays a fundamental role. The standard divergence is different from the canonical divergence.

1 Introduction: Divergence and Dual Geometry

A divergence function $D[p : q]$ is a differentiable function of two points p and q in a manifold M. It satisfies the non-negativity condition

$$D[p : q] \geq 0 \tag{1}$$

with equality when and only when $p = q$. But it is not a distance in the mathematical sense and it can be an asymmetric function of p and q. When a coordinate system $\boldsymbol{\xi}$ is given in M, we pose one condition that, for two nearby points $p = (\boldsymbol{\xi})$ and $q = (\boldsymbol{\xi} + d\boldsymbol{\xi})$, D is expanded as

$$D[p : q] = \frac{1}{2}g_{ij}(\boldsymbol{\xi})d\xi^i d\xi^j + O\left(|d\boldsymbol{\xi}|^3\right) \tag{2}$$

and $g_{ij}(\boldsymbol{\xi})$ is a positive definite matrix.

A Riemannian metric $G = (g_{ij})$ is defined from the divergence by (2). A pair of dual affine connections are also introduced from it (Eguchi 1983). We use a simplified notation of differentiation such as

$$\partial_i = \frac{\partial}{\partial \xi^i}. \tag{3}$$

We also use the following notations of differentiation with respect to coordinates of $p = (\boldsymbol{\xi})$ and $q = (\boldsymbol{\xi}')$ in $D[\boldsymbol{\xi} : \boldsymbol{\xi}']$ as

$$\partial_i = \frac{\partial}{\partial \xi^i}, \quad \partial_i' = \frac{\partial}{\partial \xi'^i}. \tag{4}$$

© Springer International Publishing Switzerland 2015
F. Nielsen and F. Barbaresco (Eds.): GSI 2015, LNCS 9389, pp. 320–325, 2015.
DOI: 10.1007/978-3-319-25040-3_35

Then, the Riemannian metric is written as

$$g_{ij} = -\partial_i \partial'_j D\left[\boldsymbol{\xi} : \boldsymbol{\xi}'\right]_{\boldsymbol{\xi}'=\boldsymbol{\xi}} = \partial_i \partial_j D\left[\boldsymbol{\xi} : \boldsymbol{\xi}'\right]_{\boldsymbol{\xi}'=\boldsymbol{\xi}} = \partial'_i \partial'_j D\left[\boldsymbol{\xi} : \boldsymbol{\xi}'\right]_{\boldsymbol{\xi}'=\boldsymbol{\xi}}. \qquad (5)$$

The two quantities

$$\Gamma_{ijk} = -\partial_i \partial_j \partial'_k D\left[\boldsymbol{\xi} : \boldsymbol{\xi}'\right]_{\boldsymbol{\xi}'=\boldsymbol{\xi}}, \qquad (6)$$

$$\Gamma^*_{ijk} = -\partial'_i \partial'_j \partial_k D\left[\boldsymbol{\xi} : \boldsymbol{\xi}'\right]_{\boldsymbol{\xi}'=\boldsymbol{\xi}} \qquad (7)$$

give coefficients of a pair of dual affine connections without torsion. They define two covariant derivatives ∇ and ∇^*. They are dual with respect to the Riemannian metric, since they satisfy the duality condition

$$X\langle Y, Z\rangle = \langle \nabla_X Y, Z\rangle + \langle Y, \nabla^*_X Z\rangle \qquad (8)$$

for three vector fields X, Y and Z, where $\langle \, , \, \rangle$ is the inner product (Amari and Nagaoka 2000).

The inverse problem is to find a divergence $D[p : q]$ which generates a given geometrical structure $\{M, \nabla, \nabla^*\}$. Matumoto (1993) showed that a divergence exists for any such manifold. However, it is not unique and there are infinitely many divergences which give the same geometrical structure. It is an interesting problem to define a standard divergence which is uniquely determined from the geometrical structure. The present paper gives an answer to it, by using the magnitude (Riemannian length) of the inverse exponential map, where we use the α-connection with $\alpha = -(1/3)$.

When a manifold is dually flat, a canonical divergence was introduced by Amari and Nagaoka (2000), which is a Bregman divergence. It has nice properties such as the generalized Pythagorean theorem and projection theorem. A standard divergence is different from it even in the dually flat case. It will be an interesting problem to search for their relations in a more general framework.

2   Exponential Map

A geodesic $\boldsymbol{\xi}(t)$ passing through $p = (\boldsymbol{\xi}_p)$ satisfies the geodesic equation,

$$\ddot{\xi}^i(t) = -\Gamma_{jk}{}^i \{\boldsymbol{\xi}(t)\} \dot{\xi}^j \dot{\xi}^k, \qquad (9)$$

$$\boldsymbol{\xi}(0) = \boldsymbol{\xi}_p, \qquad (10)$$

$$\dot{\boldsymbol{\xi}}(0) = X, \qquad (11)$$

where X is a tangent vector belonging to the tangent space at p, $\dot{}$ denotes differentiation d/dt and Γ^i_{jk} is the contravariant representation of Γ_{jki}. We fix p and consider the exponential map, denoted by $\exp_p(X)$, which maps X to q as

$$q = \boldsymbol{\xi}(1). \qquad (12)$$

We denote it as

$$q = \exp_p(X). \qquad (13)$$

We consider a neighborhood of M in which there exists a unique geodesic connecting two points.

A dual geometry is characterized by a triple $\{M, G, T\}$, where T is a third-order symmetric tensor defined by

$$T_{ijk} = \Gamma^*_{ijk} - \Gamma_{ijk}, \tag{14}$$

$$\Gamma_{ijk} = \Gamma^{(0)}_{ijk} - \frac{1}{2} T_{ijk}, \tag{15}$$

$$\Gamma^*_{ijk} = \Gamma^{(0)}_{ijk} + \frac{1}{2} T_{ijk}, \tag{16}$$

where $\Gamma^{(0)}_{ijk}$ denotes the Levi-Civita connection (Amari and Nagaoka, 2000). The α-geometry is defined by $\{M, G, \alpha T\}$, where α is a scalar parameter. We have dual connections called the $\pm\alpha$-connection,

$$\Gamma^{(\pm\alpha)}_{ijk} = \Gamma^{(0)}_{ijk} \mp \frac{\alpha}{2} T_{ijk}. \tag{17}$$

The α-connection and $-\alpha$-connection are dually coupled with respect to G.

The α-exponential map is defined in M by using the α-geodesic defined by the α-connection and is denoted by $\exp^{(\alpha)}_p(X)$.

The inverse of the exponential map or the log map is defined in a neighborhood of p, which maps q to a tangent vector X at p:

$$X(p; q) = \log_p(q). \tag{18}$$

When the α-connection is used, we call it the α-inverse exponential map or α-log map, denoting it by

$$X_\alpha(p; q) = \log^{(\alpha)}_p(q). \tag{19}$$

The square of the magnitude of $X(p; q)$ is defined by

$$\|X_\alpha(p; q)\|^2 = \frac{1}{2} g_{ij}(p) X^i_\alpha(p; q) X^j_\alpha(p; q). \tag{20}$$

It satisfies the conditions of divergence, so by denoting it as

$$D_\alpha[p : q] = \|X_\alpha(p; q)\|^2, \tag{21}$$

we call it the α-inverse-exponential divergence (α-log divergence in short).

3 α-Divergence Derived from Inverse Exponential Map

We study the dual geometry given rise to by the inverse-exponential divergence. An important question is if it recovers the original geometry of M or not. Since the geometrical quantities are derived from a divergence by its derivatives at diagonal elements $p = q$, it suffices to study the exponential map when p and q are close. We use the Taylor expansion. By expanding geodesic $\xi(t)$

$$\xi^i(t) = \xi^i_p + t a^i + \frac{t^2}{2} b^i + O\left(t^3\right), \tag{22}$$

we have

$$a^i = \dot{\xi}^i(0) = X^i, \tag{23}$$
$$b^i = \ddot{\xi}^i(0) = -\Gamma_{jk}{}^i X^j X^k. \tag{24}$$

Hence, the inverse exponential map satisfies

$$X^i(p; q) = \Delta z^i + \frac{1}{2}\Gamma_{jk}{}^i \Delta z^j \Delta z^k + O\left(|\Delta|^3\right), \tag{25}$$

where

$$\Delta z^i = \xi_q^i - \xi_p^i \tag{26}$$

and

$$\xi_q = \exp_p(tX). \tag{27}$$

For the log map, we have

$$D[p : q] = \|X\|^2 = \frac{1}{2}g_{ij}(p)\Delta z^i \Delta z^j + \frac{1}{2}\Gamma_{ijk}(\Delta z)^i \left(\Delta z^j\right)\Delta z^k. \tag{28}$$

Theorem 1. *The α-log divergence recovers the original Riemannian metric.*

Proof. From (28), by differentiation, we easily have

$$\partial_i' \partial_j' D_\alpha\left[\boldsymbol{\xi} : \boldsymbol{\xi}'\right]\big|_{\boldsymbol{\xi}=\boldsymbol{\xi}'=\boldsymbol{\xi}_p} = g_{ij}\left(\boldsymbol{\xi}_p\right) \tag{29}$$

for any α connection in which $\Gamma_{ijk}^{(\alpha)}$ is used in (28).

Theorem 2. *The log divergence of M induces $\alpha = -3$ geometry.*

Proof. We calculate the dual affine connection $\Gamma_{ijk}^{*(X)}$ induced from $D\left[\boldsymbol{\xi} : \boldsymbol{\xi}'\right] = \|X\|^2$. It is given by

$$\Gamma_{ijk}^{*(X)} = -\partial_i' \partial_j' \partial_k D\left[\boldsymbol{\xi} : \boldsymbol{\xi}'\right]_{\boldsymbol{\xi}'=\boldsymbol{\xi}} \tag{30}$$

which is proved to be

$$\Gamma_{ijk}^{*(X)} = -\partial_k g_{ij} + 3\Gamma_{(ijk)}, \tag{31}$$

where (i, j, k) implies symmetrization with respect to three indices. From the identity

$$\partial_k g_{ij} = \Gamma_{kij} + \Gamma_{kji} + T_{ijk} \tag{32}$$
$$= \Gamma_{kji}^* + \Gamma_{kji}^* - T_{ijk} \tag{33}$$

derived from

$$\nabla_k g_{ij} = T_{kij}, \tag{34}$$

we have

$$\Gamma_{ijk}^{*(X)} = \Gamma_{ijk} - T_{ijk} = \Gamma_{ijk}^{(0)} - \frac{3}{2}T_{ijk} \tag{35}$$

and hence

$$\Gamma_{ijk}^{(X)} = \Gamma_{ijk}^{(0)} + \frac{3}{2}T_{ijk}. \tag{36}$$

This is confirmed by calculating $\partial_i \partial_j \partial_k' D\left[\boldsymbol{\xi} : \boldsymbol{\xi}'\right]$ directly. They show that the derived geometry is $\alpha = -3$ geometry.

4 Standard Divergence

From theorem 2, we see that when we use $-(1/3)$-connection to define the log map, it gives the original connections. Hence, we have the following theorem.

Theorem 3. *The* $-(1/3)$ *log map divergence recovers the original geometry.*

We call it a standard divergence. We can derive it uniquely for any manifold of dual connections. In the case of Riemannian geometry where $T_{ijk} = 0$, the log divergence is a symmetric divergence. The α-divergences are equal for any α. It gives a half of the square of the Riemannian distance. In the general case, it does not coincide with the canonical divergence unfortunately when M is dually flat.

Note that there are other ways of defining a standard divergence $D[p : q]$ that recovers the original dual geometry of M. We have obtained some of such standard divergence.

5 Divergence and Projection Theorem

Given a smooth submanifold $R \subset M$ and a point p outside R, we often search for the point $\hat{p} \in R$ that minimizes the divergence from p to R,

$$\hat{p} = \arg\min_{q \in R} D[p : q]. \tag{37}$$

When the geodesic connecting p and $\hat{p}$ is orthogonal to R, the minimizer $\hat{p}$ is given by the orthogonal geodesic projection of p to R.

This is a desired property of a divergence. We see that the canonical divergence in a dually flat M has this property.

When we consider a sphere centered at p and radius t measured by a divergence,

$$S_t = \{s \,|\, D[p : s] = t\}, \tag{38}$$

q is the point that first touches R as t increases. Therefore, for X satisfying $|X|^2 = 1$, let t_0 be

$$t_0 = \arg\min_t \left\{ \exp_p(tX) \in R \right\} \tag{39}$$

and let the minimum be attained by X_0. Then, we have

$$q = \exp_p (t_0 X_0) . \tag{40}$$

Therefore, for our standard divergence, the minimizing q is given by the $\alpha = -(1/3)$ geodesic in direction X_0.

However, it is not the orthogonal projection of p to R, because the $\alpha = -(1/3)$ geodesic is not orthogonal to S_t in general. This is shown as follows. We consider two geodesics

$$\ddot{\xi}(t, X) = -\Gamma_{jk}{}^i(x)\dot{\xi}^j\dot{\xi}^k \tag{41}$$

$$\ddot{\xi}(t, X + \delta X) = -\Gamma_{jk}{}^i(\xi + \delta\xi)\dot{\xi}^j\dot{\xi}^k, \tag{42}$$

where
$$\delta \xi^i = \xi^i(t, X + dX) - \xi^i(t, X). \tag{43}$$

Since $\delta \boldsymbol{\xi}$ represents tangent directions of S_t, the angle between $\delta \boldsymbol{\xi}$ and $\dot{\boldsymbol{\xi}}$ is given by their inner product. However, we have

$$\frac{d}{dt} \langle \delta \boldsymbol{\xi}, \dot{\boldsymbol{\xi}} \rangle = \langle \nabla^*_{\dot{\boldsymbol{\xi}}} \delta \boldsymbol{\xi}, \dot{\boldsymbol{\xi}} \rangle + \langle \delta \boldsymbol{\xi}, \nabla_{\dot{\boldsymbol{\xi}}} \dot{\boldsymbol{\xi}} \rangle \tag{44}$$

$$= \langle \nabla^*_{\dot{\boldsymbol{\xi}}} \delta \boldsymbol{\xi}, \dot{\boldsymbol{\xi}} \rangle. \tag{45}$$

Since $\langle \nabla^*_{\dot{\boldsymbol{\xi}}} \delta \boldsymbol{\xi}, \dot{\boldsymbol{\xi}} \rangle$ does not vanish in general, $\delta \boldsymbol{\xi}$ and $\dot{\boldsymbol{\xi}}$ is not orthogonal, although they are orthogonal at $t = 0$.

6 Conclusions

We have proposed the problem of obtaining a standard divergence in a dual manifold M that recovers the original geometry. We derived a unique standard divergence in a manifold of dual affine connections. However, it is not fully satisfactory.

Added Notes: We have succeeded to derive a better divergence that coincides with the canonical divergence in the dually flat case and with the α-divergence when the α-connection is used. A necessary condition for the projection theorem is also derived. These results will be published in future.

References

Amari, S., Nagaoka, H.: Methods of Information Geometry. American Mathematical Society and Oxford University Press, Providence (2000)

Eguchi, S.: Second order efficiency of minimum contrast estimators in a curved exponential family. Ann. Stat. **11**, 793–803 (1983)

Matumoto, T.: Any statistical manifold has a contrast function - On the C^3-functions taking the minimum at the diagonal of the product manifold. Hiroshima Math. J. **23**, 327–332 (1993)

Transformations and Coupling Relations for Affine Connections

James Tao[1] and Jun Zhang[2]([✉])

[1] Harvard University, Cambridge, MA 02138, USA
[2] University of Michigan, Ann Arbor, MI 49109, USA
junz@umich.edu

Abstract. The statistical structure on a manifold $\mathfrak{M}$ is predicated upon a special kind of coupling between the Riemannian metric g and a torsion-free affine connection ∇ on the $T\mathfrak{M}$, such that ∇g is totally symmetric, forming, by definition, a "Codazzi pair" $\{\nabla, g\}$. In this paper, we first investigate various transformations of affine connections, including additive translation (by an arbitrary (1,2)-tensor K), multiplicative perturbation (through an arbitrary invertible operator L on $T\mathfrak{M}$), and conjugation (through a non-degenerate two-form h). We then study the Codazzi coupling of ∇ with h and its coupling with L, and the link between these two couplings. We introduce, as special cases of K-translations, various transformations that generalize traditional projective and dual-projective transformations, and study their commutativity with L-perturbation and h-conjugation transformations. Our derivations allow affine connections to carry torsion, and we investigate conditions under which torsions are preserved by the various transformations mentioned above. Our systematic approach establishes a general setting for the study of Information Geometry based on transformations and coupling relations of affine connections – in particular, we provide a generalization of conformal-projective transformation.

1 Introduction

On the tangent bundle $T\mathfrak{M}$ of a differentiable manifold $\mathfrak{M}$, one can introduce two separate structures: affine connection ∇ and Riemannian metric g. The coupling of these two structures has been of great interest to, say, affine geometers and information geometers. When coupled, $\{\nabla, g\}$ is called a Codazzi pair e.g., [14,17], which is an important concept in PDEs and affine hypersurface theory [7,8,10–12,15,16]. Codazzi coupling of a metric and an affine connection is a defining characteristics of "statistical structure" [6] of the manifold of the probability functions [1], where the metric-connection pair arises from a general construction of divergence ("contrast") functions [20,21] or pre-contrast functions [2]. To investigate the robustness of the Codazzi structure, one would perturb the metric and perturb the affine connection, and examine whether, after perturbation, the resulting metric and connection will still maintain Codazzi coupling [13].

© Springer International Publishing Switzerland 2015
F. Nielsen and F. Barbaresco (Eds.): GSI 2015, LNCS 9389, pp. 326–339, 2015.
DOI: 10.1007/978-3-319-25040-3_36

Codazzi transform is a useful concept in coupling projective transform of a connection and conformal transformation of the Riemannian metric: the pair $\{\nabla, g\}$ is jointly transformed in such a way that Codazzi coupling is preserved, see [17]. This is done through an arbitrary function that transforms both the metric and connection. A natural question to ask is whether there is a more general transformation of the metric and of the connection that preserves the Codazzi coupling, such that Codazzi transform (with the freedom of one function) is a special case. In this paper, we provide a positive answer to this question. The second goal of this paper is to investigate the role of torsion in affine connections and their transformations. Research on this topic is isolated, and the general importance has not been appreciated.

In this paper, we will collect various results on transformations on affine connection and classify them through one of the three classes, L-perturbation, h-conjugation, and the more general K-translation. They correspond to transforming ∇ via a (1,1)-tensor, (0,2)-tensor, or (1,2)-tensor. We will investigate the interactions between these transformations, based on known results but generalizing them to more arbitrary and less restrictive conditions. We will show how a general transformation of a non-degenerate two-form and a certain transformation of the connection are coupled; here transformation of a connection can be through L-perturbation, h-conjugation, and K-translation which specialzes to various projective-like transformations. We will show how they are linked in the case when they are Codazzi coupled to a same connection ∇. The outcome are depicted in commutative diagrams as well as stated as Theorems. Finally, our paper will provide a generalization of the conformal-projective transformation mentioned above, with an additional degree of freedom, and specify the conditions under which such transformation preserves Codazzi pairing of g and ∇. Due to page limit, all proofs are omitted.

2   Transformations of Affine Connections

2.1   Affine Connections

An affine (linear) connection ∇ is an endomorphism of $T\mathfrak{M}$: $\nabla : (X, Y) \in T\mathfrak{M} \times T\mathfrak{M} \mapsto \nabla_X Y \in T\mathfrak{M}$ that is bilinear in the vector fields X, Y and that satisfies the *Leibniz rule*

$$\nabla_X(\phi Y) = X(\phi)Y + \phi \nabla_X Y,$$

for any smooth function ϕ on $\mathfrak{M}$. An affine connection specifies the manner parallel transport of tangent vectors is performed on a manifold. Associated with any affine connection is a system of auto-parallel curves (also called geodesics): the family of auto-parallel curves passing through any point on the manifold is called the *geodesic spray*. The torsion of a connection ∇ is characterized by the (1,2)-tensor *torsion tensor*

$$T^{\nabla}(X, Y) = \nabla_X Y - \nabla_Y X - [X, Y],$$

which characterizes how tangent spaces twist about a curve when they are parallel transported.

2.2 Three Kinds of Transformations

The space of affine connections is convex in the following sense: if $\nabla, \widetilde{\nabla}$ are affine connections, then so is $\alpha\nabla + \beta\widetilde{\nabla}$ for any $\alpha, \beta \in \mathfrak{R}$ so long as $\alpha + \beta = 1$. This normalization condition is needed to ensure that the Leibniz rule holds. For example, $\frac{1}{2}\nabla_X Y + \frac{1}{2}\widetilde{\nabla}_X Y$ is a connection, whereas $\frac{1}{2}\nabla_X Y + \frac{1}{2}\nabla_Y X$ is not; both are bilinear forms of X, Y.

Definition 1. *A transformation of affine connections is an arbitrary map from the set $\mathfrak{D}$ of affine connections ∇ of some differentiable manifold $\mathfrak{M}$ to $\mathfrak{D}$ itself.*

In this following, we investigate three kinds of transformations of affine connections:

(i) *translation* by a (1,2)-tensor;
(ii) *perturbation* by an invertible operator or (1,1)-tensor;
(iii) *conjugation* by a non-degenerate two-form or (0,2)-tensor.

Additive Transformation: K-translation

Proposition 1. *Given two affine connections ∇ and $\widetilde{\nabla}$, then their difference $K(X,Y) := \widetilde{\nabla}_X Y - \nabla_X Y$ is a $(1,2)$-tensor. Conversely, any affine connection $\widetilde{\nabla}$ arises this way as an additive transformation by a $(1,2)$-tensor $K(X,Y)$ from ∇:*

$$\widetilde{\nabla}_X Y = \nabla_X Y + K(X,Y).$$

It follows that a transformation T of affine connections is equivalent to a choice of $(1,2)$-tensor T_∇ for every affine connection ∇. Stated otherwise, given a connection, any other connection can be obtained in this way, i.e. by adding an appropriate (1,2)-tensor, which may or may not depend on ∇. When the (1,2)-tensor $K(X,Y)$ is independent of ∇, we say that $\widetilde{\nabla}_X Y$ is a K-*translation* of ∇.

Additive transformations obviously commute with each other, since tensor addition is commutative. So additive transformations from a given affine connection form a group.

For any two connections ∇ and $\widetilde{\nabla}$, their difference tensor $K(X,Y)$ decomposes in general as $\frac{1}{2}A(X,Y) + \frac{1}{2}B(X,Y)$ where A is symmetric and B is anti-symmetric. Since the difference between the torsion tensors of $\widetilde{\nabla}$ and ∇ is given by

$$T^{\widetilde{\nabla}}(X,Y) - T^\nabla(X,Y) = K(X,Y) - K(Y,X) = B(X,Y),$$

we have the following:

Proposition 2. *K-translaton of an affine connection preserves torsion if and only if K is symmetric: $K(X,Y) = K(Y,X)$.*

The symmetric part, $A(X,Y)$, of the difference tensor $K(X,Y)$ reflects a difference in the geodesic spray associated with each affine connection: $\widetilde{\nabla}$ and ∇ have the same families of geodesic spray if and only if $A(X,Y) = 0$.

The following examples are K-translations that will be discussed in great length later on:

(i) $\mathsf{P}^\vee(\tau) : \nabla_X Y \mapsto \nabla_X Y + \tau(X)Y$, called $\mathsf{P}^\vee$-*transformation*;
(ii) $\mathsf{P}(\tau) : \nabla_X Y \mapsto \nabla_X Y + \tau(Y)X$, called P-*transformation*;
(iii) $\mathsf{Proj}(\tau) : \nabla_X Y \mapsto \nabla_X Y + \tau(Y)X + \tau(X)Y$, called *projective transformation*;
(iv) $\mathsf{D}(h,\xi) : \nabla_X Y \mapsto \nabla_X Y - h(Y,X)\xi$, called D-*transformation*, or dual-projective transformation.

Here, τ is an arbitrary one-form or $(0,1)$-tensor, h is a non-degenerate two-form or $(0,2)$-tensor, X, Y, ξ are all vector fields. From Proposition 2, $\mathsf{Proj}(\tau)$ is always torsion-preserving, while $\mathsf{D}(h,\xi)$ is torsion-preserving when h is symmetric.

It is obvious that $\mathsf{Proj}(\tau)$ is the composition of $\mathsf{P}(\tau)$ and $\mathsf{P}^\vee(\tau)$ for any τ. This may be viewed as follows: the P-transformation introduces torsion in one direction, i.e. it adds $B(X,Y) := \tau(Y)X - \tau(X)Y$ to the torsion tensor, but the $\mathsf{P}^\vee$ transformation cancels out this torsion, by adding $-B(X,Y) = \tau(X)Y - \tau(Y)X$, resulting in a torsion-preserving transformation of $\mathsf{Proj}(\tau)$.

Any affine connection ∇ on $T\mathfrak{M}$ induces an action on $T^*\mathfrak{M}$. The action of ∇ on a one-form ω is defined as:

$$(\nabla_X \omega)(Y) = X(\omega(Y)) - \omega(\nabla_X Y).$$

When ∇ undergoes a K-translation, $\nabla_X Y \mapsto \nabla_X Y + K(X,Y)$ for a $(1,2)$-tensor K, then

$$(\nabla_X \omega)(Y) \mapsto (\nabla_X \omega)(Y) - \omega(K(X,Y)).$$

In particular, the transformation $\mathsf{P}^\vee(\tau)$ of a connection acting on $T\mathfrak{M}$ induces a change of $\mathsf{P}^\vee(-\tau)$ when the connection acts on $T^*\mathfrak{M}$.

Multiplicative Transformation: L-pertubation. Complementing the additive transformation, we define a "multiplicative" transformation of affine connections through an invertible operator $L : T\mathfrak{M} \to T\mathfrak{M}$.

Proposition 3. *([13]) Given an affine connection ∇ and an invertible operator L on $T\mathfrak{M}$, then $L^{-1}(\nabla_X(L(Y)))$ is also an affine connection.*

Definition 2. *Given a connection ∇, the L-perturbation of ∇, denoted variously ∇^L, $L(\nabla)$, or $\Gamma_L(\nabla)$, is an endomorphism of $T\mathfrak{M}$ defined as:*

$$\Gamma_L(\nabla) \equiv L(\nabla) \equiv \nabla^L_X Y \equiv L^{-1}(\nabla_X LY).$$

Proposition 4. *([13]) The L-perturbations form a group such that group composition is simply operator concatenation: $\Gamma_K \circ \Gamma_L = \Gamma_{LK}$ for invertible operators K and L.*

Conjugation Transformation by h. If h is any non-degenerate $(0,2)$-tensor, it induces isomorphisms $h(X,-)$ and $h(-,X)$ from vector fields X to one-forms. When h is not symmetric, these two isomorphisms are different. Given an affine connection ∇, we can take the covariant derivative of the one-form $h(Y,-)$ with respect to X, and obtain a corresponding one-form ω such that, when fixing Y,

$$\omega_X(Z) = X(h(Y,Z)) - h(Y, \nabla_X Z).$$

Since h is non-degenerate, there exists a U such that $\omega_X = h(U,-)$ as one-forms, so that

$$X(h(Y,Z)) = h(U(X,Y),Z) + h(Y, \nabla_X Z).$$

Defining $D(X,Y) := U(X,Y)$ gives a map from $T\mathfrak{M} \times T\mathfrak{M} \to T\mathfrak{M}$.

Proposition 5. *Taking $\widetilde{\nabla}_X Y := D(X,Y)$ gives an affine connection $\widetilde{\nabla}$ as induced from ∇.*

Definition 3. *This $\widetilde{\nabla}$ is called the* left-conjugate *of ∇ with respect to h. The map taking ∇ to $\widetilde{\nabla}$ will be denoted* Left(h). *Similarly, we have a right-conjugate of ∇ and an associated map* Right(h).

If $\tilde{h}(X,Y) := h(Y,X)$, then exchanging the first and second arguments of each h in the above derivation shows that Left(h) = Right$(\tilde{h})$ and Right(h) = Left$(\tilde{h})$. When h is symmetric or anti-symmetric, the left- and right-conjugates are equal; both reduce to the special case of the usual conjugate connection ∇^* with respect to h. In this case, conjugation is involutive: $(\nabla^*)^* = \nabla$.

For a non-degenerate (but not necessarily symmetric or anti-symmetric) h, if there exists a ∇ such that

$$Z(h(X,Y)) = h(\nabla_Z X, Y) + h(X, \nabla_Z Y),$$

then ∇ = Left$(h)(\nabla)$ = Right$(h)(\nabla)$; in this case, ∇ is said to be parallel to the two-form h. Because in this case, ∇ = Left$(\tilde{h})(\nabla)$ = Right$(\tilde{h})(\nabla)$, ∇ is also parallel to the two-form $\tilde{h}$.

2.3   Codazzi Coupling and Torsion Preservation

Codazzi Coupling of ∇ with Operator L. Let L be an isomorphism of the tangent bundle $T\mathfrak{M}$ of a smooth manifold $\mathfrak{M}$, i.e. L is a smooth section of the bundle End$(T\mathfrak{M})$ such that it is invertible everywhere, i.e. an invertible $(1,1)$-tensor.

Definition 4. *Let L be an operator, and ∇ an affine connection. We call $\{\nabla, L\}$ a* Codazzi pair *if $(\nabla_X L)Y$ is symmetric in X and Y. In other words, the following identity holds*

$$(\nabla_X L)Y = (\nabla_Y L)X. \tag{1}$$

Here $(\nabla_X L)Y$ is, by definition,

$$(\nabla_X L)Y = \nabla_X(L(Y)) - L(\nabla_X Y).$$

We have the following characterization of Codazzi relations between an invertible operator and a connection:

Proposition 6. *([13]) Let ∇ and $\widetilde{\nabla}$ be arbitrary affine connections, and L an invertible operator. Then the following statements are equivalent:*

1. *$\{\nabla, L\}$ is a Codazzi pair.*
2. *∇ and $\Gamma_L(\nabla)$ have equal torsions.*
3. *$\{\Gamma_L(\nabla), L^{-1}\}$ is a Codazzi pair.*

Proposition 7. *Let $\{\nabla, L\}$ be a Codazzi pair. Let A be a symmetric $(1,2)$-tensor, and $\widetilde{\nabla} = \nabla + A$. Then $\{\widetilde{\nabla}, L\}$ forms a Codazzi pair if and only if L is self-adjoint with respect to A:*

$$A(L(X), Y) = A(X, L(Y))$$

for all vector fields X and Y.

In other words, A-translation preserves the Codazzi pair relationship of ∇ with L iff L is a *self-adjoint* operator with respect to A.

   Therefore, for a fixed operator L, the Codazzi coupling relation can be interpreted as a quality of *equivalence classes* of connections modulo translations by symmetric $(1,2)$-tensors A with respect to which L is self-adjoint.

Codazzi Coupling of ∇ with $(0,2)$-tensor h. Now we investigate Codazzi coupling of ∇ with a non-degenerate $(0,2)$-tensor h. We introduce the $(0,3)$-tensor C defined by:

$$C(X, Y, Z) \equiv (\nabla_Z h)(X, Y) = Z(h(X, Y)) - h(\nabla_Z X, Y) - h(X, \nabla_Z Y). \quad (2)$$

The tensor C is called the *cubic form* associated with $\{\nabla, h\}$ pair. When $C = 0$, then we say that h is parallel with respect to ∇.

   Recall the definition of left-conjugate $\overset{*}{\nabla}$ with respect to a non-degenerate two-form h:

$$Z(h(X, Y)) = h(\overset{*}{\nabla}_Z X, Y) + h(X, \nabla_Z Y). \quad (3)$$

Using this relation in (2) gives

$$\begin{aligned} C(X, Y, Z) &\equiv (h(\overset{*}{\nabla}_Z X, Y) + h(X, \nabla_Z Y)) - h(\nabla_Z X, Y) - h(X, \nabla_Z Y) \\ &= h((\overset{*}{\nabla} - \nabla)_Z X, Y), \end{aligned}$$

so that

$$C(X, Y, Z) - C(Z, Y, X) = h(T^{\overset{*}{\nabla}}(Z, X) - T^{\nabla}(Z, X), Y),$$

or

$$(\nabla_Z h)(X, Y) - (\nabla_X h)(Z, Y) = h(T^{\widetilde{\nabla}}(Z, X) - T^{\nabla}(Z, X), Y).$$

The non-degeneracy of h implies that $C(X, Y, Z) = C(Z, Y, X)$ if and only if ∇ and $\widetilde{\nabla}$ have equal torsions. This motivates the following definition, in analogy with the previous subsection.

Definition 5. *Let h be a two-form, and ∇ an affine connection. We call $\{\nabla, h\}$ a Codazzi pair if $(\nabla_Z h)(X, Y)$ is symmetric in X and Z.*

The cubic form associated with the pair $\{\widetilde{\nabla}, h\}$, denoted as $\widetilde{C}$, is:

$$\widetilde{C}(X, Y, Z) \equiv (\widetilde{\nabla}_Z h)(X, Y) = Z(h(X, Y)) - h(\widetilde{\nabla}_Z X, Y) - h(X, \widetilde{\nabla}_Z Y).$$

We derive, analogously,

$$\widetilde{C}(X, Y, Z) = h(X, (\nabla - \widetilde{\nabla})_Z Y),$$

from which we obtain

$$\widetilde{C}(X, Y, Z) - \widetilde{C}(Z, Y, X) = h(X, T^{\nabla}(Y, Z) - T^{\widetilde{\nabla}}(Y, Z)).$$

Summarizing the above results, we have, in analogy with Proposition 6:

Proposition 8. *Let ∇ be an arbitrary affine connection, h be an arbitrary non-degenerate two-form, and $\widetilde{\nabla}$ denotes the left-conjugate of ∇ with respect to h. Then the following statements are equivalent:*

1. *$\{\nabla, h\}$ is a Codazzi pair.*
2. *∇ and $\widetilde{\nabla}$ have equal torsions.*
3. *$\{\widetilde{\nabla}, h\}$ is a Codazzi pair.*

This proposition says that an arbitrary affine connection ∇ and an arbitrary non-degenerate two-form h form a Codazzi pair precisely when ∇ and its left-conjugate $\widetilde{\nabla}$ with respect to h have equal torsions.

Note that the definition of Codazzi pairing of ∇ with h is with respect to the first slot of h, left-conjugate is a more useful concept. The left- and right-conjugate of a connection ∇ with respect to h can become one and the same, when (i) h is symmetric; or (ii) ∇ is parallel to h: $\nabla h = 0$. These scenarios will be discussed next.

From the definition of the cubic form (2), it holds that

$$(\nabla_Z \tilde{h})(X, Y) = C(Y, X, Z) = (\nabla_Z h)(Y, X)$$

where $\tilde{h}(X, Y) = h(Y, X)$. So $C(X, Y, Z) = C(Y, X, Z)$ holds for any vector fields X, Y, Z if and only if $h = \tilde{h}$, that is, h is symmetric.

Proposition 9. *For a non-degenerate two-form h, $\nabla h = 0$ if and only if ∇ equals its left (equivalently, right) conjugate with respect to h.*

Note that in this proposition, we do not require h to be symmetric.
The following standard definition is a special case:

Definition 6. *If g is a Riemannian metric, and ∇ an affine connection, the conjugate connection ∇^* is the left-conjugate (or equivalently, right-conjugate) of ∇ with respect to g. Denote $\mathsf{C}(g)$ as the involutive map that sends ∇ to ∇^*.*

This leads to the well-known result:

Corollary 7. *$\{\nabla, g\}$ is a Codazzi pair if and only if $\mathsf{C}(g)$ preserves the torsion of ∇.*

2.4   Linking Two Codazzi Couplings

In order to relate these two notions of Codazzi pairs, one involving perturbations via a operator L, and one involving conjugation with respect to a two-form h, we need the following definition:

Definition 8. *The left L-perturbation of a $(0,2)$-tensor h is the $(0,2)$-tensor $h_L(X,Y) := h(L(X),Y)$. Similarly, the right L-perturbation is given by $h^L(X,Y) := h(X,L(Y))$.*

Proposition 10. *Let h be a non-degenerate $(0,2)$-tensor. If $\widetilde{\nabla}$ is the left-conjugate of ∇ with respect to h, then the left-conjugate of ∇ with respect to h_L is $\Gamma_L(\widetilde{\nabla})$. Analogously, if $\widehat{\nabla}$ is the right-conjugate of ∇ with respect to h, then the right-conjugate of ∇ with respect to h^L is $\Gamma_L(\widehat{\nabla})$.*

Corollary 9. *Let $\widetilde{\nabla}$ be the left-conjugate of ∇ with respect to h. If $\{\nabla, h\}$ and $\{\widetilde{\nabla}, L\}$ are Codazzi pairs, then $\{\nabla, h_L\}$ is a Codazzi pair.*

The following result describes how L-perturbation of a two-form (i.e., a $(0,2)$-tensor) induces a corresponding "L-perturbation" on the cubic form $C(X,Y,Z)$ as defined in the previous subsubsection.

Proposition 11. *Let $h(X,Y)$ be a non-degenerate two-form and L be an invertible operator. Write $f := h_L$ for notational convenience. Then, for any connection ∇,*

$$C_f(X,Y,Z) = C_h(L(X),Y,Z) + h((\nabla_Z L)X,Y),$$

where C_f and C_h are the cubic tensors of ∇ with respect to f and h.

With the notion of L-perturbation of a two-form, we can now state our main theorem describing the relation between L-perturbation of an affine connection and h-conjugation of that connection.

Theorem 10. *Fix a non-degenerate $(0,2)$-tensor h, denote its L-perturbations $h_L(X,Y) = h(L(X),Y)$ and $h^L(X,Y) = h(X,L(Y))$ as before. For an arbitrary connection ∇, denote its left-conjugate (respectively, right-conjugate) of ∇ with respect to h as $\widetilde{\nabla}$ (respectively, $\widehat{\nabla}$). Then:*

(i) $\nabla h_L = 0$ *if and only if* $\Gamma_L(\widetilde{\nabla}) = \nabla$.
(ii) $\nabla h^L = 0$ *if and only if* $\Gamma_L(\widehat{\nabla}) = \nabla$.

This Theorem means that ∇ is parallel to h_L (respectively, h^L) if and only if the left (respectively, right) h-conjugate of the L-perturbation of ∇ is ∇ itself. In this case, L-perturbation of ∇ and h-conjugation of ∇ can be coupled to render the perturbed two-form parallel with respect to ∇. Note that in the above Theorem, there is no torsion-free assumption about ∇, no symmetry assumption about h, and no Codazzi pairing assumption of $\{\nabla, h\}$.

2.5   Commutation Relations Between Transformations

Definition 11. *Given a one-form τ, we define the following transformation of an affine connection ∇:*

(i) $\mathsf{P}^\vee$*-transformation, denoted* $\mathsf{P}^\vee(\tau) : \nabla_X Y \mapsto \nabla_X Y + \tau(X)Y$;
(ii) P*-transformation, denoted* $\mathsf{P}(\tau) : \nabla_X Y \mapsto \nabla_X Y + \tau(Y)X$.
(iii) *projective transformation, denoted* $\mathsf{Proj}(\tau) : \nabla_X Y \mapsto \nabla_X Y + \tau(Y)X + \tau(X)Y$.

All these are "translations" of an affine connection (see Sect. 2). The first two transformations, (i) and (ii), are "half" of the projective transformation in (iii). While the projective transformation Proj of ∇ preserves its torsion, both $\mathsf{P}^\vee$-transformation and P-transformation introduce torsion (in opposite amounts).

Definition 12. *Given a vector field ξ and a non-degenerate 2-form h, we define the* D-*transformation of an affine connection ∇ as*

$$\mathsf{D}(h, \xi) : \nabla_X Y \mapsto \nabla_X Y - h(Y, X)\xi.$$

Furthermore, the transformation $\widetilde{\mathsf{D}}(h, \xi)$ *is defined to be* $\mathsf{D}(\tilde{h}, \xi)$.

These transformations behave very nicely with respect to left and right h-conjugation, as well as L-perturbation. More precisely, we make the following definition:

Definition 13. *We call left (respectively right) h-image of a transformation of a connection the induced transformation on the left (respectively right) h-conjugate of that connection. Similarly, we call L-image of a transformation of a connection the induced transformation on the L-perturbation of that connection.*

Proposition 12. *The left and right h-images of $\mathsf{P}^\vee(\tau)$ are both $\mathsf{P}^\vee(-\tau)$.*

Proposition 13. *The L-image of $\mathsf{P}^\vee(\tau)$ is $\mathsf{P}^\vee(\tau)$ itself.*

Proposition 14. *If V is a vector field, so that $h(V, -)$ is a one-form, then the left h-image of $\mathsf{P}(h(V, -))$ is $\mathsf{D}(h, V)$, while the right h-image of $\mathsf{P}(h(-, V))$ is $\widetilde{\mathsf{D}}(h, V)$.*

Proposition 15. *The L-image of* $D(h, V)$ *is* $D(h_L, L^{-1}(V))$, *whereas the L-image of* $\tilde{D}(h, V)$ *is* $\tilde{D}(h^L, L^{-1}(V))$.

We summarize the above results in the following commutative prisms.

Theorem 14. *Let* h *be a non-degenerate two-form,* L *be an invertible operator,* Z *be a vector field, and* τ *be a one-form. Then we have four commutative prisms:*

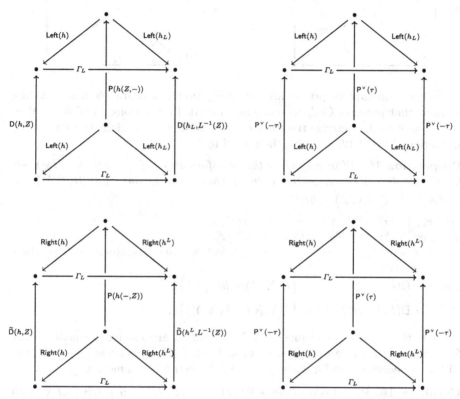

Corollary 15. *With respect to a Riemannian metric* g, *an invertible operator* L, *and an arbitrary one-form* τ, *we have the following commutative prisms:*
Each • *represents the space of affine connections of some differentiable manifold* $\mathfrak{M}$.

These commutative prisms are extremely useful in characterizing transformations that preserve Codazzi coupling. Indeed, Propositions 6 and 8 say that it is enough to characterize the torsion introduced by the various translations in Definitions 11 and 12. We have the following:

Proposition 16. *With respect to the transformation of connections:* $\nabla_X Y \mapsto \widetilde{\nabla}_X Y$, *let* $I(X,Y)$ *denote the induced change in torsion, i.e.* $B(X,Y) := T^{\widetilde{\nabla}}(X,Y) - T^{\nabla}(X,Y)$. *Then*

(i) *For* $\mathsf{P}^\vee(\tau)$: $B(X,Y) = \tau(X)Y - \tau(Y)X$.
(ii) *For* $\mathsf{P}(\tau)$: $B(X,Y) = \tau(Y)X - \tau(X)Y$.
(iii) *For* $\mathsf{Proj}(\tau)$: $B(X,Y) = 0$. *Projective transformations are torsion-preserving.*
(iv) *For* $\mathsf{D}(h,\xi)$: $B(X,Y) = \Big(h(X,Y) - h(Y,X)\Big)\xi$.
(v) *For* $\widetilde{\mathsf{D}}(h,\xi)$: $B(X,Y) = \Big(h(Y,X) - h(X,Y)\Big)\xi$.

Note that the torsion change $B(X,Y)$ is same in amount but opposite in sign for cases (i) and (ii), and for cases (iv) and (v). $B(X,Y)$ is always zero for case (iii), and becomes zero for cases (iv) and (v) when h is symmetric.

Corollary 16. $\mathsf{P}^\vee$-*transformations* $\mathsf{P}^\vee(\tau)$ *preserve Codazzi pairing of* ∇ *with* L: *For arbitrary one-form* τ, *if* $\{\nabla, L\}$ *is a Codazzi pair, then* $\{\mathsf{P}^\vee(\tau)\nabla, L\}$ *is a Codazzi pair.*

Corollary 17. D-*transformations* $\mathsf{D}(g,\xi)$ *preserve Codazzi pairing of* ∇ *with* L: *For any symmetric two-form* g, *vector field* ξ, *and operator* L *that is self-adjoint with respect to* g, *if* $\{\nabla, L\}$ *is a Codazzi pair, then* $\{\mathsf{D}(g,\xi)\nabla, L\}$ *is a Codazzi pair.*

2.6   The Conformal-Projective Transformation

The Codazzi transformation for a metric g and affine connection ∇ has been defined as

$$g(X,Y) \mapsto e^\phi g(X,Y)$$
$$\nabla_X Y \mapsto \nabla_X Y + X(\phi)Y + Y(\phi)X$$

for any smooth function ϕ. It is a known result that this preserves Codazzi pairs $\{\nabla, g\}$. In our language, this transformation can be described as follows: it is a (torsion-preserving) projective transformation $\mathsf{P}(d\phi)\mathsf{P}^\vee(d\phi)$ applied to ∇, and an L-perturbation $g \mapsto g_{e^\phi}$ applied to the metric, where e^ϕ is viewed as an invertible operator.

As generalizations to Codazzi transformation, researchers have introduced, progressively, the notions of 1-conformal transformation and α-conformal transformation in general [4], dual-projective transformation which is essentially (-1)-conformal transformation [3], and conformal-projective transformation [5], which encompass all previous cases.

Two statistical manifolds $(\mathfrak{M}, \nabla, g)$ and $(\mathfrak{M}, \nabla', g')$ are said to be *conformally-projectively equivalent* [5] if there exist two functions ϕ and ψ such that

$$\bar{g}(u, v) = e^{\phi + \psi} g(u, v),$$
$$\nabla'_u v = \nabla_u v - g(u, v)\mathrm{grad}_g \psi + \{d\phi(u)v + d\phi(v)u\}.$$

Note: $\phi = \psi$ yields conformal equivalency; $\phi = const$ yields 1-conformal (i.e., dual projective) equivalency; $\psi = const$ yields (-1)-conformal (i.e., projective) equivalency. It is shown [9] that when two statistical manifolds $(\mathfrak{M}, \nabla, g)$ and $(\mathfrak{M}, \nabla', g')$ are conformally-projectively equivalent, then $(\mathfrak{M}, \nabla^{(\alpha)}, g)$ and $(\mathfrak{M}, \nabla'^{(\alpha)}, g')$ are also conformally-projectively equivalent, with inducing functions $\phi^{(\alpha)} = \frac{1+\alpha}{2}\phi + \frac{1-\alpha}{2}\psi, \psi^{(\alpha)} = \frac{1-\alpha}{2}\psi + \frac{1+\alpha}{2}\phi$.

In our framework, we see that this transformation can be expressed as follows: it is an $e^{\psi+\phi}$-perturbation of the metric g, along with the affine connection transformation

$$\mathsf{D}(g, \mathrm{grad}_g \psi)\mathsf{Proj}(d\phi) = \mathsf{D}(g, \mathrm{grad}_g \psi)\mathsf{P}(d\phi)\mathsf{P}^\vee(d\phi).$$

The induced transformation on the conjugate connection will be

$$\Gamma_{e^{\psi+\phi}}\mathsf{P}(d\psi)\mathsf{D}(g, \mathrm{grad}_g \phi)\mathsf{P}^\vee(-d\phi) = P^\vee(d\phi + d\psi)\mathsf{P}(d\psi)\mathsf{D}(g, \mathrm{grad}_g \phi)\mathsf{P}^\vee(-d\phi)$$
$$= \mathsf{P}^\vee(d\psi)\mathsf{P}(d\psi)\mathsf{D}(g, \mathrm{grad}_g \phi)$$
$$= \mathsf{D}(g, \mathrm{grad}_g \phi)\mathsf{Proj}(d\psi),$$

which is a translation of the same form as before, but with ψ and ϕ exchanged. (The additional $\Gamma_{e^{\psi+\phi}}$ in front is induced by the $e^{\psi+\phi}$-perturbation of the metric.) In particular, this is a torsion-preserving transformation, because D and Proj are, which shows that conformal-projective transformations preserve Codazzi pairs $\{\nabla, g\}$. We can generalize the notion of conformal-projective transformation in the following way:

Definition 18. *Let V and W be vector fields, and L an invertible operator. A generalized conformal-projective transformation $\mathsf{CP}(V, W, L)$ consists of an L-perturbation of the metric g along with a torsion-preserving transformation $\mathsf{D}(g, W)\mathsf{Proj}(\tilde{V})$ of the connection, where $\tilde{V}$ is the one-form given by $\tilde{V}(X) := g(V, X)$ for any vector field X.*

Proposition 17. *A generalized conformal-projective transformation* $\mathsf{CP}(V, W, L)$ *induces the transformation* $\Gamma_L \mathsf{P}(\tilde{W})\mathsf{D}(g, V)\mathsf{P}^\vee(-\tilde{V})$ *on the conjugate connection.*

Proposition 18. *A generalized conformal-projective transformation* $\mathsf{CP}(V, W, L)$ *preserves Codazzi pairs* $\{\nabla, g\}$ *precisely when the torsion introduced by* Γ_L *cancels with that introduced by* $\mathsf{P}(\tilde{W})\mathsf{P}^\vee(-\tilde{V})$, *i.e.*

$$L^{-1}(\nabla_X(L(Y))) - L^{-1}(\nabla_Y(L(X))) - \nabla_X Y + \nabla_Y X = (\tilde{W} + \tilde{V})(X)Y + (\tilde{W} + \tilde{V})(Y)X.$$

Theorem 19. *A generalized conformal-projective transformation* $\mathsf{CP}(V, W, L)$ *preserves Codazzi pairs* $\{\nabla, g\}$ *if and only if* $L = e^f$ *for some smooth function* f, *and* $\tilde{V} + \tilde{W} = df$. *(The "only if" direction requires* $\dim \mathfrak{M} \geq 4$.*)*

This class of transformations is *strictly larger* than the class of conformal-projective transformations, since we may take $\tilde{V}$ to be an arbitrary one-form, not necessarily closed, and $\tilde{W} := df - \tilde{V}$ for some fixed smooth function f. The conformal-projective transformations result when f is itself the sum of two functions ϕ and ψ, in which case $df = d\phi + d\psi$ is a natural decomposition. Theorem 19 shows that the conformal-projective transformation admits interesting generalizations that preserve Codazzi pairs, by virtue of having an additional degree of freedom. This generalization demonstrates the utility of our "building block" transformations P, $\mathsf{P}^\vee$, D, and Γ_L in investigating Codazzi pairing relationships under general transformations of affine connections. Furthermore, this analysis shows that even torsion-free transformations may be effectively studied by decomposing them into elementary transformations that induce nontrivial torsions.

Acknowledgment. This work was completed while the second author (J.Z.) was on sabbatical visit at the Center for Mathematical Sciences and Applications at Harvard University under the auspices and support of Prof. S.T. Yau.

References

1. Amari, S., Nagaoka, H.: Method of Information Geometry. Oxford University Press, AMS Monograph, Providence, RI (2000)
2. Masayuki, H., Matsuzoe, H.: Geometry of precontrast functions and nonconservative estimating functions. In: International Workshop on Complex Structures, Integrability and Vector Fields, vol. 1340(1), AIP Publishing (2011)
3. Ivanov, S.: On dual-projectively flat affine connections. J. Geom. **53**(1–2), 89–99 (1995)
4. Kurose, T.: On the divergences of 1-conformally flat statistical manifolds. Tohoku Math. J. Second Ser. **46**(3), 427–433 (1994)
5. Kurose, T.: Conformal-projective geometry of statistical manifolds. Interdisc. Inf. Sci. **8**(1), 89–100 (2002)
6. Lauritzen, S.: Statistical manifolds. In: Amari, S., Barndorff-Nielsen, O., Kass, R., Lauritzen, S., Rao, C.R. (eds.) Differential Geometry in Statistical Inference, vol. 10, pp. 163–216. IMS Lecture Notes, Hayward, CA (1987)

7. Liu, H.L., Simon, U., Wang, C.P.: Codazzi tensors and the topology of surfaces. Ann. Glob. Anal. Geom. **16**(2), 189–202 (1998)
8. Liu, H.L., Simon, U., Wang, C.P.: Higher order Codazzi tensors on conformally flat spaces. Contrib. Algebra Geom. **39**(2), 329–348 (1998)
9. Matsuzoe, H.: On realization of conformally-projectively flat statistical manifolds and the divergences. Hokkaido Math. J. **27**, 409–421 (1998)
10. Nomizu, K., Sasaki, T.: Affine Differential Geometry - Geometry of Affine Immersions. Cambridge University Press, Cambridge (1994)
11. Pinkall, U., Schwenk-Schellschmidt, A., Simon, U.: Geometric methods for solving Codazzi and Monge-Ampere equations. Math. Ann. **298**(1), 89–100 (1994)
12. Schwenk-Schellschmidt, Angela, Udo Simon, and Martin Wiehe.: Generating higher order Codazzi tensors by functions. TU, Fachbereich Mathematik (3), 1998
13. Schwenk-Schellschmidt, A., Simon, U.: Codazzi-equivalent affine connections. RM **56**(1–4), 211–229 (2009)
14. Simon, U.: Codazzi tensors. In: Ferus, D., Ktihnel, W., Simon, U., Wegner, B.: Global Differential Geometry and Global Analysis, vol. 838, pp. 289–296. Springer, Berlin Heidelberg (1981)
15. Simon, U., Schwenk-Schellschmidt, A., Viesel, H.: Introduction to the Affine Differential Geometry of Hypersurfaces. Lecture Notes Science University of Tokyo, Tokyo (1991)
16. Simon, U.: Transformation techniques for partial differential equations on projectively flat manifolds. RM **27**(1–2), 160–187 (1995)
17. Simon, U.: Affine differential geometry. In: Dillen, F., Verstraelen, L. (eds.) Handbook of Differential Geometry, vol. 1, pp. 905–961. Elsevier Science (2000)
18. Simon, U.: Affine hypersurface theory revisited: gauge invariant structures. Russian Mathematics c/c of Izvstiia-Vysshie Uchebnye Zavedeniia Matematika **48**(11), 48 (2004)
19. Uohashi, K.: On α-conformal equivalence of statistical manifolds. J. Geom. **75**, 179–184 (2002)
20. Zhang, J.: Divergence function, duality, and convex analysis. Neural Comput. **16**, 159–195 (2004)
21. Zhang, J.: Nonparametric information geometry: from divergence function to referential-representational biduality on statistical manifolds. Entropy **15**, 5384–5418 (2013)

Online k-MLE for Mixture Modeling
with Exponential Families

Christophe Saint-Jean[1]([✉]) and Frank Nielsen[2]

[1] Mathématiques, Image, Applications (MIA), Université de La Rochelle,
La Rochelle, France
christophe.saint-jean@univ-lr.fr
[2] LIX, École Polytechnique, Palaiseau, France

Abstract. This paper address the problem of online learning finite
statistical mixtures of exponential families. A short review of the
Expectation-Maximization (EM) algorithm and its online extensions is
done. From these extensions and the description of the k-Maximum Likelihood Estimator (k-MLE), three online extensions are proposed for this
latter. To illustrate them, we consider the case of mixtures of Wishart
distributions by giving details and providing some experiments.

Keywords: Mixture modeling · Online learning · k-MLE · Wishart
distribution

1 Introduction

Mixture models are a powerful and flexible tool to model an unknown smooth
probability density function as a weighted sum of parametric density functions
$f_j(x; \theta_j)$:

$$f(x; \theta) = \sum_{j=1}^{K} w_j f_j(x; \theta_j), \text{ with } w_j > 0 \text{ and } \sum_{j=1}^{K} w_j = 1, \tag{1}$$

where K is the number of components of the mixture. The maximum likelihood
principle is a popular approach to find the unknown parameters $\theta = \{(w_j, \theta_j)\}_j$
of f. Given $\chi = \{x_i\}_{i=1}^{N}$ a set of N independent and identically distributed
observations, the maximum likelihood estimator $\hat{\theta}^{(N)}$ is defined as the maximizer
of the likelihood, or equivalently of the average log-likelihood:

$$\bar{l}(\theta; \chi) = N^{-1} \sum_{i=1}^{N} \log \sum_{j=1}^{K} w_j f_j(x_i; \theta_j). \tag{2}$$

For $K > 1$, the sum of terms appearing inside a logarithm makes this optimization quite difficult.

    The goal of this paper is to propose such a kind of estimator but for the online
setting, that is when observations x_i are available one after another. This case

© Springer International Publishing Switzerland 2015
F. Nielsen and F. Barbaresco (Eds.): GSI 2015, LNCS 9389, pp. 340–348, 2015.
DOI: 10.1007/978-3-319-25040-3_37

appears when dealing with data streams or when data sets are large enough not to fit in memory. Ideally, online methods aim to get same convergence properties as batch ones while having a single pass over the dataset. This topic receives increasing attention due to the recent challenges associated to massive datasets.

The paper is organized as follows: Sect. 2 recalls the basics of Expectation-Maximization (EM) algorithm and some of its online extensions. Section 3 describes the k-MLE technique which is derived from the formalism of EM. In the same section, two online versions of k-MLE are proposed and detailed. Section 4 gives an example of the mixture of Wishart distributions and provides some experiments before concluding in Sect. 5.

2 A Short Review of Online Mixture Learning

Before reviewing some online methods, one has to recall the basics of mixture modeling with the Expectation-Maximization (EM) algorithm [1] in the batch setting.

2.1 EM for Mixture Learning

Let Z_i be a categorical random variable over $1, ..., K$ whose parameters are $\{w_j\}_j$, that is, $Z_i \sim \mathrm{Cat}_K(\{w_j\}_j)$. Also, assuming that $X_i | Z_i = j \sim f_j(\cdot; \theta_j)$, the unconditional mixture distribution f in Eq. 1 is recovered by marginalizing their joint distribution over Z_i. Obviously, Z_i is a latent (unobservable) variable so that the realizations x_i of X_i (resp. (x_i, z_i) of (X_i, Z_i)) is often viewed as an incomplete (resp. complete) data observation. For convenience, we consider in the following that Z_i is a random vector $[Z_{i,1}, Z_{i,2}, \ldots, Z_{i,k}]$ where $Z_{i,j} = 1$ iff. X_i arises from the j-th component of the mixture and 0 otherwise[1]. Similarly to Eq. 2, the average complete log-likelihood function can be written as:

$$\bar{l}_c(\theta; \chi_c) = N^{-1} \sum_{i=1}^{N} \log \prod_{j=1}^{K} (w_j f_j(x_i; \theta_j))^{z_{i,j}},$$

$$= N^{-1} \sum_{i=1}^{N} \sum_{j=1}^{K} z_{i,j} \log(w_j f_j(x_i; \theta_j)), \tag{3}$$

where $\chi_c = \{(x_i, z_i)\}_{i=1}^{N}$, is the set of complete data observations. Here comes the EM algorithm which optimizes $\bar{l}(\theta; \chi)$ (proof in [1]) by repeating two steps until convergence. For iteration t:

E-Step Compute $\mathcal{Q}(\theta; \hat{\theta}^{(t)}, \chi) = \mathbb{E}_{\hat{\theta}^{(t)}}[\bar{l}_c(\theta; \chi_c) | \chi]$. Since $\bar{l}_c$ is linear in $z_{i,j}$, this step amounts to compute:

$$\hat{z}_{i,j}^{(t)} = \mathbb{E}_{\hat{\theta}^{(t)}}[Z_{i,j} = 1 | X_i = x_i] = \frac{\hat{w}_j^{(t)} f_j(x_i; \hat{\theta}_j^{(t)})}{\sum_{j'} \hat{w}_{j'}^{(t)} f_{j'}(x_i; \hat{\theta}_{j'}^{(t)})}. \tag{4}$$

---

[1] Thus, Z_i is distributed according to the multinomial law $\mathcal{M}_K(1, \{w_j\}_j)$.

M-Step Update mixture parameters by maximizing $\mathcal{Q}$ over θ (*i.e.*, Eq. 3 where hidden values $z_{i,j}$ are replaced by $\hat{z}_{i,j}^{(t)}$).

$$\hat{w}_j^{(t+1)} = \frac{\sum_{i=1}^{N} \hat{z}_{i,j}^{(t)}}{N}, \quad \hat{\theta}_j^{(t+1)} = \underset{\theta_j \in \Theta_j}{\arg\max} \sum_{i=1}^{N} \hat{z}_{i,j}^{(t)} \log\left(f_j(x_i; \theta_j)\right) \quad (5)$$

Remark that while $\hat{w}_j^{(t+1)}$ is always known in closed-form whatever f_j are, $\hat{\theta}_j^{(t+1)}$ are obtained by component-wise specific optimization involving **all** observations.

More generally, the improvement of $\bar{l}(\theta; \chi)$ is guaranteed whatever the increase of $\mathcal{Q}$ is in the M-Step. This leads to the Generalized EM algorithm (GEM) when partial maximization is performed.

2.2   Online Extensions

For the online setting, it is now more appropriate to denote $\hat{\theta}^{(N)}$ the current parameter estimate instead of $\hat{\theta}^{(t)}$. In the literature, we mainly distinguish two approaches according to whether the initial structure of EM (alternate optimization) is kept or not.

The first online algorithm, due to Titterington [2], corresponds to the direct optimization of $\mathcal{Q}(\theta; \hat{\theta}^{(N)}, \chi)$ using a second-order stochastic gradient ascent:

$$\hat{\theta}^{(N+1)} = \hat{\theta}^{(N)} + \gamma^{(N+1)} I_c^{-1}(\hat{\theta}^{(N)}) \nabla_\theta \log f(x_{N+1}; \hat{\theta}^{(N)}), \quad (6)$$

where $\{\gamma_N\}$ is a decreasing sequence of positive step sizes ($\gamma_N = N^{-1}$ in the original paper) and the hessian $\nabla^2 \mathcal{Q}$ of $\mathcal{Q}$ is approximated by the Fisher Information matrix I_c for the complete data ($I_c(\hat{\theta}^{(N)}) = -\mathbb{E}_{\hat{\theta}_j^{(N)}}[\frac{\log p(x,z;\theta)}{\partial \theta^t \partial \theta}]$). A major issue with that method is that $\hat{\theta}^{(N)}$ does not necessarily follow the parameters constraints.

This problem is coped by the approach of Cappé and Moulines [3] who proposed to replace the E-Step by a stochastic approximation step:

$$\hat{\mathcal{Q}}^{(N+1)}(\theta; \hat{\theta}^{(N)}, \chi^{(N+1)}) = \hat{\mathcal{Q}}^{(N)}(\theta; \hat{\theta}^{(N)}, \chi^{(N)})$$
$$+ \gamma_{N+1}(\mathbb{E}_{\hat{\theta}^{(N)}}[\bar{l}_c(\theta; \{x_{N+1}, z_{N+1}\})|x_{N+1}] - \hat{\mathcal{Q}}^{(N)}(\theta; \hat{\theta}^{(N)}, \chi^{(N)})). \quad (7)$$

Since the M-Step remains unchanged (maximizing the function $\theta \mapsto \hat{\mathcal{Q}}^{(N+1)}(\theta)$), the constrains on parameters are automatically satisfied. This method is the starting point of our proposals. One may also mention the "Incremental EM" [4] which is not detailed here. Note that previous formalisms are not limited to mixture models.

3 Online k-Maximum Likelihood Estimator

3.1 k-MLE for Mixture Learning

In this section, we describe the k-MLE algorithm, a faster alternative to EM, as introduced in [5]. The goal is now to maximize directly $\bar{l}_c(\theta; \chi_c)$. In the above description of EM, value $\hat{z}_{i,j}^{(t)}$ may be interpreted as a soft membership of x_i to the j-th component of the mixture. More generally, all values $\hat{z}_{i,j}^{(t)}$ represent a soft partition of χ which may be denoted by $\hat{Z}^{(t)}$. For fixed values of θ, the partition which maximizes $\bar{l}_c$ is a strict one:

$$\max_Z \bar{l}_c(\theta; \chi_c) = N^{-1} \sum_{i=1}^{N} \max_{j=1}^{K} \log \left(w_j f_j(x_i; \theta_j) \right). \tag{8}$$

Doing such a maximization (also called C-Step in Classification EM algorithm [6]) after the E-Step in EM induces a split of χ into K subsets ($\chi = \bigsqcup_{j=1}^{K} \hat{\chi}_j^{(t)}$). Later on, note $\tilde{z}_{i,j}^{(t)}$ the hard membership of x_i at iteration t. Then, for a fixed optimal partition, the M-step is simpler:

$$\hat{w}_j^{(t+1)} = \frac{|\hat{\chi}_j^{(t)}|}{N}, \quad \hat{\theta}_j^{(t+1)} = \arg\max_{\theta_j \in \Theta_j} \sum_{x \in \hat{\chi}_j^{(t)}} \log f_j(x; \theta_j) \tag{9}$$

The gain in computation time is obvious since a weighted MLE involving all observations is replaced by an unweighted MLE for each subset $\hat{\chi}_j^{(t)}$. The algorithm is described in Algorithm 1.

Algorithm 1. k-MLE (Lloyd's batch method)

Input: A sample $\chi = \{x_1, x_2, ..., x_N\}$
Output: Estimate $\hat{\theta}$ of mixture parameters
1 A good initialization for $\hat{\theta}^{(0)}$ (see [5]);   t = 0;
2 **repeat**
3 | Partition $\chi = \bigsqcup_{j=1}^{K} \hat{\chi}_j^{(t)}$ according to $\log \hat{w}_j^{(t)} f_j(x_i; \hat{\theta}_j^{(t)})$;   // max. w.r.t. Z
  | **foreach** $\chi_j^{(t)}$ **do**
4 | | $\hat{w}_j^{(t+1)} = N^{-1} |\hat{\chi}_j^{(t)}|$;   // max. w.r.t. w_j's
  | | $\hat{\theta}_j^{(t+1)} = \arg\max_{\theta_j \in \Theta_j} \sum_{x \in \hat{\chi}_j^{(t)}} \log f_j(x; \theta_j)$;   // max. w.r.t θ_j's
5 | t = t + 1;
6 **until** *Convergence of the complete likelihood*;

3.2 Proposed Online Extensions

In order keep ideas from online EM (stochastic E-Step) and from k-MLE (hard partition), the only possible modifications concern the assignment $z_{N+1}^{(N)}$ of the new observation x_{N+1}.

1. *Online k-MLE:* The most obvious heuristic is to maximize the complete log-likelihood for x_{N+1}. Indeed, unless all data is kept in memory, previous assignments for past observations are fixed. Note that these assignments are computed in order with mixture parameters $\theta^{(0)}, \theta^{(1)}, \ldots, \theta^{(N-1)}$. This leads to the following rule:

$$\tilde{z}_{i,j}^{(i)} = 1 \text{ if } j = \arg\max_{j'=1..K} \log(\hat{w}_{j'}^{(i-1)} f_{j'}(x_i; \hat{\theta}_{j'}^{(i-1)})) \text{ and } 0 \text{ otherwise.} \qquad (10)$$

Clearly, this choice leads to a method which is similar to the Online CEM algorithm [7]. Under the assumption that components are modeled by isotropic gaussian, the MacQueens single-point iterative k-means [8] is also recovered.

2. *Online Stochastic k-MLE:* It is well-known that the strict partitioning can give poor results in the batch setting when mixture components are not well separated. This suggests to relax the strict maximisation and replace it by a sampling from the multinomial distribution

$$\tilde{z}_i^{(i)} \text{ sampled from } \mathcal{M}_K(1, \{p_j = \log(\hat{w}_j^{(i-1)} f_j(x_i; \hat{\theta}_j^{(i-1)}))\}_j). \qquad (11)$$

Same kind of strategy was used in the Stochastic EM algorithm [6].

3. *Online Hartigan k-MLE:* Analogously to the Hartigan's version of k-MLE [9], one can select among all possible assignments of x_{N+1} the one which maximizes its complete likelihood **after** the M-step:

$$\tilde{z}_{i,j}^{(i)} = 1 \text{ if } j = \arg\max_{j'=1..K} \log(\hat{w}_{j'}^{(i)} f_{j'}(x_i; \theta_{j'}^{(i)})) \text{ and } 0 \text{ otherwise.} \qquad (12)$$

Obviously, this heuristic induces a computational overhead since K M-Steps have to be done.

To be useful, these methods require to be able to efficiently compute the MLE for components parameters. In the following, we give details for the case where these components belong to a (regular) exponential family (EF):

$$f_j(x; \theta_j) = \exp\{\langle t(x), \theta_j \rangle + k(x) - F(\theta_j)\},$$

with $t(x)$ the sufficient statistic, θ_j the natural parameter, $k(\cdot)$ the carrier measure and F the log-normalizer [10]. Under this assumption, the probability density function $p(x, z; \theta)$ is an EF^2 which can be written for the i-th observation as:

---

[2] The multinomial distribution is also an exponential family.

$$\log p(x_i, z_i; \theta) = \sum_{j=1}^{K} \langle z_{i,j}, \log w_j \rangle + \sum_{j=1}^{K} \langle z_{i,j} t(x_i), \theta_j \rangle$$

$$+ \sum_{j=1}^{K} z_{i,j} k(x_i) - \sum_{j=1}^{K} z_{i,j} F(\theta_j) \quad (13)$$

Taking into account the summation constraint for w_j's, the M-Step reduces to simple update formulas:

$$\hat{w}_j^{(N+1)} = (N+1)^{-1} \sum_{i=1}^{N+1} \tilde{z}_{i,j}^{(i-1)}, \quad (14)$$

$$\hat{\eta}_j^{(N+1)} = (\sum_{i=1}^{N+1} \tilde{z}_{i,j}^{(i-1)})^{-1} \sum_{i=1}^{N+1} \tilde{z}_{i,j}^{(i-1)} t(x_i), \quad (15)$$

where $\eta_j = \nabla F(\theta_j)$ is the expectation parameter for the j-th component (see details in [10]). Remark that these formulas can be easily turned into recursive ones:

$$\hat{w}_j^{(N+1)} = \hat{w}_j^{(N)} + (N+1)^{-1} \left(\tilde{z}_{N+1,j}^{(N)} - \hat{w}_j^{(N)} \right), \quad (16)$$

$$\hat{\eta}_j^{(N+1)} = \hat{\eta}_j^{(N)} + (N+1)^{-1} \left(\tilde{z}_{N+1,j}^{(N)} t(x_{N+1}) - \hat{\eta}_j^{(N)} \right). \quad (17)$$

Clearly, one can recognize a step of the stochastic gradient ascent method in the expectation parameter space. Note that the functional reciprocal ∇F^{-1} must be computable to get back into natural parameter space. Algorithm 2 summarizes the Online Stochastic k-MLE.

4 Example: Mixture of Wishart Distributions

4.1 Wishart Distribution is a Canonical (Curved) Exponential Family

The (central) Wishart distribution [11] is the multidimensional version of the chi-square distribution and it characterizes empirical scatter matrix estimator for the multivariate gaussian distribution $\mathcal{N}_d(\mathbf{0}, S)$. Its density function can be decomposed as

$$\mathcal{W}_d(X; \theta_n, \theta_S) = \exp\left\{ \langle \theta_n, \log|X| \rangle_{\mathbb{R}} + \langle \theta_S, -\frac{1}{2}X \rangle_F + k(X) - F(\theta_n, \theta_S) \right\} \quad (18)$$

where $(\theta_n, \theta_S) = (\frac{n-d-1}{2}, S^{-1})$, $t(X) = (\log|X|, -\frac{1}{2}X)$, $k(X) = 0$ and

$$F(\theta_n, \theta_S) = \left(\theta_n + \frac{(d+1)}{2} \right) (d\log(2) - \log|\theta_S|) + \log \Gamma_d \left(\theta_n + \frac{(d+1)}{2} \right),$$

where $\Gamma_d(y) = \pi^{\frac{d(d-1)}{4}} \prod_{j=1}^{d} \Gamma\left(y - \frac{j-1}{2}\right)$ is the multivariate gamma function defined on $\mathbb{R}_{>0}$. $\langle a, b \rangle_{\mathbb{R}} = a^\top b$ denotes the scalar product and $\langle A, B \rangle_F = \text{tr}(A^\top B)$ the Fröbenius inner product (with tr the matrix trace operator). Note that this decomposition is not unique.

Algorithm 2. Online Stochastic k-MLE for (curved) exponential families

---

Input: A sample generator $G = x_1, x_2, \ldots$ yielding a data stream of observations, a batch algorithm B for the same problem, N_w a positive integer

Output: For each observation x_{N+1} with $N \geq N_w$ an estimate $\hat{\theta}^{(N+1)}$ of mixture parameters is yielded

`// Warm-Up-Step`

1 Get $\hat{\theta}^{(N)} = \{\hat{w}_j^{(N)}, \hat{\theta}_j^{(N)}\}_j$ from B with the N_w first observations of G;

2 $N = N_w$;

3 **foreach** *component j in mixture* **do** $\hat{\eta}_j^{(N)} = \nabla F(\hat{\theta}_j^{(N)})$ **foreach** *new value x_{N+1} from G* **do**

4       $\tilde{z}_{N+1}^{(N)}$ sampled from $\mathcal{M}_K(1, \{p_j = \log(\hat{w}_j^{(N)} f_j(x_{N+1}; \hat{\theta}_j^{(N)}))\}_j)$;

5   **foreach** *component j in mixture* **do**

6           $\hat{w}_j^{(N+1)} = \hat{w}_j^{(N)} + (N+1)^{-1}\left(\tilde{z}_{N+1,j}^{(N)} - \hat{w}_j^{(N)}\right)$;

7           $\hat{\eta}_j^{(N+1)} = \hat{\eta}_j^{(N)} + (N+1)^{-1}\left(\tilde{z}_{N+1,j}^{(N)} t(x_{N+1}) - \hat{\eta}_j^{(N)}\right)$;

8   yield mixture parameters $\hat{\theta}^{(N+1)} = \{\hat{w}_j^{(N+1)}, \hat{\theta}_j^{(N+1)} = (\nabla F)^{-1}(\hat{\eta}_j^{(N+1)})\}_j$;

9       $N = N + 1$;

4.2   Details for the M-Step

Recall that find the MLE amounts to compute $(\nabla F)^{-1}$ on the average of sufficient statistics. In this specific case, the following system has to be inverted to get values of (θ_n, θ_S) given (η_n, η_S):

$$d\log(2) - \log|\theta_S| + \Psi_d\left(\theta_n + \frac{(d+1)}{2}\right) = \eta_n, \tag{19a}$$

$$-\left(\theta_n + \frac{(d+1)}{2}\right)\theta_S^{-1} = \eta_S, \tag{19b}$$

where Ψ_d the derivative of the log Γ_d should be inverted. As far as we know, no closed-form solution exists but it can be easily solve numerically:

- Isolate θ_S in Eq. 19b: $\theta_S = \left(\theta_n + \frac{(d+1)}{2}\right)(-\eta_S)^{-1}$
- Plug it in Eq. 19a and solve numerically the following one dimensional problem:

$$d\log(2) - d\log\left(\theta_n + \frac{(d+1)}{2}\right) + \log|-\eta_S| + \Psi_d\left(\theta_n + \frac{(d+1)}{2}\right) - \eta_n = 0 \tag{20}$$

with any root-finding method on $]d-1, +\infty[$.
- Substitute the solution into Eq. 19b and solve the value for θ_S.

Whole process gives the $(\nabla F)^{-1}$ function mentioned in line 8 of Algorithm 2.

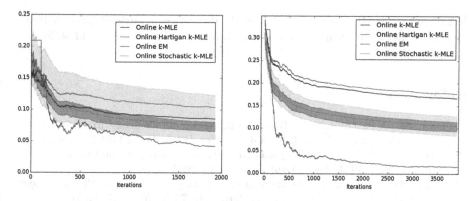

Fig. 1. $\mathrm{KL}(f(\cdot;\theta^{true}) \| f(\cdot;\hat{\theta}^{(N)}))$ for $K = 3$ (left) and $K = 10$ (right)

Line 8 of Algorithm 2 amounts to compute the following update formulas:

$$\hat{\eta}_{n_j}^{(N+1)} = \hat{\eta}_{n_j}^{(N)} + (N+1)^{-1}\left(\tilde{z}_{N+1,j}^{(N)} \log |X_{N+1}| - \hat{\eta}_{n_j}^{(N)}\right), \qquad (21)$$

$$\hat{\eta}_{S_j}^{(N+1)} = \hat{\eta}_{S_j}^{(N)} - (N+1)^{-1}\left(\tilde{z}_{N+1,j}^{(N)} \tfrac{1}{2} X_{N+1} + \hat{\eta}_{S_j}^{(N)}\right). \qquad (22)$$

4.3 Experiment on Synthetic Data-Sets

In this section, we provide a preliminary empirical analysis of our proposed methods. The protocol is the following: pick a random Wishart mixture for $K = 3$ components (left) or $K = 10$ components (right), compute the Kullback-Leibler divergence between the "true" mixture and the one yielded every iteration using a Monte Carlo approximation (10^4 samples). The initialization mixture $\hat{\theta}^{(0)}$ is computed with k-MLE for the first 100 observations. The simulations are repeated 30 times for the Online Stochastic k-MLE so that it is possible to compute mean, min, max and the first and third quartiles. Also, results of online EM are reported.

From Fig. 1, one can observe a clear hierarchy between the algorithms especially when $K = 10$. One may guess that this dataset corresponds to the case when clusters components are overlapping more. Thus, the soft assignment in online EM outperforms other methods with an additional computational cost (i.e. all sufficient statistics and cluster parameters have to be updated). The proof of convergence of Online Stochastic k-MLE still remain to be done while the reader may refer to the Sect. 3.5 of the article [7] for Online k-MLE.

5   Conclusion

This paper addresses the problem of online learning of finite statistical mixtures with a special focus on curved exponential families. The proposed methods are fast since they require only one pass over the data stream. Further speed increase may be achieved by using distributed computing for partial sums of sufficient statistics (see [12]).

References

1. Dempster, A.P., Laird, N.M., Rubin, D.B.: Maximum likelihood from incomplete data via the EM algorithm. J. R. Stat. Soc. Ser. B (Methodol.) **39**(1), 1–38 (1977)
2. Titterington, D.M.: Recursive parameter estimation using incomplete data. J. R. Stat. Soc. Ser. B (Methodol.) **46**(2), 257–267 (1984)
3. Cappé, O., Moulines, E.: On-line expectation-maximization algorithm for latent data models. J. R. Stat. Soc. Ser. B (Methodol.) **71**(3), 593–613 (2009)
4. Neal, R.M., Hinton, G.E.: A view of the EM algorithm that justifies incremental, sparse, and other variants. In: Jordan, M.I. (ed.) Learning in Graphical Models, pp. 355–368. MIT Press, Cambridge (1999)
5. Nielsen, F.: On learning statistical mixtures maximizing the complete likelihood. In: Bayesian Inference and Maximum Entropy Methods in Science and Engineering (MaxEnt 2014), AIP Conference Proceedings Publishing, 1641, pp. 238–245 (2014)
6. Celeux, G., Govaert, G.: A classification EM algorithm for clustering and two stochastic versions. Comput. Stat. Data Anal. **14**(3), 315–332 (1992)
7. Samé, A., Ambroise, C., Govaert, G.: An online classification EM algorithm based on the mixture model. Stat. Comput. **17**(3), 209–218 (2007)
8. MacQueen, J.: Some methods for classification and analysis of multivariate observations. In: Proceedings of the Fifth Berkeley Symposium on Mathematical Statistics and Probability, 1(14) (1967)
9. Saint-Jean, C., Nielsen, F.: Hartigan's method for k-MLE : mixture modeling with Wishart distributions and its application to motion retrieval. In: Nielsen, F. (ed.) Geometric Theory of Information. Signals and Communication Technology, pp. 301–330. Springer, Switzerland (2014)
10. Nielsen, F., Garcia, V.: Statistical exponential families: a digest with flash cards, November 2009. http://arxiv.org/abs/0911.4863
11. Wishart, J.: The generalised product moment distribution in samples from a normal multivariate population. Biometrika **20**(1/2), 32–52 (1928)
12. Liu, Q., Ihler, A.T.: Distributed estimation, information loss and exponential families. In: Ghahramani, Z., Welling, M., Cortes, C., Lawrence, N.D., Weinberger, K.Q. (eds.) Advances in Neural Information Processing Systems 27, pp. 1098–1106. MIT Press, Cambridge (2014)

Second-Order Optimization over the Multivariate Gaussian Distribution

Luigi Malagò[1][(⊠)] and Giovanni Pistone[2]

[1] Shinshu University & Inria Saclay – Île-de-France,
4-17-1 Wakasato, Nagano 380-8553, Japan
malago@shinshu-u.ac.jp
[2] Collegio Carlo Alberto, Via Real Collegio, 30, 10024 Moncalieri, Italy
giovanni.pistone@carloalberto.org

Abstract. We discuss the optimization of the stochastic relaxation of a real-valued function, i.e., we introduce a new search space given by a statistical model and we optimize the expected value of the original function with respect to a distribution in the model. From the point of view of Information Geometry, statistical models are Riemannian manifolds of distributions endowed with the Fisher information metric, thus the stochastic relaxation can be seen as a continuous optimization problem defined over a differentiable manifold. In this paper we explore the second-order geometry of the exponential family, with applications to the multivariate Gaussian distributions, to generalize second-order optimization methods. Besides the Riemannian Hessian, we introduce the exponential and the mixture Hessians, which come from the dually flat structure of an exponential family. This allows us to obtain different Taylor formulæ according to the choice of the Hessian and of the geodesic used, and thus different approaches to the design of second-order methods, such as the Newton method.

In this paper we study the optimization of a real-valued function by means of its *Stochastic Relaxation* (SR), i.e., we search for the optimum of the function by optimizing the expected value of the function itself over a statistical model. This approach in optimization is very general and it has been developed in many different fields, from statistical physics and random-search methods, e.g., the Gibbs sampler in optimization [1], simulated annealing and the cross-entropy method [2]; to black-box optimization in evolutionary computation, e.g., Estimation of Distribution Algorithms [3] and evolutionary strategies [4–7]; going through well known techniques in polynomial optimization, such as the method of the moments [8].

By optimizing the SR of a function, we move from the original search space to a new search space given by a statistical model, i.e., a set of probability densities. Once we introduce a parameterization for the statistical model, the parameters of the model become the new variables of the relaxed problem. Notice that the notion of stochastic relaxation differs from the common notion of relaxation in optimization, indeed the minimum of the relaxed problem does not provide a

© Springer International Publishing Switzerland 2015
F. Nielsen and F. Barbaresco (Eds.): GSI 2015, LNCS 9389, pp. 349–358, 2015.
DOI: 10.1007/978-3-319-25040-3_38

lower bound for the minimum of the original problem, since the expected value of a function is always greater of equal to the minimum of the function. The term stochastic relaxation has been borrowed from [1], and used in the context of optimization for the first time in [9]. In the original work Geman and Geman introduced the Gibbs sampler, which is described as a stochastic relaxation technique to sample a joint probability distribution, and that, combined with an annealing schedule, can be used as a maximization tool as well.

The choice of the statistical model in the SR plays a fundamental role, indeed there is a tradeoff between the complexity of the statistical model, expressed for instance by its dimension, and the difficulty of the relaxed problem, expressed for example in terms of the non-linearities which appear in the formula of the expected value of the function. For instance, consider the case of a finite search space. One could be tempted to define a relaxation over the whole probability simplex, so that the SR would become linear in the probabilities, and thus easy to optimize. However, the dimension of relaxed problem would equal that of the search space, and there would be no advantage in moving the search over a statistical model. Instead, it is more reasonable to choose a lower-dimensional statistical model in the search for the optimum. For finite search spaces this would correspond to constraining the search to a subset of the probability simplex. In this work we focus on the SR of a continous function with respect to a statistical model in the multivariate Gaussian distributions, however the theory we use applies in the general case of exponential families, with either finite, discrete or continuous sample space.

In solving the SR, we are looking for an optimal density in a statistical model. This corresponds to the distribution that in the discrete case concentrate the probability mass over an optimal solutions of the original function, while in the continous case it is more appropriate to talk about concentration of the probability density in a neighborhood of the optimal solution in the original search space. The optimization of the SR can be performed according to different paradigms. In particular, a common approach in the family of first-order methods is given by gradient descent. However, it is well known in statistics that the geometry of a statistical model is not Euclidean, indeed it was first shown by Rao [10] that the set of positive distributions on a finite state space is a Riemannian manifold endowed with the Fisher information metric. Follows that the gradient of the stochastic relaxation should be evaluated with respect to the Fisher information metric, which leads us to the definition of natural gradient introduced by Amari [11]. Natural gradient has been proved to be efficient in different contexts besides the optimization of the SR [5–7], such as the training of neural networks [12] and, more recently, in deep learning [13]; policy gradients in reinforcement learning [14]; and last but not least variational inference techniques, e.g., [15].

In this paper we follow a geometric approach based on Information Geometry [16–19] to study the first and second-order geometry of the exponential family. The purpose of this analysis is to introduce the proper tools to define second-order optimization methods over a statistical model, and in particular the notion of Riemannian Hessian which is required when the geometry of the space is not Euclidean. Notice that despite second-order methods over manifolds are widely used, as in the case of matrix manifolds [20], they appear to be new in the context of statistical manifolds. As we already mentioned, exponential families of distributions have an intrinsic Riemannian geometry, where the Fisher Information matrix plays the role of metric tensor. However, it was pointed out by Amari [16,18] that besides the Riemannian geometry there are two other relevant dually-flat affine geometries of Hessian type for an exponential family: the exponential and the mixture one. The existence of (at least) three geometries provides three definitions of connections for an exponential family, three types of geodesics and, as we will see in the following, three types of Hessian, which are at the basis of the study of second-order optimization methods over a statistical manifold.

In the first part of the paper we review the first-order geometry of the exponential family. Next we move to second-order calculus, by introducing the notion of covariant derivative, and we provide formulæ for the Riemannian, exponential, and mixture Hessians over a statistical manifold. This analysis allows us to generalize the Newton algorithm to the optimization over a statistical manifold. We conclude the paper with some remarks about the case of multivariate Gaussian distributions. A preliminary version of this paper has been presented as a poster [21] at the NIPS 2014 Workshop on Optimization for Machine Learning (OPT2014).

1   Geometry of the Exponential Family

Given a real-valued function $f : \Omega \to \mathbb{R}$ to be minimized, and a statistical model $\mathcal{M}$, the Stochastic Relaxation (SR) F of f is defined as the expected value of the function itself with respect to p in $\mathcal{M}$, i.e., $F(p) = \mathbb{E}_p[f]$. Under some regularity conditions over the choice of $\mathcal{M}$, F is a continuous function independently from the nature of the sample space Ω, which can be either finite, discrete or continuous.

We are interested in developing second-order optimization methods for the SR of f based on the Gaussian distribution. However, the approach we present is more general and can be applied to any exponential family, thus in the following we will use the formalism of the exponential family and we will come back to the Gaussian distribution in the last part of the paper. In the first part of this section we review some general properties of the exponential family and we refer to the monograph [22]. Consider the exponential family $\mathcal{E}$:

$$p(\boldsymbol{x}; \boldsymbol{\theta}) = \exp\left(\sum_{i=1}^{d} \theta_i T_i(\boldsymbol{x}) - \psi(\boldsymbol{\theta}) \right), \tag{1}$$

with $\boldsymbol{\theta} \in B$, where B is an open convex set in $\mathbb{R}^d$. The real-valued functions $T_1, \ldots, T_k$, are the sufficient statistics of the exponential family, and $\psi(\boldsymbol{\theta})$ is the log-partition function, i.e., $\psi(\boldsymbol{\theta}) = \log \int_{\Omega} \exp\left(\sum_{i=1}^{d} \theta_i T_i(\boldsymbol{x})\right) \, \mathrm{d}x$. The exponential family also admits a dual parameterization based on the expectation parameters $\boldsymbol{\eta}$ with $\boldsymbol{\eta} = \mathbb{E}_{\boldsymbol{\theta}}[\boldsymbol{T}] = \nabla \psi(\boldsymbol{\theta})$.

First and second-order methods to optimize a function F defined over an exponential family require the evaluation of the gradient and of the Hessian of F. The evaluation of such quantities depend on the geometry of the space, which is known to be non-Euclidean in the case of statistical models. To better understand the nature of $\mathcal{E}$, we refer to notions from Information Geometry [16,18], which studies the geometry of statistical models and of the exponential family from the point of view of differential geometry [25]. Statistical models are considered as manifolds of distributions endowed with a Riemannian metric, given by the Fisher information metric.

In the following we denote with $\mathrm{T}_p\mathcal{E}$ the tangent space of $\mathcal{E}$ at p, i.e., the space of the tangent vectors to any curve $p(t)$ in $\mathcal{E}$ that goes through p. Rao showed that the tangent vector to $p(t)$ can be evaluated as $\frac{\mathrm{d}}{\mathrm{d}t} \log p(t)$, so that the tangent space $\mathrm{T}_p\mathcal{E}$ can be equivalently characterized as the space of all random variable centered in p, with the canonical basis given by the centered sufficient statistics $T_i - \mathbb{E}_p[T_i]$. Given two tangent vector U, V in $\mathrm{T}_p\mathcal{E}$, the tangent space is endowed with the inner product given by $g(U,V)(p) = \mathbb{E}_p[UV]$. In the basis of the sufficient statistics we have $\mathbb{E}_p[UV] = \sum_{ij} U_i \mathbb{E}_p[(T_i - \mathbb{E}_p[T_i])(T_j - \mathbb{E}_p[T_j])]V_j$, where ${}^eI(p) = E_p[(T_i - \mathbb{E}_p[T_i])(T_j - \mathbb{E}_p[T_j])]_{ij} = [\mathrm{Cov}\ (T_i, T_j)]_{ij}$ is the Fisher information matrix.

Given an exponential family $\mathcal{E}$, a function $F : \mathcal{E} \to \mathbb{R}$ and the metric g for $\mathcal{E}$, which in our case is the Fisher information metric, the Riemannian gradient grad F is the unique vector such that for any direction identified by the vector $X \in \mathrm{T}_p\mathcal{E}$, we have:

$$g(\mathrm{grad}\, F, X)(p) = \mathrm{D}_X F(p), \tag{2}$$

i.e., grad F is defined as the unique vector such that the inner product with respect to the metric between grad F and an arbitrarily direction X, evaluated at $p \in \mathcal{E}$, is the directional derivative $\mathrm{D}_X F(p)$ of F along X in p. The previous definition of Riemannian gradient is coordinate independent. If we consider a parameterization for the exponential family, and we choose a basis for the tangent space, we can write a formula for the components of the Riemannian gradient. In the exponential family, the *natural gradient* gives the components of the Riemannian gradient evaluated with respect to the Fisher information matrix ${}^eI(\boldsymbol{\theta})$, expressed in the basis of the centered sufficient statistics:

$$\widetilde{\nabla} F(\boldsymbol{\theta}) = {}^e I(\boldsymbol{\theta})^{-1} \nabla F(\boldsymbol{\theta}). \tag{3}$$

Due to the properties of the exponential family ${}^eI(\boldsymbol{\theta})$ can be obtained as the Hessian of $\psi(\boldsymbol{\theta})$, i.e., the matrix of second-order partial derivatives $[\partial_i \partial_j \psi(\boldsymbol{\theta})]_{ij}$, and $\nabla F(\boldsymbol{\theta}) = (\partial_i F(\boldsymbol{\theta}))_i$ is the vector of first-order partial derivatives. Here ∂_i denotes the partial derivative with respect to θ_i, i.e., $\partial_i = \frac{\partial}{\partial \theta_i}$. We denote the natural gradient with $\widetilde{\nabla} F$ to distinguish it from ∇F, which corresponds to the components of the gradient evaluated with respect to the Euclidean metric.

In order to move to second-order calculus, we need a definition of Hessian of the function F over a manifold, which generalizes the Euclidean case. In the following we refer to [20], where second-order methods have been applied to the optimization over manifolds, cf. [23] for a similar approach. We study the second-order geometry of the exponential family in a general way, similar to what has been done in [24], where the focus was on applications to binary optimization. For basic notions of differential geometry, we refer to the standard book [25].

The first step in the geometric construction of the Riemannian Hessian, which is required to write a second-order Taylor approximation of the function in a neighborhood of a point, is the generalization to a manifold of the concept of directional derivative of a vector field. Indeed, differently from the Euclidean case, a definition based on the derivation of a vector field along a curve is not possible, since in each point of the curve tangent vectors belong to different tangent spaces, and without a correspondence between tangent spaces, no comparison is possible. The notion of affine connection provides a way to define such correspondence.

A *connection* ∇ over a manifold $\mathcal{M}$ is an operator $\nabla : T\mathcal{M} \times T\mathcal{M} \to T\mathcal{M}$ which given two vector fields X and Y defined over $\mathcal{M}$ returns a new vector field $\nabla_X Y$ given by the directional derivative $D_X Y$ of the Y in the direction X. The vector field $\nabla_X Y$ is called the *covariant derivative* of Y with respect to X for the given affine connection ∇. Notice that in general a manifold admits infinitely many connections. Each connection can be specified by d^2 vector fields which represent the covariant derivate $\nabla_{E_i} E_i$ where E_i and E_j are the coordinate vector fields. Then, a connection can be fully determined by d^3 symbols, called the Christoffel symbols Γ_{ij}^k, which represent the components of $\nabla_{E_i} E_j$ in the basis $E_1, \ldots, E_d$, i.e., $\nabla_{E_i} E_j = \sum_k \Gamma_{ij}^k E_k$.

Among all possible connections, there is a unique connection, called Riemannian or Levi-Civita connection, denoted by $^0\nabla$, which satisfies the properties of being symmetric and invariant with respect to the Riemannian metric. The Christoffel symbols $^0\Gamma_{ij}^k$, with $i, j, k = 1, \ldots, d$, for the Levi-Civita connection can be derived from the metric, using the formula $^0\Gamma_{ij}^k = \sum_l g^{kl}\,^0\Gamma_{ijl}$, with $^0\Gamma_{ijk} = \frac{1}{2}\left(\partial_i g_{jk} + \partial_j g_{ik} - \partial_k g_{ij}\right)$. The symbols $\Gamma_{ijk} = \sum_l g_{il}\Gamma_{jk}^l$ are called the Christoffel symbols of the first type, to distinguish them from $\Gamma_{jk}^k = \sum_l g^{kl}\Gamma_{ij}^l$ which are sometimes referred as Christoffel symbols of the second type. Here the g^{ij}'s denote the entries of the inverse Fisher information matrix, i.e., $[g^{ij}] = [g_{ij}]^{-1}$. Notice that when g can be expressed as the Hessian of a function for a given parameterization, then by symmetry we have $^0\Gamma_{ijk} = \frac{1}{2}\partial_i g_{jk}$.

As pointed out previously, besides the Riemannian connection, two other affine geometries, namely the exponential and the mixture geometry, play an important role for the exponential family. Amari [18] introduced the following family of α-connections, given by the Christoffel symbols:

$$^\alpha\Gamma_{ijk}(\boldsymbol{\xi}) = \mathrm{E}_{\boldsymbol{\xi}}\left[\left(\partial_i\partial_j \log p(\boldsymbol{x}; \boldsymbol{\xi}) + \frac{1-\alpha}{2}\partial_i \log p(\boldsymbol{x}; \boldsymbol{\xi})\partial_j \log p(\boldsymbol{x}; \boldsymbol{\xi})\right)\partial_k \log p(\boldsymbol{x}; \boldsymbol{\xi})\right]$$

For $\alpha = 0$ we recover the Christoffel symbols of the Levi-Civita connection $^0\Gamma_{ijk}(\boldsymbol{\xi})$, while for $\alpha = \pm 1$ we obtain a characterization for the exponential and mixture connection. In particular, for an exponential family parametrized by $\boldsymbol{\theta}$, it is easy to show that the Christoffel symbols of the exponential connection $^e\Gamma_{ijk}(\boldsymbol{\theta})$, for $\alpha = 1$, are identically equal to zero, i.e., the exponential family is e-flat. Similarly, once the exponential family is parametrized by $\boldsymbol{\eta}$, it turns out that the Christoffel symbols of the mixture connection $^m\Gamma_{ijk}(\boldsymbol{\eta})$, for $\alpha = -1$, are identically zero, i.e., the exponential family is m-flat as well. This is a consequence of the duality between the exponential and mixture geometry of the exponential family. It follows that we can introduce at least two alternative definitions of covariant derivative, based on the exponential and mixture geometries, which we call exponential and mixture covariant derivatives. Given the connection through its Christoffel symbols, the covariant derivative can be evaluated by the following formula:

$$\nabla_X Y = \sum_{ij} X^j \left(\sum_k Y^k \Gamma_{jk}^i + \partial_j Y^i \right) E_i. \tag{4}$$

The introduction of a connection over the manifold allows to define the notion of acceleration along a curve, which is based on the differentiation of tangent vectors along the curve itself. Thus, we can introduce a *geodesic* between two points as the curve with zero acceleration. Different definitions of covariant derivatives produce different geodesics between two points.

We can now introduce the *Riemannian Hessian* of a function defined over a manifold. In the following we interpret the Hessian as an operator which is applied to a vector field X and returns a vector field D_X grad F given by the directional derivative of the Riemannian gradient along the direction identified by X. On a Riemannian manifold $\mathcal{M}$ endowed with the metric g, the Riemannian Hessian of F is the linear mapping ^{0}Hess $F(p) : T_p\mathcal{M} \to T_p\mathcal{M}$ such that ^{0}Hess $F(p)[X(p)] = \nabla_{X(p)}$ grad $F(p)$, where $^0\nabla$ is the Levi-Civita connection associated to g on $\mathcal{M}$. The coordinate representation of the Riemannian Hessian in the basis of the centered sufficient statistics [24] is given by:

$$^0\text{Hess } F(p)[X(p)] = {}^e I(p)^{-1} \left(\text{Hess } F(p) - \frac{1}{2} \sum_k \partial_k {}^e I(p)(\widetilde{\nabla} F(p))_k \right) X(p), \tag{5}$$

where Hess $F(p)$, with no arguments, denotes the Euclidean Hessian of F in p, i.e., the matrix of second-order partial derivatives. Notice that in the natural parameters, and more in general for any Hessian manifolds, since $^e I(\boldsymbol{\theta}) = $ Hess $\psi(\boldsymbol{\theta})$, then $^0\Gamma_{ijk}(\boldsymbol{\theta}) = \frac{1}{2}\partial_i\partial_j\partial_k\psi(\boldsymbol{\theta})$ becomes symmetric with respect to the three indices. Equation (5) can be derived from Eq. (4), where the Christoffel symbols of the second type are given by the tensor contraction $\frac{1}{2}{}^e I(p)^{-1}\partial^e I(p)$. By choosing different Christoffel symbols associated to the exponential and mixture connections, we can obtain similar formulæ for ^{e}Hess $F(p)[X(p)]$ and ^{m}Hess $F(p)[X(p)]$.

2 Second-Order Optimization: The Newton Method

The Newton method is an optimization method which generates a sequence of distributions $\{p_t\}$, $t \geq 0$, in $\mathcal{M}$ which converges towards a stationary point of F, i.e., a critical point of the vector field $p \mapsto \operatorname{grad} F(p)$. At the basis of this optimization technique there is a Taylor expansion $F(p)$ which provides a second-order approximation of the function over the manifold.

Let $t \mapsto p(t)$ be a Riemannian geodesic connecting $p = p(0)$ to $q = p(1)$ in $\mathcal{E}$, and $\mathrm{D}p(t)$ denote the tangent velocity vector $\frac{\mathrm{d}}{\mathrm{d}t} \log p(t)$, then the following Taylor formula holds:

$$F(q) \approx F(p) + \langle \operatorname{grad} F(p), \mathrm{D}p(0) \rangle_p + \frac{1}{2} \left\langle {}^0\!\operatorname{Hess} F(p)[\mathrm{D}p(0)], \mathrm{D}p(0) \right\rangle_p. \quad (6)$$

However, this is not the only possible second-order approximation of F. Two similar formula can be obtained by consider the exponential geodesic connecting p and q together with the mixture Hessian ${}^m\!\operatorname{Hess} F(p)[\mathrm{D}p(0)]$, and dually, using the mixture geodesic and the exponential Hessian. Proofs are omitted due to space limitation, however they are based on the duality between covariant derivatives in terms of preserving inner products with respect to the metric, and the fact that the acceleration along the corresponding geodesic is zero.

In order to determine the next point at each iteration, the Newton method is based on the idea of choosing the step in such a way that the Taylor expansion attains its minimum in the new point. This step can be found by ensuring that the derivative of the approximation is equal to zero in the new point. This requires to solve in $X(p) \in \mathrm{T}_p\mathcal{M}$ the following Newton equation:

$$\operatorname{Hess} F(p)[X(p)] = -\operatorname{grad} F(p). \quad (7)$$

Once the previous equation has been solved, the last step consists in finding a point over the manifold along the geodesic starting from the current point with initial velocity given by the Newton step. This last step is required for any first or second-order optimization method over a manifold to find a correspondence between tangent vectors in a point and the neighborhood of the point itself in the manifold. The computation of a geodesic determined by the Newton step can be an expensive task in general, for instance when the geometry is not flat, however this step could be relaxed and approximated by the notion of *retraction*. The retraction over a manifold [20] is a mapping between the tangent space in a point and the manifold, with local rigidity conditions which preserves gradients at the point where it is evaluated.

3 Applications to the Gaussian Distribution

In this section we give some details about the application of the general theory of second-order calculus over an exponential family to the case of the Gaussian distribution. In the first part we recall some results about exponential families.

Due to the properties of the exponential family, the Fisher information matrix, the Euclidean gradient, and thus the natural gradient can be evaluated in terms of covariances, indeed we have $\widetilde{\nabla} F(\boldsymbol{\theta}) = \mathrm{Cov}_{\boldsymbol{\theta}}(\boldsymbol{T}, \boldsymbol{T})^{-1} \mathrm{Cov}_{\boldsymbol{\theta}}(\boldsymbol{T}, f)$. As remarked above, since the exponential family parameterized by $\boldsymbol{\theta}$ is a Hessian manifold, it follows that $\partial I(\boldsymbol{\theta}) = [\partial_i \partial_j \partial_k \psi(\boldsymbol{\theta})] = \mathrm{Cov}_{\boldsymbol{\theta}}(\boldsymbol{T}, \boldsymbol{T}, \boldsymbol{T}) = \mathbb{E}_{\boldsymbol{\theta}}[(\boldsymbol{T} - \mathbb{E}_{\boldsymbol{\theta}}[\boldsymbol{T}])(\boldsymbol{T} - \mathbb{E}_{\boldsymbol{\theta}}[\boldsymbol{T}])(\boldsymbol{T} - \mathbb{E}_{\boldsymbol{\theta}}[\boldsymbol{T}])]$, and Hess $F(p) = [\partial_i \partial_j F(\boldsymbol{\theta})] = (\mathrm{Cov}_{\boldsymbol{\theta}}(\boldsymbol{T}, \boldsymbol{T}, f)$. The Riemannian Hessian ${}^0\mathrm{Hess}\, F(\boldsymbol{\theta})[X(\boldsymbol{\theta})]$ can then be written in coordinates:

$$\mathrm{Cov}_{\boldsymbol{\theta}}(\boldsymbol{T}, \boldsymbol{T})^{-1} \left(\mathrm{Cov}_{\boldsymbol{\theta}}(\boldsymbol{T}, \boldsymbol{T}, f) - \frac{1}{2} \sum_k \mathrm{Cov}_{\boldsymbol{\theta}}(\boldsymbol{T}, \boldsymbol{T}, T_k)(\widetilde{\nabla} F(\boldsymbol{\theta}))_k \right) X(\boldsymbol{\theta}). \quad (8)$$

The implementation of an optimization algorithm for the SR based on the exponential family requires the evaluation of the covariances among the sufficient statistics and between the sufficient statistics and the function to be optimized. In the general case, to determine these quantities exactly can be computationally unfeasible, for this reason it is a common approach to replace the exact value with Monte Carlo estimations of the covariances based on the current sample.

We have now all the elements to write explicitly an updating formula in the natural parameters for the Newton method, where the sequence of distributions generated is identified by a corresponding sequence of parameter vectors $\{\boldsymbol{\theta}_t\}$, $t \geq 0$. The iterative formula for the Newton method can be written as:

$$\boldsymbol{\theta}_{t+1} = \boldsymbol{\theta}_t - R_{\boldsymbol{\theta}_t}(\lambda \, \mathrm{Hess}\, F(\boldsymbol{\theta}_t)^{-1} \widetilde{\nabla} F(\boldsymbol{\theta}_t)), \quad (9)$$

where the function $R_{\boldsymbol{\theta}}$ returns the coordinates of the image of the retraction, which is a mapping from the tangent space to the manifold that identifies a point along the direction specified by the vector given as an argument, which in our case is the Newton step. The parameter $\lambda > 0$ is used to control the step size and thus the convergence to a critical point of F.

We conclude this section with some comments about the application to the Gaussian case. We refer to [27] as a standard reference for the geometry of the Gaussian distribution, and to our paper [26] for a presentation of the different parameterizations of the Gaussian distribution in view of the SR. The Gaussian distribution is one of the special cases in the exponential family, where the computation of the transformation between natural parameters and expectation parameters can be done in an efficient way, through the inversion of the covariance matrix. Indeed, the natural parameters of the Gaussian distribution are a function of the inverse covariance matrix and of the mean vector, while the expectation parameters correspond to a function of covariance matrix and mean vector. This suggests an implementation of the Newton method based on the exponential Hessian in the natural parameters, for which the Christoffel symbols vanish, combined with a retraction based on the mixture geodesic, which can be evaluated efficiently in the expectation parameters.

4   Discussion and Future Work

In this paper we studied the second-order geometry of a Riemannian manifold, in the special case of exponential statistical models. We extended the analysis carried out in [24], by defining not only the Riemannian Hessian, but also the exponential and the mixture Hessians. The three Hessians we introduced, which are associated to the three privileged geometries of an exponential family, allow to derive three different Taylor formulæ and thus three alternative generalizations of the updating rule of the Newton method over an exponential family.

The alternative approaches we proposed appear to be equally well motivated from a theoretical perspective, however they are not equivalent in practice, indeed they are based on the computation of different covariant derivates and different geodesics. Moreover we expect different computational costs in the evaluation of the Newton step according to the choice of the parameterization and of the connection, as well as the type of geodesic which needs to be computed. An experimental comparison is required in order to investigate the advantages and disadvantages of the different approaches we proposed, for instance in terms of computational complexity and speed of convergence.

We conclude the paper with a remark about second-order optimization techniques. Indeed, even if the Newton method and more in general second-order methods are very popular and well-known for their quadratic local convergence properties, in practice a number of issues has to be taken into account. The Newton step does not always points in the direction of the natural gradient, and close enough to a saddle point of the function the Newton step will tend to converge to it. In order to obtain a direction of descent for the function to be optimized, the Hessian must be negative-definite, i.e., its eigenvalues must be strictly negative. In order to overcome these issues, different methods have been proposed in the literature, such as quasi-Newton methods, where the update vector is obtained using a modified Hessian which is guaranteed to be negative definite. Finally, a number of other issues has to be taken into account in the design of an algorithm, such as the uncertainty in the estimation of the Hessian and of the gradient, when they are estimated from a sample, and the choice of other parameters of the algorithm, such as the step size.

Acknowledgements. Giovanni Pistone is supported by de Castro Statistics, Collegio Carlo Alberto, Moncalieri, and he is a member of GNAMPA-INDAM.

References

1. Geman, S., Geman, D.: Stochastic relaxation, Gibbs distributions, and the Bayesian restoration of images. IEEE Trans. PAMI **6**, 721–741 (1984)
2. Rubinstein, R.: The cross-entropy method for combinatorial and continuous optimization. Methodol. Comput. Appl. Probab. **1**, 127–190 (1999)
3. Larrañaga, P., Lozano, J.A. (eds.): Estimation of Distribution Algoritms: A New Tool for Evolutionary Computation. Springer, New York (2001)
4. Hansen, N., Ostermeier, A.: Completely derandomized self-adaptation in evolution strategies. Evol. Comput. **9**, 159–195 (2001)
5. Wierstra, D., Schaul, T., Glasmachers, T., Sun, Y., Peters, J., Schmidhuber, J.: Natural evolution strategies. JMLR **15**, 949–980 (2014)
6. Malagò, L., Matteucci, M., Pistone, G.: Towards the geometry of estimation of distribution algorithms based on the exponential family. In: Proceedings of FOGA 2011, pp. 230–242, ACM (2011)

7. Ollivier, Y., Arnold, L., Auger, A., Hansen, N.: Information-geometric optimization algorithms: a unifying picture via invariance principles. arXiv:1106.3708 (2011)
8. Lasserre, J.B.: Global optimization with polynomials and the problem of moments. SIAM J. Optim. **11**, 796–817 (2001)
9. Malagò, L., Matteucci, M., Pistone, G.: Stochastic relaxation as a unifying approach in 0/1 programming. In: NIPS 2009 Workshop on Discrete Optimization in Machine Learning: Submodularity, Sparsity & Polyhedra (DISCML) (2009)
10. Rao, R.C.: Information and the accuracy attainable in the estimation of statistical parameters. Bull. Calcutta Math. Soc. **37**, 81–91 (1945)
11. Amari, S.: Natural gradient works efficiently in learning. Neural Comput. **10**, 251–276 (1998)
12. Amari, S.: Neural learning in structured parameter spaces - natural Riemannian gradient. In: NIPS 1997, pp. 127–133. MIT Press (1997)
13. Pascanu, R., Bengio, Y.: Revisiting natural gradient for deep networks. In: Proceedings of ICLR 2014 (2014)
14. Kakade, S.: A natural policy gradient. In: Dietterich, T.G., Becker, S., Ghahramani, Z. (eds.) NIPS 2001, pp. 1531–1538. MIT Press (2001)
15. Kuusela, M., Raiko, T., Honkela, A., Karhunen, J.: A gradient-based algorithm competitive with variational Bayesian EM for mixture of Gaussians. In: Neural Networks (IJCNN 2009), pp. 1688–1695 (2009)
16. Amari, S.: Differential-Geometrical Methods in Statistics. Lecture Notes in Statistics, vol. 28. Springer, New York (1985)
17. Amari, S.-I., Barndorff-Nielsen, O.E., Kass, R.E., Lauritzen, S.L., Rao, C.R.: Chapter 4: Statistical manifolds. Differential geometry in statistical inference. Institute of Mathematical Statistics, Hayward, CA (1987)
18. Amari, S., Nagaoka, H.: Methods of Information Geometry. American Mathematical Society, Providence (2000)
19. Pistone, G.: Algebraic varieties vs differentiable manifolds in statistical models. In: Gibilisco, P., Riccomagno, E., Rogantin, M.P., Wynn, H.P. (eds.) Algebraic and Geometric Methods in Statistics, pp. 339–363. Cambridge University Press, Cambridge (2009)
20. Absil, P.A., Mahony, R., Sepulchre, R.: Optimization Algorithms on Matrix Manifolds. Princeton University Press, Princeton (2008)
21. Malagò, L., Pistone, G.: Stochastic relaxation over the exponential family: second-order geometry. In: NIPS 2014 Workshop on Optimization for Machine Learning (OPT 2014), Montreal, Canada, 12 December 2014 (2014)
22. Brown, L.D.: Fundamentals of Statistical Exponential Families with Applications in Statistical Decision Theory. Lecture Notes - Monograph Series, vol. 9. Institute of Mathematical Statistics, Hayward (1986)
23. Manton, J.H.: A framework for generalising the Newton method and other iterative methods from euclidean space to manifolds. arXiv:1106.3708 (2012v1; 2014v2)
24. Malagò, L., Pistone, G.: Combinatorial optimization with information geometry: the Newton method. Entropy **16**, 4260–4289 (2014)
25. do Carmo, M.P.: Riemannian Geometry, Mathematics: Theory & Applications. Birkhäuser Boston Inc., Boston (1992)
26. Malagò, L., Pistone, G.: Information geometry of the Gaussian distribution in view of stochastic optimization. In: Proceedings of FOGA 2015, pp. 150–162 (2015)
27. Skovgaard, L.T.: A Riemannian geometry of the multivariate normal model. Scand. J. Stat. **11**, 211–223 (1984)

The Information Geometry of Mirror Descent

Garvesh Raskutti[1,2]([✉]) and Sayan Mukherjee[3,4]

[1] Department of Statistics and Computer Science, University of Wisconsin-Madison,
Madison, USA
raskutti@stat.wisc.edu
[2] Wisconsin Institute of Discovery, Optimization Group, Madison, USA
[3] Departments of Statistical Science, Computer Science, and Mathematics,
Duke University, Durham, USA
[4] Institute for Genome Sciences & Policy, Duke University, Durham, USA

Abstract. We prove the equivalence of two online learning algorithms, mirror descent and natural gradient descent. Both mirror descent and natural gradient descent are generalizations of online gradient descent when the parameter of interest lies on a non-Euclidean manifold. Natural gradient descent selects the steepest descent direction along a Riemannian manifold by multiplying the standard gradient by the inverse of the metric tensor. Mirror descent induces non-Euclidean structure by solving iterative optimization problems using different proximity functions. In this paper, we prove that mirror descent induced by a Bregman divergence proximity functions is equivalent to the *natural* gradient descent algorithm on the Riemannian manifold in the *dual* coordinate system. We use techniques from convex analysis and connections between Riemannian manifolds, Bregman divergences and convexity to prove this result. This equivalence between natural gradient descent and mirror descent, implies that (1) mirror descent is the steepest descent direction along the Riemannian manifold corresponding to the choice of Bregman divergence and (2) mirror descent with log-likelihood loss applied to parameter estimation in exponential families asymptotically achieves the classical Cramér-Rao lower bound.

1 Introduction

Recently there has been great interest in online learning both in terms of algorithms as well as in terms of convergence properties. Given a convex differentiable cost function, $f : \Theta \to \mathbb{R}$, with parameter in a convex set, $\theta \in \Theta \subseteq \mathbb{R}^p$, an online learning algorithm predicts a sequence of parameters $\{\theta_t\}_{t=1}^{\infty}$ which incur a loss $f(\theta_t)$ at each iterate t. The goal in online learning is to construct a sequence that minimizes the *regret* at a time T, $\sum_{t=1}^{T} f(\theta_t)$.

The most common approach to construct a sequence $\{\theta_t\}_{t=1}^{\infty}$ is based on online or stochastic gradient descent. The online gradient descent update is:

$$\theta_{t+1} = \theta_t - \alpha_t \nabla f(\theta_t), \tag{1}$$

© Springer International Publishing Switzerland 2015
F. Nielsen and F. Barbaresco (Eds.): GSI 2015, LNCS 9389, pp. 359–368, 2015.
DOI: 10.1007/978-3-319-25040-3_39

where $(\alpha_t)_{t=0}^{\infty}$ denotes a sequence of step-sizes. Gradient descent is the direction of steepest descent if the parameters θ_t belong to a Euclidean space. However in many applications, parameters lie on non-Euclidean manifolds (e.g. mean parameters for Poisson families, mean parameters for Bernoulli families and other exponential families). In such scenarios gradient descent in the ambient space is not the direction of steepest descent, since the parameter is restricted to a manifold. Consequently generalizations of gradient decent that incorporate non-Euclidean structure have been developed.

1.1   Riemannian Manifolds and Natural Gradient Descent

One generalization of gradient descent is *natural gradient descent* developed by Amari [1]. Natural gradient descent assumes the parameter of interest lies on a Riemannian manifold and selects the steepest descent direction along that manifold. Let $(\mathcal{M}, \mathcal{H})$ be a p-dimensional Riemannian manifold with metric tensor $\mathcal{H} = (h_{jk})$ and $\mathcal{M} \subseteq \mathbb{R}^p$. A well-known statistical example of Riemannian manifolds are manifolds induced by the Fisher information of parametric families. In particular given a parametric family $\{p(x; \mu)\}$ where $\mu \in \mathcal{M} \subseteq \mathbb{R}^p$, let $\{\mathcal{I}(\mu)\}$ for each $\mu \in \mathcal{M}$ denote the $p \times p$ Fisher information matrices. Then $(\mathcal{M}, \mathcal{I}(\mu))$ denotes a p-dimensional Riemannian manifold. Table 1 provides examples of statistical manifolds induced by parametric families (see e.g. [3,11,18] for details).

When $\mathcal{I}(\mu) = I_{p \times p}$, the Riemannian manifold corresponds to standard Euclidean space. For a short introduction to Riemannian manifolds, see [10].

Given a function $\tilde{f}$ on the Riemannian manifold $\tilde{f} : \mathcal{M} \to \mathbb{R}$, the *natural* gradient descent step is:

$$\mu_{t+1} = \mu_t - \alpha_t \mathcal{H}^{-1}(\mu_t) \nabla \tilde{f}(\mu_t), \qquad (2)$$

where $\mathcal{H}^{-1}$ is the inverse of the Riemannian metric $\mathcal{H} = (h_{jk})$ and μ is the parameter of interest. If $(\mathcal{M}, \mathcal{H}) = (\mathbb{R}^p, I_{p \times p})$, the natural gradient step corresponds to the standard gradient descent step (1). Theorem 1 in [1] proves that the natural gradient algorithm steps in the direction of steepest descent along the Riemannian manifold $(\mathcal{M}, \mathcal{H})$. Hence the name natural gradient descent.

1.2   Mirror Descent with Bregman Divergences

Another generalization of online gradient descent is mirror descent developed by Nemirovski and Yudin [16]. Mirror descent induces non-Euclidean geometry by

Table 1. Statistical manifold examples

Family	$\mathcal{M}$	$\mathcal{I}(\mu)$
$\mathcal{N}(\theta, I_{p \times p})$	$\mathbb{R}^p$	$I_{p \times p}$
Bernoulli(p)	$[0, 1]$	$\frac{1}{p(1-p)}$
Poisson(λ)	$[0, \infty)$	$\frac{1}{\lambda}$

re-writing the gradient descent update as an iterative ℓ_2-penalized optimization problem and selecting a proximity function different from squared ℓ_2 error. Note that the online gradient descent step (1) can alternatively be expressed as:

$$\theta_{t+1} = \arg\min_{\theta \in \Theta} \left\{ \langle \theta, \nabla f(\theta_t) \rangle + \frac{1}{2\alpha_t} \| \theta - \theta_t \|_2^2 \right\},$$

where $\Theta \subseteq \mathbb{R}^p$. By re-expressing the stochastic gradient step in this way, [16] introduced a generalization of gradient descent as follows: Denote the *proximity* function $\Psi : \mathbb{R}^p \times \mathbb{R}^p \to \mathbb{R}^+$, strictly convex in the first argument, then define the *mirror* descent step as:

$$\theta_{t+1} = \arg\min_{\theta \in \Theta} \left\{ \langle \theta, \nabla f(\theta_t) \rangle + \frac{1}{\alpha_t} \Psi(\theta, \theta_t) \right\}. \tag{3}$$

Setting $\Psi(\theta, \theta') = \frac{1}{2} \| \theta - \theta' \|_2^2$ yields the standard gradient descent update, hence (3) is a generalization of online gradient descent.

A standard choice for proximity function Ψ is the so-called *Bregman divergence* since it corresponds to the Kullback-Leibler divergence for different exponential families. See (e.g. [4,5]) for a detailed introduction to the connection and equivalence between Bregman divergences and exponential families.

In particular, let $G : \Theta \to \mathbb{R}$ denote a strictly convex twice-differentiable function, the divergence introduced by [8] $B_G : \Theta \times \Theta \to \mathbb{R}^+$ is:

$$B_G(\theta, \theta') = G(\theta) - G(\theta') - \langle \nabla G(\theta'), \theta - \theta' \rangle.$$

Bregman divergences are widely used in statistical inference, optimization, machine learning, and information geometry (see e.g. [2,5,15,20]). Letting $\Psi(\cdot, \cdot) = B_G(\cdot, \cdot)$, the mirror descent step defined is:

$$\theta_{t+1} = \arg\min_{\theta} \left\{ \langle \theta, \nabla f(\theta_t) \rangle + \frac{1}{\alpha_t} B_G(\theta, \theta_t) \right\}. \tag{4}$$

There is a one-to-one correspondence between Bregman divergences and exponential families [4,5] since both are defined by strictly convex functions and we exploit this connection later when we discuss estimation in exponential families. Examples of G, exponential families and the induced Bregman divergences are listed in Table 2. For a more extensive list, see [5].

1.3 Our Contribution

In this paper, we prove that the mirror descent update with Bregman divergence step (4) is equivalent to the natural gradient step (2) along the *dual* Riemannian manifold which we introduce later. The proof of equivalence uses concepts in convex analysis combined with connections between Bregman divergences and Riemannian manifolds developed in [2]. Using the equivalence of the two algorithms, we provide a new perspective and statistical optimality results for mirror descent. In particular natural gradient descent is known to be the direction of steepest descent along a Riemannian manifold and is Fisher efficient for parameter estimation in exponential families [1], neither of which are known for mirror descent.

Table 2. Bregman divergence examples

Family	$G(\theta)$	$B_G(\theta, \theta')$
$\mathcal{N}(\theta, I_{p \times p})$	$\frac{1}{2}\|\theta\|_2^2$	$\frac{1}{2}\|\theta - \theta'\|_2^2$
Poisson(e^θ)	$\exp(\theta)$	$\exp(\theta) - \exp(\theta') - \langle \exp(\theta'), \theta - \theta' \rangle$
Bernoulli($\frac{1}{1+e^{-\theta}}$)	$\log(1 + \exp(\theta))$	$\log\left(\frac{1+e^\theta}{1+e^{\theta'}}\right) - \langle \frac{e^{\theta'}}{1+e^{\theta'}}, \theta - \theta' \rangle$

2 Equivalence Through Dual Co-ordinates

In this section we prove the equivalence of natural gradient descent (2) and mirror descent (4). The key to the proof involves concepts in convex analysis, in particular the convex conjugate function and connections between Bregman divergences, convex functions and Riemannian manifolds.

2.1 Bregman Divergences and Convex Duality

The concept of convex conjugate functions is central to the main result in the paper. The convex conjugate function for a function G is defined to be:

$$H(\mu) := \sup_{\theta \in \Theta} \{\langle \theta, \mu \rangle - G(\theta)\}.$$

If G is lower semi-continuous, G is the convex conjugate of H, implying a dual relationship between G and H. Further, if we assume G is strictly convex and twice differentiable, then so is H. Note also that if $g = \nabla G$ and $h = \nabla H$, $g = h^{-1}$. For additional properties and motivation for the convex conjugate function, see [21].

Let $\mu = g(\theta) \in \Phi$ be the point at which the supremum for the dual function is attained represent the *dual* co-ordinate system to θ. The dual Bregman divergence $B_H : \Phi \times \Phi \to \mathbb{R}^+$ induced by the strictly convex differentiable function H is:

$$B_H(\mu, \mu') = H(\mu) - H(\mu') - \langle \nabla H(\mu'), \mu - \mu' \rangle.$$

Using the dual co-ordinate relationship, it is straightforward to show that $B_H(\mu, \mu') = B_G(h(\mu'), h(\mu))$ and $B_G(\theta, \theta') = B_H(g(\theta'), g(\theta))$. Dual functions and Bregman divergences for examples in Table 2 are presented in Table 3. For more examples see [5].

2.2 Bregman Divergences and Riemannian Manifolds

Now we explain how every Bregman divergence induces a Riemannian manifold as explained in [2]. Let $\mathcal{M}$ be a Riemannian manifold. For the Bregman divergence $B_G : \Theta \times \Theta \to \mathbb{R}^+$ induced by the convex function G, define the Riemannian metric on $\Theta \subseteq \mathcal{M}$, $\mathcal{G} = \nabla^2 G$ (i.e. the Hessian matrix). Since G is a strictly convex twice differentiable function, $\nabla^2 G(\theta)$ is a positive definite matrix for all $\theta \in \Theta$. Hence $B_G(\cdot, \cdot)$ induces the Riemannian manifold $(\mathcal{M}, \theta, \nabla^2 G)$ in the θ co-ordinates. Now let $\Phi \subseteq \mathcal{M}$ be the image of Θ under the continuous map $g = \nabla G$. $B_H : \Phi \times \Phi \to \mathbb{R}^+$ is the dual Bregman divergence and $(\mathcal{M}, \phi, \mathcal{H})$, where $\mathcal{H} = \nabla^2 H$ is the Riemannian manifold in terms of the *dual* co-ordinate system ϕ. As shown in Theorem 1 in [2] and Eq. (30) in Sect. 1.4 [17], $\nabla^2 G(\theta) = \nabla^2 H^{-1}(\mu)$.

For example, for the Gaussian statistical family defined on Table 1, $\Theta = \Phi = \mathbb{R}^p$ and $\nabla^2 G = \nabla^2 H = I_{p \times p}$ (i.e. the primal and dual manifolds are the same). On the other hand, for the Bernoulli(p) family in Table 1, the mean parameter is $0 \le p \le 1$ whereas the natural parameter is $\theta = \log p - \log(1 - p)$ and $G(\theta) = \log(1 + e^\theta)$. Consequently $(\Theta, \nabla^2 G) = (\mathbb{R}, \frac{e^{-\theta}}{(1+e^{-\theta})^2})$ and $(\Phi, \nabla^2 H) = ([0, 1], \frac{1}{p(1-p)})$ which is consistent with Table 1. For a more thorough introduction to the connection between Bregman divergences and Riemanninan manifolds see [2].

2.3 Main Result

In this section we present our main result, the equivalence of mirror descent and natural gradient descent. We also discuss consequences and implications.

Theorem 1. *The mirror descent step* (4) *with Bregman divergence defined by G applied to function f in the $\theta \in \Theta$ co-ordinate system is equivalent to the natural gradient step* (2) *in the dual co-ordiante system $\phi \in \Phi$.*

The proof follows by stating mirror descent in the dual Riemannian manifold and simple applications of the chain rule.

Proof. Recall that the mirror descent update (3) is:

$$\theta_{t+1} = \arg\min_\theta \left\{ \langle \theta, \nabla f(\theta_t) \rangle + \frac{1}{\alpha_t} B_G(\theta, \theta_t) \right\}.$$

Finding the minimum by differentiation yields the step:

$$g(\theta_{t+1}) = g(\theta_t) - \alpha_t \nabla_\theta f(\theta_t),$$

where $g = \nabla G$. In terms of the dual variable $\mu = g(\theta)$ and noting that $\theta = h(\mu) = \nabla H(\mu)$,

$$\mu_{t+1} = \mu_t - \alpha_t \nabla_\theta f(h(\mu_t)).$$

Applying the chain rule to $\nabla_\mu f(h(\mu)) = \nabla_\mu h(\mu) \nabla_\theta f(h(\mu))$ implies that

$$\nabla_\theta f(h(\mu_t)) = [\nabla_\mu h(\mu_t)]^{-1} \nabla_\mu f(h(\mu_t)).$$

Therefore

$$\mu_{t+1} = \mu_t - \alpha_t [\nabla^2 H(\mu_t)]^{-1} \nabla_\mu f(h(\mu_t)),$$

which corresponds to the natural gradient descent step (2). This completes the proof.

3 Consequences

In this section, we discuss how this connection directly yields optimal efficiency results for mirror descent and discuss connections to other online algorithm on Riemannian manifolds.

By Theorem 1 in Amari [1], natural gradient descent along the Riemannian manifold $(\mathcal{M}, \Phi, \nabla^2 H)$ follows the direction of steepest descent along that manifold. As an immediate consequence, mirror descent with Bregman divergence induced by G follows the direction of steepest descent along the Riemannain manifold $(\mathcal{M}, \Phi, \nabla^2 H)$ where H is the convex conjugate for G. As far as we are aware, an interpretation in terms of Riemannian manifolds had not been provided for mirror descent.

Next we explain how using existing theoretical results in Amari [1], we can prove that mirror descent achieves Fisher efficiency.

3.1 Efficient Parameter Estimation in Exponential Families

In this section we exploit the connection between mirror descent and natural gradient descent to study the efficiency of mirror descent from a statistical perspective. Prior work on the statistical theory of mirror descent has largely focussed on regret analysis and we are not aware of analysis of second-order properties such as statistical efficiency. We will see that asymptotic Fisher efficiency [12,14,19] for mirror descent which is an optimality criterion on the covariance of a parameter estimate is an immediate consequence of the equivalence between mirror descent and natural gradient descent.

The statistical problem we consider is parameter estimation in exponential families. Consider a *natural parameter* exponential family with density:

$$p(y \mid \theta) = h(y) \exp(\langle \theta, y \rangle - G(\theta)),$$

where $\theta \in \mathbb{R}^p$ and $G : \mathbb{R}^p \to \mathbb{R}$ is a strictly convex differentiable function. The probability density function can be re-expressed in terms of the Bregman divergence $B_G(\cdot, \cdot)$ as follows:

$$p(y \mid \theta) = \tilde{h}(y) \exp(-B_G(\theta, h(y))),$$

where recall that $h = \nabla H$ and H is the conjugate dual function of G. The distribution can be expressed in terms of the *mean parameter* $\mu = g(\theta)$ and the dual Bregman divergence $B_H(\cdot, \cdot)$:

$$p(y \mid \eta) = \tilde{h}(y) \exp(-B_H(y, \mu)).$$

The natural and mean parameterizations and their relationship through convex conjugates is widely known (see e.g. [6,9,13]) and this dual parameterization has been applied in many applications (see e.g. [4,5,22]).

Consider the mirror descent update for the natural parameter θ with proximity function $B_G(\cdot, \cdot)$ when the function to be minimized is the standard log loss:

$$f(\theta; y_t) = -\log p(y_t \mid \theta) = B_G(\theta, h(y_t)).$$

Note that the dependence on t for f_t is through the noisy data point y_t which varies for different t.

Then the mirror descent step is:

$$\theta_{t+1} = \arg\min_{\theta} \left\{ \langle \theta, \nabla_{\theta} B_G(\theta, h(y_t))|_{\theta=\theta_t} \rangle + \frac{1}{\alpha_t} B_G(\theta, \theta_t) \rangle \right\}. \tag{5}$$

Now if we consider the natural gradient descent step for the mean parameter μ, the function to be minimized is again the standard log-loss in the μ co-ordinates:

$$\tilde{f}_t(\mu; y_t) = -\log p(y_t \mid \mu) = B_H(y_t, \mu).$$

Using Theorem 1 (or by showing it directly), the natural gradient step is:

$$\mu_{t+1} = \mu_t - \alpha_t [\nabla^2 H(\mu_t)]^{-1} \nabla B_H(y_t, \mu_t). \tag{6}$$

A parallel argument holds if the mirror descent step was expressed in terms of the mean parameter and the natural gradient step in terms of the natural parameter.

By explicitly calculating $\nabla B_H(y_t, \mu_t)$, Eq. (6) reduces to the very simple linear update:

$$\mu_{t+1} = \mu_t - \alpha_t(\mu_t - y_t),$$

and hence is very straightforward to implement as natural gradient descent. The fact that the curvature of the loss function $B_H(y_t, \mu_t)$ perfectly matches the curvature due to the metric tensor $\nabla^2 H(\mu)$ is why the mirror descent and natural gradient descent updates reduce to this simple linear update equation. Hence in this setting mirror descent applied to the natural parameter results in the simple linear update of the mean parameter through natural gradient descent.

Now we use Theorem 2 in [1] to prove that mirror descent yields an asymptotically Fisher efficient for μ. The Cramér-Rao theorem states that any unbiased estimator based on T independent samples $y_1, y_2, ..., y_T$ of μ, which we denote by $\hat{\mu}_T$ satisfies the following lower bound:

$$\mathbb{E}[(\hat{\mu}_T - \mu)(\hat{\mu}_T - \mu)^T] \succeq \frac{1}{T} \nabla^2 H,$$

where $\succeq$ refers to the standard matrix inequality. A sequence of estimators $(\hat{\mu}_t)_{t=1}^{\infty}$ is asymptotically Fisher efficient if:

$$\lim_{T \to \infty} T\mathbb{E}[(\hat{\mu}_T - \mu)(\hat{\mu}_T - \mu)^T] \to \nabla^2 H.$$

Now by using Theorem 2 in [1] for natural gradient descent and the equivalence of natural gradient descent and mirror descent (Theorem 1), it follows that mirror descent is Fisher efficient.

Corollary 1. *The mirror descent step applied to the log loss* (5) *with step-sizes* $\alpha_t = \frac{1}{t}$ *asymptotically achieves the Cramér-Rao lower bound.*

Theorem 2 in Amari [1] applies more generally to neural network and regression models where the update along the Riemannian manifold is not exactly linear but almost linear with a vanishing remainder term. For a more detailed discussion on the statistical properties of natural gradient see [1]. Here we have illustrated how the equivalence between mirror descent with Bregman divergences and natural gradient descent gives second-order optimality properties of mirror descent.

3.2   Connection to Other Online Methods on Riemannian Manifolds

The point in using the natural gradient is to the parameter of interest in the direction of the gradient on the manifold rather than the gradient in the ambient space. Note however that any non-infinitesimal step in the direction of the gradient of the manifold will move one off the manifold, for any curved manifold. This observation has motivated algorithms [7] in which the update step is constrained to remain on the manifold.

In this section, we discuss the relation between natural gradient descent, mirror descent, and gradient based methods that along a Riemannian manifold [7]. To define the online steepest descent step used in [7], we need to define the exponential map and differentiation in curved spaces.

The *exponential map* at a point $\mu \in \mathcal{M}$ is a map $\exp_\mu : T_\mu \mathcal{M} \to \mathcal{M}$ where $T_\mu \mathcal{M}$ is the tangent space at each point $\mu \in \mathcal{M}$. The idea of an exponential map is starting at a point μ with tangent vector $v \in T_\mu$ if one starts at point μ and "flows" along the manifold in direction v for a fixed (unit) time interval at constant velocity one reaches a new point on the manifold $\exp_\mu(v)$. This idea is usually stated in terms of geodesic curves on the manifold, consider the geodesic curve $\gamma : [0,1] \to \mathcal{M}$, with $\gamma(0) = \mu$ and $\dot{\gamma}(0) = v$, where $v \in T_\mu \mathcal{M}$ then $\exp_\mu(v) = \gamma(1)$. Again, in words, $\exp_\mu(\cdot)$ is the end-point of a curve that lies along the manifold $\mathcal{M}$ that begins at μ with initial velocity $v = \dot{\gamma}(0)$ that travels one time unit. A more thorough introduction to the exponential map is provided in [10].

Now we define differentiation along a manifold. Let $f : \mathcal{M} \to \mathbb{R}$ be a differentiable function on $\mathcal{M}$. The gradient vector field $\nabla_{\mathcal{M}} f$ takes the form $\nabla_{\mathcal{M}} f(\mu) = \nabla_v(f(\exp_\mu(v)))|_{v=0}$ noting that $f(\exp_\mu(v))$ is a smooth function on $T_\mu \mathcal{M}$.

For the function f where $f : \mathcal{M} \to \mathbb{R}$ the online steepest descent step analyzed in [7] is:

$$\mu_{t+1} = \exp_{\mu_t}(-\alpha_t \nabla_{\mathcal{M}} f(\mu_t)). \tag{7}$$

The key reason why the update (7) is the standard gradient descent step instead of the natural gradient descent step introduced by Amari is that μ_{t+1} is always guaranteed to lie on the manifold $\mathcal{M}$ for (7), but not for the natural gradient descent step. Unfortunately, the exponential map is extremely difficult to evaluate in general since it is the solution of a system of second-order differential equations [10].

Consequently a standard strategy is to use a computable *retraction* R_μ: $T_\mu \mathcal{M} \to \mathbb{R}^p$ of the exponential map which yields the approximate gradient descent step:

$$\mu_{t+1} = R_{\mu_t}(-\alpha_t \nabla_{\mathcal{M}} f(\mu_t)). \tag{8}$$

The retraction $R_\mu(v) = \mu + v$ corresponds to the first-order Taylor approximation of the exponential map and yields the natural gradient descent step in [1]. Therefore as pointed out in [7], natural gradient descent can be cast as an approximation to gradient descent for Riemannian manifolds. Consequently mirror descent can be viewed as an easily computable first-order approximation to steepest descent for any Riemannian manifold induced by a Bregman divergence.

4 Conclusion

In this paper we prove that mirror descent with proximity function Ψ equal to a Bregman divergence is equivalent to the natural gradient descent algorithm in the dual co-ordinate system. Based on this equivalence, we use results developed in [1] to conclude that mirror descent is the direction of steepest in the corresponding Riemannian space and for parameter estimation in exponential families with the associated Bregman divergence, mirror descent achieves the Cramér-Rao lower bound.

Following on from this connection, there are a number of interesting and open directions. Firstly, one of the important issues for any online learning algorithm is choice of step-size. Using the connection between mirror descent and natural gradient, it would be interesting to determine whether adaptive choices of step-sizes proposed in [1] that exploit the Riemannian structure can improve performance of mirror descent. It would also be useful to determine a precise characterization of the geometry of mirror descent for other proximity functions such as ℓ_p-norms and explore links online algorithms such as projected gradient descent.

Acknowledgements. GR was partially supported by the NSF under Grant DMS-1127914 to the Statistical and Applied Mathematical Sciences Institute. SM was supported by grants: NIH (Systems Biology): 5P50-GM081883, AFOSR: FA9550-10-1-0436, and NSF CCF-1049290.

References

1. Amari, S.: Natural gradient works efficiently in learning. Neural Comput. **10**(2), 251–276 (1998)
2. Amari, S., Cichocki, A.: Information geometry of divergence functions. Bull. Pol. Acad. Sci. Tech. Sci. **58**(1), 183–195 (2010)
3. Amari, S.-I., Barndoff-Nielsen, O.E., Kass, R.E., Lauritzen, S.L., Rao, C.R.: Differential Geometry in Statistical Inference. IMS Lecture Notes - Monograph Series. Institute of Mathematical Statistic, Hayward (1987)

4. Azoury, K.S., Warmuth, M.K.: Relative loss bounds for on-line density estimation with the exponential family of dsitributions. Mach. Learn. **43**(3), 211–246 (2001)
5. Banerjee, A., Merugu, S., Dhillon, I.S., Ghosh, J.: Clustering with Bregman divergences. J. Mach. Learn. Res. **6**, 1705–1749 (2005)
6. Barndorff-Nielson, O.E.: Information and Exponential Families. Wiley, Chichester (1978)
7. Bonnabel, S.: Stochastic gradient descent on Riemannian manifiolds. Technical report, Mines Paris Tech (2011)
8. Bregman, L.M.: The relaxation method for finding the common point of convex sets and its application to the solution of problems in convex programming. USSR Comput. Math. Math. Phys. **7**, 191–204 (1967)
9. Brown, L.D.: Fundamentals of Statistical Exponential Families. Institute of Mathematical Statistics, Hayward (1986)
10. DoCarmo, M.P.: Riemannian Geometry. Springer Series in Statistics. Birkhauser, Boston (1992)
11. Cramér, H.: Mathematical Methods of Statistics. Princeton University Press, Princeton (1946)
12. Efron, B.: Defining the curvature of a statistical problem (with applications to second order efficiency). Ann. Stat. **3**(6), 1189–1242 (1975)
13. Efron, B.: The geometry of exponential families. Ann. Stat. **6**, 362–376 (1978)
14. Fisher, R.A.: Theory of statistical estimation. Math. Proc. Cambridge Philos. Soc. **22**, 700–725 (1925)
15. Lafferty, J.: Additive models, boosting, and inference for generalized divergences. In: COLT (1999)
16. Nemirovski, A., Yudin, D.: Problem Complexity and Method Efficiency in Optimization. Wiley, New York (1983)
17. Nielsen, F., Garcia, V.: Statistical exponential families: a digest with flash cards. Technical report, École Polytechnique (2011)
18. Rao, C.R.: Information and accuracy obtainable in the estimation of statistical parameters. Bull. Calcutta Math. Soc. **37**, 81–91 (1945)
19. Rao, C.R.: Asymptotic efficiency and limiting information. In: Proceedings of the Fourth Berkeley Symposium on Mathematical Statistics and Probability, vol. 1, pp. 531–546 (1961)
20. Reid, M.D., Williamson, R.C.: Information, divergence and risk for binary experiments. J. Mach. Learn. Res. **12**, 731–817 (2011)
21. Rockafeller, R.T.: Convex Analysis. Princeton University Press, Princeton (1970)
22. Wainwright, M.J., Jordan, M.I.: A variational principle for graphical models. In: New Directions in Statistical Signal Processing. MIT Press, Cambridge, MA (2006)

Information Geometry in Image Analysis

Texture Classification Using Rao's Distance on the Space of Covariance Matrices

Salem Said, Lionel Bombrun[(⊠)], and Yannick Berthoumieu

Laboratoire IMS, CNRS - UMR 5218, Université de Bordeaux, Bordeaux, France
{salem.said,lionel.bombrun,yannick.berthoumieu}@ims-bordeaux.fr

Abstract. The current paper introduces new prior distributions on the zero-mean multivariate Gaussian model, with the aim of applying them to the classification of covariance matrices populations. These new prior distributions are entirely based on the Riemannian geometry of the multivariate Gaussian model. More precisely, the proposed Riemannian Gaussian distribution has two parameters, the centre of mass $\bar{Y}$ and the dispersion parameter σ. Its density with respect to Riemannian volume is proportional to $\exp(-d^2(Y;\bar{Y}))$, where $d^2(Y;\bar{Y})$ is the square of Rao's Riemannian distance. We derive its maximum likelihood estimators and propose an experiment on the VisTex database for the classification of texture images.

Keywords: Texture classification · Information geometry · Riemannian centre of mass · Mixture estimation · EM algorithm

1 Introduction

In information geometry, a parametric family of probability densities is considered as a Riemannian manifold [1]. Precisely, the role of Riemannian metric is played by the Fisher metric, and that of Riemannian distance by Rao's distance. Rao's distance has been widely used for several statistical applications including object detection and tracking, shape classification, and image segmentation [2–4]. Nevertheless, none of them have formulated it as a probabilistic approach to clustering on Riemannian manifolds, which is the main contribution of the paper.

More precisely, this paper introduces new Riemannian prior (denoted $G(\bar{Y},\sigma)$) as Gaussian distributions on the zero-mean multivariate Gaussian model. These distributions have a unique mode $\bar{Y}$ (the unique Riemannian centre of mass), and its dispersion away from $\bar{Y}$ is given by σ. In order to improve upon the performance obtained in [5], the present paper uses mixtures of Riemannian priors as prior distributions for classification. This allows for clustering analysis to be carried out using an expectation-maximisation, or EM, algorithm, instead of the essentially deterministic k-means approach of existing works, (e.g. [2–4]).

The paper is structured as follows. Section 2 recalls some definitions concerning the Riemannian geometry of covariance matrices. Section 3 introduces the

© Springer International Publishing Switzerland 2015
F. Nielsen and F. Barbaresco (Eds.): GSI 2015, LNCS 9389, pp. 371–378, 2015.
DOI: 10.1007/978-3-319-25040-3_40

proposed Riemannian Gaussian distributions. After having presented its maximum likelihood estimators and its extension to mixture models in Sect. 4, an experiment on the VisTex database is proposed in Sect. 5 to evaluate the potential of the proposed prior for the classification of texture images. Due to the restriction length, all the mathematical proofs cannot be detailed here and will be given in a forthcoming journal paper.

2   Riemannian Geometry of Covariance Matrices

Let $\mathcal{P}_m$ denote the space of all $m \times m$ real matrices Y which are symmetric and strictly positive definite,

$$Y^\dagger - Y = 0 \qquad x^\dagger Y x > 0 \text{ for all } x \in \mathbb{R}^m \tag{1}$$

where $\dagger$ denotes the transpose. In many applications [6], $\mathcal{P}_m$ arises as a space of tensors, such as structure tensors in image processing, or diffusion tensors in medical imaging, (in these examples, $m = 2, 3$). In general, $\mathcal{P}_m$ may also be thought of as the space of non-degenerate covariance matrices [7].

When thinking of the elements Y of $\mathcal{P}_m$ as covariance matrices, it is most suitable to do so within the framework of the normal covariance model [7,8]. This model associates to $Y \in \mathcal{P}_m$ the normal probability density function $P(x|Y)$ on $\mathbb{R}^m$, with mean $0 \in \mathbb{R}^m$ and covariance Y. Recall that $\log P(x|Y) = \ell(Y)$, where

$$\ell(Y) = -\frac{1}{2} \log\left[\det(2\pi Y)\right] - \frac{1}{2} x^\dagger Y^{-1} x. \tag{2}$$

Let us now recall the definition of the Fisher information matrix [8]. Let $p = m(m+1)/2$, the dimension of $\mathcal{P}_m$, and Θ an open subset of $\mathbb{R}^p$. Assume $\theta \mapsto Y(\theta)$ is a differentiable mapping from Θ to $\mathcal{P}_m$, which is a diffeomorphism. One refers to the mapping $\theta \mapsto Y(\theta)$ as a parameterisation of $\mathcal{P}_m$, with parameters $\theta = (\theta^a ; a = 1, \ldots, p)$. Let $\ell(\theta)$ stand for $\ell(Y(\theta))$ where $\ell(Y)$ is the function defined in (2). The Fisher information matrix $I(\theta)$ has matrix elements

$$I_{ab}(\theta) = \mathbb{E}_\theta \left[\frac{\partial \ell(\theta)}{\partial \theta^a} \times \frac{\partial \ell(\theta)}{\partial \theta^b} \right], \tag{3}$$

where $\mathbb{E}_\theta$ denotes expectation with respect to the normal probability density function $p(x|Y(\theta))$.

A Riemannian metric on $\mathcal{P}_m$ is a quadratic form $ds^2(Y)$ which measures the *squared length of a small displacement* dY, separating two elements $Y \in \mathcal{P}_m$ and $Y + dY \in \mathcal{P}_m$. Here, dY is a symmetric matrix, since Y and $Y + dY$ are symmetric, by (1). The Rao-Fisher metric is the following [8,9],

$$ds^2(Y) = \text{tr}\left([Y^{-1} dY]^2 \right), \tag{4}$$

where $\text{tr}()$ denotes the trace.

The Rao-Fisher metric, like any other Riemannian metric on $\mathcal{P}_m$, defines a Riemannian distance $d : \mathcal{P}_m \times \mathcal{P}_m \to \mathbb{R}_+$. This is called Rao's distance, and is defined as follows [9,10]. Let $Y, Z \in \mathcal{P}_m$ and $c : [0,1] \to \mathcal{P}_m$ be a differentiable curve with $c(0) = Y$ and $c(1) = Z$. The length $L(c)$ of c is defined by

$$L(c) = \int_0^1 ds(c(t)) = \int_0^1 \|\dot{c}(t)\| \, dt, \tag{5}$$

where $\dot{c}(t) = \frac{dc}{dt}$. Rao's distance $d(Y, Z)$ is the infimum of $L(c)$ taken over all differentiable curves c as above.

A major property of the Rao-Fisher metric is the following. When equipped with the Rao-Fisher metric, the space $\mathcal{P}_m$ is a Riemannian manifold of negative sectional curvature. One implication of this property, (called the Cartan-Hadamard theorem [10]), is that the infimum of $L(c)$ is realised by a unique curve γ, known as the geodesic connecting Y and Z. The equation of this curve is the following [9],

$$\gamma(t) = Y^{1/2} \, (Y^{-1/2} Z Y^{-1/2})^t \, Y^{1/2}. \tag{6}$$

Given the expression (6), it is possible to compute $L(\gamma)$ from (5). This is precisely Rao's distance $d(Y, Z)$. It turns out,

$$d^2(Y, Z) = \mathrm{tr} \, [\log(Y^{-1/2} Z Y^{-1/2})]^2. \tag{7}$$

Since the Rao-Fisher metric gives a mean of measuring length, it can also be used to measure volume. Indeed, (based on the elementary fact that the "volume of a cube is the product of the lengths of its sides"), the Riemannian volume element associated to the Rao-Fisher metric is defined to be [9]

$$dv(Y) = \det(Y)^{-\frac{m+1}{2}} \prod_{i \le j} dY_{ij}. \tag{8}$$

All matrix functions appearing in (6) and (7), (square root, power and logarithm), should be understood as symmetric positive definite matrix functions.

3   Riemannian Gaussian Distributions

The main theoretical contribution of the present paper is to give an original exact formulation of Riemannian Gaussian distributions. These are probability distributions on $\mathcal{P}_m$, whose probability density function, with respect to the Riemannian volume element (8), is of the form,

$$p(Y | \bar{Y}, \sigma) = \frac{1}{Z(\sigma)} \exp\left[-\frac{d^2(Y, \bar{Y})}{2\sigma^2} \right], \tag{9}$$

where $\bar{Y} \in \mathcal{P}_m$ and $\sigma > 0$ are parameters, and where $d(Y, \bar{Y})$ is Rao's distance, given by (7). For brevity, a Riemannian Gaussian distribution, with probability density function (9), will be called a Gaussian distribution, and denoted $G(\bar{Y}, \sigma)$.

The parameter $\bar{Y}$ is called the centre of mass, and σ is called the dispersion, of the distribution $G(\bar{Y}, \sigma)$.

Distributions of the form (9) were considered by Pennec [11], defined on general Riemannian manifolds. However, in existing literature, their treatment remains incomplete, as it is based on asymptotic formulae, valid only in the limit where the parameter σ is small, (see [11] (Theorem 5., Page 140) and [12] (Theorem 3.1.1., Page 434)). In addition to being only approximations, such formulae are quite difficult, both to evaluate and to apply. These issues, (lack of an exact expression and difficulty of application), are fully overcome in the following.

Note also that a more sophisticated description by means of a concentration matrix instead of a scalar dispersion parameter σ is possible. This approach has notably been introduced in [11].

3.1 Maximum Likelihood Estimation

Let $Y_1, \ldots, Y_N$ be N independent samples from a Gaussian distribution $G(\bar{Y}, \sigma)$. Based on these samples, the maximum likelihood estimate of the parameter $\bar{Y}$ is the empirical Riemannian centre of mass $\hat{Y}_N$ of $Y_1, \ldots, Y_N$ defined as the unique global minimiser $\hat{Y}_N$ of $\mathcal{E}_N : \mathcal{P}_m \to \mathbb{R}$,

$$\mathcal{E}_N(Y) = \frac{1}{N} \sum_{n=1}^{N} d^2(Y, Y_n). \tag{10}$$

Moreover, the maximum likelihood estimate of the parameter σ is the solution $\hat{\sigma}_N$ of the equation, (for unknown σ),

$$\sigma^3 \times \frac{d}{d\sigma} \log Z(\sigma) = \mathcal{E}_N(\hat{Y}_N). \tag{11}$$

Both $\hat{Y}_N$ and $\hat{\sigma}_N$ exist and are unique for any realisation of the samples $Y_1, \ldots, Y_N$. In practice, $\bar{Y}$ is first estimated according to (10) then the estimation of σ is proceed by (11).

3.2 Application to $\mathcal{P}_2$

In (9), the normalising factor $Z(\sigma)$ can be expressed under an integral form as

$$Z(\sigma) = \int_{\mathcal{P}_m} f(Y \,|\, \bar{Y}, \sigma) \, dv(Y), \tag{12}$$

where $dv(Y)$ is the Riemannian volume element (8). It is interesting to note that $Z(\sigma)$ is independent from the centre of mass $\bar{Y}$. For the space of 2×2 covariance matrices (i.e. $m = 2$), the normalising factor admits the following close form expression:

$$Z(\sigma) = 4\pi^2 \sigma^2 \, \exp(\sigma^2/4) \, \mathrm{erf}(\sigma/2), \tag{13}$$

where erf() is the error function.

4 EM Algorithm for Mixture Estimation

While successful in application to specific data sets, the Bayesian approach of [5] summarised in the previous section fails to take into account the presence of within-class diversity. Precisely, this approach assumes that the given learning sequence is immediately subdivided into clusters, whose members display "homogeneous" properties, in the sense that they can be faithfully modelled as belonging to the same Riemannian prior. Clearly, this is a restrictive assumption. In the presence of within-class diversity, a learning sequence should be subdivided into classes, whose members display "heterogeneous" properties, in the sense that they may belong, within the same class, to different clusters, each corresponding to a different Riemannian prior.

Here, this situation is formulated as follows. If a class $\mathcal{C}$, whose members are points $Y_1, \ldots, Y_N \in \mathcal{P}_m$, is expected to contain K clusters, respectively corresponding to Riemannian priors $G(\bar{Y}_a, \sigma_a)$, where $a = 1, \ldots, K$, then $\mathcal{C}$ is modelled as a sample of size N, drawn from the mixture of Riemannian priors

$$p(Y|\Theta) = \sum_{a=1}^{K} \varpi_a\, p(Y|\bar{Y}_a, \sigma_a) \tag{14}$$

where $\varpi_1, \ldots, \varpi_K$ are positive weights, with $\sum_{a=1}^{K} \varpi_a = 1$, and each density $p(Y|\bar{Y}_a, \sigma_a)$ is given by (9).

Now, assume a training sequence is subdivided into classes, each containing a known numbers of clusters. In order to implement a decision rule which associates any test object, described by $Y_t \in \mathcal{P}_m$, to the most likely cluster within the training sequence, it is necessary, for each class $\mathcal{C}$, modelled by (14), to find maximum likelihood estimates of the mixture parameters $\vartheta = (\varpi_a, \bar{Y}, \sigma_a)$. Here, this task is realised using an expectation-maximisation (EM) algorithm. Following [13], the starting point for the EM algorithm is the introduction of the following quantities

$$\omega_a(Y_j) \propto \varpi_a \times p(Y_j|\bar{Y}_a, \sigma_a) \qquad n_a = \sum_{j=1}^{N} \omega_a(Y_j) \tag{15}$$

where, $\propto$ denotes proportionality, so that $\sum_a \omega_a(Y_j) = 1$. To emphasise the fact that $\omega_a(Y_j)$ and n_a are computed for a given value of $\vartheta = (\varpi_a, \bar{Y}, \sigma_a)$, they shall be denoted $\omega_a(Y_j, \vartheta)$ and $n_a(\vartheta)$. The algorithm iteratively updates $\hat{\vartheta} = (\hat{\varpi}_a, \hat{Y}_a, \hat{\sigma}_a)$, an approximation of the maximum likelihood estimate of $\vartheta = (\varpi_a, \bar{Y}_a, \sigma_a)$. Precisely, the update rules for $\hat{\varpi}_a$, $\hat{Y}_a$, and $\hat{\sigma}_a$ are repeated as long as this introduces a sensible change in the values of $\hat{\varpi}_a$, $\hat{Y}_a$, and $\hat{\sigma}_a$. As this is a non convex problem optimization, we reach a local stationary point. It is hence useful to run the algorithm several times, with different initialisations to reach the global optimum. The update rules are the following,

▶ **Update for $\hat{\varpi}_a$:** Based on the current value of $\hat{\vartheta}$, assign to $\hat{\varpi}_a$ the new value

$$\hat{\varpi}_a^{\text{new}} = \frac{n_a(\hat{\vartheta})}{\sum_{a=1}^{K} n_a(\hat{\vartheta})}. \tag{16}$$

▶ **Update for $\hat{Y}_a$:** Based on the current value of $\hat{\vartheta}$, compute $\hat{Y}_a$ to be the global minimiser of the following function,

$$V(Y|\hat{\vartheta}) = \frac{1}{2} \sum_{j=1}^{N} \omega_a(Y_j, \hat{\vartheta}) \times d^2(Y_j, Y).$$ (17)

$\hat{Y}_a$ is the empirical Riemannian centre of mass which may be estimated by a Riemannian gradient descent algorithm (See [12] for more details).

▶ **Update for $\hat{\sigma}_a$:** Based on the current value of $\hat{\vartheta}$, compute $\hat{\sigma}_a$ to be the solution of the following equation, for unknown σ,

$$F(\sigma) = \frac{1}{2n_a(\hat{\vartheta})} V(\hat{Y}_a|\hat{\vartheta})$$ (18)

where $F(\sigma) = \sigma^3 \times \frac{d}{d\sigma} \log Z(\sigma)$. Practically, a Newton-Raphson procedure is employed to solve (18).

These three update rules should be performed in the above given order. Therefore, the "current value of $\hat{\vartheta} = (\hat{\varpi}_a, \hat{Y}_a, \hat{\sigma}_a)$" is different, in each one of them. For instance, in the update rule of $\hat{\sigma}_a$, the current value of $\hat{Y}_a$ is found from the minimisation of (17), just before.

5   Application to Texture Image Classification

The present section proposes a new decision rule, for the classification of covariance matrices, and applies it to texture classification, using the VisTex database [14]. The following numerical experiment was carried out. Half of the database was used for training, and the other half for testing. Each training image was subdivided into 169 patches of 128×128 pixels, with a 32 pixel overlap. For each training patch, 6 wavelet subbands were computed using the stationary wavelet decomposition (with 2 scale) with Daubechies' filter db4. In texture classification, multivariate models were found very effective for modelling the spatial dependency of wavelet coefficients. Hence, two spatial neighborhoods (horizontal dH and vertical dV) of one pixel were considered. Each subband s of patch n gives two bivariate normal populations $\Pi_{s,n,dH}$ and $\Pi_{s,n,dV}$, represented respectively by a point $Y_{s,n,dH}$ and $Y_{s,n,dV} \in \mathcal{P}_2$. The size of the feature space is hence $F = 12$ (6 subbands times 2 spatial supports). For the sake of simplicity, let say that the training patch n is represented by a set of F covariance matrices denoted $Y_{f,n}$.

For each training class, a set of $N = 84$ "arrays" are extracted. These arrays Y_j are a collection of F covariance matrices and are considered as multivariate realisations of a mixture distribution (14), with independent components $Y_{f,n}$ since wavelet subbands are assumed independent. Each class is assumed to contain the same number K of clusters, and is modelled as a sample drawn from a mixture distribution (14). First, the EM algorithm of Sect. 4 is applied to each class, leading to maximum likelihood estimates $(\hat{\varpi}_a, \hat{Y}_{f,a}, \hat{\sigma}_a)$, for $a = 1, \ldots, K$ and $f = 1, \ldots, F$.

Table 1. Classification performance on the VisTex database.

Prior	Overall Accuracy
Riemannian prior on $\mathcal{P}_2$ (K=1) (9)	86.27 ± 0.45 %
Mixture prior on $\mathcal{P}_2$ (EM, K=3, (14))	**94.31 ± 0.42 %**
Mixture prior on $\mathcal{P}_2$ (K-means, K=3) [15]	92.40 ± 0.46 %
Riemannian prior on $\mathcal{H}$ [5]	83.29 ± 0.51 %
Mixture prior on $\mathcal{H}$ [17]	88.50 ± 0.88 %
Conjugate prior on $\mathcal{H}$	83.48 ± 0.53 %

Each triple of such estimates defines a cluster within the training sequence. Denote the total number of clusters defined in this way L, and the corresponding maximum likelihood estimates $(\hat{\varpi}_c, \hat{Y}_{f,c}, \hat{\sigma}_c)$, for $c = 1, \ldots, L$ and $f = 1, \ldots, F$. Then, a test population represented by $Y_t \in \mathcal{P}_2$ is associated to the class of the cluster C_*, realising the minimum over c of,

$$- \log \hat{\varpi}_c + \log Z(\hat{\sigma}_c) + \frac{1}{2\hat{\sigma}_c^2} \sum_{f=1}^{F} d^2(Y_t, \hat{Y}_{f,c}). \tag{19}$$

This is the new decision rule, proposed for use with the mixture model (14). Note that the case $K = 1$ reduces to a Bayesian classifier with the proposed Riemannian Gaussian distribution.

Table 1 displays the classification performance in terms of overall accuracy on the VisTex database. The first two lines correspond to the proposed Riemannian prior (9) on $\mathcal{P}_2$ with respectively $K = 1$ and $K = 3$. The third line corresponds to a nearest centre of mass classifier classically employed in literature [15]. In such case, the centres of mass $\hat{Y}_{f,c}$ are estimated by using a K-means algorithm. Some comparisons are also carried out with univariate normal populations where the mean and standard deviation are computed on Gabor energy subbands (see [16] for more details). In such case, a Riemannian prior on the Poincaré upper half-plane $\mathcal{H}$ has been introduced in [5] and further extended to mixture models [17]. A conjugate normal-inverse gamma prior on $\mathcal{H}$ is also displayed on the last line of Table 1

As observed in Table 1, the proposed Riemannian prior on $\mathcal{P}_2$ based on a mixture model displays much better performance than other prior. A significant gain of respectively 2 % and 6 % is observed when compared to a nearest centre of mass classifier [15] and to a mixture prior on the Poincaré upper half-plane $\mathcal{H}$.

6 Conclusion

This paper has addressed the problem of classification using Rao's distance on the space of covariance matrices. To this aim, a Riemannian Gaussian distributions has been introduced. Analogous to the classical multivariate Gaussian distribution, the proposed Riemannian Gaussian distribution has two parameters, the centre of mass $\bar{Y}$ and the dispersion parameter σ. The main difference relies on the use of the Riemannian distance in the exponential of the pdf instead of the Mahalanobis distance. After having presented its maximum likelihood estimators and its extension to mixture models, an experiment on the VisTex database have shown the potential of the proposed model for the classification of texture images.

References

1. Amari, S., Nagaoka, H.: Methods of Information Geometry. American Mathematical Society, Providence (2000)
2. Porikli, F., Tuzel, O., Meer, P.: Covariance tracking using model update based means on Riemannian manifolds. In: IEEE Conference on Computer Vision and Pattern Recognition (CVPR), pp. 728–735 (2006)
3. Kurtek, S., Klassen, E., Ding, Z., Avison, M.J., Srivastava, A.: Parameterization-invariant shape statistics and probabilistic classification of anatomical surfaces. In: Székely, G., Hahn, H.K. (eds.) IPMI 2011. LNCS, vol. 6801, pp. 147–158. Springer, Heidelberg (2011)
4. Gu, X., Deng, J., Purvis, M.: Improving superpixel-based image segmentation by incorporating color covariance matrix manifolds. In: International Conference on Image Processing (ICIP), pp. 4403–4406 (2014)
5. Said, S., Bombrun, L., Berthoumieu, Y.: New Riemannian priors on the univariate normal model. Entropy **16**(7), 4015–4031 (2014)
6. Weickert, J., Hagen, H. (eds.): Visualization and Processing of Tensor Fields. Mathematics Visualization. Springer, Heidelberg (2006)
7. Muirhead, R.J.: Aspects of Multivariate Statistical Theory. Wiley, New York (1982)
8. Atkinson, C., Mitchell, A.: Rao's distance measure. Sankhya Ser. A **43**, 345–365 (1981)
9. Terras, A.: Harmonic Analysis on Symmetric Spaces and Applications II. Springer, New York (1988)
10. Helgason, S.: Differential Geometry, Lie Groups, and Symmetric Spaces. Americal Mathematical Society, Providence (2001)
11. Pennec, X.: Intrinsic statistics on Riemannian manifolds: Basic tools for geometric measurements. J. Math. Imaging Vis. **25**(1), 127–154 (2006)
12. Lenglet, C., Rousson, M., Deriche, R., Faugeras, O.: Statistics on the manifold of multivariate normal distributions. J. Math. Imaging Vis. **25**(3), 423–444 (2006)
13. Mengersen, K., Robert, C., Titterington, M.: Mixtures : Estimation and Applications. Wiley, Chichester (2011)
14. Saravanan, T.: MIT Vision and Modeling Group. Vision Texture. VisTex : Vision Texture Database. http://vismod.media.mit.edu/pub/vistex
15. Barachant, A., Bonnet, S., Congedo, M., Jutten, C.: Multiclass brain-computer interface classification by Riemannian geometry. IEEE Trans. Biomed. Eng. **59**(4), 920–928 (2012)
16. Grigorescu, S., Petkov, N., Kruizinga, P.: Comparison of texture features based on Gabor filters. IEEE Trans. Im. Proc. **11**(10), 1160–1167 (2002)
17. Said, S., Bombrun, L., Berthoumieu, Y.: Texture classification using Rao's distance: an EM algorithm on the Poincaré half plane. In: International Conference on Image Processing (ICIP) (2015)

Color Texture Discrimination Using the Principal Geodesic Distance on a Multivariate Generalized Gaussian Manifold

Geert Verdoolaege[1,2](✉) and Aqsa Shabbir[1,3]

[1] Department of Applied Physics, Ghent University, 9000 Ghent, Belgium
`geert.verdoolaege@ugent.be`
[2] Laboratory for Plasma Physics – Royal Military Academy (LPP–ERM/KMS),
1000 Brussels, Belgium
[3] Max-Planck-Institut Für Plasmaphysik, 85748 Garching, Germany

Abstract. We present a new texture discrimination method for textured color images in the wavelet domain. In each wavelet subband, the correlation between the color bands is modeled by a multivariate generalized Gaussian distribution with fixed shape parameter (Gaussian, Laplacian). On the corresponding Riemannian manifold, the shape of texture clusters is characterized by means of principal geodesic analysis, specifically by the principal geodesic along which the cluster exhibits its largest variance. Then, the similarity of a texture to a class is defined in terms of the Rao geodesic distance on the manifold from the texture's distribution to its projection on the principal geodesic of that class. This similarity measure is used in a classification scheme, referred to as *principal geodesic classification* (PGC). It is shown to perform significantly better than several other classifiers.

Keywords: Texture classification · Rao geodesic distance · Principal geodesic analysis

1 Introduction

Texture discrimination is an essential task in various image processing applications, such as image retrieval and image segmentation. Texture is often characterized by means of the distribution of filter responses. In [1] the Rao geodesic distance (GD) based on the Fisher-Rao metric tensor was proposed as a similarity measure between multivariate generalized Gaussian distributions (MGGDs) characterizing the wavelet detail features of color textures. Among other advantages, it turns out that, for fixed shape parameter, an analytic expression exists for the GD on the MGGD submanifold, in contrast to the Kullback-Leibler divergence (KLD), barring the two-dimensional case [2]. Moreover, in [2] it was shown that, compared to the KLD, the GD provides consistently superior performance in its application to various texture classification and retrieval experiments.

© Springer International Publishing Switzerland 2015
F. Nielsen and F. Barbaresco (Eds.): GSI 2015, LNCS 9389, pp. 379–386, 2015.
DOI: 10.1007/978-3-319-25040-3_41

Texture discrimination techniques frequently compute the distance between the unlabeled (query) texture image and one or several of its nearest neighbors in the training set. However, they seldom take into account the underlying shape or variability of the class. When the features consist of distribution parameters, this may be done by characterizing the shape of the cluster on the corresponding probabilistic manifold. Provided the clusters are compact, the class centroid yields a convenient summary of the cluster, which may be sufficient to discriminate between the various classes, as was done in [3]. On the other hand, for non-compact clusters a more sophisticated measure of cluster shape is required. For this reason in [4] texture classes were modeled by multiple centroids in an eigenspace of distance matrices.

In this paper we take a different approach which hinges on the observation that clusters of MGGD dispersion matrices form elongated structures on the manifold. The elongation is typically very pronounced along one or a few directions at most. Therefore we choose to characterize the cluster shape intrinsically in terms of the cluster's geodesic subspaces obtained by principal geodesic analysis. We present a new scheme for texture discrimination on the zero-mean MGGD manifold with fixed shape. It is based on the geodesic distance between the unlabeled texture and its projection on the principal geodesic corresponding to the largest eigenvalue for each class. Using data from a challenging color texture database, we compare the performance of our proposed scheme, which we refer to as *principal geodesic classification*, with the performance of the GD-based k-nearest neighbour classifier and another strategy based on the GD to a single cluster centroid ('distance-to-centroid'). This paper builds on our earlier work in [5], but here we present more mathematical details about the method and the experiments have been extended significantly.

2   The Manifold of Multivariate Generalized Gaussian Distributions

In our application the wavelet detail coefficients of color textures are modeled by means of a zero-mean MGGD, considering the dependence between the color bands. The wavelet subbands are assumed to be mutually independent. We first introduce the MGGD model and then we discuss the geodesics, the exponential map and the Fréchet mean on the MGGD manifold. We assume that, where necessary, existence and uniqueness conditions are fulfilled.

2.1   The Multivariate Generalized Gaussian Distribution

We adopt the definition of the zero-mean MGGD (or multivariate exponential power distribution) provided in [1], with the following density function for the vector x:

$$f(x|\Sigma, \beta) = \frac{\Gamma\left(\frac{m}{2}\right)}{\pi^{\frac{m}{2}}\Gamma\left(\frac{m}{2\beta}\right)2^{\frac{m}{2\beta}}}\frac{\beta}{|\Sigma|^{\frac{1}{2}}}\exp\left[-\frac{1}{2}\left(x'\Sigma^{-1}x\right)^{\beta}\right]. \tag{1}$$

Here, m is the dimensionality of the probability space, e.g. $m = 3$ for three-band color images. Also, $\Gamma(.)$ denotes the Gamma function and Σ is the dispersion matrix. β is the shape parameter which controls the fall-off rate of the distribution. The multivariate Gaussian case is retrieved for $\beta = 1$, while we refer to the case $\beta = 1/2$ as the multivariate Laplace distribution. Owing to its heavier tails, the Laplace distribution is expected to provide a better model for wavelet statistics; a fact that was confirmed in earlier classification experiments [2]. In the experiments below, the parameters of the probability models were estimated via the method of moments, followed by an optimization through maximum likelihood estimation [2].

2.2 Geodesic Distance

The geodesics for the zero-mean MGGD were derived in [1]. We here only consider the case with fixed shape parameter β, corresponding to a set of submanifolds, each parameterized by the dispersion matrix Σ. The dimensionality of each submanifold is given by $N = m(m + 1)/2$, resulting in $N = 6$ dimensions for three-band color images. However, it turns out that the metric and geodesics assume a particularly simple form in another parameterization, obtained as follows [1]. First, we consider the geodesic between two specific dispersion matrices Σ_1 and Σ_2. Then, we calculate the regular matrix K that simultaneously diagonalizes Σ_1 and Σ_2, sending Σ_1 to the unit matrix I_m and Σ_2 to a diagonal matrix Φ_2:

$$K'\Sigma_1 K = I_m, \qquad K'\Sigma_2 K = \Phi_2.$$

The diagonal elements of Φ_2 are the eigenvalues λ_2^i of $\Sigma_1^{-1}\Sigma_2$ ($i = 1,\ldots,m$). With a final coordinate transformation to $r_2^i \equiv \ln \lambda_2^i$, the metric elements g_{ij} are constants given by

$$g_{ii} = 3b_h - \frac{1}{4},$$
$$g_{ij} = b_h - \frac{1}{4}, \quad i \neq j, \qquad \text{where} \quad b_h \equiv \frac{1}{4}\frac{m + 2\beta}{m + 2}.$$

In fact it can be proved that K diagonalizes all matrices $\Sigma(t)$ on the geodesic between Σ_1 and Σ_2, parameterized by t ($0 \leq t \leq 1$). As such, K reduces $\Sigma(t)$ to $\Phi(t)$, a diagonal matrix with elements the eigenvalues $\lambda^i(t)$ of $\Sigma_1^{-1}\Sigma(t)$, where $\lambda_2^i \equiv \lambda^i(1)$. The geodesic between Σ_1 and Σ_2 is then simply a straight line:

$$r^i(t) = \ln(\lambda_2^i)\, t, \tag{2}$$

where $r^i(t) \equiv \ln[\lambda^i(t)]$. As a result, the geodesic distance between the two distributions becomes [1]

$$\mathrm{GD}(\Sigma_1, \Sigma_2) = \left[\left(3b_h - \frac{1}{4}\right)\sum_i (r_2^i)^2 + 2\left(b_h - \frac{1}{4}\right)\sum_{i<j} r_2^i r_2^j\right]^{1/2}. \tag{3}$$

2.3   Exponential Map

The exponential map, sending tangent vectors to points on the manifold, as well as its inverse, will be needed for subsequent calculations. A tangent vector in the starting point ($t = 0$) of a geodesic provides a 'velocity vector' for that geodesic. In terms of the matrices $\Phi(t)$ the tangent vectors T are given by

$$T = \left.\frac{\mathrm{d}\Sigma(t)}{\mathrm{d}t}\right|_{t=0} = (K')^{-1} \left.\frac{\mathrm{d}\Phi(t)}{\mathrm{d}t}\right|_{t=0} K^{-1} = (K')^{-1}\ln(\Phi_2)K^{-1}, \qquad (4)$$

where the last equality follows from (2). Clearly the tangent vector T is also diagonalized by the same matrix K, resulting in the matrix $\ln(\Phi_2)$. Consequently, given a tangent vector and a point of application Σ_1 on the manifold, to find the exponential map of the tangent vector T we merely need to calculate the matrix K that diagonalizes both Σ_1 and T, sending Σ_1 to the unit matrix. The only remaining operation is the normalization, since we have to find the point on the geodesic that lies at a geodesic distance $\|T\|$ from Σ_1. From (3) and (4) it follows that rescaling the logarithmic eigenvalues $\ln\lambda_2^i$ by a factor k also rescales the GD by the same factor. Therefore, after calculating the GD, resulting in a 'temporary' value of, say GD_0, corresponding to the logarithmic eigenvalues $\ln\lambda_2^i$ obtained by diagonalization of Σ_1 and T, we simply need to rescale $\ln\lambda_2^i$ by a factor $\|T\|/\mathrm{GD}_0$. Then, the result of the exponential map Σ_2 applied to T is given by

$$\Sigma_2 = (K')^{-1}\ln(\Phi_2)\frac{\|T\|}{\mathrm{GD}_0}K^{-1}.$$

Conversely, to find the result of the inverse exponential map, or logarithmic map, taking a point Σ_2 to the tangent vector T in Σ_1, with $\|T\| = \mathrm{GD}(\Sigma_1, \Sigma_2)$, we first calculate the following 'temporary' tangent vector:

$$T_0 = (K')^{-1}\ln(\Phi_2)K^{-1}.$$

This still needs to be rescaled, resulting in the final image T under the logarithmic map:

$$T = T_0\frac{\mathrm{GD}(\Sigma_1, \Sigma_2)}{\|T_0\|}.$$

2.4   Fréchet Mean

The Fréchet or Kärcher mean provides a generalization to the manifold setting of the centroid of a cluster of points in a Euclidean space. Given a set of n points Σ_j on the fixed-shape zero-mean MGGD manifold, the centroid Σ_c is obtained through the following minimization:

$$\Sigma_c = \underset{\Sigma}{\mathrm{ArgMin}} \sum_{j=1}^{n} \mathrm{GD}^2(\Sigma, \Sigma_j). \qquad (5)$$

This poses an optimization problem on the manifold. Assuming that a solution exists and that it is unique, we solve the problem iteratively by projecting the points Σ_j on the tangent space at the current approximation to the centroid (initialized by that Σ_j which minimizes the criterion (5)). Then we calculate their Euclidean mean on the tangent space and project the result back to the manifold, as illustrated in Fig. 1(a). This is basically a gradient descent algorithm on the manifold, which was derived in [6].

3 Principal Geodesic Classification

The proposed principal geodesic classifier on the MGGD manifold is based on principal geodesic analysis (PGA). We briefly describe PGA in this section, followed by an outline of the principal geodesic classification (PGC) algorithm.

3.1 Principal Geodesic Analysis

Since a geodesic is in a sense a generalization of a straight line in a Euclidean space, PGA for a cluster of points on a manifold was proposed as a natural generalization of principal component analysis (PCA) [7]. PGA yields a set of nested submanifolds, on which the projected elongation or variance of the cluster is maximal. Approximating the projection on the subspaces by the inner product in the tangent space at the centroid, PGA can be carried out through PCA in the tangent space (exact PGA would computationally be too demanding). The resulting tangent vectors, which are the eigenvectors of the covariance matrix in the tangent space, uniquely define a set of geodesic subspaces of the manifold.

It is important to note that PGA yields an (approximately) intrinsic characterization of the cluster, which is certainly to be preferred over tangent space approximations in the case of elongated structures. For instance, in our experiments we noted that a classifier based on the Mahalonobis distance in the tangent space at each cluster centroid, did not yield satisfactory results. Although the issue was not studied in detail, it is possible that the reason lies in the distortion that occurs through the projection on the tangent space. Indeed, on geometrical grounds it is clear that, as a result of the distortion, the error on the Mahalanobis distance is generally larger for more elongated clusters.

3.2 PGC Training and Testing

The PGC training phase consists of providing the model for each class by means of PGA. In the experiments below, we retain only the first principal geodesic, characterizing the direction along which the cluster has its largest elongation or variance.

In the testing phase, each texture in the database is considered one after the other. Such a test (or query) texture is then projected on the first principal geodesic of each class. This in itself is an optimization problem, as it involves finding the point on the geodesic that has the shortest GD to the test point. As with PGA, the projected point could be approximated by performing the projection in the tangent space and taking the image under the exponential map.

However, it will be shown that this noticeably reduces the overall performance of the classifier. Therefore it is better to carry out the exact projection through optimization in terms of the parameter t along the principal geodesic. Subsequently, the GD is calculated between the test point and its projection on the principal geodesic. This *principal geodesic distance* is defined as our similarity measure between the test point and a class. This is illustrated in Fig. 1(b).

As the wavelet subbands are considered to be mutually independent, this procedure can be carried out in each individual subband. The total squared GD between the test point and the class is then taken as the sum of squared GDs in each individual subband. Finally, the test point is assigned to the class to which its total principal geodesic distance is the smallest.

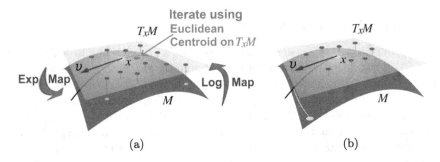

(a)   (b)

Fig. 1. (a) Principle of the iterative algorithm to calculate the centroid of a cluster on the MGGD manifold. (b) Illustration of classification of a test texture by PGC. For each class, the distance is calculated of the test texture to its projection on the first principal geodesic of that class.

4   Classification Experiments

4.1   Experimental Setup

An experiment was set up using data from the Columbia-Utrecht Reflectance and Texture Database (CUReT). It is characterized by a relatively large within-class variability, leading to a highly challenging classification task. A subset of cropped 200×200 RGB images was chosen, belonging to 61 classes. Each class is made up of a single texture, imaged under varying illumination conditions and viewpoints [8]. As such, each class consists of 92 images, resulting in a database of 5612 images to be classified.

The class features were calculated as follows. Every color component of each image was individually normalized to zero mean and unit standard deviation. Then, a discrete wavelet transform with three levels and three orientations (nine subbands) was applied individually on every color component using the Daubechies filters of length eight. The wavelet detail coefficients of every subband were then modeled jointly over the three color components by an MGGD with $\beta = 1$ (Gaussian) or $\beta = 1/2$ (Laplacian). The resulting dispersion matrices constitute the feature set for a single image. Next, in all wavelet subbands the first principal geodesic was computed for each class of 92 images.

Finally, the classification was carried out based on the principal geodesic distance to each class. The classification performance was measured by the success rate using the leave-one-out strategy. We performed a comparison with a 'distance-to-centroid' (DtC) classifier that simply calculates the GD of the test texture to the centroid of each cluster. Another comparison was made with a k-nearest neighbor (kNN) classifier. Here, $k = 91$ was chosen since ideally the other 91 subimages should be the nearest neighbors of a test texture.

4.2   Experimental Results

The results of the classification experiments are presented in Table 1. The highest classification accuracy is achieved with our proposed principal geodesic classifier, compared to the DtC classifier and kNN. This indicates that accommodating the intrinsic variability of the texture classes on the MGGD manifold potentially leads to a performance improvement.

In addition, the Laplace distribution, for which the GD takes on a closed form, is indeed seen to perform better than the Gaussian distribution in most tests, and for PGC in particular.

It is also worth noting that the performance of the DtC classifier is inferior to that of kNN. Furthermore, it is the only case where the Laplace distribution performs worse than the Gaussian. This could indicate that, for this particular database, the characterization of the classes by means of a single centroid entails an excessive loss of information. We should also mention here that in earlier results on another database, the DtC scheme did yield considerably better results than the kNN classifier [3]. This remains a matter for further investigation.

Finally, PGC also offers a significant computational advantage over kNN. Indeed, although the training phase of PGC is more demanding, during classification kNN requires a distance calculation to each image in the database, while PGC merely needs the principal geodesic distance to each class. The computational advantage becomes even more pronounced when the approximation is employed whereby the projection onto the principal geodesic is performed in the tangent space.

Table 1. Classification success rates (SR), based on Gaussian (G) and Laplace (L) models of 5612 CUReT color textures for three wavelet scales, using PGC (exact projection and approximation in the tangent space). This is compared to a distance-to-centroid (DtC) and a k-nearest neighbor (kNN) classifier.

Classifier	PGC exact		PGC approx.		DtC		kNN	
Model	G	L	G	L	G	L	G	L
SR	80.6	82.5	76.6	77.6	72.3	69.8	73.5	75.7

5 Conclusion

We have presented a new classification scheme for color textures on a probabilistic manifold, exploiting the redundancy of the information in the parameters of the distribution to characterize the variability of texture classes. The multivariate generalized Gaussian distribution remains an interesting model for multiband wavelet features, particularly in view of the existence of an analytic expression for the Rao geodesic distance in the case of a fixed shape parameter. Our proposed principal geodesic classifier exhibits superior performance in a classification task on the CUReT texture database, in comparison with a distance-to-centroid and a k-nearest neighbor classifier.

Various avenues for future research have been identified, starting with existence and uniqueness conditions for the cluster centroids, projection onto the principal geodesic, etc. The weaker performance of the distance-to-centroid classifier in the present experiments is another issue to be investigated. Finally, projection on multiple geodesic subspaces along interesting directions would be a logical next development of the principal geodesic classifier.

References

1. Verdoolaege, G., Scheunders, P.: On the geometry of multivariate generalized Gaussian models. J. Math. Imaging Vis. **43**(3), 180–193 (2011)
2. Verdoolaege, G., Scheunders, P.: Geodesics on the manifold of multivariate generalized Gaussian distributions with an application to multicomponent texture discrimination. Int. J. Comput. Vis. **95**(3), 265–286 (2011)
3. Shabbir, A., Verdoolaege, G., Van Oost, G.: Multivariate texture discrimination based on geodesics to class centroids on a generalized Gaussian manifold. In: Nielsen, F., Barbaresco, F. (eds.) GSI 2013. LNCS, vol. 8085, pp. 853–860. Springer, Heidelberg (2013)
4. Schutz, A., Bombrun, L., Berthoumieu, Y.: K-Centroids-Based supervised classification of texture images using the SIRV modeling. In: Nielsen, F., Barbaresco, F. (eds.) GSI 2013. LNCS, vol. 8085, pp. 140–148. Springer, Heidelberg (2013)
5. Shabbir, A., Verdoolaege, G.: Multivariate texture discrimination using a principal geodesic classifier. In: IEEE International Conference on Image Processing, Québec City, Canada, September 2015
6. Pennec, X.: Intrinsic statistics on Riemannian manifolds: basic tools for geometric measurements. J. Math. Imaging Vis. **25**(1), 127–154 (2006)
7. Fletcher, P.T., Lu, C.L., Pizer, S.A., Joshi, S.: Principal geodesic analysis for the study of nonlinear statistics of shape. IEEE Trans. Med. Imag. **23**(8), 995–1005 (2004)
8. Online at http://www.robots.ox.ac.uk/~vgg/research/texclass (2008)

Bag-of-Components: An Online Algorithm for Batch Learning of Mixture Models

Olivier Schwander[1,2]([✉]) and Frank Nielsen[3]

[1] Laboratoire des Signaux et Systèmes (L2S, UMR CNRS 8506),
CentraleSupélec-CNRS-Université Paris-Sud, Orsay, France
olivier.schwander@ens-lyon.org
[2] Viper Group, Computer Vision and Multimedia Laboratory,
University of Geneva, Geneva, Switzerland
[3] Laboratoire D'Informatique (LIX, UMR 7161), École Polytechnique,
Palaiseau, France

Abstract. Practical estimation of mixture models may be problematic when a large number of observations are involved: for such cases, online versions of Expectation-Maximization may be preferred, avoiding the need to store all the observations before running the algorithms. We introduce a new online method well-suited when both the number of observations is large and lots of mixture models need to be learned from different sets of points. Inspired by dictionary methods, our algorithm begins with a training step which is used to build a dictionary of components. The next step, which can be done online, amounts to populating the weights of the components given each arriving observation. The usage of the dictionary of components shows all its interest when lots of mixtures need to be learned using the same dictionary in order to maximize the return on investment of the training step. We evaluate the proposed method on an artificial dataset built from random Gaussian mixture models.

1 Introduction and Motivation

The problem of estimating the probability density function of an unknown probability law is old and well-studied and among all the techniques used mixture models are particularly widespread in practical applications. A lot of work is thus devoted to the improvement of the speed of the algorithms for mixture parameters estimation, which is of particular interest in real-time applications such as object tracking in videos [4,9,10].

The most common axes of research for mixture models can be divided into three main categories. First, the goal may be to reduce the computational burden of the algorithm itself: for example k-MLE [7] and cEM [3] are fast variants of EM where the slow step of soft assignment is replaced by a fast step of hard assignment. Second, a work on the input data may be done: in [5], coresets are used to reduce the number of points needed to build the model. Third, online algorithms can be designed to deal more easily with large datasets [2,10], avoiding the need to store all the content of the dataset.

© Springer International Publishing Switzerland 2015
F. Nielsen and F. Barbaresco (Eds.): GSI 2015, LNCS 9389, pp. 387–395, 2015.
DOI: 10.1007/978-3-319-25040-3_42

We take here a slightly different point of view: we address both the massive data problem and the online constraint in the case where a large number of different mixtures from quite similar sets of points are needed. As such, our new algorithm is divided into two steps: a first training step (which can be slow but it does not really matter since it will be done only once) is used to build a dictionary of components (where atoms are the parameters of the distributions), and a second step uses a nearest-neighbor search to associate each incoming observation to the most probable component, thus incrementally populating the vectors of weights of the mixture. This learning step is obviously online, since the processing can be done observation by observation, and is faster than Expectation-Maximization (EM), since the nearest-neighbor search is rather simple compared to a full-blown EM.

We believe that the separation between a training step and learning step for mixture model can be very useful in numerous applications. For example, in a video analysis application (on a MPEG compressed stream), the dictionary can be built on a key-frame and inter-frames can be modeled using the dictionary of the corresponding key-frame: the dictionary learned on a key-frame will be well suited for the following images but a new one will be needed if a too different scene appears.

Our contributions are the following: first we define the co-mixture concept and present an Expectation-Maximization based algorithm, called *co-EM*, to estimate the parameters; then we introduce an online algorithm, called *Bag of Components* , which relies on a co-mixture to learn a dictionary of components and uses a nearest-neighbor search to estimate a new mixture from observations arriving one by one.

This article is organized as follows: the first part describes the co-mixtures and the algorithm co-EM; the second part introduces the Bag of Components and the online algorithm; the next part discusses some improvements over the basic algorithm; and the last part shows some experimental results.

2   Co-mixture Models

We formally define a co-mixture of exponential families as a set of S mixture models sharing the same parameters for theirs components:

$$
\begin{cases}
m_1(x; \omega_1^{(1)} \ldots \omega_K^{(1)}) = \sum_{i=1}^{K} \omega_i^{(1)} p_F(x; \theta_i) \\
m_2(x; \omega_1^{(2)} \ldots \omega_K^{(2)}) = \sum_{i=1}^{K} \omega_i^{(2)} p_F(x; \theta_i) \\
\ldots \\
m_S(x; \omega_1^{(S)} \ldots \omega_K^{(S)}) = \sum_{i=1}^{K} \omega_i^{(S)} p_F(x; \theta_i)
\end{cases}
\tag{1}
$$

where p_F is the exponential family with log-normalizer F and $\theta_1 \ldots \theta_K$ are the parameters of the components and are shared between all the individual mixtures of the co-mixture; the S vectors $\omega_1^{(s)} \ldots \omega_K^{(s)}$ are the vectors of weights (thus positive and normalized to 1).

In the previous expressions, all the mixtures have the same number of components but since the weight associated to a component may be zero, it is not a limitation.

In order to build such a set of mixtures from a dataset made of S sets of point $\mathcal{X}^{(l)} = \{x_1^{(l)}, \ldots, x_{n_l}^{(l)}\}$ (where n_l is the number of observations in the set of points $\mathcal{X}^{(l)}$), we design an EM-based iterative algorithm, called co-EM. For clarity, we write a generic version working for any exponential family: it is a variant of Bregman Soft Clustering [1] for which the maximization is simply an arithmetic mean in the expectation parameters space (which is in bijection with the usual parameters). It can be described by three main steps:

- Expectation step,
- Maximization step (set of points by set of points),
- Maximization step (aggregation).

Expectation Step. We compute S responsibility matrices $p^{(1)}, \ldots, p^{(S)}$: the coefficient $p^{(l)}(i, j)$ measures the likelihood for the observation $x_i^{(l)}$ from the set of points $\mathcal{X}^{(l)}$ to come from the j-th component of the mixture m_l given the current estimate of the parameters $\eta_1, \ldots, \eta_k$ and of the weights for the l-th mixture $\omega_1^{(l)}, \ldots, \omega_k^{(l)}$. In short, we have:

$$p^{(l)}(i, j) = \frac{\omega_j^{(l)} p_F(x_i^{(l)}, \eta_j)}{m(x_i^{(l)} | \omega^{(l)}, \eta)} \tag{2}$$

Maximization Step (Set of Points by Set of Points). In the first part of the maximization step S partial estimates $(\eta_1^{(1)}, \ldots, \eta_K^{(1)}), \ldots, (\eta_1^{(S)}, \ldots, \eta_k^{(S)})$ are made, one for each individual mixture of the comixture.

The new estimates $(\eta_1^{(l)}, \ldots, \eta_K^{(l)})$ for the l-th set of points are computed using the observations for $\mathcal{X}^{(l)}$ and the l-th responsibility matrix:

$$\eta_j^{(l)} = \sum_i \frac{p^{(l)}(i, j)}{\sum_u p^{(l)}(u, j)} t(x_i^{(l)}) \tag{3}$$

And the weights of each individual mixtures are updated with:

$$\omega_j^{(l)} = \frac{1}{n_l} \sum_{i=1}^{n_l} p^{(l)}(i, j) \tag{4}$$

Maximization Step (Aggregation). All these partial estimates are then aggregated into the new estimate of the parameters $\eta_1, \ldots, \eta_K$.

For the component j, the new estimate of η_j is computed with a Bregman barycenter of all the $\eta_j^{(1)}, \ldots, \eta_j^{(S)}$:

$$\eta_j = \frac{1}{S} \sum_{l=1}^{S} \eta_j^{(l)} \tag{5}$$

This aggregation step gives the same weight to all the set of points, no matter the number of components inside, allowing to remove the influence of various set of points sizes.

Fig. 1. Segmentation with regular EM and co-EM using a 5D RGBxy description of the images.

The algorithm co-EM converges to the average of the log-likelihoods on all the individual mixtures of the co-mixture and can be used independently of the Bag of Components: Fig. 1 shows an image segmentation application.

3   Bag of Components

This online algorithm is inspired by dictionary methods. As such, the training step amounts to building a dictionary of components (the atoms of the dictionary) and the learning step amounts to computing the activation of each atom given the observations.

The dictionary can be directly extracted from the output of co-EM (or from the output of any algorithm building a co-mixture). Given a co-mixture, the dictionary is the set of parameters:

$$\mathcal{D} = \{\theta_1, \ldots, \theta_K\} \tag{6}$$

Due to the need to build a co-mixture, the training step is potentially expensive but this cost is counterbalanced by two points. First it is made only once and the results are reused during the learning step. Second, there is no overload if the set used to build the co-mixtures is a subset of the interesting sets of points: in this case, since it is not more costly to build a co-mixture of size S with co-EM than to learn S mixtures with EM, the global cost of the training and learning steps is still smaller than the cost of doing an EM on all the dataset.

The learning step can be done online: we do not need to work on the entire input points but we can rather update the model parameters each time we see a new observation. This step amounts to a hard-assignment step: given a new observation, we search the most probable component among the atoms of the dictionary (using a naive linear search):

$$\hat{\imath} = \arg\max_{\theta \in \mathcal{D}} p_F(x_i, \theta) \tag{7}$$

We then increment the value in the bin $\hat{\imath}$ of the histogram which counts how many observations have been associated to each atom. At the end of the processing, it is straightforward to go from the histogram to a real vector of weights by dividing by the total number of observations.

4   Improvements

The previous maximization problem can be rewritten as a nearest-neighbor search using the bijection between exponential families and Bregman divergences [1]:

$$\hat{\imath} = \arg\min_{\theta \in \mathcal{D}} B_{F^*}(t(x_i)\|\eta(\theta)) \tag{8}$$

where F^* is the Legendre dual of the log-normalizer F of the exponential family and $\eta(\theta)$ is the transformation of the natural parameter θ into the space of expectation parameters.

As such, it is possible to improve the linear time search described previously by using appropriate nearest-neighbor techniques and data structures such as Bregman ball tree [8] and to go below the linear time search.

Another possible variant is to enforce the sparsity of the weights: after the computation of the vector of weights, we are likely to have some components with a very low weight and thus carrying nearly no information. We assume we can remove these components by thresholding and renormalizing the weights. Another choice may be to clusterize the mixture using the k-medoids [6] algorithm to concentrate weights on most important components.

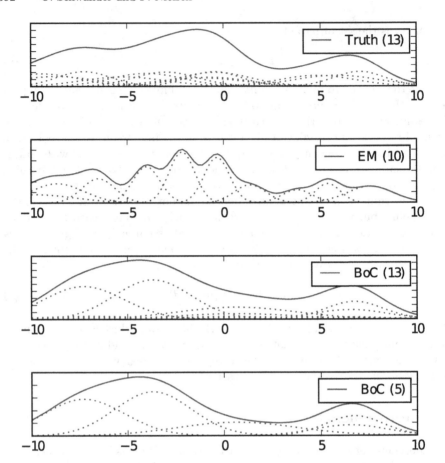

Fig. 2. From top to bottom: original mixture, Expectation-Maximization, raw Bag of Components, Bag of Components with weights thresholded. Between parentheses is the number of components with non-zero weights.

5   Experiments

We evaluate the Bag of Components algorithm on artificially generated mixture models. In order to generate mixture models sufficiently similar to use with a dictionary-based method , we first generate a dictionary of multivariate Gaussian distributions (the covariance matrices are generated from a LDL^T decomposition, where L is a triangular unit matrix and D a diagonal matrix with positive coefficients). We then generate mixtures by randomly drawing the weights, imposing that only a small fraction of the components has a non-zero weight (to enforce some diversity between the random mixtures).

In all the following experiments, the random mixtures are generated from a dictionary of size 30 with only 30 % of non-zero weights. co-EM builds a 30

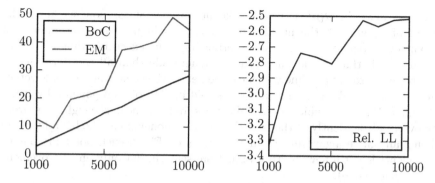

Fig. 3. Computation time for EM and BoC (left) and relative log-likelihood (in percent, right) with respect to the number of observations during the learning step (from 1000 to 10000, in dimension 5).

components co-mixture from 10 sets of 1000 points. The components of this co-mixture are used as a dictionary for Bag of Components.

The goal of the first experiment is to visually check the quality of a 1D mixture built with Bag of Components (from 1000 observations). Figure 2 compares the original mixture (first curve) with the output of EM (second curve, with 10 components) and the output of Bag of Components (third curve). On the third curve, some components have clearly a low weight compared to the most prominent Gaussians so in the fourth curve weights are thresholded under 0.06 in order to keep only 5 components: in this particular case, most of the information seem to be preserved.

A second experiment in Fig. 3 compares the execution time (left) and the quality (right) of the output of Bag of Components and of an EM (10 components) with respect to the number of points in the input set (from 1000 to 10000 points, in dimension 5). Given a dictionary built with co-EM during a preprocessing offline step, we build a mixture with the Bag of Components method from a new input set of points (not present in the dataset used for the dictionary learning step) and compare the output mixture to the result of a classical EM.

The quality of the mixtures from the two algorithms is compared using the relative log-likelihood $\frac{ll_{BoC} - ll_{EM}}{ll_{BoC}}$ so a negative value means Bag of Components produces worse mixtures than EM: on the explored range, the two algorithms produce mixtures of similar quality, with roughly between -4% and -2% of relative difference.

The left part of Fig. 3 measures the execution time of Bag of Components (without the dictionary building step, since this step is made offline): not surprisingly, it is perfectly linear with a speed-up between 1.2 and 4 compared to EM (which has a very irregular execution time).

The speed-up from EM to Bag of Components is real but not so high. Indeed, even if the learning step of Bag of Components is made in $O(nK)$ (where n is the number of observations and K the number of atoms of the dictionary) and the EM is made in $O(nki)$ (where k is the number of components and i the number of iterations), the number of atoms K is higher than the number of components k (three times in the experiments). The execution time of Bag of Components is thus of the same order of magnitude than all the iterations of EM, giving an execution time which can be nearly the same when EM converges in few iterations. We may increase the speed of Bag of Components by using a Bregman Ball Tree which would allow a sub-linear nearest-neighbor search. Moreover, independently of the time, Bag of Components has the big advantage of being an online algorithm (so in the curves on Fig. 3, each point for Bag of Components is not a new mixture built from scratch, but only an improvement of the previous one).

6   Conclusion

We described the notion of co-mixtures along with the algorithm co-EM. It is used as a basis to design a new algorithm for mixture model learning, called Bag of Components: this new algorithm works online and allows to build a mixture faster than Expectation-Maximization. It is well suited when a lot of mixtures from related or similar sets of points are needed: in such a case, it is worth building a dictionary on a subset of the sets of points and apply Bag of Components on the remaining sets of points. It is also interesting when only a few sets of points are available at a time: the available sets can be used to learn the dictionary of components and new mixtures can be built on new sets of points at soon as they become available.

There is room for lots of improvements both on the speed, by using efficient nearest-neighbor or approximate nearest-neighbor techniques, and for the sparsity of the weights, by evaluating the need and the interest of removing low weight components. Furthermore, we leave for future work to validate co-EM and Bag of Components on a real application instead of artificial mixtures.

References

1. Banerjee, A., Merugu, S., Dhillon, I.S., Ghosh, J.: Clustering with Bregman divergences. J. Mach. Learn. Res. **6**, 1705–1749 (2005)
2. Cappé, O., Moulines, E.: On-line expectation-maximization algorithm for latent data models. J. R. Stat. Soc.: Ser. B (Stat. Methodol.) **71**(3), 593–613 (2009)
3. Celeux, G., Govaert, G.: A classification EM algorithm for clustering and two stochastic versions. Comput. Stat. Data Anal. **14**(3), 315–332 (1992). doi:10.1016/0167-9473(92)90042-E
4. Chen, G., Yu, Z., Wen, Q., Yu, Y.: Improved Gaussian mixture model for moving object detection. In: Deng, H., Miao, D., Lei, J., Wang, F.L. (eds.) AICI 2011, Part I. LNCS, vol. 7002, pp. 179–186. Springer, Heidelberg (2011)
5. Feldman, D., Faulkner, M., Krause, A.: Scalable training of mixture models via coresets. In: Advances in Neural Information Processing Systems, pp. 2142–2150 (2011)

6. Kaufman, L., Rousseeuw, P.: Clustering by means of medoids. North-Holland, Amsterdam (1987)

7. Nielsen, F.: k-MLE: a fast algorithm for learning statistical mixture models. CoRR (2012)

8. Nielsen, F., Piro, P., Barlaud, M.: Tailored bregman ball trees for effective nearest neighbors. In: Proceedings of the 25th European Workshop on Computational Geometry (EuroCG), pp. 29–32 (2009)

9. Sicre, R., Nicolas, H.: Improved Gaussian mixture model for the task of object tracking. In: Real, P., Diaz-Pernil, D., Molina-Abril, H., Berciano, A., Kropatsch, W. (eds.) CAIP 2011, Part II. LNCS, vol. 6855, pp. 389–396. Springer, Heidelberg (2011). http://dx.doi.org/10.1007/978-3-642-23678-5_46

10. Stauffer, C., Grimson, W.E.L.: Adaptive background mixture models for real-time tracking. In: IEEE Computer Society Conference on Computer Vision and Pattern Recognition 1999, vol. 2, IEEE (1999)

Statistical Gaussian Model of Image Regions in Stochastic Watershed Segmentation

Jesús Angulo[✉]

MINES ParisTech, CMM-Centre de Morphologie Mathématique,
PSL-Research University, Paris, France
jesus.angulo@mines-paristech.fr

Abstract. Stochastic watershed is an image segmentation technique based on mathematical morphology which produces a probability density function of image contours. Estimated probabilities depend mainly on local distances between pixels. This paper introduces a variant of stochastic watershed where the probabilities of contours are computed from a gaussian model of image regions. In this framework, the basic ingredient is the distance between pairs of regions, hence a distance between normal distributions. Hence several alternatives of statistical distances for normal distributions are compared, namely Bhattacharyya distance, Hellinger metric distance and Wasserstein metric distance.

1 Introduction

Image segmentation is one of the most studied and relevant problems in low level computer vision. Indeed, the state-of-the-art is vast and rich in multiple paradigms. We are interested here on approaches based on statistical modeling of pixels and regions. Examples of methods fitting in such a paradigm and having excellent performance are mean shift [9] and statistical region merging [15]. Hierarchical contour detection is another successful paradigm with approaches based for instance on machine learned edge detection [4] or on watershed transform [6].

Instead of dealing with a determinist set of contours, the idea of the stochastic watershed [2] (SW) is to estimate a probability density function (pdf) of contours by MonteCarlo simulations. Some variants included multiscale framework [3], bagging framework [11], robust framework [5], etc. It was shown in [16] that the corresponding pdf obtained by SW can be calculated in closed form without simulation by using graph algorithms, for more recent results see also [17]. Nevertheless, we focuss here on an approach working on simulations. In particular, our contribution is in the line of [12], where the estimation of the probability of each contour is based on a regional model of each region, the model being in [12] the mean color. In the present work, the approach is pushed forward such that each region should be modeled as a multivariate normal distribution. The basic ingredient will be a distance between pairs of regions, hence a distance between normal distributions. In this context, several alternatives of statistical distances for normal distributions are compared.

© Springer International Publishing Switzerland 2015
F. Nielsen and F. Barbaresco (Eds.): GSI 2015, LNCS 9389, pp. 396–405, 2015.
DOI: 10.1007/978-3-319-25040-3_43

The rest of the paper is organized as follows. In Sect. 2, we remind the Mon-teCarlo framework of stochastic watershed and in particular the simulation of regionalized random germs. Section 3 introduces the contribution of the paper: first, distances between multivariate normal distributions are reviewed; second, computation of probability density of contours based on a normal region model is formulated. Results are discussed and compared to statistical region merging segmentation [15].

2   Remind on Stochastic Watershed

Regionalized Poisson Points. We first consider the notion regionalized random points as well as the algorithm used to simulate a realization of N random germs associated to a spatial density.

A rather natural way to introduce uniform random germs is to generate realizations of a Poisson point process with a constant intensity θ (i.e., average number of points per unit area). It is well known that the random number of points $N(D)$ falling in a domain D, which is considered a bounded Borel set, with area $|D|$, follows a Poisson distribution with parameter $\theta|D|$, i.e., $\Pr\{N(D) = n\} = e^{-\theta|D|}\frac{(-\theta|D|)^n}{n!}$. In addition, conditionally to the fact that $N(D) = n$, the n points are independently and uniformly distributed over D, and the average number of points in D is $\theta|D|$ (i.e., the mean and variance of a Poisson distribution is its parameter).

More generally, we can suppose that the density θ is not constant; but considered as measurable function, defined in $\mathbb{R}^d$, with positive values. For simplicity, let us write $\theta(D) = \int \theta(\mathbf{x})d\mathbf{x}$. It is also known [13] that the number of points falling in a Borel set B according to a regionalized density function θ follows a Poisson distribution of parameter $\theta(D)$, i.e., $\Pr\{N(D) = n\} = e^{-\theta(D)}\frac{(-\theta(D))^n}{n!}$. In such a case, if $N(D) = n$, the n are independently distributed over D with the probability density function $\widehat{\theta}(\mathbf{x}) = \theta(\mathbf{x})/\theta(D)$. In practice, in order to simulate a realization of N independent random germs distributed over the image with the pdf $\pi_k(\mathbf{x})$ we propose to use an inverse transform sampling method. More precisely, the algorithm to generate N random germs in an image $m : E \to \{0,1\}$ according to density $\widehat{\theta}(x)$ is as follows:

1. Initialization: $m(x_i) = 0 \forall x_i \in E;\ P = \mathrm{Card}(E)$
2. Compute cumulative distribution function: $cdf(x_i) = \frac{\sum_{k<i} \widehat{\theta}(x_k)}{\sum_{k=1}^{P} \widehat{\theta}(x_k)}$
3. for $j = 1$ to N
4. $r_j \sim \mathcal{U}(1, P)$
5.   Find the value s_j such that $r_j \leq cdf(x_{s_j})$.
6. $m(x_{s_j}) = 1$

Marker-Driven Watershed Transform. Let $g(x)$ and $mrk(x)$ be respectively a (norm of) gradient image and a marker image. Intuitively, the associated watershed transformation [6], $WS(g, mrk)(x)$, produces a binary image with the contours of regions "marked" by the image mrk according to the strength of contour

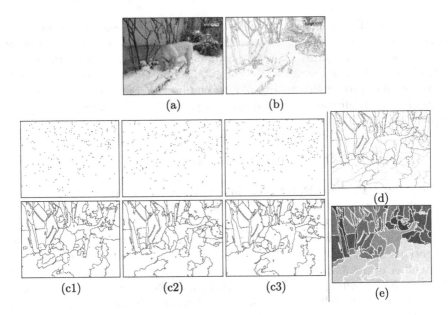

Fig. 1. Stochastic watershed segmentation of "Custard" image: (a) color image f, (b) its color gradient g, (c) top, three realizations of random markers mrk_n regionalized from g, bottom, corresponding watershed lines $WS(g, mrk_n)$, (d) estimated density of contours pdf, (e) segmentation from d. In (c) and (d) images in negative for better visualization (Color figure online).

given by the gradient image g. The classical paradigm of watershed segmentation lays on the appropriate choice of markers, which are the seeds to initiate the flooding procedure.

Probability Density of Contours Using MonteCarlo Simulations of Watershed. In the stochastic watershed (SW) approach [2], an opposite direction is followed, by spreading random germs for markers on the watershed segmentation. This arbitrary choice will be balanced by the use of a given number M of realizations, in order to filter out non significant fluctuations. Each piece of contour may then be assigned the number of times it appears during the various simulations in order to estimate a probability density function (pdf) of contours. In the case of uniformly distributed random germs, large regions will be sampled more frequently than smaller regions and will be selected more often. Using $g(x)$ as density for regionalization of random germs involves sampling high contrasted image areas and it has been proved to be an appropriate choice [2]. In this case, the probability of selecting a contour will offer a trade-off between strength of the contours and size of the adjacent regions.

More precisely, given a color image f the associated SW pdf of contours is obtained as follows. Let $\{mrk_n(x)\}_{n=1}^{M}$ be a series of M realizations of N spatially distributed random markers according to its gradient image g. Each realization of random germs is considered as the marker image for a watershed segmentation of gradient image g in order to obtain the binary image

$$WS(g, mrk_n)(x) = \begin{cases} 1 & \text{if } x \in \text{Watershed lines} \\ 0 & \text{if } x \notin \text{Watershed lines} \end{cases}$$

Consequently, a set of M realizations of segmentation is computed, i.e., $\{WS(g, mrk_n)\}_{1 \leq n \leq M}$. Note that in each realization the number of points determines the number of regions obtained (i.e., essential property of watershed transformation). Then, the probability density function of contours is computed by the kernel density estimation method as follows:

$$pdf(x) = \frac{1}{M} \sum_{n=1}^{M} WS(g, mrk_n)(x) * K_\sigma(\mathbf{x}). \tag{1}$$

Typically, the kernel $K_\sigma(\mathbf{x})$ is a spatial Gaussian function of width σ, which determines the smoothing effect.

Then, the image $pdf(x)$ can be segmented by selecting the contours of probability higher than a given contrast [2]. Figure 1 depicts an example of color image segmentation using SW. As we can note from this example, which includes large homogenous areas, well contrasted objects as well as textured zones, the pdf and corresponding segmentation produces relatively satisfactory results. However, large homogeneous areas are oversegmented and textured zones are not always well contoured. Obviously, low contrasted areas (e.g., boundary between dog head and wall) are not properly segmented. All those are well known drawbacks of SW which have been addressed by the robust stochastic watershed (RSW) [5] or by the regional regularized stochastic watershed [12]. We adopt here an approach related to the latter one, based on a statistical model of regions.

Let us remind the principle of the RSW [5] since it will be also used in the results of next section. The fundamental property of watershed is the insensitivity to the placement of seed points, which usually enables the SW segmentation to find reliably relevant boundaries, but, in the case of "false boundaries" it works to our disadvantage. The idea of RSW is to introduce a perturbation ϵ_n (i.e., small amount of noise) into the flooding function g (i.e., gradient) at each realization n, in order to reduce the number of times that a "false boundary" will appear. More precisely, the kernel density estimator (1) becomes now

$$pdf(x) = \frac{1}{M} \sum_{n=1}^{M} WS(g + \epsilon_n, mrk_n)(x) * K_\sigma(\mathbf{x}), \tag{2}$$

where $\epsilon_n(x)$ is in our experiments a zero-mean Gaussian white noise with an intensity-dependent variance from $g(x)$.

3   Multivariate Gaussian Model of Regions in SW

As we just discussed, the watershed segmentation of g from N markers (imposed minima) produces a set of thin lines dividing the image domain E into N disjoint

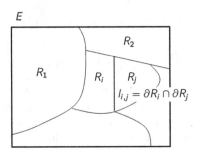

Fig. 2. Tessellation τ of E from watershed $WS(x)$.

regions, denoted $\{R_r\}_{1 \leq r \leq N}$. This structure is called tessellation τ of E, see Fig. 2, defined as a (finite) family of disjoint open sets (or classes, or regions)

$$\tau = \{R_r\}_{1 \leq r \leq N}, \quad \text{with } i \neq j \Rightarrow R_i \cap R_j = \emptyset,$$

such that

$$E = \cup_r R_r \bigcup WS(x) \Leftrightarrow WS(x) = E \setminus \cup_r R_r = \cup l_{i,j}.$$

The watershed lines $WS(x)$ can be decomposed into the curves that separates the regions. More precisely, let us denote by $l_{i,j}$ the curve (or irregular arc segment) defined as the boundary between regions R_i and R_j, i.e.,

$$l_{i,j} = \partial R_i \cap \partial R_j.$$

We obviously have $WS = \cup l_{i,j}$, but we also note that in the case of three (or more adjacent) regions, their boundary segments intersect at their junctions (or triple points).

The color image values restricted to each region of the partition, $P_i = f(R_i)$, can be modeled by different statistical distributions. Here we focuss on a multivariate normal model.

3.1   Distances for Multivariate Normal Distributions

Let us consider a family of multivariate normal distributions P_i of mean $\boldsymbol{\mu}_i$ and covariance matrix $\boldsymbol{\Sigma}_i$, i.e., $P_i \sim \mathcal{N}(\boldsymbol{\mu}_i, \boldsymbol{\Sigma}_i)$. Different distances are defined in the space of P_i.

Bhattacharyya Distance and Hellinger Metric Distance. The Bhattacharyya distance $D_B(P_1, P_2)$ measures the similarity of two discrete or continuous probability distributions P_1 and P_2. More precisely, it computes the amount of overlap between the two statistical populations, i.e., $D_B(P_1, P_2) = -\log \int \sqrt{P_1(x) P_2(x)} dx$.

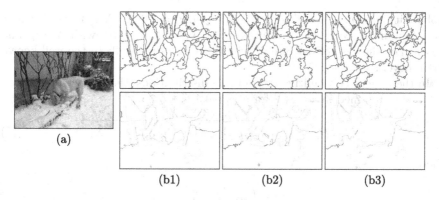

Fig. 3. Given the "Custard" color image (a), (b1)–(b3) are, on the top, three realizations of SW lines $WS(x, n)$, at the bottom, the corresponding probability maps $Pr(x, n)$ using the Bhattacharyya distance (Color figure online).

For multivariate normal distributions the Bhattacharyya distance $D_B(P_1, P_2)$ is given by

$$D_B(P_1, P_2) = \frac{1}{8}(\mu_1 - \mu_2)^T \Sigma^{-1}(\mu_1 - \mu_2) + \frac{1}{2} \log \left(\frac{\det \Sigma}{\sqrt{\det \Sigma_1 \det \Sigma_2}} \right), \quad (3)$$

where

$$\Sigma = \frac{\Sigma_1 + \Sigma_2}{2}.$$

Note that the first term in the Bhattacharyya distance is related to the Mahalanobis distance, both are the same when the covariance of both distributions is the same.

We have $0 \leq D_B \leq \infty$ and it is symmetric $D_B(P_1, P_2)$, it is not a metric. But D_B does not obey the triangle inequality and therefore it is not a metric. Nevertheless, it can be metrized by transforming it into to the following Hellinger metric distance $D_H(P_1, P_2)$ given by $D_H(P_1, P_2) = \sqrt{1 - \exp(-D_B(P_1, P_2))}$, such that

$$D_H(P_1, P_2) = \sqrt{1 - \left(\frac{\det \Sigma}{\sqrt{\det \Sigma_1 \det \Sigma_2}} \right)^{-1/2} e^{\left(-\frac{1}{4}(\mu_1 - \mu_2)^T (\Sigma_1 + \Sigma_2)^{-1}(\mu_1 - \mu_2)\right)}}. \quad (4)$$

Hellinger distance is an α-divergence [1], which corresponds to the case $\alpha = 0$ and it is the solely being a metric distance. Hellinger distance can be related to measure theory and asymptotic statistics. For more details on Bhattacharyya and Hellinger distances, see for instance [14].

Wasserstein Metric Distance. The Wasserstein metric is a distance function defined between probability measures on a given metric space based on the notion optimal transport [20]. Namely, the W_2 Wasserstein distance between probability measures μ and ν on $\mathbb{R}^n$ is $W_2(\mu, \nu) = \inf \mathbb{E}(\|X - Y\|^2)^{1/2}$, where the infimum runs over all random vectors $(X, Y) \in \mathbb{R}^n \times \mathbb{R}^n$ with $X \sim \mu$ and $Y \sim \nu$. For the case of discrete distributions, it corresponds to the well-known earth mover's distance [7,18].

In the case of two multivariate normal distributions, the Wasserstein metric distance is obtained in a closed form as [10,19]:

$$D_W(P_1, P_2) = \sqrt{\|\boldsymbol{\mu}_1 - \boldsymbol{\mu}_2\|^2 + \mathrm{Tr}(\boldsymbol{\Sigma}_1 + \boldsymbol{\Sigma}_2 - 2\boldsymbol{\Sigma}_{1,2})}, \tag{5}$$

where

$$\boldsymbol{\Sigma}_{1,2} = \left(\boldsymbol{\Sigma}_1^{1/2} \boldsymbol{\Sigma}_2 \boldsymbol{\Sigma}_1^{1/2}\right)^{1/2}.$$

We note in particular that in the commutative case $\boldsymbol{\Sigma}_1 \boldsymbol{\Sigma}_2 = \boldsymbol{\Sigma}_2 \boldsymbol{\Sigma}_1$ we have

$$D_W(P_1, P_2)^2 = \|\boldsymbol{\mu}_1 - \boldsymbol{\mu}_2\|^2 + \|\boldsymbol{\Sigma}_1^{1/2} - \boldsymbol{\Sigma}_2^{1/2}\|_F^2.$$

3.2   Probability Density Function Estimation

We have now the ingredients to compute by MonteCarlo simulations the probability density function from a color image. The idea is to assign to each piece of contour $l_{i,j}$ between regions R_i and R_j the statistical distance between the color gaussian distributions P_i and P_j:

$$\pi_{i,j} = \frac{D(P_i, P_j)}{\sum_{l_{k,l} \in WS} D(P_k, P_l)}, \tag{6}$$

where $D(P_i, P_j)$ is any of the distances discussed above. Thus, for any realization n of SW, denoted $WS(x, n)$, one can compute the following image of weighted contours:

$$Pr(x, n) = \begin{cases} \pi_{i,j} & \text{if } x \in l_{i,j}^n \\ 0 & \text{if } x \notin l_{i,j}^n \end{cases}$$

where $l_{i,j}^n$ is the boundary between regions R_i and R_j from $WS(x, n)$. Figure 3 gives an example of three realizations of SW lines $WS(x, n)$ and the corresponding probability maps $Pr(x, n)$ using the Bhattacharyya distance.

Finally, integrating across the M realizations, the MonteCarlo estimate of the probability density function of contours is given by

$$pdf(x) = \frac{1}{M} \sum_{n=1}^{M} Pr(x, n) * K_\sigma(x). \tag{7}$$

Obviously, this approach is compatible with the robust stochastic watershed (RSW) variant discussed in previous section, since each realization n of the RSW produces also a tessellation of E since the image of weighted contours $Pr(x, n)$ can be computed.

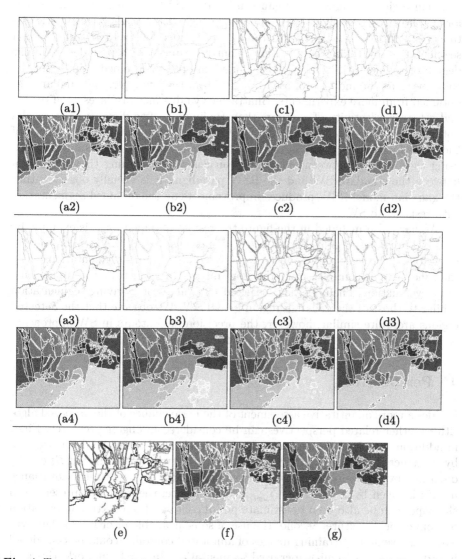

Fig. 4. Top, image segmentation of "Custard" using gaussian region model SW, where the first row is $pdf(x)$ and second row the segmentation obtained at a given probability contrast; middle, gaussian region model RSW: (a) using only color mean [12], (b) Bhattacharyya distance, (c) Hellinger metric distance, (d) Wasserstein metric distance. The number of realization is $M = 50$ and the number of random markers at each realization is $N = 200$ germs. Bottom, image segmentation of "Custard" using statistical region merging (SRM) [15]: (e) sum of contours obtained from SRM using nine values of Q (256, 128, 64, 32, 16, 8, 4, 2, 1), (f) segmentation for $Q = 128$ and (g) for $Q = 32$.

In Fig. 4 is depicted a comparison of image segmentation of "Custard" using gaussian region model SW and gaussian model RSW. In particular, the $pdf(x)$ for the three considered distances (Bhattacharyya distance, Hellinger metric distance and Wasserstein metric distance) is provided, as well as the obtained segmentation by taking a probability contrast value which provides a similar degree of segmentation. It is also included the result obtained using only the color mean as the model proposed in [12]. From this example, and similar ones obtained from more experiments, Bhattacharyya distance produces good results and a better selectivity of contours than Hellinger metric distance. The results obtained by Wasserstein metric distance are also relevant but the influence of the covariance matrix is less significant, thus being closer to the results obtained using the distance between only the mean colors. Concerning the comparison between the standard SW and the RSW paradigm, it is visually observed that the gaussian model RSW produces improvements on the obtained segmentation with respect to SW.

We have also included in the figure the results obtained for this example by statistical region merging (SRM) [15], computed using MATLAB code provided in [8]. Using the scale parameter Q, we have computed nine segmentations, such that the sum of contours from the nine images can be viewed as a contours saliency function. Then, we have selected two values of Q giving segmentation similar to those of the gaussian region model SW. We observe that the detected regions are quite similar, however, the precision of contours in SW-approaches is qualitatively better for such kind of images.

4   Perspectives

Besides a more quantitative assessment of the performance of the proposed algorithms, other related perspectives can be considered in ongoing research. First, in addition to color information, texture at each pixel x can be described also by its structure tensor $T(x) \in SPD(2)$. Thus, texture at each region R_i can be described by a zero-mean gaussian distribution $\mathcal{N}(\mathbf{0}, \Sigma_i)$, where the covariance matrix is given by $\Sigma_i = |R_i|^{-1} \sum_{x \in R_i} T(x)$. Hence, the approach presented in the paper can be also used to estimate pdf of contours from texture information or from color + texture. Second, the use of some available prior knowledge, typically represented by training images of annotated contours, could be considered in order to have a (semi-)supervised segmentation. In our framework, this goal can be formulated as a problem of distance learning in the space of multivariate normal distributions.

References

1. Amari, S., Nagaoka, H.: Methods of Information Geometry. Translations of Mathematical Monographs, vol. 191. AMS and Oxford University Press, Providence (2000)

2. Angulo, J., Jeulin, D.: Stochastic watershed segmentation. In: Proceedings of the ISMM 2007, pp. 265–276 (2007)
3. Angulo, J., Velasco-Forero, S., Chanussot, J.: Multiscale stochastic watershed for unsupervised hyperspectral image segmentation. In: Proceedings of IEEE IGARSS 2009, vol. 3, pp. 93–96 (2009)
4. Arbelaez, P., Maire, M., Fowlkes, C., Malik, J.: Contour detection and hierarchical image segmentation. IEEE Trans. PAMI **33**(5), 898–916 (2011)
5. Bernander, K.B., Gustavsson, K., Selig, B., Sintorn, I.M., Luengo Hendriks, C.L.: Improving the stochastic watershed. Pattern Recogn. Lett. **34**(9), 993–1000 (2013)
6. Beucher, S., Meyer, F.: The morphological approach to segmentation: the watershed transformation. In: Dougherty, E.R. (ed.) Mathematical Morphology in Image Processing, pp. 433–481. Marcel Dekker, New York (1992)
7. Boltz, S., Nielsen, F., Soatto, S.: Earth mover distance on superpixels. In: Proceedings of IEEE ICIP 2010, pp. 4597–4600 (2010)
8. Boltz, S.: Image segmentation using statistical region merging. http://www.mathworks.com/matlabcentral/fileexchange/25619-image-segmentation-using-statistical-region-merging. Accessed December 2014
9. Comaniciu, D., Meer, P.: Mean shift analysis and applications. In: Proceedings of ICCV 1999, vol. 2, pp. 1197–1203 (1999)
10. Dowson, D.C., Landau, B.V.: The Fréchet distance between multivariate normal distributions. J. Multivar. Anal. **12**(3), 450–455 (1982)
11. Franchi, G., Angulo, J.: Bagging stochastic watershed on natural color image segmentation. In: Proceedings of the ISMM 2015 (2015)
12. López-Mir, F., Naranjo, V., Morales, S., Angulo, J.: Probability density function of object contours using regional regularized stochastic watershed. In: Proceedings of IEEE ICIP 2014, pp. 4762–4766 (2014)
13. Matheron, G.: Random Sets and Integral Geometry. Wiley, New York (1975)
14. Nielsen, F., Boltz, S.: The Burbea-Rao and Bhattacharyya centroids. IEEE Trans. Inf. Theor. **57**(8), 5455–5466 (2010)
15. Nock, R., Nielsen, F.: Statistical region merging. IEEE Trans. PAMI **26**(11), 1452–1458 (2004)
16. Meyer, F., Stawiaski, J.: A stochastic evaluation of the contour strength. In: Goesele, M., Roth, S., Kuijper, A., Schiele, B., Schindler, K. (eds.) Pattern Recognition. LNCS, vol. 6376, pp. 513–522. Springer, Heidelberg (2010)
17. Meyer, F.: Stochastic watershed hierarchies. In: Proceedings of ICAPR 2015 (2015)
18. Rubner, Y., Tomasi, C., Guibas, L.J.: A metric for distributions with applications to image databases. In: Proceedings of ICCV 1998, pp. 59–66 (1998)
19. Takatsu, A.: Wasserstein geometry of Gaussian measures. Osaka J. Math. **48**(4), 1005–1026 (2011)
20. Villani, C.: Optimal Transport: Old and New. Grundlehren der Mathematischen Wissenschaften [Fundamental Principles of Mathematical Sciences], vol. 338. Springer, Berlin (2009)

Quantization of Hyperspectral Image Manifold Using Probabilistic Distances

Gianni Franchi and Jesús Angulo[✉]

MINES ParisTech, CMM-Centre de Morphologie Mathématique,
PSL-Research University, Paris, France
{gianni.franchi,jesus.angulo}@mines-paristech.fr

Abstract. A technique of spatial-spectral quantization of hyperspectral images is introduced. Thus a quantized hyperspectral image is just summarized by K spectra which represent the spatial and spectral structures of the image. The proposed technique is based on α-connected components on a region adjacency graph. The main ingredient is a dissimilarity metric. In order to choose the metric that best fit the hyperspectral data manifold, a comparison of different probabilistic dissimilarity measures is achieved.

Keywords: Quantization · Hyperspectral images · Information geometry · Probabilistic distances · Mathematical morphology

1 Introduction

Gray-level images are pictures where each pixel is a scalar value which usually is quantized between 0 and 255. Color and hyperspectral images are images where each pixel can be considered as a vector, such that each coordinate of the vector corresponds to the intensity of the pixel at a certain wavelength. In addition, pixels values on spectral images are most of the time different. Quantization of images can be seen as a way to remove this excess of variability by reducing the number of (spectral) vectors and therefore to address the curse of dimensionality [3]. The problem can be addressed using dictionary learning techniques [2]. In these learning methods, a dictionary composed of k atoms is learned on a set of vectors. Then each vector is represented by a (low) number of atoms of the dictionary. Among these methods, VQ [7] is a well known technique that quantizes a set of vectors, since each vector is represented by just one atom of the dictionary. We consider in this paper a vector quantization method for hyperspectral images, halfway between manifold learning and dictionary learning methods, since we explore a quantization of the image manifold. First, we review existing similarity metrics on hyperspectral imaging and consider some probabilistic distance less known on this domain. Finally we quantize hyperspectral images using the best distances.

© Springer International Publishing Switzerland 2015
F. Nielsen and F. Barbaresco (Eds.): GSI 2015, LNCS 9389, pp. 406–414, 2015.
DOI: 10.1007/978-3-319-25040-3_44

2   Quantization of Hyperspectral Images

Background on Hyperspectral Images. Let us consider a hyperspectral image where each pixel value v_i is a spectral vector of dimension $d \in \mathbb{N}$, such that $v_i \in \mathbb{R}^d$. Because of the curse of dimensionality [3], spectral variability is added to the vectors. First, let us consider a simple case where $v_i \in [0, 1]^d$ and $d = 1$; in this case it is easy to calculate that with 100 points, one get an interval between points of around 10^{-2}. However, if we consider a space of dimension $d = 10$ and if one wants the points to be separated by a ball of radius 10^{-2}, we can see that this time we need 10^{20} points. Indeed, in the case of hyperspectral images, d is usually about one or two hundred, which shows how sparse the sampled manifold is (or how empty the whole space is), and how crucial is the question about similarity between spectra. Moreover one can see a hyperspectral image as a set of spectral classes. If we consider for instance three classes: "road", "water", or "forest", then each pixel belongs to one of these three classes. Due to the high dimensional space, each class has a high variability with respect to the Euclidean distance, and thus this distance will not be discriminative enough to separate objects from different classes. A solution could be to reduce the dimension of the manifold or to find another low-dimensional space to embed data, and then to use Euclidean distance in this space. Actually, we do not focus here on this kind of classical approach. We assess the interest of metrics on the original manifold on the d-dimensional space that are more invariant to spectral variability [1,12,15]. Thanks to this kind of similarities we can expect to improve hyperspectral image quantization.

Spectral/Spatial Image Quantization. Quantization is the process which allows to approach a signal with large set of values by a signal on a smaller set. Images are signals on a spatial domain, so their quantization should takes into account the expected spatial coherence. To achieve this goal, we choose to use α-connected components representation [18,20,21], that produces an image partition into homogenous spatial classes. Two pixels belongs to the same α-connected component if there is a path linking these pixels such that the similarity between successive pixels of this path is lower than α. However, α-connected components algorithm often produces inadequate image partitions, since it fails to respect image contours. A solution is to first use an initial partition algorithm that would produce "superpixels" on our image and that must follows main image contours. Then, the superpixels are connected between then by their region adjacency graph (RAG). It is a graph where each node is a superpixel, and edges represent the dissimilarity between superpixels. Here edges are weighted by the dissimilarity between centroids of superpixels. In our case, the superpixels are obtained by computing the classical watershed on the image. Then the notion of α-connected components can be extended to RAG [18]. Moreover, by comparing nodes and thus regions, our quantization is more robust to noise. For this purpose on each superpixel SP_i the value of the original pixels is replaced by the value of its centroid C_i. Finally the choice of α is done in order to have a fixed number of centroids, and thus of different spectra.

Assessment of Image Quantization. Our quantization depends on the RAG and therefore, on the choice of distance. We focus on the potential interest of probabilistic distances on spectral pixels. In this context, our assessment is separated into two steps.

Step 1: Evaluation of the probabilistic distances. We consider that a good probabilistic distance is a dissimilarity measure that has an invariant behaviour on each spectral class in a high dimension space. During the process of assessment, hyperspectral images with a ground truth spectral classification are used. Thus the class of each pixel is known. Moreover for each class the centroid for a dissimilarity measure can be computed: it corresponds to the vector which minimizes the cumulated distance to the other vectors of the class. The centroid represents the class. Figure 1 represents two scatter-plots from a three-classes hyperspectral image. These scatter-plots depict the distance of each spectrum to the centroids of each class. A good dissimilarity measure should concentrates the data of each class around its centroid. In this example, Kullback-Leibler divergence performs better than L_2 norm from this viewpoint. To quantitatively assess such property, the image is classified by calculating the distance of each image spectrum to the class centroids. Then, the class of the nearest centroid is given to each spectrum. After having classified the image, we evaluate the Overall Accuracy (OA), and the Average Accuracy (AA), that are two measures commonly used in hyperspectral imaging [10]. In addition, the rank of classification is also computed for each pixel. For example, let us consider that we have a spectrum X and three classes C_1, C_2, C_3, and the spectrum X belongs to cluster C_1. After computing the distance of X to each class centroid, denoted receptively $D1, D2, D3$, we obtain $D1 > D3 > D2$. Obviously the classifier would consider that the class of X is C_2. Since in this case $D1$ is the highest distance, the classification mistake is very significant. The rank of good classification is in the present case 3. Hence, the rank of classification (Rank) is obtained as the mean of rank of good classification over all the pixels, divided by the number of classes. The more the rank of classification is near to one, the worst a dissimilarity measure is.

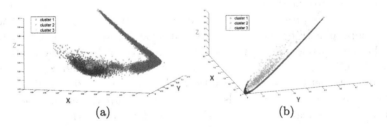

(a)   (b)

Fig. 1. Scatter plot of distances of pixels (of Pavia image) of cluster 1 (in blue) to the centroid of cluster 1 in X, to the centroid of cluster 2 in Y, to the centroid of cluster 3 in Z. Similarly, in red for cluster 2, and in green for cluster 3. In (a) we use the L_2 norm as dissimilarity measure, in (b) we used the Kullback-Leibler divergence (Color figure online).

Step 2: Evaluation of the quantization results. Quantization is a simplification operation that can destroy some relevant information. This introduces an error between the quantized signal and the original signal. This error is generally called quantization noise (or distortion). Let us consider an image f, and its quantized version $\hat{f}$. The distortion is measured as: $\|f - \hat{f}\|_2$. Thus this distortion is measured by the L_2 norm and this point can be problematic for use since we compare different measures of dissimilarity. To overcome this problem, we used as measure the SNR that is also used on signal quantization. Nevertheless, this metric might not be adapted to images since the spatial distortion is not taken into account. We propose to introduce an additional measure of distortion adapted to structured signals. The pattern spectrum (PS) [4] corresponds to the probability density function (pdf) of the granulometric decomposition, a multi-scale morphological image decomposition. We consider in fact the cumulative distribution function of the difference image $|f - \hat{f}|$. A good quantization schema filters out small spatial structures and keeps most of the large spatial objects of the image $|f - \hat{f}|$. A quantized image $\hat{f}_A$ is better than a quantized image $\hat{f}_B$ if the cumulative pattern spectrum of image $|f - \hat{f}_A|$ stochastically dominates [9] the one of $|f - \hat{f}_B|$.

3   Probabilistic Distances on Hyperspectral Images

We focus on spectral metrics which model the data uncertainty as resulting from randomness. Thus we consider each pixel as a random variable with the probability distribution obtained by normalizing the vector. Given two vectors $X = (x_1, \ldots, x_d) \in \mathbb{R}^d$, and $Y = (y_1, \ldots, y_d) \in \mathbb{R}^d$, they are represented respectively by their probability distribution function (pdf) as $P_x = (\frac{x_1}{\sum_i x_i}, \ldots, \frac{x_d}{\sum_i x_i}) \in \mathbb{R}^d$ and $P_y = (\frac{y_1}{\sum_i y_i}, \ldots, \frac{y_d}{\sum_i y_i}) \in \mathbb{R}^d$. Besides the metric presented here, we also considered (on [22]) L_p Minkowski norms, Hellinger distance.

Spectral Angle Mapper (SAM). Let us consider a spectrum $X \in \mathbb{R}^d$ of an hyperspectral image. It can be considered as a set of n random tests, where $n = \sum_i X_i$, and each event X_i is independent and provides a binary outcome. Thus, each image spectrum can be seen as following a multinomial distribution of parameter P_x. It is then possible to define a Fisher-Rao distance between X and Y, represented by their pdf P_x, P_y. This notion corresponds to the spherical distance [16]: $D_{Spher}(X, Y) = 2 \arccos(\sum \sqrt{P_{x,i}P_{y,i}})$. It is called the spherical distance since it represents the geodesic distance between distributions that are embedded on a $(d - 1)-$unit sphere of equation: $\sum_{i=1}^d \sqrt{P_{x,i}}\sqrt{P_{x,i}} = 1$. A distance classically used on hyperspectral images is the spectral angle mapper (SAM) [1,6], which for X, Y is defined between as: $D_{SAM}(X, Y) = \arccos(\frac{\sum X_i Y_i}{\sqrt{\sum X_i X_i}\sqrt{\sum Y_i Y_i}}) = \frac{1}{2}D_{Spherical}(X^2, Y^2)$. This metric is invariant to spectral multiplication since $D_{SAM}(\alpha X, Y) = D_{SAM}(X, Y), \forall \alpha \in \mathbb{R}^*$. So it is invariant to illumination changes, which can be problematic on remote sensing.

χ^2 **Distance.** The χ^2 distance is a similarity measure obtained from a statistical test that takes as input two distributions P_x, P_Y. This distance is defined as $D_{\chi^2}(X, Y) = \sum_{i=1}^{d} \frac{(P_{x,i} - m_i)^2}{m_i}$ with $m_i = \frac{P_{x,i} + P_{y,i}}{2}$, and measures the level of "adequacy" of pair P_x, P_Y [11]. More precisely, it corresponds to the probability that data of P_x follows the law P_y.

Kullback-Leibler Divergence. The Kullback-Leibler divergence [16] is a distinguishability measure between two distributions P_x and P_y. Divergence $S(P_x \| P_y)$ tells us how much the expected lengths of a code change, when the coding is optimal but made under the assumption that X follows P_y. It is defined as: $S(P_x \| P_y) = \sum_i P_{x,i} \log \frac{P_{x,i}}{P_{y,i}}$. The use of an asymmetric divergence may be problematic for some applications. This is why we try to symmetrize the Kullback-Leibler divergence and to use the Spectral Information Divergence (SID) [1]: $SID(P_x \| P_y) = S(P_x \| P_y) + S(P_y \| P_x)$.

Rényi Divergences. The Shannon entropy is a function that corresponds to the amount of information contained or send by a source of information. Somehow, the more the source emits different information, the more the entropy is large. Let us consider a pdf P_x with d possible outputs. Then, Shannon defined that the amount of information produced by knowing that an event of probability $P_{x,i}$ took place can be approximate by [16,19]: $I(P_{x,i}) = -\log(P_{x,i})$. The Shannon entropy is a mean of amount of information over P_x. It fulfils the postulate of additivity of information, which states that the information of two independent events is the sum of each information. However more general measures can be defined. Rényi [19] proved that to be able to verify the postulate of additivity of information, it is necessary to have entropy of the form: $H(P_x) = g^{-1}\left(\sum_{i=1}^{d} P_{x,i} g(I(P_{x,i}))\right)$, with $g(x) = cx$ or $g(x) = c2^{(1-\alpha)x}$. The first case leads to the Shannon entropy, whereas the second leads to other functional class called the Rényi entropy, which is defined by: $H_\alpha(P_x) = \frac{1}{1-\alpha} \log\left(\sum_{i=1}^{d} P_{x,i}^\alpha\right)$, Thus the Rényi entropy is a more flexible measure of uncertainty, where the parameter α allows different notions of information. The case $\alpha = 1$ leads to the Shannon entropy. Similarly to the Shannon entropy, it is possible to introduce divergence, that are called the Rényi divergence of order α, $\alpha > 0$ of a distribution P_x from a distribution P_y: $S_\alpha(P\|Q) = \frac{1}{\alpha-1} \log \sum_i P_{x,i}^\alpha P_{y,i}^{1-\alpha}$. Parameter α involves a family divergences, where $S_{\alpha \to 1}(P_x \| P_y) = S(P_x \| P_y)$. There are two other special cases: $\alpha = 1/2$, which is: $S_{\alpha=1/2}(P_x \| P_y) = -2\log(1 - D_{Hellinger}(X, Y)/2)$; and $\alpha = 2$ which leads to the quadratic Rényi divergence, that is a function mostly used on finance that check $S_{\alpha=2}(P_x \| P_y) = \log(1 + D_{\chi^2}(X, Y))$.

Mahalanobis Distance. We have considered above that each spectrum follows a multinomial distribution. It might be possible to consider alternatively that the spectra follow normal distributions. This model is often used on hyperspectral imaging. In our case, we assume that each spectrum X follows a multivariate normal of mean itself and with a fixed covariance for all the spectra. Hence we have that $X \sim \mathcal{N}(X, \Sigma)$. It turns out that the Fisher-Rao distance between X, Y

corresponds to the Mahalanobis distance [13,14] defined as: $D_{\text{Mahal}}(X,Y) = (X-Y)^T \Sigma^{-1}(X-Y)$. Endowed with this distance, the hyperspectral space is a submanifold of the manifold of multivariate normal distributions [14]. Moreover this kind of metric depends on the estimation of the covariance matrix. We have considered two ways to estimate the covariance matrix. The first one is the biased empirical covariance estimator: $\Sigma = \frac{1}{d}\sum_{i=1}^{d}(X_i - \overline{X})(X_i - \overline{X})^T$, where $\overline{X} = \frac{1}{d}\sum_{i=1}^{d} X_i$ is the empirical mean. The second one is inspired from the work [18], where thanks to random projections they succeed to have a good estimator of the Mahalanobis distance. We will denote respectively the corresponding distance D_{Mahal1} and D_{Mahal2}.

Kolmogorov Distance. Spectra X of a hyperspectral image can be represented by a pdf, but they can also be represented by their cumulative distribution function (cdf), denoted C_X, where $C_{x,i} = \sum_{k=1}^{i} P_{x,k}$. This function is smoother and less subject to high noise variation. Then, on can define the Kolmogorov-Smirnov distance [11] for two spectra X and Y as the maximal difference between their cumulative distribution functions: $D_{\text{Kolmo}} = \max(|P_{x,1} - P_{y,1}|, \ldots, |P_{x,2} - P_{y,2}|)$. This distance happens to be the L_∞ norm applied to the cdf's.

Earth Mover's Distance. Let us consider two spectra X and Y represented by their respective pdf P_x and P_y. Their Earth Mover's Distance [11] can be defined as $D_{\text{EMD}} = \min_{\alpha_{i,j} \in \mathcal{M}}\left(\sum_{i=1}^{d}\sum_{j=1}^{d}\alpha_{i,j}C(i,j)\right)$, where $\mathcal{M} = \{\alpha_{i,j} \geq 0; \sum_{i=1}^{d}\alpha_{i,j} = P_{y,j}; \sum_{j=1}^{d}\alpha_{i,j} == P_{x,i}\}$ and C is the cost function. Different choices of cost functions have been considered. We adopt here two different cost functions. The first one can be defined as: $C_1(i,j) = \frac{1}{d}|i-j|$. In such a case, the Earth Mover's Distance is given by [17]: $D_{\text{EMD1}} = \frac{1}{d}\|CP_x - CP_y\|_1$ where CP_x is the cumulative distribution function of X, similarly for CP_y and Y. One can also use a thresholded version of the cost function [17] to increase the computation speed: $C_2(i,j) = \begin{cases} |i-j| \text{ if } |i-j| \leq s \\ s \text{ otherwise} \end{cases}$ where s is the value of the threshold. We will write this distance D_{EMD2}.

Table 1. Comparison of probabilistic distances on hyperspectral images.

	L_1	L_2	L_∞	D_{Spher}	D_{SAM}	D_{Helli}	$D_{\chi 2}$	S	$S_{\alpha=1/2}$	$S_{\alpha=2}$	SID	D_{Mahal1}	D_{Mahal2}	D_{Kolmo}	D_{EMD1}	D_{EMD2}
Results on "Pavia" image																
OA	0.001	0.001	0.003	0.012	0.056	0.001	0.023	0.51	0.50	0.50	0.012	0.009	0.057	0.006	0.008	0.006
AA	0.001	0.001	0.013	0.085	0.12	0.001	0.11	0.25	0.22	0.22	0.089	0.003	0.22	0.09	0.1	0.08
Rank	0.93	0.93	0.91	0.81	0.62	0.46	0.33	0.21	0.47	0.22	0.36	0.58	0.37	0.90	0.90	0.41
SNR	22.82	22.96	22.89	22.72	14.90	21.92	23.90	21.88	21.97	21.20	21.26		21.4	22.0	22.47	
Results on "Indian Pines" image																
OA	0.016	0.016	0.011	0.016	0.09	0.016	0.30	0.010	0.010	0.010	0.010	0.019	0.068	0.0162	0.029	0.0162
AA	0.012	0.0022	0.014	0.016	0.13	0.016	0.24	0.09	0.09	0.09	0.09	0.016	0.13	0.0019	0.069	0.016
Rank	0.44	0.45	0.45	0.45	0.22	0.46	0.18	0.46	0.44	0.45	0.46	0.39	0.34	0.45	0.45	0.42
SNR	12.86	12.74	12.73	12.73	5.59	12.69	14.01	12.88	12.81	12.86	12.73		12.74	11.01	12.55	

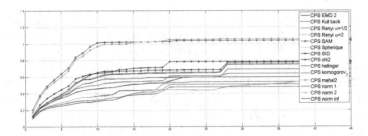

Fig. 2. Representation of the cumulative Pattern Spectrum of Pavia images for different similarity measures (Color figure online).

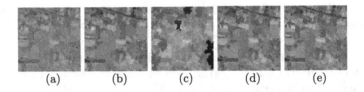

(a)   (b)   (c)   (d)   (e)

Fig. 3. (a) False RGB color image (using three spectral bands) of Indian Pines hyperspectral image. False RGB color image of the quantized hyperspectral image $K = 1500$ thanks to in (b) the norm 2, (c) the SAM, (d) the χ^2 distance, (e) the EMD.

4   Results on Hyperspectral Images

The question of the evaluation of the metrics is crucial. We first use two images conventionally used in hyperspectral image processing: (i) the Pavia image, which represents the campus of Pavia university (urban scene), of size 610×340 pixels and $d = 103$ spectral bands, and is composed of 9 classes; (ii) the Indian Pines image, test site in North-western Indiana composed for two thirds of agriculture,

(a)   (b)   (c)   (d)   (e)

Fig. 4. (a) False RGB color image (using three spectral bands) of Pavia hyperspectral image. False RGB color image of the quantized hyperspectral image $K = 3000$ thanks to in (b) the norm 2, (c) the SAM, (d) the χ^2 distance, (e) the EMD.

and one-third of forest, of 145×145 pixels and $d = 224$ spectral bands, and is composed of 16 classes. On theses images, we did a quantification, for Indian Pines we chose $K = 1500$, and for Pavia university we chose $K = 3000$. Then, we calculate on Table 1, the evaluation measures introduced in Sect. 2. We also plotted the cumulative pattern spectra, on Fig. 2, and examples of quantized Indian Pines image on Fig. 3, and of Pavia image on Fig. 4. From this study we can deduce that Mahalanobis distance may have good results, however the quality of the results of Mahalanobis distance depends on the estimation of the covariance matrix. We can also deduce that χ^2 distance can have excellent results, and seems to be quite robust to the curse of dimensionality, using this distance in dimensionality reduction [8] might improve the extracted feature.

5 Conclusion

We have done a systematic study to compare and assess different probabilistic distances in the context of hyperspectral image quantization. Our results are consistent to those previously published [1, 12, 15] for some of the distances. We infer from our study the importance of using appropriate distances to address the curse of dimensionality in hyperspectral imaging. A distance that seems to be rather efficient is the χ^2 distance. However, each dissimilarity measure has its disadvantages and benefits. A potentially interesting approach would be to take a dissimilarity measure as a linear combination of dissimilarity measures, taking advantage of the discriminatory power of each of them. This kind of approach is link with the multiple kernel learning [23]. For the particular problem of image quantization, which involves computation of centroids, advanced methods to compute centroid from divergences [5] can improve the results.

References

1. Chang, C.I. (ed.): Hyperspectral Imaging: Techniques for Spectral Detection and Classification, vol. 1. Springer Science & Business Media, New York (2003)
2. Tosic, I., Frossard, P.: Dictionary learning. IEEE Signal Process. Mag. **28**(2), 27–38 (2011)
3. Bellman, R.E.: Adaptive Control Processes. Princeton University Press, Princeton (1961)
4. Maragos, P.: Pattern spectrum and multiscale shape representation. IEEE Trans. Pattern Anal. Mach. Intell. **11**(7), 701–716 (1989)
5. Nielsen, F., Nock, R.: Sided and symmetrized Bregman centroids. IEEE Trans. Inf. Theor. **55**(6), 2882–2904 (2009)
6. Kass, R.E., Vos, P.W.: Geometrical Foundations of Asymptotic Inference, vol. 908. Wiley, New York (2011)
7. Schmid-Saugeon, P., Zakhor, A.: Dictionary design for matching pursuit and application to motion-compensated video coding. IEEE Trans. Circuits Syst. Video Technol. **14**(6), 880–886 (2004)
8. Noyel, G., Angulo, J., Jeulin, D.: Morphological segmentation of hyperspectral images. Image Anal. Stereol. **26**(3), 101–109 (2007)

9. Saporta, G.: Probabilites, analyse des donnees et statistique. Editions Technip (2011)

10. Fauvel, M.: Spectral and spatial methods for the classification of urban remote sensing data. Doctoral dissertation, Institut National Polytechnique de Grenoble-INPG, Universit d'Islande (2007)

11. Rubner, Y., Tomasi, C., Guibas, L.J.: The earth mover's distance as a metric for image retrieval. Int. J. Comput. Vis. **40**(2), 99–121 (2000)

12. Robila, S.: An investigation of spectral metrics in hyperspectral image preprocessing for classification. In: Geospatial Goes Global: From Your Neighborhood to the Whole Planet. ASPRS Annual Conference, Baltimore, Maryland, March 2005

13. Mahalanobis, P.C.: On the generalized distance in statistics. Proc. Nat. Inst. Sci. (Calcutta) **2**, 49–55 (1936)

14. Strapasson, J.E., Porto, J.P., Costa, S.I.: On bounds for the Fisher-Rao distance between multivariate normal distributions. In: MAXENT 2014, vol. 1641, pp. 313–320. AIP Publishing, January 2015

15. Paclik, P., Duin, R.P.: Dissimilarity-based classification of spectra: computational issues. Real-Time Imaging **9**(4), 237–244 (2003)

16. Bengtsson, I., Zyczkowski, K.: Geometry of Quantum States: An Introduction to Quantum Entanglement. Cambridge University Press, Cambridge (2006)

17. Pele, O., Werman, M.: Fast and robust earth mover's distances. In: 2009 IEEE 12th International Conference Computer Vision, September 2009

18. Gueguen, L., Velasco-Forero, S., Soille, P.: Local mutual information for dissimilarity-based image segmentation. J. Math. Imaging Vis. **48**(3), 625–644 (2014)

19. Renyi, A.: On measures of entropy and information. In: Fourth Berkeley Symposium on Mathematical Statistics and Probability, vol. 1, pp. 547–561 (1961)

20. Meyer, F., Maragos, P.: Nonlinear scale-space representation with morphological levelings. J. Vis. Commun. Image Represent. **11**(2), 245–265 (2000)

21. Soille, P.: Constrained connectivity for hierarchical image partitioning and simplification. IEEE Trans. Pattern Anal. Mach. Intell. **30**(7), 1132–1145 (2008)

22. Franchi, G., Angulo, J. Quantization of hyperspectral image manifold using probabilistic distances. 2015. hal-01121104

23. Bach, F.R., Lanckriet, G.R., Jordan, M.I.: Multiple kernel learning, conic duality, and the SMO algorithm. In: Proceedings of the Twenty-first International Conference on Machine Learning, p. 6. ACM, July 2004

Divergence Geometry

Generalized EM Algorithms for Minimum Divergence Estimation

Diaa Al Mohamad$^{(\boxtimes)}$ and Michel Broniatowski$^{(\boxtimes)}$

Laboratoire de Statistique Théorique et Appliquée, Université Pierre et Marie Curie,
4 Place Jussieu, 75005 Paris, France
diaa.al-mohamad@etu.upmc.fr

Abstract. Minimum divergence estimators are derived through the dual form of the divergence in parametric models. These estimators generalize the classical maximum likelihood ones. Models with unobserved data, as mixture models, can be estimated with EM algorithms, which are proved to converge to stationary points of the likelihood function under general assumptions. This paper presents an extension of the EM algorithm based on minimization of the dual approximation of the divergence between the empirical measure and the model using a proximal-type algorithm. The algorithm converges to the stationary points of the empirical criterion under general conditions pertaining to the divergence and the model. Robustness properties of this algorithm are also presented. We provide another proof of convergence of the EM algorithm in a two-component gaussian mixture. Simulations on Gaussian and Weibull mixtures are performed to compare the results with the MLE.

Introduction

The EM algorithm is a well known method for calculating the maximum likelihood estimator of a model where incomplete data is considered. For example, when working with mixture models in the context of clustering, the labels or classes of observations are unknown during the training phase. Several variants of the EM algorithm were proposed, see [11]. Another way to look at the EM algorithm is as a proximal point problem, see [5,15]. Indeed, one may rewrite the conditional expectation of the complete log-likelihood as a sum of the log-likelihood function and a distance-like function over the conditional densities of the labels provided an observation. Generally, the proximal term has a regularization effect in the sense that a proximal point algorithm is more stable and frequently outperforms classical optimization algorithms, see [9]. Chrétien and Hero [4] prove superlinear convergence of a proximal point algorithm derived by the EM algorithm. Notice that EM-type algorithms usually enjoy no more than linear convergence.

Taking into consideration the need for robust estimators, and the fact that the MLE is the least robust estimator among the class of divergence-type estimators which we present below, we generalize the EM algorithm (and the version in [15]) by replacing the log-likelihood function by an estimator of a φ-divergence

© Springer International Publishing Switzerland 2015
F. Nielsen and F. Barbaresco (Eds.): GSI 2015, LNCS 9389, pp. 417–426, 2015.
DOI: 10.1007/978-3-319-25040-3_45

between the *true distribution* of the data and the model. A φ-divergence in the sense of Csiszár [7] is defined in the same way as [3] by:

$$D_\varphi(Q, P) = \int \varphi\left(\frac{dQ}{dP}(y)\right) dP(y),$$

where φ is a nonnegative strictly convex function. Examples of such divergences are: the Kullback-Leibler (KL) divergence for $\varphi(t) = t\log(t) - t + 1$, the modified KL divergence for $\varphi(t) = -\log(t) + t - 1$, the hellinger distance for $\varphi(t) = \frac{1}{2}(\sqrt{t} - 1)$ among others. All these well-known divergences belong to the class of Cressie-Read functions defined by $\varphi_\gamma(t) = \frac{x^\gamma - \gamma x + \gamma - 1}{\gamma(\gamma-1)}$ for $\gamma \in \mathbb{R} \setminus \{0, 1\}$.

Since the φ-divergence calculus uses the unknown true distribution, we need to estimate it. We consider the dual estimator of the divergence introduced independently by [2, 10]. The use of this estimator is motivated by many reasons. Its minimum coincides with the MLE for $\varphi(t) = -\log(t) + t - 1$. Besides, it has the same form for discrete and continuous models, and does not consider any partitioning or smoothing.

Let $(P_\phi)_{\phi \in \Phi}$ be a parametric model with $\Phi \subset \mathbb{R}^d$, and denote ϕ^T the *true* set of parameters. Let dy be the Lebesgue measure defined on $\mathbb{R}$. Suppose that $\forall \phi \in \Phi$, the probability measure P_ϕ is absolutely continuous with respect to dy and denote p_ϕ the corresponding probability density. The dual estimator of the φ-divergence given an n-sample $y_1, \cdots, y_n$ is given by:

$$\hat{D}_\varphi(p_\phi, p_{\phi_T}) = \sup_{\alpha \in \Phi} \int \varphi'\left(\frac{p_\phi}{p_\alpha}\right)(x)p_\phi(x)dx - \frac{1}{n}\sum_{i=1}^n \varphi^\#\left(\frac{p_\phi}{p_\alpha}\right)(y_i), \quad (1)$$

with $\varphi^\#(t) = t\varphi'(t) - \varphi(t)$. AL Mohamad [1] argues that this formula works well under the model, however, when we are not, this quantity largely underestimates the divergence between the true distribution and the model, and proposes following modification:

$$\tilde{D}_\varphi(p_\phi, p_{\phi_T}) = \int \varphi'\left(\frac{p_\phi}{K_{n,w}}\right)(x)p_\phi(x)dx - \frac{1}{n}\sum_{i=1}^n \varphi^\#\left(\frac{p_\phi}{K_{n,w}}\right)(y_i), \quad (2)$$

where $K_{n,w}$ is the Rosenblatt-Parzen kernel estimate with window parameter w. Whether it is $\hat{D}_\varphi$, or $\tilde{D}_\varphi$, the minimum dual φ-divergence estimator (MDφDE) is defined as the argument of the infimum of the dual approximation:

$$\hat{\phi}_n = \arg\inf_{\phi \in \Phi} \hat{D}_\varphi(p_\phi, p_{\phi_T}), \quad (3)$$

$$\tilde{\phi}_n = \arg\inf_{\phi \in \Phi} \tilde{D}_\varphi(p_\phi, p_{\phi_T}). \quad (4)$$

Asymptotic properties and consistency of these two estimators can be found in [1, 3]. Robustness properties were also studied using the influence function approach in [1, 14]. The kernel-based MDφDE (4) seems to be a *better* estimator than the classical MDφDE (3) in the sense that the former is robust whereas the later is generally not. Under the model, the estimator given by (3) is, however, more efficient especially when the true density of the data is unbounded[1], see [1] for a brief comparison.

Here in this paper, we propose to calculate the MDφDE using an iterative procedure based on the work of [15] on the log-likelihood function. This procedure has the form of a proximal point algorithm, and extends the EM algorithm. Our convergence proof demands some regularity of the estimated divergence with respect to the parameter vector which is not simply checked using (1). Recent results in the book of [13] provide sufficient conditions to solve this problem. Differentiability with respect to ϕ still remains a very hard task, therefore, our results cover cases when the objective function is not differentiable.

The paper is organized as follows: In Sect. 1, we present the general context. We also present the derivation of our algorithm from the EM algorithm and passing by Tseng's generalization. In Sect. 2, we present some convergence properties. We discuss in Sect. 3 the two-gaussian mixture model and a convergence proof of the EM algorithm in the spirit of our approach. Finally, Sect. 4 contains simulations confirming our claim about the efficiency and the robustness of our approach in comparison with the MLE.

1 A Description of the Algorithm

1.1 General Context and Notations

Let (X, Y) be a couple of random variables with joint probability density function $f(x, y|\phi)$ parametrized by a vector of parameters $\phi \in \Phi \subset \mathbb{R}^d$. Let $(X_1, Y_1), \cdots,$ (X_n, Y_n) be n copies of (X, Y) independently and identically distributed. Finally, let $(x_1, y_1), \cdots, (x_n, y_n)$ be n realizations of the n copies of (X, Y). The x_i's are the unobserved data (labels) and the y_i's are the observations. The vector of parameters ϕ is unknown and need to be estimated. The observed data y_i are supposed to be real numbers, and the labels x_i belong to a space $\mathcal{X}$ not necessarily finite unless mentioned otherwise. The marginal density of the observed data is given by $p_\phi(y) = \int f(x, y|\phi) dx$.

For a parametrized function f with a parameter a, we write $f(x|a)$. We use the notation ϕ^k for sequences with the index above. Derivatives of a real valued function ψ defined on $\mathbb{R}$ are written as ψ', ψ'', etc. We use ∇f for the gradient of a real function f defined on $\mathbb{R}^d$, and J_f for the matrix of second order partial derivatives. If the function has two (vectorial) arguments $D(\phi|\theta)$, then $\nabla_1 D(\phi|\theta)$ denotes the gradient with respect to the first (vectorial) variable. Finally, for any set A, we use $int(A)$ to denote the interior of A.

1.2 EM Algorithm and Tseng's Generalization

The EM algorithm estimates the unknown parameter vector by (see [8]):

$$\phi^{k+1} = \arg\max_{\Phi} \mathbb{E}\left[\log(f(\mathbf{X}, \mathbf{Y}|\phi)) \,\middle|\, \mathbf{Y} = \mathbf{y}, \phi^k\right].$$

---

[1] More investigation is needed here since we may use asymmetric kernels to overcome this difficulty.

where $\mathbf{X} = (X_1, \cdots, X_n)$, $\mathbf{Y} = (Y_1, \cdots, Y_n)$ and $\mathbf{y} = (y_1, \cdots, y_n)$. By independence between the couples (X_i, Y_i)'s, previous iteration may be written as:

$$
\begin{aligned}
\phi^{k+1} &= \arg\max_{\Phi} \sum_{i=1}^{n} \mathbb{E}\left[\log(f(X_i, Y_i|\phi)) \,|\, Y_i = y_i, \phi^k\right] \\
&= \arg\max_{\Phi} \sum_{i=1}^{n} \int_{\mathcal{X}} \log(f(x, y_i|\phi)) h_i(x|\phi^k) dx,
\end{aligned}
\tag{5}
$$

where $h_i(x|\phi^k) = \frac{f(x,y_i|\phi^k)}{p_{\phi^k}(y_i)}$ is the conditional density of the labels (at step k) provided y_i which we suppose to be positive dx-almost everywhere. It is well-known that the EM iterations can be rewritten as a difference between the log-likelihood and a *Kullback-Liebler* distance-like function.

$$
\phi^{k+1} = \arg\max_{\Phi} \sum_{i=1}^{n} \log\left(g(y_i|\phi)\right) + \sum_{i=1}^{n} \int_{\mathcal{X}} \log\left(\frac{h_i(x|\phi)}{h_i(x|\phi^k)}\right) h_i(x|\phi^k) dx.
$$

The previous iteration has the form of a proximal point maximization of the log-likelihood, i.e. a perturbation of the log-likelihood by a distance-like function defined on the conditional densities of the labels. Tseng [15] generalizes this iteration by allowing any nonnegative convex function ψ to replace the $t \mapsto -\log(t)$ function. Tseng's recurrence is defined by:

$$
\phi^{k+1} = \arg\sup_{\phi} J(\phi) - D_\psi(\phi, \phi^k),
\tag{6}
$$

where J is the log-likelihood function and D_ψ is given by:

$$
D_\psi(\phi, \phi^k) = \sum_{i=1}^{n} \int_{\mathcal{X}} \psi\left(\frac{h_i(x|\phi)}{h_i(x|\phi^k)}\right) h_i(x|\phi^k) dx,
\tag{7}
$$

for any real nonnegative convex function ψ such that $\psi(1) = \psi'(1) = 0$. $D_\psi(\phi_1, \phi_2)$ is nonnegative, and $D_\psi(\phi_1, \phi_2) = 0$ if and only if $\forall i, h_i(x|\phi_1) = h_i(x|\phi_2)$ dx-almost everywhere.

1.3   Generalization of Tseng's Algorithm

We use the relation between maximizing the log-likelihood and minimizing the Kullback-Liebler divergence to generalize the previous algorithm. We, therefore, replace the log-likelihood function by an estimate of a φ-divergence D_φ between the true distribution and the model. We use the dual estimator of the divergence presented earlier in the introduction (1) or (2) which we denote in the same manner $\hat{D}_\varphi$ unless mentioned otherwise. Our new algorithm is defined by:

$$
\phi^{k+1} = \arg\inf_{\phi} \hat{D}_\varphi(p_\phi, p_{\phi_T}) + \frac{1}{n} D_\psi(\phi, \phi^k),
\tag{8}
$$

where $D_\psi(\phi, \phi^k)$ is defined by (7). When $\varphi(t) = -\log(t) + t - 1$, it is easy to see that we get recurrence (6). Indeed, for the case of (1) we have:

$$\hat{D}_\varphi(p_\phi, p_{\phi_T}) = \sup_\alpha \frac{1}{n} \sum_{i=1}^n \log(p_\alpha(y_i)) - \frac{1}{n} \sum_{i=1}^n \log(p_\phi(y_i)).$$

Using the fact that the first term in $\hat{D}_\varphi(p_\phi, p_{\phi_T})$ does not depend on ϕ, so it does not count in the arg inf defining ϕ^{k+1}, we easily get (6). The same applies for the case of (2). For notational simplicity, from now on, we redefine D_ψ with a normalization by n. Hence, our set of algorithms is redefined by:

$$\phi^{k+1} = \arg\inf_\phi \hat{D}_\varphi(p_\phi, p_{\phi_T}) + D_\psi(\phi, \phi^k). \tag{9}$$

We will see later that this iteration forces the divergence to decrease and that under suitable conditions, it converges to a (local) minimum of $\hat{D}_\varphi(p_\phi, p_{\phi_T})$. It results that, algorithm (9) is a way to calculate the MDφDE.

2 Some Convergence Properties of ϕ^k

We show here how, according to some possible situations, one may prove convergence of the algorithm defined by (9). Let ϕ^0 be a given initialization, and let $\Phi^0 := \{\phi \in \Phi : \hat{D}_\varphi(p_\phi, p_{\phi_T}) \le \hat{D}_\varphi(p_{\phi^0}, p_{\phi_T})\}$ which we suppose to be a subset of $int(\Phi)$. We will be using the following assumptions:

A0. Functions $\phi \mapsto \hat{D}_\varphi(p_\phi|p_{\phi_T}), D_\psi$ are lower semicontinuous.

A1. Functions $\phi \mapsto \hat{D}_\varphi(p_\phi|p_{\phi_T}), D_\psi$ and $\nabla_1 D_\psi$ are defined and continuous on, respectively, $\Phi, \Phi \times \Phi$ and $\Phi \times \Phi$;

AC. $\phi \mapsto \nabla \hat{D}_\varphi(p_\phi|p_{\phi_T})$ is defined and continuous on Φ

A2. Φ^0 is a compact subset of $int(\Phi)$;

A3. $D_\psi(\phi, \bar\phi) > 0$ for all $\bar\phi \ne \phi \in \Phi$.

Recall also that we suppose that $h_i(x|\phi) > 0, dx - a.e.$ We relax the convexity assumption of function ψ. We only suppose that ψ is nonnegative and $\psi(t) = 0$ iff $t = 1$. Besides $\psi'(t) = 0$ if $t = 1$.

Continuity and differentiability assumptions of function $\phi \mapsto \hat{D}_\varphi(p_\phi|p_{\phi_T})$ for the case of (2) can be easily checked using Lebesgue theorems. Continuity assumption for the case of (1) can be checked using Theorem 1.17 or Corollary 10.14 in [13]. Differentiability can also be checked using Corollary 10.14 or Theorem 10.31 in the same book. In what concerns D_ψ, continuity and differentiability can be obtained merely by fulfilling Lebesgue theorems conditions. When working with mixture models, we only need the continuity and differentiability of ψ and functions h_i. The later is easily deduced from regularity assumptions on the model. For assumption A2, there is no universal method, see paragraph (3) for an example. Assumption A3 can be checked using Lemma 2 in [15].

We start the convergence properties by proving the decrease of the objective function $\hat{D}_\varphi(p_\phi|p_{\phi_T})$, and a possible set of conditions for the existence of the sequence $(\phi^k)_k$. The proofs of Propositions 1–3 are adaptations of the proofs given in [15].

Proposition 1. (a) *Assume that the sequence* $(\phi^k)_k$ *is well defined in* Φ, *then* $\hat{D}_\varphi(p_{\phi^{k+1}}|p_{\phi_T}) \leq \hat{D}_\varphi(p_{\phi^k}|p_{\phi_T})$, *and* $\forall k, \phi^k \in \Phi^0$. (b) *Assume A0 and A2 are verified, then the sequence* $(\phi^k)_k$ *is defined and bounded. Moreover, the sequence* $(\hat{D}_\varphi(p_{\phi^k}|p_{\phi_T}))_k$ *converges.*

The convergence of the sequence $(\hat{D}_\varphi(\phi^k|\phi_T))_k$ is an interesting property, since in general there is no theoretical guarantee, or it is difficult to prove that the whole sequence $(\phi^k)_k$ converges. It may also continue to fluctuate around a minimum. The decrease of the error criterion $\hat{D}_\varphi(\phi^k|\phi_T)$ between two iterations helps us decide when to stop the iterative procedure.

Proposition 2. *Suppose A1 verified,* Φ^0 *is closed and* $\{\phi^{k+1} - \phi^k\} \to 0$.

(a) *If AC is verified, then any limit point of* $(\phi^k)_k$ *is a stationary point of* $\phi \mapsto \hat{D}_\varphi(p_\phi|p_{\phi_T})$;

(b) *If AC is dropped, then any limit point of* $(\phi^k)_k$ *is a "generalized" stationary point of* $\phi \mapsto \hat{D}_\varphi(p_\phi|p_{\phi_T})$, *i.e. zero belongs to the subgradient of* $\phi \mapsto \hat{D}_\varphi(p_\phi|p_{\phi_T})$ *calculated at the limit point.*

Assumption $\{\phi^{k+1} - \phi^k\} \to 0$ used in Proposition 2 is not easy to be checked unless one has a close formula of ϕ^k. The following proposition gives a method to prove such assumption. This method seems simpler, but it is not verified in many mixture models, see Sect. 3 for a counter example.

Proposition 3. *Assume that A1, A2 and A3 are verified, then* $\{\phi^{k+1} - \phi^k\} \to 0$. *Thus, by Proposition 2 (according to whether AC is verified or not) any limit point of the sequence* ϕ^k *is a (generalized) stationary point of* $\hat{D}_\varphi(.|\phi_T)$.

Corollary 1. *Under assumptions of Proposition 3, the set of accumulation points of* $(\phi^k)_k$ *is a connected compact set. Moreover, if* $\phi \mapsto \hat{D}(p_\phi, p_{\phi_T})$ *is strictly convex in a neighborhood of a limit point of the sequence* $(\phi^k)_k$, *then the whole sequence* $(\phi^k)_k$ *converges to a local minimum of* $\hat{D}(p_\phi, p_{\phi_T})$.

Proof of Corollary 1 is based on Theorem 28.1 in [12]. Proposition 3 and Corollary 1 describe what we may hope to get of the sequence ϕ^k. Convergence of the whole sequence is bound by a local convexity assumption in a neighborhood of a limit point. Although simple, this assumption remains difficult to be checked since we do not know where might be the limit points. Besides, assumption A3 is very restrictive, and is not verified in mixture models.

Propositions 2 and 3 were developed for the likelihood function in the paper of [15]. Similar results for a general class of functions replacing $\hat{D}_\varphi$ and D_ψ which may not be differentiable (but still continuous) are presented in [5]. In these results, assumption A3 is essential. Although [6] overcomes this problem, their approach demands that the log-likelihood has $-\infty$ limit as $\|\phi\| \to \infty$. This is simply not verified for mixture models. We present a similar method to [6] based on the idea of [15] of using the set Φ^0 which is valid for mixtures. We lose, however, the guarantee of consecutive decrease of the sequence.

Proposition 4. *Assume A1, AC and A2 verified. Any limit point of the sequence* $(\phi^k)_k$ *is a stationary point of* $\phi \to \hat{D}(p_\phi, p_{\phi_T})$. *If AC is dropped, then 0 belongs to the subgradient of* $\phi \mapsto \hat{D}(p_\phi, p_{\phi_T})$ *calculated at the limit point.*

The proof is very similar to the proof of Proposition 2. The key idea is to use the sequence of conditional densities $h_i(x|\phi^k)$ instead of the sequence ϕ^k. According to the application, one may be interested only in Proposition 1 or in Propositions 2–4. If one is interested in the parameters, Propositions 2 to 4 are needed, since we need a stable limit of $(\phi^k)_k$. If we are only interested in minimizing an error criterion $\hat{D}_\varphi(p_\phi, p_{\phi_T})$ between the estimated distribution and the true one, Proposition 1 should be sufficient.

3   Case Study: The Two-Component Gaussian Mixture

We suppose that the model $(p_\phi)_{\phi \in \Phi}$ is a mixture of two gaussian densities, and that we are only interested in estimating the means $\mu = (\mu_1, \mu_2) \in \mathbb{R}^2$ and the proportion $\lambda \in [\eta, 1 - \eta]$. The use of η is to avoid cancellation of any of the two components, and to keep the hypothesis $h_i(x|\phi) > 0$ for $x = 1, 2$ verified. We also suppose that the components variances are reduced ($\sigma_i = 1$). The model takes the form

$$p_{\lambda,\mu}(x) = \frac{\lambda}{\sqrt{2\pi}} e^{-\frac{1}{2}(x-\mu_1)^2} + \frac{1-\lambda}{\sqrt{2\pi}} e^{-\frac{1}{2}(x-\mu_2)^2}.$$

In the case of $\varphi(t) = -\log(t)+t-1$, the set Φ^0 is given by $\Phi^0 = J^{-1}\left([J(\phi^0), +\infty)\right)$. The log-likelihood function J is clearly of class $\mathcal{C}^1(\text{int}(\Phi))$, where $\Phi = [\eta, 1-\eta] \times \mathbb{R}^2$. The regularization term D_ψ is defined by (7) where:

$$h_i(1|\phi) = \frac{\lambda e^{-\frac{1}{2}(y_i-\mu_1)^2}}{\lambda e^{-\frac{1}{2}(y_i-\mu_1)^2} + (1-\lambda)e^{-\frac{1}{2}(y_i-\mu_2)^2}}, \quad h_i(2|\phi) = 1 - h_i(1|\phi).$$

Functions h_i are clearly of class $\mathcal{C}^1(\text{int}(\Phi))$, hence, assumptions A1 and AC are verified. We prove that Φ^0 is closed and bounded which is sufficient to conclude its compactness, since the space $[\eta, 1 - \eta] \times \mathbb{R}^2$ provided with the euclidean distance is complete.

Closedness is clear since J is continuous and $[\eta, 1-\eta] \times \mathbb{R}^2$ is complete. **Boundedness.** By contradiction, suppose that Φ^0 is unbounded, then there exists a sequence $(\phi^l)_l$ of elements of Φ^0 which tends to infinity. Since $\lambda^l \in [\eta, 1 - \eta]$, then either of μ_1^l or μ_2^l tends to infinity. Suppose that both μ_1^l and μ_2^l tend to infinity, we then have $J(\phi^l) \to -\infty$. Any finite initialization ϕ^0 will imply that $J(\phi^0) > -\infty$ so that $\forall \phi \in \Phi^0, J(\phi) \geq J(\phi^0) > -\infty$. Suppose that $\mu_1^l \to \infty$, and that μ_2^l converges[2] to μ_2. The limit of the likelihood is $L(\lambda, \infty, \mu_2) = \prod_{i=1}^n \frac{(1-\lambda)}{\sqrt{2\pi}} e^{-\frac{1}{2}(y_i-\mu_2)^2}$ which is bounded by its value for $\lambda = 0$ and $\mu_2 =$

---

[2] Normally, μ_2^l is bounded; still, we can extract a subsequence which converges.

$\frac{1}{n}\sum_{i=1}^{n} y_i$. Thus, it suffices to choose ϕ^0 such that $J(\phi^0) > J\left(0, \infty, \frac{1}{n}\sum_{i=1}^{n} y_i\right)$. By symmetry, if ϕ^0 verify:

$$J(\phi^0) > \max\left[J\left(0, \infty, \frac{1}{n}\sum_{i=1}^{n} y_i\right), J\left(1, \frac{1}{n}\sum_{i=1}^{n} y_i, \infty\right)\right], \quad (10)$$

the set Φ^0 becomes bounded. Condition (10) is very natural and means that we need to start at a likelihood higher than the likelihood of having one component. Now that Φ^0 is compact, part (b) of Proposition 1 is verified and the sequence $(\phi^k)_k$ generated by (9) is well defined and bounded. Moreover, the sequence $(J(\phi^k))_k$ converges. However, Proposition 3 cannot be applied since A3 is not fulfilled. Take for example $\mu_1 = 0, \mu_2 = 1, \lambda = \frac{2}{3}, \mu'_1 = \frac{1}{2}, \mu'_2 = \frac{3}{2}, \lambda' = \frac{1}{2}$. We get that $D_\psi((\lambda, \mu_1, \mu_2), (\lambda', \mu'_1, \mu'_2)) = 0$ although both arguments are different. Still, Proposition 4 ensures that, since A1, AC and A2 are verified, all limit points of the sequence $(\phi^k)_k$ generated by (9) are stationary points of the likelihood. But we have no information about the difference between consecutive terms of the sequence. Note that the case of $\psi(t) = \varphi(t) = -\log(t) + t - 1$, we get the classical EM, and [15] proved its convergence through Proposition 2.

4   Simulation Study

We summarize the results of 100 experiments on 100-samples by giving the average of the estimates and the error committed, and the corresponding standard deviation. The criterion error is the total variation distance which is calculated using the $L1$ distance by the Shceffé lemma. We consider the Hellinger divergence for estimators based on φ-divergences which corresponds to $\varphi(t) = \frac{1}{2}(\sqrt{t} - 1)^2$. D_ψ is calculated with $\psi(t) = \frac{1}{2}(t - 1)^2$. The kernel-based MD$\varphi$DE is calculated using the gaussian kernel, and the window is calculated using Silverman's rule. Simulations from two mixture models are given below. MLE was calculated using EM algorithm.

1. A two component gaussian mixture with unknown parameters $\lambda = 0.35, \mu_1 = -2, \mu_2 = 1.5$ and known variances equal to 1. Contamination was done by adding in the original sample to the 5 lowest values random observations from the uniform distribution $\mathcal{U}[-5, -2]$. We also added to the 5 largest values random observations from the uniform distribution $\mathcal{U}[2, 5]$. Results are summarized in Table 1.
2. A two component Weibull mixture with unknown shapes $\nu_1 = 1.2, \nu_2 = 2$ and a proportion $\lambda = 0.35$. The scales are known an equal to $\sigma_1 = 0.5, \sigma_2 = 2$. Contamination was done by replacing 10 observations of each sample chosen randomly by 10 i.i.d. observations drawn from a Weibull distribution with shape $\nu = 0.9$ and scale $\sigma = 3$. Results are summarized in Table 2.

In what concerns our simulation results. The total variation of all three estimation methods is very close when we are under the model. When we added outliers, the classical MDφDE was as sensitive as the maximum likelihood estimator. The error was doubled. The kernel-based MDφDE is clearly robust since the total variation under contamination has slightly increased. Differences in the Weibull mixture are less apparent. Indeed, this is caused by the fact that symmetric kernels suffer from a bias on the boundary. Thus, the bias slightly influence our kernel-based MDφDE. In more complicated situations, the use of bias-correction methods or other kernels which are free of the boundary effect is needed. This is not discussed here for lack of space and is let for future investigations.

Table 1. The mean and the standard deviation of the estimates and the errors committed in a 100-run experiment of a two-component gaussian mixture. The true set of parameters is $\lambda = 0.35, \mu_1 = -2, \mu_2 = 1.5$.

Estimation method	λ	sd(λ)	μ_1	sd(μ_1)	μ_2	sd(μ_2)	TVD	sd(TVD)
Without outliers								
Classical MDφDE	0.349	0.049	−1.989	0.207	1.511	0.151	0.061	0.029
New MDφDE - Silverman	0.349	0.049	−1.987	0.208	1.520	0.155	0.062	0.029
EM (MLE)	0.360	0.054	−1.989	0.204	1.493	0.136	0.064	0.025
With 10 % outliers								
Classical MDφDE	0.357	0.022	−2.629	0.094	1.734	0.111	0.146	0.034
New MDφDE - Silverman	0.352	0.057	−1.756	0.224	1.358	0.132	0.087	0.033
EM (MLE)	0.342	0.064	−2.617	0.288	1.713	0.172	0.150	0.034

Table 2. The mean and the standard deviation of the estimates and the errors committed in a 100-run experiment of a two-component Weibull mixture. The true set of parameter is $\lambda = 0.35, \nu_1 = 1.2, \nu_2 = 2$.

Estimation method	λ	sd(λ)	ν_1	sd(ν_1)	ν_2	sd(ν_2)	TVD	sd(TVD)
Without outliers								
Classical MDφDE	0.356	0.066	1.245	0.228	2.055	0.237	0.052	0.025
New MDφDE - Silverman	0.387	0.067	1.229	0.241	2.145	0.289	0.058	0.029
EM (MLE)	0.355	0.066	1.245	0.228	2.054	0.237	0.052	0.025
With 10 % outliers								
Classical MDφDE	0.250	0.085	1.089	0.300	1.470	0.335	0.092	0.037
New MDφDE - Silverman	0.349	0.076	1.122	0.252	1.824	0.324	0.067	0.034
EM (MLE)	0.259	0.095	0.941	0.368	1.565	0.325	0.095	0.035

References

1. Al Mohamad, D.: Towards a better understanding of the dual representation of phi divergences. ArXiv e-prints (2015)
2. Broniatowski, M., Keziou, A.: Minimization of divergences on sets of signed measures. Studia Sci. Math. Hungar. **43**(4), 403–442 (2006)
3. Broniatowski, M., Keziou, A.: Parametric estimation and tests through divergences and the duality technique. J. Multivar. Anal. **100**(1), 16–36 (2009)
4. Chretien, S., Hero, A.O.: Acceleration of the EM algorithm via proximal point iterations. In: Proceedings of the 1998 IEEE International Symposium on Information Theory, 1998, p. 444 (1998)
5. Chrétien, S., Hero, A.O.: Generalized proximal point algorithms and bundle implementations. Technical report, Department of Electrical Engineering and Computer Science, The University of Michigan (1998)
6. Chrétien, S., Hero, A.O.: On EM algorithms and their proximal generalizations. ESAIM: Probab. Stat. **12**, 308–326 (2008)
7. Csiszár, I.: Eine informationstheoretische Ungleichung und ihre anwendung auf den Beweis der ergodizität von Markoffschen Ketten. Publications of the Mathematical Institute of Hungarian Academy of Sciences **8**, 95–108 (1963)
8. Dempster, A.P., Laird, N.M., Rubin, D.B.: Maximum likelihood from incomplete data via the EM algorithm. J. Roy. Stat. Soc. Ser. B **39**(1), 1–38 (1977)
9. Goldstein, A.A., Russak, I.B.: How good are the proximal point algorithms? Numer. Funct. Anal. Optim. **9**(7–8), 709–724 (1987)
10. Liese, F., Vajda, I.: On divergences and informations in statistics and information theory. IEEE Trans. Inf. Theor. **52**(10), 4394–4412 (2006)
11. McLachlan, G., Krishnan, T.: The EM Algorithm and Extensions. Wiley Series in Probability and Statistics. Wiley, Hoboken (2007)
12. Ostrowski, A.M.: Solution of Equations and Systems of Equations. Pure and Applied Mathematics. Academic Press, New York (1966)
13. Rockafellar, R.T., Wets, R.J.-B.: Variational Analysis. Die Grundlehren der mathematischen Wissenschaften in Einzeldarstellungen, 3rd edn. Springer, Heidelberg (1998)
14. Toma, A., Broniatowski, M.: Dual divergence estimators and tests: robustness results. J. Multivar. Anal. **102**(1), 20–36 (2011)
15. Tseng, P.: An analysis of the EM algorithm and entropy-like proximal point methods. Math. Oper. Res. **29**(1), 27–44 (2004)

Extension of Information Geometry
to Non-statistical Systems: Some Examples

Jan Naudts[(✉)] and Ben Anthonis

Universiteit Antwerpen, Antwerp, Belgium
Jan.Naudts@uantwerpen.be, Ben.Anthonis@gmail.com

Abstract. Our goal is to extend information geometry to situations where statistical modeling is not obvious. The setting is that of modeling experimental data. Quite often the data are not of a statistical nature. Sometimes also the model is not a statistical manifold. An example of the former is the description of the Bose gas in the grand canonical ensemble. An example of the latter is the modeling of quantum systems with density matrices. Conditional expectations in the quantum context are reviewed. The border problem is discussed: through conditioning the model point shifts to the border of the differentiable manifold.

1 Introduction

One of the goals of information geometry [1] is the study of the geometry of a statistical manifold $\mathbb{M}$. A tool suited for this study is the divergence function $D(p||q)$, called relative entropy in the physics literature. It compares two probability distributions p and q. It cannot be negative and vanishes if and only if $p = q$. In our recent works [2–4] we have stressed the importance of considering the statistical manifold $\mathbb{M}$ as embedded in the set of all probability distributions. In particular, the divergence function $D(p||q)$, with p not belonging to $\mathbb{M}$, can be used to characterize exponential families. We also stressed that it is not a strict necessity that the first argument of the divergence is a probability distribution. The first example given in this paper is an illustration of this point.

In quantum probability [5,6] both arguments of the divergence function are replaced by density matrices, which are the quantum analogues of probability distributions. A renewed interest in quantum probability comes from quantum information theory (see for instance [6,7]). The theoretical developments are accompanied by a large number of novel experiments. Some of them are mentioned below [8–10]. They confirm the validity of quantum mechanics but challenge our understanding of nature. The present paper tries to situate some recent insights in the context of quantum conditional expectations. In particular, weak measurement theory [11,12] is considered.

2 The Ideal Bose Gas

The result of a thought experiment on the ideal Bose gas is a sequence of non-negative integers $n_1, n_2, \cdots$ with finite sum $\sum_{j=1}^{\infty} n_j < +\infty$. A model for these data is a two-parameter family of probability distributions

© Springer International Publishing Switzerland 2015
F. Nielsen and F. Barbaresco (Eds.): GSI 2015, LNCS 9389, pp. 427–434, 2015.
DOI: 10.1007/978-3-319-25040-3_46

$$p_{\beta,\mu}(n) = \frac{1}{Z(\beta,\mu)} \exp(-\beta \sum_j \epsilon_j n_j + \beta\mu \sum_j n_j), \tag{1}$$

with normalization given by

$$Z(\beta,\mu) = \prod_j \frac{1}{1 - \exp(-\beta(\epsilon_j - \mu))}. \tag{2}$$

The numbers ϵ_j are supposed to be known and to increase fast enough with the index j to guarantee the convergence of the infinite product. The parameters β and μ are real. By assumption is $\beta > 0$ and $\mu < \epsilon_j$ for all j.

The model space is the statistical manifold $\mathbb{M}$ formed by the probability distributions $p_{\beta,\mu}$. However, the data produced by the measurement are strictly spoken not of a stochastic origin. Indeed, a more detailed modeling of the experiment involves quantum mechanics and quantum measurement theory. The latter is much debated since the introduction of so-called weak measurements [11]. Hence, we cannot speculate about a possible stochastic origin of the data.

If all we know about the experiment is that it produces the sequence n then the simplest modeling we can do is to fit (1) to the data with a method which can be used also outside the conventional settings of statistics. Our proposal is to do the fitting by minimization of a divergence function.

The Kullback-Leibler divergence between two probability measures can be easily generalized to a divergence between a sequence of integers n and a point (β,μ) of the statistical manifold $\mathbb{M}$. Our ansatz is

$$D(n||\beta,\mu) = \ln Z(\beta,\mu) - \sum_j n_j(-\beta\epsilon_j + \beta\mu). \tag{3}$$

Minimizing this divergence produces a best fit for the data. If such a best fit β, μ exists we say that it is the orthogonal projection of n on $\mathbb{M}$. The reference to orthogonality is justified by the knowledge that a Pythagorean theorem holds for the divergence (3) — see [4]. Let us analyze in what follows the properties of this minimization procedure.

Derivatives of (3) w.r.t. β and μ can be calculated easily. It follows that, if the minimization procedure has a solution, then it satisfies the pair of equations

$$\sum_j n_j \epsilon_j = \sum_j \frac{\epsilon_j}{\exp(\beta(\epsilon_j - \mu)) - 1}, \tag{4}$$

$$\sum_j n_j = \sum_j \frac{1}{\exp(\beta(\epsilon_j - \mu)) - 1}. \tag{5}$$

These expressions are well-known in statistical physics, see for instance [13], Chap. 10.

A metric tensor $g(\beta,\mu)$ is given by the matrix of second derivatives of $D(n||\beta,\mu)$, evaluated at the minimum. One finds

$$g(\beta,\mu) = \sum_j \frac{\exp(\beta((\epsilon_j - \mu))}{[\exp(\beta(\epsilon_j - \mu)) - 1]^2} \begin{pmatrix} (\epsilon_j - \mu)^2 & -\beta(\epsilon_j - \mu) \\ -\beta(\epsilon_j - \mu) & \beta^2 \end{pmatrix}. \tag{6}$$

It is positive definite and does not depend on the choice of coordinates β, μ of the statistical manifold $\mathbb{M}$, as it should be.

The next step is the introduction of covariant derivatives ∇_a, $a = \beta, \mu$ such that for all n the Hessian of the divergence $D(n||\beta, \mu)$, evaluated at the projection point (β, μ) of n on $\mathbb{M}$, equals the Hessian of a potential $\Phi(\beta, \mu)$. The corresponding connection ω satisfies

$$\nabla_a \partial_b = \omega^c_{ab} \partial_c. \tag{7}$$

The existence of this connection ω shows that the Hessian $\nabla_a \nabla_b D(n||\beta, \mu)$ is constant on the set of all n which project on the point (β, μ) and equals the metric tensor g. This gives the inverse of g the meaning of a Fisher information.

A method for calculating ω is given in [4]. It turns out that all coefficients of ω vanish except $\omega^\mu_{\beta\mu} = 1/\beta$.

3 Quantum Measurements

The quantum analogue of a probability distribution is a density matrix. In the finite-dimensional case this is a positive-definite matrix whose trace equals 1. Its eigenvalues λ_j satisfy $\lambda_j \geq 0$ and $\sum_j \lambda_j = 1$. Hence they can be interpreted as probabilities.

On the other hand the state of the quantum system, in the most simple case, is described by a wave function ψ. This is a normalized element of a Hilbert space $\mathcal{H}$. Let $|\psi\rangle\langle\psi|$ denote the orthogonal projection onto the subspace $\mathbb{C}\psi$. This is a density matrix of rank 1. It is generally believed that a measurement on the quantum system with wave function ψ yields the diagonal part of the matrix $|\psi\rangle\langle\psi|$ in an orthonormal basis the choice of which is dictated by the experimental setup. Let $(e_j)_j$ denote this basis. Then the measured quantities are the numbers $|\langle e_j|\psi\rangle|^2$ (here $\langle\cdot|\cdot\rangle$ is the inner product of the Hilbert space). These are the diagonal elements of the matrix $|\psi\rangle\langle\psi|$. The diagonal part of this projection operator is again a density matrix, which we denote diag($|\psi\rangle\langle\psi|$). It is the result of the experiment.

The map

$$E : |\psi\rangle\langle\psi| \to \text{diag}(|\psi\rangle\langle\psi|) \tag{8}$$

can be seen as a conditioning which is introduced by the experimental setup. Indeed, E is a quantum conditional expectation in the terminology of Petz (Chap. 9 of [6]). See the Appendix A). Petz gives an overview of quantum probability theory as it is known today. The part on conditional expectations originated with the work of Accardi and Cecchini [14] and relies on Tomita-Takesaki theory.

We are interested in the question how the conditioning interferes with the modeling of experimental data using a divergence function. The quantum analogue of the Kullback-Leibler divergence (also called the relative entropy) has density matrices as its arguments. It is given by

$$D(\sigma||\rho) = \text{Tr}\,\sigma \ln \sigma - \text{Tr}\,\sigma \ln \rho. \tag{9}$$

Let $\mathbb{X}$ denote the set of density matrices σ for which $\sigma \ln \sigma$ is trace-class. Let V_σ denote the set of density matrices ρ such that $\mathcal{R}(\sigma) \subset \mathcal{R}(\rho)$ and $\sigma \ln \rho$ is a trace-class operator. The domain of D is then

$$\mathbb{D} = \{(\sigma, \rho) : \sigma \in \mathbb{X}, \rho \in V_\sigma\}. \tag{10}$$

For the sake of completeness we repeat here the following well-known result (see Theorem 5.5 of [15])

Theorem 1. $D(\sigma \| \rho) \geq 0$, with equality if and only if $\sigma = \rho$.

Fix now a model manifold $\mathbb{M}$. It is tradition to work with the quantum analogue of a Boltzmann-Gibbs distribution, which is a probability distribution belonging to the exponential family (see for instance [16]). The parametrized density matrix $\rho_\theta \in \mathbb{M}$ is of the form

$$\rho_\theta = \frac{1}{Z(\theta)} e^{-\theta^k H_k}, \tag{11}$$

with normalization

$$Z(\theta) = \operatorname{Tr} e^{-\theta^k H_k}. \tag{12}$$

The operators H_k are self-adjoint. Together they form the Hamiltonian of the system under consideration. The parameters $\theta^1, \theta^2, \cdots, \theta^n$ are real numbers. Note that Einstein's summation convention is used.

The estimation problem is the question about the optimal choice of the parameters θ^k given the result σ of the experiment. The proposal of information geometry is to use the divergence function (9) to calculate the orthogonal projection ρ_σ of σ onto the model manifold $\mathbb{M} = \{\rho_\theta : \theta \in \Theta\}$. The projection is said to be orthogonal because the following Pythagorean relation holds

$$D(\sigma \| \rho_\theta) = D(\sigma \| \rho_\sigma) + D(\rho_\sigma \| \rho_\theta) \tag{13}$$

holds for all θ in Θ.

Assume now that an experiment is done in the basis $(\psi_n)_n$ in which the elements of $\mathbb{M}$ are diagonal. Let $\sigma_c \equiv \operatorname{diag}(\sigma)$ as before. Then it follows from Theorem 9.3 of [6] that

$$D(\sigma \| \rho) = D(\sigma \| \sigma_c) + D(\sigma_c \| \rho) \quad \text{for all } \rho \in \mathbb{M}. \tag{14}$$

Now fitting the result σ_c of the experiment with elements of $\mathbb{M}$ is equivalent with fitting the unknown density matrix σ because the difference of the two divergences is constant, equal to $D(\sigma \| \sigma_c)$.

4   Weak Measurements

In many recent experiments the actual state of the system, which is described by the density matrix σ, is measured in a basis $(\psi_n)_n$ in which σ is far from diagonal. Many of these experiments involve so-called quantum entangled particles.

They confirm [8] that the Bell inequalities, which are derived using probabilistic arguments (see for instance [17]), can be violated.

Because one knows that the actual state σ is not diagonal one tries to fit a model which is not diagonal as well. In such a case the above argument based on (14) cannot be used. Instead, the conditioning implied by the experimental setup should be included in the modeling of the experiment.

Introduce a conditional manifold

$$\mathbb{M}_c = \{\rho_c : \rho \in \mathbb{M} \text{ and } \rho > 0\}, \quad \text{where } \rho_c \equiv \text{diag}(\rho). \tag{15}$$

The relation (14) then shows that the optimal ρ_c, minimizing the divergence $D(\sigma_c \| \rho_c)$, also minimizes $D(\sigma \| \rho_c)$.

It can happen[1] that $\mathbb{M}_c$ is in the border region of the manifold of positive-definite matrices, where the value of the function $\rho_c \to D(\sigma \| \rho_c)$ can become very large. This is similar to the effect exploited in weak measurements [11], namely that the denominator of the so-called weak value can become very small. See the Appendix B. This suggests that weak measurements can be understood by the behavior of the divergence function $\rho_c \to D(\sigma \| \rho_c)$ in the border region. This idea requires further exploration.

In the more common von Neumann type of experiments the measurement disturbs the quantum system in such a strong manner that the conditioning of the experimental setup also changes the state of the quantum system. Repeating the experiment then reproduces the same outcome as that of the first measurement. This is called the collapse of the wave function. If the outcome ρ_c of the experiment is very sensitive to small changes in the state σ of the quantum system then one can afford to make the interaction between quantum system and measurement apparatus so weak that repeated measurements become feasible. In recent experiments [10] thousands of consecutive measurements were feasible. They reveal a gradual change of the quantum state σ of the system as a consequence of the measurements.

5 Summary

Two situations are described where a divergence function is used with arguments which are *not* probability distributions. In the example of the ideal Bose gas the experimental data are sequences $n = (n_j)_j$ of non-negative integers. It is more natural to take the sequence n as the first argument of the divergence function rather than to introduce an empirical measure concentrating on the data points. In the example of quantum mechanics the arguments are density matrices. The use of density matrices as the arguments of the divergence has been studied extensively in the context of quantum probability.

In the final part of the paper we investigate the use of divergences in the theory of quantum measurements. Our point of view is that any quantum measurement necessarily induces a quantum condition on the range of experimental

---

[1] If $\mathbb{M}_c$ is empty there is not much to tell.

outcomes. The mathematical notion of a quantum conditional expectation is used — see the Appendix A. The recent development of weak quantum measurements is cast into this terminology. The distinction is made between the conditioning of the experimental outcomes, which is unavoidable, and the conditioning of the actual state of the system, which is avoided by the weak measurements. The eventual importance of the border of the manifold of positive definite density matrices is pointed out.

A   Quantum Conditional Expectations

Following Petz (see Chap. 9 of [6]) a **conditional expectation** consists of a subalgebra $\mathcal{A}$ of the algebra $\mathcal{B}$ of bounded linear operators in the Hilbert space $\mathcal{H}$ together with a linear map $E : \mathcal{B} \to \mathcal{A}$. They should satisfy

- $\mathbb{I}$ belongs to $\mathcal{A}$ and $E(\mathbb{I}) = \mathbb{I}$.
- If $A \in \mathcal{A}$ then also $A^\dagger \in \mathcal{A}$.
- If B is positive then also $E(B)$ is positive.
- $E(AB) = AE(B)$ for all $A \in \mathcal{A}$ and $B \in \mathcal{B}$.

Take $B = \mathbb{I}$ in the latter to find that $E(A) = A$ for all A in $\mathcal{A}$.

In the terminology of [6] a density matrix ρ is **preserved** by the conditional expectation $\mathcal{A}, E$ if

$$\mathrm{Tr}\,\rho B = \mathrm{Tr}\,\rho E(B) \tag{16}$$

holds for all b in $\mathcal{B}$.

Now, let be given an orthonormal basis $(\psi_n)_n$ of $\mathcal{H}$. Then any bounded operator B has matrix elements $(\langle \psi_m | B \psi_n \rangle)_{m,n}$. The diagonal part of the operator B is then defined by linear extension of

$$\mathrm{diag}(B)\psi_n = \langle \psi_n | B \psi_n \rangle \psi_n. \tag{17}$$

The map $B \to \mathrm{diag}(B)$, together with the algebra of all diagonal operators is a conditional expectation. In addition, for any density matrix ρ the diagonal part $\rho_c \equiv \mathrm{diag}(\rho)$ is again a density matrix and it is preserved by this conditional expectation. Indeed, one has for all ρ and B

$$\mathrm{Tr}\,\rho_c B = \mathrm{Tr}\,\rho_c \mathrm{diag}(B). \tag{18}$$

B   Weak Measurement Theory

In the seminal paper [11] about weak measurements an experimental setup is proposed. The quantum system contains two parts. The first part is the system of interest. It is weakly coupled to the second part. On the latter von Neumann type measurements are performed to collect data. The subsequent experimental implementations follow the same scheme. See for instance [9,10]. In the present paper only the first part of the experimental setup is considered as the quantum system. The remainder is then considered to be part of the measuring apparatus.

Ref. [12] discusses the notions of pre and post selected states. The density matrix $\sigma = |\psi\rangle \langle\psi|$ of the present paper describes the preselected state. It is the initial state of the experiment transported forward in time using the Schrödinger equation. In a von Neumann type of measurement the post selected state is the state $|\psi_f\rangle \langle\psi_f|$ obtained after the collapse of the wave function, transported backwards in time to the point where it meets the preselected state. The claim of [12] is that the result of the measurement is a so-called weak value of an operator C, which is the operator of the quantum system to which the measurement apparatus couples. This weak value is given by

$$\langle C\rangle = \frac{\langle\psi_f|C\psi\rangle}{\langle\psi_f|\psi\rangle}. \tag{19}$$

It can become arbitrary large by setting up the experiment in such a way that the overlap $|\langle\psi_f|\psi\rangle|^2$ of the pre and post selected states is very small. This theory of weak measurements has been criticized in the literature (see the references in [12]). Additional experiments are needed for its validation.

In the terminology of the present paper the coupling via the operator C induces a conditioning on the outcomes of the experiment.

References

1. Amari, S., Nagaoka, H.: Methods of Information Geometry. Translations of Mathematical Monographs. Oxford University Press, Oxford (2000). Originally in Japanese (Iwanami Shoten, Tokyo, 1993)
2. Naudts, J., Anthonis, B.: Data set models and exponential families in statistical physics and beyond. Mod. Phys. Lett. B **26**, 1250062 (2012)
3. Naudts, J., Anthonis, B.: The exponential family in abstract information theory. In: Nielsen, F., Barbaresco, F. (eds.) GSI 2013. LNCS, vol. 8085, pp. 265–272. Springer, Heidelberg (2013)
4. Anthonis, B.: Extension of information geometry for modeling non-statistical systems. Ph.d thesis. University of Antwerp (2014). arXiv:1501.00853
5. Accardi, L.: Topics in quantum probability. Phys. Rep. **77**, 169–192 (1981)
6. Petz, D.: Quantum Information Theory and Quantum Statistics. Theoretical and Mathematical Physics. Springer, Heidelberg (2008)
7. Nielsen, M.A., Chuang, I.L.: Quantum Computation and Quantum Information. Cambridge University Press, Cambridge (2000)
8. Aspect, A., Grangier, P., Roger, G.: Experimental realization of Einstein-Podolsky-Rosen-Bohm Gedankenexperiment: a new violation of Bell's inequalities. Phys. Rev. Lett. **49**, 91–94 (1982)
9. Ghose, P., Majumdar, A.S., Guhab, S.: Bohmian trajectories for photons. Phys. Lett. A **290**, 205–213 (2001)
10. Guerlin, C., Bernu, J., Deléglise, S., Sayrin, C., Gleyzes, S., Kuhr, S., Brune, M., Raimond, J.-M., Haroche, S.: Progressive field-state collapse and quantum non-demolition photon counting. Nature **448**, 06057 (2007)

11. Aharonov, Y., Albert, D.Z., Vaidman, L.: How the result of a measurement of a component of the spin of a spin-1/2 particle can turn out to be 100. Phys. Rev. Lett. **60**, 1351–1354 (1988)
12. Aharonov, Y., Vaidman, L.: The two-state vector formalism of quantum mechanics an updated review. Lect. Notes Phys. **734**, 399–447 (2007). arXiv:quant-ph/0105101
13. Huang, K.: Introduction to Statical Physics. Taylor & Francis, Boca Raton (2001)
14. Accardi, L., Cecchini, C.: Conditional expectations in von Neumann algebras and a theorem of Takesaki. J. Funct. Anal. **45**, 245–273 (1982)
15. Ohya, M., Petz, D.: Quantum Entropy and Its Use. Springer, Heidelberg (1993). Theoretical and Mathematical Physics
16. Naudts, J.: Generalised Thermostatistics. Springer, Heidelberg (2011)
17. Clauser, J.F., Horne, M.A., Shimony, A., Holt, R.A.: Proposed experiment to test local hidden-variable theories. Phys. Rev. Lett. **23**, 880–884 (1969)

An Information Geometry Problem in Mathematical Finance

Imre Csiszár[1]([⊠]) and Thomas Breuer[2]

[1] MTA Rényi Institute of Mathematics, Budapest, Hungary
csiszar.imre@renyi.mta.hu
[2] PPE Research Centre, FH Vorarlberg, Dornbirn, Austria
tb@fhv.at

Abstract. Familiar approaches to risk and preferences involve minimizing the expectation $E_{\mathbb{P}}(X)$ of a payoff function X over a family Γ of plausible risk factor distributions $\mathbb{P}$. We consider Γ determined by a bound on a convex integral functional of the density of $\mathbb{P}$, thus Γ may be an I-divergence (relative entropy) ball or some other f-divergence ball or Bregman distance ball around a default distribution $\mathbb{P}_0$. Using a Pythagorean identity we show that whether or not a worst case distribution exists (minimizing $E_{\mathbb{P}}(X)$ subject to $\mathbb{P} \in \Gamma$), the almost worst case distributions cluster around an explicitly specified, perhaps incomplete distribution. When Γ is an f-divergence ball, a worst case distribution either exists for any radius, or it does/does not exist for radius less/larger than a critical value. It remains open how far the latter result extends beyond f-divergence balls.

Keywords: Convex integral functional · Bregman distance · f-divergence · I-divergence · Payoff function · Pythagorean identity · Risk measure · Worst case density · Almost worst case densities

1 Preliminaries

Let Ω be any set equipped with a (finite or σ-finite) measure μ on a σ-algebra not mentioned in the sequel. The notation $\mathbb{P}$ will always mean a *distribution* (probability measure) $\mathbb{P} \ll \mu$, with *density* $p = d\mathbb{P}/d\mu$, but p will also denote any nonnegative (measurable) function on Ω. Equality of functions on Ω is meant in the μ-almost everywhere (μ-a.e.) sense.

Let $\mathcal{B}$ denote the family of functions $\beta(\omega, s)$ on $\Omega \times \mathbb{R}$, measurable in ω for each $s \in \mathbb{R}$, strictly convex and differentiable in s on $(0, +\infty)$ for each $\omega \in \Omega$, and satisfying

$$\beta(\omega, 0) = \lim_{s \downarrow 0} \beta(\omega, s), \quad \beta(\omega, s) := +\infty \text{ if } s < 0. \tag{1}$$

ICs acknowledges support by the Hungarian National Foundation for Scientific Research OTKA, grant No. K105840. TB acknowledges support by the Josef Ressel Centre for Applied Scientific Computing.

© Springer International Publishing Switzerland 2015
F. Nielsen and F. Barbaresco (Eds.): GSI 2015, LNCS 9389, pp. 435–443, 2015.
DOI: 10.1007/978-3-319-25040-3_47

Note that the functions $\beta \in \mathcal{B}$ are (convex) normal integrands, which takes care of measurability questions not entered below, see [13].

For $\beta \in \mathcal{B}$ define the integral functional

$$H(p) = H_\beta(p) := \int_\Omega \beta(\omega, p(\omega))\mu(d\omega). \tag{2}$$

Assume that for a *default distribution* $\mathbb{P}_0$ with density p_0

$$H(p) \geq H(p_0) = 0 \quad \text{whenever} \int pd\mu = 1. \tag{3}$$

Fix a measurable function X such that

$$E_{\mathbb{P}_0}(X) = \int_\Omega X(\omega)p_0(\omega)\mu(d\omega) =: b_0 \text{ exists}, \ m < b_0 < M, \tag{4}$$

where m and M denote the μ-ess inf and μ-ess sup of X.

We are interested in minimizing $E_{\mathbb{P}}(X)$ subject to $\mathbb{P} \in \Gamma$ when

$$\Gamma = \{\mathbb{P} : H(p) \leq k\}. \tag{5}$$

Thus, we address the problem

$$V(k) := \inf_{p:\int pd\mu=1, H(p)\leq k} \int Xpd\mu. \tag{6}$$

In mathematical finance, X is a *payoff function* depending on a collection $\omega \in \Omega$ of random risk factors. The risk factor distribution is unknown but assumed to belong to a known family Γ of plausible distributions. Then $\inf_{\mathbb{P} \in \Gamma} E_{\mathbb{P}}(X)$ measures (the negative of) the risk of a financial position with payoff function X. The same expression arises also in the theory of ambiguity averse preferences. For details, including axiomatic considerations, we refer to Föllmer and Schied [8], Hansen and Sargent [11], or Gilboa [9].

It is natural to consider those distributions $\mathbb{P}$ plausible that do not deviate much from the default distribution $\mathbb{P}_0$. This still admits many choices for Γ, according to what measure of deviation is used. The setting (5), introduced in Breuer and Csiszár [4], appears to cover most choices of interest.

Example 1. Take $\mu = \mathbb{P}_0$, thus $p_0 \equiv 1$, and let $\beta(\omega, s) = f(s)$ be an autonomous convex integrand, with $f(1) = 0$ to ensure (3). Then $H(p)$ in (2) for $p = \frac{d\mathbb{P}}{d\mu}$ is the f-*divergence* $D_f(\mathbb{P} \| \mathbb{P}_0)$, introduced in Csiszár [5], and (5) gives the f-divergence ball $\{\mathbb{P} : D_f(\mathbb{P} \| \mathbb{P}_0) \leq k\}$.

Example 2. Let f be any strictly convex and differentiable function on $(0, +\infty)$, and for $s \geq 0$ let $\beta(\omega, s) = \Delta_f(s, p_0(\omega))$. Here

$$\Delta_f(s, t) := f(s) - f(t) - f'(t)(s - t), \tag{7}$$

where $f(0)$ and $f'(0)$ are defined as limits, and if $f(0) = +\infty$, we set $\Delta_f(s,0) := 0$ for $s = 0$ and $\Delta_f(s,0) := \infty$ otherwise.

In this example $\mathbb{P}_0 \ll \mu$ is arbitrary, except that in case $f'(0) = -\infty$ we assume that $p_0 > 0$ μ-a.e.. Then $H(p)$ equals the Bregman distance [3]

$$B_{f,\mu}(p, p_0) := \int_\Omega \Delta_f(p(\omega), p_0(\omega)) \mu(d\omega), \tag{8}$$

and (5) gives the Bregman ball of radius k around $\mathbb{P}_0$.

In the special case $f(s) = s \log s$ both examples give as $H(p)$ for $p = d\mathbb{P}/d\mu$ the I-divergence (relative entropy) $D(\mathbb{P} \| \mathbb{P}_0) := \int p \log \frac{p}{p_0} d\mu$.

In the context of risk and preferences, I-divergence balls were used perhaps first by Hansen and Sargent [10]. Ahmadi-Javid [1] showed the corresponding risk measure, which he called *entropic value at risk*, preferable to others from the point of view of computability, and Strzalecki [14] distinguished it axiomatically. General f-divergences are used in Maccheroni *et al.*[12] and Ben-Tal and Teboulle [2], see also references in [2]. Bregman distances could be used similarly but to this we do not have references.

2 Basic Facts

Define the function

$$F(b) := \inf_{p: \int p d\mu = 1, \int X p d\mu = b} H(p). \tag{9}$$

It is convex, with minimum 0 attained at $b = b_0$, see (3) and (4).

We adopt as standing assumption, in addition to (3) and (4), that

$$k_{\max} := \lim_{b \downarrow m} F(b) > 0. \tag{10}$$

The problem (6) is nontrivial only if (10) holds and $k \in (0, k_{\max})$. Note that $k_{\max} = +\infty$ if $m = -\infty$ (subject to (10)), and $k_{\max} \le F(m)$ if m is finite (strict inequality is possible). See also Remark 1 later in this Section.

Lemma 1. *[4, Proposition 3.1] To each $k \in (0, k_{\max})$ there exists a unique $b \in (m, b_0)$ with $F(b) = k$, and then $V(k) = b$. A density p attains the minimum in (6) if and only if it attains that in (9) (for the above b)* (Fig. 1).

Lemma 1 relates problem (6) to the problem extensively studied in information geometry of minimizing a convex integral functional under moment constraints, specifically with moment mapping $\phi(\omega) := (1, X(\omega))$. We will rely upon results in Csiszár and Matúš [7] specified for this moment mapping. For proofs omitted below, see the full version of this paper [6].

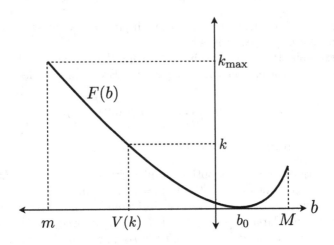

Fig. 1. Lemma 1 relates the solution of the risk problem (6) to the solution of the information geometry problem (9): $F(V(k)) = k$.

For $\phi(\omega)$ above, the value function in [7] is the convex function

$$J(a,b) := \inf_{p:\int pd\mu=a,\int Xpd\mu=b} H(p), \tag{11}$$

thus $F(b) = J(1,b)$.

A crucial fact is the instance of [7, Theorem 1.1] that the convex conjugate $J^*(\theta_1,\theta_2) := \sup_{a,b}[\theta_1 a + \theta_2 b - J(a,b)]$ of J in (11) equals

$$K(\theta_1,\theta_2) := \int \beta^*(\omega, \theta_1 + \theta_2 X(\omega))\mu(d\omega), \tag{12}$$

$$\beta^*(\omega,r) := \sup_{s\in\mathbb{R}} (sr - \beta(\omega,s)). \tag{13}$$

Since J may differ from $J^{**} = K^*$ only on the boundary of its effective domain, it follows that at least for $b \notin \{m, M\}$

$$F(b) = J(1,b) = K^*(1,b) = \sup_{\theta_1,\theta_2}[\theta_1 + \theta_2 b - K(\theta_1,\theta_2)], \tag{14}$$

or equivalently

$$F(b) = G^*(b) \quad \text{where} \quad G(\theta_2) := \inf_{\theta_1} [K(\theta_1,\theta_2) - \theta_1]. \tag{15}$$

The following family of non-negative functions on Ω plays a key role like exponential families do for I-divergence minimization:

$$p_{\theta_1,\theta_2}(\omega) := (\beta^*)'(\omega, \theta_1 + \theta_2 X(\omega)), \quad (\theta_1,\theta_2) \in \Theta; \tag{16}$$

$$\Theta := \{(\theta_1,\theta_2) \in \mathrm{dom}\, K : \theta_1 + \theta_2 X(\omega) < \beta'(\omega, +\infty) \quad \mu\text{-a.e.}\}. \tag{17}$$

The derivatives of β and β^* are by the second variable. Note that $\beta^*(\omega, r)$ equals $+\infty$ if $r > \beta'(\omega, +\infty)$, while it is finite and differentiable if $r < \beta'(\omega, +\infty)$. Its derivative (for fixed ω) equals 0 if $r \leq \beta'(\omega, 0)$ and is positive, growing to $+\infty$ for $r \in (\beta'(\omega, 0), (\beta'(\omega, +\infty))$.

The projection to the θ_2-axis of the set Θ in (17) coincides with that of dom $K := \{(\theta_1, \theta_2) : K(\theta_1, \theta_2) < +\infty\}$. This projection is an interval, it will be denoted by Θ_2, and its left endpoint by $\theta_{\min}$ (which may be $-\infty$).

Remark 1. As shown in [4], the interval Θ_2 contains the origin, and the default density p_0 belongs to the family (16) with $\theta_2 = 0$. Moreover, the standing assumption $k_{\max} > 0$ is equivalent to $\theta_{\min} < 0$. This implies, in turn, that $F(b) > 0$ for each $b < b_0$.

The following lemma gives relevant information about evaluating the function G in (15). In turn, from G one can calculate the desired $V(k)$ in (6), as the next Theorem shows. Clearly, dom $G := \{\theta_2 : G(\theta_2) < +\infty\} = \Theta_2$.

Lemma 2. *To each $\theta_2 \in \Theta_2$ there exists a unique $\theta_1 = \theta_1(\theta_2)$ with $K(\theta_1, \theta_2) - \theta_1 = G(\theta_2)$. Either $\tilde{\theta}_1 := \sup\{\theta_1 : (\theta_1, \theta_2) \in \text{dom } K\}$ is finite and satisfies $(\tilde{\theta}_1, \theta_2) \in \Theta$, $\int p_{\tilde{\theta}_1, \theta_2} d\mu < 1$, then $\theta_1(\theta_2) = \tilde{\theta}_1$, or else a unique θ_1 satisfies $(\theta_1, \theta_2) \in \Theta$, $\int p_{\theta_1, \theta_2} d\mu = 1$, then $\theta_1(\theta_2)$ equals this θ_1.*

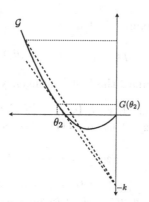

Fig. 2. Among straight lines passing through $(0, -k)$ and some point $(\theta_2, G(\theta_2))$ with $\theta_2 < 0$, the supporting line to the graph of G has maximum slope $(k + G(\theta_2))/\theta_2$. By Theorem 1, the infimum $V(k)$ in (6) is equal to the slope of this supporting line.

Theorem 1. *For $k \in (0, k_{\max})$*

$$V(k) = \max_{\theta_2 < 0} \frac{k + G(\theta_2)}{\theta_2}. \tag{18}$$

A maximizer in (18) is equivalently a maximizer of $\theta_2 b - G(\theta_2)$ where $b = V(k)$, and (θ_1, θ_2) with $\theta_1 = \theta_1(\theta_2)$ as in Lemma 2 is a maximizer in (14) (Fig. 2).

An identity equivalent to (18) appears, for autonomous integrands and bounded payoff functions, in Ahmadi-Javid [1, Theorem 5.1].

3 Main Results

For the problem (6), call a density p an $(\epsilon\text{-}\gamma)$-*Almost-Worst-Case-Density* (AWCD), where $\epsilon \geq 0$, $\gamma \geq 0$, if

$$H(p) \leq k + \gamma \quad \text{and} \quad \int X p d\mu \leq V(k) + \epsilon. \tag{19}$$

A *worst case density* (WCD) attaining the minimum in (6) is a (0-0)-AWCD.

Theorem 2 below establishes a clustering property of the $(\epsilon\text{-}\gamma)$-AWCDs. From a practical point of view, this may be relevant for efficient hedging against the almost worst scenarios, but this issue is not entered here.

An extension of the concept of Bregman distance is needed. For any $\beta \in \mathcal{B}$ define $\Delta_{\beta(\omega,\cdot)}(s,t)$ as in Example 2, with $\beta(\omega,\cdot) : s \mapsto \beta(\omega, s)$ in the role of f. Then, denoting $\Delta_{\beta(\omega,\cdot)}(p(\omega), q(\omega))$ briefly by $\Delta_\beta(p,q)$, define

$$B(p,q) = B_\beta(p,q) := \int \Delta_\beta(p,q)d\mu. \tag{20}$$

For $\theta_2 \in \Theta_2$ and $\theta_1(\theta_2)$ in Lemma 2, denote

$$q_{\theta_2} := p_{\theta_1,\theta_2}, \qquad \theta_1 = \theta_1(\theta_2). \tag{21}$$

It is a density unless in Lemma 2 the first contingency takes place.

Theorem 2. *For* $k \in (0, k_{\max})$, *each* $(\epsilon\text{-}\gamma)$-*AWCD* p *belongs to the Bregman neighborhood of radius* $(\gamma - \theta_2 \epsilon)$ *of* q_{θ_2} *in* (21) *with* $\theta_2 < 0$ *attaining the maximum in* (18), *i.e.*,

$$B(p, q_{\theta_2}) \leq \gamma - \theta_2 \epsilon \quad \text{if } p \text{ is an } (\epsilon\text{-}\gamma)\text{-AWCD}. \tag{22}$$

Corollary 1. *Each sequence* $\{p_n\}$ *of* $(\epsilon_n\text{-}\gamma_n)$-*AWCDs with* $\epsilon_n \to 0$, $\gamma_n \to 0$ *converges to* q_{θ_2} *locally in measure.*[1] *In particular,* q_{V_2} *is uniquely determined.*

Proof. By [7, Lemma 4.15], combined with [7, Remark 4.13], we have for each density p with $\int X p d\mu$ finite, and each $(\theta_1, \theta_2) \in \Theta$,

$$H(p) = \theta_1 + \theta_2 \int X p d\mu - K(\theta_1, \theta_2) + B(p, p_{\theta_1,\theta_2})$$

$$+ \int |\beta'(\omega, 0) - \theta_1 - \theta_2 X(\omega)|_+ p(\omega)\mu(d\omega). \tag{23}$$

---

[1] This means that $\mu(\{\omega \in C : |p_n(\omega) - q_{\theta_2}(\omega)| > \epsilon\}) \to 0$ for each $C \subset \Omega$ with $\mu(C)$ finite, and any $\epsilon > 0$. If μ is a finite measure, this is equivalent to standard (global) convergence in measure.

For θ_2 in Theorem 2 and $\theta_1 = \theta_1(\theta_2)$, here $p_{\theta_1,\theta_2} = q_{\theta_2}$ and $K(\theta_1,\theta_2) - \theta_1 = G(\theta_2) = \theta_2 V(k) - k$. see Theorem 1. Substituting this into (23) gives

$$H(p) = k + \theta_2 \left(\int Xp d\mu - V(k) \right) + B(p, q_{V_2})$$

$$+ \int |\beta'(\omega,0) - \theta_1 - \theta_2 X(\omega)|_+ p(\omega) \mu(d\omega). \tag{24}$$

Clearly, (24) and (19) imply (22).

The Corollary follows since $B(p_n, q_{\theta_2}) \to 0$ implies convergence of p_n to q_{V_2} locally in measure [7, Corollary 2.14].

The function $q_{\theta_2} = p_{\theta_1,\theta_2}$ in Theorem 2, uniquely determined by k according to Corollary 1, will be denoted below by $\hat{q}_k$.

Remark 2. Corollary 1 extends the known result that $\hat{q}_k$ is a *generalized solution* of problem (9) in the sense [7] that densities p_n with $\int Xp_n d\mu = b = V(k)$, $H(p_n) \to F(b) = k$ converge to $\hat{q}_k$ locally in measure, and also establishes its (new) counterpart for problem (6). The above proof uses the idea of [7], the key identity (24) is an extension of the *generalized Pythagorean identity* [7, Lemma 4.16] (in case of the currently addressed moment mapping ϕ) to densities p that need not satisfy $\int Xp d\mu = b$.

By Theorem 2, if p is a WCD then $p = q_{\theta_2} = \hat{q}_k$. Hence, necessary conditions of existence of a WCD are $\int p_{\theta_1,\theta_2} d\mu = 1$ and $\int Xp_{\theta_1,\theta_2} d\mu = V(k)$ for (θ_1,θ_2) with $p_{\theta_1,\theta_2} = \hat{q}_k$. Of course, if these conditions hold for some $(\theta_1,\theta_2) \in \Theta$ then p_{θ_1,θ_2} is a WCD. The mentioned conditions can be seen to hold for $\hat{q}_k = p_{\theta_1,\theta_2}$ if $(\theta_1,\theta_2) \in \text{int dom } K$, but if (θ_1,θ_2) is on the boundary of dom K then $\hat{q}_k = p_{\theta_1,\theta_2}$ is typically not a WCD.

Theorem 3. *(i) If $\hat{q}_k$ is a density then it is also a WCD for k, except when $\theta_{\min}$ and $G'(\theta_{\min})$ are finite and*

$$k > k_{\text{cr}} := -G(\theta_{\min}) + \theta_{\min} G'(\theta_{\min}). \tag{25}$$

In the latter case $\hat{q}_k = \hat{q}_{k_{\text{cr}}}$ and no WCD exists for k.

(ii) A sufficient condition for $\hat{q}_k$ to be a density for each $k \in (0, k_{\max})$ is the finiteness of $K(\theta_1, 0) = \int \beta^*(\omega, \theta_1) \mu(d\omega)$ for each $\theta_1 \in \mathbb{R}$.

The condition in (ii) holds, e.g., when $H(p)$ is an f-divergence, see Example 1, with *cofinite* f, i.e., $f'(+\infty) = +\infty$. Hence, for f-divergences with cofinite f, a WCD either exists for each $k \in (0, k_{\max})$ or it exists/does not exist for k not exceeding/exceeding a critical value. In the case of I-divergence ($f(s) = s \log s$) this phenomenon has been pointed out in [4].

Our final theorem establishes a similar phenomenon for non-cofinite f (but then $\hat{q}_k$ may fail to be a density, and may depend on k also above the critical value). It remains open how far this phenomenon extends beyond f-divergences, but see Example 4 below.

For $\beta(\omega, s) = f(s)$ with non-cofinite f, the standing assumption $k_{\max} > 0$ (equivalent to $\theta_{\min} < 0$) holds if and only if $m > -\infty$. With no loss of generality, assume that $m = 0$. Then $K(\theta_1, \theta_2) = \int f^*(\theta_1 + \theta_2 X) d\mu$ for $\theta_2 < 0$ is finite/infinite if θ_1 is less/larger than $f'(+\infty)$, thus $\tilde{\theta}_1$ in Lemma 2 is equal to $f'(+\infty)$. Hence, see (21), $q_{\theta_2} = p_{\theta_1, \theta_2}$ is a density if and only if

$$g(\theta_2) := \int (f^*)'(c + \theta_2 X) d\mu \geq 1, \quad c := f'(+\infty), \tag{26}$$

and Lemma 2 also gives that

$$q_{\theta_2} = p_{c, \theta_2} = (f^*)'(c + \theta_2 X) \quad \text{if} \quad g(\theta_2) \leq 1. \tag{27}$$

Theorem 4. *When $H(p)$ is an f-divergence with $f'(+\infty) = c < +\infty$, and $m = 0$, the WCD exists for all $k \in (0, k_{\max})$ if $g(\theta_2) = +\infty$ for each $\theta_2 < 0$. Otherwise, denote*

$$\tilde{\theta}_{\min} := \inf\{\theta_2 : g(\theta_2) \geq 1\}, \tag{28}$$

$$\tilde{k}_{\mathrm{cr}} := \tilde{\theta}_{\min} G'_+(\tilde{\theta}_{\min}) - G(\tilde{\theta}_{\min}). \tag{29}$$

Then $\tilde{\theta}_{\min} \in (-\infty, 0)$, $\tilde{k}_{\mathrm{cr}} \in (0, k_{\max})$, and for $k < \tilde{k}_{\mathrm{cr}}$ the WCD exists. For $k \geq \tilde{k}_{\mathrm{cr}}$ the function $\hat{q}_k$ is of form (27), and it is not a density if $k > \tilde{k}_{\mathrm{cr}}$,

Example 3. In the setting of Example 1 with $\mu = \mathbb{P}_0$, take $f(s) = -\log s$. Then $H(p) = D(\mathbb{P}_0 \| \mathbb{P})$ for $p = d\mathbb{P}/d\mathbb{P}_0$.

Take specifically $\Omega = (0, 1)$, $X(\omega) = \omega$, and let $\mu = \mathbb{P}_0$ be the distribution with Lebesgue density 2ω. As $f^*(r) = -1 - \log(-r)$ $(r < 0)$, then $K(\theta_1, \theta_2) = \int_0^1 [-1 - \log(-\theta_1 - \theta_2 \omega)] 2\omega d\omega$ and $p_{\theta_1, \theta_2}(\omega) = 1/(-\theta_1 - \theta_2 \omega)$ for $(\theta_1, \theta_2) \in \Theta = \mathrm{dom}\, K = \{(\theta_1, \theta_2) : \theta_1 \leq 0, \theta_1 + \theta_2 < 0\}$. Simple calculus shows that in Lemma 2 the second contingency holds if $-2 \leq \theta_2 < 0$, hence in that case $q_{\theta_2} = p_{\theta_1, \theta_2}$ with $\theta_1 = \theta_1(\theta_2)$ is a density, but this and $G(\theta_2)$ can not be given explicitly. For $\theta_2 < -2$, in Lemma 2 the first contingency holds, thus $\theta_1(\theta_2) = 0$, $q_{\theta_2}(\omega) = (-\theta_2 \omega)^{-1}$, and $G(\theta_2) = -\frac{1}{2} - \log(-\theta_2)$. One sees that $H(q_{\theta_2})$ ranges from 0 to $\log 2 - 1/2$ as θ_2 ranges from 0 to -2. Hence the function $q_{\theta_2}(\omega) = \hat{q}_k(\omega)$ in Theorem 2 equals the WCD if $0 < k \leq \log 2 - 1/2$, while if $k \geq \log 2 - 1/2 = \tilde{k}_{\mathrm{cr}}$ then q_{θ_2} equals $(-\theta_2 \omega)^{-1}$, with $\theta_2 \leq -2$ attaining $V(k) = \max_{\theta_2 < 0} [k + G(\theta_2)]/\theta_2$. Calculus gives that this maximum is attained by $\theta_2 = -e^{k+1/2}$ and equals $V(k) = e^{-(k+1/2)}$, and one sees that $\hat{q}_k(\omega) = e^{-(k+1/2)} \omega^{-1}$ is not a density unless $k = \log 2 - 1/2$.

Example 4. Let Ω, X, μ be as in Example 3, but this time let the default distribution $\mathbb{P}_0$ be the uniform distribution whose μ-density is $p_0(\omega) = \frac{1}{2\omega}$. Take $H(p) = B_{f, \mu}(p, p_0)$, see (8), with $f(s) = -\log s$, i.e., the integral functional (2) with $\beta(\omega, s) = \Delta_f(s, p_0(\omega)) = -\log s - \log(2\omega) + 2\omega(s - \frac{1}{2\omega})$. Then $\beta^*(\omega, r) = \log 2\omega - \log(-r + 2\omega)$, $(\beta^*)'(\omega, r) = 1/(-r + 2\omega)$, $r < 2\omega$. The set Θ (equal to $\mathrm{dom}\, K$) of this example consists of those (θ_1, θ_2) for which $(\theta_1, \theta_2 - \theta_1)$ belongs to the set $\Theta = \mathrm{dom}\, K$ of Example 3. Moreover, for such (θ_1, θ_2) the function $p_{\theta_1, \theta_2}(\omega) = 1/[-\theta_1 - (\theta_2 - 2)\omega]$ coincides with the function $p_{\theta_1, \theta_2 - 2}$ of Example 3, which can not be a density if $\theta_2 < 0$. Hence, in the present Example no WCD exists for any $k > 0$.

References

1. Ahmadi-Javid, A.: Entropic value at risk: a new coherent risk measure. J. Optim. Theor. Appl. **155**(3), 1105–1123 (2011)
2. Ben-Tal, A., Teboulle, M.: An old-new concept of convex risk measures: the optimized certainty equivalent. Math. Finance **17**, 449–476 (2007)
3. Bregman, L.M.: The relaxation method of finding the common point of convex sets and its application to the solution of problems in convex programming. USSR Comput. Math. Math. Phys. **7**, 200–217 (1967)
4. Breuer, T., Csiszár, I.: Measuring distribution model risk. Math. Finance (2013). doi:10.1111/mafi.12050
5. Csiszár, I.: Eine informationstheoretische Ungleichung und ihre Anwendung auf den Beweis der Ergodizität von Markoffschen Ketten. Publ. Math. Insti. Hung. Acad. Sci. **8**, 85–108 (1963)
6. Imre Csiszár and Thomas Breuer. Almost worst case distributions in multiple priors models. http://arxiv.org/abs/1506.01619 (2015)
7. Csiszár, I., Matúš, F.: On minimization of entropy functionals under moment constraints. Kybernetika **48**, 637–689 (2012)
8. Föllmer, H., Schied, A.: Stochastic Finance: An Introduction in Discrete Time. de Gruyter Studies in Mathematics, vol. 27, 2nd edn. Walter de Gruyter, Berlin (2004)
9. Gilboa, I.: Theory of Decision Under Uncertainty. Econometric Society Monographs, vol. 45. Cambridge University Press, Cambridge (2009)
10. Hansen, L.P., Sargent, T.: Robust control and model uncertainty. Am. Econ. Rev. **91**, 60–66 (2001)
11. Hansen, L.P., Sargent, T.: Robustness. Princeton University Press, Princeton (2008)
12. Maccheroni, F., Marinacci, M., Rustichini, A.: Ambiguity aversion, robustness, and the variational representation of preferences. Econometrica **74**, 1447–1498 (2006)
13. Rockafellar, R.T., Wets, R.J.-B.: Variational Analysis. Grundlehren der Mathematischen Wissenschaften, vol. 317. Springer, Heidelberg (1998)
14. Strzalecki, T.: Axiomatic foundations of multiplier preferences. Econometrica **79**, 47–73 (2011)

Multivariate Divergences with Application in Multisample Density Ratio Models

Amor Keziou$^{(\boxtimes)}$

Laboratoire de Mathématiques de Reims, EA 4535, UFR Sciences Exactes et
Naturelles, Moulin de la Housse, BP1039, 51687 Remis Cedex, France
amor.keziou@univ-reims.fr

Abstract. We introduce what we will call multivariate divergences
between K, $K \geq 1$, signed finite measures $(Q_1, \ldots, Q_K)$ and a given
reference probability measure P on a σ-field $(\mathcal{X}, \mathcal{B})$, extending the well
known divergences between two measures, a signed finite measure Q_1
and a given probability distribution P. We investigate the Fenchel duality theory for the introduced multivariate divergences viewed as convex
functionals on well chosen topological vector spaces of signed finite measures. We obtain new dual representations of these criteria, which we
will use to define new family of estimates and test statistics with multiple samples under multiple semiparametric density ratio models. This
family contains the estimate and test statistic obtained through empirical
likelihood. Moreover, the present approach allows obtaining the asymptotic properties of the estimates and test statistics both under the model
and under misspecification. This leads to accurate approximations of the
power function for any used criterion, including the empirical likelihood
one, which is of its own interest. Moreover, the proposed multivariate
divergences can be used, in the context of multiple samples in density
ratio models, to define new criteria for model selection and multi-group
classification.

1 Introduction, Notation and Motivation

On a measurable space $(\mathcal{X}, \mathcal{B})$, let P be a given reference probability measure
(p.m.). Denote by $\mathcal{M}$ the real vector space of all signed finite measures (s.f.m.)
on $(\mathcal{X}, \mathcal{B})$. Let

$$\psi(\cdot) : \mathbf{x} := (x_1, \ldots, x_K)^\top \in \mathbb{R}^K \mapsto \psi(\mathbf{x}) \in [0, +\infty] \tag{1}$$

be a nonnegative closed (equivalently lower semi-continuous) convex function on
$\mathbb{R}^K$, satisfying $\psi(\mathbf{1}_K) = 0$, where $\mathbf{1}_K := (1, \ldots, 1)^\top \in \mathbb{R}^K$, and such that (s.t.)
$\mathbf{1}_K$ is an interior point of its domain $\mathrm{dom}_\psi := \{\mathbf{x} \in \mathbb{R}^K \text{ s.t. } |\psi(\mathbf{x})| < \infty\}$. We
define the ψ-divergence between K-signed finite measures $\mathbf{Q} := (Q_1, \ldots, Q_K)$
and a given p.m. P through

$$D_\psi(\mathbf{Q}; P) := \int_{\mathcal{X}} \psi\left(\frac{dQ_1}{dP}(x), \ldots, \frac{dQ_K}{dP}(x)\right) dP(x), \tag{2}$$

© Springer International Publishing Switzerland 2015
F. Nielsen and F. Barbaresco (Eds.): GSI 2015, LNCS 9389, pp. 444–453, 2015.
DOI: 10.1007/978-3-319-25040-3_48

if $Q_1, \ldots, Q_K$ are absolutely continuous (a.c.) with respect to (w.r.t.) P; If there exists Q_k which is not a.c.w.r.t. P, we set $D_\psi(\mathbf{Q}; P) = +\infty$. In the above integral, $\frac{dQ_k}{dP}(\cdot)$ is the Radon-Nikodym derivative of Q_k w.r.t. P. Obviously, if $K = 1$, we obtain the classical definition of φ-divergences between two measures (a signed finite measure Q_1 and a probability measure P). It is clear that any ψ-divergence is nonnegative, and that the functional $\mathbf{Q} \in \mathcal{M}^K \mapsto D_\psi(\mathbf{Q}; P)$ is convex on the product vector space $\mathcal{M}^K$, for any given p.m. P. Moreover, if the function $\psi(\cdot)$ is strictly convex on a neighborhood of the vector $\mathbf{1}_K$, then we can prove that

$$D_\psi(\mathbf{Q}; P) = 0 \text{ if and only if } Q_1 = \cdots = Q_K = P. \tag{3}$$

In this paper, we investigate statistical inference for multiple density ratio models (DRM) of [1]. Suppose we have $1 + K$ random samples

$$(X_{0,1}, \ldots, X_{0,n_0}), (X_{1,1}, \ldots, X_{1,n_1}), \ldots, (X_{K,1}, \ldots, X_{K,n_K}), \tag{4}$$

of $1 + K$ random variables (with values in $\mathbb{R}^m, m \geq 1$) $X_0, X_1, \ldots, X_K$, with distributions $P, Q_1, \ldots, Q_K$, respectively. The DRM postulates that

$$\log\left(\frac{dQ_k}{dP}(x)\right) = \boldsymbol{\theta}_k^\top \boldsymbol{h}_k(x), \quad \text{for all} \quad k = 1, \ldots, K, \tag{5}$$

for some known vector valued function $\boldsymbol{h}_k(x) := (h_{k,0}(x), h_{k,1}(x), \ldots, h_{k,d_K}(x))^\top$, $x \in \mathbb{R}^m$, with values in $\mathbb{R}^{1+d_K}$, and corresponding unknown vector-valued parameters $\boldsymbol{\theta}_k := (\alpha_k, \boldsymbol{\beta}_k^\top)^\top \in \Theta_k \subset \mathbb{R}^{1+d_k}$. We require, for all $k = 1, \ldots, K$, the identifiability of the parameter $\boldsymbol{\theta}_k$, in the sense that

$$\text{if} \quad \boldsymbol{\theta}_k^\top \boldsymbol{h}_k(x) = \boldsymbol{\theta}_k'^\top \boldsymbol{h}_k(x) \quad \forall x - P\text{-a.s.,} \quad \text{then} \quad \boldsymbol{\theta}_k = \boldsymbol{\theta}_k'. \tag{6}$$

We will denote then by $\boldsymbol{\theta}_T := \left(\boldsymbol{\theta}_{1T}^\top, \ldots, \boldsymbol{\theta}_{KT}^\top\right)^\top$ the true value of the parameter $\boldsymbol{\theta} := \left(\boldsymbol{\theta}_1^\top, \ldots, \boldsymbol{\theta}_K^\top\right)^\top$. We require also the first component of $\boldsymbol{h}_k(\cdot)$, for each k, to be the constant one, i.e., $h_{k,0}(\cdot) := \mathbb{1}_{\mathbb{R}^m}(\cdot)$, the indicator function of $\mathbb{R}^m$. This makes the corresponding first real true value α_{kT}, of the vector $\boldsymbol{\theta}_{kT} := (\alpha_{kT}, \boldsymbol{\beta}_{kT}^\top)^\top$, a normalization parameter, for all $k = 1 \ldots, K$. In particular, we have $\alpha_{kT} = 0$, whenever $\boldsymbol{\beta}_{kT} = \mathbf{0}$. We will assume that, for all $k = 1, \ldots, K$, the interior $\text{int}(\Theta_k)$ of the parameter space Θ_k in $\mathbb{R}^{1+d_k}$ is not void. The function $\boldsymbol{h}_k(\cdot)$ in (5) may depend on k, with possibly different dimensions $d_1, \ldots, d_K$. Observe also that, for the DRM, the distribution P is unspecified. Only the density ratios are specified through the vector valued functions $\boldsymbol{h}_k(\cdot)$. The components of $\boldsymbol{h}_k(\cdot)$, for each k, may be chosen, for instance, to be the first $1 + d_k$ elements of some basis functions. For notational convenience, we will denote $q_k(\boldsymbol{\theta}_k, x) := \exp\left\{\boldsymbol{\theta}_k^\top \boldsymbol{h}_k(x)\right\}, \forall x \in \mathbb{R}^m, \forall \boldsymbol{\theta}_k \in \Theta_k, \forall k = 1, \ldots, K$. The DRM writes then $\frac{dQ_k}{dP}(x) = q_k(\boldsymbol{\theta}_k, x), x \in \mathbb{R}^m, \boldsymbol{\theta}_k := (\alpha_k, \boldsymbol{\beta}_K^\top)^\top \in \Theta_k \subset \mathbb{R}^{1+d_k}, k = 1, \ldots, K$. Consider the test problem of the null hypothesis

$$\mathcal{H}_0 : Q_1 = \cdots = Q_K = P \quad \text{against the alternative} \quad \mathcal{H}_1 : \exists Q_k \neq P, \tag{7}$$

which is equivalent, in the context of the DRM, to testing $\mathcal{H}_0 : \theta_{1T} = \cdots = \theta_{KT} = \mathbf{0}$ against the alternative $\mathcal{H}_1 : \exists \theta_{kT} \neq \mathbf{0}$. In view of the basic property (3), in order to build a statistical test of the above hypotheses, we can use an estimate of $D_\psi(Q; P)$, and decide to reject the null hypothesis $\mathcal{H}_0$ when the obtained estimate takes large value. The plug-in estimate of $D_\psi(Q; P)$ writes $\widetilde{D}_\psi(Q; P) := D_\psi(\widehat{Q}_1, \ldots, \widehat{Q}_K; \widehat{P})$, obtained from the definition (2) of the ψ-divergence by replacing the distribution $P, Q_1, \ldots, Q_K$ by their corresponding empirical measures

$$\widehat{P} := \frac{1}{n_0} \sum_{i=1}^{X_{0,i}} \delta_{X_{0,i}}, \widehat{Q}_1 := \frac{1}{n_1} \sum_{i=1}^{n_1} \delta_{X_{1,i}}, \ldots, \widehat{Q}_K := \frac{1}{n_K} \sum_{i=1}^{n_K} \delta_{X_{K,i}}, \quad (8)$$

associated to the K samples (4). Unfortunately, the above estimate is not generally well defined, since, the absolute continuity condition is not satisfied (even for large sample sizes, if the distributions are continuous, or discrete with unbounded support). Moreover, the plug-in estimate does not use the semiparametric form of the DRM. In the following section, we will give a new dual representation for the ψ-divergences, extending to multivariate case the dual representations obtained in [2,6], and we will show how to use the obtained dual representation to give well defined estimates using the form of the DRM, in the present multiple sample context. We will use the Fenchel duality theory for the convex functionals $Q \mapsto D_\psi(Q; P)$ on well chosen product topological vector space of signed measures to be specified in the following sections. For more details and proofs, we refer to [7].

2   Fenchel Duality of Convex Functions on Finite Dimensional Vector Spaces

We can refer to [11] for the proof of the following duality theorem on $\mathbb{R}^k$.

Theorem 1. *Let* $g(\cdot) : \mathbf{x} \in \mathbb{R}^K \mapsto g(\mathbf{x}) \in] - \infty, +\infty]$ *be a proper convex function. Denote by* $g^*(\cdot)$ *its convex conjugate, called also Fenchel-Legendre transform, defined by* $g^*(\cdot) : \mathbf{t} \in \mathbb{R}^K \mapsto g^*(\mathbf{t}) := \sup_{\mathbf{x} \in \mathbb{R}^K} \{ \langle \mathbf{t}, \mathbf{x} \rangle - g(\mathbf{x}) \} \in]-\infty, +\infty]$, *where* $\langle \mathbf{t}, \mathbf{x} \rangle := \mathbf{t}^\top \mathbf{x} := \sum_{i=1}^K t_i x_i$ *is the inner product on* $\mathbb{R}^K$. *Consider the convex conjugate of the proper convex function* $g^*(\cdot)$, *i.e., the function* $g^{**}(\cdot) :$ $\mathbf{x} \in \mathbb{R}^K \mapsto g^{**}(\mathbf{x}) := \sup_{\mathbf{t} \in \mathbb{R}^K} \{ \langle \mathbf{t}, \mathbf{x} \rangle - g^*(\mathbf{t}) \} \in] - \infty, +\infty]$. *If* $g(\cdot)$ *is lower semi-continuous (l.s.c.), then we have for all* $\mathbf{x} \in \mathbb{R}^K$, $g(\mathbf{x}) = g^{**}(\mathbf{x})$, *namely,* $g(\mathbf{x}) = \sup_{\mathbf{t} \in \mathbb{R}^K} \{ \langle \mathbf{t}, \mathbf{x} \rangle - g^*(\mathbf{t}) \}$, *for all* $\mathbf{x} \in \mathbb{R}^K$.

The above theorem means that any proper convex function on $\mathbb{R}^K$, if it is l.s.c. (or equivalently closed), can be represented as the convex conjugate of its convex conjugate. By applying the above theorem, we obtain for the function $\psi(\cdot) : \psi(\mathbf{x}) = \sup_{\mathbf{t} \in \mathbb{R}^K} \{ \langle \mathbf{x}, \mathbf{t} \rangle - \psi^*(\mathbf{t}) \}$, for all $\mathbf{x} \in \mathbb{R}^K$, where $\forall \mathbf{t} := (t_1, \ldots, t_K)^\top \in \mathbb{R}^K, \psi^*(\mathbf{t}) := \sup_{\mathbf{x} \in \mathbb{R}^K} \{ \langle \mathbf{t}, \mathbf{x} \rangle - \psi(\mathbf{x}) \}$. We give in the following lines, some properties of the convex conjugate $\psi^*(\cdot)$ of $\psi(\cdot)$. For the proofs, we can extend, to the multivariate case, the arguments in Sect. 26 of [12]. One can show that $\psi^*(\cdot)$, in turn, is convex, closed, and proper, since $\psi^*(\mathbf{t}) > -\infty$, for all $\mathbf{t} \in \mathbb{R}^K$, and $\psi^*(\mathbf{0}) = \mathbf{0}$. It takes its values in $] -\infty, +\infty]$. Moreover, the strict convexity of $\psi(\cdot)$ on its domain dom_ψ is equivalent to the condition that $\psi^*(\cdot)$ is "essentially smooth", i.e., differentiable on $\mathrm{int}(\mathrm{dom}_{\psi^*})$ and, if the boundary point set of dom_{ψ^*} is not void, the directional derivatives satisfy, for all boundary point $\mathbf{b}$ of dom_{ψ^*}, $\lim_{\mathbf{y} \to \mathbf{b}, \mathbf{y} \in \mathrm{dom}_{\psi^*}} \psi^{*\prime}(\mathbf{y}; \mathbf{x} - \mathbf{y}) :=$ $\lim_{\mathbf{y} \to \mathbf{b}, \mathbf{y} \in \mathrm{dom}_{\psi^*}} \lim_{\lambda \downarrow 0} \frac{\psi^*(\mathbf{y} + \lambda(\mathbf{x} - \mathbf{y})) - \psi^*(\mathbf{y})}{\lambda} = +\infty$, $\forall \mathbf{x} \in \mathrm{dom}_{\psi^*}$. In the above display, $\psi^{*\prime}(\mathbf{y}; \mathbf{x} - \mathbf{y})$ denotes the derivative of $\psi^*(\cdot)$ at the point $\mathbf{y}$ in the direction $\mathbf{x} - \mathbf{y}$, which always exists (may be infinite) since $\psi^*(\cdot)$ is convex. Conversely, $\psi(\cdot)$ is essentially smooth if and only if $\psi^*(\cdot)$ is strictly convex on its domain dom_{ψ^*}. Assume that ψ is differentiable on $\mathrm{int}(\mathrm{dom}_\psi)$, and denote by $\psi'_{x_k}(\cdot)$ the partial derivative $\frac{\partial \psi}{\partial x_k}(\cdot)$, for all $k = 1, \ldots, K$. Then, we can give the explicit expression of $\psi^*(\mathbf{t})$ for all $\mathbf{t} \in \{(\psi'_{x_1}(\mathbf{x}), \ldots, \psi'_{x_K}(\mathbf{x})) \in \mathbb{R}^K \text{ s.t. } \mathbf{x} \in \mathrm{int}(\mathrm{dom}_\psi)\}$. Indeed, one can show that, $\forall \mathbf{x} \in \mathrm{int}(\mathrm{dom}_\psi)$, $\psi^*(\psi'_{x_1}(\mathbf{x}), \ldots, \psi'_{x_K}(\mathbf{x})) = \sum_{k=1}^K x_k \psi'_{x_k}(\mathbf{x}) - \psi(\mathbf{x})$.

3 Fenchel Duality of Convex Functionals on Infinite Dimensional Vector Spaces

We refer to [4], for the proof of the following duality theorem, for convex functionals on more general vector spaces, with possibly infinite dimension.

Theorem 2. *Let $\mathcal{S}$ be a locally convex Hausdorff topological vector space, and $g(\cdot) : \mathcal{S} \mapsto] -\infty, +\infty]$ be a proper convex function. Denote by $g^*(\cdot)$ the convex conjugate of $g(\cdot)$, i.e., the proper convex function on $\mathcal{S}^*$ (the topological dual of $\mathcal{S}$) with values in $] -\infty, +\infty]$, defined by $g^*(\cdot) : \ell \in \mathcal{S}^* \mapsto g^*(\ell) := \sup_{x \in \mathcal{S}} \{\ell(x) - g(x)\}$. If $g(\cdot)$ is l.s.c. on $\mathcal{S}$, then we have the following dual representation for $g(\cdot) : g(x) = \sup_{\ell \in \mathcal{S}^*} \{\ell(x) - g^*(\ell)\}, \forall x \in \mathcal{S}$.*

We will apply the above theorem in order to obtain dual representations for ψ-divergences. We will look at ψ-divergences as convex functionals $\mathbf{Q} \mapsto D_\psi(\mathbf{Q}; P)$ on the product vector space of K well chosen vector spaces of signed finite measures. Here, P is a fixed reference probability measure. The question is how to chose the vector spaces $\mathcal{S}$ (of signed finite measures) and the topology, on which the ψ-divergence functionals satisfy all conditions of the above duality theorem? At the same time, we need to determine the corresponding topological dual, in order to get sufficiently explicit representations. We give answers to these questions in the following. Let $\boldsymbol{\mathcal{F}} := \mathcal{F}_1 \times \cdots \times \mathcal{F}_K$ be a collection of measurable functions defined on $\mathcal{X}$ with values in $\mathbb{R}^K$. We will denote by $f_1, \ldots, f_K$ the K components of any function $\boldsymbol{f} \in \boldsymbol{\mathcal{F}}$. The spaces $\mathcal{X}$ and $\mathbb{R}^K$ are endowed, respectively, by the σ-field $\mathcal{B}$ and the Borel σ-field $\mathcal{B}(\mathbb{R}^K)$, induced by the usual topology of $\mathbb{R}^K$. We assume that all elements $\boldsymbol{f} := (f_1, \ldots, f_K)^\top$ of $\mathcal{F}$ satisfy the condition

$$\int_{\mathcal{X}} |f_k(x)| \, dP(x) < \infty, \forall f_k \in \mathcal{F}_k, \forall k = 1, \ldots, K. \tag{9}$$

Denote by $\mathcal{F}_b$ the real vector space of all bounded measurable functions on $\mathcal{X}$ with values in $\mathbb{R}^K$. Define the real (product) vector space $\mathcal{M}_{\mathcal{F}} := \left\{ Q \in \mathcal{M}^K : \int_{\mathcal{X}} |f_k(x)| \, d|Q_k(x)| < +\infty, \forall k = 1, \ldots, K, \forall f \in \mathcal{F} \right\}$, and denote by $\langle \mathcal{F} \cup \mathcal{F}_b \rangle$ the real vector space induced by the collection $\mathcal{F} \cup \mathcal{F}_b$. Note that the space $\mathcal{M}_{\mathcal{F}}$ is not void, since it contains the "K-dimension" measure vector $(P, \ldots, P)$, by assumption (9). We equip $\mathcal{M}_{\mathcal{F}}$ with the $\tau_{\mathcal{F}}$-topology, the weakest one that makes any mapping $Q \in \mathcal{M}_{\mathcal{F}} \mapsto \sum_{k=1}^{K} \int_{\mathcal{X}} f_k(x) \, dQ_k(x)$ continuous, for all $f := (f_1, \ldots, f_K)^\top \in \langle \mathcal{F} \cup \mathcal{F}_b \rangle$. Then, we have

Proposition 1. *(1) The real vector space* $\mathcal{M}_{\mathcal{F}}$, *equipped with the* $\tau_{\mathcal{F}}$-*topology, is locally convex Hausdorff topological vector space; and the topological dual of* $\mathcal{M}_{\mathcal{F}}$ *is given by*

$$\mathcal{M}_{\mathcal{F}}^* = \left\{ Q \in \mathcal{M}_{\mathcal{F}} \mapsto \sum_{k=1}^{K} \int_{\mathcal{X}} f_k(x) \, dQ_k(x); \ f \in \langle \mathcal{F} \cup \mathcal{F}_b \rangle \right\};$$

(2) The proper convex function $Q \in \mathcal{M}_{\mathcal{F}} \mapsto D_\psi(Q; P)$ *is lower semi-continuous.*

In view of the above theorem, the Fenchel-Legendre transform of the proper convex function $D_\psi(\cdot; P) : Q \in \mathcal{M}_{\mathcal{F}} \mapsto D_\psi(Q; P)$ is defined by

$$D_\psi^*(f) := \sup_{Q \in \mathcal{M}_{\mathcal{F}}} \left\{ \sum_{k=1}^{K} \int_{\mathcal{X}} f_k(x) \, dQ_k(x) - D_\psi(Q; P) \right\}, f \in \langle \mathcal{F} \cup \mathcal{F}_b \rangle.$$

In the following proposition, we give explicit formula for the convex conjugate $D_\psi^*(\cdot)$. Moreover, we prove that $D_\psi^*(\cdot) \equiv D_{\psi^*}(\cdot)$, in the sense that $D_\psi^*(f) = \int_{\mathcal{X}} \psi^* (f(x)) \, dP(x) =: D_{\psi^*}(f), \forall f := (f_1, \ldots, f_K)^\top \in \langle \mathcal{F} \cup \mathcal{F}_b \rangle$.

Proposition 2. *Assume that,* $\psi(\cdot)$ *is differentiable on the interior of its domain* dom_ψ, *and denote by* $\psi'_{x_k}(\cdot), k = 1, \ldots, K$, *the partial derivatives* $\frac{\partial \psi}{\partial x_k}(\cdot), k = 1 \ldots, K$, *respectively. Then, for all* $Q \in \mathcal{M}_{\mathcal{F}}$ *such that* $D_\psi(Q; P) < \infty$ *and* $\left(\psi'_{x_1}(dQ/dP), \ldots, \psi'_{x_K}(dQ/P) \right) \in \langle \mathcal{F} \cup \mathcal{F}_b \rangle$, *the* ψ-*divergence* $D_\psi(Q; P)$ *admits the dual representation* $D_\psi(Q; P) = \sup_{f \in \langle \mathcal{F} \cup \mathcal{F}_b \rangle}$ $\left\{ \sum_{k=1}^{K} \int_{\mathcal{X}} f_k(x) \, dQ_k(x) - \int_{\mathcal{X}} \psi^* (f(x)) \, dP(x) \right\}$, *and the function* $\overline{f}(x) := \left(\psi'_{x_1} \left(\frac{dQ}{dP}(x) \right), \ldots, \psi'_{x_K} \left(\frac{dQ}{dP}(x) \right) \right), x \in \mathcal{X}$, *is a dual optimal solution. Moreover, if* ψ *is essentially smooth, then* $\overline{f}$ *is unique* P-*a.s.*

The above proposition remains valid if the vector space $\langle \mathcal{F} \cup \mathcal{F}_b \rangle$ is restricted to the class of functions $\mathcal{F}$. We state this result in the following theorem.

Theorem 3. *Assume that,* $\psi(\cdot)$ *is differentiable on the interior of its domain* dom_ψ. *Then, for all* $Q \in \mathcal{M}_{\mathcal{F}}$ *such that* $D_\psi(Q; P) < \infty$ *and* $\left(\psi'_{x_1}(dQ/dP), \ldots, \psi'_{x_K}(dQ/P) \right) \in \mathcal{F}$, *the* ψ-*divergence* $D_\psi(Q; P)$ *admits the dual representation* $D_\psi(Q; P) = \sup_{f \in \mathcal{F}} \left\{ \sum_{k=1}^{K} \int_{\mathcal{X}} f_k(x) \, dQ_k(x) - \int_{\mathcal{X}} \psi^* (f(x)) \, dP(x) \right\}$, *and the function* $\overline{f}(x) := \left(\psi'_{x_1} \left(\frac{dQ}{dP}(x) \right), \ldots, \psi'_{x_K} \left(\frac{dQ}{dP}(x) \right) \right), x \in \mathcal{X}$, *is a dual optimal solution. Moreover, if* ψ *is essentially smooth, then* $\overline{f}$ *is unique* P-*a.s.*

4 Inference with Multiple Samples in DRM Through Dual Representations of ψ-divergences

Here, we consider the Borel σ-filed $(\mathcal{X}, \mathcal{B}) = (\mathbb{R}^m, \mathcal{B}(\mathbb{R}^m))$, and we will use the notation $\boldsymbol{q}(\boldsymbol{\theta}, x) := (q(\boldsymbol{\theta}_1, x), \ldots, q(\boldsymbol{\theta}_K, x))$, where $\boldsymbol{\theta} := (\boldsymbol{\theta}_1^\top, \ldots, \boldsymbol{\theta}_K^\top)^\top$. Taking into account the form of the DRM (5), and choosing the class of functions

$$\mathcal{F} := \prod_{k=1}^K \mathcal{F}_k := \prod_{k=1}^K \left\{ x \in \mathbb{R}^m \mapsto \psi'_{x_k}\left(q(\boldsymbol{\theta}_1, x), \ldots, q(\boldsymbol{\theta}_K, x)\right); \boldsymbol{\theta} \in \prod_{k=1}^K \Theta_k \right\},$$

we obtain from Theorem 3, the following dual formula

$$D_\psi(\boldsymbol{Q}; P) = \sup_{\boldsymbol{\theta} \in \prod_{k=1}^K \Theta_k} \left\{ \sum_{k=1}^K \int_{\mathbb{R}^m} \psi'_{x_k}\left(\boldsymbol{q}(\boldsymbol{\theta}, x)\right) dQ_k(x) \right.$$
$$\left. - \int_{\mathbb{R}^m} \psi^*\left(\psi'_{x_1}(\boldsymbol{q}(\boldsymbol{\theta}, x)), \ldots, \psi'_{x_K}(\boldsymbol{q}(\boldsymbol{\theta}, x))\right) dP(x) \right\}. \quad (10)$$

Furthermore, the supremum in the above display is unique and achieved in $\boldsymbol{\theta}_T := \left(\boldsymbol{\theta}_{1T}^\top, \ldots, \boldsymbol{\theta}_{KT}^\top\right)^\top$:

$$\boldsymbol{\theta}_T = \operatorname*{argsup}_{\boldsymbol{\theta} \in \prod_{k=1}^K \Theta_k} \left\{ \sum_{k=1}^K \int_{\mathbb{R}^m} \psi'_{x_k}\left(\boldsymbol{q}(\boldsymbol{\theta}, x)\right) dQ_k(x) \right.$$
$$\left. - \int_{\mathbb{R}^m} \psi^*\left(\psi'_{x_1}(\boldsymbol{q}(\boldsymbol{\theta}, x)), \ldots, \psi'_{x_K}(\boldsymbol{q}(\boldsymbol{\theta}, x))\right) dP(x) \right\}. \quad (11)$$

The uniqueness of the dual solution $\boldsymbol{\theta}_T$ comes from the uniqueness of the dual solution $\overline{\boldsymbol{f}}(\cdot)$ in the above theorem and the identifiability assumption (6). Furthermore, we have, for the function in the second integral in (10) and (11), the explicit expression

$$\psi^*\left(\psi'_{x_1}(\boldsymbol{q}(\boldsymbol{\theta}, x)), \ldots, \psi'_{x_K}(\boldsymbol{q}(\boldsymbol{\theta}, x))\right) = \sum_{k=1}^K q(\boldsymbol{\theta}_k, x) \psi'_{x_k}(\boldsymbol{q}(\boldsymbol{\theta}, x)) - \psi\left(\boldsymbol{q}(\boldsymbol{\theta}, x)\right).$$

For notational simplicity, for all $\boldsymbol{\theta} \in \prod_{k=1}^K \Theta_k$, denote by $f_k(\boldsymbol{\theta}, \cdot)$ and $g(\boldsymbol{\theta}, \cdot)$, respectively, the real valued functions

$$f_k(\boldsymbol{\theta}, \cdot) : x \in \mathbb{R}^m \mapsto f_k(\boldsymbol{\theta}, x) := \psi'_{x_k}\left(\boldsymbol{q}(\boldsymbol{\theta}, x)\right), k = 1, \ldots, K, \quad (12)$$

and

$$g(\boldsymbol{\theta}, \cdot) : x \in \mathbb{R}^m \mapsto g(\boldsymbol{\theta}, x) := \sum_{k=1}^K q(\boldsymbol{\theta}_k, x) \psi'_{x_k}(\boldsymbol{q}(\boldsymbol{\theta}, x)) - \psi\left(\boldsymbol{q}(\boldsymbol{\theta}, x)\right). \quad (13)$$

We may write also Qf, instead of $\int_{\mathcal{X}} f(x)\,dQ(x)$, for any function f and any measure Q. Whence, the above formulas (10) and (11) can be written, respectively, under the form

$$D_\psi(Q_1,\ldots,Q_K;P) = \sup_{\boldsymbol{\theta}\in\prod_{k=1}^K \Theta_k}\left\{\sum_{k=1}^K Q_k f_k(\boldsymbol{\theta}) - Pg(\boldsymbol{\theta})\right\} \qquad (14)$$

and

$$\boldsymbol{\theta}_T = \operatorname*{argsup}_{\boldsymbol{\theta}\in\prod_{k=1}^K \Theta_k}\left\{\sum_{k=1}^K Q_k f_k(\boldsymbol{\theta}) - Pg(\boldsymbol{\theta})\right\} =: \operatorname*{argsup}_{\boldsymbol{\theta}\in\prod_{k=1}^K \Theta_k}\ell_\psi(\boldsymbol{\theta}). \qquad (15)$$

Whence, by replacing, in the above dual formula, the distributions $P, Q_1,\ldots,Q_K$ by their empirical ones (8), we obtain the following "dual" plug-in estimate of the ψ-divergence $D_\psi(Q_1,\ldots,Q_K;P)$

$$\widehat{D}_\psi(Q_1,\ldots,Q_K;P) := \sup_{\boldsymbol{\theta}\in\prod_{k=1}^K \Theta_k}\left\{\sum_{k=1}^K \widehat{Q}_k f_k(\boldsymbol{\theta}) - \widehat{P}g(\boldsymbol{\theta})\right\} \qquad (16)$$

and the following dual ψ-divergence estimates of the vector parameters $\boldsymbol{\theta}_T$

$$\widehat{\boldsymbol{\theta}}_\psi := \operatorname*{argsup}_{\boldsymbol{\theta}\in\prod_{k=1}^K \Theta_k}\left\{\sum_{k=1}^K \widehat{Q}_k f_k(\boldsymbol{\theta}) - \widehat{P}g(\boldsymbol{\theta})\right\} =: \operatorname*{argsup}_{\boldsymbol{\theta}\in\prod_{k=1}^K \Theta_k}\widehat{\ell}_\psi(\boldsymbol{\theta}). \qquad (17)$$

5   Empirical Likelihood and ψ-divergences Under the DRM

We give a brief introduction of empirical likelihood (EL) inference under the DRM; for more information see [3,5,8–10]. For any $\boldsymbol{\theta}\in\prod_{k=1}^K \Theta_k$, the EL associated to the independent samples $(X_{k,i},\ i=1,\ldots,n_k)$, $k=0,1,\ldots,K$, under the DRM (5), is $L(\boldsymbol{\theta},p) = \prod_{k=0}^K \prod_{i=1}^{n_k} p(X_{k,i})q(\boldsymbol{\theta}_k,X_{k,i})$. For $k=0$, we set $\boldsymbol{\theta}_0 = \mathbf{0}$, and then $q(\boldsymbol{\theta}_0,x) = q(\mathbf{0},x) = 1, \forall x$. In the EL approach, we consider the probability distribution P as if it is discrete, and $p(X_{k,i})$ is the mass induced by P on the observation $X_{k,i}$, for all $k=1,\ldots,K$, and all $i=1,\ldots,n_k$. Hence, p can be seen as a vector in $\mathbb{R}_+^{*\,n}$, where $n := n_0+n_1+\cdots+n_K$ is the total sample size, and such that the sum of its components is equal to one. The log-EL is $\ell(\boldsymbol{\theta},p) :=$ $\log\prod_{k=0}^K \prod_{i=1}^{n_k} p(X_{k,i})q(\boldsymbol{\theta}_k,X_{k,i}) = \sum_{k=0}^K \sum_{i=1}^{n_k}(\log p(X_k,i) + \log q(\boldsymbol{\theta}_k,X_{k,i}))$. The model assumption (5) implies that $\int_{\mathbb{R}^m} q(\boldsymbol{\theta}_r,x)\,dP(x) = 1, \forall r=0,1,\ldots,K$. Thus, $\sum_{k=0}^K \sum_{i=1}^{n_k} p(X_{k,i})q(\boldsymbol{\theta}_r,X_{k,i}) = 1, \forall r=0,1,\ldots,K$, for the empirical version. The profil log-EL (in $\boldsymbol{\theta}$) is then defined by

$$\ell(\boldsymbol{\theta}) = \sup_{p\in\mathcal{C}_\theta}\ell(\boldsymbol{\theta},p) = \sup_{p\in\mathcal{C}_\theta}\sum_{k=0}^K \sum_{i=1}^{n_k}(\log p(X_{k,i}) + \log q(\boldsymbol{\theta}_k,X_{k,i})), \qquad (18)$$

where p, due to the above constraints, is constrained to the set $\mathcal{C}_{\boldsymbol{\theta}}$ of all $p \in \mathbb{R}_+^{*\,n}$ s.t. $\sum_{k=0}^{K} \sum_{i=1}^{n_k} p(X_{k,i}) = 1$ and $\sum_{k=0}^{K} \sum_{i=1}^{n_k} p(X_{k,i}) q(\boldsymbol{\theta}_r, X_{k,i}) = 1, \forall r = 1, \ldots, K$. The EL estimate $\widehat{\boldsymbol{\theta}}_{EL}$ of $\boldsymbol{\theta}_T$ is then $\widehat{\boldsymbol{\theta}}_{EL} := \underset{\boldsymbol{\theta} \in \prod_{k=1}^{K} \Theta_k}{\operatorname{argsup}} \ell(\boldsymbol{\theta})$. For a given $\boldsymbol{\theta} \in \prod_{k=1}^{K} \Theta_k$, if the condition $\exists p \in \mathcal{C}_{\boldsymbol{\theta}}$ such that $|\ell(\boldsymbol{\theta}, p)| < \infty$ is satisfied, then by Lagrange multipliers theorem, the supremum over p, in (18), is attained when $p(X_{k,i}) = \dfrac{1}{n\{1 + \sum_{s=1}^{K} \widehat{\lambda}_s [q(\boldsymbol{\theta}_s, X_{k,i}) - 1]\}}, \forall k = 0, 1, \ldots, K, \forall i = 1, \ldots, n_k$, where $(\widehat{\lambda}_1, \ldots, \widehat{\lambda}_K)$ is the solution, which depends on $\boldsymbol{\theta}$ and on the data, of the system of equations $\sum_{k=0}^{K} \sum_{i=1}^{n_k} p(X_{k,i}) q(\boldsymbol{\theta}_r, X_{k,i}) = 1, \forall r = 1, \ldots, K$. Given $\widehat{\boldsymbol{\theta}}_{EL}$. Interestingly, by differentiating $\ell(\boldsymbol{\theta})$ w.r.t. $\boldsymbol{\theta}$ at the point $\widehat{\boldsymbol{\theta}}_{EL}$, and solving the system of equations $(\partial/\partial \alpha_k)\ell(\widehat{\boldsymbol{\theta}}_{EL}) = 0$, $(\partial/\partial \beta_k)\ell(\widehat{\boldsymbol{\theta}}_{EL}) = \mathbf{0}$, $\forall k = 1, \ldots, K$, one can show that the corresponding Lagrange multipliers $\widehat{\lambda}_1, \ldots, \widehat{\lambda}_K$ depend only on the sample sizes, and we always have $\widehat{\lambda}_s = n_s/n, \forall s = 1, \ldots, K$. We obtain then,

$$\widehat{\boldsymbol{\theta}}_{EL} := \underset{\boldsymbol{\theta} \in \prod_{k=1}^{K} \Theta_k}{\operatorname{argsup}} \ell(\boldsymbol{\theta}) = \underset{\boldsymbol{\theta} \in \prod_{k=1}^{K} \Theta_k}{\operatorname{argsup}} \sup_{p \in \mathcal{C}_{\boldsymbol{\theta}}} \ell(\boldsymbol{\theta}, p) = \underset{\boldsymbol{\theta} \in \prod_{k=1}^{K} \Theta_k}{\operatorname{argsup}} \ell_d(\boldsymbol{\theta}), \text{ where}$$

$$\ell_d(\boldsymbol{\theta}) := -n \log n - \sum_{k=0}^{K} \sum_{i=1}^{n_k} \log \left(\widehat{\lambda}_0 + \sum_{s=1}^{K} \widehat{\lambda}_s q(\boldsymbol{\theta}_s, X_{k,i}) \right) + \sum_{k=1}^{K} \sum_{i=1}^{n_k} \log q(\boldsymbol{\theta}_k, X_{k,i}),$$

with $\widehat{\lambda}_s = n_s/n, \forall s = 0, 1, \ldots, K$. The function $\ell_d(\cdot)$ has the same maximum value and point as the profile log-EL $\ell(\cdot)$. Moreover, the function $\ell_d(\cdot)$ is parametric and simpler than $\ell(\cdot)$. Therefore, in the literature, $\ell_d(\cdot)$ is regarded as the profile log-EL instead of $\ell(\cdot)$. Consider now, in the context of DRM, the test problem of the null hypothesis $\mathcal{H}_0 : Q_1 = \cdots = Q_K = P$ against the alternative $\mathcal{H}_1 : \exists Q_k \neq P$, or equivalently $\mathcal{H}_0 : \boldsymbol{\theta}_{1T} = \cdots = \boldsymbol{\theta}_{KT} = \mathbf{0}$ against the alternative $\mathcal{H}_1 : \exists \boldsymbol{\theta}_{kT} \neq \mathbf{0}$. The corresponding EL ratio statistic can be written as

$$S_n := 2 \log \frac{\sup_{\boldsymbol{\theta} \in \prod_{k=1}^{K} \Theta_k} \sup_{p \in \mathcal{C}_{\boldsymbol{\theta}}} \ell(\boldsymbol{\theta}, p)}{\sup_{p \in \mathcal{C}_0} \prod_{k=0}^{K} \prod_{i=1}^{n_k} p(X_{k,i})} = 2 \left(\sup_{\boldsymbol{\theta} \in \prod_{k=1}^{K} \Theta_k} \ell_d(\boldsymbol{\theta}) + n \log n \right)$$

$$= 2n \widehat{D}_{\psi_{EL}}(Q_1, \ldots, Q_K; P),$$ where $\psi_{EL}(\cdot)$ is the particular nonnegative proper closed convex function on $\mathbb{R}^k$ defined by

$$\psi_{EL}(x_1, \ldots, x_K) = \left(\sum_{s=1}^{K} \widehat{\rho}_s \varphi_1(x_s) \right) - \varphi_1 \left(\widehat{\rho}_0 + \sum_{s=1}^{K} \widehat{\rho}_s x_s \right), \tag{19}$$

where φ_1 is the nonnegative proper closed convex function on $\mathbb{R}$ defined by

$$\varphi_1(x) := (x \log x - x + 1) \mathbb{1}_{\mathbb{R}_+}(x) + (+\infty) \mathbb{1}_{]-\infty, 0[}(x). \tag{20}$$

Moreover, we have $\psi_{EL}(\mathbf{1}_K) = 0$, and it is straightforward to see that $\psi_{EL}(\cdot)$ is a member of the ψ-functions considered in (1). Whence, the empirical likelihood ratio statistic S_n is equal to $2n$ times the dual estimate of the particular ψ_{EL}-divergence defined by the convex function (19). Moreover, the EL estimate $\widehat{\boldsymbol{\theta}}_{EL}$ is equal to the corresponding dual maximum ψ_{EL}-divergence estimate, namely, $\widehat{\boldsymbol{\theta}}_{EL} = \underset{\boldsymbol{\theta} \in \prod_{k=1}^{K} \Theta_k}{\operatorname{argsup}} \widehat{\ell}_{\psi_{EL}}(\boldsymbol{\theta}).$

6   Asymptotic Behavior of the Estimates and Test Statistics

In this section, we give the asymptotic properties of the maximum ψ-divergence estimates (17) of the parameter $\boldsymbol{\theta}_T$, and the dual plug-in estimates (16) of the ψ-divergences $D_\psi(Q_1, \ldots, Q_K; P)$, when ψ is of the form

$$\psi : \boldsymbol{x} \in \mathbb{R}^K \mapsto \psi(\boldsymbol{x}) = \sum_{s=1}^{K} \widehat{\rho}_s \varphi(x_s) - \varphi\left(\widehat{\rho}_0 + \sum_{s=1}^{K} \widehat{\rho}_s x_s\right) \in [0, \infty], \qquad (21)$$

where $\varphi : x \in \mathbb{R} \mapsto \varphi(x) \in [0, +\infty]$ is any nonnegative proper closed convex function, satisfying $\varphi(1) = 0$, and such that 1 is an interior point of its domain $\mathrm{dom}_\varphi := \{x \in \mathbb{R} \text{ s.t. } |\varphi(x)| < \infty\}$. Note that all functions in (21) are members of the ψ-functions considered in (1). We assume also that $\varphi'(1) = 0$, and, without loss of generally, that $\varphi''(1) = 1$. Recall that the particular ψ_{EL}-function is obtained for the particular φ-function $\varphi(\cdot) = \varphi_1(\cdot)$ given in (20). Whence, the EL estimate and test statistic will be treated in this section as special case. Assume that the (1+K) sample sizes, n_k, $k = 0, 1, \ldots, K$, satisfy the condition $\frac{n_k}{n} \to \rho_k$ for some real $\rho_k \in]0, 1[$, when the total sample size $n \to +\infty$. We will use the following assumptions.

(A.1) There exists a neighborhood $N(\boldsymbol{\theta}_T)$ of $\boldsymbol{\theta}_T$ such that (a) for each k, the third order partial derivative functions $\{x \in \mathbb{R}^m \mapsto (f_k'''(\boldsymbol{\theta}, x))_{i,j,k}; \boldsymbol{\theta} \in N(\boldsymbol{\theta}_T)\}$ are dominated by some Q_k-integrable function; (b) the third order partial derivative functions $\{x \in \mathbb{R}^m \mapsto (g'''(\boldsymbol{\theta}, x))_{i,j,k}; \boldsymbol{\theta} \in N(\boldsymbol{\theta}_T)\}$ are dominated by some P-integrable function;

(A.2) The integrals $\int_{\mathbb{R}^m} |f_k'(\boldsymbol{\theta}_T, x)|^2 \, dQ_k(x)$, $\int_{\mathbb{R}^m} |f_k''(\boldsymbol{\theta}_T, x)| \, dQ_k(x)$, $k = 1, \ldots, K$, $\int_{\mathbb{R}^m} |g'(\boldsymbol{\theta}_T, x)|^2 \, dP(x)$, $\int_{\mathbb{R}^m} |g''(\boldsymbol{\theta}_T, x)| \, dP(x)$ are finite, and the matrix $D := -\sum_{k=1}^{K} Q_k f_k''(\boldsymbol{\theta}_T) + Pg''(\boldsymbol{\theta}_T)$ is nonsingular.

Theorem 4. *Assume that assumptions (A.1) and (A.2) hold.*

(1) Let $B(\boldsymbol{\theta}_T, n^{-1/3}) := \{\boldsymbol{\theta} \in \Theta : |\boldsymbol{\theta} - \boldsymbol{\theta}_T| \le n^{-1/3}\}$. Then, as $n \to \infty$, with probability one, $\widehat{\ell}_\psi(\boldsymbol{\theta})$ attains its maximum value at some point $\widehat{\boldsymbol{\theta}}_\psi$ in the interior of the ball $B(\boldsymbol{\theta}_T, n^{-1/3})$, satisfying $\widehat{\ell}_\psi'\left(\widehat{\boldsymbol{\theta}}_\psi\right) = 0$;

(2) $\sqrt{n}\left(\widehat{\boldsymbol{\theta}}_\psi - \boldsymbol{\theta}_T\right)$ converges in distribution to a centered normal random vector with explicit covariance matrix;

(3) Under the null hypothesis $\mathcal{H}_0$, the statistic $2n\widehat{D}_\psi(\boldsymbol{Q}; P)$ convergences in distribution to a $\chi^2(d \times K)$ random variable;

(4) Under the alternative hypothesis $\mathcal{H}_1$, $\sqrt{n}\left(\widehat{D}_\psi(\boldsymbol{Q}; P) - D_\psi(\boldsymbol{Q}; P)\right)$ convergences in distribution to a centered normal random variable with explicit variance.

7 Conclusion

We have introduced multivariate divergences between signed finite measure vectors and a given reference probability distribution. We have considered Fenchel duality theory of such divergences viewed as convex functionals on appropriate topological vector spaces of signed finite measure vectors and $\mathbb{R}^K$-valued functions. Dual representations of multivariate divergences have been obtained, and applied for estimating and testing in multi-sample density ratio models. This approach recovers the dual empirical likelihood one and allows to obtain the asymptotic properties of the proposed estimates and test statistics, including the empirical likelihood ones, both under the model and under misspecification. In testing context, the obtained asymptotic distributions under alternative hypotheses lead to accurate approximation of the power functions. Moreover, the proposed dual estimates of the introduced multivariate divergences can be used as criteria for model selection, in multi-sample density ratio models, and for multi-group classification.

References

1. Anderson, J.A.: Multivariate logistic compounds. Biometrika **66**(1), 17–26 (1979). http://dx.doi.org/10.1093/biomet/66.1.17
2. Broniatowski, M., Keziou, A.: Minimization of ϕ-divergences on sets of signed measures. Studia Sci. Math. Hung. **43**(4), 403–442 (2006). arXiv:1003.5457
3. Chen, J., Liu, Y.: Quantile and quantile-function estimations under density ratio model. Ann. Stat. **41**(3), 1669–1692 (2013). http://dx.doi.org/10.1214/13-AOS1129
4. Dembo, A., Zeitouni, O.: Large Deviations Techniques and Applications, 2nd edn. Springer-Verlag, New York (1998)
5. Fokianos, K., Kedem, B., Qin, J., Short, D.A.: A semiparametric approach to the one-way layout. Technometrics **43**(1), 56–65 (2001)
6. Keziou, A.: Dual representation of ϕ-divergences and applications. C. R. Math. Acad. Sci. Paris **336**(10), 857–862 (2003)
7. Keziou, A.: Mulivariate divergences: applications in statistical inference for multiple samples under density ratio models. Manuscript in preparation (2015)
8. Keziou, A., Leoni-Aubin, S.: On empirical likelihood for semiparametric two-sample density ratio models. J. Stat. Plan. Infer. **138**(4), 915–928 (2008). http://dx.doi.org/10.1016/j.jspi.2007.02.009
9. Qin, J.: Inferences for case-control and semiparametric two-sample density ratio models. Biometrika **85**(3), 619–630 (1998)
10. Qin, J., Zhang, B.: A goodness-of-fit test for logistic regression models based on case-control data. Biometrika **84**(3), 609–618 (1997). http://dx.doi.org/10.1093/biomet/84.3.609
11. Rockafellar, R.T.: Integrals which are convex functionals. II. Pac. J. Math. **39**, 439–469 (1971)
12. Rockafellar, R.T.: Convex Analysis. Princeton University Press, Princeton (1970)

Generalized Mutual-Information Based Independence Tests

Amor Keziou and Philippe Regnault[✉]

Laboratoire de Mathématiques de Reims, EA 4535, UFR Sciences Exactes
et Naturelles, Moulin de la Housse, BP1039, 51687 Remis Cedex, France
{amor.keziou,philippe.regnault}@univ-reims.fr

Abstract. We derive independence tests by means of dependence measures thresholding in a semiparametric context. Precisely, estimates of mutual information associated to φ-divergences are derived through the dual representations of φ-divergences. The asymptotic properties of the estimates are established, including consistency, asymptotic distribution and large deviations principle. The related tests of independence are compared through their relative asymptotic Bahadur efficiency and numerical simulations.

1 Introduction

Measuring the dependence between random variables has been a central aim of probability theory since its earliest developments. Classical examples of dependence measures are correlation measures, such as Pearson's correlation or Kendall's and Spearman's correlations. While the first one focuses on linear relationship between real random variables, the two second ones measure the monotonic relationship between variables taking values in ordered sets. Pure-independence measures between variables X and Y taking values in general measurable sets $(\mathcal{X}, \mathcal{A}_\mathcal{X})$ and $(\mathcal{Y}, \mathcal{A}_\mathcal{Y})$ can be defined by considering any divergence between the joint distribution $\mathbb{P}$ of (X, Y) and the product of its margins $\mathbb{P}^\perp := \mathbb{P}_X \otimes \mathbb{P}_Y$. The most outstanding example of such a dependence measure is the χ^2-divergence, for a finite distribution $\mathbb{P} := (p_{x,y})_{(x,y)}$, given by

$$\chi^2(\mathbb{P}, \mathbb{P}^\perp) := \frac{1}{2} \int_{\mathcal{X} \times \mathcal{Y}} \left(\frac{d\mathbb{P}}{d\mathbb{P}^\perp} - 1 \right)^2 d\mathbb{P}^\perp = \frac{1}{2} \sum_{(x,y) \in \mathcal{X} \times \mathcal{Y}} \frac{(p_{x,y} - p_x p_y)^2}{p_x p_y}, \quad (1)$$

where $d\mathbb{P}/d\mathbb{P}^\perp$ denotes the density of $\mathbb{P}$ w.r.t. $\mathbb{P}^\perp$. Another classical example is the mutual information (MI) associated to Kullback-Leibler (KL) divergence – denote it by KL-MI, defined by

$$I_{KL}(\mathbb{P}) := \mathbb{K}(\mathbb{P}, \mathbb{P}^\perp) := \int_{\mathcal{X} \times \mathcal{Y}} \frac{d\mathbb{P}}{d\mathbb{P}^\perp} \log \frac{d\mathbb{P}}{d\mathbb{P}^\perp} d\mathbb{P}^\perp. \quad (2)$$

When dealing with observations of two random variables, we may test the null hypothesis that these variables are independent by means of estimating such a

© Springer International Publishing Switzerland 2015
F. Nielsen and F. Barbaresco (Eds.): GSI 2015, LNCS 9389, pp. 454–463, 2015.
DOI: 10.1007/978-3-319-25040-3_49

dependence measure and deciding to reject the independence hypothesis if the estimate is too far away from zero — the classical χ^2-independence test is such a procedure using the test statistic

$$2n\chi^2\left(\widehat{\mathbb{P}}, \widehat{\mathbb{P}}^\perp\right) = n \sum_{(x,y)} \frac{(\widehat{p}_{x,y} - \widehat{p}_x\widehat{p}_y)^2}{\widehat{p}_x\widehat{p}_y}, \tag{3}$$

where $\widehat{\mathbb{P}}$ is the empirical distribution associated to the observations. The dependence measure can also be the KL-MI or any other divergence measure between $\mathbb{P}$ and $\mathbb{P}^\perp$. The tests based on such dependence measures have been extensively studied when $\mathcal{X}$ and $\mathcal{Y}$ are finite sets; see e.g. [8], Chap. 8. When dealing with real or multidimensional continuous variables, numerous nonparametric estimates of (2) built from i.i.d. samples $(X_1, Y_1), \ldots, (X_n, Y_n)$ exist in the literature. See e.g. [6] for an overview and numerical comparisons of existing estimates. Unfortunately, their (asymptotic) distributions remain inaccessible. Hence, testing independence from these estimates requires Monte-Carlo or Bootstrap approximations of the related p-values.

This paper introduces semiparametric estimates of φ-mutual information (φ-MI), i.e., independence measures associated to φ-divergence functionals. These estimates are obtained by making use of a dual representation of φ-MI, presented in Sect. 2.2. Their asymptotic properties are presented in Sect. 3. This leads to explicit independence tests, whose Bahadur efficiencies are compared in Sect. 4; the most efficient test is shown to be the one based on KL-MI (2). Finally, the power of φ-MI based tests is compared to classical non-correlation tests in the Gaussian setting in Sect. 5.

2 Dual Representation of φ-Mutual Informations

In this Section, we first define φ-mutual informations, then specify the semiparametric modeling of the ratio $d\mathbb{P}/d\mathbb{P}^\perp$, and finally make use of the so-called dual representation of φ-divergences (see [1,4]) to get a dual representation of φ-MI.

2.1 Introducing φ-mutual Informations

Let (X, Y) be a random vector taking values in $(\mathcal{X} \times \mathcal{Y}, \mathcal{A}_\mathcal{X} \otimes \mathcal{A}_\mathcal{Y})$, with joint distribution $\mathbb{P}$ and margins $\mathbb{P}_X$ and $\mathbb{P}_Y$. Let $\varphi : \mathbb{R} \to [0, +\infty]$ be some non negative closed proper convex three times differentiable function such that $\varphi(1) = 0$, $\varphi'(1) = 0$ and $\varphi''(1) = 1$. Let us introduce the φ-mutual information (φ-MI) $I_\varphi(X, Y) = I_\varphi(\mathbb{P})$ of (X, Y), defined as the φ-divergence from $\mathbb{P}$ to $\mathbb{P}^\perp := \mathbb{P}_X \otimes \mathbb{P}_Y$; precisely

$$I_\varphi(\mathbb{P}) := \mathbb{D}_\varphi(\mathbb{P}, \mathbb{P}^\perp) = \int_{\mathcal{X} \times \mathcal{Y}} \varphi\left(\frac{d\mathbb{P}}{d\mathbb{P}^\perp}\right) d\mathbb{P}^\perp. \tag{4}$$

Note that $I_\varphi(\mathbb{P}) \geq 0$; moreover, if φ is strictly convex on some neighborhood of 1, we have

$$I_\varphi(\mathbb{P}) = 0 \quad \text{iff} \quad \mathbb{P} = \mathbb{P}^\perp,$$

i.e., the φ-MI of (X, Y) is null if and only if X and Y are independent. KL-MI (2) is obtained for

$$\varphi_1(x) := x \log x - x + 1, \quad x > 0, \tag{5}$$

while χ^2 independence measure (1) is obtained for $\varphi_2(x) := (x-1)^2/2$.

2.2 Semiparametric Modeling of the Ratio $dP/dP^\perp$

Denote the set of probability measures on $(\mathcal{X} \times \mathcal{Y}, \mathcal{A}_\mathcal{X} \otimes \mathcal{A}_\mathcal{Y})$ by $\mathcal{M}_1$. For $P \in \mathcal{M}_1$, let $P^\perp$ denote the product distribution $P^\perp := P_1 \otimes P_2$ of the margins P_1 and P_2 of P.

In the following, the distribution $\mathbb{P}$ of (X, Y) is assumed to belong to the semiparametric model

$$\mathcal{M}_\Theta := \left\{ P \in \mathcal{M}_1 \text{ such that } \frac{dP}{dP^\perp} =: h_\theta; \theta \in \Theta \right\}, \tag{6}$$

where $\Theta \subseteq \mathbb{R}^{1+d}$, with $d \geq 1$, is the set of parameters, and $h_\theta(.,.)$ is some specified real-valued function, indexed by the $(1+d)$-dimensional parameter θ. The following assumptions on the model $\mathcal{M}_\Theta$ will be of use in the sequel.

Assumptions
A1: $(h_\theta(x, y) = h_{\theta'}(x, y), \forall (x, y) \in \mathcal{X} \times \mathcal{Y}) \Rightarrow (\theta = \theta')$ *(identifiability)*;
A2: *There exists (a unique)* $\theta_0 \in \text{int}(\Theta)$ *satisfying* $h_{\theta_0}(x, y) = 1, \forall (x, y) \in \mathcal{X} \times \mathcal{Y}$ *(independence is covered by the model).*

Denote by θ_T the true unknown value of the parameter, namely, the unique value satisfying

$$\frac{d\mathbb{P}}{dP^\perp}(x, y) = h_{\theta_T}(x, y), \quad \forall (x, y) \in \mathcal{X} \times \mathcal{Y}.$$

Examples 1. *1. If (X, Y) is a 2-dimensional Gaussian vector with nondegenerate covariance matrix, then straightforward computations show that the ratio $d\mathbb{P}/dP^\perp$ can be written under the form of the model (6) with*

$$h_\theta(x, y) = \exp\left\{ \alpha + \beta_1 x + \beta_2 y + \beta_3 x^2 + \beta_4 y^2 + \beta_5 xy \right\}, \tag{7}$$

and $\theta = (\alpha, \beta_1, \beta_2, \beta_3, \beta_4, \beta_5)^\top$. Note that the number of free parameters in $\theta_T := (\alpha_T, \beta_T^\top)^\top$ is $d = 5$, and that α_T is considered as a normalizing parameter due to the constraint $\int_{\mathcal{X} \times \mathcal{Y}} h_{\theta_T}(x, y) \, dP^\perp(x, y) = \int_{\mathcal{X} \times \mathcal{Y}} d\mathbb{P}(x, y) = 1$ since $\mathbb{P}$ is a probability distribution.

2. Assume that $\mathcal{X} \times \mathcal{Y}$ is a finite set; denote by $(p_{x,y})_{(x,y)}$ the density of $\mathbb{P}$ with respect to the counting measure on $\mathcal{X} \times \mathcal{Y}$. Then

$$\frac{d\mathbb{P}}{dP^\perp}(x, y) = \frac{p_{x,y}}{p_x p_y} = \exp\left(\sum_{(a,b) \in \mathcal{X} \times \mathcal{Y}} \theta_{a,b} \delta_{(a,b)}(x, y) \right) =: h_\theta(x, y), \tag{8}$$

where

$$\theta_{a,b} = \log \frac{p_{a,b}}{p_a p_b}, \quad (a, b) \in \mathcal{X} \times \mathcal{Y}.$$

3. Assume that (X, Y) is a 2-dimensional random vector such that $\mathbb{P}$ is dominated by the Lebesgue measure on $\mathbb{R}^2$. Denote by $c(\cdot, \cdot)$ the copula density associated to (X, Y), i.e., the density of the copula

$$C(u, v) := F(F_X^{-1}(u), F_Y^{-1}(v)), \ (u, v) \in]0, 1[^2,$$

where F, F_X and F_Y are, respectively, the cumulative distribution functions of (X, Y), X and Y. Assume that c belongs to some parametric model of copula densities $\{c_\beta, \beta \in \mathbb{R}^d\}$ and F_X and F_Y belong to some parametric models, see e.g. [3, 7] for examples of such models. We then have the relation

$$\frac{d\mathbb{P}}{d\mathbb{P}^\perp}(x, y) = c_\beta \left(F_X^{\gamma_1}(x), F_Y^{\gamma_2}(y) \right).$$

Numerous examples of models (6) can be then obtained by considering

$$h_\theta(x, y) = c_\beta(F_X^{\gamma_1}(x), F_Y^{\gamma_2}(y)) \quad \text{with} \quad \theta = (\beta, \gamma_1, \gamma_2).$$

We can also deal with semiparametric models induced by parametric models of copula densities, with nonparametric unknown continuous marginal distribution functions $F_1(\cdot)$ and $F_2(\cdot)$,

$$h_\theta(x, y) := h_\theta(x, y, F_1(x), F_2(x)) = c_\theta(F_1(x), F_2(y)); \ \theta \in \Theta \subset \mathbb{R}^d.$$

Other examples of models for $d\mathbb{P}/d\mathbb{P}^\perp$ can be found in [5].

2.3 Dual Representation of φ-mutual Informations

Provided that Assumptions **A1** and **A3–4** below are fulfilled, a direct application of the dual representation results for φ-divergences obtained in [1,4], yields

$$I_\varphi(\mathbb{P}) = \sup_{\theta \in \Theta} \left\{ \int_{\mathcal{X} \times \mathcal{Y}} \varphi'(h_\theta) \, d\mathbb{P} - \int_{\mathcal{X} \times \mathcal{Y}} \varphi^* \left(\varphi'(h_\theta) \right) \, d\mathbb{P}^\perp \right\}, \tag{9}$$

where φ^* denotes the convex conjugate of φ, namely, the real function defined by

$$\varphi^*(t) := \sup_{x \in \mathbb{R}_+} (tx - \varphi(x)), \quad t \in \mathbb{R}.$$

Moreover, the supremum in (9) is unique and achieved at $\theta = \theta_T$.

Assumptions
A3: *The φ-mutual information is finite: $I_\varphi(\mathbb{P}) < \infty$.*
A4: *For all $\theta \in \Theta$, we have $\int_{\mathcal{X} \times \mathcal{Y}} |\varphi'(h_\theta(x, y))| \, d\mathbb{P}(x, y) < \infty$.*

3 Estimation of φ-MI and Tests of Independence

From an i.i.d. sample, $(X_1, Y_1), \ldots, (X_n, Y_n)$, of (X, Y), we aim at testing the null hypothesis of independence of the margins X and Y, and estimating the parameter of interest θ_T, in the context of the model (6). Hence, we consider the test problem

$$\mathcal{H}_0 : \mathbb{P} = \mathbb{P}^\perp \quad \text{against} \quad \mathcal{H}_1 : \mathbb{P} \neq \mathbb{P}^\perp,$$

which can be reformulated in the context of the present paper as

$$I_\varphi(\mathbb{P}) = 0 \quad \text{against} \quad I_\varphi(\mathbb{P}) > 0, \qquad (10)$$

which requires to estimate $I_\varphi(\mathbb{P})$.

3.1   Dual Estimation of φ-MI

A natural attempt to estimate the φ-MI of (X, Y) consists in considering the plug-in estimate of $I_\varphi(\mathbb{P})$ obtained by replacing $\mathbb{P}$ by its empirical counter-part $\widehat{\mathbb{P}} := \frac{1}{n} \sum_{i=1}^n \delta_{(X_i, Y_i)}$. Unfortunately, by doing so, we do not take advantage of the information carried by the model (6) and we only measure dependence of the contingency table associated to the sample. When dealing with variables X and Y absolutely continuous with respect to Lebesgue measure, the contingency table is a $n \times n$ table with all coefficients but diagonal ones equal to zero; particularly, variables X and Y appear (misleadingly) purely dependent.

An alternative consists in taking advantage of the dual representation (9) by introducing the semiparametric estimates – say dual estimates – of $I_\varphi(\mathbb{P})$ and θ_T given by

$$\widehat{I}_\varphi := \sup_{\theta \in \Theta} \left\{ \int_{\mathcal{X} \times \mathcal{Y}} \varphi'(h_\theta) \, d\widehat{\mathbb{P}} - \int_{\mathcal{X} \times \mathcal{Y}} \varphi^*(\varphi'(h_\theta)) \, d\widehat{\mathbb{P}}^\perp \right\} \qquad (11)$$

$$= \sup_{\theta \in \Theta} \left\{ \frac{1}{n} \sum_{i=1}^n \varphi'(h_\theta(X_i, Y_i)) - \frac{1}{n^2} \sum_{i=1}^n \sum_{j=1}^n \varphi^*(\varphi'(h_\theta(X_i, Y_j))) \right\}$$

$$\widehat{\theta}_\varphi := \arg \sup_{\theta \in \Theta} \left\{ \frac{1}{n} \sum_{i=1}^n \varphi'(h_\theta(X_i, Y_i)) - \frac{1}{n^2} \sum_{i=1}^n \sum_{j=1}^n \varphi^*(\varphi'(h_\theta(X_i, Y_j))) \right\}. \, (12)$$

Note that if $\mathcal{X} \times \mathcal{Y}$ is a finite set, then the dual estimate $\widehat{I}_\varphi$, computed in the context of the model described in Example 1.2, is equal to the plug–in estimate $I_\varphi(\widehat{\mathbb{P}})$. Particularly, for $\varphi(x) = (x-1)^2/2$, we obtain that $\widehat{I}_\varphi$ is equal (up to a factor $1/2n$) to the classical χ^2 statistic (3).

3.2   Asymptotic Behavior of Dual Estimates

The asymptotic properties of estimates (11) and (12) stated in Propositions 1 and 2 to follow are obtained by classical techniques from M-estimation theory; they require some regularity and integrability assumptions on h_θ, stated below.

Assumptions
A5: the parameter space Θ is a compact subset of $\mathbb{R} \times \mathbb{R}^d$.
A6: $\int \sup_{\theta \in \Theta} |\varphi'(h_\theta)| \, d\mathbb{P} < \infty$.
A7: $\int \sup_{\theta \in \Theta} \varphi^*(\varphi'(h_\theta))^2 \, d\mathbb{P}^\perp < \infty$.

Note that Assumptions **A6–7** imply **A3–4**.

Proposition 1. *Assume that **A1, 5–7** hold. Then, the estimates $\widehat{I}_\varphi$ of $I_\varphi(\mathbb{P})$ defined by (11) and the estimates $\widehat{\theta}_\varphi$ of θ_T defined by (12) are consistent. Precisely, as $n \to \infty$, the following convergences in probability hold*

$$\widehat{I}_\varphi \to I_\varphi(\mathbb{P}) \quad and \quad \widehat{\theta}_\varphi \to \theta_T.$$

We now derive the asymptotic distribution of the estimate of KL-MI I_{φ_1} (where φ_1 is given by (5)) under the null hypothesis of independence, for models $\mathcal{M}_\Theta$ of the specific form

$$h_\theta(x,y) = \exp\left(\alpha + m_\beta(x,y)\right) \quad \text{with} \quad m_\beta(x,y) := \sum_{k=1}^{d} \beta_k \xi_k(x)\zeta_k(y), \qquad (13)$$

for some specified measurable real valued functions ξ_k and ζ_k, $k = 1,\ldots,d$, defined, respectively, on $\mathcal{X}$ and $\mathcal{Y}$ and $\theta = (\alpha, \beta^\top)^\top \in \Theta \subseteq \mathbb{R}^{1+d}$.

Assumptions
A8: *The third order partial derivatives $(\partial^3/\partial\theta^3)\varphi_1'(h_\theta)$ (resp. $(\partial^3/\partial\theta^3)\varphi_1^*$ $(\varphi_1'(h_\theta)))$ are dominated, on some neighborhood of θ_T, by some $\mathbb{P}$-integrable function (resp. some $\mathbb{P}^\perp$-square-integrable function);*
A9: *The integrals $\mathbb{P}\left\|(\partial/\partial\theta)\varphi_1'(h_{\theta_T})\right\|^2$, $\mathbb{P}^\perp\left\|(\partial/\partial\theta)\varphi_1^*(\varphi_1'(h_{\theta_T}))\right\|^2$, $\mathbb{P}\left\|(\partial^2/\partial\theta^2)\varphi_1'(h_{\theta_T})\right\|$, $\mathbb{P}^\perp\left\|(\partial^2/\partial\theta^2)\varphi_1^*(\varphi_1'(h_{\theta_T}))\right\|^2$ are finite, and the matrix*

$$\mathbb{P}(\partial^2/\partial\theta^2)\varphi_1'(h_{\theta_T}) - \mathbb{P}^\perp(\partial^2/\partial\theta^2)\varphi_1^*(\varphi_1'(h_{\theta_T}))$$

is nonsingular.

Proposition 2. *Assume that conditions **A1–2, 5–9** hold and that $\mathbb{P} = \mathbb{P}^\perp$, in the context of the model (13) (i.e., $\theta_T = \theta_0 = \mathbf{0}$). Then,*

(1) $\sqrt{n}\widehat{\theta}_{\varphi_1}$ converges in distribution to a $(1+d)$-dimensional centered Gaussian vector with explicit covariance matrix;
(2) $2n\widehat{I}_{\varphi_1}$ converges in distribution to the random variable Z^tZ, where Z is a $(1+d)$-dimensional centered Gaussian vector with explicit covariance matrix.

See [5] for explicit expressions of the asymptotic covariance matrices of $\sqrt{n}\widehat{\theta}_{\varphi_1}$ and Z, as well as the proofs of Propositions 1 and 2.

The decision rule for the test (10) for a signification level $\alpha \in (0,1)$, consists in rejecting the independence hypothesis if the estimate $\widehat{I}_\varphi$ is greater than some critical value, namely, the quantile $q_{1-\alpha}$ of the distribution of $\widehat{I}_\varphi$. If $\varphi = \varphi_1$ and the sample size is large, the distribution of $2n\widehat{I}_{\varphi_1}$ can be replaced by the asymptotic distribution given in Proposition 2.

4   Bahadur Asymptotic Efficiency of φ-MI-Based Tests

Given $(\widehat{I}_{\varphi_1})_n$ and $(\widehat{I}_{\varphi_2})_n$ two sequences of statistics for the test problem (10), numbers $\alpha \in (0,1)$, $\gamma \in (0,1)$ and an alternative $P \in \mathcal{M}_\Theta$, we define $n_i(\alpha, \gamma, P)$, for $i \in \{1, 2\}$, respectively, as the minimal number of observations needed for the test based on $\widehat{I}_{\varphi_i}$ to have signification level α and power level γ. Then, Bahadur asymptotic relative efficiency of $(\widehat{I}_{\varphi_1})_n$ with respect to $(\widehat{I}_{\varphi_2})_n$ is defined as (if the limit exists) $\lim_{\alpha \to 0} n_2(\alpha, \gamma, P)/n_1(\alpha, \gamma, P)$.

It is well known (see for example [9], Chap. 14), that if both sequences $(\widehat{I}_{\varphi_1})_n$ and $(\widehat{I}_{\varphi_2})_n$ satisfy a large deviations principle (LDP) under the null hypothesis (with good rate functions $e_{\varphi_1}(\cdot)$ and $e_{\varphi_2}(\cdot)$) and also a law of large number under the alternative hypothesis, with asymptotic means $\mu_{\varphi_1}(P)$ and $\mu_{\varphi_2}(P)$, respectively, then their Bahadur asymptotic relative efficiency equals $e_{\varphi_1}(\mu_{\varphi_1}(P))/e_{\varphi_2}(\mu_{\varphi_2}(P))$. Particularly, the most efficient test maximizes Bahadur slope $e_\varphi(\mu_\varphi(P))$.

4.1   A Large Deviations Principle for Dual Estimates of φ-MI

The following theorem establishes a LDP for $(\widehat{I}_\varphi)_n$ under the null hypothesis for specific models described by Assumptions **A11.a** or **A11.b** to follow. It relies on some generalization due to [2] of classical Sanov's theorem to finer topologies – requiring the finiteness of some exponential moments, as stated in Assumption **A10** to follow – and the contraction principle.

Let $\mathcal{F}$ be the set of measurable functions from $\mathcal{X} \times \mathcal{Y}$ into $\mathbb{R}$, given by

$$\mathcal{F} := \mathcal{B} \cup \{\varphi'(h_\theta); \theta \in \Theta\} \cup \{\varphi^*(\varphi'(h_\theta)); \theta \in \Theta\},$$

where $\mathcal{B}$ is the set of measurable bounded functions from $\mathcal{X} \times \mathcal{Y}$ into $\mathbb{R}$. Let us introduce the subset

$$M_\mathcal{F} := \left\{ P \in \mathcal{M}_1 : \int_{\mathcal{X} \times \mathcal{Y}} |f|\, \mathrm{d}P < \infty \text{ and } \int_{\mathcal{X} \times \mathcal{Y}} |f|\, \mathrm{d}P^\perp < \infty, \forall f \in \mathcal{F} \right\}.$$

Define on $M_\mathcal{F}$ the $\tau_\mathcal{F}$-topology as the coarsest one that makes applications $P \in M_\mathcal{F} \mapsto \int f\, \mathrm{d}P$ and $P \in M_\mathcal{F} \mapsto \int f\, \mathrm{d}P^\perp$ continuous, for all $f \in \mathcal{F}$.

Assumptions
A10: *for all $f \in \mathcal{F}$, for all $a > 0$, $\int \exp(a|f|)\, \mathrm{d}\mathbb{P} < \infty$.*
A11.a: *The model $\{h_\theta(\cdot, \cdot); \theta = (\alpha, \beta^\top)^\top \in \Theta\}$ is of the form*

$$h_\theta(x, y) = \exp(\alpha + m_\beta(x, y))$$

with the condition that, for any constant c and any β, we have

$$\mathbb{P}^\perp (m_\beta(X, Y) = c) \neq 0 \quad \text{iff} \quad \beta = (0, \ldots, 0)^\top, \alpha = 0 \text{ and } c = 0.$$

A11.b: *(X, Y) is finite-discrete, supported by $\mathcal{X} \times \mathcal{Y}$.*

Proposition 3. *Let (X, Y) be a couple of independent random variables with joint distribution* $\mathbb{P} = \mathbb{P}^{\perp} \in \mathcal{M}_{\Theta} \cap M_{\mathcal{F}}$.

(1) Assume that $A1$–2, 5, 10, $11.a$ hold. Then, the sequence $(\widehat{I}_{\varphi})_n$ of estimates, of $I_{\varphi}(\mathbb{P}) = 0$, given by (11), satisfies the following large deviation principle

$$\frac{1}{n} \log \mathbb{P}^{\perp} \left(\widehat{I}_{\varphi} > d \right) \overset{n \to \infty}{\longrightarrow} -e_{\varphi}(d), \quad d > 0, \tag{14}$$

where $e_{\varphi}(d) := \inf\{\mathbb{K}(\mathbb{Q}, \mathbb{P}^{\perp}) : \mathcal{D}_{\varphi}(\mathbb{Q}|\mathbb{Q}^{\perp}) > d\}$, with

$$\mathcal{D}_{\varphi}(\mathbb{Q}, \mathbb{Q}^{\perp}) := \sup_{\theta \in \Theta} \left\{ \int \varphi'(h_{\theta}) \, d\mathbb{Q} - \int \varphi^* \left(\varphi'(h_{\theta}) \right) d\mathbb{Q}^{\perp} \right\}, \quad \mathbb{Q} \in M_{\mathcal{F}}.$$

(2) Assume that $A1$–2, 5, $11.b$ hold. Then the above statement holds if $M_{\mathcal{F}}$ is replaced by the set of all discrete-finite distributions with the same finite support $\mathcal{X} \times \mathcal{Y}$.

Again, see [5] for the proof and also some sufficient conditions on $\mathcal{M}_{\Theta}$ for **A10** to hold.

4.2 KL-MI Test Is the Most Efficient Test

From the law of large number stated in Proposition 1 and the above LDP, we can compute and compare Bahadur slopes of φ-MI based independence tests, the final result being stated in the following Theorem.

Theorem 1. *Let (X, Y) be a couple of random variables with joint distribution $\mathbb{P} \in \mathcal{M}_{\Theta} \cap M_{\mathcal{F}}$. Suppose that either $A1$–2, 5, 10, $11.a$ or $A1$–2, 5, $11.b$ hold. For the test problem (10), the test based on the estimate $\widehat{I}_{\varphi_1}$, see (11), of the Kullback-Leibler mutual information, is uniformly (i.e., whatever be the alternative $\mathbb{P} \neq \mathbb{P}^{\perp}$) the most efficient test, in Bahadur sense, among all $\widehat{I}_{\varphi}$-based tests, including the classical χ^2-independence one.*

If $\mathcal{X} \times \mathcal{Y}$ is finite and $\mathcal{M}_{\Theta}$ is given by (8), assumptions $A1$–2, 5 and $11.b$ are obviously fulfilled. We then obtain that KL-MI-based independence test is more efficient than classical χ^2 independence test. This result was already stated in the context of goodness-of-fit testing, for e.g. in [9], Chap. 14. The above Theorem extends it to independence testing and to more general probability distributions.

5   Numerical Comparison of Independence Tests

We have compared the powers of KL-MI and χ^2-MI independence tests with non-correlation tests for samples drawn according to bi-variate centered normal distributions with marginal variances equal to 1 and covariance ρ varying from 0 to 1. We have fixed a signification level $\alpha = 0.05$ and computed the critical values of φ-MI-based tests whether using the asymptotic distribution of $\widehat{I}_{\varphi_1}$

stated in Proposition 2 if $\varphi = \varphi_1$ (i.e., for KL-MI) or by means of Monte-Carlo simulations of the distribution of $\widehat{I}_{\varphi_2}$ (i.e., for χ^2-MI). Then we have estimated the power of these tests as well as of non-correlation tests, still by Monte-Carlo methods. Figure 1 presents the power curves for KL-MI (plain black curve), χ^2-MI (dotted black curve) independence tests and Pearson (dashed red curve), Kendall and Spearman (mixed dashed and dotted red and blue curves) correlation tests, obtained from $N = 1000$ samples of size $n = 50$ of bi-variate Gaussian distributions. For this setting, KL-MI independence test is almost as powerful as the most uniformly powerful independence test (Pearson). χ^2-MI, Spearman and Kendall tests have comparable powers, slightly lower than KL-MI and Pearson's ones.

Other Monte-Carlo based power comparisons in the Gaussian setting are available in [5], as well as power comparisons for semiparametric copula models described in Example 1.3, for which critical values are estimated by using bootstrap methods.

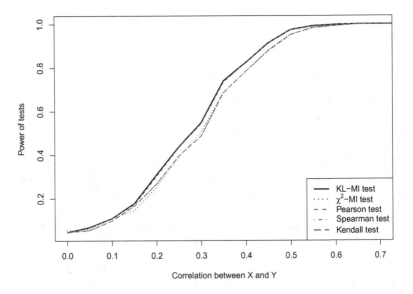

Fig. 1. Comparison of powers of φ-MI and χ^2-MI tests with non-correlation tests of Pearson, Spearman and Kendall tests.

6   Conclusion and Discussion

In this paper, we have defined and studied estimates of φ-mutual informations, based on the dual representation of φ-divergences and a semiparametric modeling of the density ratio between the joint distribution of the couple and the product distribution of its margins. The consistency of these estimates has been established assuming some classical regularity conditions on the model;

the asymptotic normality has been established for classical Kullback-Leibler mutual information (KL-MI) and specific models. A class of independence tests has been derived from these estimates, recovering as a particular case, the classical χ^2-independence test. For a large variety of situations including finite-discrete random couples, the most efficient test is based on the proposed KL-MI estimates, outperforming the classical χ^2-independence one.

References

1. Broniatowski, M., Keziou, A.: Minimization of ϕ-divergences on sets of signed measures. Studia Sci. Math. Hung. **43**(4), 403–442 (2006). arXiv:1003.5457
2. Eichelsbacher, P., Schmock, U.: Large deviations of U-empirical measures in strong topologies and applications. Ann. Inst. Henri Poincaré Probab. Stat. **38**(5), 779–797 (2002)
3. Joe, H.: Multivariate Models and Dependence Concepts. Monographs on Statistics and Applied Probability, vol. 73. Chapman & Hall, London (1997). http://dx.doi.org/10.1201/b13150
4. Keziou, A.: Dual representation of ϕ-divergences and applications. C. R. Math. Acad. Sci. Paris **336**(10), 857–862 (2003)
5. Keziou, A., Regnault, P.: Semiparametric estimation of mutual information and related criteria: Optimal test of independence. Manuscript in preparation (2015)
6. Khan, S., Bandyopadhyay, S., Ganguly, A.R., Saigal, S., Erickson, D.J., Protopopescu, V., Ostrouchov, G.: Relative performance of mutual information estimation methods for quantifying the dependence among short and noisy data. Phys. Rev. E **76**, 026209 (2007). http://link.aps.org/doi/10.1103/PhysRevE.76.026209
7. Nelsen, R.B.: An Introduction to Copulas. Springer Series in Statistics, Second edn. Springer, New York (2006)
8. Pardo, L.: Statistical Inference Based on Divergence Measures. Statistics: Textbooks and Monographs, vol. 185. Chapman & Hall/CRC, Boca Raton (2006)
9. van der Vaart, A.W.: Asymptotic Statistics. Cambridge Series in Statistical and Probabilistic Mathematics. Cambridge University Press, Cambridge (1998)

Optimization on Manifold

Riemannian Trust Regions with Finite-Difference Hessian Approximations are Globally Convergent

Nicolas Boumal[(⊠)]

Inria & D.I., UMR 8548, Ecole Normale Supérieure, Paris, France
nicolasboumal@gmail.com

Abstract. The Riemannian trust-region algorithm (RTR) is designed to optimize differentiable cost functions on Riemannian manifolds. It proceeds by iteratively optimizing local models of the cost function. When these models are exact up to second order, RTR boasts a quadratic convergence rate to critical points. In practice, building such models requires computing the Riemannian Hessian, which may be challenging. A simple idea to alleviate this difficulty is to approximate the Hessian using finite differences of the gradient. Unfortunately, this is a nonlinear approximation, which breaks the known convergence results for RTR.

We propose RTR-FD: a modification of RTR which retains global convergence when the Hessian is approximated using finite differences. Importantly, RTR-FD reduces gracefully to RTR if a linear approximation is used. This algorithm is available in the Manopt toolbox.

Keywords: RTR-FD · Optimization on manifolds · Convergence · Manopt

1 Introduction

The Riemannian trust-region method (RTR) is a popular algorithm designed to minimize differentiable cost functions f over Riemannian manifolds $\mathcal{M}$ [1,2]. That is, RTR attempts to compute $\min_{x \in \mathcal{M}} f(x)$. Starting with a given initial guess $x_0 \in \mathcal{M}$, it iteratively reduces the cost $f(x_k)$ along a sequence $x_0, x_1, \ldots$

Under conditions we explicit later, the sequences of iterates produced by RTR converge to critical points regardless of the initial guess (this is called *global convergence*). A critical point $x \in \mathcal{M}$ is such that $\operatorname{grad} f(x) = 0$, where $\operatorname{grad} f(x)$ is the Riemannian gradient of f at x. Since all global optimizers are critical points, this property is highly desirable.

RTR proceeds as follows. At the current iterate $x_k \in \mathcal{M}$, it produces a candidate next iterate, $x_k^+ \in \mathcal{M}$, by (approximately) minimizing a local *model* m_k of f in a neighborhood of x_k, called a *trust region*—because this is where we trust the model. This procedure always reduces the model cost m_k, but of course, the aim is to reduce the actual cost f. RTR then computes the actual cost improvement and decides to accept or reject the proposed step x_k^+ accordingly.

© Springer International Publishing Switzerland 2015
F. Nielsen and F. Barbaresco (Eds.): GSI 2015, LNCS 9389, pp. 467–475, 2015.
DOI: 10.1007/978-3-319-25040-3_50

Furthermore, depending on how accurately the actual cost improvement was predicted by the model, the size Δ_k of the trust region is reduced, increased or left unchanged for the next iteration. See Algorithm 1.

To be precise, the *inner problem* at iteration k takes the following form:

$$\min_{\eta \in T_{x_k}\mathcal{M}, \|\eta\|_{P_k^{-1}} \leq \Delta_k} m_k(\eta) := f(x_k) + \langle \eta, \mathrm{grad} f(x_k) \rangle_{x_k} + \frac{1}{2} \langle \eta, H_k[\eta] \rangle_{x_k}, \quad (1)$$

where $T_x\mathcal{M}$ is the tangent space to $\mathcal{M}$ at x, $\langle \cdot, \cdot \rangle_x$ is the Riemannian metric on $T_x\mathcal{M}$, $H_k \colon T_{x_k}\mathcal{M} \to T_{x_k}\mathcal{M}$ is an operator (conditions on H_k are the topic of this paper), $P_k \colon T_{x_k}\mathcal{M} \to T_{x_k}\mathcal{M}$ is a symmetric *positive definite* preconditioner, $\|\eta\|_{P_k^{-1}}^2 := \langle \eta, P_k^{-1}\eta \rangle_{x_k}$ defines a norm and Δ_k is the size of the trust region at iteration k. Ideally, P_k is a cheap, positive approximation of the inverse of the Hessian of f at x_k. For a first read, it is safe to assume $P_k = \mathrm{Id}$ (identity).

An approximate solution η_k to the inner problem is computed, and the candidate next iterate is obtained as $x_k^+ = \mathrm{Retr}_{x_k}\eta_k$, where $\mathrm{Retr}_x \colon T_x\mathcal{M} \to \mathcal{M}$ is a *retraction* on $\mathcal{M}$ [2, def. 4.1.1]: a relaxation of the differential geometric notion of *exponential*. For all x, it satisfies $\mathrm{Retr}_x(0) = x$, and the derivative of $t \mapsto \mathrm{Retr}_x t\eta$ at $t = 0$ equals η, for all tangent η. Furthermore, $\mathrm{Retr}_x\eta$ is smooth in both x and η. If $\mathcal{M}$ is a Euclidean space such as $\mathbb{R}^n$, the classical choice is $\mathrm{Retr}_x\eta = x + \eta$.

A remarkable feature of RTR is that it guarantees global convergence under very lax conditions on both the H_k's and how well (1) is solved [2, thm. 7.4.4]. Essentially two things are required: (a) that the H_k's be uniformly bounded,

---

Algorithm 1. RTR : preconditioned Riemannian trust-region method

1: **Given:** $x_0 \in \mathcal{M}$, $0 < \Delta_0 \leq \bar{\Delta}$ and $0 < \rho' < 1/4$
2: **Init:** $k = 0$
3: **repeat**
4:     $\eta_k = \mathrm{tCG}(x_k, \Delta_k, H_k, P_k)$ ▷ solve inner problem (1) (approximately)
5:     $x_k^+ = \mathrm{Retr}_{x_k}(\eta_k)$ ▷ candidate next iterate
6:     $\rho_1 = f(x_k) - f(x_k^+)$ ▷ actual improvement
7:     $\rho_2 = m_k(0) - m_k(\eta_k)$ ▷ model improvement
8:     **if** $\rho_1/\rho_2 < 1/4$ **then**   ▷ if the model made a poor prediction
9:         $\Delta_{k+1} = \Delta_k/4$ ▷ reduce the trust region radius
                    ▷ if the model is good but the region is too small
10:    **else if** $\rho_1/\rho_2 > 3/4$ **and** tCG hit the boundary **then**
11:        $\Delta_{k+1} = \min(2\Delta_k, \bar{\Delta})$ ▷ enlarge the radius
12:    **else**
13:        $\Delta_{k+1} = \Delta_k$
14:    **end if**
15:    **if** $\rho_1/\rho_2 > \rho'$ **then**   ▷ if the relative decrease is sufficient
16:        $x_{k+1} = x_k^+$ ▷ accept the step
17:    **else**   ▷ otherwise
18:        $x_{k+1} = x_k$ ▷ reject it
19:    **end if**
20:    $k = k + 1$
21: **until** a stopping criterion triggers

symmetric *linear* operators, and (b) that the approximate model minimizers η_k produce at least the following decrease in the *model* cost at each iteration [2, eq. (7.14)]:

$$m_k(0) - m_k(\eta_k) \geq c_1 \|\mathrm{grad} f(x_k)\|_{x_k} \min\left(\Delta_k, c_2 \|\mathrm{grad} f(x_k)\|_{x_k}\right), \qquad (2)$$

where $\|\eta\|_{x_k}^2 := \langle \eta, \eta \rangle_{x_k}$ and $c_1, c_2 > 0$ are constants.

Then, two things are known: (a) if η_k is produced by the *Steihaug-Toint truncated conjugate-gradients algorithm* (tCG, Algorithm 2), sufficient decrease is attained; and (b) still using tCG, if H_k is a sufficiently good approximation of the Riemannian Hessian of f at x_k, RTR achieves a *superlinear local convergence rate* [2, thm. 7.4.11], i.e., close to an isolated local minimizer, convergence is fast.

As computing the Riemannian Hessian can be cumbersome (at best), there is a need for good, generic approximations of it. Linear approximations based on finite differences of the gradient have been proposed [2, §8.2.1], but they are impractical, since they require the computation of a full operator H_k expanded in a basis of $\mathrm{T}_{x_k}\mathcal{M}$: this is an issue if the dimension of $\mathcal{M}$ is large or if it is difficult to define natural bases of the tangent spaces, that is, in most cases. Alternatives based on transporting an approximate Hessian from tangent space to tangent space may constitute a good solution, especially in low-dimension, even if they are arguably delicate to implement and typically require extra memory [6,7].

On the other hand, it is quite natural to propose a *nonlinear* approximation of the Hessian at x_k based on finite differences, as $H_k^{\mathrm{FD}}[0] = 0$ and

$$H_k^{\mathrm{FD}}[\eta] = \frac{\mathrm{Transp}_{x_k \leftarrow y} \mathrm{grad} f(y) - \mathrm{grad} f(x_k)}{c}, \quad \text{with} \quad \begin{cases} c = \alpha/\|\eta\|_{x_k}, \\ y = \mathrm{Retr}_{x_k} c\eta, \end{cases} \qquad (3)$$

where $\alpha > 0$ is a small constant (more on this later) and $\mathrm{Transp}_{x \leftarrow y}$ is a *transporter*, i.e., a linear operator from $\mathrm{T}_y\mathcal{M}$ to $\mathrm{T}_x\mathcal{M}$ whose dependence on x and y is jointly continuous and such that $\mathrm{Transp}_{x \leftarrow x} = \mathrm{Id}$ for all x.[1] Transporters allow comparing vectors in different tangent spaces. In this respect, they are loose relaxations of the concept of *parallel transport* in differential geometry. For $\mathcal{M}$ a Euclidean space, the classical choice is $\mathrm{Transp}_{x \leftarrow y} = \mathrm{Id}$ since $\mathrm{T}_x\mathcal{M} \equiv \mathrm{T}_y\mathcal{M}$.

H_k^{FD} is cheap and simple to compute: it essentially requires a single extra gradient evaluation. Unfortunately, because it is nonlinear, the known global convergence theory for RTR does not apply as is. In this paper, we show how a tiny modification to the tCG algorithm makes it possible to retain global convergence even if H_k is only *radially linear*, by which we mean:

$$\forall \eta \in \mathrm{T}_{x_k}\mathcal{M}, \forall a \geq 0, \quad H_k[a\eta] = aH_k[\eta]. \qquad (4)$$

Since H_k^{FD} is radially linear, this is a good first step. We then show that H_k^{FD} satisfies the other important condition, namely, uniform boundedness, under mild extra assumptions. Lastly, we note that the modification of tCG is innocuous if H_k is linear, so that it is safe to use the modified version for all purposes.

---

[1] Transporters [7, §4.3] are mostly equivalent to *vector transports* [2, def. 8.1.1].

Algorithm 2. tCG(x, Δ, H, P) : modified Steihaug-Toint truncated CG method. It is obtained from the classical tCG by adding the highlighted instructions. See Section 4 for details on how to evaluate $m(\eta)$ and P^{-1}-norms.

1: **Given:** $x \in \mathcal{M}$ and $\Delta, \theta, \kappa > 0$, $H, P \colon T_x\mathcal{M} \to T_x\mathcal{M}$,
        H radially linear (4), P symmetric positive definite.
2: **Init:** $\eta^0 = 0 \in T_x\mathcal{M}, r_0 = \mathrm{grad}f(x), z_0 = P[r_0], \delta_0 = -z_0$
3: **for** $j = 0 \ldots$ max inner iterations $- 1$ **do**
4:     $\kappa_j = \langle \delta_j, H[\delta_j] \rangle_x$
5:     $\alpha_j = \langle z_j, r_j \rangle_x / \kappa_j$
6:     **if** $\kappa_j \leq 0$ **or** $\|\eta^j + \alpha_j \delta_j\|_{P^{-1}} \geq \Delta$ **then**
                    ▷ the model Hessian has negative curvature or TR exceeded:
7:        Set τ_j to be the positive root of $\|\eta^j + \tau_j \delta_j\|_{P^{-1}}^2 = \Delta^2$
8:        $\eta^{j+1} = \eta^j + \tau_j \delta_j$ ▷ hit the boundary

> 9:   **if** $m(\eta^{j+1}) \geq m(\eta^j)$ **then**   ▷ never triggers if H is linear or $j = 0$
> 10:   **return** η^j ▷ η^j is sure to decrease the model cost
> 11:   **end if**

12:        **return** η^{j+1}
13:     **end if**
14:     $\eta^{j+1} = \eta^j + \alpha_j \delta_j$

> 15:   **if** $m(\eta^{j+1}) \geq m(\eta^j)$ **then**   ▷ idem
> 16:   **return** η^j
> 17:   **end if**

18:     $r_{j+1} = r_j + \alpha_j H[\delta_j]$
19:     **if** $\|r_{j+1}\|_x \leq \|r_0\|_x \cdot \min(\|r_0\|_x^\theta, \kappa)$ **then**
20:        **return** η^{j+1} ▷ this approximate solution is good enough
21:     **end if**
22:     $z_{j+1} = P[r_{j+1}]$
23:     $\beta_j = \langle z_{j+1}, r_{j+1} \rangle_x / \langle z_j, r_j \rangle_x$
24:     $\delta_{j+1} = -z_{j+1} + \beta_j \delta_j$
25: **end for**
26: **return** η^{last}

We name RTR with the modified tCG algorithm and the finite-difference Hessian approximation $H_k := H_k^{\mathrm{FD}}$ the *RTR-FD algorithm*. The Manopt toolbox [4] implements RTR-FD as a default fall-back in case the user does not specify the Hessian. Experience shows it performs well in practice (see for example [3]).

2   Global Convergence with Bounded, Radially Linear H_k

Let $\mathcal{M}$ be a finite-dimensional Riemannian manifold and $f \colon \mathcal{M} \to \mathbb{R}$ be a scalar field on $\mathcal{M}$. We use the notation dist(x, y) to denote the *Riemannian distance* between two points x and y on $\mathcal{M}$. The *injectivity radius* of $\mathcal{M}$ is defined as

$$i(\mathcal{M}) := \inf_{x \in \mathcal{M}} \sup\{\varepsilon > 0 : \text{Exp}_x|_{\{\eta \in T_x \mathcal{M}: \|\eta\|_x < \varepsilon\}} \text{ is a diffeomorphism}\},$$

where $\text{Exp}_x : T_x \mathcal{M} \to \mathcal{M}$ is the (geometric) *exponential map* at x; loosely, the operator that generates geodesics. In other words, for all x, y such that $\text{dist}(x, y) < i(\mathcal{M})$, there exists a unique minimizing geodesic joining x to y. In particular, $i(\mathbb{R}^n) = \infty$. For such close points, there is a unique, privileged transporter $\text{PTransp}_{x \leftarrow y}$, called the *parallel transporter* [2, p. 148]. Assuming $i(\mathcal{M}) > 0$, we say f is *Lipschitz continuously differentiable* [2, def. 7.4.3] if it is differentiable and there exists β_1 such that, for all x, y with $\text{dist}(x, y) < i(\mathcal{M})$,

$$\|\text{PTransp}_{x \leftarrow y} \text{grad} f(y) - \text{grad} f(x)\|_x \leq \beta_1 \text{dist}(x, y). \tag{5}$$

We make the following assumptions. They differ from the standard assumptions in only two ways: (a) the H_k's are allowed to be radially linear rather than linear, which requires a slight modification of the tCG algorithm, but no modification of the proofs; and (b) preconditioners are explicitly allowed.

Assumption 1 $\mathcal{M}$ has a positive injectivity radius, $i(\mathcal{M}) > 0$.

Assumption 2 f is Lipschitz continuously differentiable (5) and $f \circ \text{Retr}$ is radially Lipschitz continuously differentiable [2, def. 7.4.1].

Assumption 3 f is bounded below , that is, $\inf_{x \in \mathcal{M}} f(x) > -\infty$.

Assumption 4 The H_k's are radially linear (4) and bounded, i.e., there exists $\beta < \infty$ such that $\|H_k\|_{\text{op}} := \max\{\|H_k[\eta]\|_{x_k} : \eta \in T_{x_k}\mathcal{M}, \|\eta\|_{x_k} = 1\} \leq \beta, \forall k$.

Assumption 5 There exist $\beta_P, \beta_{P^{-1}}$ such that, for all k, $\|P_k\|_{\text{op}} \leq \beta_P < \infty$ and $1/\|P_k^{-1}\|_{\text{op}} \geq \beta_{P^{-1}} > 0$.

Assumption 6 There exist $\mu, \delta_\mu > 0$ such that for all $x \in \mathcal{M}$ and for all $\eta \in T_x \mathcal{M}$ with $\|\eta\|_x \leq \delta_\mu$, the retraction satisfies: $\text{dist}(x, \text{Retr}_x \eta) \leq \|\eta\|_x / \mu$.

Theorem 1. *Under assumptions 1–5, the sequence $x_0, x_1, x_2, \ldots$ generated by the modified RTR-tCG algorithm (Algorithms 1–2) satisfies:*
$$\liminf_{k \to \infty} \|\text{grad} f(x_k)\|_{x_k} = 0.$$

Proof. This is essentially Theorem 7.4.2 in [2], with H_k's allowed to be radially linear rather than linear, and with the possibility to use a preconditioner P_k. Without preconditioner ($P_k = \text{Id}$), the proof in [2] turns out to apply verbatim. In the more general case, it can be verified that the first step η^1 computed by the tCG algorithm at iteration k is the *preconditioned Cauchy step*:

$$\eta^1 = \underset{\eta = -\tau P_k \text{grad} f(x_k)}{\text{argmin}} m_k(\eta), \text{ subject to: } \|\eta\|_{P_k^{-1}} \leq \Delta_k \text{ and } \tau > 0. \tag{6}$$

To verify this, execute the first step of tCG by hand (it is oblivious to the fact that H_k is only radially linear), and compare the results to the solution of (6).

The latter is simple to solve since it is a quadratic in τ, to be minimized on an interval. Using the analytic expression for η^1, it can be seen that

$$m_k(0) - m_k(\eta^1) \geq \frac{1}{2}\|\operatorname{grad}f(x_k)\|_{P_k} \min\left(\Delta_k, \frac{\|\operatorname{grad}f(x_k)\|_{P_k}}{\|P_k^{1/2} \circ H_k \circ P_k^{1/2}\|_{\mathrm{op}}}\right),$$

where $\|\eta\|_{P_k}^2 = \langle \eta, P_k\eta\rangle_{x_k}$. By submultiplicativity of the operator norm,

$$\|P_k^{1/2} \circ H_k \circ P_k^{1/2}\|_{\mathrm{op}} \leq \|P_k\|_{\mathrm{op}} \|H_k\|_{\mathrm{op}} \leq \beta\beta_P.$$

Furthermore, $\|\operatorname{grad}f(x_k)\|_{P_k} \geq \beta_{P-1}^{1/2}\|\operatorname{grad}f(x_k)\|_{x_k}$. Thus, the sufficient decrease condition (2) is fulfilled by η^1. If H_k is linear, then tCG guarantees $m_k(\eta^{j+1}) < m_k(\eta^j)$ [2, Prop. 7.3.2], so that if η^1 is a sufficiently good approximate solution to the inner problem (which it is), then certainly the solution tCG returns, η_k, is too. For nonlinear H_k, this is not guaranteed anymore, hence the proposed modified tCG, which ensures that, if η^{j+1} is worse than η^j (as per the model), then η^j is returned. The latter is at least as good as η^1, hence (2) holds. $\square$

The proof that all accumulations points are critical points holds verbatim, even though we allow the H_k's to be merely radially linear [2, Thm. 7.4.4]:

Theorem 2. *Under assumptions 1–6, the sequence $x_0, x_1, x_2, \ldots$ generated by the modified RTR-tCG algorithm (Algorithms 1–2) satisfies:* $\lim_{k\to\infty} \operatorname{grad}f(x_k) = 0.$

3 H_k^{FD} is bounded and radially linear

We now show that setting $H_k := H_k^{\mathrm{FD}}$ (3) to approximate the Hessian of f using finite differences fulfills Assumption 4, under these mild additional assumptions:

Assumption 7. There exist $\mu', \delta_{\mu'} > 0$ such that, for all x, y with $\mathrm{dist}(x,y) \leq \delta_{\mu'}$, the transporter satisfies $\|\operatorname{Transp}_{x\leftarrow y}\|_{\mathrm{op}} \leq \mu'$.

Assumption 8. There exist $\beta_2, \delta_{\beta_2} > 0$ such that, for all x with $f(x) \leq f(x_0)$ and y with $\mathrm{dist}(x,y) \leq \delta_{\beta_2}$, it holds that $\|\operatorname{grad}f(y)\|_y \leq \beta_2$.

Assumption 7 is inconsequential, since ideal transporters are close to isometries: it would make little sense to violate it. Assumption 8 should be easily achieved, given that f is already assumed Lipschitz continuously differentiable. These assumptions allow to make the following statement:

Theorem 3. *Under assumptions 1–3 and 5–8, with $0 < \alpha < \min(\delta_\mu, \mu\delta_{\mu'}, \mu\delta_{\beta_2})$, the operators H_k^{FD} satisfy Assumption 4, so that the sequence $x_0, x_1, x_2, \ldots$ generated by RTR-FD satisfies:*

$$\lim_{k\to\infty} \operatorname{grad}f(x_k) = 0.$$

Proof. It is clear that H_k^{FD} is radially linear. Let us show that it is also uniformly bounded. For all $\eta \in \mathrm{T}_{x_k}\mathcal{M}$ with $\|\eta\|_{x_k} = 1$ and $y = \operatorname{Retr}_{x_k}(\alpha\eta)$, Assumption 6 ensures that $\alpha < \delta_\mu$ implies $\mathrm{dist}(x_k, y) \leq \alpha/\mu < \min(\delta_{\mu'}, \delta_{\beta_2})$, so that:

$$\|H_k^{\mathrm{FD}}[\eta]\|_{x_k} = \frac{1}{\alpha} \left\| \mathrm{Transp}_{x_k \leftarrow y} \mathrm{grad} f(y) - \mathrm{grad} f(x_k) \right\|_{x_k}$$

$$\leq \frac{1}{\alpha} \left(\left\| \mathrm{Transp}_{x_k \leftarrow y} \right\|_{\mathrm{op}} \left\| \mathrm{grad} f(y) \right\|_y + \left\| \mathrm{grad} f(x_k) \right\|_{x_k} \right)$$

$$\leq \frac{(1 + \mu')\beta_2}{\alpha} =: \beta.$$

Note: The dependence on $1/\alpha$ is likely artificial and might be removed. One potential start is to argue that $g(y) = \left\| \mathrm{Transp}_{x \leftarrow y} - \mathrm{PTransp}_{x \leftarrow y} \right\|_{\mathrm{op}}$ cannot grow faster than $c_x \cdot \mathrm{dist}(x, y)$ for some constant c_x (since $g(x) = 0$ and g is continuous), and then to use Lipschitz continuous differentiability of f. □

Corollary 1. *If $\mathcal{M}$ is a Euclidean space (for example, $\mathbb{R}^n$), equipped with the standard tools* $\mathrm{Retr}_x \eta = x + \eta$ *and* $\mathrm{Transp}_{x \leftarrow y} = \mathrm{Id}$, *under assumptions 2, 3 and 5, the sequence $x_0, x_1, x_2, \ldots$ generated by RTR-FD with $\alpha > 0$ satisfies:*

$$\lim_{k \to \infty} \mathrm{grad} f(x_k) = 0.$$

Proof. Assumptions 1 and 6 are clearly fulfilled, with $i(\mathcal{M}) = \delta_\mu = \infty$ and $\mu = 1$. Assumptions 7 and 8 are not necessary, since, by Assumption 2, for $\eta \neq 0$,

$$\|H_k^{\mathrm{FD}}[\eta]\|_{x_k} / \|\eta\|_{x_k} = \frac{1}{\alpha} \left\| \mathrm{grad} f(x_k + c\eta) - \mathrm{grad} f(x_k) \right\|_{x_k} \leq \beta_1 =: \beta.$$

□

Corollary 2. *If $\mathcal{M}$ is a compact manifold and f is twice continuously differentiable, under Assumption 5 and with the same constraint on α as in Theorem 3, the sequence $x_0, x_1, x_2, \ldots$ generated by RTR-FD satisfies:* $\lim_{k \to \infty} \mathrm{grad} f(x_k) = 0$.

Proof. $\mathcal{M}$ compact implies assumptions 1, 6 and 7. f twice continuously differentiable with $\mathcal{M}$ compact implies assumptions 2, 3 and 8. See [2, Cor. 7.4.6]. □

4   A Technical Point for Computational Efficiency

Proposition 7.3.2 in [2] ensures that, provided the operator H is linear, then the model cost strictly decreases at each iteration of tCG: $m(\eta^{j+1}) < m(\eta^j)$. This notably means that there is no need to track $m(\eta^j)$. Allowing for nonlinear H's, this property is lost. The proposed fix (the modified tCG, Algorithm) tracks the model cost and safely terminates if a violation (a non-decrease) is witnessed.

A direct implementation of the modified tCG algorithm evaluates the model cost $f(x) + \langle \eta^j, \mathrm{grad} f(x) \rangle_x + 1/2 \langle \eta^j, H[\eta^j] \rangle_x$ at each iteration. This is not advisable, because it requires computing $H[\eta^j]$ whereas only $H[\delta_j]$ is readily available. If H were linear, then it would hold that $H[\eta^{j+1}] = H[\eta^j + c_j \delta_j] = H[\eta^j] + c_j H[\delta_j]$, with c_j either equal to τ_j or to α_j, as prescribed by the algorithm. This suggests a recurrence to evaluate the model cost without requiring additional applications of H, which is what we use in practice. The sequence $\zeta_0, \zeta_1, \ldots$

defined by $\zeta_0 = 0$ and $\zeta_{j+1} = \zeta_j + c_j H[\delta_j]$ coincides with $H[\eta^0], H[\eta^1], \dots$ when H is linear. The model cost at η^j is evaluated as $f(x) + \langle \eta^j, \operatorname{grad} f(x) \rangle_x + 1/2 \langle \eta^j, \zeta_j \rangle_x$.

Of course, for nonlinear H, this does not correspond to the original model. But the convergence result still holds if it corresponds to a model using $\tilde{H}$, where the $\tilde{H}$'s are still radially linear and uniformly bounded.

Conceptually run tCG a first time as described above. Then, define $\tilde{H}$ such that it is radially linear, satisfies $\tilde{H}[\eta^j] = \zeta_j$ and $\tilde{H}[\delta_j] = H[\delta_j]$, and coincides with H otherwise. ($\tilde{H}$ is never constructed in practice; it merely serves the argument.) This is well defined as long as no two vectors among $\delta_1, \delta_2, \dots, \delta_{\text{last}}$ and $\eta^1, \eta^2, \dots, \eta^{\text{last}}$ are aligned on the same (positive) ray (δ_0 and η^1 are aligned by construction, in a compatible fashion). We do not prove that this property holds, but we note that it seems unlikely that it would not, in practical instances. Then, the operators $\tilde{H}$ remain uniformly bounded provided the $\|\zeta_j\|/\|\eta^j\|$'s are uniformly bounded. If so, RTR with the modified tCG behaves exactly as if the models were defined using the $\tilde{H}$'s which satisfy Assumption 4, with true evaluation of the model. This would ensure global convergence.

In the same spirit, the tCG algorithm requires computations of P^{-1}-norms, to ensure iterates remain in the trust region. Since the preconditioner is often only available as a black box P, these P^{-1}-norms are typically computed via recurrences that only involve applying P—see [5, eqs.(7.5.5–7)]. These recurrences make use of the fact that, for linear H, r_{j+1} is orthogonal to $\delta_0, \dots, \delta_j$. This may not be the case for nonlinear H, so that, in general, using these recurrences may lead to iterates leaving the trust region. One possible fix is to modify the recurrences so that they do not assume the aforementioned orthogonality, but we refrain from doing so in practice, for it does not appear to affect performance.

5   Conclusion

From extensive experience, it seems that RTR-FD achieves a superlinear local convergence rate, which is expected since H_k^{FD} is "close" to the true Hessian. See for example [3]. Unfortunately, the existing local convergence analyses rely deeply on the linearity of H_k. We do not expect that a simple modification of the argument would suffice to establish superlinear convergence of RTR-FD. A possible starting point in that direction would be work by Huang et al. on Riemannian trust regions with approximate Hessians [6,7].

Acknowledgment. The author thanks P.-A. Absil for numerous helpful discussions.

References

1. Absil, P.A., Baker, C.G., Gallivan, K.A.: Trust-region methods on Riemannian manifolds. Found. Comput. Math. **7**(3), 303–330 (2007)
2. Absil, P.A., Mahony, R., Sepulchre, R.: Optimization Algorithms on Matrix Manifolds. Princeton University Press, Princeton (2008)

3. Boumal, N.: Interpolation and regression of rotation matrices. In: Nielsen, F., Barbaresco, F. (eds.) GSI 2013. LNCS, vol. 8085, pp. 345–352. Springer, Heidelberg (2013)
4. Boumal, N., Mishra, B., Absil, P.A., Sepulchre, R.: Manopt, a Matlab toolbox for optimization on manifolds. J. Mach. Learn. Res. **15**, 1455–1459 (2014). http://www.manopt.org
5. Conn, A., Gould, N., Toint, P.: Trust-region methods. MPS-SIAM Series on Optimization. Society for Industrial and Applied Mathematics, Philadelphia (2000)
6. Huang, W., Absil, P.A., Gallivan, K.: A Riemannian symmetric rank-one trust-region method. Math. Program. **150**(2), 179–216 (2014)
7. Huang, W., Gallivan, K., Absil, P.A.: A Broyden class of quasi-Newton methods for Riemannian optimization. Technical report UCL-INMA-2014.01, Université catholique de Louvain (2015)

Block-Jacobi Methods with Newton-Steps and Non-unitary Joint Matrix Diagonalization

Martin Kleinsteuber and Hao Shen[(⊠)]

Department of Electrical and Computer Engineering, Technische Universität
München, Arcisstr. 21, 80333 Munich, Germany
{kleinsteuber,hao.shen}@tum.de
http://www.gol.ei.tum.de

Abstract. In this work, we consider block-Jacobi methods with Newton steps in each subspace search and prove their local quadratic convergence to a local minimum with non-degenerate Hessian under some orthogonality assumptions on the search directions. Moreover, such a method is exemplified for non-unitary joint matrix diagonalization, where we present a block-Jacobi-type method on the oblique manifold with guaranteed local quadratic convergence.

Keywords: Jacobi algorithms · Local convergence properties · Manifold optimization · Signal separation

1 Introduction

Jacobi-type methods have been very successful in numerical linear algebra for the task of diagonalizing matrices, dating back to the seminal work of Carl Gustaf Jacob Jacobi from 1846 [1], where a scheme is presented to iteratively find the eigenvalue decomposition of a Hermitian matrix. A characteristic feature of many Jacobi-type methods is that they act to minimize the distance to diagonality while preserving some predefined constraints, cf. [2] and the references therein. Furthermore, their inherent parallelizability makes them also useful for large scale matrix computations.

In the context of jointly diagonalizing a set of Hermitian matrices, eg. JADE [3,4] is a very prominent example that borrows the idea of iteratively minimizing the distance to joint diagonality by means of unitary similarity transforms. Moreover, Jacobi-type algorithms have also demonstrated their promising performance in solving the problem of blind source separation, such as [5,6].

Block Jacobi-type procedures were developed as a generalization of standard Jacobi method in terms of grouped variables for solving symmetric eigenvalue problems or singular value problems [7]. In each Jacobi sweep, it is required to solve a sequence of sub-optimization problems, which can be computationally expensive or infeasible. To cope with this problem, we propose to employ a Newton-step that approximates a solution to this sub problem. We show that under this setting, if the predefined search directions are orthogonal with respect

© Springer International Publishing Switzerland 2015
F. Nielsen and F. Barbaresco (Eds.): GSI 2015, LNCS 9389, pp. 476–483, 2015.
DOI: 10.1007/978-3-319-25040-3_51

to the non degenerate Hessian of the minimum, then the Jacobi method converges locally to that minimum with quadratic, hence superlinear rate. We exemplify these insights in order to develop an efficient method for the problem of non-unitary joint matrix diagonalization (JMD), which arises in the context of Blind Source Separation.

2   Block-Jacobi-type Methods on Manifolds

From a viewpoint of geometric optimization, Jacobi-type methods can be considered as a generalization of coordinate descent methods to the manifold setting. Given some point on a manifold, Jacobi-type methods optimize a cost function along some predefined directions in the tangent space in order to find the next iterate on the manifold. In practice, this requires two more algorithmic ingredients: (i) a map from the tangent space to the manifold, which is traditionally done via the Riemannian exponential, but also more general concepts like retraction have been introduced for general line search method adaptions to the manifold setting; (ii) a practical step-size selection rule that approximates the search for a minimizer of the restricted cost function. In the following, we provide a formal setup and introduce some notations that help to make these concepts more concrete.

Let M be an n-dimensional smooth manifold and consider the problem of minimizing a smooth function $f \colon M \to \mathbb{R}$. Let the map $\mu \colon \mathbb{R}^n \times M \to M$ be smooth and fulfill the property that $\mu_x \colon \mathbb{R}^n \to M$, $v \mapsto \mu(v, x)$ is a local parametrization around x with $\mu_x(0) = x$[1]. Actually, it would suffice for μ_x to be defined only in an appropriate neighbourhood of 0, but we omit this detail for the sake of readability. In order to explain the predefined directions on the manifold for the purpose of optimization, let $\mathbb{R}^n = \oplus_i V^{(i)}$ be a vector space decomposition of $\mathbb{R}^n$ into a direct sum of N subspaces V_i, with $\dim V_i = \ell_i$. By a slight abuse of notation, we denote by $V_i \in \mathbb{R}^{n \times \ell_i}$ also basis of these vector spaces. Since μ_x is a local diffeomorphism, its differential map at 0

$$T_0 \mu_x \colon \mathbb{R}^n \to T_x M \tag{1}$$

is bijective, hence the images of V_i under this map, namely $\mathcal{V}_x^{(i)} := T_0 \mu_x(V_i)$, form a direct vector space decomposition of the tangent space $T_x M$, i.e.

$$T_x M = \oplus_i \mathcal{V}_x^{(i)}. \tag{2}$$

Note, that the restrictions of μ_x to subspace V_i, i.e. $\mu_x(V_i)$ for all $i = 1, \ldots, N$, are often referred to as basic transformations. Our main result in this paper states that Jacobi-type methods are locally quadratically convergent to a local minimum of f, if the Hessian at this minimum is non-degenerate and if the above decomposition of the tangent space is orthogonal w.r.t. this Hessian at the local minimum.

---

[1] That is, μ_x^{-1} is a coordinate chart around x.

In the following, we first consider the case where the $\mathcal{V}_x^{(i)}$ are one-dimensional, leading to (one-dimensional) predefined directions in the tangent-space along which optimization is performed in each step. We then consider the generalization to higher dimensions, leading to a manifold adaption of block-coordinate descent methods on manifolds.

2.1   Coordinate Descent on Manifolds

Let us consider the case where the dimension of the V_i's is equal to one, i.e. $\ell_i = 1$ for all $i = 1, \ldots n$. A straightforward adaption of coordinate descent methods to manifolds for minimizing a smooth cost function f is now as follows. It consists of iterating *sweeps*, where one sweep sequentially works off all the initially predetermined directions V_i. That is, starting from some point $x \in M$, we determine the local minimum[2] that is closest to zero of the restricted cost function $t \mapsto f \circ \mu_x(V_1 t)$. This minimum t^* then delivers the initial point $x_{\text{new}} = \mu_x(V_1 t^*)$ for a subsequent minimization along the next predetermined direction $\mu_{x_{\text{new}}}(V_2 t)$. This procedure is repeated until all directions V_i are worked off. The Jacobi-sweep is visualized in Fig. 1 and concretized in Algorithm 1.

Algorithm 1. *Jacobi-Sweep on a manifold M*

INPUT: initial point $x^{(0)} \in M$ and and directions V_i, $i = 1 \ldots n$
FOR $i = 1, \ldots, n$ DO
  STEP 1. Compute the local minimum t^* with smallest absolute value of

$$\varphi \colon \mathbb{R} \to \mathbb{R}, \quad t \mapsto f \circ \mu_{x^{(i-1)}}(V_i t) \tag{3}$$

  STEP 2. Set $x^{(i)} := \mu_{x^{(i-1)}}(V_i t^*)$
  STEP 3. Increase i

It can be shown that a *Jacobi method*, that is iterating Algorithm 1, leads to a locally quadratic convergent algorithm, if the descent directions are orthogonal with respect to the non-degenerated Hessian at a local minimum of f.

Theorem 1 ([8]). *Let M be an n-dimensional manifold and let x^* be a local minimum of the smooth cost function $f \colon M \to \mathbb{R}$ with nondegenerate Hessian $\mathsf{H}_f(x^*)$. If the $\mathcal{V}_i := T_0 \mu_{x^*}(V_i) \in T_{x^*} M$ are orthogonal with respect to $\mathsf{H}_f(x^*)$, then the Jacobi method is locally quadratic convergent to x^*.*

In practice, the search for a local minimum in STEP 1 of Algorithm 1 is often infeasible. We therefore follow a different approach that is based on a one dimensional Newton optimisation step. Similar approximations of the optimal

---

[2] The reason why we choose a local and not a global minimum here is that for the convergence analysis, this choice is needed to be smooth around a minimizer of the cost function. This can only be guaranteed by choosing the nearest local minimum along basic transformations.

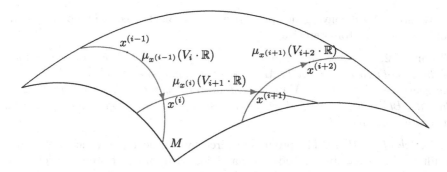

Fig. 1. Illustration of Jacobi-type method on smooth manifolds.

step size have already been used in [9,10]. The idea is to replace STEP 1 by the approximation

$$t^* := -\left(\frac{\mathrm{d}^2}{\mathrm{d}t^2}\varphi(t)\big|_{t=0}\right)^{-1}\frac{\mathrm{d}}{\mathrm{d}t}\varphi(t)\big|_{t=0}. \tag{4}$$

We show in a more general result on block-coordinate descent methods in the next section that this approximation maintains the local quadratic convergence property.

2.2 Approximate Block-Coordinate Descent on Manifolds

With the established setting, it is quite easy to generalize the above line-search methods to so-called block-Jacobi-type methods. The idea is to extend the search for a new iterate to higher dimensional subspaces, i.e. we drop the assumption that $\ell_i = 1$. To search for a new iterate, we propose to extend the one-dimensional Newton step (4) to higher dimensions. More precisely, for the i-th iteration within one sweep, we consider the restricted cost function

$$\varphi: \mathbb{R}^{\ell_i} \to \mathbb{R}, \quad t \mapsto f \circ \mu_{x^{(i-1)}}(V_i t). \tag{5}$$

Note, that now $t \in \mathbb{R}^{\ell_i}$ and $V_i \in \mathbb{R}^{n \times \ell_i}$. We denote the Hessian matrix of φ at 0 as $\mathsf{H}_\varphi(0)$ and the standard gradient as $\nabla_\varphi(0)$. The ℓ_i-dimensional Newton step

$$t^* := -\mathsf{H}_\varphi(0)^{-1}\nabla_\varphi(0) \tag{6}$$

is then a straightforward generalization of (4) to higher dimensions. The sweep for a block-Jacobi method with Newton-step size is summarized in Algorithm 2.

Algorithm 2. *Block-Jacobi-Sweep with Newton-step size on a manifold M*

INPUT: initial point $x^{(0)} \in M$ and and matrices $V_i \in \mathbb{R}^{n \times \ell_i}$, $i = 1...N$
FOR $i = 1,\ldots,N$ DO
    STEP 1. Compute the $t^* \in \mathbb{R}_i^\ell$ according to (6)
    STEP 2. Set $x^{(i)} := \mu_{x^{(i-1)}}(V_i t^*)$

We study local convergence properties of the Block-Jacobi-Sweep on a manifold M with Newton-step size.

Theorem 2. *Let M be an n-dimensional manifold and let x^* be a local minimum of the smooth cost function $f: M \to \mathbb{R}$ with nondegenerate Hessian $\mathsf{H}_f(x^*)$. If the subspaces $\mathcal{V}_i := T_0 \mu_{x^*}(V_i) \subset T_{x^*}M$ are orthogonal with respect to $\mathsf{H}_f(x^*)$, then the Block-Jacobi method which consists of iterating Algorithm 2 is locally quadratic convergent to x^*.*

Sketch of the Proof. The proof is inspired by the local quadratic convergence result for block-Jacobi methods on manifolds with an exact step-size selection rule in [11]. It consists of essentially three steps. First, one has to show that a local minimum x^* is a fixed-point of the algorithm. Second, we formulate one sweep as a map $s: M \mapsto M$ and compute its derivative at x^*. Third, we show that this derivative vanishes if the subspaces $\mathcal{V}_i$ are orthogonal with respect to the Hessian at x^*. Finally, using the Taylor Series expansion, we can then conclude that

$$\|s(x) - x^*\| \leq C\|x - x^*\|^2 \tag{7}$$

for all x being close enough to x^*, which ensures local quadratic convergence of the algorithm.

The fixed-point condition holds because $Df(x^*) = 0$ implies by use of the chain rule, that $\nabla_\varphi(0) = 0$, and hence $t^*(x^*) = 0$ for all directions. We now consider one step within a sweep given by

$$r: M \to M, \quad x \mapsto \mu(V_i t^*(x), x). \tag{8}$$

Using $t^*(x^*) = 0$, the derivative of r at x^* applied to a tangent vector $\xi \in T_{x^*}M$ is

$$Dr(x)|_{x=x^*}\xi = D_1\mu(V_i(Dt^*(x)|_{x=x^*}\xi), x^*) + D_2\mu(0, x^*)\xi \tag{9}$$

$$= D_1\mu(V_i(Dt^*(x)|_{x=x^*}\xi), x^*) + \xi \tag{10}$$

where D_l denotes the derivative w.r.t. the l-th argument. So the next step is to calculate $Dt^*(x)|_{x=x^*}\xi$. Let $e_k \in \mathbb{R}^{\ell_i}$ be the k-th standard basis vector and denote

$$\xi_k := D_1\mu(V_i e_k, x) \in T_x M. \tag{11}$$

Then the (k, l) entry of $\mathsf{H}_\varphi(0)$ is $\mathsf{H}_f(x^*)(\xi_k, \xi_l)$ and the k-th entry of $\nabla_\varphi(0)$ is $Df(x)\xi_k$. Using the product rule, we have

$$Dt^*(x)|_{x=x^*}\xi = -\big(D\mathsf{H}_\varphi(0)^{-1}\xi\big)\nabla_\varphi(0)|_{x=x^*} - \mathsf{H}_\varphi(0)^{-1}\big(D\nabla_\varphi(0)|_{x=x^*}\xi\big) \tag{12}$$

The first summand vanishes since $\nabla_\varphi(0)|_{x=x^*} = 0$. The entries of the vector on the right-hand side are $D(Df(x)|_{x=x^*}\xi_k)\xi = \mathsf{H}_f(x^*)(\xi_k, \xi)$. It follows that if $\xi^{(i)} \in \mathcal{V}_i$, i.e. if $\xi^{(i)} = \sum_k h_k \xi_k$ is a linear combination of the ξ_k, then

$$Dt^*(x)|_{x=x^*}\xi^{(i)} = -\mathsf{H}_\varphi(0)^{-1}\mathsf{H}_\varphi(0)|_{x=x^*}\begin{bmatrix} h_1, \\ \vdots \\ h_{\ell_i} \end{bmatrix} = -\begin{bmatrix} h_1, \\ \vdots \\ h_{\ell_i} \end{bmatrix}, \tag{13}$$

so that, by using (10) and the fact that $D_1\mu(V_ih, x^*) = \xi^{(i)}$, the derivative $Dr(x)|_{x=x^*}$ annihilates the V_i-component of ξ. On the other hand, if $\xi^{(i),\perp}$ is such that $H_f(x^*)(\xi_k, \xi^{(i),\perp}) = 0$ for all k, then $Dt^*(x)|_{x=x^*}\xi^{(i),\perp} = 0$ and thus

$$Dr(x)|_{x=x^*}\xi^{(i),\perp} = \xi^{(i),\perp}. \tag{14}$$

This shows that the derivative of the i-th step in the Jacobi-Sweep is an orthogonal projection with respect to $H_f(x^*)$ onto $V_i^\perp$. Therefore, using the fixed-point property of x^* and the chain rule, we conclude that

$$Ds(x)|_{x=x^*}\xi = 0, \tag{15}$$

which concludes the proof of Theorem 2.

Remark 1. It is worthwhile to notice that convergence of Jacobi-type algorithms is strongly dependent on the construction of basic transformations. Local quadratic convergence can only be attained, when the subspaces in the tangent space specified by the basic transformations are orthogonal with respect to the Hessian at a critical point. Unfortunately, both characterization of the Hessian at critical points and construction of computationally light basic transformations are non-trivial tasks in general.

3   Applications in Signal Separation

In order to investigate performance of the theoretical results presented in the last section, we employ the problem of joint matrix diagonalization as an illustrative and important example. Given a set of $m \times m$ real symmetric matrices $\{C_i\}_{i=1}^n$, constructed by $C_i = A\Lambda_i A^\top$, for $i = 1, \ldots, n$, where $\Lambda_i = \text{diag}\left(\lambda_{i1}, \ldots, \lambda_{im}\right) \in \mathbb{R}^{m \times m}$ with $\lambda_{ij} \neq 0$ for $j = 1, \ldots, m$ and $A \in Gl(m)$. The problem of estimating the matrix A given only the set $\{C_i\}_{i=1}^n$ leads to finding an $X \in Gl(m)$ such that the matrices $Y_i = X^\top C_i X$ are simultaneously diagonalised. In a generic situation, a joint diagonalizer X can only be determined up to column-wise permutation and scaling, i.e. if X is a diagonalizer, so is any XDP where D is an $m \times m$ invertible diagonal matrix and P an $m \times m$ permutation matrix, cf. [12] for a uniqueness analysis of non-unitary JMD. To deal with the scaling ambiguity, we restrict the solutions to the oblique manifold, i.e.

$$\mathcal{OB}(m) := \left\{X \in \mathbb{R}^{m \times m} \mid \text{ddiag}(X^\top X) = I_m, \text{rank}\, X = m\right\}, \tag{16}$$

where $\text{ddiag}(Z)$ forms a diagonal matrix, whose diagonal entries are just those of Z, and I_m is the $m \times m$ identity matrix.

We employ the popular off-norm function for measuring the diagonality of matrices, i.e.

$$f: \mathcal{OB}(m) \to \mathbb{R}, \quad X \mapsto \frac{1}{4}\sum_{i=1}^n \left\|\text{off}(X^\top C_i X)\right\|_{\text{F}}^2, \tag{17}$$

where $\mathrm{off}(Z) = Z - \mathrm{ddiag}(Z)$ is a matrix by setting the diagonal entries of Z to zero, and $\|\cdot\|_F$ is the Frobenius norm. In order to develop a block Jacobi algorithm to minimise the cost function f, we recall firstly a local parameterisation on $\mathcal{OB}(m)$. Let us denote the set of all $m \times m$ matrices with all diagonal entries equal to zero by

$$\mathrm{off}(m) = \{Z \in \mathbb{R}^{m \times m} | z_{ii} = 0, \text{ for } i = 1, \dots, m\}, \qquad (18)$$

then, for every point $X \in \mathcal{OB}(m)$, the following map

$$\mu_X : \mathrm{off}(m) \to \mathcal{OB}(m), \quad Z \mapsto X(I_m + Z) \mathrm{diag}\left\{\frac{1}{\|X(e_1 + z_1)\|}, \dots, \frac{1}{\|X(e_m + z_m)\|}\right\}, \quad (19)$$

where $Z = [z_1, \dots, z_m] \in \mathrm{off}(m)$ and e_i is the i-th standard basis vector of $\mathbb{R}^m$, is a local and smooth parameterisation around X. Let us define the set of matrices, whose entries are all zero except the (i,j) and (j,i) position, as

$$V_{ij} := \{Z = (z_{ij}) \in \mathbb{R}^{m \times m} | z_{pq} = 0, \text{ for } (p,q) \notin \{(i,j),(j,i)\}\}, \qquad (20)$$

with $\oplus_{i \neq j} V_{ij} = \mathrm{off}(m)$. We denote

$$\mathcal{V}_{ij}(X) := \{\tfrac{\mathrm{d}}{\mathrm{d}t} \mu_X(t \cdot Z)|_{t=0} | Z \in V_{ij}\}, \qquad (21)$$

being a predefined vector space decomposition of the tangent space $T_X\mathcal{OB}(m)$, i.e. $T_X\mathcal{OB}(m) = \oplus_{i \neq j} \mathcal{V}_{ij}(X)$.

The results in [4] have shown that the subspaces $\mathcal{V}_{ij}(X^*)$ are orthogonal with respect to $\mathsf{H}_f(X^*)$, hence validate the feasibility of construction of a block Jacobi algorithm with Newton step size selection that is locally quadratically convergent to an exact joint diagonalizer.

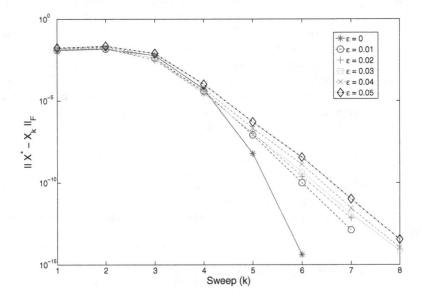

Fig. 2. Convergence properties of the proposed block Jacobi algorithm.

The task of our experiment is to jointly diagonalize a set of symmetric matrices $\{\widetilde{C}_i\}_{i=1}^n$, constructed by

$$\widetilde{C}_i = A\Lambda_i A^\top + \varepsilon E_i, \qquad i = 1, \ldots, n, \tag{22}$$

where $A \in \mathbb{R}^{m \times m}$ is a randomly picked matrix in $\mathcal{OB}(m)$, diagonal entries of Λ_i are drawn from a uniform distribution on the interval $(9, 11)$, $E_i \in \mathbb{R}^{m \times m}$ is the symmetric part of an $m \times m$ matrix, whose entries are generated from a uniform distribution on the unit interval $(-0.5, 0.5)$, representing additive noise, and $\varepsilon \in \mathbb{R}$ is the noise level. We set $m = 5$, $n = 20$, and run six tests in accordance with increasing noise, by using $\varepsilon = d \times 10^{-2}$ where $d = 0, \ldots, 5$.

The convergence of algorithms is measured by the distance of the accumulation point $X^* \in \mathcal{OB}(m)$ to the current iterate $X_k \in \mathcal{OB}(m)$, i.e. by $\|X_k - X^*\|_F$. According to Fig. 2, it is clear that our proposed algorithm converges locally quadratically fast to a joint diagonalizer under the exact nonunitary JMD setting, i.e. $\varepsilon = 0$, while with presence of noise, the algorithm seems to converge only linearly.

References

1. Jacobi, C.G.J.: Über ein leichtes verfahren, die in der theorie der säcularstörungen vorkommenden gleichungen numerisch aufzulösen. Crelle's J. für die reine und angewandte Mathematik 30, pp. 51–94 (1846)
2. Kleinsteuber, M.: A sort-Jacobi algorithm for semisimple Lie algebras. Linear Algebra Appl. **430**(1), 155–173 (2009)
3. Cardoso, J.F.: High-order contrasts for independent component analysis. Neural Comput. **11**(1), 157–192 (1999)
4. Shen, H., Hüper, K.: Block Jacobi-type methods for non-orthogonal joint diagonalisation. In: Proceedings of the 34th IEEE International Conference on Acoustics, Speech, and Signal Processing, pp. 3285–3288 (2009)
5. Shen, H.: Geometric algorithms for whitened linear independent component analysis. Ph.D. thesis, The Australian National University, Australia (2008)
6. Shen, H., Kleinsteuber, M., Hüper, K.: Block Jacobi-type methods for log-likelihood based linear independent subspace analysis. In: Proceedings of the 17th IEEE International Workshop on Machine Learning for Signal Processing (MLSP 2007), Thessaloniki, Greece, pp. 133–138 (2007)
7. Golub, G., van Loan, C.F.: Matrix Computations, 4th edn. The Johns Hopkins University Press, Baltimore (2013)
8. Kleinsteuber, M.: Jacobi-Type methods on semisimple Lie algebras - a Lie algebraic approach to the symmetric eigenvalue problem. Ph.D. thesis, Bayerische Julius-Maximilians-Universität Würzburg, Germany (2006)
9. Götze, J., Paul, S., Sauer, M.: An efficient Jacobi-like algorithm for parallel eigenvalue computation. IEEE Trans. Comput. **42**(9), 1058–1065 (1993)
10. Modi, J.J., Pryce, J.D.: Efficient implementation of Jacobi's diagonalization method on the DAP. Numer. Math. **46**(3), 443–454 (1985)
11. Hüper, K.: A calculus approach to matrix eigenvalue algorithms. Ph.D. thesis, University of Würzburg, Germany (2002)
12. Kleinsteuber, M., Shen, H.: Uniqueness analysis of non-unitary matrix joint diagonalization. IEEE Trans. Sig. Proc. **61**(7), 1786–1796 (2013)

Weakly Correlated Sparse Components with Nearly Orthonormal Loadings

Matthieu Genicot[1]([✉]), Wen Huang[1], and Nickolay T. Trendafilov[2]

[1] Université Catholique de Louvain, Louvain-la-Neuve, Belgium
matthieu.genicot@uclouvain.be
[2] Department of Mathematics and Statistics, Open University, Milton Keynes, UK

Abstract. There is already a great number of highly efficient methods producing components with sparse loadings which significantly facilitates the interpretation of principal component analysis (PCA). However, they produce either only orthonormal loadings, or only uncorrelated components, or, most frequently, neither of them. To overcome this weakness, we introduce a new approach to define sparse PCA similar to the Dantzig selector idea already employed for regression problems. In contrast to the existing methods, the new approach makes it possible to achieve simultaneously nearly uncorrelated sparse components with nearly orthonormal loadings. The performance of the new method is illustrated on real data sets. It is demonstrated that the new method outperforms one of the most popular available methods for sparse PCA in terms of preservation of principal components properties.

Keywords: Dantzig selector · LASSO · Orthonormal and oblique component loadings matrices · Optimization on matrix manifolds

1 Introduction

Modern data analysis deals with high-dimensional data sets, which number of variables is too large to allow an efficient representation. Hence, the first step in data analysis is traditionally some kind of low-dimensional data representation, e.g., performing principal component analysis (PCA) [4], and then, followed by interpretation of the initial variables contributions to the component, namely the component loadings. However, the classical PCA results involve all p input variables which complicates the interpretation. For problems with thousands of variables it is natural to look for solutions using fairly limited part of them, which, in other words, calls for sparseness. This can be achieved only by modifying the standard PCA to produce sparse component loadings. Let $\|a\|_0$ denote the cardinality of $a \in \mathbb{R}^p$, i.e., the number of its non-zero entries. Then a is called

M. Genicot—Supported by a FRIA grant from F.R.S.-FNRS, Belgium.
W. Huang—Supported by grant FNRS PDR T.0173.13.
N. T. Trendafilov—Supported by a grant RPG-2013-211 from The Leverhulme Trust, UK.

© Springer International Publishing Switzerland 2015
F. Nielsen and F. Barbaresco (Eds.): GSI 2015, LNCS 9389, pp. 484–490, 2015.
DOI: 10.1007/978-3-319-25040-3_52

sparse if $\|a\|_0 \ll p$. Usually it is more convenient to work with $\|a\|_1$, which also promotes sparseness in a, but avoids using discrete variables.

There exist a great number of methods for sparse PCA, and their number continues to increase. A recent review of different approaches and methods for sparse PCA is available in a paper of Trendafilov [11]. The existing methods for sparse PCA have a common weakness: they produce sparse matrices of component loadings that are not completely orthonormal, or that the corresponding components are correlated, or both. Hence, they do not preserve the properties of the original principal components. Only SCoTLASS [5] and the method recently proposed by Qi et al. [8] are capable to produce either orthonormal sparse loadings or uncorrelated components.

Recently, Lu and Zhang [7] consider a novel type of sparse PCA explicitly controlling the orthonormality of the loadings and the correlations among components. They produce nearly uncorrelated components and nearly orthogonal sparse loadings. Such sparse PCA approximates better the optimal features of the standard PCA. In their work, the correlation between components is controlled by means of inequality constraints. In this paper we propose another natural and simple approach to achieve the same goal.

The paper is organized as follows. In the next Sect. 2, the existing methods for sparse PCA are classified according to the form of the objective function and the constraints involved in their definitions. It will be seen that there is one way to define sparse PCA, known as *function-constrained form*, which is not explored so far. Section 3 is dedicated to follow this way and define a new sparse PCA. Finally, the performance of the new method is illustrated on a real large data set. It is demonstrated that the proposed method outperforms in many occasions one of the best available methods for sparse PCA, in that it produces solution with properties very close to the original principal components.

2   Taxonomy of PCA Subject to ℓ_1 Constraint (LASSO)

Wright [12] proposed the following taxonomy of problems seeking for sparse minimizers a of $f(a)$ through the ℓ_1 norm:

- Weighted form: $\min f(a) + \tau\|a\|_1$, for some $\tau > 0$;
- ℓ_1-constrained form (variable selection): $\min f(a)$ subject to $\|a\|_1 \leq \tau$;
- Function-constrained form: $\min \|a\|_1$ subject to $f(a) \leq \bar{f}$.

All three options were explored for regression type of problems when $f(a) = \|Xa-b\|$ and X is a given data matrix, e.g., [3,9]. This taxonomy can be restated accordingly for sparse PCA. For a given $p \times p$ correlation matrix R, let $f(a) = a^\top R a$ and consider finding a vector of loadings a, $(\|a\|_2 = 1)$, by solving one of the following:

- Weighted form: $\max a^\top R a - \tau\|a\|_1$, for some $\tau > 0$.
- ℓ_1-constrained form (variable selection): $\max a^\top R a$ subject to $\|a\|_1 \leq \tau$, $\tau \in [1, \sqrt{p}]$.

– Function-constrained form: min $\|a\|_1$ subject to $a^\top Ra \geq \lambda_{max} - \epsilon$, for some $\epsilon > 0$ and with λ the largest eigenvalue of R.

The first two forms were explored in a number of papers. For example, SCoT-LASS is in the ℓ_1-constrained form (variable selection), while SPCA [14] uses the weighted form to define the sparsification problem. It is interesting that the function-constrained form has never been used to address PCA sparsification.

Trendafilov [11] adopted the function-constrained approach and considered several possible definitions of sparse PCA aiming to approximate different features of the original PCs. One of these sparse PCA definitions has been identified to have particularly interesting features, and is considered in details in the next Sect. 3.

3 Weakly Correlated Sparse Components with Oblique Loadings

Consider the set of all $p \times r$ matrices A with unit length columns, i.e., $\text{diag}(A^\top A) = I_r$, where $\text{diag}(.)$ of a square matrix produces diagonal matrix with the same main diagonal. We call this set the oblique manifold $\mathcal{OB}(p, r)$ [1, 10]. We consider solving the following matrix optimization problem:

$$\min_{\mathcal{OB}(p,r)} \|A\|_1 + \mu\|A^\top RA - D^2\|_F^2, \tag{1}$$

where $R = X^\top X$, D^2 is diagonal matrix whose diagonal entries are dominant eigenvalues of R, $\|A\|_1 = \sum_{i=1}^{p}\sum_{j=1}^{r}|A_{ij}|$, and A_{ij} is i-th row j-th column entry of the matrix of component loadings A. The first term in (1) encourages the sparseness, while the second term, weighted by a parameter μ, promotes a solution as close to standard PCA as possible (i.e., the second term promotes components explaining as much variance as possible).

The components obtained from (1) can be uncorrelated only approximately. However, this sparse PCA formulation is very interesting, because the resulting loadings A stay nearly orthonormal and $A^\top RA$ nearly diagonal. Indeed, let us consider the Riemannian gradient of $h(A) = \|A^\top RA - D^2\|_F^2, A \in \mathcal{OB}(p, r)$ with respect to the Euclidean metric $\langle U, V \rangle = \text{tr}(U^\top V)$. It follows that the Riemannian gradient is

$$\text{grad}h(A) = \nabla h(A) - A\,\text{diag}(A^\top \nabla h(A)), \tag{2}$$

where $\nabla h(A) = 4RA(A^\top RA - D^2)$ is the Euclidean gradient of $h(A)$.

By the first order optimality condition $\text{grad}h(A) = 0$ and (2), we have $RA(A^\top RA - D^2) = A\,\text{diag}(A^\top RA(A^\top RA - D^2))$. It follows that

$$A^\top RA(A^\top RA - D^2) = (A^\top A)\,\text{diag}[A^\top RA(A^\top RA - D^2)]. \tag{3}$$

It can be shown that, if at the minimum $A^\top RA \approx D^2$, then $A^\top A \approx I_r$, where '$\approx$' means approximate with respect to $\|.\|_F$. Since $A^\top RA = D^2$ is only possible for exact eigenvectors, it cannot be a solution to (1) with the presence of the $\|A\|_1$ term.

These considerations show that the minimization of $\|A^\top RA - D^2\|_F$ subject to $\mathrm{diag}(A^\top A) = I_r$ leads to diagonalization of R and orthonormality of A.

In this paper, problem (1) is solved by smoothing its cost function and using some standard solvers, e.g., Manopt [2]. Specifically, $\|A\|_1$ is replaced by the pseudo-Huber loss approximation $\sum_{ij} \left(\sqrt{A_{ij}^2 + \epsilon^2} - \epsilon \right)$, for some small positive ϵ. The parameter ϵ determines the trade-off between the smoothness of the cost function and the goodness of the approximation, i.e., $\sum_{ij} \left(\sqrt{A_{ij}^2 + \epsilon^2} - \epsilon \right) \rightarrow \|A\|_1$ as $\epsilon \rightarrow 0$. The final problem to solve thus becomes:

$$\min_{\mathcal{OB}(p,r)} \sum_{ij} \left(\sqrt{A_{ij}^2 + \epsilon^2} - \epsilon \right) + \mu \|A^\top RA - D^2\|_F^2, \tag{4}$$

In the rest of this work, ϵ has been fixed to 10^{-6}.

4 Numerical Example

Tests are performed on real DNA methylation data sets using Manopt, a Matlab toolbox for optimization on manifolds [2]. The data set used is available online on the NCBI website with the reference number GSE32393 [13]. As a preprocessing step, 2000 genes are randomly selected and standardized, such that their mean is equal to 0 and their standard deviation to 1. Three measures are of interest to evaluate the performances of our method: the variance explained, the orthogonality of the loading factors and the uncorrelation of the resulting components. Each of these measures is analyzed in regard to the sparseness achieved. For each run, 10 components were computed, such that A is a 2000×10 matrix. While many runs have been performed, results were so constant that only the mean of these runs is shown. Results are compared to the method of Journee et al. [6], which is one of the state of the art methods to perform sparse PCA.

A drawback of our method is that it does not produce exact zero values but only values that are close to zero. Hence, a threshold th has to be fixed, for the values below this threshold being reduced to 0. To take this modification into account, the matrix A was then column-normalized, such that $\mathrm{diag}(A^\top A) = I$.

First, the sparseness that can be achieved with respect to μ and th is illustrated in Fig. 1. As stated above, a large value of μ increases the importance of the second term in (4) and thus decreases the sparseness.

The 'trade-off' curve between the variance explained by the components and the sparseness of the loading factors is shown in Fig. 2. While the variance of the components can simply be computed as $\mathrm{tr}(A^\top X^\top XA)$ for the original PCA, this measure is not suitable when the new components are correlated, as noted by Zou et al. [14]. Indeed, since the components are correlated, they 'share' some common variance with the measure described above. An adjusted measure,

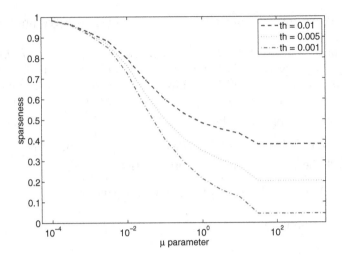

Fig. 1. The sparseness of the loading factors, computed as $\frac{\#\text{elements}<\text{th}}{\#\text{elements}}$, is shown with respect to μ, for different thresholds (`th`).

proposed by Zou et al. [14], is thus used here. Using the QR decomposition of the components $XA = QR$, this measure is computed as $\text{tr}(R^2)$. Results for both the non-adjusted (Fig. 2(a)) and the adjusted variance (Fig. 2(b)) are shown. While the results of Journée's method look better with the non-adjusted variance, the difference is small for the adjusted variance. Indeed, because of a higher correlation between the components for Journée's method (see below), more variance explained with the non-adjusted measure is redundant and so discarded when computing the adjusted variance.

Figure 3 reveals the strengths of our method. First, in Fig. 3(a), the 'non-orthogonality' of the loading factors is computed as $\| A^\top A - \text{diag}(A^\top A) \|_F$ / $\| \text{diag}(A^\top A) \|_F$. A large value indicates that the components are far from orthogonal. It can be seen that a high degree of orthogonality is maintained through any level of sparseness, in contrast to Journée's method that produces more non-orthogonal components for high degrees of sparseness. The correlation of the resulting components is then computed as $\| A^\top XX^\top A - \text{diag}(A^\top XX^\top A) \|_F$ / $\| \text{diag}(A^\top XX^\top A) \|_F$ (Fig. 3(b)). A high value indicates that the resulting components are highly correlated. Our method produces also more uncorrelated components in comparison to Journée's method.

More research is needed to figure out a more elaborate stopping rule that relates the unknowns' convergence to the model parameters. Indeed, since both the cost function and the gradient depend on μ, it is not straightforward to determine a standard stopping rule. While some simulations are very fast (< 1 s), more research should determine whether such a fast computational time can be reached for every value of μ.

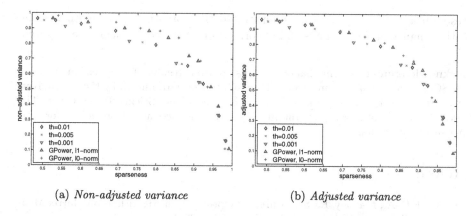

(a) *Non-adjusted variance*   (b) *Adjusted variance*

Fig. 2. The variance of the components, non-adjusted (a) and adjusted (b) plotted in regards to the sparseness for different thresholds (th). The variance is normalized by the variance of the components resulting from standard PCA. Results are compared to the variance of the components of Journée's method for various degree of sparseness, both using the l_1 and the l_0 norm (see [6]).

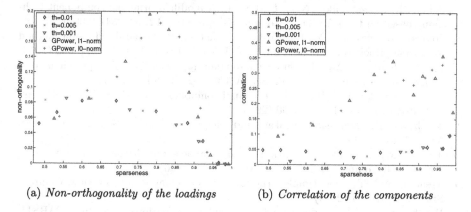

(a) *Non-orthogonality of the loadings*   (b) *Correlation of the components*

Fig. 3. The 'non-orthogonality' of the loadings (a) and the correlation of the components (b) are shown with respect to the sparseness of the components. Results are displayed for different thresholds th for our method, and both with the l_1 and the l_0 norm for Journée's method.

5   Conclusion

A new approach to solve the sparse PCA problem has been introduced. The trade-off curve has shown that our method is able to explain a significant part of the variance even with a high degree of sparseness. However, its strengths lie in the preservation of the characteristics of the original PCA. Indeed, unlike most of the existing methods used until now, our method preserves both a high level of orthogonality between the loading factors and a high uncorrelation between the resulting components. While experiments at a larger scale are still required, tests on a real data set have demonstrated the potential of our method when orthogonality of the loadings and uncorrelation of the components are required.

Acknowledgements. This paper presents research results of the Belgian Network DYSCO (Dynamical Systems, Control, and Optimization), funded by the Interuniversity Attraction Poles Programme initiated by the Belgian Science Policy Office. The authors thank Professor Pierre-Antoine Absil for his advices and the reviewers for their pertinent comments.

References

1. Absil, P.-A., Mahony, R., Sepulchre, R.: Optimization Algorithms on Matrix Manifolds. Princeton University Press, Princeton (2008)
2. Boumal, N., Mishra, B., Absil, P.-A., Sepulchre, R.: Manopt, a Matlab toolbox for optimization on manifolds. J. Mach. Learn. Res. **15**, 1455–1459 (2014)
3. Candès, E.J., Tao, T.: The Dantzig selector: statistical estimation when p is much larger than n. Ann. Stat. **35**, 2313–2351 (2007)
4. Jolliffe, I.T.: Principal Component Analysis, 2nd edn. Springer, New York (2002)
5. Jolliffe, I.T., Trendafilov, N.T., Uddin, M.: A modified principal component technique based on the LASSO. J. Comput. Graph. Stat. **12**, 531–547 (2003)
6. Journée, M., Nesterov, Y., Richtárik, P., Sepulchre, R.: Generalized power method for sparse principal component analysis. J. Mach. Learn. Res. **11**, 517–553 (2010)
7. Lu, Z., Zhang, Y.: An augmented Lagrangian approach for sparse principal component analysis. Math. Program. Ser. A **135**(1–2), 149–193 (2012)
8. Qi, X., Luo, R., Zhao, H.: Sparse principal component analysis by choice of norm. J. Multivar. Anal. **114**, 127–160 (2013)
9. Tibshirani, R.: Regression shrinkage and selection via the LASSO. J. R. Stat. Soc. **58**, 267–288 (1996)
10. Trendafilov, N.T.: The dynamical system approach to multivariate data analysis, a review. J. Comput. Graph. Stat. **50**, 628–650 (2006)
11. Trendafilov, N.T.: From simple structure to sparse components: a review. Comput. Stat. **29**, 431–454 (2014)
12. Wright, S.: Gradient algorithms for regularized optimization. In: SPARS11, Edinburgh, Scotland (2011). http://pages.cs.wisc.edu/-swright/
13. Zhuang, J., Jones, A., Lee, S.-H., Ng, E., Fiegl, H., et al.: The dynamics and prognostic potential of DNA methylation changes at stem cell gene loci in women's cancer. PLoS Genet. **8**(2), e1002517 (2012)
14. Zou, H., Hastie, T., Tibshirani, R.: Sparse principal component analysis. J. Comput. Graph. Stat. **15**, 265–286 (2006)

Fitting Smooth Paths on Riemannian Manifolds: Endometrial Surface Reconstruction and Preoperative MRI-Based Navigation

Antoine Arnould[1], Pierre-Yves Gousenbourger[1]([⊠]), Chafik Samir[2], Pierre-Antoine Absil[1], and Michel Canis[2]

[1] ICTEAM Institute, Université Catholique de Louvain,
1348 Louvain-la-Neuve, Belgium
`pierre-yves.gousenbourger@uclouvain.be`
[2] ISIT UMR 6284 CNRS, Clermont University, Clermont-Ferrand, France
`chafik.samir@udamail.fr`

Abstract. We present a new method to fit smooth paths to a given set of points on Riemannian manifolds using C^1 piecewise-Bézier functions. A property of the method is that, when the manifold reduces to a Euclidean space, the control points minimize the mean square acceleration of the path. As an application, we focus on data observations that evolve on certain nonlinear manifolds of importance in medical imaging: the shape manifold for endometrial surface reconstruction; the special orthogonal group $SO(3)$ and the special Euclidean group $SE(3)$ for preoperative MRI-based navigation. Results on real data show that our method succeeds in meeting the clinical goal: combining different modalities to improve the localization of the endometrial lesions.

Keywords: Path fitting on Riemannian manifolds · Bézier functions · Optimization on manifolds · MRI-based navigation · Endometrial surface reconstruction

1 Introduction

Surface reconstruction problem has been widely studied because of its importance in different applications such as medical imaging, computer graphics, mechanical simulations, virtual reality, etc. Particularly, the reconstruction of surfaces from given $3D$ point clouds is important since they are frequently used in medical imaging and computer graphics [4]. For example, one can use a continuous formulation using PDEs and compute the solution as an implicit surface, which is usually the zero level set of a sufficiently smooth function. Therefore, one can control the resulting surface by adding physics-inspired constraints depending on geometry or external forces [17]. However, when the given data is a set of curves one needs to find an optimal fitting between them by taking into account their parametrization and the non-linearity of their spatial evolution. In this

© Springer International Publishing Switzerland 2015
F. Nielsen and F. Barbaresco (Eds.): GSI 2015, LNCS 9389, pp. 491–498, 2015.
DOI: 10.1007/978-3-319-25040-3_53

work, we formulate the problem of reconstructing a surface from a given set of curves as a smooth path fitting on the space of curves.

Path fitting on manifolds has been addressed in the literature with different approaches and for various purposes. Generic path fitting methods on manifolds include splines on manifolds [8], rolling procedures [6], subdivision schemes [10], gradient descent [13], and geodesic finite elements [15]. Interpolation of rotations (where the manifold $\mathcal{M}$ is the special orthogonal group $SO(3)$) is useful in robotics for motion planning of rigid bodies and in computer graphics for the animation of 3D objects [11]. More related to our work, morphing between shapes can be tackled as an interpolation problem on shape space [5].

With an estimated prevalence of 10 %, endometriosis is one of the most common clinical problems affecting women of reproductive age [16]. Various structures can be affected by endometriosis, including the uterosacral ligaments, rectosigmoid colon, vagina, uterus, and bladder [9]. Generally, the absence of an accurate preoperative diagnosis leads to unnecessary surgeries even with possible complications. To reduce the risk of such failure, surgeons need to know exactly the number, the size, the locations and the depth of infiltration of endometrial cysts before the intervention. A preoperative mapping of the lesions is crucial for managing the disease. This mapping can be well defined through medical imaging techniques such as Magnetic Resonance Imaging (MRI) and 2D Transvaginal Ultrasound (TVUS) [3].

When using these two modalities separately, there is an increased risk of false negative and false positive, due to their different advantages and inconvenients [1]. Moreover, they complement each other excellently, as lacking information from one modality can be provided by the other in terms of spatial, contrast or temporal resolution. For instance, lesions hard to detect in MRI are revealed at TVUS acquisition (due to an approximate distance between the probe and the tissue with relatively free movement). That is why registration is helpful here in order to show TVUS with lesions into MRI volume.

After TVUS-MRI registration [14], the position p of the ultrasound probe and its orientation n can be precisely determined within the $3D$ MRI volume and around the TVUS curve. They form a certain plane $\Pi_{n,p}$. The MRI *views* corresponding to the intersections of the MRI *volume* with the registered planes $\Pi_{n,p}$ are used to reproduce the probe movement. As a result, the clinician is able to explore the MRI volume to search for very close and clear views including the pelvic organs. We illustrate this idea by fitting a smooth path γ to different key positions of $\Pi_{n,p}$ viewed as points on Riemannian manifolds like $SO(3)$ (rotations) or $SE(3)$ (rotations and translations). For simplicity, we will refer to the resulting path γ (in both cases) as a preoperative MRI-based navigation to locate and characterize lesions.

Clinical Context. In practice, TVUS is done on women presenting symptoms corresponding to the presence of endometrial tissues. When 2D TVUS does not provide enough information to confirm the diagnosis, MRI is performed. Given TVUS and MRI, practicians select a set of corresponding landmarks to define surrounding organ boundaries in both images, manually. These anatomical correspondences between structures on MRI and TVUS are then used to measure and locate lesions, separately, which is still a challenging task.

The rest of this paper is organized as follows. Section 2 describes the formulation of our path fitting method. Section 3 presents two applications of this method: (i) MRI surface reconstruction as a path on shape manifold and (ii) navigation in the MRI volume as a path on $SE(3)$. We close this paper with a brief summary in Sect. 4.

2   Problem Formulation

Given a finite set of points $p_0, \ldots, p_n$ at time instants $t_0, \ldots, t_n$ on a Riemannian manifold $\mathcal{M}$, our goal is to construct a smooth path γ that interpolates p_i at t_i for $i = 0, \ldots, n$. When the manifold $\mathcal{M}$ reduces to the Euclidean space $\mathbb{R}^m$, we propose a method that generates a piecewise-Bézier C^1 path with minimal mean squared acceleration. This method improves on the technique recently proposed in [5], where the choice of the path velocity direction at the interpolation points was suboptimal even in the Euclidean case. Let the function $t \mapsto \beta_k(t; b_0, \ldots, b_k)$ denote the Bézier curve of order k and $b_0, \ldots, b_k$ the control points. For simplicity, we set time instants $t_i = i$ with straightforward extension to general timestamps. In this work, we only use Bézier curves of degree 2 and 3, expressed in $\mathbb{R}^m$ as:

$$\beta_2(t; b_0, b_1, b_2) = b_0(1-t)^2 + 2b_1 t(1-t) + b_2 t^2 \tag{1}$$

$$\beta_3(t; b_0, b_1, b_2, b_3) = b_0(1-t)^3 + 3b_1 t(1-t)^2 + 3b_2 t^2(1-t) + b_3 t^3. \tag{2}$$

Using these polynomials, we construct a C^1 curve $\gamma : [0, n] \to \mathbb{R}^m$ (we call it a *piecewise-Bézier curve*) consisting of Bézier curves of degree 2 for the extremal segments and of degree 3 for the others:

$$\gamma(t) = \begin{cases} \beta_2(t; p_0, b_1^-, p_1) & \text{if } t \in [0, 1] \\ \beta_3(t - (i - 1); p_{i-1}, b_{i-1}^+, b_i^-, p_i) & \text{if } t \in [i-1, i], \ i=2,\ldots,n-1 \\ \beta_2(t - (n - 1); p_{n-1}, b_{n-1}^+, p_n) & \text{if } t \in [n-1, n], \end{cases} \tag{3}$$

where b_i^+ and b_i^- are the control points respectively on the right and left hand side of the interpolation point p_i. One can observe that this formulation satisfies the interpolation conditions $\gamma(t_i) = p_i$. The differentiability condition of the curve is ensured by imposing velocities to be equal on the left and right of the interpolation points p_is, which allows us to express b_i^+ in terms of b_i^-:

$$\begin{aligned} b_1^+ &= \tfrac{5}{3}p_1 - \tfrac{2}{3}b_1^-, \\ b_i^+ &= 2p_i - b_i^- \quad i = 2, \ldots, n-2, \\ b_{n-1}^+ &= \tfrac{5}{2}p_{n-1} - \tfrac{3}{2}b_{n-1}^-. \end{aligned} \tag{4}$$

The resulting optimization problem is an unconstrained minimization with b_i^- as variables and the mean square acceleration of the piecewise-Bézier curve as the objective function, defined as follows:

$$\int_0^1 \|\ddot{\beta}_2(t; p_0, b_1^-, p_1)\|^2 dt + \sum_{i=1}^{n-2} \int_0^1 \|\ddot{\beta}_3(t; p_{i-1}, b_{i-1}^+, b_i^-, p_i)\|^2 dt$$

$$+ \int_0^1 \|\ddot{\beta}_2(t; p_{n-1}, b_{n-1}^+, p_n)\|^2 dt. \quad (5)$$

As the Bézier segments are linear functions of the control points, the objective function is quadratic. The optimal solution is then computed as a critical point of the gradient, which gives rise to a linear system of the form: $AX = CP$ where $X = \begin{bmatrix} b_1^- \dots b_{n-1}^- \end{bmatrix}^T \in \mathbb{R}^{n-1 \times m}$, $P = \begin{bmatrix} p_0 \dots p_n \end{bmatrix}^T \in \mathbb{R}^{n+1 \times m}$, $A \in \mathbb{R}^{n-1 \times n-1}$ and $C \in \mathbb{R}^{n-1 \times n+1}$ are tridiagonal matrices with coefficients:

$$A_{(1,1:2)} = \begin{bmatrix} 64 & 24 \end{bmatrix}, \quad (6)$$

$$A_{(2,1:3)} = \begin{bmatrix} 24 & 144 & 36 \end{bmatrix}, \quad (7)$$

$$A_{(i,i-1:i+1)} = \begin{bmatrix} 36 & 144 & 36 \end{bmatrix}, i = 3 : n - 2 \quad (8)$$

$$A_{(n-1,n-2:n-1)} = \begin{bmatrix} 36 & 144 \end{bmatrix} \quad (9)$$

and

$$C_{(1,1:2)} = \begin{bmatrix} 16 & 72 \end{bmatrix}, \quad (10)$$

$$C_{(2,2:3)} = \begin{bmatrix} 60 & 144 \end{bmatrix}, \quad (11)$$

$$C_{(i,i:i+1)} = \begin{bmatrix} 72 & 144 \end{bmatrix}, i = 3 : n - 2 \quad (12)$$

$$C_{(n-1,n-1:n+1)} = \begin{bmatrix} 72 & 132 & -24 \end{bmatrix}. \quad (13)$$

Since A is invertible, the unique solution is given by:

$$X = A^{-1}CP = DP \quad \text{with} \quad \sum_{j=0}^n D_{ij} = 1, \forall i. \quad (14)$$

We generalize this result on a Riemannian submanifold $\mathcal{M}$ embedded in a Euclidean space $\mathcal{E}$. In order to make this possible for a Riemannian manifold $\mathcal{M}$, one needs the tangent space $T_p(\mathcal{M})$ of $\mathcal{M}$ at a given point p, the Riemannian exponential map Exp_p, and its inverse Log_p (see [2,12, Sect. 4] for a formal definition on specific manifolds). Bézier curves (1) and (2) are generalized by means of the Riemannian De Casteljau's algorithm (see, e.g., [5] for the literature). Conditions (4), of the form $b_i^+ = p_i + \alpha(b_i^- - p_i)$ generalize to $b_i = \mathrm{Exp}_{p_i}(\alpha \mathrm{Log}_{p_i}(b_i^-))$ and ensure that γ on $\mathcal{M}$ is $\mathcal{C}^1$. It then remains to generalize (14) for which we propose two approaches:

1. **method 1:** In order to solve the fitting problem on a linear space as for the Euclidean case, we proceed as follows. We initially choose an arbitrary point among the p_is that we will call a root point, e.g., p_0. Next we map the rest of the given data points to $T_{p_0}(\mathcal{M})$ as $\tilde{p}_i = \mathrm{Log}_{p_0}(p_i)$. Then we solve the linear system $\widetilde{X} = D\widetilde{P}$. Finally, we project the solution $\widetilde{X}$ back to $\mathcal{M}$ as $b_i^- = \mathrm{Exp}_{p_0}(\tilde{x}_i)$. Numerically, the choice of the root point may affect the quality of the solution.

2. **method 2:** In order to avoid the dependance on the choice of a single root point, an alternative is to choose p_i as the root point for row i of (14). Thus, for each $i = 1 \ldots n - 1$ we map the rest of data points into the tangent space $T_{p_i}(\mathcal{M})$ using the logarithmic map Log_{p_i}. The mapped data are then given by $\widetilde{P} = \left[\text{Log}_{p_i}(p_0) \ldots \text{Log}_{p_i}(p_n)\right]$ and the solution is given by $\widetilde{x}_i = \text{Log}_{p_i}(b_i^-) = \sum_{j=0}^{n} D_{ij} \text{Log}_{p_i}(p_j)$. Therefore, each $\widetilde{x}_i$ is mapped back to $\mathcal{M}$ as $b_i^- = \text{Exp}_{p_i}(\widetilde{x}_i)$.

As stated earlier, our method minimizes the mean square acceleration objective when $\mathcal{M}$ is a Euclidean space. This follows from the fact that, when $\mathcal{M}$ is a Euclidean space, we have $\text{Log}_p(b) = b - p$; this, along with the property $\sum_{j=0}^{n} D_{ij} = 1$, makes the root point cancel out and recover (14). Even if they seem to provide good solutions, the proposed generalizations of (14) do not guarantee a minimal mean square acceleration when $\mathcal{M}$ is nonlinear. Nevertheless, we will use the second method for the experiments as it was observed to be more efficient than the first one, at least when $\mathcal{M}$ is the unit sphere S^2, the special orthogonal group $SO(3)$, or the special Euclidean group $SE(3)$.

3   Experimental Results

In this section, we present two different applications of our framework. On both cases, results are given using real data images obtained from patients with endometriosis characterized by different localizations and depths of infiltration. Figure 1(a,b) shows examples of corresponding landmark curves (uterus, rectum, lesions) in TVUS and MR images. The curves in TVUS have been deformed along during the exam due to the transducer's pressure as shown in Fig. 2(a).

3.1   Endometrial Surface Reconstruction

As a first application, we performed our path fitting method to reconstruct the endometrial surface S_{MRI} from curves in three steps. First, a radiologist was asked to select different slices (from 4 to 7) and segment curves as boundaries of an interest zone on each slice. Second, we represented each curve as a point on the shape manifold. Note that we aligned and fixed the starting point of each curve (Fig. 1(f)). As given time indexes that have spatial meaning in this case, we used the $z - values$ for each curve from its corresponding slice. Third, we used a modified version of [7] to compute a geodesic path between any two points on shape space. Finally, we applied our method as detailed in Sect. 2 to construct S_{MRI} (Fig. 1(c) and (h)) as a C^1-fitting path between curves (see $\|\dot{\gamma}(t)\|$ in Fig. 1(d)). To give an idea about the quality of the reconstructed surface, we show an example of S_{MRI} constructed from a set of 3D curves (f) using a linear interpolation between them (g) and our method (h).

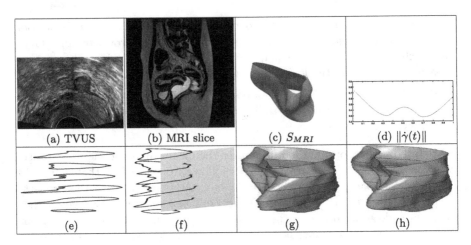

(a) TVUS	(b) MRI slice	(c) S_{MRI}	(d) $\|\dot{\gamma}(t)\|$
(e)	(f)	(g)	(h)

Fig. 1. Application 1: from TVUS and MR images (a,b), we reconstruct the MRI surface (c) as a path γ interpolating 4 key curves extracted on MRI slices. The velocity of γ is continuous (d). From the 6 key curves (e) with fixed starting points (f) we reconstruct (g) as a linear interpolation and (h) with our method.

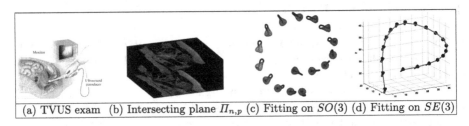

(a) TVUS exam	(b) Intersecting plane $\Pi_{n,p}$	(c) Fitting on $SO(3)$	(d) Fitting on $SE(3)$

Fig. 2. (a) An illustration of TVUS movement during the exam and (b) is an example of the intersection between MRI volume and a plane $\Pi_{n,p}$ by means of interpolation. (c) and (d) Are two examples of fitting paths on $SO(3)$ and $SE(3)$, respectively.

3.2   MRI-based Navigation

The problem of 2D-3D TVUS-MRI registration was recently addressed by Samir et al. in [14]. The basic idea of their method is a follows. First, they manually segment the cylindrical endometrial tissue surface S_{MRI} from the MRI image and the planar contour from the corresponding TVUS image. This registration provides a one-to-one correspondence of curves between TVUS and MRI. We will refer to the resulting intersecting plane by $\Pi_{n,p}$ (Fig. 2(b)).

To look for a very close and more clear views including the pelvic organs in the MRI volume, one has to consider the movement of the probe (rotations only or rotations and translations) to search around $\Pi_{n,p}$. We illustrate this idea by fitting smooth paths of different key positions of $\Pi_{n,p}$ on $SO(3)$ (rotations: Fig. 2(c)) and on $SE(3)$ (rotations and translations: Fig. 2(d)). In both cases, we consider the resulting path γ as a preoperative MRI-based navigation. It is

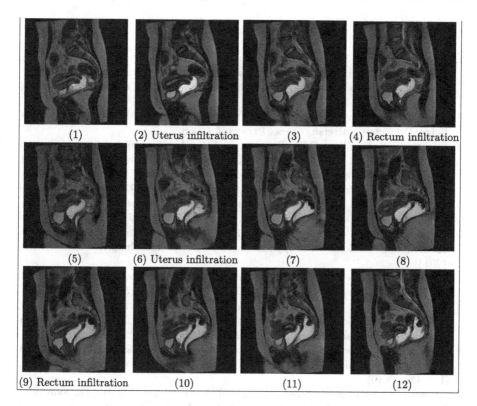

(1)   (2) Uterus infiltration   (3)   (4) Rectum infiltration

(5)   (6) Uterus infiltration   (7)   (8)

(9) Rectum infiltration   (10)   (11)   (12)

Fig. 3. Application 2: examples of uniformly sampled frames from MRI navigation obtained as a fitting on $SE(3)$.

clear from Fig. 3 that such navigation has more chances to locate the extent of lesions than TVUS and MRI when used separately. This idea is illustrated in Fig. 3(frames $(2, 4, 6, 9)$) where new views as points from γ provide more accurate characterization of the lesion. In this case, the red coloured areas denotes the region of interest (delineated by the expert) which were not clear enough or non visible on sagittal, coronal, and axial views.

4   Summary

In this work, we have proposed a new Riemannian framework for a MR-based navigation system to locate and characterize endometrial tissues in order to improve the preoperative diagnosis. The information is available in the form of landmark curves (extended to surfaces) in the 3D MRI and curves in the 2D TVUS images. Our approach embeds the TVUS intersecting plane into MRI and use a new path fitting method to construct an MRI-based navigation. This way, we have reached very close and more clear views including the pelvic organs in the MRI volume.

References

1. Abrao, M.S., da C. Goncalves, M.O., Dias, J.A., Podgaec, J.S., Chamie, L.P., Blasbalg, R.: Comparison between clinical examination, transvaginal sonography and magnetic resonance imaging for the diagnosis of deep endometriosis. Hum. Reprod. **22**, 3092–3097 (2007)
2. Absil, P.A., Mahony, R., Sepulchre, R.: Optimization Algorithms on Matrix Manifolds. Princeton University Press, Princeton (2008)
3. Bagaria, S.J., Rasalkar, D.D., Paunipagar, B.K.: Imaging tools for ndometriosis: Role of Ultrasound, MRI and Other Imaging Modalities in Diagnosis and Planning Intervention. In: Chaudhury, K. (ed.) Endometriosis - Basic Concepts and Current Research Trends, InTech (2012). doi:10.5772/29063, ISBN: 978-953-51-0524-4
4. Bajaj, C.L., Xu, G.L., Zhang, Q.: Bio-molecule surfaces construction via a higher-order level-set method. J. Comput. Sci. Technol. **23**(6), 1026–1036 (2008)
5. Gousenbourger, P.Y., Samir, C., Absil, P.A.: Piecewise-Bézier C^1 interpolation on Riemannian manifolds with application to 2D shape morphing. In: IEEE ICPR (2014)
6. Hüper, K., Silva Leite, F.: On the geometry of rolling and interpolation curves on S^n, SO_n, and Grassmann manifolds. J. Dyn. Control Syst. **13**(4), 467–502 (2007)
7. Joshi, S., Jermyn, I., Klassen, E., Srivastava, A.: An efficient representation for computing geodesics between n-dimensional elastic shapes. IEEE Conference on Computer Vision and Pattern Recognition (CVPR) (2007)
8. Machado, L., Leite, F.S., Krakowski, K.: Higher-order smoothing splines versus least squares problems on Riemannian manifolds. J. Dyn. Control Syst. **16**(1), 121–148 (2010)
9. Massein, A., Petit, E., Darchen, M., Loriau, J., Oberlin, O., Marty, O., Sauvanet, E., Afriat, R., Girard, F., Molini, V., Duchatelle, V., Zins, M.: Imaging of intestinal involvement in endometriosis. Diagn. Interv. Imaging **94**(3), 281–291 (2013)
10. Nira, D.: Linear and nonlinear subdivision schemes in geometric modeling. In: Foundations of computational mathematics, Hong Kong 2008. London Math. Soc. Lecture Note Ser., vol. 363, pp. 68–92. Cambridge Univ. Press, Cambridge (2009)
11. Park, J.: Interpolation and tracking of rigid body orientations. In: ICCAS, pp. 668–673 (2010)
12. Rentmeesters, Q.: A gradient method for geodesic data fitting on some symmetric Riemannian manifolds. In: 2011 50th IEEE Conference on Decision and Control and European Control Conference (CDC-ECC), pp. 7141–7146 (2011)
13. Samir, C., Absil, P.A., Srivastava, A., Klassen, E.: A gradient-descent method for curve fitting on Riemannian manifolds. Found. Comput. Math. **12**, 49–73 (2012)
14. Samir, C., Kurtek, S., Srivastava, A., Canis, M.: Elastic shape analysis of cylindrical surfaces for 3D/2D registration in endometrial tissue characterization. IEEE Trans. MI **33**, 1035–1043 (2014)
15. Sander, O.: Geodesic finite elements of higher order. Technical report 356 (2013)
16. Umaria, N., Olliff, J.: Imaging features of pelvic endometriosis. Br. J. Radiol. **74**, 556–562 (2001)
17. Zhao, H.K., Osher, S., Fedkiw, R.: Fast surface reconstruction using the level set method. In: Proceedings of the IEEE Workshop on Variational and Level Set Methods in Computer Vision, pp. 194–201 (2001)

PDE Constrained Shape Optimization as Optimization on Shape Manifolds

Volker H. Schulz, Martin Siebenborn, and Kathrin Welker$^{(\boxtimes)}$

Department of Mathematics, University of Trier, 54296 Trier, Germany
{volker.schulz,siebenborn,welker}@uni-trier.de

Abstract. The novel Riemannian view on shape optimization introduced in [14] is extended to a Lagrange–Newton as well as a quasi–Newton approach for PDE constrained shape optimization problems.

Keywords: Shape optimization · Riemannian manifold · Newton method · Quasi–Newton method · Limited memory BFGS

1 Introduction

Shape optimization problems arise frequently in technological processes which are modeled in the form of partial differential equations as for example in [2,8]. In many practical circumstances, the shape under investigation is parametrized by finitely many parameters, which on the one hand allows the application of standard optimization approaches, but on the other hand limits the space of reachable shapes unnecessarily. Shape calculus which has been the subject of several monographs [5,10,17] presents a way out of that dilemma. However, so far it is mainly applied in the form of gradient descent methods, which can be shown to converge. The major difference between shape optimization and the standard PDE constrained optimization framework is the lack of the linear space structure in shape spaces. If one cannot use a linear space structure, then the next best structure is the Riemannian manifold structure as discussed for shape spaces for example in [9]. The publication [14] makes a link between shape calculus and shape manifolds and thus enables the usage of optimization techniques on manifolds in the context of shape optimization.

In this paper we consider problems of finding the interfaces of two subdomains. It is organized in the following way. In the first part of this paper we consider an elliptic (Sect. 2) and in the second part a parabolic interface shape optimization problem (Sect. 3). Section 2 presents a vector bundle framework [15] based on the Riemannian framework established in [14], which enables the discussion of Lagrange–Newton methods within the shape calculus framework for PDE constrained shape optimization. Newton–type methods have been used in shape optimization since many years, e.g. [7,12]. Often, the unknown diffusivity in diffusive processes is structured by piecewise constant patches. Section 3 is devoted to efficient methods for the determination of such structured diffusion parameters by exploiting shape calculus. Quasi–Newton methods on general

© Springer International Publishing Switzerland 2015
F. Nielsen and F. Barbaresco (Eds.): GSI 2015, LNCS 9389, pp. 499–508, 2015.
DOI: 10.1007/978-3-319-25040-3_54

manifolds have already been discussed in [1,6,13]. Here, we specify them for the particular case of shape manifolds. Finally, Sect. 4 discusses numerical results for the inverse problem of finding the interfaces of two subdomains.

The methodology and algorithm derived in this paper applies for example to the problem of inversely determining cell shapes in the human skin as investigated in [11].

2   A Lagrange–Newton Approach on Riemannian Manifolds

In the first part of this section we formulate a Lagrange–Newton approach for PDE constrained shape optimiziation problems which is based on Riemannian vector space bundles. Then, we exemplify these theoretical discussions to an example which is motivated by electrical impedance tomography.

2.1   Riemannian Vector Bundle Framework

Let (M, g) denote a Riemannian manifold and $\Omega(u)$ the interior of a shape $u \in M$. We consider the following equality constrained optimization problem

$$\min_{(y,u) \in E} J(y,u), \; J: E \to \mathbb{R} \tag{1}$$

$$\text{s.t. } a_u(y,p) = b_u(p), \; \forall p \in H \tag{2}$$

where H is a given Hilbert space $H(u)$ such that $E := \{(H(u), u) : u \in M\}$ is the total space of a vector bundle (E, π, M), $a_u(\cdot, \cdot)$ is a bilinear form and $b_u(\cdot)$ a linear form defined on each fiber H.

The Lagrangian of the minimization problem (1–2) is defined as

$$\mathscr{L}(y,u,p) := J(y,u) + a_u(y,p) - b_u(p) \tag{3}$$

where $(y,u,p) \in F := \{(H(u), u, H(u)) \,|\, u \in M\}$. Its Gradient grad$\mathscr{L}$ is based on the scalar product on TF with $T_{(y,u,p)}F \cong H(u) \times T_y M \times H(u)$ and its Hessian Hess$\mathscr{L}$ on the Riemannian connection on F. For more details we refer the reader to [15].

Let $(\hat{y}, \hat{u}) \in E$ solve the optimization problem (1–2). Then, the variational problem which we get by differentiating $\mathscr{L}$ with respect to y is given by

$$a_{\hat{u}}(z,p) = -\frac{\partial}{\partial y} J(\hat{y}, \hat{u})z, \; \forall z \in H(\hat{u}) \tag{4}$$

and the design problem which we get by differentiating $\mathscr{L}$ with respect to u by

$$\frac{\partial}{\partial u}\bigg|_{u=\hat{u}} [J(\hat{y}, u) + a_u(\hat{y}, \hat{p}) - b_u(\hat{p})] \, w = 0, \; \forall w \in T_{\hat{u}} M \tag{5}$$

where $\hat{p} \in H$ solves (4). If we differentiate $\mathscr{L}$ with respect to p, we get the state equation. These three (KKT) conditions could be collected in the following condition:

$$DL(\hat{y}, \hat{u}, \hat{p})h = 0, \ \forall h \in T_{(y,u,p)}F \tag{6}$$

The scalar product on TF can be used to define the gradient of the Lagrangian $\mathrm{grad}\mathscr{L} \in TF$ as a Riesz representation by the condition

$$\langle \mathrm{grad}\mathscr{L}, h \rangle_{T_{(y,u,p)}F} := DL(y, u, p)h, \ \forall h \in T_{(y,u,p)}F. \tag{7}$$

Now, similar to standard nonlinear programming we can solve the problem of finding $(y, u, p) \in F$ with $\mathrm{grad}\mathscr{L}(y, u, p) = 0$ as a means to find solutions to the optimization problem (1–2). This nonlinear problem has exactly the form of the root finding problems discussed in [14]. Exploiting the Riemannian structure on TF, we can formulate a Newton iteration involving the Riemannian Hessian which is based on the resulting Riemannian connection. In each iteration we have to compute the increment $\Delta\xi = (z, w, q)^T$ as solution of

$$\mathrm{Hess}\mathscr{L}(y, u, p)\Delta\xi = -\mathrm{grad}\mathscr{L}(y, u, p). \tag{8}$$

It is advantageous to solve Eq. (8) in a weak formulation, i.e., in detail, the following equations have to be satisfied for all $h = (\bar{z}, \bar{w}, \bar{q})^T \in T_{(y,u,p)}F$:

$$H_{11}(z, \bar{z}) + H_{12}(w, \bar{z}) + H_{13}(q, \bar{z}) = -a_u(\bar{z}, p) - \frac{\partial}{\partial y}J(y, u)\bar{z} \tag{9}$$

$$H_{21}(z, \bar{w}) + H_{22}(w, \bar{w}) + H_{23}(q, \bar{w}) = -\frac{\partial}{\partial u}[J(y, u) + a_u(y, p) - b_u(p)]\bar{w} \tag{10}$$

$$H_{31}(z, \bar{q}) + H_{32}(w, \bar{q}) + H_{33}(q, \bar{q}) = -a_u(y, \bar{q}) + b_u(\bar{q}) \tag{11}$$

where the expressions H_{ij} ($i = 1, 2, 3$) can be found in [15]. A key observation in [14] is that the expression $H_{22}(w, \bar{w})$ is symmetric in the solution of the shape optimization problem. This motivates a shape–SQP method, where away from the solution only expressions in $H_{22}(w, \bar{w})$ are used which are nonzero at the solution. Its basis is the following observation:

If the term $H_{22}(w, \bar{w})$ is replaced by an approximation $\hat{H}_{22}(w, \bar{w})$, which omits all terms in $H_{22}(w, \bar{w})$, which are zero at the solution and if the reduced Hessian of $\mathrm{Hess}\mathscr{L}$ built with this approximation is coercive, Eq. (8) or Eqs. (9–11) are equivalent to the linear–quadratic problem

$$\min_{(z,w)} \frac{1}{2}\left(H_{11}(z, z) + 2H_{12}(w, z) + \hat{H}_{22}(w, w)\right) + \frac{\partial}{\partial y}J(y, u)z + \frac{\partial}{\partial u}J(y, u)w \tag{12}$$

$$\text{s.t.} \ a_u(z, \bar{q}) + \frac{\partial}{\partial u}[a_u(y, \bar{q}) - b_u(\bar{q})]w = -a_u(y, \bar{q}) + b_u(\bar{q}), \ \forall \bar{q} \in H(u) \tag{13}$$

where the adjoint variable to the constraint (13) is just $p + q$. We also omit terms in H_{11} and H_{12}, which are zero, when evaluated at the solution of the optimization problems. Nevertheless, quadratic convergence of the resulting SQP method is to be expected and indeed observed in Sect. 4.

2.2   Application of the Riemannian Vector Bundle Framework

We consider a domain $\Omega(u) := (0,1)^2 \subset \mathbb{R}^2$ which is split into the two subdomains $\Omega_1(u), \Omega_2(u) \subset \Omega(u)$ such that $\partial\Omega_1(u) \cap \partial\Omega_2(U) = u$ and $\Omega_1(u) \cup u \cup \Omega_2(u) = \Omega(u)$ where $\cup$ denotes the disjoint union. The interface u is an element of the manifold

$$B_e^0([0,1],\mathbb{R}^2) := \mathrm{Emb}^0([0,1],\mathbb{R}^2)/\mathrm{Diff}^0([0,1]) \tag{14}$$

i.e., an element of the set of all equivalence classes of the set of embeddings

$$\mathrm{Emb}^0([0,1],\mathbb{R}^2) := \{\phi \in C^\infty([0,1],\mathbb{R}^2) \mid \phi(0) = (0.5,0), \phi(1) = (0.5,1), \atop \phi \text{ injective immersion}\} \tag{15}$$

where the equivalence relation is defined by the set of all C^∞ re–parametrizations, i.e., by the set of all diffeomorphisms

$$\mathrm{Diff}^0([0,1],\mathbb{R}^2) := \{\phi\colon [0,1] \to \mathbb{R}^2 \mid \phi(0) = (0.5,0), \phi(1) = (0.5,1), \atop \phi \text{ diffeomorphism}\}. \tag{16}$$

The construction of the domain $\Omega(u)$ from the interface $u \in B_e^0([0,1],\mathbb{R}^2)$ is illustrated in Fig. 1.

The PDE constrained shape optimization problem is given in strong form by

$$\min_u \; J(y,u) \equiv \frac{1}{2} \int_{\Omega(u)} (y - \bar{y})^2 dx + \mu \int_u 1 ds \tag{17}$$

$$\text{s.t.} \; -\triangle y = f \quad \text{in } \Omega(u) \tag{18}$$

$$y = 0 \quad \text{on } \partial\Omega(u) \tag{19}$$

where $f \equiv f_1 = \text{const. in } \Omega_1(u)$, $f \equiv f_2 = \text{const. in } \Omega_2(u)$. The perimeter regularization with $\mu > 0$ in the objective (17) is a frequently used means to overcome ill–posedness of the optimization problem (e.g. [3]). Let n be the unit outer normal to $\Omega_1(u)$ at u. Furthermore, we have interface conditions at the interface u. We formulate explicitly the continuity of the state and of the flux at the boundary u as

$$[\![y]\!] = 0, \quad \left[\!\left[\frac{\partial y}{\partial n}\right]\!\right] = 0 \quad \text{on } u \tag{20}$$

where the jump symbol $[\![\cdot]\!]$ denotes the discontinuity across the interface u and is defined by $[\![v]\!] := v_1 - v_2$ where $v_1 := v\big|_{\Omega_1}$ and $v_2 := v\big|_{\Omega_2}$.

The boundary value problem (18–20) is written in weak form as

$$a_u(y,p) = b_u(p), \quad \forall p \in H_0^1(\Omega(u)) \tag{21}$$

where

$$a_u(y,p) := \int_{\Omega(u)} \nabla y^T \nabla p\, dx - \int_u \left[\!\left[\frac{\partial y}{\partial n} p\right]\!\right] ds \tag{22}$$

$$b_u(p) := \int_{\Omega(u)} fp\, dx. \tag{23}$$

Now, F from the previous subsection takes the specific form

$$F := \left\{ (H_0^1(\Omega(u)), u, H_0^1(\Omega(u))) \mid u \in B_e^0([0,1], \mathbb{R}^2) \right\}. \tag{24}$$

For the convenience of the reader we only state the quadratic problem to the minimization problem (17–19). Its derivation is very technical and achieved by an application of the theorem of Correa and Seeger [4, Theorem 2.1]. We refer the reader for its derivation to [15]. Let κ denote the curvature corresponding to n and $\frac{\partial}{\partial \tau}$ the derivative tangential to u. If the optimal solution is a straight line connection of two fixed endpoints, then the QP (12, 13) is given strong form as the following optimal control problem:

$$\min_{(z,w)} F(z, w, y, p) := \int_{\Omega(u)} \frac{z^2}{2} + (y - \bar{y})z \, dx + \int_u \mu \kappa w - [\![f]\!] pw \, ds$$

$$+ \frac{1}{2} \int_u \mu \left(\frac{\partial w}{\partial \tau} \right)^2 - [\![f]\!] \kappa p w^2 \, ds \tag{25}$$

$$\text{s.t.} \quad -\Delta z = \Delta y + f \quad \text{in } \Omega(u) \tag{26}$$

$$\frac{\partial z}{\partial n} = f_1 w, \quad -\frac{\partial z}{\partial n} = f_2 w \quad \text{on } u \tag{27}$$

$$z = 0 \quad \text{on } \partial\Omega(u) \tag{28}$$

The adjoint problem to this optimal control problem is given by

$$-\Delta q = -z - (y - \bar{y}) \quad \text{in } \Omega(u) \tag{29}$$

$$q = 0 \quad \text{on } \partial\Omega(u) \tag{30}$$

and the resulting design equation for the optimal control problem (25–28) by

$$0 = -[\![f]\!] (p + \kappa p w + q) + \mu \kappa - \mu \frac{\partial^2 w}{\partial \tau^2} \quad \text{on } u. \tag{31}$$

3   A Quasi–Newton Approach

In this section we first formulate the minimization problem, a parabolic interface problem, and its shape derivative. The second part of this section presents a limited memory BFGS quasi–Newton technique in the shape space.

3.1   Problem Formulation and Shape Derivative

We will denote by u the interior boundary of a bounded domain $X(u) \subset \mathbb{R}^2$ with fixed Lipschitz boundary $\Gamma_{\text{out}} := \partial X(u)$. The interior boundary u is assumed to be smooth and variable. Moreover, let this domain $X(u)$ split into two disjoint subdomains $X_1(u), X_2(u) \subset X(u)$ such that $X_1(u) \,\dot\cup\, u \,\dot\cup\, X_2(u) = X(u)$, $\Gamma_{\text{bottom}} \,\dot\cup\, \Gamma_{\text{left}} \,\dot\cup\, \Gamma_{\text{right}} \,\dot\cup\, \Gamma_{\text{top}} = \partial X(u) \; (=: \Gamma_{\text{out}})$ and $\partial X_1(u) \cap \partial X_2(u) = u$ where $\dot\cup$ denotes the disjoint union. An example of such a domain is illustrated in Fig. 1.

The parabolic PDE constrained shape optimization problem is given by

$$\min_{u} J(y,u) := \int_{X(u)} \int_{0}^{T} (y - \bar{y})^2 dt\, dx + \mu \int_{u} 1\, ds \tag{32}$$

$$\text{s.t. } \frac{\partial y}{\partial t} - \operatorname{div}(k\nabla y) = f \quad \text{in } X(u) \times (0,T] \tag{33}$$

$$y = 1 \quad \text{on } \Gamma_{\text{top}} \times (0,T] \tag{34}$$

$$\frac{\partial y}{\partial n} = 0 \quad \text{on } (\Gamma_{\text{bottom}} \cup \Gamma_{\text{left}} \cup \Gamma_{\text{right}}) \times (0,T] \tag{35}$$

$$y = y_0 \quad \text{in } X(u) \times \{0\} \tag{36}$$

where $k \equiv k_1 = $ const. in $X_1(u) \times (0,T]$, $k \equiv k_2 = $ const. in $X_2(u) \times (0,T]$ and n denotes the unit outer normal to $X_2(u)$ at u. As mentioned in the previous section, the perimeter regularization with $\mu > 0$ in the objective (32) is used to overcome ill–posedness of the optimization problem. We formulate explicitly the continuity of the state and the flux at u as

$$[\![y]\!] = 0, \quad \left[\!\!\left[k\frac{\partial y}{\partial n}\right]\!\!\right] = 0 \quad \text{on } u \times (0,T] \tag{37}$$

where the jump $[\![\cdot]\!]$ is defined as in Sect. 2. The shape derivative of J in the direction of a continuous vector field V is given by

$$dJ(y,u)[V] = \int_{u} \left[\int_{0}^{T} \langle V, n\rangle \left[\!\!\left[-2k\frac{\partial y}{\partial n}\frac{\partial p}{\partial n} + k\nabla y^T \nabla p \right]\!\!\right] dt + \langle V, n\rangle \mu\kappa \right] ds \tag{38}$$

where κ denotes the curvature corresponding to the normal n. Its derivation is very technical. As in the elliptic case considered in the previous section, it is achieved by an application of the theorem of Correa and Seeger [4, Theorem 2.1]. For its derivation we refer the reader to [16].

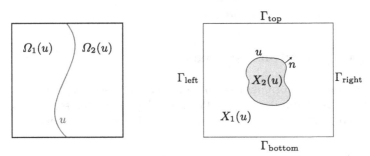

Fig. 1. Example of the domains $\Omega(u)$ and $X(u)$

3.2 Riemannian Limited Memory BFGS Update

As pointed out in [14], shape optimization can be viewed as optimization on Riemannian shape manifolds and resulting optimization methods can be constructed and analyzed within this framework which combines algorithmic ideas from [1] with the differential geometric point of view established in [9]. As in [14], we study connected and compact subsets $X(u)$ of $\mathbb{R}^2$ with $X(u) \neq \emptyset$ and C^∞ boundary $c = u$. In [9], this set of smooth boundary curves c is characterized by $B_e(S^1, \mathbb{R}^2) := \mathrm{Emb}(S^1, \mathbb{R}^2)/\mathrm{Diff}(S^1)$, where $\mathrm{Emb}(S^1, \mathbb{R}^2)$ denotes the set of all C^∞ embeddings of S^1 into the plane and $\mathrm{Diff}(S^1)$ the set of all diffeomorphisms of S^1 into itself. A particular point on the manifold $B_e(S^1, \mathbb{R}^2)$ is represented by a curve $c \colon S^1 \ni \theta \mapsto c(\theta) \in \mathbb{R}^2$. Because of the equivalence relation $(\mathrm{Diff}(S^1))$, the tangent space is isomorphic to the set of all normal C^∞ vector fields along c, i.e., $T_c B_e \cong \{h \mid h = \alpha n, \, \alpha \in C^\infty(S^1, \mathbb{R})\}$ where n is the unit exterior normal field of the shape defined by the boundary $u = c$ such that $n(\theta) \perp c'$ for all $\theta \in S^1$ and c' denotes the circumferential derivative as in [9]. For our discussion, we pick the following Sobolev metric family of [9] which defines a Riemannian metric on B_e for $A > 0$:

$$g^1 \colon T_c B_e \times T_c B_e \to \mathbb{R}, \, (h, k) \mapsto \int_c \alpha\beta + A\alpha'\beta' ds = ((I - A\triangle_c)\alpha, \beta)_{L^2(c)} \quad (39)$$

where $\triangle_c$ denotes the Laplace–Beltrami operator on the surface c.

The application of quasi–Newton methods is based on the secant condition, which can be formulated on the Riemannian manifold B_e analogously to [1]. In standard formulation, update formulas require the storage of the whole convergence history up to the current iteration. Limited memory update techniques have been developed, in order to reduce the amount of storage. In our situation, a BFGS update H_k of $\mathrm{Hess}J(c_k)$ can be formulated in the following way:

$\rho_k \leftarrow \frac{1}{g(w_k, s_k)}, \, q \leftarrow \mathrm{grad}\, J(c_k)$
for $i = k - 1, \ldots, k - m$ **do**
  $s_i \leftarrow Ts_i, \, w_i \leftarrow Tw_i, \, \alpha_i \leftarrow \rho_i\, g(s_i, q), \, q \leftarrow q - \alpha_i d_i$
end for
$z \leftarrow \mathrm{grad}J(c_k), \, q \leftarrow \frac{g(w_{k-1}, s_{k-1})}{g(w_{k-1}, w_{k-1})}\, \mathrm{grad}\, J(c_k)$
for $i = k - m, \ldots, k - 1$ **do**
  $\beta_i \leftarrow \rho_i\, g(w_i, z), \, q \leftarrow q + (\alpha_i - \beta_i)\, s_i$
end for
return $q = H_k^{-1} \mathrm{grad}J(c_k)$

where s_k denotes the distance between iterated shapes, w_k the difference of iterated Riemannian shape gradients, T the vector transport of elements in tangential space to updated shape and the memory contains only m shape gradients. One should note however, that the computation of each gradient involves the solution of an elliptic equation on the surface c due to the Sobolev metric. I.e., if the shape derivative is given in the form $dJ[V] = \int_c \gamma \langle V, n \rangle\, ds$, then the Riemannian gradient is given as the normal vector field $\mathrm{grad}J = gn$ with $(id - A\triangle_c)g = \gamma$.

4 Numerical Results and Implementation Details

The numerical results to the example of the Lagrange–Newton approach of Sect. 2 are given in Sect. 4.1. The second part of this section discusses numerical results for the inverse problem of Sect. 3.

4.1 Lagrange–Newton Approach

We solve the optimal control problem (25–28) by employing a CG–iteration for the reduced problem (31), i.e., we iterate over the variable w. Each time the CG–iteration needs a residual of Eq. (31) from w^k. We compute the state variable z^k from (26–28) and then the adjoint variable q^k from (29, 30), which enables the evaluation of the residual

$$r^k := - \llbracket f \rrbracket \left(p + \kappa p w^k + q^k \right) + \mu \kappa - \mu \frac{\partial^2 w^k}{\partial \tau^2}. \tag{40}$$

In this way we create an iterative solution technique very similar to SQP techniques known from linear spaces by using the QP (25–28) away from the optimal solution as a means to determine the step in the shape normal direction.

The particular values for the parameters are chosen as $f_1 = 1000$ and $f_2 = 1$ and the regularization parameter as $\mu = 10$. The data $\bar{y}$ are generated from a solution of the state Eqs. (18, 19) with u being the straight line connection of the points $(0.5, 0)$ and $(0.5, 1)$. The starting point of our iterations is described by a B–spline defined by the two control points $(0.6, 0.7)$ and $(0.4, 0.3)$. We built three unstructured tetrahedral grids Ω_h^1, Ω_h^2 and Ω_h^3 with roughly 6000, 24000 and 98000 triangles. In each iteration, the volume mesh is deformed according to the elasticity equation. The following table gives the distances of each shape to the solution approximated by $\mathrm{dist}(u^k, u^*) := \int_{u^*} |\langle u^k, e_1 \rangle - \frac{1}{2}| \, ds$ where u^* denotes the solution shape and $e_1 = (1, 0)$ is the first unit vector: This table

It.–No.	Ω_h^1	Ω_h^2	Ω_h^3
0	0.0705945	0.070637	0.0706476
1	0.0043115	0.004104	0.0040465
2	0.0003941	0.000104	0.0000645

demonstrates that quadratic convergence can be observed on the finest mesh, but also that the mesh resolution has a strong influence on the convergence properties revealed.

4.2 Quasi–Newton Approach

We test the algorithm developed in Sect. 3 with the problem (17–19) in the domain $X(u) = [-1, 1]^2$. First, we build artificial data $\bar{y}$, by solving the state equation for the setting $\bar{X}_2(u) := \{x \colon \|x\|_2 \leq 0.5\}$. Afterwards, we choose another initial domain $X_1(u)$ and $X_2(u)$. Figure 2 illustrates the interior boundary u around the initial domain $X_2(u)$ and the target domain $\bar{X}_2(u)$.

We choose the values for the parameters as $k_1 = 1$ and $k_2 = 0.001$ and a regularization parameter of $\mu = 0.0001$. The final time of simulation is $T = 20$. In order to solve the boundary value problem (33–36) we use standard linear

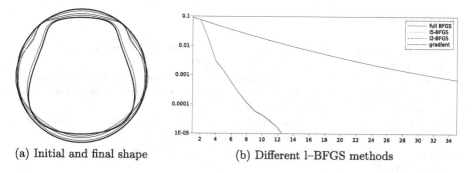

(a) Initial and final shape   (b) Different l–BFGS methods

Fig. 2. Initial and final shape geometry as well as the convergence history of different BFGS strategies compared to a pure gradient method

finite elements. Furthermore, we choose the implicit Euler method for the temporal discretization. The interval $[0, T]$ is therefore divided by 30 equidistantly distributed time steps.

The measurements of convergence rates ideally has to be performed in terms of the geodesic distance which is a highly expensive operation. In the discrete setting we therefore compute for each node of the iterated shape c_j the shortest distance to the optimal solution $\hat{c}$ in normal direction. We then form the L^2–norm of this distance field over $\hat{c}$, which is used to measure the convergence. It should be mentioned that the cost of this operation is quadratic with respect to the number of nodes on the surface. Starting in one node on c_j in normal direction, the determination of a point of intersection with $\hat{c}$ requires to check all boundary segments. This is the reason why we restrict our numerical results to 2D computations. Following this approach, Fig. 2 visualizes the convergence history of different BFGS strategies compared to a pure gradient method. It can clearly be seen that the BFGS methods are superior to the gradient based method. Furthermore, we partly obtain superlinear convergence in the BFGS case. It is yet surprising that, in this particular test case, there is hardly any difference between the number of stored gradients in the limited memory BFGS.

5   Conclusions

The novelties of this paper lie in the generalizations of the Riemannian shape calculus framework in [14] to Lagrange–Newton and quasi–Newton approaches for PDE constrained shape optimization problems. It is shown that this approaches are viable and lead to computational methods with superior convergence properties, when compared to only linearly converging standard steepest descent methods. We observe very fast convergence to the level of the approximation error – and this without any line–search. These promising results are to be extended to more practically challenging problems in a large–scale framework in subsequent papers. Moreover, as observed during the computations, the shape deformation sometimes leads to shapes, where normal vectors can no longer be reliably evaluated. Provisions for those cases have to be developed.

References

1. Absil, P.-A., Mahony, R., Sepulchre, R.: Optimization Algorithms on Matrix Manifolds. Princeton University Press, Princeton (2008)
2. Arian, E., Vatsa, V.N.: A preconditioning method for shape optimization governed by the Euler equations. Technical report, Institute for Computer Applications in Science and Engineering (ICASE), pp. 98–14 (1998)
3. Ameur, H.B., Burger, M., Hackl, B.: Level set methods for geometric inverse problems in linear elasticity. Inverse Prob. **20**, 673–696 (2004)
4. Correa, R., Seeger, A.: Directional derivative of a minmax function. Nonlinear Anal. **9**(1), 13–22 (1985)
5. Delfour, M.C., Zolésio, J.-P.: Shapes and Geometries: Analysis, Differential Calculus, and Optimization. Advances in Design and Control. SIAM Philadelphia, Philadelphia (2001)
6. Gabay, D.: Minimizing a differentiable function over a differential manifold. J. Optim. Theor. Appl. **37**(2), 177–219 (1982)
7. Eppler, K., Harbrecht, H.: A regularized Newton method in electrical impedance tomography using shape Hessian information. Control Cybern. **34**(1), 203–225 (2005)
8. Eppler, K., Schmidt, S., Schulz, V.H., Ilic, C.: Preconditioning the pressure tracking in fluid dynamics by shape Hessian information. J. Optim. Theory Appl. **141**(3), 513–531 (2009)
9. Michor, P.W., Mumford, D.: Riemannian geometries on spaces of plane curves. J. Eur. Math. Soc. (JEMS) **8**, 1–48 (2006)
10. Mohammadi, B., Pironneau, O.: Applied Shape Optimization for Fluids, Numerical Mathematics and Scientific Computation. Clarendon Press, Oxford (2001)
11. Nägel, A., Schulz, V.H., Siebenborn, M., Wittum, G.: Scalable shape optimization methods for structured inverse modeling in 3D diffusive processes. Comput. Vis. Sci. (2015). doi10.1007/s00791-015-0248-9
12. Novruzi, A., Roche, J.R.: Newton's method in shape optimisation: a three-dimensional case. BIT Numer. Math. **40**, 102–120 (2000)
13. Ring, W., Wirth, B.: Optimization methods on Riemannian manifolds and their application to shape space. SIAM J. Optim. **22**, 596–627 (2012)
14. Schulz, V.H.: A Riemannian view on shape optimization. Found. Comput. Math. **14**, 483–501 (2014)
15. Schulz, V.H., Siebenborn, M., Welker, K.: Towards a Lagrange-Newton approach for PDE constrained shape optimization. In: Leugering, G., et al., (eds.) New Trends in Shape Optimization. International Series of Numerical Mathematics, vol. 166. Springer (2015). doi:10.1007/978-3-319-17563-8
16. Schulz, V.H., Siebenborn, M., Welker, K.: Structured inverse modeling in parabolic diffusion problems. SICON (2014, submitted to). arXiv:1409.3464
17. Sokolowski, J., Zolésio, J.-P.: Introduction to Shape Optimization: Shape Sensitivity Analysis. Springer Series in Computational Mathematics, vol. 16. Springer, Heidelberg (1992)

Lie Groups and Geometric Mechanics/Thermodynamics

Poincaré Equations for Cosserat Shells: Application to Cephalopod Locomotion

Frederic Boyer[1](✉) and Federico Renda[2]

[1] EMN, IRCCyN, La Chantrerie 4, Rue Alfred Kastler, B.P. 20722,
44307 Nantes Cedex 3, France
frederic.boyer@emn.fr
[2] KUSTAR, KURI, Abu Dhabi Campus, 127788 Abu Dhabi, UAE
federico.renda@kustar.ac.ae

Abstract. In 1901 Henri Poincaré proposed a new set of equations for mechanics. These equations are a generalization of Lagrange equations to a system whose configuration space is a Lie group, which is not necessarily commutative. Since then, this result has been extensively refined by the Lagrangian reduction theory. In this article, we show the relations between these equations and continuous Cosserat media, i.e. media for which the conventional model of point particle is replaced by a rigid body of small volume named microstructure. In particular, we will see that the usual shell balance equations of nonlinear structural dynamics can be easily derived from the Poincaré's result. This framework is illustrated through the simulation of a simplified model of cephalopod swimming.

1 Introduction

The aim of this article is to contribute to establish the relations between the model of Cosserat media [1] and a set of equations first introduced by H. Poincaré [2] which became over time the corner stone of the Lagrangian reduction theory [3,4]. Though they may seem abstract at first, these equations are in fact a powerful tool for deriving blindly the balance equations of Cosserat media. Furthermore, by revealing the intrinsically geometric nature of these media, they are of great assistance in the implementation of a numerical "geometrically exact approach" of finite elements [5]. We established in [6] the relation between Cosserat beams and Poincaré equations. This requires their extension to field theory, as proposed in [7,8]. Here we show that these equations can be applied to two-dimensional Cosserat media, to recover the so called geometrically exact balance equations of shells in finite transformations and small strains. In a second step, we will briefly illustrate the general Poincaré-Cosserat picture by applying it to an axi-symmetric shell modelling the *mantle* of cephalopods, i.e., the open soft cavity that these animals contract and dilate for jet propulsion, a topic which is of great interest in the field of soft bio-robotics [9].

2 Poincaré Equations of a Classical Mechanical System

We consider a mechanical system with a configuration group G of transformations g. If the system is free, its lagrangian $L(g, \dot{g})$ (with $dg/dt = \dot{g}$) is invariant

© Springer International Publishing Switzerland 2015
F. Nielsen and F. Barbaresco (Eds.): GSI 2015, LNCS 9389, pp. 511–518, 2015.
DOI: 10.1007/978-3-319-25040-3_55

under any change of inertial frame, and $L(g, \dot{g}) = L(g^{-1}g, g^{-1}\dot{g}) = l(\eta)$ with $\eta \in \mathfrak{g}$ and l the reduced Lagrangian in the Lie algebra $\mathfrak{g}$. Poincaré applies the Hamilton principle to the action $\int_{t_1}^{t_2} l(\eta)dt$ with no use of the group's parameters, but rather by constraining the variations $\delta\zeta = g^{-1}\delta g \in \mathfrak{g}$ to satisfy $\delta\eta = d(\delta\zeta)/dt + ad_\eta(\delta\zeta)$. Then, applying the usual variational calculus leads to the equations:

$$\frac{d}{dt}\left(\frac{\partial l}{\partial \eta}\right) - ad_\eta^*\left(\frac{\partial l}{\partial \eta}\right) = 0. \tag{1}$$

These equations named "Poincaré" or "Euler-Poincaré" represent a balance of conjugate momenta in $\mathfrak{g}^*$ and have to be supplemented with the reconstruction equations $\dot{g} = g\eta$.

3   Poincaré Equations of a Cosserat Medium

Let us consider a three-dimensional body $\mathcal{B}$, i.e. a set of material points X of $\mathbb{R}^3$ labelled by 3 parameters $\{X^I\}_{I=1,2,3}$ in a Cartesian frame (O, E_1, E_2, E_3) we call the material frame. A configuration of $\mathcal{B}$ is an orientation preserving embedding $\Phi : \mathcal{B} \to \mathcal{E} = \mathbb{R}^3$. Using simplified notations, we call $\Phi(\mathcal{B}) = \{x = \Phi(X)/X \in \mathcal{B}\}$ the current configuration of the body. The ambient space $\mathcal{E}$ is endowed with an orthonormal inertial (o, e_1, e_2, e_3) relative to which a vector $x \in \mathcal{E}$ has the expansion $x = x^i e_i$. Among all the possible configurations of $\mathcal{B}$, we distinguish one of them $\Phi_o(\mathcal{B})$, called the reference configuration, in which $\mathcal{B}$ is internally at rest. In practise we take $(O, E_1, E_2, E_3) = (o, e_1, e_2, e_3)$ and will interchangeably speak of an inertial or material frame, depending on the context. In the subsequent developments, $\mathcal{B}$ is a Cosserat medium, i.e. a medium for which we can define a material p-dimensional subspace $\mathcal{D}$ of $\mathcal{B}$, which can be $\mathcal{B}$ itself, in each point of which, a Lie group G of rigid body mechanics ($SO(3)$, $SE(2)$, $SE(3)$...) acts on a subspace $\mathcal{M}$ of $\mathcal{B}$, named "micro-structure", to generate all the configurations $\Phi(\mathcal{B})$. The parameterization of $\mathcal{B}$ is chosen in such a manner that $\mathcal{D}$ is coordinatized by $\{X^\alpha\}_{\alpha=1,2..p}$, with $1 \leq p \leq 3$. As a result, the configuration space of such a medium, can be intrinsically defined as the set of maps:

$$\mathcal{C} = \{g : (X^1, X^2, ...X^p) \in \mathcal{D} \mapsto g(X^1, X^2, ...X^p) \in G\}. \tag{2}$$

Following Poincaré's approach, a Cosserat medium $\mathcal{B}$ submitted to a set of external forces is governed by the extended Hamilton's principle, which can be stated directly on the definition (2) of $\mathcal{C}$ (α running from 1 to p), as:

$$\delta \int_{t_1}^{t_2} \int \int ... \int_\mathcal{D} \mathcal{L}\left(g, \frac{\partial g}{\partial t}, \frac{\partial g}{\partial X^\alpha}\right)\sqrt{|h_o|}dX^1 dX^2...dX^p dt = -\int_{t_1}^{t_2} \delta W_{ext}dt, \tag{3}$$

for any $\delta g = g\delta\zeta$, where $\delta\zeta \in \mathfrak{g}$ is a field of material variation of g, achieved while t and all the X^α are maintained fixed and such that $\delta\zeta(t_1) = \delta\zeta(t_2) = 0$. In (3), $\mathcal{L}$ is the density of the Lagrangian of $\mathcal{B}$ per unit of metric volume $\sqrt{|h_o|}dX^1 dX^2...dX^p$ of $\Phi_o(\mathcal{D})$, with $|h_o|$, the determinant of the Euclidean metric in the natural basis $\{\partial\Phi_o/\partial X^\alpha(X)\}_{\alpha=1,2..p}$. δW_{ext} models the virtual work of the external forces. Physically, the $\partial g/\partial t$-dependency stands for the kinetic energy of $\mathcal{B}$ while that with respect to the $\partial g/\partial X^\alpha$'s models the internal strain energy of the material that will be assumed to be elastic. Remarkably, the Lagrangian is left-invariant in the sense that substituting g by hg with h a constant transformation in G, does not change the Lagrangian. Physically, this reflects the fact that the kinetic energy is unchanged by a change of inertial frame while the internal strain energy is not affected by a rigid overall displacement by virtue of its material objectivity. As a result, taking $h = g^{-1}$, allows (3) to be changed into:

$$\delta \int_{t_1}^{t_2} \int \int \cdots \int_{\mathcal{D}} \mathfrak{L}\left(\eta, \xi_\alpha\right) \sqrt{|h_o|} dX^1 dX^2 ... dX^p dt = - \int_{t_1}^{t_2} \delta W_{ext} dt, \qquad (4)$$

where $\mathfrak{L}$ is the reduced Lagrangian density, while $\eta = g^{-1}(\partial g/\partial t)$ and the $\xi_\alpha = g^{-1}(\partial g/\partial X^\alpha)$ expressions denote left-invariant vector fields in the Lie algebra $\mathfrak{g}$ of G. Now, let us invoke the constraints of variation at fixed time and material labels:

$$\delta \frac{\partial g}{\partial t} = \frac{\partial \delta g}{\partial t}, \ \delta \frac{\partial g}{\partial X_\alpha} = \frac{\partial \delta g}{\partial X_\alpha}, \text{ for } \alpha = 1, 2..p. \qquad (5)$$

Then inserting "$\delta g = g\delta\zeta$" into (5) gives the following relations, which play a key role in the variational calculus on Lie groups:

$$\delta\eta = \frac{\partial \delta\zeta}{\partial t} + ad_\eta(\delta\zeta), \ \delta\xi = \frac{\partial \delta\zeta}{\partial X^\alpha} + ad_{\xi_\alpha}(\delta\zeta). \qquad (6)$$

Finally, using (6) in (4), before the usual by part integration in time, and the divergence theorem in the chart $\{X^\alpha\}_{\alpha=1,2..p}$, gives the Poincaré equations of a Cosserat medium in the material frame (we use summation convention on repeated indices α):

$$\frac{\partial}{\partial t}\left(\frac{\partial \mathfrak{L}}{\partial \eta}\right) - ad_\eta^*\left(\frac{\partial \mathfrak{L}}{\partial \eta}\right) + \frac{1}{\sqrt{|h_o|}}\left(\frac{\partial}{\partial X^\alpha}\left(\sqrt{|h_o|}\frac{\partial \mathfrak{L}}{\partial \xi_\alpha}\right) - ad_{\xi_\alpha}^*\left(\sqrt{|h_o|}\frac{\partial \mathfrak{L}}{\partial \xi_\alpha}\right)\right) = F_{ext},$$

$$\left(\frac{|h_o|}{|\overline{h}_o|}\right)^{1/2}\frac{\partial \mathfrak{L}}{\partial \xi_\alpha}\nu_\alpha = -\overline{F}_{ext}, \qquad (7)$$

where $\nu = \nu_\alpha E^\alpha$ is the unit outward normal to the tangent planes of $\partial\Phi_o(\mathcal{D})$ pulled-back by the deformation of $\mathcal{D}$, and $|\overline{h}_o|^{1/2}dX^1 dX^2$ is its metric volume element. All the terms of the balance (7) define densities of vectors in $\mathfrak{g}^*$ related to the metric volume of $\Phi_o(\mathcal{D})$ (top), and $\partial\Phi_o(\partial\mathcal{D})$ (bottom). Alternatively, using (3) with a Lagrangian density per unit of metric volume $\sqrt{|h|}dX^1 dX^2 ... dX^p$ of $\Phi(\mathcal{D})$, gives a set of equations on the current configuration that will be detailed in the case of shells. Finally, all the conjugate momenta and external forces of (7) are reduced in $\mathfrak{g}^*$ and removing the X^α-dependence leads to the equations of a single rigid microstructure (body) given by (1).

4 Application to Cosserat Shells

While the case of beams has been addressed in [6]; we here focus on shells, which will be assumed to endure small strains but finite transformations in the Lie group $G = SE(3)$ represented by the transformations $g = (R, r)$, with $R \in SO(3)$ and $r \in \mathbb{R}^3$ the rotation and translation components.

4.1 Geometrically Exact Shells

Kinematics. In the case of shells, the Cosserat micro-structures $\mathcal{M}$ models rigid material lines, or "directors" (supported by E_3), which traverse the shell's material mid-surface $\mathcal{D}$ (supported by (E_1, E_2)). The directors are labelled by (X^1, X^2), two parameters which define a set of (generally curvilinear) coordinates of $\Phi_o(\mathcal{D})$. The configuration space of one such rigid microstructure, say the (X^1, X^2)-director is : $g(X^1, X^2) \in SE(3)$; and the whole configuration space of the shell is the space of (X^1, X^2)-parameterized surfaces in $SE(3)$:

$$C := \{g : (X^1, X^2) \in [0, 1]^2 \mapsto g(X^1, X^2) \in SE(3)\}. \tag{8}$$

Note that this definition appears in the basic Cosserat shells kinematics in $\mathbb{R}^3$:

$$\Phi(X^1, X^2, X^3) = r(X^1, X^2) + R(X^1, X^2).(X^3 E_3), \tag{9}$$

where, let us recall that (O, E_1, E_2, E_3) interchangeably defines the material or inertial frame to which r and R are related. With this definition of the shell configuration space, the left-invariant fields of the general construction are:

$$\eta = g^{-1} \frac{\partial g}{\partial t} = \begin{pmatrix} \Omega & V \\ 0 & 0 \end{pmatrix}, \ \xi_1 = g^{-1} \frac{\partial g}{\partial X^1} = \begin{pmatrix} K_1 & \Gamma_1 \\ 0 & 0 \end{pmatrix}, \ \xi_2 = g^{-1} \frac{\partial g}{\partial X^2} = \begin{pmatrix} K_2 & \Gamma_2 \\ 0 & 0 \end{pmatrix}, \tag{10}$$

where $\Omega(X^1, X^2)$ and $V(X^1, X^2)$ denote the linear and angular velocities of the (X^1, X^2)-director pulled back in the material frame, while (K_1, K_2) and (Γ_1, Γ_2) are in the same frame, and appear in the shell strain measurements of [5]:

$$\gamma_\alpha = E_3.(\Gamma_\alpha - \Gamma_\alpha^o), \ \rho_{\alpha\beta} = E_3.((K_\alpha \times \Gamma_\beta) - (K_\alpha^o \times \Gamma_\beta^o)), \ \epsilon_{\alpha\beta} = \Gamma_\alpha.\Gamma_\beta - \Gamma_\alpha^o.\Gamma_\beta^o, \tag{11}$$

where γ_α, $\rho_{\alpha\beta}$ and $\epsilon_{\alpha\beta}$ measure the transverse shearing, the curvature and the membrane stretching and shearing of the shell in the two material directions with respect to the reference configuration (denoted with the upper index o). Let us note that these measures, being dependent only on ξ_1 and ξ_2, satisfy de facto the frame-indifference requirement of nonlinear structural mechanics.

Geometrically Exact Force Balance of Shells. Based on the previous kinematics, the reduced shell's Lagrangian density of (4) takes the form:

$$\mathfrak{L}(\eta, \xi_1, \xi_2) = \mathfrak{T}(\eta) - \mathfrak{U}(\xi_1, \xi_2), \tag{12}$$

which, once introduced in the general Cosserat equations (7), cause to appear the density fields of internal wrenches per unit of metric area of $\Phi_o(\mathcal{D})$:

$$\frac{\partial \mathfrak{U}}{\partial \xi_1} = \begin{pmatrix} \frac{\partial \mathfrak{U}}{\partial K_1} \\ \frac{\partial \mathfrak{U}}{\partial \Gamma_1} \end{pmatrix} = \begin{pmatrix} M^1 \\ N^1 \end{pmatrix} , \frac{\partial \mathfrak{U}}{\partial \xi_2} = \begin{pmatrix} \frac{\partial \mathfrak{U}}{\partial K_2} \\ \frac{\partial \mathfrak{U}}{\partial \Gamma_2} \end{pmatrix} = \begin{pmatrix} M^2 \\ N^2 \end{pmatrix} . \tag{13}$$

Once muliplied by $\sqrt{|h_o|}$, N^1 and M^1 (respectively N^2 and M^2) define the resultant and the momentum of stress forces exerted across the shell cross section X^1 =constant (respectively X^2 =constant) per unit of X^2-length (X^1-length). Let us give the density of kinetic energy of a shell of mass density ρ_o in $\Phi_o(\mathcal{D})$:

$$\mathfrak{T}(\eta) = \frac{1}{2}(\Omega^T, V^T) \begin{pmatrix} \rho_o J & 0 \\ 0 & \rho_o l \delta \end{pmatrix} \begin{pmatrix} \Omega \\ V \end{pmatrix}, \tag{14}$$

where we assumed that the director frames are centered on the mass center of the shell's directors while l and $J = j_\perp(E_1 \otimes E^1 + E_2 \otimes E^2) + j_\parallel E_3 \otimes E^3$ denote the length and the geometric inertia tensor of the director respectively. Using the expression of ad^* on $se(3) \cong \mathbb{R}^6$, the field equations of (7) become for a shell:

$$\rho_o l(\frac{\partial V}{\partial t} + \Omega \times V) = \frac{1}{\sqrt{|h_o|}}(\frac{\partial \sqrt{|h_o|}N^\alpha}{\partial X^\alpha} + K_\alpha \times \sqrt{|h_o|}N^\alpha) + N_{ext}, \tag{15}$$

$$\rho_o J\frac{\partial \Omega}{\partial t} + \Omega \times \rho_o J\Omega = \frac{1}{\sqrt{|h_o|}}(\frac{\partial \sqrt{|h_o|}M^\alpha}{\partial X^\alpha} + K_\alpha \times \sqrt{|h_o|}M^\alpha + \Gamma_\alpha \times \sqrt{|h_o|}N^\alpha) + M_{ext}.$$

In (15), all the terms are densities related to the metric volume element $\sqrt{|h_o|}dX^1$ dX^2 of $\Phi_o(\mathcal{D})$ and can be equivalently related to $\sqrt{|h|}dX^1dX^2$, the metric volume element of $\Phi(\mathcal{D})$, if in our general picture, we start with a density of Lagrangian (3) related to the deformed configuration $\Phi(\mathcal{D})$. In this latter case, using the mass conservation $\rho_o\sqrt{|h_o|} = \rho\sqrt{|h|}$ and pushing forward the balance from the material frame to the directors' frames allows one to rewrite (15) in a form known in the shell literature [5], where they are derived from Newton laws:

$$\begin{cases} \rho l\frac{\partial v}{\partial t} = \frac{1}{\sqrt{|h|}} \left(\frac{\partial \sqrt{|h|}n^\alpha}{\partial X^\alpha} \right) + n_{ext}, \\ \rho\frac{\partial I\omega}{\partial t} = \frac{1}{\sqrt{|h|}} \left(\frac{\partial \sqrt{|h|}m^\alpha}{\partial X^\alpha} \right) + \frac{\partial r}{\partial X^\alpha} \times n^\alpha + m_{ext}. \end{cases} \tag{16}$$

From the Lie group point of view (16) stands for the Poincaré -Cosserat equations in the spatial setting where we have $\omega = R\Omega$, $v = RV$, $n^\alpha = RM^\alpha$, $m^\alpha = RM^\alpha$, $n_{ext} = RM_{ext}$, $m_{ext} = RM_{ext}$ and $I = RJR^T$. In spite of their elegance, these equations are not practically usable in general. In fact, they would hold perfectly for a 2-D Cosserat medium whose the micro-structures have a full rank inertia tensor J, i.e. with non negligible spin and couple stress around the directors, here modelled by $j_\parallel$ and the wrench components M^{31} and M^{32} of (15). By contrast, the usual shells have no such features and Eqs. (15) and (16) need to be modified further to cope with what appears as an artifactual effect, often named "drill rotation", of the nominal Cosserat model. A canonical way to achieve this is to replace $SE(3)$ in the definition of the configuration space of the shell by $S^2 \times \mathbb{R}^3$ where S^2 stands for the configuration space of each of the director alone with no reference to a director frame. Alternatively, one can follow the usual procedure leading to the symmetry of the Cauchy stress tensor of non-Cosserat 3D media and neglect M^{31} and M^{32} in (15) along with the kinetic momenta around the directors (i.e. $j_{\parallel} = 0$), which changes the third row of the dynamics (15) into the static constitutive constraint:

$$E_3.(\Gamma_\alpha \times N^\alpha + (K_\alpha \times E_3) \times (M^\alpha \times E_3)) = 0. \tag{17}$$

Then, rewriting (17) in the field of convected basis on the mid-surface pulled back in the material frame, i.e. in $(\Gamma_1, \Gamma_2, E_3)$, allows one to introduce the effective stress couple $\mathcal{M}^\alpha = M^\alpha \times E_3 = \mathcal{M}^{\alpha\beta}\Gamma_\alpha \otimes \Gamma_\beta$ and the effective stress tensor $\mathcal{N} = \mathcal{N}^{\alpha\beta}\Gamma_\alpha \otimes \Gamma_\beta$, which is symmetric, as is the Cauchy stress tensor field, in a conventional (not Cosserat) medium. In the subsequent developments, the motions will be restricted in such a manner that the "drill rotation" effect does not express and the full Cosserat-Poincaré Eqs. (15, 16) will hold. However, the constitutive laws will handle the effective stress and couple stress tensors and will implicitly take the above constitutive restriction into account.

Constitutive Equations. For isotropic elastic materials in small strains, the constitutive equations of shells can be stated in the so called reduced Hooke's law of shells [5]:

$$\mathcal{N}^{\alpha\beta} = H_d^{\alpha\beta\lambda\mu}\epsilon_{\lambda\mu}, \ \mathcal{M}^{\alpha\beta} = H_r^{\alpha\beta\lambda\mu}\rho_{\lambda\mu}, \ \mathcal{Q}^\alpha = H_s^{\alpha\beta}\gamma_\beta, \tag{18}$$

where $\mathcal{N}^{\alpha\beta}$, $\mathcal{M}^{\alpha\beta}$ and $\mathcal{Q}^\alpha$ are the effective internal stresses which can be defined as the dual of the strains measurements (11). By contrast to the case of beams [6], the Cosserat internal stress wrenches of the general construction cannot be identified to these effective stresses but are rather related to them through the relation:

$$\frac{\partial \mathcal{U}}{\partial \xi_\alpha} = \begin{pmatrix} \frac{\partial \mathcal{U}}{\partial K_\alpha} \\ \frac{\partial \mathcal{U}}{\partial \Gamma_\alpha} \end{pmatrix} = \begin{pmatrix} M^\alpha \\ N^\alpha \end{pmatrix} = \begin{pmatrix} (E_3 \times \Gamma_\beta)\mathcal{M}^{\alpha\beta} \\ (K_\beta \times E_3)\mathcal{M}^{\beta\alpha} + \Gamma_\beta\mathcal{N}^{\alpha\beta} + E_3\mathcal{Q}^\alpha \end{pmatrix}. \tag{19}$$

Finally, by substituting (11) into (18) and the result into (19), we obtain the constitutive laws relating the left-invariant fields of (10) with their dual (13). It is worth noting here that the constitutive law (18), as it is detailed in [5], does force (17), along with $M^{31} = M^{32} = 0$.

5   Application to Cephalopod Swimming

In this section, we apply the previous general picture to an axisymmetric cephalopod's shell-like mantle $\mathcal{B}$ undergoing a net translation and axi-symmetric shape deformations along the (o, e_3) direction of an inertial frame (o, e_1, e_2, e_3) with no rotation around it. As a result, the directors have no drill motion, and the previous entire Cosserat-Poincaré picture holds. According to the problem symmetry, the inertial frame (o, e_1, e_2, e_3) is endowed with a chart of cylindrical coordinates (r, ϕ, z) of local basis (e_r, e_ϕ, e_3). The material space $\mathcal{B}$, of material frame $(O, E_1, E_2, E_3) = (o, e_1, e_2, e_3)$, is identified to $\mathcal{D} \times \mathcal{M}$, with $\mathcal{D}$ the mantle's material mid-surface supported by (E_1, E_2), and $\mathcal{M}$ its director supported by E_3. The reference configuration is $\Phi_o(\mathcal{B})$. Its symmetry axis is (O, e_3) and the cylindrical coordinates of its points are noted (r_o, z_o, ϕ_o). In this context, $\mathcal{D}$ is covered with the polar material chart $\{X^1, X^2\} = \{X, \phi\}$ centered on O, where X is the metric length along the meridians of $\Phi_o(\mathcal{D})$. In any configuration $\Phi(\mathcal{B})$ of the mantle, any cross section fiber crossing $\mathcal{D}$ in (X, ϕ), is supported by the third unit vector of a director frame deduced from (O, E_1, E_2, E_3) through a transformation of $SE(3)$ of the form:

$$g(X, \phi) = \begin{pmatrix} \exp(\phi \hat{e}_3) & 0 \\ 0 & 1 \end{pmatrix} \begin{pmatrix} \exp(-\theta \hat{e}_\phi) & re_r + ze_3 \\ 0 & 1 \end{pmatrix}, \qquad (20)$$

with θ, the angle parameterizing the local rotation of the director with respect to the z-axis. Using (20) in (10), gives, with $se(3) \cong \mathbb{R}^6$:

$$\eta = \begin{pmatrix} 0 \\ \Omega_2 \\ 0 \\ V_1 \\ 0 \\ V_3 \end{pmatrix}, \xi_X = \begin{pmatrix} 0 \\ K_{2X} \\ 0 \\ \Gamma_{1X} \\ 0 \\ \Gamma_{3X} \end{pmatrix} = \begin{pmatrix} 0 \\ -\theta' \\ 0 \\ r' \cos\theta + z' \sin\theta \\ 0 \\ z' \cos\theta - r' \sin\theta \end{pmatrix}, \xi_\phi = \begin{pmatrix} K_{1\phi} \\ 0 \\ K_{3\phi} \\ 0 \\ \Gamma_{2\phi} \\ 0 \end{pmatrix} = \begin{pmatrix} \sin\theta \\ 0 \\ \cos\theta \\ 0 \\ r \\ 0 \end{pmatrix},$$

where $'$ denotes $\partial./\partial X$. These expressions define Γ_α and K_α ($\alpha = X, \phi$), which once inserted in (11), give the expressions of strains. Applying Eq. (15) to our mantle with $\sqrt{|h|} = \Gamma_{2\phi} \sqrt{\Gamma_{1X}^2 + \Gamma_{3X}^2}$ and no ϕ-dependency of the Lagrangian density, gives the following three scalar equations:

$$\begin{cases} \rho l \left(\frac{\partial V_1}{\partial t} - V_3 \Omega_2 \right) = \frac{1}{\sqrt{|h|}} \frac{\partial \sqrt{|h|} N^{1X}}{\partial X} + K_{2X} N^{3X} - K_{3\phi} N^{2\phi} + N_{ext}^1, \\ \rho l \left(\frac{\partial V_2}{\partial t} + V_1 \Omega_2 \right) = \frac{1}{\sqrt{|h|}} \frac{\partial \sqrt{|h|} N^{3X}}{\partial X} - K_{2X} N^{1X} + K_{1\phi} N^{2\phi} + N_{ext}^2, \quad (21) \\ \rho j \frac{\partial \Omega_2}{\partial t} = \frac{1}{\sqrt{|h|}} \frac{\partial \sqrt{|h|} M^{2X}}{\partial X} - \Gamma_{1X} N^{3X} + \Gamma_{3X} N^{1X} + K_{3\phi} M^{1\phi} + M_{ext}^2. \end{cases}$$

These balance Eq. (21) which involve densities of wrenches per unit of metric volume of $\Phi(\mathcal{D})$, have to be supplemented with the expressions of η, ξ_X and ξ_ϕ, the constitutive law (18), the reconstruction equation $\partial g / \partial t = g\eta$, and a model of hydrodynamic forces $(N_{ext}^1, N_{ext}^2, M_{ext}^2)$. Figure 1 illustrates a simulation result obtained with the mantle described in [9]. The hydrodynamic forces are modelled by distributing uniformly along the mantle the thrust exerted on a rigid rocket of same current shape whose expelled mass is deduced from the time-variations of the mantle's volume. With 1.7 as drag coefficient and 1.1 for the added mass coefficient, the snapshots above illustrate a forward motion of average speed $\simeq 0.14$ m/s and maximum internal volume reduction $\simeq 39$ %.

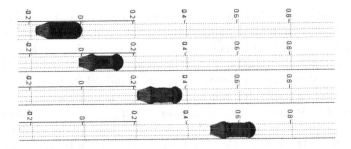

Fig. 1. Some snapshots of a simulated jet-propelled shell-like mantle.

6 Conclusion

In this article we proposed a general picture that permits the derivation of the balance equations of a Cosserat medium from a unique Lagrangian density. The approach is systematic and requires no phenomenological inputs. Based on the Poincaré equations, it allows the recover of the usual formulations of non-linear shell's theory, both in the reference and current configuration. It can be used to model the cephalopod mantle. In future, this model will be used to study hydrodynamic forces, and extended to the case of an octopus with a mantle prolonged by eight arms modelled by Cosserat beams.

References

1. Cosserat, E., Cosserat, F.: Théorie des corps déformables. A. Hermann et fils, Hermann, Paris (1909)
2. Poincaré, H.: Sur une forme nouvelle des équations de la mécanique. Compte Rendu de l'Académie des Sciences de Paris **132**, 369–371 (1901)
3. Marsden, J.E., Montgomery, R., Ratiu, T.S.: Reduction, symmetry, and phases in mechanics. Memoire AMS **436**, 1–110 (1990)
4. Marle, C.-M.: On Henri Poincaré's note: sur une forme nouvelle des equations de la mécanique. J. Geom. Symmetry Phys. **29**, 1–38 (2013)
5. Simo, J.C., Fox, D.D.: On stress resultant geometrically exact shell model. part i: formulation and optimal parametrization. Comput. Methods Appl. Mech. Eng. **72**(3), 267–304 (1989)
6. Boyer, F., Primault, D.: The Poincaré-Chetayev equations and flexible multibody systems. J. Appl. Math. Mech. **69**(6), 925–942, 2005. http://hal.archives-ouvertes.fr/hal-00672477
7. Castrillón-López, M., Ratiu, T.S., Shkoller, S.: Reduction in principal fiber bundles: covariant Euler-Poincaré equations. Proc. Amer. Math. Soc. **128**(7), 2155–2164 (2000)
8. Pommaret, J.F.: Partial Differential Equations and Group Theory. Kluwer Academic Publishers, Dordrecht (1994)
9. Renda, F., Serchi, F.G., Boyer, F., Laschi, C.: Propulsion and elastodynamic model of an underwater shell-like soft robot. In: Proceedings of IEEE International Conference on Robotics and Automation (ICRA), Seattle, 26–30 May 2015 (Accepted)

Entropy and Structure
of the Thermodynamical Systems

Géry de Saxcé[(⊠)]

Laboratoire de Mécanique de Lille (UMR CNRS 8107), Villeneuve d'Ascq, France
gery.desaxce@univ-lille1.fr

Abstract. With respect to the concept of affine tensor, we analyse in this work the underlying geometric structure of the theories of Lie group statistical mechanics and relativistic thermodynamics of continua, formulated by Souriau independently one of each other. We reveal the link between these ones in the classical Galilean context. These geometric structures of the thermodynamics are rich and we think they might be source of inspiration for the geometric theory of information based on the concept of entropy.

1 Affine Tensors

Points of an Affine Space. Let $A\mathcal{T}$ be an affine space associated to a linear space $\mathcal{T}$ of finite dimension n. By the choice of an affine frame f composed of a basis of $\mathcal{T}$ and an origin a_0, we can associate to each point a a set of n (affine) components V^i gathered in the n-column $V \in \mathbb{R}^n$. For a change of affine frames, the transformation law for the components of a point reads:

$$V = C + PV', \qquad (1)$$

which is an affine representation of the affine group of $\mathbb{R}^n$ denoted $\mathbb{A}ff(n)$. It is clearly different from the usual transformation law of vectors $V = PV'$.

Affine Forms. The affine maps $\boldsymbol{\Psi}$ from $A\mathcal{T}$ into $\mathbb{R}$ are called affine forms and their set is denoted $A^*\mathcal{T}$. In an affine frame, $\boldsymbol{\Psi}$ is represented by an affine function Ψ from $\mathbb{R}^n$ into $\mathbb{R}$. Hence, it holds:

$$\boldsymbol{\Psi}(a) = \Psi(V) = \chi + \Phi V,$$

where $\chi = \Psi(0) = \boldsymbol{\Psi}(a_0)$ and $\Phi = lin(\Psi)$ is a n-row. We call $\Phi_1, \Phi_2, \cdots, \Phi_n, \chi$ the components of $\boldsymbol{\Psi}$ or, equivalently, the couple of χ and the row Φ collecting the Φ_α. The set $A^*\mathcal{T}$ is a linear space of dimension $(n+1)$ called the vector dual of $A\mathcal{T}$. If we change the affine frame, the components of an affine form are modified according to the induced action of $\mathbb{A}ff(n)$, that leads to, taking into account (1):

$$\chi' = \chi - \Phi P^{-1}C, \qquad \Phi' = \Phi P^{-1}, \qquad (2)$$

which is a linear representation of $\mathbb{A}ff(n)$.

© Springer International Publishing Switzerland 2015
F. Nielsen and F. Barbaresco (Eds.): GSI 2015, LNCS 9389, pp. 519–528, 2015.
DOI: 10.1007/978-3-319-25040-3_56

Affine Tensors. We can generalize this construction and define an affine tensor as an object:

- that assigns a set of components to each affine frame f of an affine space AT of finite dimension n,
- with a transformation law, when changing of frames, which is an affine or a linear representation of $\mathbb{A}ff(n)$.

With this definition, the affine tensors are a natural generalization of the classical tensors that we shall call linear tensors, these last ones being trivial affine tensors for which the affine transformation $a = (C, P)$ acts through its linear part $P = lin(a)$. An affine tensor can be constructed as a map which is affine or linear with respect to each of its arguments. As the linear tensors, the affine ones can be classified in three families: covariants, contravariant and mixed. The most simple affine tensors are the points which are 1-contravariant and the affine forms which are 1-covariant but we can construct more complex ones having a strong physical meaning, the torsors proposed in [2] and the momenta introduced later. For more details on the affine dual space, affine tensor product, affine wedge product and affine tangent bundles, the reader interested in this topic is referred to the so-called AV-differential geometry [13].

G-Tensors. A subgroup G of $\mathbb{A}ff(n)$ naturally acts onto the affine tensors by restriction to G of their transformation law. Let F_G be a set of affine frames of which G is a transformation group. The elements of F_G are called G-frames. A G-tensor is an object:

- that assigns a set of components to each G-frame f,
- with a transformation law, when changing of frames, which is an affine or a linear representation of G.

Hence each G-tensor can be identified with an orbit of G within the space of the tensor components.

2   Momentum as Affine Tensor

Let $\mathcal{M}$ be a differential manifold of dimension n and G a Lie subgroup of $\mathbb{A}ff(n)$. In the applications to Physics, $\mathcal{M}$ will be for us typically the space-time and G a subgroup of $\mathbb{A}ff(n)$ with a physical meaning in the framework of classical mechanics (Galileo's group) or relativity (Poincaré's group). The points of the space-time $\mathcal{M}$ are events of which the coordinate X^0 is the time t and $X^i = x^i$ for i running from 1 to 3 gives the position.

The tangent space to $\mathcal{M}$ at the point $\boldsymbol{X}$ equipped with a structure of affine space is called the affine tangent space and is denoted $AT_{\boldsymbol{X}}\mathcal{M}$. Its elements are called tangent points at $\boldsymbol{X}$. The set of affine forms on the affine tangent space is denoted $A^*T_{\boldsymbol{X}}\mathcal{M}$. We call momentum a bilinear map $\boldsymbol{\mu}$:

$$\boldsymbol{\mu} : T_{\boldsymbol{X}}\mathcal{M} \times A^*T_{\boldsymbol{X}}\mathcal{M} \to \mathbb{R} : (\overrightarrow{\boldsymbol{V}}, \boldsymbol{\Psi}) \mapsto \boldsymbol{\mu}(\overrightarrow{\boldsymbol{V}}, \boldsymbol{\Psi})$$

It is a mixed 1-covariant and 1-contravariant affine tensor. Taking into account the bilinearity, it is represented in an affine frame f by:

$$\mu(\overrightarrow{V}, \Psi) = (\chi F_\beta + \Phi_\alpha L_\beta^\alpha) V^\beta$$

where F_β and L_β^α are the components of μ in the affine frame f or, equivalently, the couple $\mu = (F, L)$ of the row F collecting the F_β and the $n \times n$ matrix L of elements L_β^α. Owing to (2), the transformation law is given by the induced action of $\mathbb{A}ff(n)$:

$$F' = F P^{-1}, \qquad L' = (P L + C F) P^{-1} \tag{3}$$

If the action is restricted to the subgroup G, the momentum μ is a G-tensor. Identifying the space of the momentum components μ to the dual $\mathfrak{g}^*$ of the Lie algebra of G thanks to the dual pairing:

$$\mu Z = \mu \, da = (F, L)' (dC, dP) = F \, dC + Tr(L \, dP) \tag{4}$$

it is noteworthy to observe that the transformation law (3) of momenta *is nothing else* the coadjoint action:

$$\mu = Ad^*(a) \, \mu'.$$

However, this mathematical construction is not relevant for all considered physical applications and we need to extend it by considering a map θ from G into $\mathfrak{g}^*$ and a generalized transformation law:

$$\mu = a \cdot \mu' = Ad^*(a) \, \mu' + \theta(a), \tag{5}$$

where θ eventually depends on an invariant of the orbit. It is an affine representation of G in $\mathfrak{g}^*$ (because we wish the momentum to be an affine tensor) provided:

$$\forall a, b \in G, \qquad \theta(ab) = \theta(a) + Ad^*(a) \, \theta(b) \tag{6}$$

Remark: this action induces a *structure of affine space* on the set of momentum tensors. Let $\pi : \mathcal{F} \to \mathcal{M}$ be a G-principal bundle of affine frames with the free action $(a, f) \mapsto f' = a \cdot f$ on each fiber. Then we can build the associated G-principal bundle:

$$\hat{\pi} : \mathfrak{g}^* \times \mathcal{F} \to (\mathfrak{g}^* \times \mathcal{F})/G : (\mu, f) \mapsto \boldsymbol{\mu} = orb(\mu, f)$$

for the free action:

$$(a, (\mu, f)) \mapsto (\mu', f') = a \cdot (\mu, f) = (a \cdot \mu, a \cdot f)$$

where the action on $\mathfrak{g}^*$ is (5). Clearly, the orbite $\boldsymbol{\mu} = orb(\mu, f)$ can be identified to the momentum G-tensor $\boldsymbol{\mu}$ of components μ in the G-frame f.

3   Symplectic Action and Momentum Map

Let $(\mathcal{N}, \omega)$ be a symplectic manifold. A Lie group G smoothly left acting on $\mathcal{N}$ and preserving the symplectic form ω is said to be symplectic. The interior product of a vector $\overrightarrow{V}$ and a p-form ω is denoted $\iota(\overrightarrow{V})\omega$. A map $\psi : \mathcal{N} \to \mathfrak{g}^*$ such that:

$$\forall \eta \in \mathcal{N}, \quad \forall Z \in \mathfrak{g}, \quad \iota(Z \cdot \eta)\,\omega = -d(\psi(\eta)Z),$$

is called a momentum map of G. It is the quantity involved in Noether's theorem that claims ψ is constant on each leaf of $\mathcal{N}$. In ([8] (Theorem (11.17), p. 109, or its English translation [11]), Souriau proved there exists a smooth map θ from G into $\mathfrak{g}^*$:

$$\theta(a) = \psi(a \cdot \eta) - Ad^*(a)\,\psi(\eta), \tag{7}$$

which is a symplectic cocycle, that is a map $\theta : G \to \mathfrak{g}$ verifying the identity (6) and such that $(D\theta)(e)$ is a 2-form. An important result called Kirillov-Kostant-Souriau theorem reveals the orbit symplectic structure [8] (Theorem (11.34), pp. 116–118). Let G be a Lie group and an orbit of the coadjoint representation $orb\,(\mu) \subset \mathfrak{g}^*$. Then the orbit $orb\,(\mu)$ is a symplectic manifold, G is a symplectic group and any $\mu \in \mathfrak{g}^*$ is its own momentum.

Remark 1: replacing η by $a^{-1} \cdot \eta$ in (7), this formula reads:

$$\psi(\eta) = Ad^*(a)\,\psi'(\eta) + \theta(a),$$

where $\psi \mapsto \psi' = a \cdot \psi$ is the induced action of the one of G on $\mathcal{N}$. It is worth observing it is just (5) with $\mu = \psi(\eta)$ and $\mu' = \psi'(\eta)$. In this sence, the values of the momentum map *are just* the momentum G-tensors defined in the previous Section.

Remark 2: we saw at Remark of Sect. 2 that the momentum G-tensor μ is identified to the orbit $\mu = orb(\mu, f)$ and, disregarding the frames for simplification, we can identify μ to the component orbit $orb(\mu)$.

4   Lie Group Statistical Mechanics

In order to discover the underlined geometric structure of the statistical mechanics, we are interested by the affine maps Θ on the affine space of momentum tensors. In an affine frame, Θ is represented by an affine function Θ from $\mathfrak{g}^*$ into $\mathbb{R}$:

$$\boldsymbol{\Theta}(\boldsymbol{\mu}) = \Theta(\mu) = z + \mu Z,$$

where $z = \Theta(0) = \boldsymbol{\Theta}(\boldsymbol{\mu}_0)$ and $Z = lin(\Theta) \in \mathfrak{g}$ are the affine components of $\boldsymbol{\Theta}$. If the components of the momentum tensors are modified according to (5), the change of affine components of $\boldsymbol{\Theta}$ is given by the induced action:

$$z = z' - \theta(a)\,Ad(a)\,Z', \qquad Z = Ad(a)\,Z'. \tag{8}$$

Then $\boldsymbol{\Theta}$ is a G-tensors. In [8,11], Souriau proposed a statistical mechanics model using geometric tools. Let $d\lambda$ be a measure on $\boldsymbol{\mu} = orb\,(\mu)$ and a Gibbs probability measure $p\,d\lambda$ with:

$$p = e^{-(\mu)} = e^{-(z+\mu\,Z)}.$$

The normalization condition $\int_{orb(\mu)} p\,d\lambda = 1$ links the components of $\boldsymbol{Z}$, allowing to express z in terms of $\boldsymbol{Z}$:

$$z(Z) = \ln \int_{orb(\mu)} e^{-\mu\,Z}\,d\lambda. \tag{9}$$

The corresponding entropy and mean momenta are:

$$s = -\int_{orb(\mu)} p\,\ln p\,d\lambda = z + M\,Z, \qquad M = \int_{orb(\mu)} \mu\,p\,d\lambda = -\frac{\partial z}{\partial Z}. \tag{10}$$

satisfying the same transformation law as the one (5) of μ. Hence M are the components of a momentum tensor $\boldsymbol{M}$ which can be identified to the orbit $orb(M)$, that defines a map $\boldsymbol{\mu} \mapsto \boldsymbol{M}$ i.e. a correspondance between two orbits. This construction is formal and, for reasons of integrability, the integrals will be performed only on a subset of the orbit according to an heuristic way explained latter on.

People generally consider that the definition of the entropy is relevant for applications insofar the number of particles in the system is very huge. For instance, the number of atoms contained in one mole is Avogadro's number equal to 6×10^{23}. It is worth to notice that Vallée and Lerintiu proposed a generalization of the ideal gas law based on convex analysis and a definition of entropy which does not require the classical approximations (Stirling's Formula) [15].

5   Relativistic Thermodynamics of Continua

Independently of his statistical mechanics, Souriau proposed in [9,10] a thermodynamics of continua compatible with general relativity. In his footstep, one can quote the works by Iglesias [5] and Vallée [14]. In his Ph.D thesis, Vallée studied the invariant form of constitutive laws in the context of special relativity where the gravitation effects are neglected. In [3], the author and Vallée proposed a Galilean version of this theory of which we recal the cornerstone results.

G being Galileo's group and $\mathcal{M}$ being the space-time equipped with a G-connection ∇ representing the gravitation, the matter and its evolution is characterized by a line bundle $\pi_0 : \mathcal{M} \mapsto \mathcal{M}_0$. The trajectory of the particle $X_0 \in \mathcal{M}_0$ is the corresponding fiber $\pi_0^{-1}(X_0)$. In local charts, X_0 is represented by $s' \in \mathbb{R}^3$ and its position x at time t is given by a map $x = \varphi(t,s')$. The 4-velocity $\overrightarrow{U} = \overrightarrow{dX}/dt$ is the tangent vector to the fiber parameterized by the time. β being the reciprocal temperature, that is $1/k_B T$ where k_B is Boltzmann's constant and T the absolute temperature, there are five basic tensor fields defined on the space-time $\mathcal{M}$:

- the 4-flux of mass $\vec{N} = \rho \vec{U}$ where ρ is the density,
- the 4-flux of entropy $\vec{S} = \rho s \, \vec{U} = s \, \vec{N}$ where s is the specific entropy,
- Planck's temperature vector $\vec{W} = \beta \, \vec{U}$,
- its gradient $\mathbf{f} = \nabla \vec{W}$ called friction tensor,
- the momentum tensor of a continuum T, a linear map from $T_X \mathcal{M}$ into itself.

In local charts, they are respectively represented by two 4-columns N, W and two 4×4 matrices f and T. Then we proved in [3] the following result characterizing the reversible processes:

Theorem 1. *If Planck's potential ζ smoothly depends on s', W and $F = \partial x / \partial s'$ through right Cauchy strains $C = F^T F$, then:*

$$T = U \, \Pi_R + \begin{pmatrix} 0 & 0 \\ -\sigma v & \sigma \end{pmatrix} \tag{11}$$

with

$$\Pi = -\rho \, \frac{\partial \zeta}{\partial W}, \qquad \sigma = -\frac{2\rho}{\beta} F \frac{\partial \zeta}{\partial C} F^T, \tag{12}$$

represents the momentum tensor of the continuum and is such that:

$$(\nabla \zeta) \, N = -Tr \, (T \, f),$$

Combining this result with the geometric version of the first principle of thermodynamics:

$$\boldsymbol{Div\, T = 0}, \qquad \boldsymbol{Div\, \vec{N} = 0}, \tag{13}$$

it can be proved that the 4-flux of specific entropy:

$$\vec{S} = \boldsymbol{T} \, \vec{W} + \zeta \, \vec{N},$$

is divergence free and the specific entropy s is an integral of the motion.

6   Planck's Potential of a Continuum

Now, let us reveal the link between the previous relativistic thermodynamics of continua and Lie group statistical mechanics in the classical Galilean context and, to simplify, in absence of gravitation. In other words, how to deduce T from M and ζ from z? We work in five steps:

- **Step 1: parameterizing the orbit.** In order to calculate the integral (9), the orbit is parameterized thanks to a momentum map. Galileo's group G is a subgroup of the affine group $\mathbb{GA}(4)$, collecting the Galilean transformations, that is the affine transformations $X = P \, X' + C$ such that:

$$C = \begin{pmatrix} \tau_0 \\ k \end{pmatrix}, \qquad P = \begin{pmatrix} 1 & 0 \\ u & R \end{pmatrix},$$

where $u \in \mathbb{R}^3$ is a Galilean boost, $R \in \mathbb{SO}(3)$ is a rotation, $k \in \mathbb{R}^3$ is a spatial translation and $\tau_0 \in \mathbb{R}$ is a clock change. Hence Galileo's group is a Lie group of dimension 10. Considering its infinitesimal generators $Z = (dC, dP)$:

$$dC = \begin{pmatrix} d\tau_0 \\ dk \end{pmatrix}, \qquad dP = \begin{pmatrix} 0 & 0 \\ du & j(d\varpi) \end{pmatrix},$$

where $j(d\varpi)\, v = d\varpi \times v$, the dual pairing (4) reads:

$$\mu Z = l \cdot d\varpi - q \cdot du + p \cdot dk - e\, d\tau_0.$$

The most general form of the action (5) itemizes in:

$$p = R\, p' + m\, u, \qquad q = R\, (q' - \tau_0\, p') + m\, (k - \tau_0\, u), \tag{14}$$

$$l = R\, l' - u \times (R\, q') + k \times (R\, p') + m\, k \times u, \tag{15}$$

$$e = e' + u \cdot (R\, p') + \frac{1}{2} m \parallel u \parallel^2. \tag{16}$$

where the orbit invariant m occuring in the symplectic cocycle θ is physically interpreted as the particle mass. This transformation law reveals the physical meaning of the momentum tensor components as the linear momentum p, the passage q, the angular momentum l and the energy e. For the generic orbits, the dimension of the isotropy group of μ is 2, hence $dim(orb(\mu)) = dim\, G - 2 = 8$. In the 10 dimensional space $\mathfrak{g}^*$ of the momentum components, the orbit is defined by $dim\, \mathfrak{g}^* - 8 = 2$ equations which are invariant by galilean transformations:

$$s_0 = \parallel l_0 \parallel = C^{te}, \qquad e_0 = e - \frac{1}{2m} \parallel p \parallel^2 = C^{te}. \tag{17}$$

where occurs the spin angular momentum $l_0 = l - q \times p/m$. If the particle has an internal structure, introducing the moment of inertia matrix $\mathcal{J}$ and the spin ϖ, we have, according to König's theorem:

$$l_0 = \mathcal{J}\, \varpi, \qquad e_0 = \frac{1}{2} \varpi \cdot (\mathcal{J}\, \varpi).$$

Hence each orbit defines a particle of mass m, spin s_0, inertia $\mathcal{J}$ and can be parameterized by 8 coordinates, the 3 components of q, the 3 components of p and the 2 components of the unit vector n defining the spin direction, thanks to the momentum map $\mathbb{R}^3 \times \mathbb{R}^3 \times \mathbb{S}^2 \to \mathfrak{g}^* : (q, p, n) \mapsto \mu = \psi(q, p, n)$ such that:

$$l = \frac{1}{m} q \times p + s_0 n, \qquad e = \frac{1}{2m} \parallel p \parallel^2 + \frac{s_0^2}{2} n \cdot (\mathcal{J}^{-1} n).$$

The corresponding measure is $d\lambda = d^3q\, d^3p\, d^2n$. For simplicity, we consider further only a spinless particle, being at rest in a coordinate system X', then characterized by null components. In another coordinate system $X = P X' + C$ with a Galilean boost u and a translation of the origin at $k = x$ (hence $\tau_0 = 0$ and $R = 1_{\mathbb{R}^3}$), the new components are determined by (14), (15) and (16):

$$p = m u, \qquad q = m x, \qquad l = m x \times u, \qquad e = \frac{1}{2} m \parallel u \parallel^2 = \frac{1}{2 m} \parallel p \parallel^2 .$$

- **Step 2: modelling the deformation.** Let us consider N identical particles contained in a box of finite volume V, large with respect to the particles but representing the volume element of the continuum thermodynamics. For a change of coordinate $t = t'$, $x = \varphi(t', s')$, the jacobean matrix reads:

$$\frac{\partial X}{\partial X'} = P = \begin{pmatrix} 1 & 0 \\ v & F \end{pmatrix} \tag{18}$$

Besides, we suppose that the box of initial volume V_0 is at rest in the considered coordinate system ($v = 0$) and the deformation gradient F is uniform in the box, then $dx = F \, ds'$. According to (3), the linear momentum is transformed according to $p = F^{-T} p'$. The measure becomes $d\lambda = m^3 d^3 x \, d^3 p \, d^2 n = m^3 d^3 s' \, d^3 p' \, d^2 n$. Replacing the orbit by the subset $V_0 \times \mathbb{R}^3 \times \mathbb{S}^2$ and integrating, (9) gives for a particle after leaving out the primes:

$$z = \frac{1}{2} \ln(\det(C)) - \frac{3}{2} \ln \beta + C^{te}, \tag{19}$$

where the value of the constant is not relevant in the sequel.

As pointed out by Barbaresco [1], there is a puzzling analogy between the integral occuring in (9) and Koszul-Vinberg characteristic function [6,16]:

$$\psi_\Omega(Z) = \int_{\Omega^*} e^{-\mu Z} d\lambda,$$

where Ω is a sharp open convex cone and Ω^* is the set of linear strictly positive forms on $\bar{\Omega} - \{0\}$. Considering Galileo's group, it is worth to remark the cone of future directed timelike vectors (i.e. such that $\beta > 0$) [7] is preserved by linear Galilean transformations. The momentum orbits are contained in Ω^* but the integral does not converge neither on the orbits either all the more on Ω^*.

- **Step 3: identification.** We claim that $Z = (-W, 0)$ where $W^T = (\beta, w^T)$ is the transposed of the 4-column gathering the components of Planck's temperature vector. This identification is suggested by the fact that the transformation law $W = P W'$ of vectors *is nothing else* (8) where $Z = (-W, 0)$. For the box at rest in the coordinate system X, we put $W^T = (\beta, 0)$.
- **Step 4: boost method.** A new coordinate system $\bar{X}$ in which the box has the velocity v can be deduced from $X = P \bar{X} + C$ by applying a boost $u = -v$ (hence $k = 0$, $\tau_0 = 0$ and $R = 1_{\mathbb{R}^3}$). The transformation law of vectors gives the new components $\bar{W}^T = (\beta, \beta v^T)$ and (8) leads to:

$$\bar{z} = z + \frac{m \beta}{2} \parallel v \parallel^2 = z + \frac{m}{2 \beta} \parallel w \parallel^2 .$$

Taking into account (19) and leaving out the bars:

$$z = \frac{1}{2} \ln(\det(C)) - \frac{3}{2} \ln \beta + \frac{m}{2\beta} \parallel w \parallel^2 + C^{te}. \tag{20}$$

It is clear from (10) that s is Legendre conjugate of $-z$, then, introducing the internal energy (which is nothing else the Galilean invariant e_0 of (17)):

$$e_{int} = e - \frac{1}{2m} \parallel p \parallel^2,$$

the entropy is:

$$s = \frac{3}{2} \ln e_{int} + \frac{1}{2} \ln(\det(C)) + C^{te},$$

and, by $Z = \partial s / \partial M$, we derive the corresponding momenta:

$$\beta = \frac{\partial s}{\partial e} = \frac{3}{2 e_{int}}, \qquad w = -grad_p \, s = \frac{3}{2 e_{int}} \frac{p}{m}.$$

– **Step 5: link between z and ζ.** As z is an extensive quantity, its value for N identical particles is $z_N = N z$. Planck's potential ζ being a specific quantity, we claim that:

$$\zeta = \frac{z_N}{N m} = \frac{z}{m} = \frac{1}{2m} \ln(\det(C)) - \frac{3}{2m} \ln \beta + \frac{1}{2\beta} \parallel w \parallel^2 + C^{te}.$$

By (11) and (12), we obtain the linear 4-momentum $\Pi = (\mathcal{H}, -p^T)$ and Cauchy's stresses:

$$\mathcal{H} = \rho \left(\frac{3}{2} \frac{k_B T}{m} + \frac{1}{2} \parallel v \parallel^2 \right), \qquad p = \rho v, \qquad \sigma = -q \, 1_{\mathbb{R}^3},$$

where, by the expression of the pressure, we recover the *ideal gas law*:

$$q = \frac{\rho}{m} k_B T = \frac{N}{V} k_B T.$$

The first principle of thermodynamics (13) reads:

$$\frac{\partial \mathcal{H}}{\partial t} + div \, (\mathcal{H} v - \sigma v) = 0, \qquad \rho \frac{dv}{dt} = -grad \, q, \qquad \frac{\partial \rho}{\partial t} + div \, (\rho v) = 0.$$

We recognize the balance of energy, linear momentum and mass.

Remark: the Hessian matrix I of $-z$, considered as function of W through Z, is positive definite [8]. It is *Fisher metric* of the Information Geometry. For the expression (20), it is easy to verify it:

$$-\delta M \, \delta Z = \frac{1}{\beta} \left(e_{int} (\delta \beta)^2 + m \parallel \delta w - \frac{\delta \beta}{m} p \parallel^2 \right) > 0,$$

for any non vanishing δZ taking into account $\beta > 0, e_{int} > 0$ and $m > 0$. On this ground, we can construct a *thermodynamic length* of a path $t \mapsto X(t)$ [4]:

$$\mathcal{L} = \int_{t_0}^{t_1} \sqrt{(\delta W(t))^T I(t) \, \delta W(t)} dt,$$

where $\delta W(t)$ is the perturbation of the temperature vector, tangent to the spacetime at $X(t)$. We can also define a related quantity, *Jensen-Shannon divergence* of the path:

$$\mathcal{J} = (t_1 - t_0) \int_{t_0}^{t_1} (\delta W(t))^T I(t) \, \delta W(t) dt.$$

References

1. Barbaresco, F.: Koszul information geometry and Souriau geometric temperature/capacity of lie group thermodynamics. Entropy **16**, 4521–4565 (2014)
2. de Saxcé, G., Vallée, C.: Affine tensors in mechanics of freely falling particles and rigid bodies. Math. Mech. Solid J. **17**(4), 413–430 (2011)
3. de Saxcé, G., Vallée, C.: Bargmann group, momentum tensor and Galilean invariance of Clausius-Duhem inequality. Int. J. Eng. Sci. **50**, 216–232 (2012)
4. Crooks, G.E.: Measuring thermodynamic length. Phys. Rev. Lett. **99**, 100602 pp. (2009). doi:10.1103/PhysRevLett.99.100602
5. Iglesias, P.: Essai de thermodynamique rationnelle des milieux continus. Ann. Inst. Henri Poincaré **34**, 1–24 (1981)
6. Koszul, J.L.: Ouverts convexes homogènes des espaces affines. Math. Z. **79**, 254–259 (1962)
7. Künzle, H.P.: Galilei and Lorentz structures on space-time: comparison of the corresponding geometry and physics. Ann. l'Institut Henri Poincaré Sect. A **17**(4), 337–362 (1972)
8. Souriau, J.-M.: Structure des Systèmes Dynamiques. Dunod, Paris (1970)
9. Souriau, J.-M.: Thermodynamique et géométrie. In: Bleuler, K., Reetz, A., Petry, H.R. (eds.) Differential Geometrical Methods in Mathematical Physics II. Lecture Notes in Mathematics, vol. 676, pp. 369–397. Springer, Heidelberg (1976)
10. Souriau, J.-M.: Thermodynamique relativiste des fluides. Rend. Sem. Mat. Univ. Politech. Torino **35**, 21–34 (1978)
11. Souriau, J.-M.: Structure of Dynamical Systems, a Symplectic View of Physics. Birkhäuser Verlag, New York (1997)
12. Souriau, J.-M.: Milieux continus de dimension 1, 2 ou 3: statique et dynamique. Actes du 13ème Congrés Français de Mécanique, Poitiers-Futuroscope, pp. 41–53 (1997)
13. Tulczyjew, W., Urbański, P., Grabowski, J.: A pseudocategory of principal bundles. Atti della Reale Accademia delle Scienze di Torino **122**, 66–72 (1988)
14. Vallée, C.: Relativistic thermodynamics of continua. Int. J. Eng. Sci. **19**(5), 589–601 (1981)
15. Vallée, C., Lerintiu, C.: Convex analysis and entropy calculation in statistical mechanics. Proc. A. Razmadze Math. Inst. **137**, 111–129 (2005)
16. Vinberg, E.B.: Structure of the group of automorphisms of a homogeneous convex cone. Trudy Moskovskogo Matematicheskogo Obshchestva **13**, 56–83 (1965)

Symplectic Structure of Information Geometry: Fisher Metric and Euler-Poincaré Equation of Souriau Lie Group Thermodynamics

Frédéric Barbaresco[(✉)]

Thales Land & Air Systems, Voie Pierre-Gilles de Gennes,
F91470 Limours, France
frederic.barbaresco@thalesgroup.com

Abstract. We introduce the Symplectic Structure of Information Geometry based on Souriau's Lie Group Thermodynamics model, with a covariant definition of Gibbs equilibrium via invariances through co-adjoint action of a group on its momentum space, defining physical observables like energy, heat, and momentum as pure geometrical objects. Using Geometric (Planck) Temperature of Souriau model and Symplectic cocycle notion, the Fisher metric is identified as a Souriau Geometric Heat Capacity. In the framework of Lie Group Thermodynamics, an Euler-Poincaré equation is elaborated with respect to thermodynamic variables, and a new variational principal for thermodynamics is built through an invariant Poincaré-Cartan-Souriau integral. Finally, we conclude on Balian Gauge theory of Thermodynamics compatible with Souriau's Model.

Keywords: Information Geometry · Symplectic Geometry · Momentum Map · Cartan-Poincaré Integral Invariant · Lie Group Thermodynamics · Geometric Mechanics · Euler-Poincaré Equation · Gibbs Equilibrium · Fisher Metric · Maximum Entropy · Gauge Theory

1 Souriau Symplectic Geometry of Statistical Physics

In 1970, Souriau introduced the concept of co-adjoint action of a group on its momentum space (or "*moment map*": mapping induced by symplectic manifold symmetries), based on the orbit method works, that allows to define physical observables like energy, heat and momentum as pure geometrical objects (the moment map takes its values in a space determined by the group of symmetries: the dual space of its Lie algebra). The moment map is a constant of the motion and is associated to symplectic cohomology (assignment of algebraic invariants to a topological space that arises from the algebraic dualization of the homology construction). Souriau has observed that Gibbs equilibrium states are not covariant by dynamical groups (Galileo or Poincaré groups) and then he has developed a covariant model that he called "*Lie Group Thermodynamics*", where equilibriums are indexed by a "*geometric (planck) temperature*", given by a vector β that lies in the Lie algebra of the dynamical group. For Souriau, all the details of classical mechanics appear as geometric

© Springer International Publishing Switzerland 2015
F. Nielsen and F. Barbaresco (Eds.): GSI 2015, LNCS 9389, pp. 529–540, 2015.
DOI: 10.1007/978-3-319-25040-3_57

necessities (e.g., mass is the measure of the symplectic cohomology of the action of a Galileo group). Based on this new covariant model of thermodynamic Gibbs equilibrium, Souriau has formulated statistical mechanics and thermodynamics in the framework of Symplectic Geometry by use of symplectic moments and distribution-tensor concepts, giving a geometric status for temperature, heat and entropy. This work has been extended by Claude Vallée and Gery de Saxcé [11, 12]. More recently, M. Kapranov (arXiv:1108.3472) has also given a thermodynamical interpretation of the moment map for toric varieties. The conservation of the moment of a Hamiltonian action was called by Souriau the "*Symplectic or Geometric Noether theorem*". Considering phases space as symplectic manifold, cotangent fiber of configuration space with canonical symplectic form, if Hamiltonian has Lie algebra, moment map is constant along system integral curves. Noether theorem is obtained by considering independently each component of moment map.

We will enlighten Souriau's Model with Koszul Information Geometry, recently studied in [10], where we have shown that this last Geometry is founded on the notion of Koszul-Vinberg Characteristic function $\psi_\Omega(x) = \int_{\Omega^*} e^{-\langle x, \xi \rangle} d\xi, \forall x \in \Omega$ where Ω is a convex cone and Ω^* the dual cone with respect to Cartan-Killing inner product $\langle x, y \rangle = -B(x, \theta(y))$ invariant by automorphisms of Ω, with $B(.,.)$ the Killing form and $\theta(.)$the Cartan involution. This characteristic function is at the cornerstone of modern concept of Information Geometry, defining Koszul density by Solution of Maximum Koszul-Shannon Entropy $\Phi^*(\bar\xi) = - \int_{\Omega^*} p_{\bar\xi}(\xi) \log p_{\bar\xi}(\xi) d\xi$:

$$\underset{p}{Max} \left[- \int_{\Omega^*} p_{\bar\xi}(\xi) \log p_{\bar\xi}(\xi) d\xi \right] \text{ such that } \int_{\Omega^*} p_{\bar\xi}(\xi) d\xi = 1 \text{ and } \int_{\Omega^*} \xi . p_{\bar\xi}(\xi) d\xi = \bar\xi \tag{1}$$

$$p_{\bar\xi}(\xi) = \frac{e^{-\langle \xi, \Theta^{-1}(\bar\xi) \rangle}}{\int_{\Omega^*} e^{-\langle \xi, \Theta^{-1}(\bar\xi) \rangle} d\xi} \text{ with } \bar\xi = \Theta(x) = \frac{\partial \Phi(x)}{\partial x} \text{ where } \Phi(x) = -\log \int_{\Omega^*} e^{-\langle x, \xi \rangle} d\xi \tag{2}$$

The inversion $\Theta^{-1}(\bar\xi)$ is given by the Legendre transform based on the property that the Koszul-Shannon Entropy is given by the Legendre transform of minus the logarithm of the characteristic function:

$$\Phi^*(x^*) = \langle x, x^* \rangle - \Phi(x) \text{ with } \Phi(x) = -\log \int_{\Omega^*} e^{-\langle \xi, x \rangle} d\xi \, \forall x \in \Omega \text{ and } \forall x^* \in \Omega^* \tag{3}$$

We can observe the fundamental property that $E[\Phi^*(\xi)] = \Phi^*(E[\xi]), \xi \in \Omega^*$, and also as observed by Maurice Fréchet [18] that "distinguished functions" (densities with estimator reaching the Fréchet-Darmois bound) are solutions of the *Alexis Clairaut Equation* introduced by Clairaut in 1734:

$$\Phi^*(x^*) = \langle \Theta^{-1}(x^*), x^* \rangle - \Phi[\Theta^{-1}(x^*)] \; \forall x^* \in \{\Theta(x)/x \in \Omega\} \tag{4}$$

In this structure, the Fisher metric $I(x)$ makes appear naturally a *Koszul hessian geometry*, if we observe that $\log p_x(\xi) = -\langle x, \xi \rangle + \Phi(x) \Rightarrow \frac{\partial^2 \log p_x(\xi)}{\partial x^2} = \frac{\partial^2 \Phi(x)}{\partial x^2}$.

$$I(x) = -E_\xi \left[\frac{\partial^2 \log p_x(\xi)}{\partial x^2} \right] = -\frac{\partial^2 \Phi(x)}{\partial x^2} = \frac{\partial^2 \log \psi_\Omega(x)}{\partial x^2} = E_\xi[\xi^2] - E_\xi[\xi]^2 = Var(\xi) \tag{5}$$

with Crouzeix relation established in 1977, $\frac{\partial^2 \Phi}{\partial x^2} = \left[\frac{\partial^2 \Phi^*}{\partial x^{*2}} \right]^{-1}$ giving the dual metric, in dual space, where Entropy Φ^* and (minus) logarithm of characteristic function$\Phi$are dual potential functions.

We will see hereafter that Souriau has generalized this Fisher metric for Lie Group Thermodynamics, and interpreted the Fisher Metric as a Geometric Heat Capacity.

2 Souriau Model of Lie Group Thermodynamics

Souriau has defined Gibbs canonical ensemble on symplectic manifold M for a Lie group action on M. In classical statistical mechanics, a state is given by the solution of Liouville equation on the phase space, the partition function. As symplectic manifolds have a completely continuous measure, invariant by diffeomorphisms, the Liouville measure λ, all statistical states will be the product of Liouville measure by the scalar function given by the generalized partition function $e^{\Phi(\beta) - \langle \beta, U(\xi) \rangle}$ defined by the energy U (defined in dual of Lie Algebra of this dynamical group) and the geometric temperature β, where Φ is a normalizing constant such the mass of probability is equal to 1, $\Phi(\beta) = -\log \int_M e^{-\langle \beta, U(\xi) \rangle} d\lambda$. Jean-Marie Souriau then generalizes the Gibbs equilibrium state to all symplectic manifolds that have a dynamical group. To ensure that all integrals, that will be defined, could converge, *the canonical Gibbs ensemble is the largest open proper subset (in Lie algebra) where these integrals are convergent. This canonical Gibbs ensemble is convex.* The derivative of Φ, $Q = \frac{\partial \Phi}{\partial \beta}$ (thermodynamic heat) is equal to the mean value of the energy U. The minus derivative of this generalized heat$Q$, $K = -\frac{\partial Q}{\partial \beta}$ is symmetric and positive (this is a geometric heat capacity). Entropy s is then defined by Legendre transform of Φ, $s = \langle \beta, Q \rangle - \Phi$. If this approach is applied for the group of time translation, this is the classical thermodynamic theory. But *Souriau has observed that if we apply this theory for non-commutative group (Galileo or Poincaré groups), the symmetry has been broken. Classical Gibbs equilibrium states are no longer invariant by this group.* This symmetry breaking provides new equations, discovered by Souriau [2–6].

For each temperature β, Souriau has introduced a tensor $\tilde{\Theta}_\beta$, equal to the sum of the cocycle $\tilde{\Theta}$ and the Heat coboundary (with [.,.] Lie bracket):

$$\tilde{\Theta}_\beta(Z_1, Z_2) = \tilde{\Theta}(Z_1, Z_2) + \langle Q, ad_{Z_1}(Z_2) \rangle \text{ with } ad_{Z_1}(Z_2) = [Z_1, Z_2] \qquad (6)$$

This tensor $\tilde{\Theta}_\beta$ has the following properties:

- $\tilde{\Theta}(X, Y) = \langle \Theta(X), Y \rangle$ where the map Θ is the one-cocycle of the Lie algebra g with values in g^*, with $\Theta(X) = T_e\theta(X(e))$ where θ the one-cocycle of the Lie group G. $\tilde{\Theta}(X, Y)$ is constant on M and the map $\tilde{\Theta}(X, Y) : g \times g \to \Re$ is a skew-symmetric bilinear form, and is called the **Symplectic Cocycle of Lie algebra** g associated to the **momentum map** J, with the following properties:

$$\tilde{\Theta}(X, Y) = J_{[X,Y]} - \{J_X, J_Y\} \text{ with } \{.,.\} \text{ Poisson Bracket and } J \text{ the Moment Map} \qquad (7)$$

$$\tilde{\Theta}([X, Y], Z) + \tilde{\Theta}([Y, Z], X) + \tilde{\Theta}([Z, X], Y) = 0 \qquad (8)$$

where J_X linear application from g to differential function on M : $\begin{array}{c} g \to C^\infty(M, R) \\ X \to J_X \end{array}$

and the associated differentiable application J, called moment(um) map:

$$J : M \to g^* \text{ with } x \mapsto J(x) \text{ such that } J_X(x) = \langle J(x), X \rangle, X \in g \qquad (9)$$

If instead of J we take the following momentum map: $J'(x) = J(x) + Q, x \in M$ where $Q \in g^*$ is constant, the symplectic cocycle θ is replaced by $\theta'(g) = \theta(g) + Q - Ad_g^*Q$ where $\theta' - \theta = Q - Ad_g^*Q$ is one-coboundary of G with values in g^*. We have also properties $\theta(g_1g_2) = Ad_{g_1}^*\theta(g_2) + \theta(g_1)$ and $\theta(e) = 0$.

- $$\beta \in Ker \tilde{\Theta}_\beta, \text{ such that } \tilde{\Theta}_\beta(\beta, \beta) = 0, \forall \beta \in g \qquad (10)$$

- The following symmetric tensor g_β, defined on all values of $ad_\beta(.) = [\beta, .]$ is positive definite

$$g_\beta([\beta, Z_1], [\beta, Z_2]) = \tilde{\Theta}_\beta(Z_1, [\beta, Z_2]) \qquad (11)$$

$$g_\beta([\beta, Z_1], Z_2) = \tilde{\Theta}_\beta(Z_1, Z_2), \forall Z_1 \in g, \forall Z_2 \in Im(ad_\beta(.)) \qquad (12)$$

$$g_\beta(Z_1, Z_2) \geq 0, \forall Z_1, Z_2 \in Im(ad_\beta(.)) \qquad (13)$$

where the linear map $ad_X \in gl(g)$ is the adjoint representation of the Lie algebra g defined by $X, Y \in g (= T_eG) \mapsto ad_X(Y) = [X, Y]$, and the co-adjoint representation of the Lie algebra g the linear map $ad_X^* \in gl(g^*)$ which satisfies, for each $\xi \in g^*$ and $X, Y \in g : \langle ad_X^*(\xi), Y \rangle = \langle \xi, -ad_X(Y) \rangle$

These equations are universal, because they are not dependent of the symplectic manifold but only of the dynamical group G, the symplectic cocycle Θ, the temperature β and the heat Q. Souriau called this model *"Lie Groups Thermodynamics"*.

We will give the main theorem of Souriau for this "Lie Group Thermodynamics":

■ *Souriau Theorem of Lie Group Thermodynamics:*

Let Ω be the largest open proper subset of g, the fundamental equations of Souriau $\int_M e^{-\langle \beta, U(\xi) \rangle} d\lambda$ and $\int_M \xi.e^{-\langle \beta, U(\xi) \rangle} d\lambda$ are convergent integrals, this set Ω is convex and is invariant under every transformation $Ad_g(.)$, where $g \mapsto Ad_g(.)$ is the adjoint representation of G, such that $Ad_g = T_e i_g$ with $i_g : h \mapsto ghg^{-1}$. Let $a : G \times g^* \to g^*$ a unique affine action a such that linear part is coadjoint representation of G, that is the contragradient of the adjoint representation. It associates to each $g \in G$ the linear isomorphism $Ad_g^* \in GL(g^*)$, satisfying, for each:

$$\xi \in g^* \text{ and } X \in g : \left\langle Ad_g^*(\xi), X \right\rangle = \left\langle \xi, Ad_{g^{-1}}(X) \right\rangle$$

Then, the fundamental equations of Souriau Lie Group Thermodynamics are:

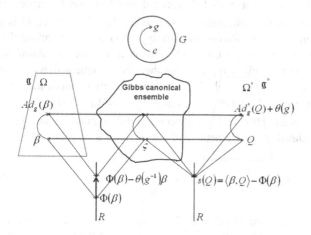

- $$\beta \to Ad_g(\beta) \tag{14}$$

- $$\Phi \to \Phi - \theta(g^{-1})\beta \tag{15}$$

- $$s \to s \tag{16}$$

- $$Q \to a(g, Q) = Ad_g^*(Q) + \theta(g) \tag{17}$$

For Hamiltonian, actions of a Lie group on a connected symplectic manifold, the equivariance of the momentum map with respect to an affine action of the group on the dual of its Lie algebra has been studied by C.M. Marle & P. Libermann [7, 9]:

■ *Marle's Formula between cocycles*:

Let G be a connected and simply connected Lie group, $R : G \to GL(E)$be a linear representation of G in a finite-dimensional vector space E, and $r : g \to gl(E)$be the associated linear representation of its Lie algebra g. For any one-cocycle $\Theta : g \to E$of the Lie algebra g for the linear representation r, there exists a unique one-cocycle $\theta : G \to E$ of the Lie group G for the linear representation R such that $\Theta(X) = T_e\theta(X(e))$, which has Θ as associated Lie algebra one-cocycle. The Lie group one-cocycle θ is a Lie group one-coboundary if and only if the Lie algebra one-cocycle Θ is a Lie algebra one-coboundary.

Let G be a Lie group whose Lie algebra is g. The skew-symmetric bilinear form $\tilde{\Theta}$on $g = T_eG$ can be extended into a closed differential two-form on G, since the identity on $\tilde{\Theta}$ means that its exterior differential $d\tilde{\Theta}$ vanishes. In other words, $\tilde{\Theta}$ is a 2-cocycle for the restriction of the de Rham cohomology of G to left invariant differential forms. In the framework of Lie Group Action on a Symplectic Manifold, equivariance of moment could be studied to prove that there is a unique action $a(.,.)$ of the Lie group G on the dual g^* of its Lie algebra for which the momentum map J is equivariant, that means for each $x \in M$:$J(\Phi_g(x)) = a(g, J(x)) = Ad_g^*(J(x)) + \theta(g)$

where $\Phi : G \times M \to M$ is an action of Lie Group G on differentiable manifold M, the fundamental field associated to an element X of Lie algebra g of group G is the vectors field X_M on M:

$$X_M(x) = \frac{d}{dt}\Phi_{\exp(-tX)}(x)\Big|_{t=0} \tag{18}$$

with $\Phi_{g_1}(\Phi_{g_2}(x)) = \Phi_{g_1g_2}(x)$ and $\Phi_e(x) = x$. Φ is hamiltonian on a Symplectic Manifold M, if Φ is symplectic and if for all $X \in g$, the fundamental field X_M is globally Hamiltonian. The cohomology class of the symplectic cocycle θ only depends on the Hamiltonian action Φ, and not on J.

3   Fisher Metric of Souriau Lie Group Thermodynamics

If we differentiate this relation of Souriau theorem $Q(Ad_g(\beta)) = Ad_g^*(Q) + \theta(g)$,
This relation occurs:

$$\frac{\partial Q}{\partial \beta}(-[Z_1, \beta], .) = \tilde{\Theta}(Z_1, [\beta, .]) + \langle Q, Ad_{.Z_1}([\beta, .]) \rangle = \tilde{\Theta}_\beta(Z_1, [\beta, .]) \tag{19}$$

$$-\frac{\partial Q}{\partial \beta}([Z_1, \beta], Z_2.) = \tilde{\Theta}(Z_1, [\beta, Z_2]) + \langle Q, Ad_{Z_1}([\beta, Z_2])\rangle = \tilde{\Theta}_\beta(Z_1, [\beta, Z_2]) \Rightarrow -\frac{\partial Q}{\partial \beta}$$

$$= g_\beta([\beta, Z_1], [\beta, Z_2])$$

We observe that the Fisher Metric $I(\beta) = -\frac{\partial Q}{\partial \beta}$ is exactly the Souriau Metric defined through Symplectic cocycle:

$$I(\beta) = \tilde{\Theta}_\beta(Z_1, [\beta, Z_2]) = g_\beta([\beta, Z_1], [\beta, Z_2]) \qquad (20)$$

The Fisher Metric $I(\beta) = -\frac{\partial^2 \Phi(\beta)}{\partial \beta^2} = -\frac{\partial Q}{\partial \beta}$ has been considered by Souriau as a *generalization of "Heat Capacity"*. Souriau called it K the *"Geometric Capacity"*. For $\beta = \frac{1}{kT}$, $K = -\frac{\partial Q}{\partial \beta} = -\frac{\partial Q}{\partial T}\left(\frac{\partial(1/kT)}{\partial T}\right)^{-1} = kT^2\frac{\partial Q}{\partial T}$ linking the geometric capacity to calorific capacity, then Fisher metric can be introduced in Fourier heat equation:

$$\frac{\partial T}{\partial t} = \frac{\kappa}{C.D}\Delta T \text{ with } \frac{\partial Q}{\partial T} = C.D \Rightarrow \frac{\partial \beta^{-1}}{\partial t} = \kappa\left[(\beta^2/k).I_{Fisher}(\beta)\right]^{-1}\Delta\beta^{-1} \qquad (21)$$

We can also observe that Q is related to the mean, and K to the variance of U:

$$K = I(\beta) = -\frac{\partial Q}{\partial \beta} = var(U) = \int_M U(\xi)^2.p_\beta(\xi)d\omega - \left(\int_M U(\xi).p_\beta(\xi)d\omega\right)^2 \qquad (22)$$

We observe that the entropy s is unchanged, and Φ is changed but with linear dependence to β, with consequence that Fisher Souriau metric is invariant:

$$s\left[Q\left(Ad_g(\beta)\right)\right] = s(Q(\beta)) \text{ and } I\left(Ad_g(\beta)\right) = \frac{\partial^2(\Phi - \theta(g^{-1})\beta)}{\partial \beta^2} = \frac{\partial^2 \Phi}{\partial \beta^2} = I(\beta) \qquad (23)$$

General definition of Heat Capacity has also been introduced by Pierre Duhem [17].

4   Euler-Poincaré Equation of Lie Group Thermodynamics

When a Lie algebra acts locally transitively on the configuration space of a Lagrangian mechanical system, Henri Poincaré proved that the Euler-Lagrange equations are equivalent to a new system of differential equations defined on the product of the configuration space with the Lie algebra. C.M. Marle [8] has written the Euler-Poincaré equations, under an intrinsic form, without any reference to a particular system of local coordinates, proving that they can be conveniently expressed in terms of the Legendre and momentum maps of the lift to the cotangent bundle of the Lie algebra action on the configuration space. The Lagrangian is a smooth real valued function L defined on the tangent bundle TM. To each parameterized continuous, piecewise smooth curve $\gamma : [t_0, t_1] \to M$, defined on a closed interval $[t_0, t_1]$, with values in M, one associates the value at γ of the action integral:

$$I(\gamma) = \int_{t_0}^{t_1} L\left(\frac{d\gamma(t)}{dt}\right) dt \tag{24}$$

The partial differential of the function $L : M \times g \to \Re$ with respect to its second variable $d_2\bar{L}$, which plays an important part in the Euler-Poincaré equation, can be expressed in terms of the momentum and Legendre maps: $d_2\bar{L} = p_{g^*} \circ \phi^t \circ L \circ \phi$ with $J = p_{g^*} \circ \phi^t (\Rightarrow d_2\bar{L} = J \circ L \circ \phi)$ the moment map, $p_{g^*} : M \times g^* \to g^*$ the canonical projection on the second factor, $L : TM \to T^*M$ the Legendre transform, with $\phi : M \times g \to TM/\phi(x, X) = X_M(x)$ and $\phi^t : T^*M \to M \times g^*/\phi^t(\xi) = (\pi_M(\xi), J(\xi))$

The Euler-Poincaré equation can therefore be written under the form:

$$\left(\frac{d}{dt} - ad^*_{V(t)}\right)(J \circ L \circ \phi(\gamma(t), V(t))) = J \circ d_1\bar{L}(\gamma(t), V(t)) \text{ with } \frac{d\gamma(t)}{dt} = \phi(\gamma(t), V(t))$$

With $H(\xi) = \langle \xi, L^{-1}(\xi) \rangle - L(L^{-1}(\xi)), \xi \in T^*M, L : TM \to T^*M, H : T^*M \to R$

Following the remark made by Poincaré at the end of his note, the most interesting case is when the map $\bar{L} : M \times g \to R$ only depends on its second variable $X \in g$. The Euler-Poincaré equation becomes:

$$\left(\frac{d}{dt} - ad^*_{V(t)}\right)(d\bar{L}(V(t))) = 0 \tag{25}$$

We can use analogy of structure when the convex Gibbs ensemble is homogeneous [13]. We can then apply Euler-Poincaré equation for Lie Group Thermodynamics. Considering Clairaut equation:

$$s(Q) = \langle \beta, Q \rangle - \Phi(\beta) = \langle \Theta^{-1}(Q), Q \rangle - \Phi(\Theta^{-1}(Q)) \tag{26}$$

with $Q = \Theta(\beta) = \frac{\partial \Phi}{\partial \beta} \in g^*$, $\beta = \Theta^{-1}(Q) \in g$, a ***Souriau-Euler-Poincaré equation*** can be elaborated for Souriau Lie Group Thermodynamics:

$$\frac{dQ}{dt} = ad^*_\beta Q \text{ or } \frac{d}{dt}\left(Ad^*_g Q\right) = 0 \tag{27}$$

An associated equation on Entropy is: $\frac{ds}{dt} = \left\langle \frac{d\beta}{dt}, Q \right\rangle + \left\langle \beta, ad^*_\beta Q \right\rangle - \frac{d\Phi}{dt}$ that reduces to $\frac{ds}{dt} = \left\langle \frac{d\beta}{dt}, Q \right\rangle - \frac{d\Phi}{dt}$ due to $\langle \xi, ad_V X \rangle = -\langle ad^*_V \xi, X \rangle \Rightarrow \left\langle \beta, ad^*_\beta Q \right\rangle = \langle Q, ad_\beta \beta \rangle = 0$.

5 Poincaré-Cartan Integral Invariant of Thermodynamics

We will define the Poincaré-Cartan Integral Invariant [1] for Lie Group Thermodynamics. Classically in mechanics, the Pfaffian form $\omega = p.dq - H.dt$ is related to Poincaré-Cartan integral invariant. P. Dedecker has observed, based on the relation $\omega = \partial_{\dot{q}}L.dq - (\partial_{\dot{q}}L.\dot{q} - L).dt = L.dt + \partial_{\dot{q}}L\varpi$ with $\varpi = dq - \dot{q}.dt$, that the property that among all forms $\chi \equiv L.dt$ mod ϖ the form $\omega = p.dq - H.dt$ is the only one satisfying $d\chi \equiv 0$ mod ϖ, is a particular case of more general T. Lepage congruence.

Analogies between Geometric Mechanics & Geometric Lie Group Thermodynamics, provides the following similarities of structures:

$$\begin{cases} \dot{q} \leftrightarrow \beta \\ p \leftrightarrow Q \end{cases}, \begin{cases} L(\dot{q}) \leftrightarrow \Phi(\beta) \\ H(p) \leftrightarrow s(Q) \\ H = p.\dot{q} - L \leftrightarrow s = \langle Q, \beta \rangle - \Phi \end{cases} \text{ and } \begin{cases} \dot{q} = \dfrac{dq}{dt} = \dfrac{\partial H}{\partial p} \leftrightarrow \beta = \dfrac{\partial s}{\partial Q} \\ p = \dfrac{\partial L}{\partial \dot{q}} \leftrightarrow Q = \dfrac{\partial \Phi}{\partial \beta} \end{cases} \tag{28}$$

We can then consider a similar **Poincaré-Cartan-Souriau Pfaffian form**:

$$\omega = p.dq - H.dt \leftrightarrow \omega = \langle Q, (\beta.dt) \rangle - s.dt = (\langle Q, \beta \rangle - s).dt = \Phi(\beta).dt \tag{29}$$

This analogy provides an associated **Poincaré-Cartan-Souriau Integral Invariant**:

$$\int_{C_a} p.dq - H.dt = \int_{C_b} p.dq - H.dt \text{ is transformed in } \int_{C_a} \Phi(\beta).dt = \int_{C_b} \Phi(\beta).dt \tag{30}$$

We can then deduce an **Euler-Poincaré-Souriau Variational Principle** for Thermodynamics: *The Variational Principle holds on g, for variations* $\delta\beta = \dot{\eta} + [\beta, \eta]$, *where* $\eta(t)$ *is an arbitrary path that vanishes at the endpoints,*

$$\eta(a) = \eta(b) = 0 : \delta \int_{t_0}^{t_1} \Phi(\beta(t)).dt = 0 \tag{31}$$

6 Compatible Balian Gauge Theory of Thermodynamics

Supported by TOTAL group, Roger Balian has introduced in [15] a Gauge Theory of Thermodynamics. Balian has observed that the Entropy S (we use Balian notation, contrary with previous chapter where we use $-S$ as neg-Entropy) can be regarded as an extensive variable $q^0 = S(q^1, \ldots, q^n)$, with q^i $(i = 1, \ldots, n)$, n independent quantities, usually extensive and conservative, characterizing the system. The n intensive variables γ_i are defined as the partial derivatives:

$$\gamma_i = \frac{\partial S(q^1, \ldots, q^n)}{\partial q^i} \tag{32}$$

Balian has introduced a non-vanishing gauge variable p_0, without physical relevance, which multiplies all the intensive variables, defining a new set of variables:

$$p_i = -p_0 . \gamma_i, i = 1, \ldots, n \tag{33}$$

The $2n+1$-dimensional space is thereby extended into a $2n+2$-dimensional thermodynamic space T spanned by the variables p_i, q^i with $i = 0, 1, \ldots, n$, where the physical system is associated with a n+1-dimensional manifold M in T, parameterized for instance by the coordinates $q^1, \ldots, q^n$ and p_0. A gauge transformation which changes the extra variable p_0 while keeping the ratios $p_i/p_0 = -\gamma_i$ invariant **is not observable**, so that a state of the system is represented by any point of a one-dimensional ray lying in M, along which the physical variables $q^0, \ldots, q^n, \gamma_1, \ldots, \gamma_n$ are fixed. Then, the relation between contact and canonical transformations is a direct outcome of this gauge invariance: the contact structure $\tilde{\omega} = dq^0 - \sum_{i=1}^{n} \gamma_i . dq^i$ in $2n+1$ dimension can be embedded into a symplectic structure in $2n +2$ dimension, with 1-form:

$$\omega = \sum_{i=0}^{n} p_i . dq^i \tag{34}$$

as **symplectization**, with geometric interpretation in the theory of fibre bundles.

The $n +1$-dimensional thermodynamic manifolds M are characterized by the vanishing of this form $\omega = 0$. The 1-form induces then **a symplectic structure on** T:

$$d\omega = \sum_{i=0}^{n} dp_i \wedge dq^i \tag{35}$$

Any thermodynamic manifold M belongs to the set of the so-called Lagrangian manifolds in T, which are the integral submanifolds of $d\omega$ with maximum dimension (n +1). Moreover, M **is gauge invariant**, which is implied by $\omega = 0$. The extensivity of the entropy function $S(q^1, \ldots, q^n)$ is expressed by the Gibbs-Duhem relation $S = \sum_{i=1}^{n} q^i \frac{\partial S}{\partial q^i}$, rewritten with previous relation $\sum_{i=0}^{n} p_i q^i = 0$, defining a $2n+1$-dimensional extensivity sheet in T, where the thermodynamic manifolds M should lie. Considering an infinitesimal canonical transformation, generated by the Hamiltonian $h(q^0, q^1, \ldots, q^n, p_0, p_1, \ldots, p_n)$, $\dot{q}_i = \frac{\partial h}{\partial p_i}$ and $\dot{p}_i = \frac{\partial h}{\partial q^i}$, the Hamilton's equations are given by Poisson bracket:

$$\dot{g} = \{g, h\} = \sum_{i=0}^{n} \frac{\partial g}{\partial q^i} \frac{\partial h}{\partial p_i} - \frac{\partial h}{\partial q_i} \frac{\partial g}{\partial p_i} \tag{36}$$

The concavity of the entropy $S(q^1, \ldots, q^n)$, as function of the extensive variables, expresses the stability of equilibrium states. This property produces constraints on the physical manifolds M in the $2n+2$-dimensional space. It entails the existence of a metric structure [14, 16] in the n-dimensional space q_i relying on the quadratic form:

$$ds^2 = -d^2 S = - \sum_{i,j=1}^{n} \frac{\partial^2 S}{\partial q^i \partial q^j} dq^i dq^j \tag{37}$$

which defines a distance between two neighboring thermodynamic states.

As $d\gamma_i = \sum_{j=1}^{n} \frac{\partial^2 S}{\partial q^i \partial q^j} dq^j$, then : $ds^2 = - \sum_{i=1}^{n} d\gamma_i dq_i = \frac{1}{p_0} \sum_{i=0}^{n} dp_i dq^i \tag{38}$

The factor $1/p_0$ ensures gauge invariance. In a continuous transformation generated by h, the metric evolves according to:

$$\frac{d}{d\tau}(ds^2) = \frac{1}{p_0} \frac{\partial h}{\partial q^0} ds^2 + \frac{1}{p_0} \sum_{i,j=0}^{n} \left(\frac{\partial^2 h}{\partial q^i \partial p_j} dp_i dp_j - \frac{\partial^2 h}{\partial q^i \partial q^j} dq^i dq^j \right) \tag{39}$$

We can observe that *this Gauge Theory of Thermodynamics is compatible with Souriau Lie Group Thermodynamics*, where we have to consider the Souriau vector

$$\beta = \begin{bmatrix} \gamma_1 \\ \vdots \\ \gamma_n \end{bmatrix}, \text{ transformed in a new vector } p_i = -p_0 \cdot \gamma_i, \ p = \begin{bmatrix} -p_0 \gamma_1 \\ \vdots \\ -p_0 \gamma_n \end{bmatrix} = -p_0 \cdot \beta \tag{40}$$

"La Physique mathématique, en incorporant à sa base la notion de groupe, marque la suprématie rationnelle... Chaque géométrie - et sans doute plus généralement chaque organisation mathématique de l'expérience - est caractérisée par un groupe spécial de transformations... Le groupe apporte la preuve d'une mathématique fermée sur elle-même. Sa découverte clôt l'ère des conventions, plus ou moins indépendantes, plus ou moins cohérentes" **G. BACHELARD, Le nouvel esprit scientifique, 1934**

References

1. Cartan, E.: Leçons sur les Invariants Intégraux. Hermann, Paris (1922)
2. Souriau, J.M.: Définition covariante des équilibres thermodynamiques. Suppl. Nuov. Cimento **1**, 203–216 (1966)

3. Souriau, J.M.: Géométrie de l'espace de phases. Comm. Math. Phys. **374**, 1–30 (1966)
4. Souriau, J.M.: Structure des systèmes dynamiques. Ed Dunod, Paris (1970)
5. Souriau, J.M.: Thermodynamique et géométrie. In: Bleuler, K., Reetz, A., Petry, H.R. (eds.) Differential Geometrical Methods in Mathematical Physics II, pp. 369–397. Springer, Berlin (1978)
6. Souriau, J.M.: Mécanique classique et géométrie symplectique. CNRS Marseille. Cent. Phys. Théor., Report ref. CPT-84/PE-1695 (1984)
7. Libermann, P., Marle, C.M.: Symplectic Geometry and Analytical Mechanics. Springer, Heidelberg (1987)
8. Marle, C.M.: On Henri Poincaré's note "Sur une forme nouvelle des équations de la mécanique". J. Geom. Symmetry Phys. **29**, 1–38 (2013)
9. Marle, C.M.: Symmetries of Hamiltonian systems on symplectic and poisson manifolds. In: 15th International Conference on Geometry, Integration and Quantitative., Sofia, pp.1-86 (2014)
10. Barbaresco, F.: Koszul information geometry and Souriau geometric temperature/capacity of lie group thermodynamics. Entropy **16**, 4521–4565 (2014)
11. de Saxcé, G., Vallée, C.: Bargmann group, Momentum tensor and Galilean invariance of Clausius-Duhem inequality. Int. J. Eng. Sci. **50**, 216–232 (2012)
12. Vallée, C., Lerintiu, C.: Convex analysis and entropy calculation in statistical mechanics. Proc. A. Razmadze Math. Inst. **137**, 111–129 (2005)
13. Sternberg, S.: Symplectic Homogeneous Spaces. Trans. Am. Math. Soc. **212**, 113–130 (1975)
14. Balian, R., Alhassid, Y., Reinhardt, H.: Dissipation in many-body systems: a geometric approach based on information theory. Phys. Rep. **131**(1), 1–146 (1986)
15. Balian, R., Valentin, P.: Hamiltonian structure of thermodynamics with gauge. Eur. Phys. J. B **21**, 269–282 (2001)
16. Balian, R.: The entropy-based quantum metric. Entropy **16**(7), 3878–3888 (2014)
17. Duhem, P.: Sur les équations générales de la thermodynamique. In: Annales scientifiques de l'ENS, $3^{\text{ème}}$ série, tome 9, pp. 231–266 (1891)
18. Fréchet, M.: Sur l'extension de certaines évaluations statistiques au cas de petits échantillons. Revue de l'Institut International de Statistique **11**(3/4), 182–205 (1943)

Pontryagin Calculus in Riemannian Geometry

François Dubois[1]([⊠]), Danielle Fortuné[2], Juan Antonio Rojas Quintero[3],
and Claude Vallée[1,2,3]

[1] Department of Mathematics, University Paris Sud and CNAM Paris, Structural
Mechanics and Coupled Systems Laboratory, Orsay, France
francois.dubois@math.u-psud.fr
[2] 21, Rue du Hameau du Cherpe, 86280 Saint Benoît, France
danielle.fortune123@orange.fr
[3] School of Mechanics and Engineering, Southwest Jiaotong University,
Chengdu, China
jrojasqu@yahoo.com

Abstract. In this contribution, we study systems with a finite number
of degrees of freedom as in robotics. A key idea is to consider the mass
tensor associated to the kinetic energy as a metric in a Riemannian configuration space. We apply Pontryagin's framework to derive an optimal
evolution of the control forces and torques applied to the mechanical
system. This equation under covariant form uses explicitly the Riemann
curvature tensor.

Keywords: Robotics · Euler-Lagrange · Riemann curvature tensor

AMS classification: 49S05, 51P05, 53A35.

1 Introduction

As part of studies based on the calculus of variations, the choice of a Lagrangian
or a Hamiltonian is essential. When we study the dynamics of articulated systems, the choice of the Lagrangian is directly linked to the conservation of energy.
The Euler-Lagrange methodology is applied to the kinetic and potential energies.
This establishes a system of second order ordinary differential equations for the
movement. These equations are identical to those deduced from the fundamental principle of dynamics. The choice of configuration parameters does not affect
the energy value. Because the kinetic energy is a positive definite quadratic form
with respect to the configuration parameters derivatives, its coefficients are ideal
candidates to define and create a Riemannian metric structure on the configuration space. The Euler Lagrange equations have a contravariant tensorial nature
and highlight the covariant derivatives with respect to time with the introduction
of the Christoffel symbols.

• For the control of articulated robot choosing a Hamiltonian and a cost function are delicate. Here the presence of the Riemann structure is sound. It enables

---

This contribution is dedicated to the memory of Claude Vallée.

© Springer International Publishing Switzerland 2015
F. Nielsen and F. Barbaresco (Eds.): GSI 2015, LNCS 9389, pp. 541–549, 2015.
DOI: 10.1007/978-3-319-25040-3_58

a cost function invariant when coordinates change. The application of the Pontryagin method from the optimal Hamiltonian leads to a system of second order differential equations for the control variables. Its tensorial nature is covariant and the Riemann-Christoffel curvature tensor is naturally revealed. In this development, the adjoint variables are directly interpreted and have a physical sense.

(1)   Pontryagin Framework for Differential Equations

We study a dynamical system, where the state vector $y(t\,; \lambda(\bullet))$ is a function of time. This system is controlled by a set of variables $\lambda(t)$ and satisfies a first order ordinary differential equation:

$$\frac{dy}{dt} = f(y(t),\, \lambda(t),\, t). \tag{1}$$

We suppose also given an initial condition:

$$y(0\,; \lambda(\bullet)) = x. \tag{2}$$

We search an optimal solution associated to the optimal control $t \longmapsto \lambda(t)$ in order to minimize the following cost function J:

$$J(\lambda(\bullet)) \equiv \int_0^T g\big(y(t),\, \lambda(t),\, t\big)\, dt, \tag{3}$$

where $g(\bullet,\, \bullet,\, \bullet)$ is a given real valued function.

• Pontryagin's main idea [3] is to consider the differential equation (1) as a *constraint* satisfied by the variable y. Then he introduces a Lagrange multiplier p associated to the constraint (1). This new variable, due to the continuous nature of the constraint (1), is a covariant vector function of time: $p = p(t)$. A global Lagrangian functional can be considered:

$$\mathcal{L}(y,\, \lambda,\, p) \equiv \int_0^T g(y,\, \lambda,\, t)\, dt \; + \; \int_0^T p\left(\frac{dy}{dt} - f(y,\, \lambda,\, t)\right) dt.$$

• **Proposition 1. Adjoint Equations.** If the Lagrange multiplier $p(t)$ satisfies the so-called *adjoint equations*,

$$\frac{dp}{dt} + p\,\frac{\partial f}{\partial y} - \frac{\partial g}{\partial y} = 0 \tag{4}$$

and the so-called *final condition*,

$$p(T) = 0, \tag{5}$$

then the variation of the cost function for a given variation $\delta\lambda$ of the paramater is given by the simple relation

$$\delta J = \int_0^T \left[\frac{\partial g}{\partial \lambda} - p\,\frac{\partial f}{\partial \lambda}\right] \delta\lambda(t)\, dt.$$

At the optimum this variation is identically null and we find the so-called Pontryagin optimality condition:

$$\frac{\partial g}{\partial \lambda} - p\frac{\partial f}{\partial \lambda} = 0. \tag{6}$$

Proof of Proposition 1. We write in a general way the variation of the Lagrangian $\mathcal{L}(y, \lambda, p)$ in a variation δy, $\delta \lambda$ and δp of the variables y, λ, and p respectively. We use classical calculus rules as

$$\delta\left(\int_0^T g\,dt\right) = \int_0^T \delta g\,dt, \quad \delta\left(\frac{dy}{dt}\right) = \frac{d}{dt}(\delta y),$$

and we integrate by parts. We get

$$\delta\mathcal{L} = \int_0^T \left[\frac{\partial g}{\partial y}\delta y + \frac{\partial g}{\partial \lambda}\delta\lambda\right]dt + \int_0^T p\left(\frac{d\delta y}{dt} - \frac{\partial f}{\partial y}\delta y - \frac{\partial f}{\partial \lambda}\delta\lambda\right)dt + \int_0^T \delta p\left(\frac{dy}{dt} - f(y, \lambda, t)\right)dt$$

$$= \int_0^T \left[\frac{\partial g}{\partial y} - p\frac{\partial f}{\partial y}\right]\delta y\,dt + \int_0^T \left[\frac{\partial g}{\partial \lambda} - p\frac{\partial f}{\partial \lambda}\right]\delta\lambda\,dt + \left[p\,\delta y\right]_0^T - \int_0^T \frac{dp}{dt}\delta y\,dt$$

$$= p(T)\,\delta y(T) - \int_0^T \left[\frac{dp}{dt} + p\frac{\partial f}{\partial y} - \frac{\partial g}{\partial y}\right]\delta y\,dt + \int_0^T \left[\frac{\partial g}{\partial \lambda} - p\frac{\partial f}{\partial \lambda}\right]\delta\lambda\,dt$$

because $\delta y(0) = 0$ taking fixed the initial condition (2). By canceling the first two terms of the right hand side of the previous relation, we find the adjoint equation (4) giving the evolution of the Lagrange multiplier and the associated final condition (5). The third term allows to calculate the change in the functional $J(\bullet)$ for a variation $\delta\lambda$ of control. $\qquad\square$

(2)   Pontryagin Hamiltonian

We introduce the Hamiltonian

$$\mathcal{H}(p, y, \lambda) \equiv pf - g \tag{7}$$

and the optimal Hamiltonian

$$H(p, y) \equiv \mathcal{H}(p, y, \lambda^*)$$

for $\lambda(t) = \lambda^*(t)$ equal to the optimal value associated to the optimal condition (6).

• **Proposition 2. Symplectic form of the Dynamic Equations.** With the notations introduced previously, the "forward" differential equation (1) and the "backward" adjoint differential equation (4) take the following symplectic form:

$$\frac{dy}{dt} = \frac{\partial H}{\partial p}, \quad \frac{dp}{dt} = -\frac{\partial H}{\partial y}. \tag{8}$$

Proof of Proposition 2. Since $\frac{\partial \mathcal{H}}{\partial \lambda} = 0$ at the optimum, we have $\frac{\partial H}{\partial p} = \frac{\partial \mathcal{H}}{\partial p} = f$ and the first relation of (8) is proven. On the other hand, $\frac{\partial H}{\partial y} = \frac{\partial \mathcal{H}}{\partial y} = p\frac{\partial f}{\partial y} - \frac{\partial g}{\partial y}$ and the property is established. $\qquad\square$

(3)   Riemanian Metric

We consider now a dynamical system parameterized by a finite number of functions $q^j(t)$ like a poly-articulated system for robotics applications, as developed previously in [1,4,5,7]. The set of all states $q \equiv \{q^j\}$ is denoted by Q. The kinetic energy K is a positive definite quadratic form of the time derivatives $\dot{q}^j$ for each state $q \in Q$. The coefficients of this quadratic form define a so-called *mass tensor* $M(q)$. The mass tensor is composed by *a priori* a nonlinear regular function of the state $q \in Q$. We have

$$K(q, \dot{q}) \equiv \frac{1}{2} \sum_{k\,\ell} M_{k\ell}(q)\, \dot{q}^k\, \dot{q}^\ell. \tag{9}$$

The mass tensor $M(q)$ in (9) is symmetric and positive definite for each state q. It contains the mechanical caracteristics of mass, inertia of the articulated system. With Lazrak and Vallée [1] and Siebert [6], we consider the Riemannian metric g defined by the mass tensor M. We set:

$$g_{k\ell}(q) \equiv M_{k\ell}(q).$$

• With this framework, the space of states Q has now a structure of *Riemannian manifold*. Therefore all classical geometrical tools of Riemannian geometry can be used (see *e.g.* the book [2]):

covariant space derivation $\partial_j \equiv \frac{\partial}{\partial q^j}$

contravariant space derivation ∂^j: $<\partial^j,\, \partial_k> = \delta^j_k$

component j, ℓ of the inverse mass tensor M^{-1}: $M^{j\ell},\; M_{ij}\, M^{j\ell} = \delta^\ell_i$

connection $\Gamma^j_{ik} = \frac{1}{2} M^{j\ell} \left(\partial_i M_{\ell k} + \partial_k M_{\ell i} - \partial_\ell M_{ik} \right),\; \Gamma^j_{ki} = \Gamma^j_{ik}$,

$\mathrm{d}\,\partial_j = \Gamma^\ell_{jk}\, \mathrm{d}q^k\, \partial_\ell,\; \mathrm{d}\,\partial^j = -\Gamma^j_{k\ell}\, \mathrm{d}q^k\, \partial^\ell$,

relations between covariant components φ_j and contravariant components φ^k of a vector field: $\varphi_j = M_{jk}\, \varphi^k,\; \varphi^k = M^{kj}\, \varphi_j$

covariant derivation of a vector field $\varphi \equiv \varphi^j\, \partial_j$: $\mathrm{d}\varphi = \left(\partial_\ell \varphi^j + \Gamma^j_{\ell k}\, \varphi^k \right) \mathrm{d}q^\ell\, \partial_j$

covariant derivation of a covector field $\varphi \equiv \varphi_\ell\, \partial^\ell$:

$\mathrm{d}\varphi = \left(\partial_k \varphi_\ell - \Gamma^j_{k\ell}\, \varphi_j \right) \mathrm{d}q^k\, \partial^\ell$

Ricci identities:

$$\begin{cases} \partial_j M_{k\ell} = \Gamma^p_{jk}\, M_{\ell p} + \Gamma^p_{j\ell}\, M_{kp}, \\ \partial_j M^{k\ell} = -\Gamma^k_{jp}\, M^{p\ell} - \Gamma^\ell_{jp}\, M^{pk} \end{cases} \tag{10}$$

gradient of a scalar field: $\mathrm{d}V = \partial_\ell V\, \mathrm{d}q^\ell = <\nabla V,\, \mathrm{d}q^j\, \partial_j>$ and

$\nabla V = \partial_\ell V\, \partial^\ell$

gradient of a covector field $\varphi = \varphi_\ell\, \partial^\ell$: $\mathrm{d}\varphi \equiv <\nabla \varphi,\, \mathrm{d}q^j\, \partial_j>$ and

$\nabla \varphi = \left(\partial_k \varphi_\ell - \Gamma^j_{k\ell}\, \varphi_j \right) \partial^k\, \partial^\ell$

second order gradient of a scalar field V: $\nabla^2 V = \nabla(\nabla V)$ and

$$\nabla^2 V = \left(\partial_k \partial_\ell V - \Gamma^j_{k\ell}\, \partial_j V \right) \partial^k\, \partial^\ell \tag{11}$$

components $R^j_{ik\ell}$ of the Riemann tensor:

$$R^j_{ik\ell} \equiv \partial_\ell \Gamma^j_{ik} - \partial_k \Gamma^j_{i\ell} + \Gamma^p_{ik}\, \Gamma^j_{p\ell} - \Gamma^p_{i\ell}\, \Gamma^j_{pk} \tag{12}$$

anti-symmetry of the Riemann tensor: $R^j_{ik\ell} = -R^j_{i\ell k}$.

- **Proposition 3. Riemanian form of the Euler-Lagrange Equations.**
With the previous framework, in the presence of an external potential $V = V(q)$, the Lagrangian $L(q, \dot{q}) = K(q, \dot{q}) - V(q)$ allows to write the equations of motion in the classical Euler-Lagrange form:

$$\frac{\mathrm{d}}{\mathrm{d}t}\left(\frac{\partial L}{\partial \dot{q}^i}\right) = \frac{\partial L}{\partial q^i} \tag{13}$$

These equations take also the Riemannian form:

$$M_{k\ell}\left(\ddot{q}^\ell + \Gamma^\ell_{ij}\,\dot{q}^i\,\dot{q}^j\right) + \partial_k V = 0. \tag{14}$$

Proof of Proposition 3. The proof is presented in the references [5,7]. We detail it to be complete. We have, due to (9),

$$\frac{\partial K}{\partial \dot{q}^k} = M_{k\ell}\,\dot{q}^\ell.$$

We have also the following calculus:

$$\frac{\mathrm{d}}{\mathrm{d}t}\left(\frac{\partial L}{\partial \dot{q}^k}\right) - \frac{\partial L}{\partial q^k} = \frac{\mathrm{d}}{\mathrm{d}t}\left(\frac{\partial K}{\partial \dot{q}^k}\right) - \partial_k(K - V(q))$$

$$= \frac{\mathrm{d}}{\mathrm{d}t}\left(M_{k\ell}\,\dot{q}^\ell\right) - \partial_k\left(\frac{1}{2}\,M_{ij}\,\dot{q}^i\,\dot{q}^j\right) + \partial_k V$$

$$= \left(\partial_j M_{k\ell}\right)\dot{q}^j\,\dot{q}^\ell + M_{k\ell}\,\ddot{q}^\ell - \frac{1}{2}\left(\partial_k M_{ij}\right)\dot{q}^i\,\dot{q}^j + \partial_k V$$

$$= \left(M_{ks}\,\Gamma^s_{j\ell} + M_{\ell s}\,\Gamma^s_{jk}\right)\dot{q}^j\,\dot{q}^\ell - \frac{1}{2}\left(M_{is}\,\Gamma^s_{kj} + M_{js}\,\Gamma^s_{ki}\right)\dot{q}^i\,\dot{q}^j + M_{k\ell}\,\ddot{q}^\ell + \partial_k V$$

due to the first Ricci identity (10)

$$= \left(M_{ks}\,\Gamma^s_{ji} + M_{is}\,\Gamma^s_{jk} - \frac{1}{2}\,M_{is}\,\Gamma^s_{kj} - \frac{1}{2}\,M_{js}\,\Gamma^s_{ki}\right)\dot{q}^i\,\dot{q}^j + M_{k\ell}\,\ddot{q}^\ell + \partial_k V$$

$$= M_{k\ell}\left(\ddot{q}^\ell + \Gamma^\ell_{ij}\,\dot{q}^i\,\dot{q}^j\right) + \partial_k V$$

and the relation (14) is established. $\square$

- When a mechanical forcing control u is present (forces and torques typically), the equations of motion can be formulated as follows:

$$\ddot{q}^j + \Gamma^j_{k\ell}\,\dot{q}^k\,\dot{q}^\ell + M^{j\ell}\,\partial_\ell V = u^j. \tag{15}$$

We observe that with this form (15) of the equations of motion, the contravariant components of the control u have to be considered in the right hand side of the dynamical equations.

(4)   Optimal Dynamics

In this section, we follow the ideas proposed in [4,5,7]. The space of states Q has a natural Riemannian structure. Therefore, it is natural to choose a cost function that is intrinsic and invariant, and in consequence non sensible to the change of coordinates. Following Rojas Qinteros's thesis [4], we introduce a particular cost function to control the dynamics (15):

$$J(u) = \frac{1}{2}\int_0^T M_{k\ell}(q)\,u^k\,u^\ell\,\mathrm{d}t. \tag{16}$$

The controlled system (15) (16) is of type (1) (3) with

$$
\begin{cases}
Y = \{q^j,\, \dot{q}^j\},\ f = \{Y_2^j, -\Gamma_{k\ell}^j\, \dot{q}^k\, \dot{q}^\ell - M^{j\ell}\, \partial_\ell V + u^j\}, \\
\lambda = \{u^k\}, \qquad g = \tfrac{1}{2} M_{k\ell}(Y_1)\, u^k\, u^\ell.
\end{cases}
\tag{17}
$$

The Pontryagin method introduces Lagrangre multipliers (or adjoint states) p_j and ξ_j to form the Hamiltonian $\mathcal{H}(Y, P, \lambda)$ function of state Y defined in (17) and adjoint P obtained by combining the two adjoint states:

$$
P = \{p_j,\, \xi_j\}
\tag{18}
$$

and $\lambda = \{u^k\}$ as proposed in (17). Taking into account (7), (17) and (18), we have:

$$
\mathcal{H}(Y, P, \lambda) = p_j\, \dot{q}^j + \xi_j \left[- \Gamma_{k\ell}^j\, \dot{q}^k\, \dot{q}^\ell - M^{j\ell}\, \partial_\ell V + u^j \right] - \frac{1}{2} M_{k\ell}(Y_1)\, u^k\, u^\ell.
\tag{19}
$$

• **Proposition 4. Interpretation of One Adjoint State.** When the cost function J defined in (16) is stationary, the adjoint state ξ^j is exactly equal to the applied force (and torque!) u^j in the right hand side of the dynamic equation (15):

$$
\xi^j = u^j.
\tag{20}
$$

Proof of Proposition 4. Due to the expression (19) of the Hamiltonian function, the optimality condition $\frac{\partial \mathcal{H}}{\partial \lambda} = 0$ takes the simple form

$$
\xi_j = M_{j\ell}\, \xi^\ell.
$$

This relation is equivalent to the condition (20). □

• The reduced Hamiltonian $H(Y, P)$ at the optimum can be explicited without difficulty. We just replace the control force u_j by the adjoint state ξ_j:

$$
H(Y, P) = p_j\, \dot{q}^j + \xi_j \left[- \Gamma_{k\ell}^j\, \dot{q}^k\, \dot{q}^\ell - M^{j\ell}(Y_1)\, \partial_\ell V \right] + \frac{1}{2} M^{k\ell}(Y_1)\, \xi_k\, \xi_\ell.
$$

The symplectic dynamics (8) can be written simply:

$$
\dot{q}^j = \frac{\partial H}{\partial p_j}, \quad \ddot{q}^j = \frac{\partial H}{\partial \xi_j}, \quad \dot{p}_j = -\frac{\partial H}{\partial q^j}, \quad \dot{\xi}_j = -\frac{\partial H}{\partial \dot{q}^j}.
\tag{21}
$$

The two first equations of (21) give the initial controlled dynamics (15). We have also

$$
\begin{cases}
\dfrac{\partial H}{\partial q^j} = -(\partial_j \Gamma_{k\ell}^i)\, \dot{q}^k\, \dot{q}^\ell\, \xi_i - \partial_j(M^{i\ell}\, \partial_\ell V)\, \xi_i + \dfrac{1}{2}(\partial_j M^{k\ell})\, \xi_k\, \xi_\ell \\
\dfrac{\partial H}{\partial \dot{q}^j} = p_j - 2\,\Gamma_{kj}^i\, \dot{q}^k\, \xi_i.
\end{cases}
$$

We deduce the developed form of the two last equations of (21):

$$\dot{p}_j = \left(\partial_j \Gamma^i_{k\ell}\right) \dot{q}^k \dot{q}^\ell \, \xi_i + \partial_j \left(M^{i\ell} \, \partial_\ell V\right) \xi_i - \frac{1}{2} \left(\partial_j M^{k\ell}\right) \xi_k \, \xi_\ell \qquad (22)$$

$$\dot{\xi}_j = 2\, \Gamma^i_{kj} \, \dot{q}^k \, \xi_i - p_j. \qquad (23)$$

(5)   Intrinsic Evolution of the Generalized Force

We introduce the covector ξ according to its covariant coordinates: $\xi = \xi_j \, \partial^j$. We have the following result, first established in [4,7]:

- **Proposition 5. Covariant Evolution Equation of the Optimal Force.** With the above notations and hypotheses, the forces and torques u satisfy the following time evolution:

$$\left(\frac{d^2 u}{dt^2}\right)_j + R^i_{k\ell j}\, \dot{q}^k \, \dot{q}^\ell \, u_i + \left(\nabla^2_{jk} V\right) u^k = 0. \qquad (24)$$

Proof of Proposition 5. The time covariant derivative of the covector ξ is given by

$$\frac{d\xi}{dt} = \left(\dot{\xi}_j - \Gamma^i_{jk} \, \dot{q}^k \, \xi_i\right) \partial^j$$

that is $\left(\dfrac{d\xi}{dt}\right)_j = \dot{\xi}_j - \Gamma^i_{jk} \, \dot{q}^k \, \xi_i$. We report this expression in (23):

$$p_j = \Gamma^i_{jk} \, \dot{q}^k \, \xi_i - \left(\frac{d\xi}{dt}\right)_j. \qquad (25)$$

We wish to differentiate relative to time the expression p_j given in (25). The covariant derivatives of the covector p can be evaluated as follows:

$$\left(\frac{dp}{dt}\right)_j = \dot{p}_j - \Gamma^\ell_{jk} \, \dot{q}^k \, p_\ell.$$

Then we have, taking into account again the relation (25):

$$\dot{p}_j = \frac{d}{dt}\left[\Gamma^i_{jk} \, \dot{q}^k \, \xi_i - \left(\frac{d\xi}{dt}\right)_j\right] + \Gamma^\ell_{jk} \, \dot{q}^k \left[\Gamma^i_{s\ell} \, \dot{q}^s \, \xi_i - \left(\frac{d\xi}{dt}\right)_\ell\right]$$

$$= \partial_\ell\left(\Gamma^i_{jk}\right) \dot{q}^k \, \dot{q}^\ell \, \xi_i + \Gamma^i_{jk} \, \ddot{q}^k \, \xi_i + \Gamma^i_{jk} \, \dot{q}^k \left(\frac{d\xi}{dt}\right)_i - \left(\frac{d^2\xi}{dt^2}\right)_j$$

$$\quad + \Gamma^\ell_{jk} \, \Gamma^i_{s\ell} \, \dot{q}^k \, \dot{q}^s \, \xi_i - \Gamma^\ell_{jk} \, \dot{q}^k \left(\frac{d\xi}{dt}\right)_\ell$$

$$= \partial_\ell\left(\Gamma^i_{jk}\right) \dot{q}^k \, \dot{q}^\ell \, \xi_i + \Gamma^s_{kj} \, \Gamma^i_{s\ell} \, \dot{q}^k \, \dot{q}^\ell \, \xi_i - \left(\frac{d^2\xi}{dt^2}\right)_j + \Gamma^i_{kj} \, \ddot{q}^k \, \xi_i$$

$$\qquad\qquad\qquad\qquad\qquad \text{due to the simplification of two terms}$$

$$= \partial_\ell\left(\Gamma^i_{jk}\right) \dot{q}^k \, \dot{q}^\ell \, \xi_i + \Gamma^s_{kj} \, \Gamma^i_{s\ell} \, \dot{q}^k \, \dot{q}^\ell \, \xi_i - \left(\frac{d^2\xi}{dt^2}\right)_j$$

$$\quad + \Gamma^i_{kj}\left(-\Gamma^k_{\ell s} \, \dot{q}^s \, \dot{q}^\ell - M^{k\ell} \, \partial_\ell V + \xi^k\right) \xi_i \qquad\qquad \text{due to (15)}$$

$$= \left(\partial_\ell \Gamma^i_{jk} + \Gamma^s_{jk} \, \Gamma^i_{s\ell} - \Gamma^s_{k\ell} \, \Gamma^i_{sj}\right) \dot{q}^k \, \dot{q}^\ell \, \xi_i - \left(\frac{d^2\xi}{dt^2}\right)_j - \Gamma^i_{kj} \, M^{k\ell} \, \partial_\ell V \, \xi_i + \Gamma^i_{kj} \, \xi^k \, \xi_i$$

and

$$\dot{p}_j = \left(R^i_{kj\ell} + \partial_j \Gamma^i_{k\ell}\right) \dot{q}^k \dot{q}^\ell \xi_i - \left(\frac{d^2\xi}{dt^2}\right)_j - \Gamma^i_{kj} M^{k\ell} \partial_\ell V \xi_i + \Gamma^i_{kj} \xi^k \xi_i \qquad (26)$$

taking into account the expression (12) of the Riemann tensor. We confront the relations (26) and (22). We deduce

$$\begin{cases} R^i_{kj\ell} \dot{q}^k \dot{q}^\ell \xi_i - \left(\frac{d^2\xi}{dt^2}\right)_j - \Gamma^i_{kj} M^{k\ell} \partial_\ell V \xi_i + \Gamma^i_{kj} \xi^k \xi_i \\ \qquad = \partial_j \left(M^{i\ell} \partial_\ell V\right) \xi_i - \frac{1}{2}\left(\partial_j M^{k\ell}\right) \xi_k \xi_\ell. \end{cases} \qquad (27)$$

We take into account the second Ricci identity (10). It comes

$$\left(\partial_j M^{k\ell}\right) \xi_k \xi_\ell = -\Gamma^k_{js} \xi_k \xi^s - \Gamma^\ell_{js} \xi_\ell \xi^s = -2\,\Gamma^k_{j\ell} \xi_k \xi^\ell = -2\,\Gamma^i_{jk} \xi^k \xi_i\,.$$

Then we can write the relation (27) in a simpler way:

$$\begin{aligned} R^i_{kj\ell} \dot{q}^k \dot{q}^\ell \xi_i - \left(\frac{d^2\xi}{dt^2}\right)_j &= \left[\Gamma^i_{kj} M^{k\ell} \partial_\ell V + \partial_j\left(M^{i\ell}\partial_\ell V\right)\right] \xi_i \\ &= \left[\Gamma^i_{kj} M^{k\ell} \partial_\ell V - \Gamma^i_{js} M^{s\ell} \partial_\ell V - \Gamma^\ell_{js} M^{is} \partial_\ell V + M^{i\ell} \partial_\ell \partial_j V\right] \xi_i \\ &= \left(\partial_\ell \partial_j V - \Gamma^s_{j\ell} \partial_s V\right) M^{i\ell} \xi_i = \left(\nabla^2_{j\ell} V\right) \xi^\ell \end{aligned}$$

due to the expression (11) of the second gradient of a scalar field. We have established the following evolution equation

$$\left(\frac{d^2\xi}{dt^2}\right)_j + \left(\nabla^2_{jk} V\right) \xi^k = R^i_{kj\ell} \dot{q}^k \dot{q}^\ell \xi_i$$

and the relation (24) is a simple consequence of the anti-symmetry of the Riemann tensor and of the identity (20).   □

2 Conclusion

We have established that the methods of Euler-Lagrange and Pontryagin conduct to two second order differential systems that couples state and control variables. The choice of a Riemannian metric allows the two systems to be in a well-defined tensorial nature: contravariant for the equation of motion and covariant for the equation of the control variables. The study of a robotic system, of which we try to optimize the control, shows how important is the introduction of an appropriate geometric structure. Riemannian geometry selected on the configuration parameter space favors the metric directly related to the mass tensor as suggested by the expression of the kinetic energy. An undeniable impact is the choice of an invariant cost function with respect to the choice of parameters, this is a stabilizing factor for numerical developments. Pontryagin's principle applied to contravariant equation of motion associated with the cost function conducts to a mechanical interpretation of adjoint states.

The adjoint control equation is established in a condensed form by the introduction of second order covariant derivatives and shows the Riemann curvature tensor. Moreover, this framework exhibits a numerically stable method when discretization is considered. The resolution of the coupled system gives a direct access to control variables without any additional calculation. Thus, future numerical developments will have to juggle between two coupled systems of second-order ordinary differential equations: the equation of motion and the equation for the control.

References

1. Lazrak, M., Vallée, C.: Commande de robots en temps minimal. Revue d'Automatique et de Productique Appliquées (RAPA) 8(2–3), 217–222 (1995)
2. Lovelock, D., Rund, H.: Tensors, Differential Forms and Variational Principles. Wiley, New York (1975)
3. Pontryagin, L.S., Boltyanskii, V.G., Gamkrelidze, R.V., Mishchenko, E.F.: The Mathematical Theory of Optimal Processes (english translation). Interscience. Wiley, New York (1962)
4. Rojas Quintero, J.A.: Contribution à la manipulation dextre dynamique pour les aspects conceptuels et de commande en ligne optimale. Thesis Poitiers University, 31 October 2013
5. Rojas Quintero, J.A., Vallée, C., Gazeau, J.P., Seguin, P., Arsicault, M.: An alternative to Pontryagin's principle for the optimal control of jointed arm robots. Congrès Français de Mécanique, Bordeaux, 26–30 August 2013
6. Siebert, R.: Mechanical integrators for the optimal control in multibody dynamics. Dissertation, Department Maschinenbau, Universität Siegen (2012)
7. Vallée, C.,. Rojas Quintero, J.A, Fortuné, D., Gazeau, J.P.: Covariant formulation of optimal control of jointed arm robots: an alternative to Pontryagin's principle. arXiv:1305.6517, 28 May 2013

Rolling Symmetric Spaces

Krzysztof A. Krakowski[1,2], Luís Machado[1,3]([✉]), and Fátima Silva Leite[1,4]

[1] Institute of Systems and Robotics, University of Coimbra,
3030-290 Coimbra, Portugal
[2] Faculty of Mathematics and Natural Sciences,
Cardinal Stefan Wyszyński University, 01-815 Warsaw, Poland
k.krakowski@uksw.edu.pl
[3] Department of Mathematics, University of Trás-os-Montes and Alto Douro
(UTAD), 5001-801 Vila Real, Portugal
lmiguel@utad.pt
[4] Department of Mathematics, University of Coimbra, 3001-454 Coimbra, Portugal
fleite@mat.uc.pt

Abstract. Riemannian symmetric spaces play an important role in many areas that are interrelated to information geometry. For instance, in image processing one of the most elementary tasks is image interpolation. Since a set of images may be represented by a point in the Graßmann manifold, image interpolation can be formulated as an interpolation problem on that symmetric space. It turns out that rolling motions, subject to nonholonomic constraints of no-slip and no-twist, provide efficient algorithms to generate interpolating curves on certain Riemannian manifolds, in particular on symmetric spaces. The main goal of this paper is to study rolling motions on symmetric spaces. It is shown that the natural decomposition of the Lie algebra associated to a symmetric space provides the structure of the kinematic equations that describe the rolling motion of that space upon its affine tangent space at a point. This generalizes what can be observed in all the particular cases that are known to the authors. Some of these cases illustrate the general results.

Keywords: Rolling · Isometry · Graßmann manifold · Symmetric spaces · Lie algebra

1 Introduction

Techniques from Riemannian geometry have become increasingly successful and important in image processing, machine learning and data analysis. This is due to the fact that data representation is usually more realistic on manifolds rather then on vector spaces. Examples of manifolds that became popular in computer vision are: the Graßmann manifold, each point of which is associated to a set of images; the essential manifold, which parameterizes the epipolar constraint encoding the relation between correspondences across two images of the same

This work was developed under FCT project PTDC/EEA-CRO/122812/2010.

© Springer International Publishing Switzerland 2015
F. Nielsen and F. Barbaresco (Eds.): GSI 2015, LNCS 9389, pp. 550–557, 2015.
DOI: 10.1007/978-3-319-25040-3_59

scene taken from two different locations; and the manifold of symmetric and positive-definite matrices (SPD), typically identified with the space of diffusion tensors. These examples do not exhaust the list of manifolds that play an important role in areas that are interrelated to information geometry, but serve as motivations for the problems studied here since they are particular cases of Riemannian symmetric spaces.

Averaging, regression and interpolation problems are often employed in pattern recognition and other related areas of computer vision. Image interpolation is one of the most elementary image processing tasks. Many image interpolation techniques have been proposed in the literature. When the data is represented on some manifolds, one approach that is quite effective is based on the notion of rolling motions of a manifold over another one. In which concerns the representation of images by points on the Graßmann, it is known that this correspondence is many to one. For that reason, results for interpolation on that manifold cannot be uniquely turned back to a set of interpolating images.

An algorithm to generate interpolating curves on spheres, on the orthogonal group and on the Graßmann manifold was proposed in [6]. This algorithm was also implemented in [9] for the essential manifold. Another interesting aspect of rolling, in the context of computer vision, is that it can be used to solve multi-class classification problems, as explained in [2]. The main idea behind these algorithms is to use rolling motions to project the data from the manifold to a simpler space where classical methods can be applied and then roll back the solution in order to solve the initial problem on the manifold. It turns out that these algorithms can be adapted to other manifolds, such as the ellipsoid with a left-invariant metric [8] or other Riemannian or even pseudo-Riemannian manifolds, in particular to Riemannian symmetric spaces. This is one of the main motivations behind the research presented in this paper. Other optimization problems and methods to solve them on these manifolds can be found, for instance, in [1].

In this paper, we concentrate on the rolling motion of symmetric homogeneous spaces over the affine tangent space at a point. Section 2 introduces the necessary background. It recalls the definition of rolling subject to the nonholonomic constraints of *"no-slip"* and *"no-twist"*, and contains the fundamentals of homogeneous symmetric spaces. Our results are given in Sect. 3. Theorem 5 proves a strong relationship between rolling maps and the structure of a Lie algebra associated to the symmetric space. Theorem 6 shows how to generate left-invariant parallel vector fields on symmetric spaces from rolling maps. These results are illustrated by a few examples. Example 1 shows how the Lie algebra forces the structure of the kinematic equations. This is also illustrated with the Graßmann manifold in Example 2 and with the Lorentzian sphere, a pseudo-Riemannian manifold, in Example 3. We finish with a few concluding remarks.

2   Preliminaries

We are interested in submanifolds of a Riemannian manifold $\widetilde{M}$. Typically, $\widetilde{M}$ will be the Euclidean space $\mathbb{R}^m$.

2.1 Rolling Maps

In this section we introduce a rolling map of submanifolds isometrically embedded in a Riemannian manifold. The definition of rolling is a generalisation of that given in [10, Appendix B] applicable to a general situation, where the embedding space $\mathbb{R}^m$ is replaced by an orientable Riemannian manifold $\widetilde{M}$, cf. [5]. We assume here and in the remainder of this paper that all manifolds are connected and orientable.

Let $\widetilde{M}$ be a Riemannian complete m-dimensional manifold and let $\widetilde{\mathfrak{G}}$ be the group of isometries on $\widetilde{M}$. Let $I \subset \mathbb{R}$ be a closed interval. From now on, we closely follow the notations used in [5].

Definition 1. Let M and M_0 be two n-manifolds isometrically embedded in an m-dimensional Riemannian manifold $\widetilde{M}$. Then a *rolling* of M on M_0 without slipping or twisting is a map $\chi \colon I \to \widetilde{\mathfrak{G}}$ satisfying the following conditions.

Rolling. *There is a piecewise smooth rolling curve on M given by $\sigma \colon I \to M$ such that:*

  (a) $\chi(t) \cdot \sigma(t) \in M_0$, *and*
  (b) $\mathbf{T}_{\chi(t) \cdot \sigma(t)}(\chi(t)(M)) = \mathbf{T}_{\chi(t) \cdot \sigma(t)} M_0$, *for all $t \in I$.*
  These properties imply that at each point of contact, both manifolds, M_0 and $\chi(t)(M)$, have the same tangent space. This is identified as a subspace of the tangent space of $\widetilde{M}$ at the considered point. The curve $\sigma_0 \colon I \to M_0$ defined by $\sigma_0(t) := \chi(t) \cdot \sigma(t)$ is called the development curve of σ.

No-slip. $\dot{\sigma}_0(t) = \chi(t)_* \cdot \dot{\sigma}(t)$, *for almost all $t \in I$. This condition expresses the fact that the two curves have the same velocity at the point of contact.*

No-twist. *the two complementary conditions:*

  tangential $\left(\dot{\chi}(t)\, \chi(t)^{-1} \right)_* \left(\mathbf{T}_{\sigma_0(t)} M_0 \right) \subset \mathbf{T}_{\sigma_0(t)}^{\perp} M_0$, *and*
  normal $\left(\dot{\chi}(t)\, \chi(t)^{-1} \right)_* \left(\mathbf{T}_{\sigma_0(t)}^{\perp} M_0 \right) \subset \mathbf{T}_{\sigma_0(t)} M_0$, *for almost all $t \in I$.*

We conclude this part by a crucial observation about the operator $\left(\dot{\chi}\, \chi^{-1} \right)_*$ made by Sharpe in [10, p. 379], when $\widetilde{M}$ is the Euclidean space, and in [5] in a more general setting. If χ is a rolling map of M upon M_0, then in suitable coordinates in a neighbourhood of $p \in M_0$ we may choose orthonormal basis in $\mathbf{T}_p \widetilde{M} = \mathbf{T}_p M_0 \oplus \mathbf{T}_p^{\perp} M_0$ so that the operator $\left(\dot{\chi}\, \chi^{-1} \right)_*$ has the matrix form ($m = n + r$)

$$\left(\dot{\chi}(t)\, \chi(t)^{-1} \right)_* = \left[\begin{array}{c|c} 0 & X_{n \times r} \\ \hline -X_{n \times r}^{\mathrm{T}} & 0 \end{array} \right] \begin{array}{l} \mathbf{T}_p M_0 \\ \mathbf{T}_p^{\perp} M_0 \end{array} \tag{1}$$
$$ \mathbf{T}_p M_0 \quad \mathbf{T}_p^{\perp} M_0$$

In essence, our main result, Theorem 5 captures the structure of $\left(\dot{\chi}\, \chi^{-1} \right)_*$ given by (1), that is carried from the Lie algebra of the symmetry acting transitively on M.

2.2   Symmetric Riemannian Homogeneous Spaces

This section gives a very brief introduction to symmetric Riemannian homogeneous spaces. For more details we refer to [4].

Let $\mathfrak{G}$ be a connected Lie group with Lie algebra $\mathfrak{g}$. Suppose $\mathfrak{G}$ acts transitively on a Riemannian manifold M, i.e., there is a smooth map $\mathfrak{G} \times M \to M$, denoted by $(a, p) \mapsto a \cdot p$, such that, for any $p \in M$: $a \cdot (b \cdot p) = (ab) \cdot p$, for any $a, b \in \mathfrak{G}$; $e \cdot p = p$, where e is the identity element of $\mathfrak{G}$; for any $q \in M$ there exists an element $a \in \mathfrak{G}$ such that $q = a \cdot p$. For an arbitrary fixed point $p_0 \in M$ the closed subgroup

$$H := \{ a \in \mathfrak{G} \ : \ a \cdot p_0 = p_0 \}$$

is an isotropy group of $\mathfrak{G}$ at p_0. Then M is diffeomorphic to the space $\mathfrak{G}/H$ of left cosets aH, with $p \mapsto aH$, where $a \in \mathfrak{G}$ is such that $p = a \cdot p_0$. Let the metric on M be invariant under $\mathfrak{G}$, i.e., for any $x \in \mathfrak{G}$ the mapping $\tau(x) \colon aH \mapsto xaH$ of $\mathfrak{G}/H$ onto $\mathfrak{G}/H$ is an isometry. We will assume further that the homogeneous space $\mathfrak{G}/H$ is *reductive*, i.e., there exists a decomposition $\mathfrak{g} = \mathfrak{h} \oplus \mathfrak{p}$, invariant under $\mathrm{Ad}(H)$. The natural projection $\pi \colon \mathfrak{G} \to M \cong \mathfrak{G}/H$ induces the linear surjection $\pi_* \colon \mathbf{T}_e\mathfrak{G} \to \mathbf{T}_{p_0}M$ and we have the following isomorphisms

$$\mathbf{T}_{p_0}M \cong \mathbf{T}_e\mathfrak{G}/\ker \pi_* \cong \mathfrak{g}/\mathfrak{h} \cong \mathfrak{p}.$$

The space $M \cong \mathfrak{G}/H$ is called a *symmetric Riemannian homogeneous space* (*symmetric space* for short) if the above vector subspace $\mathfrak{p}$ satisfies $[\mathfrak{p}, \mathfrak{p}] \subset \mathfrak{h}$. For such spaces we have the following relations

$$\mathfrak{g} = \mathfrak{h} \oplus \mathfrak{p}, \quad [\mathfrak{p}, \mathfrak{p}] \subset \mathfrak{h}, \quad [\mathfrak{p}, \mathfrak{h}] \subset \mathfrak{p} \quad \text{and} \quad [\mathfrak{h}, \mathfrak{h}] \subset \mathfrak{h}.$$

3   Rolling Riemannian Symmetric Spaces

In the remainder of this paper we assume that a manifold M, isometrically embedded in the ambient space $\widetilde{M}$, is rolling upon its affine tangent space at a point p_0. Let $\widetilde{\mathfrak{G}} = \mathfrak{G} \ltimes V$ be the group of isometries preserving orientation of $\widetilde{M}$. For instance, if the ambient space is $\mathbb{R}^m$, its isometry group is the special Euclidean group $\mathbf{SE}(m) = \mathbf{SO}(m) \ltimes \mathbb{R}^m$. If $\chi = (g, s)$ is a rolling map then χ acts as follows

$$I \times \widetilde{M} \ \xrightarrow{\ \chi\ } \ \widetilde{M} \qquad (t, p) \ \xrightarrow{\ \chi\ } \ g(t) \cdot p + s(t)$$

$$I \times \mathbf{T}_{p_0}\widetilde{M} \ \xrightarrow{\ \chi_*\ } \ \mathbf{T}_{p_0}\widetilde{M} \qquad (t, V) \ \xrightarrow{\ \chi_*\ } \ g(t)_* \cdot V$$

We shall assume that M is the symmetric space $\mathfrak{G}/H$, that is $M \cong \mathfrak{G}/H$, so that the subgroup $\mathfrak{G} \subset \widetilde{\mathfrak{G}}$ acts transitively on M and H is the isotropy group of $p_0 \in M$. We identify elements of Lie algebra $\mathfrak{g}$ of $\mathfrak{G}$ with the vector space of linear maps from $\mathbf{T}_{p_0}M$ to itself. Let μ denoting the group action on M then the above relationships can be illustrated with the following diagrams.

$$\mathfrak{G} \times \widetilde{\mathbf{M}} \xrightarrow{\ \mu\ } \widetilde{\mathbf{M}} \qquad\qquad (g,p) \xrightarrow{\ \mu\ } g \cdot p$$

$$\exp\uparrow\exp \qquad\qquad \uparrow\exp \qquad\qquad \exp\uparrow\exp \qquad\qquad \uparrow\exp$$

$$\mathfrak{g} \times \mathbf{T}_{p_0}\widetilde{\mathbf{M}} \xrightarrow{\ \mu_*\ } \mathbf{T}_{p_0}\widetilde{\mathbf{M}} \qquad\qquad (X,V) \xrightarrow{\ \mu_*\ } X \cdot V$$

With these assumptions, if $\chi = (g,s)$ is a rolling map of $\mathbf{M}$ upon its affine tangent space at p_0, then $\sigma(t) = g^{-1}(t) \cdot p_0$ is the rolling curve.

Proposition 2. *Let $\mathfrak{h}$ be the Lie algebra of the isotropy group H of $p_0 \in \mathbf{M}$. Then $\mathfrak{h}(\mathbf{T}_{p_0}\mathbf{M}) \subset \mathbf{T}_{p_0}\mathbf{M}$ and $\mathfrak{h}(\mathbf{T}_{p_0}^{\perp}\mathbf{M}) \subset \mathbf{T}_{p_0}^{\perp}\mathbf{M}$.*

Proof. Let $g\colon (-\varepsilon,\varepsilon) \to H$ be a differentiable curve in the isotropy group H such that $g(0)$ is the identity. Moreover, let $\gamma\colon (-\delta,\delta) \to \widetilde{\mathbf{M}}$ be a differentiable curve in the ambient manifold, with $\gamma(0) = p_0$. Then $c(t,s) := g(t) \cdot \gamma(s)$ is a smooth map from $(-\varepsilon,\varepsilon) \times (-\delta,\delta)$ to $\widetilde{\mathbf{M}}$ such that $c(t,0) = p_0$, for all $t \in (-\varepsilon,\varepsilon)$. The derivative of c with respect to s is

$$\partial_s c(t,0) = g(t)_* \cdot \dot{\gamma}(0),$$

therefore $g(t)_*$ is a map from $\mathbf{T}_{p_0}\widetilde{\mathbf{M}}$ to itself. Since H is also a subgroup of a Lie group $\mathfrak{G}$, that acts transitively on $\mathbf{M}$, then, by restricting γ to $\mathbf{M}$, $\partial_s c(t,0) = g(t)_* \cdot V$, where $V \in \mathbf{T}_{p_0}\mathbf{M}$, is a curve in the tangent space $\mathbf{T}_{p_0}\mathbf{M}$. Similarly $g(t)_* \cdot \Lambda$, where $\Lambda \in \mathbf{T}_{p_0}^{\perp}\mathbf{M}$ is a curve in the normal space $\mathbf{T}_{p_0}^{\perp}\mathbf{M}$, because H is an isometry. Taking derivative with respect to t, noting that $g(0) = e$, yields $\dot{g}(0)_* \cdot \dot{\gamma}(0) = \partial_t \partial_s c(0,0)$, where $\dot{g}(0)_* \in \mathfrak{h}$. The proof is now complete. $\qquad\square$

The following proposition has been proved in [3].

Proposition 3. *Assume that $\widetilde{\mathbf{M}}$ is Euclidean and let $\mathfrak{p} = \mathfrak{g}/\mathfrak{h}$, where $\mathfrak{h}$ is the Lie algebra of the isotropy group H of $p_0 \in \mathbf{M}$. Then $\mathfrak{p}(\mathbf{T}_{p_0}\mathbf{M}) \subset \mathbf{T}_{p_0}^{\perp}\mathbf{M}$ and $\mathfrak{p}(\mathbf{T}_{p_0}^{\perp}\mathbf{M}) \subset \mathbf{T}_{p_0}\mathbf{M}$.*

Remark 4. We are strongly convinced that Proposition 3 is true for general Riemannian manifolds, although we have not been able to produce a complete proof yet. Our believe is based on all the cases that we have analyzed including some of the examples that appear later.

Theorem 5. *Let $\mathfrak{p}$ be as above and χ be a rolling map of a symmetric space $\mathbf{M} \cong \mathfrak{G}/H$ embedded in Euclidean space. Then $\left(\dot{\chi}\,\chi^{-1}\right)_*$ is an element of $\mathfrak{p}$.*

Proof. Denote $\left(\dot{\chi}\,\chi^{-1}\right)_*$ by $u \in \mathfrak{g}$. Let $u = u_{\mathfrak{h}} + u_{\mathfrak{p}}$ be a decomposition of u into components in $\mathfrak{h}$ and $\mathfrak{p}$, respectively. For any vector $V \in \mathbf{T}_{p_0}\mathbf{M}$ there is

$$u \cdot V = (u_{\mathfrak{h}} + u_{\mathfrak{p}}) \cdot V = u_{\mathfrak{h}} \cdot V + u_{\mathfrak{p}} \cdot V,$$

where $u_{\mathfrak{h}} \cdot V \in \mathbf{T}_{p_0}\mathbf{M}$ and $u_{\mathfrak{p}} \cdot V \in \mathbf{T}_{p_0}^{\perp}\mathbf{M}$, by Propositions 2 and 3, respectively. From the tangential part of the *"no-twist"* conditions $u \cdot V \in \mathbf{T}_{p_0}^{\perp}\mathbf{M}$ then it follows that $u_{\mathfrak{h}} \cdot V$ is zero, for all $V \in \mathbf{T}_{p_0}\mathbf{M}$. By a similar reasoning with the normal part of the *"no-twist"* conditions one shows that also $u_{\mathfrak{h}} \cdot V = 0$, for all $V \in \mathbf{T}_{p_0}^{\perp}\mathbf{M}$. Therefore $u_{\mathfrak{h}} \equiv 0$ hence $u = u_{\mathfrak{p}} \in \mathfrak{p}$. This completes the proof. $\qquad\square$

Theorem 6. *Let $\chi = (g, s)$ be a rolling map of a symmetric space $\mathbf{M} \cong \mathfrak{G}/H$ and $\sigma(t) = g^{-1}(t) \cdot p_0$ be the corresponding rolling curve. For any $V_0 \in \mathbf{T}_{p_0}\mathbf{M}$ define a vector field along σ by*

$$V(t) := g^{-1}(t)_* \cdot V_0.$$

Then V is a left-invariant parallel vector field along σ.

Proof. Clearly $V(t) \in \mathbf{T}_{\sigma(t)}\mathbf{M}$. We show first that V is left-invariant. Let L_a denote the left translation by $a \in \mathfrak{G}$ then $V = \left(L_{g^{-1}}\right)_* \cdot V_0$ and

$$V(f \circ L_g) = \left(\left(L_{g^{-1}}\right)_* \cdot V_0\right)(f \circ L_g) = V_0(f \circ L_g \circ L_{g^{-1}}) = V_0(f),$$

for any differentiable f on $\mathbf{M}$. Hence $(L_g)_* V = V_0 = V(0)$ and V is left invariant.

The rolling map χ generates vector field $\widetilde{V}$ along development curve $\sigma_0(t) = \chi(t) \cdot \sigma(t)$ and since rolling maps preserve covariant differentiation, *cf.* [5], then $D_t V = \widetilde{D}_t \widetilde{V}$, where $\widetilde{D}_t$ is the covariant derivative on the affine tangent space. Because $\widetilde{V}(t) = \chi(t)_* \cdot V(t) = \left(g(t)_* \, g^{-1}(t)_*\right) \cdot V_0 = V_0$ is constant therefore $D_t V = 0$, what was to show. $\qquad\square$

Examples. Here we give a few examples of rolling symmetric spaces on their respective affine tangent spaces. These examples illustrate the main ideas behind the structure of the rolling maps and decomposition of a Lie algebra.

Example 1 (The Sphere). Consider the well studied problem of rolling the sphere $\mathbf{S}^n$ on its affine tangent space. Since $\mathbf{S}^n = \mathbf{SO}(n+1)/\mathbf{SO}(n)$ is homogeneous space, take any $p_0 \in \mathbf{S}^n$, then $H = \mathbf{SO}(n)$ is an isotropy group leaving p_0 fixed.

To be more precise, take $p_0 = (0, \ldots, 0, -1)$ be the "south pole" of $\mathbf{S}^n$. The Lie algebra $\mathfrak{g} = \mathfrak{so}(n+1)$ splits into the direct sum $\mathfrak{p} \oplus \mathfrak{h}$, where

$$\mathfrak{h} = \left\{ x \in \mathfrak{so}(n+1) \; : \; x = \begin{bmatrix} A & 0 \\ 0 & 0 \end{bmatrix} \quad \text{and} \quad A \in \mathfrak{so}(n) \right\}$$

and $\mathfrak{p} = \mathfrak{h}^\perp$ is given by

$$\mathfrak{p} = \left\{ x \in \mathfrak{so}(n+1) \; : \; x = \begin{bmatrix} 0 & m \\ -m^{\mathrm{T}} & 0 \end{bmatrix} \quad \text{and} \quad m \in \mathbb{R}^{n \times 1} \right\} \cong \mathbf{T}_{p_0}\mathbf{S}^n.$$

It is easy to see that $\mathfrak{p} \cdot p_0 = \mathbf{T}_{p_0}\mathbf{S}^n$ and $\mathfrak{h} \cdot p_0 = 0$. Note that $\mathrm{span}(p_0) = \mathbf{T}_{p_0}^\perp\mathbf{S}^n$. Let χ be the rolling map and let $u = \left(\dot{\chi}\,\chi^{-1}\right)_*$ then $u \in \mathfrak{g}$ and $\langle u \cdot (\mathfrak{p} \cdot p_0), p_0 \rangle = -\langle \mathfrak{p} \cdot p_0, u \cdot p_0 \rangle$. From the tangential part of the *"no-twist"* condition it follows that $u \in \mathfrak{p}$.

Example 2 (The Graßmann Manifold). We now look at the Graßmann manifold rolling on its affine tangent space, *cf.* [6]. The Graßmann manifold $\mathbf{Gr}_{k,n}$ is defined by $\mathbf{Gr}_{k,n} := \left\{ P \in \mathfrak{s}(n) \; : \; P^2 = P \text{ and } \mathrm{rank}(P) = k \right\}$ and considered embedded in $\mathfrak{s}(n)$, where $\mathfrak{s}(n)$ is the set of $n \times n$ symmetric matrices. Group

$\mathfrak{G} = \mathbf{SO}(n)$ acts transitively on $\mathbf{Gr}_{k,n}$ by $(X, P) \mapsto X \cdot P \cdot X^{\mathrm{T}}$. This action induces Lie algebra action $(a, V) \mapsto a \cdot V + V \cdot a^{\mathrm{T}}$. Take $P_0 = \begin{bmatrix} \mathbf{1}_k & 0 \\ 0 & 0 \end{bmatrix}$ and let $H \subset \mathfrak{G}$ be the isotropy group leaving P_0 fixed. Then

$$H = \left\{ \begin{bmatrix} H_1 & 0 \\ 0 & H_2 \end{bmatrix} : H_1 \in \mathbf{SO}(k) \quad \text{and} \quad H_2 \in \mathbf{SO}(n-k) \right\}.$$

Then Lie algebra $\mathfrak{h}$ of the group H is

$$\mathfrak{h} = \left\{ \begin{bmatrix} h_1 & 0 \\ 0 & h_2 \end{bmatrix} : h_1 \in \mathfrak{so}(k) \quad \text{and} \quad h_2 \in \mathfrak{so}(n-k) \right\}.$$

The orthogonal complement $\mathfrak{p} = \mathfrak{h}^{\perp}$ is therefore

$$\mathfrak{p} = \left\{ \begin{bmatrix} 0 & m \\ -m^{\mathrm{T}} & 0 \end{bmatrix} : m \in \mathbb{R}^{k \times (n-k)} \right\}.$$

The tangent and normal spaces at P_0 are given by

$$\mathbf{T}_{P_0} \mathbf{Gr}_{k,n} = \left\{ \begin{bmatrix} 0 & Z \\ Z^{\mathrm{T}} & 0 \end{bmatrix} : Z \in \mathbb{R}^{k \times (n-k)} \right\} \quad \text{and}$$

$$\mathbf{T}_{P_0}^{\perp} \mathbf{Gr}_{k,n} = \left\{ \begin{bmatrix} S_1 & 0 \\ 0 & S_2 \end{bmatrix} : S_1 \in \mathfrak{s}(k), \quad S_2 \in \mathfrak{s}(n-k) \right\}.$$

The normal part of the *"no-twist"* conditions

$$\begin{bmatrix} u_1 & u_2 \\ -u_2^{\mathrm{T}} & u_3 \end{bmatrix} \begin{bmatrix} S_1 & 0 \\ 0 & S_2 \end{bmatrix} + \begin{bmatrix} S_1 & 0 \\ 0 & S_2 \end{bmatrix} \begin{bmatrix} u_1^{\mathrm{T}} & -u_2 \\ u_2^{\mathrm{T}} & u_3^{\mathrm{T}} \end{bmatrix} = \begin{bmatrix} [u_1, S_1] & u_2 \cdot S_2 - S_1 \cdot u_2 \\ -u_2^{\mathrm{T}} \cdot S_1 + S_2 \cdot u_2^{\mathrm{T}} & [u_3, S_2] \end{bmatrix}$$

yields $[u_1, S_1] = 0$ and $[u_3, S_2] = 0$, for any symmetric S_1 and S_2. This is only possible when $u_1 = 0$ and $u_3 = 0$, hence $u \in \mathfrak{p}$, as expected.

Example 3 (the Lorentzian sphere). We now look at the pseudo-Riemannian case, cf. [7]. The embedding space is $\mathbb{R}^{n+1}$ endowed with the Minkowski metric with the signature $(n, 1)$, denoted by J. Let $\mathbf{S}^{n,1}$ be the surface defined by

$$\mathbf{S}^{n,1} := \left\{ x \in \mathbb{R}^{n+1} : \langle x, x \rangle_J = 1 \right\}.$$

Surface $\mathbf{S}^{n,1}$ is called the *Lorentzian sphere* also known as *de Sitter space*. The symmetry group acting transitively on $\mathbf{S}^{n,1}$ is $\mathbf{SO}(n, 1)$ defined as

$$\mathbf{SO}(n, 1) := \left\{ X \in \mathbb{R}^{(n+1) \times (n+1)} : X^{\mathrm{T}} J X = J \quad \text{and} \quad \det X = 1 \right\},$$

with its Lie algebra $\mathfrak{so}(n, 1) := \left\{ \Omega \in \mathbb{R}^{(n+1) \times (n+1)} : \Omega^{\mathrm{T}} J = -J \Omega \right\}$. It is known that $\mathbf{S}^{n,1} = \mathbf{SO}(n, 1)/\mathbf{SO}(n-1, 1)$ is a symmetric space. Choose $p_0 = (1, 0, \ldots, 0)$ and $n > 1$ then the isotropy group becomes

$$H = \left\{ X \in \mathbf{SO}(n, 1) : X = \begin{bmatrix} 1 & 0 \\ 0 & \mathbf{SO}(n-1, 1) \end{bmatrix} \right\}.$$

Its Lie algebra is therefore

$$\mathfrak{h} = \left\{ x \in \mathfrak{so}(n,1) \;:\; x = \begin{bmatrix} 0 & 0 \\ 0 & \mathfrak{so}(n-1,1) \end{bmatrix} \right\}$$

and its orthogonal complement is

$$\mathfrak{p} = \left\{ x \in \mathfrak{so}(n,1) \;:\; x = J \cdot \begin{bmatrix} 0 & -u^{\mathrm{T}} \\ u & 0 \end{bmatrix} \quad \text{and} \quad u \in \mathbb{R}^{n \times 1} \right\}.$$

This is consistent with the results in [7].

Cases of symmetric spaces like the *essential manifold* [9] and the *ellipsoid* embedded in a space with a left-invariant metric [8] arise naturally from the above three examples.

4   Final Remarks

We have proven that the natural decomposition of the Lie algebra associated to a symmetric space embedded in a Euclidean space provides the structure of the kinematic equations that describe the rolling motion of that space upon its affine tangent space at a point. Several examples have been provided to illustrate the general results.

References

1. Absil, P.A., Mahony, R., Sepulchre, R.: Optimization Algorithms on Matrix Manifolds. Princeton University Press, Princeton (2008)
2. Caseiro, R., Martins, P., Henriques, J., Silva Leite, F., Batista, J.: Rolling Riemannian manifolds to solve the multi-class classification problem. In: IEEE Conference on Computer Vision and Pattern Recognition (CVPR), pp. 41–48, June 2013
3. Eschenburg, J., Heintze, E.: Extrinsic symmetric spaces and orbits of s-representations. Manuscripta Math. **88**, 517–524 (1995)
4. Helgason, S.: Differential Geometry, Lie Groups and Symmetric Spaces. Academic Press, London (1978)
5. Hüper, K., Krakowski, K.A., Silva Leite, F.: Rolling Maps in a Riemannian Framework, Textos de Matemática, vol. 43, pp. 15–30. Department of Mathematics, University of Coimbra, Portugal (2011)
6. Hüper, K., Silva Leite, F.: On the geometry of rolling and interpolation curves on S^n, SO_n and Graßmann manifolds. J. Dyn. Control Syst. **13**(4), 467–502 (2007)
7. Korolko, A., Silva Leite, F.: Kinematics for rolling a Lorentzian sphere. In: 2011 50th IEEE Conference on Decision and Control and European Control Conference (CDC-ECC), pp. 6522–6527, December 2011
8. Krakowski, K.A., Silva Leite, F.: An algorithm based on rolling to generate smooth interpolating curves on ellipsoids. Kybernetika **50**(4), 544–562 (2014)
9. Machado, L., Pina, F., Silva Leite, F.: Rolling maps for the Essential Manifold, Chap. 21. In: Bourguignon, J.-P., Jeltsch, R., Pinto, A.A., Viana, M. (eds.) Dynamics, Games and Science. CIM Series in Mathematical Sciences, pp. 399–415. Springer, Cham (2015)
10. Sharpe, R.W.: Differential Geometry: Cartan's Generalization of Klein's Erlangen Program. Graduate Texts in Mathematics, vol. 166. Springer, New York (1997)

Enlargement, Geodesics, and Collectives

Eric W. Justh[2] and P. S. Krishnaprasad[1]([✉])

[1] Department of Electrical and Computer Engineering,
Institute for Systems Research, University of Maryland, College Park 20742, USA
krishna@umd.edu
[2] Naval Research Laboratory, Washington, DC 20375, USA

Abstract. We investigate optimal control of systems of particles on matrix Lie groups coupled through graphs of interaction, and characterize the limit of strong coupling. Following Brockett, we use an enlargement approach to obtain a convenient form of the optimal controls. In the setting of drift-free particle dynamics, the coupling terms in the cost functionals lead to a novel class of problems in subriemannian geometry of product Lie groups.

Keywords: Subriemannian geometry · Subriemannian geodesics · Non-holonomic integrator · Lie-poisson reduction · Collective behavior

1 Introduction

Consider the drift-free left-invariant system on the Heisenberg group $H(3)$ given by

$$\dot{g} = g\xi = g(u_1 X_1 + u_2 X_2), \quad g \in H(3), \; \xi \in \mathfrak{h}(3), \text{ the Lie algebra of } H(3), \quad (1)$$

where

$$X_1 = \begin{bmatrix} 0 & 1 & 0 \\ 0 & 0 & 0 \\ 0 & 0 & 0 \end{bmatrix}, \; X_2 = \begin{bmatrix} 0 & 0 & 0 \\ 0 & 0 & 1 \\ 0 & 0 & 0 \end{bmatrix}, \; X_3 = \begin{bmatrix} 0 & 0 & 1 \\ 0 & 0 & 0 \\ 0 & 0 & 0 \end{bmatrix}, \quad (2)$$

is a basis for $\mathfrak{h}(3)$, with $[X_1, X_2] = X_3$, $[X_1, X_3] = 0$, and $[X_2, X_3] = 0$, where

$$g = \begin{bmatrix} 1 & g_{12} & g_{13} \\ 0 & 1 & g_{23} \\ 0 & 0 & 1 \end{bmatrix}. \quad (3)$$

The elements of g are real-valued, as are the controls u_1 and u_2. The Magnus expansion (single exponential representation) relates the solution g of (1) to the logarithmic coordinates defined by

$$g = \exp(Z) = \exp(z_1 X_1 + z_2 X_2 + z_3 X_3), \quad (4)$$

This research was supported in part by the Air Force Office of Scientific Research under AFOSR Grant No. FA9550-10-1-0250, the ARL/ARO MURI Program Grant No. W911NF-13-1-0390, and by the Office of Naval Research.

where $Z \in \mathfrak{h}(3)$. Due to the nilpotency of $\mathfrak{h}(3)$, the Magnus expansion terminates, so that defining $U = (u_1 X_1 + u_2 X_2)$, the system

$$\dot{Z} = U + \frac{1}{2}[Z,U] + \frac{1}{12}[Z,[Z,U]] - \frac{1}{720}[Z,[Z,[Z,U]]] \pm \cdots = U + \frac{1}{2}[Z,U], \quad (5)$$

or equivalently, as in [9],

$$\dot{z}_1 = u_1, \quad \dot{z}_2 = u_2, \quad \dot{z}_3 = \frac{1}{2}(z_1 u_2 - z_2 u_1), \quad (6)$$

is a globally valid coordinate representation for the dynamics on $H(3)$ given by (1). The system (6) is the nonholonomic integrator, a prototype in the study of subriemannian geometry [4,5]. Other such prototypes may be seen in [1,2].

Although (1) evolves on the Lie group $H(3)$, it can be viewed as a special case of a dynamics defined on the Lie group $Gl(3)$, the general linear group of 3×3 matrices, with initial condition $g(0) \in H(3)$, or even as a special case of a dynamics defined on Mat(3), the vector space of 3×3 matrices (with the same restriction on initial conditions). This latter *extrinsic* viewpoint is adopted in the present paper, and while we use the group $H(3)$ as an illustrative example, the results apply to the Mat(n) setting, $n \geq 3$.

We study control of collective dynamics, where the elementary units (particles) are coupled through a communication graph, and controls are found by extremizing a particular form of cost functional. In prior work, we used Lie-Poisson reduction to study such problems, where reduction was made possible by symmetry of the cost functional, [8] (see also [7]). Here we show that an extrinsic (enlargement) approach offers the advantage of a convenient form of the necessary conditions of optimality. This method was pioneered by Brockett [3] who also developed subriemannian geometry with the nonholonomic integrator (6) and its generalization in [4,5] as model problems. For a penetrating view of subriemannian geometry see the work of Mikhael Gromov [6].

This paper is organized as follows. In Sect. 2 we formulate an optimal control problem for a collective of particles, each of which evolves according to the same right-invariant control-affine dynamics. The Maximum Principle is then applied to find optimal controls, and through this process, a symmetry of the corresponding hamilton's equations is observed. In Sect. 3 this symmetry is shown to yield the Kirillov-Kostant-Souriau (KKS) bracket, and the reduced dynamics are presented. The limiting case of "strong coupling," analogous to [8] is also discussed. Finally, in Sect. 6 we return to the special case of the Heisenberg group in concluding remarks.

2   Enlargement

Here we use an enlargement approach to formulate and analyze the dynamics of a collective of particles evolving on copies of a matrix Lie group (such as $H(3)$). The basic idea is to formulate a higher-dimensional system more amenable to analysis, which specializes to the dynamics of interest under appropriately constrained initial conditions. As we show by explicit calculation, the optimality conditions for the enlarged system admit reduction, and the interpretation of this reduction is provided in the subsequent section.

2.1 Collective Dynamics

For N (identical) particles $Q_i : \mathbb{R}^+ \to \text{Mat}(n)$, $t \mapsto Q_i(t)$, i.e., Q_i an $n \times n$ matrix with real-valued elements, we consider the right-invariant dynamics

$$\dot{Q}_i = \left(A + \sum_{j=1}^m u_i^j B_j\right) Q_i, \quad i = 1, \ldots, N, \tag{7}$$

for some positive integer m, where $u_i^j : \mathbb{R}^+ \to \mathbb{R}$ are controls, and $A, B_j \in \text{Mat}(n)$, $j = 1, \ldots, m$.

For the special case of the nonholonomic integrator (in right-invariant form), $A = 0$ (i.e., the system is drift-free), $m = 2$, $B_1 = -X_1$, $B_2 = -X_2$, and $Q \in H(3)$. In general, the drift-free restriction is required for the subriemannian setting. But we postpone this specialization, because certain key results hold in the more general formulation of (7).

2.2 Cost Functional and Lagrangian

The cost functional we seek to minimize is

$$\mathcal{L} = \int_0^T L(u_1(t), \ldots, u_N(t)) dt, \tag{8}$$

with fixed endpoints $t = 0$ and $t = T$, where $u_i \in \mathbb{R}^m$ is the column vector of controls applied to the ith particle, $i = 1, \ldots, N$. The Lagrangian

$$L = \frac{1}{2}\left(\sum_{k=1}^N |u_i|^2 + \chi \sum_{k=1}^N \sum_{i=1}^N \alpha_{ki}|u_k - u_i|^2\right) \tag{9}$$

penalizes (steering) control "energy" and (steering) control differences, with a nonnegative coupling constant χ determining the relative contribution of these two terms. In (9), the α_{ki} are elements of the adjacency matrix corresponding to a communication graph, which is assumed to be connected, undirected, and without self loops. The corresponding graph Laplacian is denoted by $\beta = D - \alpha$, where D is the (diagonal) degree matrix. Here, $|\cdot|$ denotes the norm associated with a vector inner product $\langle \cdot, \cdot \rangle$ on $\mathbb{R}^m$. A calculation incorporating these assumptions on the interaction graph shows that we can rewrite (9) as

$$L = \frac{1}{2}\sum_{i=1}^N |u_i|^2 + \chi \sum_{k=1}^N \left\langle u_k, \sum_{i=1}^N \beta_{ki} u_i \right\rangle. \tag{10}$$

2.3   Pre-hamiltonian

Because the Lagrangian (10) only involves the controls u_i, $i = 1, \ldots, N$, the adjoint system to (7) is

$$\dot{P}_i = - \left(A + \sum_{j=1}^{m} u_i^j B_j \right)^T P_i, \quad i = 1, \ldots, N, \tag{11}$$

where $P_i : \mathbb{R}^+ \to \mathrm{Mat}(n)$.

The Pontryagin pre-hamiltonian is defined as

$$H = \sum_{i=1}^{N} \left\langle\!\left\langle \dot{Q}_i, P_i \right\rangle\!\right\rangle - L, \tag{12}$$

where $\langle\!\langle \cdot, \cdot \rangle\!\rangle$ denotes a suitable inner product, here the trace inner product, yielding

$$H = \sum_{i=1}^{N} \mathrm{tr}\left(\dot{Q}_i P_i^T \right) - L = \sum_{i=1}^{N} \mathrm{tr}\left(\left(A + \sum_{j=1}^{m} u_i^j B_j \right) Q_i P_i^T \right) - L$$

$$= \sum_{i=1}^{N} \mathrm{tr}\left(\left(A + \sum_{j=1}^{m} u_i^j B_j \right) K_i \right) - L, \tag{13}$$

where we have defined

$$K_i = Q_i P_i^T, \quad i = 1, \ldots, N. \tag{14}$$

With this definition of K_i we then have

$$\dot{K}_i = \dot{Q}_i P_i^T + Q_i \dot{P}_i^T = \left[\left(A + \sum_{j=1}^{m} u_i^j B_j \right), K_i \right]. \tag{15}$$

Thus, (15) and (13) form a self-contained system involving K_i and u_i, $i = 1, \ldots, N$.

2.4   Application of the Maximum Principle

We note that the cost functional $\mathcal{L}$ given by (8) with Lagrangian L given by (10) depends only on the controls $u_1(t), \ldots, u_N(t)$ and L is convex. The Maximum Principle in the present setting can be stated as follows.

Theorem (Maximum Principle): If $u_1(t), \ldots, u_N(t)$ are optimal controls for $\mathcal{L}$, $(Q_1, \ldots, Q_N)$ denotes the corresponding optimal trajectory in $GL(n)^N$, and the only extremals of $\mathcal{L}$ are regular extremals, then

(a) there exist $P_1(t), \ldots, P_N(t)$ such that

$$H(P_1(t), \ldots, P_N(t), Q_1(t), \ldots, Q_N(t), u_1(t), \ldots, u_N(t))$$
$$= \sup_{v_k \in \mathbb{R}^m, \, k=1,\ldots,N} H(P_1(t), \ldots, P_N(t), Q_1(t), \ldots, Q_N(t), v_1, \ldots, v_N),$$

$$(16)$$

for a.e. $t \in [0, T]$; and
(b) defining

$$\mathrm{H}(P_1(t), \ldots, P_N(t), Q_1(t), \ldots, Q_N(t))$$
$$= \sup_{v_k \in \mathbb{R}^m, \, k=1,\ldots,N} H(P_1(t), \ldots, P_N(t), Q_1(t), \ldots, Q_N(t), v_1, \ldots, v_N),$$

$$(17)$$

we have that $P_1(t), \ldots, P_N(t), Q_1(t), \ldots, Q_N(t)$ satisfy hamilton's equations for the hamiltonian H.

Applying the Maximum Principle to the specific Lagrangian (10), the first-order necessary condition for (17) is $\partial H/\partial u_i^j = 0$, $i = 1, \ldots, N$, $j = 1, \ldots, m$, and differentiating (13) with respect to u_i^j yields

$$\frac{\partial H}{\partial u_i^j} = \mathrm{tr}\,(B_j K_i) - u_i^j - 2\chi \sum_{k=1}^N \beta_{ki} u_k^j = 0, \tag{18}$$

so that $\mathrm{tr}\,(B_j K_i) = u_i^j + 2\chi \sum_{k=1}^N \beta_{ik} u_k^j$, where we have used the symmetry of the graph Laplacian β.

We define

$$\mu_i^j \triangleq \mathrm{tr}\,(B_j K_i), \quad i = 1, \ldots, N, \; j = 1, \ldots, m, \tag{19}$$

so that $\mu_i^j = u_i^j + 2\chi \sum_{k=1}^N \beta_{ik} u_k^j$, or, equivalently,

$$\mu_i = u_i + 2\chi \sum_{k=1}^N \beta_{ik} u_k, \tag{20}$$

where $\mu_i \in \mathbb{R}^m$, $i = 1, \ldots, N$. This can be rewritten, via Kronecker products, as

$$\begin{bmatrix} \mu_1 \\ \vdots \\ \mu_N \end{bmatrix} = ((\mathbb{I}_N + 2\chi\beta) \otimes \mathbb{I}_m) \begin{bmatrix} u_1 \\ \vdots \\ u_N \end{bmatrix}. \tag{21}$$

Defining $\Psi \triangleq ((\mathbb{I}_N + 2\chi\beta) \otimes \mathbb{I}_m)^{-1} = (\mathbb{I}_N + 2\chi\beta)^{-1} \otimes \mathbb{I}_m$, and noting that all eigenvalues of β are real and nonnegative (including exactly one zero eigenvalue), we see that Ψ is guaranteed to exist for all $\chi \geq 0$. We then have

$$\begin{bmatrix} u_1 \\ \vdots \\ u_N \end{bmatrix} = \Psi \begin{bmatrix} \mu_1 \\ \vdots \\ \mu_N \end{bmatrix}, \tag{22}$$

which we can substitute back into the hamiltonian. From (13) and (10), with the optimal controls substituted in, we find (after some calculation)

$$h = \sum_{i=1}^{N} \mathrm{tr}\,(AK_i) + \frac{1}{2}\, [\,\mu_1^T \cdots \mu_N^T\,]\, \Psi \begin{bmatrix} \mu_1 \\ \vdots \\ \mu_N \end{bmatrix}. \tag{23}$$

We have used the notation "h" (rather than "H") to denote the hamiltonian because it depends on Q_i and P_i only through $K_i = Q_i P_i^T$, $i = 1, \ldots, N$, so it has the interpretation of a reduced hamiltonian on a reduced space. We make this precise in the next section.

3   Reduction and the Kirillov-Kostant-Souriau Bracket

It can be shown that the passage from the hamiltonian system defined in terms of Q_i, P_i to the system involving only K_i, $i = 1, \ldots, N$, is Lie-Poisson reduction, and a KKS bracket is obtained:

$$\{\phi, \psi\}\,(K_1, \ldots, K_N) = -\sum_{i=1}^{N} \left\langle \left\langle \left[\frac{\delta\phi}{\delta K_i}, \frac{\delta\psi}{\delta K_i} \right], K_i \right\rangle \right\rangle, \tag{24}$$

where $\phi, \psi : (\mathrm{Mat}(n))^N \to \mathbb{R}$. Although the KKS bracket (24) is decoupled, we note that coupling among the K_i, $i = 1, \ldots, N$, remains present in the reduced dynamics,

$$\dot{K}_i = \left[A + \sum_{j=1}^{m} \left([\,\mu_1^T \cdots \mu_N^T\,]\, \Psi \begin{bmatrix} \delta_{i1}e^j \\ \vdots \\ \delta_{iN}e^j \end{bmatrix} \right) B_j, K_i \right], \quad i = 1, \ldots, N, \tag{25}$$

where e^j is an m-vector with the jth element equal to one and all other elements zero, and δ_{il} is the Kronecker delta (i.e., one for $i = l$ and zero otherwise). The derivative of the reduced hamiltonian (23), which is incorporated into (25), is given by

$$\frac{\delta h}{\delta K_i} = A^T + \sum_{j=1}^{m} \left([\,\mu_1^T \cdots \mu_N^T\,]\, \Psi \begin{bmatrix} \delta_{i1}e^j \\ \vdots \\ \delta_{iN}e^j \end{bmatrix} \right) B_j^T, \tag{26}$$

and the coupling present in the dynamics (25) is thus seen to enter through the gradient of the hamiltonian (via Ψ, which depends on the interaction graph).

The hamiltonian h is constant along trajectories. Furthermore, $\mathrm{tr}(K_i^k)$ is a Casimir (commutes with any h under the bracket (24)) for any $k > 0$ (and $i = 1, \ldots, N$). Using the Cayley-Hamilton Theorem, we thus have n functionally independent Casimirs for each $i = 1, \ldots, N$ - a total of nN Casimirs.

4   Strong Coupling Limit

The limit of strong coupling is defined as $\chi \to \infty$, and in (25), this limit corresponds to [8]

$$\Psi_\infty = \lim_{\chi \to \infty} \Psi = \frac{1}{N} \mathbf{1}_N \mathbf{1}_N^T \otimes \mathbb{I}_m, \tag{27}$$

where $\mathbf{1}_N = [1 \ 1 \ \cdots \ 1]^T$, resulting in $\dot{K}_i = \left[A + \sum_{j=1}^m \mathrm{tr}\left(B_j \bar{K} \right) B_j, K_i \right]$, for $\bar{K} = \frac{1}{N} \sum_{i=1}^N K_i$, and in fact

$$\dot{\bar{K}} = \left[A + \sum_{j=1}^m \mathrm{tr}\left(B_j \bar{K} \right) B_j, \bar{K} \right], \tag{28}$$

which is equivalent to the single-particle ($N = 1$) solution. Along trajectories, h and the nN Casimirs (n associated with each particle) are constant. In the $\chi \to \infty$ limit, we have n additional constants associated with $\mathrm{tr}(\bar{K}^k)$, $k > 0$: these correspond to "hidden symmetries."

5   Subriemannian Geodesics

The enlargement process just described can be specialized to the subriemannian geodesic problem for collectives of particles on matrix Lie groups and metrics of the form (8) and (9). The value function is the optimal cost of passing between two configurations. In this context, we require drift-free dynamics, i.e., $A = 0$, because the value function shouldn't depend on which configuration is "initial," and which is "final." The enlarged version of the subriemannian geodesic dynamics are thus

$$\dot{Q}_i = \left(\sum_{j=1}^m u_i^j B_j \right) Q_i, \quad i = 1, \dots, N, \tag{29}$$

i.e., (7) with $A = 0$, and the optimal controls (used to compute the value function) are found using (22), (19) and

$$\dot{K}_i = \sum_{j=1}^m \left(\begin{bmatrix} \mu_1^T & \cdots & \mu_N^T \end{bmatrix} \Psi \begin{bmatrix} \delta_{i1} e^j \\ \vdots \\ \delta_{iN} e^j \end{bmatrix} \right) [B_j, K_i], \quad i = 1, \dots, N \tag{30}$$

(i.e., (25) with $A = 0$). Constancy of h and the Casimirs, along with the limiting behavior as $\chi \to \infty$, may be useful in the process of calculating (or numerically approximating) the value function, e.g., to exhibit geodesic spheres, as in Fig. 1 of [4].

6   Concluding Remarks

A class of optimal control problems on a product of matrix Lie groups is interpreted in terms of cost-coupled interacting particles. The interactions are governed by a connected graph and a coupling constant. Using the method of enlargement due to Brockett, necessary conditions for optimality are derived and examined in the strong coupling limit. For drift-free dynamics and appropriate controllability conditions the setup leads to novel problems of subriemannian geometry, and convenient forms for associated geodesic equations. For the Heisenberg group (and the nonholonomic integrator), the calculus of variations can be used for expressions equivalent to (29) and (30) for computing geodesics and corresponding value functions (metric distances). In the context of the Heisenberg group, working out the enlargement route for low-dimensional examples (e.g., $N = 2$ or $N = 3$ with various interaction graphs), can help make contact with these more direct calculations. These details are omitted for space limitations. The utility of the enlargement approach is in providing a general technique for a novel class of subriemannian geometries on matrix Lie groups.

Acknowledgment. It is a pleasure to acknowledge stimulating discussions with Roger Brockett on the subject of this paper

References

1. Baillieul, J.: Some optimization problems in geometric control theory. Ph.D. thesis, Harvard University, Cambridge, MA (1975)
2. Baillieul, J.: Geometric methods for nonlinear optimal control problems. J. Opt. Theor. Appl. **25**(4), 519–548 (1978)
3. Brockett, R.W.: Lie theory and control systems defined on spheres. SIAM J. Appl. Math. **25**(2), 213–225 (1973)
4. Brockett, R.W.: Control theory and singular riemannian geometry. In: Hilton, P.J., Young, G.S. (eds.) New Directions in Applied Mathematics: Papers Presented April 25/26, 1980, on the Occasion of the Case Centennial Celebration, pp. 11–27. Springer, New York (1982)
5. Brockett, R.W.: Nonlinear control theory and differential geometry. In: Ciesielski, Z., Olech, C. (eds.) Proceedings of the International Congress of Mathematicians, Warszawa, 16–24 August 1983, vol. 2, sect. 14, pp. 1357–1378. Polish Scientific Publishers, Warszawa (1984)
6. Gromov, M.: Carnot-Carathéodory spaces seen from within. In: Bellaiche, A., Risler, J.-J. (eds.) Sub-Riemannian Geometry. Progress in Mathematics, vol. 144, pp. 79–323. Birkhuser-Verlag, Basel (1996)
7. Justh, E.W., Krishnaprasad, P.S.: Optimal natural frames. Comm. Inf. Syst. **11**(1), 17–34 (2011)
8. Justh, E.W., Krishnaprasad, P.S.: Optimality, reduction, and collective motion. Proc. R. Soc. A **471**, 20140606 (2015). http://dx.doi.org/10.1098/rspa.2014.0606
9. Struemper, H.: Motion control for nonholonomic systems on matrix Lie groups. Ph.D. thesis, University of Maryland, College Park, MD (1998)

Computational Information Geometry

Geometry of Goodness-of-Fit Testing in High Dimensional Low Sample Size Modelling

Paul Marriott[1], Radka Sabolova[2][(✉)], Germain Van Bever[2], and Frank Critchley[2]

[1] University of Waterloo, Waterloo, ON, Canada
[2] The Open University, Milton Keynes, UK
radka.sabolova@open.ac.uk

Abstract. We introduce a new approach to goodness-of-fit testing in the high dimensional, sparse extended multinomial context. The paper takes a computational information geometric approach, extending classical higher order asymptotic theory. We show why the Wald – equivalently, the Pearson χ^2 and score statistics – are unworkable in this context, but that the deviance has a simple, accurate and tractable sampling distribution even for moderate sample sizes. Issues of uniformity of asymptotic approximations across model space are discussed. A variety of important applications and extensions are noted.

1 Introduction

A major contribution of classical information geometry to statistics is the geometric analysis of higher order asymptotic theory, see the seminal work [2] and for example [5]. It has excellent tools for constructing higher order corrections to approximations of sampling distributions, an example being the work on the geometry of Edgeworth expansions in [2, Chapter 4]. These expressions use curvature terms to correct for skewness and other higher order moment (cumulant) issues and provide good, operational corrections to sampling distributions, such as those in Fig. 3(b) and (c) below. However, as discussed in [3,6], these curvature terms grow unboundedly as the boundary of the probability simplex is approached. Since this region plays a key role in modelling in the sparse setting – the MLE often being on the boundary – extensions to the classical theory are needed. This paper starts such a development.

Independently, there has been increased interest in categorical, (hierarchical) log-linear and graphical models. See, in particular, [6–8,10]. As stated by [7] '[their] statistical properties under sparse settings are still very poorly understood. As a result, [analysis of such data] remains exceptionally difficult'.

This paper is an introduction to a novel approach which combines and extends these two areas. The extension comes from using approximations based on the asymptotics of high dimensionality (k-asymptotics) rather than the more familiar sample size approach (N-asymptotics). This is connected to, but distinct from, the landmark paper by [12] and related work. In particular, for a practical

© Springer International Publishing Switzerland 2015
F. Nielsen and F. Barbaresco (Eds.): GSI 2015, LNCS 9389, pp. 569–576, 2015.
DOI: 10.1007/978-3-319-25040-3_61

example of so-called sparse-data asymptotics, see [1, Sect. 6.3]. Computational information geometry – in all its forms: see, for example, [4,6,11,13] – has been a significant recent development, and this paper is a further contribution to it.

We address the challenging problems which arise in the high dimensional sparse extended multinomial context where the dimension k of the underlying probability simplex, one less than the number of categories or cells, is much more than the number of observations N, so that boundary effects necessarily occur, see [3]. In particular, arbitrarily small (possibly, zero) expected cell frequencies must be accommodated. Hence we work with *extended* multinomial models thus taking us out of the manifold structure of classical information geometry, [4].

For practical relevance, our primary focus is on (a) accurate, finite sample and dimension approximation, rather than asymptotic limiting results *per se*; and (b) performance at or near the boundary, rather than (as in earlier studies) the centre of the simplex.

Section 2.1 shows why the Wald statistic – identical, here, to the Pearson χ^2 or score statistic – is unworkable in this context. In contrast analysis and simulation exercises (Sect. 2.2) indicate that the same is not true of the deviance D. We demonstrate that a simple normal (or shifted χ^2) approximation to the distribution of D is accurate and tractable even as the boundary is approached. In contrast to other approaches, this appears to hold effectively *uniformly* across the simplex. The worst place is at its centre (where all cells are equiprobable), due to discretisation effects. However, further theory shows that, even here, the accuracy of approximation improves without limit when $N, k \to \infty$ with $N/k \to c > 0$.

Section 3 considers the uniformity of asymptotic approximations. Its three subsections address issues associated with the boundary, higher moments and discreteness, respectively.

2   Analysis

2.1   Why the Wald Statistic is Unworkable

With i ranging over $\{0, 1, ..., k\}$, let $n = (n_i) \sim$ Multinomial $(N, (\pi_i))$, where here each $\pi_i > 0$. In this context the Wald, Pearson's χ^2, and score statistics all coincide, their common value, W, being

$$W := \sum_{i=0}^{k} \frac{(\pi_i - n_i/N)^2}{\pi_i} \equiv \frac{1}{N^2} \sum_{i=0}^{k} \frac{n_i^2}{\pi_i} - 1.$$

Defining $\pi^{(\alpha)} := \sum_i \pi_i^\alpha$ we note the inequality, for each $m \geq 1$,

$$\left\{ \pi^{(-m)} - (k+1)^{m+1} \right\} \geq 0,$$

in which equality holds if and only if $\pi_i \equiv 1/(k+1)$ – i.e. iff (π_i) is uniform. We then have the following theorem, which establishes that the statistic W is unworkable as $\pi_{\min} := \min(\pi_i) \to 0$ for fixed k and N.

Theorem 1. *For $k > 1$ and $N \geq 6$, the first three moments of W are:*

$$E(W) = \frac{k}{N}, var(W) = \frac{\left\{\pi^{(-1)} - (k+1)^2\right\} + 2k(N-1)}{N^3}$$

and $E[\{W - E(W)\}^3]$ given by

$$\frac{\left\{\pi^{(-2)} - (k+1)^3\right\} - (3k + 25 - 22N)\left\{\pi^{(-1)} - (k+1)^2\right\} + g(k,N)}{N^5}$$

where $g(k,N) = 4(N-1)k(k + 2N - 5) > 0$.
In particular, for fixed k and N, as $\pi_{\min} \to 0$

$$var(W) \to \infty \text{ and } \gamma(W) \to +\infty$$

where $\gamma(W) := E[\{W - E(W)\}^3]/\{var(W)\}^{3/2}$.

2.2   The Deviance Statistic

Unlike the triumvirate of statistics above, the deviance has a workable distribution in the same limit: that is, for fixed N and k as we approach the boundary of the probability simplex. The paper [3] demonstrated the lack of uniformity across this simplex of standard first order N-asymptotic approximations. In sharp contrast to this we see the very stable and workable behaviour of the k-asymptotic approximation to the distribution of the deviance.
Define the deviance D via

$$D/2 = \sum_{\{0 \leq i \leq k : n_i > 0\}} n_i \log(n_i/N) - \sum_{i=0}^{k} n_i \log(\pi_i)$$

$$= \sum_{\{0 \leq i \leq k : n_i > 0\}} n_i \log(n_i/\mu_i),$$

where $\mu_i := E(n_i) = N\pi_i$. We will exploit the characterisation that the multinomial random vector n has the same distribution as a vector of independent Poisson random variables conditioned on their sum. Specifically, let the elements of (n_i^*) be *independently* distributed as Poisson $Po(\mu_i)$. Then, $N^* := \sum_{i=0}^{k} n_i^* \sim Po(N)$, while $(n_i) := (n_i^* | N^* = N) \sim$ Multinomial$(N, (\pi_i))$. Define

$$S^* := \binom{N^*}{D^*/2} = \sum_{i=0}^{k} \binom{n_i^*}{n_i^* \log(n_i^*/\mu_i)}$$

where D^* is defined implicitly and $0 \log 0 := 0$. The terms ν, τ and ρ are defined by the first two moments of S^* via

$$\binom{N}{\nu} := E(S^*) = \binom{N}{\sum_{i=0}^{k} E(n_i^* \log\{n_i^*/\mu_i\})},$$

$$\begin{pmatrix} N & \rho\tau\sqrt{N} \\ \cdot & \tau^2 \end{pmatrix} := cov(S^*) = \begin{pmatrix} N & \sum_{i=0}^{k} C_i \\ \cdot & \sum_{i=0}^{k} V_i \end{pmatrix},$$

where $C_i := Cov(n_i^*, n_i^* \log(n_i^*/\mu_i))$ and $V_i := Var(n_i^* \log(n_i^*/\mu_i))$. Careful analysis gives:

Theorem 2. *Each of these terms ν, τ and ρ are bounded as $\pi_{\min} \to 0$ and hence the distribution of the deviance is stable in this limit.*

Moreover, these terms can be easily and accurately approximated using standard truncate and bound computational methods, exploited below.

Under standard Lindeberg conditions, multivariate central limit theorem (CLT) gives for large k and N that S^* is approximately distributed as a bivariate normal $N_2(E(S^*), cov(S^*))$. Mild conditions (see [12]) ensuring uniform equicontinuity of the conditional characteristic functions $D^*/2|\{N^* = N\}$ (see [14]) then gives, in the same limit,

$$D/2 = D^*/2|\{N^* = N\} \sim N_1(\nu, \tau^2(1 - \rho^2)). \tag{1}$$

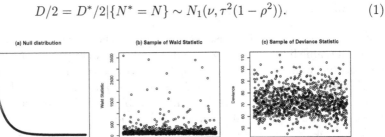

Fig. 1. Stability of the sampling distributions

3   Uniformity of Asymptotic Approximations

3.1   Uniformity Near the Boundary

In general asymptotic approximations are not uniformly accurate as is shown in [3]. Consider the consequences of Theorem 1 when $\pi_{\min}$ is close to zero as illustrated in Fig. 1. This shows, in panel (a), the distribution, π, where we see that $\pi_{\min}$ is indeed very small. Here, and throughout, we plot the distributions in rank order without loss since all sampling distributions considered are invariant to permutation of the labels of the multinomial cells. Panel (b) shows a sample of 1000 values of W drawn from its distribution when there are $N = 50$ observations in dimension $k = 200$. The extreme non-normality, and hence the failure of the standard N-asymptotic approximation, is evident. In contrast, consider panel (c), which shows 1000 replicates of D for the same (N, k) values. The much greater stability, which is implied by Approximation (1), is extremely clear in this case.

The performance of Approximation (1) can, in fact, be improved by simple adjustments. Here we show a couple of examples in Fig. 2. Panel (a) shows a QQ-plot of the deviance, against the normal, in the case where the underlying distribution is shown in Fig. 1(a) – one that is very close to the boundary. We see the normal approximation is good but shows some skewness. Panel (b) shows a scaled χ^2-approximation, designed to correct for skewness in the sampling distribution, while panel (c) shows a symmetrised version of the deviance statistic, defined by randomising across upper and lower tails of the test statistic, which, if it is used for testing against a two tailed alternative, is a valid procedure. Both these simple corrections show excellent performance.

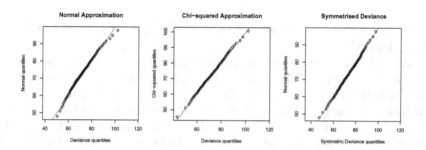

Fig. 2. Evaluation of the quality of k-asymptotic approximations

Having seen that the N-asymptotic approximation does not hold uniformly across the simplex, it is natural to investigate the uniformity of the k-asymptotic approximation given by (1). This approximation exploited a bivariate normal approximation to the distribution of $S^* = (N^*, D^*/2)^T$ and it is sufficient to check the normal approximation to any linear function of N^* and $D^*/2$. In particular, initially, we focus on the component D^*. We note that we can express $D^*/2$ via

$$D^*/2 = \sum_{\{0 \leq i \leq k : n_i^* > 0\}} n_i^* \log(n_i^*/\mu_i) = \Gamma^* + \Delta^* \qquad (2)$$

where

$$\Gamma^* := \sum_{i=0}^{k} \alpha_i n_i^* \text{ and } \Delta^* := \sum_{\{0 \leq i \leq k : n_i^* > 1\}} n_i^* \log n_i^* \geq 0$$

and $\alpha_i := -\log \mu_i$. It is insightful to consider the terms Γ^* and Δ^* separately.

3.2   Uniformity and Higher Moments

One way of assessing the quality of the k-asymptotic approximation for the distribution of Γ^* would be based on how well the moment generating function

of the (standardised) Γ^* is approximated by that of a (standard) normal. Writing
the moment generating function as

$$M_\gamma(t) = \exp\left(-\frac{E(\Gamma^*)}{\sqrt{Var(\Gamma^*)}}\right) \exp\left[\sum_{i=0}^{k}\left\{\sum_{h=1}^{\infty}\frac{(-1)^h}{h!}\mu_i(\log\mu_i)^h\left(\frac{t}{\sqrt{Var(\Gamma^*)}}\right)^h\right\}\right]$$

then, when analysing where the approximation would break down, it is natural
to make the third order term (i.e. the skewness)

$$\sum_{i=0}^{k}\mu_i(\log\mu_i)^3$$

as large as possible for fixed mean $E(\Gamma^*) = -\sum_{i=0}^{k}\mu_i\log(\mu_i)$ and $Var(\Gamma^*) = \sum_{i=0}^{k}\mu_i(\log\mu_i)^2$.

Solving this optimisation problem gives a distribution with three distinct values for μ_i. An example of this is shown in Fig. 3, where $k = 200$. Panels (b) and
(c) are histograms for a sample of 1000 values of W and D, respectively, drawn
from their distribution when $N = 30$. In this example, we see both the Wald
and deviance statistics are close to normal but with significant skewness which
disappears with a larger sample size. This is to be expected from the analysis
of [9,12] who look at the behaviour of deviance, when bounded away from the
boundary of the simplex, when both N and k tend to infinity together. In particular [9] shows the accuracy of this normal approximation improves without
limit when $N, k \to \infty$ with $N/k \to c > 0$. Symmetrising the deviance would
of course reduce this skewness, but in this example would hide the underlying
geometric structure and only works in the two tailed testing problem.

3.3   Uniformity and Discreteness

In fact the hardest cases for the normal approximation (1) to the distribution of
the deviance are in complementary parts of the simplex to the hardest cases from
the Wald statistic. For W, it is the boundary where there are problems, while
for (1) the worst place is the centre of the simplex, i.e. the uniform distribution.
The difficulties there are not due to large higher order moments, but rather to
discreteness.

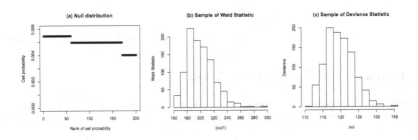

Fig. 3. Worst case solution for normality of Γ^*

In this analysis consider again decomposition (2). Note that Γ^* is completely degenerate here, while there are never any contributions to the Δ^* term from cells for which n_i is 0 or 1. However, for $k >> N$, we would expect that for all i, $n_i^* \in \{0, 1\}$, with high probability, hence, after conditioning on $N^* = N$ there is no variability in D – it has a completely degenerate (singular) distribution. In the general case all the variability comes from the cases where $n_i^* > 1$ and these events can have a very discrete distribution – so the approximation given by the continuous normal must be poor.

We illustrate this 'granular' behaviour in Fig. 4. Panel (a) shows the uniform distribution when $k = 200$, panel (b) displays 1000 realisations of D when $N = 30$. The discreteness of the distribution is very clear here, and is also illustrated in panel (c) which shows a QQ-plot of the sample against a normal distribution. Note that any given quantile is not far from the normal, but the discreteness of the underlying distribution means that not all quantiles can be attained. This may, or may not, be a problem in a goodness-of-fit testing situation.

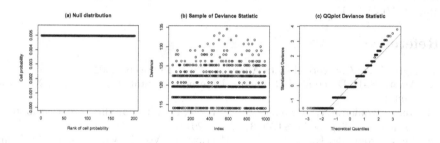

Fig. 4. Behaviour at the centre of the simplex, N=30

Again following the analysis of [9] this behaviour will disappear as N gets larger relative to k. This is shown in Fig. 5 where the N is now 60 – twice what it was in Fig. 4. The marked drop in granularity of panel (b) between Figs. 4 and 5 is due to the much greater variability in the maximum observed value of n_i^* as N increases. Clearly, for the distribution of any discrete random variable to be well approximated by a continuous one, it is necessary that it have a large number of support points, close together. The good news here is that, for the deviance, this condition appears also to be sufficient.

4 Discussion

Overall, we have seen that the deviance remains stable and eminently useable in high-dimensional, sparse contexts – of accelerating practical importance. Discreteness issues are rare, predictable and well-understood, while simple modifications are available to deal with any higher moment concerns, such as skewness. When using the deviance, computational information geometry can be used to gain insight into the power of the implicit likelihood ratio test, exploiting the

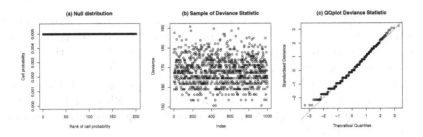

Fig. 5. Behaviour at the centre of the simplex, N=60

fact that D is constant on high-dimensional affine subspaces in the mean parameterisation, [6], while both its null and alternative approximating distributions depend only on a few low-order moments, inducing a pivotal foliation.

Acknowledgements. The authors acknowledge with gratitude the support of EPSRC grant EP/L010429/1.

References

1. Agresti, A.: Categorical Data Analysis. Wiley, Hoboken (2002)
2. Amari, S.-I.: Differential-Geometrical Methods in Statistics. Springer, New York (1985)
3. Anaya-Izquierdo, K., Critchley, F., Marriott, P.: When are first order asymptotics adequate? a diagnostic. STAT **3**, 17–22 (2014)
4. Anaya-Izquierdo, K., Critchley, F., Marriott, P., Vos, P.: Computational information geometry in statistics: foundations. In: Nielsen, F., Barbaresco, F. (eds.) GSI 2013. LNCS, vol. 8085, pp. 311–318. Springer, Heidelberg (2013)
5. Barndorff-Nielsen, O.E., Cox, D.R.: Inference and Asymptotics. Chapman & Hall, London (1994)
6. Critchley, F., Marriott, P.: Computational information geometry in statistics. Entropy **16**, 2454–2471 (2014)
7. Fienberg, S.E., Rinaldo, A.: Maximum likelihood estimation in log-linear models. Ann. Stat. **40**, 996–1023 (2012)
8. Geyer, C.J.: Likelihood inference in exponential families and directions of recession. Electron. J. Stat. **3**, 259–289 (2009)
9. Holst, L.: Asymptotic normality and efficiency for certain goodness-of-fit tests. Biometrika **59**, 137–145 (1972)
10. Lauritzen, S.L.: Graphical Models. Clarendon Press, Oxford (1996)
11. Liu, M., Vemuri, B.C., Amari, S.-I., Nielsen, F.: Shape retrieval using heirarchical total Bregman soft clustering. IEEE Trans. Pattern Anal. Mach. Intell. **34**, 2407–2419 (2012)
12. Morris, C.: Central limit theorems for multinomial sums. Ann. Stat. **3**, 165–188 (1975)
13. Nielsen, F., Nock, N.: Optimal interval clustering: application to Bregman clustering and statistical mixture learning. IEEE Trans. Pattern Anal. Mach. Intell. **21**(10), 1289–1292 (2014)
14. Steck, G.P.: Limit Theorems for Conditional Distributions. University of California Press, Berkeley (1957)

Computing Boundaries in Local Mixture Models

Vahed Maroufy[(⊠)] and Paul Marriott

Department of Statistics and Actuarial Science,
University of Waterloo, Waterloo N2L 3G1, Canada
vmaroufy@uwaterloo.ca

Abstract. Local mixture models give an inferentially tractable but still
flexible alternative to general mixture models. Their parameter space
naturally includes boundaries; near these the behaviour of the likelihood
is not standard. This paper shows how convex and differential geometries
help in characterising these boundaries. In particular the geometry of
polytopes, ruled and developable surfaces is exploited to develop efficient
inferential algorithms.

Keywords: Computing boundaries · Computational information geometry · Embedded manifolds · Local mixture models · Polytopes · Ruled
and developable surfaces

1   Introduction

Often, in statistical inference, the parameter space of a model includes a boundary, which can affect the maximum likelihood estimator (MLE) and its asymptotic properties. Important examples include the (extended) multinomial family,
logistic regression models, contingency tables, graphical models and log-linear
models, all of which are commonly used in statistical modelling, see [2]. In
[12,22,23], it is shown that the MLE in a log-linear model exists, if and only
if, the observed sufficient statistic lies in the interior of the marginal polyhedron
i.e. away from the boundary. The paper [14] studies the influence of the nonexistence of the MLE on asymptotic theory, confidence intervals and hypothesis
testing for a binomial, a logistic regression model, and a contingency table. Further, in [2] a diagnostic criterion is provided for the MLE which defines how
far it is required to be from the boundary so that first order asymptotics are
adequate.

Boundary computation is, in general, a hard problem, [13,14]. Although it is
insufficiently explored in statistics, there are numerous mathematical and computational results in other literatures. Their focus are on (i) approximating a
convex closed subspace by a polytope, see [5,7,10,17] and (ii) approximating a
polytope by a smooth manifold, see [6,15,16].

While the general problem of computing boundaries is difficult, in this paper
we show some new results about computing them for local mixture models
(LMM), introduced in [19] and studied further in [3]. The parameter space of

© Springer International Publishing Switzerland 2015
F. Nielsen and F. Barbaresco (Eds.): GSI 2015, LNCS 9389, pp. 577–585, 2015.
DOI: 10.1007/978-3-319-25040-3_62

a LMM includes two forms of boundary: the hard and soft. Here we consider a continuous and a discrete LMM: based on the normal and Poisson distributions respectively. We show here that the boundary can have both discrete and smooth aspects, and provide novel geometric methods for computing the boundaries.

Section 2 is a brief review of LMM's and their geometry, while Sect. 3 introduces some explicit, and new, results on the structure of the fibre of a local mixture in important examples and uses the classical geometric notions of ruled surfaces in the computations. Section 4 concludes with discussion and future directions.

2   Local Mixture Models

The theory of local mixture models is motivated by a number of different statistical modelling situations which share a common structure. Suppose that there is a baseline statistical model which describes the majority of the observed variation, but there remains appreciable residual variation that is not consistent with the baseline model. These situations include over-dispersion in binomial and Poisson regression models, frailty analysis in lifetime data analysis [4] and measurement errors in covariates in regression models [20]. Other applications include local influence analysis [9] and the analysis of predictive distributions [19].

The geometric complexity of the space of general mixture models means that undertaking inference in this class is a hard problem. It has issues of identification, singularity and multi-modality in the likelihood function, interpretability problems and non-standard asymptotic expansions.

The key identification and multi-modality problem comes from the general observation that if a set of densities $f(x; \theta)$ lies uniformly close to an low-dimensional -1-affine space – as defined by [1]– then all mixtures of that model would also lie close to that space. Hence the space of mixtures is much lower dimensional than might be expected. The local mixture model, [3,19], is designed to have the 'correct' dimension by restricting the class of mixing distributions to so-called localising distributions. This allows a much more tractable geometry and corresponding inference theory. The restriction often comes only at a small cost in modelling terms. The local mixture model is, in geometric terms, closely related to a fibre-bundle over the baseline model, and has the elegant information geometric properties, described formally in Theorem 1, that (i) inference on the 'interest parameters' of the baseline model only weakly depends on the values of the nuisance parameters of the fibres because of orthogonality, (ii) the log-likelihood on the fibre has only a single mode due to convexity (iii) the local mixture model is a higher order approximation to the actual mixture.

As defined in [19] a LMM is a union of -1-convex subsets of -1-affine subspaces of the set of densities, in the information geometry of Amari, [1]. Here -1 refers to the $\alpha = -1$ or mixture connection.

Definition 1. *Let S be a common sample space. The local mixture, of order k, of a regular exponential family $f(x; \mu)$ in its mean parameterization, μ, is defined as*

$$g(x; \lambda, \mu) = f(x; \mu) + \lambda_2 f^{(2)}(x; \mu) + \cdots + \lambda_k f^{(k)}(x; \mu), \quad \lambda \in \Lambda_\mu \subset \mathbb{R}^{k-1} \quad (1)$$

where $\lambda = (\lambda_2, \cdots, \lambda_k)$ *and* $f^{(j)}(x;\mu) = \frac{\partial^j f}{\partial \mu^j}(x;\mu)$. *Also* $q_j(x;\mu) := \frac{f^{(j)}(x;\mu)}{f(x;\mu)}$, *then for any fixed* μ,

$$\Lambda_\mu = \left\{ \lambda | 1 + \sum_{j=2}^{k} \lambda_j\, q_j(x;\mu) \geq 0, \forall x \in S \right\},$$

is a convex subspace obtained by intersection of half-spaces. Its boundary is called the hard boundary *and corresponds to a positivity condition on* $g(x;\lambda,\mu)$.

A local mixture model has a structure similar to that of a fibre bundle and for each fixed μ_0 the subfamily, $g(x;\lambda,\mu_0)$, is called a fibre – although more strictly it is a convex subset of the full fibre. The paper [3] shows that LMMs have the following excellent statistical properties.

Theorem 1. *(i) The set* $\{g(x;\lambda,\mu_0) - f(x;\mu_0)\}$ *is* -1*-flat and Fisher orthogonal to the score of* $f(x;\mu)$ *at* μ_0. *Thus* μ *and* λ *are orthogonal parameters.*
(ii) On each fibre the log-likelihood function is concave - though not necessarily strictly concave.
(iii) A continuous mixture model $\int f(x;\mu)\, dQ(\mu)$ *can be approximated by a LMM to an arbitrary order if* Q *satisfies the properties of a localizing distribution defined in [19].*

In such an approximation the parameter vector λ represents the mixing distribution Q through its moments; however, for some values of λ a LMM can have moments not attainable by a mixture model of the form $\int f(x;\mu)\, dQ(\mu)$. A true LMM, defined in [3], is a LMM which behaves similarly to a mixture model, in terms of a finite set of moments. For a true LMM, additional to hard boundary, there is another type of restricting boundary, called soft boundary, and characterized by following definition.

Definition 2. *For a density function* $f(x;\mu)$ *with* k *finite moments let,*

$$\mathcal{M}_k(f) := (E_f(X), E_f(X^2), \cdots, E_f(X^k)).$$

Then $g(x;\mu,\lambda)$, *defined in Definition 1, is called a* true local mixture, *if and only if, for each* μ *in a compact subset* I, $\mathcal{M}_k(g)$ *lies inside the convex hull of* $\{\mathcal{M}_r(f)|\mu \in I\}$.
The boundary of the convex hull is called the soft boundary.

Inferentially Model (1) might be used for marginal inference about μ where λ is treated as a nuisance parameter in, for example, random effect or frailty models, see [18,19]. The properties of Theorem 1 on the (μ,λ)-parameterization guarantees asymptotic independence of $\hat{\mu}$ and $\hat{\lambda}$ and simplifies determination of $(\hat{\mu},\hat{\lambda})$, [8]. Therefore, the profile likelihood method would be expected to be a promising approach for marginal inference about μ when λ is away from boundaries. This intuition is confirmed by simulation exercises. To use such an approach in general it is necessary that the analyst can compute the inferential effect of the boundary. The rest of this paper explores the geometric structure of the boundaries of LMMs and the computational consequences of such a structure.

3   Computing the Boundaries

In this section we compute the hard and soft boundaries for LMMs of order $k = 4$, as lower order LMMs have trivial boundaries and typically LMMs with $k > 4$ do not add greatly to modelling performance, see [21].

3.1   Hard Boundary for the LMM of Poisson Distribution

Consider the following LMM of the Poison probability mass function $p(x; \mu)$,

$$g(x; \mu, \lambda) = p(x; \mu) + \sum_{j=2}^{4} \lambda_j \, p^{(j)}(x; \mu), \qquad \lambda \in \Lambda_\mu \subset R^3. \qquad (2)$$

It is straightforward to show that

$$E_g(X) = E_p(X) = \mu, \qquad Var_g(X) = Var_p(X) + 2\lambda_2, \qquad (3)$$

illustrating that the λ parametrization of LMMs is tractable and intuitive as the model in Eq. (2) produces higher (lower) dispersion compared to $p(x; \mu)$. Furthermore, as shown in [3], the other parameters also have interpretable moment based meanings.

For model (2), the hard boundary is obtained by analysing half spaces defined, for fixed μ, by

$$S_x = \left\{ \lambda \,\middle|\, A_2(x)\,\lambda_2 + A_3(x)\lambda_3 + A_4(x)\,\lambda_4 + 1 \geq 0, \forall x \in \mathbb{Z}^+ \right\},$$

where $A_j(x)$'s are polynomials of x defined by Definition 1. The space Λ_μ, for fixed μ, will be the countable intersection of such half spaces over $x \in \{0, 1, \dots\}$ i.e., we can write $\Lambda_\mu = \bigcap_{x \in \mathbb{Z}^+} S_x$. In fact, as we show in Proposition 1, the space can be arbitrarily well approximated by a polytope. In this paper all proofs are omitted due to space constraints.

Proposition 1. *For a LMM of a Poisson distribution, for each μ, the space Λ_μ can be arbitrarily well approximated, as measured by volume for example, by a finite polytope.*

Figure 1 shows some issues related to this proposition. It shows two slices through the space Λ_μ by fixing a value of λ_2 (left panel) and λ_3 (right panel). The shaded polytope is a subset of Λ_μ in both cases. The lines are sets of the form

$$A_2(x)\,\lambda_2 + A_3(x)\lambda_3 + A_4(x)\,\lambda_4 + 1 = 0,$$

for different values of $x \in \{0, 1, 2, \cdots\}$, with solid lines being support lines and dashed lines representing redundant constraints. In $\mathbb{R}^3$ a finite number of such planes will define a polytope which is arbitrarily close to Λ_μ. A second feature, which is clear from Fig. 1, is that parts of the boundary look like they can be well approximated by a smooth curve, ([6,15,16]), which has the potential to simplify computational aspects of the problem.

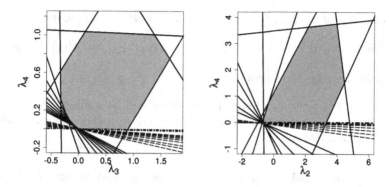

Fig. 1. Left: slice through $\lambda_2 = -0.1$; Right: slice through $\lambda_3 = 0.3$. Solid lines represent active and dashed lines redundant constraints. For our model $\lambda_4 > 0$ is a necessary condition for positivity.

3.2   Hard Boundary for the LMM of Normal

In the previous example the boundary was defined by a countable intersection of half-spaces. Now we look at an example, the LMM of normal distributions, where we have an uncountably infinite intersection of half spaces and we observe a smooth, manifold like boundary with lower dimensional sets of singularities. First, we need to review some classical differential geometry.

Ruled and Developable Surfaces

Definition 3. *A ruled surface is a surface generated by a smooth curve, $\alpha(x) \subset \mathbb{R}^3$, and a set of vectors, $\beta(x) \subset \mathbb{R}^3$, with following parameterization*

$$\Gamma(x, \gamma) = \alpha(x) + \gamma \cdot \beta(x), \quad x \in I \subset \mathbb{R}, \gamma \in \mathbb{R}^3.$$

If, in addition, at each x the three vectors $\beta(x)$, $\beta'(x)$ and $\alpha'(x)$ are coplanar then $\Gamma(x, \gamma)$ is called a developable surface ([11, p.188]).

Cylinders and cones are simple ruled surfaces which are also developable. Another way of constructing a developable surface is by finding the *envelope* of a one-parameter family of planes, see [24, Sects. 1, 2, 3, 4 and 5].

Definition 4. *Let $\lambda = (\lambda_2, \lambda_3, \lambda_4) \in \mathbb{R}^3$ and $a(x) = (a_1(x), a_2(x), a_3(x))$ where $d(x)$ and $a_j(x)$, $j = 1, 2, 3$ are differentiable functions of x. The family of planes, $\mathcal{A} = \{\lambda \in \mathbb{R}^3 |\ a(x) \cdot \lambda + d(x) = 0, x \in \mathbb{R}\}$, each determined by an $x \in \mathbb{R}$, is called a one-parameter infinite family of planes. Each element of the set*

$$\{\lambda \in \mathbb{R}^3 | a(x) \cdot \lambda + d(x) = 0, a'(x) \cdot \lambda + d'(x) = 0, x \in \mathbb{R}\}$$

is called a characteristic line of the surface at x and the union is called the envelope of the family.

An envelope of the set of characteristic lines is the set of points

$$\{\lambda \in \mathbb{R}^3 | a(x) \cdot \lambda + d(x) = 0, a'(x) \cdot \lambda + d'(x) = 0, a''(x) \cdot \lambda + d''(x) = 0, x \in \mathbb{R}\}$$

and is called the edge of recession.

Under general regularity conditions the envelope will be a ruled (often a developable) surface, and we will use such a construction in this paper to find the boundary of an LMM.

Application to LMMs. Consider the LMM of a normal distribution $N(\mu, \sigma)$, where $\sigma > 0$ is fixed and known, and for which, without loss of generality, we assume $\sigma = 1$. Let $y = x - \mu$, then

$$\Lambda_\mu = \{\lambda \, | (y^2 - 1)\lambda_2 + (y^3 - 3y)\lambda_3 + (y^4 - 6y^2 + 3)\lambda_4 + 1 \geq 0, \forall y \in \mathbb{R}\}. \quad (4)$$

is the intersection of infinite set of half-spaces in R^3.

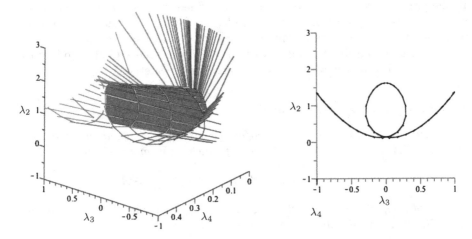

Fig. 2. Left: The hard boundary for the normal LMM (shaded) as a subset of a self intersecting ruled surface (unshaded); Right: slice through $\lambda_4 = 0.2$.

To understand the boundary of Λ_μ we first use Definition 4 to define the envelope of a one parameter set of planes in $\mathbb{R}^3$. These planes, in λ-space, are parameterised by $y \in \mathbb{R}$, and are the solutions of

$$(y^2 - 1)\lambda_2 + (y^3 - 3y)\lambda_3 + (y^4 - 6y^2 + 3)\lambda_4 + 1 = 0.$$

The envelope of this family forms a ruled surface, and can be thought of as a self-intersecting surface in $\mathbb{R}^3$. The surface partitions $\mathbb{R}^3$ into disconnected regions and one of these – the one containing the origin $(0, 0, 0)$ – is the set Λ_μ. Figure 2 shows the self-intersecting surface and the shaded region is the subset which is the boundary of Λ_μ.

While the boundary of Λ_μ will have large regions which are smooth, it also has singular lines and points. These are the self-intersection points of the envelope and it is at these points where the boundary fails to be an immersed manifold-but is still locally smooth. The general structure of the boundary is a non-smooth union of a finite number of smooth components.

3.3 Soft Boundary Calculations

The previous section looks at issues associated with the hard boundary calculations for LMMs. In this section we look at similar issues connected with computing soft boundaries, Definition 2, in moment spaces for true LMMs.

For visualization purposes, consider $k = 3$ and we use the normal example from the previous section. The moment maps are given by

$$\mathcal{M}_3(f) = (\mu, \mu^2 + \sigma^2, \mu^3 + 3\mu\sigma^2),$$
$$\mathcal{M}_3(g) = (\mu, \mu^2 + \sigma^2 + 2\lambda_2, \mu^3 + 3\mu\sigma^2 + 6\mu\lambda_2 + 6\lambda_3).$$

Suppose $I = [a, b]$, then $\mathcal{M}_3(f)$ defines a smooth space curve, $\varphi : [a, b] \to \mathbb{R}^3$. To construct the convex hull, denoted by $convh\{\mathcal{M}_3(f), \ \mu \in [a, b]\}$, all the lines between $\varphi(a)$ and $\varphi(\mu)$ and all the lines between $\varphi(\mu)$ and $\varphi(b)$, for $\mu \in [a, b]$, are required. Each of the two family of lines are attached to the curve and construct a surface in $\mathbb{R}^3$. Hence, we have two surfaces each formed by a smooth curve and a set of straight lines (Fig. 3, right). Thus we have the following two ruled surfaces,

$$\begin{cases} \gamma_a(\mu, u) = \varphi(\mu) + u\, L_a(\mu), \text{ surface } a, \\ \gamma_b(\mu, u) = \varphi(\mu) + u\, L_b(\mu), \text{ surface } b, \end{cases}$$

where $u \in [0, 1]$, and for each $\mu \in [a, b]$, $L_a(\mu)$ is the line connecting $\varphi(\mu)$ to $\varphi(a)$, and similarly $L_b(\mu)$ is the line between $\varphi(b)$ and $\varphi(\mu)$.

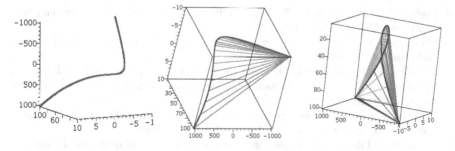

Fig. 3. Left: the 3-D curve $\varphi(\mu)$; Middle: the bounding ruled surface $\gamma_a(\mu, u)$; Right: the convex subspace restricted to soft boundary.

The soft boundary of the Poisson model can be characterized similarly.

4   Discussion and Future Work

This paper gives an introduction to some of the issues associated with computing the boundaries of local mixture models. Understanding these boundaries is important if we want to exploit the nice statistical properties of LMM, given by Theorem 1. The 'cost' associated with these properties is that boundaries will potentially play a role in inference giving, typically, non-standard results. The boundaries described in this paper have both discrete aspects, (i.e. the ability to be approximated by polytopes), and smooth aspects (i.e. regions where the boundaries are exactly or approximately smooth). It is an interesting and important open research area to develop computational information geometric tools which can efficiently deal with such geometric objects.

References

1. Amari, S.I.: Differential-Geometrical Methods in Statisitics: Lecture Notes in Statisitics, 2nd edn. Springer, New York (1990)
2. Anaya-Izquierdo, K., Critchley, F., Marriott, P.: When are first order asymptotics adequate? Diagn.Stat. **3**(1), 17–22 (2013)
3. Anaya-Izquierdo, K., Marriott, P.: Local mixture models of exponential families. Bernoulli **13**, 623–640 (2007)
4. Anaya-Izquierdo, K., Marriott, P.: Local mixtures of the exponential distribution. Ann. Inst. Stat. Math. **59**(1), 111–134 (2007)
5. Barvinok, A.: Thrifty approximations of convex bodies by polytopes. Int. Math. Res. Not. rnt078 (2013)
6. Batyrev, V.V.: Toric varieties and smooth convex approximations of a polytope. RIMS Kokyuroku **776**, 20 (1992)
7. Boroczky, K., Fodor, F.: Approximating 3-dimensional convex bodies by polytopes with a restricted number of edges. Contrib. Algebra Geom. **49**(1), 177–193 (2008)
8. Cox, D.R., Reid, N.: Parameter orthogonality and approximate conditional inference. J. R. Stat. Soc. **49**(1), 1–39 (1987)
9. Critchley, F., Marriott, P.: Data-informed influence analysis. Biometrika **91**(1), 125–140 (2004)
10. Dieker, A.B., Vempala, S.: Stochastic billiards for sampling form boundary of a convex set (2014). arXiv:1410.5775
11. Do Carmo, M.P.: Differential Geometry of Curves and Surfaces, vol. 2. Prentice-Hall, Englewood (1976)
12. Eriksson, N., Fienberg, S.E., Rinaldo, A., Sullivant, S.: Polyhedral conditions for the nonexistence of the mle for hierarchical log-linear models. J. Symbolic Comput. **41**, 222–233 (2006)
13. Fukuda, K.: From the zonotope construction to the minkowski addition of convex polytopes. J. Symbolic Comput. **38**(4), 1261–1272 (2004)
14. Geyer, C.J.: Likelihood inference in exponential familes and direction of recession. Electron. J. Stat. **3**, 259–289 (2009)
15. Ghomi, M.: Strictly convex submanifolds and hypersurfaces of positive curvature. J. Differ. Geom. **57**(2), 239–271 (2001)
16. Ghomi, M.: Optimal smoothing for convex polytopes. Bull. London Math. Soc. **36**(4), 483–492 (2004)

17. Lopez, M., Reisner, S.: Hausdorff approximation of 3d convex polytopes. Inf. Process. Lett. **107**(2), 76–82 (2008)
18. Maroufy, V., Marriott, P.: Generalizing the frailty assumptions in survival analysis. Preprint (2014)
19. Marriott, P.: On the local geometry of mixture models. Biometrika **89**, 77–93 (2002)
20. Marriott, P.: On the geometry of measurement error models. Biometrika **90**(3), 567–576 (2003)
21. Marriott, P.: Extending local mixture models. AISM **59**, 95–110 (2006)
22. Rinaldo, A., Fienberg, S.E., Zhou, Y.: On the geometry of discrete exponential families with application to exponential random graph models. Electro. J. Stat. **3**, 446–484 (2009)
23. Sontag, D., Jaakkola, T.: New outer bounds on the marginal polytopes. Adv. Nat. Inf. Process. **20**, 1393–1400 (2007)
24. Struik, D.J.: Lectures on Classical Differential Geometry. Dover Publications, New York (1988)

Approximating Covering and Minimum Enclosing Balls in Hyperbolic Geometry

Frank Nielsen[1] and Gaëtan Hadjeres[2]([⊠])

[1] École Polytechnique, Palaiseau, France
Frank.Nielsen@acm.org
[2] Sony Computer Science Laboratories, Paris, France
gaetan.hadjeres@etu.upmc.fr

Abstract. We generalize the $O(\frac{dn}{\epsilon^2})$-time $(1 + \epsilon)$-approximation algorithm for the smallest enclosing Euclidean ball [2,10] to point sets in hyperbolic geometry of *arbitrary* dimension. We guarantee a $O\left(1/\epsilon^2\right)$ convergence time by using a closed-form formula to compute the geodesic α-midpoint between any two points. Those results allow us to apply the hyperbolic k-center clustering for statistical location-scale families or for multivariate spherical normal distributions by using their Fisher information matrix as the underlying Riemannian hyperbolic metric.

1 Introduction and Prior Work

Given a metric space $(X, d_X(.,.))$, fitting the smallest enclosing ball of a point set $P = \{p_1, \ldots, p_N\}$ consists in finding the circumcenter $c \in X$ minimizing $\max_{p \in P} d_X(c, p)$. In practice, this non-differentiable problem is computationally intractable as the dimension increases, and has thus to be approximated. A simple algorithm was proposed in [2] for euclidean spaces and generalized in [7] to dually flat manifolds.

In this article, we consider the case of the hyperbolic Poincaré conformal ball model $\mathbb{B}^d$ which is a model of d-dimensional geometry [8]. Even if balls in this hyperbolic model can be interpreted as euclidean balls with *shifted* centers [9], we cannot transpose directly results obtained in the euclidean case to the hyperbolic one because the euclidean enclosing balls may intersect the boundary ball $\partial \mathbb{B}^d$ (and are thus not proper hyperbolic balls, see Fig. 1).

An exact solution for the hyperbolic Poincaré ball was proposed in [6] as a LP-type problem, but such an approach cannot be used in practice in high dimensions. A generic Riemannian approximation algorithm was studied by Arnaudon and Nielsen [1] but no explicit bounds were reported in hyperbolic geometry besides convergence, and moreover the heuristic assumed to be able to precisely cut geodesics and that step was approximated in [1].

We propose an intrinsic solution based on a closed-form formula making explicit the computation of geodesic α-midpoints (generalization of barycenter between two points) in hyperbolic geometry. We derive a $(1 + \epsilon)$-approximation algorithm for computing and enclosing ball in hyperbolic geometry in arbitrary

© Springer International Publishing Switzerland 2015
F. Nielsen and F. Barbaresco (Eds.): GSI 2015, LNCS 9389, pp. 586–594, 2015.
DOI: 10.1007/978-3-319-25040-3_63

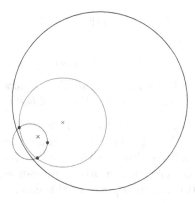

Fig. 1. Difference between euclidean MEB (in blue) and hyperbolic MEB (in red) for the set of blue points in hyperbolic Poincaré disk (in black). The red cross is the hyperbolic center of the red circle while the pink one is its euclidean center (Color figure online).

dimension in $O(\frac{dn}{\epsilon^2})$. This is all the more interesting from a machine learning perspective when dealing with data whose underlying geometry is hyperbolic. As an example, we illustrate our results on location-scale families or on multivariate spherical normal distributions. In the reminder, we assume the reader familiar with the basis of differential and Riemannian geometry, and recommend the textbook [8], otherwise.

The paper is organized as follows: Sect. 2 presents the exact computation of the α-midpoint between any arbitrary pair of points. Section 3 describes and analyzes the approximation algorithms for (i) fixed-radius covering balls and (ii) minimum enclosing balls. Section 4 presents the experimental results and discusses on k-center clustering applications.

2   Geodesic α-midpoints in the Hyperbolic Poincaré ball model

Let $\langle \cdot, \cdot \rangle$ and $\|x\| = \sqrt{\langle x, x \rangle}$ denote the usual scalar product and norm on the euclidean space $\mathbb{R}^d$. The Poincaré conformal ball model of dimension d is defined as the d-dimensional open unit ball $\mathbb{B}^d = \{x \in \mathbb{R}^d : \|x\| < 1\}$ together with the hyperbolic metric distance $\rho(.,.)$ given by:

$$\rho(p,q) = \operatorname{arcosh}\left(1 + \frac{2\|p-q\|^2}{(1 - \|p\|^2)(1 - \|q\|^2)}\right), \quad \forall p,q \in \mathbb{B}^d.$$

This distance induces on $\mathbb{B}^d$ a Riemannian structure.

Definition 1. *Let $p, q \in \mathbb{B}^d$ and $\gamma_{p,q}$ the unique geodesic joining p to q in the hyperbolic Poincaré model. For $\alpha \in [0,1]$, we define the α-midpoint, $p\#_\alpha q$,* between p and q as the point $m_\alpha \in \gamma_{p,q}([0,1]) \subset \mathbb{B}^d$ on the geodesic $\gamma_{p,q}$ such that

$$\rho(p, m_\alpha) = \alpha\rho(p, q).$$

Lemma 1. *For all $\alpha \in [0,1]$, we can compute the α-midpoint $p\#_\alpha q$ between two points p, q in the d-dimensional hyperbolic Poincaré ball model in constant time.*

Proof. We first consider the case where one of the point, say p, is equal to the origin $(0,\dots,0)$ of the unit ball. In this case, the only geodesic running through p and q is the straight euclidean line. As the distance ρ on the hyperbolic ball is invariant under rotation around the origin, we can assume without loss of generality that $q = (x_q, 0, 0, \dots, 0)$, $x_q \geq 0$. In this case, we have:

$$\rho(p, q) = \operatorname{arcosh}\left(1 + \frac{2\|q\|^2}{1 - \|q\|^2}\right) = \log\left(\frac{1 + \|q\|}{1 - \|q\|}\right) = \log\left(\frac{1 + x_q}{1 - x_q}\right), \quad (1)$$

using $\operatorname{arcosh}(x) = \log\left(x + \sqrt{x^2 - 1}\right)$. The α-midpoint $p\#_\alpha q$ has coordinates $(x_\alpha, 0, \dots, 0)$, $x_\alpha \geq 0$, which satisfies $\rho(p, p\#_\alpha q) = \alpha\rho(p, q)$. By (1), this is equivalent to solving, after exponentiating, $\frac{1+x_\alpha}{1-x_\alpha} = \left(\frac{1+x_q}{1-x_q}\right)^\alpha$. It follows that:

$$x_\alpha = \frac{c_{\alpha,q} - 1}{c_{\alpha,q} + 1}, \quad \text{where} \quad c_{\alpha,q} := e^{\alpha\rho(p,q)} \left(= \left(\frac{1 + x_q}{1 - x_q}\right)^\alpha\right). \quad (2)$$

For $p = (0, \dots, 0)$ and $q \neq p$ arbitrary, we have $p\#_\alpha q = \frac{\|x_\alpha\|}{\|q\|} q$, taking (2) as a definition for x_α with $c_{\alpha,q} = e^{\alpha\rho(p,q)}$.

Now, for arbitrary p and q, we first perform a *hyperbolic translation* T_{-p} of vector $-p$ to both p and q in order to resort to the preceding case, then compute the α-midpoint and translate it using the inverse hyperbolic translation T_p. The translation of $x \in \mathbb{B}^d$ by a vector $p \in \mathbb{B}^d$ of the hyperbolic Poincaré conformal ball model is given by (see [8], page 124 formula (4.5.5)):

$$T_p(x) = \frac{\left(1 - \|p\|^2\right) x + \left(\|x\|^2 + 2\langle x, p\rangle + 1\right) p}{\|p\|^2 \|x\|^2 + 2\langle x, p\rangle + 1}, \quad (3)$$

Since hyperbolic translations preserve the hyperbolic distance, using Definition 1, we have indeed

$$T_p\left(T_{-p}(p) \#_\alpha T_{-p}(q)\right) = p\#_\alpha q.$$

Note that those computations can be made *exactly* without numerical loss since they involve only rationals and square root operations, see [3].

3   Approximation Algorithms: Enclosing Balls and Minimum Enclosing Balls

In the following, we will denote by $P = \{p_1, \dots, p_N\}$ a set of N points of the hyperbolic Poincaré ball model. For $q \in \mathbb{B}^d$, we write $\rho(q, P) := \max_{p \in P} \rho(q, p)$.

Definition 2. *Let* $r > 0$. *A point* $c \in \mathbb{B}^d$ *is called the center of a hyperbolic enclosing ball of* P *of radius* r *(abbreviated* EHB (P, r)*) if*

$$\rho(c, P) \leq r. \tag{4}$$

If c *is the center of a* EHB *of minimal radius among all hyperbolic enclosing balls of* P, *then* c *is called the center of a minimum hyperbolic enclosing ball of* P *(abbreviated* MEHB (P)*).*

As the MEHB is unique [1], let R^* denote its radius and c^* its center. In practice, computing the MEHB (P) is intractable in high dimensions, we will focus on approximation algorithms by modifying (4).

Definition 3. *The point* $c \in \mathbb{B}^d$ *is the center of an* $(1 + \epsilon)$-*approximation of* EHB (P, r) *if* $\rho(c, P) \leq (1 + \epsilon) r$ *and the center of an* $(1 + \epsilon)$-*approximation of* MEHB (P) *if* $\rho(c, P) \leq (1 + \epsilon) R^*$.

3.1 A $(1 + \epsilon)$-approximation of an Enclosing Ball of Fixed Radius

We generalize the EHB (P, r) approximation introduced in [10], Algorithm 2 Sect. 3.1, to point clouds in the hyperbolic Poincaré ball model. Given P, $r > R^*$ and ϵ, this algorithm returns the center of a $(1 + \epsilon)$-approximation of EHB (P, r).

Algorithm 1. $(1 + \epsilon)$-approximation of EHB (P, r)

1: $c_0 := p_1$
2: $t := 0$
3: **while** $\exists p \in P$ such that $p \notin B(c_t, (1 + \epsilon) r)$ **do**
4:     let $p \in P$ be such a point
5:     $\alpha := \frac{\rho(c_t, p) - r}{\rho(c_t, p)}$
6:     $c_{t+1} := c_t \# _\alpha p$
7:     $t := t + 1$
8: **end while**
9: **return** c_t

As in [1] or [7], we took into consideration the fact that this geometry is not euclidean. The update move (step 5 and 6) consists in taking a point c_{t+1} on the geodesic from c_t to p such that the ball $B(c_{t+1}, r)$ "touches" p (i.e. such that $\rho(c_{t+1}, p) = r$).

Proposition 1. *Algorithm 1 returns the center of an* $(1 + \epsilon)$-*approximation of* EHB (P, r) *in* $O(1/\epsilon^2)$ *iterations (exactly less than* $4/\epsilon^2$ *iterations).*

Proof. Let $\rho_t := \rho(c_t, c^*)$. Figure 2 illustrates the update of c_t, straight lines represent geodesic between points. From step 5 and 6, we have $\rho(c_{t+1}, p_t) = r$ which implies $\rho(c_{t+1}, c_t) > \epsilon r$. Since $B(c^*, R^*)$ is a MEHB, $\rho(c^*, p_t) \leq R^*$. Denote the angle $\angle c^* c_{t+1} c_t$ by θ and the distance $\rho(c^*, p_t)$ by r'. The hyperbolic law of cosines gives: $\cos(\theta)\operatorname{sh}(\rho_{t+1})\operatorname{sh}(r) = \operatorname{ch}(\rho_{t+1})\operatorname{ch}(r) - \operatorname{ch}(r')$, so that $\cos(\theta) \geq 0$ since $\operatorname{ch}(r') \leq \operatorname{ch}(r)$ and $\operatorname{ch}(\rho_{t+1}) \geq 1$.

Let θ' be the angle $\angle c_t c_{t+1} c^*$, it follows that $\cos(\theta') \leq 0$. Let h be the distance between c_t and c_{t+1}, the hyperbolic law of cosines gives $0 \leq \cos(\theta')\operatorname{sh}(h)\operatorname{sh}(\rho_{t+1}) = \operatorname{ch}(h)\operatorname{ch}(\rho_{t+1}) - \operatorname{ch}(\rho_t)$. Thus $\operatorname{ch}(\rho_t) \geq \operatorname{ch}(h)\operatorname{ch}(\rho_{t+1})$. After T iterations, we have the following inequality:

$$\operatorname{ch}(\rho_1) \geq \frac{\operatorname{ch}(\rho_1)}{\operatorname{ch}(\rho_T)} \geq \operatorname{ch}(h)^T \geq \operatorname{ch}(\epsilon r)^T, \tag{5}$$

which proves that the algorithm converges since $\operatorname{ch}(\epsilon r) > 1$. We can rewrite (5) as:

$$T \leq \frac{\log(\operatorname{ch}(\rho_1))}{\log(\operatorname{ch}(\epsilon r))} \leq \frac{\log(\operatorname{ch}(2r))}{\log(\operatorname{ch}(\epsilon r))} \leq \frac{4}{\epsilon^2} \tag{6}$$

using the fact that $\rho_1 \leq 2r$ and that $f := r \mapsto \frac{\log(\operatorname{ch}(2r))}{\log(\operatorname{ch}(\epsilon r))}$ is a decreasing function from $[0, +\infty[$ to $]0, +\infty[$ with $\lim_{r=0+} f(r) = 4/\epsilon^2$.

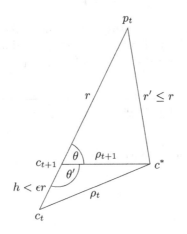

Fig. 2. Update of c_t

We now show how far from the real center c^* of the MEHB is the center of an $(1 + \epsilon)$-approximation of EHB (P, r). We need the following lemma which generalizes Lemma 2 from [5]. For this, denote by $\langle ., .\rangle_p$ the scalar product given by the Riemannian metric on the tangent space $T_p \mathbb{B}^d$ in $p \in \mathbb{B}^d$ and by $\exp_p : T_p \mathbb{B}^d \to \mathbb{B}^d$ the exponential map.

Lemma 2. *For every tangent vector $v \in T_{c^*}\mathbb{B}^d$, there exists $p \in P \cap \mathcal{H}_v$ such that $\rho(c^*, p) = R^*$ where we denoted by*

$$\mathcal{H}_v = \{\exp_{c^*}(u) \in \mathbb{B}^d, \quad u \in T_{c^*}\mathbb{B}^d, \quad \langle v, u\rangle_{c^*} \geq 0\} \tag{7}$$

the points in $\mathbb{B}^d$ obtained by following geodesics whose tangent vector at point c^* lie in the half-space defined by v.

Proof. Assume it exists $v \in T_{c^*} \mathbb{B}^d$ such that for all $p \in P \cap \mathcal{H}_v$, $\rho(c^*, p) < R^*$. We will show that "moving" c^* in the direction $-v$ results in a new center c whose distance to P is strictly less than R^*, contradicting the fact that c^* is the center of MEHB (P).

For each point $q \in P$ not in $\mathcal{H}_v$ we have

$$\frac{d}{dt} \rho\left(\exp_{c^*}(-tv), q\right)\bigg|_{t=0} = \left\langle \frac{-\exp_{c^*}^{-1}(q)}{\rho(c^*, q)}, -v \right\rangle_{c^*} < 0 \tag{8}$$

by (7). So we can find $t > 0$ small enough to obtain $\rho\left(\exp_{c^*}(-tv), p\right) < R^*$ for all $p \in P$ since there is only a finite number of points in P.

Proposition 2. *Let c be the center of an $(1 + \epsilon)$-approximation of EHB (P, r). We have the following inequality:*

$$\rho(c, c^*) \le \operatorname{arcosh}\left(\frac{\operatorname{ch}((1 + \epsilon) r)}{\operatorname{ch}(R^*)}\right) \tag{9}$$

where c^* and R^* are respectively the center and radius of the MEHB (P).

Proof. We can assume that $c \ne c^*$, otherwise (9) is true. Let $d := \rho(c, d^*)$. Consider the geodesic $\gamma : [0, d] \rightarrow M$ from c c^*. By applying the preceding lemma with $v := \dot{\gamma}(d)$, we obtain a point $p \in P \cap \mathcal{H}_v$ such that $\rho(c^*, p) = R^*$. By definition, the angle $\theta := \angle cc^*p$ is obtuse.

We name h the distance $\rho(c, p)$ and apply the hyperbolic law of cosines to obtain $0 \ge \cos(\theta) \operatorname{sh}(d) \operatorname{sh}(R^*) = \operatorname{ch}(d) \operatorname{ch}(R^*) - \operatorname{ch}(h)$. Since c is the center of an $(1+\epsilon)$-approximation de EHB (P, r), $h \le (1 + \epsilon) r$. we deduce $\operatorname{ch}((1 + \epsilon) r) \ge \operatorname{ch}(h) \ge \operatorname{ch}(d) \operatorname{ch}(R^*)$ from which we derive (9).

3.2   A $(1 + \epsilon + \epsilon^2/4)$-approximation of MEHB (P)

We can use the previous results to derive an algorithm computing the MEHB (P) in hyperbolic geometry of arbitrary dimension. The proposed algorithm (Algorithm 2) consists in a dichotomic search of the radius of the MEHB (P). Indeed, we can discard a radius smaller than R^* using (6) and use inequality (9) in order to obtain a tighter bound.

Proposition 3. *Algorithm 2 returns the center of an $(1+\epsilon+\frac{\epsilon^2}{4})$-approximation of MEHB (P) in $O\left(\frac{1}{\epsilon^2} \log\left(\frac{1}{\epsilon}\right)\right)$ iterations.*

Proof. In Algorithm 2, as $\rho(p_1, P) > R^*$, the first call to Algorithm 1 returns an $(1 + \epsilon/2)$-approximation of EHB $(\rho(p_1, P), P)$. The fact that c is the center of an $(1 + \epsilon/2)$ approximation of EHB $(P, r_{\max})$ becomes a loop invariant. We also ensure that at each loop

$$r_{\min} \le R^* \le r_{\max}, \tag{10}$$

so that the maximum number T of iterations of Alg1 $(P, r, \epsilon/2)$ can be bounded by

$$T \le \frac{\log(\operatorname{ch}(\rho(c_1, c^*)))}{\log(\operatorname{ch}(\epsilon r/2))} \le \frac{\log(\operatorname{ch}(1 + \epsilon/2) r) - \log(\operatorname{ch}(r_{\min}))}{\log(\operatorname{ch}(r\epsilon/2))} \tag{11}$$

Algorithm 2. $(1 + \epsilon)$-approximation of EHB (P)

1: $c := p_1$
2: $r_{\max} := \rho\,(c, P);\ r_{\min} = \frac{r_{\max}}{2};\ t_{\max} := +\infty$
3: $r := r_{\max};$
4: **repeat**
5:     $c_{\text{temp}} := \text{Alg1}\left(P, r, \frac{\epsilon}{2}\right),$ interrupt if $t > t_{\max}$ in Alg1
6:     **if** call of Alg1 has been interrupted **then**
7:         $r_{\min} := r$
8:     **else**
9:         $r_{\max} := r\ ;\ c := c_{\text{temp}}$
10:    **end if**
11:    $dr := \frac{r_{\max} - r_{\min}}{2}\ ;\ r := r_{\min} + dr\ ;\ t_{\max} := \frac{\log(\text{ch}(1 + \epsilon/2)r) - \log(\text{ch}(r_{\min}))}{\log(\text{ch}(r\epsilon/2))}$
12: **until** $2dr < r_{\min}\frac{\epsilon}{2}$
13: **return** c

---

using (9) and the left side of (6) and (10). At the end of the repeat-until loop, we know that $r_{\max} \leq R^* + dr$ and that c is the center of an $(1 + \epsilon/2)$ approximation of EHB $(P, r_{\max})$. So

$$\rho\,(c, P) \leq \left(1 + \frac{\epsilon}{2}\right) r_{\max} \leq \left(1 + \frac{\epsilon}{2}\right)\left(R^* + r_{\min}\frac{\epsilon}{2}\right) \leq \left(1 + \epsilon + \frac{\epsilon^2}{4}\right) R^*. \quad (12)$$

This approximation is obtained in precisely $O\left(\frac{N}{\epsilon^2}\log\left(\frac{1}{\epsilon}\right)\right)$ since after T iterations of the main loop, $dr \approx \frac{R^*}{2^T}$.

4   Experimental Results

4.1   Performance

To evaluate the performance of Algorithm 2, we computed MEHB centers for a point cloud of $N = 200$ points for different values of the dimension d and the precision parameter ϵ. For each test, the point cloud was sampled uniformly (euclidean sampling) in the unit ball of dimension d. In order to check the relevance of our theoretical bounds, we plotted in Fig. 3 the *average number* of α-midpoints calculations and the mean execution time as a function of ϵ for different values of d. We evaluated convergence comparing the returned values of c to a value c^* computed with high precision. The algorithms have been implemented in Java using the arbitrary-precision arithmetic library Apfloat.

4.2   One-Class Clustering in Some Subfamilies of Multivariate Distributions

One-class clustering consists, given a set P of points, to sum up the information contained in P while minimizing a measure of distortion. In our case, we associate to a point set P a point c minimizing $\rho\,(c, P)$, i.e. the center

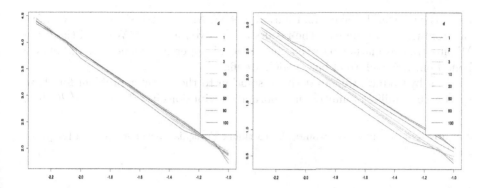

Fig. 3. Number of α-midpoint calculations (left) as a function of ϵ and execution time (right) as a function of ϵ, both in logarithmic scale for different values of d. We observe that the number of iterations does not depend on d, and that the running time is approximately $O(\frac{dn}{\epsilon^2})$ (vertical translation in logarithmic scale).

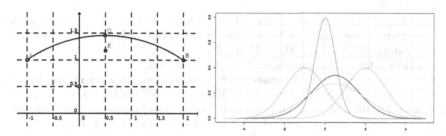

Fig. 4. (Best viewed in color). Graphical representation of the center of the MEHB, in the (μ, σ) superior half-plane (left), by showing corresponding probability density functions (right). In red (point E) is represented the center of MEHB (A,B,C). In pink (point D) is the 1/2-midpoint between A and B. The geodesic joining A to B is also displayed (Color figure online)

of the MEHB (P). This is particularly relevant when applied to parameterizations of subfamilies of multivariate normal distributions. Indeed, a sufficiently smooth family of probability distributions can be seen as a statistical manifold (a Riemannian manifold whose metric is given by the Fisher information matrix, see [4]). As proved in [4],

- the family $\mathcal{N}(\mu, \sigma^2 I_d)$ of d-variate normal distributions with scalar covariance matrix (I_d is the $d \times d$ identity matrix)
- the family $\mathcal{N}(\mu, \mathrm{diag}\left(\sigma_1^2, \ldots, \sigma_n^2\right))$ of n-variate normal distributions with diagonal covariance matrix
- the family $\mathcal{N}(\mu_0, \Sigma)$ of d-variate normal distributions with fixed mean μ_0 and arbitrary positive definite covariance matrix Σ

all induce a hyperbolic metric on their respective parameter spaces. We can thus apply Algorithm 2 to those subfamilies in order to perform one-class clustering using their natural Fisher information metric. An example showing how different from the euclidean case the results are is given in Fig. 4. We used the usual Möbius transformation between the Poincaré upper half-plane and the hyperbolic Poincaré conformal ball model, see [6].

As a byproduct, we can derive a solution to the k-center problem for those specific subfamilies of multivariate normal distributions in $2^{O(k \log k \log(1/\epsilon)/\epsilon^2)} dn$.

Acknowledgements. We would like to thank François Pachet for his kind support and advice.

References

1. Arnaudon, M., Nielsen, F.: On approximating the riemannian 1-center. Comput. Geom. **46**(1), 93–104 (2013)
2. Badoiu, M., Clarkson, K.L.: Optimal core-sets for balls. Comput. Geom. **40**(1), 14–22 (2008)
3. Burnikel, C., Fleischer, R., Mehlhorn, K., Schirra, S.: Efficient exact geometric computation made easy. In: Proceedings of the Fifteenth Annual Symposium on Computational Geometry, Miami Beach, Florida, USA, 13–16 June 1999, pp. 341–350 (1999). http://doi.acm.org/10.1145/304893.304988
4. Costa, S.I., Santos, S.A., Strapasson, J.E.: Fisher information matrix and hyperbolic geometry. In: Information Theory Workshop, 2005 IEEE, p. 3. IEEE (2005)
5. Kumar, P., Mitchell, J.S., Yildirim, E.A.: Approximate minimum enclosing balls in high dimensions using core-sets. J. Exp. Algorithmics (JEA) **8**, 1–1 (2003)
6. Nielsen, F., Nock, R.: Hyperbolic Voronoi diagrams made easy. In: 2010 International Conference on Computational Science and Its Applications (ICCSA), pp. 74–80, IEEE (2010)
7. Nock, R., Nielsen, F.: Fitting the smallest enclosing bregman ball. In: Gama, J., Camacho, R., Brazdil, P.B., Jorge, A.M., Torgo, L. (eds.) ECML 2005. LNCS (LNAI), vol. 3720, pp. 649–656. Springer, Heidelberg (2005)
8. Ratcliffe, J.: Foundations of Hyperbolic Manifolds, vol. 149. Springer Science & Business Media, New York (2006)
9. Tanuma, T., Imai, H., Moriyama, S.: Revisiting hyperbolic Voronoi diagrams from theoretical, applied and generalized viewpoints. In: 2010 International Symposium on Voronoi Diagrams in Science and Engineering (ISVD), pp. 23–32. IEEE (2010)
10. Tsang, I.W., Kocsor, A., Kwok, J.T.: Simpler core vector machines with enclosing balls. In: Proceedings of the 24th International Conference on Machine Learning, pp. 911–918. ACM (2007)

From Euclidean to Riemannian Means: Information Geometry for SSVEP Classification

Emmanuel K. Kalunga[1,2]($\boxtimes$), Sylvain Chevallier[2], Quentin Barthélemy[3], Karim Djouani[1], Yskandar Hamam[1], and Eric Monacelli[2]

[1] Department of Electrical Engineering/F'SATI, Tshwane University of Technology, Pretoria 0001, South Africa
emmanuelkalunga.k@gmail.com, djouanik@tut.ac.za, yskandar@hamam.ws
[2] Laboratoire d'Ingénierie des Systèmes de Versailles, Université de Versailles Saint-Quentin, 78140 Velizy, France
{sylvain.chevallier,eric.monacelli}@uvsq.fr
[3] Mensia Technologies, ICM, Hôpital de la Pitié-Salpêtrière, 75013 Paris, France
qb@mensiatech.com

Abstract. Brain Computer Interfaces (BCI) based on electroencephalography (EEG) rely on multichannel brain signal processing. Most of the state-of-the-art approaches deal with covariance matrices, and indeed Riemannian geometry has provided a substantial framework for developing new algorithms. Most notably, a straightforward algorithm such as Minimum Distance to Mean yields competitive results when applied with a Riemannian distance. This applicative contribution aims at assessing the impact of several distances on real EEG dataset, as the invariances embedded in those distances have an influence on the classification accuracy. Euclidean and Riemannian distances and means are compared both in term of quality of results and of computational load.

Keywords: Information geometry · Riemannian means · Brain-Computer Interfaces · Steady State Visually Evoked Potentials

1 Introduction

Brain-Computer Interfaces (BCI) allow interaction with a computer or a machine without relying on the user's motor capabilities. In rehabilitation and assistive technologies, BCI offer promising solutions to compensate for physical disabilities. To record brain signals in BCI systems, the most common choice is to rely on electroencephalography (EEG) [15], as the recording systems are smaller and less expensive than other brain imaging technologies (such as MEG or fMRI). BCI systems rely on different brain signals, such as event-related desynchronization or evoked potentials. The former is observed in the premotor cortex when the subject imagines moving some part of his own body (also known as Motor Imagery paradigm) and the latter qualifies the brain response to a specific sensory stimulation, usually visual or auditory. This contribution focuses on

© Springer International Publishing Switzerland 2015
F. Nielsen and F. Barbaresco (Eds.): GSI 2015, LNCS 9389, pp. 595–604, 2015.
DOI: 10.1007/978-3-319-25040-3_64

Steady-State Visually Evoked Potentials (SSVEP), which are potentials emerging when a subject concentrates his attention on a stimulus blinking at a given frequency. Shortly after the user concentrates on this stimulus, brain waves in visual cortex could be observed with matching frequencies. To date, BCI still faces challenges and a major limitation is the EEG poor spatial resolution. This limitation is due to the volume conductance effect [15], as the skull bones act as a non-linear low pass filter, mixing the brain source signals and thus reducing the signal-to-noise ratio.

Consequently, spatial filtering methods are used, such as xDAWN [17], Independent Component Analysis (ICA) [20], Common Spatial Pattern (CSP) [10] and Canonical Correlation Analysis (CCA) [12]. Spatial filters, obtained by diagonalization of data covariance matrices, enhance the differences between variances of signals of different classes/tasks. They are efficient on clean datasets obtained from strongly constrained environment. However they are sensitive to artifacts and outliers [13,19]. Working directly on covariance matrices is advantageous: it simplifies the whole BCI system [21], avoiding the alignment of two learning steps (spatial filters and classifiers) that might lead to overfitting. Covariance matrices being Symmetric and Positive-Definite (SPD), they are best handled by tools provided by Riemannian geometry. Classification in the space of SPD matrices eliminates the need of spatial filters and improves the system robustness [5,7,21].

A classification technique referred to as *minimum distance to Riemannian mean* (MDRM) has been recently introduced to EEG classification [5]. It entirely relies on covariance matrices and the fact that they belong to the manifold of SPD matrices. New EEG trials are assigned to the class whose average covariance matrix is the closest to the trial covariance matrix according to the affine-invariant Riemannian metric [14]. It is a simple, yet robust classification scheme outperforming complex and highly parametrized state-of-the-art classifiers. The limitations of using Euclidean metrics in the computation of distances between SPD matrices and their means have been demonstrated [3]. Using information geometry, a number of Riemannian distances have been developed and appropriately used on SPD matrices [1,3]. The present work applies some of these distances to SSVEP data, providing a practical analysis and a comparison with Euclidean distance.

Moreover, most applications of Riemannian geometry to BCI are thus far focusing only on Motor Imagery (MI) paradigm. Riemannian BCI is well suited for MI experiment as the spatial information linked with synchronization are directly embedded in covariance matrices obtained from multichannel recordings. However, for BCI that rely on evoked potential such as SSVEP or event-related potential, as P300, frequency or temporal information are needed. In [7], the authors propose a rearrangement of the covariance matrices that embed the timing or the frequency information, thus allowing a direct application of the Riemannian framework. This contribution relies on this rearrangement to apply MDRM on covariance matrices of SSVEP signals. The signals are recorded in an application of assistive robotics where an SSVEP-based BCI is used in tandem with a 3D touchless interface based on IR-sensors as a multimodal system to control an arm exoskeleton [11].

The paper is organized as follows: Sect. 2 describes the framework for the classification of covariance matrices. The distances and means considered for this study are presented. In Sect. 3, the classification results obtained on real EEG dataset are presented and discussed. Section 4 concludes this paper.

2   Classification of Covariance Matrices for SSVEP

A SSVEP classifier based on covariance matrices is presented. The computation of means of training covariance matrices is crucial to the classifier performance.

2.1   Means for Covariance Matrices

In the following, we will consider covariance matrices belonging to the manifold $\mathcal{M}_C$ of the $C \times C$ symmetric positive definite matrices, defined as:

$$\mathcal{M}_C = \left\{ \Sigma \in \mathbb{R}^{C \times C} : \Sigma = \Sigma^{\mathsf{T}} \text{ and } x^{\mathsf{T}} \Sigma x > 0, \forall x \in \mathbb{R}^C \backslash 0 \right\}.$$

Given a set of covariance matrices $\{\Sigma_i\}_{i=1,\dots,I}$, we consider the mean matrix $\bar{\Sigma}$ of the set, which is a covariance matrix that minimizes the sum of the squared distances to matrices Σ_i:

$$\bar{\Sigma} = \mu(\Sigma_1, \dots, \Sigma_I) = \arg \min_{\Sigma \in \mathcal{M}_C} \sum_{i=1}^{I} d^m(\Sigma_i, \Sigma), \tag{1}$$

where $m = 1$ when $d(\cdot, \cdot)$ is a divergence (i.e. a generalization of squared distance), and $m = 2$ when $d(\cdot, \cdot)$ is a distance.

From Eq. (1), several means can be defined and those considered in this study are indicated in Table 1. We consider the Euclidean distance d_{E}, as a baseline, yielding the arithmetic mean. The first considered Riemannian distance is the Log-Euclidean d_{LE} distance. Its mean is expressed explicitly [3]. The second is the Affine-Invariant d_{AI} [14]. Unlike the d_{LE}, it does not have an explicit expression for the mean. It could be efficiently computed with the gradient-based iterative algorithm proposed in [8]. The two last distances considered in this study are the log-determinant α-divergence [6] and the Bhattacharyya distance [16], the later being a special case of the former with $\alpha = 0$. Since the α-divergence is not symmetrical, its right version is used in this work [6].

2.2   Minimum Distance to Mean Classifier for SSVEP

The considered classifier is referred to as Minimum Distance to Mean (MDM), and is inspired from [5] where it is limited to Riemannian mean. Covariance matrices of EEG trials are classified based on their distance to the centers of the classes, equal to means. To embed frequency information in the covariance

Table 1. Distances, divergences and means considered in the experimental study.

	Distance/Divergence	Mean	References
Euclidean	$d_E(\Sigma_1, \Sigma_2) = \|\Sigma_1 - \Sigma_2\|_F$	$\bar{\Sigma}_E = \frac{1}{I}\sum_{i=1}^I \Sigma_i$	
Log-Euclidean	$d_{LE}(\Sigma_1, \Sigma_2) = \|\log(\Sigma_1) - \log(\Sigma_2)\|_F$	$\bar{\Sigma}_{LE} = \exp\left(\sum_{i=1}^I \log(\Sigma_i)\right)$	[2,3]
Affine-invariant	$d_{AI}(\Sigma_1, \Sigma_2) = \|\log(\Sigma_1^{-1}\Sigma_2)\|_F$	Algorithm 3 in [8]	[8,14]
α-divergence[a]	$d_{\alpha D}(\Sigma_1, \Sigma_2) = \frac{4}{1-\alpha^2}\log\dfrac{\det(\frac{1-\alpha}{2}\Sigma_1+\frac{1+\alpha}{2}\Sigma_2)}{\det(\Sigma_1)^{\frac{1-\alpha}{2}}\det(\Sigma_2)^{\frac{1+\alpha}{2}}}$ $-1<\alpha<1$	Algorithm 1 in [6]	[6]
Bhattacharyya	$d_B(\Sigma_1, \Sigma_2) = \left(\log\dfrac{\det\frac{1}{2}(\Sigma_1+\Sigma_2)}{(\det(\Sigma_1)\det(\Sigma_2))^{1/2}}\right)^{1/2}$	Algorithm 1 in [6]	[6,16]

[a] For $\alpha = -1$, the log-determinant α-divergence is defined as: $\mathrm{tr}(\Sigma_1^{-1}\Sigma_2 - I) - \log\det(\Sigma_1^{-1}\Sigma_2)$, and for $\alpha = 1$: $\mathrm{tr}(\Sigma_2^{-1}\Sigma_1 - I) - \log\det(\Sigma_2^{-1}\Sigma_1)$ [6]

matrices, we use a construction of matrices proposed in [7]. Let $X \in \mathbb{R}^{C \times N}$ be an EEG trial measured on C channels and N samples in a SSVEP experiment with F stimulus blinking at different frequencies. The covariance matrices are estimated from a modified version of the input signal X:

$$X \in \mathbb{R}^{C \times N} \rightarrow \begin{bmatrix} X_{\mathrm{freq}_1} \\ \vdots \\ X_{\mathrm{freq}_F} \end{bmatrix} \in \mathbb{R}^{FC \times N}, \tag{2}$$

where X_{freq_f} is the input signal X band-pass filtered around frequency freq_f, $f = 1, \ldots, F$. Henceforth, all EEG signals will be considered as filtered and modified by Eq. (2). The associated covariance matrix $\Sigma \in \mathcal{M}_{FC}$ is estimated using the Schäfer shrinkage estimator [18].

For SSVEP classification, $K = F + 1$ classes are considered: one class for each target frequency, and one for the resting state. As described in Algorithm 1, from I labelled training trials $\{X_i\}_{i=1}^I$ recorded per subject, K centers of classes $\bar{\Sigma}^{(k)}$ are estimated (step 3). In this step, outlier matrices are removed to have a reliable mean estimation, using an offline *Riemannian potato* [4]. A new unlabeled test trial Y is predicted to belong to the class whose mean $\bar{\Sigma}^{(k)}$ is the closest to the trial covariance matrix, w.r.t. one of the distances from Table 1 (step 5). Remark that test trial has to be finished before being classified: in this paper, there is no early classification.

3 Experimental Results

This section presents experimental results obtained applying Euclidean and Riemannian distances in SSVEP classification task. The first part of this section describes the data used and the second part provides the assessment of the classification for the considered distances and divergences.

3.1 SSVEP Dataset

The experimental study is conducted on multichannel EEG signals recorded during an SSVEP-based BCI experiment [11]. EEG are recorded on $C = 8$

Algorithm 1. Minimum Distance to Mean Classifier

Inputs: $X_i \in \mathbb{R}^{FC \times N}$, for $i = 1, \ldots, I$, a set of labelled EEG trials.
Inputs: $\mathcal{I}(k)$, a set of indices of trials belonging to class k.
Input: $Y \in \mathbb{R}^{FC \times N}$, an unlabeled test EEG trial.
Output: k^*, the predicted label of Y.
1: Compute covariance matrices Σ_i of X_i
2: **for** $k = 1$ **to** K **do**
3:    Compute center of class : $\bar{\Sigma}^{(k)} = \mu(\Sigma_i : i \in \mathcal{I}(k))$
4: **end**
5: Compute covariance matrix Σ of Y, and classify it : $k^* = \arg\min_k d(\Sigma, \bar{\Sigma}^{(k)})$
6: **return** k^*

---

channels (i.e. Oz, O1, O2, PO3, POz, PO7, PO8, PO4) from 12 subjects. The subjects are presented with $F = 3$ visual target stimuli blinking respectively at 13 Hz, 17 Hz and 21 Hz. It is a $K = 4$ classes setup combining $F = 3$ stimulus classes and one resting class (no-SSVEP). In a session, 32 trials are recorded: 8 for each visual stimulus and 8 for the resting class. The number of sessions recorded per subject varies from 2 to 5. For each subject, a test set is made of 32 trials while the remaining trials (which might vary from 32 to 128) make up for the training set.

3.2   Results and Discussion

The MDM classifier is simple. Once the covariance matrices have been estimated, the only major calculations involved are the mean and distance computations. The covariance matrices obtained from SSVEP data extended with Eq. (2) have interesting features, allowing the discrimination between signals of identical sources but with different frequencies. Figure 1 shows the K classes mean covariance matrices $\bar{\Sigma}^{(k)}$ from subjects with the highest (a) and lowest (b) classification accuracies. The three 8×8 diagonal blocks hold the covariance matrices of the $F = 3$ target frequencies. Inter-frequencies covariances blocks are almost null. In each mean covariance matrix, the block holding the covariance of the target frequency has the largest values. For the resting class, all F blocks tend to have similar and small values. These features are more visible in the subject with the highest classification accuracy, and less visible in the one with lowest classification accuracy. Contrary to discriminative classifiers classically used in BCI, such as LDA or SVM [9] which can appear as *black-boxes* with difficult interpretation, it is very interesting to see that the presented covariance based classifier uses features with a simple representation, and thus allows for an intuitive understanding. The observed covariance matrices have a physiological meaning and interpretation. In this framework, EEG processing complexity is encoded by a dedicated distance and not by a machine learning algorithm.

Based on those covariance matrices, the different distances and means of Table 1 are compared in terms of classification accuracy and average CPU time elapsed on a trial classification, which involves the computation of 4 means of

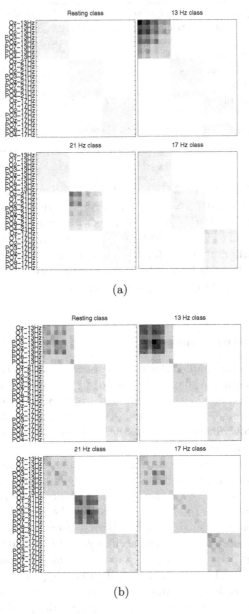

Fig. 1. Representation of covariance matrices: each image is the covariance matrix mean $\bar{\Sigma}^{(k)}$ of the class k, for one session of the recording. The diagonal blocks show the covariance in different frequency bands, i.e. 13 Hz in the upper-left block, 21 Hz in the middle, and 17 Hz in the bottom-right. The two chosen subjects are those with the highest (a) and the lowest (b) BCI performance.

Table 2. Subject classification accuracies (acc(%)) and average CPU time (time(s)) elapsed for the classification of a single trial. Classification is performed with MDM using either Euclidean or Riemannian means (see Table 1). Results obtained with a state-of-the-art method using CCA and SVM [12] are included.

Sub	CCA+SVM [12] acc (%)	Euclidean arithmetic acc (%)	time(s)	Riemannian Log-Euclidean acc (%)	time(s)	Affine-Invariant acc (%)	time(s)	α-divergence acc (%)	time(s)	Bhattacharyya acc (%)	time(s)
1	54.68	53.12	0.025	71.88	**0.150**	**73.44**	0.194	59.37	0.155	68.75	0.225
2	37.50	43.75	0.020	78.13	0.160	79.69	0.190	79.69	0.200	**81.25**	**0.065**
3	89.06	67.19	0.020	85.94	0.120	85.93	0.205	**95.31**	0.155	85.94	**0.100**
4	79.69	54.68	0.030	84.38	0.225	87.50	0.315	**89.07**	0.250	85.94	**0.100**
5	50.00	37.50	0.020	62.50	**0.115**	68.75	0.290	**73.44**	0.140	65.62	0.125
6	87.50	34.37	0.015	84.38	0.120	85.94	0.210	**87.50**	0.145	82.81	**0.100**
7	77.08	60.42	0.027	87.50	0.267	88.54	0.410	**91.66**	0.417	86.46	**0.137**
8	73.44	67.19	0.035	90.63	0.215	**92.19**	0.290	**92.19**	0.290	**92.19**	0.125
9	60.94	57.81	0.035	70.31	0.275	70.31	0.380	**75.00**	0.300	67.19	**0.134**
10	67.97	38.28	0.035	75.00	0.254	80.47	0.514	**82.03**	0.510	78.13	**0.160**
11	71.88	48.44	0.025	60.94	0.144	65.63	0.235	57.81	0.150	**75.00**	**0.105**
12	95.63	71.25	0.032	96.25	**0.292**	96.69	0.534	95.62	0.634	**96.88**	0.300
Avg.	70.45	52.83	0.027	78.98	0.194	81.27	0.314	**81.56**	0.279	80.51	**0.140**

class and a distance to each mean. One can note that for an optimal implementation, the 4 means are computed only once. Table 2 summarizes results obtained for each subject and each distance/divergence. Results obtained with a state-of-the-art method are also included, combining CCA and SVM [12].

Euclidean distance yields drastically low accuracy. This supports the fact that using Euclidean distance and arithmetic mean on SPD matrices is not appropriate. This is generally attributed to the invariance under inversion that is not guaranteed (i.e. $\bar{\Sigma}(\Sigma_i) \neq \bar{\Sigma}^{-1}(\Sigma_i^{-1})$) and the fact that the determinant of the arithmetic mean of SPD matrices can be larger than the determinant of its parts; it is referred to as the *swelling effect*. Since the value of the determinant is a direct measure of dispersion of the multivariate variables (i.e. EEG channels and frequency bands), it leads to poor discrimination in the classification task. The swelling effect of arithmetic mean is shown in Fig. 2: the determinant of the arithmetic mean is strictly larger than other means, the Log-Euclidean, Affine-Invariant and Bhattacharyya ones yielding similar determinants, close to trials values.

Using Riemannian distances significantly improves classification performances, with regards to state-of-the-art method (70.45 %) and Euclidean distance. The α-divergence yields the best results (81.56 %). The value of α was set to 0.6 through cross-validation. This procedure lasted 225.42 s and makes α-divergence the most costly method, due to the optimization of its parameter α. Log-Euclidean yields lower classification accuracy (average 78.98 %) but could be computed faster than α-divergence or Affine-Invariant distance. However, the Bhattacharyya distance has the lowest computational cost of the considered Riemannian distances (average CPU time 0.140 s), with a higher average accuracy

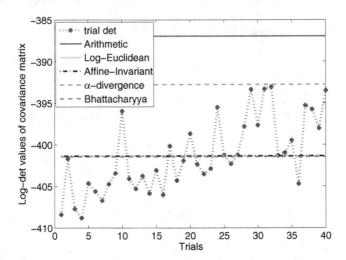

Fig. 2. (a): Swelling effect of arithmetic mean shown through log-determinant values. Training trials are taken from the 13 Hz class of the subject with the highest BCI performance. Log-determinant values are given for each trial covariance (points), and for means of Table 1 (horizontal lines).

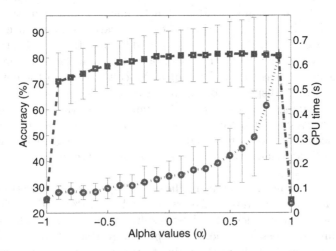

Fig. 3. Classification accuracy and CPU time obtained for the log-determinant α-divergence, with $-1 \leqslant \alpha \leqslant 1$. The values are averaged across all sessions and subjects.

of 80.51 %. So, it is good trade-off between efficiency and speed. The accuracies and CPU time of the α-divergence at different values of α are shown in Fig. 3. It is seen that for $\alpha = \pm 1$, where α-divergence represents a Bregman divergence associated with the log-determinant function, the classification accuracy are drastically low (25 %). For the rest, the accuracy varies smoothly with changes in α, with the highest accuracy scored while α is positive.

This experiment on real EEG data shows that it is crucial to process covariance matrices with dedicated Riemannian tools, impacting the efficiency of the classification.

4   Conclusion

Riemannian approaches have been successfully applied on EEG signals for brain computer interfaces. Straightforward algorithms, such as Minimum Distance to Mean, provide competitive results with state-of-the-art methods, without requiring meticulous parametrization or optimization. Working on covariance matrices in Riemannian spaces offers a wide choice of distances, embedding desirable invariances: it is thus possible to avoid the computation of user-specific spatial filters which are sensitive to artifacts and outliers. Nonetheless, the estimation of the Riemannian geometric mean has a strong impact on the classifier accuracy. This study investigates the performance of several distances and divergence on a real EEG dataset in the context of BCI based on the SSVEP paradigm. The experimental results indicate that the α-divergence yields the best accuracy after the selection of the best α values, but the Bhattacharyya distance has the lowest computational cost while providing decent accuracies.

References

1. Amari, S.I.: α-divergence is unique, belonging to both f-divergence and Bregman divergence classes. IEEE Trans. Inf. Theor. **55**(11), 4925–4931 (2009)
2. Arsigny, V., Fillard, P., Pennec, X., Ayache, N.: Log-Euclidean metrics for fast and simple calculus on diffusion tensors. Magn. Reson. Med. **56**(2), 411–421 (2006)
3. Arsigny, V., Fillard, P., Pennec, X., Ayache, N.: Geometric means in a novel vector space structure on symmetric positive-definite matrices. SIAM J. Matrix Anal. Appl. **29**(1), 328–347 (2007)
4. Barachant, A., Andreev, A., Congedo, M., et al.: The Riemannian potato: an automatic and adaptive artifact detection method for online experiments using Riemannian geometry. In: Proceedings of TOBI Workshop IV, pp. 19–20 (2013)
5. Barachant, A., Bonnet, S., Congedo, M., Jutten, C.: Multiclass brain-computer interface classification by Riemannian geometry. IEEE Trans. Biomed. Eng. **59**(4), 920–928 (2012)
6. Chebbi, Z., Moakher, M.: Means of Hermitian positive-definite matrices based on the log-determinant α-divergence function. Linear Algebra Appl. **436**(7), 1872–1889 (2012)
7. Congedo, M., Barachant, A., Andreev, A.: A new generation of brain-computer interface based on Riemannian geometry. arXiv preprint arXiv:1310.8115 (2013)

8. Fletcher, P.T., Joshi, S.: Principal geodesic analysis on symmetric spaces: statistics of diffusion tensors. In: Sonka, M., Kakadiaris, I.A., Kybic, J. (eds.) CVAMIA/MMBIA 2004. LNCS, vol. 3117, pp. 87–98. Springer, Heidelberg (2004)
9. Hastie, T., Tibshirani, R., Friedman, J., Hastie, T., Friedman, J., Tibshirani, R.: The Elements of Statistical Learning, vol. 2. Springer, New York (2009)
10. Johannes, M.G., Pfurtscheller, G., Flyvbjerg, H.: Designing optimal spatial filters for single-trial EEG classification in a movement task. Clin. Neurophysiol. **110**(5), 787–798 (1999)
11. Kalunga, E.K., Chevallier, S., Rabreau, O., Monacelli, E.: Hybrid interface: integrating BCI in multimodal human-machine interfaces. In: 2014 IEEE/ASME International Conference on Advanced Intelligent Mechatronics (AIM), pp. 530–535. IEEE (2014)
12. Kalunga, E.K., Djouani, K., Hamam, Y., Chevallier, S., Monacelli, E.: SSVEP enhancement based on canonical correlation analysis to improve BCI performances. In: AFRICON 2013, pp. 1–5. IEEE (2013)
13. Lotte, F., Guan, C.: Regularizing common spatial patterns to improve BCI designs: unified theory and new algorithms. IEEE Trans. Biomed. Eng. **58**(2), 355–362 (2011)
14. Moakher, M.: A differential geometric approach to the geometric mean of symmetric positive-definite matrices. SIAM J. Matrix Anal. Appl. **26**(3), 735–747 (2005)
15. Niedermeyer, E., da Silva, F.L.: Electroencephalography: Basic Principles, Clinical Applications, and Related Fields, 5th edn. Lippincott Williams & Wilkins, Baltimore (2004)
16. Nielsen, F., Bhatia, R.: Matrix Information Geometry. Springer Publishing Company Incorporated, Heidelberg (2012)
17. Rivet, B., Souloumiac, A., Attina, V., Gibert, G.: xDAWN algorithm to enhance evoked potentials: application to brain-computer interface. IEEE Trans. Biomed. Eng. **56**(8), 2035–2043 (2009)
18. Schäfer, J., Strimmer, K.: A shrinkage approach to large-scale covariance matrix estimation and implications for functional genomics. Stat. Appl. Genet. Mol. Biol. **4**(1) (2005)
19. Tomioka, R., Aihara, K., Müller, K.R.: Logistic regression for single trial EEG classification. In: NIPS, vol. 19, pp. 1377–1384 (2007)
20. Wang, S., James, C.J.: Enhancing evoked responses for BCI through advanced ICA techniques. In: MEDSIP, pp. 1–4 (2006)
21. Yger, F.: A review of kernels on covariance matrices for BCI applications. In: 2013 IEEE International Workshop on Machine Learning for Signal Processing (MLSP), pp. 1–6. IEEE (2013)

Group Theoretical Study on Geodesics for the Elliptical Models

Hiroto Inoue$^{(\boxtimes)}$

Graduate School of Mathematics, Kyushu University, Fukuoka, Japan
hi-inoue@math.kyushu-u.ac.jp

Abstract. We consider the geodesic equation on the elliptical model, which is a generalization of the normal model. More precisely, we characterize this manifold from the group theoretical view point and formulate Eriksen's procedure to obtain geodesics on normal model and give an alternative proof for it.

1 Introduction

The study of statistical manifolds began from a geometrical consideration on statistical estimation (Ref. Amari, Nagaoka 2001). Especially, the geodesics on statistical manifolds has been investigated since Rao 1945. In this article, we consider a special type of statistical manifold called the elliptical model, which is defined by each family of the elliptical distributions. The elliptical distributions are known to be important probability distributions in the multivariate statistics. It includes the multivariate normal distribution, Student-t and Cauchy distribution as special case (Ref. Calvo, Oller 1982; Muirhead 2001). Then the Normal model is a special case of the elliptical model.

Let $P_n(\mathbb{R})$ be the set of n-dimensional positive-definite symmetric matrices. The normal model is a Riemannian manifold with one coordinate system $\{(\mu, \Sigma) \in \mathbb{R}^n \times P_n(\mathbb{R})\}$ equipped with the metric

$$ds^2 = ({}^t d\mu)\Sigma^{-1}(d\mu) + \frac{1}{2}\mathrm{tr}((\Sigma^{-1}d\Sigma)^2).$$

This metric is defined by the Fisher information matrix. It is known that the geodesic equation on normal model is given as

$$\begin{cases} \ddot{\mu} - \dot{\Sigma}\Sigma^{-1}\dot{\mu} = 0, \\ \ddot{\Sigma} + \dot{\mu}{}^t\dot{\mu} - \dot{\Sigma}\Sigma^{-1}\dot{\Sigma} = 0. \end{cases} \tag{1}$$

The solution of this geodesic equation was firstly obtained by Eriksen. He obtained the geodesic by extracting a block matrix in a matrix exponential as follows:

H. Inoue—Research Fellow of Japan Society for the Promotion of Science.

© Springer International Publishing Switzerland 2015
F. Nielsen and F. Barbaresco (Eds.): GSI 2015, LNCS 9389, pp. 605–614, 2015.
DOI: 10.1007/978-3-319-25040-3_65

Theorem 1 (Eriksen 1987). *For any* $x \in \mathbb{R}^n, B \in \mathrm{Sym}_n(\mathbb{R})$, *define a matrix exponential* $\Lambda(t)$ *by*

$$\Lambda(-t) := \exp(-tA) =: \begin{pmatrix} \Delta & \delta & \Phi \\ {}^t\delta & \epsilon & {}^t\gamma \\ {}^t\Phi & \gamma & \Gamma \end{pmatrix}, \quad A := \begin{pmatrix} B & x & 0 \\ {}^tx & 0 & -{}^tx \\ 0 & -x & -B \end{pmatrix} \in \mathrm{Mat}_{2n+1}. \quad (2)$$

Then, the curve $(\mu(t), \Sigma(t)) := (-\Delta^{-1}\delta, \Delta^{-1})$ *is the solution of the geodesic Eq.* (1) *on* N_n *satsfying the initial condition*

$$(\mu(0), \Sigma(0)) = (0, I_n), \quad (\dot{\mu}(0), \dot{\Sigma}(0)) = (x, B).$$

$\square$

After that, in Calvo, Oller 1991, the explicit formula for geodesics on normal model was obtained by solving general differential equation that includes (1). Recently, Imai, Takaesu, Wakayama 2011 concretely calculated the exponential matrix given in (2), and reproved the result of Calvo, Oller 1991.

However, the question remains how to give a geometrical explanation for the result about the geodesic by Eriksen 1987. On the other hand, Eriksen pointed out that the matrix A in (2) is regarded as an element of the $\mathfrak{p}$-part of the Cartan decomposition $\mathfrak{so}(n+1, n) = \mathfrak{p} \oplus \mathfrak{k}$. And it is well known that geodesics on a symmetric spaces are obtained from matrix exponential. Considering these facts, it is suggested that there is a similar aspect for normal model with symmetric spaces. Indeed, the 1-dimensional normal model is a symmetric isometric to the Poincaré upper half plane. The result of Eriksen was geometrically explained in Imai, Takaesu, Wakayama 2011.

The purpose of this article is as follows. First, to give the geodesic on elliptical model explicitly. And secondly, to attempt a group theoretical interpretation to the result of Eriksen 1987 for the geodesic on more than 1-dimensional normal model. In Sect. 2, using a result of Calvo, Oller 2001, we see that the elliptical models are simultaneously embedded into a symmetric space $P_{n+1}(\mathbb{R}) \simeq GL(n+1)/O(n+1)$ as Riemannian manifolds, except a small class. In particular, the normal model is regarded as a Riemannian submanifold of $P_{n+1}(\mathbb{R})$. In Sect. 3, we rewrite the geodesic equation on normal model as a differential equation of a curve on $P_{n+1}(\mathbb{R})$. In other words, we express the geodesic equation on normal model as the linear projection from the geodesic equation on $P_{n+1}(\mathbb{R})$. Taking advantage of this expression, we rewrite the geodesic equation on elliptical model in the same manner. In Sect. 4, we formulate the procedure to obtain the geodesic by Theorem 1. For this purpose, we use the geometric structure of the positive-definite symmetric matrices based on the group actions. And we give an alternative proof of Theorem 1.

2   Elliptical Model and its Embedding

In this section, we consider the elliptical model, that includes the normal model for its special case.

2.1   Elliptical Model $E_n(\alpha)$

Put $d_\alpha := (n+1)\alpha^2 + 2\alpha$ ($\alpha \in \mathbb{C}$). For $\alpha \in \mathbb{C}$ which satisfies $d_\alpha > -1/n$, we define an $n(n+3)/2$-dimensional Riemannian manifold $E_n(\alpha) = (M, ds^2)$ by

$$M = \{(\mu, \Sigma) \in \mathbb{R}^n \times P_n(\mathbb{R})\},$$

$$ds^2 = ({}^t d\mu)\Sigma^{-1}(d\mu) + \frac{1}{2}\mathrm{tr}((\Sigma^{-1}d\Sigma)^2) + \frac{d_\alpha}{2}\mathrm{tr}^2\left(\Sigma^{-1}d\Sigma\right). \tag{3}$$

Here, (3) is a scale-transformed metric of the Fisher metric defined from a family of elliptical distributions $\{f(x; \mu, \Sigma)\}_{\mu, \Sigma}$ (see Calvo, Oller 2001). Then, each elliptical model is isometric to $E_n(\alpha)$. Hereafter, we call $E_n(\alpha)$ an elliptical model. Notice that for $\alpha_1, \alpha_2 \in \mathbb{C}$, $\alpha_1 \neq \alpha_2$ that satisfy $d_{\alpha_1} = d_{\alpha_2}$, we have $E_n(\alpha_1) \simeq E_n(\alpha_2)$ as Riemannian manifolds. And in particular, $E_n(0)$ is the normal model, and we write $E_n(0) = N_n$.

Example 1. The metric of $N_1 = \{(\mu, \sigma^2)\}$ is

$$ds^2 = \sigma^{-2}(d\mu^2 + 2d\sigma^2).$$

Then, N_1 is isometric to the Poincaré H, and is a symmetric space. We have following isomorphisms (see Imai, Takaesu, Wakayama 2011):

$$\begin{array}{ccccc}
N_1 & \xrightarrow{\sim} & H & \cong \mathrm{SL}(2,\mathbb{R})/\mathrm{SO}(2) \cong \mathrm{SO}_o(2,1)/\mathrm{S}(\mathrm{O}(2) \times \mathrm{O}(1)) \\
(\mu, \sigma^2) & \mapsto & \frac{1}{\sqrt{2}}\mu + i\sigma = g.i \leftarrow & gK & \mapsto & \mathrm{Ad}(g)
\end{array}$$
$$\tag{4}$$

where $\mathrm{Ad}(g) \in \mathrm{End}(\mathfrak{sl}_2(\mathbb{R}))$ is the adjoint action defined by $\mathrm{Ad}(g) : X \mapsto gXg^{-1}$.

2.2   Embedding of $E_n(\alpha)$ into $P_{n+1}(\mathbb{R})$

By using a result of Calvo, Oller 2001, we see that all the elliptical models $E_n(\alpha)$ for $\alpha \in \mathbb{R}$ are simultaneously embedded into a symmetric space $P_{n+1}(\mathbb{R}) \simeq \mathrm{GL}(n+1)/\mathrm{O}(n+1)$ as Riemannian manifolds.

The Symmetric Space $P_n(\mathbb{R})$. Consider the set $P_n(\mathbb{R})$ of n-dimensional positive-definite symmetric matrices. The tangent space of $P_n(\mathbb{R})$ is $\mathrm{Sym}_n(\mathbb{R})$. The following bijection is known:

$$\begin{array}{ccc}
\mathrm{GL}(n,\mathbb{R})/\mathrm{O}(n) & \cong & P_n(\mathbb{R}) \\
gK & \mapsto & \Psi = gI_n{}^t g.
\end{array}$$

The $\mathrm{GL}(n,\mathbb{R})$-invariant Riemannian metric on $P_n(\mathbb{R})$ is defined by $ds^2 = \frac{1}{2}\mathrm{tr}\left((\Psi^{-1}d\Psi)^2\right)$. Moreover, $P_n(\mathbb{R}) \simeq \mathrm{GL}(n)/\mathrm{O}(n)$ with this metric is known to be a Riemannian symmetric space, Ref. Helgason 1962.

The geodesic equation on $P_n(\mathbb{R})$ and its solution $\Psi(s)$ that satisfies $\Psi(0) = \Psi_1$ are given by

$$\dot{\Psi}(s) = \Psi(s)H, \quad \Psi(s) = \exp(Hs)\Psi_1.$$

Here H is a constant matrix that commutes with Ψ_1. For two matrices A, B, define $\langle A, B \rangle := \operatorname{tr}(A^t B)$, $\|A\| := \langle A, A \rangle^{1/2}$. Then, the Riemannian distance between $\Psi_1, \Psi_2 \in P_n(\mathbb{R})$ is written as

$$d(\Psi_1, \Psi_2) = \sqrt{\frac{1}{2}} \| \ln(\Psi_1^{-1/2} \Psi_2 \Psi_1^{-1/2}) \| = \left(\frac{1}{2} \sum_{i=1}^n \ln^2 \lambda_i \right)^{1/2},$$

where λ_i denotes an eigenvalue of $\Psi_1^{-1} \Psi_2$.

Embedding and Homogeneity of $E_n(\alpha)$. For our purpose, we rewrite the embedding of elliptical models into $P_{n+1}(\mathbb{R})$ by Calvo, Oller 2001 as follows. Put $|A| := \det A$.

Lemma 1. *For each $\alpha \in \mathbb{R}$, we have the embedding of $E_n(\alpha)$ into $P_{n+1}(\mathbb{R})$ as a Riemannian manifold:*

$$f_\alpha : E_n(\alpha) \hookrightarrow P_{n+1}(\mathbb{R})$$

$$(\mu, \Sigma) \mapsto |\Sigma|^\alpha \begin{pmatrix} \Sigma + \mu^t \mu & \mu \\ {}^t\mu & 1 \end{pmatrix}.$$

$f_\alpha(E_n(\alpha)) \simeq E_n(\alpha)$ *is a non-geodesic submanifold of $P_{n+1}(\mathbb{R})$.* $\square$

Then as a corollary of this result, elliptical models $E_n(\alpha)$ $(\alpha \in \mathbb{R})$ are simultaneously embedded into a symmetric space $P_{n+1}(\mathbb{R})$.

Proposition 1. *1. $P_{n+1}(\mathbb{R})$ is decomposed as follows:*

$$P_{n+1}(\mathbb{R}) = \left\{ \begin{pmatrix} \Sigma + \mu^t \mu & \mu \\ {}^t\mu & 1 \end{pmatrix} \,\Big|\, \det \Sigma = 1 \right\} \sqcup \bigsqcup_{\alpha \in \mathbb{R}} \left\{ |\Sigma|^\alpha \begin{pmatrix} \Sigma + \mu^t \mu & \mu \\ {}^t\mu & 1 \end{pmatrix} \,\Big|\, \det \Sigma \neq 1 \right\}$$

$$=: Q \sqcup \bigsqcup_{\alpha \in \mathbb{R}} R_\alpha. \tag{5}$$

For each $\alpha \in \mathbb{R}$, we have the isometry

$$E_n(\alpha) \xrightarrow{\sim} Q \sqcup R_\alpha \subset P_{n+1}(\mathbb{R})$$

$$(\mu, \Sigma) \mapsto \begin{pmatrix} \Sigma + \mu^t \mu & \mu \\ {}^t\mu & 1 \end{pmatrix}. \tag{6}$$

2. Consider the action $\Psi \mapsto g \Psi^t g$ of $\operatorname{GL}(n+1, \mathbb{R})$ on $P_{n+1}(\mathbb{R}) \simeq \operatorname{GL}(n+1)/\operatorname{O}(n+1)$. Then, for each $\alpha \in \mathbb{R}$, the subgroup

$$G_A(\alpha) := \left\{ |P|^\alpha \begin{pmatrix} P & \delta \\ 0 & 1 \end{pmatrix} \,\Big|\, P \in \operatorname{GL}(n, \mathbb{R}), \delta \in \mathbb{R}^n \right\} \quad (\alpha \in \mathbb{R})$$

of $\operatorname{GL}(n+1, \mathbb{R})$ acts transitively on $Q \sqcup R_\alpha \simeq E_n(\alpha)$.

3. *For any $\Psi \in P_{n+1}(\mathbb{R})$, put $\Psi = \begin{pmatrix} A & B \\ {}^tB & D \end{pmatrix}$, $A \in P_n(\mathbb{R}), B \in M_{n,1}(\mathbb{R}), D > 0$.*
   Then, we have
$$\Psi \in Q \Leftrightarrow \det \left(D^{-1}A - D^{-2}B^tB \right) = 1. \tag{7}$$
   And provided that $\Psi \notin Q$,
$$\Psi \in R_\alpha \Leftrightarrow \alpha = \frac{\log D}{\log |D^{-1}A - D^{-2}(B^tB)|}. \tag{8}$$

We have seen that each elliptical model $E_n(\alpha) \simeq Q \sqcup R_\alpha$ is identified as the orbit of the subgroup $G_A(\alpha) \subset \mathrm{GL}(n+1)$ through $I_{n+1} \in P_{n+1}(\mathbb{R})$, equipped with the metric induced from the metric on $P_{n+1}(\mathbb{R})$.

$$\mathrm{GL}(n) \quad \subset \quad G_A(\alpha) \quad \subset \quad \mathrm{GL}(n+1)$$

$$P_n(\mathbb{R}) \quad \subset \quad E_n(\alpha) \simeq Q \sqcup R_\alpha \quad \subset \quad P_{n+1}(\mathbb{R})$$

$$\left\{ |\Sigma|^\alpha \begin{pmatrix} \Sigma & 0 \\ 0 & 1 \end{pmatrix} \right\} \subset \left\{ |\Sigma|^\alpha \begin{pmatrix} \Sigma + \mu^t\mu & \mu \\ {}^t\mu & 1 \end{pmatrix} \right\} \subset \left\{ \Psi = \begin{pmatrix} A & B \\ {}^tB & D \end{pmatrix} \right\}$$

In particular, $E_n(\alpha)$ is a homogeneous space. Then, in the arguments of geodesics on $E_n(\alpha)$, we only have to consider the geodesics that go through $I_{n+1} \in P_{n+1}(\mathbb{R})$ without loss of generality.

We can also embed isometric elliptical models into $P_n(\mathbb{R})$ simultaneously.

Proposition 2. *1. Assume that $n \geq 2$. Then, for any $\alpha \in \mathbb{R}$, $\alpha \neq -1/n$, we have the following decomposition:*

$$P_{n+1}(\mathbb{R}) = \bigsqcup_{\beta>0} \left\{ |\Sigma|^\alpha \begin{pmatrix} \Sigma + \beta\mu^t\mu & \beta\mu \\ \beta^t\mu & \beta \end{pmatrix} \right\} =: \bigsqcup_{\beta>0} R'_\beta. \tag{9}$$

The subgroup $G_A(\alpha)$ acts transitively on each R'_β ($\beta > 0$) by $\Psi \mapsto g\Psi^tg$. And each R'_β is isometric to $E_n(\alpha)$ by the map

$$E_n(\alpha) \xrightarrow{\sim} R'_\beta \subset P_{n+1}(\mathbb{R})$$
$$(\mu, \Sigma) \mapsto \begin{pmatrix} \Sigma + \beta^2\mu^t\mu & \beta^{3/2}\mu \\ \beta^{3/2t}\mu & \beta \end{pmatrix}. \tag{10}$$

2. *For any $\Psi \in P_{n+1}(\mathbb{R})$, put $\Psi = \begin{pmatrix} A & B \\ {}^tB & D \end{pmatrix}$, $A \in P_n(\mathbb{R}), B \in M_{n,1}(\mathbb{R}), D > 0$.*
   Then, $\beta > 0$ such that $\Psi \in R'_\beta$ is uniquely determined by

$$\beta = \left(\frac{D}{|D^{-1}A - D^{-2}B^tB|^\alpha} \right)^{1/(n\alpha+1)}. \tag{11}$$

By the above proposition, we get the orbit decomposition

$$\mathrm{GL}(n+1,\mathbb{R}) = \bigsqcup_{\beta>0} G_A(\alpha)\, g_\beta\, O(n+1) \quad (g_\beta \in R'_\beta, \alpha \in \mathbb{R}, \alpha \neq -1/n).$$

The complete invariant $\beta > 0$ on each orbit is given by (11).

3 Geodesics on Elliptical Model

In this section, we regard the normal model N_n as a submanifold of $P_{n+1}(\mathbb{R})$. And we rewrite the geodesic equation on N_n as a system of linear equation. We also rewrite the geodesic equation on elliptical model $E_n(\alpha)$ in the same manner.

3.1 Geodesic Equation on N_n

The geodesic Eq. (1) on N_n can be partially integrated as

$$\begin{cases} \dot{\mu} = \Sigma x, \\ \dot{\Sigma} = \Sigma(B - x^t \mu), \end{cases} \tag{12}$$

where $(\dot{\mu}(0), \dot{\Sigma}(0)) = (x, B) \in \mathbb{R}^n \times \mathrm{Sym}_n(\mathbb{R})$ are the integration constants. We express $(\mu(t), \Sigma(t))$ as a curve $S(t)$ on $N_n \subset P_{n+1}(\mathbb{R})$.

Lemma 2. *The Eq. (12) is written as the geodesic equation on $N_n \simeq Q \sqcup R_0$:*

$$(I_n, 0)S^{-1}\dot{S} = (B, x), \quad S(t) = \begin{pmatrix} \Sigma + \mu^t \mu & \mu \\ {}^t\mu & 1 \end{pmatrix}. \tag{13}$$

We notice that geodesics on the symmetric space $P_{n+1}(\mathbb{R})$ are given by matrix exponentials. In other words, the geodesic equation is given by

$$S^{-1}\dot{S} = C, \tag{14}$$

where $C \in M_{n+1}(\mathbb{R})$ is a constant matrix. We get the geodesic equation on N_n formally by the linear projection of (14). As we see in Theorem 1, geodesic equation on normal model has been solved. We will give an alternative proof for it.

3.2 Geodesic Equation on $E_n(\alpha)$ and its Solution

The geodesic equation on elliptical model $E_n(\alpha)$ is given by

$$\begin{cases} \ddot{\mu} - \dot{\Sigma}\Sigma^{-1}\dot{\mu} = 0, \\ \ddot{\Sigma} + \dot{\mu}^t\dot{\mu} - \dot{\Sigma}\Sigma^{-1}\dot{\Sigma} - \dfrac{d_\alpha}{nd_\alpha + 1}{}^t\dot{\mu}\Sigma^{-1}\dot{\mu}\Sigma = 0, \end{cases} \quad (d_\alpha := (n+1)\alpha^2 + 2\alpha) \tag{15}$$

(cf. Calvo, Oller 2001). Integrating (15) by part, we get

$$\begin{cases} \dot{\mu} = \Sigma x, \\ \dot{\Sigma} = \Sigma\left(B - x^t\mu + \dfrac{d_\alpha}{nd_\alpha + 1}({}^t\mu x)I_n\right), \end{cases} \tag{16}$$

where $(\dot{\mu}(0), \dot{\Sigma}(0)) = (x, B) \in \mathbb{R}^n \times \mathrm{Sym}_n(\mathbb{R})$ are the integration constants. We can assume $(\mu(0), \Sigma(0)) = (0, I_n)$ because of the homogeneity of $E_n(\alpha)$. Hereafter, we assume that $\alpha \in \mathbb{R}$ and regard $E_n(\alpha) \simeq Q \sqcup R_\alpha$ as a submanifold in $P_{n+1}(\mathbb{R})$.

Proposition 3. *The Eq. (16) is written as the geodesic equation on $E_n(\alpha) \simeq Q \sqcup R_\alpha$:*

$$(I_n, 0)S_\alpha^{-1}\dot{S}_\alpha = (B, x) + \frac{d}{dt}F_\alpha(t)(I_n, 0), \quad S_\alpha(t) = |\Sigma|^\alpha \begin{pmatrix} \Sigma + \mu^t\mu & \mu \\ {}^t\mu & 1 \end{pmatrix}, \quad (17)$$

$$F_\alpha(t) = -\alpha \log |S_\alpha| + d_\alpha \mathrm{tr} B \cdot t.$$

Corollary 1. *The equations in (17) are written as*

$$(I_n, 0)\left(T_\alpha^{-1}\right)' = (-B, -x)\, T_\alpha^{-1},$$

$$T_\alpha = e^{-d_\alpha \mathrm{tr} B \cdot t} |S_\alpha|^\alpha S_\alpha = |e^{-tB}\,\Sigma|^{d_\alpha} \begin{pmatrix} \Sigma + \mu^t\mu & \mu \\ {}^t\mu & 1 \end{pmatrix}.$$

Remark 1. Let $S \mapsto S^* = S^{-1}$ be the involution of $P_{n+1}(\mathbb{R}) \simeq \mathrm{GL}(n+1)/O(n+1)$. The geodesic equation on $P_{n+1}(\mathbb{R})$ is written by using the dual coordinates: $\ddot{S}^* = S^{-1}\ddot{S}S^{-1}$. In general, such form of equations are explicitly written in Barbaresco 2014 as the geodesic equations for the Koszul Hessian metric. Equation (17) is obtained by adding the submanifold-constrain condition:

$$\ddot{S}^* = S^{-1}\ddot{S}S^{-1} + \lambda(t)\frac{\partial g}{\partial S},$$

where $g(S) = \alpha \log |S| - ((n+1)\alpha + 1) \log |{}^t e_{n+1} S e_{n+1}|$ and $\lambda(t)$ is the Lagrange multiplier.

4   Geometric Description of Theorem 1

In this section, we formulate Theorem 1. First, we observe the geometric structure of $P_{2n+1}(\mathbb{R})$. On the basis of these observations, we give a group theoretical interpretation for the matrix A and the extraction of a block matrix from $\Lambda(-t)$ in (2).

4.1   Geometric Structure of $P_{2n+1}(\mathbb{R})$

We extend the embedding in Calvo, Oller 2001.

Lemma 3. *Let $P_{m+n}(\mathbb{R})$ be the set of $(m + n)$-dimensional positive-definite symmetric matrices. Then, any $\Lambda \in P_{m+n}(\mathbb{R})$ can be uniquely written as follows:*

$$\Lambda = \begin{pmatrix} S + c\Gamma^t c & c\Gamma \\ \Gamma^t c & \Gamma \end{pmatrix}, \quad S \in P_m(\mathbb{R}), \ \Gamma \in P_n(\mathbb{R}), \ c \in M_{m,n}. \quad (18)$$

From above formula, we have

$$\mathrm{tr}\left((\Lambda^{-1}d\Lambda)^2\right) = \mathrm{tr}\left((S^{-1}dS)^2\right) + 2\mathrm{tr}\left(\Gamma^t dc S^{-1}dc\right) + \mathrm{tr}\left((\Gamma^{-1}d\Gamma)^2\right). \quad (19)$$

Moreover, taking $\tilde{S} \in \mathrm{GL}(m)/O(m)$, $\tilde{\Gamma} \in \mathrm{GL}(n)/O(n)$ such that $S = \tilde{S}^t\tilde{S}$, $\Gamma = \tilde{\Gamma}^t\tilde{\Gamma}$, we have

$$\Lambda = g^t g, \quad g = \begin{pmatrix} \tilde{S} & c\tilde{\Gamma} \\ 0 & \tilde{\Gamma} \end{pmatrix}. \quad (20)$$

Remark 2. (19) includes a result of Calvo, Oller 2001 with $n = 1$. Equation (20) is a weak form of Cholesky decomposition.

Proposition 4.   *1. Define the Lie group $H = \mathrm{SO}(n+1, n)$ by*

$$H := \left\{ h \in \mathrm{GL}(2n+1, \mathbb{R}) \mid {}^t h J h = J \right\}, \quad J := \begin{pmatrix} & & I_n \\ & 1 & \\ I_n & & \end{pmatrix}.$$

Let $\mathfrak{h}$ be the Lie algebra of H and $\mathfrak{h} = \mathfrak{p} \oplus \mathfrak{k}$ be the Cartan decomposition. Then, we have

$$\mathfrak{p} = \mathrm{sym}_{2n+1} \cap \mathfrak{h} = \left\{ X = \begin{pmatrix} B & x & D \\ {}^t x & 0 & -{}^t x \\ {}^t D & -x & -B \end{pmatrix} ; \ B, D \in \mathrm{M}_n, \ {}^t B = B, \ {}^t D = -D \right\}.$$

$$(21)$$

2. Consider the map $\pi' : P_{2n+1}(\mathbb{R}) \ni \Lambda \mapsto S \in P_{n+1}(\mathbb{R})$ defined by the decomposition (18) and its restriction $\pi := \pi'|_{P_{2n+1}(\mathbb{R}) \cap H}$. Then, the image of π coincides with N_n;

$$\pi : P_{2n+1}(\mathbb{R}) \cap H \twoheadrightarrow Q \sqcup R_0 \simeq N_n.$$

3. For any $X \in \mathrm{sym}_{2n+1} \cap \mathfrak{h}$, define $\Lambda(t) := \exp(tX)$. Then, $S(t) := \pi(\Lambda(t)) \in P_{n+1}(\mathbb{R})$ is a curve on $Q \sqcup R_0 \simeq N_n$. In particular, If $D = 0$ for X in the form of (21), $S(t)$ satisfies the equation

$$(I_n, 0) S^{-1} \dot{S}(t) = (B, x). \tag{22}$$

In other words, defining $(\mu(t), \Sigma(t))$ by $S(t) =: \begin{pmatrix} \Sigma + \mu {}^t \mu & \mu \\ {}^t \mu & 1 \end{pmatrix}$, (22) is written as the geodesic equation on normal model for (μ, Σ).

In this proposition, we see that the matrix A in (2) is an element of $\mathfrak{p} = \mathrm{sym}_{2n+1} \cap \mathfrak{h}$, which appears in the Cartan decomposition $\mathfrak{h} = \mathfrak{p} \oplus \mathfrak{k}$. And we regard the extraction of a block matrix in (2) as the projection π to an element in Cholesky decomposition. Based on these description, we can give an alternative proof for Theorem 1. Our argument can be summarized as follows.

$$\mathrm{sym}_{2n+1} \cap \mathfrak{h} \rightarrow P_{2n+1}(\mathbb{R}) \cap H \twoheadrightarrow N_n \subset P_{n+1}(\mathbb{R})$$
$$A \quad \mapsto \quad e^{tA} = \Lambda(t) \quad \mapsto \pi(\Lambda(t)) = S(t) : \text{geodesic.}$$

Proof.   1. The Cartan decomposition $\mathfrak{h} = \mathfrak{p} \oplus \mathfrak{k}$ gives the eigenspace decomposition of $\mathfrak{h}$ for the transpose map $X \mapsto {}^t X$ (Ref. Helgason 1962). Thus, the first equation of (21) holds. By the definition of the Lie algebra $\mathfrak{h}$, we see that

$$\mathfrak{h} = \left\{ Y \mid J {}^t Y J = -Y \right\}.$$

Then, any $X \in \mathrm{sym}_{2n+1} \cap \mathfrak{h}$ is the form of (21).

2. We show that $\pi(\Lambda) \in N_n$ for any $\Lambda = g^t g \in P_{2n+1}(\mathbb{R}) \cap H$. Put

$$g = \begin{pmatrix} \tilde{S} & * \\ 0 & \tilde{\Gamma} \end{pmatrix}, \quad \tilde{S} = \begin{pmatrix} P & \mu \\ 0 & a \end{pmatrix}, \quad P, \tilde{\Gamma} \in GL(n, \mathbb{R}), \quad \mu \in \mathbb{R}^n, \quad a \in \mathbb{R}.$$

Since $\Lambda = g^t g \in H$, g satisfies $J(g^t g)J = {}^t g^{-1} g^{-1}$. Then we have

$$\tilde{\Gamma}^t \tilde{\Gamma} = {}^t P^{-1} P^{-1}.$$

Therefore,

$$1 = \det \Lambda = (\det g)^2 = a^2 (\det P)^2 (\det \tilde{\Gamma})^2 = a^2,$$

so we have

$$\pi(\Lambda) = \tilde{S}^t \tilde{S} = \begin{pmatrix} P & \mu \\ 0 & a \end{pmatrix} \begin{pmatrix} {}^t P & 0 \\ {}^t \mu & a \end{pmatrix} \in Q \sqcup R_0 \simeq N_n.$$

Here we can take $P \in GL(n, \mathbb{R})$, $\mu \in \mathbb{R}^n$ arbitrarily, so we see that π is surjective.

3. We give the proof in next subsection as an alternative proof for Theorem 1.

4.2   Alternative Proof for Theorem 1

If $D = 0$, $\Lambda(t) = \exp(tX)$ is identified with the matrix exponential in (2). We will proof (22) by calculating the sub-block in the differential equation of $\Lambda(t)$. We set the matrix-valued functions $S(t), \mathbf{b}(t)$ by

$$S^{-1}(t) := \begin{pmatrix} \Delta & \delta \\ {}^t\delta & \epsilon \end{pmatrix}, \quad \mathbf{b}(t) := \begin{pmatrix} \Phi \\ {}^t\gamma \end{pmatrix}.$$

We notice that $S(t) = \pi(\Lambda(t))$. First, we can see that

$$\Lambda(t) = \begin{pmatrix} S(I_{n+1} + \mathbf{b}\Delta^t \mathbf{b}S) & -S\mathbf{b}\Delta \\ -\Delta^t \mathbf{b}S & \Delta \end{pmatrix}, \quad \Lambda^{-1}(t) = \begin{pmatrix} S^{-1} & \mathbf{b} \\ {}^t\mathbf{b} & \Delta^{-1} + {}^t\mathbf{b}S\mathbf{b} \end{pmatrix}.$$

From the differential equaiton $\dot{\Lambda}(t) = X\Lambda(t)$, we get

$$\begin{pmatrix} -B & -x & * \\ -{}^tx & 0 & * \\ * & * & * \end{pmatrix} = (\Lambda^{-1}(t))'\Lambda(t) = \begin{pmatrix} (S^{-1})'S(I_{n+1} + \mathbf{b}\Delta^t \mathbf{b}S) - \dot{\mathbf{b}}\Delta^t \mathbf{b}S & * \\ * & * \end{pmatrix},$$

and

$$\dot{\mathbf{b}} = \begin{pmatrix} -B & -x \\ -{}^tx & 0 \end{pmatrix} \mathbf{b} + \begin{pmatrix} 0 \\ {}^tx\Gamma \end{pmatrix}.$$

Combining above two equations, we get

$$\left\{ (S^{-1})'S - \begin{pmatrix} -B & -x \\ -{}^tx & 0 \end{pmatrix} \right\} \{I_{n+1} + \mathbf{b}\Delta^t \mathbf{b}S\} + \begin{pmatrix} 0 \\ {}^tx\Gamma \end{pmatrix} \Delta^t \mathbf{b}S = 0.$$

Then, multiplying $(I_n, 0) \in M_{n,n+1}(\mathbb{R})$ to both sides from the left, we get

$$\{(I_n, 0)(S^{-1})'S - (-B, -x)\} \{I_{n+1} + \mathbf{b}\Delta^t \mathbf{b}S\} = 0.$$

Since $\Lambda(0) = I_{2n+1}$, we see that $I_{n+1} + \mathbf{b}\Delta^t\mathbf{b}S$ is contained in a neighbor of I_{n+1} for any sufficient small $t \in \mathbb{R}$. Therore, for such t, we have

$$(I_n, 0)(S^{-1})'S = (-B, -x),$$

so that $S(t)$ satisfies the geodesic Eq. (13) on $Q \sqcup R_0 \simeq N_n$. $\square$

References

Amari, S., Nagaoka, H.: Methods of Information Geometry. Translations of Mathematical Monographs. American Mathematical Society and Oxford University Press, Oxford (2001)

Barbaresco, F.: Koszul information geometry and Souriau geometric temperature/capacity of Lie group thermodynamics. Entropy **16**, 4521–4565 (2014). doi:10.3390/e16084521

Calvo, M., Oller, J.M.: A distance between elliptical distributions based in an embedding into the Siegel group. J. Comput. Appl. Math. **145**, 319–334 (2002)

Calvo, M., Oller, J.M.: An explicit solution of information geodesic equations for the multivariate normal model. Stat. Decis. **9**, 119–138 (1991)

Eriksen, P.S.: Geodesics connected with the Fisher metric on the multivariate normal manifold, pp. 225–229. Proceedings of the GST Workshop, Lancaster (1987)

Helgason, S.: Differential Geometry and Symmetric Spaces. Academic Press, New York and London (1962)

Imai, T., Takaesu, A., Wakayama, M.: Remarks on geodesics for multivariate normal models. J. Math-for-Ind. **3**(2011B–6), 125–130 (2011B)

Muirhead, R.J.: Aspects of Multivariate Statistical Theory. Wiley, New York (1982)

Rao, C.R.: Information and the accuracy attainable in the estimation of statistical parameters. Bull. Calcutta Math. Soc. **37**, 81–91 (1945)

Path Connectedness on a Space of Probability Density Functions

Shinto Eguchi[1] and Osamu Komori[2]([⊠])

[1] Institute of Statistical Mathematics, Tachikawa, Tokyo 190-8562, Japan
eguchi@ism.ac.jp
[2] University of Fukui, Fukui 910-8507, Japan
komori0926@gmail.com

Abstract. We introduce a class of paths or one-parameter models connecting arbitrary two probability density functions (pdf's). The class is derived by employing the Kolmogorov-Nagumo average between the two pdf's. There is a variety of such path connectedness on the space of pdf's since the Kolmogorov-Nagumo average is applicable for any convex and strictly increasing function. The information geometric insight is provided for understanding probabilistic properties for statistical methods associated with the path connectedness. The one-parameter model is extended to a multidimensional model, on which the statistical inference is characterized by sufficient statistics.

1 Introduction

Information geometry has provided intrinsic understandings in a variety of mathematical sciences such as statistics, information science, statistical physics and machine learning [1, 2, 9]. We view the parameter describing objects and natures in the science as a coordinate expressing the geometric space in which such a viewpoint gives clear and strong intuition to understand the interactive mechanism for backgrounds of the science applied. A typical example is a parametric family of pdf's including the Gaussian distribution model. Information geometry directly reveals the central role of the parameter to give the uncertainty and probability excluding apparent properties depending on the choice of the parameter. The nonparametric formulation for information geometry is discussed from a point of infinite dimensional manifold or functional manifold such as Hilbert and Banach manifolds [3, 18, 21–23, 27]. There are several difficulties unless some integrability assumptions are imposed.

We employ the Kolmogorov-Nagumo average to avoid such a technical problem mentioned above, see [15–17] for the detailed discussion on the Kolmogorov-Nagumo average from statistical physics. This average is also called quasi-arithmetic mean as in [19], where a generalized upper bounds on Bayes error using quasi-arithmetic means is discussed.

We view the Kolmogorov-Nagumo average between two pdf's as a path connecting the pdf's. A convexity property guarantees the existence of such a path without such integrability assumptions. We can make a freehand drawing for the

© Springer International Publishing Switzerland 2015
F. Nielsen and F. Barbaresco (Eds.): GSI 2015, LNCS 9389, pp. 615–624, 2015.
DOI: 10.1007/978-3-319-25040-3_66

path on the infinite-dimensional space. The path is characterized as a geodesic curve in a dualistic pair of linear connections. In particular, we consider the path in a class of information divergence which the pair of linear connections are associated with, see [6,8,9,12] for the divergence geometry.

The paper is organized as follows. Section 2 introduce ϕ-path employing the Kolmogorov-Nagumo average with respect to a generator function ϕ. Typical examples are discussed. In Sect. 3 we consider a pair of parallel transport and the geodesic associated. We show that a ϕ-path is a geodesic curve with respect to a parallel transport associated with the generator function ϕ. Section 4 provides the dualistic pair of linear connections associated with information divergence. We show that the ϕ-path is explicitly given in the framework. Finally we overview the discussion here with existing literature,

2 ϕ-path Connectedness

Let $\mathcal{F}_P$ be the space of all P-a.s. strictly positive pdf's with respect to a probability measure P of a data space $\mathcal{X}$. For example, P is taken as a standard Gaussian measure and the uniform probability measure corresponding to continuous and discrete random variables, respectively. Thus a pdf f of $\mathcal{F}_P$ associates with a probability measure $P_f(B) = \int_B f(x)dP(x)$. If we take another measure Q that is mutually absolute continuous with P, then $\mathcal{F}_P$ is changed to $\mathcal{F}_Q$ defined by $\{(\partial P/\partial Q)f : f \in \mathcal{F}_P\}$. Let ϕ be a monotone increasing and concave function defined on $\mathbb{R}_+$. Then there exists the inverse function ϕ^{-1} that is monotone increasing and convex in $\mathbb{R}$. The Kolmogorov-Nagumo average for f and g in $\mathcal{F}_P$ is defined by

$$\phi^{-1}\Big((1-t)\phi(f(x)) + t\phi(g(x))\Big). \tag{1}$$

for $t, 0 \le t \le 1$. From the convexity of ϕ^{-1} we observe that

$$0 \le \phi^{-1}\Big((1-t)\phi(f(x)) + t\phi(g(x))\Big) \le (1-t)f(x) + tg(x),$$

which implies that

$$0 \le \int_{\mathcal{X}} \phi^{-1}\Big((1-t)\phi(f(x)) + t\phi(g(x))\Big)dP(x) \le 1.$$

We define κ_t to satisfy that

$$\int_{\mathcal{X}} \phi^{-1}\Big((1-t)\phi(f(x)) + t\phi(g(x)) - \kappa_t\Big)dP(x) = 1 \tag{2}$$

and we write

$$f_t(x, \phi) = \phi^{-1}\Big((1-t)\phi(f(x)) + t\phi(g(x)) - \kappa_t\Big), \tag{3}$$

which we call the ϕ-path connecting f and g. By the definition of κ_t, we know $\kappa_t \le 0$ with equality if $t = 0$ or $t = 1$. The simplest example is $\phi(x) = x$, in which the ϕ-path is noting but the mixture geodesic, or $f_t(x, \phi) = (1-t)f(x) + tg(x)$ with $\kappa_t = 0$. We note the existence for κ_t for all $t \in [0, 1]$ as follows.

Theorem 1. *There uniquely exists* κ_t *satisfying* (2).

Proof. Let $B_\delta = \{x \in \mathcal{X} : f(x) \geq \phi^{-1}(\delta), g(x) \geq \phi^{-1}(\delta)\}$. Then we see that

$$\int_{\mathcal{X}} \phi^{-1}((1-t)\phi(f(x)) + t\phi(g(x)) + c)dP(x) \geq \phi^{-1}(c+\delta)P(B_\delta).$$

We observe that $\lim_{c \to \infty} \phi^{-1}(c+\delta) = +\infty$ since ϕ^{-1} is convex and monotone increasing from the assumption of ϕ. Also $P(B_\delta) > 0$ for sufficiently small δ. This guarantees the existence of such a κ_t to satisfy (2) and the uniqueness is trivial.

Accordingly we see that the space $\mathcal{F}_P$ is path-connected for any monotone increasing and concave function ϕ since we can define $F^{(\phi)} : [0,1] \to \mathcal{F}_P$ by $F^{(\phi)}(t) = f_t(\cdot, \phi)$, where $f_t(x, \phi)$ is defined in (3). There is a variety of path-connectedness in $\mathcal{F}_P$ because the choice for ϕ is rather arbitrary. Let us consider a few of examples as follows.

Example 1. (i). $\phi_0(x) = \log(x)$. The ϕ_0-path is given by

$$f_t(x, \phi_0) = \exp((1-t)\log f(x) + t\log g(x) - \kappa_t),$$

where

$$\kappa_t = \log \int \exp((1-t)\log f(x) + t\log g(x))dP(x).$$

(ii). $\phi_\eta(x) = \log(x+\eta)$ with $\eta \geq 0$. The ϕ_η-path is given by

$$f_t(x, \phi_\eta) = \exp\left[(1-t)\log\{f(x)+\eta\} + t\log\{g(x)+\eta\} - \kappa_t\right],$$

where

$$\kappa_t = \log\left[\int \exp\{(1-t)\log\{f(x)+\eta\} + t\log\{g(x)+\eta\}\}dP(x) - \eta\right].$$

We note that κ_t is well defined because

$$\exp\left[(1-t)\log\{f(x)+\eta\} + t\log\{g(x)+\eta\}\right] \geq \eta.$$

(iii). $\phi_\beta(x) = (x^\beta - 1)/\beta$ with $\beta \leq 1$. The ϕ_β-path is given by

$$f_t(x, \phi_\beta) = \{(1-t)f(x)^\beta + tg(x)^\beta - \kappa_t\}^{\frac{1}{\beta}},$$

where κ_t is the normalizing factor. We observe that $-1 \leq \kappa_t \leq 0$ since

$$\int_{\mathcal{X}} \{(1-t)f(x)^\beta + tg(x)^\beta - \kappa\}^{\frac{1}{\beta}}dP(x) \geq (-\kappa)^{\frac{1}{\beta}}$$

for any $\kappa < 0$. We note that the explicit form of κ_t is not known in general.

On the other hand, there is another way to normalize (1) as

$$g_t(x, \phi) = z_t^{-1} \phi^{-1}((1-t)\phi(f(x)) + t\phi(g(x))),$$

where

$$z_t = \int_{\mathcal{X}} \phi^{-1}((1-t)\phi(f(x)) + t\phi(g(x))dP(x).$$

In Examples (i) and (ii) $\kappa_t = \log z_t$; in Example (iii) there is no explicit relation of κ_t with z_t.

3 A Pair of Parallel Transports

We discuss a geometric characterization for the ϕ-path $C^{(\phi)} = \{f_t(x, \phi) : 0 \leq t \leq 1\}$ connecting f and g. We introduce an extended expectation using the generator function ϕ as

$$E_f^{(\phi)}\{a(X)\} = \frac{\displaystyle\int_{\mathcal{X}} \frac{1}{\phi'(f(x))} a(x) dP(x)}{\displaystyle\int_{\mathcal{X}} \frac{1}{\phi'(f(x))} dP(x)}.$$

We note that
(i). $E_f^{(\phi)}(c) = c$ for any constant c.
(ii). $E_f^{(\phi)}\{ca(X)\} = cE_f^{(\phi)}\{a(X)\}$ for any constant c.
(iii). $E_f^{(\phi)}\{a(X) + b(X)\} = E_f^{(\phi)}\{a(X)\} + E_f^{(\phi)}\{b(X)\}$.
(iv). $E_f^{(\phi)}\{a(X)^2\} \geq 0$ with equality if and only if $a(x) = 0$ for P-almost everywhere x in $\mathcal{X}$.
We note that, if $\phi(t) = \log t$, then $E^{(\phi)}$ reduces to the usual expectation. Let H_f be a Hilbert space with the inner product defined by $\langle a, b \rangle_f = E_f^{(\phi)}\{a(X)b(X)\}$, and the tangent space

$$T_f = \{a \in H_f : \langle a, 1 \rangle_f = 0\}.$$

For a statistical model $M = \{f_\theta(x)\}_{\theta \in \Theta}$

$$E_{f_\theta}^{(\phi)}\{\partial_i \phi(f_\theta(X))\} = 0$$

for all θ of Θ, where $\partial_i = \partial/\partial\theta_i$ with $\theta = (\theta_i)_{i=1,\cdots,p}$. Further,

$$E_{f_\theta}^{(\phi)}\{\partial_i \partial_j \phi(f_\theta(X))\} = E_{f_\theta}^{(\phi)}\left\{ \frac{\phi''(f_\theta(X))}{\phi'(f_\theta(X))^2} \partial_i \phi(f_\theta(X)) \partial_i \phi(f_\theta(X)) \right\}. \qquad (4)$$

If $\phi(t) = \log t$, then (4) is nothing but the Bartlett identity. Here we have to remark that the existence of the generalized expectation and these two identities are not always guaranteed, so we assume the existence and differentiability.

The geometric meaning of the ϕ-path is important to understand the statistical applications via ϕ-path. Let C be a path connecting f and g such that C is parametrized as $f_t(x)$, where $f_0 = f$ and $f_1 = g$. Define $A_t^{(\phi)}(x)$ of T_{f_t} by the solution for a differential equation

$$\dot{A}_t^{(\phi)}(x) - \mathrm{E}_{f_t}^{(\phi)}\left\{A_t^{(\phi)} \dot{f}_t \frac{\phi''(f_t)}{\phi'(f_t)}\right\} + \mathrm{E}_{f_t}^{(\phi)}\{A_t^{(\phi)}(X)\}\mathrm{E}_{f_t}^{(\phi)}\left\{\frac{\phi''(f_t(X))}{\phi'(f_t(X))}\dot{f}_{t(X)}\right\} = 0.$$

(5)

Then we confirm that

$$\frac{d}{dt}\mathrm{E}_{f_t}^{(\phi)}\{A_t^{(\phi)}(X)\} = 0.$$

The idea of parallel transport provides a characterization of ϕ-path defined in Sect. 2.

Theorem 2. *The geodesic curve $\{f_t\}_{0\leq t\leq 1}$ by the parallel displacement $A^{(\phi)}$ is the ϕ-path defined in (3).*

Proof. We apply the parallel displacement to $A_t^{(\phi)}(x) = (d/dt)\phi(f_t(x))$ as

$$\frac{d^2}{dt^2}\phi(f_t(x)) - \mathrm{E}_{f_t}^{(\phi)}\left\{\frac{d}{dt}\phi(f_t(x))\dot{f}_t \frac{\phi''(f_t)}{\phi'(f_t)}\right\} + \mathrm{E}_{f_t}^{(\phi)}\left\{\frac{d}{dt}\phi(f_t(x))\right\}\mathrm{E}_{f_t}^{(\phi)}\left\{\dot{f}_t \frac{\phi''(f_t)}{\phi'(f_t)}\right\} = 0,$$

which is written by

$$\frac{d^2}{dt^2}\phi(f_t(x)) - \mathrm{E}_{f_t}^{(\phi)}\left\{\frac{d^2}{dt^2}\phi(f_t(x))\right\} = 0.$$

(6)

Hence we can solve this differential equation as

$$\phi(f_t(x)) = (1 - t)\phi(f(x)) + t\phi(g(x)) - \kappa_t,$$

or equivalently $f_t(x)$ is nothing but $f_t^{(\phi)}(x)$ as defined in (3). This completes the proof.

Similarly we define $B_t^{(\phi)}(x)$ of T_{f_t} by the solution for a differential equation

$$\dot{B}_t^{(\phi)}(x) - B_t^{(\phi)}\dot{f}_t \frac{\phi''(f_t)}{\phi'(f_t)} + B_t^{(\phi)}\mathrm{E}_{f_t}^{(\phi)}\left\{\dot{f}_t \frac{\phi''(f_t)}{\phi'(f_t)}\right\} = 0.$$

(7)

We observe the following by an argument similar to that just before Theorem 2.

$$\frac{d}{dt}\mathrm{E}_{f_t}^{(\phi)}\{B_t^{(\phi)}(X)\} = 0.$$

Applying to $B_t^{(\phi)}(x) = (d/dt)\phi(f_t(x))$ we find that

$$\phi'(f_t(x))\ddot{f}_t(x) = 0,$$

(8)

of which the solution is given by $f_t(x) = (1-t)f(x) + tg(x)$. Hence the geodesic curve by the parallel displacement $B^{(\phi)}$ is always the mixture curve, of which the discussion is independent of ϕ.

Basically the ϕ-path can connect between any f and g in $\mathcal{F}_P$. Let $f_0, \cdots, f_K$ be distinct in $\mathcal{F}_P$. Then the ϕ-surface is defined by

$$M^{(\phi)} = \{f_\pi(x) := \phi^{-1}\Big(\sum_{k=0}^{K} \pi_k \phi(f_k(x)) - \kappa_\pi\Big) : \pi \in \mathbb{S}_K\}$$

where $\pi = (\pi_0, \cdots, \pi_K)$ and $\mathbb{S}_K$ denotes a K-dimensional simplex $\{\pi : \pi_k \geq 0, \sum_{k=0}^{K} \pi_k = 1\}$. In this way ϕ-surface $M^{(\phi)}$ connects among any K pdf's f_k's. Further, $M^{(\phi)}$ is total geodesic.

Theorem 3. Let $\{f_t(x, \phi)\}_{0 \leq t \leq 1}$ be the ϕ-path connecting $f_{\pi_{(0)}}$ and $f_{\pi_{(1)}}$ of $M^{(\phi)}$. Then, $f_t(\cdot, \phi) \in M^{(\phi)}$ for any $t \in (0, 1)$.

Proof. By definition,

$$f_t(x, \phi) = \phi^{-1}((1-t)\phi(f_{\pi_{(0)}}(x)) + t\phi(f_{\pi_{(1)}}(x)) - \kappa_t)$$

which is nothing but $f_{\pi_t}(x)$, where $\pi_t = (1-t)\pi_{(0)} + t\pi_{(1)}$ We conclude that $\{f_t(x)\}_{0 \leq t \leq 1}$ is embedded in $M^{(\phi)}$ since $\{\pi_t\}_{0 \leq t \leq 1}$ is a curve in the simplex $\mathbb{S}_K$. The proof is complete.

4 Minimum Divergence Geometry

4.1 U-divergence and the Geodesic Associated

Assume that $U(s)$ is a convex and increasing function and let $\xi(t) = \mathrm{argmax}_s\{st - U(s)\}$. Then the U-divergence is defined

$$D_U(f, g) = \int \{U(\xi(g)) - f\xi(g)\}dP - \int \{U(\xi(f)) - f\xi(f)\}dP.$$

In fact, U-divergence is the difference of the cross entropy $C_U(f, g)$ with the diagonal entropy $C_U(f, f)$, where $C_U(f, g) = \int \{U(\xi(g)) - f\xi(g)\}dP$. See [8,9] for the detailed discussion.

We review the geometry generated by the U-divergence $D_U(f, g)$ based on a manifold of finite dimension $M = \{f_\theta(x) : \theta \in \Theta\}$ and vector fields X and Y on M as follows. The Riemannian metric is

$$G^{(U)}(X, Y)(f) = -\int Xf \, Y\xi(f)dP$$

for $f \in M$ and linear connections $\nabla^{(U)}$ and $^*\nabla^{(U)}$ are

$$G^{(U)}(\nabla_X^{(U)}Y, Z)(f) = -\int XYf \, Z\xi(f)dP \tag{9}$$

and

$$G^{(U)}(^*\nabla_X^{(U)}Y, Z)(f) = -\int Zf \, XY\xi(f)dP \tag{10}$$

for $f \in M$. The components are given by

$$G_{ij}^{(U)}(\theta) = -\mathrm{E}_{f_\theta}^{(\xi)}\left\{\partial_i\xi(f_\theta)\partial_j\xi(f_\theta)\right\}, \tag{11}$$

$$G^{(U)}(\nabla_{\partial_i}^{(U)}\partial_j, \partial_k)(\theta) = -\mathrm{E}_{f_\theta}^{(\xi)}\left\{\left(\partial_i\partial_j\xi(f_\theta) - \frac{\xi''(f_\theta)}{\xi'(f_\theta)^2}\partial_i\xi(f_\theta)\partial_j\xi(f_\theta)\right)\partial_k\xi(f_\theta)\right\} \tag{12}$$

and

$$G^{(U)}(^*\nabla_{\partial_i}^{(U)}\partial_j, \partial_k)(\theta) = -\mathrm{E}_{f_\theta}^{(\xi)}\left\{\partial_k\xi(f_\theta)\partial_i\partial_j\xi(f_\theta)\right\} \tag{13}$$

for $\theta \in \Theta$, where $\partial_i = \partial/\partial\theta_i$.

Let C be a one-parameter model parametrized by $C = \{f_t(x) : 0 \le t \le 1\}$. We consider a two-parameter model to include C as

$$M_1(h) = \{f_{(t,s)}(x) = (1-s)f_t(x) + sh(x) : 0 \le t \le 1, 0 \le s \le 1\}, \tag{14}$$

where h is arbitrarily fixed in $\mathcal{F}_P$. Apply $M_1(h)$ to the formula (12), and thus

$$G^{(U)}(\nabla_{\partial_1}^{(U)}\partial_1, \partial_2)(t, s)$$
$$= -\mathrm{E}_{f_{(t,s)}}^{(\xi)}\left\{\left(\partial_1\partial_1\xi(f_{(t,s)}) - \frac{\xi''(f_{(t,s)})}{\xi'(f_{(t,s)})^2}\partial_1\xi(f_{(t,s)})\partial_1\xi(f_{(t,s)})\right)\partial_2\xi(f_{(t,s)})\right\}.$$

Here we confirm that

$$\partial_2\xi(f_{(t,s)}) = \xi'(f_{(t,s)})(h - f_t)$$

and

$$\partial_1\partial_1\xi(f_{(t,s)}) - \frac{\xi''(f_{(t,s)})}{\xi'(f_{(t,s)})^2}\partial_1\xi(f_{(t,s)})\partial_1\xi(f_{(t,s)}) = (1-s)\xi'(f_{(t,s)})\partial_1\partial_1 f_{(t,s)}$$

Therefore,

$$G^{(U)}(\nabla_{\partial_1}^{(U)}\partial_1, \partial_2)(t, 0) = -\mathrm{E}_{f_t}^{(\xi)}\{(\xi'(f_t))^2\partial_1\partial_1 f_t(h - f_t)\}. \tag{15}$$

Let $\{f_t(x)\}_{0 \le t \le 1}$ is a $\nabla^{(U)}$-geodesic with an affine parameter t. By definition for any Riemannian metric G,

$$G(\nabla_{\dot{f}_t}^{(U)}\dot{f}_t, \eta_t) = 0 \tag{16}$$

for any η_t of T_{f_t}. We observe that if $G = G^{(U)}$ and $\eta_t = \partial_2$, then

$$G(\nabla_{\dot{f}_t}^{(U)}\dot{f}_t, \eta_t) = G^{(U)}(\nabla_{\partial_1}^{(U)}\partial_1, \partial_2)(t, 0)$$

which means that

$$\mathrm{E}_{f_t}^{(\xi)}\left\{(\xi'(f_t))^2\frac{d^2}{dt^2}f_t(h-f_t)\right\} = 0 \tag{17}$$

for any h of $\mathcal{F}_P$ according to $f_{(t,0)}(x) = f_t(x)$. Because (17) holds for any h in $\mathcal{F}_P$, we conclude that

$$\frac{d^2}{dt^2}f_t = c_1(t)\{\xi'(f_t)\}^{-1} \tag{18}$$

for almost everywhere x in the sense of P, where $c_1(t)$ is a constant in x. Immediately we know $c_1(t) = 0$ since the integration for both sides in (18) leads to $c_1(t)\int(\xi'(f_t))^{-1}dP = 0$. This implies that

$$f_t(x) = (1-t)f(x) + tg(x).$$

There is another discussion parallel to that for M_1. Consider another two-parameter model to include C, which is parametrized by

$$M_2(h) = \{f_{(t,s)}^*(x) = \xi^{-1}((1-s)\xi(f_t(x)) + s\xi(h(x)) - \kappa(t,s)) : 0 \le t \le 1, 0 \le s \le 1\},$$

where $h \in \mathcal{F}_P$ and $\kappa(t,s)$ is the normalizing factor. We note that $\kappa(t,0) = 0$. Applying $M_2(h)$ to the formula (13), an argument similar to that above yields that

$$G^{(U)}(^*\nabla_{\partial_1}^{(U)}\partial_1,\partial_2)(t,0) = -\mathrm{E}_{f_t}^{(\xi)}\left[\{-\xi(f_t) + \xi(h) - \partial_2\kappa(t,s)|_{s=0}\}\{\partial_1^2\xi(f_t) - \partial_1^2\kappa(t,0)\}\right]. \tag{19}$$

Assume that $f_t(x)$ is a $^*\nabla^{(U)}$-geodesic with an affine parameter t. Then,

$$G^{(U)}(^*\nabla_{\partial_1}^{(U)}\partial_1,\partial_2)(t,0) = 0$$

for any h of $\mathcal{F}_P$. It follows from (19) that $(d^2/dt^2)\xi(f_t(x)) = (d^2/dt^2)\kappa(t,0)$ for almost everywhere x in the sense of P, where $(d^2/dt^2)\kappa(t,0)$ is a constant in x. This implies that

$$f_t(x) = \xi^{-1}((1-t)\xi(f(x)) + t\xi(g(x)) - \kappa_t),$$

where $\kappa_t = \kappa(t,0)$. Now we can summarize our discussion as follows.

Theorem 4. *Let $\nabla^{(U)}$ and $^*\nabla^{(U)}$ be linear connections associated with U-divergence D_U as defined in (9) and (10) and let $C^{(\phi)}$ be the ϕ path connecting f and g of $\mathcal{F}_P$ as defined in (3). Then a ∇-geodesic curve connecting f and g is equal to $C^{(\mathrm{id})}$, where id denotes the identity function; while a $^*\nabla$-geodesic curve C_t^* connecting f and g is equal to $C^{(\xi)}$, where*

$$\xi(t) = \mathrm{argmax}_s\{st - U(s)\}.$$

5 Conclusion

We cast a new light on the Kolmogorov-Nagumo average in which a ϕ-path is defined in a simple manner via the generator function ϕ. The idea of ϕ-path is robust and generative, so that we can define the ϕ-path connecting between arbitrary two pdf's without any integrability assumptions. Thus the geometry associated with ϕ-path is appropriate to object a space of all pdf's. The approach to the infinite-dimensional space by Orlicz space is rigorous with a framework of Banach manifold, cf. [21]. However, the theory does not have a direct connection with the nonparametric statistics which has been established in statistical literature. The framework of Hilbert space is more feasible to apply to discussion of statistics and machine learning based on an empirical datasets [18]. In particular, this is compatible with the framework on semiparametric inference in which the statistical model has parametric and nonparametric components. However, it is necessary to assume some integrability conditions such as finite entropy condition. In this sense the approach by ϕ-path is more robust.

We have discussed a specific property of ϕ-path with respect to the geometry associated with U-divergence. There are several applications of statistics and machine learning using U-divergence for robust statistics and boosting learning algorithm, see [10,11,13,14,20,24]. In the class of U-divergence a extensive perspective is given in [4,5]. The U-divergence is closely related with Tsallis entropy [25,26] if U is taken by a power exponential function. We will discuss these applications in the ϕ-path geometry as a future work. The application for semiparametric model and inference will be promising from this viewpoint.

Acknowledgments. Authors were supported by Japan Science and Technology Agency (JST), Core Research for Evolutionary Science and Technology (CREST), and express sincere gratitude to the reviewers for their helpful comments and suggestions for improving the original manuscript.

References

1. Amari, S.: Differential-Geometrical Methods in Statistics. Lecture notes in Statistics, vol. 28. Springer, New York (1985)
2. Amari, S., Nagaoka, H.: Methods of Information Geometry. Oxford University Press, Oxford (2000)
3. Amari, S.-I.: Information geometry of positive measures and positive-definite matrices: decomposable dually flat structure. Entropy **16**, 2131–2145 (2014)
4. Cichocki, A., Amari, S.I.: Families of alpha-beta-and gamma-divergences: flexible and robust measures of similarities. Entropy **12**, 1532–1568 (2010)
5. Cichocki, A., Cruces, S., Amari, S.: Generalized alpha-beta divergences and their application to robust nonnegative matrix factorization. Entropy **13**, 134–170 (2011)
6. Eguchi, S.: Second order efficiency of minimum contrast estimators in a curved exponential family. Ann. Stat. **11**, 793–803 (1983)
7. Eguchi, S.: Geometry of minimum contrast. Hiroshima Math. J. **22**, 631–647 (1992)
8. Eguchi, S.: Information geometry and statistical pattern recognition. Sugaku Expositions Amer. Math. Soc. **19**, 197–216 (2006)

9. Eguchi, S.: Information divergence geometry and the application to statistical machine learning. In: Emmert-Streib, F., Dehmer, M. (eds.) Information Theory and Statistical Learning, pp. 309–332. Springer, US (2008)

10. Eguchi, S., Kato, S.: Entropy and divergence associated with power function and the statistical application. Entropy **12**, 262–274 (2010)

11. Eguchi, S., Komori, O., Kato, S.: Projective power entropy and maximum tsallis entropy distributions. Entropy **13**, 1746–1764 (2011)

12. Eguchi, S., Komori, O., Ohara, A.: Duality of maximum entropy and minimum divergence. Entropy **16**(7), 3552–3572 (2014)

13. Fujisawa, H., Eguchi, S.: Robust parameter estimation with a small bias against heavy contamination. J. Multivar. Anal. **99**, 2053–2081 (2008)

14. Mihoko, M., Eguchi, S.: Robust blind source separation by beta divergence. Neural Comput. **14**(8), 1859–1886 (2002)

15. Naudts, J.: The q-exponential family in statistical physics. Cent. Eur. J. Phys. **7**, 405–413 (2009)

16. Naudts, J.: Generalized exponential families and associated entropy functions. Entropy **10**, 131–149 (2008)

17. Naudts, J.: Generalized Thermostatistics. Springer, London (2011)

18. Newton, N.: An infinite dimensional statistical manifold modelled on Hilbert space. J. Funct. Anal. **263**, 1661–1681 (2012)

19. Nielsen, F.: Generalized Bhattacharyya and Chernoff upper bounds on Bayes error using quasi-arithmetic means. Pattern Recogn. Lett. **42**, 25–34 (2014)

20. Notsu, A., Komori, O., Eguchi, S.: Spontaneous clustering via minimum gamma-divergence. Neural Comput. **26**, 421–448 (2014)

21. Pistone, G., Sempi, C.: An infinite-dimensional geometric structure on the space of all the probability measures equivalent to a given one. Ann. Stat. **23**, 1543–1561 (1995)

22. Pistone, G., Rogantin, M.: The exponential statistical manifold: mean parameters, orthogonality and space transformations. Bernoulli **5**, 721–760 (1999)

23. Santacroce, M., Siri, P., Trivellato, B.: new results on mixture and exponential models by Orlicz spaces. Bernoulli (2015, to appear)

24. Takenouchi, T., Eguchi, S.: Robustifying AdaBoost by adding the naive error rate. Neural Comput. **16**, 767–787 (2004)

25. Tsallis, C.: Possible generalization of Boltzmann-Gibbs statistics. J. Stat. Phys. **52**, 479–487 (1988)

26. Tsallis, C.: Introduction to Nonextensive Statistical Mechanics. Springer, New York (2009)

27. Zhang, J.: Nonparametric information geometry: from divergence function to referential-representational biduality on Statistical Manifolds. Entropy **15**, 5384–5418 (2013)

Lie Groups: Novel Statistical and Computational Frontiers

Image Processing in the Semidiscrete Group of Rototranslations

Dario Prandi[1]($\boxtimes$), Ugo Boscain[2,3], and Jean-Paul Gauthier[3,4]

[1] CNRS, CEREMADE, Univ. Paris-Dauphine, Paris, France
prandi@ceremade.dauphine.fr
[2] CNRS, CMAP, Ecole Polytechnique, Route de Saclay, 91128 Saclay, France
[3] INRIA Team GECO, Saclay, France
[4] LSIS, Université de Toulon USTV, 83957 La Garde Cedex, France

Abstract. It is well-known, since [12], that cells in the primary visual cortex V1 do much more than merely signaling position in the visual field: most cortical cells signal the local orientation of a contrast edge or bar – they are tuned to a particular local orientation. This orientation tuning has been given a mathematical interpretation in a sub-Riemannian model by Petitot, Citti, and Sarti [6,14]. According to this model, the primary visual cortex V1 lifts grey-scale images, given as functions $f : \mathbb{R}^2 \to [0,1]$, to functions Lf defined on the projectivized tangent bundle of the plane $PT\mathbb{R}^2 = \mathbb{R}^2 \times \mathbb{P}^1$. Recently, in [1], the authors presented a promising semidiscrete variant of this model where the Euclidean group of roto-translations $SE(2)$, which is the double covering of $PT\mathbb{R}^2$, is replaced by $SE(2,N)$, the group of translations and discrete rotations. In particular, in [15], an implementation of this model allowed for state-of-the-art image inpaintings.

In this work, we review the inpainting results and introduce an application of the semidiscrete model to image recognition. We remark that both these applications deeply exploit the Moore structure of $SE(2,N)$ that guarantees that its unitary representations behaves similarly to those of a compact group. This allows for nice properties of the Fourier transform on $SE(2,N)$ exploiting which one obtains numerical advantages.

1 The Semi-discrete Model

The starting point of our work is the sub-Riemannian model of the primary visual cortex V1 [6,14], and our recent contributions [1–4]. This model has also been deeply studied in [8,11]. In the sub-Riemannian model, V1 is modeled as the projective tangent bundle $PT\mathbb{R}^2 \cong \mathbb{R}^2 \times \mathbb{P}^1$, whose double covering is the roto-translation group $SE(2) = \mathbb{R}^2 \rtimes \mathbb{S}^1$, endowed with a left-invariant sub-Riemannian structure that mimics the connections between neurons. In particular, grayscale visual stimuli $f : \mathbb{R}^2 \to [0,1]$ feeds V1 neurons $N = (x, \theta) \in PT\mathbb{R}^2$

This research has been supported by the European Research Council, ERC StG 2009 "GeCoMethods", contract number 239748 and by the iCODE institute, research project of the Idex Paris-Saclay

© Springer International Publishing Switzerland 2015
F. Nielsen and F. Barbaresco (Eds.): GSI 2015, LNCS 9389, pp. 627–634, 2015.
DOI: 10.1007/978-3-319-25040-3_67

with an extracellular voltage $\mathcal{L}f(\xi)$ that is widely accepted to be given by $\mathcal{L}f(\xi) = \langle f, \Psi_\xi \rangle$. The functions $\{\Psi_\xi\}_{\xi \in PT\mathbb{R}^2}$ are the *receptive fields*. A good fit is $\Psi_{(x,\theta)} = \pi(x,\theta)\Psi$ where Ψ is the Gabor filter (a sinusoidal multiplied by a Gaussian function) and $\pi(x,\theta)\Psi(y) := \Psi(R_{-\theta}(x-y))$.

In this work we consider a slightly different setting, by assuming that *neurons are sensible only to a finite (small) number of orientations*. This assumption is based on the observation of the organization of the visual cortex in pinwheels: we conjecture that there are topological constraints that prevent the possibility of detecting a continuum of directions even when sending the distance between pinwheels to zero. This assumption leads us to consider the group of translations and discrete rotations $SE(2, N) = \mathbb{R}^2 \rtimes \mathbb{Z}_N$, for some $N \in \mathbb{N}$, where the action of $k \in \mathbb{Z}_N$ on $\mathbb{R}^2$ is the rotation of angle $2\pi k/N$.

To be more precise, our model is based on the following assumptions:

1. Grayscale visual stimuli coming from the retina are modeled as functions $f \in L^2(\mathbb{R}^2)$;
2. The primary visual cortex is modeled as $SE(2, N)$ and its activation patterns as $\varphi \in L^2(SE(2, N))$;
3. There exists a linear function $\mathcal{L} : L^2(\mathbb{R}^2) \to L^2(SE(2, N))$ that lifts visual stimuli to activation patterns in the primary visual cortex, of the form $\mathcal{L}f(x,k) = \langle f, \pi(x,k)\Psi \rangle$ for some $\Psi \in L^2(\mathbb{R}^2)$.
4. An excited neuron activate neighboring neurons according to the SDE

$$dA_t = X_1 dW_t + d\Theta_t, \tag{1}$$

where $X_1(x,k) = \cos(2\pi k/N)\partial_{x_1} + \sin(2\pi k/N)\partial_{x_2}$, W is a Wiener process and Θ is a Poisson jump process on $\mathbb{Z}_N$ with jump probability equal to $1/2$ on both sides.

Remark 1. Observe that the lift operator $\mathcal{L}$ respects the *shift-twist* symmetry of V1. (See e.g. [5].) That is, letting Λ be the left regular representation of $SE(2, N)$ in $L^2(SE(2, N))$ (i.e., $[\Lambda(x,k)\varphi](y,h) = \varphi((x,k)^{-1}(y,h))$) and π the quasi-regular representation of $SE(2, N)$ in $L^2(\mathbb{R}^2)$ (i.e., $[\pi(x,k)f](y) = f(R_{-k}(y-x)))$, it holds $\Lambda(x,k)\mathcal{L}f = L(\pi(x,k)f)$.

2   Image Inpainting

The algorithm we now present for image inpainting is inspired by the neurophysiological process of amodal completion, that is, the perception of a shape even when it is not actually drawn. A famous example is that of the Kanizsa triangle. Our working assumption is that amodal completion is caused by the following neurophysiological principle: *Corrupted images are reconstructed by the natural diffusion in V1, induced by the SDE (4), which for small times follows the less expensive paths (geodesics) to activate unexcited neurons.*

From the practical point of view, images are reconstructed through the following algorithm. For details see [1].

Algorithm 1. Image inpainting algorithm

Data: Input (corrupted) image $f \in L^2(\mathbb{R}^2)$. The corrupted points are assumed to be those x's such that $f(x) = 0$.
Result: Inpainted image $\tilde{f} \in L^2(\mathbb{R}^2)$.

1 $h \leftarrow$ GaussianFilter(f)
2 $\mathcal{L}h \leftarrow$ Lift(h)
3 $\mathcal{L}h \leftarrow$ EvolveDiffusion($\mathcal{L}h$)
4 **return** Project($\mathcal{L}h$)

A description of the 4 functions used in Algorithm 1 follows.

1. GaussianFilter: Smooths the input via a Gaussian filter. As explained in [2], the result of this procedure is generically a Morse function (i.e. with isolated non-degenerate critical points only).
2. Lift: Given a Morse function h lifts it to Lh defined on $SE(2, N)$, obtained as follows. We let $\theta(x) \in [0, \pi)$ to be orientation of $\nabla h(x)$, when it is well defined. Then, we define $Lh(x, k) = h(x)$ if $k \cong \theta(x)$ and 0 otherwise Here, the formulation $k \cong \theta(x)$ means that $2\pi k/N$ is the nearest point to $\theta(x)$ among $\{2\pi\ell/N \mid \ell \in \mathbb{Z}_N\}$. Since h is a Morse function, $\theta(x)$ is not well defined on isolated points. In this case, we let $Lh(x, h) := h(x)/N$ for any $h \in \mathbb{Z}_N$.
3. EvolveDiffusion: Given a function Lh on $SE(2, N)$ evolves it according to (4). An efficient way to compute this diffusion is presented in [1], and recalled in Algorithm 2.
4. Project: Given a function φ on $SE(2, N)$ returns its projection on $\mathbb{R}^2$ defined as $P\varphi(x) := \max_{k \in \mathbb{Z}_N} \varphi(x, k)$.

Remark 2. The Lift procedure detailed above is not obtained via a convolution with an oriented wavelet as it is supposed to be the case in V1. However, it can be seen as the limit when the support of the wavelet tends to zero and experiments have shown that it yields more precise reconstructions.

Algorithm 2. Evolution of the diffusion, as explained in [1]

---

1 **Function** EvolveDiffusion:
    Data: A function φ on $SE(2, N)$
    Result: The evolved function $\tilde{\varphi}$
2   For $k = 0, \ldots, N - 1$ let $\hat{\varphi}_k \leftarrow \mathcal{F}(\varphi(\cdot, k))$
3   For $x \in \mathbb{R}^2$ let $\{\hat{\psi}_k(x)\}_k \leftarrow$ Solution of an ODE with datum $\{\hat{\varphi}_k(x)\}_k$
4   For $k = 0, \ldots, N - 1$ let $\psi(\cdot, k) \leftarrow \mathcal{F}^{-1}(\hat{\psi}_k)$
5   **return** ψ
6 **end**

In Fig. 1 we present two different inpainting results. While the first one is obtained using Algorithm 1, to produce the second one we added some heuristic

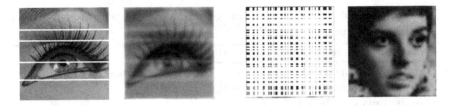

Fig. 1. Two inpaintings.

procedure (detailed in [1,15]) in order to prevent the diffusion from modifying the non-corrupted part of the image.

3   Image Recognition

The fact that images are lifted to V1, which has the (group) structure of $SE(2, N)$ allows for a natural description of the process of invariant image recognition. (That is, recognizing images under the roto-translation action of $SE(2, N)$.) Namely, we propose to use the *bispectrum* as an invariant under the action of $SE(2, N)$. These invariants are well established in statistical signal processing [7] and have been introduced and studied in the context of $SE(2, N)$ and of compact groups in [18]. We mention also [13], devoted to the bispectrum on homogeneous spaces of compact groups.

Let us introduce some generalities on the (generalized) Fourier transform on $SE(2, N)$. Since this group is a non-commutative unimodular semi-direct product, computing the Fourier transform of an $L^2(SE(2, N))$ requires the knowledge of the (continuous) irreducible unitary representations T^λ of $SE(2, N)$. Here, λ is an index taking values in the dual object of $SE(2, N)$, which is denoted by $\widehat{SE(2, N)}$ and is the set of equivalence classes of irreducible unitary representations. (See, e.g., [10].) Exploiting the semi-direct product structure of $SE(2, N)$, by Mackey machinery this dual can be shown to be the union of the slice $\mathcal{S} \subset \mathbb{R}^2$, which in polar coordinates is $\mathbb{R}_+^* \times [0, 2\pi/N)$, to which we glue $\mathbb{Z}_N$ on 0. Since it is possible to show that to invert the Fourier transform it is enough to consider representations parametrized by $\mathcal{S}$, we will henceforth ignore the $\mathbb{Z}_N$ part of the dual. **A crucial fact** for the following is that $SE(2, N)$ is a Moore group, that is, all the T^λ act on finite-dimensional spaces, that is $\mathbb{C}^N$ for $\lambda \in \mathcal{S}$. This is **not** true for the roto-translation group $SE(2)$ and is indeed one of the main theoretical advantages of the semi-discrete model.

The matrix-valued Fourier coefficient of a function $\varphi \in L^2(SE(2, N)) \cap L^1(SE(2, N))$ for $\lambda \in \widehat{SE(2, N)}$ is $\hat{\varphi}(T^\lambda) = \int_{SE(2,N)} f(a) T^\lambda(a^{-1}) \, da$. This is essentially the same formula for the Fourier transform on $\mathbb{R}$, which is a scalar and is obtained using the representations $T^\lambda(x) = e^{2\pi i x\lambda}$. As usual, the above formula can be extended to a linear isometry $\mathcal{F} : L^2(SE(2, N)) \to L^2(\widehat{SE(2, N)})$. The **bispectrum** of φ is then the quantity

$$B_\varphi(\lambda_1, \lambda_2) = \hat{\varphi}(T^{\lambda_1}) \otimes \hat{\varphi}(T^{\lambda_2}) \circ \hat{\varphi}(T^{\lambda_1} \otimes T^{\lambda_2})^* \qquad \forall (\lambda_1, \lambda_2) \in \mathcal{S}.$$

This quantity can be interpreted as the Fourier transform of the triple correlation function, see [13].

In a forthcoming paper we will present (in a more general setting) the following result.

Theorem 1. *The bispectral invariants discriminate on the set $\mathcal{G}$ of functions $\varphi \in L^2(SE(2,N))$ such that the matrices $\hat{\varphi}(T^\lambda)$ are invertible for a.e. $\lambda \in \mathcal{S}$. That is, $\varphi_1, \varphi_2 \in \mathcal{G}$ are such that $B_{\varphi_1} = B_{\varphi_2}$ if and only if $\varphi_1 = \Lambda(x,k)\varphi_2$ for some $(x,k) \in SE(2,N)$.*

Unfortunately, when considering the lifts of visual stimuli $f \in L^2(\mathbb{R}^2)$ under lifts, an easy computation shows that $\widehat{\mathcal{L}f}(T^\lambda) = \overline{\omega_f(\lambda)} \otimes \overline{\omega_\Psi(\lambda)}$ where $\omega_f(\lambda) = (\hat{f}(R_{-k}\lambda))_{k \in \mathbb{Z}_N} \in \mathbb{C}^N$. This immediately implies that rank $\widehat{\mathcal{L}f}(T^\lambda) \leq 1$ and hence that range $\mathcal{L} \cap \mathcal{G} = \varnothing$.

Using the previous formula for the Fourier transform of lifted functions and under mild assumptions on the wavelet Ψ one can show that the bispectrum $B(\lambda_1, \lambda_2)$ is completely determined by the quantity

$$I_f^2(\lambda_1, \lambda_2) = \langle \omega_f(\lambda_1) \odot \omega_f(\lambda_2), \omega_f(\lambda_1 + \lambda_2) \rangle. \tag{2}$$

It is still an open question (although we conjecture it to be true) whether the bispectrum discriminates on a "big" set of range $\mathcal{L}$.

To bypass the difficulty posed by the non-invertibility of the Fourier transform for lifted functions, we are led to consider the **rotational bispectrum**:

$$\tilde{B}_\varphi(\lambda_1, \lambda_2, k) := \hat{\varphi}(T^{R_h \lambda_1}) \otimes \hat{\varphi}(T^{\lambda_2}) \circ \hat{\varphi}(T^{\lambda_1} \otimes T^{\lambda_2})^* \qquad \forall (\lambda_1, \lambda_2) \in \mathcal{S}, \forall h \in \mathbb{Z}_N.$$

Observe that the rotational bispectrum is invariant only under the action of $\mathbb{Z}_N \subset SE(2,N)$ but not under translations. To avoid this problem, let us consider the set $\mathcal{A} \subset L^2(\mathbb{R}^2)$ of compactly supported functions with non-zero average[1]. We can then define the barycenter $c_f \in \mathbb{R}^2$ of $f \in \mathcal{A}$ as

$$c_f = \frac{1}{\text{avg } f} \left(\int_{\mathbb{R}^2} x_1 f(x) \, dx, \int_{\mathbb{R}^2} x_2 f(x) \, dx \right), \qquad j = 1,2,$$

and the centering operator $\Phi : \mathcal{A} \to \mathcal{A}$ as $\Phi f(x) := f(x - c_f)$. Then, considering the lift $\mathcal{L}_c = \mathcal{L} \circ \Phi$, we have that $\mathcal{L}_c f = \mathcal{L}_c g$ if and only if g is a translate of f.

Finally, we have the following.

Theorem 2. *Let $\mathcal{R} \subset L^2(\mathbb{R}^2)$ be the set of compactly supported functions f such that*

1. $\hat{f}(\lambda) \neq 0$ for a.e. $\lambda \in \mathbb{R}^2$;
2. the circulant matrix associated with $\omega_f(\lambda)$ is invertible for a.e. $\lambda \in \mathbb{R}^2$.

Then, if $\Psi \in \mathcal{R}$ the rotational bispectrum discriminates on $\mathcal{L}_c(\mathcal{R} \cap \mathcal{A})$. That is, for any $f, g \in \mathcal{R} \cap \mathcal{A}$ it holds that $\tilde{B}_{\mathcal{L}_c f} = \tilde{B}_{\mathcal{L}_c g}$ if and only if $f = \pi(x,k)g$ for some $(x,k) \in SE(2,N)$.

Moreover, since set $\mathcal{R}$ is residual in the compactly supported functions $L^2(\mathbb{R}^2)$, the rotational bispectrum is generically discriminating on the compactly supported functions of $L^2(\mathbb{R}^2)$.

---

[1] Recall that the average of $f \in L^1(\mathbb{R}^2)$ is avg $f = \int_{\mathbb{R}^2} f(x) \, dx$.

Let us observe that, if $\Psi \in \mathcal{R}$, then $\tilde{B}_{\mathcal{L}_c f}(\lambda_1, \lambda_2, k)$ is completely determined by the quantities

$$I_f^2(\lambda_1, \lambda_2, k) = \langle \omega_{\Phi f}(R_k \lambda_1) \odot \omega_{\Phi f}(\lambda_2), \omega_{\Phi f}(\lambda_1 + \lambda_2) \rangle.$$

In particular, computing the rotational bispectrum requires N times more operations than computing the bispectrum.

3.1   Implementation and Numerical Experiments

We now describe how to efficiently compute the bispectrum invariants[2]. The same method, with the obvious modifications, also works for the computation of the rotational bispectrum. We then show that the difference in norm of the bispectrum and the rotational bispectrum strongly separates images of different objects. The next natural step, that we will tackle in a forthcoming paper, is to use these invariants in machine learning algorithms as SVM's or AdaBoosts.

As previously remarked, to compute the bispectrum invariants it is enough to compute the quantities $I_f^2(\lambda_1, \lambda_2)$ given in (3). Thus, the main obstacle is to efficiently and precisely compute the vectors $\omega_f(\lambda)$ for a given λ. This vector is obtained by evaluating the Fourier transform of f on the orbit of λ under the action of the rotations $R_{\frac{2\pi k}{N}}$ for $k \in \mathbb{Z}_N$.

Since the Fourier transform $\hat{f}$ of an image is given as a discrete matrix, the usual way to proceed would be to implement rotations as functions on the plane and then evaluate $\hat{f}(R_{\frac{2\pi k}{N}} \lambda)$ by bilinear interpolation on the values of $\hat{f}$. However, this requires a lot of matrix products and, especially for values of λ very near to 0 where most of the information for natural images is contained, is prone to errors.

We thus chose to consider only $N = 6$ and to work with images composed of hexagonal pixels. This choice was motivated by the following reasons:

- It is well-known that retinal cells are distributed in an hexagonal grid.
- Hexagonal grids are invariant under the action of $\mathbb{Z}_6$ and discretized translations, which is the most we can get in the line of the invariance w.r.t. $SE(2,6)$.
- We can exploit the Spiral Architecture introduced by Sheridan [16,17]. This is a way to index hexagons of the grid with only one index which allows to introduce an operation, *spiral multiplication*, that, with the same complexity of a normal multiplication, computes rotations by multiples $\pi/3$.
- There exist efficient methods [9] to simulate hexagonal pixels by oversampling the image by a ratio of 7 and then using so-called *hyperpels* composed of 56 pixels to approximate an hexagonal pixel.

Indeed, once the spiral addressing described in [9] has been implemented in the function `SpiralAddr` and the spiral multiplication in `SpiralMult`, to evaluate $\omega_f(\lambda)$ it suffices to apply Algorithm 3.

---

[2] The iPython notebook with the code is available at http://nbviewer.ipython.org/github/dprn/GSI15/blob/master/Invariants-computation.ipynb.

Algorithm 3. Evaluation of $\omega_f(\lambda)$

---

1 **Function** Omega(F,λ):

 Data: F: FFT of the input image f oversampled by a factor of 7
 λ: The spiral address of the hexagon where to compute $\omega_f(\lambda)$
 Result: The vector $\omega_f(\lambda)$

2   For $k = 0,\ldots,N-1$ let $\omega_f(\lambda)_k \leftarrow$ SpiralAddr(F,SpiralMult(λ,k))

3   **return** $\omega_f(\lambda)$

4 **end**

| | bisp. | | rot. bisp. | |
	Max.	Min	Max.	Min
Triangle	9.2×10^{10}	7.0×10^{12}	2.2×10^{11}	1.7×10^{13}
Rectangle	7.9×10^{10}	8.2×10^{12}	1.9×10^{11}	2.0×10^{13}
Ellipse	7.4×10^{10}	7.1×10^{12}	1.8×10^{11}	1.7×10^{13}
Star	7.2×10^{10}	5.5×10^{12}	1.7×10^{11}	1.3×10^{13}
Diamond	3.7×10^{10}	5.4×10^{12}	9.2×10^{10}	1.3×10^{13}

Fig. 2. Hexagons used in the computation of the invariants. The covered area corresponds to roughly 11 square pixels.

Fig. 3. Results of the comparisons of the invariants.

To test the invariants, we built a library composed of 5 geometrical figures rotated of angles $\pi k/3$, $k \in \mathbb{Z}_6$, and 14 natural images and computed the invariants corresponding to λ_1 and λ_2 chosen between the subset of central hexagons of the grid shown in Fig. 2, obtaining a vector I_f^2 of invariants with 49 elements. Then, for each geometrical figure f, we computed the difference in norm $\|I_f^2 - I_g^2\|$ between its invariants and those of another image g, for all the images. In the second and third column of Table 3 we reported the maximal difference w.r.t. the rotated of the same image and the minimal difference w.r.t. the other images. In particular, since the difference between these two values is at least in the order of 10^2, we observe that already simply using the norm seems to be a good discriminating factor for these simple images.

We then repeated the same test with the rotational bispectrum, whose results are reported in the fourth and fifth column of Fig. 3. We point out that, accordingly to our conjecture regarding the completeness of the bispectrum, considering the rotational invariants do not seem to add discriminating power.

4   Conclusions

In this work we presented a framework for image reconstruction and invariant recognition. We remark that the numerical work for the image recognition part has just started. Presently, we are testing the bispectrum as a source of invariant for different machine learning algorithms. In particular, the AdaBoost algorithm seems very promising and well adapted to the problem.

References

1. Boscain, U., Chertovskih, R., Gauthier, J.P., Remizov, A.: Hypoelliptic diffusion and human vision: a semi-discrete new twist on the petitot theory. SIAM J. Imaging Sci. **7**(2), 669–695 (2014)
2. Boscain, U., Duplaix, J., Gauthier, J.P., Rossi, F.: Anthropomorphic image reconstruction via hypoelliptic diffusion. SIAM J. Control Optim. **46**, 1–25 (2012)
3. Boscain, U., Gauthier, J.P., Prandi, D., Remizov, A.: Image reconstruction via non-isotropic diffusion in Dubins/Reed-Shepp-like control systems. In: 53rd IEEE Conference on Decision and Control, pp. 4278–4283 (2014)
4. Boscain, U., Charlot, G., Rossi, F.: Existence of planar curves minimizing length and curvature. Tr. Mat. Inst. Steklova **270**, 49–61 (2010)
5. Bressloff, P.C., Cowan, J.D., Golubitsky, M., Thomas, P.J., Wiener, M.C.: Geometric visual hallucinations, Euclidean symmetry and the functional architecture of striate cortex. Philos. Trans. R. Soc. Lond. B. Biol. Sci. **356**, 299–330 (2001)
6. Citti, G., Sarti, A.: A cortical based model of perceptual completion in the Roto-Translation space. J. Math. Imaging Vis. **24**(3), 307–326 (2006)
7. Dubnov, S., Tishby, N., Cohen, D.: Polyspectra as measures of sound texture and timbre. J. New Music Res. **26**(4), 277–314 (1997)
8. Duits, R., Franken, E.: Left-invariant parabolic evolutions on SE(2) and contour enhancement via invertible orientation scores. Part I: linear left-invariant diffusion equations on SE(2). Q. Appl. Math. **68**(2), 255–292 (2010)
9. He, X., Hintz, T., Wu, Q., Wang, H., Jia, W.: A new simulation of spiral architecture. In: Proceedings of the 2006 International Conference Image Processing Computer Vision, Pattern Recognition, Las Vegas, Nevada, USA, 26–29 June 2006
10. Hewitt, E., Ross, K.: Abstract Harmonic Analysis, vol. 1. Springer, New York (1963)
11. Hladky, R.K., Pauls, S.D.: Minimal surfaces in the roto-translation group with applications to a neuro-biological image completion model. J. Math. Imaging Vis. **36**(1), 1–27 (2010)
12. Hubel, D., Wiesel, T.: Receptive fields of single neurones in the cat's striate cortex. J. Physiol. **148**, 574–591 (1959)
13. Kakarala, R.: The bispectrum as a source of phase-sensitive invariants for fourier descriptors: a group-theoretic approach. J. Math. Imaging Vis. **44**(3), 341–353 (2012)
14. Petitot, J.: Neurogéométrie de la vision - Modèles mathématiques et physiques des architectures fonctionnelles. Les Éditions de l'École Polytechnique (2008)
15. Prandi, D., Remizov, A., Chertovskih, R., Boscain, U., Gauthier, J.P.: Highly corrupted image inpainting through hypoelliptic diffusion (2015). (arXiv:1502.07331)
16. Sheridan, P.: Spiral architecture for machine vision. Ph.D. thesis (1996)
17. Sheridan, P., Hintz, T., Alexander, D.: Pseudo-invariant image transformations on a hexagonal lattice. Image Vis. Comput. **18**, 907–917 (2000)
18. Smach, F., Lemaître, C., Gauthier, J.P., Miteran, J., Atri, M.: Generalized fourier descriptors with applications to objects recognition in SVM context. J. Math. Imaging Vis. **30**, 43–71 (2008)

Universal, Non-asymptotic Confidence Sets for Circular Means

Thomas Hotz$^{(\boxtimes)}$, Florian Kelma, and Johannes Wieditz

Institute of Mathematics, Technische Universität Ilmenau, 98684 Ilmenau, Germany
{thomas.hotz,florian.kelma,johannes.wieditz}@tu-ilmenau.de

Abstract. Based on Hoeffding's mass concentration inequalities, non-asymptotic confidence sets for circular means are constructed which are universal in the sense that they require no distributional assumptions. These are then compared with asymptotic confidence sets in simulations and for a real data set.

1 Confidence Sets for Means of Circular Data

In applications, data assuming values on the circle, i.e. *circular data*, arise frequently, examples being measurements of wind directions, or time of the day patients are admitted to a hospital unit. We refer to the literature for an overview of statistical methods for circular data, e.g. [5,9,10].

Here, we will concern ourselves with the arguably simplest statistic, the *mean*. But, given that a circle does not carry a vector space structure, i.e. there is neither a natural addition of points on the circle nor can one divide them by a natural number, what should the meaning of "mean" be?

In order to simplify the exposition, we specifically consider the unit circle in the complex plane, $\mathbb{S}^1 = \{z \in \mathbb{C} : |z| = 1\}$, and we assume the data can be modelled as independent random variables $Z_1, \ldots, Z_n$ which are identically distributed as the random variable Z taking values in $\mathbb{S}^1$.

Of course, $\mathbb{C}$ is a Euclidean (i.e. real) vector space, so the *Euclidean sample mean* $\bar{Z}_n = \frac{1}{n} \sum_{k=1}^{n} Z_k \in \mathbb{C}$ is well-defined. However, unless all Z_k take identical values, it will (by the strict convexity of the closed unit disc) lie inside the circle, i.e. its *modulus* $|\bar{Z}_n|$ will be less than 1. Though $\bar{Z}_n$ cannot be taken as a mean on the circle, if $\bar{Z}_n \neq 0$ one might say that it specifies a direction; this leads to the idea of calling $\bar{Z}_n/|\bar{Z}_n|$ the circular sample mean of the data.

Observing that the Euclidean sample mean is the minimiser of the sum of squared distances, this can be put in the more general framework of *Fréchet means* [6]: define the *set of circular sample means* to be

$$\hat{\mu}_n = \operatorname*{argmin}_{\zeta \in \mathbb{S}^1} \sum_{k=1}^{n} |Z_k - \zeta|^2, \tag{1}$$

T. Hotz—wishes to thank Stephan Huckemann from the Georgia Augusta University of Göttingen for fruitful discussions concerning the first construction of confidence regions described in Sect. 2.

F. Nielsen and F. Barbaresco (Eds.): GSI 2015, LNCS 9389, pp. 635–642, 2015.
DOI: 10.1007/978-3-319-25040-3_68

and analoguously define the *set of circular population means* of the random variable Z to be

$$\mu = \underset{\zeta \in \mathbb{S}^1}{\operatorname{argmin}} \, \mathbf{E}|Z - \zeta|^2. \tag{2}$$

Then, as usual, the circular sample means are the circular population means with respect to the empirical distribution of $Z_1, \ldots, Z_n$.

The circular population mean can be related to the Euclidean population mean $\mathbf{E}Z$ by noting that $\mathbf{E}|Z - \zeta|^2 = \mathbf{E}|Z - \mathbf{E}Z|^2 + |\mathbf{E}Z - \zeta|^2$ (in statistics, this is called the *bias-variance decomposition*), so that

$$\mu = \underset{\zeta \in \mathbb{S}^1}{\operatorname{argmin}} \, |\mathbf{E}Z - \zeta|^2 \tag{3}$$

is the set of points on the circle closest to $\mathbf{E}Z$. It follows that μ is unique if and only if $\mathbf{E}Z \neq 0$ in which case it is given by $\mu = \mathbf{E}Z/|\mathbf{E}Z|$, the *orthogonal projection* of $\mathbf{E}Z$ onto the circle; otherwise, i.e. if $\mathbf{E}Z = 0$, the set of circular population means is all of $\mathbb{S}^1$. Analogously, $\hat{\mu}_n$ is either all of $\mathbb{S}^1$ or uniquely given by $\bar{Z}_n/|\bar{Z}_n|$ according as $\bar{Z}_n$ is 0 or not.

The expected squared distances minimised in (2) are given by the metric inherited from the ambient space $\mathbb{C}$; therefore μ is also called the set of *extrinsic* population means. Had we measured distances intrinsically along the circle, i.e. using arc-length instead of chordal distance, we had obtained what is called the set of *intrinsic* population means. We will not consider the latter in the following, see e.g. [8] for a comparison and [1,2] for generalizations of these concepts.

Our aim is to construct *confidence sets* for the circular population mean μ which form a superset of μ with a certain (so-called) *coverage probability* which is required to be not less than some pre-specified signifance level $1 - \alpha \in (0,1)$.

The classical approach is to construct an *asymptotic confidence interval* where the coverage probability converges to $1 - \alpha$ when n tends to infinity. This can be done as follows: since Z is a bounded random variable, $\sqrt{n}(\bar{Z}_n - \mathbf{E}Z)$ converges to a bivariate normal distribution when identifying $\mathbb{C}$ with $\mathbb{R}^2$. Now, assume $\mathbf{E}Z \neq 0$ so μ is unique. Then the orthogonal projection is differentiable in a neighbourhood of $\mathbf{E}Z$, so the δ-method can be applied and one easily obtains

$$\sqrt{n} \, \mathbf{Arg}(\mu^{-1}\hat{\mu}_n) \xrightarrow{\mathcal{D}} \mathcal{N}\left(0, \frac{\mathbf{E}(\mathrm{Im}(\mu^{-1}Z))^2}{|\mathbf{E}Z|^2}\right) \tag{4}$$

where $\mathbf{Arg} : \mathbb{C} \setminus \{0\} \to (-\pi, \pi] \subset \mathbb{R}$ denotes the *argument* of a complex number (it is defined arbitrarily at $0 \in \mathbb{C}$), while multiplying with μ^{-1} rotates such that $\mathbf{E}Z = \mu$ is mapped to $0 \in (-\pi, \pi]$, see e.g. [9, Proposition 3.1] or [8, Theorem 5]. Estimating the asymptotic variance and applying Slutsky's lemma, one arrives at the asymptotic confidence set $C_A = \{\zeta \in \mathbb{S}^1 : |\mathbf{Arg}(\zeta^{-1}\hat{\mu}_n)| < \delta_A\}$ provided $\hat{\mu}_n$ is unique, where the angle determining the interval is given by

$$\delta_A = \frac{q_{1-\frac{\alpha}{2}}}{n|\bar{Z}_n|} \sqrt{\sum_{k=1}^{n} (\mathrm{Im}(\hat{\mu}_n^{-1}Z_k))^2} \tag{5}$$

with $q_{1-\frac{\alpha}{2}}$ denoting the $(1 - \frac{\alpha}{2})$-quantile of the standard normal distribution.

There are two major drawbacks to the use of asymptotic confidence intervals: firstly, by definition they do not guarantee a coverage probability of at least $1 - \alpha$ for finite n, so the coverage probability for a fixed distribution and sample size may be much smaller; we will demonstrate this for simulated data in Sect. 4. Secondly, they *assume* that $\mathbf{E}Z \neq 0$, so they are not applicable to all distributions on the circle. Of course, one could test the hypothesis $\mathbf{E}Z = 0$, possibly again by an asymptotic test, and construct the confidence set conditioned on this hypothesis having been rejected, setting $C_A = \mathbb{S}^1$ otherwise. But this sequential procedure would require some adaptation for mutiple testing – we come back to this point in Sect. 5 – and it is not commonly implemented in practice.

We therefore aim to construct *non-asymptotic* confidence sets for μ which are *universal* in the sense that they do not make any distributional assumptions about the circular data besides them being independent and identically distributed. It has been shown in [8] that this is possible; however, the confidence sets that were constructed there were far too large to be of use in practice. Nonetheless, we start by varying that construction in Sect. 2 but using Hoeffding's inequality instead of Chebyshev's as in [8]. Considerable improvements are possible if one takes the variance $\mathbf{E}(\mathbf{Im}(\mu^{-1}Z))^2$ "perpendicular to $\mathbf{E}Z$" into account; this is achieved by a second construction in Sect. 3. We illustrate and compare those confidence sets for simulations and for an application to real data in Sect. 4, discussing the results obtained in Sect. 5.

2   Construction Using Hoeffding's Inequality

We modify the construction in [8, Sect. 6] by replacing Chebyshev's inequality – which is too conservative here – by three applications of Hoeffding's inequality [7, Theorem 1]: if $U_1, \ldots, U_n$ are independent random variables taking values in the bounded interval $[a, b]$ with $-\infty < a < b < \infty$ then $\bar{U}_n = \frac{1}{n} \sum_{k=1}^{n} U_k$ with $\mathbf{E}\bar{U}_n = \nu$ fulfills

$$\mathbf{P}(\bar{U}_n - \nu \geq t) \leq \left[\left(\frac{\nu - a}{\nu - a + t} \right)^{\nu - a + t} \left(\frac{b - \nu}{b - \nu - t} \right)^{b - \nu - t} \right]^{\frac{n}{b-a}} \tag{6}$$

for any $t \in (0, b - \nu)$. Denoting the bound on the right hand side by $\beta(t)$, an elementary calculation shows that it is (as expected) strictly decreasing in t with $\beta(0) = 1$ and $\lim_{t \to b-\nu} \beta(t) = \left(\frac{\nu-a}{b-a} \right)^n$. By continuity, we conclude that the equation $\beta(t) = \gamma$ has a unique solution $t = t(\gamma, \nu, a, b)$ for any $\gamma \in \left(\left(\frac{\nu-a}{b-a} \right)^n, 1 \right)$; t is strictly decreasing in γ and another elementary calculation shows that $\nu + t(\gamma, \nu, a, b)$ is strictly increasing in ν (which is also to be expected). While there is no closed form expression for $t(\gamma, \nu, a, b)$, it can without difficulty be determined numerically.

Note that the estimate

$$\beta(t) \leq \exp(-2nt^2/(b-a)^2) \tag{7}$$

is often used and called Hoeffding's inequality [7]. While this would allow to solve explicitly for t, we prefer to work with β as it is sharper, especially for ν close to b as well as for large t. Nonetheless, it shows that the tail bound $\beta(t)$ tends to zero as fast as if using the central limit theorem which is why it is widely applied for bounded variables, see e.g. [3].

We construct our first confidence set as the acceptance region of a series of tests: for any $\zeta \in \mathbb{S}^1$ we will test the hypothesis that ζ is a circular population mean. Well, this hypothesis is equivalent to saying that there is some $\lambda \in [0,1]$ such that $\mathbf{E}Z = \lambda\zeta$. Multiplication by ζ^{-1} then rotates $\mathbf{E}Z$ onto the non-negative real axis: $\mathbf{E}\zeta^{-1}Z = \lambda \geq 0$.

Now fix ζ and consider $X_k = \mathbf{Re}(\zeta^{-1}Z_k)$, $Y_k = \mathbf{Im}(\zeta^{-1}Z_k)$ for $k = 1, \ldots, n$ which may be viewed as the projection of $Z_1, \ldots, Z_k$ onto the line in the direction of ζ and onto the line perpendicular to it. Both are sequences of independent random variables with values in $[-1,1]$ and $\mathbf{E}X_k = \lambda$, $\mathbf{E}Y_k = 0$ under the hypothesis. They thus fulfill the conditions for Hoeffding's inequality with $a = -1$, $b = 1$ and $\nu = \lambda$ or 0, respectively.

We will first consider the case of non-uniqueness, $\mu = \mathbb{S}^1$, or equivalently $\lambda = 0$. Then, assuming $\frac{\alpha}{4} > 2^{-n}$, the critical value $s_0 = t(\frac{\alpha}{4}, 0, -1, 1)$ is well-defined and we get $\mathbf{P}(\bar{X}_n \geq s_0) \leq \frac{\alpha}{4}$, and also, by considering $-X_1, \ldots, -X_n$, that $\mathbf{P}(-\bar{X}_n \geq s_0) \leq \frac{\alpha}{4}$. Analogously, $\mathbf{P}(|\bar{Y}_n| \geq s_0) \leq 2\frac{\alpha}{4} = \frac{\alpha}{2}$. We conclude that $\mathbf{P}(|\bar{Z}_n| \geq \sqrt{2}s_0) = \mathbf{P}(|\bar{X}_n|^2 + |\bar{Y}_n|^2 \geq 2s_0^2) \leq \mathbf{P}(|\bar{X}_n|^2 \geq s_0^2) + \mathbf{P}(|\bar{Y}_n|^2 \geq s_0^2) \leq \alpha$. Rejecting the hypothesis $\mu = \mathbb{S}^1$, i.e. $\mathbf{E}Z = 0$, if $|\bar{Z}_n| \geq \sqrt{2}s_0$ thus leads to a test whose probability of false rejection is at most α.

In case $\lambda > 0$, i.e. under the hypothesis that ζ is the unique circular population mean, we use the monotonicity of $\nu + t(\gamma, \nu, a, b)$ in ν and get $\mathbf{P}(\bar{X}_n \leq -s_0) = \mathbf{P}(-\bar{X}_n \geq s_0) \leq \mathbf{P}(-\bar{X}_n \geq -\lambda + t(\frac{\alpha}{4}, -\lambda, -1, 1)) \leq \frac{\alpha}{4}$ as well. For the perpendicular direction, however, we may now work with $\frac{3}{8}\alpha$, so consider $s_p = t(\frac{3}{8}\alpha, 0, -1, 1)$ – which is also well-defined since $\frac{3}{8}\alpha \geq \frac{\alpha}{4} > 2^{-n}$ by the assumption made above. Rejecting if $\bar{X}_n \leq -s_0$ or $|\bar{Y}_n| \geq s_p$ then will happen with probability at most $\frac{\alpha}{4} + 2\frac{3}{8}\alpha = \alpha$ under this hypothesis. In case we already rejected the hypothesis $\mu = \mathbb{S}^1$, i.e. if $|\bar{Z}_n| \geq \sqrt{2}s_0$, ζ will not be rejected if and only if $\bar{X}_n > s_0 > 0$ and $|\bar{Y}_n| < s_p < s_0$ which is then equivalent to $|\mathbf{Arg}(\zeta^{-1}\bar{Z}_n)| = \arcsin(|\bar{Y}_n|/|\bar{Z}_n|) < \arcsin(s_p/|\bar{Z}_n|)$.

Letting the confidence set C_H comprise all ζ which we could not reject means

$$\zeta \in C_H \text{ if } \alpha \leq 2^{-n+2} \text{ or } |\bar{Z}_n| \leq \sqrt{2}s_0 \text{ or } |\mathbf{Arg}(\zeta^{-1}\hat{\mu}_n)| < \arcsin\frac{s_p}{|\bar{Z}_n|}, \quad (8)$$

so that a non-trivial confidence set $C_H \neq \mathbb{S}^1$ is given by directions at an angle less than $\delta_H = \arcsin(s_p/|\bar{Z}_n|)$ from the circular sample mean $\hat{\mu}_n$. The above discussion shows that the coverage probability of C_H is then – for any sample size n and any distribution of Z – guaranteed to be $1 - \alpha$.

From (7), we obtain the estimate $\alpha \leq \exp(-ns_0^2/2)$ which implies that s_0 is of the order $n^{-\frac{1}{2}}$. Now, if μ is unique consider $\zeta = -\mu$ which has $\tau = \mathbf{E}\bar{X}_n < 0$ to see that the probability of obtaining the trivial confidence set $C_H = \mathbb{S}^1$ is eventually bounded by $\mathbf{P}(\zeta \in C_H) \leq \mathbf{P}(\bar{X}_n > -s_0) \leq \mathbf{P}(\bar{X}_n > \frac{\tau}{2}) = \mathbf{P}(\bar{X}_n - \mathbf{E}\bar{X}_n >$

$-\frac{\tau}{2}) \leq \exp(-n\tau^2/8)$, and hence will go to zero exponentially fast as n tends to infinity (note that eventually $\alpha > 2^{-n+2}$). Moreover, since $\bar{Z}_n \to \mathbf{E}Z$, with $s_p < s_0$ also δ_H is of the order $n^{-\frac{1}{2}}$, just like the angle δ_A for the asymptotic confidence interval.

3   Estimating the Variance

From the central limit theorem for $\hat{\mu}_n$ in case of unique μ, cf. (4), we see that the aymptotic variance of $\hat{\mu}_n$ gets small if $|\mathbf{E}Z|$ is close to 1 (then $\mathbf{E}Z$ is close to the boundary $\mathbb{S}^1$ of the unit disc which is possible only if the distribution is very concentrated) or if the variance $\mathbf{E}(\mathbf{Im}(\mu^{-1}Z))^2$ in the direction perpendicular to μ is small (if the distribution were concentrated on $\pm\mu$ this variance would be zero and $\hat{\mu}_n$ would equal μ with large probability). While δ_H ($|\bar{Z}_n|$ being the denominator of its sine) takes the former into account, the latter has not been exploited yet. To do so, we need to estimate $\mathbf{E}(\mathbf{Im}(\mu^{-1}Z))^2$.

Now, varying the construction in Sect. 2, consider $V_n = \frac{1}{n}\sum_{k=1}^n Y_k^2$ which under the hypothesis that the corresponding ζ is the unique circular population mean has expectation $\sigma^2 = \mathbf{Var}(Y_k)$. Once again we will apply (6), this time to $1 - V_n = \frac{1}{n}\sum_{k=1}^n (1 - Y_k^2)$ which then is the mean of n independent random variables taking values in $[0,1]$, having expectation $1-\sigma^2$. We thus obtain $\mathbf{P}(\sigma^2 \geq V_n + t) = \mathbf{P}(1 - V_n \geq 1 - \sigma^2 + t) \leq \frac{\alpha}{4}$ for $t = t(\frac{\alpha}{4}, 1 - \sigma^2, 0, 1)$, the latter existing if $\frac{\alpha}{4} > (1 - \sigma^2)^n$. Since $1 - \sigma^2 + t(\frac{\alpha}{4}, 1 - \sigma^2, 0, 1)$ increases with $1 - \sigma^2$, there is a minimal σ^2 for which $1 - V_n \geq 1 - \sigma^2 + t(\frac{\alpha}{4}, 1 - \sigma^2, 0, 1)$ holds and becomes an equality; we denote it by $\widehat{\sigma^2} = V_n + t(\frac{\alpha}{4}, 1 - \widehat{\sigma^2}, 0, 1)$. Inserting into (6), it by construction fulfills

$$\frac{\alpha}{4} = \left[\left(\frac{1 - \widehat{\sigma^2}}{1 - V_n} \right)^{1 - V_n} \left(\frac{\widehat{\sigma^2}}{V_n} \right)^{V_n} \right]^n . \tag{9}$$

It is easy to see that the right hand side depends continuously on and is strictly decreasing in $\widehat{\sigma^2} \in [V_n, 1]$, thereby traversing the interval $[0, 1]$ so that one can again solve the equation numerically. We then may, with an error probability of at most $\frac{\alpha}{4}$, use $\widehat{\sigma^2}$ as an upper bound for σ^2; comparing the condition $(1 - \widehat{\sigma^2})^n < \frac{\alpha}{4}$ with (9), $\widehat{\sigma^2} > V_n$ exists for any $V_n < 1$ and else $\widehat{\sigma^2} = 1$ is the trivial upper bound.

With such an upper bound on its variance, we now can get a better estimate for $\mathbf{P}(\bar{Y}_n > t)$. Indeed, one may use another inequality by Hoeffding [7, Theorem 3]: the mean $\bar{W}_n = \frac{1}{n}\sum_{k=1}^n W_k$ of a sequence $W_1, \ldots, W_n$ of independent random variables taking values in $(-\infty, 1]$, each having zero expectation as well as variance ρ^2 fulfills for any $w \in (0, 1)$

$$\mathbf{P}(\bar{W}_n \geq w) \leq \left[\left(1 + \frac{w}{\rho^2} \right)^{-\rho^2 - w} \left(1 - w \right)^{w-1} \right]^{\frac{n}{1+\rho^2}} . \tag{10}$$

Again, an elementary calculation shows that the right hand side is strictly decreasing in w, continuously ranging between 1 and $\left(\frac{\rho^2}{1+\rho^2} \right)^n$ as w varies in

$(0,1)$, so that there exists a unique $w = w(\gamma, \rho^2)$ for which the right hand side equals γ, provided $\gamma \in \left(\left(\frac{\rho^2}{1+\rho^2} \right)^n, 1 \right)$. Moreover, the right hand side increases with ρ^2 (as expected), so that $w(\gamma, \rho^2)$ is increasing in ρ^2, too.

Therefore, under the hypothesis that the corresponding ζ is the unique circular population mean, $\mathbf{P}(|\bar{Y}_n| \geq w(\frac{\alpha}{4}, \sigma^2)) \leq 2\frac{\alpha}{4} = \frac{\alpha}{2}$. Now, since $\mathbf{P}(w(\frac{\alpha}{4}, \sigma^2) \geq w(\frac{\alpha}{4}, \widehat{\sigma^2})) = \mathbf{P}(\sigma^2 \geq \widehat{\sigma^2}) \leq \frac{\alpha}{4}$, setting $s_v = w(\frac{\alpha}{4}, \widehat{\sigma^2})$ we get $\mathbf{P}(|\bar{Y}_n| \geq s_v) \leq \frac{3}{4}\alpha$. Note that $\frac{\rho^2}{1+\rho^2}$ increases with ρ^2, so in case s_0 exists $\widehat{\sigma^2} \leq 1$ implies $\frac{\alpha}{4} > 2^{-n} \geq \left(\frac{\widehat{\sigma^2}}{1+\widehat{\sigma^2}} \right)^n$, i.e. the existence of s_v.

Because the bound in (10) for $\rho^2 = 1$ agrees with the bound in (6) for $a = -1$, $b = 1$ and $\nu = 0$, we have $s_v \leq s_0$. So we can argue as for C_H before – which allows us to construct a confidence set C_V for μ with coverage probability at least $1 - \alpha$ by letting $\zeta \in C_V$ if $\alpha \leq 2^{-n+2}$ or $|\bar{Z}_n| \leq \sqrt{2}s_0$ or $|\mathbf{Arg}(\zeta^{-1}\hat{\mu}_n)| < \arcsin\frac{s_v}{|\bar{Z}_n|}$.

Consequently, if C_V is non-trivial – which happens if and only if C_H is non-trivial – then it is given by all $\zeta \in \mathbb{S}^1$ forming an angle less than $\arcsin(s_v/|\bar{Z}_n|)$ with the circular sample mean $\hat{\mu}_n$. This angle depends via $\widehat{\sigma^2}$ on ζ, though, so we set $\delta_V = \sup_{\zeta \in C_V} |\mathbf{Arg}(\zeta^{-1}\hat{\mu}_n)|$. Since $s_v \leq s_0$, it nonetheless follows that the asymptotic behaviour of C_V is qualitatively the same as that of C_H.

4   Simulation and Application to Real Data

We will compare the asymptotic confidence set C_A, the confidence set C_H constructed directly using Hoeffding's inequality in Sect. 2, and the last confidence set C_V taking the variance perpendicular to the circular population mean into account by reporting their corresponding opening angles δ_A, δ_H, and δ_V in degrees ($°$) as well as their coverage frequencies in simulations.

Implementation. All computations have been performed using our own code based on the software package R [11]. In order to determine δ_V, all $\zeta \in \mathbb{S}^1$ with $\frac{1800}{\pi}\mathbf{Arg}\,\zeta \in \mathbb{Z}$ have been considered which corresponds to a grid of angles with a spacing of $0.1°$.

Simulation 1: two points of equal mass at $\pm 10°$. First, we consider a rather favourable situation: $n = 400$ independent draws from the distribution with $\mathbf{P}(Z = \exp(10\pi i/180)) = \mathbf{P}(Z = \exp(-10\pi i/180)) = \frac{1}{2}$. Then we have $|\mathbf{E}Z| = \mathbf{E}Z = \cos(10\pi i/180) \approx 0.985$, implying that the data are highly concentrated, $\mu = 1$ is unique, and the variance of Z in the direction of μ is 0; there is only variation perpendicular to μ, i.e. in the direction of the imaginary axis.

Table 1 shows the results based on 10,000 repetitions for a nominal coverage probability of $1 - \alpha = 95\%$: the average δ_H is about 3.5 times larger than δ_V which is about twice as large as δ_A. As expected, the asymptotics are rather precise in this situation: C_A did cover the true mean in about 95% of the cases which implies that the other confidence sets are quite conservative; indeed C_H and C_V covered the true mean in all repetitions. One may also note that the angles varied only little between repetitions, δ_V a little more than the others.

Table 1. Results for simulation 1 (two points of equal mass at $\pm 10°$) based on 10,000 repetitions with $n = 400$ observations each: average observed δ_H, δ_V, and δ_A (with corresponding standard deviation), as well as frequency (with corresponding standard error) with which $\mu = 1$ was covered by C_H, C_V, and C_A, respectively; the nominal coverage probability was $1 - \alpha = 95\%$.

confidence set	mean δ	($\pm$ s.d.)	coverage frequency	($\pm$ s.e.)
C_H	8.2°	($\pm 0.0005°$)	100.0%	($\pm 0.0\%$)
C_V	2.3°	($\pm 0.0252°$)	100.0%	($\pm 0.0\%$)
C_A	1.0°	($\pm 0.0019°$)	94.8%	($\pm 0.2\%$)

Table 2. Results for simulation 2 (three points placed asymmetrically) based on 10,000 repetitions with $n = 100$ observations each: average observed δ_H, δ_V, and δ_A (with corresponding standard deviation), as well as frequency (with corresponding standard error) with which $\mu = 1$ was covered by C_H, C_V, and C_A, respectively; the nominal coverage probability was $1 - \alpha = 90\%$.

confidence set	mean δ	($\pm$ s.d.)	coverage frequency	($\pm$ s.e.)
C_H	16.5°	($\pm 0.8502°$)	100.0%	($\pm 0.0\%$)
C_V	5.0°	($\pm 0.3740°$)	100.0%	($\pm 0.0\%$)
C_A	0.4°	($\pm 0.2813°$)	62.8%	($\pm 0.5\%$)

Simulation 2: three points placed asymmetrically. Secondly, we consider a situation which has been designed to show that even a considerably large sample size ($n = 100$) does not guarantee approximate coverage for the asymptotic confidence set C_A: the distribution of Z is concentrated on three points, $\xi_j = \exp(\theta_j \pi i / 180)$, $j = 1, 2, 3$ with weights $\omega_j = \mathbf{P}(Z = \xi_j)$ chosen such that $\mathbf{E}Z = |\mathbf{E}Z| = 0.9$ (implying a small variance and $\mu = 1$), $\omega_1 = 1\%$ and $\mathbf{Arg}\,\xi_1 > 0$ while $\mathbf{Arg}\,\xi_2, \mathbf{Arg}\,\xi_3 < 0$. In numbers, $\theta_1 \approx 25.8$, $\theta_2 \approx -0.3$, and $\theta_3 \approx -179.7$ (in °) while $\omega_2 \approx 94\%$, and $\omega_3 \approx 5\%$.

The results based on 10,000 repetitions are shown in Table 2 where a nominal coverage probability of $1 - \alpha = 90\%$ was prescribed. Clearly, C_A with its coverage probability of less than 64% performs quite poorly while the others are conservative; $\delta_V \approx 5°$ still appears small enough to be useful in practice, though.

Real data: movements of ants. N. I. Fisher [5, Example 4.4] describes a data set of the directions 100 ants took in response to an illuminated target placed at 180° for which it may be of interest to know whether the ants indeed (on average) move towards that target. The data set is available as Ants_radians within the R package CircNNTSR [4].

The circular sample mean for this data set is about $-176.9°$; for a nominal coverage probability of $1 - \alpha = 95\%$ one gets $\delta_H \approx 27.3°$, $\delta_V \approx 20.4°$, and $\delta_A \approx 9.6°$ so that all confidence sets contain $\pm 180°$. The data set's concentration is not very high, however, so that the circular population mean could – according to C_V – also be $-156.5°$ or $162.7°$.

5 Discussion

We have derived two confidence sets, C_H and C_V, for the set of circular sample means. Both guarantee coverage for any finite sample size – at the cost of potentially being quite conservative – without making any assumptions on the distribution of the data (besides that they are independent and identically distributed): they are non-asymptotic and universal in this sense. Judging from the simulations and the real data set, C_V – which estimates the variance perpendicular to the mean direction – appears to be preferable over C_H (as expected) and small enough to be useful in practice.

While the asymptotic confidence set's opening angle appears to be less than half of the one for C_V, it has the drawback that even for a sample size of $n = 100$ it may fail to give a coverage probability close to the nominal one; also, one has to assume that the circular population mean is unique. Of course, one could also devise an asymptotically justified test for the latter but this would entail a correction for multiple testing (for example working with $\frac{\alpha}{2}$ each time) which would also render the asymptotic confidence set conservative.

Further improvements would require sharper "universal" mass concentration inequalities taking the first or the first two moments into account; that, however, is beyond the scope of this article.

References

1. Afsari, B.: Riemannian L^p center of mass: existence, uniqueness, and convexity. Proc. Am. Math. Soc. **139**(2), 655–673 (2011)
2. Arnaudon, M., Miclo, L.: A stochastic algorithm finding p-means on the circle. Bernoulli (to appear, available online). http://www.e-publications.org/ims/submission/BEJ/user/submissionFile/16932?confirm=239c7c43
3. Bousquet, O., Boucheron, S., Lugosi, G.: Introduction to statistical learning theory. In: Bousquet, O., von Luxburg, U., Rätsch, G. (eds.) Machine Learning 2003. LNCS (LNAI), vol. 3176, pp. 169–207. Springer, Heidelberg (2004)
4. Fernandez-Duran, J.J., Gregorio-Dominguez, M.M.: CircNNTSR: CircNNTSR: An R package for the statistical analysis of circular data using nonnegative trigonometric sums (NNTS) models (2013). http://CRAN.R-project.org/package=CircNNTSR, R package version 2.1
5. Fisher, N.I.: Statistical Analysis of Circular Data. Cambridge University Press, Cambridge (1993)
6. Fréchet, M.: Les éléments aléatoires de nature quelconque dans un espace distancié. Annales de l'Institut Henri Poincaré, vol. 10(4), pp. 215–310 (1948)
7. Hoeffding, W.: Probability inequalities for sums of bounded random variables. J. Am. Stat. Assoc. **58**(301), 13–30 (1963)
8. Hotz, T.: Extrinsic vs intrinsic means on the circle. In: Nielsen, F., Barbaresco, F. (eds.) GSI 2013. LNCS, vol. 8085, pp. 433–440. Springer, Heidelberg (2013)
9. Jammalamadaka, S.R., SenGupta, A.: Topics in Circular Statistics. Series on Multivariate Analysis, vol. 5. World Scientific, Singapore (2001)
10. Mardia, K.V., Jupp, P.E.: Directional Statistics. Wiley, New York (2000)
11. R Core Team.: R: A Language and Environment for Statistical Computing. R Foundation for Statistical Computing, Vienna (2013). http://www.R-project.org/

A Methodology for Deblurring and Recovering Conformational States of Biomolecular Complexes from Single Particle Electron Microscopy

Bijan Afsari[✉] and Gregory S. Chirikjian

Johns Hopkins University, Baltimore, USA
bijan@cis.jhu.edu, gchirik1@jhu.edu

Abstract. In this paper we study two forms of blurring effects that may appear in the reconstruction of 3D Electron Microscopy (EM), specifically in single particle reconstruction from random orientations of large multi-unit biomolecular complexes. We model the blurring effects as being due to independent contributions from: (1) variations in the conformation of the biomolecular complex; and (2) errors accumulated in the reconstruction process. Under the assumption that these effects can be separated and treated independently, we show that the overall blurring effect can be expressed as a special form of a convolution operation of the 3D density with a kernel defined on $SE(3)$, the Lie group of rigid body motions in 3D. We call this form of convolution mixed spatial-motional convolution. We discuss the ill-conditioned nature of the deconvolution needed to deblur the reconstructed 3D density in terms of parameters associated with the unknown probability in $SE(3)$. We provide an algorithm for recovering the conformational information of large multi-unit biomolecular complexes (essentially deblurring) under certain biologically plausible prior structural knowledge about the subunits of the complex in the case the blurring kernel has a special form.

1 Introduction

Reconstructing three dimensional densities associated with large biomolecular complexes using single particle 3D Electron Microscopy (EM) has proved very promising in structural biology and other biological applications. The reader is referred to [4] for general introduction and extensive references.

At the core of single particle reconstruction lies the problem of reconstruction of a 3D volume from thousands of very noisy 2D projections of the volume formed along *random (unknown)* projection directions relative to the body-fixed frame of the biomolecular complex. What makes this problem different from standard tomography is exactly the fact that the projection directions are unknown and need to be determined before one can apply a standard 3D reconstruction such as weighted back-projection. In addition, due to certain biological restrictions the signal to noise ratio in a single projection image is extremely low (e.g., typically

© Springer International Publishing Switzerland 2015
F. Nielsen and F. Barbaresco (Eds.): GSI 2015, LNCS 9389, pp. 643–653, 2015.
DOI: 10.1007/978-3-319-25040-3_69

at the order of $1/100$). The reason one has to deal with such random projections is that in single particle EM imaging (specifically cryo-EM) one essentially takes a 2D image of a layer of a frozen sample containing a large number of copies or instances of a biomolecular complex lying at random positions and orientations within the sample. The output of a reconstruction algorithm is a *blurred* 3D (so-called) *density* map representing the Coulomb potentials of the atoms of the biomolecular complex under experiment improve our algorithm [4].

In this paper (in Sect. 2) we study two sources of blurring: the first one is due to variations in the structure or conformational states of the biomolecular complex in the sample (i.e., not all instances of the biomolecular complex are exactly the same). The second source of blurring is due to errors introduced in the process of reconstructing the 3D density from the collected images. We first show that each of these blurring effects can be modeled as a specific form of averaging or convolution of the ground truth 3D volume with probability density (kernel) defined on $SE(3)$, the group of rigid body motions. The associated *blind deblurring* or *deconvolution* is severely ill-posed and requires prior information or information (fusion) from other imaging modalities to yield a well-posed problem. In certain cases of dealing with large multi-unit complexes, however, one may have information about the shape of the subunits and the problem recovering the shape of the complex basically boils down to recovering the relative positions of the subunits. In Sect. 3 we derive a set of equations describing blurring of a rigid body model under a $SE(3)$ kernel in terms of the body parameters and the parameters of the kernel (in particular its Lie-algebraic $SE(3)$ *mean* and *covariance*). We also derive a simple algorithm for recovering conformational information under the assumption of *isotropic* blurring and we show the application of this algorithm to simulated data; and we conclude the paper in Sect. 4. We mention that closely related works include [6] and [7], where, respectively, Eculidean convolution and spherical convolution have been employed to model the blurring effects.

2 Blurring as Mixed Spatial-Motional Convolution

In this section we study two sources of blurring effects in 3D single particle EM, both of which can be modeled using probability densities on $SE(3)$. The first effect is conformational blurring within a biomolecular complex due to internal motions. The second is blurring during the process of reconstructing 3D densities from an ensemble of noisy 2D projections. As an idealization, we assume that these effects can be treated independently.

The preparation of the sample for single particle EM usually starts with a solution containing the designated biomolecular complex, each consisting of multiple macromolecules, and freezing the solution in the form of a very thin layer. For various reasons the instances of the biomolecular complex in the sample may not have exactly the same shape. For example, they may be at different conformational states (e.g., open or close) or their subunits might have been displaced in the freezing process. Let us consider a biomolecular complex with 3D density ρ consisting of N macromolecular subunits

$$\rho(\mathbf{r}) = \sum_{i=1}^{N} \rho_i(\mathbf{r}), \tag{1}$$

where $\rho_i : \mathbb{R}^3 \to \mathbb{R}$ is the 3D density of subunit i. Often in large biomolecular complexes we may model ρ_i as a rigid body. In this case, we model the effect of conformational states or motions as the ensemble or average of the *action* of $SE(3)$ on the rigid bodies. Specifically, let $\cdot$ denote the standard $SE(3)$ action in $\mathbb{R}^3$

$$r \mapsto g \cdot \mathbf{r} = R\,\mathbf{r} + t, \tag{2}$$

where each $g \in SE(3)$ is represented with the rotation-translation pair $(R, t) \in SO(3) \times \mathbb{R}^3$, and the group operation for $SE(3)$ is $g_1 \circ g_2 = (R_1 R_2, g_1 \cdot t_2)$. Then a copy (or instance) of subunit $i \geq 2$ with density ρ_i might be under transformation g relative to the subunit $i = 1$, which can be described in the global (lab) coordinates as $\rho_i(g^{-1} \cdot \mathbf{r})$. Throughout the sample the copies might go through different transformations which we model by a $SE(3)$ probability density $f_i : SE(3) \to \mathbb{R}$ and the ensemble average of such motional or conformational variations can be modeled as

$$\tilde{\rho}_i(\mathbf{r}) = (f_i \star \rho_i)(\mathbf{r}) := \int_{SE(3)} f_i(g) \rho_i(g^{-1} \cdot \mathbf{r}) dg \tag{3}$$

where dg is the Haar measure for $SE(3)$. The above operation may be called *mixed spatial-motional* convolution. The operation resembles convolution on $SE(3)$,

$$(k * f_i)(g) := \int_{SE(3)} k(h) f_i(h^{-1} \circ g) dh,$$

which we denote with an asterisk $*$ rather than a $\star$, but $f_i \star \rho_i$ is not a convolution since the functions under operation have different domains.

The total conformationally blurred 3D density with body 1 fixed can then can be expressed as

$$\tilde{\rho}(\mathbf{r}) = \sum_{i=1}^{N} \tilde{\rho}_i(\mathbf{r}) = \sum_{i=1}^{N} (f_i \star \rho_i)(\mathbf{r}). \tag{4}$$

In the above f_1 is assumed to be the Dirac delta function at the identity of $SE(3)$, denoted by $\delta(g)$. As far as cryo-EM imaging is concerned, (3) and hence (4) show non-physical ensemble averages, since they are not directly measured. This is in contrast to Small-Angle-X-ray-Scattering (SAXS) measurements in which the ensemble average is measured directly [3]. Here, the actual averaging or superposition of the different (continuum of) conformational states is to happen in the reconstruction process (algorithm). Specifically, in the imaging step, many copies of the biomolecular complex in each conformational state, positioned and oriented randomly throughout the sample, are imaged *separately*.

Then, these 2D images are fed to a 3D reconstruction algorithm to reconstruct a 3D density. Therefore, one expects that the ensemble averaging should happen in the reconstruction process. However, this also means that what a specific algorithm does may matter. We first consider an idealized algorithm (meaning that the algorithm introduces no errors). We also assume that we have an algorithm designed to deal with homogeneous samples. Most commonly used algorithms are such and they assume a single conformational state of the biomolecular complex in the sample. To be compatible with this assumption we also assume that the probability densities associated with conformational state variation (f_i's) are unimodal and concentrated enough (i.e., small conformational variations within the sample). Before proceeding further, we mention that the problem of heterogeneity of data is a challenging problem in single particle reconstruction, which in reality limits the accuracy of these methods [4, p. 266], [5]. Source of heterogeneity could range from impurity in the sample to presence of ligands and different conformational states. The latter is our main focus here. Here, we have distinguished between large and small variation in conformational states. The presence of large deviations in conformational states essentially is equivalent to a multi-modal or non-concentrated distribution f_i. The existence of such modes or classes makes the 3D reconstruction problem much more difficult. Specifically, the step of classification of the images will be very hard for heterogeneous data due to the intermingling between variation in pose and conformational state as portrayed on the 2D projections ([5], and see below). Nevertheless, specific algorithms for heterogeneous data have been developed (see e.g., [11]), but the subject is still in its fancy [5].

In the rest of discussion for convenience we consider a biomolecular complex comprised of only two subunits $\rho(\mathbf{r}) = \rho_1(\mathbf{r}) + \rho_2(\mathbf{r})$, and we assume that its conformational states are determined only by a single copy of $SE(3)$, i.e., the total density under a conformational state change $g \in SE(3)$ is $\rho_g(\mathbf{r}) = \rho_1(\mathbf{r}) + \rho_2(g^{-1} \cdot \mathbf{r})$. We assume that g has the $SE(3)$ probability density f. A typical biomolecular complex in the frozen sample will be $\rho_g(h^{-1} \cdot \mathbf{r})$, where $h = (R_h, t_h) \in SE(3)$ denotes a random orientation (pose) and position of the biomolecular complex in the sample. Henceforth we use the following notation:

$$\rho_g^h(\mathbf{r}) := \rho_g(h^{-1} \cdot \mathbf{r}) = \rho(g^{-1} \cdot (h^{-1} \cdot \mathbf{r})) = \rho((h \circ g)^{-1} \cdot \mathbf{r}).$$

In the imaging process an image from each copy of $\rho_g^h(\mathbf{r})$ is formed by the projection operation (along the z axis)

$$p_g^h(x, y) = \int \rho_g^h(\mathbf{r}) dz \qquad (5)$$

where $\mathbf{r} = [x, y, z]^\top$. A typical (homogeneous) 3D reconstruction algorithm first brings all the 2D images to a common origin, which we assume is the origin of the lab frame. Due to the large amount of noise in these images, they are class averaged. This process can be described as an in-plane $SE(2)$ blurring [8,9]. A class is meant to correspond to the biomolecular complex being imaged along similar directions (ideally exactly the same direction). This means that a class roughly corresponds to images from the copies of the biomolecular complex in the sample that are at the same *orientation*, i.e., a full 3D rotation modulo an in-plane rotation in the $x - y$ plane.

The next step is finding the actual projection direction for each class relative to the body-fixed frame of the biomolecular complex. This is known as *angular reconstitution* (see [12, 14] and references therein for related methods). Assuming the angles are found correctly, the actual 3D reconstruction is the standard tomographic reconstruction. Often the *weighted backprojection* algorithm is used, which given enough number of sampled projection directions can reconstruct the 3D volume without any aliasing [4]. In our case this means that the ensemble average $\tilde{\rho}(\mathbf{r})$ is reconstructed. Hence, although, as mentioned before, $\tilde{\rho}(\mathbf{r})$ in (4) is a non-physical ensemble average, it can be realized in the 3D reconstruct due to the fact that the 2D images are averaged from many of copies of the biomolecular complex at different conformational states and also the fact that the steps involved in the reconstruction are linear operations. Notice, however, that this is under the assumption of the steps of centering the 2D projection, classification, and angular reconstitution are error free. In reality, all these steps are highly prone to error due to the extremely high level noise in the image formation process. Additionally, notice that a possible interplay between g and h can result in complications in the classification step. However, also note that whether such an interplay (and the ensued misclassification) necessarily results in reconstruction errors also depends on the structure of the complex (e.g., if certain symmetries exist then the misclassification won't be harmful).

We postulate that the output of the 3D reconstruction is a version of the conformationally blurred density $\tilde{\rho}$, where an additional $SE(3)$ blurring kernel includes both motional blurring due to class averaging and reconstruction errors. That is, the contribution to the blurred density of the biomolecular complex from the ith macromolecular subunit will be of the form

$$\tilde{\tilde{\rho}}_i(\mathbf{r}) = (k \star \tilde{\rho}_i)(\mathbf{r}) = (k \star (f_i \star \rho_i))(\mathbf{r}) = ((k * f_i) \star \rho_i)(\mathbf{r}).$$

Here $k : SE(3) \rightarrow \mathbb{R}$ is the reconstruction blurring kernel that contains contributions from both class averaging effects and 3D reconstruction.

Of course, we state this under certain assumptions most notably that conformational states and projections orientations do not interplay and that error kernels are independent of the poses. Both assumptions are plausible under small conformational variation and if many different poses are available. We also add that in many image processing applications modeling blurring using a convolution is a viable and common approach (independent of the source and mechanism of the blurring which could be highly nonlinear). However, the more challenging part is the fact that the kernel is unknown and hence one has to resort to blind de-convolution methods.

3 Recovering Conformational Information Based on Moment Matching

Blind deconvolution or deblurring, in general, without prior information is ill-posed and difficult. In certain biological applications the goal is to understand the conformational state of a large biomolecular complex comprised of subunits, while the structure of each of subunit is a-priori *known*, and the goal is to find the relative position (pose) of the subunits with respect to each other. For example, given a complex comprised of two subunits the goal might be to decide whether it is in close or open configuration or to find the relative position of the two subunits. We assume that each subunit can be modeled by a rigid body, in particular, an ellipsoid itself modeled by a Gaussian in $\mathbb{R}^3$. This, in particular, means that in (1) ρ_i's are assumed to be known up to a rotation and translation. Furthermore, we assume that upon reconstruction we can separate the reconstructed subunits $\tilde{\rho}_i$ from each other. This may be done through a 3D segmentation algorithm, manually, or using a clustering algorithm such k-means. The extent to which this assumption is practical or valid depends on the problem and needs further verification. Assuming these simplifications, in the following we will consider blurring a 3D Gaussian distribution with an $SE(3)$ kernel and find the mean and covariance of the blurred density in terms of the parameters (mean and covariance) of the kernel and the density.

Parameterization of $SE(3)$ Kernel. Let $\mathfrak{se}(3)$ denote the Lie algebra of $SE(3)$. Also let exp : $\mathfrak{se}(3) \rightarrow SE(3)$ denote the matrix exponential and $\log : SE(3) \rightarrow \mathfrak{se}(3)$ its inverse. Recall that an element $\Omega \in \mathfrak{se}(3)$ can be represented as $\Omega = \begin{bmatrix} \Omega_R & \omega_t \\ \hline 0 & 0 \end{bmatrix}$, where Ω_R is a 3×3 skew-symemtri matrix and $\omega_t \in \mathbb{R}^3$. We will need the following well-known fact which gives a closed form expression for the logarithm map (see e.g., [10] for a proof):

Proposition 1. *Let* $\Omega = \begin{bmatrix} \Omega_R & \omega_t \\ \hline 0 & 0 \end{bmatrix} \in \mathfrak{se}(3)$. *Then* $e^\Omega = \begin{bmatrix} e^{\Omega_R} & \frac{e^u - 1}{u}|_{u=\Omega_R} \, \omega_t \\ \hline 0 & 1 \end{bmatrix}$.
Conversely if $g = \begin{bmatrix} R & t \\ \hline 0 & 1 \end{bmatrix} \in SE(3)$, *then* $\log(g) = \begin{bmatrix} \log(R) & \frac{u}{e^u - 1}|_{u=\log(R)} \, t \\ \hline 0 & 0 \end{bmatrix} \in \mathfrak{se}(3)$. *This result holds if all the eigenvalues of* Ω *are less than* π *in absolute value or equivalently* g *has no eigenvalue of* -1.

We now define the notion of Lie-algebraic mean [13] (also known as bi-invariant mean [1,10]) and covariance [2] for $SE(3)$-valued random variables:

Definition 1. *Let* **g** *be an* $SE(3)$-*valued random variable with probability density* $f : SE(3) \rightarrow \mathbb{R}$. *Then we define a mean* $\mu_{\mathbf{g}}$ *of* **g** *as a solution to*[1]

$$\mathbb{E}\{\log(\mu_{\mathbf{g}}^{-1}g)\} = \int_{SE(3)} \log(\mu_{\mathbf{g}}^{-1}g)f(g)dg = 0 \tag{6}$$

---

[1] Here $\mu_{\mathbf{g}}$ and $\Sigma_{\mathbf{g}}$ are not functions of g, but are properties of the random variable **g** that has distribution $f(g)$.

and the associated covariance $\Sigma_{\mathbf{g}}$

$$\Sigma_{\mathbf{g}} := \mathbb{E}\{\text{vec}(\Omega_g)\text{vec}(\Omega_g)^\top\} = \int_{SE(3)} \text{vec}(\Omega_g)\text{vec}(\Omega_g)^\top f(g)dg \qquad (7)$$

where $\text{vec} : \mathfrak{se}(n) \to \mathbb{R}^6$ is an isomorphism between $\mathfrak{se}(3)$ and $\mathbb{R}^6$ and $\Omega_g = \log(g\mu_{\mathbf{g}}^{-1})$.

Due to topological constraints the Eq. (6) for mean has always at least two solutions on $SE(3)$. However, it can be shown that if f is concentrated in a small enough region, then there exists a unique mean in that region [10]. To our knowledge stronger results are not known. The covariance $\Sigma_{\mathbf{g}}$ depends on the isomorphism used. We use the standard isomorphism induced by the basis

$$E_1 = \begin{bmatrix} 0&0&0&0 \\ 0&0&-1&0 \\ 0&1&0&0 \\ 0&0&0&0 \end{bmatrix}, \; E_2 = \begin{bmatrix} 0&0&1&0 \\ 0&0&0&0 \\ -1&0&0&0 \\ 0&0&0&0 \end{bmatrix}, \; E_3 = \begin{bmatrix} 0&-1&0&0 \\ 1&0&0&0 \\ 0&0&0&0 \\ 0&0&0&0 \end{bmatrix}, \qquad (8a)$$

$$E_4 = \begin{bmatrix} 0&0&0&1 \\ 0&0&0&0 \\ 0&0&0&0 \\ 0&0&0&0 \end{bmatrix}, \; E_5 = \begin{bmatrix} 0&0&0&0 \\ 0&0&0&1 \\ 0&0&0&0 \\ 0&0&0&0 \end{bmatrix}, \; E_6 = \begin{bmatrix} 0&0&0&0 \\ 0&0&0&0 \\ 0&0&0&1 \\ 0&0&0&0 \end{bmatrix}. \qquad (8b)$$

Thus, if $\Omega = \sum_{i=1}^N \omega_i E_i$, then we have $\text{vec}(\Omega) = (\omega_1, \cdots, \omega_6)^\top \in \mathbb{R}^6$.

Mean and Covariance of the Blurred 3D Density. Consider the model:

$$\mathbf{y} = \mathbf{R}\mathbf{r} + \mathbf{t}, \; \mathbf{r} \in \mathbb{R}^3, \; \mathbb{E}\{\mathbf{r}\} = 0, \mathbb{E}\{\mathbf{r}\mathbf{r}^\top\} = C_{\mathbf{r}}, g = \begin{bmatrix} \mathbf{R}&\mathbf{t} \\ 0&1 \end{bmatrix} \in SE(3), \quad (9)$$

with $\mathbf{g}$ and $\mathbf{r}$ being statistically independent. This model corresponds to the mixed spatial-motional convolution (3). The goal is to express the Euclidean mean and covariance matrix of $\mathbf{y}$ (which we assumed can be estimated from blurry 3D reconstruction) in terms of covariance $C_{\mathbf{r}}$ (which we assumed is given) and $SE(3)$ mean and covariance of $\mathbf{g}$ which are to be estimated. Denote the $SE(3)$-mean of $\mathbf{g}$ by $\mu_{\mathbf{g}}$, where $\mu_{\mathbf{g}} = \begin{bmatrix} \mu_{\mathbf{R}}&\mu_{\mathbf{t}} \\ 0&1 \end{bmatrix}$. Note that $\int \log(\mu_{\mathbf{g}}^{-1}g)f(g)dg = 0$ implies that $\int \mu_{\mathbf{g}}^{-1}\log(g\mu_{\mathbf{g}}^{-1})\mu_{\mathbf{g}}f(g)dg = 0$ and $\int \log(g\mu_{\mathbf{g}}^{-1})f(g)dg = 0$, hence

$$g = e^{\log(g\mu_{\mathbf{g}}^{-1})}\mu_{\mathbf{g}} = e^{\Omega_g}\mu_{\mathbf{g}}, \; \mathbb{E}\{\Omega_g\} = 0, \text{ where } \Omega_g = \log(g\mu_{\mathbf{g}}^{-1}) = \begin{bmatrix} \Omega_{\mathbf{R}}&\omega_{\mathbf{t}} \\ 0&0 \end{bmatrix} \in \mathfrak{se}(3).$$

$$(10)$$

The following proposition gives the first two moments of $\mathbf{y}$ up to 2nd order terms in terms of those of $\mathbf{r}$ and $\mathbf{g}$. The proof is straightforward using Proposition 1 and some algebraic manipulation.

Proposition 2. *Under statistical independence of rotation and translation at the Lie algebra (i.e., independence of Ω_R and ω_t in (10)) and statistical independence of g and* **r** *the forward equations for the mean and covariance of* **y** *in (9) up to second order are:*

$$\mathbb{E}\{\mathbf{y}\} = \mathbb{E}\{\mathbf{t}\} \overset{2nd}{=} (I + \frac{1}{2}\mathbb{E}\{\Omega_R^2\})\mu_t \tag{11a}$$

$$C_\mathbf{y} = \mathbb{E}\{RC_\mathbf{r}R^\top\} + C_\mathbf{t} \overset{2nd}{=} \tilde{C}_\mathbf{r} + \mathbb{E}\{\Omega_R\tilde{C}_\mathbf{r}\Omega_R^\top\} + \frac{1}{2}\mathbb{E}\{\Omega_R^2\}\tilde{C}_\mathbf{r} + \frac{1}{2}\tilde{C}_\mathbf{r}\mathbb{E}\{\Omega_R^2\}$$
$$+ \mathbb{E}\{\Omega_R\mu_t\mu_t^\top\Omega_R^\top\} + \mathbb{E}\{\omega_t\omega_t^\top\} \tag{11b}$$

where $\tilde{C}_\mathbf{r} = \mu_R C_\mathbf{r}\mu_R^\top$ and the expectations of quantities quadratic in Ω_R and ω_t can be expressed in terms of the $SE(3)$ covariance of g, i.e., $\Sigma_\mathbf{g}$ in (7).

Simplified Equations Under Isotropic Blurring. The unknowns in (11) are the 6×6 covariance matrix $\Sigma_\mathbf{g}$ and the 6×1 vector $\mu_\mathbf{g}$, which in general amounts to 27 unknowns, whereas the number of independent equations is 9. However, if we assume that blurring is isotropic in translational and rotational directions, i.e., $\Sigma_\mathbf{g}$ is diagonal and variances along E_1, E_2 and E_3 are equal to σ_R^2 and along E_4, E_5 and E_6 are σ_t^2, then the number of unknowns will be 8. Thus, we have

$$\mathbb{E}\{\mathbf{y}\} = \mathbb{E}\{\mathbf{t}\} \overset{2nd}{=} (1 - \sigma_R^2)\mu_t \tag{12a}$$

$$C_\mathbf{y} \overset{2nd}{=} \mu_\mathbf{R}C_\mathbf{r}\mu_\mathbf{R}^\top + \sigma_\mathbf{R}^2\left(\text{tr}(C_\mathbf{r})I_3 - 3\mu_\mathbf{R}C_\mathbf{r}\mu_\mathbf{R}^\top\right) + \sigma_\mathbf{R}^2\left(\|\mu_\mathbf{t}\|^2 I_3 - \mu_\mathbf{t}\mu_\mathbf{t}^\top\right) + \sigma_t^2 I_3, \tag{12b}$$

where I_3 is the 3×3 identity matrix. The interesting point here is that if μ_t is large (even for small rotational noise σ_R^2) $C_\mathbf{y}$ can become large merely due to large translational mean. Considering our argument about blurring due to 3D reconstruction errors the assumption of isotropic blurring might not be justified, nevertheless, as a starting point to solve the inverse problem in Proposition 2 we choose this assumption. Figure 1a shows the blurring effect of an istropic $SE(3)$ kernel with mean $\mu_\mathbf{g} = I_4$, $\sigma_\mathbf{R} = \pi/10$ and $\sigma_t = \frac{1}{10}$ applied to a unit vector along the z-direction in $\mathbb{R}^3$.

(a) Example of blurring by an isotropic $SE(3)$ kernel. (b) The right panel shows the blurred version of right configuration in our numerical simulation.

Fig. 1. Examples of blurring under $SE(3)$ kernels.

Algorithm. The two equations in (12) are coupled and nonlinear in the unknowns; however, by fixing σ_R^2 in (12a) and μ_t in (12b) they decouple. Thus, in the first step, we find μ_t from (12a) (fixing σ_R^2) and in the next step $\mu_R, \sigma_R^2, \sigma_t^2$ from (12b) using min-square fitting, and iterate these steps. Specifically, based on (12b) we consider the cost function

$$F(\mu_R, \sigma_R^2, \sigma_t^2)$$
$$= \|\mu_R C_{\mathbf{r}} \mu_R^\top + \sigma_R^2 (\mathrm{tr}(C_{\mathbf{r}}) I_3 - 3\mu_R C_{\mathbf{r}} \mu_R^\top) + \sigma_R^2 (\|\mu_t\|^2 I_3 - \mu_t \mu_t^\top) + \sigma_t^2 I_3 - C_y\|_F^2, \quad (13)$$

where $\|\cdot\|_F$ is the Frobenius norm. We solve the regularized minimization

$$\min_{\mu_R \in SO(3), \sigma_R^2, \sigma_t^2} F_r(\mu_R, \sigma_R^2, \sigma_t^2; \lambda_R, \lambda_t) \quad (14)$$

where $F_r(\mu_R, \sigma_R^2, \sigma_t^2; \lambda_R, \lambda_t) = F(\mu_R, \sigma_R^2, \sigma_t^2) + \lambda_R(\sigma_R^2)^2 + \lambda_t(\sigma_t^2)^2$ and $\lambda_R, \lambda_t > 0$ are small regularization weights. Our experiments show that although the number of unknowns is more than the number of equations in (12a) and (12b), still sensitivity can be high; thus we add the regularization terms in this minimization. Solving (14) in an *alternative* minimization fashion results in simple (closed-form) eigendecomposition-based solution for μ_R and scalar min-square solution with thresholding to enforce $\sigma_R^2, \sigma_t^2 \geq 0$.

Numerical Simulations. We simulate a complex with two subunits ρ_1 and ρ_2 modeled with two Gaussians $\mathbf{r}_1$ and $\mathbf{r}_2$ with covariances $C_{\mathbf{r}_1} = \mathrm{diag}(3,2,1)$ and $C_{\mathbf{r}_2} = \mathrm{diag}(4,3,5)$, respectively. We consider two $SE(3)$ blurring kernels with

$$\mu_{\mathbf{g}_1} = \begin{bmatrix} 0.6063 & 0.3861 & -0.6952 & 5.0000 \\ -0.7453 & -0.5807 & 0.3275 & 5.0000 \\ -0.2773 & 0.7167 & 0.6399 & 5.0000 \\ 0 & 0 & 0 & 1.0000 \end{bmatrix}, \mu_{\mathbf{g}_2} = \begin{bmatrix} -0.6196 & -0.3585 & -0.6983 & -1.0000 \\ -0.3601 & -0.6607 & 0.6587 & -3.0000 \\ -0.6975 & 0.6595 & 0.2802 & -2.0000 \\ 0 & 0 & 0 & 1.0000 \end{bmatrix} \quad (15)$$

and with variances $(\sigma_{R_1}^2, \sigma_{t_1}^2) = (0.2, .02)$ and $(\sigma_{R_2}^2, \sigma_{t_2}^2) = (.1, .01)$. We generate $T = 2000$ i.i.d. samples of $\mathbf{r}_i, \mathbf{g}_i$ $(i = 1, 2)$ and then $\mathbf{y}_i$ according to (9). The left panel in Fig. 1b shows the original configuartion and the right panel shows the blurred configuration, in which the subunits appear bloated (blue (or $\cdot$) and black (or $*$) correpond to $\mathbf{r}_1$ and $\mathbf{r}_2$, respectively). We run a k-means algorithm to separate the two clouds (subunits). Using the above algorithm with $\lambda_R = \lambda_t = 1$ to get the estimates:

$$\hat{\mu}_{\mathbf{g}_1} = \begin{bmatrix} -0.7361 & 0.4145 & -0.5351 & 4.7676 \\ -0.3606 & -0.9092 & -0.2082 & 4.8539 \\ -0.5728 & 0.0397 & 0.8187 & 4.7737 \\ 0 & 0 & 0 & 1.0000 \end{bmatrix}, \hat{\mu}_{\mathbf{g}_2} = \begin{bmatrix} -0.6596 & -0.5253 & -0.5375 & -0.9976 \\ -0.2048 & -0.5625 & 0.8010 & -3.0366 \\ -0.7232 & 0.6384 & 0.2635 & -2.0821 \\ 0 & 0 & 0 & 1.0000 \end{bmatrix} \quad (16)$$

and $\hat{\sigma}_{R_1}^2 = 0.15$, $\hat{\sigma}_{t_1}^2 = 0.89$, $\hat{\sigma}_{R_2}^2 = 0.11$, and $\hat{\sigma}_{t_2}^2 = 0.09$. There is an indeterminacy in estimating $\mu_{\mathbf{g}}$ in the form of a rotation by π, i.e., a factor of the form $\Pi = \begin{bmatrix} -1 & 0 & 0 \\ 0 & -1 & 0 \\ 0 & 0 & 1 \end{bmatrix}$ and its permutations. After fixing the indeterminacy, we get $d(\mu_{R_1}, \hat{\mu}_{R_1}) = 0.2491\pi$ and $d(\mu_{R_2}, \hat{\mu}_{R_2}) = 0.0751\pi$, where $d(\cdot, \cdot)$ is the standard Riemannian distance on $SO(3)$. Thus, the error in estimating $\mu_{\mathbf{g}}$ is low; however, estimating σ_R^2 and σ_t^2 is more difficult. Nevertheless, note that $\mu_{\mathbf{g}}$ is the more important or informative variable in determining relative configurations.

4   Conclusions

In this paper we reproted preliminary studies for the modeling of blurring effects in 3D reconstruction of densities in single particle EM using $SE(3)$ blurring kernels. We derived a set of blurring equations relating the parameters of the original 3D density and the blurring kernel to quantities which can be calculated from the reconstructed density. The equations are highly ill-posed to invert. However, in the case of a multi-unit complex one might have prior knowledge about the shape of the subunits. We examined this in the case of isotropic blurring and derived a simple regularized minimization algorithm to find conformational information of the complex (i.e., the relative positions of subunits). We plan to improve our algorithm e.g., by using more prior information and better regularizations.

Acknowledgements. Research reported in this publication was supported by the National Institute of General Medical Sciences of the National Institutes of Health under award number R01GM113240.

References

1. Arsigny, V., Pennec, X., Ayache, N.: Bi-invariant means in Lie groups. application to left-invariant polyaffine transformations (2006)
2. Chirikjian, G.S.: Stochastic Models, Information Theory, and Lie Groups. Springer, Boston (2012)
3. Feigin, L., Svergun, D.I., Taylor, G.W.: Structure Analysis by Small-angle X-ray and Neutron Scattering. Springer, Berlin (1987)
4. Frank, J.: Three-dimensional electron microscopy of macromolecular assemblies: Visualization of biological molecules in their native (2006)
5. Frank, J.: Story in a sample-the potential (and limitations) of cryo-electron microscopy applied to molecular machines. Biopolymers **99**(11), 832–836 (2013)
6. Hirsch, M., Schölkopf, B., Habeck, M.: A blind deconvolution approach for improving the resolution of cryo-EM density maps. J. Comput. Biol. **18**(3), 335–346 (2011)
7. Kishchenko, G.P., Leith, A.: Spherical deconvolution improves quality of single particle reconstruction. J. Struct. Biol. **187**(1), 84–92 (2014)
8. Park, W., Chirikjian, G.S.: An assembly automation approach to alignment of non-circular projections in electron microscopy. IEEE Trans. Autom. Sci. Eng. **11**(3), 668–679 (2014)
9. Park, W., Midgett, C.R., Madden, D.R., Chirikjian, G.S.: A stochastic kinematic model of class averaging in single-particle electron microscopy. Int. J. Rob. Res. **30**(6), 730–754 (2011)
10. Pennec, X., Arsigny, V.: Exponential barycenters of the canonical cartan connection and invariant means on Lie groups. In: Nielsen, F., Bhatia, R. (eds.) Matrix Information Geometry, pp. 123–166. Springer, Heidelberg (2013)
11. Scheres, S.H., Gao, H., Valle, M., Herman, G.T., Eggermont, P.P., Frank, J., Carazo, J.-M.: Disentangling conformational states of macromolecules in 3D-EM through likelihood optimization. Nature Methods **4**(1), 27–29 (2007)
12. van Heel, M., Gowen, B., Matadeen, R., Orlova, E.V., Finn, R., Pape, T., Cohen, D., Stark, H., Schmidt, R., Schatz, M., et al.: Single-particle electron cryo-microscopy: towards atomic resolution. Q. Rev. Biophys. **33**(04), 307–369 (2000)

13. Wang, Y., Chirikjian, G.S.: Error propagation on the Euclidean group with applications to manipulator kinematics. IEEE Trans. Robot. **22**(4), 591–602 (2006)
14. Zhao, Z., Singer, A.: Rotationally invariant image representation for viewing direction classification in cryo-EM. Journal of structural biology **186**(1), 153–166 (2014)

Nonlinear Operators on Graphs via Stacks

Santiago Velasco-Forero[✉] and Jesús Angulo

MINES ParisTech - PSL-Research University - Centre de Morphologie
Mathématique, Paris, France
jesus.angulo@mines-paristech.fr

Abstract. We consider a framework for nonlinear operators on functions evaluated on graphs via stacks of level sets. We investigate a family of transformations on functions evaluated on graph which includes adaptive flat and non-flat erosions and dilations in the sense of mathematical morphology. Additionally, the connection to mean motion curvature on graphs is noted. Proposed operators are illustrated in the cases of functions on graphs, textured meshes and graphs of images.

1 Introduction

Recent years have witnessed an enormous growth of interest in the description and analysis of problems via similarities or dependencies between data elements. A common way to represent this structure is to use graphs, so that data elements are indexed by graph nodes, and the strength of dependences between pairs of elements is represented by corresponding weighted graph edges. In this paper, we analyze nonlinear (morphological) operators in the context of *discrete signal processing on graphs* [27]. Our framework is used to extend the traditional adaptive (non-flat) morphology on images to more complex structures as sets of images, meshes, point clouds [3] and so on. In graph-based modeling, digital images are a particular case, where the pixel information is represented by the two-dimensional rectangular grid, and pixels correspond to graph nodes related by links according to the four or eight adjacent neighborhood. On the one hand, we note that in the literature, one can find some works about nonlinear filters on graphs and hypergraphs, particularly mathematical morphology operators in the algebraic sense [6,14,21,32], where the couple of nonlinear operator (dilation/erosion) are maps from two different lattices, i.e., they are maps "from nodes to edges" or "edges to nodes". On the other hand, some regularization techniques and nonlinear operators have been introduced for functions evaluated on graph via directional derivative [8,28,29] or discrete version of the p-Laplacian [9]. We adopt a different viewpoint, our approach is inspired from the signal processing approach on graphs [12,24–26]. Thus, we firstly review graph signal decomposition by upper level sets, convolution and diffusion on graphs, and then we present a general formulation of flat and non-flat morphology on graphs, a family of nonlinear transformations and its connection to mean curvature motion on graphs [4,11,12]. Finally, we include some examples to illustrate the interest of our method.

© Springer International Publishing Switzerland 2015
F. Nielsen and F. Barbaresco (Eds.): GSI 2015, LNCS 9389, pp. 654–663, 2015.
DOI: 10.1007/978-3-319-25040-3_70

2 Convolution and Morphology by Stacks on Graphs

We start by introducing the notation used throughout this paper. The objects under study are considered as the nodes (or vertices) of the graph $\mathcal{G}$. A simple, connected, undirected, and weighted graph $\mathcal{G} = (\mathcal{V}, E)$ consists of a set of nodes $\mathcal{V} = \{v_1, v_2, \ldots, v_N\}$ and edges $E = \{(v_n, v_m, w_{nm})\}, v_n, v_m \in \mathcal{V}$, where (v_n, v_m, w_{nm}) denoted an edge of weight w_{nm} between node v_n and v_m. For ease of exposition and to avoid tedious notation, in the sequel we use only subscripts to denote the vertices in the graph, i.e., $\mathcal{V} = \{1, 2, \ldots, N\}$. The degree d_n of a node n is the sum of the edge-weights connected to node n, and the degree matrix of the graph consists of degrees of all nodes arranged in a diagonal matrix $\mathbf{D} = diag\{d_1, d_2, \ldots, d_N\}$. Denote the maximal and minimal degrees by $d_+ = \max_{i \in \mathcal{V}} d_i := \bigvee_{i \in \mathcal{V}} d_i$ and $d_- = \min_{i \in \mathcal{V}} d_i := \bigwedge_{i \in \mathcal{V}} d_i$. The adjacency matrix $\mathbf{W}$ of the graph is an $N \times N$ matrix with $\mathbf{W}(n, m) = w_{nm}$, the *combinatorial Laplacian* matrix is $\mathbf{L} = \mathbf{D} - \mathbf{W}$ and the *graph Laplacian* $\mathcal{L} = \mathbf{I} - \mathbf{D}^{-1/2}\mathbf{W}\mathbf{D}^{-1/2}$ is a generalizations of the Laplacian on the grid, where frequency and smoothness are relative to $\mathbf{W}$ and interrelated through these operators [5]. A graph signal is defined as a scalar valued discrete mapping $\mathbf{f} : \mathcal{V} \to \mathbb{R}$, such that $\mathbf{f}(n)$ is the value of the signal on node n. Thus a graph signal can also be represented as a vector $\mathbf{f}$ in the space of functions from $\mathcal{V}$ to $\mathbb{R}$, denoted by V, with indices corresponding to the nodes in the graph. Additionally, we often analyze operators transforming signals evaluated on graphs, for instance $\phi : \mathbf{f} \to \phi(\mathbf{f})$, in this case we say that $\phi \in \mathsf{V} \times \mathsf{V}$. Finally, the *graph Fourier transform* $\hat{\mathbf{f}}$ of a function $\mathbf{f} \in \mathsf{V}$ is the expansion of $\mathbf{f}$ in terms of the eigenvectors of the graph Laplacian, denoted by Λ_l, with $l = 1, \cdots, N$. More precisely, it is defined in [26] by $\hat{\mathbf{f}}(l) := \langle \mathbf{f}, \Lambda_l \rangle = \sum_{n=1}^{N} \Lambda_l^*(n)\mathbf{f}(n)$, by using the conjugate matrix in the definition.[1]

Definition 1. *The* upper level set *(ULS) of* $\mathbf{f} \in \mathsf{V}$ *at level* $\lambda \in \mathbb{R}$ *is defined by* $\chi(\mathbf{f}, \lambda) = \{n \in \mathcal{V} : \mathbf{f}(n) \geq \lambda\}.$

The set of ULS constitutes a family of decreasing sets: $\lambda \geq \mu \Rightarrow \chi(\mathbf{f}, \lambda) \subseteq \chi(\mathbf{f}, \mu)$ and $\chi(\mathbf{f}, \lambda) = \cap\{\chi(\mathbf{f}, \mu), \mu < \lambda\}$. Any graph signal $\mathbf{f} \in \mathsf{V}$ can be viewed as a unique stack of its cross-sections, which leads to the following superposition description.

Definition 2. *The* threshold-max superposition *of* $\mathbf{f} \in \mathsf{V}$ *is defined by:* $\mathbf{f}(n) = \bigvee\{\lambda \in \mathbb{R} : n \in \chi(\mathbf{f}, \lambda)\}.$

This definition say that for each node n, the signal $\mathbf{f}$ can be recomposed from the ULS finding the largest value of λ where the predicate $n \in \chi(\mathbf{f}, \lambda)$ is valid. Similar to the image description as a topographic surface in [17,18,33], we consider here the alternative stacking reconstruction using a numerical sum of the characteristic function of upper level sets.

---

[1] For a given matrix $\mathbf{W} = [\mathbf{W}(i, j)]$, the *conjugate* of $\mathbf{W}$ to be $\mathbf{W}^* = [-\mathbf{W}(j, i)]$, i.e., $\mathbf{W}^*$ is derived from $\mathbf{W}$ by transposing and negating.

Definition 3. *The* threshold-linear superposition *of* $\mathbf{f} \in V$ *is defined by:* $\mathbf{f}(n) = \int_0^{+\infty} \chi(\mathbf{f}, \lambda)(n) d\lambda$

In the particular case of discrete range, $\mathcal{T} = \{c_1, c_2, \ldots, c_{|\mathcal{T}|}\}$, the signal $\mathbf{f}$ can be reconstructed from the discrete stack $\chi(\mathbf{f}, \lambda)$ via addition, i.e., $\mathbf{f}(n) = \sum_{\lambda \in \mathcal{T}} \chi(\mathbf{f}, \lambda)(n)$.

Definition 4. *We shall say that an operator* $\phi \in V \times V$ *commutes with thresholding if* $\phi(\chi(\mathbf{f}, \lambda)) = \chi(\phi(\mathbf{f}), \lambda)$ *for any signal* $\mathbf{f} \in V$ *and any value* $\lambda \in \mathbb{R}$.

In other words, if an operator ϕ commutes with thresholding, processing by ϕ the upper level set at λ gives the same result as processing first the signal $\mathbf{f}$ byϕ and then thresholding $\phi(\mathbf{f})$ at level λ.

Definition 5. *We shall say that an operator* ϕ *obeys the threshold-linear superposition provided that* $\phi(\mathbf{f}) = \int_0^{+\infty} \phi(\chi(\mathbf{f}, \lambda)) d\lambda$ *for any signal* $\mathbf{f} \in V \times \mathbb{R}^+$.

As it was pointed out in [18], for grey scale images, the threshold-max superposition is more general than the thresholded-linear superposition since the latter applies only to nonnegative input signals, while the former applies to any real-valued input signal. But alternatively, the max-superposition can be applied only when $\phi(\chi(\mathbf{f}, \lambda))$ are binary signals, an assumption not needed by the linear superposition. In fact, the threshold sum/integral ties well also with linear systems. In our case, we assume that $\mathbf{f} \in V \times \mathbb{R}^+$ is continuous and nonnegative so we will consider the reconstruction formula given by (5).

Proposition 1. *The class of operators* $\phi \in V \times V$ *that obey the threshold-linear superposition: (a) is closed under minimum, maximum and composition. (b) It forms a* vector space *over the field of real numbers under vector addition* $(\phi_1 + \phi_2)(\mathbf{f}) := \phi_1(\mathbf{f}) + \phi_2(\mathbf{f})$ *and the scalar multiplication* $(c\phi)(\mathbf{f}) := c\phi(\mathbf{f})$ *with* $c \in \mathbb{R}$.

2.1   Convolution on Graphs

For signals $\mathbf{f}, \mathbf{g} \in L^2(\mathbb{R})$, the convolution product $\mathbf{h} = \mathbf{f} * \mathbf{g}$ satisfies

$$\mathbf{h} = (\mathbf{f} * \mathbf{g}) = \int_{\mathbb{R}} \hat{\mathbf{h}}(\xi) \exp\{2\pi i \xi t\} d\xi = \int_{\mathbb{R}} \hat{\mathbf{f}}(\xi) \hat{\mathbf{g}}(\xi) \exp\{2\pi i \xi t\} d\xi. \qquad (1)$$

By replacing the complex exponentials in (1) with the graph Fourier transform, i.e., the graph Laplacian eigenvectors Λ_l, in [26] has defined a *generalized convolution* of signals $\mathbf{f}, \mathbf{g} \in V$ by $(\mathbf{f} * \mathbf{g}) := \sum_{l=1}^{N} \left[\hat{\mathbf{f}}(l) \hat{\mathbf{g}}(l) \Lambda_l \right] = \sum_{l=1}^{N} \left[(\Lambda_l^* \mathbf{f})(\Lambda_l^* \mathbf{g}) \Lambda_l \right] = \sum_{l=1}^{N} \left[\Lambda_l^* (\mathbf{f}\mathbf{g}) \Lambda_l \right] = \sum_{l=1}^{N} \mathbf{f}(l) \mathbf{g}(l)$.

Proposition 2. *The linear operator associated to the convolution signal function* $\mathbf{g}$ *commutes with the stacking of cross-sections according to the threshold-linear superposition, i.e.,* $(\mathbf{f} * \mathbf{g}) = \int_0^{+\infty} (\chi(\mathbf{f}, \lambda) * \mathbf{g}) d\lambda$

2.2   Diffusion on Graphs

Consider an arbitrary graph $\mathcal{G} = (\mathcal{V}, E, \mathbf{W})$ with Laplacian matrix $\mathbf{L}$ and a signal $\mathbf{f} \in \mathcal{V} \to \mathbb{R}^N$. For a given constant $\sigma > 0$, define the time-varying vector $\mathbf{f}_{\sigma,t} \in \mathbb{R}^N$ as the solution of the linear differential equation:

$$\frac{\partial \mathbf{f}_{\sigma,t}}{\partial t} = -\sigma \mathbf{L} \mathbf{f}_{\sigma,t}, \quad \mathbf{f}_{\sigma,0} = \mathbf{f}, \tag{2}$$

where σ is the thermal conductivity [25] and controls the heat diffusion rate. The differential equation in (2) represents the heat diffusion process on the graph $\mathcal{G}$ due to the fact that $-\mathbf{L}$ can be shown to be the discrete approximation of the continuous Laplacian operator used to characterize the heat diffusion in physics [15,25]. The general solution of the heat equation on $\mathbb{R}^N$ is obtained by convolution [10]. However, the solution of (2) denoted by $\mathbf{f}_{\sigma,t}^{\mathbf{L}} \in \mathsf{V} \times \mathsf{V}$, is given by the matrix exponential as follows

$$\mathbf{f}_{\sigma,t}^{\mathbf{L}} := \exp\left(-\sigma \mathbf{L} t\right) \mathbf{f} \tag{3}$$

which can be verified by direct substitution in (2). It is important to note that for a given time t, the n-th element of $\mathbf{f}_{\sigma,t}(n)$ is $\frac{\partial \mathbf{f}_{\sigma,t}^{\mathbf{L}}(n)}{\partial t} = \sum_{k \in \mathcal{N}(n)} \sigma \mathbf{W}(n,k)(\mathbf{f}_{\sigma,t}^{\mathbf{L}}(k) - \mathbf{f}_{\sigma,t}^{\mathbf{L}}(n))$ where $\mathcal{N}(n)$ is the neighborhood of n, i.e., the set of k such that $\mathbf{W}(i,k) > 0$. Thus, the heat flow on an edge grows proportionally with both the "temperature differential" $\mathbf{f}_{\sigma,t}^{\mathbf{L}}(k) - \mathbf{f}_{\sigma,t}^{\mathbf{L}}(n)$ and the weight $\mathbf{W}(n,k)$. Now in Proposition 3, we see the behavior of the graph diffusion in the stack of cross-sections.

Proposition 3. *The operator $\mathbf{f}_{\sigma,t}^{\mathbf{L}}$ in Eq.(3) associated to the diffusion of a graph signal $\mathbf{f}$, commutes with the stacking of cross-sections according to the threshold-linear superposition, i.e., $\mathbf{f}_{\sigma,t}^{\mathbf{L}}(n) = \int_0^{+\infty} (\chi(\mathbf{f}, \lambda))_{\sigma,t}^{\mathbf{L}}(n) d\lambda$.*

Note that the right part of the equality means that the graph diffusion is applied in each upper level set of the graph signal function $\mathbf{f}$. The proof of Proposition 3 is straightforward by means of Taylor series expansion of the graph heat equation and interchanging summation and integration, i.e., $\mathbf{f}_{\sigma,t}^{\mathbf{L}}(n) = \sum_{k=0}^{\infty} \frac{(-\sigma \mathbf{L} t)^k \mathbf{f}}{k!} = \int_0^{+\infty} \sum_{k=0}^{\infty} \frac{(-\sigma \mathbf{L} t)^k}{k!} \chi(\mathbf{f}, \lambda)(n) d\lambda = \int_0^{+\infty} (\chi(\mathbf{f}, \lambda))_{\sigma,t}^{\mathbf{L}}(n) d\lambda$. At this point, we should highlight that the behavior of the diffusion in (3) is controlled by the choice of the Laplacian matrix, i.e., therefore expression (3) includes isotropic and anisotropic diffusion.

2.3   Morphological Operators in Graphs

In the case of a graph value function $\mathbf{f} \in \mathsf{V}$, we can have the following counterparts of dilation and erosion of numerical functions [7,13] viewed as a convolution in max-plus algebra (and its adjoint/dual algebra).

Definition 6. *The matrix $\mathbf{W}$ is a morphological weight matrix if $-\infty \leq \mathbf{W}(n,m) \leq 0$, for all n, m.*

Note that is a really simple characterization of the weight matrix because in later definition do not require symmetry ($\mathbf{W}(n,m) \neq \mathbf{W}(m,n)$) neither zero-diagonal ($\mathbf{W}(i,i) \neq 0$) as in [31].

Definition 7. *The dilation of a signal function* $\mathbf{f}$ *on a graph* $\mathcal{G} = (\mathcal{V}, E)$ *is defined by* $\delta_{\mathbf{W}}(\mathbf{f})(n) = \bigvee_{m=1}^{N}(\mathbf{f}(m) + \mathbf{W}(n,m)) := \mathbf{W} \oplus \mathbf{f}(n)$ *and the dual adjoint erosion is given by* $\varepsilon_{\mathbf{W}}\mathbf{f}(n) := \bigwedge_{m=1}^{N}(\mathbf{f}(m) + \mathbf{W}^*(n,m)) = \bigwedge_{m=1}^{N}(\mathbf{f}(m) - \mathbf{W}(m,n)) := \mathbf{W}^* \ominus \mathbf{f}(n)$

We remark that, $\varepsilon_{\mathbf{W}}, \delta_{\mathbf{W}}$ are both in $\mathcal{V} \times \mathcal{V}$ and include morphological transformations by flat, non-flat, adaptive [7] and nonlocal structuring elements [31]. Additionally, Theorem 1 do not require the symmetry in the matrix $\mathbf{W}$ as in [31]. A crucial point is the existence of a *Galois adjunction theorem* [13,23] for graph valued signals.

Theorem 1. *Given a* $\mathbf{W}$ *morphological weight matrix and, the pair of operators* $(\varepsilon_{\mathbf{W}}, \delta_{\mathbf{W}})$ *defines an Galois adjunction, i.e., for all* $\mathbf{f}, \mathbf{g}$ *in* $\mathcal{V} \to \mathbb{R}$, *we have* $\mathbf{W} \oplus \mathbf{f} \leq \mathbf{g} \iff \mathbf{f} \leq \mathbf{W}^* \ominus \mathbf{g}$.

Proof. $\mathbf{W} \oplus \mathbf{f} \leq \mathbf{g} \iff \forall n, \quad \bigvee_{m=1}^{N}(\mathbf{f}(m) + \mathbf{W}(n,m)) \leq \mathbf{g}(n)$
$\iff \forall n, \forall m, \mathbf{f}(m) + \mathbf{W}(n,m) \leq \mathbf{g}(n),$
$\iff \forall n, \forall m, \mathbf{f}(m) \leq \mathbf{g}(n) - \mathbf{W}(n,m), \iff \forall m, \mathbf{f}(m) \leq \bigwedge_{n=1}^{N} \mathbf{g}(n) - \mathbf{W}(n,m),$
$\iff \forall m, \mathbf{f}(m) \leq \bigwedge_{n=1}^{N} \mathbf{g}(n) + \mathbf{W}^*(m,n), \iff \forall m, \mathbf{f} \leq \mathbf{W}^* \ominus \mathbf{g}.$

However, we do not have an order between the original signal $\mathbf{f}$ and its dilation or erosion, *i.e.*, we have $\mathbf{W}^* \ominus \mathbf{f} \leq \mathbf{W} \oplus \mathbf{f}$, but $\mathbf{f} \nleq \mathbf{W} \oplus \mathbf{f}$ neither $\mathbf{W} \ominus \nleq \mathbf{f}$.

Definition 8. *A morphological weight matrix* $\mathbf{W}$ *is called* conservative *if* $\mathbf{W}(i,i) = 0$ *for all* $i \in 1 \dots, N$.

Proposition 4. *If* $\mathbf{W}$ *is a conservative morphological weight matrix then* $\mathbf{W}^* \ominus \mathbf{f} \leq \mathbf{f} \leq \mathbf{W} \oplus \mathbf{f}$ *for every* $\mathbf{f}$ *in* $\mathcal{V} \to \mathbb{R}$.

Thanks to Theorem 1, we can have a large set of morphological filters such as openings, closings, alternate sequential filters, leveling and so on, because they are defined by combination of dilations and erosions [16,22].

Definition 9. *A morphological weight matrix* $\mathbf{B}$ *is called* flat *if* $\mathbf{B}(i,j) = 0$ *or* $\mathbf{B}(i,j) = -\infty$ *for all* $i,j \in 1 \dots, N$.

Proposition 5. *The flat dilation and erosion obey both the threshold-max superposition and the threshold linear superposition, i.e.,* $\delta_{\mathbf{B}}(\mathbf{f}) = \bigvee_{\lambda=0}^{+\infty}\{\delta_{\mathbf{B}}(\chi(\mathbf{f}, \lambda)) = 1\} = \int_{0}^{+\infty} \delta_{\mathbf{B}}(\chi(\mathbf{f}, \lambda))d\lambda$

Since by Propositions 1 and 5, we can directly have that the class of operators $\phi \in \mathcal{V} \times \mathcal{V}$ that obey the threshold-linear superposition also contain morphological gradients (difference between dilation and erosion), opening and closing (composition of dilation and erosion), top-hat transformation, granulometries, reconstruction operators, leveling, additive morphological decompositions [30] and skeleton transformation (based on generalized Lantuejoul formula).

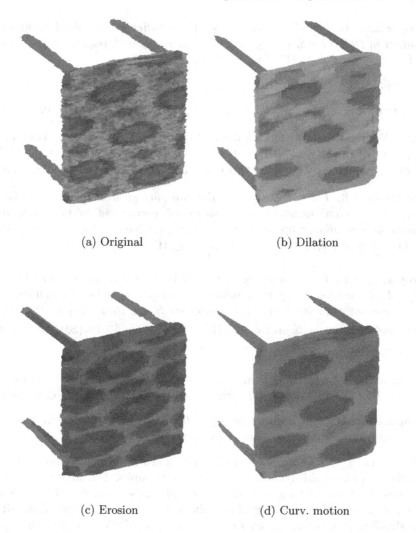

(a) Original   (b) Dilation

(c) Erosion   (d) Curv. motion

Fig. 1. Illustration of nonlinear filters on textured mesh. The textured mesh is obtained from [2]. Note that both colors and mesh coordinates have been modified in the processing. (d) Curvature motion is obtained by iterating $\psi_{\mathbf{L},.5}$ (Color figure online).

2.4 Morphological Operators via Convolution on Graph

Definition 10. *For a graph signal value* $\mathbf{f} \in \mathsf{V}$, *the convolution-thresholding nonlinear operator associated to the heat diffusion* $\mathbf{W}$ *of conductivity* σ *at scale* t, *and the threshold* τ, *with* $\tau \in [0,1]$, *is the mapping* $\mathcal{F}(\mathcal{V}, \mathbb{R}) \to \mathcal{F}(\mathcal{V}, \mathbb{R})$ *defined by*

$$\psi_{\mathbf{L},\,\tau}(\mathbf{f}) = \int_0^{+\infty} \left[(\chi(\mathbf{f}, \lambda))_{\sigma,t}^{\mathbf{L}} \geq \tau \right] d\lambda \tag{4}$$

The next proposition is easy to prove.

Proposition 6. *For all* $\mathbf{f} \in \mathcal{V}$: *(a)* $\psi_{\mathbf{L}, \tau}(\mathbf{f})$ *satisfies the threshold-linear super-position in Definition 3. (b)* $\psi_{\mathbf{L}, \tau}(\mathbf{f})$ *is monotonous with respect to the choice of* τ, *i.e.,* $\tau_1 \leq \tau_2 \Rightarrow \psi_{\mathbf{L}, \tau_1}(\mathbf{f}) \leq \psi_{\mathbf{L}, \tau_2}(\mathbf{f})$.

We can also prove that (4) is increasing.

Proposition 7. *If* $\mathbf{f}_1 \leq \mathbf{f}_2$, *then* $\psi_{\mathbf{L}, \tau}(\mathbf{f}_1) \leq \psi_{\mathbf{L}, \tau}(\mathbf{f}_2)$ *for all* $\sigma, t \geq 0$.

Proof. We note that (3) follows the called *comparison principle*, (Lemma 2.6, property (d) in [12]), i.e. if $\mathbf{f}_1 \leq \mathbf{f}_2$, then $(\mathbf{f}_1)_{\sigma,t}^{\mathbf{L}} \leq (\mathbf{f}_2)_{\sigma,t}^{\mathbf{L}}$ for all $\sigma, t \geq 0$. The proof is completed by applying this result in each ULS and integrating in λ.

Proposition 8. *For the case of a flat morphological weight matrix* $\mathbf{B}$, *the morphological flat operators in Proposition 5 correspond to the convolution-thresholding nonlinear operator in* (4) *with particular values of* τ *as follows:* $\delta_{\mathbf{B}}(\mathbf{f}) = \lim_{\tau \to 0^+} \psi_{\mathbf{B}, \tau}(\mathbf{f})$ *and* $\varepsilon_{\mathbf{B}}(\mathbf{f}) = \lim_{\tau \to 1^-} \psi_{\mathbf{B}, \tau}(\mathbf{f})$.

Proposition 9. *For a binary signal* $S \in \mathcal{V} \times \{0,1\}$, *the set of measures* $Vol(S) = \sum_{i \in S} d_i$. *Let* $\rho(\mathbf{L})$ *be the spectral radius of the graph Laplacian,* $\mathbf{L}$, *then iterations in Proposition 8 on the graph with initial set* S *are stationary if either of the two conditions are satisfied:* $\sigma \leq \rho(\mathbf{L})^{-1} \log\left(1 + \tau d_-^{r/2}(Vol(S))^{-1/2}\right)$ *or* $\sigma \leq \frac{\tau}{||\mathbf{L}\chi(S)||_{\mathcal{V}, \infty}}$

The proof is direct by using Lemma 2.2 and Theorem 4.2 in [12]. Now, we point out a link of the operator in (4) with motion by mean curvature on a graph.

Proposition 10. *The operator in* (4) *in the case of* $\tau = .5$ *is an iteration with of the approximate motion by mean curvature on a graph.*

Firstly, the $\psi_{\mathbf{L}, .5}$ is an iteration of the well-known Merriman, Bence and Osher (MBO) [20] threshold dynamics algorithms on graphs. The *MBO algorithm* is obtained by time splitting the *Allen-Cahn phase-field equation* for motion by mean curvature.[2] The resulting scheme alternates two steps, diffusion, and simple thresholding [12]. Secondly, several papers use the MBO algorithm on a graph to approximate motion by mean curvature [12,19]. It is important to note that the curvature is defined by means of the isotropic total variation [12] instead of the one-Laplacian as it is the case in [8,9,28,29].

---

[2] The semi-linear heat equation called the *Allen-Cahn equation* is a reaction-diffusion equation of mathematical physics of the form:

$$\frac{\partial u}{\partial t} - \delta u + \frac{W'(u)}{\epsilon^2} = 0 \in \mathbb{R}^N \times (0, \infty) \tag{5}$$

which was introduced by S.M. Allen and J.W. Can (1979) [1] to describe the process of phase separation in iron alloys, including order-disorder transitions. Here W is a function that has only two equal minima; its typical form is $W(v) = \frac{(v^2-1)^2}{2}$, W' denotes the derivative and ϵ is a positive parameter.

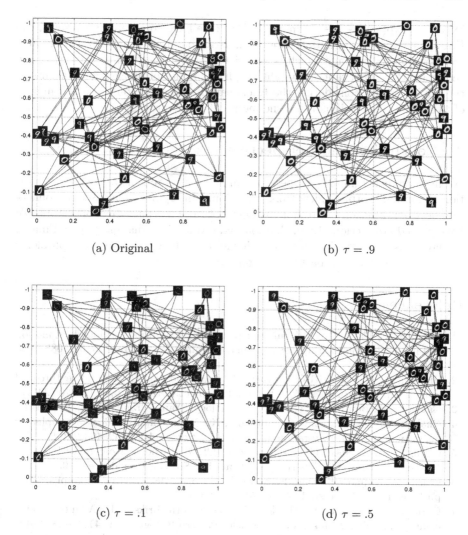

(a) Original

(b) $\tau = .9$

(c) $\tau = .1$

(d) $\tau = .5$

Fig. 2. $\mathcal{G}$ is the five-nearest neighbors with $\mathbf{W}$ the Euclidean distance between pairs of images. Lines are linking images that where $\mathbf{W}(i,j)$ is not zero. Diffusion parameters ($t = 20$ and $\sigma = .03$). Note that digits tend to be similar in (d) (after ten iterations).

2.5 Examples of Applications

The family of filters proposed can be used to analyze any function defined on the vertices of a graph. We provide some illustrations of the results in color data on mesh in Fig. 1 and images on a graph of images in Fig. 2. We first constructed a graph from these datasets by treating the nodes in the graph to be the sample points in the dataset and the edges weight to be the similarity between the features of the different samples. Edge weights were determined via the Radial Bases Function (RBF) kernel with σ^2 set to the variance in the respective dataset

$\mathbf{W}(i,j) = \exp(-\frac{||\mathbf{x}_i - \mathbf{x}_j||^2}{2\sigma^2})$. Finally, in Fig. 2 we have considered a subset of the USPS handwritten digit database for illustration. Each image is a digit in a 28×28 grey scale image which is considered in our approach as a multivariate vector in $\mathbb{R}^{784}$. This random weight graph has nodes on images and weights in the 5-KNN graph considering the Euclidean distance. We only use 200 images of two digits (0,9) to illustrated the merits of our approach.

3 Conclusion and Perspectives

We have analyzed nonlinear operators on stack of graphs as a discrete adaptation of non-flat morphological transformation. This approach is based on connection between diffusion+ thresholding operators and morphological operators. We have proved adjunction of pair dilation and erosion in signal on graphs. Finally, we have illustrated the interest and behavior of such operators in some problems of image processing and pattern recognition.

References

1. Allen, S.M., Cahn, J.W.: A microscopic theory for antiphase boundary motion and its application to antiphase domain coarsening. Acta Metall. **27**(6), 1085–1095 (1979)
2. Biasotti, S., et al.: SHREC 14 track: Retrieval and classification on textured 3d models. In: Eurographics Workshop on 3D Object Retrieval, pp. 111–120 (2014)
3. Calderon, S., Boubekeur, T.: Point morphology. Trans. Graph. **33**(4), 45 (2014)
4. Chakik, A.E., Elmoataz, A., Desquesnes, X.: Mean curvature flow on graphs for image and manifold restoration and enh. Sig. Process. **105**, 449–463 (2014)
5. Chung, F.R.K.: Spectral Graph Theory (CBMS Regional Conference Series in Mathematics, No. 92). American Mathematical Society, Providence (1996)
6. Cousty, J., Najman, L., Dias, F., Serra, J.: Morphological filtering on graphs. Comput. Vis. Image Underst. **117**(4), 370–385 (2013)
7. Ćurić, V., Landström, A., Thurley, M.J., Luengo Hendriks, C.L.: Adaptive mathematical morphology-a survey of the field. Pattern Recogn. Lett. **47**, 18–28 (2014)
8. Elmoataz, A., et al.: Nonlocal discrete regularization on weighted graphs: a framework for image and manifold processing. Trans. Image Proc. **17**(7), 1047–1060 (2008)
9. Elmoataz, A., et al.: Non-local morphological PDEs and p-Laplacian equation on graphs with applications in image processing and machine learning. J. Sel. Top. Sig. Process. **6**(7), 764–779 (2012)
10. Evans, L.C.: Partial Differential Equations. American Mathematical Society, Providence (1998)
11. Garcia-Cardona, C., et al.: Multiclass data segmentation using diffuse interface methods on graphs. Trans. Patt. Ana. Mach. Lear. **36**(8), 1600–1613 (2014)
12. van Gennip, Y., et al.: Mean curvature, threshold dynamics, and phase field theory on finite graphs. Milan J. Math. **82**(1), 3–65 (2014)
13. Heijmans, H.J., Ronse, C.: The algebraic basis of mathematical morphology i. dilations and erosions. CVGIP **50**(3), 245–295 (1990)

14. Heijmans, H., Nacken, P., Toet, A., Vincent, L.: Graph morphology. J. Vis. Commun. Image Represent. **3**(1), 24–38 (1992)
15. Kondor, R.I., Lafferty, J.: Diff. kernels on graphs. In: ICML, pp. 315–322 (2002)
16. Maragos, P.: Lattice image processing: a unification of morphological and fuzzy algebraic systems. J. Math. Imaging Vis. **22**(2–3), 333–353 (2005)
17. Maragos, P., Schafer, R.W.: Morphological filters-part II. IEEE Trans. Acoust. Speech Sig. Process. **35**(8), 1170–1184 (1987)
18. Maragos, P., Ziff, R.D.: Threshold superposition in morphological image analysis systems. Trans. Pattern Ana. Mach. Lear. **12**(5), 498–504 (1990)
19. Merkurjev, E., Kostic, T., Bertozzi, A.L.: An MBO scheme on graphs for segmentation and image processing. SIAM J. Imaging Sci. **6**(4), 1903–1930 (2013)
20. Merriman, B., Bence, J.K., Osher, S.: Diffusion generated motion by mean curvature. Department of Mathematics, University of California, Los Angeles (1992)
21. Meyer, F.: Watersheds on weighted graphs. Pattern Recogn. Lett. **47**(1), 72–79 (2014)
22. Ronse, C.: Why math. morph. needs complete lattices. Sig. Process. **21**(2), 129–154 (1990)
23. Ronse, C., Heijmans, H.J.: The algebraic basis of mathematical morphology: II. openings and closings. CVGIP **54**(1), 74–97 (1991)
24. Sandryhaila, A., Moura, J.: Big data analysis with signal processing on graphs. IEEE Sig. Process. Mag. **31**(5), 80–90 (2014)
25. Segarra, S., Huang, W., Ribeiro, A.: Diffusion and superposition distances for signals supported on networks. CoRR abs/1411.7443 (2014)
26. Shuman, D.I., et al.: A windowed graph fourier transform. In: 2012 IEEE Statistical Signal Processing Workshop (SSP), pp. 133–136 (2012)
27. Shuman, D.I., et al.: The emerging field of signal processing on graphs. IEEE Sig. Process. Mag. **30**(3), 83–98 (2013)
28. Ta, V.T., Elmoataz, A., Lézoray, O.: Nonlocal pdes-based morphology on weighted graphs for image and data processing. Trans. Imaging Proc. **20**(6), 1504–1516 (2011)
29. Ta, V.-T., Elmoataz, A., Lézoray, O.: Partial difference equations over graphs: morphological processing of arbitrary discrete data. In: Forsyth, D., Torr, P., Zisserman, A. (eds.) ECCV 2008, Part III. LNCS, vol. 5304, pp. 668–680. Springer, Heidelberg (2008)
30. Velasco-Forero, S., Angulo, J.: Classification of hyperspectral images by tensor modeling and additive morph. decom. Pattern Recogn. **46**(2), 566–577 (2013)
31. Velasco-Forero, S., Angulo, J.: On nonlocal mathematical morphology. In: Hendriks, C.L.L., Borgefors, G., Strand, R. (eds.) ISMM 2013. LNCS, vol. 7883, pp. 219–230. Springer, Heidelberg (2013)
32. Vincent, L.: Graphs and mathematical morphology. Sig. Proc. **16**(4), 365–388 (1989)
33. Wendt, P., et al.: Stack filters. Trans. Acoust. Speech Sig. **34**(4), 898–911 (1986)

An Intrinsic Cramér-Rao Bound on Lie Groups

Silvère Bonnabel and Axel Barrau[⊠]

Centre for Robotics, MINES ParisTech, PSL Research University,
60 Bd St Michel, 75006 Paris, France
{silvere.bonnabel,axel.barrau}@mines-paristech.fr

Abstract. In his 2005 paper, S.T. Smith proposed an intrinsic Cramér-Rao bound on the variance of estimators of a parameter defined on a Riemannian manifold. In the present technical note, we consider the special case where the parameter lives in a Lie group. In this case, by choosing, e.g., the right invariant metric, parallel transport becomes very simple, which allows a more straightforward and natural derivation of the bound in terms of Lie bracket, albeit for a slightly different definition of the estimation error. For bi-invariant metrics, the Lie group exponential map we use to define the estimation error, and the Riemannian exponential map used by S.T. Smith coincide, and we prove in this case that both results are identical indeed.

1 Introduction

The Cramér-Rao bound is a lower bound on the achievable precision of any unbiased estimator of a vector θ which parametrizes a family of probability distributions $p(X|\theta)$, from a sample $X_1, \cdots, X_n$. This bound is standard in classical estimation theory. Differential geometry considerations in statistics can be traced back to equivariant estimation [10] (see also [6] and references therein for a more recent exposure) and of course to the work by Fisher on the Information metric, and all the works that followed, notably in information geometry. The paper [11] proposes to derive an intrinsic Cramér-Rao bound for the case where the parameter lives in a Riemannian manifold. The two examples given are subspace estimation (that pertains to the Grassman manifold) and covariance matrix estimation (that pertains to the cone of positive definite matrices), both examples being related to signal processing applications. See also the nice extensions proposed since then by N. Boumal [8,9], and our gentle introduction to the subject [2] for more details.

Our motivating example is the so-called Wahba's problem [13], named after Grace Wahba, which is an optimization problem where the parameter is a rotation matrix, but which can also be viewed as the search for a maximum likelihood rotation estimator. The application invoked in [13] is satellite attitude determination. The derivation of more sophisticated attitude estimators has been the subject of a lot of research over the last decade, mainly driven by the burst of mini UAVs (unmanned aerial vehicles), especially quadrotors. The reader is referred to, e.g., [4] for examples.

© Springer International Publishing Switzerland 2015
F. Nielsen and F. Barbaresco (Eds.): GSI 2015, LNCS 9389, pp. 664–672, 2015.
DOI: 10.1007/978-3-319-25040-3_71

In the present paper we propose a general derivation of an intrinsic Cramér-Rao lower bound on Lie groups, that is similar to the one proposed by S.T. Smith on manifolds, that is, we retain terms up to the second order in the estimation error. The discrepancy between the estimation and the true parameter is naturally defined in terms of group operation (which makes it intrinsic). Thus, the bound differs from the Euclidean one because of the non-commutativity of the group operation, yielding some additional terms that are expressed thanks to the Lie bracket (or alternatively structure constants). It is interesting to note our result coincides with the result of S.T. Smith in the case where the metric is bi-invariant, as in $SO(3)$. However, both formula disagree in the general case, as the definition of estimation error in terms of group multiplication differs from the intrinsic estimation error based on the Riemannian exponential proposed by S.T. Smith.

2 An Intrinsic Cramér-Rao Bound on Lie Groups

We compute here the Intrinsic Cramér-Rao Lower Bound (ICRLB) on a Lie group G, up to the second order terms in the estimation error $\log(g\hat{g}^{-1})$, where $g \in G$ is the true value of the parameter and $\hat{g} \in G$ the estimate.

2.1 Preliminaries

Let G be a Lie group of dimension n. To simplify notations we assume G is a matrix Lie group. The tangent space at the Identity element Id, denoted $\mathfrak{g}$, is called the Lie Algebra of G and can be identified as $\mathbb{R}^n$, that is

$$\mathfrak{g} \approx \mathbb{R}^n.$$

The (group) exponential map

$$\exp : \mathfrak{g} \mapsto G$$
$$\xi \to \exp(\xi),$$

provides a local diffeomorphism in a neighborhood of Id. The (group) logarithmic map

$$\log : G \mapsto \mathfrak{g},$$

is defined as the principal inverse of exp. For any estimator $\hat{g}$ of a parameter $g \in G$, it allows to measure the mean quadratic estimation error projected onto the Lie algebra (the error being intrinsically defined in terms of group operation, where the group multiplication replaces the usual addition in $\mathbb{R}^n$)

$$\mathbb{E}_g \left(\log(g\hat{g}(X)^{-1}) \right)$$

where $\mathbb{E}$ denotes the expectation assuming X is sampled from $\mathbb{P}(X|g)$. The logarithmic map allows also to define a covariance matrix of the estimation error:

$$P = \mathbb{E}_g \left(\log(g\hat{g}(X)^{-1}) \log(g\hat{g}(X)^{-1})^T \right) \in \mathbb{R}^{n \times n}. \tag{1}$$

2.2 Main Result

Consider a family of densities parameterized by elements of G

$$p(X|g),\ X \in \mathbb{R}^k,\ g \in G.$$

Using the exponential map, the intrinsic information matrix $J(g)$ can be defined in a right-invariant basis as follows: for any $\xi \in \mathbb{R}^n$,

$$\xi^T J(g)\xi = \int \left(\frac{d}{dt}\Big|_{t=0} \log p\left(X \mid \exp\left(t\xi\right)g\right)\right)^T \left(\frac{d}{dt}\Big|_{t=0} \log p\left(X \mid \exp\left(t\xi\right)g\right)\right) p(X \mid g)dX, \quad (2)$$

and then $J(g)$ can be recovered using the standard polarization formulas

$$\xi^T J(g)\nu = \frac{1}{2}\left((\xi+\nu)^T J(g)(\xi+\nu) - \xi^T J(g)\xi - \nu^T J\nu\right).$$

Besides, using the fact that $\int p\left(X \mid \exp\left(t\xi\right)g\right)dX$ is constant (equal to 1), which implies

$$0 = \frac{d}{dt}\int p\left(X \mid \exp\left(t\xi\right)g\right)dX = \int \left(\frac{d}{dt}\log p\left(X \mid \exp\left(t\xi\right)g\right)\right) p\left(X \mid \exp\left(t\xi\right)g\right)dX, \tag{3}$$

we have, differentiating equality (3) a second time w.r.t t and reusing that $\frac{d}{dt}p = p\frac{d}{dt}\log p$:

$$0 = \int \left(\frac{d^2}{dt^2}\log p\left(X \mid \exp\left(t\xi\right)g\right)\right) p\left(X \mid g\right)dX$$
$$+ \int \left(\frac{d}{dt}\Big|_{t=0} \log p\left(X \mid \exp\left(t\xi\right)g\right)\right)\left(\frac{d}{dt}\Big|_{t=0} \log p\left(X \mid \exp\left(t\xi\right)g\right)\right) p(X|g)dX,$$

allowing to recover an intrinsic version of the classical result according to which the information matrix can be also defined using a second order derivative

$$\boxed{\xi^T J(g)\xi = -\mathbb{E}_g \left(\frac{d^2}{dt^2}\log p\left(X \mid \exp\left(t\xi\right)g\right)\right).}$$

Let $\hat{g}$ be an unbiased estimator of g in the sense of the intrinsic (right invariant) error $g\hat{g}^{-1}$, that is,

$$\int_X \log\left(g\hat{g}\left(X\right)^{-1}\right) p\left(X|g\right)dX = 0.$$

Let P be the covariance matrix of the estimation error as defined in (1). Our main result of this section is as follows

$$\boxed{P \succeq \left(Id + \frac{1}{12}P.H\right) J(g)^{-1} \left(Id + \frac{1}{12}P.H\right)^T,} \tag{4}$$

where we have neglected terms of order $\mathbb{E}_g\left(\left\|\log\left(g\hat{g}\left(X\right)^{-1}\right)\right\|^3\right)$, and where H is the (1,3)-structure tensor defined by

$$H\left(X,Y,Z\right) := [X,[Y,Z]],$$

and where $P.H$ is the tensor contraction of P and H on the two first lower indices of H, defined by $(P.H)_{kl} = \sum_{ij} P^{ij} H^l_{ijk}$. Using the structure constants of G defined by

$$[e_i, e_j] : \sum_k c^k_{ij} e_k, \quad (ad_{e_i})^j_k = c^k_{ij}, \tag{5}$$

note that the components of H can be expressed by the equality

$$H^m_{ijk} = \sum_l c^m_{il} c^l_{jk}. \tag{6}$$

The latter result is totally intrinsic, that is, it is independent of the choice of the metric in the Lie algebra $\mathfrak{g}$.

For small errors, we can neglect the terms in P on the right hand side (curvature terms) yielding the approximation which reminds the Euclidean case

$$P = \int_X \log\left(g\hat{g}\left(X\right)^{-1}\right) \log\left(g\hat{g}\left(X\right)^{-1}\right)^T p\left(X|g\right) dX \succeq J\left(g\right)^{-1} + \text{curvature terms}.$$

2.3 Proof of the Result

Let $\hat{g}$ be an unbiased estimator of g in the sense of the intrinsic (right invariant) error $g\hat{g}^{-1}$, that is,

$$\int_X \log\left(g\hat{g}\left(X\right)^{-1}\right) p\left(X|g\right) dX = 0.$$

If we let ξ be any vector of the Lie algebra and $t \in \mathbb{R}$, the latter formula holds with g replaced by $\exp\left(t\xi\right) g$ and X sampled from $p(X|\exp\left(t\xi\right) g)$. Thus we have $\mathbb{E}_{\exp(t\xi)g}\left(\log\left[\exp\left(t\xi\right) g\hat{g}\left(X\right)^{-1}\right]\right) = 0$ for any $t \in \mathbb{R}$. Differentiating this equality written as an integral over X we get

$$\frac{d}{dt} \int_X \log\left[\exp\left(t\xi\right) g\hat{g}\left(X\right)^{-1}\right] p\left(X|\exp\left(t\xi\right) g\right) dX = 0.$$

Formally, this implies at $t = 0$

$$\int_X \left(D\log|_{g\hat{g}(X)^{-1}}\left[\xi g\hat{g}(X)^{-1}\right] p\left(X|g\right) + \log\left(g\hat{g}(X)^{-1}\right) Dp\left(X\mid\cdot\right)|_g\left[\xi g\right]\right) dX = 0. \tag{7}$$

For any linear form $u(\cdot)$ of the Lie algebra $\mathfrak{g}$ we have thus:

$$-\int_X u\left(D\log|_{g\hat{g}(X)^{-1}}\left[\xi g\hat{g}\left(X\right)^{-1}\right]\right)p\left(X|g\right)dX$$

$$=\int_X u\left(\log\left(g\hat{g}(X)^{-1}\right)Dp(X\mid\cdot)|_g[\xi g]\right)dX$$

$$\leqslant\sqrt{\left(\int_X u\left(\log\left(g\hat{g}(X)^{-1}\right)\right)^2 p(X|g)dX\right)\left(\int_X\left(D\log p(X\mid\cdot)|_g[\xi g]\right)^2 p(X|g)\right)dX},$$

$$(8)$$

where we used the Cauchy Schwarz inequality and the relationships

$$Dp=pD\log p,\quad\text{and then}\quad p=(\sqrt{p})^2.$$

We then introduce a basis of $\mathfrak{g}$ and the vector $\tilde{A}(X)=\log(g\hat{g}(X)^{-1})$. According to (2) the right-hand integral in (8) is $J(g)$, which yields (u being assimilated to a vector of $\mathfrak{g}$):

$$\left(u^T\int_X D\log|_{g\hat{g}(X)^{-1}}[\xi_\times g\hat{g}(X)^{-1}]p(X|g)dX\right)^2$$

$$\leqslant\left(u^T\left[\int_X\tilde{A}(X)\tilde{A}(X)^T p(X|g)dX\right]u\right)\left(\xi^T J(g)\xi\right),$$

$$(9)$$

where we added the subscript $\times$ to ξ, distinguishing the element $\xi_\times$ of $\mathfrak{g}$ and the column vector ξ containing its coordinates in the chosen basis. Now we compute a second-order expansion (in the estimation error) of the left-hand term. To do so, we note from the Baker-Campbell-Hausdorff (BCH) formula retaining only terms proportional to t

$$\log\left[\exp\left(t\xi\right)Q\right]=\log\left[\exp\left(t\xi\right)\exp\left(\log(Q)\right)\right]$$

$$=t\xi-\frac{1}{2}\left[\log(Q),t\xi\right]+\frac{1}{12}\left[\log(Q),[\log(Q),t\xi]\right]+O\left(\|\log(Q)\|^3\right)\right]t\xi.$$

This gives the formula below: the second-order expansion in $\log(Q)$ of the derivative of the left-hand term w.r.t to t. Note that this approximation will be integrated over Q in Eq. (11) and therefore a rigorous reasoning should prove the density p is small where the higher-order terms become larger. Here we assume p is sufficiently peaked for this approximation to be valid.

$$D\log_Q[(\xi)_\times Q]=[I-\frac{1}{2}ad_{\log(Q)}+\frac{1}{12}ad^2_{\log(Q)}]\xi.\qquad(10)$$

Moreover we have by linearity (ξ is here deterministic):

$$\mathbb{E}_g\left(\frac{1}{2}\left[\log\left(g\hat{g}(X)^{-1}\right),t\xi\right]\right)=\frac{1}{2}\left[\mathbb{E}_g\left(\log\left(g\hat{g}(X)^{-1}\right)\right),t\xi\right]=0,$$

and also

$$\mathbb{E}_g[x,[x,\xi]]=\sum H^l_{ijk}\mathbb{E}_g(x_i x_j)\xi_k e_l=G^0\xi,$$

where G^0 is a matrix whose entries are functions of $\mathbb{E}_g(xx^T)$ and the structure constants (see (5)): G^0 is defined by $G^0 = P.H$, i.e. $G^0_{k,l} = \sum_{ij} P^{ij} H^l_{ijk}$ with H defined as in (6). We introduce the latter second-order expansion in the error in Eq. (9):

$$\left[u^T \int_X \left[I + \frac{1}{12} ad^2_{\tilde{A}(X)} \right] p(X|g) dX \xi \right]^2 \leqslant \left(u^T \left[\int_X \tilde{A}(X)\tilde{A}(X)^T p(X|g) dX \right] u \right) \left(\xi^T J(g) \xi \right).$$
(11)

Letting $P = \int_X \tilde{A}(X)\tilde{A}(X)^T p(X|g) dX$ we get:

$$\left[u^T \left(I + \frac{1}{12} G^0(P) \right) \xi \right]^2 \leqslant (u^T P u) \left(\xi^T J(g) \xi \right).$$

Replacing ξ with the variable $\xi = J(g)^{-1} \left(I + \frac{1}{12} G^0(P) \right) u$ directly allows to prove that

$$P \succeq \left(1 + \frac{1}{12} G^0(P) \right) J(g)^{-1} \left(1 + \frac{1}{12} G^0(P) \right)^T,$$
(12)

where

$$G^0(P) = \mathbb{E}_g \left[\log \left(g\hat{g}^{-1} \right), \left[\log \left(g\hat{g}^{-1} \right), \cdot \right] \right].$$
(13)

Remark 1. *If the model is equivariant, i.e., verifies $\forall h \in G, p\left(hX|gh^{-1}\right) = p\left(X|g\right)$ (see [2, 6]), the study can be restricted to equivariant estimators (estimators verifying $\hat{g}(hX) = \hat{g}(X)h^{-1}$). In this case Eq. (13) simplifies:*

$$G^0(P) = \int_X \left[\log \left(g\hat{g}(X)^{-1} \right), \left[\log \left(g\hat{g}(X)^{-1} \right), \cdot \right] \right] p(X|g) dX$$

$$= \int_X \left[\log \left(\hat{g}(gX)^{-1} \right), \left[\log \left(\hat{g}(gX)^{-1} \right), \cdot \right] \right] p(gX|Id) dX$$

$$= \int_{X'} \left[\log \left(\hat{g}(X')^{-1} \right), \left[\log \left(\hat{g}(X')^{-1} \right), \cdot \right] \right] p(X'|Id) dX',$$

where the integration variable X has been replaced by $X' = gX$ in the latter equality. Thus if the model is equivariant, the Cramér-Rao bound is constant over the Lie group.

3   Links with the More General Riemannian Manifolds Case

In the paper [11], the author derives the following intrinsic Cramér-Rao bound (see Corollary 2). Assume θ lives in a Riemannian manifold and $\hat{\theta}$ is an unbiased estimator, i.e. $\mathbb{E}_g(\exp_\theta^{-1} \hat{\theta}) = 0$ where exp is the geodesic exponential map at point θ associated with the chosen metric. The proposed ICRLB writes (up to higher order terms)

$$P := \mathbb{E}_g \left(\exp_\theta^{-1} \hat{\theta} \right) \left(\exp_\theta^{-1} \hat{\theta} \right)^T \succeq J(\theta)^{-1} - \frac{1}{3} \left(R_m(P)J(\theta)^{-1} + J(\theta)^{-1} R_m(P) \right),$$

where for sufficiently small covariance norm the matrix $R_m(P)$ can be approximated by the quadratic form

$$\Omega \rightarrow \langle R_m(P)\Omega, \Omega \rangle = \mathbb{E}_g \left\langle R \left(\exp_\theta^{-1} \hat{\theta}, \Omega \right) \Omega, \exp_\theta^{-1} \hat{\theta} \right\rangle,$$

where R is the Riemannian curvature tensor at θ.

The considered error $\log(g\hat{g}^{-1})$ being right-invariant, we can assume $g = Id$ to compare our result to the latter. We then see, that up to third order terms, formula (12) coincides indeed with the result of [11] **if**

$$G^0(P) = -4R_m(P),$$

where

$$\langle R_m(P)\xi, \xi \rangle = \mathbb{E}_g \left\langle R \left(\log \left(g\hat{g}(X)^{-1} \right), \xi \right) \xi, \log \left(g\hat{g}(X)^{-1} \right) \right\rangle,$$

$$\text{and } \left\langle G^0(P)\xi, \xi \right\rangle = \mathbb{E}_g \left\langle \left[\log \left(g\hat{g}(X)^{-1} \right), \left[\log \left(g\hat{g}(X)^{-1} \right), \xi \right] \right], \xi \right\rangle.$$

If the metric is bi-invariant, as is the case for $G = SO(3)$, we recover the result of [11]. Indeed, for bi-invariant metrics on Lie groups, the Riemannian curvature tensor satisfies for right-invariant vector fields (see, e.g., [1])

$$R(X,Y)Z = -\frac{1}{4}[[X,Y],Z]. \tag{14}$$

The question of comparing both formulas boils down to proving for Z random vector s.t. $E(Z) = 0$ that $\mathbb{E}_g\langle \xi, [Z,[Z,\xi]] \rangle = -4\mathbb{E}_g \langle R(Z,\xi)\xi, Z \rangle$. It can be verified as follows:

$$-4\mathbb{E}_g\langle R(Z,\xi)\xi, Z \rangle = \mathbb{E}_g\langle [[Z,\xi],\xi], Z \rangle = \mathbb{E}_g \langle [Z,[Z,\xi]], \xi \rangle = \mathbb{E}_g\langle \xi, [Z,[Z,\xi]] \rangle$$

where the second equality stems from a property of the bi-invariant case (e.g., [1]) that generalizes the mixed product on $SO(3)$ property, namely $\langle [X,Y], Z \rangle = \langle [Z,X], Y \rangle$.

3.1   Differences

If the metric is not bi-invariant the results are different. This is merely because then the Lie group exponential map and the Riemannian exponential map are not the same. Thus, our definition of the estimation error differs, so it is logical that the results be different.

4   Conclusion

In this paper, we have proposed an intrinsic lower bound for estimation of a parameter that lives in a Lie group. The main difference with the Euclidean case is that the estimation error between the estimate and the true parameter underlying the data is measured in terms of group multiplication (that is, the error is an element of the group). This is a much more natural way to measure estimation errors, and it has been used in countless works on statistical estimation and filtering on manifolds. But it comes at a price of additional terms in the bound, that are not easy to interpret. The bound is not closed-form, but by retaining only first and second order terms in the estimation error, we ended up with a closed form bound, that is the covariance of any unbiased estimator is lower bounded by the inverse of the intrinsic Fisher information matrix, as in the Euclidean case, plus additional terms that are functions of the covariance. Moreover, the nice structure of Lie groups allows straightforward calculations, and the bound is expressed with the help of the Lie bracket, this, in turn, being related to the sectional curvature at the identity, helping to draw a link with the general result of S.T. Smith on Riemannian manifolds [11].

This note generalizes previous calculations [3] obtained on SO(3), in the context of attitude filtering for a dynamical rigid body in space, the latter Cramér-Rao bound being compared to the covariance yielded by the intrinsic Kalman filter of [4]. It would be interesting to apply the obtained general bound to pose averaging (that is on SE(3)), as, e.g., proposed in [12], which could be then attacked by means of intrinsic stochastic approximation as in, e.g., [5] or [7].

Another future route could be to derive an intrinsic Cramér-Rao bound on homogeneous spaces. It is interesting to note that both examples of [11] are homogeneous spaces. Moreover, the results on manifolds are bound to be local, but we can hope for results on Lie groups and homogeneous spaces with a large domain of validity (if not global).

Acknowledgments. We thank Y. Ollivier for his help on some Riemannian geometry matters, N. Boumal and P.A. Absil for the time they took to discuss with us about the subject of intrinsic Cramér-Rao bounds on manifolds, and J. Jakubowicz for kindly inviting us to submit this paper.

References

1. Arnol'd, V.I.: Sur la géométrie différentielle des groupes de Lie de dimension infinie et ses applications à l'hydrodynamique des fluides parfaits. Ann. Inst. Fourier **16**, 319–361 (1966)
2. Barrau, A., Bonnabel, S.: A note on the intrinsic Cramer-Rao bound. In: Nielsen, F., Barbaresco, F. (eds.) GSI 2013. LNCS, vol. 8085, pp. 377–386. Springer, Heidelberg (2013)
3. Barrau, A., Bonnabel, S.: An intrinsic Cramér-Rao bound on SO(3) for (dynamic) attitude filtering. In: IEEE Conference on Decision and Control (2015)
4. Barrau, A., Bonnabel, S.: Intrinsic filtering on Lie groups with applications to attitude estimation. IEEE Trans. Autom. Control **60**(2), 436–449 (2015)
5. Bellachehab, A., Jakubowicz, J.: Random pairwise gossip on CAT(0) metric spaces. In: IEEE Conference on Decision and Control, pp. 5593–5598 (2014)
6. Berger, J.O.: Statistical Decision Theory and Bayesian Analysis. Springer, New York (1985)
7. Bonnabel, S.: Stochastic gradient descent on Riemannian manifolds. IEEE Trans. Autom. Control **58**(9), 2217–2229 (2013)

8. Boumal, N.: On intrinsic Cramér-Rao bounds for Riemannian submanifolds and quotient manifolds. IEEE Trans. Sig. Process. **61**(5–8), 1809–1821 (2013)

9. Boumal, N., Singer, A., Absil, P.-A., Blondel, V.D.: Cramér-Rao bounds for synchronization of rotations. Inf. Infer. **3**(1), 1–39 (2014)

10. Pitman, E.J.G.: The estimation of the location and scale parameters of a continuous population of any given form. Biometrika **30**(3–4), 391–421 (1939)

11. Smith, S.T.: Covariance, subspace, and intrinsic Cramér-Rao bounds. IEEE Trans. Sig. Process. **53**(5), 1610–1629 (2005)

12. Tron, R., Vidal, R., Terzis, A.: Distributed pose averaging in camera networks via consensus on SE(3). In: Second ACM/IEEE International Conference on Distributed Smart Cameras, ICDSC 2008, pp. 1–10. IEEE (2008)

13. Wahba, G.: A least squares estimate of satellite attitude. SIAM Rev. **7**(3), 409–409 (1965)

Geometry of Time Series and Linear Dynamical systems

Clustering Random Walk Time Series

Gautier Marti[1,2]([✉]), Frank Nielsen[2], Philippe Very[1], and Philippe Donnat[1]

[1] Hellebore Capital Management, Paris, France
{gautier.marti,philippe.very}@helleborecapital.com
[2] Ecole Polytechnique, Palaiseau, France

Abstract. We present in this paper a novel non-parametric approach useful for clustering independent identically distributed stochastic processes. We introduce a pre-processing step consisting in mapping multivariate independent and identically distributed samples from random variables to a generic non-parametric representation which factorizes dependency and marginal distribution apart without losing any information. An associated metric is defined where the balance between random variables dependency and distribution information is controlled by a single parameter. This mixing parameter can be learned or played with by a practitioner, such use is illustrated on the case of clustering financial time series. Experiments, implementation and results obtained on public financial time series are online on a web portal http://www.datagrapple.com.

1 Introduction

Random walks are sometimes used to perform data clustering [13] or can be a point of view on spectral clustering [21,27]. In this paper, we consider the original converse problem: clustering random walks. These stochastic processes are an important mathematical formalization used to model, for instance, the path of molecules travelling in gas, or financial market prices as stated in the random walk hypothesis [3] and the efficient-market hypothesis [12].

1.1 Clustering Time Series

Partition-based clustering is the task of grouping a set of objects in such a way that objects in the same group (cluster) are more similar to each other than those in different groups. This task leverages a representation of the dataset and a distance between objects. In practice, such semantic representation and distance are unknown and the ones used are motivated by some heuristics.

When working with time series, researchers have considered, for instance, L^p metrics or Dynamic Time Warping (DTW) [7] for comparing them, and wavelets [23] or SAX [19] as means of representation. These approaches were found useful to detect anomalies in time series [16] with a strong focus on pattern recognition [15].

© Springer International Publishing Switzerland 2015
F. Nielsen and F. Barbaresco (Eds.): GSI 2015, LNCS 9389, pp. 675–684, 2015.
DOI: 10.1007/978-3-319-25040-3_72

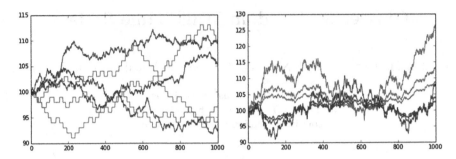

Fig. 1. Different criteria (apparently signal shape and homothetic scaling) are used for grouping these random walks in the two examples

1.2 Shortcomings of a Standard Approach

To understand why a standard approach fails to properly cluster random walks, we have to give a close look at the definition: a random walk is the sum $\sum_i X_i$ of a series of independent and identically distributed (i.i.d.) random variables X_i. So, there are no temporal patterns and thus approaches looking for them such as using a distance DTW and compressing time series using patterns as a way of representation are useless here. Note also that all information is carried by the increments X_i, it is therefore the underlying time series to study. By using a L^p metric between the increment time series, we may capture similarity in co-movements but, informally, we observe that we lose information of the random walk "shape", a criterion to take into account to cluster random walks as we can see it in the left panel of Fig. 1. Moreover, since increments are independent and identically distributed, time does not matter in these time series and we actually consider equivalence classes of random walks consisting in all the permutations of the X_i. To cluster this special kind of time series, one cannot simply use the standard machinery of machine learning on time series. Common normalizations do not make sense either. So, this work is a first step to study the problem of clustering random walks with application to financial time series in mind [20].

To alleviate the shortcomings of a standard approach, this paper propounds in Sect. 2.1 a proper random walk representation capturing all information which is leveraged by a relevant distance. In Sect. 2.3, the approach is validated on synthetic datasets. In-depth results using the presented workflow on real and public financial time series from the credit default swap market, and implementations for reproducible research are available online (http://www.datagrapple.com).

2   Generic Non-parametric Representation

We explain in this section our approach to represent and then cluster N random walks using a pre-processing we dubbed TS-GNPR for Generic Non-Parametric Representation of random walk Time Series.

2.1   Representation and Distance

Let $(\Omega, \mathcal{F}, \mathbf{P})$ be a probability space. Ω is the sample space, $\mathcal{F}$ is the σ-algebra of events, and $\mathbf{P}$ is the probability measure. Let $\mathcal{V}$ be the space of all continuous real valued random variables defined on $(\Omega, \mathcal{F}, \mathbf{P})$. Let $\mathcal{U}$ be the space of random variables following a uniform distribution on $[0,1]$ and $\mathcal{G}$ be the space of absolutely continuous cumulative distribution functions (cdf). We define the following representation of random vectors, that actually splits the joint behaviours of the marginal variables from their distributional information:

Let $\mathcal{T}$ be a mapping which transforms a random vector $X = (X_1, \ldots, X_N)$ into its TS-GNPR, an element of $\mathcal{U}^N \times \mathcal{G}^N$ representing X, defined as

$$\mathcal{T} : \mathcal{V}^N \to \mathcal{U}^N \times \mathcal{G}^N \tag{1}$$
$$X \mapsto (G_X(X), G_X)$$

where $G_X = (G_{X_1}, \ldots, G_{X_N})$, and G_{X_i} being the cumulative distribution function of X_i. $\mathcal{T}$ is a bijection, and thus preserves the whole information. Note that it replicates Sklar's theorem [26], seminal result of copula theory.

Statistical distances (or non-metric divergences) have been intensively studied [4] for data processing. One important class of divergences is f-divergences that ensures the property of information monotonicity [1]. Informally, information monotonicity guarantees that the divergence between coarse-binned histograms is less than fine-binned histograms as some information are lost due to the binning process.

In our setting, which actually does not require the copula theory framework, using the generic *non-parametric* representation, we introduce artificially a separable divergence as follows: we leverage TS-GNPR by defining a distance d_θ between random variables taking into account both distributional forms and joint behaviours.

Let $(X, Y) \in \mathcal{V}^2$. Let G_X, G_Y be vectors of marginal cdf.
We define the following distance depending on $\theta \in [0,1]$:

$$d_\theta^2(X, Y) = \theta d_1^2(G_X(X), G_Y(Y)) + (1 - \theta) d_0^2(G_X, G_Y),$$

where

$$d_1^2(G_X(X), G_Y(Y)) = 3\mathbf{E}[|G_X(X) - G_Y(Y)|^2], \tag{2}$$

and

$$d_0^2(G_X, G_Y) = \frac{1}{2} \int_{\mathbf{R}} \left(\sqrt{\frac{dG_X}{d\lambda}} - \sqrt{\frac{dG_Y}{d\lambda}} \right)^2 d\lambda. \tag{3}$$

As particular cases, d_0 is the Hellinger distance, a particular f-divergence, quantifying the similarity between two probability distributions, and the distance $d_1 = \sqrt{(1 - \rho_S)/2}$ is a distance correlation measuring statistical dependence between two random variables, where ρ_S is the Spearman's correlation between X and Y.

We can notice that for $\theta \in [0,1]$, $0 \le d_\theta \le 1$ and for $0 < \theta < 1$, d_θ is a metric. For $\theta = 0$ or $\theta = 1$, the separation axiom of metrics does not hold. This distance d_θ is invariant under diffeomorphism, i.e. let $h : \mathcal{V} \to \mathcal{V}$ be a diffeomorphism, let $(X,Y) \in \mathcal{V}^2$, we have $d_\theta(h(X), h(Y)) = d_\theta(X,Y)$. It is a desirable property as it ensures to be insensitive to scaling (e.g. choice of units) or measurement scheme (e.g. device, mathematical modelling) of the underlying phenomenon.

To apply the proposed distance on sampled data, we define a statistical estimate of d_θ: distance d_1 working with continuous uniform distributions can be approximated by rank statistics yielding to discrete uniform distributions, in fact coordinates of the multivariate empirical copula [9]; distance d_0 can be approximated using its discrete form working with estimates of marginal densities obtained from a basic kernel density estimator. For computing d_1, we need a bijective ranking function and since we consider application to time series, it is natural to choose arrival order to break ties. Let $(X_i)_{i=1}^M$ be M realizations of $X \in \mathcal{V}$. Let S_M be the permutation group of $\{1, \ldots, M\}$ and let $\sigma \in S_M$ be any fixed permutation, say $\sigma = Id_{\{1, \ldots, M\}}$. A bijective ranking function for $(X_i)_{i=1}^M$ can be defined as a function

$$\mathrm{rk}^X : \{1, \ldots, M\} \to \{1, \ldots, M\} \tag{4}$$
$$i \mapsto \#\{k \in \{1, \ldots, M\} \mid \mathcal{P}_\sigma\}$$

where $\mathcal{P}_\sigma \equiv (X_k < X_i) \vee (X_k = X_i \wedge \sigma(k) \le \sigma(i))$.

Let $(X_i)_{i=1}^M$ and $(Y_i)_{i=1}^M$ be M realizations of random variables $X, Y \in \mathcal{V}$. An empirical distance between realizations of random variables can be defined by

$$\tilde{d}_\theta^2 \left((X_i)_{i=1}^M, (Y_i)_{i=1}^M \right) \overset{a.s.}{=} \theta \tilde{d}_1^2 + (1 - \theta)\tilde{d}_0^2, \tag{5}$$

where

$$\tilde{d}_1^2 = \frac{3}{M^2(M-1)} \sum_{i=1}^M \left(\mathrm{rk}^X(i) - \mathrm{rk}^Y(i) \right)^2 \tag{6}$$

and

$$\tilde{d}_0^2 = \frac{1}{2} \sum_{k=-\infty}^{+\infty} \left(\sqrt{g_X^h(hk)} - \sqrt{g_Y^h(hk)} \right)^2, \tag{7}$$

h being a suitable bandwidth, and $g_X^h(x) = \frac{1}{M} \sum_{i=1}^M \mathbf{1}\{\lfloor \frac{x}{h} \rfloor h \le X_i < (\lfloor \frac{x}{h} \rfloor + 1)h\}$ being a density histogram estimating the probability density function g_X from $(X_i)_{i=1}^M$, M realizations of random variable $X \in \mathcal{V}$.

2.2 Parameter Selection Using Clustering Stability

To effectively use d_θ it boils down to select a particular value for θ. For instance, this value can be chosen by an expert who intends to give more weight on joint behaviours rather than distribution information, or the converse if one focuses on marginals. To aggregate both information in a balanced data-driven manner, we suggest using stability principles. Several researchers [6,8,18,25] advocate that stability of some kind is a desirable property of clustering, i.e. partitions obtained should be similar while data undergo small perturbations, yet some critics have arose [5,17] warning about the pitfalls of using stability as a method for clustering validation and model selection. In [24], authors conclude that stability is still a relevant criterion over finite samples.

Similarity Between Partitions. To measure clustering stability, we first have to define a similarity measure between clusters, and then partitions. We consider a correlation-flavoured similarity which can be seen as the scalar product of representation vectors [10]. Given two clusters C_1 and C_2, their similarity s_C is defined by

$$s_C(C_1, C_2) = \frac{\#(C_1 \cap C_2)}{\sqrt{\#(C_1)\#(C_2)}} \tag{8}$$

where $\#(C)$ is the number of elements in a cluster C. Given two partitions $\mathcal{P}_1$ and $\mathcal{P}_2$, with the same size K, of a dataset $\mathcal{X}$, i.e. $\mathcal{P}_i = \{C_i^k\}_{k=1}^K$ for $i \in \{1,2\}$, and $\mathcal{X} = \biguplus_{k=1}^K C_i^k$, we define a similarity s_P between $\mathcal{P}_1$ and $\mathcal{P}_2$ by averaging the pairwise similarities between clusters from $\mathcal{P}_1$ and $\mathcal{P}_2$, where each cluster in $\mathcal{P}_1$ is optimally assigned to a cluster in $\mathcal{P}_2$ with respect to maximizing the average cluster similarity, i.e.

$$s_P(P_1, P_2) = \max_{\sigma \in S_K} \frac{1}{K} \sum_{k=1}^K s_C(C_1^k, C_2^{\sigma(k)}) \tag{9}$$

Hungarian algorithm [22] is used to find the best assignment σ between the clusters from $\mathcal{P}_1$ and $\mathcal{P}_2$.

Time Stability of a Clustering. Many different kind of data perturbations can be considered, a popular one being the bootstrap [11] as it preserves the statistical properties of the initial sample. In the context of time series context, it seems more natural to consider perturbations due to a time-sliding window. In a steady regime, practitioners want their model stable with respect to passing time. Since increments of the random walks are i.i.d. this perturbation also preserves the data statistical properties.

To define time stability, we suggest to apply a clustering algorithm at different periods and compute the partition similarities between the resulting clusterings. More precisely, we propose to apply the same clustering algorithm to a sliding window, compute all the similarities between partitions of two successive windows and finally average all of them.

Let $X = (x_1, \ldots, x_N)^\top$ be a matrix describing N time series, where each x_i is a vector in $\mathbf{R}^T$ and T is the time horizon under focus. Given a window of width H, we note $\mathcal{P}_H^K(t)$ the partition computed by a given clustering algorithm on the window $]t - H, t]$. Given a number of cluster K, a window width H, and a time step δt, the stability index is defined by

$$S_I(X, K, H, \delta t) = \frac{1}{W} \sum_{t=H}^{T} s_{\mathcal{P}}(\mathcal{P}_H^K(t), \mathcal{P}_H^K(t + \delta t)) \tag{10}$$

where $W = \lfloor \frac{T-H}{\delta t} \rfloor + 1$ is the number of slidings.

Stability Index for Model Selection. We present a simple example where time series are aggregated using a one level factorial model:

$\forall i \in [1, N], \forall t \in [1, T],$

$$x_i(t) = \sqrt{\rho_m} m(t) + \sqrt{\rho_k} f_{k(i)}(t) + \sqrt{\rho_s} \epsilon_i(t) \tag{11}$$

where m, $(f_k)_{k=1}^K$ and $(\epsilon_i)_{i=1}^N$ are multivariate uncorrelated Gaussian noises, $\rho_m, \rho_k \geq 0$, such that $\rho_m + \rho_k \leq 1$, and $\rho_s = 1 - \rho_m - \rho_k$.

In economical terms, m is a systemic factor that correlates all the x_i together whereas $(f_k)_{k=1}^K$ are sectoral factors that lead to the grouping of the time series in K clusters. Finally, $(\epsilon_i)_{i=1}^N$ are residual noises that decrease pairwise correlations. Two series x_i and x_j belong to the same clusters if they share the same sectoral factor, that is if $k(i) = k(j)$.

Here we choose $K = 10$ clusters among $N = 100$ time series, for an horizon $T = 500$. Time series are evenly distributed among the factors, forming clusters of size $\frac{N}{K} = 10$. We choose $\rho_m = 40\%$ and $\rho_k = 30\%$. We compute our stability index with a window of size $H = \frac{T}{2} = 250$ and a time step $\delta t = 5$ and obtain the results shown in Fig. 2. We see that the stability index is equal to 1 for degenerated cases $K = 1$ and $K = N$ but also for the ground truth $K = 10$ clusters. This stability index usefulness depends on the signal-to-noise ratio $\sqrt{\rho_k/(1 - \rho_k)}$, usually small in applications, and the length of the time series, usually finite horizon in applications, to obtain a good estimate. The mentioned bias which is obvious for $K = 1$ or $K = N$ exists for all values of K. We look for an estimate of this bias by computing the stability score on purely Gaussian noise and obtain the following stability curve plotted on Fig. 3.

We thus propose the following adjusted stability index by subtracting this estimate. Given a set of time series X and a multivariate Gaussian noise $\mathcal{N}$, we define the adjusted stability index by

$$AS_I(X, K, H, \delta t) = \frac{S_I(X, K, H, \delta t) - S_I(\mathcal{N}, K, H, \delta t)}{1 - S_I(\mathcal{N}, K, H, \delta t)} \tag{12}$$

$\theta^\star$ can be estimated similarly with this stability index.

2.3   Validation and Experiments

To benchmark our approach, we use the following generative model that generalizes the one presented in Sect. 2.2. Let $S \in \mathbf{N}$. Let $(K_1, \ldots, K_S) \in \mathbf{N}^S$. Let

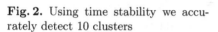
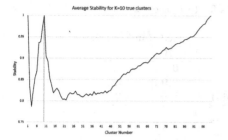
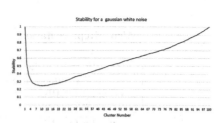

Fig. 2. Using time stability we accurately detect 10 clusters

Fig. 3. Estimate of the stability index bias obtained on purely Gaussian noise

$(Y_k^s)_{k=1}^{K_s}$, $1 \leq s \leq S$, be i.i.d. random variables following a standard normal distribution. Let $p, K \in \mathbf{N}$. Let $N = pK \prod_{s=1}^{S} K_s$. Let $(Z_k^i)_{k=1}^{K}$, $1 \leq i \leq N$, be independent random variables. For $1 \leq i \leq N$, we define

$$X_i = \sum_{s=1}^{S} \sum_{k=1}^{K_s} \beta_{k,i}^s Y_k^s + \sum_{k=1}^{K} \alpha_{k,i} Z_k^i, \tag{13}$$

where $\alpha_{k,i} = 1$, if $i \equiv k - 1 \pmod{K}$, 0 otherwise; $\beta_k^s \in [0,1]$, $\beta_{k,i}^s = \beta_k^s$, if $\lceil iK_s/N \rceil = k$, 0 otherwise. $(X_i)_{i=1}^N$ are partitioned into $Q = K \prod_{s=1}^{S} K_s$ clusters of p random variables each. Playing with the model parameters, we define in Table 2 some interesting test case datasets to study distribution clustering, dependence clustering or a mix of both. For clarity, we set $S \leq 2$ and $K \leq 4$, and use the following notations as a shorthand: $\mathcal{N} := \mathcal{N}(0,1)$; $\mathcal{J} := \sum_{n \geq 0} (-1)^n \mathbf{1}_{\{t=T_n\}}$, with $T_n = \sum_{i=1}^{n} X_i$ and $X_i \sim \text{Pois}(\lambda)$ are i.i.d., with $\lambda = 5$; $\mathcal{L} := \text{Laplace}(0, 1/\sqrt{2})$; $\mathcal{S} := \text{t-distribution}(3)/\sqrt{3}$.

Our approach is essentially not algorithm dependent as can be seen in Table 1 where k-means++ [2] and Ward, a hierarchical clustering, algorithms have the same behaviour on datasets A, B and C which are described in Table 2. As expected algorithms working on standard representation, and TS-GNPR with $\theta = 1$ (working only on rank correlations) cannot retrieve distribution information which is the only information present in dataset A, whereas TS-GNPR with $\theta = 0$ (working only on distributions) or estimated $\theta^\star$ (working on an optimal mix of co-movements and distributions) can. On dataset B containing only co-movements information, all approaches but expectedly TS-GNPR with $\theta = 0$ perform accurately. Nonetheless, when distribution and dependence information are mixed (dataset C), only TS-GNPR with $\theta^\star$ can recover the ground truth. Notice that TS-GNPR with $\theta = 1$ achieves a much better Adjusted Rand Index (ARI) [14] than the standard representations (0.72 against 0.45) which shows that working on a proper representation, even if only a part of the total information is available, is a better practice than working directly on the time series where heavy-tailed distributions can obfuscate the dependence relations between them.

Table 1. Comparative results for test case datasets

Algo.	Representation	Adjusted Rand Index		
		A	B	C
Ward	X	0	**0.94**	0.42
	$(X - \mu_X)/\sigma_X$	0	**0.94**	0.42
	$(X - \min)/(\max - \min)$	0	0.48	0.45
	TS-GNPR $\theta = 0$	1	0	0.47
	TS-GNPR $\theta = 1$	0	**0.91**	0.72
	TS-GNPR θ^*	1	**0.92**	1
k-m++	X	0	**0.90**	0.44
	$(X - \mu_X)/\sigma_X$	0	**0.91**	0.45
	$(X - \min)/(\max - \min)$	0.11	0.55	0.47
	TS-GNPR $\theta = 0$	1	0	0.53
	TS-GNPR $\theta = 1$	0.06	**0.99**	0.80
	TS-GNPR θ^*	1	**0.99**	1

Table 2. Model parameters for some interesting test case datasets

Dataset	Clustering	N	M	Q	K_1	β_k^1	K_2	β_k^2	Z_1^i	Z_2^i	Z_3^i	Z_4^i
A	Distribution	400	10000	4	0	0	0	0	$\mathcal{N}$	$\mathcal{J}$	$\mathcal{L}$	$\mathcal{S}$
B	Dependence	300	500	30	3	0.1	10	0.1	$\mathcal{N}$	$\mathcal{N}$	$\mathcal{N}$	$\mathcal{N}$
C	Mix	100	1000	20	0	0	10	0.1	$\mathcal{N}$	$\mathcal{N}$	$\mathcal{J}$	$\mathcal{J}$

3 Discussion

The aim over the long term is to design a full framework for the study of random walks in finite samples which will tackle multivariate inference and outlier detection based on clustering dynamics. The presented work was but a first step toward this goal by allowing us to have a proper data representation with a model selection based on a criterion that is dear to practitioners in finance, i.e. time stability. To complete this work, it remains to show that clustering using TS-GNPR could achieve consistency in simple factorial models where correlation matrices are slightly perturbed. We might also wish to improve the distance working on the TS-GNPR representation as we may want to compare distributions differently by taking into account, for instance, tail dependence.

References

1. Amari, S.I., Cichocki, A.: Information geometry of divergence functions. Bull. Pol. Acad. Sci. Tech. Sci. **58**(1), 183–195 (2010)
2. Arthur, D., Vassilvitskii, S.: k-means++: the advantages of careful seeding. In: Proceedings of the Eighteenth Annual ACM-SIAM Symposium on Discrete Algorithms, pp. 1027–1035. Society for Industrial and Applied Mathematics (2007)
3. Bachelier, L.: Théorie de la spéculation. Gauthier-Villars (1900)

4. Basseville, M.: Divergence measures for statistical data processing. Sig. Process. **93**(4), 621–633 (2013)
5. Ben-David, S., Von Luxburg, U., Pál, D.: A sober look at clustering stability. In: Lugosi, G., Simon, H.U. (eds.) COLT 2006. LNCS (LNAI), vol. 4005, pp. 5–19. Springer, Heidelberg (2006)
6. Ben-Hur, A., Elisseeff, A., Guyon, I.: A stability based method for discovering structure in clustered data. In: Pacific Symposium on Biocomputing, vol. 7, pp. 6–17 (2001)
7. Berndt, D.J., Clifford, J.: Using dynamic time warping to find patterns in time series. In: KDD Workshop, Seattle, WA, vol. 10, pp. 359–370 (1994)
8. Carlsson, G., Mémoli, F.: Characterization, stability and convergence of hierarchical clustering methods. J. Mach. Learn. Res. **11**, 1425–1470 (2010)
9. Deheuvels, P.: La fonction de dépendance empirique et ses propriétés. Un test non paramétrique d'indépendance. Acad. Roy. Belg. Bull. Cl. Sci. (5) 65(6), 274–292 (1979)
10. Ding, C., He, X.: K-means clustering via principal component analysis. In: Proceedings of the Twenty-First International Conference on Machine Learning, p. 29. ACM (2004)
11. Efron, B.: Bootstrap methods: another look at the jackknife. Ann. Stat. **7**, 1–26 (1979)
12. Fama, E.F.: The behavior of stock-market prices. J. Bus. **38**, 34–105 (1965)
13. Harel, D., Koren, Y.: On clustering using random walks. In: Hariharan, R., Mukund, M., Vinay, V. (eds.) FSTTCS 2001. LNCS, vol. 2245, pp. 18–41. Springer, Heidelberg (2001)
14. Hubert, L., Arabie, P.: Comparing partitions. J. Classif. **2**(1), 193–218 (1985)
15. Ivanov, P.C., Rosenblum, M.G., Peng, C., Mietus, J., Havlin, S., Stanley, H., Goldberger, A.L.: Scaling behaviour of heartbeat intervals obtained by wavelet-based time-series analysis. Nature **383**(6598), 323–327 (1996)
16. Keogh, E., Lin, J., Fu, A.: Hot sax: efficiently finding the most unusual time series subsequence. In: Fifth IEEE International Conference on Data Mining, pp. 8-pp. IEEE (2005)
17. Krieger, A.M., Green, P.E.: A cautionary note on using internal cross validation to select the number of clusters. Psychometrika **64**(3), 341–353 (1999)
18. Lange, T., Roth, V., Braun, M.L., Buhmann, J.M.: Stability-based validation of clustering solutions. Neural Comput. **16**(6), 1299–1323 (2004)
19. Lin, J., Keogh, E., Lonardi, S., Chiu, B.: A symbolic representation of time series, with implications for streaming algorithms. In: Proceedings of the 8th ACM SIGMOD Workshop on Research Issues in Data Mining and Knowledge Discovery, pp. 2–11. ACM (2003)
20. Marti, G., Very, P., Donnat, P.: Toward a generic representation of random variables for machine learning (2015). arXiv preprint arXiv:1506.00976
21. Meila, M., Shi, J.: A random walks view of spectral segmentation. In: AI and STATISTICS (AISTATS) (2001)
22. Munkres, J.: Algorithms for the assignment and transportation problems. J. Soc. Ind. Appl. Math. **5**(1), 32–38 (1957)
23. Percival, D.B., Walden, A.T.: Wavelet Methods for Time Series Analysis, vol. 4. Cambridge University Press, Cambridge (2006)
24. Shamir, O., Tishby, N.: Cluster stability for finite samples. In: NIPS (2007)
25. Shamir, O., Tishby, N.: Model selection and stability in k-means clustering. In: Learning Theory (2008)

26. Sklar, A.: Fonctions de répartition à n dimensions et leurs marges. Université Paris 8 (1959)
27. Von Luxburg, U.: A tutorial on spectral clustering. Stat. Comput. **17**(4), 395–416 (2007)

A Common Symmetrization Framework for Iterative (Linear) Maps

Alain Sarlette[✉]

QUANTIC, INRIA, Paris Sciences Lettres University, Rocquencourt, France
alain.sarlette@alumni.ulg.ac.be

Abstract. This paper highlights some more examples of maps that follow a recently introduced "symmetrization" structure behind the average consensus algorithm. We review among others some generalized consensus settings and coordinate descent optimization.

1 Introduction

The linear consensus algorithm [1]

$$x_k(t+1) = x_k(t) + \sum_{j=1}^{N} a_{jk}(x_j(t) - x_k(t)), \quad k = 1, 2, ..., N, \qquad (1)$$

with $a_j k \geq 0$, or equivalently for $x_k \in \mathbb{R}$, $x(t+1) = A(t)x(t)$ where the matrix A has off-diagonal components $A_{jk} = a_{jk}$ and diagonal component $A_{kk} = 1 - \sum_{j=1}^{N} a_{jk}$, features several geometric structures that could serve as a basis for generalization. In previous work we have considered the structure of (1) on a cone space, associated to the Hilbert metric, which generalizes (1) e.g. towards "non-commutative consensus" and the Kraus maps of quantum dynamics [2]. More recently [3], we have explored a structure related to the network interaction and generalized (1) into a "symmetrization" class of iterative maps, which is restricted to linear procedures but allows to cover consensus on (quantum) probability spaces and some geometric control design algorithms.

In the present paper, we want to give a few more examples of applications covered by the former approach. In particular:

- In Sect. 3 we show how symmetrization readily applies to the modeling of some direct variants of (1).
- In Sect. 4 we slightly relax the symmetrization assumptions to model gradient and coordinate descent (for a quadratic cost function to stay in the linear context).

The intended message of this paper is in showing the possibility to model a set of a priori different maps as one class of dynamics.

We denote by I the identity matrix and by e the identity element of a group; when numbering group elements we use the convention $g_1 = e$.

F. Nielsen and F. Barbaresco (Eds.): GSI 2015, LNCS 9389, pp. 685–692, 2015.
DOI: 10.1007/978-3-319-25040-3_73

2 Symmetrization

Let us first recall the symmetrization framework developed in [3]. The main point is to write an iterative map

$$x(t+1) = \Phi(x(t), t) \tag{2}$$

as a stochastic combination of actions of a discrete group $\mathcal{G}$, i.e.

$$\Phi(\cdot, t) = \sum_{i=1}^{|\mathcal{G}|} w_i(t) a(g_i, \cdot) \tag{3}$$

where $g_i \in \mathcal{G}$ are the elements of the group, $a(g_i, \cdot)$ is a linear action associated to g_i on the space $\mathcal{X} \ni x$, and the w_i satisfy: $w_i(t) \geq 0$ for all i, $\sum_{i=1}^{|\mathcal{G}|} w_i(t) = 1$.

The iterative map (2) can then be lifted, possibly non-uniquely, to dynamics on the stochastic weights of group elements. I.e. we can write

$$x(t) = \sum_{i=1}^{|\mathcal{G}|} p_i(t) \, a(g_i, x(0))$$

and describe the evolution of the $p_i(t)$. Explicitly, defining $\pi_i(k)$ such that $g_{\pi_i(k)} = g_i^{-1} g_k$ in the group sense, we have

$$p_k(t+1) = \sum_{i=1}^{|\mathcal{G}|} w_i(t) \, p_{\pi_i(k)}(t), \quad k = 1, 2, ..., |\mathcal{G}|, \tag{4}$$

starting with $p_1(0) = 1$ for $g_1 = e$ and $p_k(0) = 0$ for $k \neq 1$. In general, $|\mathcal{G}|$ can be much larger than the dimension of $\mathcal{X}$. Yet the interesting point is that (4) is completely independent of $\mathcal{X}$ and of the particular action $a(g_i, \cdot)$. This has allowed us to cover with one single model, consensus on variables belonging to vector spaces, so-called "symmetric consensus" on probabilities over a discrete set and similarly "symmetric quantum consensus". In the two latter applications, not only marginal probabilities but also all multi-partite correlation probabilities converge to the same values for all agents.

It is not difficult to see that the $|\mathcal{G}| \times |\mathcal{G}|$ state matrix describing (4) is doubly stochastic. It is hence stable and features as a particular stationary point the uniform weight distribution $p = \bar{p}$ where $\bar{p}_k = 1/|\mathcal{G}|$ for all k. For this stationary point, in the $\mathcal{X}$ space, we have

$$\bar{x} = \frac{1}{|\mathcal{G}|} \sum_{i=1}^{|\mathcal{G}|} a(g_i, x(0)).$$

We hence call $\bar{x}$ the *symmetrization* of x_0 with respect to $(\mathcal{G}, a(\cdot, \cdot))$, and *a fortiori* if Φ can be written as (3) then $\bar{x}$ is a stationary point of Φ.

In general the goal of Φ is to make x converge to $\bar{x}$. A sufficient condition for this is that the associated lift (4) makes p converge to $\bar{p}$. In some situations, especially with time-varying Φ, the lift might provide more insights, e.g.:

Proposition 1. *Given $\delta > 0$, $T > 0$, denote $\mathcal{S}(t) \subseteq \mathcal{G}$ the set of g_i for which $\sum_{t'=t}^{t+T} w_i(t') > \delta$. If there exist $\delta > 0$, $T > 0$ and $\mathcal{H} \subseteq \mathcal{G}$ generating $\mathcal{G}$ such that*
• $g_1 = e \in \mathcal{S}(t)$ for each t and • $\mathcal{H} \subseteq \mathcal{S}(t)$ for each t,
then the dynamics (4) makes p converge to $\bar{p}$.

With different choices of the lift, Proposition 1 can prove convergence of several different linear iterative maps at once, instead of having to repeat a "similar" proof argument for each specific map separately. In [3] a few examples are given. In the following we present a few others. Note that each of these examples has been efficiently solved on its own in the respective literature; our point here is that a common framework is established for all of them.

3   Variations on Consensus

Let us first review how (1) can be cast into the symmetrization framework, at least for the usual case where A is doubly stochastic, implying a.o. $\sum_j a_{jk} \leq 1$ for all k. Take $\bar{\mathcal{G}} = \text{Perm}_N$ the group of permutations on N elements. The easiest case is "gossip", where at each t a single pair (m, n) of agents interacts with $a_{mn} = a_{nm} \neq 0$: we can then write $w_1 = 1 - a_{m,n}$ for $g_1 = e$ and $w_i = a_{m,n}$ for g_i the pairwise permutation of (m, n), all other w_i zero. Several variations that allow several interactions at once, admit a similarly direct treatment.

For a general situation with A doubly stochastic, we use a result by Birkhoff [4] to decompose Φ into a convex sum of permutations. If there exists $\alpha > 0$ such that $a_{kk} > \alpha$ for all k, then $\tilde{A} = \frac{1}{1-\alpha}(A - \alpha I)$ is also doubly stochastic and by applying the Birkhoff result to $\tilde{A}$, we obtain a decomposition of $A = \alpha I + (1-\alpha)\tilde{A}$ into permutation matrices weighted by w_i and where $w_1 \geq \alpha$. Except for special time dependencies of the $a_{jk}(t)$, generally an element for which $w_i > \delta$ at some t satisfies $\sum_{t'=t}^{t+T} w_i(t') > \delta$ for all t, for T large enough. We then say that all elements are recurrent. In this case, Proposition 1 ensures that $p(t)$ converges to the uniform distribution $\bar{p}$ over all elements of the group $\mathcal{G} \subseteq \bar{\mathcal{G}}$ generated by the elements that appear in the decomposition. Concretely, for instance:

- For pairwise gossip, if the interaction graph corresponding to the $\bar{a}_{jk} = \sum_{t'=t}^{t+T} a_{jk}(t')$ is connected, then the associated pairwise permutations generate all of $\bar{\mathcal{G}} = \text{Perm}_N$. A generalization to the case where the interaction graph is composed of several connected components is immediate.
- When $a_{jk} = \beta/N$ for all $j \neq k$ and some $\beta \in (0, 1)$, an associated lift would be $w_1 = 1 - \beta$ and $w_i = \beta/N$ for all g_i which are a power 1 to $N - 1$ of the circular permutation $g_c : (1, 2, 3, ..., N) \rightarrow (N, 1, 2, ..., N - 1)$. These form a closed subgroup $\mathcal{G} \subset \text{Perm}_N$. But still, having p converge to the uniform distribution over these N elements is sufficient to have the x_k converge to $1/N \sum_{k=1}^{N} x_k(0)$ for each k. Problems with the lift could arise if an element of $\text{Perm}_N \backslash \mathcal{G}$ appears *rarely*, i.e. without belonging to the set $\mathcal{S}(t)$ of Proposition 1 for any large T. This seems unlikely in practical situations.

In this setting, whether x_k is a scalar, vector or other object of a linear space $\mathcal{X}$ does not matter, it just affects the group actions. Another action allows to apply the same group dynamics to the symmetrization of joint probability distributions, both classical and quantum, over a finite set [3]. We now turn to different variants of (1) for $x_k \in \mathbb{R}$.

3.1 Consensus with Antagonistic Interactions

A variant of consensus proposed e.g. in [5] considers that agents might be attracted to the *opposite* of the values of some neighbors, i.e. the dynamics become

$$x_k(t+1) = a_{kk}\, x_k(t) + a_{jk}x_j(t) \qquad (5)$$

with $a_{kk} = 1 - \sum_{j \neq k} |a_{jk}| > 0$, but possibly some negative a_{jk} for $j \neq k$.

To cover this possibility, we take $\tilde{\mathcal{G}}$ the group corresponding to all permutation matrices with arbitrary sign ± 1 on each component; we call this "generalized" permutations. This can be viewed as a particular product of the group of permutations with the group $\{-1, 1\}^N$. We then lift (5) as follows.

- Take the Birkhoff decomposition associated to the corresponding standard consensus algorithm, where each a_{jk} in (5) is replaced by $|a_{jk}|$. This gives nonzero weights $\tilde{w}_i$ on the $g_i \in \mathcal{G}$ corresponding to permutation matrices.
- Then sequentially consider each pair (j, k) for which $a_{jk} < 0$ and swap the weights attributed to the generalized permutation matrices which are equal up to the sign of their element (j, k).

Convergence of (5) is efficiently characterized a.o. in [6]. We just provide an informal summary on the basis of $\tilde{\mathcal{G}}$. As for standard consensus, we may generally assume that all the elements appearing in the lift are recurrent. The generated group $\mathcal{G}$ over which we get $p = \bar{p}$ at convergence then depends (only) on the group generated by the appearing elements. Three main situations can occur:

- The interaction graph is not connected.
- The component (k, k) takes the same sign in the matrices associated to all the generated group elements. By group properties, this implies that all components (j, j) have the same sign in all the generated group elements, provided the interaction graph is connected. Then in terms of x_k, the agents split into two communities with opposite component values as $t \to \infty$.
- The component (k, k) can take opposite signs in the generated group elements. Then, thanks to group properties, for each j, k there should be an equal number of elements with negative component (j, k) as with positive component (j, k) in the generated group $\mathcal{G}$. This implies that when p converges to $\bar{p}$ uniform over $\mathcal{G}$, the corresponding x_k converge to 0.

Our lift can involve many elements but still a finite number, i.e. it abstracts away the exact values of the $a_{j,k}(t)$ in (5). While our framework may appear more cumbersome than the graph analysis of [6], it can also yield more precise results in some cases. Indeed, we can exactly cover the following example which [6] provides as an inconclusive case of their graph-conditions-based theorem:

$$x_k(t+1) = 1/2 \; (x_1(t) + x_2(t)), \quad k = 1, 2; \; t \text{ even}$$
$$x_k(t+1) = (-1)^k/2 \; (x_2(t) - x_1(t)), \quad k = 1, 2; \; t \text{ uneven}.$$

The dynamics can be explicitly solved and converges to $x_1 = x_2 = 0$ after iterations $t = 0$, $t = 1$, yet the theorem of [6] cannot tell this. In terms of our lift, we have three recurrent elements

$$g_1 = \begin{pmatrix} 1 & 0 \\ 0 & 1 \end{pmatrix}; \ g_2 = \begin{pmatrix} 0 & 1 \\ 1 & 0 \end{pmatrix}; \ g_3 = \begin{pmatrix} 0 & -1 \\ -1 & 0 \end{pmatrix}$$

with weights $w_1 = 1/2$, $w_2(t) = 1/4 + (-1/4)^t$, $w_3(t) = 1/4 - (-1/4)^t$. These weights are in fact irrelevant, we only need to compute that g_1, g_2, g_3 generates a 4-group element, together with $g_4 = -g_1$. We then readily conclude that the system converges to $\bar{x} = \frac{1}{4}\sum_{i=1}^{4} g_i x = 0$.

At this stage we do not claim that the group lift approach is efficiently scalable to larger problems. Yet it might suggest e.g. how to adapt the graph analysis argument of [6] towards covering the above example.

3.2 Doubly Sub-stochastic Matrices

Consensus is a particular linear dynamics which features a one-dimensional invariant space. In iterative maps, the most common situation is when there is only one target invariant point. This happens with "smaller than stochastic" A. Shifting the invariant point to $x = 0$, we define the following class of systems:

$$x_k(t+1) = \sum_{j=1}^{N} a_{jk}(t)\, x_j(t) \ \text{ with } \quad \begin{aligned} 1 &\geq \sum_j |a_{jk}(t)| \ \text{ for all } k, t \\ 1 &\geq \sum_k |a_{jk}(t)| \ \text{ for all } j, t. \end{aligned} \tag{6}$$

Such system can be straightforwardly rewritten as consensus with (possibly antagonistic interactions and) a virtual leader $x_0(t) = 0$, hence proving convergence is not a big issue. Again our intended contribution is more to show that this dynamics, as well, can be cast into the class of symmetrization algorithms. For simplicity of notations we describe the following assuming $a_{jk} \geq 0$ for all j, k.

Take the same group $\tilde{\mathcal{G}}$ as in Sect. 3.1. We first define $b_{jk} = a_{jk} + c_{jk}$ with $c_{jk} \geq 0$ and such that $\sum_{j=1}^{N} b_{jk} = \sum_{j=1}^{N} b_{kj} = 1$ for all k, i.e. the state matrix associated to the b_{jk} is doubly stochastic. Such a construction is always possible for a sub-stochastic matrix, e.g. as follows:

1. Let $a'_{jk} := a_{jk}$ for all j, k.
2. Compute $a^c_k = \sum_{j=1}^{N} a'_{jk}$ and $a^r_k = \sum_{j=1}^{N} a'_{kj}$ for all k.
3. Since $\sum_k a^c_k = \sum_{j,k} a'_{jk} = \sum_k a^r_k$: if $a^c_k < 1$ for some k then there exists k' such that $a^r_{k'} < 1$, and vice versa. Else $a^c_k = a^r_k = 1$ for all k, i.e. the a'_{jk} matrix is already doubly stochastic and we can stop.
4. Find $a^r_m = \max(a^r_k : a^r_k < 1)$ and $a^c_n = \max(a^c_k : a^c_k < 1)$. Then let $b_{mn} = a'_{mn} + c_{mn}$ with $c_{mn} = 1 - \max(a^r_m, a^c_n)$, and $b_{jk} = a'_{jk}$ for all $(j, k) \neq (m, n)$.
5. Define $a'_{jk} := b_{jk}$ for all j, k and iterate the procedure starting again from 1.

This procedure converges to a doubly stochastic matrix. Indeed, at each iteration: if the a'_{jk} are doubly sub-stochastic then also the b_{jk} are; and the b_{jk} matrix has at least one more row or column whose elements sum to 1 than the a'_{jk} matrix.

Define by $\{\tilde{w}_i\}$ a lift of the dynamics associated to the standard consensus $x_k(t+1) = \sum_{j=1}^{N} b_{jk}(t)\, x_j(t)$. We then sequentially go through all (j, k) for which $c_{jk} > 0$ and transform the weights as follows.

1. Let $w_i' = \tilde{w}_i$ for all i and choose some (j, k) for which $c_{jk} > 0$.
2. Denote $f_{jk}^-(m) = n$ such that g_n is identical to g_m except that the matrix element (j, k) associated to g_m has an opposite sign in g_n. Then for all i for which g_i has a positive element (j, k), apply

$$w_i = (1 - \tfrac{c_{jk}}{2b_{jk}}) w_i', \qquad w_{f_{jk}^-(i)} = \tfrac{c_{jk}}{2b_{jk}} w_i'.$$

3. Let $w_i' = w_i$, choose a different (j, k) with $c_{jk} > 0$ and iterate from point 1.

The weight modification in point 1 above preserves the total weight, $\sum_i w_i = \sum_i w_i'$. The w_i remain non-negative, since $c_{jk} \geq 0$, $b_{jk} \geq 0$, $a_{jk} \geq 0$ and $1 - \tfrac{c_{jk}}{2b_{jk}} = 1 - \tfrac{c_{jk}}{2c_{jk}+2a_{jk}} > 0$. Summing up the effect on $\sum_i w_i a(g_i, \cdot)$, we see that when point 1 treats (j, k), the matrix components different from (j, k) undergo no change while the component on (j, k) is modified:

$$\text{from} \quad \sum_{i \in \mathcal{F}_{j,k}} w_i' =: B_1 \qquad \text{to} \qquad \sum_{i \in \mathcal{F}_{j,k}} w_i' \left((1 - \tfrac{c_{jk}}{2b_{jk}}) - \tfrac{c_{jk}}{2b_{jk}} \right) =: B_2,$$

where $\mathcal{F}_{j,k} \subset \tilde{\mathcal{G}}$ is the subset of group elements whose associated matrix have a coefficient $+1$ at position (j, k). By definition of the lift, we have $B_1 = b_{jk}$. We then get $B_2 = B_1(1 - \tfrac{c_{jk}}{b_{jk}}) = a_{jk}$, thus indeed the matrix that we had to model.

From there, we can go on and analyze convergence of the resulting group dynamics. By construction, if $\mathcal{G}$ contains g_i then also contains $g_{f_{jk}^-(i)}$ for each (j, k) for which $c_{jk} > 0$. Hence if $p(t)$ converges to $\bar{p}$ uniform over $\mathcal{G}$, then the corresponding (j, k) elements will be zero in the matrix associated to $\tfrac{1}{|G|} \sum_{i=1}^{|\mathcal{G}|} a(g_i, \cdot)$.

4   Gradient Descent and Coordinate Descent

In optimization, the gradient descent method consists in iteratively searching for $\min f(x)$ by applying

$$x(t + 1) = x(t) - \alpha \operatorname{grad}_x f(x(t)), \quad \alpha > 0, \ x \in \mathbb{R}^N. \tag{7}$$

For a local quadratic expansion $f = \tfrac{1}{2} x^T A x$ around the minimum, A symmetric positive definite, approximating (7) by a linear map, there is a trivial symmetrization viewpoint on the resulting $x(t + 1) = x(t) - \alpha A x(t)$.

In the eigenbasis of A, (7) becomes $x_k(t + 1) = (1 - \alpha a_k) x_k(t)$ with a_k the associated eigenvalues. For stable algorithms, $(1 - \alpha a_k) \in (-1, 1)$ such that the associated state transition matrix is (diagonal and) doubly substochastic. A simplified version of Sect. 3.2 then rewrites this iteration as symmetrization over the group $\mathcal{G}$ represented by diagonal $N \times N$ matrices with elements ± 1. This abstract definition does not require to actually compute the eigenbasis – only assume it exists and the a_k are small enough. Yet this viewpoint must be adapted to address the following.

4.1   Coordinate Descent

Coordinate descent selects some $k \in \{1, 2, ..., N\}$ at each step and applies gradient descent to minimize $f(x)$ along a line where only x_k varies, assuming fixed all x_j with $j \neq k$. This comes closer to a (possibly stochastically) time-varying map as in consensus. It links to symmetrization as follows, e.g. for $N = 2$: define

$$y = \begin{pmatrix} \sqrt{a_1} & 0 \\ 0 & \sqrt{a_2} \end{pmatrix} \begin{pmatrix} \cos\theta & \sin\theta \\ -\sin\theta & \cos\theta \end{pmatrix} x$$

such that $f(x) = \frac{1}{2} x^T A x = \frac{1}{2} y^T y$. This simplifies the cost function, but coordinate descent now implies iterative optimization along two arbitrary lines, of slopes $\tan\phi_1 = \tan\theta\, a_2/a_1$ and $\tan\phi_2 = -\cot\theta\, a_2/a_1$. In general those lines are not orthogonal, but f will still be quadratic along each of them. It is not difficult to see that we can then write, if $k \in \{1, 2\}$ is chosen at time t:

$$y(t+1) = (1 - \lambda)\, y(t) + \lambda \,(\text{reflection of } y(t) \text{ around axis of slope } \cot\alpha\, \phi_k)$$
$$= (1 - \lambda)\, y(t) + \lambda \begin{pmatrix} -\cos 2\phi_k & \sin 2\phi_k \\ \sin 2\phi_k & \cos 2\phi_k \end{pmatrix} y(t),$$

for some $\lambda \in (0, 1)$. In general, $(1 - \lambda) + \lambda(\cos + \sin) > 1$ and the associated transition matrix is not sub-stochastic. Nevertheless, we were able to write this as symmetrization with respect to the 8-element group $\mathcal{G} =$

$$\left\{ g_1^{\pm} = \pm \begin{pmatrix} 1 & 0 \\ 0 & 1 \end{pmatrix}, \; g_2^{\pm} = \pm \begin{pmatrix} 1 & 0 \\ 0 & -1 \end{pmatrix}, \; g_3^{\pm} = \pm \begin{pmatrix} 0 & 1 \\ 1 & 0 \end{pmatrix}, \; g_4^{\pm} = \pm \begin{pmatrix} 0 & 1 \\ -1 & 0 \end{pmatrix} \right\}.$$

Explicitly, $y(t+1) = (w_{1+} g_1^+ + w_{2-} g_2^- + w_{2+} g_2^+ + w_{3-} g_3^- + w_{3+} g_3^+)\, y(t)$ with

$$w_{1+} = 1 - \lambda; \quad w_{2-} = \lambda\,(\tfrac{1}{4} + \tfrac{\cos\phi_k}{2}); \quad w_{3+} = \lambda\,(\tfrac{1}{4} + \tfrac{\sin\phi_k}{2})$$
$$w_{2+} = \lambda\,(\tfrac{1}{4} - \tfrac{\cos\phi_k}{2}); \quad w_{3-} = \lambda\,(\tfrac{1}{4} - \tfrac{\sin\phi_k}{2}).$$

By construction $\sum_{g_i \in \mathcal{G}} w_i = 1$. However we do not have $w_i > 0$ for all i. This requires to slightly generalize the class of dynamics accepted on the group. In the present case:

The state transition matrix for the weights $p_i(t)$ on the group $\mathcal{G}$ is symmetric and doubly stochastic, up to having possibly negative off-diagonal components. Hence it still features $\bar{p}$ as a stationary point. Its convergence is not as straightforward to analyze as standard consensus. However thanks to symmetry, in a time-varying context it suffices for convergence to examine the common Lyapunov function $V(t) = \sum_i (p_i(t))^2$.

For this particular example, i.e. a two-dimensional x, the state transition matrices for the $p_i(t)$ further admit an efficient tensor product decomposition such that convergence conditions can be easily formalized. At this point we cannot claim that the symmetrization framework is efficient to study convergence for high-dimensional x. Yet, the possibility to model the system as symmetrization, with generalized (possibly negative) weights, could suggest a way to include coordinate descent in versatile theoretical formalizations that exploit the symmetrization structure.

5   Conclusion

We have abstractly defined a "symmetrization" class of linear iterative maps. This class comprises a great variety of algorithms, as illustrated in [3]. The present paper highlights some further elements of the class. Although each instance encountered so far, admits a relatively easy specific convergence proof, the common symmetrization framework allows to treat them all at once, modulo a finite group analysis step. In the future we hope that further properties of this symmetrization class can be exploited in general analyzes.

Acknowledgments. I thank F.Ticozzi and L.Mazzarella for sharing ideas on the symmetrization approach and R.Sepulchre for encouraging to address these particular applications.

References

1. Tsitsiklis, J.N.: Problems in decentralized decision making and computation. Ph.D. thesis, MIT (1984)
2. Sepulchre, R., Sarlette, A., Rouchon, P.: Consensus in non-commutative spaces. In: Proceedings of 49th IEEE Conference on Decision and Control, pp. 5596–6601 (2010)
3. Mazzarella, L., Sarlette, A., Ticozzi, F.: Extending robustness and randomization from consensus to symmetrization algorithms, SIAM J. Control Optim. (2014, in press). http://arxiv.org/abs/1311.3364
4. Birkhoff, G.: Three observations on linear algebra. Univ. Nac. Tucuan. Rev. A **5**, 147–151 (1946)
5. Meng, Z., Shi, G., Johansson, K.-H., Cao, M., Yiguang, H.: Modulus consensus over networks with antagonistic interactions and switching topologies (2014). Preprint http://arxiv.org/abs/1402.2766
6. Hendrickx, J.: A lifting approach to models of opinion dynamics with antagonisms. In: Proceedings of 53rd IEEE Conference on Decision and Control (2014)

New Model Search for Nonlinear Recursive Models, Regressions and Autoregressions

Anna-Lena Kißlinger[1] and Wolfgang Stummer[2,3]($\boxtimes$)

[1] Chair of Statistics and Econometrics, University of Erlangen–Nürnberg,
Lange Gasse 20, 90403 Nürnberg, Germany
[2] Department of Mathematics, University of Erlangen–Nürnberg, Cauerstrasse 11,
91058 Erlangen, Germany
stummer@math.fau.de
[3] Affiliated Faculty Member of the School of Business and Economics,
University of Erlangen–Nürnberg, Lange Gasse 20, 90403 Nürnberg, Germany

Abstract. Scaled Bregman distances SBD have turned out to be useful tools for simultaneous estimation and goodness-of-fit-testing in parametric models of random data (streams, clouds). We show how SBD can additionally be used for model preselection (structure detection), i.e. for finding appropriate candidates of model (sub)classes in order to support a desired decision under uncertainty. For this, we exemplarily concentrate on the context of nonlinear recursive models with additional exogenous inputs; as special cases we include nonlinear regressions, linear autoregressive models (e.g. AR, ARIMA, SARIMA time series), and nonlinear autoregressive models with exogenous inputs (NARX). In particular, we outline a corresponding information-geometric 3D computer-graphical selection procedure. Some sample-size asymptotics is given as well.

Keywords: Scaled Bregman distances · Model selection · Nonlinear regression · AR · SARIMA · NARX · Autorecursions · 3D score surface

1 Introduction and Results

Especially within the last two decades, some distances (divergences, (dis)similarity measures, discrepancy measures) between probability distributions have been successfully used for (minimum distance) parameter estimation and goodness-of-fit testing. Amongst them, let us mention exemplarily the ϕ-divergences of Csiszar [5] and Ali & Silvey [1], as well as the classical Bregman distances (see e.g. Pardo & Vajda [12]) which also include the density power divergences of Basu et al. [2]. Some comprehensive coverages can e.g. be found in Pardo [11] and Basu et al. [3]. Recently, Stummer and Vajda [15] (cf. also Stummer [14]) introduced the concept of scaled Bregman distances (SBD), which cover all the above-mentioned distances as special cases; see Kißlinger & Stummer [6,7] for some applications of SBD to simultaneous parameter estimation and goodness-of-fit investigations, and Kißlinger & Stummer [8] for utilizations in robust change point detections. In this paper, we would like to indicate how SBD can be employed for model search

© Springer International Publishing Switzerland 2015
F. Nielsen and F. Barbaresco (Eds.): GSI 2015, LNCS 9389, pp. 693–701, 2015.
DOI: 10.1007/978-3-319-25040-3_74

within a context which covers many widely used intertemporal respectively inter-spatial models:

Let $k \in \mathbb{Z} \cup \{-\infty\}$ be either an integer or $-\infty$. For the set $\tau := \{m \in \mathbb{Z} : m \geq k\}$ let the generation of the m-th data point be represented by the random variable X_m which takes values in some space $\mathscr{X}^1$, e.g. $\mathscr{X}$ is a finite or infinite set/space of d-dimensional vectors or $\mathscr{X}$ is a set/space of special functions. We consider collections $(X_m, m \in \tau)$ of data which are *autorecursions* in the sense of

$$F_{\gamma_{m+1}}\Big(m+1, X_{m+1}, X_m, X_{m-1}, \ldots, X_k, Z_{k-}, a_{m+1}, a_m, a_{m-1}, \ldots, a_k\Big) = \varepsilon_{m+1},$$
$$m \geq k, \qquad (1)$$

where $(\varepsilon_{m+1})_{m \geq k}$ is a family of independent and identically distributed (i.i.d.) random variables on some space $\mathscr{Y}^2$ having parametric distribution Q_θ ($\theta \in \Theta$). The nonlinear function F is parametrized by $\gamma_{m+1} \in \Gamma$; notice that Γ and Θ can be any sets but typically consist of numbers or vectors (of numbers). Furthermore, the $(a_k)_{m \geq k}$ are independent variables which we assume to be non-stochastic (deterministic). For a comfortable unified treatment of subclasses with different lags, we use the "backlog-input" Z_{k-} to denote additional input on X and a before k needed to get the recursion started. Technically, without loss of generality we suppose zero expectation $E_{Q_\theta}[\varepsilon_{m+1}] = 0$ ($m \geq k$). Moreover, for $k \neq -\infty$ we assume the initial data X_k as well as the backlog-input Z_{k-} to be deterministic, for the sake of brevity; the case of random X_k, Z_{k-} can be treated analogously.

Notice that in (1), the recursive appearance of the "next" data point X_{m+1} is given only implicitly; this suffices for our purposes. However, in most applications one can equivalently isolate X_{m+1} from the "other input" in terms of, say,

$$X_{m+1} = g\Big(f_{\gamma_{m+1}}(m+1, X_m, X_{m-1}, \ldots, X_k, Z_{k-}, a_{m+1}, a_m, a_{m-1}, \ldots, a_k), \varepsilon_{m+1}\Big)$$
$$m \geq k, \qquad (2)$$

for some appropriate functions $f_{\gamma_{m+1}}$ and g. This is in particular helpful for explicit forecasting, with which we do not deal here; the outer function $g(\cdot, \cdot)$ often describes a (possibly vector-component wise) addition $g(z_1, z_2) = z_1 + z_2$ or (possibly vector-component wise) multiplication $g(z_1, z_2) = z_1 \cdot z_2$, or some weighted variant thereof; in the light of this, $(\varepsilon_{m+1})_{m \geq k}$ can be interpreted as "randomness-driving innovations (noise)".

The general context (1) allows for a broad spectrum of applications; for instance, the following well-known, very widely used models are covered:

(I) nonlinear regressions with deterministic independent variables:

$$X_{m+1} = f_{\gamma_{m+1}}(a_{m+1}, a_m, a_{m-1}, \ldots, a_k, Z_{k-}) + \varepsilon_{m+1}, \qquad m \geq k, \qquad (3)$$

---

[1] Equipped with some σ-algebra $\mathscr{A}$.
[2] Equipped with some σ-algebra $\mathscr{B}$.

with the special case of fixed lag $r \in \mathbb{N}_0$ in the independent variables

$$X_{m+1} = f_\psi(a_{m+1}, \ldots, a_{m-r+1}) + \varepsilon_{m+1}, \quad m \geq k,$$

where all necessary a_j with $j < k$ are interpreted as part of Z_{k-}. The most prominent context is linear regression where f_ψ is an affine-linear function $f_\psi(a_{m+1}, a_m, a_{m-1}, \ldots, a_{m-r+1}) = \psi_{-1} + \psi_0 \cdot a_{m+1} + \ldots + \psi_r \cdot a_{m-r+1}$ for some vector of real numbers $\psi = (\psi_{-1}, \psi_0, \ldots, \psi_r)$ with $\psi_r \neq 0$; altogether, one has constant parameter vector $\gamma_{m+1} = (r, \psi) = (r, \psi_{-1}, \psi_0, \ldots, \psi_r)$ for all $m \geq k$. The case $r = 0$ gives $X_{m+1} = \psi_{-1} + \psi_0 \cdot a_{m+1} + \varepsilon_{m+1}$ $(m \geq k)$.

(II) AR(r): univariate linear autoregressive models (time series) of order $r \in \mathbb{N}$:

$$X_{m+1} = f_\psi(X_m, X_{m-1}, \ldots, X_{m-r+1}) + \varepsilon_{m+1}, \quad m \geq k, \tag{4}$$

where $f_\psi(X_m, X_{m-1}, \ldots, X_{m-r+1}) := \psi_1 \cdot X_m + \ldots + \psi_r \cdot X_{m-r+1}$ for some vector of real numbers $\psi = (\psi_1, \ldots \psi_r)$ with $\psi_r \neq 0^3$. All necessary X_j with $j < k$ are interpreted as part of Z_{k-}. In terms of the backshift operator B defined by $B X_m := X_{m-1}$, the corresponding r-polynomial $\psi_1 \cdot B + \psi_2 \cdot B^2 + \ldots + \psi_r \cdot B^r$, and the identity operator $\mathbb{1}$ given by $\mathbb{1}X_m := X_m$, the Eq. (4) can be rewritten as

$$\left(\mathbb{1} - \sum_{j=1}^{r} \psi_j B^j\right) X_{m+1} = \varepsilon_{m+1}, \quad m \geq k,$$

which is a special case of (1) with $F_{\gamma_{m+1}}\left(X_{m+1}, X_m, X_{m-1}, \ldots, X_k, Z_{k-}\right) = \left(\mathbb{1} - \sum_{j=1}^{r} \psi_j B^j\right) X_{m+1}$ with constant parameter vector $\gamma_{m+1} = (r, \psi)$ for all $m \geq k$. As an explicit example, the AR(2) amounts to

$$X_{m+1} - \psi_1 \cdot X_m - \psi_2 \cdot X_{m-1} = \varepsilon_{m+1}, \quad m \geq k.$$

We also define AR(0) by $X_{m+1} = \varepsilon_{m+1}$ $(m \geq k)$.

(III) ARIMA(r, d, 0): linear autoregressive integrated models (time series) of order $r \in \mathbb{N}_0$ and $d \in \mathbb{N}_0$. By definition, this means that the transformed sequence $(Y_{m+1})_{m \geq k} := ((\mathbb{1}-B)^d X_{m+1})_{m \geq k}$ is an AR(r) model; notice that for $d = 1$ one has $Y_{m+1} := X_{m+1} - X_m$ and for general d, Y_{m+1} is nothing but the d-fold backward difference built from the original-data extraction $\{X_{m+1}, \ldots, X_{m-d+1}\}$. Hence, an ARIMA(r, d, 0) model is described by

$$\left(\mathbb{1} - \sum_{j=1}^{r} \psi_j B^j\right)\left(\mathbb{1} - B\right)^d X_{m+1} = \varepsilon_{m+1}, \quad m \geq k; \tag{5}$$

where all necessary X_j with $j < k$ are interpreted as part of Z_{k-}. With the help of the binomial expansion $(\mathbb{1} - B)^d = \mathbb{1} + \sum_{j=1}^{d} \binom{d}{j}(-1)^j \cdot B^j$

---

[3] Notice that, here, for the definition of AR models we do not assume the stationarity of $(X_m)_{m \geq k}$.

one can easily rewrite (5) in terms of (2) with corresponding function f_γ with parameter vector $\gamma = (r, d, \psi_1, \ldots, \psi_r)$ as well as $g(z_1, z_2) = z_1 + z_2$. Clearly, ARIMA$(r, 0, 0)$ coincides with AR(r), and e.g. ARIMA$(2, 1, 0)$ is given by

$$X_{m+1} = (1 + \psi_1) \cdot X_m + (\psi_2 - \psi_1) \cdot X_{m-1} - \psi_2 \cdot X_{m-2} + \varepsilon_{m+1}, \quad m \geq k,$$

which formally looks like an AR(3) model with some special structure on the corresponding parameter vector $\overline{\psi} = (\overline{\psi}_1, \overline{\psi}_2, \overline{\psi}_3) := (1 + \psi_1, \psi_2 - \psi_1, -\psi_2)$.

(IV) SARIMA$(r, d, 0)(R, D, 0)^s$: linear seasonal autoregressive integrated models (time series) of order $d \in \mathbb{N}_0$ of non-seasonal differencing, order $r \in \mathbb{N}_0$ of the non-seasonal AR-part, length $s \in \mathbb{N}_0$ of a season, order $D \in \mathbb{N}_0$ of seasonal differencing and order $R \in \mathbb{N}_0$ of the seasonal AR-part. This means

$$\left(1 - \sum_{j=1}^{r} \gamma_j B^j\right)\left(1 - \sum_{i=1}^{R} \overline{\gamma}_i B^{s \cdot i}\right)\left(1 - B\right)^d \left(1 - B^s\right)^D X_{m+1} = \varepsilon_{m+1}, \quad m \geq k,$$

$$(6)$$

where all necessary X_j with $j < k$ are interpreted as part of Z_{k-}. One can straightforwardly rewrite (6) equivalently in terms of (2) with a corresponding function f_γ and $g(z_1, z_2) = z_1 + z_2$, too. Notice that SARIMA$(r, d, 0)$ $(0, 0, 0)^0$ is the same as ARIMA(r, d), and e.g. SARIMA$(0, 1, 0)(0, 1, 0)^{365}$ is given by

$$X_{m+1} = X_m + X_{m-364} - X_{m-365} + \varepsilon_{m+1}, \quad m \geq k.$$

(V) NARX: nonlinear autoregressive models with exogenous input:

$$X_{m+1} = f_\gamma\left(X_m, X_{m-1}, \ldots, X_k, Z_{k-}, a_{m+1}, a_m, a_{m-1}, \ldots, a_k\right) + \varepsilon_{m+1}, \quad m \geq k,$$

with some nonlinear function f_γ (including a fixed lag for the involved variables) and deterministic independent variables $(a_m)_{m \geq k}$. Accordingly, (1) holds with $F_\gamma(X_{m+1}, \ldots) = X_{m+1} - f_\gamma(\ldots)$. As a special case, one gets nonlinear autoregression models of order r in the sense that (4) for a nonlinear function f_γ holds. Furthermore, (3) is covered as well. For a survey on NARX methods in time, frequency and spatio-temporal systems, see Billings [4].

This finishes the overview of covered special cases of (1) which are widely applicable. Within this underlying general autorecursion framework (1), we introduce stepwise a universal information-geometric toolbox for model-search, in the following. To start with, let us first recall that under a[4] correct model $((F_{\gamma_{m+1}^0})_{m \geq k}, Q_{\theta_0})$ the sample $(X_{k+N}, X_{k+N-1}, \ldots)$ behaves in a way such that the derived quantities

$$\left\{F_{\gamma_{k+i}^0}\left(k + i, X_{k+i}, X_{k+i-1}, \ldots, X_k, Z_{k-}, a_{k+i}, a_{k+i-1}, \ldots, a_k\right)\right\}_{i=1,\ldots,N}$$

---

[4] The use of the indefinite article reflects the possible non-uniqueness.

behave like a sample of size N from an iid sequence under the distribution Q_{θ_0}. This means that under a correct model the corresponding empirical distribution

$$P_N^{y^0}[\,\cdot\,] := \frac{1}{N} \cdot \sum_{i=1}^{N} \delta_{F_{\gamma_{k+i}^0}\left(k+i, X_{k+i}, X_{k+i-1}, \ldots, X_k, Z_{k-}, a_{k+i}, a_{k+i-1}, \ldots, a_k\right)}[\,\cdot\,] \qquad (7)$$

converges to Q_{θ_0} as the sample size N tends to ∞, and thus

$$D_{\alpha,\beta}(P_N^{y^0}, Q_{\theta_0}) \xrightarrow[N\to\infty]{} 0 \qquad (8)$$

for a very broad spectrum $\mathscr{D} := \{D_{\alpha,\beta}(\cdot,\cdot) : \alpha \in [\underline{\alpha},\overline{\alpha}], \beta \in [\underline{\beta},\overline{\beta}]\}$ of distances (divergences, (dis)similarity measures, discrepancy measures) $D_{\alpha,\beta}(P,Q)$ between probability distributions P,Q. In (7), we have used $y^0 := (\gamma_{k+i}^0)_{i\geq 1}$ and δ_y for Dirac's one-point distribution at y (i.e. $\delta_y[A] = 1$ iff $y \in A$ and $\delta_y[A] = 0$ else); thus, P_N^y is nothing else but the histogram-according probability distribution where the probability mass function (pmf) p_N^y corresponds to the relative frequencies[5]

$$p_N^y(y) = \frac{1}{N} \cdot \#\Big\{i \in \{1,\ldots,N\} :$$
$$F_{\gamma_{k+i}}\left(k+i, X_{k+i}, X_{k+i-1}, \ldots, X_k, Z_{k-}, a_{k+i}, a_{k+i-1}, \ldots, a_k\right) = y\Big\} \qquad (9)$$

for all $y \in \mathscr{Y}$. As an example, for an AR(2) model the pmf (9) simplifies to

$$p_N^y(y) = \frac{1}{N} \cdot \#\Big\{i \in \{1,\ldots,N\} : X_{k+i} - \gamma_1 \cdot X_{k+i-1} - \gamma_2 \cdot X_{k+i-2} = y\Big\}.$$

Hence, in the light of (8), preselection is to find "good" candidates $((F_{\gamma_{m+1}})_{m\geq k}, Q_\theta)$ for an unknown correct model $((F_{\gamma_{m+1}^0})_{m\geq k}, Q_{\theta_0})$ in the sense that they fulfill

$$D_{\breve{\alpha},\breve{\beta}}(P_N^y, Q_\theta) \approx 0 \qquad \text{for large enough sample size } N, \qquad (10)$$

for some arbitrarily fixed probability distance $D_{\breve{\alpha},\breve{\beta}}(\cdot,\cdot) \in \mathscr{D}$, $(\breve{\alpha},\breve{\beta}) \in [\underline{\alpha},\overline{\alpha}] \times [\underline{\beta},\overline{\beta}]$. Because of the unavoidable imprecision $\approx$ in (10), the amount of preselected models may be quite large; to narrow this down it makes sense to replace the criterion (10) by

$$D_{\alpha,\beta}(P_N^y, Q_\theta) \approx 0 \qquad \text{for large enough sample size } N$$
$$\text{and all } (\alpha,\beta) \in [\underline{\alpha},\overline{\alpha}] \times [\underline{\beta},\overline{\beta}]. \qquad (11)$$

Accordingly, we propose the following procedure which we call *universal model-search by probability distance (UMSPD)*: choose $(F_{\gamma_{m+1}})_{m\geq k}$ from a principal model class (according to fundamental insights from the situation-based context)

---

[5] One can also take variants which (according to some principle) "synthetically un-zero-ize" the empirical probability mass of non-appearing outcomes.

– e.g. one out of (I) to (V) – and choose Q_θ from some prefixed principal class of parametric distributions (e.g. Gaussian distributions, binomial distributions[6]), and find $\gamma_{m+1} \in \Gamma$ $(m \geq k)$ as well as $\theta \in \Theta$ such that (11) holds where the concrete data are plugged into P_N^γ. To implement this graphically, one can plot the corresponding 3D surface $\mathscr{S} := \{(\alpha, \beta, D_{\alpha,\beta}(P_N^\gamma, Q_\theta) : \alpha \in [\underline{\alpha}, \overline{\alpha}], \beta \in [\underline{\beta}, \overline{\beta}]\}$; if (the z-coordinate of) $\mathscr{S}$ is (over a longer period of sampling) smaller than some threshold, say T[7], then the model $((F_{\gamma_{m+1}})_{m \geq k}, Q_\theta)$ is preselected in the sense of short-listed. Since the concrete data set was realized with a certain probability, the threshold should be taken in a way to reflect the preselection decision with appropriately high probabilistic confidence. Of course, the search for good $\gamma_{m+1} \in \Gamma$ $(m \geq k)$ as well as $\theta \in \Theta$ can be operationalized in a fast interactive (semiautomatic) way, especially for the many models which have constant parameter $\gamma_{m+1} \equiv \gamma$, or $\gamma_{m+1} = h(m+1, \gamma)$. The advantage of this approach is that after the preselection process one can continue to work with $D_{\alpha,\beta}(\cdot, \cdot)$ in order to perform amongst all preselected candidate models a statistically sound inference in terms of *simultaneous exact* parameter-estimation (of $(\gamma_{m+1})_{m \geq k}, \theta)$) and goodness-of-fit. With UMSPD, we have settled these tasks approximately (with maybe good precision already).

Two issues must be discussed particularly: (VI) the choice of the distance family $\mathscr{D}$, and (VII) the choice of the threshold T. Concerning (VI), we propose to use $D_{\alpha,\beta}(P_N^\gamma, Q_\theta) := B_{\phi_\alpha}(P_N^\gamma, Q_\theta \| M_\beta)$[8] derived from the following general concept of Stummer & Vajda [15] (see also Stummer [14], Kißlinger & Stummer [6]):

Definition 1. *Let $\phi : (0, \infty) \mapsto \mathbb{R}$ be a finite convex function, continuously extended to $t = 0$, with right-hand derivative ϕ'_+. Furthermore, let P, Q be two probability distributions on $\mathscr{Y}$ (with $|\mathscr{Y}| \geq 2$) and M an arbitrary distribution (measure) on $\mathscr{Y}$ having densities $p = \frac{dP}{d\lambda}$, $q = \frac{dQ}{d\lambda}$, $m = \frac{dM}{d\lambda}$ w.r.t. a σ-finite distribution λ. Then the Bregman distance of P, Q scaled by M is defined by*

$$0 \leq B_\phi(P, Q \| M)$$
$$= \int_{\mathscr{Y}} \left[\phi\left(\frac{p(y)}{m(y)}\right) - \phi\left(\frac{q(y)}{m(y)}\right) - \phi'_+\left(\frac{q(y)}{m(y)}\right) \cdot \left(\frac{p(y)}{m(y)} - \frac{q(y)}{m(y)}\right) \right] dM(y)$$
$$\tag{12}$$
$$= \int_{\mathscr{Y}} \left[m(y) \cdot \left\{ \phi\left(\frac{p(y)}{m(y)}\right) - \phi\left(\frac{q(y)}{m(y)}\right) \right\} - \phi'_+\left(\frac{q(y)}{m(y)}\right) \cdot (p(y) - q(y)) \right] d\lambda(y).$$
$$\tag{13}$$

---

[6] In the sense of putting probability mass $\binom{c}{j} \theta^j (1-\theta)^{c-j}$ $(\theta \in]0,1[)$ on the j-th point y_j of a finite set $\mathscr{Y} = \{y_1, \ldots, y_c\}$.

[7] Which may vary in α, β.

[8] Notice that P_N^γ is a discrete distribution and hence the reference distribution λ is typically the counting distribution (attributing the value 1 to each possible outcome); if Q_θ has a different reference distribution $\overline{\lambda}$ of completely different type (e.g. Q_θ is a classical (absolutely) continuous distribution, say Gaussian, and accordingly $\overline{\lambda}$ is the Lebesgue measure), then one can e.g. smooth the histogram and hence P_N^γ, or "discretize" Q_θ by appropriately partitioning its support.

To guarantee the existence of the integrals in (12), (13) (with possibly infinite values), the zeros of p, q, m have to be combined by proper conventions; if $\mathscr{Y}$ is an infinite space, the integrals may take the infinite value ∞.

If $p(y) > 0$, $q(y) > 0$ for all $y \in \mathscr{X}$ and the function ϕ is convex on $[0, \infty)$, continuous on $(0, \infty)$ as well as strictly convex at 1 with $\phi(1) = 0$, then (13) leads to the special case (cf. Stummer [14], Stummer and Vajda [15])

$$B_\phi\left(P, Q \,\|\, Q\right) = \int_{\mathscr{Y}} q(y) \cdot \phi\left(\frac{p(y)}{q(y)}\right) \, \mathrm{d}\lambda(y) =: D_\phi^{CAS}\left(P, Q\right) \qquad (14)$$

which is nothing but the well-known Csiszar-Ali-Silvey ϕ-divergence between P and Q [1,5]; for $\phi(t) = \phi_1^{\sim}(t) := t \log t + 1 - t \geq 0$ $(t > 0)$ one ends up with the *KL divergence* $D_{\phi_1^{\sim}}^{CAS}\left(P, Q\right)$, for $\phi(t) = \phi_0^{\sim}(t) := -\log t + t - 1 \geq 0$ with the *reversed KL divergence* $D_{\phi_0^{\sim}}^{CAS}\left(P, Q\right) = D_{\phi_1^{\sim}}^{CAS}\left(Q, P\right)$, and for $\phi(t) = \phi_\alpha^{\sim}(t) := \frac{t^\alpha - 1}{\alpha(\alpha-1)} - \frac{t-1}{\alpha-1} \geq 0$ $(\alpha \in \mathbb{R}\backslash\{0,1\})$ with the *power divergences* $D_{\phi_\alpha^{\sim}}^{CAS}\left(P, Q\right)$ (cf. Liese & Vajda [9], Read & Cressie [13]) with Pearson's chi-square divergence as the most prominent special case $\alpha = 2$. Moreover, $B_\phi\left(P, Q \,\|\, \mathbb{1}\right) =: D_\phi^{CBD}\left(P, Q\right)$ corresponds to the *classical Bregman distances* where $M = \mathbb{1}$ stands for the unscaled case $m(x) \equiv 1$ (see e.g. Pardo & Vajda [12] for finite probability spaces, and e.g. Nock et al. [10] for applications in the framework of universal nearest neighbor classification); the special case $D_{\phi_\alpha^{\sim}}^{CBD}\left(P, Q\right)$ coincides with the *density power divergences* of Basu et al. [2].

Returning back to the central model-search task, for the rest of this paper we confine ourselves to the general context of *finite* state space $\mathscr{Y}$ with $|\mathscr{Y}| < \infty$, $\phi = \phi_\alpha$ and $M_\beta = W_\beta(P_N^{\mathscr{Y}}, Q_\theta)$ in the sense that $m_\beta(y) = w_\beta(p_N^{\mathscr{Y}}(y), q_\theta(y)) \geq 0$ for some β-family of "scale-connectors" $w_\beta : [0,1] \times [0,1] \mapsto [0, \infty]$ between the probability mass functions $p_N^{\mathscr{Y}}(y)$ and $q_\theta(y)$. Accordingly, one obtains the new class of model-preselection scores $D_{\alpha,\beta}(P_N^{\mathscr{Y}}, Q_\theta) := B_{\phi_\alpha}(P_N^{\mathscr{Y}}, Q_\theta \,\|\, W_\beta(P_N^{\mathscr{Y}}, Q_\theta))$ in terms of the scaled Bregman distances[9,10]

$$0 \leq B_{\phi_\alpha}(P_N^{\mathscr{Y}}, Q_\theta \,\|\, W_\beta(P_N^{\mathscr{Y}}, Q_\theta))$$
$$= \sum_{y \in \mathscr{Y}} w_\beta(p_N^{\mathscr{Y}}(y), q_\theta(y)) \cdot \left[\phi_\alpha\left(\frac{p_N^{\mathscr{Y}}(y)}{w_\beta(p_N^{\mathscr{Y}}(y), q_\theta(y))}\right) - \phi_\alpha\left(\frac{q_\theta(y)}{w_\beta(p_N^{\mathscr{Y}}(y), q_\theta(y))}\right) \right.$$
$$\left. - \phi_{\alpha+}'\left(\frac{q_\theta(y)}{w_\beta(p_N^{\mathscr{Y}}(y), q_\theta(y))}\right) \cdot \left(\frac{p_N^{\mathscr{Y}}(y)}{w_\beta(p(y), q_\theta(y))} - \frac{q_\theta(y)}{w_\beta(p_N^{\mathscr{Y}}(y), q_\theta(y))}\right) \right]. \quad (15)$$

---

[9] Choosing the counting distribution for λ; one can use (15) also for non-probability contexts (e.g. general nonnegative vectors) with $\sum_{x \in \mathscr{X}} p(x) \neq 1$, $\sum_{x \in \mathscr{X}} q(x) \neq 1$.

[10] In case of $w(u, v) = w(v, u)$ for all (u, v), one can easily produce symmetric preselection-score versions by means of either $B_{\phi, W}^{new}(P, Q) + B_{\phi, W}^{new}(Q, P)$, $\max\{B_{\phi, W}^{new}(P, Q), B_{\phi, W}^{new}(Q, P)\}$, $\min\{B_{\phi, W}^{new}(P, Q), B_{\phi, W}^{new}(Q, P)\}$; this also works for $\phi(t) = \phi_1(t)$ together with arbitrary scale-connectors w.

The corresponding 3D preselection-score surface[11] $\mathscr{S} := \{(\alpha, \beta, B_{\phi_\alpha}(P_N^{\mathbf{y}}, Q_\theta\|M_\beta) : \alpha \in [\underline{\alpha}, \overline{\alpha}], \beta \in [\underline{\beta}, \overline{\beta}]\}$ may look like the one in Fig. 1(a). The scale connector family $(w_\beta)_\beta$ may be chosen in a way to favor those models for which the data-derived quantities do not appear as outliers[12] and inliers[13] in a probabilistic sense; for instance, for the choice $\phi_\alpha(t) = \phi_2^{\sim}(t)$ we get from (15)

$$B_{\phi_2^{\sim}}(P_N^{\mathbf{y}}, Q_\theta \| W_\beta(P_N^{\mathbf{y}}, Q_\theta)) = \sum_{y \in \mathscr{Y}} \frac{(p_N^{\mathbf{y}}(y) - q_\theta(y))^2}{w_\beta(p_N^{\mathbf{y}}(y), q_\theta(y))} \; ;$$

hence, the scale connector $w_\beta(u, v)$ should be small for (u, v) "close" to $(1, 0)$ (outliers) at least for some β and also small for (u, v) "close" to $(0, 1)$ (inliers) at least for some β (and maybe reflect also further features of situation-based importance). For example, this can be achieved with the choice $w_\beta(u, v) = w_\beta^{\sim}(u, v) := u^\beta \cdot v^{1-\beta}$ ($\beta \in [0, 1]$, with $\beta = 0$ corresponding to the Csiszar-Ali-Silvey ϕ_α-divergences); see Fig. 1(b), (c) for exemplary parameter choices. Other scale connectors can be found in Kißlinger & Stummer [7, 8]. Finally, concerning the above-mentioned task (VII) we exemplarily show how to quantify the above-mentioned preselection criterion "the 3D surface $\mathscr{S}$ should be smaller than a threshold T" by some sound asymptotic analysis for the above special choices $\phi_\alpha(t) = \phi_\alpha^{\sim}(t)$ and $w_\beta(u, v) = w_\beta^{\sim}(u, v)$. The cornerstone is the following assertion under the true model $((F_{\gamma_{m+1}}^0)_{m \geq k}, Q_{\theta^0})$ to be traced:

Theorem 1. *Let Q_{θ_0} be a finite discrete distribution with $c := |\mathscr{Y}| \geq 2$ possible outcomes and strictly positive densities $q_{\theta_0}(y) > 0$ for all $y \in \mathscr{Y}$. Then for each $\alpha > 0$, $\alpha \neq 1$ and each $\beta \in [0, 1[$ the random scaled Bregman power distance*

$$2N \cdot B_{\phi_\alpha^{\sim}}\left(P_N^{\mathbf{y}_0}, Q_{\theta_0} \mid (P_N^{\mathbf{y}_0})^\beta \cdot Q_{\theta_0}^{1-\beta}\right) =: 2N \cdot B(\alpha, \beta; \mathbf{y}_0, \theta_0; N)$$

is asymptotically chi-squared distributed in the sense that

$$2N \cdot B(\alpha, \beta; \mathbf{y}_0, \theta_0; N) \xrightarrow[N \to \infty]{\mathscr{L}} \chi_{c-1}^2.$$

In terms of the corresponding χ_{c-1}^2-quantiles, one can derive the threshold T which the 3D preselection-score surface $\mathscr{S}$ has to (partially) exceed in order to believe with appropriate level of confidence that the investigated model $((F_{\gamma_{m+1}})_{m \geq k}, Q_\theta)$ is not good enough to be preselected. The proof of Theorem (1) will appear elsewhere, as part of a complete treatment for simultaneous model preselection, parameter estimation and goodness-of-fit tests for general ϕ and w.

---

[11] Goodness-of-approximation score surface.

[12] Quantities which modelwise should be rare but actually appear much more often.

[13] Quantities which modelwise should be very frequent but actually appear much less often.

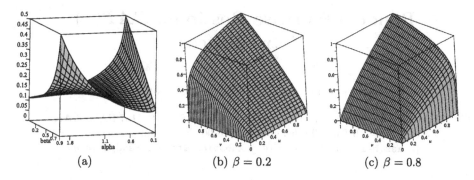

(a)   (b) $\beta = 0.2$ (c) $\beta = 0.8$

Fig. 1. (a): 3D score surface for model preselection;   (b), (c): scale-connectors $w_{\tilde{\beta}}(u, v) = u^{\beta} \cdot v^{1-\beta}$

References

1. Ali, M.S., Silvey, D.: A general class of coefficients of divergence of one distribution from another. J. Roy. Statist. Soc. **B-28**, 131140 (1966)
2. Basu, A., Harris, I.R., Hjort, N.L., Jones, M.C.: Robust and efficient estimation by minimising a density power divergence. Biometrika **B-85**, 549–559 (1998)
3. Basu, A., Shioya, H., Park, C.: Statistical Inference: The Minimum Distance Approach. CRC Press, Boca Raton (2011)
4. Billings, S.A.: Nonlinear System Identification. Wiley, Chichester (2013)
5. Csiszar, I.: Eine informationstheoretische Ungleichung und ihre Anwendung auf den Beweis der Ergodizität von Markoffschen Ketten. Publ. Math. Inst. Hungar. Acad. Sci. **A-8**, 85–108 (1963)
6. Kißlinger, A.-L., Stummer, W.: Some decision procedures based on scaled bregman distance surfaces. In: Nielsen, F., Barbaresco, F. (eds.) GSI 2013. LNCS, vol. 8085, pp. 479–486. Springer, Heidelberg (2013)
7. Kißlinger, A.-L., Stummer, W.: Robust statistical engineering by means of scaled Bregman divergences (to appear)
8. Kißlinger, A.-L., Stummer, W.: A new information-geometric method of change detection. Preprint
9. Liese, F., Vajda, I.: Convex Statistical Distances. Teubner, Leipzig (1987)
10. Nock, R., Piro, P., Nielsen, F., Ali, W.B.H., Barlaud, M.: Boosting k-NN for categorization of natural sciences. Int. J. Comput. Vis. **100**, 294–314 (2012)
11. Pardo, L.: Statistical Inference Based on Divergence Measures. Chapman & Hall, Boca Raton (2006)
12. Pardo, M.C., Vajda, I.: On asymptotic properties of information-theoretic divergences. IEEE Trans. Inf. Theor. **49**(7), 1860–1868 (2003)
13. Read, T.R.C., Cressie, N.A.C.: Goodness-of-Fit Statistics for Discrete Multivariate Data. Springer, New York (1988)
14. Stummer, W.: Some Bregman distances between financial diffusion processes. Proc. Appl. Math. Mech. **7**(1), 1050503–1050504 (2007)
15. Stummer, W., Vajda, I.: On Bregman distances and divergences of probability measures. IEEE Trans. Inf. Theor. **58**(3), 1277–1288 (2012)

Random Pairwise Gossip on $CAT(\kappa)$ Metric Spaces

Anass Bellachehab[✉] and Jérémie Jakubowicz

Institut Mines-Télécom Télécom SudParis CNRS UMR 5157 SAMOVAR,
9 Rue Charles Fourier, 91000 Evry, France
{anass.bellachehab,jeremie.jakubowicz}@telecom-sudparis.eu

Abstract. In the context of sensor networks, gossip algorithms are a popular, well established technique, for achieving consensus when sensor data are encoded in linear spaces. Gossip algorithms also have several extensions to non linear data spaces. Most of these extensions deal with Riemannian manifolds and use Riemannian gradient descent. This paper, instead, studies gossip in a broader $CAT(\kappa)$ metric setting, encompassing, but not restricted to, several interesting cases of Riemannian manifolds. As it turns out, convergence can be guaranteed as soon as the data lie in a small enough ball of a mere $CAT(\kappa)$ metric space. We also study convergence speed in this setting and establish linear rates of convergence.

1 Introduction

In the field of distributed algorithms, the consensus problem has attracted much attention. Whether in database management [Bur06], clock synchronization [SG07] or signal estimation in wireless sensor networks [SRG08]. Consider a network of sensors in which each sensor is able to make local measurements. Because of resource constraints, each sensor can only communicate with its neighbors in terms of network topology. The sensors seek to reach a consensus on their measurements. If the measurement data can be encoded in a vector space (for example temperatures or velocities); gossip algorithms [BGPS06] are known efficient candidates that converge with exponential speed towards a consensus state, assuming the network is connected.

Prior Work. There has been significant work in order to extend the gossip algorithms to cases where data cannot be encoded in a Euclidean space (*e.g.* space of 3-D rotations, Grassmannians...). Recently [Bon13] used the framework of stochastic gradient descent to cast the asynchronous consensus problem as a problem of minimizing a cost function (the sum of pairwise distances squared); the author then proposes a gossip algorithm analogous to that of [BGPS06] in the case of manifolds of nonpositive curvature. In [BJ] we have provided an adaption of the Random Pairwise Gossip (RPG) algorithm [BGPS06] to the purely metric setting of $CAT(0)$ spaces; convergence with linear rate was established.

Paper Contribution. In this paper, we extend the previous result to the case of positive curvature, the $CAT(\kappa)$ metric space with $\kappa > 0$. The main result is

© Springer International Publishing Switzerland 2015
F. Nielsen and F. Barbaresco (Eds.): GSI 2015, LNCS 9389, pp. 702–709, 2015.
DOI: 10.1007/978-3-319-25040-3_75

that provided the initial set of data is located inside a compact of sufficiently diameter, we can ensure the same results, (convergence at linear rate) as the nonpostive curvature case.

2 Framework

2.1 Notation

Assume V is some finite set. We denote by $\mathcal{P}_2(V)$ the set of *pairs* of elements in V: $\mathcal{P}_2(V) = \{\{v, w\} : v \neq w\}$. Notice that, by definition, for $v \neq w$, $\{v, w\} = \{w, v\}$ whereas $(v, w) \neq (w, v)$. Throughout the paper, $\mathcal{M}$ will denote a metric space, equipped with metric d. Associated with any subset $S \subset \mathcal{M}$, we define its *diameter* $\mathrm{diam}(S) = \sup\{d(s, s') : s, s' \in S\}$. We also define (closed) balls $B(x, r) = \{y \in \mathcal{M} : d(x, y) \leq r\}$. Random variables are denoted by upper-case letters (*e.g.*, X, ...) while their realizations are denoted by lower-case letters (*e.g.* x, ...) Without any further notice, random variables are assumed to be functions from a probability space Ω equipped with its σ-field $\mathcal{F}$ and probability measure $\mathbb{P}$; $x = X(\omega)$ denotes the realization associated to $\omega \in \Omega$. For any set S and any subset A, $\delta\{A\}$ denotes the indicator function that takes value 1 on A and 0 otherwise.

2.2 Network

We consider a network of N agents represented by a graph $G = (V, E)$, where $V = \{1, \ldots, N\}$ stands for the set of agents and E denotes the set of available communication links between agents. A link $e \in E$ is given by a pair $\{v, w\} \in \mathcal{P}_2(V)$ where v and w are two distinct agents in the network that are able to communicate directly. Note that the graph is assumed undirected, meaning that whenever agent v is able to communicate with agent w, the reciprocal communication is also assumed feasible. This assumption makes sense when communication speed is fast compared to agents movements speed. When a communication link $e = \{v, w\}$ exists between two agents, both agents are said to be neighbors and the link is denoted $v \sim w$. We denote by $\mathcal{N}(v)$ the set of all neighbors of the agent $v \in V$. The number of elements in $\mathcal{N}(v)$ is referred to as the *degree* of v and denoted $\deg(v)$.

2.3 Time

As in [BGPS06], we assume that the time model is asynchronous, *i.e.* that each agent has its own Poisson clock that ticks with a common intensity λ (the clocks are identically made), and moreover, each clock is independent from the other clocks. When an agent clock ticks, the agent is able to perform some computations and wake up some neighboring agents. This time model has the same probability distribution than a global single clock ticking with intensity $N\lambda$ and selecting uniformly randomly a single agent at each tick. This equivalence is described, *e.g.* in [BGPS06]. From now on, we represent time by the set of integers: for such an integer k, time k stands for the time at which the kth event occurred.

2.4   Communication

At a given time k, we denote by V_k the agent whose clock ticked and by W_k the neighbor that was in turn awaken. Therefore, at time k, the only communicating agents in the whole network are V_k and W_k. A single link is then active at each time, hence, at a given time, most links are not used. The previously described time model implies that (V_k, W_k)'s are independent and identically distributed.

It is going to turn out convenient to consider directly the link $\{V_k, W_k\}$, forgetting which node was the first to wake up and which node was second. In this case $\mathbb{P}[\{V_k, W_k\} = \{v, w\}]$ is of course symmetric in (v, w). One has:

$$\mathbb{P}[\{V_k, W_k\} = \{v, w\}] = \begin{cases} \frac{1}{N}\left(\frac{1}{\deg(v)} + \frac{1}{\deg(w)}\right) & \text{if } v \sim w \\ 0 & \text{otherwise} \end{cases}$$

The communication framework considered here is standard [BGPS06].

2.5   Data

Each node $v \in V$ stores data represented as an element x_v belonging to some space $\mathcal{M}$. Initially each node v has a value $x_v(0)$ and $X_0 = (x_1(0), \ldots, x_N(0))$ is the tuple of initial values of the network. We focus on iterative algorithms that tend to drive the network to a *consensus state*; meaning a state of the form $X_\infty = (x_\infty, \ldots, x_\infty)$ with: $x_\infty \in \mathcal{M}$. We denote by $x_v(k)$ the value stored by the agent $v \in V$ at the k-th iteration of the algorithm, and $X_k = (x_1(k), \ldots, x_N(k))$ the global state of the network at instant k. The general scheme is as follows: network is in some state X_{k-1}; agents V_k and W_k wake up, communicate and perform some computation to lead the network to state X_k.

3 $CAT(\kappa)$ Metric Spaces

3.1   Preliminary Definitions

Definition 1 (Geodesic, Segments). *A path* $c : [0, l] \to \mathcal{M}$, $l \geq 0$ *is said to be a* geodesic *if* $d(c(t), c(t')) = |t - t'|$, *for all* $(t, t') \in [0, l]^2$; $x = c(0)$ *and* $y = c(1)$ *are the* endpoints *of the geodesic and* $l = d(x, y)$ *is the* length *of the geodesic. The image of* c *is called a* geodesic segment *with endpoints* x *and* y. *If there is a single segment with endpoints* x *and* y, *it is denoted* $[x, y]$.

Definition 2 (Midpoint). *The* midpoint *of segment* $[x, y]$ *is denoted* $\langle \frac{x+y}{2} \rangle$, *it is defined as the unique point* m *such that* $d(x, m) = d(y, m) = d(x, y)/2$.

In what follows we also use notation: $D_\kappa = \frac{\pi}{\sqrt{\kappa}}$ and: $r_\kappa = \frac{D_\kappa}{2}$.

Convexity can have several meaning in the context of metric spaces (*cf.*, *e.g.* [Cha06, p.403]).

Definition 3 (Convexity). *A subset S of $\mathcal{M}$ is said* convex *when for every couple of points $(x, y) \in S^2$, every geodesic segment γ joining x and y in $(\mathcal{M}, d)$ is such that $\gamma \subset S$.*

3.2 The $CAT(\kappa)$ Inequality

Definition 4 ($CAT(\kappa)$ Inequality). *Assume $(\mathcal{M}, d)$ is a metric space and $\Delta = (c_0, c_1, c_2)$ is a geodesic triangle with vertices $p = c_0(0)$, $q = c_1(0)$ and $r = c_2(0)$ and with perimeter strictly less than $2D_\kappa$. Let $\bar{\Delta} = (\bar{p}, \bar{q}, \bar{r})$ denote a comparison triangle in $\mathcal{M}_\kappa^2$. Δ is said to satisfy the $CAT(\kappa)$ inequality if for any $x = c_0(t)$ and $y = c_2(t')$, one has:*

$$d(x, y) \leq \bar{d}(\bar{x}, \bar{y})$$

where $\bar{x}$ is the unique point of $[\bar{p}, \bar{q}]$ such that $d(p, x) = \bar{d}(\bar{p}, \bar{x})$ and $\bar{y}$ on $[\bar{p}, \bar{r}]$ such that $d(p, y) = \bar{d}(\bar{p}, \bar{y})$.

Definition 5 ($CAT(\kappa)$ Metric Space). *A metric space $(\mathcal{M}, d)$ is said $CAT(\kappa)$ if every geodesic triangle of $\mathcal{M}$ satisfies the $CAT(\kappa)$ inequality.*

3.3 Further Proprieties of $CAT(\kappa)$ Spaces

Proposition 1 ([BH99] [prop. II.1.4]). *Let $\mathcal{M}$ denote a $CAT(\kappa)$ metric space.*

1. *If x and y in $\mathcal{M}$ are such that $d(x, y) < D_\kappa$, there exists a unique geodesic $[x, y]$ joining them.*
2. *For any $x \in \mathcal{M}$, the ball $B_{x,r}$ with $r < r_\kappa$ is convex.*

For notational convenience, define the functions: $C_\kappa(t) = \cos(\sqrt{\kappa}t)$, $S_\kappa(t) = \frac{\sin(\sqrt{\kappa}t)}{\sqrt{\kappa}}$, $\chi_\kappa(t) = 1 - C_\kappa(t)$.

4 Algorithm

At each count of the virtual global clock one node v is selected uniformly randomly from the set of agents V; the node v then randomly selects a node w from $\mathcal{N}(v)$. Both nodes v and w then compute and update their value to $\langle \frac{x_v + x_w}{2} \rangle$.

Remark 1. *Please note that given two points $x, y \in \mathcal{M}$, their midpoint is not necessarily well-defined when $\kappa > 0$. However if $d(x, y) < D_\kappa$ then according to a result in $CAT(\kappa)$ metric space [BH99, p.160] their midpoint exists and is unique. If one can ensure that the convex hull of the initial set of points is of diameter $< D_\kappa$, then by convexity the points will remain in that hull and the algorithm will be well defined.*

Algorithm Random Pairwise Midpoint

Input: a graph $G = (V, E)$ and the initial nodes configuration $X_v(0), v \in V$
for all $k > 0$ **do**

At instant k, uniformly randomly choose a node V_k from V and a node W_k uniformly randomly from $\mathcal{N}(V_k)$.

Update:

$$X_{V_k}(k) = \left\langle \frac{X_{V_k}(k-1) + X_{W_k}(k-1)}{2} \right\rangle$$

$$X_{W_k}(k) = \left\langle \frac{X_{V_k}(k-1) + X_{W_k}(k-1)}{2} \right\rangle$$

$$X_v(k) = X_v(k-1) \text{ for } v \notin \{V_k, W_k\}$$

end for

---

5 Convergence Results

In order to study convergence we recall the following assumptions, already explained in Sect. 2.

Assumption 1.

1. $G = (V, E)$ *is connected*
2. $(V_k, W_k)_{k \geq 0}$ *are i.i.d random variables, such that:*
    (a) (V_k, W_k) is independent from $X_0, \dots, X_{k-1}, (V_0, W_0), \dots, (V_{k-1}, W_{k-1})$,
    (b) $\mathbb{P}[\{V_0, W_0\} = \{v, w\}] = \frac{1}{N}(\deg^{-1}(v) + \deg^{-1}(w))\delta\{v \sim w\}$
3. $\kappa > 0$
4. $(\mathcal{M}, d)$ *is a complete* $CAT(\kappa)$ *metric space.*
5. $diam(\{X_v(0) : v \in V\}) < r_\kappa$

We are ensured that Algorithm Random Pairwise Midpoint is well-defined. Since, by convexity of balls with radius smaller than r_κ [BH99, p.167] points will remain within distance less than r_κ of each other. Moreover midpoints are well-defined and unique since $r_\kappa < D_\kappa$.

We now define the *disagreement function*:

Definition 6. *For* $x \in \mathcal{M}^n$ *define:*

- $\Delta_\kappa(x) = \frac{1}{2} \sum_{\substack{v \sim w \\ \{v, w\} \in E}} \frac{1}{N}(\deg(v)^{-1} + \deg(w)^{-1})\chi_\kappa(d(x_v, x_w))$
- $\sigma_\kappa^2(x) = \frac{2}{N} \sum_{\{v, w\} \in \mathcal{P}_2(V)} \chi_\kappa(d(x_v, x_w))$

One can remark that for all $k \geq 0$, $(v, w) \in V^2$: $\sigma_\kappa^2(X_k) \geq 0$ and $\Delta_\kappa(X_k) \geq 0$. Notice that $\sigma_\kappa^2(x) = 0$ implies that for all $\{v, w\} \in \mathcal{P}_2(V)$: $\chi_\kappa(d(v, w)) = 0$; and, since $0 \leq d(v, w) \leq \frac{\pi}{2\sqrt{\kappa}}$, it implies that $d(v, w) = 0$, hence the system is in a consensus state. Moreover, when $\kappa \to 0$, $\chi_\kappa(d(v, w)) \to d^2(v, w)$.

Using the following well known lemma in positive curvature trigonometry.

Lemma 1 (Law of Cosines). *Given a complete manifold* $\mathcal{M}_\kappa^n$ *with constant sectional curvature* κ *and a geodesic triangle* $\Delta(pqr)$ *in* $\mathcal{M}_\kappa^n$, *assume* $\max\{d(p, r), d(q, r), d(p, q)\} < r_\kappa$ *and let* $\alpha := \widehat{prq}$. *We have:*

$$C_\kappa(d(p, q)) = C_\kappa(d(p, r))C_\kappa(d(q, r)) + S_\kappa(d(p, r))S_\kappa(d(q, r))\cos(\alpha)$$

We prove following proposition.

Proposition 2. *Under Assumption 1, for any triangle $\Delta(pqr)$ in C where m is the midpoint of $[p,q]$ we have:*

$$\chi_\kappa(d(m,r)) \leq \frac{\chi_\kappa(d(p,r)) + \chi_\kappa(d(q,r))}{2}$$

Proof. Using the trig we get:

$C_\kappa(d(p,r)) + C_\kappa(d(q,r)) - 2C_\kappa(d(m,r)) \leq 2C_\kappa(d(m,r))C_\kappa(d(p,q)) - 2C_\kappa(d(m,r))$

Since $\max\{d(m,r), d(p,q)\} < \frac{\pi}{2\sqrt{\kappa}}$ we have: $0 \leq C_\kappa(d(p,q)) \leq 1$ and $0 \leq C_\kappa(d(m,r)) \leq 1$ Which means that: $2C_\kappa(d(m,r))C_\kappa(d(p,q)) - 2C_\kappa(d(m,r)) \leq 0$. And thus:

$2\chi_\kappa(d(m,r)) \leq \chi_\kappa(d(p,r)) + \chi_\kappa(d(q,r))$

Proposition 3.

$$\mathbb{E}[\sigma_\kappa^2(X_{k+1}) - \sigma_\kappa^2(X_k)] \leq -\frac{1}{N}\mathbb{E}\Delta_\kappa(X_k)$$

Proof. At round k, two nodes woke up with indices V_k and W_k, it follows that:

$N(\sigma_\kappa^2(X_k) - \sigma_\kappa^2(X_{k-1})) = -\chi_\kappa(d(X_{V_k}(k-1), X_{W_k}(k-1))) + \sum_{\substack{u \in V \\ u \neq V_k, u \neq W_k}} T_\kappa(V_k, W_k, u)$

Where T_κ is defined as:

$T_\kappa(V_k, W_k, u) = 2\chi_\kappa(d(X_u(k), M_k)) - \chi_\kappa(d(X_u(k), X_{V_k}(k-1))) - \chi_\kappa(d(X_u(k), X_{W_k}(k-1)))$.

Now, using the inequality of Proposition 2, one gets that $T_\kappa(V_k, W_k, u) \leq 0$ and:

$N(\sigma_\kappa^2(X_k) - \sigma_\kappa^2(X_{k-1})) \leq \chi_\kappa(d(X_{V_k}(k-1), X_{W_k}(k-1)))$.

Taking expectations on both sides and dividing by N gives:

$\mathbb{E}[\sigma_\kappa^2(X_k) - \sigma_\kappa^2(X_{k-1})] \leq -\frac{1}{N}\mathbb{E}[\chi_\kappa(d(X_{V_k}(k-1), X_{W_k}(k-1)))]$

Recalling that $\mathbb{P}[\{V_k, W_k\} = \{u,v\}] = \frac{1}{N}(\frac{1}{\deg u} + \frac{1}{\deg v})$ when $u \sim v$ and 0 otherwise, and that (V_k, W_k) are independent from X_{k-1}, one can deduce:

$\mathbb{E}[\chi_\kappa(d(X_{V_k}(k-1), X_{W_k}(k-1)))] = \mathbb{E}[\Delta_\kappa(X_{k-1})]$

Which yields:

$\mathbb{E}[\sigma_\kappa^2(X_{k+1}) - \sigma_\kappa^2(X_k)] \leq -\frac{1}{N}\mathbb{E}\Delta_\kappa(X_k)$

Proposition 4. *Assume $G = (V, E)$ is an undirected connected graph, there exists a constant C_κ depending on the graph only such that:*

$$\forall x \in \mathcal{M}^N, \quad \frac{\kappa}{\pi^2}\Delta_\kappa(x) \leq \sigma_\kappa^2(x) \leq C_\kappa\Delta_\kappa(x)$$

Proof. First when $\kappa \to 0$:

$$\Delta_0(x) = \sum_{v \sim w} \frac{1}{N}(\deg(v)^{-1} + \deg(w)^{-1})d^2(x_v, x_w)$$

$$\leq \frac{2}{N} \sum_{v \sim w} d^2(x_v, x_w)$$

$$\leq \frac{2}{N} \sum_{\{v,w\} \in \mathcal{P}_2(V)} d^2(x_v, x_w) = 2\sigma_0^2(x)$$

For the second inequality, consider $v \neq w$ two vertices in V, not necessarily adjacent. Since G is connected, there exists a path $u_0 = v, \ldots, u_l = w$ such that $u_i \sim u_{i+1}$. Then, using Cauchy-Schwartz inequality:

$$d(x_v, x_w)^2 \leq l \sum_{i=0}^{l-1} d^2(x_{u_i}, x_{u_{i+1}}) \leq \frac{\deg(G)}{2} \operatorname{diam}(G) \sum_{i=0}^{l-1} (\deg(u_i)^{-1} + \deg(u_{i+1})^{-1})d^2(x_{u_i}, x_{u_{i+1}})$$

where $\deg(G)$ denotes the *maximum degree* $\max\{\deg(v) : v \in V\}$ and $\operatorname{diam}(G)$ the *diameter* of G. Hence taking $C_G = (N-1)\frac{\deg(G)}{2}\operatorname{diam}(G) \geq 1$, one recover the sought inequality for $\kappa \to 0$.

One has: $\frac{2\kappa}{\pi^2}x^2 \leq \chi_\kappa(x) \leq \frac{\kappa}{2}x^2$ when $0 \leq x < \frac{\pi}{2\sqrt{\kappa}}$. Hence, under Assumption 1, χ_κ and d are equivalent.

Lemma 2. *Assume a_n is a sequence of nonnegative numbers such that $a_{n+1} - a_n \leq -\beta a_n$ with $\beta \in (0,1)$. Then,*

$$\forall n \geq 0, \quad a_n \leq a_0 \exp(-\beta n)$$

Proof. Indeed if $l_n = \log a_n$, then $l_{n+1} - l_n \leq \log(1-\beta) \leq -\beta$. Hence $l_n \leq l_0 - \beta n$. Taking exponential on both side gives the expected result.

Theorem 1. *Let $X_k = (x_1(k), \ldots, x_N(k))$ denote the sequence of random variables generated by Algorithm Random Pairwise Midpoint; under Assumption 1, there exists $L \in (-1, 0)$ such that,*

$$\mathbb{E}\Delta_\kappa(X_k) \leq \exp(Lk)$$

Proof. Denote by $a_n = \mathbb{E}\sigma_\kappa^2(X_k)$. From Propositions 4 and 3, we know that the constant $L = -\frac{1}{2C_G}$ verifies $a_{n+1} - a_n \leq La_n$ with $L \in (-1, 0)$ since $C_G \geq 1$. We conclude using Lemma 2.

The proof of the following theorem is the same as the one for the nonpositive curvature case, see [BJ].

Theorem 2. *Let $X_k = (X_1(k), \ldots, X_N(k))$ denote the sequence generated by Algorithm Random Pairwise Midpoint, then under Assumption 1, there exists a random variable X_∞ taking values in the consensus subspace, such that X_k tends to X_∞ almost surely.*

References

[BGPS06] Boyd, S., Ghosh, A., Prabhakar, B., Shah, D.: Randomized gossip algorithms. IEEE Trans. Inf. Theory **52**(6), 2508–2530 (2006)

[BH99] Bridson, M., Haefliger, A.: Metric Spaces of Non-Positive Curvature.
Springer, Heidelberg (1999)

[BJ] Bellachehab, A., Jakubowicz, J.: Random pairwise gossip on CAT (0) metric
spaces

[Bon13] Bonnabel, S.: Stochastic gradient descent on Riemannian manifolds. IEEE
Trans. Autom. Control **58**(9), 2217–2229 (2013)

[Bur06] Burrows, M.: The chubby lock service for loosely-coupled distributed systems. In: Proceedings of the 7th Symposium on Operating Systems Design
and Implementation, pp. 335–350 (2006)

[Cha06] Chavel, I.: Riemannian Geometry, 2nd edn. Cambridge University Press,
New York (2006)

[SG07] Schenato, L., Gamba, G.: A distributed consensus protocol for clock synchronization in wireless sensor network. In: IEEE Conference on Decision
and Control, pp. 2289–2294 (2007)

[SRG08] Schizas, I., Alejandro, R., Georgios, G.: Consensus in ad hoc WSNs with
noisy links-part I: distributed estimation of deterministic signals. IEEE
Trans. Signal Process **56**(1), 350–364 (2008)

Bayesian and Information Geometry
for Inverse Problems

Stochastic PDE Projection on Manifolds: Assumed-Density and Galerkin Filters

John Armstrong[1]([📧]) and Damiano Brigo[2]

[1] Department of Mathematics, King's College London, London, UK
john.1.armstrong@kcl.ac.uk
[2] Department of Mathematics, Imperial College London,
180 Queen's Gate, London SW7 2AZ, UK
damiano.brigo@imperial.ac.uk

Abstract. We review the manifold projection method for stochastic nonlinear filtering in a more general setting than in our previous paper in Geometric Science of Information 2013. We still use a Hilbert space structure on a space of probability densities to project the infinite dimensional stochastic partial differential equation for the optimal filter onto a finite dimensional exponential or mixture family, respectively, with two different metrics, the Hellinger distance and the L^2 direct metric. This reduces the problem to finite dimensional stochastic differential equations. In this paper we summarize a previous equivalence result between Assumed Density Filters (ADF) and Hellinger/Exponential projection filters, and introduce a new equivalence between Galerkin method based filters and Direct metric/Mixture projection filters. This result allows us to give a rigorous geometric interpretation to ADF and Galerkin filters. We also discuss the different finite-dimensional filters obtained when projecting the stochastic partial differential equation for either the normalized (Kushner-Stratonovich) or a specific unnormalized (Zakai) density of the optimal filter.

1 The Filtering Problem in Continuous Time

The state of a system X evolves over time according to some stochastic process driven by a noise W. We cannot observe the state directly but we make an imperfect measurement Y which is also perturbed stochastically by random noise V. In a diffusion setting this problem is formulated as

$$dX_t = f_t(X_t)\, dt + \sigma_t(X_t)\, dW_t, \quad X_0, \quad dY_t = b_t(X_t)\, dt + dV_t, \quad Y_0 = 0. \quad (1)$$

In these equations the unobserved state process $\{X_t, t \geq 0\}$ takes values in $\mathbb{R}^n$, the observation $\{Y_t, t \geq 0\}$ takes values in $\mathbb{R}^d$ and the noise processes $\{W_t, t \geq 0\}$ and $\{V_t, t \geq 0\}$ are two Brownian motions. The nonlinear filtering problem consists in finding the conditional probability distribution π_t of the state X_t given the observations up to time t and the prior distribution π_0 for X_0. Let us assume that X_0, and the two Brownian motions are independent. Let us also assume that the covariance matrix for V_t is invertible. We can then assume

© Springer International Publishing Switzerland 2015
F. Nielsen and F. Barbaresco (Eds.): GSI 2015, LNCS 9389, pp. 713–722, 2015.
DOI: 10.1007/978-3-319-25040-3_76

without any further loss of generality that its covariance matrix is the identity. We introduce a variable a_t defined by $a_t = \sigma_t \sigma_t^T$. With these preliminaries, and a number of rather more technical conditions for which we refer to [9], one can show that π_t satisfies the Kushner–Stratonovich equation. We further suppose that the measure π_t is determined by a probability density p_t. A formal calculation then gives the following Stratonovich calculus version of the optimal filter stochastic PDE (SPDE) for the evolution of p:

$$dp_t = \mathcal{L}_t^* p_t \, dt - \frac{1}{2} p_t \left[|b_t|^2 - E_{p_t}\{|b_t|^2\} \right] dt + \sum_{k=1}^{d} p_t \left[b_t^k - E_{p_t}\{b_t^k\} \right] \circ dY_t^k. \quad (2)$$

We use Stratonovich calculus because we need the formal chain rule to hold when identifying the projected evolution from the projected right hand side of the equation, as we hint below after Eq. (7).

Here $\mathcal{L}^*$ is the formal adjoint of $\mathcal{L}$ – the so-called forward diffusion operator for X, where the backward diffusion operator is defined by:

$$\mathcal{L}_t = \sum_{i=1}^{n} f_t^i \frac{\partial}{\partial x_i} + \frac{1}{2} \sum_{i,j=1}^{n} a_t^{ij} \frac{\partial^2}{\partial x_i \partial x_j}, \quad \mathcal{L}_t^* \phi = - \sum_{i=1}^{n} \frac{\partial}{\partial x_i}[f_t^i \phi] + \frac{1}{2} \sum_{i,j=1}^{n} \frac{\partial^2}{\partial x_i \partial x_j}[a_t^{ij} \phi].$$

If the coefficients f and b are linear, σ is deterministic (and does not depend on X), and if the prior distribution is normal, this equation can be solved analytically to give the so-called Kalman Filter, where p is a Gaussian density. This Kalman filter reduces the problem to a vector SDE for the mean and a matrix SDE for the covariance matrix of the normal distribution. However, in general the optimal filter is not finite dimensional. We should point out that, in the general case, the preferred SPDE for the optimal filter is a SPDE for an unnormalized version q of the optimal filter density p. The Zakai equation for a specific unnormalized density $q_t(x)$ of the optimal filter reads, in Stratonovich form

$$dq_t = \mathcal{L}_t^* q_t \, dt - \frac{1}{2} q_t \, |b_t|^2 \, dt + \sum_{k=1}^{d} q_t \left[b_t^k \right] \circ dY_t^k \, , \quad q_0 = p_0,$$

see for example Eq. 14.31 in [1]. This is a linear Stochastic PDE and as such it is more tractable than the KS Equation. The reason why we still resort to KS will be clarified when we introduce the projection filter below. A general advantgage of the Zakai version is the possibility to derive a robust non-stochastic PDE for the optimal filter, see for example [10].

2   The Projection Method: From PDEs to ODEs

As we summarized previously in [3], the projection method can be understood abstractly as a technique to approximate the solution of a differential equation on a Riemannian manifold M. Given a vector field $\mathcal{X}$ defined on M, we wish to find the trajectory of a particle p as it flows along $\mathcal{X}$. We attempt to approximate this trajectory by choosing a submanifold Σ of M and using the Riemannian metric on M to project $\mathcal{X}$ applied to the current approximation p' in Σ onto the tangent space of Σ at p'. This gives rise to a tangent vector $\mathcal{X}'$ on Σ that is closest to the original L^2 tangent vector $\mathcal{X}$ in p'. One hopes that the trajectories of $\mathcal{X}'$ will be a good approximation for the trajectories of $\mathcal{X}$.

The approach becomes interesting for the filtering problem when one considers an infinite dimensional Hilbert manifold M where the exact stochastic PDE solution for the optimal filter evolves (in Stratonovich calculus), and a finite dimensional Σ where the projected stochastic ODE for the approximate filter will evolve. This idea was first sketched by Hanzon in [11] and fully developed in [2,8,9]. This is not only interesting for filtering. Indeed many standard approaches to the numerical solution of PDE's can be re-interpreted geometrically this way. Thus we will attempt to numerically solve the filtering problem by mapping the space of probability distributions into a Hilbert manifold and then projecting onto a finite dimensional submanifold. In fact the Hilbert manifolds we use will simply be Hilbert spaces.

3   Choice of Hilbert Space Structure

There are two obvious ways of embedding the state of our system as belonging in a Hilbert space. One can consider $\sqrt{p}$ which lies inside $L^2(\mathbb{R})$ or one can assume that p is itself square integrable and so lies inside $L^2(\mathbb{R})$. These two approaches give two different metrics on the space of probability distributions. The former yields the Hellinger metric, the latter we will call the direct L^2 metric.

Since the first approach requires no further assumptions on the integrability of p than p being integrable to one, the Hellinger metric immediately seems more attractive from a theoretical standpoint. Moreover, its definition can be extended to probability measures via their densities and it is invariant with respect to the base measure used to express densities of the two measures.

The direct L^2 metric is only defined on square integrable distributions and is not invariant under reparameterizations. However, it has one distinct advantage over the Hellinger metric: it is defined in terms of p rather than $\sqrt{p}$. Since the metric is bilinear in p, using the L^2 metric gives more convenient formulae for particular manifolds like mixture distributions than does the Hellinger metric, as we shall observe explicitly later. In [3] we observed that the direct metric also offers numerical advantages for mixture manifolds.

We should finally point out that the space of probability distributions is not a submanifold of $L^2(\mathbb{R})$. Fortunately we can view the stochastic PDE we wish to solve as an equation on the whole of $L^2(\mathbb{R})$ and so avoid the thorny question of defining a manifold structure on the space of probability measures, which is solved in [9] by introducing an enveloping manifold for the exponential case. A discussion on whether (the above Zakai version of) the optimal filter equation can be seen as a functional equation in L^2 is in the monograph [1]. More generally, the study of the infinite dimensional geometry for spaces of probability distributions is a broad field that has received increased attention over the last two decades, and we refer for example to [12,14].

4   Exponential and Mixture Submanifolds

Earlier research in [5,6,8,9] illustrated in detail how the Hellinger distance and the metric it induces on a finite dimensional exponential family, namely the Fisher metric, are ideal tools when using the projection onto exponential families of densities. The above references illustrate this by applying the above framework to the infinite dimensional stochastic PDE describing the optimal solution of the nonlinear filtering problem. The use of exponential families allows the correction step in the filtering algorithm to become exact, so that only the prediction step is approximated. Furthermore, and independently from the filtering application, exponential families and the Fisher metric are known to interact well thanks to a number of properties we will explain shortly.

We give now a summary of why the Fisher metric/Hellinger distance works well with exponential families and a summary of the Fisher-metric based projection filter. Section 4.2 will deal with the direct metric and mixtures families.

4.1   Exponential Families

We use the following equivalent notations for multiple partial differentiation:

$$\frac{\partial^k}{\partial \theta_{i_1} \cdots \partial \theta_{i_k}} = \partial^k_{i_1, \cdots, i_k}.$$

Let $\{c_1, \cdots, c_m\}$ be scalar functions $c_i : \mathbb{R}^n \to \mathbb{R}$, $i = 1, 2, \cdots, m$ such that $\{1, c_1, \cdots, c_m\}$ are *linearly independent*, and assume that the convex set

$$\Theta_0 := \{\theta \in \mathbb{R}^m : \psi(\theta) = \log \int \exp[\theta^T c(x)]\, dx < \infty\},$$

has *non–empty interior*. Then

$$p(x, \theta) := \exp[\theta^T c(x) - \psi(\theta)],$$

where $\Theta \subseteq \Theta_0$ is open, is called an exponential family of probability densities. We define $E_\theta[\varphi] := \int \varphi(x) p(x, \theta) dx$. An important role in exponential families is played by differentiation of ψ. In fact for an exponential family ψ is infinitely differentiable in Θ and

$$E_\theta\{c_i\} = \partial_i \psi(\theta) =: \eta^i(\theta), \quad E_\theta\{c_i c_j\} = \partial^2_{ij}\psi(\theta) + \partial_i \psi(\theta)\, \partial_j \psi(\theta),$$

and more generally

$$E_\theta\{c_{i_1} \cdots c_{i_k}\} = \exp[-\psi(\theta)] \frac{\partial^k \exp[\psi(\theta)]}{\partial \theta_{i_1} \cdots \partial \theta_{i_k}}.$$

The Fisher information matrix satisfies

$$g_{ij}(\theta) = \partial^2_{ij}\psi(\theta) = \partial_i \eta^j(\theta) .$$

The quantities

$$(\eta^1, \cdots, \eta^m) \in \mathcal{E} = \eta(\Theta) \subset \mathbb{R}^m$$

form a coordinate system for the given exponential family. The two coordinate systems θ (canonical parameters) and η (expectation parameters) are related by diffeomorphism, and according to the above results the Jacobian matrix of the transformation $\eta = \eta(\theta)$ is the Fisher information matrix. The canonical parameters and the expectation parameters are *biorthogonal* w.r.t. the Fisher information metric.

We can now look at the particular shape taken by the Fisher metric projection for exponential families. We obtain

$$\Pi_\theta[v] = \sum_{i=1}^{m} [\sum_{j=1}^{m} g^{ij}(\theta) \, (E_\theta[vc_j] - E_\theta[v]E_\theta[c_j])] \, (c_i(\cdot) - E_\theta[c_i])p(\cdot, \theta). \quad (3)$$

The Fisher metric projection amounts to take covariance expectations of the function to be projected with the family exponents. The Fisher metric works well with exponential families essentially because in case of exponential families the square root amounts simply to add a $1/2$ factor into the exponent of the family of density, and then differentiation of exponential functions is easy and regular.

4.2 Mixture Families

Besides exponential families, there is another general framework that is powerful in modeling probability densities, and this is the mixture family. Mixture distributions are ubiquitous in statistics and may account for important stylized features such as skewness, multi-modality and fat tails.

We define a mixture family as follows. Suppose we are given $m+1$ fixed squared integrable probability densities, say $q = [q_1, q_2, \ldots, q_{m+1}]^T$. Suppose we define the following space of probability densities:

$$p(x, \theta) = \theta_1 q_1(x) + \theta_2 q_2(x) + \cdots + \theta_m q_m(x) + (1 - \theta_1 - \cdots - \theta_m)q_{m+1}(x), \quad (4)$$

$$\theta \in \Theta, \quad \Theta = \{\theta : \theta_i \geq 0 \text{ for all } i, \quad \theta_1 + \cdots + \theta_m < 1\}.$$

For convenience, define $\hat{\theta}(\theta) := [\theta_1, \theta_2, \ldots, \theta_m, 1 - \theta_1 - \theta_2 - \cdots - \theta_m]^T$. With this definition, $p(x, \theta) = \hat{\theta}(\theta)^T q(x)$. If we consider the direct L^2 distance, the metric $h(\theta)$ that is induced on $p(x, \theta)$ and the related projection become very simple. Indeed, one can immediately check from the definition of h that for the mixture family we have tangent vectors and metric

$$\frac{\partial p(\cdot, \theta)}{\partial \theta_i} = q_i - q_{m+1}, \quad h_{ij}(\theta) = \int (q_i(x) - q_m(x))(q_j(x) - q_m(x))dx =: h_{ij}$$

i.e., the L^2 direct metric (and matrix) does not depend on the specific point θ of the manifold. The same holds for the tangent space as we just saw. The L^2 projection is thus particularly simple:

$$\Pi_\theta[v] = \sum_{i=1}^{m} [\sum_{j=1}^{m} h^{ij} \langle v, q_j - q_{m+1} \rangle] (q_i - q_{m+1}).$$ (5)

We conclude by observing that, from the above calculations, the manifold for the direct metric that simplifies our projection equations drastically is the mixture choice. We analyzed numerical studies of the projection filter for the quadratic sensor with this direct metric - mixture setup and for the Hellinger-exponential setting in [3] under the special case of a scalar system for X, Y and with "q_i normal with mean μ_i and variance σ_i^2" and with "$c_i(x) = x^i$".

More generally, the motivation for considering these particular submanifolds is that, even in low dimensions, they allow us to reproduce many of the qualitative phenomena seen in the filtering problem. In particular we can produce highly skewed distributions and multi modal distributions. Many other possible choices of submanifold are worth consideration and are being investigated. In this respect, it is worth mentioning that in general there is no strict algorithmic method to select the manifold for a specific filtering problem, and this turns out to be a case by case matter. For example, in the quadratic sensor case one may expect a bimodal conditional density for the optimal filter, so one knows one will probably need about five parameters (two means, two standard deviations and a mixing parameter). As a general recipe, one can try a specific projection filter with a small manifold based on qualitative or heuristic considerations, as in the above-mentioned quadratic sensor case. Once this is done, one measures the L^2 norm of the projection residuals, and if this is large one may increase the manifold dimension until a sufficiently small projection residual norm is attained.

5   The Projected Equation

We now derive the direct L^2 projection filter for a general manifold M. Let M be an m dimensional submanifold of L^2 parameterized by $\theta = (\theta_1, \theta_2, \ldots, \theta_m)$. Define $v_i = \frac{\partial p}{\partial \theta^i}$ so that $\{v_1, v_2, \ldots v_m\}$ gives a basis for the tangent space of M at a point θ.

The direct L^2 metric induces a Riemannian metric $h_{ij}(\theta)$ on M. By projecting both sides of the Stratonovich equation for the evolution of p_t given above, we can obtain a stochastic differential for the evolution of the parameter θ.

To simplify the result, we introduce the following notation:

$$\gamma_t^0(p) := \frac{1}{2} [|b_t|^2 - E_p\{|b_t|^2\}] \, p, \quad \gamma_t^k(p) := [b_t^k - E_p\{b_t^k\}] p,$$ (6)

for $k = 1, \cdots, d$. Using the chain rule

$$dp(\theta_t) = \sum_{i=1}^{m} \frac{\partial p(\theta_t)}{\partial \theta^i} \circ d\theta_t^i$$ (7)

one can then show [2] that the projected equation for θ is equivalent to the stochastic differential equation:

$$d\theta^i = \sum_{j=1}^m h^{ij} \left\{ \langle p(\theta), \mathcal{L}v_j \rangle dt - \langle \gamma^0(p(\theta)), v_j \rangle dt + \sum_{k=1}^d \langle \gamma^k(p(\theta)), v_j \rangle \circ dY^k \right\}. \quad (8)$$

Here $\langle \cdot, \cdot \rangle$ denotes the direct L^2 inner product. If preferred, one could instead project the Kushner–Stratonovich equation using the Hellinger metric instead. This yields the following stochastic differential equation [9]:

$$d\theta^i = \sum_{j=1}^m g^{ij} \left(\langle \frac{\mathcal{L}^* p(\theta)}{p(\theta)}, v_j \rangle dt - \langle \frac{1}{2}|b|^2, v_j \rangle dt + \sum_{k=1}^d \langle b^k, v_j \rangle \circ dY^k \right) \quad (9)$$

It is now possible to explain why we resorted to the Kushner-Stratonovich (KS) Equation rather than the unnormalized but linear Zakai equation in deriving the projection filter. Consider the nonlinear terms in the KS Eq. (2), namely

$$\frac{1}{2} p_t E_{p_t}\{|b_t|^2\} dt, \quad \sum_{k=1}^d p_t \left[-E_{p_t}\{b_t^k\} \right] \circ dY_t^k.$$

Consider first the Hellinger projection filter (9). By inspection, we see that there is no impact of the nonlinear terms in the projected equation. Therefore, projecting the Zakai equation would result in the same Hellinger projection filter.

Proposition 1. *The Hellinger projection takes care of dimensionality reduction and adds normalization as a bonus, without further approximation. Hellinger projection of either KS or the Zakai Eq. leads to the same projection filter given by Eq. (9).*

This equivalence between KS and Zakai projection, however, is broken when we project according to the L^2 direct metric, obtaining the projection filter (8). For this filter we do have an impact of the nonlinear terms. In fact, it is easy to adapt the derivation of the L^2 direct filter to the Zakai equation, which leads to the filter

$$d\theta^i = \sum_{j=1}^m h^{ij} \left\{ \langle p(\theta), \mathcal{L}v_j \rangle dt - \langle \frac{1}{2}|b_t|^2 p(\theta), v_j \rangle dt + \sum_{k=1}^d \langle b_t^k p(\theta), v_j \rangle \circ dY^k \right\}$$
$$(10)$$

which is clearly different from (8).

Proposition 2. *For the L^2 direct metric projection filter, the dimensionality reduction approximation coming with the projection does not take care of normalization and we obtain two different projection filters depending on whether we project the normalized KS Equation or the unnormalized Zakai Equation, leading to Eqs. (8) and (10) respectively.*

Since we aim at studying mostly the pure dimensionality reduction approximation, we use KS rather than Zakai, meaning that for the L^2 direct metric projection filter we will consider Eq. (8) rather than (10). A numerical comparison of the two projection filters for the cubic sensor is under investigation.

6 Equivalence with ADF and Galerkin Filters

The projection filter with specific metrics and manifolds can be shown to be equivalent to earlier filtering algorithms. We summarize the equivalence results here, starting from

ADF = ProjectionFilter(Hellinger, Exponential) (full details in [9]).

By computing the c-moments of the optimal filter $\hat{\eta}_i(t) = E[c_i(X_t)|\mathcal{Y}_t] = \int c_i(x)p_t(x)dx$ with p the optimal filter (2), one can write an equation for the $d\hat{\eta}_i(t)$ vector driven by dY. This will not be a closed vector differential equation, in that the right hand side will depend on the whole filter p_t and not just on its moments $\hat{\eta}_t$. However, if we *replace* the optimal filter p_t in the right hand side of this equation for $d\hat{\eta}(t)$ with the exponential density in the family with exponent c characterized by the expectation parameters $\hat{\eta}$, then we can close the differential equation and obtain a finite dimensional filter. This will not be the optimal filter but just an approximation, as the replacement is based on an arbitrary assumption. This approximation is called exponential assumed density filter (E-ADF). The resulting equation is

$$d\eta_t^i = E_{\eta_t}\{\mathcal{L}_t\,c_i\}\,dt - \tfrac{1}{2}\,[\,E_{\eta_t}\{|b_t|^2\,c_i\} - E_{\eta_t}\{|b_t|^2\}\,\eta_t^i\,]\,dt$$

$$+ \sum_{k=1}^{d}[\,E_{\eta_t}\{b_t^k\,c_i\} - E_{\eta_t}\{b_t^k\}\,\eta_t^i\,]\circ dY_t^k, \qquad i = 1,\cdots,m. \tag{11}$$

Recall from our earlier section on exponential families that η's are an alternative coordinate system to θ in the exponential manifold, so that the above equation for η can be seen as evolving in the exponential manifold. In fact, we can say more. In [9] we proved the following

Theorem 1. *The E-ADF (11) and the projection filter (9) on the same exponential family coincide. Forcing an exponential density on the right-hand-side of the exponent-moments equation results in the same filter as projecting the optimal filter onto the exponential family in Hellinger distance.*

This result is important because it shows that a heuristic approximation like the E-ADF can be justified in rigorous geometric terms by resorting to the Hellinger distance.

We now move to our second equivalence result:

Galerkin Filter = ProjectionFilter(Direct, Mixture) ([2] for details).

Our second equivalence result is that the projection filter in direct metric for simple mixture families is equivalent to an approximated filter derived via a Galerkin method, as first noticed in the preprint [7].

The basic Galerkin approximation is obtained by approximating the exact solution of the filtering SPDE (8) with a linear combination of basis functions $\phi_i(x)$, namely

$$\tilde{p}_t(x) := \sum_{i=1}^{\ell} \alpha_i(t)\phi_i(x). \tag{12}$$

Ideally, the ϕ_i can be extended to indices $\ell+1, \ell+2, \ldots, +\infty$ so as form a basis of L^2. The method can be sketched intuitively as follows. We could write the optimal filtering Eq. (2) as

$$\langle -dp_t + \mathcal{L}_t^* \, p_t \, dt - \gamma_t^0(p_t) \, dt + \sum_{k=1}^d \gamma_t^k(p_t) \circ dY_t^k \,, \xi \rangle = 0$$

for all smooth L^2 test functions ξ such that the inner product exists.

We replace this equation with the equation where p_t is replaced by $\tilde{p}_t$ in (12) and ξ is given by $\phi_1, \ldots, \phi_\ell$. Using the linearity of the inner product in each argument and integration by parts we obtain easily a stochastic ODE for the combinators $\alpha(t)$. We call this equation the Galerkin filter for ϕ.

Consider now the projection filter with the manifold (4) and the direct metric. The projection filter Eq. (8) specializes to an equation that can be shown, by inspection, to be identical to the equation for the $d\alpha(t)$ coming from the Galerkin method if one sets

$\ell = m+1$, $\alpha_i = \theta_i$ and $\phi_i = q_i - q_{m+1}$ for $i = 1, \ldots, m$, and $\alpha_{m+1} = 1$, $\phi_{m+1} = q_{m+1}$.

The choice of the simple mixture is related to a choice of the L^2 basis in the Galerkin method. A typical choice could be based on Gaussian radial basis functions, see for example [13]. We have thus sketched the proof of the following

Theorem 2. *For simple mixture families (4), the direct-metric projection filter (8) coincides with a Galerkin method where the basis functions are the mixture components q.*

However, this equivalence holds only for the simple mixture family (4). More complex mixture families, such as the one we used to analyze the quadratic sensor in [3], will not allow for a Galerkin-based filter and only the L^2 projection filter can be defined there. Note also that even in the simple case (4) our L^2 Galerkin/projection filter will be different from the Galerkin projection filter seen for example in [4], because we use Stratonovich calculus to project the Kushner-Stratonovich equation in L^2 metric.

References

1. Ahmed, N.U.: Linear and Nonlinear Filtering for Scientists and Engineers. World Scientific, Singapore (1998)
2. Armstrong, J, Brigo, D.: Stochastic filtering via L^2 projection on mixture manifolds with computer algorithms and numerical examples. Submitted, preprint available at arXiv.org (2013)
3. Armstrong, J., Brigo, D.: Stochastic filtering by projection: the example of the quadratic sensor. In: Nielsen, F., Barbaresco, F. (eds.) GSI 2013. LNCS, vol. 8085, pp. 685–692. Springer, Heidelberg (2013)
4. Beard, R., Gunther, J.: Galerkin Approximations of the Kushner Equation in Nonlinear Estimation. Working Paper, Brigham Young University (1997)

5. Brigo, D.: Diffusion processes, manifolds of exponential densities, and nonlinear filtering, In: Barndorff-Nielsen, O.E., Vedel Jensen, E.B. (ed.) Geometry in Present Day Science, World Scientific (1999)

6. Brigo, D.: On SDEs with marginal laws evolving in finite-dimensional exponential families, STAT PROBABIL LETT, vol. 49, pp. 127–134 (2000)

7. Brigo, D.: The direct L2 geometric structure on a manifold of probability densities with applications to Filtering. Available at arXiv.org (2012)

8. Brigo, D., Hanzon, B., LeGland, F.: A differential geometric approach to nonlinear filtering: The projection filter. IEEE Trans. Autom. Control **43**, 247–252 (1998)

9. Brigo, D., Hanzon, B., Le Gland, F.: Approximate nonlinear filtering by projection on exponential manifolds of densities. BERNOULLI **5**, 495–534 (1999)

10. Clark, J.M.C.: The design of robust approximations to the stochastic differential equations of nonlinear filtering. In: Skwirzynski, J.K. (ed.) Communication Systems and Random Process Theory, NATO Advanced Study Institute Series (Sijthoff and Noordhoff, Alphen aan den Rijn) (1978)

11. Hanzon, B.: A differential-geometric approach to approximate nonlinear filtering. In: Dodson, C.T.J. (ed.) Geometrization of Statistical Theory, pp. 219–223. ULMD Publications, University of Lancaster (1987)

12. Khesin, B., Lennels, J., Misiolek, G., Preston, S.C.: Geometry of Diffeomorphism Groups, Complete integrability and Geometric statistics. Geom. Funct. Anal. **23**, 334–366 (2013)

13. Kormann, K., Larsson, E.: A galerkin radial basis function method for the schroedinger equation. SIAM J. Sci. Comput. **35**(6), A2832–A2855 (2014)

14. Pistone, G., Sempi, C.: An infinite-dimensional geometric structure on the space of all probability measures equivalent to a given one. Ann. Statist. **23**, 1543–1561 (1995)

Variational Bayesian Approximation Method for Classification and Clustering with a Mixture of Student-t Model

Ali Mohammad-Djafari[✉]

Laboratoire des signaux et systèmes (L2S),
CNRS-CentraleSupélec-UNIV PARIS SUD, Gif-sur-yvette, France
djafari@lss.supelec.fr
http://djafari.free.fr

Abstract. Clustering, classification and Pattern Recognition in a set of data are between the most important tasks in statistical researches and in many applications. In this paper, we propose to use a mixture of Student-t distribution model for the data via a hierarchical graphical model and the Bayesian framework to do these tasks. The main advantages of this model is that the model accounts for the uncertainties of variances and covariances and we can use the Variational Bayesian Approximation (VBA) methods to obtain fast algorithms to be able to handle large data sets.

1 Introduction

Clustering and classification of a set of data are not trivial problems. In fact, we can consider them as ill-posed inverse problems in which the solutions are not unique. Mixture models are natural ones for classification and clustering [1–8]. The Mixture of Gaussians (MoG) models have been used very extensively [9–11]. In this paper, we propose to use a mixture of Student-t model and a Bayesian framework for these tasks. The main advantages of this model is that the model accounts for the uncertainties of variances and covariances and we can use the Variational Bayesian Approximation (VBA) methods to obtain fast algorithms as well. Even if this model may have been used before [12–26], here we propose a novel unifying presentation for all the steps: training, supervised or semi-supervised classification and clustering (non-supervised). We also use VBA framework and some simplifications to develop fast algorithms to be able to handle big data sets.

2 Mixture Models for Classification and Clustering

A mixture model is generally given as:

$$p(\boldsymbol{x}|\boldsymbol{a}, \boldsymbol{\Theta}, K) = \sum_{k=1}^{K} a_k \, p_k(\boldsymbol{x}_k|\boldsymbol{\theta}_k), \tag{1}$$

© Springer International Publishing Switzerland 2015
F. Nielsen and F. Barbaresco (Eds.): GSI 2015, LNCS 9389, pp. 723–731, 2015.
DOI: 10.1007/978-3-319-25040-3_77

where K is the number of classes, $\boldsymbol{a} = \{a_k, k = 1, \cdots, K\}$ the proportion parameters and $\boldsymbol{\Theta} = \{\boldsymbol{\theta}_k, k = 1, \cdots, K\}$ all the other parameters of the model. If we assume different classes can be modeled by the same family $p_k(\boldsymbol{x}_k | \boldsymbol{\theta}_k) = p(\boldsymbol{x}_k | \boldsymbol{\theta}_k)$ and introduce a hidden class variable $c_n \in \{1, \cdots, K\}$, then for a given sample $\boldsymbol{x}_n$ in class k we can write:

$$p(\boldsymbol{x}_n | c_n = k, \boldsymbol{\theta}_k) = p(\boldsymbol{x}_n | \boldsymbol{\theta}_k) \tag{2}$$

or

$$p(\boldsymbol{x}_n, c_n = k | a_k, \boldsymbol{\theta}_k, K) = a_k \, p(\boldsymbol{x}_n | \boldsymbol{\theta}_k). \tag{3}$$

The Mixture of Gaussians (MoG) corresponds to the case where $p(\boldsymbol{x}_n | c_n = k, \boldsymbol{\theta}_k) = \mathcal{N}(\boldsymbol{x} | \boldsymbol{\mu}_k, \boldsymbol{\Sigma}_k)$ with $\boldsymbol{\theta}_k = (\boldsymbol{\mu}_k, \boldsymbol{\Sigma}_k)$.

Now, imagine a set of data $\boldsymbol{X} = \{\boldsymbol{x}_n, n = 1, \cdots, N\}$ where each element $\boldsymbol{x}_n$ can be in one of these classes. Then, we can write:

$$p(\boldsymbol{X}_n, c_n = k | \boldsymbol{a}, \boldsymbol{\theta}) = \prod_{n=1}^{N} p(\boldsymbol{x}_n, c_n = k | \boldsymbol{a}, \boldsymbol{\theta}). \tag{4}$$

Noting by $\boldsymbol{c} = \{c_n, n = 1, N\}$ with $c_n \in \{1, \cdots, K\}$, $\boldsymbol{a} = \{a_k, k = 1, \cdots, K\}$ and $\boldsymbol{\Theta} = \{\boldsymbol{\theta}_k, k = 1, \cdots, K\}$, we have:

$$\begin{aligned} p(\boldsymbol{X}_n, \boldsymbol{c} | \boldsymbol{a}, \boldsymbol{\Theta}, K) &= \prod_{n=1}^{N} \prod_{k=1}^{K} p(c_n = k) \, p(\boldsymbol{x}_n | \boldsymbol{\theta}_k) \\ &= \prod_{n=1}^{N} \prod_{k=1}^{K} a_k \, p(\boldsymbol{x}_n | \boldsymbol{\theta}_k). \end{aligned} \tag{5}$$

The classification problems can then be summarized as follows:

Training: Given a set of (training) data $\boldsymbol{X}$ and classes $\boldsymbol{c}$, estimate the parameters $\boldsymbol{a}$ and $\boldsymbol{\Theta}$. The classical frequentist method is the Maximum Likelihood (ML) which defines the solution as

$$(\widehat{\boldsymbol{a}}, \widehat{\boldsymbol{\Theta}}) = \arg \max_{(\boldsymbol{a}, \boldsymbol{\Theta})} \{ p(\boldsymbol{X}, \boldsymbol{c} | \boldsymbol{a}, \boldsymbol{\Theta}, K) \}. \tag{6}$$

The Bayesian way is to assign priors $p(\boldsymbol{a} | K)$ and $p(\boldsymbol{\Theta} | K) = \prod_{k=1}^{K} p(\boldsymbol{\theta}_k)$, then the joint posterior laws is given by:

$$p(\boldsymbol{a}, \boldsymbol{\Theta} | \boldsymbol{X}, \boldsymbol{c}, K) = \frac{p(\boldsymbol{X}, \boldsymbol{c} | \boldsymbol{a}, \boldsymbol{\Theta}, K) \, p(\boldsymbol{a} | K) \, p(\boldsymbol{\Theta} | K)}{p(\boldsymbol{X}, \boldsymbol{c} | K)} \tag{7}$$

where

$$p(\boldsymbol{X}, \boldsymbol{c} | K) = \iint p(\boldsymbol{X}, \boldsymbol{c} | \boldsymbol{a}, \boldsymbol{\Theta} | K) p(\boldsymbol{a} | K) \, p(\boldsymbol{\Theta} | K) \, \mathrm{d}\boldsymbol{a} \, \mathrm{d}\boldsymbol{\Theta} \tag{8}$$

from which we can deduce $\widehat{\boldsymbol{a}}$ and $\{\widehat{\boldsymbol{\theta}}_k, k = 1, \cdots, K\}$ either as the Maximum A Posteriori (MAP) or Posterior Mean (PM).

Supervised Classification: For a given sample $\boldsymbol{x}_m$ and given the parameters K, $\boldsymbol{a}$ and $\boldsymbol{\Theta}$ determine

$$p(c_m = k | \boldsymbol{x}_m, \boldsymbol{a}, \boldsymbol{\Theta}, K) = \frac{p(\boldsymbol{x}_m, c_m = k | \boldsymbol{a}, \boldsymbol{\Theta}, K)}{p(\boldsymbol{x}_m | \boldsymbol{a}, \boldsymbol{\Theta}, K)} \tag{9}$$

where $p(\boldsymbol{x}_m, c_m = k | \boldsymbol{a}, \boldsymbol{\Theta}, K) = a_k p(\boldsymbol{x}_m | \boldsymbol{\theta}_k)$ and

$$p(\boldsymbol{x}_m | \boldsymbol{a}, \boldsymbol{\Theta}, K) = \sum_{k=1}^{K} a_k \, p(\boldsymbol{x}_m | \boldsymbol{\theta}_k) \tag{10}$$

and its best class k^*, for example the MAP solution:

$$k^* = \arg\max_{k} \{ p(c_m = k | \boldsymbol{x}_m, \boldsymbol{a}, \boldsymbol{\Theta}, K) \}. \tag{11}$$

Semi-supervised Classification: For a given sample $\boldsymbol{x}_m$ and given the parameters K and $\boldsymbol{\Theta}$, determine the probabilities

$$p(c_m = k | \boldsymbol{x}_m, \boldsymbol{\Theta}, K) = \frac{p(\boldsymbol{x}_m, c_m = k | \boldsymbol{\Theta}, K)}{p(\boldsymbol{x}_m | \boldsymbol{\Theta}, K)} \tag{12}$$

where

$$p(\boldsymbol{x}_m, c_m = k | \boldsymbol{\Theta}, K) = \int p(\boldsymbol{x}_m, c_m = k | \boldsymbol{a}, \boldsymbol{\Theta}, K) p(\boldsymbol{a} | K) \, \mathrm{d}\boldsymbol{a} \tag{13}$$

and

$$p(\boldsymbol{x}_m | \boldsymbol{\Theta}, K) = \sum_{k=1}^{K} p(\boldsymbol{x}_m, c_m = k | \boldsymbol{\Theta}, K) \tag{14}$$

and its best class k^*, for example the MAP solution:

$$k^* = \arg\max_{k} \{ p(c_m = k | \boldsymbol{x}_m, \boldsymbol{\Theta}, K) \}. \tag{15}$$

Clustering or Non-supervised Classification: Given a set of data $\boldsymbol{X}$, determine K and $\boldsymbol{c}$. When these are determined, we can also determine the characteristics of those classes $\boldsymbol{a}$ and $\boldsymbol{\Theta}$. To do this we need the following relations:

$$p(K = L | \boldsymbol{X}) = \frac{p(\boldsymbol{X}, K = L)}{p(\boldsymbol{X})} = \frac{p(\boldsymbol{X} | K = L) \, p(K = L)}{p(\boldsymbol{X})} \tag{16}$$

and

$$p(\boldsymbol{X}) = \sum_{L=1}^{L_0} p(K = L) \, p(\boldsymbol{X} | K = L), \tag{17}$$

where L_0 is the a priori maximum number of classes and

$$p(\boldsymbol{X} | K = L) = \int \int \prod_n \prod_{k=1}^{L} a_k p(\boldsymbol{x}_n, c_n = k | \boldsymbol{\theta}_k) p(\boldsymbol{a} | K) \, p(\boldsymbol{\Theta} | K) \, \mathrm{d}\boldsymbol{a} \, \mathrm{d}\boldsymbol{\Theta}. \tag{18}$$

As we will see later the main difficulty is the computation of these two last equations. The Variational Bayesian Approximation technics try to find upper and lower bounds for them.

3   Mixture of Student-t Model

Let us consider the following representation of the Student-t probability density function (pdf):

$$T(\boldsymbol{x}|\nu, \boldsymbol{\mu}, \boldsymbol{\Sigma}) = \int_0^\infty \mathcal{N}(\boldsymbol{x}|\boldsymbol{\mu}, z^{-1}\boldsymbol{\Sigma})\, \mathcal{G}(z|\frac{\nu}{2}, \frac{\nu}{2})\, \mathrm{d}z, \tag{19}$$

where

$$\begin{aligned}
\mathcal{N}(\boldsymbol{x}|\boldsymbol{\mu}, \boldsymbol{\Sigma}) &= |2\pi\boldsymbol{\Sigma}|^{-\frac{1}{2}} \exp\left[-\tfrac{1}{2}(\boldsymbol{x}-\boldsymbol{\mu})'\boldsymbol{\Sigma}^{-1}(\boldsymbol{x}-\boldsymbol{\mu})\right] \\
&= |2\pi\boldsymbol{\Sigma}|^{-\frac{1}{2}} \exp\left[-\tfrac{1}{2}\mathrm{Tr}\left\{(\boldsymbol{x}-\boldsymbol{\mu})\boldsymbol{\Sigma}^{-1}(\boldsymbol{x}-\boldsymbol{\mu})'\right\}\right],
\end{aligned} \tag{20}$$

and

$$\mathcal{G}(z|\alpha, \beta) = \frac{\beta^\alpha}{\Gamma(\alpha)} z^{\alpha-1} \exp\left[-\beta z\right]. \tag{21}$$

Let us also consider the finite mixture of Student-t model:

$$p(\boldsymbol{x}|\{\nu_k, a_k, \boldsymbol{\mu}_k, \boldsymbol{\Sigma}_k, k=1, \cdots, K\}, K) = \sum_{k=1}^{K} a_k\, T(\boldsymbol{x}_n|\nu_k, \boldsymbol{\mu}_k, \boldsymbol{\Sigma}_k). \tag{22}$$

Introducing the hidden variables z_{nk} this model can be written via:

$$p(\boldsymbol{x}_n, c_n = k, z_{nk}|\boldsymbol{\mu}_k, \boldsymbol{\Sigma}_k, K) = a_k\, \mathcal{N}(\boldsymbol{x}_n|\boldsymbol{\mu}_k, z_{n,k}^{-1}\boldsymbol{\Sigma}_k)\, \mathcal{G}(z_{n,k}|\frac{\nu_k}{2}, \frac{\nu_k}{2}). \tag{23}$$

Noting by: $\boldsymbol{Z} = \{z_{nk}\}$, $\boldsymbol{z}_k = \{z_{nk}, n = 1, \cdots, N\}$, $\boldsymbol{c} = \{c_n, n = 1, \cdots, N\}$, $\boldsymbol{\theta}_k = \{\nu_k, a_k, \boldsymbol{\mu}_k, \boldsymbol{\Sigma}_k\}$, $\boldsymbol{\Theta} = \{\boldsymbol{\theta}_k, k = 1, \cdots, K\}$ and assigning the priors $p(\boldsymbol{\Theta}) = \prod_k p(\boldsymbol{\theta}_k)$, we can write:

$$p(\boldsymbol{X}, \boldsymbol{c}, \boldsymbol{Z}, \boldsymbol{\Theta}|K) = \prod_n \prod_k a_k \mathcal{N}(\boldsymbol{x}_n|\boldsymbol{\mu}_k, z_{nk}^{-1}\boldsymbol{\Sigma}_k)\, \mathcal{G}(z_{nk}|\tfrac{\nu_k}{2}, \tfrac{\nu_k}{2})\, p(\boldsymbol{\theta}_k) \tag{24}$$

Then, the joint posterior law of all the unknowns $(\boldsymbol{c}, \boldsymbol{Z}, \boldsymbol{\Theta})$ given the data $\boldsymbol{X}$ and K can be written as

$$p(\boldsymbol{c}, \boldsymbol{Z}, \boldsymbol{\Theta}|\boldsymbol{X}, K) = \frac{p(\boldsymbol{X}, \boldsymbol{c}, \boldsymbol{Z}, \boldsymbol{\Theta}|K)}{p(\boldsymbol{X}|K)}. \tag{25}$$

The main task now is to propose some approximations to it in such a way that we can use it easily in all the above mentioned tasks of classification or clustering. The main idea behind the VBA technics is exactly this.

4   Variational Bayesian Approximation (VBA)

4.1   Main Idea

The main idea behind the VBA is to propose an approximation $q(\boldsymbol{c}, \boldsymbol{Z}, \boldsymbol{\Theta})$ for $p(\boldsymbol{c}, \boldsymbol{Z}, \boldsymbol{\Theta}|\boldsymbol{X}, K)$. This approximation can be in such a way that $\mathrm{KL}(q : p)$ be minimized. Interestingly, by noting that $p(\boldsymbol{c}, \boldsymbol{Z}, \boldsymbol{\Theta}|\boldsymbol{X}, K) = p(\boldsymbol{X}, \boldsymbol{c}, \boldsymbol{Z}, \boldsymbol{\Theta}|K)/p(\boldsymbol{X}|K)$, it is easy to showed that

$$KL(q:p) = -\mathcal{F}(q) + \ln p(\boldsymbol{X}|K) \qquad (26)$$

where

$$\mathcal{F}(q) = \langle -\ln p(\boldsymbol{X}, \boldsymbol{c}, \boldsymbol{Z}, \boldsymbol{\Theta}|K)\rangle_q \qquad (27)$$

is called free energy of q and we have the following properties:

- Maximizing $\mathcal{F}(q)$ or minimizing $KL(q:p)$ are equivalent and both give un upper bound to the evidence of the model $\ln p(\boldsymbol{X}|K)$.
- When the optimum q^* is obtained, $\mathcal{F}(q^*)$ can be used as a criterion for model selection.
- If p is in the exponential family, then choosing appropriate conjugate priors, the structure of q will be the same and we can obtain appropriate fast optimization algorithms.

In our case, noting that

$$p(\boldsymbol{X}, \boldsymbol{c}, \boldsymbol{Z}, \boldsymbol{\Theta}|K) = \prod_n \prod_k p(\boldsymbol{x}_n, c_n, z_{nk}|a_k, \boldsymbol{\mu}_k, \boldsymbol{\Sigma}_k, \nu_k) \\ \prod_k [p(\alpha_k)\, p(\beta_k)\, p(\boldsymbol{\mu}_k|\boldsymbol{\Sigma}_k)\, p(\boldsymbol{\Sigma}_k)] \qquad (28)$$

with

$$p(\boldsymbol{x}_n, c_n, z_{nk}|a_k, \boldsymbol{\mu}_k, \boldsymbol{\Sigma}_k, \nu_k) = \mathcal{N}(\boldsymbol{x}_n|\boldsymbol{\mu}_k, z_{n,k}^{-1}\boldsymbol{\Sigma}_k)\,\mathcal{G}(z_{nk}|\alpha_k, \beta_k) \qquad (29)$$

is separable, in one side for $[\boldsymbol{c}, \boldsymbol{Z}]$ and in other size in components of $\boldsymbol{\Theta}$, we propose to use

$$q(\boldsymbol{c}, \boldsymbol{Z}, \boldsymbol{\Theta}) = q(\boldsymbol{c}, \boldsymbol{Z})\, q(\boldsymbol{\Theta}). \qquad (30)$$

With this decomposition, the expression of the Kullback-Leibler divergence becomes:

$$KL(q_1(\boldsymbol{c}, \boldsymbol{Z})q_2(\boldsymbol{\Theta}) : p(\boldsymbol{c}, \boldsymbol{Z}, \boldsymbol{\Theta}|\boldsymbol{X}, K) \\ = \sum_c \int\int q_1(\boldsymbol{c}, \boldsymbol{Z})q_2(\boldsymbol{\Theta}) \ln \frac{q_1(\boldsymbol{c}, \boldsymbol{Z})q_2(\boldsymbol{\Theta})}{p(\boldsymbol{c}, \boldsymbol{Z}, \boldsymbol{\Theta}|\boldsymbol{X}, K)}\, d\boldsymbol{\Theta}\, d\boldsymbol{Z} \qquad (31)$$

and the expression of the Free energy becomes:

$$\mathcal{F}(q_1(\boldsymbol{c}, \boldsymbol{Z})q_2(\boldsymbol{\Theta})) = \sum_c \int\int q_1(\boldsymbol{c}, \boldsymbol{Z})q_2(\boldsymbol{\Theta}) \ln \frac{p(\boldsymbol{X}, \boldsymbol{c}, \boldsymbol{Z}|\boldsymbol{\Theta}, K)p(\boldsymbol{\Theta}|K)}{q_1(\boldsymbol{c}, \boldsymbol{Z})q_2(\boldsymbol{\Theta})}\, d\boldsymbol{\Theta}\, d\boldsymbol{Z}. \qquad (32)$$

In the following we propose appropriate priors and obtain the expressions of q and appropriate fast algorithms.

5   Proposed VBA for Mixture of Student-t Priors Model

As we discussed in previous section, here we consider the Mixture of Student-t priors model and propose appropriate conjugate priors and appropriate factorized form for the testing or approximation q and finally give the details of the parameters updating algorithm. To be able to propose conjugate priors for all the parameters, we change slightly the model by replacing ν_k in the Gamma expression $\mathcal{G}(z_{n,k}|\frac{\nu_k}{2}, \frac{\nu_k}{2})$ of the Student-t expression by two parameters $\mathcal{G}(z_{n,k}|\alpha_k, \beta_k)$:

$$p(\boldsymbol{x}_n, c_n = k, z_{nk}|a_k, \boldsymbol{\mu}_k, \boldsymbol{\Sigma}_k, \alpha_k, \beta_k, K) = a_k \, \mathcal{N}(\boldsymbol{x}_n|\boldsymbol{\mu}_k, z_{n,k}^{-1}\boldsymbol{\Sigma}_k) \, \mathcal{G}(z_{n,k}|\alpha_k, \beta_k). \tag{33}$$

The final hierarchical model that we propose is shown in the Fig. 1.

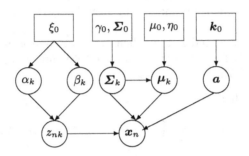

Fig. 1. Graphical representation of the model.

5.1   Conjugate Priors

In the following, noting by $\boldsymbol{\Theta} = \{(a_k, \boldsymbol{\mu}_k, \boldsymbol{\Sigma}_k, \alpha_k, \beta_k), k = 1, \cdots, K\}$, we propose to use the factorized prior laws:

$$p(\boldsymbol{\Theta}) = p(\boldsymbol{a}) \sum_k [p(\alpha_k) \, p(\beta_k) \, p(\boldsymbol{\mu}_k|\boldsymbol{\Sigma}_k) \, p(\boldsymbol{\Sigma}_k)] \tag{34}$$

with the following components:

$$\begin{cases} p(\boldsymbol{a}) = \mathcal{D}(\boldsymbol{a}|\boldsymbol{k}_0), \quad \boldsymbol{k}_0 = [k_0, \cdots, k_0] = k_0 \boldsymbol{1} \\ p(\alpha_k) = \mathcal{E}(\alpha_k|\zeta_0) = \mathcal{G}(\alpha_k|1, \zeta_0) \\ p(\beta_k) = \mathcal{E}(\beta_k|\zeta_0) = \mathcal{G}(\alpha_k|1, \zeta_0) \\ p(\boldsymbol{\mu}_k|\boldsymbol{\Sigma}_k) = \mathcal{N}(\boldsymbol{\mu}_k|\mu_0 \boldsymbol{1}, \eta_0^{-1}\boldsymbol{\Sigma}_k) \\ p(\boldsymbol{\Sigma}_k) = \mathcal{IW}(\boldsymbol{\Sigma}_k|\gamma_0, \gamma_0\boldsymbol{\Sigma}_0) \end{cases} \tag{35}$$

where

$$\mathcal{D}(\boldsymbol{a}|\boldsymbol{k}) = \frac{\Gamma(\sum_l k_k)}{\prod_l \Gamma(k_l)} \prod_l a_l^{k_l - 1} \tag{36}$$

is the Dirichlet pdf,

$$\mathcal{E}(t|\zeta_0) = \zeta_0 \exp[-\zeta_0 t] \tag{37}$$

is the Exponential pdf,

$$\mathcal{G}(t|a, b) = \frac{b^a}{\Gamma(a)} t^{a-1} \exp[-bt] \tag{38}$$

is the Gamma pdf and

$$\mathcal{IW}(\boldsymbol{\Sigma}|\gamma,\gamma\boldsymbol{\Delta}) = \frac{|\frac{1}{2}\boldsymbol{\Delta}|^{\gamma/2}\exp\left[-\frac{1}{2}\mathrm{Tr}\left\{\boldsymbol{\Delta}\boldsymbol{\Sigma}^{-1}\right\}\right]}{\Gamma_D(\gamma/2)|\boldsymbol{\Sigma}|^{\frac{\gamma+D+1}{2}}}. \tag{39}$$

is the inverse Wishart pdf.

With these prior laws and the likelihood: $p(\boldsymbol{x}_n|c(n),\boldsymbol{z}_k(n),\boldsymbol{\Theta},k)$ we can obtain the joint posterior law:

$$p_k(\boldsymbol{c},\boldsymbol{Z},\boldsymbol{\Theta}|\boldsymbol{X}) = \frac{p(\boldsymbol{X},\boldsymbol{c},\boldsymbol{Z},\boldsymbol{\Theta})}{p(\boldsymbol{X})}. \tag{40}$$

Now, we have to choose a factored form for q in such a way that we can transform the optimization of the $KL(q:p)$ or the free energy $\mathcal{F}(q)$ to the updating of the parameters of the different components of q. We propose to use the following decomposition:

$$\begin{aligned}
q(\boldsymbol{c},\boldsymbol{Z},\boldsymbol{\Theta}) &= q(\boldsymbol{c},\boldsymbol{Z})\,q(\boldsymbol{\Theta}) \\
&= \textstyle\prod_n \prod_k [q(c_n = k|z_{nk})\,q(z_{nk})] \\
&\quad \textstyle\prod_k [q(\alpha_k)\,q(\beta_k)\,q(\boldsymbol{\mu}_k|\boldsymbol{\Sigma}_k)\,q(\boldsymbol{\Sigma}_k)]\,q(\boldsymbol{a}).
\end{aligned} \tag{41}$$

with:

$$\begin{cases}
q(\boldsymbol{a}) = \mathcal{D}(\boldsymbol{a}|\widetilde{\boldsymbol{k}}), \quad \widetilde{\boldsymbol{k}} = [\widetilde{k}_1,\cdots,\widetilde{k}_K] \\
q(\alpha_k) = \mathcal{G}(\alpha_k|\widetilde{\zeta}_k,\widetilde{\eta}_k) \\
q(\beta_k) = \mathcal{G}(\beta_k|\widetilde{\zeta}_k,\widetilde{\eta}_k) \\
q(\boldsymbol{\mu}_k|\boldsymbol{\Sigma}_k) = \mathcal{N}(\boldsymbol{\mu}_k|\widetilde{\boldsymbol{\mu}},\widetilde{\eta}^{-1}\boldsymbol{\Sigma}_k) \\
q(\boldsymbol{\Sigma}_k) = \mathcal{IW}(\boldsymbol{\Sigma}_k|\widetilde{\gamma},\widetilde{\gamma}\widetilde{\boldsymbol{\Sigma}})
\end{cases} \tag{42}$$

With these choices, we have

$$\begin{aligned}
\mathcal{F}(q(\boldsymbol{c},\boldsymbol{Z},\boldsymbol{\Theta})) &= \langle \ln p(\boldsymbol{X},\boldsymbol{c},\boldsymbol{Z},\boldsymbol{\Theta}|K)\rangle_{q(\boldsymbol{c},\boldsymbol{Z},\boldsymbol{\Theta})} \\
&= \textstyle\prod_k \prod_n \mathcal{F}_{1_{kn}} + \prod_k \mathcal{F}_{2_k}
\end{aligned} \tag{43}$$

with

$$\begin{aligned}
\mathcal{F}_{1_{kn}} &= \langle \ln p(\boldsymbol{x}_n,c_n,z_{nk},\boldsymbol{\theta}_k)\rangle_{q(c_n=k|z_{nk})q(z_{nk})} \\
\mathcal{F}_{2_k} &= \langle \ln p(\boldsymbol{x}_n,c_n,z_{nk},\boldsymbol{\theta}_k)\rangle_{q(\boldsymbol{\theta}_k)}
\end{aligned} \tag{44}$$

Now, to obtain the expressions of the updating expressions of the tilded parameters, we need to go to the following three steps:

- E step: Optimizing $\mathcal{F}$ with respect to $q(\boldsymbol{c},\boldsymbol{Z})$ when keeping $q(\boldsymbol{\Theta})$ fixed, we obtain the expression of $q(c_n = k|z_{nk}) = \widetilde{a}_k$, $q(z_{nk}) = \mathcal{G}(z_{nk}|\widetilde{\alpha}_k,\widetilde{\beta}_k)$.
- M step: Optimizing $\mathcal{F}$ with respect to $q(\boldsymbol{\Theta})$ when keeping $q(\boldsymbol{c},\boldsymbol{Z})$ fixed, we obtain the expression of $q(\boldsymbol{a}) = \mathcal{D}(\boldsymbol{a}|\widetilde{\boldsymbol{k}})$, $\widetilde{\boldsymbol{k}} = [\widetilde{k}_1,\cdots,\widetilde{k}_K]$, $q(\alpha_k) = \mathcal{G}(\alpha_k|\widetilde{\zeta}_k,\widetilde{\eta}_k)$, $q(\beta_k) = \mathcal{G}(\beta_k|\widetilde{\zeta}_k,\widetilde{\eta}_k)$, $q(\boldsymbol{\mu}_k|\boldsymbol{\Sigma}_k) = \mathcal{N}(\boldsymbol{\mu}_k|\widetilde{\boldsymbol{\mu}},\widetilde{\eta}^{-1}\boldsymbol{\Sigma}_k)$, and $q(\boldsymbol{\Sigma}_k) = \mathcal{IW}(\boldsymbol{\Sigma}_k|\widetilde{\gamma},\widetilde{\gamma}\widetilde{\boldsymbol{\Sigma}})$, which gives the updating algorithm for the corresponding tilded parameters.

– $\mathcal{F}$ evaluation After each E step and M step, we can also evaluate the expression of $\mathcal{F}(q)$ which can be used for stopping rule of the iterative algorithm. Also, final value of this expression for each value of K, noted $\mathcal{F}_k$, can be used as a criterion for the model selection, i.e.; the determination of the number of clusters.

The expressions of all the tilded parameters update as well as the expression of $\mathcal{F}_K$ are easily obtained thanks to the properties of the conjugate priors. However, these expressions are cumbersome and will be given in the appendix.

6   Conclusion

Clustering and classification of a set of data are between the most important tasks in statistical researches for many applications such as data mining in biology. Mixture models and in particular Mixture of Gaussians are classical models for these tasks. In this paper, we proposed to use a mixture of Student-t distribution model for the data via a hierarchical graphical model. Then, we proposed a Bayesian framework to do these tasks. The main advantages of this model is that the model accounts for the uncertainties of variances and covariances and we can use the Variational Bayesian Approximation (VBA) methods. To obtain fast algorithms and be able to handle large data sets, we used conjugate priors everywhere it was possible. The proposed algorithm has been used for clustering, classification and discriminent analysis of some biological data, but in this paper, we only presented the main algorithm.

Acknowledgment. This work has been supported partially by the C5SYS (https://www.erasysbio.net/index.php?index=272) project of ERASYSBIO.

References

1. Forgy, E.W.: Cluster analysis of multivariate data: efficiency vs interpretability of classifications. Biometrics **21**, 768–769 (1965)
2. MacKay, D.J.C.: A practical Bayesian framework for backpropagation networks. Neural Comput. **4**, 448–472 (1992)
3. Redner, R., Walker, H.: Mixture densities, maximum likelihood and the em algorithm. SIAM Rev. **26**(2), 195–239 (1984)
4. Husmeier, D., Penny, W., Roberts, S.: Empirical evaluation of Bayesian sampling for neural classifiers. In: Niklason, M.L., Ziemke, T. (eds.): ICANN 98: Proceedings of the 8th International Conference on Artificial Neural Networks (1998)
5. Lee, T., Lewicki, M., Sejnowski, T.: Independent component analysis using an extended infomax algorithm for mixed sub-gaussian and super-gaussian sources. Neural Comput. **11**, 409–433 (1999)
6. Hyvärinen, A., Oja, E.: Independent component analysis: algorithms and applications. Neural Netw. **13**, 411–430 (2000)
7. Ma, J., Xu, L., Jordan, M.I.: Asymptotic convergence rate of the EM algorithm for Gaussian mixtures. Neural Comput. **12**, 2881–2907 (2001)

8. Nielsen, F., Nock, R.: Clustering multivariate normal distributions. In: Nielsen, F. (ed.) ETVC 2008. LNCS, vol. 5416, pp. 164–174. Springer, Heidelberg (2009)
9. Quandt, R., Ramsey, J.: Estimating mixtures of normal distributions and switching regressions. J. Am. Stat. Assoc. **73**, 2 (1978)
10. Hathaway, R.: A constrained formulation of maximum-likelihood estimation for normal mixture distributions. Ann. Stat. **13**(2), 795–800 (1985)
11. Hastie, T., Tibshirani, R.: Discriminant analysis by gaussian mixture. J. Roy. Stat. Soc. **58**, 155–176 (1996)
12. Dempster, A., Laird, N., Rubin, D.: Maximum likelihood from incomplete data via the EM algorithm. J. Roy. Stat. Soc. Ser. B **39**(1), 1–38 (1977)
13. Neal, R., Hinton, G.: A view of the EM algorithm that justifies incremental, sparse, and other variants. Learn. Graph. Models **89**, 355–368 (1998)
14. Tipping, M.E., Bishop, C.M.: Mixtures of probabilistic principal components analysis. Neural Comput. **11**, 443–482 (1999)
15. Jordan, M., Ghahramani, Z., Jaakkola, T., Saul, L.: An introduction to variational methods for graphical models. Mach. Learn. **37**, 183–233 (2006)
16. Jaakkola, T.S., Jordan, M.I.: Bayesian parameter estimation via variational methods. Stat. Comput. **10**, 25–37 (2000)
17. Friston, K., Penny, W.: Bayesian inference and posterior probability maps. In: Proceedings of the 9th International Conference on Neural Information Processing (ICONIP 2002), pp. 413–417 (2002)
18. Winn, J., Bishop, C.M., Jaakkola, T.: Variational message passing. J. Mach. Learn. Res. **6**, 661–694 (2005)
19. David, M.B., Michael, I.J.: Variational inference for the dirichlet process mixtures. Bayesian Anal. **1**(1), 121–144 (2006)
20. Beal, M.: Variational Algorithms for Approximate Bayesian Inference. Ph.D. thesis, Gatsby Computational Neuroscience Unit, University College London (2003)
21. Beal, M., Ghahramani, Z.: Variational Bayesian learning of directed graphical models with hidden variables. Bayesian Stat. **1**(4), 793–832 (2006)
22. Kim, H., Ghahramani, Z.: Bayesian gaussian process classification with the em-ep algorithm. IEEE Trans. Pattern Anal. Mach. Intell. **28**, 1948–1959 (2006)
23. Nasios, N., Bors, A.: Variational learning for gaussian mixture models. IEEE Trans. Syst. Man Cybern. Part B **36**, 849–862 (2006)
24. Ghahramani, Z., Griffiths, T., Sollich, P.: Bayesian nonparametric latent feature models. Bayesian Stat. **8**, 1–25 (2007)
25. McGrory, C., Titterington, D.: Variational approximations in Bayesian model selection for finite mixture distributions. Comput. Stat. Data Anal. **51**, 5352–5367 (2007)
26. Qiao, Z., Zhou, L., Huang, J.Z.: Sparse linear discriminant analysis with applications to high dimensional low sample size data. Int. J. Appl. Math. **39**, 48–60 (2008)

Geometric Properties of Textile Plot

Tomonari Sei[1]([✉]) and Ushio Tanaka[2]

[1] University of Tokyo, 7-3-1 Hongo, Bunkyo-ku, Tokyo 113-8656, Japan
sei@mist.i.u-tokyo.ac.jp
[2] Osaka Prefecture University, 1-1 Gakuen-cho, Naka-ku, Sakai-shi,
Osaka 599-8531, Japan

Abstract. The textile plot proposed by Kumasaka and Shibata (2008) is a method for data visualization. The method transforms a data matrix in order to draw a parallel coordinate plot. In this paper, we investigate a set of matrices induced by the textile plot, which we call the textile set, from a geometrical viewpoint. It is shown that the textile set is written as the union of two differentiable manifolds if data matrices are restricted to be full-rank.

1 Introduction

The textile plot is a method for data visualization proposed by [3]. In this section, we briefly describe the method. We define the textile set by a set of matrices induced by the textile plot. We will use bold uppercase letters for matrices and bold lowercase letters for column vectors.

Let $X = (x_1, \ldots, x_p) \in \mathbb{R}^{n \times p}$ be a data matrix of n individuals and p variates, where each x_i is a column vector. For example, imagine a data of characteristics of n students in a school. Then the first column of X may represent age, the second is height, the third is weight, and so on. Note that, in each column, every element has the common unit ([years], [cm], [kg], etc.). For simplicity, we assumed that the data matrix has no missing value and that each variate is numeric, although the original method of [3] can deal with missing and categorical values. Each column of X is assumed to be non-degenerate, i.e., take at least two distinct values. In the following, without loss of generality, we assume the data is scaled:

$$x_j' 1_n = 0 \quad \text{and} \quad \|x_j\| = 1,$$

where x' is the transpose of x, $\|x\| = (x'x)^{1/2}$ is the Euclidean norm, and $1_n = (1, \ldots, 1)' \in \mathbb{R}^n$.

The textile plot generates another matrix $Y \in \mathbb{R}^{n \times p}$ from X by location and scale transformations in the following way. For simplicity, we do not change the order of columns as opposed to [3]. The matrix $Y = (y_1, \ldots, y_p)$ is defined by

$$y_j = b_j x_j, \quad j = 1, \ldots, p,$$

where $(b_1, \ldots, b_p)'$ is the unit eigenvector corresponding to the maximum eigenvalue of the sample correlation matrix $(x_i' x_j)_{i,j=1}^p$. The resultant matrix Y is used to draw a parallel coordinate plot (e.g. [2]). Refer to [3] for details.

© Springer International Publishing Switzerland 2015
F. Nielsen and F. Barbaresco (Eds.): GSI 2015, LNCS 9389, pp. 732–739, 2015.
DOI: 10.1007/978-3-319-25040-3_78

If we replace the condition of the maximum eigenvalue with *some* eigenvalue, we have a necessary condition

$$\exists \lambda \in \mathbb{R}, \quad \forall k \in \{1,\ldots,p\}, \quad \sum_{\ell=1}^{p}(\boldsymbol{x}_k'\boldsymbol{x}_\ell)b_\ell = \lambda b_k.$$

This condition can be written in terms of $(\boldsymbol{y}_1,\ldots,\boldsymbol{y}_p)$. Indeed, multiplying b_k to both sides, we have a condition

$$\exists \lambda \in \mathbb{R}, \quad \forall i \in \{1,\ldots,p\}, \quad \sum_{j=1}^{p}\boldsymbol{y}_i'\boldsymbol{y}_j = \lambda\|\boldsymbol{y}_i\|^2. \tag{1}$$

Furthermore, the conditions $\sum_{j=1}^{p} b_j^2 = 1$ and $\|\boldsymbol{x}_j\| = 1$ imply

$$\sum_{j=1}^{p}\|\boldsymbol{y}_j\|^2 = 1. \tag{2}$$

An extra constraint $\boldsymbol{y}_j'\boldsymbol{1}_n = 0$ can be removed by the orthogonalization argument. More precisely, the constraint is represented by $\boldsymbol{y}_j = \boldsymbol{Q}\boldsymbol{z}_j$, where $\boldsymbol{Q}$ is a fixed $n \times (n-1)$ matrix such that $(\boldsymbol{1}_n/\sqrt{n}, \boldsymbol{Q})$ is an orthogonal matrix. The vector $\boldsymbol{z}_j \in \mathbb{R}^{n-1}$ has no constraint. By substituting $\boldsymbol{y}_j = \boldsymbol{Q}\boldsymbol{z}_j$ to (1) and (2), we have the same equations as (1) and (2) for $\boldsymbol{z}_j \in \mathbb{R}^{n-1}$ without extra constraints.

The textile set is defined as follows.

Definition 1 (Textile Set). *Let n and p be positive integers. Then the set of matrices $\boldsymbol{Y} = (\boldsymbol{y}_1,\ldots,\boldsymbol{y}_p) \in \mathbb{R}^{n\times p}$ that satisfy (1) and (2) is denoted by $T_{n,p} \subset \mathbb{R}^{n\times p}$, and called the textile set in this paper. The set of $\boldsymbol{Y}$ that satisfies (1) is denoted by $U_{n,p}$ and called the unnormalized textile set.*

If $\|\boldsymbol{y}_i\| \neq 0$ for all i, the condition (1) is equivalent to

$$\frac{\boldsymbol{y}_1'(\sum_k \boldsymbol{y}_k)}{\|\boldsymbol{y}_1\|^2} = \cdots = \frac{\boldsymbol{y}_p'(\sum_k \boldsymbol{y}_k)}{\|\boldsymbol{y}_p\|^2} \quad (=\lambda). \tag{3}$$

We investigate geometric properties of the textile set. It is shown that the restriction of $T_{n,p}$ to full-rank matrices consists of two differentiable submanifolds of $\mathbb{R}^{n\times p}$. The result will be needed in the future to understand, for example, probabilistic properties of $\boldsymbol{Y}$ when the original data matrix $\boldsymbol{X}$ is distributed according to multivariate normal distributions.

In Sect. 2, examples for small p or small n are given. In Sect. 3, a topological property is studied. The main result is given in Sect. 4.

2   Low-Dimensional Cases

If $p = 1$, it is easy to see that

$$U_{n,1} = \mathbb{R}^n \quad \text{and} \quad T_{n,1} = S^{n-1} = \{\boldsymbol{y} \in \mathbb{R}^n \mid \|\boldsymbol{y}\| = 1\}.$$

In particular, $T_{1,1}$ is the two-point set $\{-1, 1\}$.

If $p = 2$, then

$$\begin{aligned}
U_{n,2} &= \{(\boldsymbol{y}_1, \boldsymbol{y}_2) \mid \exists \lambda, \ (\lambda - 1)\|\boldsymbol{y}_1\|^2 = \boldsymbol{y}_1'\boldsymbol{y}_2 = (\lambda - 1)\|\boldsymbol{y}_2\|^2\} \\
&= \{(\boldsymbol{y}_1, \boldsymbol{y}_2) \mid \|\boldsymbol{y}_1\| = \|\boldsymbol{y}_2\|\} \cup \{(\boldsymbol{y}_1, \boldsymbol{y}_2) \mid \boldsymbol{y}_1'\boldsymbol{y}_2 = 0\} \\
&= \{(\boldsymbol{y}_1, \boldsymbol{y}_2) \mid \|\boldsymbol{y}_1\| = \|\boldsymbol{y}_2\|\} \cup \{(\boldsymbol{y}_1, \boldsymbol{y}_2) \mid \|\boldsymbol{y}_1 - \boldsymbol{y}_2\| = \|\boldsymbol{y}_1 + \boldsymbol{y}_2\|\}
\end{aligned}$$

and

$$T_{n,2} = \left\{(\boldsymbol{y}_1, \boldsymbol{y}_2) \ \middle| \ \|\boldsymbol{y}_1\| = \|\boldsymbol{y}_2\| = \frac{1}{\sqrt{2}}\right\} \cup \{(\boldsymbol{y}_1, \boldsymbol{y}_2) \mid \|\boldsymbol{y}_1 - \boldsymbol{y}_2\| = \|\boldsymbol{y}_1 + \boldsymbol{y}_2\| = 1\}$$

Thus $T_{n,2}$ is the union of two manifolds, each of which is diffeomorphic to $S^{n-1} \times S^{n-1}$. The intersection of the two manifolds is a non-empty set

$$\left\{(\boldsymbol{y}_1, \boldsymbol{y}_2) \ \middle| \ \|\boldsymbol{y}_1\| = \|\boldsymbol{y}_2\| = \frac{1}{\sqrt{2}}, \ \boldsymbol{y}_1'\boldsymbol{y}_2 = 0\right\},$$

which is diffeomorphic to the Stiefel manifold. In particular, $T_{n,2}$ is connected if $n \geq 2$. In contrast, $T_{1,2}$ consists of 8 points: $(\pm 1/\sqrt{2}, \pm 1/\sqrt{2})$, $(\pm 1, 0)$ and $(0, \pm 1)$.

Consider the case $n = 1$ for general p. Let $[p] = \{1, \ldots, p\}$.

Proposition 1. *If $n = 1$, then the unnormalized textile set is*

$$U_{1,p} = \left\{\sum_{i=1}^{p} y_i = 0\right\} \cup \left(\bigcup_{\emptyset \neq I \subset [p]} \{y_i = y_j \ (i, j \in I), \ y_k = 0 \ (k \notin I)\}\right),$$

which consists of a $(p-1)$-dimensional plane and $2^p - 1$ lines. Correspondingly, the textile set $T_{1,p}$ consists of a $(p-2)$-dimensional manifold

$$S^{p-1} \cap \left\{\sum_{i=1}^{p} y_i = 0\right\}$$

and $2(2^p - 1)$ points

$$\bigcup_{\emptyset \neq I \subset [p]} \bigcup_{e \in \{-1, 1\}} \left\{y_i = \frac{e}{\sqrt{|I|}} \ (i \in I), \ y_k = 0 \ (k \notin I)\right\}.$$

Proof. Fix $(y_1, \ldots, y_p) \in \mathbb{R}^{1 \times p}$ and let I be the set of i such that $y_i \neq 0$. Then the condition (1) is rewritten as

$$\exists \lambda, \quad \forall i \in I, \quad \frac{\sum_{j \in I} y_j}{y_i} = \lambda.$$

This condition is satisfied if and only if $\sum_{i \in I} y_i = 0$ or $y_i = y_j$ for any $i, j \in I$. The expressions of $U_{1,p}$ and $T_{1,p}$ then follow. $\qquad \square$

Corollary 1. *If $(y_1, \ldots, y_p) \in U_{1,p}$ and $y_i \neq 0$ for all i, then $\sum_{j=1}^{n} y_j = 0$ or $y_1 = \cdots = y_p$.*

3   The Closedness Property

In this section, we show that $U_{n,p}$ is closed, which implies $T_{n,p}$ is compact. This property is not obvious from the definition since $U_{n,p}$ is the union of an infinite number of closed subsets. In contrast, $U_{n,p}$ is easily shown to be connected. Indeed, $U_{n,p}$ is a cone since the condition (1) is invariant under the scalar multiple of $\boldsymbol{Y}$. In the following, we assume $p \geq 2$ since $U_{n,1} = \mathbb{R}^n$.

For each $1 \leq i < j \leq p$, define a set $Z_{ij} \subset \mathbb{R}^{n \times p}$ by

$$Z_{ij} = \{(\boldsymbol{y}_1, \ldots, \boldsymbol{y}_p) \mid (\|\boldsymbol{y}_j\|^2 \boldsymbol{y}_i - \|\boldsymbol{y}_i\|^2 \boldsymbol{y}_j)'(\textstyle\sum_k \boldsymbol{y}_k) = 0\}. \tag{4}$$

Since Z_{ij} is the zero set of a continuous function, it is closed.

We prove the following proposition.

Proposition 2. *Let $p \geq 2$. Then we have*

$$U_{n,p} = \bigcap_{1 \leq i < j \leq p} Z_{ij}. \tag{5}$$

In particular, $U_{n,p}$ is closed and $T_{n,p}$ is compact.

Proof. Define two vectors $\boldsymbol{v} = (v_i)_{i=1}^p$ and $\boldsymbol{w} = (w_i)_{i=1}^p$ by $v_i = \boldsymbol{y}_i'(\sum_{k=1}^p \boldsymbol{y}_k)$ and $w_i = \|\boldsymbol{y}_i\|^2$. Then the condition (1) is written as $\boldsymbol{v} = \lambda \boldsymbol{w}$ for some λ.

We first show that the condition (1) is equivalent to linear dependence of the vectors $\boldsymbol{v}$ and $\boldsymbol{w}$. If (1) is satisfied, then $\boldsymbol{v} = \lambda \boldsymbol{w}$ and the linear dependence follows. Conversely, assume $\boldsymbol{v}$ and $\boldsymbol{w}$ are linearly dependent. Then either $\boldsymbol{v} = \lambda \boldsymbol{w}$ for some λ or $\boldsymbol{w} = \boldsymbol{0}$. The former is the same as (1). If $\boldsymbol{w} = \boldsymbol{0}$, then $\|\boldsymbol{y}_i\| = 0$ for any i, which implies $\boldsymbol{v} = \boldsymbol{0}$. Hence the condition (1) is fulfilled.

The vectors $\boldsymbol{v}$ and $\boldsymbol{w}$ are dependent if and only if $v_i w_j - v_j w_i = 0$ for any pair of i and j. This equation is equivalent to $\boldsymbol{Y} \in Z_{ij}$. Hence $\boldsymbol{Y} \in U_{n,p}$ if and only if $\boldsymbol{Y} \in \bigcap_{i<j} Z_{ij}$. □

Remark 1. The Eqs. (4) and (5) show that $U_{n,p}$ is an algebraic variety.

4   Main Result

Let $n \geq p$. Denote the set of all full-rank matrices by

$$V^* = \{\boldsymbol{Y} \in \mathbb{R}^{n \times p} \mid \mathrm{rank}(\boldsymbol{Y}) = p\}.$$

The set V^* is called *the noncompact Stiefel manifold* in literature (e.g. [1]). Although V^* depends on n and p, we omit the indices just for simplicity.

The restriction of $U_{n,p}$ to V^* is denoted by

$$\begin{aligned} U_{n,p}^* &= U_{n,p} \cap V^* \\ &= \{\boldsymbol{Y} \in \mathbb{R}^{n \times p} \mid \mathrm{rank}(\boldsymbol{Y}) = p, \ \boldsymbol{Y} \text{ satisfies } (3)\}. \end{aligned}$$

Note that, for $\boldsymbol{Y} \in V^*$, the condition (1) is equivalent to (3). We can also define $T_{n,p}^*$ in the same manner. However, it is sufficient to study only $U_{n,p}^*$. Indeed, the two sets $T_{n,p}^*$ and $U_{n,p}^*$ are related to each other by the projection to the unit sphere (2). We will show that $U_{n,p}^*$ is the union of two differentiable manifolds.

Remark 2. The (compact) Stiefel manifold is defined by $\{Y \in \mathbb{R}^{n \times p} \mid Y'Y = I_p\}$, where I_p is the identity matrix. This manifold is contained in $U^*_{n,p}$ since the condition (3) is satisfied with $\lambda = 1$ if $Y'Y = I_p$.

The sets $U^*_{n,p}$ and V^* are closed under multiplication of orthogonal matrix from the left. In other words, the group $O(n)$ acts on $U^*_{n,p}$ and V^*, respectively. By the Gram-Schmidt orthonormalization, each equivalence class of the quotient space $V^*/O(n)$ is identified with an upper-triangular matrix

$$\begin{pmatrix} y_{11} & \cdots & y_{1p} \\ 0 & \ddots & \vdots \\ \vdots & \ddots & y_{pp} \\ 0 & \cdots & 0 \\ \vdots & & \vdots \\ 0 & \cdots & 0 \end{pmatrix}, \quad y_{ii} > 0 \ (1 \le i \le p). \tag{6}$$

See e.g. Example 4.1.2 of [1]. Let us call it *the canonical form*. The set of all canonical forms is denoted by

$$V^{**} = \{Y \in \mathbb{R}^{n \times p} \mid Y \text{ is written as (6)}\}.$$

Correspondingly, define

$$\begin{aligned} U^{**}_{n,p} &= U_{n,p} \cap V^{**} \\ &= U^*_{n,p} \cap V^{**} \\ &= \{Y \in \mathbb{R}^{n \times p} \mid Y \text{ is written as (6)}, \ Y \text{ satisfies (3)}\}. \end{aligned}$$

Let $\pi : V^* \to V^{**}$ be a map defined by the Gram-Schmidt orthonormalization. Since π is a submersion, $\pi^{-1}(A)$ is a submanifold of V^* for any submanifold A of V^{**}. Therefore we investigate $U^{**}_{n,p}$ instead of $U^*_{n,p} = \pi^{-1}(U^{**}_{n,p})$.

For each canonical form $Y \in V^{**}$, (3) is written as

$$\frac{y_{1i}s_1 + \cdots + y_{ii}s_i}{y_{1i}^2 + \cdots + y_{ii}^2} = \frac{s_1}{y_{11}}, \quad i = 2, \ldots, p, \tag{7}$$

where $s_j = y_{jj} + \cdots + y_{jp}$.

Example 1. If $p = 2$, (7) is

$$\frac{y_{12}(y_{11} + y_{12}) + y_{22}^2}{y_{12}^2 + y_{22}^2} = \frac{y_{11} + y_{12}}{y_{11}}.$$

This is equivalent to

$$y_{11}^2 = y_{12}^2 + y_{22}^2 \quad \text{or} \quad y_{12} = 0.$$

The two equations represent hyperbola and plane, respectively. Thus $U^{**}_{n,2}$ is the union of the two-dimensional manifolds. □

Our main theorem is stated as follows.

Theorem 1. *Let $n \geq p \geq 3$. Then we have a decomposition*

$$U_{n,p}^{**} = M_1 \cup M_2,$$

where M_1 and M_2 are differentiable manifolds with dimension

$$\dim M_1 = \frac{p(p+1)}{2} - (p-1),$$

$$\dim M_2 = \frac{p(p+1)}{2} - p,$$

respectively. The manifold M_2 is connected while M_1 may not.

Proof. Let $\boldsymbol{Y} = (\boldsymbol{y}_1, \ldots, \boldsymbol{y}_p) \in V^{**}$. We solve the Eq. (7). Recall that $s_i = y_{ii} + \cdots + y_{ip}$.

We will consider two cases, $s_1 \neq y_{11}$ and $s_1 = y_{11}$. Put

$$M_1 = U_{n,p}^{**} \cap \{s_1 \neq y_{11}\}, \quad M_2 = U_{n,p}^{**} \cap \{s_1 = y_{11}\}. \tag{8}$$

(i) First consider the case $s_1 \neq y_{11}$. We solve (7) with respect to the $(p-1)$ underlined variables in

$$\begin{pmatrix} y_{11} & y_{12} & \cdots & y_{1,p-1} & y_{1p} \\ & y_{22} & & y_{2,p-1} & \underline{y_{2p}} \\ & & \ddots & \vdots & \vdots \\ & & & y_{p-1,p-1} & \underline{y_{p-1,p}} \\ 0 & & & & \underline{y_{pp}} \end{pmatrix}.$$

In other words, the underlined variables are considered as dependent variables and the others are independent variables. Denote the independent variables by $\boldsymbol{r}$, that is,

$$\boldsymbol{r} = (y_{11}, \ldots, y_{1p}, y_{22}, \ldots, y_{2,p-1}, \ldots, y_{p-1,p-1}) \in \mathbb{R}^{p(p+1)/2 - (p-1)}.$$

First focus on y_{2p}. Let $i = 2$ in (7) to obtain

$$\frac{y_{12}s_1 + y_{22}s_2}{y_{12}^2 + y_{22}^2} = \frac{s_1}{y_{11}}.$$

In this equation, y_{2p} is contained only in s_2 and its coefficient $y_{22}/(y_{12}^2 + y_{22}^2)$ is not zero since $\boldsymbol{Y} \in V^{**}$. The other variables are contained in $\boldsymbol{r}$. Hence y_{2p} is written as a rational function of $\boldsymbol{r}$.

In the following, we use induction to show that y_{ip} is written as a rational function of $\boldsymbol{r}$ for each $2 \leq i \leq p-1$. We have already proved the claim for $i = 2$. Let $i \geq 3$ and assume that the claim is correct up to $i - 1$. In (7), the variables not in $\boldsymbol{r}$ are $y_{2p}, \ldots, y_{ip}$. Furthermore, the coefficient of y_{ip} is not zero. Hence y_{ip} is a rational function of $\boldsymbol{r}$ and $y_{2p}, \ldots, y_{i-1,p}$. By induction, y_{ip} is written as a rational function of $\boldsymbol{r}$.

Finally, we focus on y_{pp}. Let $i = p$ in (7). Then, by noting $s_p = y_{pp}$, we have

$$\frac{\sum_{k=1}^{p-1} y_{kp} s_k + y_{pp}^2}{\sum_{k=1}^{p-1} y_{kp}^2 + y_{pp}^2} = \frac{s_1}{y_{11}},$$

that is,

$$y_{pp}^2 = \frac{\sum_{k=1}^{p-1}(y_{11} y_{kp} s_k - y_{kp}^2 s_1)}{s_1 - y_{11}}, \tag{9}$$

where we used the condition $s_1 - y_{11} \neq 0$. Therefore, y_{pp} has a positive solution if and only if the right hand side of (9) is positive, in that case the solution is unique.

In summary, over the domain

$$D = \left\{ r \in \mathbb{R}^{p(p+1)/2-(p-1)} \mid s_{11} \neq y_{11}, \text{ the right hand side of (9) is positive} \right\},$$

the quantities $y_{2p}, \ldots, y_{pp}$ are uniquely written as differentiable functions (rational functions and square root). Since D is an open set, we deduce that M_1 is a differentiable manifold as long as D is not empty. We give an element of D explicitly. Let $y_{11} = \cdots = y_{p-1,p-1} = 1$, $y_{1p} = \alpha \in (0,1)$, and the other variables in r be zero. Then (7) implies $y_{ip} = \alpha$ $(2 \leq i \leq p-1)$, and (9) is

$$y_{pp}^2 = (p-1)(1-\alpha^2) > 0.$$

It was shown that M_1 is a (possibly unconnected) differentiable manifold. (ii) Next consider the case $s_1 = y_{11}$. In this case, (7) is simplified to

$$\sum_{k=1}^{i} y_{ki}(s_k - y_{ki}) = 0, \quad i = 2, \ldots, p. \tag{10}$$

We solve (10) with respect to the p underlined variables in

$$\begin{pmatrix} y_{11} \cdots & & \cdots & y_{1,p-1} & \underline{y_{1p}} \\ & \ddots & & \vdots & \vdots \\ & y_{p-3,p-3} & \cdots & y_{p-3,p-1} & \underline{y_{p-3,p}} \\ & & y_{p-2,p-2} & \underline{y_{p-2,p-1}} & \underline{y_{p-2,p}} \\ & & & y_{p-1,p-1} & \underline{y_{p-1,p}} \\ 0 & & & & \underline{y_{pp}} \end{pmatrix}.$$

The underlined variables are dependent variables and the others are independent variables. Denote the independent variables by r.

Then by the condition $s_1 = y_{11}$,

$$y_{1p} = -\sum_{j=2}^{p-1} y_{1j}.$$

In particular, y_{1p} is written as a linear function of r.

Use induction to show that y_{ip} is written as a rational function of r for each $1 \leq i \leq p-3$. We have already proved the claim for $i = 1$. Let $2 \leq i \leq p-3$ and assume the claim is correct up to $i-1$. In (10), the variables not in r are $y_{1p}, \ldots, y_{ip}$. In addition, the coefficient of y_{ip} is positive. Hence y_{ip} is a rational function of r and $(y_{1p}, \ldots, y_{i-1,p})$. By induction, y_{ip} is written as a rational function of r.

Finally, we focus on $a := y_{p-2,p-1}$, $b := y_{p-2,p}$, and $c := y_{p-1,p}$. The Eqs. (10) for $i = p-2, p-1, p$ are

$$R_1 + y_{p-2,p-2}(a + b) = 0, \tag{11}$$

$$R_2 + y_{p-2,p-2}a + ab + y_{p-1,p-1}c = 0, \tag{12}$$

$$R_3 + y_{p-2,p-2}b + ab + y_{p-1,p-1}c = 0, \tag{13}$$

where the terms without a, b, c are abbreviated to R_i $(i = 1, 2, 3)$. The difference between (12) and (13) is

$$R_4 + y_{p-2,p-2}(a - b) = 0.$$

From this and (11), we obtain a and b. Then c is determined from (12).

It was shown that M_2 is a differentiable manifold. Since r has no constraint, M_2 is connected.   □

Corollary 2. *Let $n \geq p \geq 1$. Then $U^*_{n,p}$ is the union of two differentiable manifolds:*

$$U^*_{n,p} = \pi^{-1}(M_1) \cup \pi^{-1}(M_2),$$

where π is the Gram-Schmidt orthonormalization map, and M_1 and M_2 are defined by (8).

Acknowledgment. The authors are grateful to three anonymous reviewers for their helpful comments.

References

1. Absil, P.-A., Mahony, R., Sepulchre, R.: Optimization Algorithms on Matrix Manifolds. Princeton University Press, Princeton (2008)
2. Inselberg, A.: Parallel Coordinates: VISUAL Multidimensional Geometry and its Applications. Springer, New York (2009)
3. Kumasaka, N., Shibata, R.: High-dimensional data visualisation: the textile plot. Comput. Stat. Data Anal. **52**, 3616–3644 (2008)

A Generalization of Independence and Multivariate Student's t-distributions

Monta Sakamoto[1] and Hiroshi Matsuzoe[2]([✉])

[1] Engineering School of Information and Digital Technologies, Efrei,
30-32 Avenue de la Republique, 94800 Villejuif, France
[2] Department of Computer Science and Engineering Graduate School of Engineering,
Nagoya Institute of Technology, Gokiso-cho, Showa-ku,
Nagoya 466-8555, Japan
matsuzoe@nitech.ac.jp

Abstract. In anomalous statistical physics, deformed algebraic structures are important objects. Heavily tailed probability distributions, such as Student's t-distributions, are characterized by deformed algebras. In addition, deformed algebras cause deformations of expectations and independences of random variables. Hence, a generalization of independence for multivariate Student's t-distribution is studied in this paper. Even if two random variables which follow to univariate Student's t-distributions are independent, the joint probability distribution of these two distributions is not a bivariate Student's t-distribution. It is shown that a bivariate Student's t-distribution is obtained from two univariate Student's t-distributions under q-deformed independence.

Keywords: Deformed exponential family · Deformed independence · Statistical manifold · Tsallis statistics · Information geometry

1 Introduction

In the theory of complex systems, heavily tailed probability distributions are important objects. Power law tailed probability distributions and their related probability distributions have been studied in anomalous statistical physics ([6,12,15]). One of an important probability distribution in anomalous statistical physics is a q-Gaussian distribution. It is a noteworthy fact that a q-Gaussian distribution coincides with a Student's t-distribution in statistics. Hence we can discuss Student's t-distributions from the viewpoint of anomalous statistical physics. Though Student's t-distributions have been studied by many authors (cf. [3,7]), our motivation is quite different from the others.

Heavily tailed probability distributions including Student's t-distributions are represented using deformed exponential functions (cf. [11,12]). However, these functions do not satisfy the law of exponents. Hence deformed algebraic

H. Matsuzoe—This work was partially supported by MEXT KAKENHI Grant Numbers 15K04842 and 26108003.

F. Nielsen and F. Barbaresco (Eds.): GSI 2015, LNCS 9389, pp. 740–749, 2015.
DOI: 10.1007/978-3-319-25040-3_79

structures are naturally introduced (cf. [4,6]). Once such a deformed algebra is introduced, the sample space can be regarded as a some deformed algebraic space, not the standard Euclidean space (cf. [11]). Hence it is natural to introduce suitable deformed expectations and independences of random variables. In fact, we find that the duality of exponential and logarithm can express the notion of independence of random variables. Hence we can generalize the independence using deformed exponential and deformed logarithm functions [9].

In this paper, we summarize such deformed algebraic structures, then we apply these deformed algebras to multivariate Student's t-distributions. Even if two independent random variables follow to univariate Student's t-distributions, the joint probability distribution is not a bivariate Student's t-distribution. Hence we show that a bivariate Student's t-distribution can be obtained from two univariate Student's t-distributions under q-deformed independence with a suitable normalization.

We remark that deformed algebraic structures for statistical models and generalization of independence are discussed in information geometry. (cf. [5,9,11]. See also [1].) Though normalizations of positive densities are necessary in the arguments of generalized independence, statistical manifold structures are changed by normalizations of positive densities. In particular, generalized conformal equivalence relations for statistical manifolds are needed (cf. [9,10]). Hence a statistical manifold of the set of bivariate Student's t-distributions with q-independent random variables is not equivalent to a product statistical manifold of two sets of univariate Student's t-distributions.

2   Deformed Exponential Families

In this paper, we assume that all objects are smooth for simplicity. Let us begin by reviewing the foundations of deformed exponential functions and deformed exponential families (cf. [9,12]).

Let χ be a strictly increasing function from $(0, \infty)$ to $(0, \infty)$. We define a χ-*logarithm function* or a *deformed logarithm function* by

$$\ln_\chi s := \int_1^s \frac{1}{\chi(t)} dt.$$

The inverse of $\ln_\chi s$ is called a χ-*exponential function* or a *deformed exponential function*, which is defined by

$$\exp_\chi t := 1 + \int_0^t u(s) ds,$$

where the function $u(s)$ is given by $u(\ln_\chi s) = \chi(s)$.

From now on, we suppose that χ is a power function, that is, $\chi(t) = t^q$. Then the deformed logarithm and the deformed exponential are defined by

$$\ln_q s := \frac{s^{1-q} - 1}{1 - q}, \qquad\qquad (s > 0),$$

$$\exp_q t := (1 + (1-q)t)^{\frac{1}{1-q}}, \qquad\qquad (1 + (1-q)t > 0).$$

We say that $\ln_q s$ is a *q-logarithm function* and $\exp_q t$ is a *q-exponential function*. By taking a limit $q \to 1$, these functions coincide with the standard logarithm $\ln s$ and the standard exponential $\exp t$, respectively. In this paper, we focus on q-exponential case. However, many of arguments for q-exponential family can be generalized for χ-exponential family ([9,11]).

A statistical model S_q is called a *q-exponential family* if

$$S_q = \left\{ p(x,\theta) \,\middle|\, p(x;\theta) = \exp_q\left[\sum_{i=1}^n \theta^i F_i(x) - \psi(\theta)\right], \ \theta \in \Theta \subset \mathbf{R}^n \right\},$$

where $F_1(x),\ldots,F_n(x)$ are functions on the sample space Ω, $\theta = {}^t(\theta^1,\ldots,\theta^n)$ is a parameter, and $\psi(\theta)$ is the normalization with respect to the parameter θ.

Example 1 (Student's t-distribution). Fix a number q $(1 < q < 1+2/d,\ d \in \mathbf{N})$, and set $\nu = -d - 2/(1-q)$. We define an *n-dimensional Student's t-distribution with degree of freedom ν or a q-Gaussian distribution* by

$$p_q(x;\mu,\Sigma) := \frac{\Gamma\left(\frac{1}{q-1}\right)}{(\pi\nu)^{\frac{d}{2}}\Gamma\left(\frac{\nu}{2}\right)\sqrt{\det(\Sigma)}}\left[1 + \frac{1}{\nu}{}^t(x-\mu)\Sigma^{-1}(x-\mu)\right]^{\frac{1}{1-q}},$$

where $X = {}^t(X_1,\ldots,X_d)$ is a random vector on $\mathbf{R}^d$, $\mu = {}^t(\mu^1,\ldots,\mu^d)$ is a location vector on $\mathbf{R}^d$ and Σ is a scale matrix on $\mathrm{Sym}^+(d)$. For simplicity, we assume that Σ is invertible. Otherwise, we should choose a suitable basis $\{v^\alpha\}$ on $\mathrm{Sym}^+(d)$ such that $\Sigma = \sum_\alpha w_\alpha v^\alpha$. Then, the set of all Student's t-distributions is a q-exponential family. In fact, set

$$z_q = \frac{(\pi\nu)^{\frac{d}{2}}\Gamma\left(\frac{\nu}{2}\right)\sqrt{\det(\Sigma)}}{\Gamma\left(\frac{1}{q-1}\right)}, \quad \tilde{R} = \frac{z_q^{q-1}}{(1-q)d+2}\Sigma^{-1}, \quad \text{and} \quad \theta = 2\tilde{R}\mu. \quad (1)$$

Then we have

$$p_q(x;\mu,\Sigma) = \frac{1}{z_q}\left[1 + \frac{1}{\nu}{}^t(x-\mu)\Sigma^{-1}(x-\mu)\right]^{\frac{1}{1-q}}$$

$$= \left[\left(\frac{1}{z_q}\right)^{1-q} - \frac{1-q}{(1-q)d+2}\left(\frac{1}{z_q}\right)^{1-q}{}^t(x-\mu)\Sigma^{-1}(x-\mu)\right]^{\frac{1}{1-q}}$$

$$= \exp_q\left[-{}^t(x-\mu)\tilde{R}(x-\mu) + \ln_q\frac{1}{z_q}\right]$$

$$= \exp_q\left[\sum_{i=1}^d \theta^i x_i - \sum_{i=1}^d \tilde{R}_{ii}x_i^2 - 2\sum_{i<j}\tilde{R}_{ij}x_i x_j - \frac{1}{4}{}^t\theta\tilde{R}^{-1}\theta + \ln_q\frac{1}{z_q}\right].$$

Since $\theta \in \mathbf{R}^d$ and $\tilde{R} \in \mathrm{Sym}^+(d)$, the set of all Student's t-distributions is a $d(d+3)/2$-dimensional q-exponential family. The normalization $\psi(\theta)$ is given by

$$\psi(\theta) = \frac{1}{4}{}^t\theta\tilde{R}^{-1}\theta - \ln_q\frac{1}{z_q}.$$

3 Statistical Manifold Structures Based on q-Fisher Metric

In this section we give a brief review of statistical manifold structures on a q-exponential family. We consider a q-Fisher metric in this paper. However, it is known that a q-exponential family naturally has three kinds of statistical manifold structures. See [2, 9, 11] for more details.

Let S_q be a q-exponential family. The normalization $\psi(\theta)$ on S_q is convex, but may not be strictly convex. Hence we assume that ψ is strictly convex throughout this paper. In fact, we obtain the following proposition.

Proposition 1. *Let $S_q = \{p(x;\theta)\}$ be a q-exponential family. Then the normalization function $\psi(\theta)$ is convex.*

Proof. Set $u(x) = \exp_q x$ and $\partial_i = \partial/\partial\theta^i$. Then we have

$$\partial_i p(x;\theta) = u'\left(\sum \theta^k F_k(x) - \psi(\theta)\right)(F_i(x) - \partial_i\psi(\theta)),$$

$$\partial_i\partial_j p(x;\theta) = u''\left(\sum \theta^k F_k(x) - \psi(\theta)\right)(F_i(x) - \partial_i\psi(\theta))(F_j(x) - \partial_j\psi(\theta))$$
$$- u'\left(\sum \theta^k F_k(x) - \psi(\theta)\right)\partial_i\partial_j\psi(\theta).$$

Since $\partial_i\int_\Omega p(x;\theta)dx = \int_\Omega \partial_i p(x;\theta)dx = 0$ and $\int_\Omega \partial_i\partial_j p(x;\theta)dx = 0$, we have

$$Z_q(p) = \int_\Omega \{(p(x;\theta))\}^q dx = \int_\Omega u'\left(\sum \theta^k F_k(x) - \psi(\theta)\right)dx,$$

$$\partial_i\partial_j\psi(\theta) = \frac{1}{Z_q(p)}\int_\Omega u''\left(\sum \theta^k F_k(x) - \psi(\theta)\right)$$
$$\times (F_i(x) - \partial_i\psi(\theta))(F_j(x) - \partial_j\psi(\theta))dx.$$

For an arbitrary vector $c = {}^t(c^1, c^2, \ldots .c^n) \in \mathbf{R}^n$, since $Z_q(p) > 0$ and $u''(x) > 0$, we have

$$\sum_{i,j=1}^n c^i c^j (\partial_i\partial_j\psi(\theta)) = \frac{1}{Z_q(p)}\int_\Omega u''\left(\sum_{k=1}^n \theta^k F_k(x) - \psi(\theta)\right)$$
$$\times \left\{\sum_{i=1}^n c^i(F_i(x) - \partial_i\psi(\theta))\right\}^2 dx \;\geq\; 0.$$

This implies that the Hessian matrix $(\partial_i\partial_j\psi(\theta))$ is semi-positive definite. □

From the assumption for $\psi(\theta)$, we can define the q-*Fisher metric* and the q-*cubic form* by

$$g_{ij}(\theta) = \partial_i\partial_j\psi(\theta), \qquad C_{ijk}(\theta) = \partial_i\partial_j\partial_k\psi(\theta),$$

respectively. For a fixed real number α, set

$$g\left(\nabla_X^{q(\alpha)}Y, Z\right) = g\left(\nabla_X^{q(0)}Y, Z\right) - \frac{\alpha}{2}C\left(X, Y, Z\right),$$

where $\nabla^{q(0)}$ is the Levi-Civita connection with respect to g. Since g is a Hessian metric, from standard arguments in Hessian geometry [13], $\nabla^{q(e)} := \nabla^{q(1)}$ and $\nabla^{q(m)} := \nabla^{q(-1)}$ are flat affine connections and mutually dual with respect to g. Hence the quadruplet $(S_q, g, \nabla^{q(e)}, \nabla^{q(m)})$ is a dually flat space.

Next, we consider deformed expectations for q-exponential families. We define the *escort distribution* $P_q(x; \theta)$ of $p(x; \theta) \in S_q$ and the *normalized escort distribution* $P_q^{esc}(x; \theta)$ by

$$P_q(x; \theta) = \{p(x; \theta)\}^q,$$

$$P_q^{esc}(x; \theta) = \frac{1}{Z_q(p)}\{p(x; \theta)\}^q, \quad \text{where} \quad Z_q(p) = \int_\Omega \{p(x; \theta)\}^q dx,$$

respectively. Let $f(x)$ be a function on Ω. The *q-expectation* $E_{q,p}[f(x)]$ and the *normalized q-expectation* $E_{q,p}^{esc}[f(x)]$ are defined by

$$E_{q,p}[f(x)] = \int_\Omega f(x)P_q(x; \theta)dx, \qquad E_{q,p}^{esc}[f(x)] = \int_\Omega f(x)P_q^{esc}(x; \theta)dx,$$

respectively. Under q-expectations, we have the following proposition. (cf. [8])

Proposition 2. *For S_q a q-exponential family, (1) set $\phi(\eta) = E_{q,p}^{esc}[\log_q p(x; \theta)]$, then $\phi(\eta)$ is the potential of g with respect to $\{\eta_i\}$. (2) Set $\eta_i = E_{q,p}^{esc}[F_i(x)]$. Then $\{\eta_i\}$ is a $\nabla^{q(m)}$-affine coordinate system such that*

$$g\left(\frac{\partial}{\partial\theta^i}, \frac{\partial}{\partial\eta_j}\right) = \delta_i^j.$$

$\qquad\square$

We define an *α-divergence* $D^{(\alpha)}$ with $\alpha = 1 - 2q$ and a *q-relative entropy* (or a *normalized Tsallis relative entropy*) D_q^T by

$$D^{(1-2q)}(p(x), r(x)) = \frac{1}{q}E_{q,p}[\log_q p(x) - \log_q r(x)] = \frac{1 - \int_\Omega p(x)^q r(x)^{1-q} dx}{q(1-q)},$$

$$D_q^T(p(x), r(x)) = E_{q,p}^{esc}[\log_q p(x) - \log_q r(x)] = \frac{1 - \int_\Omega p(x)^q r(x)^{1-q} dx}{(1-q)Z_q(p)},$$

respectively. It is known that the α-divergence $D^{(1-2q)}(r, p)$ induces a statistical manifold structure $(S_q, g^F, \nabla^{(2q-1)})$, where g^F is the Fisher metric on S_q and $\nabla^{(2q-1)}$ is the α-connection with $\alpha = 2q - 1$, and the q-relative entropy $D_q^T(r, p)$ induces $(S_q, g, \nabla^{q(e)})$.

Proposition 3 (cf. [10]). *For a q-exponential family S_q, the two statistical manifolds $(S_q, g^F, \nabla^{(2q-1)})$ and $(S_q, g, \nabla^{q(e)})$ are 1-conformally equivalent.* $\qquad\square$

We remark that the difference of a α-divergence and a q-relative entropy is only the normalization $q/Z_q(p)$. Hence a normalization for probability density imposes a generalized conformal change for a statistical model.

4   Generalization of Independence

In this section, we review the notions of q-deformed product and generalization of independence. For more details, see [9, 11].

Let us introduce the q-deformed algebras since q-exponential functions and q-logarithm functions do not satisfy the law of exponent. Let $\exp_q x$ be a q-exponential function and $\ln_q y$ be a q-logarithm function. For a fixed number q, we suppose that

$$1 + (1-q)x_1 > 0, \quad 1 + (1-q)x_2 > 0, \quad y_1^{1-q} + y_2^{1-q} - 1 > 0, \tag{2}$$

$$y_1 > 0, \quad y_2 > 0. \tag{3}$$

We define the q-sum $\tilde{\oplus}^q$ and the q-product $\otimes_q$ by the following formulas [4]:

$$\begin{aligned}
x_1 \tilde{\oplus}^q x_2 &:= \ln_q \left[\exp_q x_1 \cdot \exp_q x_2 \right] \\
&= x_1 + x_2 + (1-q)x_1 x_2, \\
y_1 \otimes_q y_2 &:= \exp_q \left[\ln_q y_1 + \ln_q y_2 \right] \\
&= \left[y_1^{1-q} + y_2^{1-q} - 1 \right]^{\frac{1}{1-q}}.
\end{aligned}$$

Since the base of an exponential function and the argument of a logarithm functions must be positive, conditions (2) and (3) are necessary. We then obtain q-deformed law of exponents as follows.

$$\exp_q(x_1 \tilde{\oplus}^q x_2) = \exp_q x_1 \cdot \exp_q x_2, \qquad \ln_q(y_1 \cdot y_2) = \ln_q y_1 \tilde{\oplus}^q \ln_q y_2,$$

$$\exp_q(x_1 + x_2) = \exp_q x_1 \otimes_q \exp_q x_2, \quad \ln_q(y_1 \otimes_q y_2) = \ln_q y_1 + \ln_q y_2.$$

We remark that the q-sum works on the domain of a q-exponential function and a q-product works on the target space. This implies that the domain of q-exponential function (i.e. the total sample space Ω) may not be a standard Euclidean space.

Let us recall the notion of independence of random variables. Suppose that X and Y are random variables which follow to probabilities $p_1(x)$ and $p_2(y)$, respectively. We say that two random variables are *independent* if the joint probability $p(x, y)$ is given by the product of $p_1(x)$ and $p_2(y)$:

$$p(x, y) = p_1(x) p_2(y).$$

Hence $p_1(x)$ and $p_2(y)$ are marginal distributions of $p(x, y)$. When $p_1(x) > 0$ and $p_2(y) > 0$, the independence is equivalent to the additivity of information:

$$\ln p(x, y) = \ln p_1(x) + \ln p_2(y).$$

Let us generalize the notion of independence based on q-products. Suppose that X and Y are random variables which follow to probabilities $p_1(x)$ and $p_2(y)$, respectively. We say that X and Y are *q-independent with e-normalization* (or *exponential normalization*) if a probability density $p(x, y)$ is decomposed by

$$p(x, y) = p_1(x) \otimes_q p_2(y) \otimes_q (-c),$$

where c is the normalization defined by

$$\iint_{Supp(p(x,y))\subset \Omega_X \times \Omega_Y} p_1(x) \otimes_q p_2(y) \otimes_q (-c) \, dxdy = 1.$$

We say that X and Y are *q-independent with m-normalization* (or *mixture normalization*) if a probability density $p(x, y)$ is decomposed by

$$p(x, y) = \frac{1}{Z(p_1, p_2)} p_1(x) \otimes_q p_2(y),$$

where $Z(p_1, p_2)$ is the normalization defined by

$$Z(p_1, p_2) := \iint_{Supp(p(x,y))\subset \Omega_X \times \Omega_Y} p_1(x) \otimes_q p_2(y) \, dxdy.$$

In the case of q-exponential families, including the standard exponential families, we can change normalizations from exponential type to mixture type and vice versa. (See the calculation in Example 1.) Hence we can carry out e- and m-normalization simultaneously. However, e- and m-normalizations are different in general [14].

In some problems, the normalization of probability density is not necessary. In this case, we say that X and Y are *q-independent* if a positive function $f(x, y)$ is decomposed by a q-product of two probability densities $p_1(x)$ and $p_2(y)$:

$$f(x, y) = p_1(x) \otimes_q p_2(y).$$

The function $f(x, y)$ is not necessary to be a probability density. In addition, the total integral of $f(x, y)$ may diverge.

5   q-independence and Student's t-distributions

In this section, we consider relations between univariate and bivariate Student's t-distributions

Suppose that X_1 and X_2 are random variables which follow to univariate Student's t-distributions $p_1(x_1)$ and $p_2(x_2)$, respectively. Even if X_1 and X_2 are independent, the joint probability $p_1(x_1)p_2(x_2)$ is not a bivariate Student's t-distribution [7]. We show that q-deformed algebras work for bivariate Student's t-distributions.

Theorem 1. *Suppose that X_1 and X_2 are random variables which follow to univariate Student's t-distributions $p_1(x_1)$ and $p_2(x_2)$, respectively, with same parameter q $(1 < q < 2)$. Then there exist a bivariate Student's t-distribution $p(x_1, x_2)$ such that X_1 and X_2 are q-independent with e-normalization.*

Proof. Suppose that X_1 follows to a univariate Student's t-distribution (or a q-Gaussian distribution) given by

$$p(x_1; \mu_1, \sigma_1) = \frac{\Gamma\left(\frac{1}{q-1}\right)}{\sqrt{\pi}\sqrt{\frac{3-q}{q-1}}\Gamma\left(\frac{3-q}{2(q-1)}\right)\sigma_1} \left[1 - (1-q)\frac{(x_1-\mu_1)^2}{(3-q)\sigma_1^2}\right]^{\frac{1}{1-q}},$$

where μ_1 $(-\infty < \mu < \infty)$ is a location parameter, and σ_1 $(0 < \sigma < \infty)$ is a scale parameter. Similarly, suppose that X_2 follows to $p(x_2; \mu_2, \sigma_2)$. By setting

$$z_q(\sigma_1) = \frac{\sqrt{\pi}\sqrt{\frac{3-q}{q-1}}\,\Gamma\left(\frac{3-q}{2(q-1)}\right)\sigma_1}{\Gamma\left(\frac{1}{q-1}\right)} = \sqrt{\frac{3-q}{q-1}}\,\mathrm{Beta}\left(\frac{3-q}{2(q-1)}, \frac{1}{2}\right)\sigma_1,$$

we obtain a q-exponential representation as follows:

$$p(x_1; \mu_1, \sigma_1) = \exp_q\left[\theta^1 x_1 - \theta^{11} x_1^2 - \frac{(\theta^1)^2}{4\theta^{11}} + \ln_q \frac{1}{z_q(\sigma_1)}\right],$$

where θ^1 and θ^{11} are natural parameters defined by

$$\theta^1 = \frac{2\mu_1\{z_q(\sigma_1)\}^{q-1}}{(3-q)\sigma_1^2}, \quad \theta^{11} = \frac{\{z_q(\sigma_1)\}^{q-1}}{(3-q)\sigma_1^2}.$$

We remark that the normalization $z_q(\sigma_1)$ can be determined by the parameter θ^{11}. Therefore, $p(x_1; \mu_1, \sigma_1)$ is uniquely determined from natural parameters θ^1 and θ^{11}. Set θ^2 and θ^{22} by changing parameters to μ_2 and σ_2. Then we obtain a positive density by

$$p(x_1; \mu_1, \sigma_1) \otimes_q p(x_2; \mu_2, \sigma_2)$$
$$= \exp_q\left[\theta^1 x_1 + \theta^2 x_2 - \theta^{11} x_1^2 - \theta^{22} x_2^2 - \frac{(\theta^1)^2}{4\theta^{11}} - \frac{(\theta^2)^2}{4\theta^{22}} + A(\theta)\right], \quad (4)$$

where $A(\theta)$ is given by

$$A(\theta) = \ln_q \frac{1}{z_q(\sigma_1)} + \ln_q \frac{1}{z_q(\sigma_2)}.$$

Recall that $p(x_1; \mu_1, \sigma_1) \otimes_q p(x_2; \mu_2, \sigma_2)$ is not a probability distribution. Set the e-normalization function c by

$$c = A(\theta) - \ln_q \frac{1}{z_q} = \left(\ln_q \frac{1}{z_q(\sigma_1)} + \ln_q \frac{1}{z_q(\sigma_2)}\right) - \ln_q \frac{1}{z_q}, \quad (5)$$

where z_q is the m-normalization function of bivariate Student's t-distribution. As a consequence, we have

$$p(x_1, x_2) = p(x_1; \mu_1, \sigma_1) \otimes_q p(x_2; \mu_2, \sigma_2) \otimes_q (-c)$$
$$= \exp_q\left[\theta^1 x_1 + \theta^2 x_2 - \theta^{11} x_1^2 - \theta^{22} x_2^2 - \frac{(\theta^1)^2}{4\theta^{11}} - \frac{(\theta^2)^2}{4\theta^{22}} + \ln_q \frac{1}{z_q}\right].$$

This implies that X_1 and X_2 are q-independent with e-normalization, and the joint positive measure $p(x_1, x_2)$ is a bivariate Student's t-distribution. □

Let us give the normalization function z_q in θ-coordinate, explicitly. Using a property of gamma function, we have

$$\frac{\Gamma\left(\frac{1}{1-q}\right)}{\nu\Gamma\left(\frac{\nu}{2}\right)} = \frac{\Gamma\left(\frac{\nu+2}{2}\right)}{\nu\Gamma\left(\frac{\nu}{2}\right)} = \frac{1}{2}.$$

Hence the m-normalization function of bivariate Student's t-distribution is simply given by

$$z_q = 2\pi\sqrt{\det\Sigma}.$$

From Equation (1) and (4), the constant z_q should be given by

$$z_q = 2\pi\left(\frac{4(2-q)^2}{(2\pi)^{2q-2}}\det\tilde{R}\right)^{\frac{1}{2(q-2)}} = \left(\frac{2-q}{\pi}\right)^{\frac{1}{q-2}}(\theta^{11}\theta^{22})^{\frac{1}{2q-4}}.$$

6 Concluding Remarks

In this paper, we showed that a bivariate Student's t-distribution can be obtained from two univariate Student's t-distributions using e- and m-normalizations. Recall that statistical manifold structures of statistical models are changed by their normalizations. Hence a statistical manifold structure of a bivariate Student's t-distribution does not coincide with the product manifold structure of two univariate Student's t-distributions.

References

1. Amari, S., Nagaoka, H.: Method of Information Geometry. American Mathematical Society, Providence. Oxford University Press, Oxford (2000)
2. Amari, S., Ohara, A., Matsuzoe, H.: Geometry of deformed exponential families: invariant, dually-flat and conformal geometry. Phys. A **391**, 4308–4319 (2012)
3. Berg, C., Vignat, C.: On the density of the sum of two independent Student t-random vectors. Statist. Probab. Lett. **80**, 1043–1055 (2010)
4. Borgesa, E.P.: A possible deformed algebra and calculus inspired in nonextensive thermostatistics. Phys. A **340**, 95–101 (2004)
5. Fujimoto, Y., Murata, N.: A generalization of independence in naive bayes model. In: Fyfe, C., Tino, P., Charles, D., Garcia-Osorio, C., Yin, H. (eds.) IDEAL 2010. LNCS, vol. 6283, pp. 153–161. Springer, Heidelberg (2010)
6. Kaniadakis, G.: Theoretical foundations and mathematical formalism of the power-law tailed statistical distributions. Entropy **15**, 3983–4010 (2013)
7. Kotz, S., Nadarajah, S.: Multivariate t Distributions and Their Applications. Cambridge University Press, New York (2004)
8. Matsuzoe, H.: Statistical manifolds and geometry of estimating functions. In: Prospects of Differential Geometry and its Related Fields, World Scientific Publishing, pp. 187–202 (2013)
9. Matsuzoe, H., Henmi, M.: Hessian structures and divergence functions on deformed exponential families. In: Nielsen, F. (ed.) Geometric Theory of Information. Signals and Communication Technology. Springer, Switzerland (2014)
10. Matsuzoe, H.., Ohara, A..: Geometry for q-exponential families. In: Recent Progress in Differential Geometry and its Related Fields, World Scientific Publishing, pp. 55–71 (2011)

11. Matsuzoe, H., Wada, T.: Deformed algebras and generalizations of independence on deformed exponential families. Entropy **17**, 5729–5751 (2015)
12. Naudts, J.: Generalised Thermostatistics. Springer, London (2011)
13. Shima, H.: The Geometry of Hessian Structures. World Scientific Publishing, Singapore (2007)
14. Takatsu, A.: Behaviors of φ-exponential distributions in Wasserstein geometry and an evolution equation. SIAM J. Math. Anal. **45**, 2546–2556 (2013)
15. Tsallis, C.: Introduction to Nonextensive Statistical Mechanics: Approaching a Complex World. Springer, New York (2009)

Probability Density Estimation

Probability Density Estimation on the Hyperbolic Space Applied to Radar Processing

Emmanuel Chevallier[1], Frédéric Barbaresco[2], and Jesús Angulo[1(✉)]

[1] CMM-Centre de Morphologie Mathématique, MINES ParisTech,
PSL-Research University, Fontainebleau, France
{emmanuel.chevallier,jesus.angulo}@mines-paristech.fr
[2] Thales Air Systems, Surface Radar Domain, Technical Directorate,
Advanced Developments Department, 91470 Limours, France

Abstract. The two main techniques of probability density estimation on symmetric spaces are reviewed in the hyperbolic case. For computational reasons we chose to focus on the kernel density estimation and we provide the expression of Pelletier estimator on hyperbolic space. The method is applied to density estimation of reflection coefficients derived from radar observations.

1 Introduction

The problem of probability density estimation is a vast topic. Their exists several standard methods in the Euclidean context, such as histograms, kernel methods, or the characteristic function method. These methods can sometimes be transposed to the case of Riemannian manifolds. However, the transposition often introduces additional computational efforts. This additional effort depends on the method used and the nature of the manifold. The hyperbolic space is one of the most elementary non-Euclidean spaces. It is one of the three simply connected isotropic manifolds, the two others being the sphere and the Euclidean space. The specificity of the hyperbolic space enables to adapt the different density estimation methods at a reasonable cost. Convergence rates of the density estimation using kernels and orthogonal series were progressively generalized to Riemannian manifolds, see [1–3]. More recently convergence rates for the kernel density estimation without the compact assumption have been introduced [4], which enables the use of Gaussian-type kernels [5]. One already encounters the problem of density estimation in the hyperbolic space for electrical impedance [2] and networks [6]. We are interested here in the estimation of the density of the reflection coefficients extracted from a radar signal [7]. These coefficients have a intrinsic hyperbolic structure [8,9]. For computational reasons we chose to focus our applications on the kernel density estimation. The paper begins with an introduction to the hyperbolic geometry in Sect. 2. Section 3 reviews the two main density estimation techniques on the hyperbolic space. Section 4 presents an application to radar data estimation.

© Springer International Publishing Switzerland 2015
F. Nielsen and F. Barbaresco (Eds.): GSI 2015, LNCS 9389, pp. 753–761, 2015.
DOI: 10.1007/978-3-319-25040-3_80

2   The Hyperbolic Space and the Poincaré Disk Model

The hyperbolic geometry results of a modification of the fifth Euclid's postulate on parallel lines. In two dimensions, given a line D and a point $p \notin D$, the hyperbolic geometry is an example where there are at least two lines going through p, which do not intersect D. Let us consider the unit disk of the Euclidean plane endowed with the Riemannian metric:

$$ds_{\mathbb{D}}^2 = 4\frac{dx^2 + dy^2}{(1 - x^2 - y^2)^2} \tag{1}$$

where x and y are the Cartesian coordinates. The unit disk $\mathbb{D}$ endowed with $ds_{\mathbb{D}}$ is called the Poincaré disk and is a model of the two-dimensional hyperbolic geometry $\mathcal{H}_2$. The construction is generalized to higher dimensions. It can be shown that the obtained Riemannian manifold $\mathcal{H}_n$ is homogeneous. In other words,

$$\forall p, q \in \mathcal{H}_n, \exists \varphi \in ISO(\mathcal{H}_n), \varphi(p) = q$$

where $ISO(\mathcal{H}_n)$ is the set of isometric transformations of $\mathcal{H}_n$.

In $\mathbb{R}^n$ the convolution of a function f by a kernel g consists in the integral of translated kernel weighted by f in each point p of the support space. The group law $+$ of $\mathbb{R}^n$ is an isometry that enables to transport the kernel in the whole space. In the Riemannian setting the definition of convolution $f * g$ needs some homogeneity assumption: in an homogeneous space it is possible to transport a kernel from a reference point to any other point by an isometry. Let the isotropy group of p be the set of isometries that fix p. The convolution is properly defined for kernels that are invariant under elements of the isotropy group of p, for a p in $\mathcal{H}_n$. Formally, let $K_{p_{ref}}$ be a function invariant to the isotropy group of p_{ref}. Let $K_p = K_{p_{ref}} \circ \varphi_{p,p_{ref}}$ with $\varphi_{p,p_{ref}}$ an isometry such that $\varphi_{p,p_{ref}}(p) = p_{ref}$. When it exists, since $\mathcal{H}_n$ is homogeneous, we can define the convolution of a function f by the kernel $K_{p_{ref}}$ by:

$$(f * K_{p_{ref}})(q) = \int_{p \in \mathcal{H}_n} f(p) K_p(q) dvol$$

where vol is the hyperbolic measure.

Furthermore it can be shown that for any couple of geodesic γ_1 and γ_2 starting from $p \in \mathcal{H}_n$, there exists φ in the isotropy group of p such that $\varphi(\gamma_1) = \gamma_2$. In other words, a kernel K_p invariant under the isotropy group of p looks the same in every directions at point p. For more details on the hyperbolic space, see [10].

3   Non-parametric Probability Density Estimation

Let Ω be a space endowed with a probability measure. Let X be a random variable $\Omega \to \mathcal{H}_n$. The measure on $\mathcal{H}_n$ induced by X is noted μ_X. We assume that μ_X has a density, noted f, with respect to vol, and that the support of X is a compact set noted $Supp(X)$. Let $(x_1, .., x_k) \in (\mathcal{H}_n)^k$ be a set of draws of X.

Let $\mu_k = \frac{1}{k}\sum_i \delta_{x_i}$ denote the empirical measure of the set of draws. This section presents the main techniques of estimation of f from the set of draws $(x_1, .., x_k)$. The estimated density at $x \in \mathcal{H}_n$ is noted $\hat{f}_k(x) = \hat{f}_k(x, x_1, ..., x_k)$. Observe that $\hat{f}_k(x)$ can be seen as a random variable. The relevance of density estimation technique depends on several aspects. Recall that $\mathcal{H}_n$ is isotropic, every point and direction is indiscernible. In absence of prior information on the density, the estimation technique should not privilege specific directions or locations. This results in a homogeneity and an isotropy condition. Convergence rates of the different density estimators is widely studied. Results were first obtained in the Euclidean case, and are progressively extended to the probability densities on manifold [1–4]. The last aspect, is computational. Each estimation technique has its own computational framework, which presents pro and cons given the different applications. For instance, the estimation by orthogonal series presents an initial pre-processing, but provides a fast evaluation of the estimated density in compact manifolds. These aspects are studied for the main techniques of density estimation in the remaining of the paper.

Every standard density estimation technique involves a scaling parameter. This scaling factor controls the influence of the observation x_i on the estimated density at x, depending on the distance between x and x_i. In the experiments, the scaling factor is chosen following the framework proposed in [11]: a cross validation of the likelihood of the estimator.

3.1   Characteristic Function, or Orthogonal Series

Let $U \in \mathcal{H}_n$ be such that $Supp \subset U$. Consider a generalized orthonormal basis (e_j) consisting of eigenfunctions of the Laplace operator on $L^2(U)$. Recall that

$$\langle f, g \rangle = \int_U f\overline{g}dvol,$$

where $\overline{x}$ denotes the complex conjugate of x. Eigenfunctions of the Laplace operator in $\mathcal{H}_n$ behave similarly to the eigenfunctions in the real case. Thus, if the density f is in $L^2(U)$ the Fourier-Helgason transform gives:

$$f = \sum_j \langle f, e_j \rangle e_j, \text{ or } f = \int_{\mathcal{B}} \langle f, e_j \rangle e_j de_j,$$

respectively when U is compact and non compact. $\mathcal{B}$ is a subset of the eigenfunctions of the Lapacian. See [12] for an expression of de_j and $\mathcal{B}$ when $U = \mathcal{H}_n$. The law of large number gives,

$$\langle f, e_i \rangle = \int f\overline{e_j}dvol = \mathbb{E}\left(\overline{e_j}(X)\right) \approx \frac{1}{k}\sum_{j=1}^{k}\overline{e_j}(x_i).$$

Let e_λ be such that $\Delta e_\lambda = \lambda e_\lambda$, where Δ denotes the Laplace operator. Given $T > 0$, let $\mathcal{B}_T = \{e_\lambda \in \mathcal{B}, \lambda < T\}$. The density estimator becomes:

$$\hat{f}_k = \frac{1}{k} \sum_{e_j \in \mathcal{B}_T} \left[\sum_{i=1}^{k} \overline{e_j}(x_i) \right] e_j, \text{ or } \hat{f}_k = \frac{1}{k} \int_{\mathcal{B}_T} \left[\sum_{i=1}^{k} \overline{e_j}(x_i) \right] e_j de_j.$$

The parameter T plays the role of the inverse of the scaling parameter. In the Euclidean context, this is equivalent to the characteristic function density estimator. The choice of the basis is motivated by the regularity of the eigenfunctions of the Laplace operator. Let I_T be the indicator function of $\mathcal{B}_T$ in $\mathcal{B}$. For a right $\mathcal{B}_T$, the inverse Fourier Helgason transform of I_T, $\mathcal{FH}^{-1}(I_T)$, is invariant under the isotropy group of a $p \in \mathcal{H}_n$. Then, the estimator is a convolution [13] written as

$$\hat{f}_k = \mu_k * \mathcal{FH}^{-1}(I_T),$$

In other words, the estimation does not privilege specific locations or directions. Convergence rates are provided in [2]. When U is compact, the estimation $\hat{f}$ is made through the estimation of N scalar product, that is to say kN summation operation. However, the evaluation at $x \in \mathcal{H}_n$ involves only a sum of N terms. On the other hand, when U is not compact, the evaluation of the integral requires significantly higher computational cost. Unfortunately, the eigenfunction of the Laplacian for $U \subset \mathcal{H}_n$ are only known for $U = \mathcal{H}_n$. See [1,14] for more details on orthogonal series density estimation on Riemannian manifolds.

3.2   Kernel Density Estimation

Let $K : \mathbb{R}_+ \to \mathbb{R}_+$ be a map which verifies the following properties: $\int_{\mathbb{R}^n} K(||x||)dx = 1$, $\int_{\mathbb{R}^n} xK(||x||)dx = 0$, $K(x > 1) = 0$, $sup(K(x)) = K(0)$. Given a point $p \in \mathcal{H}_n$, exp_p defines a new injective parametrization of $\mathcal{H}_n$. The Lebesgue measure of the tangent space is noted Leb_p. The function $\theta_p : \mathcal{H}_n \to \mathbb{R}_+$ defined by:

$$\theta_p : q \mapsto \theta_p(q) = \frac{dvol}{dexp^*(Leb_p)}(q), \tag{2}$$

is the density of the Riemannian measure with respect to the image of the Lebesgue measure of $T_p\mathcal{H}_n$ by exp_p. Given K and a scaling parameter λ, the estimator of f proposed by Pelletier in [3] is defined by:

$$\hat{f}_k = \frac{1}{k} \sum_i \frac{1}{\lambda^n} \frac{1}{\theta_{x_i}(x)} K\left(\frac{d(x, x_i)}{\lambda}\right). \tag{3}$$

It can be noted that this estimator is the usual kernel estimator in the case of Euclidean space. Convergence rates are provided in [3] for the case of compact manifold. In the present situation, we only have a compact hypothesis on *Supp*. The notion of the double of a manifold [20] enables to consider any compact manifold with boundaries as a sub-manifold of a compact manifold without boundaries. This enables to extend the convergence rates to the present situation. These rates are similar to those of the orthogonal series. In [3] is furthermore shown that x_i is the intrinsic mean of $\frac{1}{\theta_{x_i}(.)}K(d(., x_i)/\lambda)$. Under reasonable assumptions on the true density f, the shape of the kernel does not have a significant impact on the quality of the estimation in the Euclidean context [15]. Figure 1 experimentally confirms the result in $\mathcal{H}_2$.

Given a reference point $p_{ref} \in \mathcal{H}_n$, let

$$\tilde{K}(q) = \frac{1}{k\lambda^n} \frac{1}{\theta_{p_{ref}}(q)} K\left(\frac{d(p_{ref}, q)}{\lambda}\right). \tag{4}$$

Note first that if ϕ is an isometry of $\mathcal{H}_n$, $\theta_p(q) = \theta_{\phi(p)}(\phi(q))$. After noticing that $\tilde{K}$ is invariant under is isotropy group of p_{ref}, a few calculations lead to:

$$\hat{f}_k = \mu_k * \tilde{K}. \tag{5}$$

As the density estimator based on the eigenfunction of the Laplacian operator, the kernel density estimator is a convolution and does not privilege specific locations or directions. In order to evaluate the estimated density at $x \in \mathcal{H}_n$, one first need to determine the observations x_i such that $d(x, x_i) < \lambda$, and perform a sum over the selected observations.

One still needs to obtain an explicit expression of θ_p. Given a reference point p, the point of polar coordinates (r, α) of the hyperbolic space is defined as the point at distance r of p on the geodesic with initial direction $\alpha \in \mathbb{S}^{n-1}$. Since the $\mathcal{H}_n$ is isotropic the expression the length element in polar coordinates depends only on r. Expressed in polar coordinates the hyperbolic metric expression is [16, 17]:

$$g_{\mathcal{H}_n} = dr^2 + sinh(r)^2 g_{\mathbb{S}^{n-1}}.$$

The polar coordinates are a polar expression of the exponential map at p. In an adapted orthonormal basis of the tangent plane the metric takes then the following form:

$$G = \begin{pmatrix} 1 & 0 \\ 0 & sinh(r)^2 \frac{1}{r^2} I_{n-1} \end{pmatrix} \tag{6}$$

where G is the matrix of the metric and I_{n-1} is the identity matrix of size $n-1$. Thus, using (6), the volume element $d\mu_{exp_p^*}$ is given by

$$dvol = \sqrt{|G|}.dexp_p^*(Leb_p) = (\tfrac{1}{r}sinh(r))^{n-1} dexp_p^*(Leb_p). \tag{7}$$

where $r = d(p, q)$. From (2) and (7), one obtains

$$\theta_p(q) = (\tfrac{1}{r}sinh(r))^{n-1}. \tag{8}$$

Finally, plugging (8) into (3) gives

$$\hat{f}_k = \frac{1}{k}\sum_i \frac{1}{\lambda^n} \frac{d(x, x_i)^{n-1}}{sinh(d(x, x_i))^{n-1}} K\left(\frac{d(x, x_i)}{\lambda}\right). \tag{9}$$

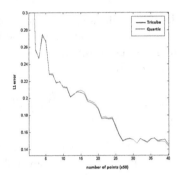

Fig. 1. Consider a law X whose density in the tangent plane at $(0,0)$ is a centered hemisphere. From a set of draws, the density is estimated using two standard kernels of the Euclidean plane, $K(||x||) = \frac{3}{\pi}(1-||x||^2)^2 \mathbf{1}_{||x||<1}$ and $K(x) = \frac{220}{81\pi}(1-||x||^3)^3 \mathbf{1}_{||x||<1}$. The L_1 distance between the estimated and the true density is plotted depending on the number of draws.

4   Application to Radar Estimation

4.1   From Radar Observations to Reflection Coefficients $\mu_k \in \mathbb{D}$

Let us discuss briefly how radar data is related to hyperbolic space via reflection coefficients, for more details see [9]. Each radar cell is a complex vector $z = (z_0, \cdots, z_{n-1})$ considered as a realization of a centered stationary Gaussian process $Z = (Z_0, \cdots, Z_{n-1})$ of covariance matrix $R_n = \mathbb{E}[ZZ^*]$. The matrix R_n has a Toeplitz structure, that is to say $(R_n)_{i,j} = (R_n)_{i+1,j+1}$ for $0 \le i,j < n$. For $1 \le k \le l \le n-1$, the k-th order autoregressive estimate of Z_l is given by $\hat{Z}_l = -\sum_{j=1}^{k} a_j^{(k)} Z_{l-j}$, where the autoregressive coefficients $a_1^{(k)} \cdots a_k^{(k)}$ are chosen such that the mean squared error $\mathbb{E}(|Z_l - \hat{Z}_l|^2)$ is minimized. In practice, reflection coefficients are estimated by regularized Burg algorithm [7]. The last autoregressive coefficient $a_k^{(k)}$ is called the k-th reflection coefficient, denoted by μ_k and which has the property $|\mu_k| < 1$. The coefficient for $k = 0$ corresponds to the power, denoted $P_0 \in \mathbb{R}_+^*$. The reflection coefficients induce a (diffeomorphic) map φ between the Toeplitz Hermitian positive definite (HPD) matrices of order n, $\mathcal{T}^n$, and reflection coefficients:

$$\varphi: \ \mathcal{T}^n \to \mathbb{R}_+^* \times \mathbb{D}^{n-1}, \ \ R_n \mapsto (P_0, \mu_1, \cdots, \mu_{n-1})$$

where $\mathbb{D} = \{\zeta \in \mathbb{C} \ : \ |\zeta| < 1\}$ is the open unit disk of the complex plane. Diffeomorphism φ is very closely related to theorems of Trench [19].

The Riemannian geometry of the space of reflection coefficients has been explored in [8] through the Hessian of Kähler potential, whose metric is

$$ds^2 = n\frac{dP_0^2}{P_0^2} + \sum_{k=1}^{n-1}(n-k)\frac{|d\mu_k|^2}{(1-|\mu_k|^2)^2}. \tag{10}$$

According to the metric (10) the space $\mathcal{T}^n$ can be seen as a product of the Riemannian manifold $\left(\mathbb{R}_+^*, ds_0^2\right)$, with $ds_0^2 = n(dP_0^2/P_0^2)$ (logarithmic metric multiplied by n), and $(n-1)$ copies of $\left(\mathbb{D}, ds_k^2\right)_{1 \leq k \leq n-1}$, with $ds_k^2 = (n-k)ds_{\mathbb{D}}^2$. $\left(\mathbb{R}_+^* \times \mathbb{D}^{n-1}, ds^2\right)$ is a Cartan–Hadamard manifold whose sectional curvatures are bounded, i.e., $-4 \leq K \leq 0$. This metric is related to information geometry and divergence functions [18]. From the product metric, closed forms of the Riemannian distance, arc-length parameterized geodesic, etc. can be obtained [8,9].

The next paragraph presents the estimations of the marginal densities coefficients μ_k.

4.2   Experimental Results

Data used in the experimental tests are radar observations from THALES X-band Radar, recorded during 2014 field trials campaign at Toulouse Blagnac Airport for European FP7 UFO study (Ultra-Fast wind sensOrs for wake-vortex hazards mitigation). Data are representative of Turbulent atmosphere monitored by radar in rainy conditions. Figure 2 illustrates the density estimation of the six reflection coefficients on the Poincaré unit disk. For each coefficient the dataset is composed of 120 draws.

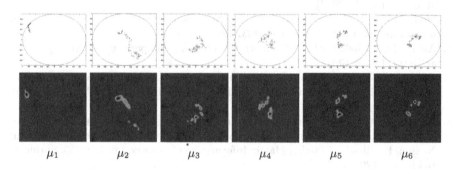

Fig. 2. Estimation of the density of the 6 first coefficients μ_k under rainy conditions. The expression of the used kernel is $K(x) = \frac{3}{\pi}(1-x^2)^2 \mathbf{1}_{x<1}$

5   Conclusion and Perspectives

We have discussed the problem of density estimation on the hyperbolic space. After having computed the volume change factor, we have adopted the approach based on kernel density estimation by Pelletier [3]. The method has been used to estimate the density of reflection coefficients from radar signals. According to the diffeomorphism between Toeplitz HPD matrices and reflection coefficients [19], densities estimated on a product of Poincaré disk can be interpreted as a probability density on the space of Toeplitz HPD matrices. Other symmetric homogenous spaces such as the Siegel disk can be addressed using similar methods.

The link between the Siegel disk and the space of Toeplitz-Block-Toeplitz HPD matrices makes of it an other interesting study case.

Alternative approaches of density estimation can be considered in future research. For data lying in a known compact symmetric subspace of the hyperbolic space, it is possible to use the orthogonal series technique, where the eigenfunctions of Laplace operator are numerically estimated. From an application viewpoint, densities from radar reflection coefficients can be used as basic ingredient in radar detection algorithms (finding modes of density and segmenting the density).

References

1. Hendriks, H.: Nonparametric estimation of a probability density on a Riemannian manifold using Fourier expansions. Ann. Stat. **18**(2), 832–849 (1990)
2. Huckeman, S., Kim, P., Koo, J., Munk, A.: Mobius deconvolution on the hyperbolic plan with application to impedance density estimation. Ann. Stat. **38**(4), 2465–2498 (2010)
3. Pelletier, B.: Kernel density estimation on Riemannian manifolds. Stat. Probab. Lett. **73**(3), 297–304 (2005)
4. Asta, D.: Kernel Density Estimation on Symmetric Spaces. arXiv preprint arXiv:1411.4040 (2014)
5. Said, S., Bombrun, L., Berthoumieu, Y.: New Riemannian priors on the univariate normal model. Entropy **16**, 4015–4031 (2014)
6. Asta, D., Shalizi, C.: Geometric network comparison. arXiv preprint arXiv:1411.1350 (2014)
7. Barbaresco, F.: Super resolution spectrum analysis regularization: burg, capon and ago antagonistic algorithms. In: Proceedings of EUSIPCO 1996, Trieste, pp. 2005–2008 (1996)
8. Barbaresco, F.: Information geometry of covariance matrix: cartan-siegel homogeneous bounded domains, mostow/berger fibration and fréchet median. In: Nielsen, F., Bhatia, R. (eds.) Matrix Information Geometry, pp. 199–255. Springer, Heidelberg (2013)
9. Arnaudon, M., Barbaresco, F., Yang, L.: Riemannian medians and means with applications to radar signal processing. IEEE J. Sel. Top. Sign. Proces. **7**(4), 595–604 (2013)
10. Cannon, J.W., Floyd, W.J., Kenyon, R., Parry, W.R.: Hyperbolic geometry. In: Levy, S. (ed.) Flavors of Geometry, vol. 31. MSRI Publications, Cambridge (1997)
11. Duin, R.P.W.: On the choice of smoothing parameters for Parzen estimators of probability density functions. IEEE Trans. Comput. **25**(11), 1175–1179 (1976)
12. Helgason, S.: Non-Euclidean Analysis, arXiv:math/0411411 [math.DG] (2004)
13. Helgason, S.: The Abel, Fourier and Radon transform on symmetric spaces. Indagationes Math. **16**(3), 531–551 (2005)
14. Kim, P., Richards, D.: Deconvolution density estimation on the space of positive definite symmetric matrices. In: Nonparametric Statistics and Mixture Models: A Festschrift in Honor of Thomas P. Hettmansperger, pp. 147–168. World Scientific Publishing (2008)
15. Silverman, B.W.: Density Estimation for Statistics and Data Analysis. Chapman and Hall, New York (1986)

16. Anker, J.-Ph., Ostellari, P.: The heat kernel on noncompact symmetric spaces. In: Lie groups and symmetric spaces, AMS Transl. Ser. 2, vol. 210, pp. 27–46 (2003)
17. Grigor'yan, A.: Heat kernel and analysis on manifolds. Am. Math. Soc. **47** (2012)
18. Barbaresco, F.: Koszul information geometry and souriau geometric temperature/capacity of lie group thermodynamics. Entropy **16**, 4521–4565 (2014)
19. Trench, W.F.: An algorithm for the inversion of finite Toeplitz matrices. J. Soc. Indust. Appl. Math. **12**, 515–522 (1964)
20. Munkres, J.R.: Elementary Diffrentaible Topology. Annals of Mathematics Studies, vol. 54. Princeton University Press, Princeton (1966)

Histograms of Images Valued in the Manifold of Colours Endowed with Perceptual Metrics

Emmanuel Chevallier[1], Ivar Farup[2], and Jesús Angulo[1](✉)

[1] CMM-Centre de Morphologie Mathématique, MINES ParisTech,
PSL-Research University, Fontainebleau, France
{emmanuel.chevallier,jesus.angulo}@mines-paristech.fr
[2] Gjøvik University College, Gjøvik, Norway

Abstract. We address here the problem of perceptual colour histograms. The Riemannian structure of perceptual distances is measured through standards sets of ellipses, such as Macadam ellipses. We propose an approach based on local Euclidean approximations that enables to take into account the Riemannian structure of perceptual distances, without introducing computational complexity during the construction of the histogram.

Keywords: Colour images · Image histograms · Riemannian metrics

1 Introduction

The histogram of the intensities is a fundamental descriptor of a grayscale image. It is one of the most important tools to address problems such as contrast enhancement (by histogram linear stretching or using advanced approaches [1]), image segmentation (by 1D clustering), texture processing [2], image retrieval [3], etc. The standard way of computing a histogram is to cut the value space into regular bins and to count the number of pixels that fall into each bin. However the obtained histogram presents important discontinuities. One thus prefers sometimes to use kernel methods which provide smoother results. One sometimes considers the color space as a part of a three dimensional Euclidean space. Under this assumption, the histogram of a color image can be built in the same way as for gray scale image. However, the distances induced on colours by the human perceptual system cannot be represented by a Euclidean space structure. Observation showed that the perceptual relation between colours is better represented in the framework of Riemannian manifolds. The local metrics of the Riemannian structure are experimentally measured by a set of ellipses, such as the Macadam ellipses [4], BFD-P [5] and RIT-DuPont [6]. This Riemannian structure makes the construction of the histogram difficult. On the one hand, except rare situations, there are no regular tilings of the space. On the other hand, kernel methods have been generalized to Riemannian manifolds in [7], but requires the knowledge of the geodesics. In this paper, we propose an approach that takes into account the Riemannian structure while keeping the computation in the

© Springer International Publishing Switzerland 2015
F. Nielsen and F. Barbaresco (Eds.): GSI 2015, LNCS 9389, pp. 762–769, 2015.
DOI: 10.1007/978-3-319-25040-3_81

Euclidean framework. Thus we propose a way of building histograms that better respects the perceptual distances than histograms built in Euclidean spaces, without increasing the computation time.

2   Image Histogram and Density Estimation

Let us consider an image I as a map:

$$I : \begin{cases} D \to V \\ p \mapsto I(p) \end{cases}$$

We have, for instance $V = \mathbb{R}$ for grey-scale images or $V = \mathbb{R}^n$ for multi-spectral images. D is the support space of pixels/voxels, typically a subset of $\mathbb{Z}^2$ or $\mathbb{Z}^3$. The set of values $\{I(p), p \in D\}$ is interpreted as a set of realizations of a random variable X. Let us assume that a reference measure μ is given on the space V. Furthermore, make the strong assumption that the law of X has a density f with respect to μ. The density f is an interesting quantity in image processing.

There are various ways of addressing the problem of probability density estimation. In the Euclidean context the most popular techniques are mainly the histograms, the kernels, and the characteristic function density estimator. The characteristic function density estimator consists in the estimation of the Fourier transform or series of the density. Each of these three techniques can be transported in most of Riemannian manifolds. However, the kernel methods become often significantly simpler than the two others. On the one hand, the histogram requires a regular tiling of the space which is a difficult problem for most of Riemannian manifolds. On the other hand the characteristic function method requires explicit expressions of the eigenfunctions of the Laplacian operator, these functions being known only in a few spaces. For its part, the kernel method only requires the knowledge of geodesic distances. In what follows, we chose to focus on the kernel method. Recall that the kernel method in the Euclidean case has the following form:

$$f(x) = \frac{1}{k} \sum_{p_i \in \{pixels\}} \frac{1}{\lambda^n} K(\frac{||x - I(p_i)||}{\lambda})$$

where λ is a scaling parameter, n the dimension of the space, k the number of pixels, and $K : \mathbb{R}_+ \to \mathbb{R}_+$ a map which obeys the following properties: $\int_{\mathbb{R}^n} K(||x||)dx = 1$, $\int_{\mathbb{R}^n} xK(||x||)dx = 0$, $sup(K(x)) = K(0)$. In this paper, we assume a supplementary condition of bounded support $K(x > 1) = 0$.

3   Basics on Riemannian Manifolds

Let $\mathcal{M}$ be a topological space, homeomorphic to an open subset of $\mathbb{R}^n$. An homeomorphism is bijective continuous map whose inverse is also continuous.

Let ϕ be an homeomorphism from $U_\phi \subset \mathbb{R}^n$ to $\mathcal{M}$. ϕ is a parametrization of $\mathcal{M}$. A Riemannian metric is a smooth field of scalar product on U_ϕ. In other words, a Riemannian metric associates a positive definite matrix to each points of U_ϕ. A smooth path is a map $\gamma : [a, b] \to \mathcal{M}$ such that $\phi^{-1} \circ \gamma$ is piece-wise C^1. Let γ be such a path. The Riemannian metric induces a notion of length on smooth path:

$$L(\gamma) = \int_a^b \sqrt{\langle (\phi^{-1} \circ \gamma)'(t), (\phi^{-1} \circ \gamma)'(t) \rangle_{(\phi^{-1} \circ \gamma)(t)}} dt$$

Where $\langle .,. \rangle_{(\phi^{-1} \circ \gamma)(t)}$ is the scalar product attached to the point $(\phi^{-1} \circ \gamma)(t)$. The notion of shortest path between two points induces a distance on $\mathcal{M}$. A shortest path is called a geodesic path and can be seen as straight segments.

The scalar product is entirely determined by its unit ball. Expressed in vector coordinates, the associated unit ball takes the form of an ellipse in two dimensions or of an ellipsoid in three dimensions. Thus, the Riemannian metric is given by a field of ellipses or ellipsoids.

4   Perceptual Metric on Colours

Already Riemann used colour as an illustration of the applicability of his geometry [8], and concrete examples of such colour geometries were developed by Helmholz [9], Schrödinger [10] and Stiles [11].

4.1   Ellipses, Local Metric

The first experimental determination of the field of ellipses describing the Riemannian metric of the colour space was performed by MacAdam [4]. The experiment consisted of about 25 000 colour matches with one observer, and the ellipses were derived from the covariance matrices of the repeated observations. Later, it has become common practice to denote ellipses obtained in this manner as JND (just noticeable difference) ellipses or ellipsoids.

Later, another type of experiment has become more commonplace. Pairs of colours that are barely perceptually different, are presented to the observer, who is given the task to estimate the magnitude of the perceptual distance using a set of standard pairs. Ellipses, ellipsoids and metrics obtained in this way are normally denoted supra-threshold ellipses. Examples of supra-threshold color difference based data include BFD-P [5] and RIT-DuPont [6].

4.2   Global Model

Data sets of measurements provide information on distances through the local metric or through distances between specific colours. A global model provides an analytic expression of the distance between two arbitrary colours. The closest the proposed expression is to the Riemannian perceptual distances, the better the model is. The more conventional procedure for going from a tristimulus space to a space closer linked to a perceptual homogeneous space typically includes the following steps. First, apply a linear transform in the tristimulus space such that the base gets close to the cone fundamentals of the retina. Secondly, perform a non-linear compression of the coordinates (e.g., logarithmic or cubic root) in order to mimic the non-linear response of the human visual system. Finally, perform a linear transformation of the resulting coordinates in order to correspond better to the perceptual attributes of colour. Typically, the first coordinate is a weighted sum of the coordinates and represent a lightness correlate, whereas the two other coordinates are weighted differences, and represent colour opponent channels such as, e.g., red–green and blue–yellow.

In order to identify the different parameters of the various transforms, different optimisation criteria are used. In the CIELAB colour space [13], the parameters were optimised such that the lightness should correspond to perceived lightness, and that the Euclidean metric in the resulting space should correspond to perceptual colour differences. For the IPT colour space [14], the parameters were optimised in order to achieve a constant perceived hue along straight radial lines in cylindrical coordinates. It is furthermore reasonably well established that in such perceptual spaces, the Euclidean metric is not the one best corresponding to the perceived colour differences, and other models have been proposed, see, e.g., Luo et al. [15] and Farup [12]. In the hyperbolic models proposed in Farup [12], histograms can be computed using adapted kernels, see [16].

5   Kernel Density Estimation Using Local Euclidean Approximations

In general Riemannian manifolds, computing the distance between two arbitrary points given the metric field is a difficult problem. Indeed, finding the distance is a minimization problem over a set of paths. However, for two close points, the local metric provides a satisfying approximation of the Riemannian distance. A probability density measures the ratio between the probability of an infinitesimal volume element and its volume. It is thus a local notion. The central idea of this section takes advantage of the fact that histograms mainly involves local phenomena. Since in a Riemannian manifold the computation of an histogram does not involve computation of long geodesics, the need of a global model that provides distances between every pairs of colours is of lower importance than in most applications.

Figure 1 shows a set of ellipses in the projective ab plane. Let us assume that these ellipses represent the local perceptual metric. Let c be a point where the metric has been measured through the ellipse $\mathcal{E}_c$. In a neighborhood of c, computing distances using the metric measured at c is a better approximation of the perceptual distance than using the canonical euclidean distance of the ab plane. At a point p where the metric is originally unknown, a metric interpolated from the neighbor points c_i has all the odds of being more relevant than the canonical Euclidean metric of the map, see Fig. 1.

Let $d_R(p,q)$ be the perceptual distance between color p and color q. $d_R(p,q)$ is the Riemannian distance induced by the field of ellipses. Let $||p-q||$ be the distance associated with the canonical scalar product of the ab plane, and $||p-q||_c$ be the distance associated with the scalar product induced by the ellipse $\mathcal{E}_c$. Let $B(c,R)$ and $B_c(c,R)$ be the respective balls of center c and radius R. The previous discussion can be formalized as follow. It can be shown that:

$$lim_{x \to c} \frac{||x-c||_c}{d_R(x,c)} = 1$$

while if $||.||_c \neq ||.||$, the equality case being exceptional,

$$lim_{x \to c} \frac{||x-c||}{d_R(x,c)} \neq 1$$

Therefore for such a c there exists $A > 0$ such that,

$$\forall R > 0, \exists x \in B(c,R), A < \left| \frac{||x-c||}{d_R(x,c)} - 1 \right|. \tag{1}$$

On the other hand there exists a real positive number $R_c = R_{c,A}$ such that,

$$\forall x \in B(c,R_c), \left| \frac{||x-c||_c}{d_R(x,c)} - 1 \right| < A. \tag{2}$$

We have

$$sup_{B(c,R_c)} \left(\left| \frac{||x-c||_c}{d_R(x,c)} - 1 \right| \right) < A < sup_{B(c,R_c)} \left(\left| \frac{||x-c||}{d_R(x,c)} - 1 \right| \right).$$

Thus for $x \in B(c,R_c)$, $||x-c||_c$ is preferred to $||x-c||$. Consider a kernel K and a scaling parameter λ such that

$$\lambda \leq R_c \text{ and } B_c(c,\lambda) \subset B(c,R_c).$$

For $x \in B(c,R_c)$, $K\left(\frac{||x-c||_c}{\lambda}\right)$ is preferred to $K\left(\frac{||x-c||}{\lambda}\right)$. For $x \notin B(c,R_c)$, $K\left(\frac{||x-c||_c}{\lambda}\right) = K\left(\frac{||x-c||}{\lambda}\right) = 0$. Thus under these assumptions on the scaling parameter λ, the histogram

$$f(x) = \frac{1}{k} \sum_{p_i \in \{pixels\}} \frac{1}{\lambda^n} K\left(\frac{||x - I(p_i)||_{I(p_i)}}{\lambda} \right) \tag{3}$$

is preferred to the classical histogram. We think that the hypothesis on λ is reasonable in practice, its validation is a subject of further research. Note that the higher the resolution of the image is, the smaller λ is and then the more the hypothesis becomes reasonable.

5.1   Metric Interpolation and Euclidean Approximation

Let $\mathcal{M}$ be topological space, homeomorphic to an open subset of $\mathbb{R}^n$. Let ϕ be an homeomorphism from $U_\phi \subset \mathbb{R}^n$ to $\mathcal{M}$. A set of scalar products G_{c_i} is given for a set of points $(c_i) \in \mathcal{M}$. We consider here the problem of interpolation of the field of metrics. Let F_1 and F_2 be two smooth metric fields that coincide with the observed ellipses at the points (c_i). Despite the intuition, if no assumption is made on ϕ regarding the Riemannian distance, there are no criteria that enables to prefer F_1 or F_2. The problem of metric tensor interpolation is thus a difficult problem. In this paper, we adopt an elementary solution. Ellipses are represented in the projective ab plane. A Delaunay triangulation with respect to the canonical Euclidean metric of the plane is performed on the set (c_i). At a point p in the triangle $c_i c_j c_k$ the parameters of the interpolated ellipse $\mathcal{E}_p$ are linearly interpolated between the parameters of $\mathcal{E}_{c_i}$, $\mathcal{E}_{c_j}$, and $\mathcal{E}_{c_k}$ with respect to the barycentric coordinates of p. If p does not belong to one of the triangles of the Delaunay triangulation, we set $\mathcal{E}_p = \mathcal{E}_q$ where q is the projection of p on the convex hull of the set of centers.

5.2   Experimental Results

The RIT-DuPont dataset [6] shows that the perceptual metric is dependent of the luminance. Nevertheless, for visualization purpose we choose to abandon the luminance information in order to work with two dimensional data. The Macadam ellipses were measured at a fixed luminance, in the CIE chromaticity diagram. The ellipses are transported in the $L = 40$ plane of the Lab space. Forgetting the luminance coordinate, one obtains then a transport of the Macadam ellipses in an ab plane. Remind that the proposed framework is independent of the dimension and can be used in three dimensional spaces with standard datasets of ellipsoids.

Figure 1 presents the transported Macadam ellipses in the projective ab plane, the Delaunay triangulation of the set of centers and the interpolation of the ellipses. Figure 2 represents the density of the Riemannian measure with respect to the Lebesgue density of the plane. Recall that the expression of the density is given by $\sqrt{det(G)}$ where G is the metric tensor derived from the ellipse.

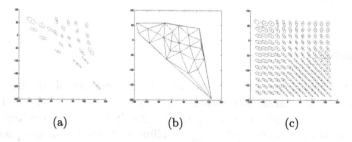

(a)   (b)   (c)

Fig. 1. (a) Macadam ellipses transported in an ab plane, (b) Delaunay triangulation, (c) ellipses interpolation

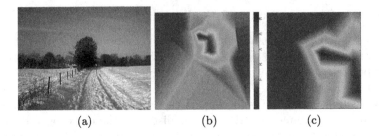

(a)   (b)   (c)

Fig. 2. (a) Color photography (b) local density change induced by the interpolated ellipses, (c) Zoom adapted to colours present in the photography

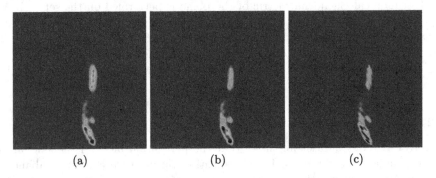

(a)   (b)   (c)

Fig. 3. The histogram of the image of Fig. 2(a) is computed using: the canonical Euclidean metric of the ab projective plane in (a), the canonical metric followed by a division by the local density of the perceptual metric in (b) and the formula (3) in (c) (Color figure online).

Figure 3 presents the histograms of the image of Fig. 2(a). Panels (b) and (c) aim a studying the density f with respect to the perceptual Riemannian volume measure. The main difference between (b) and (c) is that in (c) the shape of the kernel follows the Riemannian metric. The density with respect to the Euclidean measure is visibly different from the histogram with respect to the Riemannian measure. The amplitude of the upper spot, representing white colours, is significantly decreased when using the Riemannian measure. Perceptually, this results from the fact that the eyes have an higher sensitivity around white than around blue.

6   Conclusion

Given a set of ellipses representing the perceptual metric on colours, we proposed an approach for histogram computation that takes into account the Riemannian structure of the perceptual metric without introducing supplementary computational complexity. Indeed, the step of ellipses interpolation only has to be achieved once and does thus not introduce computational complexity. The relevance of the approach is conditioned by the relevance of the set of perceptual ellipses and the quality of the interpolation. The deep problem of metric tensor interpolation has been partially left aside and will be subject of future research. The second topic of our future research will be on the convergence of the proposed histogram to the density of the underlying random variable with respect to the interpolated Riemannian metric.

References

1. Sapiro, G., Caselles, V.: Histogram modification via differential equations. J. Differ. Equ. **135**(2), 238–268 (1997)
2. Portilla, J., Simoncelli, E.P.: A parametric texture model based on joint statistics of complex wavelet coefficients. Int. J. Comput. Vis. **40**, 47–71 (2000)
3. Gong, Y., Chuan, C.H., Xiaoyi, G.: Image indexing and retrieval using color histograms. Multimedia Tools Appl. **2**, 133–156 (1996)
4. MacAdam, D.L.: Visual sensitivities to color differences in daylight. J. Opt. Soc. Am. **32**(5), 247–274 (1942)
5. Luo, M.R., Rigg, B.: Chromaticity-discrimination ellipses for surface colours. Color Res. Appl. **11**(1), 25–42 (1986)
6. Berns, R.S., Alman, D.H., Reniff, L., Snyder, G.D., Balonon-Rosen, M.R.: Visual determination of suprathreshold color-difference tolerances using probit analysis. Color Res. Appl. **16**(5), 297–316 (1991)
7. Pelletier, B.: Kernel density estimation on Riemannian manifolds. Stat. Probab. Lett. **73**(3), 297–304 (2005)
8. Riemann, B.: Ueber die Hypothesen, welche der Geometrie zu Grunde liegen. Abh. Ge. Wiss. Gött **13**(1), 133–152 (1868)
9. von Helmholtz, H.: Versuch einer erweiterten Anwendung des Fechnerschen Gesetzes im farbensystem. Z. Psychol. Physiol. Sinnesorg. **2**, 1–30 (1891)
10. Schrödinger, E.: Grundlinien einer Theorie der Farbenmetrik im Tagessehen (III. Mitteilung). Ann. Phys. **368**(22), 481–520 (1920)
11. Stiles, W.S.: A modified Helmholtz line-element in brightness-colour space. P. Phys. Soc. **58**, 41–65 (1946)
12. Farup, I.: Hyperbolic geometry for colour metrics. Opt. Express **22**(10), 12369–12378 (2014)
13. Robertson, A.R.: The cie 1976 color-difference formulae. Color Res. Appl. **2**(1), 7–11 (1977)
14. Ebner, F., Fairchild, M.D.: Development and testing of a color space (IPT) with improved hue uniformity. In: The Sixth Color Imaging Conference: Color Science, Systems and Applications, IS&T, pp. 8–13 (1998)
15. Luo, M.R., Cui, G., Rigg, B.: The development of the CIE 2000 colour-difference formula: CIEDE2000. Color Res. Appl. **26**(5), 340–350 (2001)
16. Chevallier, E., Barbaresco, F., Angulo, J.: Probability density estimation on the hyperbolic space applied to radar processing. HAL, <hal-01121090>

Entropy Minimizing Curves with Application to Automated Flight Path Design

Stephane Puechmorel$^{(\boxtimes)}$ and Florence Nicol

Department SINA/MAIAA, Ecole Nationale de l'Aviation Civile (ENAC),
7, Avenue Edouard Belin, 31500 Toulouse, France
stephane.puechmorel@enac.fr, nicol@recherche.enac.fr

Abstract. Air traffic management (ATM) aims at providing companies with a safe and ideally optimal aircraft trajectory planning. Air traffic controllers act on flight paths in such a way that no pair of aircraft come closer than the regulatory separation norm. With the increase of traffic, it is expected that the system will reach its limits in a near future: a paradigm change in ATM is planned with the introduction of trajectory based operations. This paper investigate a mean of producing realistic air routes from the output of an automated trajectory design tool. For that purpose, an entropy associated with a system of curves is defined and a mean of iteratively minimizing it is presented. The network produced is suitable for use in a semi-automated ATM system with human in the loop.

1 Introduction

Based on recent studies [1], traffic in Europe is expected to grow on an average yearly rate of 2.6 %, yielding a net increase of 2 million flights per year at the 2020 horizon. Long term forecast gives a two fold increase in 2050 over the current traffic, pointing out the need for a paradigm change in the way flights are managed. Two major framework programs, SESAR (Single European Sky Air traffic management Research) in Europe and Nextgen in the US have been launched in order to first investigate potential solutions and to deploy them in a second phase. One of the main changes that the air traffic management (ATM) system will undergo is a switch from airspace based to trajectory based operations with a delegation of the separation task to the crews. Within this framework, trajectories become the basic object of ATM, changing the way air traffic controllers will be working. In order to alleviate the workload of controllers, trajectories will be planned several weeks in advance in such a way that close encounters are minimized and ideally removed. For that purpose, several tools are currently being developed, most of them coming from the field of robotics [6]. Unfortunately, flight path issued by these algorithms are not tractable for a human controller and need to be simplified. The purpose of the present work is to introduce an automated procedure that takes as input a set of trajectories and outputs a simplified one that can be used in an operational context. Using an entropy associated with a curves system, a gradient descent is performed in

© Springer International Publishing Switzerland 2015
F. Nielsen and F. Barbaresco (Eds.): GSI 2015, LNCS 9389, pp. 770–778, 2015.
DOI: 10.1007/978-3-319-25040-3_82

order to reduce it so as to straighten trajectories while avoiding areas with low air craft density, thus enforcing route-like behavior. This effect is related to the fact that entropy minimizing distributions favor concentration.

2   Entropy Minimizing Curves

2.1   Motivation

As previously mentioned, air traffic management of the future will make an intensive use of 4D trajectories as a basic object. Full automation is a far reaching concept that will probably not implemented before 2040–2050 and even in such a situation, it will be needed to keep humans in the loop so as to gain a wide societal acceptance of the concept. Starting from SESAR or Nextgen initial deployment and aiming towards this ultimate objective, a transition phase with human-system cooperation will take place. Since ATC controllers are used to a well structured network of routes, it is advisable to post-process the 4D trajectories issued by automated systems in order to make them as close as possible to line segments connecting beacons. To perform this task, in an automatic way, flight paths will be moved Add "iteratively" to dictionary so as to minimize an entropy criterion, that enforces avoidance of low density area and at the same time penalize length. Compared to already available bundling algorithms [3] that tend to move curves to high density areas, this new procedure generates geometrically correct curves, without excess curvature.

2.2   Entropy of a System of Curves

Let a set $\gamma_1, \ldots \gamma_N$ of smooth curves be given, that will be aircraft flight paths for the air traffic application. It will be assumed in the sequel that all curves are smooth mappings from $[0, 1]$ to a domain Ω of $\mathbb{R}^2$ with everywhere non vanishing derivatives in $]0, 1[$. This last condition allows to view trajectories as smooth immersions with boundaries and is sound from the application point of view as aircraft velocities cannot be 0 expect at the endpoints. A classical performance indicator used in ATM is the aircraft density [2], obtained from sampled positions $\gamma_i(t_j), j = 1 \ldots n_i$. It is constructed from a partition $U_k, k = 1 \ldots P$ of Ω by counting the number of samples occurring a given U_k then dividing out by the total number of samples $n = \sum_{i=1}^{N} n_i$. More formally, the density d_k in the subset U_k of Ω is:

$$d_k = n^{-1} \sum_{i=1}^{N} \sum_{j=1}^{n_i} 1_{U_k} \left(\gamma_i(t_j) \right) \tag{1}$$

with 1_{U_k} the characteristic function of the set U_k. It seems natural to extend the density obtained from samples to another one based on the trajectories themselves using an integral form:

$$d_k = \lambda^{-1} \sum_{i=1}^{N} \int_0^1 1_{U_k} \left(\gamma_i(t) \right) dt \tag{2}$$

where the normalizing constant λ is obtained as:

$$\lambda = \sum_{k=1}^{P} \sum_{i=1}^{N} \int_0^1 1_{U_k} \left(\gamma_i(t) \right) dt = \sum_{i=1}^{N} \int_0^1 \sum_{k=1}^{P} 1_{U_k} \left(\gamma_i(t) \right) dt$$

and since $U_k, k = 1 \ldots P$ is a partition:

$$\lambda = \sum_{i=1}^{N} \int_0^1 \gamma_i(t) dt \tag{3}$$

Density can be viewed as an empirical probability distribution with the U_k considered as bins in an histogram. It is thus natural to extend the above computation so as to give rise to a continuous distribution on Ω. For that purpose, a kernel function $K \colon \mathbb{R} \to \mathbb{R}^+$ is selected and a smooth version of the density [5] is defined as a mapping d from Ω to $[0,1]$:

$$d \colon x \mapsto \frac{\sum_{i=1}^{N} \int_0^1 K \left(\|x - \gamma_i(t)\| \right) dt}{\sum_{i=1}^{N} \int_{\Omega} \int_0^1 K \left(\|x - \gamma_i(t)\| \right) dt dx} \tag{4}$$

Standard choices for the K function are the ones used for non-parametric kernel estimation like the Epanechnikov function:

$$K \colon x \mapsto \left(1 - x^2 \right) 1_{[-1,1]}(x)$$

When K is compactly supported, which is the case of the Epanechnikov function and all its relatives, it comes:

$$\int_{\Omega} K \left(\|x - \gamma_i(t)\| \right) dx = \int_{\mathbb{R}^2} K \left(\|x\| \right) dx$$

provided that Ω contains the set:

$$\{ x \in \mathbb{R}^2, \inf_{i=1 \ldots N, t \in [0,1]} \|x - \gamma_i(t)\| \leq A \}$$

where the interval $[-A, A]$ contains the support of K. The case of kernels with unbounded support, like Gaussian functions, may be dealt with provided $\Omega = \mathbb{R}^2$. In the application considered, only compactly supported kernels are used, mainly to allow fast machine implementation of the density computation. Normalizing the kernel is not mandatory as the normalization occurs with the definition of d. It is nevertheless easier to consider only kernels satisfying:

$$\int_{\mathbb{R}^2} K \left(\|x\| \right) dx = 1$$

Using the polar coordinates (ρ, θ) and the rotation invariance of the integrand, the relation becomes:

$$2\pi \int_{\mathbb{R}^+} K(\rho) \rho d\rho = 1$$

Which yields a normalizing constant of $2/\pi$ for the Epanechnikov function, instead of the usual $3/4$ in the real case. When the normalization condition is fulfilled, the expression of the density simplifies to:

$$d: x \mapsto N^{-1} \sum_{i=1}^{N} \int_{0}^{1} K\left(\|x - \gamma_i(t)\|\right) dt \tag{5}$$

As an example, one day of traffic over France is considered and pictured on Fig. 1 with the corresponding density map, computed on a evenly spaced grid with a normalized Epanechnikov kernel:

Fig. 1. Traffic over France and associated density

Unfortunately, density computed this way suffers a severe flaw for the ATM application: it is not related to the shape of trajectories but also to the time behavior. Formally, it is defined on the set $\mathbf{Imm}\left([0,1], \mathbb{R}^2\right)$ of smooth immersions from $[0,1]$ to $\mathbb{R}^2$ while the space of primary interest will be the quotient by smooth diffeomorphisms of the interval $[0,1]$, $\mathbf{Imm}\left([0,1], \mathbb{R}^2\right)/\mathbf{Diff}([0,1])$. Invariance of the density under the action of $\mathbf{Diff}([0,1])$ is obtained as in [4] by adding a term related to velocity in the integrals. The new definition of d becomes:

$$\tilde{d}: x \mapsto \frac{\sum_{i=1}^{N} \int_{0}^{1} K\left(\|x - \gamma_i(t)\|\right) \|\gamma_i'(t)\| dt}{\sum_{i=1}^{N} \int_{\Omega} \int_{0}^{1} K\left(\|x - \gamma_i(t)\|\right) \|\gamma_i'(t)\| dt dx} \tag{6}$$

Assuming again a normalized kernel and letting l_i be the length of the curve γ_i, is simplifies to:

$$\tilde{d}: x \mapsto \frac{\sum_{i=1}^{N} \int_{0}^{1} K\left(\|x - \gamma_i(t)\|\right) \|\gamma_i'(t)\| dt}{\sum_{i=1}^{N} l_i} \tag{7}$$

The new **Diff**-invariant density is pictured on Fig. 2 along with the standard density. While the overall aspect of the plot is similar, one can observe that routes are more apparent the right picture and that the density peak located above Paris area is of less importance and less symmetric is due to the fact that near airports, aircraft are slowing down and this effect exaggerates the density with the non-invariant definition.

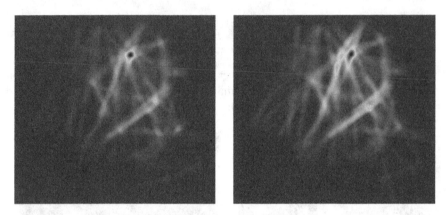

Fig. 2. Density (left) and **Diff** invariant density (right) for the 12th February 2013 traffic

Having a density at hand, the entropy of the system of curves $\gamma_1, \ldots, \gamma_N$ is defined the usual way as:

$$E(\gamma_1, \ldots, \gamma_N) = -\int_\Omega \tilde{d}(x) \log\left(\tilde{d}(x)\right) dx$$

2.3   Minimizing the Entropy

In order to fulfill the initial requirement of finding bundles of curve segments as straight as possible , one seeks after the system of curves minimizing the entropy $E(\gamma_1, \ldots, \gamma_N)$, or equivalently maximizing:

$$\int_\Omega \tilde{d}(x) \log\left(\tilde{d}(x)\right) dx$$

The reason why this criterion gives the expected behavior will become more apparent after derivation of its gradient at the end of this part. Nevertheless, when considering a single trajectory its is intuitive that the most concentrated density distribution is obtained with a straight segment connecting the endpoints.

Letting ϵ be a perturbation of the curve γ_j such that $\epsilon(0) = \epsilon(1) = 0$, the first order expansion of $-E(\gamma_1, \ldots, \gamma_N)$ will be computed in order to get a maximizing displacement field, analogous to a gradient ascent[1] in the finite dimensional setting. The notation:

$$\frac{\partial F}{\partial \gamma_j}$$

---

[1] Choice has been made to maximize the opposite of the entropy, so that the algorithm will be a gradient ascent one.

will be used in the sequel to denote the derivative of a function F of the curve γ_j in the sense that for a perturbation ϵ:

$$F(\gamma_j + \epsilon) = F(\gamma_j) + \frac{\partial F}{\partial \gamma_j}(\epsilon) + o(\|\epsilon\|_2)$$

First of all, please note that since $\tilde{d}$ has integral 1 over the domain Ω:

$$\int_\Omega \frac{\partial \tilde{d}(x)}{\partial \gamma_j}(\epsilon) dx = 0$$

so that:

$$-\frac{\partial}{\partial \gamma_j} E(\gamma_1, \ldots, \gamma_N)(\epsilon) = \int_\Omega \frac{\partial \tilde{d}(x)}{\partial \gamma_j}(\epsilon) \log\left(\tilde{d}(x)\right) dx \qquad (8)$$

Starting from the expression of $\tilde{d}$ given in Eq. 7, the first order expansion of $\tilde{d}$ with respect to the perturbation ϵ of γ_j is obtained as a sum of a term coming from the numerator:

$$\int_0^1 K\left(\|x - \gamma_j(t)\|\right) \|\gamma_j'(t)\| dt \qquad (9)$$

and a second one coming from the length of γ_j in the denominator. This last term is obtained from the usual first order variation formula of a curve length:

$$\int_{[0,1]} \|\gamma_j'(t) + \epsilon'(t)\| dt$$

$$= \int_{[0,1]} \|\gamma_j'(t)\| dt + \int_{[0,1]} \left\langle \frac{\gamma_j'(t)}{\|\gamma_j'(t)\|}, \epsilon'(t) \right\rangle dt + o(\|\epsilon\|_2)$$

Using an integration by part, the first order term can be written as:

$$\int_{[0,1]} \left\langle \frac{\gamma_j'(t)}{\|\gamma_j'(t)\|}, \epsilon'(t) \right\rangle dt \qquad (10)$$

$$= -\int_{[0,1]} \left\langle \left(\frac{\gamma_j''(t)}{\|\gamma_j'(t)\|}\right)_{\mathcal{N}}, \epsilon \right\rangle dt \qquad (11)$$

with:

$$\left(\frac{\gamma_j''(t)}{\|\gamma_j'(t)\|}\right)_{\mathcal{N}} = \frac{\gamma_j''(t)}{\|\gamma_j'(t)\|} - \frac{\gamma_j'(t)}{\|\gamma_j'(t)\|} \left\langle \frac{\gamma_j'(t)}{\|\gamma_j'(t)\|}, \frac{\gamma_j''(t)}{\|\gamma_j'(t)\|} \right\rangle$$

the normal component of:

$$\frac{\gamma_j''(t)}{\|\gamma_j'(t)\|}$$

for a curve in $\mathbb{R}^2$.

The integral in 9 can be expanded in a similar fashion. Using as above the notation $()_\mathcal{N}$ for normal components, the first order term is obtained as:

$$\int_{[0,1]} \left\langle \left(\frac{\gamma_j(t) - x}{\|\gamma_j(t) - x\|} \right)_\mathcal{N}, \epsilon \right\rangle K' \left(\|\gamma_j(t) - x\| \right) \|\gamma_j'(t)\| dt \tag{12}$$

$$- \int_{[0,1]} \left\langle \left(\frac{\gamma_j''(t)}{\|\gamma_j'(t)\|} \right)_\mathcal{N}, \epsilon \right\rangle K \left(\|\gamma_j(t) - x\| \right) dt \tag{13}$$

From the expressions in 12 and 10, the first order variation of the entropy is:

$$\frac{1}{\sum_{i=1}^N l_i} \tag{14}$$

$$\left(\int_{[0,1]} \left\langle \int_\Omega \left(\frac{\gamma_j(t) - x}{\|\gamma_j(t) - x\|} \right)_\mathcal{N} K' \left(\|\gamma_j(t) - x\| \right) \log \tilde{d}(x) dx, \epsilon \right\rangle \|\gamma_j'(t)\| dt \tag{15}$$

$$- \int_{[0,1]} \left(\int_\Omega K \left(\|\gamma_j(t) - x\| \right) \log \tilde{d}(x)) dx \right) \left\langle \left(\frac{\gamma_j''(t)}{\|\gamma_j'(t)\|} \right)_\mathcal{N}, \epsilon \right\rangle dt \tag{16}$$

$$+ \left(\int_\Omega \tilde{d}(x) \log(\tilde{d}(x)) dx \right) \int_{[0,1]} \left\langle \left(\frac{\gamma_j''(t)}{\|\gamma_j'(t)\|} \right)_\mathcal{N}, \epsilon \right\rangle dt \right) \tag{17}$$

As expected, only moves normal to the trajectory will change at first order the value of the criterion: the displacement of the curve γ_j will thus be performed at t along the normal vector $N_{\gamma_j}(t)$ and is given, up to the $(\sum_{i=1}^N l_i)^{-1}$ term by:

$$\int_\Omega \left(\frac{\gamma_j(t) - x}{\|\gamma_j(t) - x\|} \right)_\mathcal{N} K' \left(\|\gamma_j(t) - x\| \right) \log \tilde{d}(x) dx \|\gamma_j'(t)\| \tag{18}$$

$$- \left(\int_\Omega K \left(\|\gamma_j(t) - x\| \right) \log \tilde{d}(x)) dx \right) \left(\frac{\gamma_j''(t)}{\|\gamma_j'(t)\|} \right)_\mathcal{N} \tag{19}$$

$$+ \left(\int_\Omega \tilde{d}(x) \log(\tilde{d}(x)) dx \right) \left(\frac{\gamma_j''(t)}{\|\gamma_j'(t)\|} \right)_\mathcal{N} \tag{20}$$

The first term in the expression will favor moves towards areas of high density, while the second and third one are moving along normal vector and will straighten trajectory.

In practical implementation, the scaling factor in front of the whole expression is dropped and moves are made proportionally to the given vector. As usual with gradient ascent algorithms, one must carefully select the step taking in the maximizing direction in order to avoid divergence. A simple fixed step strategy was selected here and gives satisfactory results. The procedure applied to one day of traffic over France yields the picture of Fig. 3 As expected, a route-like network emerges. In such a case, since the traffic comes from an already organized situation, the recovered network is indeed a subset of the route network

Fig. 3. Initial and Bundled traffic of the 24/02/2013

in the french airspace. Please note that there is a trade-off between density concentration and minimal curvature of the recovered trajectories.

In the second example of Fig. 4, the problem of automatic conflict solving is addressed. In the initial situation, aircraft are converging to a single point, which is unsafe. Air traffic controllers will proceed in such a case by diverting aircraft from their initial flight path so as to avoid each other, but only using very simple maneuvers. An automated tool will make a full use of the available airspace and the resulting set of trajectories may fail to be manageable by a human: in the event of a system failure, no backup can be provided by controllers. The entropy minimization procedure was added to an automated conflict solver in order to end up with flight paths still tractable by humans. The final result is shown on the right part of Fig. 4, where encounters no longer exists but aircraft are bound to simple trajectories, with a merging and a splitting point. Note that since the automated planner acts on velocity, all aircraft are separated in time on the inner part.

 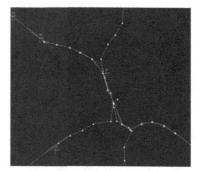

Fig. 4. Initial flight plans and final ones

3 Conclusion and Future Work

Algorithms coming from the field of shape spaces emerge as a valuable tool for applications in ATM. In this work, the foundations of a post processing procedure that may be applied after an automated flight path planner are presented. Entropy minimization makes straight segments bundle emerge, which fulfills the operational requirements. Computational efficiency has to be improved in order to release an usable building block for future ATM systems. One way to address this issue is to compute kernel density estimators using GPUs which excel in this kind of task, very similar texture manipulations. Furthermore, theoretical insights have to be gained in the next step of the work.

References

1. EUROCONTROL/NMD/STATFOR. Eurocontrol seven-year forecast, 2014. https://www.eurocontrol.int/sites/default/files/content/documents/official-docu ments/forecasts/seven-year-flights-service-units-forecast-2014-2020-sep2014.pdf
2. Harman, W.H.: Air traffic density and distribution measurements. Lincoln Laboratory, MIT, Report ATC-80 May 1979
3. Hurter, C., Ersoy, O., Telea, A.: Smooth bundling of large streaming and sequence graphs. In: Visualization Symposium (PacificVis), 2013 IEEE Pacific, pp. 41–48 Feb 2013
4. Michor, P.W., Mumford, D.: Riemannian geometries on spaces of plane curves. J. Eur. Math. Soc. (JEMS) **8**, 1–48 (2006)
5. Parzen, E.: On estimation of a probability density function and mode. Ann. Math. Statist. **33**(3), 1065–1076 (1962)
6. Roussos, G.P., Dimarogonas, D.V., Kyriakopoulos, K.J.: Distributed 3d navigation and collision avoidance for nonholonomic aircraft-like vehicles. In: Proceedings of the 2009 European Control Conference (2009)

Kernel Density Estimation on Symmetric Spaces

Dena Marie Asta[✉]

Carnegie Mellon University, Pittsburgh, PA 15217, USA
dasta@andrew.cmu.edu

Abstract. We introduce a novel kernel density estimator for a large class of symmetric spaces and prove a minimax rate of convergence as fast as the minimax rate on Euclidean space. We prove a minimax rate of convergence proven without any compactness assumptions on the space or Hölder-class assumptions on the densities. A main tool used in proving the convergence rate is the Helgason-Fourier transform, a generalization of the Fourier transform for semisimple Lie groups modulo maximal compact subgroups. This paper obtains a simplified formula in the special case when the symmetric space is the 2-dimensional hyperboloid.

1   Introduction

Data, while often expressed as collections of real numbers, are often more naturally regarded as points in non-Euclidean spaces. To take one example, radar systems can yield the data of bearings for planes and other flying objects; those bearings are naturally regarded as points on a sphere [8]. To take another example, diffusion tensor imaging (DTI) can yield information about how liquid flows through a region of the body being imaged; that three-dimensional movement can be expressed in the form of symmetric positive definite (3×3)-matrices [8]. To take yet another example, the nodes of certain hierarchical real-world networks can be regarded as having latent coordinates in a hyperboloid [2,6]. In all such examples, the spaces can be regarded as subsets of Euclidean space, but Euclidean distances between data points do not reflect the true distances between the points. Ordinary kernel density estimators applied to the sample data, expressed as points in Euclidean space, generally will not be optimal in terms of the L_2-risk with respect to the volume measure on the non-Euclidean manifold.

The literature offers some generalizations of kernel density estimation for (Riemannian) manifolds. One example for compact manifolds [7] requires that the kernels have small enough supports. Another example for compact manifolds [3] requires that the kernels generally vary at each point and the true density satisfies a Sobolev condition. A minimax rate in terms of a Sobolev parameter is proven. Moreover, that generalized kernel does not vary when the compact manifold is symmetric [3]. Another example for complete manifolds requires that analogues of kernels have to be chosen at each point [5]. A minimax convergence rate in terms of a Hölder class exponent and the differentiability of the true density is proven [5]. It is also noted in [3] that harmonic techniques, used to

© Springer International Publishing Switzerland 2015
F. Nielsen and F. Barbaresco (Eds.): GSI 2015, LNCS 9389, pp. 779–787, 2015.
DOI: 10.1007/978-3-319-25040-3_83

prove minimax convergence rates in terms of a Sobolev parameter in the compact case, extend to general symmetric spaces and in particular unify kernel density estimations on Euclidean space and compact manifolds.

The goal of this paper is to investigate kernel density estimation on a large class of symmetric spaces manifolds, describing the construction in detail and proving a minimax rate of convergence with no requirements that the space be compact, no requirements that the kernel vanish outside of a neighborhood, no Hölder class assumptions, and no requirement that the kernel be defined for each point of the space. The idea of kernel density estimation is to smooth out, or convolve, an empirical estimator with some noise K so as to obtain a smooth estimate of the true density on $\mathbb{R}^n$. Noise is a random translation of points on $\mathbb{R}^n$. Thus while K is also a density on $\mathbb{R}^n$, we should really think of K as a density on the space of translations of $\mathbb{R}^n$ (which is usually identified with $\mathbb{R}^n$ itself.) On a general Riemannian manifold $\mathbf{X}$, noise is generally a density on a more general space $\mathbf{G}$ of *symmetries* of $\mathbf{X}$ – and generally $\mathbf{G}$ cannot be identified with $\mathbf{X}$. For example, noise on a hyperboloid $\mathbb{H}_2$ of uniform curvature -1 can be regarded as a density on the space $\mathbf{SL}_2$ of (2×2)-matrices with determinant 1: each such matrix determines a distance-preserving smooth map from the hyperboloid to itself.

Thus the class of manifolds $\mathbf{X}$ on which we define our estimator are certain *symmetric spaces*, manifolds equipped with spaces of symmetries. In particular, the symmetric spaces we consider are quotients $\mathbf{X} = \mathbf{G}/\mathbf{K}$ of a semisimple Lie group $\mathbf{G}$ modulo a maximal compact subgroup $\mathbf{K}$. An example of such a symmetric space is the (non-compact) hyperboloid of uniform curvature -1, which can be regarded as the quotient $\mathbf{SL}_2/\mathbf{SO}_2$ of the group $\mathbf{SL}_2$ of (2×2)-matrices with determinant 1 by the group $\mathbf{SO}_2$ of (2×2)-rotation matrices with positive determinant.

We define a $\mathbf{G}$-*kernel density estimator*

$$\hat{f}^{(n,h,T)}_{\hat{x}_1, \hat{x}_2, \ldots, \hat{x}_n} : \mathbf{X} \to \mathbb{R},$$

an estimator for a density on the space $\mathbf{X}$ defined in terms of samples $\hat{x}_1, \hat{x}_2, \ldots, \hat{x}_n \in \mathbf{X}$, a bandwidth parameter h, and a cutoff parameter T. The $\mathbf{G}$-*kernel* used to define our estimator is a density on $\mathbf{G}$. An example of a $\mathbf{G}$-kernel is a generalized Gaussian, a solution to the heat equation on $\mathbf{G}$. Under smoothness assumptions, we bound the risk of the $\mathbf{G}$-kernel density estimator in terms of h, T, n, and the sum of the restricted roots of $\mathbf{X}$ [Theorem 1]. Optimizing h and T in terms of n, we obtain a minimax rate of

$$n^{-2\alpha/(2\alpha + \dim \mathbf{X})},$$

where α is a smoothing parameter, under natural assumptions on the density space and generalized kernel (D.1)–(D.5) [Corollary 2]. We then obtain a simplified formula (3), that can be easily implemented on a computer, for the special case where $\mathbf{X}$ is the hyperboloid.

2   Background

The main contribution of this paper is to use a specific extension of Harmonic Analysis to investigate kernel density estimation in detail for a large class of symmetric spaces, including non-compact symmetric spaces. Therefore we briefly recall the basic theory of *Helgason-Fourier Analysis* and fix some relevant notation for use throughout the paper. We refer the reader to [9] for a detailed treatment of the theory, and to [4] for an example of how the theory is applied to the deconvolution of noise.

2.1   Convolution

The convolution of densities on $\mathbb{R}^n$, necessary to deconvolve noise and smooth out empirical observations to obtain density estimators, is defined as follows. For a pair g, f of densities on $\mathbb{R}^n$ equipped with the Lebesgue measure $\mu_{\mathbb{R}^n}$, define the density $g * f$ on $\mathbb{R}^n$ by the rule

$$(g * f)(t) = \int_{\mathbb{R}^n} g(x) f(t - x) \, d\mu_{\mathbb{R}^n}.$$

Convolution, which involves the operation of subtraction, generalizes to symmetric spaces defined as follows. A *Lie group* is a manifold $\mathbf{G}$ consisting of invertible matrices such that multiplication and inversion are smooth maps

$$\cdot : \mathbf{G} \times \mathbf{G} \to \mathbf{G}, \quad (-)^{-1} : \mathbf{G} \to \mathbf{G}.$$

We take a $\mathbf{G}$-*space* $\mathbf{X}$ to be a Riemannian manifold $\mathbf{X}$ equipped with an *action*

$$\mathbf{G} \times \mathbf{X} \to \mathbf{X}$$

of a Lie group $\mathbf{G}$, a function whose restriction to a function $\mathbf{X} \to \mathbf{X}$ for each $g \in \mathbf{G}$ is an isometry. For a density g on a Lie group $\mathbf{G}$ having Haar measure $\mu_{\mathbf{G}}$ and density f on a $\mathbf{G}$-space $\mathbf{X}$ having volume measure $\mu_{\mathbf{X}}$, define the density $g * f$ on $\mathbb{R}^n$ by the rule

$$(g * f)(t) = \int_{\mathbf{G}} g(x) f(x^{-1}t) \, d\mu_{\mathbf{G}}.$$

Convolutions allow us to define ordinary kernel density estimators on $\mathbb{R}^n$ and more general $\mathbf{G}$-*kernel density estimators* on certain spaces $\mathbf{X}$ with symmetries described by a group $\mathbf{G}$. Just as Fourier analysis is useful for indirectly constructing and analyzing convolutions of densities on $\mathbb{R}^n$, a more general *Helgason-Fourier Analysis* will allow us to indirectly construct and analyze convolution of densities on more general manifolds. We define all relevant spaces, define the transform, and give an explicit formula for the inverse transform.

2.2   The Symmetric Space

We henceforth fix the following data.

1. connected noncompact semisimple Lie group $\mathbf{G}$ with finite center
2. maximal compact connected Lie subgroup $\mathbf{K}$ of $\mathbf{G}$.

The symmetric space $\mathbf{X}$ on which we wish to define a density estimator is the quotient space

$$\mathbf{X} = \mathbf{G}/\mathbf{K}.$$

Let $\mathbf{G} = \mathbf{KAN}$ be an Iwasawa decomposition of $\mathbf{G}$, a decomposition of $\mathbf{G}$ into a product of Lie subgroups such that $\mathbf{A}$ is Abelian and $\mathbf{N}$ is nilpotent. Let $\mathbf{M}$ be the centralizer of $\mathbf{A}$ in $\mathbf{K}$. Let $\mathfrak{g}, \mathfrak{k}, \mathfrak{a}, \mathfrak{n}$ represent the Lie algebras of $\mathbf{G}, \mathbf{K}, \mathbf{A}, \mathbf{N}$, respectively. Define $k_g \in \mathbf{G}$, $a_g \in \mathbf{A}$, and $n_g \in \mathbf{N}$ so that $g = k_g a_g n_g$. Let $\mu_{\mathbf{G}}$ denote the Haar measure on $\mathbf{G}$ and $\mu_{\mathbf{X}}$ denote the volume measure on $\mathbf{X}$. Let ρ be half of the sum of positive restricted roots of $\mathbf{X}$. We also write c for the *Harish-Chandra function* on $\mathfrak{a}^*$, which plays a role in defining the *Plancherel measure* on an appropriate analogue of frequency space for an appropriate analogue of Fourier Analysis; we refer the reader to [9] for suitable definitions. Let $C_c^\infty(N)$ denote the set of C^∞ real-valued functions with compact support on a manifold N.

2.3   Helgason-Fourier Transform

The *Helgason-Fourier transform* $\mathcal{H}$ sends each $f \in C_c^\infty(\mathbf{X})$ to the map

$$\mathcal{H}[f] : \mathfrak{a}^* \times \mathbf{K}/\mathbf{M} \to \mathbb{R}$$

defined as follows, where $\hat{f}$ is the lift of f to a right $\mathbf{K}$-invariant function on $\mathbf{G}$:

$$(\mathcal{H}[f])(s, k\mathbf{M}) = \int_{\mathbf{G}} \hat{f}(g) e^{(\rho - i\lambda)(\log a_{g^{-1}k^{-1}})} \, d\mu_{\mathbf{G}}, \quad (s, k\mathbf{M}) \in (\mathfrak{a}^* \otimes \mathbb{C}) \times \mathbf{K}/\mathbf{M},$$

The Helgason-Fourier transform extends to an isometry

$$L_2(\mathbf{X}, \mu_{\mathbf{X}}) \to L_2\left(\mathfrak{a}^* \times \mathbf{K}/\mathbf{M}, \frac{\mu_\lambda \otimes \mu_{\mathbf{K}/\mathbf{M}}}{|c(\lambda)^2|}\right),$$

where the measure on the analogue $\mathfrak{a}^* \times \mathbf{K}/\mathbf{M}$ of frequency space in Fourier Analysis is given in terms of standard measures as well as the Harish-Chandra c-function.

As in the classical case, for each $f \in C_c^\infty(\mathbf{X})$ we have a Plancherel identity

$$\int_{\mathbf{G}/\mathbf{K}} |f(x)|^2 d\mu_{\mathbf{X}} = \int_{\lambda \in \mathfrak{a}^*} \int_{k\mathbf{M} \in \mathbf{K}/\mathbf{M}} |\mathcal{H}f(\rho + i\lambda, k\mathbf{M})|^2 \frac{d\mu_\lambda}{|c(\lambda)^2|} d\mu_{\mathbf{K}/\mathbf{M}}. \quad (1)$$

The Helgason-Fourier transform $\mathcal{H}$ sends convolutions to products in the following sense. We have for each left and right $\mathbf{K}$-invariant $h \in C_c^\infty(\mathbf{G})$ and

$f \in C_c(\mathbf{X})$ the following identity, where $\bar{h}$ is the well-defined density on $\mathbf{X} = \mathbf{G}/\mathbf{K}$ sending each element $g\mathbf{K}$ to $h(g)$:

$$\mathcal{H}[h * f] = \mathcal{H}[\bar{h}]\mathcal{H}[f],$$

The *inverse Helgason-Fourier transform* $\mathcal{H}^{-1}$, for $f \in C_c^\infty(\mathbf{X})$, satisfies

$$f(x) = \int_{\lambda \in \mathfrak{a}^*} \int_{k\mathbf{M} \in \mathbf{K}/\mathbf{M}} \mathcal{H}f(i\lambda + \rho, k\mathbf{M}) e^{(i\lambda+\rho)(\log a_{k(x)})} \frac{d\mu_\lambda}{|c(\lambda)^2|} d\mu_{\mathbf{K}/\mathbf{M}}.$$

3 The G-kernel Density Estimator

In Euclidean space, kernel density estimation smooths out the empirical distribution around each observation. Formally, the ordinary kernel density estimator $f^{n,h}$ satisfies

$$f^{n,h} = \mathcal{F}^{-1}\left[\hat{\phi}\mathcal{F}[K_h]\right],$$

where $\mathcal{F}$ denotes the ordinary Fourier transform, $\hat{\phi}$ denotes the empirical characteristic function of the samples, K denotes a kernel, h denotes a bandwidth parameter, and $\mathcal{F}[K_h](s) = \mathcal{F}[K](hs)$.

Define the **G**-*Kernel Density Estimator* $f^{(n,T,h)}$ by

$$f^{(n,T,h)} = \mathcal{H}^{-1}\left[\hat{\phi}\mathcal{H}K_h I_{(-T,+T)}\right]$$

where we abuse notation and treat $\mathcal{H}K_h$ as the function sending $(i\lambda + \rho, k\mathbf{M})$ to $\mathcal{H}K(h(i\lambda+\rho), \mathbf{M})$, $I_{(-T,+T)}$ as the function sending $(i\lambda + \rho, k\mathbf{M})$ to 1 if $|\lambda| \leqslant T$ and 0 otherwise, and

$$\hat{\phi}(s, k\mathbf{M}) = \frac{1}{n}\sum_{i=1}^n \overline{e^{s(\log a_{k(X_i)})}} \tag{2}$$

for observed samples $X_1, \ldots, X_n \in \mathbf{X}$ and density K on $\mathbf{G}$ invariant under left and right multiplication by $\mathbf{K}$.

Let $\mathbf{X}$ denote a symmetric space such that for fixed semisimple Lie group $\mathbf{G}$ $\mathbf{X} = \mathbf{G}/\mathbf{K}$ for a maximal complete subgroup $\mathbf{K}$. Let f_X denote a density on $\mathbf{X}$ with respect to the standard volume measure $d\mu_{\mathbf{X}}$. Let K denote a density on $\mathbf{G}$ with respect to the Haar measure $d\mu_{\mathbf{G}}$.

First of all, we assume our densities are L_2.

(D.1) Assume $f_X \in L^2(\mathbf{X}, d\mu_{\mathbf{X}})$ and $K \in L^2(\mathbf{G}, d\mu_{\mathbf{G}})$.

Second of all, we need to restrict K to guarantee that its Helgason-Fourier transform is well-defined. Thus we assume K is **K**-*invariant* in the following sense.

(D.2) Assume $K(acb) = K(c)$ for $c \in \mathbf{G}$ $a, b \in \mathbf{K}$.

Third of all, we make assumptions on the smoothness of the true density f_X. The operator Δ^k defined below in terms of the Helgason-Fourier transform, generalizes the kth derivative operator from integers k to non-negative real numbers k.

(D.3) There exist $\alpha > 1$ and $\mathcal{Q} > 0$ such that

$$f_X \in \mathcal{F}_\alpha(\mathcal{Q}) = \{f_X \in L^2(\mathbf{X}, d\mu_{\mathbf{X}}) : \|\Delta^{\alpha/2} f_X\|^2 \leqslant \mathcal{Q}\},$$

where $\Delta^{\alpha/2} f_X$ denotes the unique function $h \in L^2(\mathbf{X}, d\mu_{\mathbf{X}})$ such that

$$\mathcal{H}h(s, k\mathbf{M}) = \overline{(s(s - 2\rho))}^{\alpha/2} \mathcal{H}f_X(s, k\mathbf{M})$$

Last of all, we make assumptions on the smoothness of the kernel K.

(D.4) There exist constants $\beta, \gamma, C_1, C_2 > 0$ such that

$$C_1 e^{-\frac{|s|^\beta}{\gamma}} \leqslant |\mathcal{H}K(s, k\mathbf{M})| \leqslant C_2 e^{-\frac{|s|^\beta}{\gamma}}$$

for all $s \in \mathfrak{a}^* \otimes \mathbb{C}$.

(D.5) For some $\alpha > 1$, there exist a constant $A > 0$ such that

$$\underset{s \in \mathfrak{a}^* \otimes \mathbb{C}}{ess\ sup} \frac{|\mathcal{H}K(s, k\mathbf{M}) - 1|}{|s|^\alpha} \leqslant A$$

3.1   Main Theorems

Proofs of the following main results can be found in [1].

Theorem 1. *Assume (D.1)–(D.5). Then for a density f_X on a symmetric space $\mathbf{X}$,*

$$\mathbb{E}\|f_X^{(n,T,h)} - f_X\|^2 \leqslant \mathcal{Q}A^2 h^{2\alpha} + \mathcal{Q}T^{-2\alpha} + C\frac{T^{\dim \mathbf{X}}}{n} e^{[-(2/\gamma)(h|\rho|)^\beta]}.$$

for some constant $C > 0$ not dependent on $T, \alpha, \mathcal{Q}, n$.

By choosing a smooth enough kernel density K, an optimal cutoff of T and optimal bandwidth h, we obtain the following rate of convergence.

Corollary 1. *Assume (D.1)–(D.5). Then*

$$\mathbb{E}\|f_X^{(n,T,h)} - f_X\|^2 \leqslant Cn^{-2\alpha/(2\alpha + \dim \mathbf{X})}$$

for some constant $C > 0$ not dependent on $T, \alpha, \mathcal{Q}, n$ and

$$T = \left[\frac{2\alpha \mathcal{Q}n}{\dim \mathbf{X}C} e^{[-(2/\gamma)(h|\rho|)^\beta]}\right]^{1/(2\alpha + \dim \mathbf{X})} \qquad h \in \mathcal{O}(n^{-1/(2\alpha + \dim \mathbf{X})}).$$

The convergence rate for the upper bound is matched by the lower bound, as shown below.

Theorem 2. *Assume (D.3). There exists a constant $C > 0$ such that*

$$\inf_{g^{(n)}} \sup_{f_X \in \mathcal{F}_\alpha(\mathcal{Q})} \mathbb{E}\|g^{(n)} - f_X\|^2 \geqslant Cn^{-2\alpha/(2\alpha + \dim \mathbf{X})}$$

where the infimum is taken over all estimators $g^{(n)}$.

By the previous results, we obtain our minimax rate below for our adapted kernel density estimator.

Corollary 2. *If f_X and K satisfy (D.1)–(D.5), the minimax rate for $f_X^{(n,T,h)}$ is*

$$n^{-2\alpha/(2\alpha + \dim \mathbf{X})}.$$

4   A Special Case: $\mathbb{H}_2$

Hyperbolic spaces provide a logical, generative model for real-world networks [2,6]. The simplest example of a hyperbolic space is the 2-dimensional hyperboloid $\mathbb{H}_2$ of constant curvature -1. Other examples are the nodes of hierarchical, tree-like networks under the minimum path length metric. In fact, sampling points from $\mathbb{H}_2$ according to a *node density* and assigning edges based on geodesic distances generates networks sharing strikingly similar global and local features (e.g. clusterability, power law distributions, significant clustering coefficients, exchangeability) of real-world hierarchical networks (e.g. online social networks, the Internet) [6]. The inference of a generative density on $\mathbb{H}_2$ from sample networks provides a succinct description of the large-scale geometric structure of the samples. Efficient non-parametric density estimation on $\mathbb{H}_2$ should provide important foundational tools for capturing large-scale geometric structure from a broad class of hierarchical networks [2].

4.1   2-Dimensional Hyperboloid

For this paper, we regard the hyperboloid $\mathbb{H}_2$ as the *Poincaré half-plane*

$$\mathbb{H}_2 = \{z \in \mathbb{C} \mid \operatorname{Im}(z) > 0\},$$

equipped with the Riemannian metric

$$ds^2 = y^{-2}(dx^2 + dy^2).$$

The space $\mathbb{H}_2$ is isometric to the quotient space

$$\mathbb{H}_2 = \mathbb{SL}_2/\mathbb{SO}_2.$$

Under this identification, the matrices in $\mathbb{SL}_2$ act on $\mathbb{H}_2$ by *Möbius transformations*:

$$\begin{pmatrix} a & b \\ c & d \end{pmatrix}(z) = \frac{az+b}{cz+d}.$$

Our density estimator is defined on $\mathbb{H}_2$ because $\mathbb{SL}_2$ is a semisimple Lie group admitting an Iwasawa decomposition as $\mathbb{SL}_2 = \mathbb{SO}_2 \mathbf{AN}$, where $\mathbf{A}$ is the group of diagonal (2×2)-matrices in $\mathbb{SL}_2$ with non-negative entries and $\mathbf{N}$ is the group of upper triangular (2×2)-matrices with 1's along the diagonal. For each $z \in \mathbb{H}_2$, n_z and a_z are characterized by

$$(n_z)_{1,2} = \mathrm{Re}(z), \quad (a_z)_{1,1} = \mathrm{Im}(z)^{1/2}.$$

The Harish-Chandra c-function satisfies the formula

$$c(\lambda)^{-2} = \frac{1}{8\pi^2} \lambda \tanh(\pi\lambda).$$

There exists a unique restricted root of $\mathbb{H}_2 = \mathbb{SL}_2/\mathbb{SO}_2$. Under the natural identification of $\mathbf{A}$ with the Lie group of multiplicative non-negative real numbers and hence an identification of $\mathfrak{a}^*$ with $\mathbb{R}$, we identify the unique restricted root (taking a (2×2)-matrix to the difference in its diagonal elements) with 1 and hence ρ with $1/2$.

4.2   Hypergaussian Kernel

We can also choose our kernel K to be the *Hypergaussian*, an analogue of a Gaussian density on Euclidean space defined as follows in [4]. Just as ordinary Gaussians are characterized as solutions to the heat equation, we define K to be the unique ($\mathbb{SO}_2$-invariant solution) to the heat equation on $\mathbb{H}_2$, lifted to a function on $\mathbb{SL}_2$. Concretely,

$$\mathcal{H}[K](s, kM) \propto \overline{e^{s(s-1)}}$$

and hence K satisfies the assumptions (D.1), (D.2), and (D.4) for $\beta = 2$ and $\gamma = 1$.

4.3   Simplified Formula

Under these simplifications, our $\mathbb{SL}_2$-*kernel density estimator* takes the form:

$$f^{(n,T,h)}(z) \propto \frac{1}{n}\sum_{i=1}^{n} \int_{-T}^{+T} \int_0^{2\pi} F(Z_i, \theta, \lambda)\, \frac{\lambda \tanh(\pi\lambda)}{8\pi^2}\, d\theta\, d\lambda, \qquad (3)$$

where F is defined as follows by letting k_θ denote rotation by θ:

$$F(Z, \theta, \lambda) = \mathrm{Im}(k_\theta(Z))^{\frac{1}{2}-i\lambda} e^{-(\frac{h^2}{4}+h^2\lambda^2)} (\mathrm{Im}(k_\theta(Z)))^{\frac{1}{2}+i\lambda}.$$

5   Conclusion

We have introduced a new density estimator on a large class of symmetric spaces, and have proven a minimax rate of convergence identical to the minimax rate of convergence for a Euclidean kernel density estimator. We then specialize our generalized kernel density estimator to the hyperboloid, motivated by applications to network inference. Future work will explore adaptivity, optimizations in implementation, and applications to symmetric spaces other than the hyperboloid.

Acknowledgements. This work was partially supported by an NSF Graduate Research Fellowship under grant DGE-1252522. Also, the author is grateful to her advisor, Cosma Shalizi, for his invaluable guidance and feedback throughout this research.

References

1. Asta, D.: Kernel density estimation on symmetric spaces. E-print, arXiv:1411.4040 (2014)
2. Asta, D., Shalizi, C.R.: Geometric network comparisons. E-print, arXiv:1411.1350 (2014)
3. Hendriks, H.: Nonparametric estimation of a probability density on a Riemannian manifold using Fourier expansions. Ann. Stat. **18**, 832–849 (1990)
4. Huckemann, S.F., Kim, P.T., Koo, J.-Y., Munk, A.: Möbius deconvolution on the hyperbolic plane with application toimpedance density estimation. Ann. Stat. **38**, 2465–2498 (2010). doi:10.1214/09-AOS783
5. Kim, Y.T., Park, H.S.: Geometric structures arising from kernel density estimation on Riemannian manifolds. J. Multivar. Anal. **114**, 112–126 (2013). doi:10.1016/j.jmva.2012.07.006
6. Krioukov, D., Papadopoulos, F., Kitsak, M., Vahdat, A., Boguñá, M.: Hyperbolic geometry of complex networks. Phys. Rev. E **82**, 036106 (2010). doi:10.1103/PhysRevE.82.036106
7. Pelletier, B.: Kernel density estimation on Riemannian manifolds. Stat. Probab. Lett. **73**, 297–304 (2005). doi:10.1016/j.spl.2005.04.004
8. Rahman, I.U., Drori, I., Stodden, V., Donoho, D., Schroöder, P.: Multiscale representations for manifold-valued data. Multiscale Model. Simul. **4**, 1201–1232 (2005). doi:10.1137/050622729
9. Terras, A.: Harmonic Analysis on Symmetric Spaces and Applications, vol. 2. Springer, New York (1988)

Author Index

Printed in the United States
By Bookmasters